THE SCIENCE OF Ecology

Richard Brewer SECOND EDITION

Western Michigan University

SAUNDERS COLLEGE PUBLISHING HARCOURT BRACE COLLEGE PUBLISHERS

Ft. Worth Philadelphia San Diego New York Orlando Austin San Antonio Toronto Montreal London Sydney Tokyo

Text Typeface: 10/12 New Baskerville
Compositor: University Graphics, Inc.
Executive Editor/Biology: Julie Alexander
Developmental Editor: Christine Rickoff
Managing Editor: Carol Field
Project Editor: Maureen Iannuzzi
Copy Editor: Jennifer Holness-Harze
Manager of Art and Design: Carol Bleistine
Art Developmental Editor: Leslie Ernst
Art Assistant: Sue Kinney
Text Designer: Tracy Baldwin
Cover Designer: Lou Fuiano
Text Artwork: Tasa Graphics and Ellen Palm
Director of EDP: Tim Frelick
Production Manager: Joanne Cassetti
Director of Marketing: Margie Waldron

Cover Credit: Animals Animals/Earth Scenes.

Printed in the United States of America

The Science of Ecology, Second Edition

ISBN: 0-03-096575-6

Library of Congress Catalog Card Number: 93-087006

4567 032 987654321

This book was printed on acid-free recycled content paper, containing
MORE THAN
10% POSTCONSUMER WASTE

To

E. Esther Smith (1902–1988)

Willard M. Gersbacher (1906–1989)

John W. Voigt (1920–1990)

S. Charles Kendeigh (1904–1986)

Arthur G. Vestal (1888–1964)

Atque inter silvas Academi quaerere verum.

And what is the meaning of so tiny a being as the transparent wisp of protoplasm that is a sea lace, existing for some reason inscrutable to us—a reason that demands its presence by the trillion amid the rocks and weeds of the shore? The meaning haunts and ever eludes us, and in its very pursuit we approach the ultimate mystery of Life itself.

RACHEL CARSON

Preface

Ecology is the most fascinating and most important field of human inquiry. A major reason for its compelling interest is that ecology tries to see nature as a whole, with all its interconnections. This approach yields knowledge and insights that offer humans their best chance for saving the biosphere as we know it. Any ecology text that does not convey both the intellectual fascination of the science and the necessity of its societal applications aims too low.

This is a textbook of ecology. Its framework is the principles of the science. I have tried to cover these at all levels, from the ecology of the individual to that of the ecosystem. Applied or practical aspects are discussed throughout in connection with the appropriate ecological principles, and the last three chapters deal specifically with the practical application of ecological knowledge.

This book, like the first edition and its predecessor, *Principles of Ecology,* has four broad underlying themes. It may help readers in knowing what to look for, or to look out for, if they recognize these. The first is ecology as system, the idea of the manifold consequences of any action. The second is the organizing principle of evolution. The third is the awareness of the place of humans in nature and the environment. This is two pronged. We need to approach humans as organisms with a life history, social traits, ecological relations, and an evolutionary background. We also need to realize humans' special power to alter the biosphere and special responsibility for its health. The fourth theme is science as our way of learning the truth about the universe.

Among major topics enlarged or added in this edition—taken approximately in the order they appear—are the following:

- phenotypic plasticity
- fire ecology
- population regulation with special attention to red grouse populations
- metapopulations
- mating systems and sex ratios
- genetic fitness and selection
- foraging theory
- the "ideal free" distribution
- plant defenses
- cyclic populations
- chaos
- ecology of disease
- mutualism and the exploiters of mutualism

- detritus food chains
- human expropriation of the earth's production
- nutrient limitation of primary production
- sand dune succession
- status of the climax concept
- the application of paleoecological knowledge to problems of global change
- the importance to community structure of indirect effects such as apparent competition and cascading trophic levels
- landscape ecology
- the longleaf pine ecosystem
- the old-growth temperate rain forests
- mountain zonation
- destruction and preservation of tropical rain forest
- wetlands
- the aeolian biome
- the economics of pesticide use
- pollution in the Great Lakes
- the *Exxon Valdez* oil spill
- recycling
- radical environmentalism
- global climate change
- other types of global change including depletion

There is one new chapter, Chapter 19, Conservation Biology. This emerging discipline has provided a focus for a large body of thought and data on populations, communities, and landscapes. Some of the topics this chapter deals with, such as the history of wildlife in North America, extinction, the design of nature preserves, and reasons for preserving biodiversity, were included in the previous edition. These are given extended treatment in this new edition, and much new material is included on the demographics and genetics of small populations, habitat fragmentation, corridors, restoration, and management of ecosystems and landscapes.

There has been some rearrangement of topics, notably to give mutualism and paleoecology their own chapters and to combine the short original chapter on reactions with biogeochemical cycling. And, of course, there has been updating throughout, including more than 600 new literature citations. There are many new illustrations, including numerous maps, graphs, diagrams, and photos.

Although much has changed, the purpose and approach have remained constant. The text is written for the undergraduate ecology course for majors and minors. Although it assumes no knowledge beyond that usually attained by the time freshman biology is completed, junior standing is probably optimal.

I have tried to keep the language as plain as possible for the sake of clarity. Lucidity has been one of two major goals of the writing. Interest for the reader has been the second. No textbook is likely to rival the latest John Grisham novel for most of us, but a book that draws the reader's attention, rather than repelling it, facilitates learning.

Ecology, like all science, is a human endeavor. I have tried to show this by including an expanded discussion of scientific inference, something of the history of important concepts, indications of competing hypotheses for various phenomena, and ways in which certain hypotheses have been tested. Short biographies of a few of the major figures in the development of the science are a part of this effort (there is one new biography, of Robert MacArthur). More than most disciplines, ecology is plagued with reinventions (and renamings) of the wheel; perhaps inclusion of a historical perspective is a small contribution toward a remedy.

Ecology has been called a science of the particular. That it is, though not just that. There are generalizations that apply to all of nature but also ones that apply only to organisms that are sessile or homoiothermic or populations that are predator limited or annual. There are hierarchies of principles, along several axes. One dimension is habitat and community. Some statements are true of all communities, but others apply only to forests or to temperate deciduous forests. Many generalizations about individuals and populations also have to be restricted by habitat or community. In this book, one expression of these ideas is the section on biomes. Much that is not globally general in ecology is, nevertheless, interesting and important.

Up to a point, I have favored the use of familiar ecosystems and organisms—North American vertebrates and trees, for example. For most readers, this furthers readability and clarity. But I have also paid attention to balance, taxonomically, geographically, and by habitat. In this edition more than in the past, I have made an effort to include aquatic examples throughout the book, not just in the specifically aquatic sections.

I have also tried to achieve balance in terms of the whole field of ecology. Ecology is a broad field.

We should admit it and delight in it rather than trying to ignore the parts that are currently unpopular or that do not engage us personally.

Many persons have been helpful with this book directly or indirectly, by recent or past actions, with a few incisive words or by a long, helpful association. The following lists include many of these persons, and I apologize to those I have inadvertently omitted.

Thanks to J. W. Hardy, K. D. Stewart, C. Heckrotte, W. B. Robertson, Jr., G. C. West, G. W. Cox, W. L. Gillespie, T. S. Robinson, W. J. Davis, W. L. Minckley, R. W. Olsen, C. G. Goodnight, R. W. Pippen, A. M. Laessle, J. G. Engemann, D. L. Regehr, M. L. Kaufmann, K. A. Daklberg, D. P. Cowan, J. Davey, M. T. McCann, R. J. Adams, W. B. Leak, R. E. Graber, Alan Covich, D. S. Wilson, Steve Archer, Russell Davis, Douglas Slack, Gary Mittelbach, L. E. Hurd, Lucy Sharp Brewer, Diane Stephenson, Mark Betz, John Yunger, Chad Stuart, Katja Schultz, Diane Wirt, Gail Celio, Geoff Hickok, Tim Clark, Robert McKelvey, Lee Metzgar, Richard Stemberger, Daniel Simberloff, Frances James, Joseph Travis, Thomas Miller, Todd Engstrom, Gail McPeek, and Stephen Malcolm.

I should thank separately the excellent photographers who have allowed me to use their photographs. Especially I am indebted to Bruce Moffett, Gerald and Marcella Martin, George and Betty Walker, Joseph Engemann, Richard Pippen, Larry Walkinshaw, Clayton Alway, Michael Gaule, Carl Snow, Todd Engstrom, and Janet Vail.

Several persons read portions of this new edition. For their helpful comments, I thank

Richard Stemberger, Dartmouth College
Steven Brewer, Western Michigan University
Guy Cameron, University of Houston
Al Craft, College of Misericordia
Dennis Frey, California Polytechnic State University
Thomas Jurik, Iowa State University
Ken Marion, University of Alabama at Birmingham
John Mertz, Delaware Valley College
Beth Middleton, Southern Illinois University
George Middendorf, Howard University
Lawrence Mueller, University of California, Irvine
Dennis Murphy, Stanford University
Steve Murray, California State University, Fullerton
David Polcyn, California State University, San Bernadino
Barbara Rafaill, Georgetown College
Merrill Sweet, Texas A & M University
Tad Theimer, Adams State College

I am also indebted to the people at Saunders that I have worked with over the years. Those persons, listed on the copyright page, involved with this edition were helpful indeed.

Richard Brewer
November 1993

Contents

1
Ecology as a Science 1

Ecology Defined 1
History of Ecology 2
Charles Darwin 4
Ecology Today 8
Scientific Inference 9
Ecology as System 11
Interrelationships Within the Ecosystem 12

2
Ecology of Individual Organisms: Principles 13

Relationships to the Abiotic Environment 14
Tolerance Range 14
Range of the Optimum 15
When Conditions Change 16
Phenotypic Ways of Coping with Changed Environments 17
Phenotypic Plasticity 18
Extreme Conditions 19
Daily Fluctuations 19
Limiting Factors and the Environmental Complex 20
Ecological Indicators 21
Energy Balance 22
Energy and Work 22
Energy in Organisms 23
Autotrophs 23
Heterotrophs 24
Variations 24
Energy Flow Through an Individual Organism 25
Energy Subsidies 28
Animal Behavior 28
Evolutionary Considerations 29
Fitness 29
Proximate and Ultimate Factors 29
Adaptation 30
Ecotypes 31
Habitat Selection 32
The Spread of Organisms 33
Dispersal 33
Range Expansion 36

3
Ecology of Individual Organisms: Important Abiotic Factors 40

Temperature 41
Thermal Relations of Organisms 41
Homoiotherms and Poikilotherms 42
Hibernation and Other Combinations 45
Interaction of Temperature and Wind 46
Life Forms of Plants 47
Moisture 48
Humidity 48
Vapor Pressure 49
Water Balance in Plants 49
Xerophytes and Mesophytes 50
Hydrophytes and Hydroperiod 52
Halophytes 52
Water Balance in Animals 53

Light 54
Shade Tolerance 55
Photoperiodism 57
Circadian and Other Rhythms 58
Soil 59
Soil Structure 59
Soil Texture and Fertility 60
Soil Water 63
Loams and Humus 64
Soil Names 64
Fire 65
Types of Fires 66
Effects of Fire on Plants 67
Effects of Fire on the Physical Environment 70
Effects of Fire on Animals 71
Fire Frequency 72
Fire as a Management Tool 75
Pollution 79

4
Population Ecology: Growth and Density 81

Birth and Death 81
Birth Rate and Death Rate 81
Life Tables and Longevity 82
Types of Life Tables 83
Survivorship Curves 84
Population Growth 86
Exponential Population Growth 86
Biotic Potential 89
Logistic Population Growth 91
The Logistic Curve as a Model 92
The Allee Effect 93
Mathematical Treatment of Population Growth 95
Exponential Growth 96
Logistic Growth 96
Adding Time Lags to the Logistic Equation 97
Ways of Looking at *r* 98
Intrinsic Rate of Natural Increase and Net Reproductive Rate 99
Adding the Allee Effect to the Logistic 100
Projection Matrices 101
Population Density and Population Regulation 102
Density-Dependent Factors 102
Habitat Distribution 105
Carrying Capacity 106
Population Regulation 108
Intraspecific Competition 109
Red Grouse Populations 111
Intercompensation 112
Nonequilibrium Populations 112
Sources, Sinks, and Metapopulations 114
Growth and Regulation of Human Populations 114
Thomas Robert Malthus 117

5
Population Ecology: Organization and Evolution 118

Organization 118
Spacing 118
Sociality 120
Predator Protection 121
Increased Foraging Efficiency 121
Modifying Environment 125
Other Benefits of Sociality 125
Social Organisms 126
Mating Systems 127
Cuckoldry, Pro and Anti 128
Sex Ratio 129
Age Structure 130
Age Structure in Plants 131
Evolution 133
Natural Selection 133
The Hardy-Weinberg Law 133
Genetic Fitness 135
Types of Selection 136
The Evolution of Life History Traits 137
r and *K* Selection 137
Selection of Low Reproductive Rate 138
Characteristics of *r*- and *K*-Selected Organisms 139
Annuals and Perennials 141
Salmon and Century Plants: Long-Lived Semelparous Organisms 144
Pioneer Versus Climax Species 144
Beyond the *r* and *K* Selection Model 145
Practical Aspects 148
The Evolution of Behavior 148
Heritability 148
Sociobiology 150
Altruism 151
Selfish Behavior 151
Kin Selection 152
Eusociality 154
Reciprocal Altruism 156
Human Sociobiology 157
Group Selection 157

The Ecological Theater and the Evolutionary Play 160
The Ant-Acacia System 161
Bat-Moth Coevolution 162
Extinction 165

6
The Population–Community Ecology Interface: Herbivory and Predation 167

Types of Species Interactions 167
Trophic Interactions 169
Optimal Foraging 171
Where to Look 173
Whether to Pursue 174
Central-Place Foraging 175
What Do Tests of Optimal Foraging Tell Us? 175
Optimization 176
Herbivory 177
Types of Herbivores 177
Plant Defenses Against Herbivory 178
Effects of Herbivory on Plant Distribution and Abundance 182
Frugivory 184
Seed Predation 185
Predation 187
Types of Predators 187
The Lotka-Volterra Predator-Prey Model 188
Fluctuations in Population Size 190
Cyclic Populations 192
Chaos 194
Functional and Numerical Responses and the Control of Prey Numbers by Predators 197
Mimicry and Other Kinds of Advantageous Resemblance 198
Cannibalism, Siblicide, and Intraguild Predation 203
Biological Control 202

7
The Population-Community Ecology Interface: Parasitism, Commensalism, and Saprobism 205

Parasitism 205
The Prevalence of Parasites 205
Types of Parasites 208
The Ecology of Parasites 209
Rates of Increase 211
Parasite-Host Interactions 211
Epidemics 211
Threshold Size and Human Diseases 213
AIDS 214
Effects of Parasitism on Host Numbers and Distribution 216
Evolution Within the Parasite-Host System 220
When Should Parasites Harm Their Hosts? 221
Health and Disease 222
Commensalism 222
Phoresy 223
Saprobism 223
Carrion 223
Vertebrates 224
Insects 224
Dung 225
Tumblebugs and Other Buriers 226
Dead Trees 226
Litter 228
Aquatic Habitats 228

8
The Population-Community Ecology Interface: Competition and Related Coactions 229

Interspecific Competition 229
Outcomes of Competition 229
Ecological Niches and Guilds 230
Competitive Exclusion 231
Competitive Replacement 231
Competition as a Factor in Geographical and Habitat Distribution 232
Apparent Competition 235
Competitive Exclusion in the Laboratory 238
Coexistence of Competing Species 238
Mathematical Representation of Interspecific Competition 240
The Lotka–Volterra Model 240
Adding Species to the Lotka–Volterra Model 243
Can Two Species Live on One Resource? 244
Amensalism 246
Allelopathy 247
Neutralism 248

9
The Population-Community Ecology Interface: Mutualism 250

Symbiotic Mutualism 250
Nonsymbiotic Mutualism 251

Mycorrhizae 253
Pollination 255
The Exploitation of Mutualisms 256
The Evolution of Mutualism 259
The World's Greatest Symbiosis: Evolution of the Eucaryote Cell 260

10
Community and Ecosystem Ecology: Structure and Diversity 263

Community and Ecosystem 263
Indirect Effects 264
Community Structure 267
Dominance 268
Chemical Ecology 268
Species Composition 270
Spatial Structure 271
Synusia and Guild 272
Temporal Structure 272
Day-Night Change 272
Seasonal Change 274
Phenology 275
Ecological Niche 279
The Niche as a Hypervolume 281
Partitioned Resources 282
What Produces Community Organization? 284
The Importance of Competition 285
Robert H. MacArthur 286
Are There Regularities in Species Composition Based on Competitive Exclusion? 287
The Species/Genus Ratio 287
Assembly Rules 288
Is There Evidence That Current Competition Affects the Structure of Communities? 289
Is There Evidence That Evolution Based on Reduction of Competition Has Been Widespread? 290
The Integrated Versus the Individualistic Community 293
Ecological Diversity 296
Ways of Being Diverse 296
Factors Affecting Diversity 297
Unique History 298
Time 298
Extreme Habitats 298
Resource Diversity 298
Productivity 299
Climatic Stability 299
Predation 299
Disturbance 299
Regional Diversity 300
Diversity Indices 301
What Determines the Number of Species on Islands? 302

11
Community and Ecosystem Ecology: Energy 307

Trophic Structure and the Food Chain 307
Producers 308
Consumers 308
Decomposers 309
Pyramids of Numbers and Biomass 309
Energy Flow 310
Solar Energy Input 310
Energy Flow Within the Ecosystem 311
Grazing Food Chains and Detritus Food Chains 314
Biomass 315
Productivity 316
Primary and Secondary Production 316
Energetic Steady States 318
NPP in the Biosphere 320
Relationships Between Productivity and Biomass 321
Ecological Efficiencies 324
Producer-Level Efficiencies 324
GPP/Sunlight 324
NPP/GPP 324
Harvesting Efficiency of Herbivores 325
Consumer Efficiencies 326
Within-Trophic-Level Efficiencies 326
Between-Trophic-Level Efficiencies 327
Food Chains and Food Webs 327
Difficulties With Trophic Levels 328
Chemolithoautotrophs 329
Lakes and Coastal Marine Ecosystems 329
Hydrothermal Vents 331
Regulatory Processes and Energy Quality 332
Energy in Agriculture 334
Basic Patterns 334
Energy Flow in Food Systems 334
Human Expropriation of the Earth's Primary Production 341

12

Community and Ecosystem Ecology: Reactions and Biogeochemistry 343

Reactions on Land 344
Soil Formation 344
Topography 346
Soil Moisture 348
Reactions on Air 348
Solar Radiation 348
Temperature, Humidity, Rainfall, and Wind 352
Reactions in Fresh Water, 353
Reactions in the Ocean 353
Nutrient Cycling 354
The Carbon Cycle 355
The Nitrogen Cycle 358
The Phosphorus Cycle 359
The Sulfur Cycle 361
Nutrient Limitation of NPP 361
The Hydrological Cycle 363
Watershed Studies 365
Biological Control of the Composition of the Atmosphere and Oceans 369
The Early Earth 369
The World's Greatest Reaction 369
Biological Controls 370
The Gaia Hypothesis 372

13

Community and Ecosystem Ecology: Community Change 373

Types of Community Change 373
Replacement Changes 374
Cyclic Replacement 374
Canopy Replacement in Forest 375
Fluctuation 377
The Great Drought 378
Succession 379
Types of Succession 381
Microseres 382
Causes of Succession 382
The Role of Reactions 382
Frederic E. Clements 384
An Evolutionary View of Succession 386
The Role of Coactions 387
Examples of Succession 387
Sand Dune Succession 387
Lower and Middle Beaches 388
Henry Chandler Cowles 389
Dune Formation 389
Blowouts and Wandering Dunes 392
Plant Succession 393
Water Levels 394
Animal Succession 394
Old-field Succession 396
The Climax Community 398
Monoclimax and Polyclimax 400
Is There Such a Thing as a Climax Community? 402
Stability 403

14

Community and Ecosystem Ecology: Paleoecology 406

Uniformitarianism 406
The Paleocommunity 407
Autecology 408
Population Ecology 409
Community Ecology 411
The Early Tertiary 411
The Later Tertiary 412
The Pleistocene 414
Full-Glacial Plant and Animal Distribution 416
Late-Glacial and Postglacial Changes 417
Pleistocene Changes Elsewhere 419
Extinction of the Pleistocene Megafauna 421
Applying Paleoecological Knowledge 424

15

Community and Ecosystem Ecology: The Biome System, Temperate Deciduous Forest, and Southeastern Evergreen Forest 426

Physiognomy 426
The Biome System 427
Victor E. Shelford 428
Landscape Ecology 431
Temperate Deciduous Forest 431
Location and Climate 431
Soils 432
Plants and Vegetation 432
Subdivisions of Temperate Deciduous Forest 434

E. Lucy Braum 436
Mixed Mesophytic Forest 435
Western Mesophytic Forest 435
Oak-Hickory Forest 437
Oak-Chestnut Forest 437
Oak-Pine Forest 438
Beech-Maple Forest 438
Maple-Basswood Forest 438
Hemlock-White Pine-Northern Hardwood Forest 438
Animals 439
Trees and Shade 440
Seasonality 440
Human Occupation 442
Deciduous Forest Edge 442
Southeastern Evergreen Forest 446
Vegetation 447
Upland Hardwoods 447
Pinelands 448
Hydric Communities 450
Scrub 452
Animals 452

16

Community and Ecosystem Ecology: Biomes of the High Latitudes and Elevations 459

Arctic Tundra 459
Distribution and Climate 459
Soil 461
Vegetation 461
Animals 463
Biogeographical Relations 466
Human Occupation 467
The Antarctic 468
Boreal Coniferous Forest 469
Distribution and Climate 469
Soil 470
Plants 470
Animals 472
Biogeographical Relations 475
Ecotones 476
Mountain Zonation 477
Alpine Tundra 478
Climate and Soil 479
Vegetation 481
Animals 482
Human Relations 484
Montane and Subalpine Forests 486
Appalachians 487
Rocky Mountains and the High Plateaus 488
Cascade Range and Sierra Nevada 490
Temperate Rain Forests of the Pacific Northwest 494
Animals of the Western Coniferous Forests 498
Shrublands and Woodlands 505
California Chaparral 505
Arizona Chaparral 510
Pinyon-Juniper Woodland 510

17

Community and Ecosystem Ecology: Grassland, Desert, and Tropical Biomes 514

Temperate Grassland 514
Distribution and Climate 514
Soils 515
Plants 515
Types of Grassland 519
Tallgrass Prairie 519
Mixed Prairie 520
Shortgrass Prairie 520
Bunchgrass Prairie 521
California Prairie 521
Desert Grassland 521
Animals 522
Vegetation 522
Climate 525
Effects of Animals 526
Animal Distribution 526
Oak Openings 528
Prairie Restoration and Reconstruction 528
Desert 531
Distribution and Climate 531
Topography and Soil 532
Plants 533
Cold Desert (Basin Sagebrush) 534
Warm Desert (Desert Scrub) 535
Animals 538
Animal Communities in North American Desert 539
Human Relations 542
Tropical Biomes 543
Tropical Rain Forest 543
Location and Climate 543
Soils 543
Plants 544
Animals 547
Destruction of the Rain Forest 551
Other Tropical and Subtropical Forests 554

Tropical Savanna and Grassland 554
Location and Climate 554
Soils 555
Plants and Animals 555
Loss of the African Megafauna 559

18
Community and Ecosystem Ecology: Aquatic Communities and Special Habitats 561

Lakes and Ponds 561
Origins of Lakes 562
Physical Factors in Lakes 562
Water as a Habitat 562
Thermal Stratification and Oxygen Depletion 563
Meromictic Lakes 564
Communities and Organisms 564
Pond and Lake Regions 564
Habitat Groups 565
Energy Flow 567
Succession and Eutrophication 567
Streams 569
Riffles 570
Pools 570
Downstream Drift 571
Energy Flow 571
Physiographic Succession 573
The Oceans 574
Ocean Life Zones 574
Littoral Zone 574
Bathyal, Abyssal, and Hadal Zones 587
Neritic and Oceanic Zones 587
Productivity 588
Wetlands 589
Marsh 589
Bog 590
Bog Succession 590
Habitat Conditions in Bogs 590
Bog Forest 594
Bogs as Boreal Islands 595
Fen 596
Shrub-Carr 597
Swamps 598
Wetlands Productivity 598
Caves, Phytotelmata, and Other Special Habitats 598
Caves 599
Phytotelmata 600
The Aeolian Biome 601

19
The Practical Ecologist: Conservation Biology 602

The Idea of Conservation 603
Conservation Biology 603
Wildlife Management 604
Hunting in the United States 604
Management Practices 606
Control of Hunting 606
Predator Control 607
Reservation of Land for Game 608
Game Stocking 608
Environmental Manipulation 609
A Sixth Management Stage 610
Extinction 610
Endangered and Threatened Species 610
The Process of Extinction 613
Deterministic and Stochastic Models 615
Stochasticity in Extinction 616
Demographic Stochasticity 616
Genetic Stochasticity 616
Environmental Stochasticity 618
The Threshold of Extinction 619
A Case Study of Extinction: The Black-Footed Ferret 619
Preserving Ecosystems and Landscapes 621
The Preservation of Natural Areas 621
The Design of Nature Preserves 622
Habitat Fragmentation 622
Deforestation and Other Changes on the Wintering Grounds 627
How Big is Big Enough? 629
Advantages of Multiplicity 632
Corridors 636
Landscapes 636
Preservation and Management 637
Restoration 639
Why Preserve Biodiversity? 640
Aesthetic and Practical Reasons 640
Moral and Ethical Reasons 642
Aldo Leopold 643

20
The Practical Ecologist: Pollution 646

The Touchstone Formulation for Environmental Damage, Destruction, Degradation, and Deterioriation 646

Pollution as a Global Problem 647
Types of Pollutants 649
Pesticides 649
Human Toxicity 650
Rachel Carson 651
Effects on Nontarget Organisms 651
Persistence and Accumulation 652
Problems of Spreading 652
Sublethal Effects 652
Synergistic Effects 653
Ecosystem Effects 654
Resistance to Pesticides 654
Herbicides 655
The Economics of Pesticide Use 655
Radioactive Materials 656
Nuclear War 659
Other Toxic Materials 660
Biological Magnification 661
Pollution in the Environment 663
Water Pollution 663
Pollution in the Great Lakes 665
The *Exxon Valdez* Oil Spill 669
Groundwater 671
Atmospheric Pollution 672
Acid Rain 673
The Greenhouse Effect and Global Climatic Change 679
The Hole in the Ozone 683
Genetically Engineered Organisms in the Environment 684
Recycling 686

21
The Practical Ecologist: Energy, Food, Health, Population, and Land 688

Ecology and Global Change 688
Energy 689
Nuclear Power 690
Our Energy Future 691
Food Production and Organic Farming 692
The Role of Environment in Health and Disease 694
Holistic Medicine 696
Population 698
Optimal Population Size 698
Local and National Populations 698
Immigration 700
World Populations 701
Land Use 701
Phases in Environmental Planning 702
Radical Environmentalism 703
Spaceship Earth 705

Bibliography 707
Glossary 752
Scientific Names of Species Mentioned in Text 760
Index I-1

Ecology as a Science

OUTLINE

- Ecology Defined
- History of Ecology
- Ecology Today
- Scientific Inference
- Ecology as System
- Interrelationships Within the Ecosystem

1

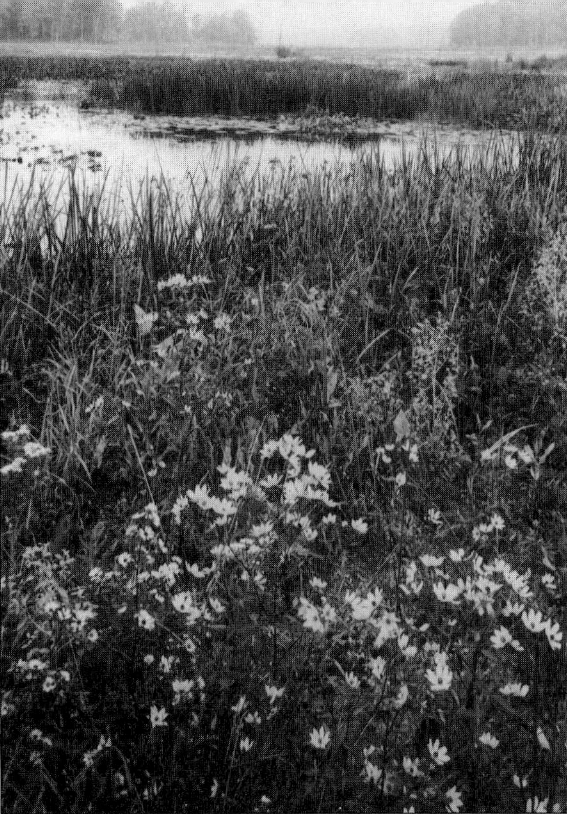

This pond in western Pennsylvania is an example of a familiar ecosystem. *(Ian J. Adams/Dembinsky Photo Associates)*

Ecology Defined

Ecology is a relatively new science. The term "ecology" was first defined in 1866 by the German zoologist Ernst Haeckel (1870). He based it on the Greek word *oikos,* meaning "home," and wrote (Allee et al. 1949):

> *By ecology we mean the body of knowledge concerning the economy of nature—the investigation of the total relations of the animal both to its inorganic and its organic environment; including, above all, its friendly and inimical relations with those animals and plants with which it comes directly or indirectly into contact—in a word, ecology is the study of all those complex interrelations referred to by Darwin as the conditions of the struggle for existence.*

Basically the same definition is used today: ecology is the study of the relationships of organisms to their environment and to one another. The key word is "relationships." Ecology is a study of interactions. A list of the plants and animals of a forest is only a first step in ecology. Ecology is knowing who eats whom, or what plants fail to grow in the forest because they can't stand the shade or because, when they do grow there, they get eaten.

History of Ecology

One important strand in the development of ecology has been natural history; in fact, the English ecologist Charles Elton (1927) gave "scientific natural history" as his definition of ecology. The 19th century was the golden age of natural history, and, as the quotation shows, Haeckel was strongly influenced by the greatest naturalist of that—or any—time, Charles Darwin. The grasp of ecological principles shown in Darwin's *The Origin of Species* seems modern today.

Another 19th-century naturalist who clearly made the transition to ecology was Stephen A. Forbes of the Illinois Natural History Survey (Brewer 1960, McIntosh 1985). Even the titles of his papers, such as "On Some Interactions of Organisms" and "The Lake as a Microcosm," both published in the 1880s, show the ecological orientation of his thinking.

Despite Forbes's important personal contributions, ecology as a distinct field of study in the United States stemmed more from the work of two slightly later biologists, both botanists. Henry Chandler Cowles at the University of Chicago and Frederic E. Clements at the University of Nebraska were strongly influenced by the work of European plant geographers. Accordingly, many of the questions they asked dealt with distribution—What recurring patterns of vegetation can be observed on the earth and what produces them?

It is traditional, if oversimplified, to date ecology in this country from Clements' *Phytogeography of Nebraska* (Pound and Clements 1898) and Cowles' "The Ecological Relations of the Vegetation of the Sand Dunes of Lake Michigan" (1899). Two early animal ecologists, both of whom were associated with Cowles and also, at the University of Illinois, with Forbes, were Charles C. Adams and Victor E. Shelford.

The development of ecology in the early part of the 20th century was rapid and especially enthusiastic in the United States. We can imagine an early ecologist setting out with a class from the University of Illinois or Chicago or Nebraska (Figure 1-1). The instructor and the students wear dark suits, ties, and hats after the fashion of the period. They travel by

Figure 1-1

A field trip 20 May 1913 led by the pioneer ecologist C. C. Adams from the University of Illinois (Western Michigan Univ. Archives, Adams Collection).

electric railroad to the station nearest their study area, and then walk. The study area may be a beech-maple forest, the beaches and dunes of Lake Michigan, or a bog. They conduct their studies, eat the sandwiches their landladies fixed for them, and at the end of the day walk back to the station for the trip home. If some of the students are planning to become ecologists, they might be needled a bit by their classmates. What difference does it make what plants grow on sand dunes? What can a study of the animals that live on the bottom of the Illinois River tell us? Why waste time studying ecology when you can work on a good solid topic like the embryology of the sea urchin or go to medical school?

Times change (Figure 1-2). If we were to visit the dunes studied by that class of 70 years ago, we would travel by expressway rather than interurban, and we might find, not dune grass, lupine, and puccoon, but steel mills. Since Cowles' and Shelford's day, the steel companies have come in, flattened the dunes, made steel and money, and fallen into decline. Steel is still made, and the air is still polluted, but buildings are discolored and go unrepaired, and many of the people who once worked here have entered an early and undesired retirement. Times change, and the science of ecology is no longer of interest only to ecologists or a subject to be studied only for its intellectual fascination.

The period around World War II was a watershed, a major transition, in ecology. Through the first 40 years of the century, most ecologists had continued to concentrate on questions of the distribution of organisms and communities. In retrospect, the book *Bio-ecology* by Clements and Shelford (1939) was a culmination of this phase of the science's development. Other ecological currents were running, but, because of the war, they were not evident for several years.

During the Vietnam war, stateside life went on undisturbed—or would have except for the efforts of the antiwar protestors. "Operation Desert Storm" had even less effect at home. World War II was much different. Many peacetime activities were curtailed; for example, gas and tires were rationed, greatly reducing travel. A high percentage of the nation's scientists were in the military or doing re-

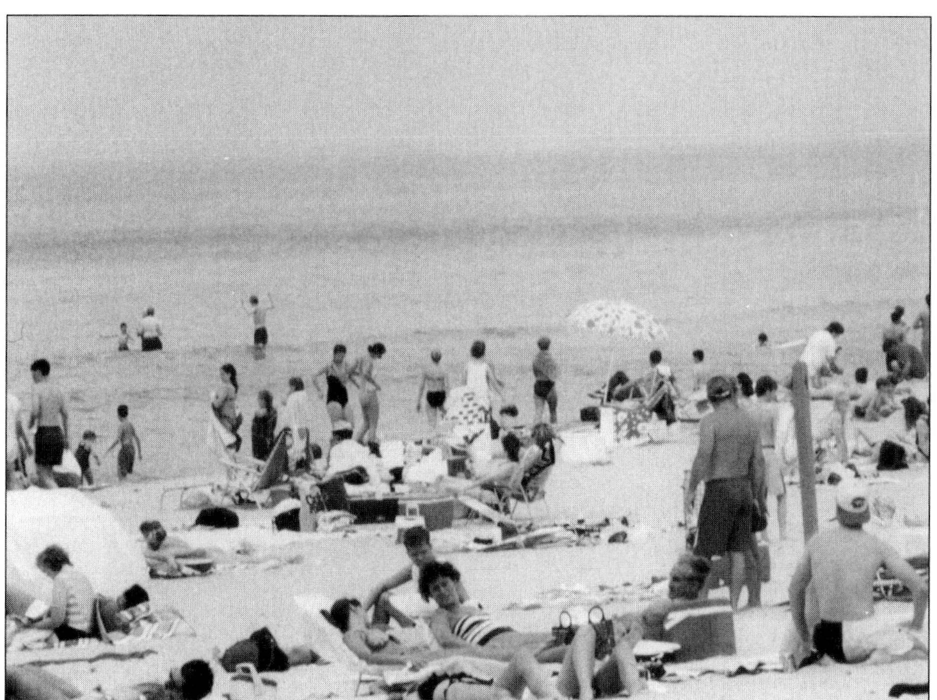

Figure 1-2

The middle beach of Lake Michigan today. (Photo by the author.)

***Charles Darwin** (English, 1809–1882)*

February 12, 1809, was a good day for humanity, the day on which both Abraham Lincoln and Charles Darwin were born. Their circumstances were different. Darwin was the son of a prominent physician in Shrewsbury in northwestern England and was sent to Shrewsbury School and later the Universities of Edinburgh and Cambridge. Darwin's approach to schooling was not particularly serious and, late in his life, he recalled that his father had once told him, "You care for nothing but shooting, dogs, and rat-catching, and you will be a disgrace to yourself and all your family." As it turned out, this was not the case, though he was fond of dogs throughout his life.

The crucial experience of Darwin's life was the four-year, nine-month voyage as a naturalist aboard H.M.S. *Beagle.* Leaving Plymouth just after Christmas 1827, the *Beagle* circumnavigated the globe, visiting both coasts of South America, the Galápagos Islands, Tahiti, New Zealand, and Australia. The specimens, observations, and ideas that Darwin accumulated during that period set the course of the rest of his life.

Although he began his first notebook on the "transmutation of species" in 1837, most of the time just after Darwin's return to England was spent reporting on geological findings of the voyage. Then for eight years, he monographed the living and fossil barnacles of the world. Finally, in 1854, he turned his full attention to the fact and the process of evolution.

Darwin published a brief statement of his views in 1858, largely because Alfred Wallace had come up with exactly the same idea, that of natural selection. Their joint publication, "On the Tendency of Species to Form Varieties; and on the Perpetuation of Varieties and Species by Natural Means of Selection," in the August issue of the *Journal of the Linnaean Society* attracted no special attention. In fact, the president of the society, in summing up the events of the year, noted that it had "not been marked by any of those striking discoveries which at once revolutionize, so to speak, the department of science on which they bear."

With the basic idea now in

search on war-related projects. Biological journals from the war years are slim, with little indication of the broadening scope of the science. The late 1940s and early 1950s, with ecologists back at their study plots, laboratories, and typewriters, were years of renewal and transformation. Among the branches of ecology that showed the greatest growth were population ecology, energetics, and evolutionary ecology.

All of these, of course, had roots that antedated World War II. Royal N. Chapman and Raymond Pearl had laid the foundation on which Thomas Park's population studies were built, but only a couple of Park's papers on grain beetles had appeared prior to 1940. The war itself gave impetus to studies of grain beetle populations in Australia; grain and flour stores had built up as the threat of the Japanese navy kept Australian cargo ships at home.

The catalytic paper for the study of ecological energy flow was Raymond L. Lindeman's 1942 "The Trophic-Dynamic Aspect of Ecology." The enormous impact of the paper was not evident until the late 1940s. The excitement of the energy-ecosystem approach to ecology was caught in the first edition (1953) of E. P. Odum's *Fundamentals of Ecology,* the most widely used textbook of the next 20-odd years. The International Biological Program, an enormous cooperative study of the world's biomes that began in 1964 (field work in 1967), was an outgrowth of the energy-ecosystem strand of ecology (Golley 1991).

The beginnings of evolutionary ecology go back to Darwin, but the immediate precursor of the modern interest in the subject was the "new synthesis" beginning about 1936. This blend of evolution, genetics, and systematics, with its focus on print, Darwin began writing in earnest, and *On the Origin of Species* was published in November 1859. The first printing was sold out on the first day. Biology, not to mention the rest of human thought, was, this time, revolutionized.

The later lives of Charles Darwin and Abraham Lincoln were, of course, also very different. Lincoln made a living as a frontier lawyer and spent his later years in the rough and tumble of politics in Springfield and, then, Washington. Darwin moved to Down House in a rural part of Kent in 1842 when he was 33 years old. Money from his father, along with his wife's dowry and allowance, allowed him to devote the rest of his life—he lived to be 73—to his researches. The fact that his health was poor also kept him away from activities that might have interrupted his work. His wife, Emma, a Wedgwood of the china Wedgwoods, guarded him from visitors and took care of him.

What caused Darwin's illness is uncertain. Social gatherings and even scientific meetings overexcited him. Afterward he would suffer from some combination of gastrointestinal pain, nausea, dizziness, sleeplessness, and lassitude. Some have suggested that all his problems were psychological, perhaps brought on by feelings of guilt about the blow he was dealing to religion. Others have claimed that his symptoms are those of Chagas' disease, related to sleeping sickness and spread by a kind of bedbug that Darwin's journal records as having bitten him one night in a small village in Argentina.

The evolutionist is the Darwin that the public knows, but he also stated many ecological concepts that we are apt to ascribe to much later authors. The first ten pages of Chapter 3 of *On the Origin of Species* contain the following ideas and methods: exponential population growth, the connection between low reproductive rate and parental care, the lack of connection between reproductive rate and carrying capacity, competitive exclusion, the role of predation or grazing in maintaining species diversity, the exclosure method, and predation as a limiting factor. The rest of the chapter and many of his other books are equally fertile.

the formation and coexistence of species, was promoted in the United States by several scientists that had left Europe ahead of the war.

The 1960s and 1970s, when ecology came to public consciousness, composed another important transition period. Thoughtful persons began to realize that many of the major problems of the world—pollution, overpopulation, and the misuse of resources—are at heart ecological problems. If we were to pick one event as the beginning of the environmental movement, it would probably be the publication of Rachel Carson's *Silent Spring,* a well-documented and utterly convincing description of the hazards of pesticide use as of 1962. The idea around which the book is built—that disregard of ecological relationships is likely to have ruinous consequences—was not new. Ecologists such as Paul Sears and Aldo Leopold and many others with less famous names had been saying the same thing since the 1930s. What was new was that people were listening.

It may not be quite so clear that this aspect of the development of ecology also has roots in World War II; but it does. At the end of the war, one of the flourishing movements in agriculture was an attempt at an environmentally sound approach to the land, exemplified in the United States by Louis Bromfield's Malabar Farm (Bromfield 1948). Faced with peacetime overcapacity, the chemical companies turned their production facilities and advertising budgets to fertilizers, pesticides, and herbicides—many of the last two the result of military research on biological weapons (e.g., Whiteside 1970). In an incredibly few years, the approach to agricultural land as a living system had all but vanished, replaced by heavy reliance on chemicals.

Landmarks in the History of Ecology

This list is arbitrary in several ways. I have omitted the earliest roots of ecology; the writings of such worthies as Aristotle, Lucretius, and Frederick II were landmarks in many fields of knowledge, but listing them tells us little about ecology. Some of the events would be landmarks in anyone's version of ecology, but judgment would differ on others. For several of the ecologists mentioned, any of several other of their books or articles could also be considered landmarks. Finally, I have listed relatively few events of the past 25 years. It has been a period of great progress but is too recent to see in perspective; besides, that is what much of this book is about. Landmarks of environmental science are listed in Chapter 20.

1798. Thomas Malthus (English): *Essay on the Principle of Population.*

1805. Alexander von Humboldt (German) recognized plant communities and related plant distribution to the physical environment in his essay on the geography of plants.

1859. Charles Darwin (English): *On the Origin of Species.*

1866. Ernst Haeckel (German) first used the term "ecology" in *Generelle Morphologie der Organismen.*

1877. In a study of oyster beds and oyster culture, Karl Möbius (German) described the interactions among members of the community, which he termed a "biocoenosis."

1887. Stephen A. Forbes (American) gave an early statement of the ecosystem concept in "The Lake as a Microcosm."

1892. F. A. Forel (Swiss) published the first volume of his monograph on the limnology of Lac Léman.

1895. Eugene Warming (Danish): *Plantesamfund,* the first strictly ecological book.

1898. R. Pound and F. E. Clements (American): *The Phytogeography of Nebraska.*

1899. Henry Chandler Cowles (American): "The Ecological Relations of the Vegetation of the Sand Dunes of Lake Michigan," like Pound and Clements' study, a pioneer study of plant succession.

1909. Under the direction of Charles C. Adams (American) an attempt was made at a thorough ecological survey of a terrestrial area, Isle Royale, Michigan.

1911. The first International Phytogeographic Excursion was held, using *Types of British Vegetation,* by Arthur G. Tansley (English), as its guidebook.

1913. Victor E. Shelford (American): *Animal Communities in Temperate America.*

1915. The Ecological Society of America was organized, with Victor E. Shelford elected first president.

1916. Frederic E. Clements (American): Publication of monograph on plant succession.

1925. Alfred J. Lotka (American) discussed demography, nutrient cycling, and energy flow in *Elements of Physical Biology.*

1926. H. A. Gleason (American) argued against a holistic approach to communities in "The Individualistic Concept of the Plant Association."

1927. Charles Elton (English): *Animal Ecology,* emphasizing regulation of population size, ecological niches, and community function as related to food chains.

1927. J. Braun-Blanquet (Swiss): *Pflanzensoziologie,* a cornerstone of the Zurich-Montpellier school of vegetation study.

1929. The first edition of *Plant Ecology* by John E. Weaver and Frederic E. Clements, the most influential textbook of the pre-World War II period, was published.

1929. W. I. Vernadsky (Russian) dealt with the earth as an ecological system and coined the term "biosphere."

1930. *The Genetical Theory of Natural Selection* by R. A. Fisher (English) set the stage for much of the fusion of ecology and evolution that has recently occurred.

1931. V. V. Stanchinsky (Russian): "On the Importance of Biomass in the Dynamic Equilibrium of the Biocenose," which anticipated much of the work on community energetics of the 1940s and 1950s.

1934. G. F. Gause (Russian) in *The Struggle for Existence* examined interspecific competition and predator-prey relations experimentally.

1939. Frederic E. Clements and Victor E. Shelford: *Bio-ecology.*

1942. "Trophic-dynamic Aspect of Ecology" by Raymond Lindeman (American) stimulated the post-World War II boom in research on energy relations in ecosystems.

1949. The encyclopedic *Principles of Animal Ecology* by W. C. Allee, Alfred E. Emerson, Orlando Park, Thomas Park, and Karl P. Schmidt (American) was published.

1950. E. Lucy Braun (American): *The Deciduous Forests of Eastern North America.*

1953. Eugene P. Odum (American): *Fundamentals of Ecology,* the most influential ecology text of the mid-20th century.

1954. Realism in the study of population ecology was emphasized in *The Distribution and Abundance of Animals* by H. G. Andrewartha and L. C. Birch (Australian).

1955. First paper by R. H. MacArthur (American), "Fluctuations of Animal Populations and a Measure of Community Stability," was one of the stimuli for studying diversity-stability relations of communities.

1957. "Concluding Remarks" by G. Evelyn Hutchinson (English) to the Cold Spring Harbor symposium on *Population Studies: Animal Ecology and Demography* set off a great deal of work on community organization.

1959. A model for the descriptive ecology of a region as well as a statement of the continuum view of vegetation was provided by John T. Curtis (American) in *The Vegetation of Wisconsin.*

1964–1974. The International Biological Program, established with the theme "The biological basis of productivity and human welfare," generated many studies of energy flow and nutrient cycling in various biomes.

1963. W. D. Hamilton's "The Evolution of Altruistic Behavior" provided a new direction for thinking about evolution, leading to sociobiology.

1964. Paul R. Ehrlich and Peter H. Raven (Americans) published "Butterflies and Plants: a Study in Coevolution," one of the roots of a greatly expanded interest in the study of coactions.

1966. In "On Optimal Use of a Patchy Environment," R. H. MacArthur and E. R. Pianka (Americans) emphasized the role of natural selection in molding an organism's use of time and energy. In the succeeding development of foraging theory papers in 1969 and 1971 by T. W. Schoener (American) were also important.

1967. "Nutrient Cycling" by F. H. Bormann and G. E. Likens (Americans) described the early results from the Hubbard Brook watershed study begun in 1962.

1977. J. L. Harper's (English) *The Population Biology of Plants* pulled together material that set the stage for a new major industry of plant demography.

1985. The Society for Conservation Biology was formed, publishing *Conservation Biology* (1987–) and providing a focus for the scientific study of biodiversity, extinction, and related topics.

The 1960s and 1970s were a period of rapid growth in both fundamental and applied ecology. The Ecological Society of America, founded 1916 to 1917 with 307 members, still had fewer than 2000 at the end of the 1950s. During the 1960s and 1970s it grew at an average rate of 250 members a year, peaking in the mid-1970s at better than 6500. Some of these were in the traditional careers for ecologists—teaching and research in universities and research and application in governmental agencies. Many, however, went into careers newly informed by the ecological viewpoint—land-use planning, for example—or into totally new fields, such as environmental law, ecological consultation, and public-interest promotion of environmental quality.

Support in the executive branch of the federal government for environmental improvement was weak from 1980 to 1992. Although opinion polls showed that the public's desire for a better environment never wavered, the highly visible and vocal public support of the 1960s and 1970s declined in the face of governmental apathy or hostility. Individuals and organizations dedicated to the environment, such as the Nature Conservancy, the Natural Resources Defense Council, and Greenpeace continued their work. Recently, the degenerating global situation—acid rain, the greenhouse effect, the loss of biotic diversity—has renewed the popular environmental movement. *Time* magazine chose not a man of the year in 1989, but a planet—Endangered Earth. The state of the earth was the subject of the June 1992 United Nations Conference on Environment and Development—"The Earth Summit"—in Rio de Janeiro.

In the past quarter-century, some environmental problems have been solved or helped, others remain, and still others loom on the horizon. Ecologists need to work to find solutions. They must also work to educate all persons about the connections between ecology and their own lives so that all of us can try to live, do business, and vote in ways that make solutions possible.

Ecology Today

"Ecology" went from a word that few people knew to one that is widely misused (Figure 1-3). Ecology is a science. It is not synonymous with environmental concern, although it has a great deal to contribute to that worthy activity. "Environmental science" refers to all the sciences that help to explore environmental relationships; besides ecology, which provides the conceptual framework, are included parts of geology, climatology, anthropology, chemistry, and physics. A good term for environmental science plus all the other fields that have an interest in human relationships with environment is "environmental studies." A list of these fields would run from economics to religion and back again.

The application of ecological thought to societal problems is one important direction in ecology today. All three of the items listed as a proposed research agenda (the **Sustainable Biosphere Initiative**) for the 1990s by the Ecological Society of America (ESA Committee 1991) are aimed in this direction. The three areas for study are ecological aspects of global change, the ecology and conservation of biological diversity, and strategies for sustainable ecological systems. Biological diversity—understanding it and preserving it—is one of the important topics of the newly defined field of conservation biology. Conceptually, all three areas of the Sustainable Biosphere Initiative emphasize complex interconnections. How will changing carbon dioxide levels in the atmosphere, for example, affect species composition of vegetation directly or by global warming, how will animal populations be affected by vegetation and climatic change and by altered interactions with one another, and what ramifying effects will these changes have on ecosystems and human lives?

Two other currently important directions in ecology as a whole are (1) continued interest in interpreting ecological phenomena in terms of evolution and (2) continued attention to scientific rigor. Scientific inference is discussed a little further in the next section.

Evolution by natural selection is the great unifying generalization of the whole of biology. Reformulation of various ecological ideas, such as succession and population regulation, in realistic evolutionary terms has allowed ecologists to throw out certain long-held but erroneous ideas and yielded clearer statements of others. The past few years have seen an enormous increase in the use of

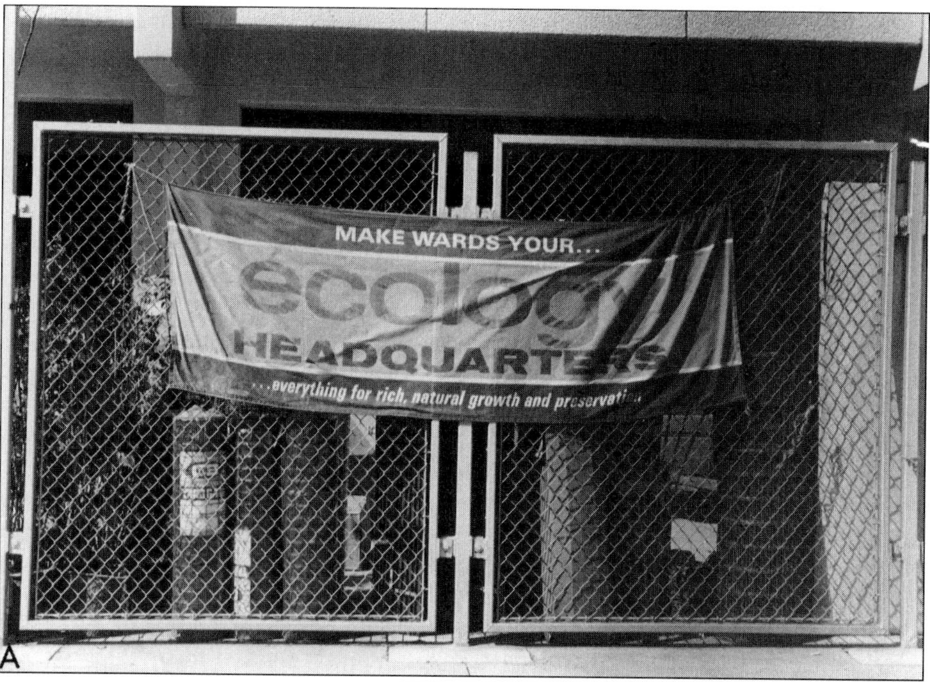

Figure 1-3

Few ecologists would agree that Montgomery Ward is anybody's ecology headquarters, but few will object to the public's casual use of the term as long as it indicates a knowledge of the earth's environmental problems and support for efforts to solve them. (Photo by the author.)

quantitative genetics to study ecologically relevant evolutionary questions.

One other trend that must be mentioned is the increased use of molecular techniques to study ecological and evolutionary problems (Chasan 1991). For example, DNA fingerprinting helps in clarifying mating systems (such as "monogamy"), constructing phylogenetic trees (increasingly important in inferring connections between environment and morphology or behavior), and genetic diversity of populations (with its relevance to extinction). Chemicals produced as a response to stress may be useful in such tasks as identifying limiting factors.

Scientific Inference

Great progress has been made in recent years in making sure that "scientific natural history" really is scientific. Science is a method of learning the nature of—that is, the truth about—the universe. Most college students have been exposed to the scientific method in one or more beginning science classes. They may think that they know what science is and probably do know enough to pass some multiple-choice questions; nevertheless, a thoroughly scientific way of thinking does not come easily to most of us.

Partly this is because science is new in human history. Although there are some glimmerings in the writings of the Greeks, very little worthy of the name science went on prior to about 1600. This is not to say that nobody knew anything before 1600 but rather that they knew a lot of silliness along with the truth, and, what is more important, they had no clear way to tell the difference.

How does science differ from other ways of knowing the truth, or gaining knowledge? If we want to know the answer to some specific question about the universe—how do birds know when to migrate south—we can approach getting the truth in various ways. We can meditate and hope for some sort of revelation. We can apply common sense (we go to Florida when it gets cold, so . . .). We can take a vote.

One of my colleagues of past years taught a general biology course in which the class, in groups of four or five, dissected a cat. He would sometimes be asked, "How do we tell if it's a boy cat or a girl cat?" He would suggest they decide by voting—three male, two female; it's a tomcat. The students would carry out the dissection and then, no matter what the vote had been, the sex would be established by direct observation. If the cat turned out to have ovaries, oviducts, and a vagina, it was a female even if the vote had been 5 to 0, male.

The point is that science tests its knowledge. It compares its ideas with reality. Ecologists come up with explanations or **hypotheses** by any of a variety of means, but they then test them **empirically.** That is, they go into the field or laboratory and make observations that can potentially disprove the explanation. The observations must be of a sort that other persons can repeat and, thereby, **verify.**

None of this implies that science is, at a given time, exactly right in its explanations. In fact, through much of the early part of science, its explanations were no better than those of magic, superstition, folklore, or common sense. What *is* implied is that science corrects itself. A wrong idea gets tested and rejected or modified, so that science gets progressively closer to the truth.

Many scientists adhere more or less consciously to a version of scientific inference outlined by Karl Popper (1968), a German philosopher of science. This is the idea that testing, hence science, proceeds by **falsifying** hypotheses. If we test various competing hypotheses for some phenomenon, the one for which no disproof is forthcoming is taken as the correct explanation.

The basis of this idea is that generalizations about nature usually cannot be directly confirmed. We could only confirm the generalization that humans have ten fingers by examining every human in the world. This, conceivably, we could almost do (though the backtracking to check new babies would probably defeat us in the end); however, we would probably be able to disprove the statement after looking at a few hundred persons and finding one with 9 or 11 fingers.

Hypotheses that are immune to falsification are of no interest. We cannot test a hypothesis that our destiny is ruled by the stars because, if this is true, the stars must control our hypothesis testing just as they do other parts of our destiny.

We cannot test the statement, "Two species that occupy the same ecological niche cannot indefinitely coexist." Ecological niches, by definition, have n dimensions, so two coexisting species may have the same niche for 50 dimensions but differ in the 51st and, consequently, have an excuse for coexisting. Also, "indefinitely" sets no time limit, so two species may coexist for a long time but one may be about ready to go extinct if we wait a little longer.

Skepticism is an important attribute of a scientist. Scientists are correctly skeptical about nontestable ideas. Scientists should place no credence in a hypothesis until it has been subjected to testing and, in fact, should continue to be cautious until the testing has been repeated. Ideas that conflict with established theory, such as spiritualism or the healing power of crystals, should be met with doubt.

This skepticism, to the point of dismissiveness of heterodox ideas, is what explains a peculiarity of how science tends to advance. Established ideas, or generally accepted models, are termed **paradigms.** Once a paradigm has been established, once a particular explanation has been adopted, the conventional scientist is highly dubious about any observations that seem to conflict with the paradigm. Only overwhelming evidence is likely to lead to a replacement of the old paradigm with a new one, a **paradigm shift** (Kuhn 1970).

One example of this phenomenon was the paradigm of the fixity of continents. Observations that seemed to conflict with this view, such as similar Permian fossils in India, Africa, Australia, and Antarctica, were explained away or considered to be mistakes. Observations best explained by continental drift continued to accumulate, a mechanism for drift (sea—floor spreading) was proposed, and over the course of about a decade, the paradigm of continental fixity was replaced by plate tectonics as a way of viewing the earth.

A paradigm that held sway during the days I was taking chemistry classes was that the noble gases (helium, xenon, etc.) were inert. When xenon tetrafluoride was synthesized, an editorial in *Science* magazine (Abelson 1962) denounced chemists of the preceding 50 years for simply accepting the dogma of the nonreactivity of the "inert" gases. "What is really needed," the editorial said, "is more healthy skepticism."

But that is exactly wrong (Brewer 1992). The chemists had plenty of skepticism; they were skeptical that elements having an outer shell of eight electrons could be reactive and highly skeptical of anyone who said anything different. The deficiency was not of skepticism but of **open-mindedness**.

Skepticism is a scientific virtue but it is not the same thing as closed-mindedness. It is just as unscientific to dismiss an idea that has not been satisfactorily tested as it is to accept it. Combining skepticism with open-mindedness is a difficult prescription, but the best science couples a free market in ideas with rigorous testing. There is more about scientific thought in ecology in Appendix 1 of *A Laboratory and Field Manual of Ecology* by Brewer and McCann (1982).

Ecology as System

The following chapters will examine the science of ecology at three levels. The ecology of the **individual organism** will be considered first. The second level deals with the ecology of groups of individuals, or **populations.** Populations of several kinds of organisms, several species, live together in **communities,** the third level. At this same level we may consider the community along with its physical setting or habitat as a single, interacting unit, the **ecosystem.** A pond is a familiar ecosystem, with the plants and animals and bacteria forming the community, and the water, the dissolved salts and gases, and the mud of the bottom being elements of the rest of the system. We can recognize communities and ecosystems of various scales, from the stomach of a cow, with its interesting populations of microorganisms, on up to the earth itself, the largest ecosystem with which we are familiar. The entire global environment supporting life from the depths of the oceans and as far down in the soil as organisms occur up to the highest part of the atmosphere occupied by organisms is termed the **biosphere.**

One of the great lessons of ecology is the interrelatedness of nature. Many good examples of such interrelatedness will show up in later pages of this book. For now, one of the earliest published examples will do. In Chapter 3 of *The Origin of Species,* Darwin told how experiments of his had shown that seed production by certain plants such as red clover are dependent on pollination by bumblebees (or humble-bees, as they are called in England); other kinds of insects do not visit them because they cannot reach the nectar.

Darwin suggested that if bumblebees were to become extinct, the numbers of red clover would probably decline. Quoting Col. Newman, "who has long attended to the habits of humble-bees," Darwin connected the numbers of bees with the numbers of field mice, which destroy the combs and nests of the bees. The number of mice, in turn, depends on the number of cats that eat the mice that destroy the nests of the bees that pollinate the clover. "Hence," wrote Darwin, "it is quite credible that the presence of a feline animal might determine the frequency of certain flowers." But the cat population is determined mainly by the number of households that keep cats, so that seed production by red clover may really depend on the number of cat lovers in the district.

Darwin is not known as a kidder, but we may suppose that he was smiling when he wrote this story connecting the commonness of red clover to the numbers of spinsters in a region. The point of the story, however, is serious. It was to show "how plants and animals, remote in the scale of nature, are bound together by a web of complex relationships." A little more than a hundred years later, Barry Commoner made the same point in his "first law of ecology" (in *The Closing Circle*). Everything, Commoner wrote, is connected to everything else. When we touch one part of any ecosystem, we eventually and in some way touch the rest.

This point is worth emphasis because of the tendency of most of us to think in terms of simple, linear, cause-and-effect relationships rather than of interacting systems. At the same time or, perhaps, as the next step, we need to realize that the effect of some connections may be strong and of others weak or negligible. We do not know whether red clover actually declines in a region of cat haters until we try the experiment. A part of the day-to-day work of ecology is assessing the strength of the multitude of possible interconnections.

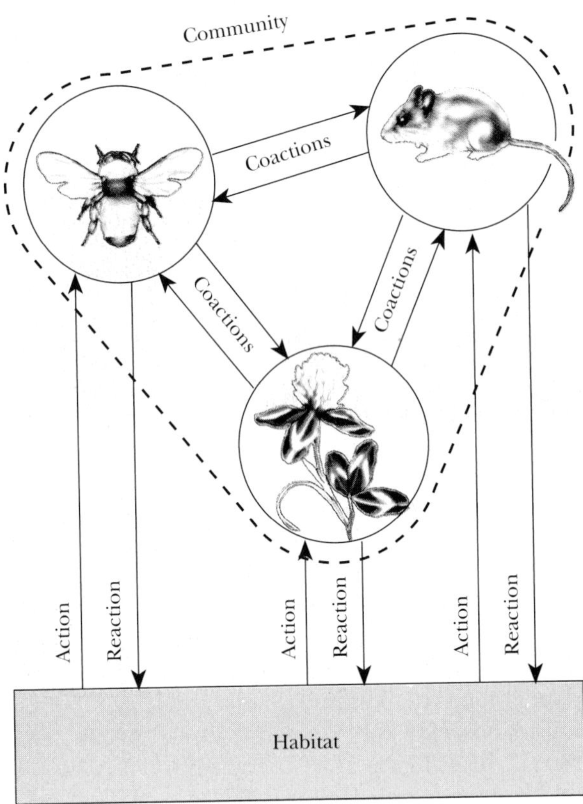

Figure 1-4

Interactions in the ecosystem.

Interrelationships Within the Ecosystem

F. E. Clements recognized three types of interactions within the ecosystem that form the bases for its structure and function (Figure 1-4). The physical environment provides conditions under which organisms function, live, and die. The effects of the physical environment on organisms are termed **actions.** Included are all the ecological influences of such factors as temperature, pH, and photoperiod, discussed in Chapters 2 and 3.

The reciprocal interactions, the effects of organisms on their environment, Clements termed **reactions.** Reactions, such as shading and soil formation, are discussed in Chapter 12. Interactions within the community of living organisms, the effects of organism on organism, are **coactions.** Coactions—herbivory, predation, competition, mutualism, etc.—are described in Chapters 6 to 10.

Ecology of Individual Organisms:
Principles

OUTLINE

- Relationships to the Abiotic Environment
- Range of the Optimum
- Energy Balance
- Animal Behavior
- Evolutionary Considerations
- Habitat Selection
- The Spread of Organisms

2

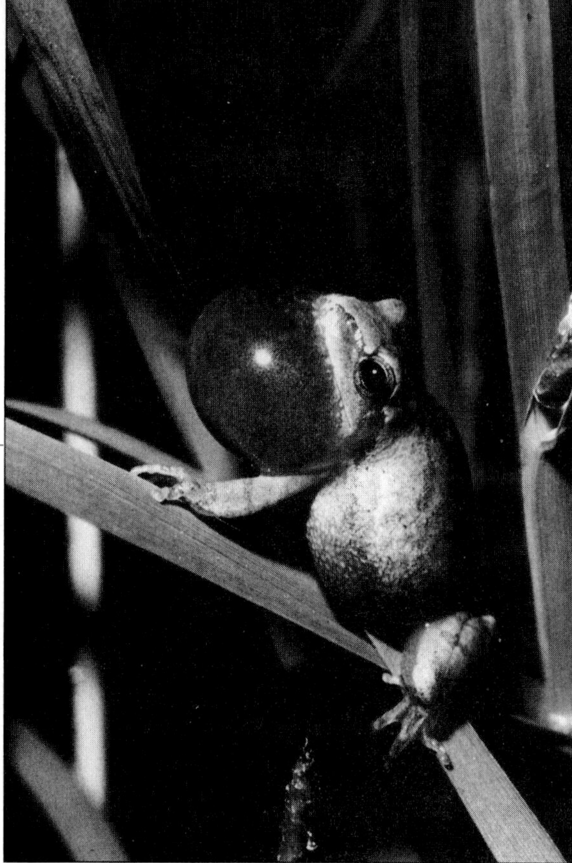

As soon as temperatures warm in the spring, spring peepers come out of hibernation and begin breeding behavior. (*John Gerlach/Dembinsky Photo Associates*)

Individual animals and plants have relationships with their physical environment and with other organisms. Individuals sprout or hatch, they disperse, they live more or less prosperously depending on the conditions they encounter. They reproduce, or not, and die.

The ecology of individuals (termed **autecology** by early plant ecologists) tends to be neglected as a subdivision of ecology for two reasons: Much of what happens to the individual is rooted in physiology; consequently, a great deal of autecology is included within the discipline of environmental physiology, or physiological ecology. Because the outcome of the interaction of individual with environment is expressed in population structure or dynamics, much autecology tends to get telescoped into the next level of the hierarchy, population ecology.

It is, nevertheless, ordinarily the individual that meets the environment. Even though most discussions in autecology lead quickly to physiology or population ecology (among other fields), individual ecology is the logical starting point for any general discourse on ecology.

Relationships to the Abiotic Environment

Tolerance Range

We refer to the surroundings of an organism as its **environment** or **habitat.** The basic subdivision is between **terrestrial,** or land, habitats and **aquatic,** or watery, habitats. A trout is an aquatic organism, living in cool streams, whereas a trout lily is terrestrial, living in moist forests. Aquatic habitats are often subdivided into **marine,** or ocean, habitats, and **fresh-water** habitats which are lakes, ponds, streams, and springs whose waters usually have low salt concentrations. Much finer subdivisions can be made—habitats may be subdivided into subhabitats and eventually **microhabitats,** such as a south-facing slope, a tree hole, or the spaces between the sand grains along a beach.

Specific features of the habitat may be studied. For the trout we can measure such factors as water temperature and the amount of oxygen dissolved in the water. For the trout lily we can measure the amount of light reaching it on the forest floor and the concentration of calcium in the soil. If we are able to determine the whole range over which the species is able to live, for one of these factors, we then will know the **range of tolerance** of that species for that factor. A goldfish, for example, might be able to live in waters in which the temperature ranges from 2 to 34°C (about 36 to 93°F), but it will die of heat or cold at temperatures above or below this range.

Field observations give us some idea of the range of tolerance of a species, but the field work must be checked by experimentation in the laboratory and in the field. One of the reasons for this is that organisms interact with other organisms. The habitat features we have mentioned so far have all been **abiotic factors,** or factors of the physical environment. The organism also exists in a world of **biotic factors,** consisting of its relationships with other organisms. These relationships are of several kinds: one organism may use another as food and at the same time serve as food for a third; two organisms may compete for food, for nesting sites, or for some other requirement. An organism may not be able to live in an area where physical factors are favorable for it due to some biotic factor.

Plants that tend to be found growing only in acid soils are called **calciphobes.** Some of these species may do so because they are unable to tolerate neutral or basic soils but this is not always the case. In his ecology lectures (I have been told), Henry Chandler Cowles described a visit to the botanical gardens in Berlin, which had an excellent collection of calciphobes and their opposite number, **calciphiles,** characteristic of basic soils. Cowles questioned one of the gardeners, asking if it wasn't difficult treating the soil in the different beds so as to maintain each at just the right pH. Not at all, the gardener is supposed to have answered; we grow them all in the same soil.

Clearly such species in nature are being restricted to just one end of their tolerance ranges. Competition from the many species that do well in average favorable conditions is probably often the cause of such restriction to extreme habitats.

This is not to say that pH or other abiotic factors do not limit plant or animal distribution. Often, the apparent physical cause of an organism's restriction is the real one. The absence of lichens from areas of air pollution was noted in the 19th century, and research since then has tended to confirm that sensitivity to sulfur dioxide is the main factor in this distribution (Anderson and Treshow 1984).

In other cases, the apparent abiotic causes are not borne out when experiments are done. An annual plant called sea rocket along the Pacific Coast of the United States grows only on open, bare sandy beaches in the salt-spray zone. When M. G. Barbour and his coworkers (1973) at the University of California at Davis planted seeds in various habitats, they found that the seedlings grew best inland in good grassland soil (Figure 2-1). This was true, however, only when the area of grassland soil was weeded of competitors; in undisturbed grassland, the sea rocket seedlings soon died. At a later stage, another biotic factor came into play; grazing by small mammals was severe on the sea rockets growing on the grassland site.

Tolerance ranges differ from one species to another. We would expect carp (goldfish), for example, to be able to tolerate warmer temperatures and lower oxygen concentrations than trout, and this is exactly what we find when we make the appropriate tests (Figure 2-2). The fact that different species have different tolerance ranges is one basic reason

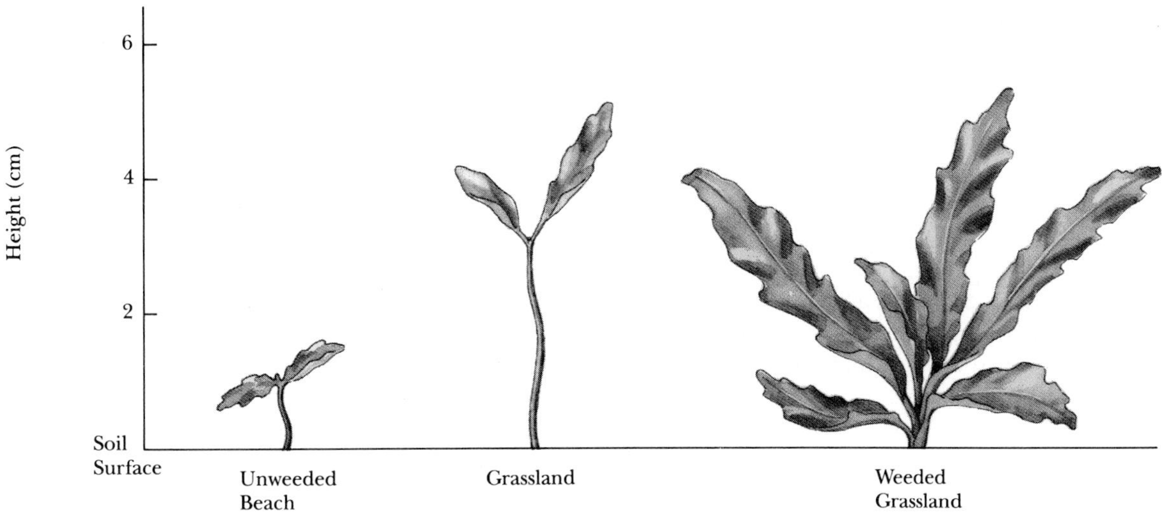

Figure 2-1

Sprout development by sea rocket seedlings six weeks after being sown in different habitats. Plots in grassland soil from which competing plants had been removed produced seedlings that were larger and healthier looking than those in the normal beach habitat. The seedlings in grassland plots from which competitors had not been removed were tall, spindly, and pale; two months later they were all dead. (Adapted from Barbour et al. 1973.)

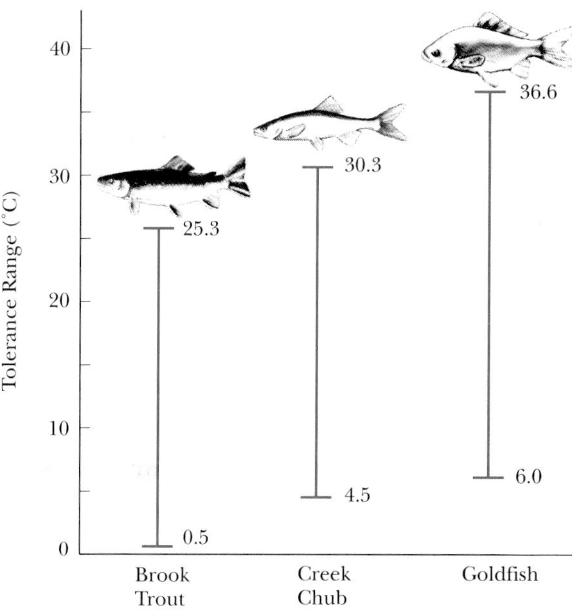

Figure 2-2

Different species have different ranges of tolerance. The goldfish is able to live in warmer waters than the brook trout but does not tolerate cold temperatures as well as the brook trout. (Adapted from Brett 1956, p. 75.)

why different habitats support different communities, and why communities vary geographically from warmer to cooler and from wetter to drier regions.

Range of the Optimum

Generally, an organism does not do equally well throughout its whole range of tolerance. Instead there is some part, usually toward the middle, where it does better (Figure 2-3). This section is referred to as the **range of the optimum.** If we look at the rate of any one process at different levels of any one factor, we can usually specify an optimum range with no trouble. For example, we may find that pea plants photosynthesize fastest around 30°C. Does this mean that this is the optimal temperature? It is for photosynthesis, but it may not be the most favorable temperature for another function, such as growth. Photosynthetic rate and respiration rate both go up with temperature, but photosynthetic rate drops at higher temperatures whereas respiration rate keeps increasing. When these two curves are combined, the temperature at which growth is greatest will be below 30°C, perhaps 25°C (Figure 2-4).

Figure 2-3

When some measure of success of an organism is plotted against different levels of an environmental factor, a bell-shaped curve often results. It rises from zero success at the lower tolerance limit to high values in the range of the optimum and then falls again to zero at the upper tolerance limit.

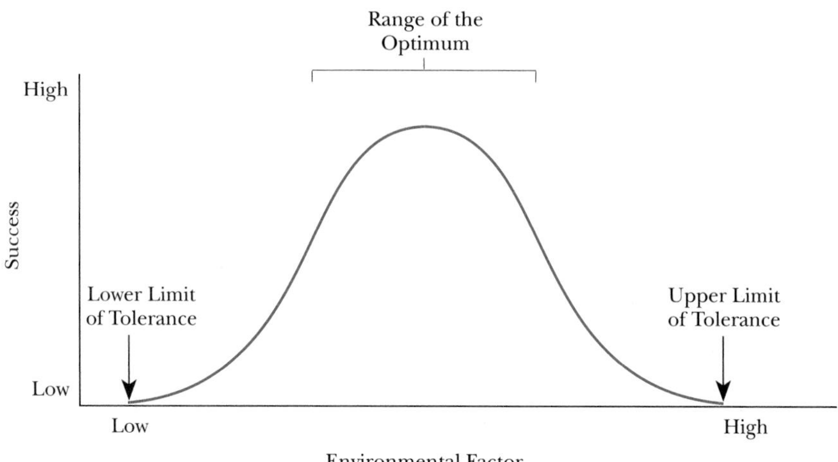

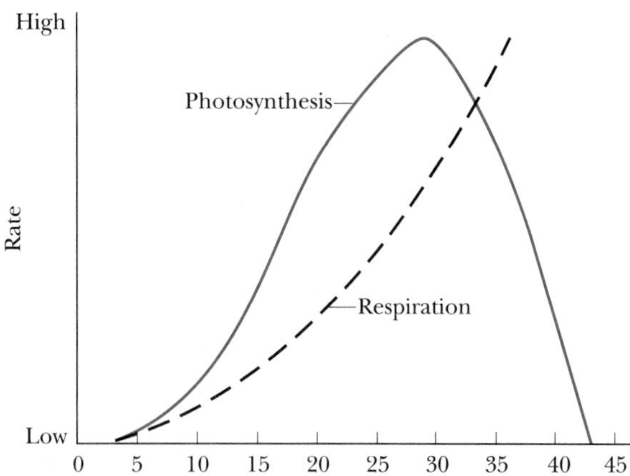

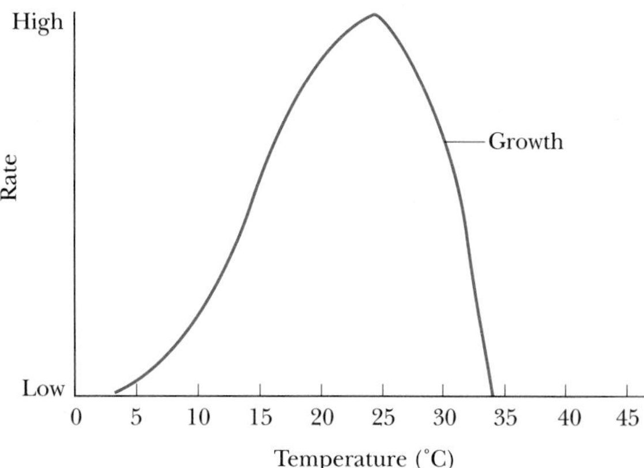

Figure 2-4

Optimal conditions may differ for different aspects of the life of organisms. The optimum temperature for photosynthesis is about 30°, shown in the top graph. Growth rate, the difference between photosynthetic rate and respiration rate at a given temperature, is shown in the bottom graph. It peaks at 25°.

So, is 25°C the optimum temperature for pea plants? They grow fastest at that temperature, but is this the temperature at which they reproduce fastest or live longest? Possibly not. Quite different temperatures may be optimal for growth and for flowering, for example. Even considering just growth rate, the example is oversimplified because the effects of day-night temperature changes (thermoperiodism) are not considered.

The ecological optimum is a useful concept but a blunt one, like a hammer. When precision is needed, some other tool may be better.

When Conditions Change

Few habitats stay the same for very long. Some, such as the ocean depths, come close, but for most habitats environmental factors change between day and night, between drought and a wet period, between summer and winter. When some important feature of the habitat changes, the organism changes in response. The change may be mainly **physiological.** As the air temperature warms up during the day, a plant and a bird both exposed to the sun will undergo physiological changes because of the heat. These changes often have the effect of keeping certain important aspects of the organisms's internal environment constant despite the changing external environment. This tendency, which is an important physiological principle, is known as *homeostasis.* The bird may respond by lowering its rate of heat production and by arranging its feathers so as to lose heat rapidly, resulting in its body temperature remaining constant.

If the sun keeps shining and the temperature keeps rising, an important difference between the plant and the bird or the trout lily and the trout may be seen. The plants are stationary, attached to one spot **(sessile),** whereas the animals can move about **(motile).** The bird or trout can move away from unfavorable conditions—the bird can go into the shade; the trout can swim to cooler water. These are **behavioral** changes that occur in addition to the purely physiological ones. The range of behavioral responses for a sessile organism is much smaller.

Phenotypic Ways of Coping With Changed Environments

If some environmental factor seasonally shifts beyond the tolerance range of an organism, it can do various things:

1. It can fail to cope, in which case it dies. In this case, the species disappears from the biota of the area unless it reinvades.
2. The adults can die but some other life history stage, such as eggs, pupae, or seeds, with broader tolerance limits survives. This strategy can provide for the regular yearly occurrence of organisms such as many insects of the forest and for annual herbs. It may also tide organisms over a longer period of unfavorability. For example, many plants produce seeds that remain viable in the soil for years or even decades (termed a **seed bank**) (Leck et al. (1989). An open-country plant may persist in the seed bank of a site made uninhabitable by the growth of trees and reappear if the trees are cut or burned.
3. It can hibernate or enter some other type of resting state (Storey and Storey 1990). It can move to a protected spot below the frost line in the soil where the temperature is low enough that the animal has a slow, energy-saving metabolism but not low enough to kill it. Many snakes, turtles, frogs, woodchucks, and invertebrates cope with cold seasons by hibernation. **Aestivation** is a similar inactive state shown by various animals in response to intolerably hot or dry conditions.
4. It can migrate. Many birds live in one geographical area in the summer and another in the winter; so do a few mammals and insects. The scale of migration varies enormously. Olive-sided flycatchers may migrate from Canada to Ecuador, whereas some soil invertebrates merely move below the frost line.
5. It can acclimate.

This last way of coping with environmental change is less a part of common knowledge than hibernation and migration. The range of temperature tolerance of a fish (or an insect or a pine tree) may vary from season to season, depending on the temperature at which it has recently been living (Figure 2-5). In midwinter, a fish may be able to live in the range from 0 to 24 °C, but in summer the same individual may have a tolerance range that extends all the way up to 33 °C but down only to 15 °C. Such basically physiological adjustments to a changed environment are known as **acclimation.** Acclimation does not occur just in temperature tolerance; organisms may also acclimate to different oxygen levels in the air (if they go to higher altitudes, for example) and to other changed condi-

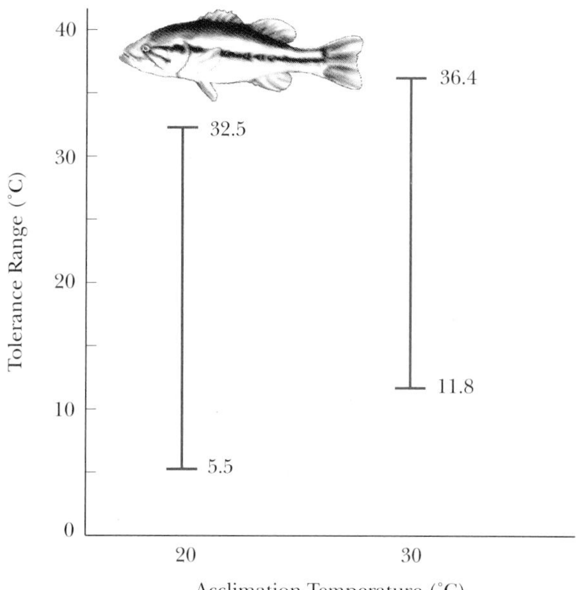

Figure 2-5

The tolerance range of organisms may be changed by acclimation. Kept at 20°C, the large-mouthed bass has an upper limit of temperature tolerance of just over 32°C. If the aquarium temperature is slowly raised to 30°C and kept there for a few days, the fishes now acclimated to 30°C are able to tolerate temperatures over 36°C. At the same time, however, they lose some of their ability to tolerate lower temperatures. (Adapted from Brett 1956.)

tions. Acclimation is of great importance in allowing organisms to exist permanently in changeable environments.

Phenotypic Plasticity

In the long term, evolutionary changes may occur in organisms as a consequence of changed environments. The first installment of our discussion of evolutionary change can be found in a later part of this chapter and the rest in the population ecology chapters. In the preceding section, we talked about one type of nonevolutionary adaptive change in individuals—the physiological changes referred to as acclimation. Acclimation is one category of a broad group of nonevolutionary changes included under the heading of **phenotypic plasticity.**

Phenotypic plasticity is defined as environmentally induced phenotypic variation (Stearns 1989). Phenotypic plasticity, then, encompasses all the sorts of differences that can be shown by an individual (or by several individuals of one genotype, such as a clone) as a result of environment.

In some cases, the resulting variation is discontinuous, such as the production of two different morphological, physiological, or life history types depending on conditions. A striking example of this type of phenotypic plasticity involves a moth feeding on oaks (Stearns 1989). Eggs hatching in the spring give rise to caterpillars that feed on catkins, encounter low levels of tannins, and develop traits that camouflage them among the catkins. Eggs hatching a little later feed on leaves, encounter higher levels of tannins, and develop traits that camouflage them on oak leaves. Evidently tannin level in the food is the developmental switch that leads to the two different morphological types.

One well-studied type of phenotypic plasticity involves the morphological changes of various planktonic animals, such as water fleas (cladocerans) in response to the presence of predators, or sometimes just to the changes in physical factors that will usually correlate with the presence of abundant predators (Dodson 1989). Under these conditions, newly hatched zooplankters grow up to have various sorts of spines or armaments that discourage capture or ingestion by the predators (such as copepods or midgefly larvae).

If the change in phenotype with environment is continuous, the relationship is often characterized using **reaction norms.** The general idea of a reaction norm is shown in Figure 2-6. An individual of genotype A develops one phenotype when it grows up under environment 1 and a different phenotype under environment 2. The environmental features we are graphing may be temperature, crowding, food, or a variety of other things. The phenotypic feature on the other axis could be size, shell thickness, time to metamorphosis, date of egg laying, etc. If we are dealing with a number of individuals, there is likely to be a range of environments at each locality and, accordingly, a range of phenotypes, as in Figure 2-7.

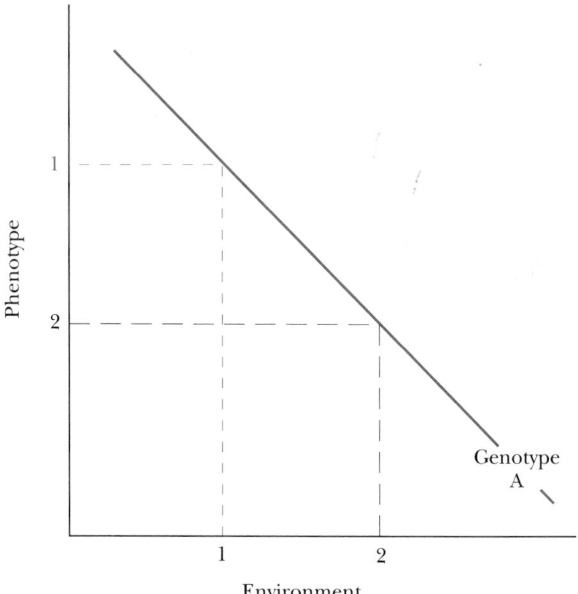

Figure 2-6

A reaction norm. An individual developing in environment 1 develops phenotype 1, whereas if it develops under environment 2, it develops a different phenotype.

If the population contains more than one genotype, there will be a bundle of reaction norm lines rather than one. The reaction norms for the different genotypes may be parallel, or some may cross, as in Figure 2-8. Where this is the case, the relative rankings of genotypes may shift depending on environment. Suppose, for example, that the phenotype in Figure 2-8 is size. In environment 1, genotype B produces the bigger organisms, but in environment 1, genotype A produces bigger ones.

Presumably much of the phenotypic plasticity seen in nature is adaptive; that is, the changes result

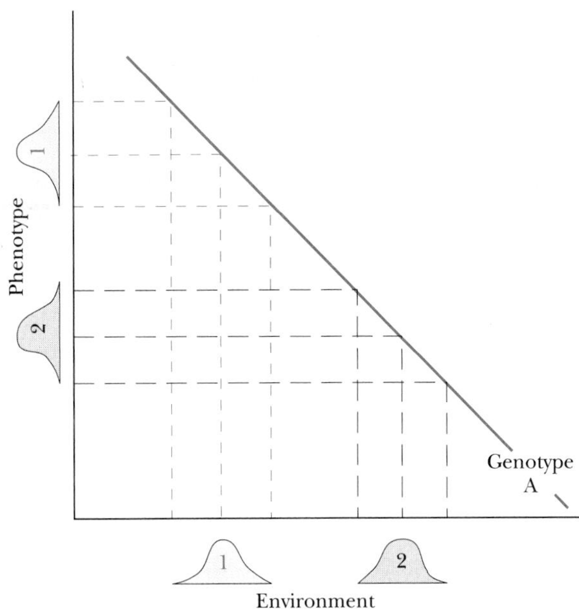

Figure 2-7

Usually a range of slightly different environments is present at any particular locality, hence the phenotypes reflect this range.

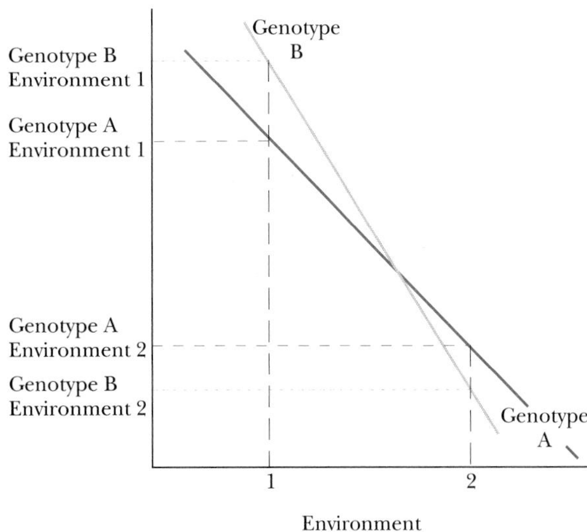

Figure 2-8

Different genotypes will have different reaction norms. If the lines for reaction norms cross, the resulting phenotypes will show these relationships.

in better survival of the organisms. This has occasionally been demonstrated and is logical for many other cases. It is likely that maintaining a broad tolerance range is costly for an organism in energetic or genetic terms. Phenotypic plasticity in the form of reaction norms may allow a given individual to develop a narrow tolerance range that is, nevertheless, adaptive for the environment it is occupying (Gabriel and Lynch 1992).

Extreme Conditions

The ability of many organisms to acclimate to temperature, the salt content of water, and various other factors is one complication that must be dealt with in going from the laboratory to the field—that is, in explaining the occurrence of organisms in nature on the basis of experimental findings. Another fact that must be remembered is that conditions need not be perpetually outside the tolerance limits of a species for the species to be absent. This was pointed out by W. P. Taylor (1934), one of the early students of wildlife management in the western United States. The point is, you only have to kill a plant or animal once. Six days a week a stream may be well within the tolerance limits of a certain fish, but if on Saturday night a factory releases a dose of toxic effluent, the fish will not be found living there. The occasional extreme condition may be more important than the average condition in determining whether or not an organism can exist permanently in a given area.

Daily Fluctuations

It is well to remember that constant conditions, such as are often maintained in the laboratory, are abnormal to most organisms. In most habitats, organisms are exposed to daily alternations of many features related to the change from day to night. This is obviously true of light but, depending on the habitat, it may also be true of such factors as temperature, humidity, pH, and oxygen concentration.

Organisms may respond differently when they experience the alternation of conditions typical of their natural habitat instead of artificial constant conditions. Growth may be faster, reproductive rate may be higher (Hoffman 1947), and a higher percentage of seeds may germinate when subjected to alternating warm and cool temperatures than when

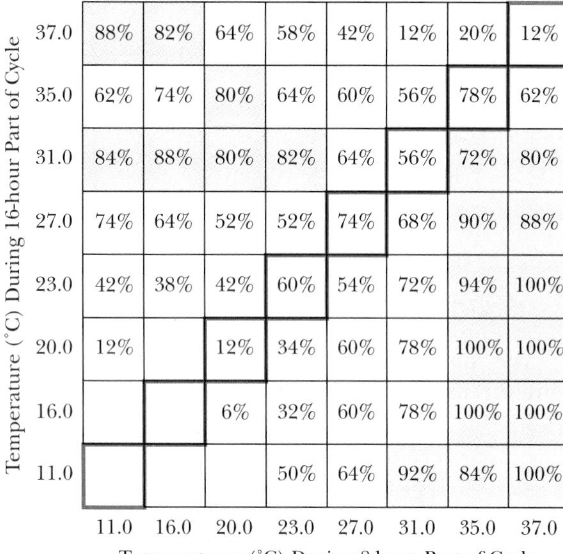

Temperature (°C) During 16-hour Part of Cycle	11.0	16.0	20.0	23.0	27.0	31.0	35.0	37.0
37.0	88%	82%	64%	58%	42%	12%	20%	12%
35.0	62%	74%	80%	64%	60%	56%	78%	62%
31.0	84%	88%	80%	82%	64%	56%	72%	80%
27.0	74%	64%	52%	52%	74%	68%	90%	88%
23.0	42%	38%	42%	60%	54%	72%	94%	100%
20.0	12%		12%	34%	60%	78%	100%	100%
16.0			6%	32%	60%	78%	100%	100%
11.0				50%	64%	92%	84%	100%

Temperatures (°C) During 8-hour Part of Cycle

Figure 2-9

Thermoperiodism in seed germination. Sand dropseed germination was tested at a variety of temperature combinations. If a constant temperature had been best, the highest germination percentage would have been somewhere along the diagonal from lower left (11°/11°) to upper right (37°/37°). Actually germination was poor along this line and the best germination was at temperatures of 35°–37° alternating with temperatures of 11°–23°. (Adapted from Sabo et al. 1979.)

kept at either of the temperatures without alternation (Sabo et al. 1979). Such a physiological or ecological response to alternating temperatures is called **thermoperiodism.** An example is given in Figure 2-9.

Some habitats, such as grasslands, deserts, and shallow ponds, show very large changes between day and night. Certain other habitats may be nearly as constant as laboratory chambers. Some of the more constant natural environments are hot springs, caves, the ocean depths, and the insides of homoiotherms (as a habitat for parasites like blood flukes).

Limiting Factors and the Environmental Complex

So far we have considered only the organism's response to a single factor of its environment, but the environment of every organism is composed of many factors. The law of limiting factors, stated by a plant physiologist named F. F. Blackman (1905), will often help us to understand a particular case. Liebig's law of the minimum is an earlier, less general statement of the same idea. The law of limiting factors states that when some process depends on several different factors, the speed of the process at a given time is determined by the "slowest" factor. The slowest or **limiting factor** may be either too little or too much of something. A process may be limited early in the morning by too little light and limited in the middle of the day by temperatures that are too high.

Good farmers and gardeners understand the law of limiting factors. They know that if their soil is deficient in calcium, adding phosphate will not make their crops grow any better; calcium is the limiting factor, and only if calcium is added will growth improve. If enough calcium is added, then some other factor will become limiting, perhaps some other mineral such as phosphate. If this is the case, then addition of that mineral will now aid growth, even though it had no effect before. The limiting factor may be something other than a mineral; for instance, there may not be enough or there may be too much moisture, or temperatures may be too low or too high.

The law of limiting factors is the basis for the ecologist's concern about phosphates being added to lakes, where phosphate is very often the limiting factor for algal growth. Because the various other materials algae require are generally present in excess (nitrate may sometimes be low), the addition to lakes of phosphates from detergents, lawn or agricultural fertilizers, or sewage results in increased growth. This has several effects that may be undesirable for some lakes: The increased growth in itself may be a nuisance; once the algae have died and fallen to the bottom, their decay depletes the oxygen supply, and this may kill some of the more desirable fishes; and the undecayed algal remains help build up the bottom of the lake so that it fills in faster. The addition of some other nonlimiting material to the lake would be of much less concern, even if that material is used as a nutrient by the algae. Adding calcium to a lake will not usually cause an algal bloom; there is already more calcium there than the algae can use.

The law of limiting factors is simple enough; it

is much the same thing as the old saying that a chain is only as strong as its weakest link. However, it does not always hold true, because of the **interaction** of different factors. The combined effects of two factors cannot always be predicted from a knowledge of their effects singly. There are cases known in which a surplus of one factor may compensate for a deficiency in another. The upper lethal temperature of the American lobster is 29°C when the oxygen content of the water in which it is living is low, but by increasing the oxygen content, the upper lethal temperature can be raised to 32°C.

Two factors may, instead, work **synergistically.** In such a situation the combined effect is greater than the sum of the separate effects (a kind of 1 plus 1 equals 5 situation). Two slightly unfavorable factors may, for example, combine to make an area uninhabitable for some organism. A blue-green alga was found to be able to grow in the presence of either DDT or salt, but when both were present, the cells did not divide—a possibly serious problem in estuaries where DDT used on the uplands is carried to salt water (Table 2-1).

One well-documented example of synergism among humans is that of oral contraceptives and smoking. Women over the age of 40 years who use birth control pills have a death rate from heart disease and blood clots slightly higher than among those women who use some other effective method of contraception. Mortality rate among pill users is 6 deaths per 100,000 women compared with 4 per 100,000 for those who use other methods (both of these are, of course, much lower than the death rate among women who become pregnant after age 40). Women who use the pill and also smoke have a death rate from heart disease and blood clots of nearly 60 per 100,000, ten times higher than among nonsmokers.

Table 2-1
Synergistic Effect of DDT and Salt (NaCl) in Inhibiting Algal Growth

The alga used was *Anacystis nidulans,* a blue-green alga.

Treatment	*Growth Rate (Number of Doublings per 24 Hours)*
No DDT, no NaCl	102
DDT (800 ppb), no NaCl	89
NaCl (1%), no DDT	26
DDT (800 ppb) + NaCl (1%)	0

Calculated from Batterton, J. C., Boush, G. M., and Matsumura, F., "DDT: inhibition of sodium chloride tolerance by the blue-green alga *Anacystis nidulans,*" *Science,* 176:1141–1143, 1972.

Ecological Indicators

"Every plant," wrote J. E. Weaver and F. E. Clements in their classic textbook *Plant Ecology,* "is a product of the conditions under which it grows and is, therefore, a measure of environment." Although this is true, in practice we are interested in plants or animals (or their responses) that indicate some specific trait of its community or the habitat. We are interested in a species or a combination of species that tells us that the soil is salty, that a pasture is overgrazed, or that a stream is polluted.

The ideal indicator, of course, would always be associated with the condition in which we are interested and would never occur in its absence. The tendency of the organism to occur only with the condition it is supposed to indicate corresponds with the plant sociologists' concept of **fidelity.** The frequency with which the indicator is associated with the condition corresponds to the phytosociological concept of **constancy.** A species might, for example, occur exclusively on one type of soil or in one kind of community (high fidelity) but be rare enough that patches of the characteristic habitat often fail to support the plant (low constancy).

The recent tendency to use ecological information for practical purposes is not new. The period from about 1945 to 1970 in which it was thought that chemical and physical technology would allow us to ignore ecological relationships was an aberration that it is hoped will not be repeated. Early in this century—and early in the development of ecology—the work of H. L. Shantz (1911) on plant communities of the Great Plains as indicators of crop capabilities was used for classifying public lands under the Stock Raising Homestead Act of 1916. Clements published a 400-page monograph on plant indicators in 1920, with most of it devoted to agricultural indicators such as indicators of crops and overgrazing.

As examples of indicator species, heavy grazing

in the dry prairies of Wisconsin used as pasture is shown by a decrease in little bluestem, an increase in annual ragweed, and the invasion of dandelion and white clover. Other indications of overgrazing, in the Midwest or elsewhere, are an increase or invasion of spiny plants such as thistles and prickly pear cactus, and also an increase in unpalatable species such as milkweed.

Lately there has been a great deal of interest in using biological indicators of pollution, sometimes now referred to as **biomonitoring.** This, too, is a revival of a subject that German and American aquatic ecologists began to study in the early 1900s. It has long been known that dense populations of certain kinds of annelid worms (''sludge worms'') or certain kinds of fly larvae on stream bottoms generally indicate pollution by organic wastes, such as sewage (Wilhm 1975). These tend to be organisms that have breathing mechanisms that allow them to stay alive under low oxygen conditions. If gill-breathing mayflies, stoneflies, and caddisflies are present, on the other hand, the water is likely to be clean and well oxygenated. Subdivisions between these extremes can also be recognized. Another approach to the problem of a pollution indicator has been to use a mathematical index based on the number of species and their relative abundances. This is discussed in a later section on diversity indices (Chapter 10).

A fundamental difference between ecological indicators of pollution and indicators of naturally occurring circumstances, such as serpentine soil, is that the association of species with pollution is a new phenomenon in their evolutionary history. Consequently, the species that are common in polluted sites also occur in specific microhabitats where pollution is absent. Sludge worms, for example, should not be looked down on because they grow in dirty rivers; they also occur, in fairly small numbers, in the muddier, slow sections of perfectly clean ones. The mere presence of sludge worms is not a sign of polluted waters, though high numbers of them coupled with the near absence of other species usually is.

As a general rule, we are less likely to make such a mistake if we use information beyond the mere presence or absence of a single species. Comparisons of abundance are helpful as can be the use of a combination of species (Cairns et al. 1985) or a community as the indicator.

The examples given so far involve the presence or absence of naturally occurring organisms. Another approach to biomonitoring is to sample these organisms, such as earthworms, bees, or European starlings, and find out what the heavy metal or pesticide load in their tissues is or how many and what kind of tumors they have (Root 1990). Still another approach is to place an organism known to be sensitive to the specific pollutant of interest in the environment. Taking canaries down into coal mines is a prescientific example of what we would now call a **bioassay.** Confining fish in cages in polluted streams and observing their survival or the rate at which they accumulate poisons, such as dioxin, is a common application of this use of ecological indicators.

Ecological indicators have been used to determine the depth of ground water, map soil types and glacial deposits, and determine types of rock formations, including prospecting for uranium (Viktorov et al. 1965). One interesting application is the use of the succession of carrion insects on corpses for forensic (legal) purposes, such as determining how long a body has been dead (Greenberg 1991).

The applications of ecological indicators do not have to be practical, of course. By the use of evidence of fire, of former cultivation, and the like, we can often reconstruct a great deal of the history of an area. There is a fascinating book, *Reading the Landscape of America* by May T. Watts (1975), devoted to this subject.

Energy Balance

Energy and Work

The functioning of every organism is based on energy. Don't worry if you don't have a thorough grasp of the concept of energy; almost no one has. You should, however, try to avoid being completely ignorant of the subject, because such ignorance has led to some of our current environmental problems. The conventional definition of **energy** is the capacity to do work. **Work** in this sense can be thought of as moving something; the something may be large like a locomotive, but it may also be small like the molecules moved around by an organism to construct a new cell or repair an old one.

In general, work is done when energy is changed from one form to another. If we pick up a ball and put it on a table, **potential energy** in the form of chemical bonds in our body is transformed in a series of steps to the energy of position of the ball (with a considerable loss of energy from our muscles as heat). If the ball rolls off the table and falls to the floor, its potential energy is converted to **kinetic energy** and is lost as heat by friction with the air and friction between molecules in the ball and floor.

The basic physical principles describing the transfer of energy are called the first and second laws of thermodynamics. It is sufficient for our purposes to say that the first law of thermodynamics indicates that energy does not vanish. If there is a certain amount of potential energy in a system, and an energy change occurs, then the work done and the heat produced will equal the amount of potential energy lost. The second law of thermodynamics states that no process is 100% efficient in converting potential energy to work; in other words, some of the original energy is always lost as heat.

Because 100% conversion of other kinds of energy to heat is possible, units of heat are convenient measures of energy. Biologists have tended to use the **calorie,** which is the amount of heat required to raise 1 gram of water 1°C, starting at 14.5° C.* Because this unit is so small, the actual unit used is the Calorie, with a capital *c,* also known as the kilocalorie, which is equal to 1000 small calories. The Calorie, or kilocalorie, is the unit most of us are familiar with from watching our diet. A quarter pound of hamburger, for example, has about 300 Calories.

* There is a trend toward using the **joule** as the only unit for expressing energy, in conformance with the "International System of Units (SI)" set forth by the *Cónference Générale des Poids et Mesures.* The joule, defined as the work done when the point of application of 1 newton is displaced a distance of 1 meter in the direction of the force, seems remote from a statement of the energy content of biomass. Insistence on its use reflects the same preference for uniformity over sense that asks us to identify a 40-acre woodlot as being 16.19 hectares. The sentiments of Oliver (1990)—"We should use whatever units are best for the job in hand"—are very much to the point. For the conformist, kilocalories can be converted to joules by the formula 1 kilocalorie = 4186 joules.

Energy in Organisms

The energy that an organism uses for its work comes from the breakdown of organic molecules (or food) within its cells. There are two broad categories of organisms called **autotrophs** and **heterotrophs,** based on how the organisms obtain their food.

Autotrophs

Autotrophs, mainly green plants, produce their own foods from simple inorganic materials in the process called **photosynthesis.** It involves the putting together of carbon dioxide and water (plus some other inorganic materials) to form complex organic molecules, especially sugars but also amino and organic acids. Oxygen is also produced. The green pigment chlorophyll and various enzymes must be present for this process to occur.

Photosynthesis is usually summarized in chemical symbols as

$$\underset{\text{Carbon dioxide}}{6CO_2} + \underset{\text{Water}}{6H_2O} \xrightarrow{\text{Light energy}} \underset{\text{Glucose}}{C_6H_{12}O_6} + \underset{\text{Oxygen}}{6O_2}$$

Such an equation tells us little more than simply stating the same thing in words. It illustrates only the synthesis of one compound, the sugar glucose; summary statements for the other compounds produced in photosynthesis would be different. Also, it tells us nothing about the actual process by which carbon dioxide and water end up as glucose or some other organic compound.

The details of photosynthesis are unnecessary for our purposes. It is enough to say that the process consists of two rather complicated steps, the light reaction and the dark reaction. In the first, radiant energy absorbed by chlorophyll is used to split water into hydrogen and oxygen. The oxygen released in photosynthesis comes from this reaction. In the dark reaction hydrogen is combined with carbon dioxide to produce an organic compound, the exact identity of which varies among different plants. This compound is then used to synthesize many different kinds of organic molecules. During the process, part of the energy absorbed by the chlorophyll is converted to the energy of the chemical bonds of the organic molecules produced.

Although we are not concerning ourselves with the physiological details of photosynthesis, they are of importance in ecological research. In the mid-1960s, plant physiologists discovered a new pathway in the dark reaction portion of photosynthesis. Plants using this pathway for making carbohydrates are referred to as **C_4 plants,** whereas those using the pathway that plant physiologists had previously worked out are known as **C_3 plants.** For several reasons, but particularly because light strongly stimulates respiration **(photorespiration)** in C_3 plants, C_4 plants produce food more efficiently. Ecologically, this means that plants having the C_4 pathway and ecosystems dominated by these plants may be more productive than plants and ecosystems lacking it. Because of the virtual absence of photorespiration in C_4 plants and their ability to use CO_2 at extremely low concentrations (continuing photosynthesis even after the stomata have been closed for a time), C_4 plants tend to do well in arid or sunny ecosystems. Sugarcane, corn, and crabgrass are examples of C_4 plants. The C_3 pathway occurs in many plants, such as wheat, sugar beets, rice, and all trees.

Heterotrophs

Heterotrophs, which include all animals and other organisms such as fungi and many bacteria, cannot make their foods directly from simple inorganic materials. They must take in food that is already formed, by eating plants or other animals that have eaten plants.

The process of **respiration** is basically the same in autotrophs and heterotrophs. Respiration occurs within all living cells. Organic molecules—carbohydrates, fats, proteins—are broken down to yield energy. In both plants and animals the energy is stored in a phosphorus-containing compound called adenosine triphosphate (ATP), the immediate source of energy for contracting muscles or doing other kinds of work.

In **aerobic respiration,** the breakdown occurs through the combination of the organic molecule with oxygen in a complicated series of steps well discussed in any beginning biology text. Aerobic respiration may be summarized as

$$\underset{\text{Glucose}}{C_6H_{12}O_6} + \underset{\text{Oxygen}}{6O_2} \rightarrow \underset{\text{Carbon dioxide}}{6CO_2} + \underset{\text{Water}}{6H_2O}$$

In this reaction energy that produces 38 molecules of ATP is released, as well as an additional amount of energy lost as heat.

Anaerobic respiration or **fermentation** begins in the same way as aerobic respiration, but it does not go as far. The carbohydrate being respired may end up as carbon dioxide and alcohol rather than as carbon dioxide and water. Because there is still energy left in the alcohol, anaerobic respiration is obviously a less efficient process than aerobic respiration. Anaerobic respiration as practiced by yeast may be shown as

$$\underset{\text{Glucose}}{C_6H_{12}O_6} \rightarrow \underset{\text{Ethyl alcohol}}{2\ C_2\ H_5OH} + \underset{\text{Carbon dioxide}}{2CO_2}$$

In this reaction energy that produces two molecules of ATP is released, as well as an additional amount of energy lost as heat. Anaerobic respiration in animal cells produces lactic acid instead of alcohol.

Variations

Not all organisms fit exactly into these descriptions of autotrophy and heterotrophy. Variations exist, especially among the bacteria and the protists. For example, bacterial photosynthesis differs from that practiced by plants. **Green bacteria** and **purple sulfur bacteria** do not use water and do not release oxygen. The hydrogen donors in bacterial photosynthesis are reduced sulfur compounds, such as hydrogen sulfide, or organic compounds. Most such bacteria are obligate anaerobes, living at the top of anaerobic sediments in lakes, marshes, and estuaries.

Nonsulfur purple bacteria carry on photosynthesis in anaerobic environments with light, but in aerobic environments in the dark, they can live as heterotrophs. Many algae have similar abilities. Generally, they behave as standard photosynthetic autotrophs, but under certain conditions, such as very low light (under snow-covered ice, for example) or heavy pollution by organic wastes, they can shift to heterotrophy.

Chemosynthetic autotrophs make sugar from carbon dioxide but do so using energy derived from the oxidation of an inorganic compound rather than the energy of sunlight. For example, the **nitrifying bacteria** convert ammonia to nitrite and nitrite to nitrate. The **sulfur bacteria** produce sulfate from a more reduced form of sulfur.

Many algae are photosynthetic but require one or more organic compounds in the same way that animals require vitamins. Vitamin B_{12}, thiamine, and biotin are the best known of these compounds. Organisms having such requirements are **auxotrophic.**

Energy Flow Through an Individual Organism

Let us follow the flow of energy through an individual animal (Figure 2-10). The food that it eats contains a certain amount of stored energy that at some time in the past originated in photosynthesis. The energy contained in its food may be called the animal's **gross energy intake.** Because the processes of digestion and assimilation are not 100% efficient, some of the energy is lost in feces. Also, some of the energy stored in compounds that are digested and absorbed is later given off in various excretions such as urine and sweat. The undigested, unassimilated, and excretory fractions are often lumped together and called simply **excretory energy.** The energy remaining, called **assimilated energy,** is essentially all the energy that the animal has to perform all its work.

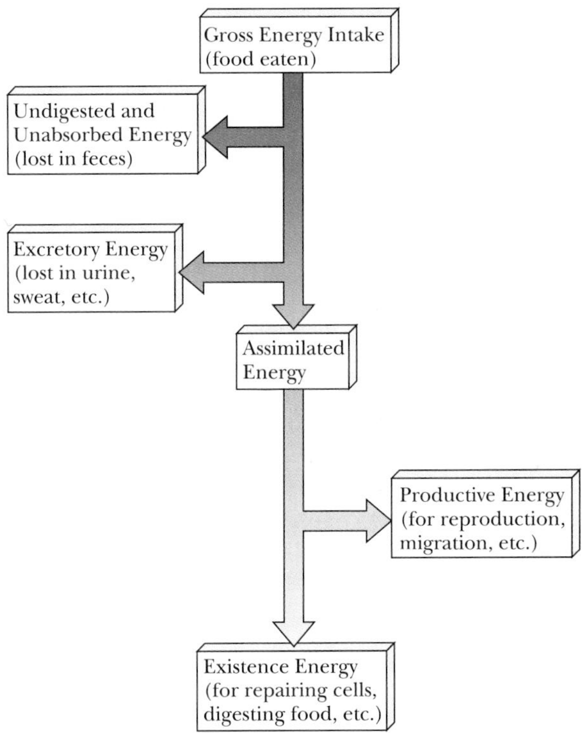

Figure 2-10

The flow of energy through an individual animal. Energy enters as food; some is lost in feces and urine. The rest is used (and eventually given off as heat) or stored (as fat or as other kinds of tissue).

A certain amount of energy is required by the animal just for existence. It is convenient to look at **existence energy** in four categories: (1) standard metabolism, (2) specific dynamic action, (3) thermoregulation, and (4) the cost of living a free existence. **Standard metabolism** is the energy used just to stay alive with minimal activity; it is the energy needed to repair cells, pump blood, and breathe. It is measured by putting an animal that has fasted for a few hours into a dark cage (so that it won't move around) that is warm but not hot, so that the animal will not have to use any energy in temperature regulation.

For birds and mammals, the main factor affecting standard metabolic rate (SMR) is size. Figure 2-11 shows the approximate relationship between SMR and body weight. Stated in words, the relationship is that SMR increases as the animal gets bigger, but the increase is not proportional. A 50-g bird uses more energy than a 10-g one, but only about three times rather than five times as much.

Although other factors may affect this relationship (Schmidt-Nielsen 1972), the main factor seems to be the relation between the mass or volume of an animal and its surface area. Heat production is related to the weight of tissue of an animal, whereas heat loss tends to be proportional to its surface area. If we compared some related animals of different sizes, we would find that mass goes up roughly as a cube of body length, whereas surface area goes up as the square (Figure 2-12). With proportionally more heat-producing tissue and less surface from which to lose heat, bigger animals can get by with proportionally less energy. Another way of thinking of the relationship is that larger animals use more energy overall than small ones, but they use less per gram of body weight.

The second category of existence energy is called **specific dynamic action.** An animal must eat to exist, but digesting and assimilating food take energy. This energy is included in specific dynamic

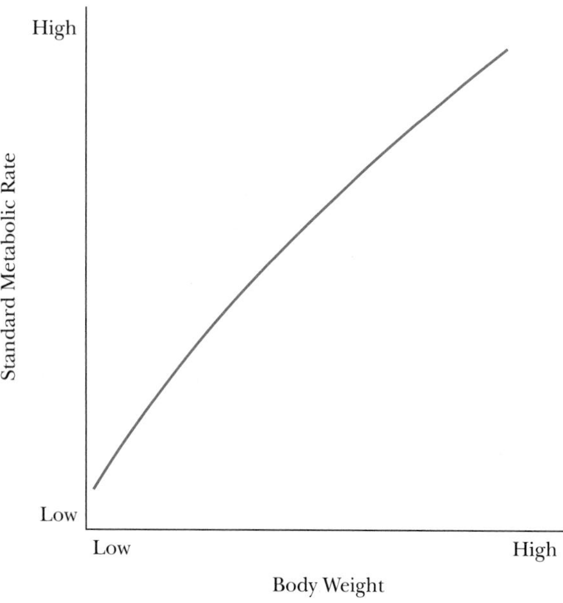

Figure 2-11

The relationship between standard metabolic rate and weight of an animal. Standard metabolic rate is energy used per animal per time period, for example, kilocalories per day. Anything that can be equated with energy usage—for example, cc of oxygen used per day or hour—could be plotted instead. An equation that gives such a curve has the form

$$SMR = a\ (\text{body weight})^b$$

For mammals, the constants are approximately $a = 68$, $b = 0.75$ (Kleiber 1975). If body weight is in kilograms, SMR is in kcal/day.

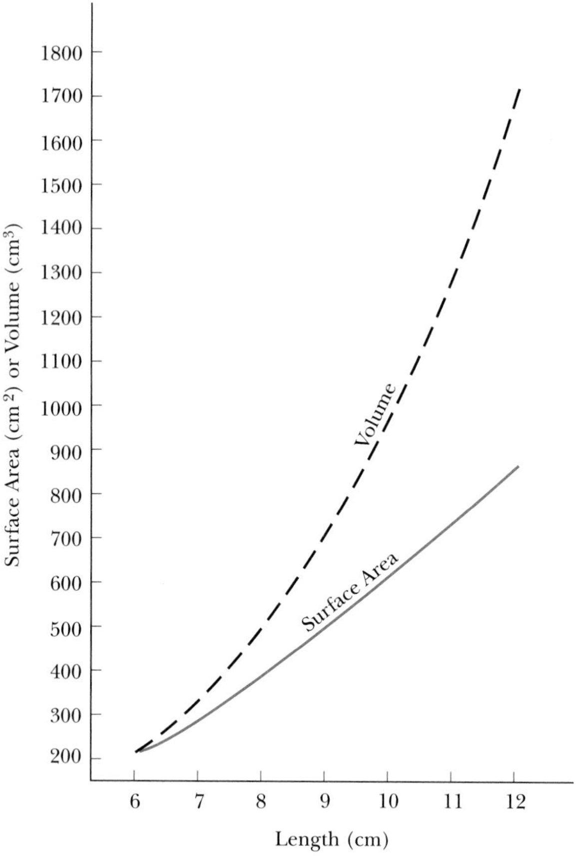

Figure 2-12

The increase in surface area is compared with the increase in volume of animals of different lengths. What is actually compared is the surface area of a cube ($6 \times \text{length}^2$) with the volume of the cube (length^3). For animals that are not cubes, details may be different but the trend will be the same: volume (or mass or weight) increases at a faster rate than does surface area.

action (and the occurrence of specific dynamic action is the reason that SMR is measured in the "postabsorptive" animal, one that has finished absorbing the food from its last meal).

The third category of existence energy is the energy needed for **thermoregulation,** that is, the energy spent to keep the body temperature in a narrow range. We will discuss thermoregulation and the differences between warm-blooded and cold-blooded animals in the next chapter in the section on temperature.

Fourth is the **added energetic cost of a free existence.** Animals that do not live in a cage with food and water supplied have additional energy-requiring activities such as traveling from burrow or roost to feeding grounds, catching food, and escaping predators. Any activity raises energy costs above SMR. For ducks, walking (or waddling) uses energy at a rate 1.7× SMR, slow swimming 2.2× SMR, and flying 12.0× SMR.

All in all, the total energy used by a warm-blooded animal (Table 2-2) living a free existence tends to be two or three times the measured or calculated SMR. This is on an average day. At certain times, the animal's energy usage could be much higher than this, for example, during very cold weather or on a day of migratory flight.

Beyond this existence energy level, the organism must be able to marshal some additional **pro-**

Table 2-2
Daily Time and Energy Budget for Free-Living Male Dickcissels

Dickcissels are small grassland birds. The figures below are generalized for a whole breeding season, but some activities obviously take more time and energy at some stages of the nesting season than others. For example, courtship is restricted to the period just after the females return to the breeding grounds. The bird's foraging seems efficient inasmuch as the bird needs to spend only about 3 hours and 2.8 kcal to gather the food that supplies all its energetic needs.

Activity	*Percent of the Day*	*Energetic Cost (kcal/Bird)*
Sleeping at night	37.5	6.5
Resting	6.5	1.1
Foraging	12.4	2.8
Singing	31.7	6.0
Courtship	0.6	0.2
Maintenance of female	4.9	1.4
Territory defense (chasing, etc.)	0.3	0.1
Interspecific aggression	<0.1	<0.1
Distant flight	6.1	6.3
Totals	100	24.4

From Schartz, R. L., and Zimmerman, J. L., "The time and energy budget of the male dickcissel *(Spiza americana)*," *Condor,* 73:65–76, 1971.

ductive energy. The amount by which gross energy intake is greater than existence plus excretory energy is the amount of productive energy available. The productive energy can be used for various kinds of work such as growing, mating, nest building, ditch digging, or tennis playing. If it is not so used, it accumulates as fat. For birds, this is a normal seasonal occurrence in preparation for migration or for surviving low winter temperatures. For humans getting fat is the result of the same process of taking in more calories than are used, but the fat rarely serves any useful function.

The importance of a favorable energy balance in the life of an individual organism is clear. A plant that does not fix more energy in photosynthesis than it uses in respiration soon dies. Such a fate may befall a seedling that sprouts beneath a dense forest canopy. A bird that is not able to generate enough productive energy in midwinter will die of starvation or from some malfunction caused by its body temperature falling too far below normal. Sparrow-sized birds do not have enough energy stored as fat to last them more than a day or so without food, depending on how low the temperature is. Circumstances that prevent their feeding for very long are deadly. In the northern part of the range of the bobwhite, storms that cover the food supply with a layer of ice or crusted snow often leave a covey dead on its roost, their stomachs empty.

Energy balance also has implications at the population level. Even if the individual can maintain an energy balance to ensure its own survival, the species will not persist in an area unless there is enough additional productive energy available to reproduce—energy to carry on courtship, fight off rivals, produce eggs, and feed the young. The University of Illinois ecologist S. C. Kendeigh (1949) pointed out that the various activities requiring heavy expenditures of productive energy tend to be spaced seasonally so that they do not overlap (Figure 2-13). In migratory birds, for example, the fall molt, in which all the feathers are replaced, is sand-

Month	Major Energy-Requiring Activities
March	Hunting for scarce food
April	Territory establishment and courtship
May	Nest-building and egg-laying
June	Incubation
July	Feeding young
August	Molting
September October	This seems to be a fairly prosperous time; Some species cache surplus food
November December January February	Surviving low winter temperatures

Figure 2-13

Seasonal spacing of the prime energy-requiring activities of a nonmigratory bird. Migratory birds need less energy in the winter for surviving low temperature but have high energy requirements for migration in fall and spring.

wiched between the end of feeding young and the beginning of laying down fat as fuel for the southward flight. There are a few kinds of birds, though, in which the young must be fed until about the time when the birds go south. Included are birds whose feeding methods require the development of considerable skill, such as flycatchers. In many of these species, the molt does not occur in the fall but instead is delayed until the birds complete migration and are on their wintering grounds (Morehouse and Brewer 1968).

Energy Subsidies

Probably all organisms receive **energy subsidies,** that is, energy from outside their own metabolism, to some extent. Gulls use energy from air currents to soar (Figure 2-14). Organisms in tidal zones use tidal energy to help bring in food and carry out wastes. Lizards and probably even some blackbirds use solar energy to help raise their body temperature. Humans in developed countries, however, are uniquely dependent on energy subsidies. One human, depending on size and activity, can get by very well on 2400 kcal per day, 100 per hour. An average color television set has a power demand of a little more than 100 watts. Since 1 watt is equal to about 0.9 kcal per hour, a color television set also requires about 100 kcal per hour. As you sit and watch television, you and your television set are each consuming almost the same amount of energy.

In fact, the average energy consumption per capita in the United States is about 8 million kcal per day (Department of Energy 1990), or about 3000 times the caloric intake a person needs to live well. Most of this energy is not expended by each person directly, of course. Much of it is used in the manufacture, transportation, and sale of things that are eaten, worn, driven, watched, and discarded. This figure is not typical of the world as a whole. North America, with 5% of the world's population, uses 27% of its oil (World Resources Institute 1986).

The functioning of the ecosystem, no less than the functioning of individuals and populations, depends on energy. In a later section we will consider energy flow at the ecosystem level.

Figure 2-14

A soaring laughing gull at Ocracoke, North Carolina, uses solar energy in the form of wind currents as an energy subsidy. (Photo by R. Bruce Moffett.)

Animal Behavior

The scientific study of animal behavior is called **ethology.** It is concerned with what animals do in their daily lives, the functions of their various acts, and the evolution of behavior. Konrad Lorenz, Niko Tinbergen, and Karl von Frisch won the Nobel prize (in physiology and medicine) in 1973 for their pioneering studies of animal behavior.

Much of the ecology of an animal is rooted in its behavior. Where it roosts may determine how cold a climate it can tolerate. Its feeding behavior is the basis of predator-prey relationships and a link in a food chain. Habitat selection, territoriality, and many other important ecological topics depend on the animal's behavior. Much behavior concerns individual ecology, such as maintenance behavior (preening, bathing, sleeping, etc.). However, a large part of the behavior of vertebrates and some invertebrates is social behavior, and properly belongs under the heading of population ecology. Included are sexual behavior (courtship and mating), **agonistic** behavior (fighting, fleeing, and related acts), and caregiving and care-soliciting behavior (parental care).

One of the great contributions of the field of ethology has been the realization that the world of an animal is not necessarily the same as a human's world. An ant's world is different from ours, and not just in the obvious sense of scale. Its perceptions based on its compound eyes, its other sense organs, and its arthropod nervous system are not the same perceptions we have, even if we lie on the ground and look through grass stems. The German word *umwelt* is sometimes used to refer to the surroundings as perceived by a particular species.

Most animals seem to react in a stereotyped way to very specific features of other organisms or their environment (Tinbergen 1951). A European robin will defend its territory not only against another male robin but against a tuft of red feathers hung up in the air. In the three-spined stickleback, a small fish, the male will court a glob of clay of almost any shape as long as it is roundish underneath (like a female ready to lay eggs) and is not red (like male sticklebacks). The features of other organisms or the environment that elicit stereotyped behavior of this sort are called **releasers.** In the last example, if the glob of clay is red, it will release territorial rather than courtship behavior.

Mammals, with their well-developed brains, have behavioral patterns in which the innate becomes less important and the learned becomes more so. (The textbook definition of "learning" is the adaptive modification of behavior by experience.) In humans, learning dominates behavior, but even in humans, behavior develops within constraints set by heredity.

Evolutionary Considerations

Fitness

On a quiz, I once asked, "How do individuals evolve?" This made me guilty of the worst of all pedagogical sins, asking a trick question, since the answer, obviously, is that individuals do not evolve, populations do. Evolution—descent with modification or, genetically, a change in gene frequencies with time—is discussed in a later chapter on populations. It is well to begin the discussion here, however, under individual ecology, with a quick discussion of fitness.

Probably the most important mechanism of evolution in most circumstances is natural selection. The catchphrase for natural selection is "survival of the fittest." **Fitness** has come to be defined as representation in the next breeding generation. If an individual leaves behind four or five offspring that take part in the next breeding generation, it has a higher fitness than another individual that leaves behind zero or one offspring that survives to breed. We can also speak of the fitness of a phenotype or a genotype.

This way of defining fitness is unfortunate because it seems to reduce "survival of the fittest" to a tautology—survival of the survivors. This is not what the phrase meant to Darwin, however. He thought of fitness in terms of features that resulted in the individual organism being successful in a particular environment: an individual with a white coat is fitter in a snowy environment than one with a yellow one. Two synonyms for the fitness of an individual—adaptive value or selective value (Falconer 1989)—are rarely used.

In the broadest sense, there are two components of fitness. One is survivorship and the other reproduction. An individual can live a short time or a long time. Although Darwin's phrase talks about survival, the main contribution of survival is that living longer allows an individual to reach reproductive age and to reproduce more often. An individual with traits that cause it to succumb to cold temperature or fail to find enough food or to fall prey to a predator before it is able to reproduce is an individual of low fitness. The measure of fitness, however—representation in the next breeding generation—uses offspring as the currency.

Proximate and Ultimate Factors

When we ask a biological question with a "why" in it—"Why do birds migrate?" or "Why do varying hares turn white in the winter?"—we are usually asking two separate questions with quite different answers. We are asking, for example, in regard to the hares, (1) What happens environmentally and physiologically that makes them lose their brown fur and grow white? But we are also asking, (2) Why are hares that turn white in winter fitter than ones that don't? For bird migration, we are asking, (1) What cues cause birds to fly south in the autumn? and (2) What makes individuals do better if they go south instead of staying north?

The answers to the first question in each case

are **proximate factors** and the answers to the second, **ultimate factors.** This terminology was first employed by the British biologist J. R. Baker (1938) in writing about another question of the same sort: "Why do birds breed when they do?"

Ultimate factors are the evolutionary advantage that the trait or activity confers, mediated by differences in survival and reproduction. For example, the most likely ultimate factor in the varying hare's changing color is camouflage. Individuals that match the snow in the winter and the ground in the summer are more successful in avoiding predators.

It is not the presence of snow or the low temperature that causes the hare to molt its brown fur and grow white; it is day length. If we take a brown hare and put it in the laboratory under a short day length of 9 hours, it will molt from brown to white, and it will do this whether the laboratory is hot or cold. If we give a white hare day lengths of 14 hours, it will begin to shed its white fur and brown will grow in. Under natural conditions, of course, the changes in day length would be associated with the change of seasons and the arrival and departure of snow cover, and, consequently, the response would synchronize the animal's color with the color of its background. Day length, or **photoperiod,** is acting as a proximate factor. A proximate factor is an environmental cue, mediated by behavior or physiology.

Adaptation

This way of approaching biological problems through analysis of both proximate and ultimate causes is so important a tool that another example may be useful. We may ask the question, Why do animals become fevered?

It is common knowledge that when people or certain other kinds of animals are sick, they develop a fever. What causes this? Many years ago physiologists discovered that fever usually indicated an infection. Observations and experiments eventually led to the following chain of events as an explanation: Disease organisms produce chemicals that the body's white blood cells respond to by producing their own chemical (interleukin-1) that acts on the hypothalamus. The specific effect is that the hypothalamus increases the body's heat production. The center in the hypothalamus that regulates heat loss is unaffected, and so the overall effect is a rise in body temperature.

This is a good proximate explanation of fever, but it is only half the answer. Is there an evolutionary reason why fever accompanies sickness? Is fever adaptive, maladaptive, or neutral?

Not long ago if a child was running a fever from a cold or similar infection, pediatricians would recommend aspirin, which is an antipyretic—that is, it makes you quit having a fever (Kiester 1984). The underlying assumption was that the fever was a harmful side effect of the disease. Some biologists with an evolutionary outlook examined the question of the adaptiveness of fever using an ingenious approach (Kluger et al. 1975).

First, they showed that poikilotherms develop a fever. They used a lizard, the desert iguana. If sick lizards are kept in a constant-temperature room at, say, 38°C, they do not develop a fever. If they are kept in a chamber that has a variety of temperatures, they select hotter temperatures when they are sick than when they are healthy. They give themselves a behavioral fever.

This result, in itself, was suggestive that elevated temperature is not some physiological mistake. It also gave a good way of testing whether fever does, or does not, aid survival, or, in other words, is or is not adaptive. If fever is merely a by-product of infection, there should be no difference in survival between sick desert iguanas with normal body temperatures and ones with fevers. Increased survival of the fevered lizards would be evidence on the side of fever as an adaptation. Accordingly, two groups of lizards were infected with a particular disease-producing bacterium. One group was placed in a chamber with a constant 38°C temperature, about the same temperature the lizards choose if they are well. The other group was put in a chamber with a temperature of 42°C, about the temperature sick lizards choose if they have a range of temperatures available.

The results of the experiment are shown in Figure 2-15. The animals with elevated temperatures survived much better. Why? That is another question and another ecological one. Bacteria, like any other organisms, have ranges of tolerance and optima for temperature and other factors. If the host organism is able to alter one or more of these factors, making it relatively unfavorable, bacterial growth may be slowed or stopped, making it easier

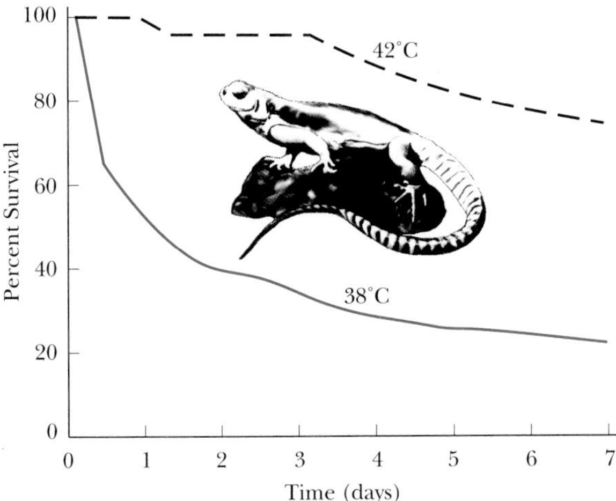

Figure 2-15

Survival of desert iguanas infected with disease bacteria and kept either at a temperature simulating their usual body temperature (38°) or at a higher temperature corresponding to a fever (42°). In a week about 75% of the lizards prevented from developing a fever had died compared with only about 20% of the fevered lizards. (Adapted from M. J. Kluger et al. 1975. Science vol. 188, pp. 166–168. Copyright 1975 by the AAAS.)

for the other infection-fighting mechanisms of the body to kill the bacteria and bring the infection under control. This may not be the complete explanation, but clearly the results point to fever as an adaptive response of the body to infection.

We need, nevertheless, to realize that not every feature of an organism has to be an adaptation. Some structures, processes, or behaviors may be merely by-products of other adaptive features. This point has been reviewed at considerable length by Gould and Lewontin (1979). For example, the color of our internal organs may not be adaptive. Our liver and lungs may be the colors they are simply because that is how they turn out, given the physiological functions they perform. Even so, if we approach every biological question with both a proximate why? and an ultimate why? in mind, we are more apt to be led to the complete truth than if we do not.

Ecotypes

The fact that two individuals belong to the same species does not guarantee that they will respond identically to some ecological factor such as temperature or light. Individuals differ, of course, and so may local populations. For example, a grass called sheep's fescue grows in England on both acid and alkaline soils; this is not the result of a wide ecological tolerance range of individuals but rather because the two soils have populations differently adapted to growth at high and low levels of calcium (Jones and Wilkins 1971). Populations having genetically based differences of ecological importance have been termed **ecotypes** (Turreson 1922), though the term now seems to be used mainly to refer to plants showing striking morphological adaptations to habitat.

Two botanists at the University of Missouri conducted an interesting experiment showing ecological variations in different portions of the geographical range of a species. Seeds of a plant called white snakeroot were gathered from widely separated localities and planted in a greenhouse (Vance and Kucera 1960). After the plants had reached a certain size, they were exposed to light equivalent to the long day lengths conducive to flowering in this species. After 120 days the plants grown from North Dakota seeds had produced mature fruits, those from South Dakota seeds had just reached full flower, and those from Kansas had not yet produced flower buds. These differences corresponded very well to the growing season (the length of time between the last frost in the spring and the first in the fall), which is 129 days in North Dakota and 195 in Kansas. Those plants having the responses of the Kansas population would probably be unable to reproduce in the short growing season found in the northern part of the range of the species.

Because the plants were all grown under similar conditions we may conclude that the differences seen are genetic or hereditary. A similar "transplant garden" approach has shown that populations of salamanders in different ponds vary in the proportions of individuals that undergo metamorphosis compared with those reproducing while still retaining the immature, larval form (Semlitsch et al. 1990). The difference is related to the ecology of the pond, such as permanency and perhaps to the types of predators. These studies and many others suggest that evolutionary fine-tuning of populations to local environments is a common phenomenon.

Habitat Selection

For some organisms, reaching a favorable environment is largely a matter of chance. A dandelion plant may produce a thousand seeds and those that reach a suitable area—bare, sunny soil—will germinate and grow. J. L. Harper (1977) has used the term **"safe site"** to describe the set of habitat conditions that favor the establishment of a seedling of a particular species, though the heuristic value of this concept is still unclear (Primack and ShiLi 1991). Other organisms, including most animals, show behavior that serves to locate them in a favorable environment, in the process known as **habitat selection.**

When we ask why these organisms occur where they do—why the chipmunk lives in forest or why the bobolink lives in grassland—we are again asking two separate questions. We are first asking what is it about grassland that makes it favorable for the bobolink. What are the ultimate factors causing (or allowing) it to live there? But we are also asking what environmental cues cause the bird to establish territories in such a habitat. In other words, what are the proximate factors causing it to live in grassland?

At the level of ultimate factors we may conclude that in this habitat the organisms find what they need to survive and prosper—the right kind of food for themselves and their young, suitable nest sites and materials for nests, cover for escape from enemies, and perches to deliver the song that will attract mates and drive away rivals. Some of these may also serve as proximate factors, but some or all of them may not be observable by the animal at the time it is selecting its habitat. The caterpillars on which many birds feed their young may still be eggs if habitat is selected in the spring, or they (those that survived) may already be hidden away as pupae if habitat is selected in the fall. The animal may then have to use some other feature of the environment, one that is associated with the essential feature, as a basis for its choice.

Female mosquitoes visit humans and other vertebrates to suck their blood, using the protein so obtained in egg production. Without this food, egg production is much lower. Vertebrate blood, constituting a suitable food, is the ultimate factor in host selection by mosquitoes. The factors that get the mosquitoes to the host, proximate factors, tend to be chemicals that are characteristic of the appropriate animal. For example, yellow fever mosquitoes are attracted by L-lactic acid in the presence of carbon dioxide (Acree et al. 1968). The carbon dioxide given off in respiration acts as an activator or locomotor stimulant and the L-lactic acid from sweat on the skin of the human as an attractant.

Two chemicals also seem involved in the settling of the free-swimming larvae of the abalone, a large marine snail. The specific chemical that induces the larvae to drop out of the plankton and begin their metamorphosis to adult form is produced by coralline red algae that coat rocks and other surfaces and that seem to be the primary food source for the young snails (Morse 1991). The chemical has proved to be a small peptide similar in structure to gamma-aminobutyric acid (GABA), a neurotransmitter in the nervous system of mammals. In experiments, the larva's sensitivity to GABA is greatly increased by the presence of lysine or similar amino acids in the seawater. It is possible that such molecules are indicators of productive areas, near which it would be advantageous for the adult abalone to be located.

Many cases of habitat selection may fit a two-stage model proposed for birds by Scandinavian ecologists G. Svärdson (1949) and O. Hildén (1965). Birds visit areas on the basis of general appearance, structure, or landscape; finer details are examined before they actually settle in the area (Figure 2-16).

In the peach aphid, a small insect, dispersal is by winged forms that launch themselves into the wind, allow themselves to be carried along, and then actively drop out and land on vegetation. An entomologist (Kennedy 1950) counted the numbers landing on two small trees, one a peach and one a euonymus. The euonymus is not a host plant for this insect.

The numbers landing on the two trees were essentially the same, with more than 10,000 aphids landing on some days. But, at the end of the dispersal season, there were dense populations of aphids on the peach and none on the euonymus. The many thousands of individuals that landed on the euonymus later moved on. The first step in habitat selection was evidently visual, a response to the sight of the small trees (or perhaps to some more abstract feature like projecting greenness), but

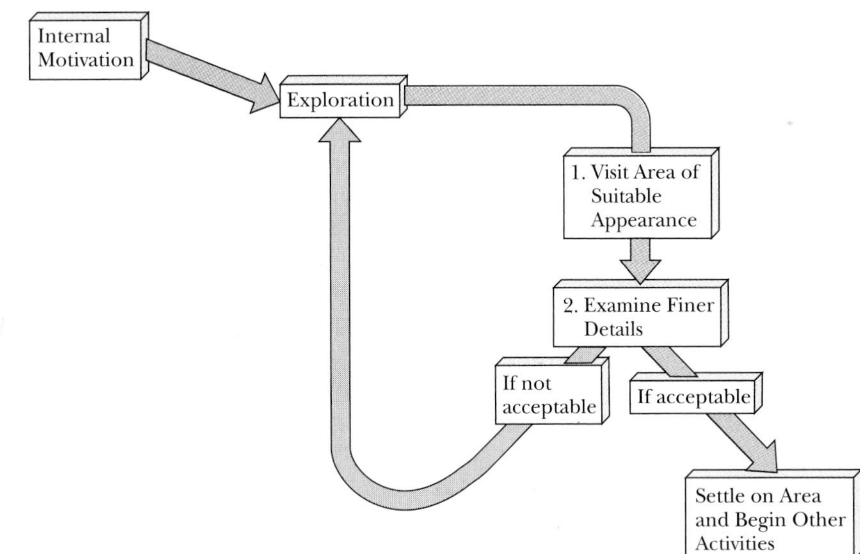

Figure 2-16

Two-stage habitat selection. If finer details of the habitat are acceptable, the animal settles and goes on to other activities, such as establishing a territory, or attracting a mate. If the habitat is unacceptable the animal returns to the first stage of visiting superficially suitable areas discovered in exploration.

there was a second stage involving a choice between peach and other trees that was, perhaps, based on some chemical factor.

One interesting question is the degree to which the preferred habitat is innate and the degree to which it is influenced by the early experience of the organism. The answer may be different for different kinds of animals. The few experiments and observations that have been done with birds and mammals suggest that a strong predisposition for a particular habitat is inherited (Table 2-3), but that this can sometimes be modified slightly by experience. Much the same statement seems to be true for insect-host plant relations (Courtney and Kibota 1989).

Table 2-3
Habitat Selection in the Pied Flycatcher

Two German ornithologists, R. Berndt and W. Winkel, studied banded pied flycatchers to see whether young birds tended to choose the kind of forest they were born in (deciduous forest is considered the typical habitat of the species). They kept track of all the cases of pied flycatchers they had banded as young that were found nesting at least 10 kilometers from where they hatched. What they found is shown below.

Kind of Forest Where Hatched	*Percentage of Young Breeding in*	
	Deciduous Forest	*Coniferous Forest*
Deciduous	69	31
Coniferous	79	21

Pied flycatchers, whether they were hatched and reared in deciduous or coniferous forest, settled if possible in deciduous forest. There seemed to be no learned preference for coniferous forest. Those in coniferous forest probably settled there initially because the deciduous forest areas were already full.

From Berndt, R., and Winkel, W., *J. Ornithol.*, 116:195, 1975.

The Spread of Organisms

Dispersal

As a statement of biogeographic logic, we can say that a species will occupy any area that (1) can be reached, (2) satisfies its habitat requirements, both proximate and ultimate, and (3) is not uninhabitable because of competitors, predators, or disease (Jordan 1928, Macan 1963). Points 2 and 3 have been discussed. Getting to the area—dispersal and range expansion—will be the next topic.

A formal definition of **dispersal** is the movement of individuals from the homesite. The means of dispersal vary from species to species. Among plants there is often a special life history stage (seeds or spores) having special structural features that seem to be adaptations for achieving dispersal.

Examples are the winged fruits of maples or the plumed fruits of dandelion that use the wind and the stick-tight fruits of beggar's-ticks and spiny fruits of cocklebur that use animals (Table 2-4, Figure 2-17). Structural adaptations for dispersal are not evident in many animals, probably because most animals can move about readily throughout their lives.

It is almost always the immature organism that disperses. This is obviously true of plants and of sessile marine animals such as barnacles (in which dispersal is by the larvae), but it is also true of many other animals. In birds the adult is clearly just as motile as the young, but it is usually the young that moves. In his classic study of the house wren S. C. Kendeigh (1941) found that 84% of the adult males and 70% of the adult females that survived the winter bred within 1000 feet of their nest of the previous year. Of the young that lived through the winter, however, only 15% nested within 1000 feet of where they had hatched. The adults tended to return each year to the site where they had settled in their first breeding season, a tendency called **philopatry** (among several other names, including **site tenacity** and **ortstreue**). The young tended to move from the immediate vicinity where they were hatched or born but stayed in the general area (of the house wren young, 70% nested in the zone be-

Table 2-4
Agents of Plant Dispersal

Below are listed environmental agents that have the potential for moving plant seeds or other plant parts around and morphological features of plants that take advantage of these forces. The classification is based on *The Dispersal of Plants Throughout the World* by H. N. Ridley.

Agent	***Plant Adaptations and Examples***
Water	*Waterproof, buoyant fruits* such as water lilies, the coconut palm, and cockleburs. *Entire floating plants* such as duckweeds and water hyacinth.
Wind	*Fruits with long hairs* such as milkweeds, many composites, willows, and cottonwood. Dandelion fruits can be kept aloft by a wind of two miles per hour. *Winged fruits or seeds* such as the samaras of maples and ashes, which spiral down a few tens of feet away from the parent tree. *Balloon-like fruits.* Inflated, spherical fruits such as those of many nightshades and the shrub bladdernut that can be rolled along the ground. *Tumbleweeds.* The whole above-ground portion of such plants as russian thistle and wild indigo breaks off and, being spherical, is rolled around by the wind. *Very small size,* such as spores and the seeds of orchids and plants of the heath family.
Animals	*Fleshy fruits with hard seeds.* Animals eat fruits such as blackberries, strawberries, and cherries, and the seeds are regurgitated or left in feces. *Sticky seeds.* Birds feed on the rind of the plant parasite mistletoe and the slimy seeds stick to their beak and may be wiped free on another plant. Seeds of some wetland plants, though not themselves sticky, are transported in the mud on the feet of water birds. *Storage without recovery.* This is probably the main method of dispersal for many nut-bearing trees such as oaks and hickories. Certain kinds of herbs (such as violets) have oily seed projections called *elaiosomes* that encourage ants to carry the seeds to their nests. *Stick-tights.* Many kinds of seeds have hairs or spines that cling to the coats of animals or even stick to their flesh. Beggar's-ticks are an example of the first, sandburs and some cacti, of the second.
Explosive fruits	Several kinds of plants have mechanisms for ejecting seeds distances up to a few meters. Examples are touch-me-not and witch hazel.
Gravity	Many kinds of fruits and seeds seem to have no special features ensuring movement of their seeds. They simply fall off the plant and lie there. Wild leek and false mermaid weed, two mesic forest plants, are examples.

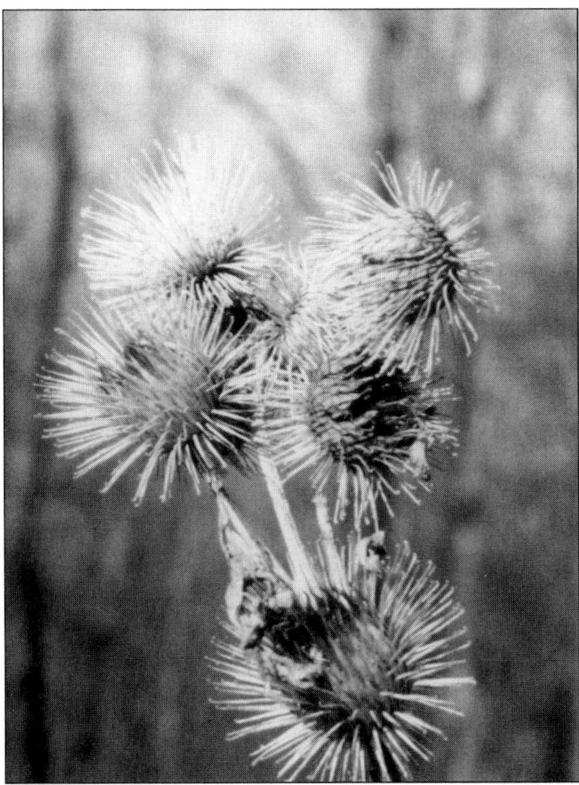

Figure 2-17

The hooks on the long bracts of these burdock heads catch on the coat of passing mammals and result in seed dispersal. (Photo by the author.)

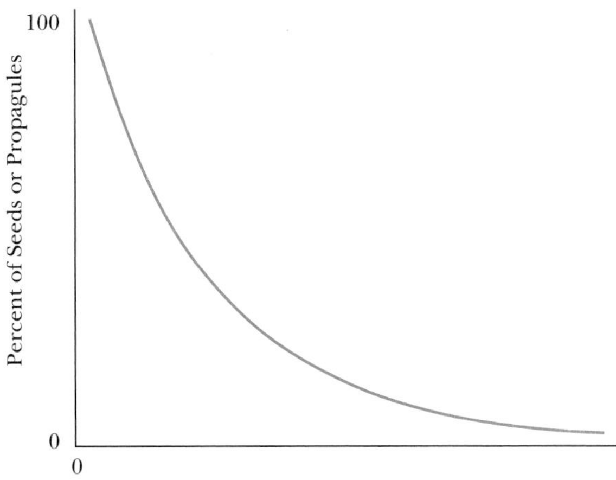

Figure 2-18

The Kettle curve, a common pattern of dispersal, especially for organisms whose dispersal is passive. The *Y* axis is the number or percent of dispersing units reaching a given distance. Some settle out in the next interval and some go on. The curve can be represented by the equation

$$Y = a\, e^{-bx}$$

where *Y* is the number of seeds or other dispersing units getting at least as far as distance *x*, *a* the number of seeds at the source (that is, at distance zero), *e* the base of natural logarithms, and *b* the rate at which seeds decline with distance.

tween 1000 and 10,000 feet of where they had hatched).

The general pattern of dispersal, as pointed out by Kettle (1951) among others, is that most of the seeds or young animals, or whatever is doing the dispersing, settle close to the homesite. There is then a rapid decline with distance, and only a few individuals reach very substantial distances. If the number reaching or going beyond a certain distance is plotted against distance, the resulting curve looks like that in Figure 2-18.

Although the dispersal of many organisms fits the Kettle curve more or less well certain kinds show consistent deviations from it. Many animals, notably birds and other terrestrial vertebrates, have been found to have more dispersers staying closer to the point and more going long distances than predicted. If we look not at the number reaching or going beyond a given distance but at the actual number settling in a given zone, the result is a curve with two (or more) peaks (Table 2-5).

The explanation for this pattern is not yet clear. It may result from different genotypes favoring different dispersal distances (Howard 1960, Johnston 1961). It may be simply a by-product of various aspects of the behavior of the species, particularly site tenacity and territoriality (Murray 1967). Habitat patchiness, resulting in individuals that fail to find a place in the patches where they hatched moving to the next favorable patch, is a third factor that can be important by itself or in combination with the other two (Brewer and Swander 1977).

There are two ways of increasing dispersal—that is, of getting more individuals farther from the homesite. One is increasing the number of seeds or other dispersing units. This is usually expensive (in the currency of energy) and not very effective. Doubling the number of seeds simply doubles the num-

Table 2-5

The Number of House Wren Yearlings Nesting at Various Distances From the Nest Box Where They Hatched

Distance (Feet)	*Number of Returning Nestlings*
0–999	27
1,000–1,999	34
2,000–2,999	21
3,000–3,999	24
4,000–4,999	12
5,000–5,999	19
6,000–6,999	3
7,000–7,999	7
8,000–8,999	4
9,000–9,999	1
10,000–10,999	2
11,000–11,999	0
12,000–12,999	0
13,000–13,999	1
14,000–14,999	0
15,000–15,999	2
More than 16,000	25

From Kendeigh, S. C., "Territorial and mating behavior of the house wren," *University of Illinois Biological Monographs*, 18(3):1–120, 1941.

ber reaching any given distance. If on the average 0.00001 seed is able to disperse a mile from the plant where it was produced, then if the plant produces twice as many seeds, 0.00002, on the average, will travel a mile.

The other method of improving dispersal is to increase dispersal ability, or **vagility.** For plants, as already mentioned, this seems to have taken the form of structural adaptations using wind, water, or animals as dispersal agents. For animals, dispersal ability seems to be improved by behavioral means that may include adult intolerance to young, forcing them to keep moving, and perhaps some greater tendency for the young to wander.

The relative importance of these two forces in animal dispersal is uncertain. No one doubts that "environmental dispersal" (Howard 1960) occurs, that is, that emigration is higher from crowded than from uncrowded areas (Greenwood and Harvey 1982). Studies on organisms from crawfish to pine voles have shown this. Such emigration is clearly an important factor in the regulation of the size of local populations.

The ultimate factors in dispersal, the things that make dispersers (or the parents of dispersers) fitter than philopatric, or nondispersing, individuals (or their parents) include the following (Johnson and Gaines 1990):

1. Changes in the habitat with time, so that a suitable habitat for the parents may not be suitable for the next generation
2. Avoiding inbreeding with its possibly deleterious genetic consequences (Packer 1985)
3. Avoiding crowding (Economou 1991)
4. Avoiding competition with relatives

For many species, the last two of these could be achieved if any genetic tendency to disperse were triggered by some level of assessed crowding. For organisms of unstable habitats, crowding would not be a particularly good trigger for dispersal (Gadgil 1985); some sort of habitat assessment would be better.

Range Expansion

Most dispersal does not produce any change in the geographical range of the species (Figure 2-19). The dispersal is within the range already occupied, perhaps into a newly created area of suitable habitat

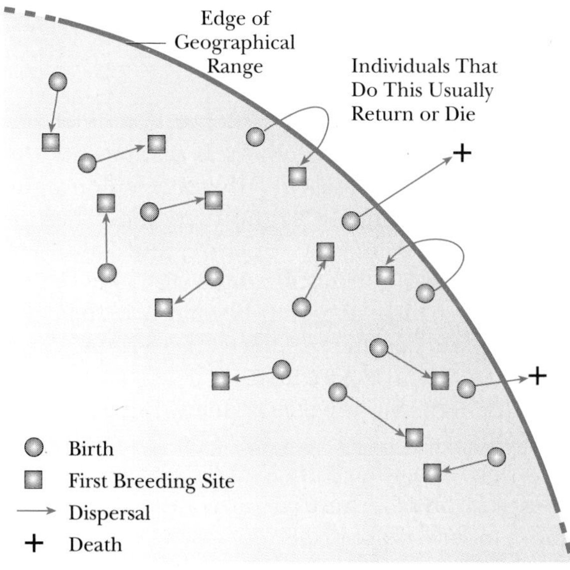

Figure 2-19

Ordinarily dispersal moves individuals but produces no change in the geographical range of the species.

(dandelions into a new lawn) or just into a spot left vacant when an older individual died. Successful dispersal into an area not formerly occupied by the species is called **range expansion.** Two primary modes of range expansion have been recognized (Brewer 1991). They are here referred to as diffusion and jump dispersal (Mundinger and Hope 1982). **Diffusion** is a gradual process in which the range boundary of a species advances by spreading into the previously unoccupied zone at the edge of the preceding year's range. **Jump dispersal** is the long-range movement of individuals or groups that colonize and reproduce at sites far from the old boundary.

Probably most range expansions have elements of both modes, for example, long-range movements followed by diffusion filling in the spaces between. Virtually all organisms have the potential to expand their geographical ranges by diffusion; an example might be the northward movement of the opossum diagrammed in Figure 2-20 or the eastward spread of the Brewer's blackbird from the grasslands of Minnesota to the open farmlands of Wisconsin and Michigan (Stepney and Power 1973).

Only organisms of high vagility are likely to spread by jump dispersal. A likely example is the spread of the cattle egret from the Old World to South America in the late 1800s (Terres 1980). The species reached North America by the 1940s and has since spread widely in the southern United States. Many cases of jump dispersal probably lead to only temporary occupation of the new territory.

The usual events allowing range expansion are (1) a barrier to the species is somehow removed or circumvented; (2) a formerly unsuitable area becomes suitable; or (3) evolutionary change in a species allows it to make use of a formerly unsuitable area.

Removal or circumvention of barriers usually occurs by one of two agents: geological processes, such as the joining and separating of continents, and human activities. The results of the removal of barriers through geological processes are difficult to study except by means of fossils. We know, for instance, that North and South America did not have a land connection such as the present Central American isthmus through much of the preceding 70 million years. During this period the two continents had almost no families of mammals in common. By glacial times, after the establishment of a connection between the two continents, there was rapid invasion in both directions. At the present time about 14 families of mammals are found in both North and South America (Marshall et al. 1982).

There are many examples of human removal or circumvention of barriers that have occurred in the lifetime of ourselves or our parents. Humans brought starlings, house sparrows, and Dutch elm disease across the Atlantic Ocean. The zebra mussel

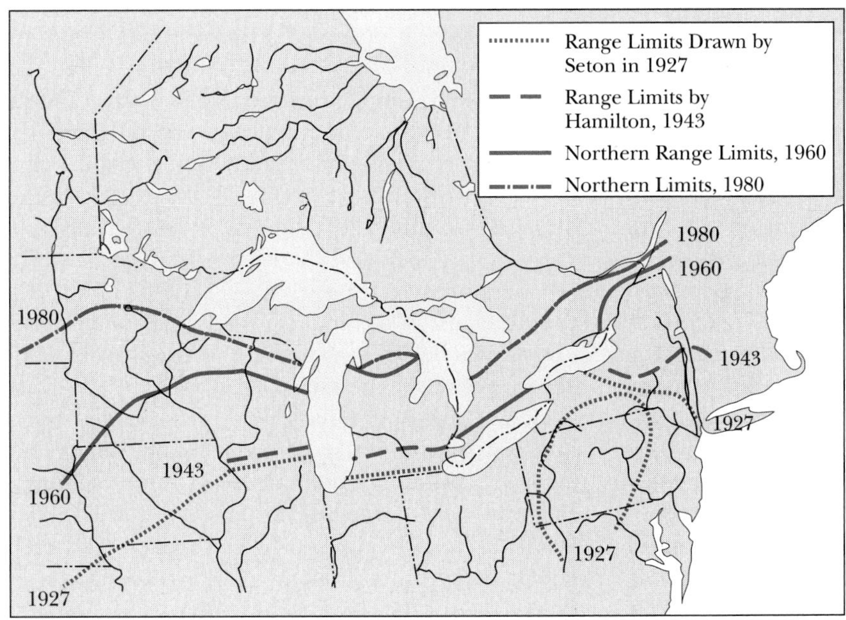

Figure 2-20

The northward spread of the opossum in the Great Lakes region. A hundred years ago, the opposum was absent from most of the northern parts of Iowa, Indiana, Ohio, and Pennsylvania. The 1927, 1943, and 1960 lines are from De Vos 1964. The line labeled 1980 is from E. R. Hall, *The Mammals of North America,* 2nd ed., New York: Wiley, 1981; J. N. Jones, Jr., D. M. Armstrong, and J. R. Choate, *Guide to the mammals of the Plains States.* Lincoln: University of Nebraska Press, 1985; and R. H. Baker, *Michigan mammals,* Detroit: Michigan State University Press, 1983.

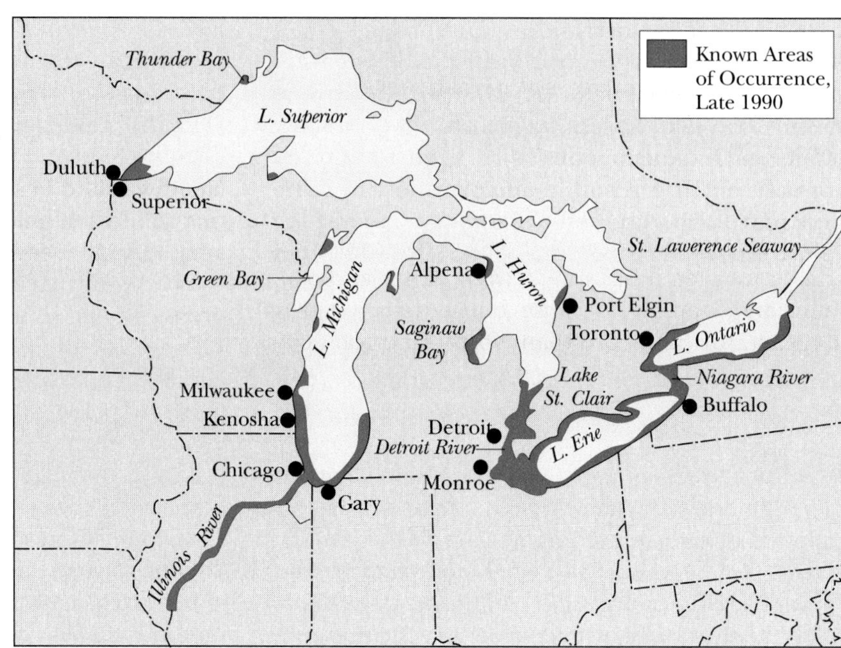

Figure 2-21

Approximate distribution of the zebra mussel in the Great Lakes region, 1991. Based on Illinois Natural History Survey Reports, October 1991, no. 310.

(Figure 2-21) is a recent addition to a long list, having been introduced into the Great Lakes about 1985 (Sparks and Marsden 1991).

Human activities have reached the magnitude of geological events. The Welland Canal was dug (1914 to 1932) between Lakes Ontario and Erie, bypassing Niagara Falls which had been a barrier to the westward spread of the sea lamprey. As a result the lamprey reached the Detroit River in the 1930s and by the 1950s had spread all the way to western Lake Superior. A proposed sea-level canal across Panama (or somewhere else in Central America) is a similar project (Aron and Smith 1971). Such a canal would allow the marine biota of the Pacific to spread into the Atlantic, and vice versa (the present Panama Canal does not do this because ships are locked up about 25 m above sea level and through two fresh-water lakes for the crossing). The devastating effect of the introduction of the sea lamprey on the Great Lakes trout population is well known; the results of range expansions following construction of a sea-level canal across Central America can only be speculated on.

In the category of environmental changes allowing range expansion, changes in climate and vegetation seem most important. The clearing of the land in the eastern and midwestern United States allowed the eastward spread of several grassland birds, including the horned lark and the Savannah sparrow (Brewer 1991). Several species of birds and mammals extended their ranges northward in the past 50 to 75 years, possibly in response to a climatic warming trend; one example is the northern cardinal. The cardinal was absent from Michigan (and comparable areas in Wisconsin and westward and in Ontario and eastward) in the late 1800s. The first known nesting in Michigan was in 1892. By 1920 it occupied a range about one third of the way up the state; by 1950 it had reached the base of the "thumb," and in a rapid expansion over the next 20 years it had reached the northern tip of the lower peninsula. Some other species that moved northward in the United States include the opossum (Figure 2-20), the nine-banded armadillo, the mockingbird, and the tufted titmouse.

Beginning in the early 1970s, the rate of northward expansion of ranges slowed, and some northern species began to expand their ranges south. Included have been such boreal species as pine siskins, brown creepers, and red-breasted nuthatches (Brewer 1981). Climatic change also seems an important ingredient in these range changes. Beginning in the 1960s, the eastern United States returned to slightly cooler summers and much cooler winters, reminiscent of the colder years of the 19th century (Diaz and Quayle 1980, Kerr

1985). Still more recently, beginning about 1980, a pronounced warming trend, perhaps related to a worldwide greenhouse effect, has been evident (Kerr 1991).

Similar patterns are known from Europe (Udvardy 1969). A hundred years ago the nightingale in Scandinavia occurred only in southern Sweden. Since then it has spread as far north as Stockholm and Uppsala. Because Europe has a much longer historical record than North America, however, we know that the nightingale was common around Stockholm and Uppsala at the time of Linnaeus in the middle 1700s. It had retreated southward in the next hundred years and then readvanced more recently. This points up the fact that geographical ranges are dynamic. Boundaries fluctuate in response to many environmental factors, and the shifts, especially for animals and short-lived plants, may be rapid.

Ecology of Individual Organisms: *Important Abiotic Factors*

OUTLINE

- Temperature
- Moisture
- Light
- Soil
- Fire
- Pollution

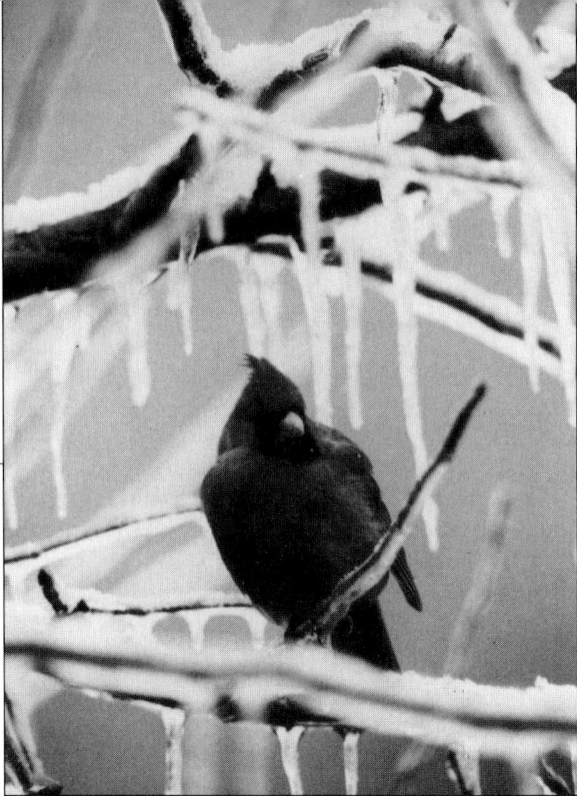

For homoiotherms like this northern cardinal, enough food to maintain body temperature is the key to surviving cold weather. *(Carl R. Sams II/Dembinsky Photo Associates)*

The list of abiotic factors having ecological effects on organisms is long. The more widely important factors include temperature, moisture, light and other kinds of radiation, texture and chemical composition of soil (for some plants and soil animals), dissolved gases and chemicals (for some aquatic organisms), gravity, pressure, and sound. The study of all the specific effects that such factors have on organisms constitutes a vast subfield of ecology, overlapping the field of physiology, which is referred to as **physiological ecology.**

Organisms affect the physical environment through their reactions, thereby altering conditions and indirectly affecting other organisms. A few reactions are discussed in this chapter, but because the reactions of individual plants or animals are usually slight compared with the combined effects in a community setting (the effects of a single tree versus those of a forest canopy, for example), most of the discussion is delayed to the section on community ecology in Chapter 12.

Only temperature, moisture, light, and fire are discussed extensively in this chapter. Additional information on various abiotic factors are included in Chapters 15 to 17 for terrestrial habitats and Chapter 18 for aquatic ones.

Temperature

Thermal Relations of Organisms

Temperature is important in controlling the rate of processes inside an organism and, thereby, its activity. In general, processes go faster at higher body temperatures. Any organism tends to have an optimal range of body temperatures, and most species show adaptations for maintaining this temperature. Such morphological, physiological, or behavioral features work by affecting the rate of heat gain or loss, or both.

When the relationship

$$\text{Heat gain} = \text{heat loss}$$

holds, the body temperature of the animal or plant will stay the same. If heat gain exceeds heat loss, body temperature will rise, and if heat loss exceeds heat gain, body temperature will drop.

We can look at the two sides of the equation in slightly more detail. Heat gain is from three main sources (Figure 3-1):

1. Solar radiation. This includes heat radiation and also shorter-wave radiation such as visible light (see Figure 3-14) which, when it is absorbed by the animal or plant, is converted to heat.
2. Infrared and heat radiation from surroundings. For example, heat is radiated from the ground or other organisms.
3. Metabolism. As the organism uses energy stored in organic compounds, this energy is released as heat in its tissues.

Heat is lost along three main avenues:

1. Loss by infrared and heat radiation. The heat contained by the organism from metabolism or other sources is lost by infrared and heat radiation from its body. The loss is to its surroundings, including other organisms, the ground, the sky, etc.
2. Convectional heat loss. The temperature of the air next to an organism may be raised by heat conducted from its body. If this air is carried away by convection and replaced by cooler air, the organism has a net loss of heat. On the other hand, if the surrounding air is warmer than the temperature of the organism's outer surface, convection can be a source of heat gain rather than loss.
3. Evaporative heat loss. When water changes from liquid to vapor, heat is lost from the evaporating surface. Plants may lose heat in the process of transpiration, animals by the evaporation of sweat or saliva. When the temperature of the air and the surfaces around an organism is equal to or higher than that of the organism, evaporation is the only avenue available for losing heat.

Organisms with a dark surface, skin, or fur absorb more solar radiation than ones with light-colored surfaces and so can heat up faster in sunlight (Cossins and Bowler 1987). Although a black bird gains more heat from sunlight than a white one, it loses no more by thermal radiation. This is because color has little effect on the exchange of infrared and heat radiation between the organism and its surroundings. The surfaces of most organisms, no matter what the color, are good at absorbing (and emitting) thermal radiation. What does influence loss of heat by radiation is temperature: increased skin temperature or decreased temperature of the surfaces to which the organism is radiating both increase radiational heat loss.

If we walk barefoot across a cold floor, we lose heat by conduction, but for most terrestrial animals heat loss by conduction is a minor factor in the balance sheet because of the very low thermal conductivity of air. The thermal conductivity of water is about 23 times higher and so for aquatic animals heat loss by conduction is an important design consideration.

Homoiotherms and Poikilotherms

Fundamental differences exist among organisms in the relationship of body temperature to environmental temperature. In **poikilotherms,** body temperature tends to match environmental temperature, so that as environmental temperatures go down, the rates of their bodily processes also go down. This is an oversimplification that we will correct shortly. In **homoiotherms** or **homeotherms,** body temperature tends to stay constant even when the environmental temperature changes. "Endotherm" and "ectotherm" are synonyms for homoiotherm and poikilotherm that focus on whether the main source of heat gain is internal, from metabolism, or external, from the environment.

Homoiotherms, or warm-blooded animals, are mainly birds and mammals. They have automatic, basically physiological, mechanisms for keeping body temperature constant despite changing outside temperatures. Keeping body temperature high at low environmental temperatures involves increasing metabolic rate (which is the same thing as increasing heat production) and increasing insulation by adding fat or fluffing up feathers or fur (to increase air space). Keeping body temperatures low at high environmental temperatures involves lowering heat production and increasing heat loss by various means, such as evaporating water from sweat glands or by panting.

Poikilotherms—cold-blooded animals and plants—have no internal physiological means of keeping body temperature constant; if you put a

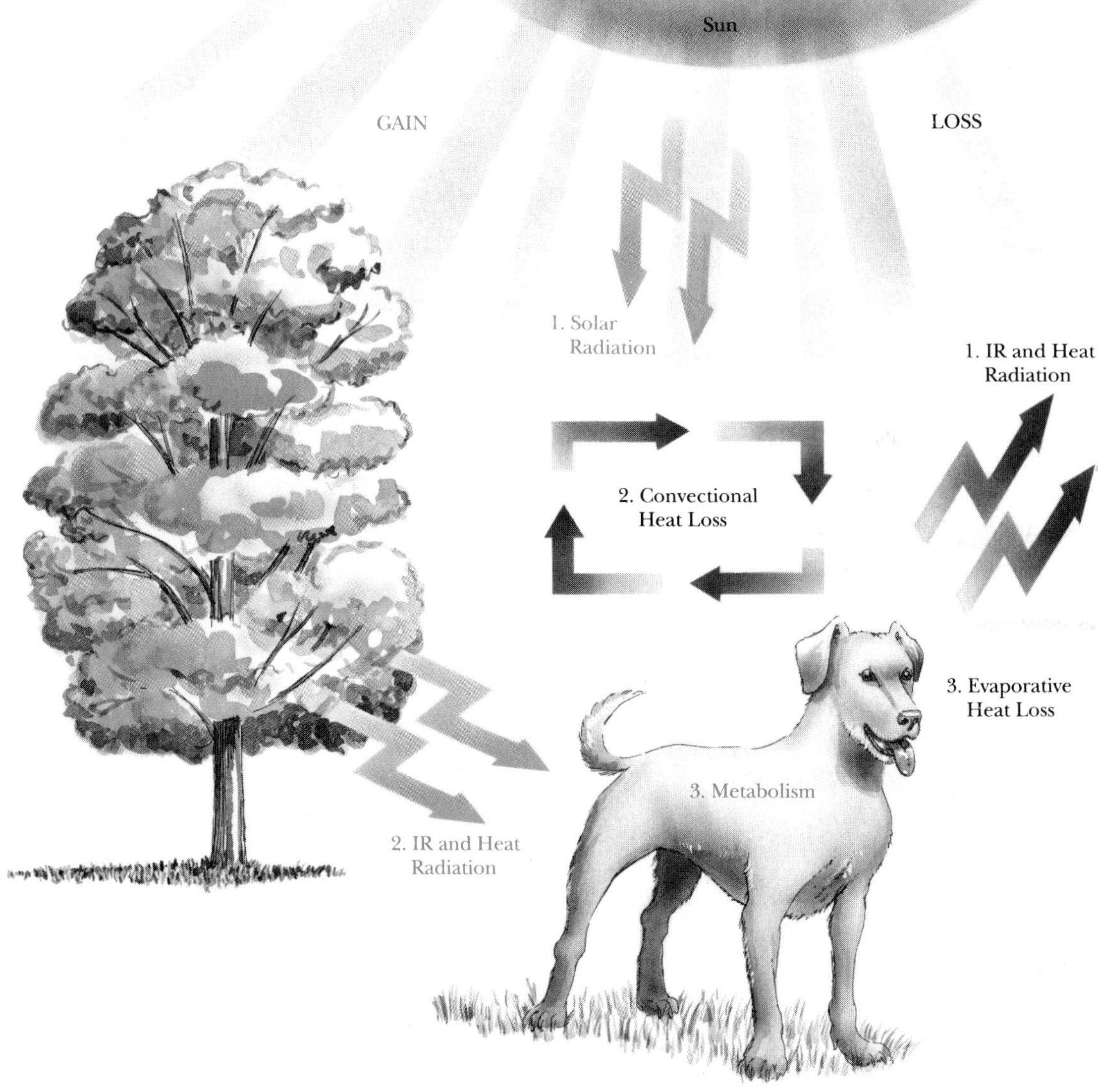

Figure 3-1

The main avenues of heat gain and loss for an organism.

lizard in a refrigerator, its body temperature goes down. Even though this is true, if you were to take the body temperature of a bird and a lizard at various times during a day while they were free to move around, you would find that the body temperature of the lizard was not very much more variable than that of the robin. The lizard, however, uses behavioral rather than physiological means of regulating its body temperature (Cowles and Bogert 1944). When the air temperature is low, the lizard sits broadside to the sun and presses its belly against warm rocks. When the air temperature is high, the lizard walks on tiptoe or stays quietly in the shade (Putnam and Wrattan 1984). If you continue taking the lizard's temperature on a cold night, you will find that its temperature drops several degrees, and if you continue as winter sets in, its temperature drops still further when it finds a hole for hibernation and settles in. During this time, you would find that the bird's temperature stays practically constant.

Figure 3-2 gives an idea of the relative constancy of body temperature that poikilotherms can show and also emphasizes the behavioral basis of it. Plotted are shell temperatures of Hermann's tortoises on a preserve in Italy (Chelazzi and Calzolai 1986). Resident animals, familiar with the area, showed body temperatures mostly between 27 and 30°C over a range of air temperatures from about 20 to 26°C. Newly imported tortoises brought in from a similar habitat 50 km away had a much more variable body temperature. Probably the imports could not thermoregulate as well because they were still unfamiliar with the preserve's microclimates such as the distribution of shady and sunny spots.

The relationship between the organism and its environment is very different in homoiotherms and poikilotherms (Figure 3-3). The way in which extremely low environmental temperatures are tolerated, for example, is different. Poikilotherms begin to become inactive as temperatures drop, though "low" is a relative term. Antarctic mites are active at subzero temperatures, though their optimal temperature for activity is considerably higher (Block 1990). Poikilotherms may seek protected microhabitats, if available, but after that their survival depends on tolerating the cold. It is useful to think of **freeze-tolerant species,** which survive ice formation in their extracellular spaces, and **freeze-susceptible** species, which do not survive freezing (Lee and

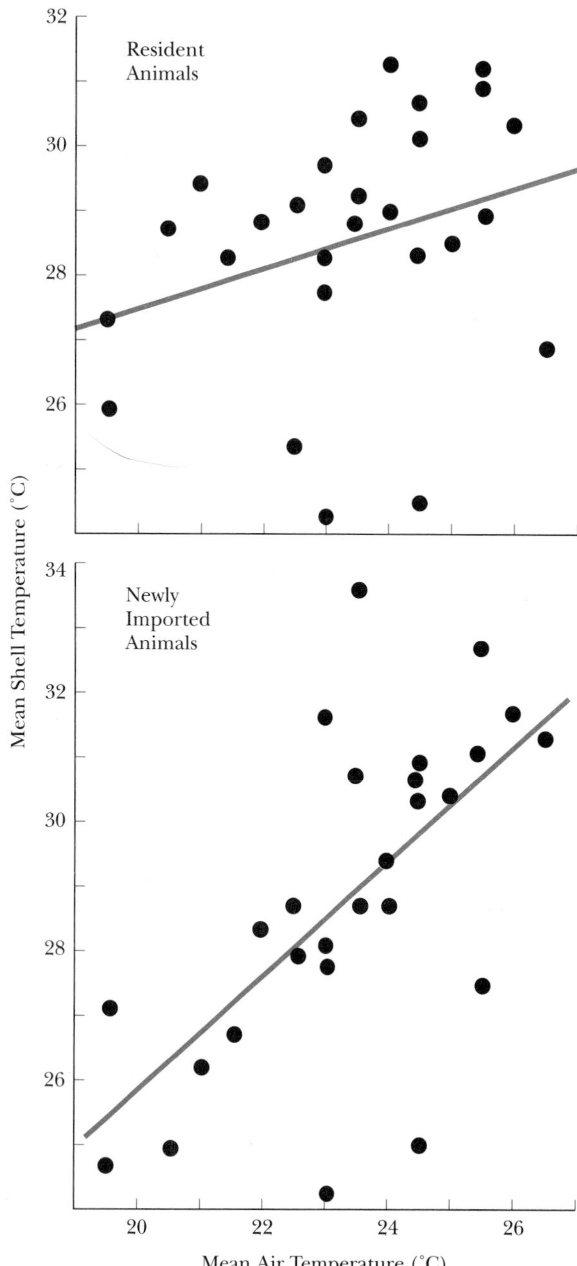

Figure 3-2

Relationship between shell temperature and air temperature of resident and newly imported land turtles on a preserve in Italy. From Chealazzo and Calzolai (1986).

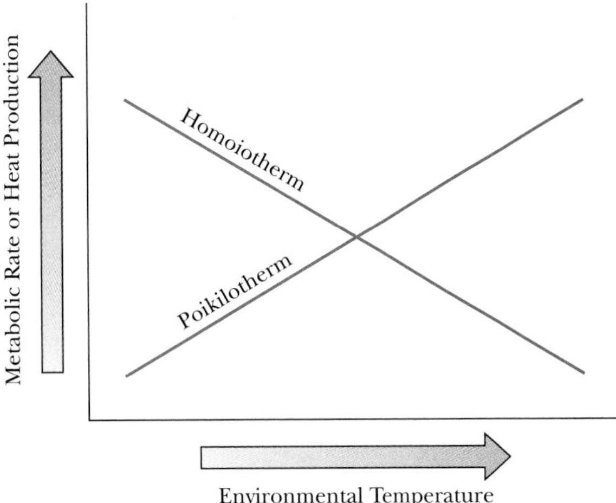

Figure 3-3

A simplified view of the relationship between metabolic rate and environmental temperature in homoiotherms and poikilotherms. In homoiotherms, over a wide range of temperatures, metabolic rate increases as environmental temperature decreases, with the result that body temperature remains about constant. In poikilotherms, decreased environmental temperature leads to decreased body temperature (unless the animal can compensate by behavioral means) with a consequent decrease in metabolic rate. The actual relationship for many homoiotherms is a little more complicated than shown. At moderate temperatures, *a thermoneutral zone* exists in which no added energy is needed for warming (or cooling) the body; for such organisms, it would be more realistic to add a flat segment on the right end of the homoiotherm line.

Denlinger 1991). Based on lower temperatures or other proximate factors, both groups tend to build up chemicals in their cells that lower the freezing point (like the antifreeze in a car radiator).

Freeze-tolerant species include many plants (Halfpenny and Ozanne 1989) and also, it is now realized, a considerable number of animals. Included are sessile or poorly motile animals of habitats that get very cold, like the intertidal zone of northern shores; barnacles and various mollusks are examples. Also freeze-tolerant are various other animals whose life histories compel them to overwinter in cold, exposed situations; known examples include tardigrades (Westh et al. 1991); several insects, mostly beetles, flies, and hymenoptera; and a few vertebrates, such as the wood frog (Cossins and Bowler 1987). These organisms have ice-nucleating agents (INAs) that encourage freezing of extracellular fluid at a moderate temperature; this presumably lowers the danger of intracellular ice formation, which is usually lethal.

Freeze-susceptible organisms include many species—tropical and temperate ones but also many with northerly ranges. The latter tolerate environmental temperatures substantially below 0°C by accumulating chemicals, such as glycerol, that lower the freezing point of their fluids and also facilitate supercooling, that is, avoiding ice formation at temperatures below the freezing point.

For homoiotherms, surviving low temperatures is approximately a matter of getting enough food to keep heat production high enough to maintain normal body temperature. If a homoiotherm cannot find enough food (for example, a bobwhite during an ice storm), its body temperature will drop and it will die quickly at a body temperature well above freezing.

Depending on your preconceptions, you may at this point be feeling sorry for poikilotherms, poor sluggish things wasting much of their time comatose in hibernation. Poikilotherms, you may think, would be homoiotherms if they could, just as the poor would be rich and the plain would be handsome. If you are of a different cast of mind, you may wonder instead what point there is to homoiothermy. The homoiotherm is condemned to a life of running around, eating continually to get enough food to metabolize the energy to keep running around. Why isn't the life of a poikilotherm better? On a chilly day, its body temperature stays too low to allow it to be active, but at the same time its bodily processes slow down so it doesn't require much food. In the winter, when food is scarce anyway, doesn't it make more sense to be snug (although comatose) with a body temperature of 4 or 5°C and a metabolic rate so slow that a few grams of fat provide enough energy to last the winter?

Poikilothermy and homoiothermy are alternative strategies. Poikilothermy is an energy-conservative, but low-power, strategy; homoiothermy is an extravagant but powerful one. In a favorable, food-rich habitat, the homoiotherm wins. This is not to say that a poikilotherm living by itself in such a habitat would not get along quite well. But in a community of homoiotherm competitors the poikilotherm's food supply gets eaten up while it is munching slowly on a leaf or sitting in its burrow.

There are, however, many situations that are not favorable to homoiotherms (Pough 1980). Poikilotherms can more readily be permanent residents of periodically unfavorable habitats. When the food or the water or the oxygen disappears for the season, poikilotherms can virtually shut down while the homoiotherm, even if it greatly restricts activity, continues to use a large amount of energy and, accordingly, to require oxygen and water continually. Aquatic environments seem to favor poikilotherms, in part because water is such a good conductor. The homoiotherms that do spend time in water conserve heat with insulating layers of feathers, blubber, or (in the case of the sea otter) fur.

Also, poikilotherms can be two things morphologically that are difficult for homoiotherms. First, they can be very small. Because of the surface-volume relationship discussed in Chapter 2, homoiotherms simply require too much energy to exist at sizes much below 5 g (Figure 3-4). In contrast, poikilotherms in great numbers exist in sizes to inhabit such microhabitats as rock crevices, gravel banks, soil, and the insides of other animals and plants.

The other morphological form that is difficult for homoiotherms is an elongate wormlike or snakelike shape. The reason is similar. This shape, with its large surface area relative to volume, leads to high energy demands compared with a more compact physique. This is true for the homoiotherm, needing to maintain its body temperature from within. The poikilotherm, without this drain, is free to be a worm or a snake and, thereby, to exploit ways of life not available to homoiotherms. Large snakes in narrow mouse burrows and the great variety of worms in the soil are examples.

Hibernation and Other Combinations

The swordfish is a large, active predator that comes to the surface waters of the ocean at night and retreats to lower depths in the day. The range of water temperature to which it's exposed in the course of a day may be as much as 19°C. The swordfish has a brain-eye heater that keeps these organs—essential for its predatory activities—nearly stable in temperature despite the great range in environmental temperature (Carey 1982, 1990). The heater consists of mitochondria-rich heat-generating tissue, a countercurrent blood supply that con-

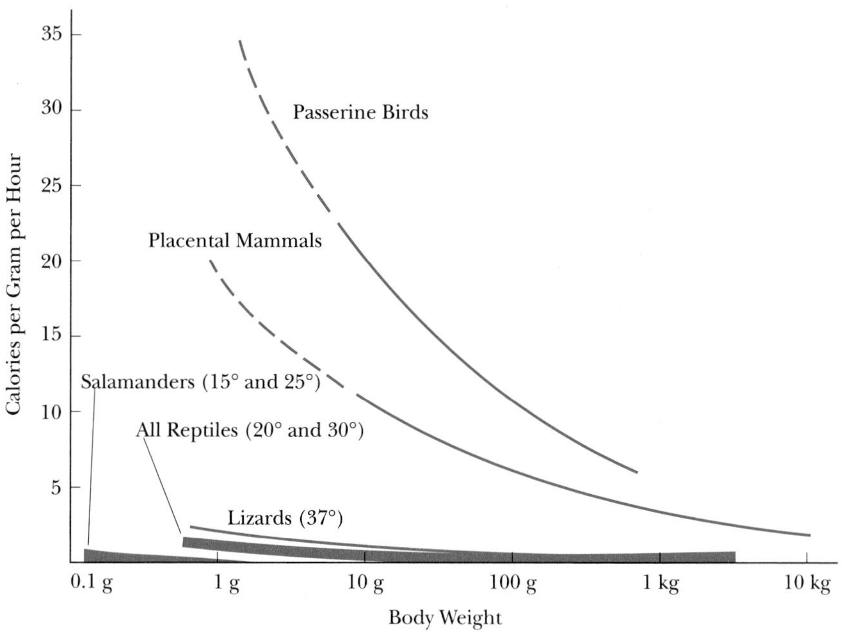

Figure 3-4

Resting metabolic rate in various groups of organisms as a function of body size. Metabolic rate is given per gram of tissue, not per animal. Note also that weight is plotted on a logarithmic scale. Passerine birds and placental animals, as homoiotherms, show higher metabolic rates and ones that go up more sharply at smaller body sizes. Temperatures are in °C. (From F. H. Pough, "The advantages of ectothermy for tetrapods." Am. Nat., vol. 115, 1980, p. 99. Copyright © University of Chicago Press. Reprinted with permission.)

serves the heat, and a thick layer of insulating fat around the brain and eye.

The point of the swordfish is that homoiothermy and poikilothermy are modes that are adaptive to a particular environment and way of life; when something less than a total commitment to one mode may be fitter, such a modification may evolve. The restriction of homoiothermy to the brain and eyes in the swordfish, saving the energy required for whole-body homoiothermy is one such case. The temporary homoiothermy while incubating eggs shown by certain reptiles is another. **Hibernation,** a much more familiar example, is a third.

The basically warm-blooded animals that hibernate in the winter are termed **heterotherms.** As winter approaches, the homoiotherms that hibernate find a place to stay and become essentially poikilothermic, so that as the temperature in their hibernaculum decreases, so does their body temperature. They save the energy they would otherwise spend in keeping their body temperature constant and in moving about. Their hibernation is, however, a more controlled state than that of poikilotherms. If the temperature in their hibernaculum gets close to freezing, their heat production begins to increase. If the temperature goes still lower, they rouse, become thoroughly homoiothermic again and, if they are lucky, find a deeper hole.

Hibernation is so much a part of folklore and common knowledge that we do not always realize that only a few mammals (and still fewer birds) use this means of coping with the low temperatures and diminished food of winter. Woodchucks, thirteen-lined ground squirrels, and jumping mice hibernate. Raccoons, possums, foxes, and squirrels do not. Not even bears and chipmunks hibernate, although they spend the winter in a burrow. They sleep much of the winter, but their body temperature does not drop much more than ours does when we sleep. The bear is a large animal with a slow metabolic rate and can store enough fat to sleep through the winter without becoming poikilothermic. The chipmunk stores food to eat when it wakes up every so often. Very small animals like shrews and most mice do not hibernate; they would have a hard time accumulating enough fat to get through the winter as hibernators and, anyway, it is not very wintery where they live under the insulation of leaves and snow.

Interaction of Temperature and Wind

A familiar example of the way factors interact is the relationship between low temperatures and wind in contributing to heat loss. The **wind-chill index** (devised by Siple and Passel in 1945), well known to all of us from weather reports, expresses this interaction. Wind, or "forced convection," removes the warm air layer on the outer surface of organisms and may even penetrate their fur, feathers, or clothes and blow away the warm air trapped there. As a consequence, heat loss is faster as a result of both lower temperatures and higher wind velocities. The usual way of expressing wind-chill effect for weather reports is to give the equivalent temperature for still air (Figure 3-5). For example, a temperature of 20°F (−7°C) when the wind is 35 mph (56 kph) is equivalent in chilling ability to a temperature of −20°F (−29°C) with no wind.

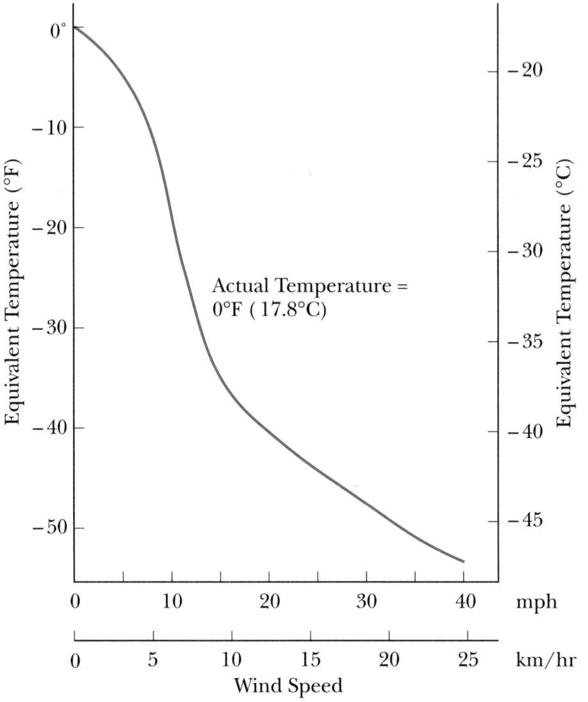

Figure 3-5

The effect of wind velocity in increasing the chilling power of the air. A temperature of 0°F (17.8°C) is used for illustration purposes. This temperature in still air is equivalent to about −40°F (also −40°C) when the wind is 20 mph.

Although the wind-chill index has been studied mostly in connection with the ability of humans to live and work in cold environments, the interaction is also important for other organisms. Quantitative data on the phainopepla, a dark, crested songbird of the desert scrub, illustrate the point. These birds, like many other species, spend the night on roosts sheltered by vegetation. Although these sites save the birds some energy by reducing their radiation loss to the night sky, the biggest saving comes through protection from the wind. Protection from forced convection provides five times the savings that stem from the improvement of radiation balance (Walsberg 1986).

Careful observers have long been aware of the importance of wind in the life of birds. Aldo Leopold (in the essay "65290" in *Sand County Almanac*) suggested that one of the commandments that chickadees learn in Sunday school is, Thou shalt not venture into windy places in winter. "In winter [the chickadee] ventures away from woods only on calm days, the distance varies inversely as the breeze. I know several wind-swept woodlots that are chickless all winter but are freely used at all other seasons. . . . To the chickadee, winter wind is the boundary of the habitable world."

Life Forms of Plants

The **life form** system of categorizing plants was devised by the Danish ecologist C. Raunkiaer (1934) and is based primarily on the methods by which plants survive the coldest season. Specifically, the main categories are based on the location of the bud from which new growth will sprout. As modified by the Swiss plant sociologist J. Braun-Blanquet and by others, the system now generally includes the broad divisions shown in Table 3-1 and Figure 3-6.

Table 3-1
Raunkiaer's Life Form System*

Life Form	*Method of Surviving Unfavorable Season*	*Examples*
Therophytes	Annual plants that survive the winter (or dry season) as seeds. Seeds are typically much more resistant to cold and drought than are growing plants.	Annual ragweed, blue-eyed mary, lettuce
Hydrophytes	Rooted water plants. The bud that produces the next year's growth is under water during the winter and is thereby insulated.	Pondweed, water lily, pickerelweed
Geophytes	Next year's bud buried in the soil and thus well insulated. Includes plants with underground stems, such as bulb plants.	Tulips, ferns, trout lily
Hemicryptophytes	Buds are located close to the surface of the ground; accordingly, they may be insulated by litter or snow but do not have the further insulation of soil that geophytes have.	Goldenrods; perennial grasses; all biennials such as carrot (wild and tame); mullein
Chamaephytes	Here the renewal buds are located above the ground but not very high (no more than 25–30 cm). The buds may sometimes be insulated by a snow cover, and they are low enough that they are not exposed to strong winds.	Trailing and creeping shrubs such as wintergreen; fruticose lichens; succulents
Phanerophytes	Mainly trees and shrubs; here the buds are located on shoots more than 25–30 cm above the ground and are thus more exposed than any other group.	Essentially all trees and shrubs; also lianas (climbing vines) such as poison ivy; in the tropics are some herbaceous phanerophytes

* In addition to these major divisions, the classification also includes phytoplankton (e.g., aquatic algae), phytoedaphon (e.g., soil algae), endophytes (mainly plants that are internal parasites), and tree epiphytes (e.g., Spanish moss and many tropical orchids).

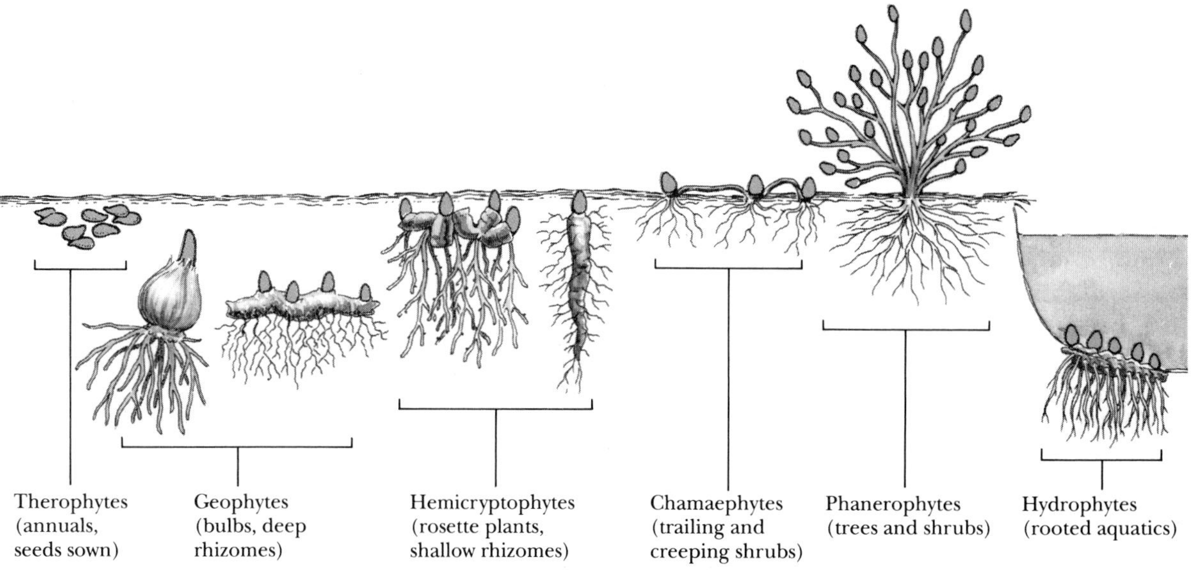

Figure 3-6

The buds or perennating structures of various plants are in blue in this diagram illustrating the major divisions in Raunkiaer's life form system. The buds are shown somewhat exaggerated in size.

The proportions of the flora in the various categories (which Raunkiaer called the biological spectrum) vary from one climate to another (Figure 3-6): deserts typically have a high proportion of therophytes; prairies and steppes have a preponderance of hemicryptophytes; alpine areas may have a high proportion of chamaephytes (although hemicryptophytes may be even more important) that are well protected through the winter by deep snow. Tropical rain forest is, of course, typical of a phanerophytic climate. Other factors besides regional climate affect the biological spectrum; in temperate deciduous forests, among the herbs, hemicryptophytes predominate in the dry oak forests and geophytes predominate in the mesic beech-maple forests.

The life-form system is an early and still-useful attempt to relate plant morphology and life history to climate. Particular plant species may, however, show much more subtle adaptations than the basic divisions of the Raunkiaer system. For example, alpine areas in the tropics are subject to low temperatures, not in winter but at night. Above 4000 m, frost occurs almost every night; one botanist referred to the climate as summer every day and winter every night. Giant rosette plants characteristic of these areas of alpine tundra form "night buds," in which the adult outer leaves fold inward over the temperature-sensitive growing points, providing insulation during the night (Beck et al. 1982).

Moisture

Whether a piece of land is wet or dry for the organisms living there depends mainly on two factors, **soil moisture** and the **evaporative power of the air** (Daubenmire 1959). Soil moisture is discussed in the next section, on soil. The evaporative power of the air depends on humidity, temperature, and wind. The evaporative power of the air is increased and an area made drier, as far as organisms are concerned, by lower humidities, higher temperatures, and more wind.

Humidity

Humidity is defined as moisture in the air in the form of vapor; it may be expressed in several ways.

It can be given as the actual mass of water in a certain volume of air, **absolute humidity.** A more familiar measure is **relative humidity,** which is the actual absolute humidity expressed as a percentage of what the absolute humidity would be if the air were saturated with water vapor at the same temperature and pressure. The amount of water vapor that air can hold increases with temperature. At −30°F (−34°C) 1 cubic foot of air can hold only about 0.1 g; at 40°F (4°C) it can hold nearly 3 g; and at 100°F (38°C) it can hold nearly 20 g. If the amount of water stays constant, lowering the temperature raises the relative humidity and raising the temperature lowers it. This is why the air in your house in the winter is very dry (unless you have a humidifier); the cold air outside may have a relative humidity of 80% but actually very little water. Bringing this air into the house and heating it to 68°F (20°C) may drop the humidity to a desertlike 10 or 20%.

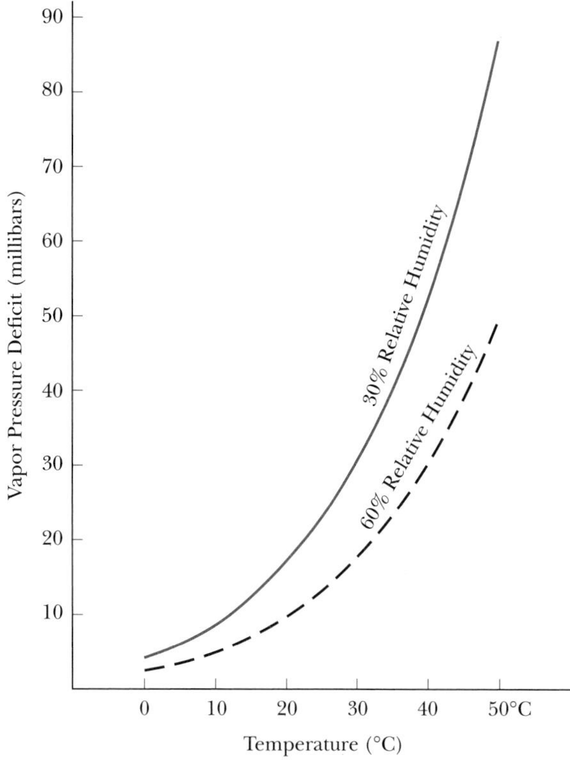

Figure 3-7

Evaporative power of the air (as measured by vapor pressure deficit) at different temperatures for two relative humidities. (Based on Table 6.1 in Fritschen & Gay 1979.)

Vapor Pressure

Humidities may also be expressed in terms of the **partial pressure** of the water vapor in the air (that is, the contribution the water molecules make to the total air pressure). This is usually expressed in millibars (mb). Relative humidity by itself gives only a crude idea of the evaporative power of the air because air with the same relative humidity has a higher evaporative power at high temperatures than at low (Figure 3-7).

That is as much as we need to know about such things for everyday ecological purposes, but it is interesting to pursue the reason a short distance into the land of physics. The rate of water loss from the evaporating surface of an organism depends on the difference in vapor pressure between the surface and the air (that is, on the **vapor pressure gradient**). Vapor pressure rises exponentially with temperature, and the rate of increase is steeper at higher relative humidities. The other thing we need to know is that the vapor pressure of an organism's evaporating surface is always taken to be the saturation vapor pressure.

When the temperature of the organism is higher than that of the air—as is nearly always the case with homoiotherms—water loss will be higher than if they were at the same temperature. In fact, water loss by evaporation will occur into air of 100% relative humidity as long as the organism's temperature is higher than the air temperature.

Water Balance in Plants

Plants and animals must maintain a favorable water balance; that is, they must take in about as much water as they lose and use. For terrestrial plants, water is absorbed from soil water through the roots. The plant uses water in photosynthesis, the hydrogen ending up in carbohydrates, and it loses water to the air through its leaves **(transpiration).** Most of the loss is not through the leaf surface but through microscopic pores called **stomata** (Figure 3-8). In most plants these are open during the daytime and closed at night; they are also closed when water balance becomes unfavorable and the plant wilts. Less than 1% of the water taken up by the roots is used

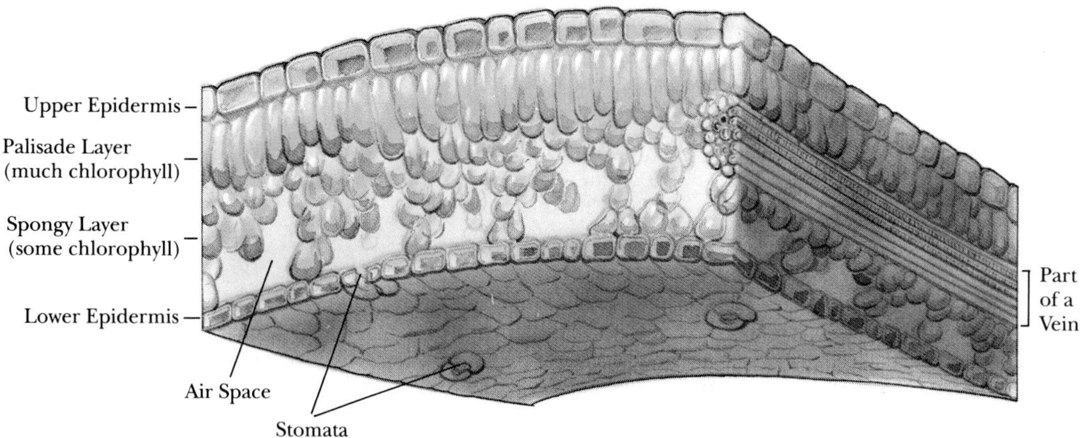

Figure 3-8

A diagrammatic view of a small part of a typical leaf blade of a mesophyte. The front consists of a cross section through the leaf. The side is a longitudinal section. The bottom is a view of the lower surface of the leaf. Water loss from the leaf is mainly through the small pores (stomata) in the lower surface. They are each surrounded by two cells (guard cells) that close the stomata at certain times (usually at night and when the plant is wilted). The stomata open into air spaces in the spongy layer. Water is transported up from the roots through the stems and into the leaves in conducting tissue like that shown in the vein at the right. (Adapted from Edmund W. Sinnott, *Botany Principles and Problems,* New York, McGraw-Hill, 1946, p.110.)

in photosynthesis; the rest is transpired. Presumably the function of the stomata is to allow carbon dioxide and oxygen from the air to enter the leaf. The water loss is a by-product of this arrangement necessary for photosynthesis and respiration. Consequently, botanists traditionally regarded transpiration as a necessary evil (or, as the University of Illinois plant ecologist A. G. Vestal is supposed to have said, a damned shame). Of course, the upward movement of water that results from transpiration brings with it needed mineral nutrients; also there is evidence that the cooling effect of transpiration is useful to plants in some circumstances (Gates 1968).

Xerophytes and Mesophytes

Xerophytes, mesophytes, and **hydrophytes** are categories of plants based on their water relationships. Xerophytes grow where it is dry *(xeric),* hydrophytes grow where it is wet *(hydric),* and mesophytes grow where there is a moderate amount of water (*mesic,* meaning medium and pronounced "mezzick"). For xerophytes (and, during drought, for mesophytes), maintaining water balance poses a problem.

There are several kinds of xerophytes, each adapted in its own way to life in a dry environment (Figure 3-9). To botanists, the **true xerophytes** are perennials, usually shrubs such as sagebrush, that have small, hard leaves which can decrease water loss greatly when the soil gets dry and which are dropped altogether during prolonged drought.

Succulents are another kind of xerophyte. Cacti are examples. They have fleshy tissue in which water, when it is available, can be stored for later use. Many succulents have uncoupled the light and dark reactions of photosynthesis in such a way that they can keep their stomata closed in the daytime, when it is hot and water loss would be high, and open them taking in carbon dioxide at night when it is cooler. The carbon dioxide is used to form organic acids that are stored in cell vacuoles in the fleshy stems or leaves. In the daylight, carbon dioxide yielded by the organic acids is processed in standard C_3 fashion. The arrangement is known as **crassulacean acid metabolism** (CAM), after the Crassulaceae, a plant family that includes such succulents as the rock-garden sedums and sempervivums.

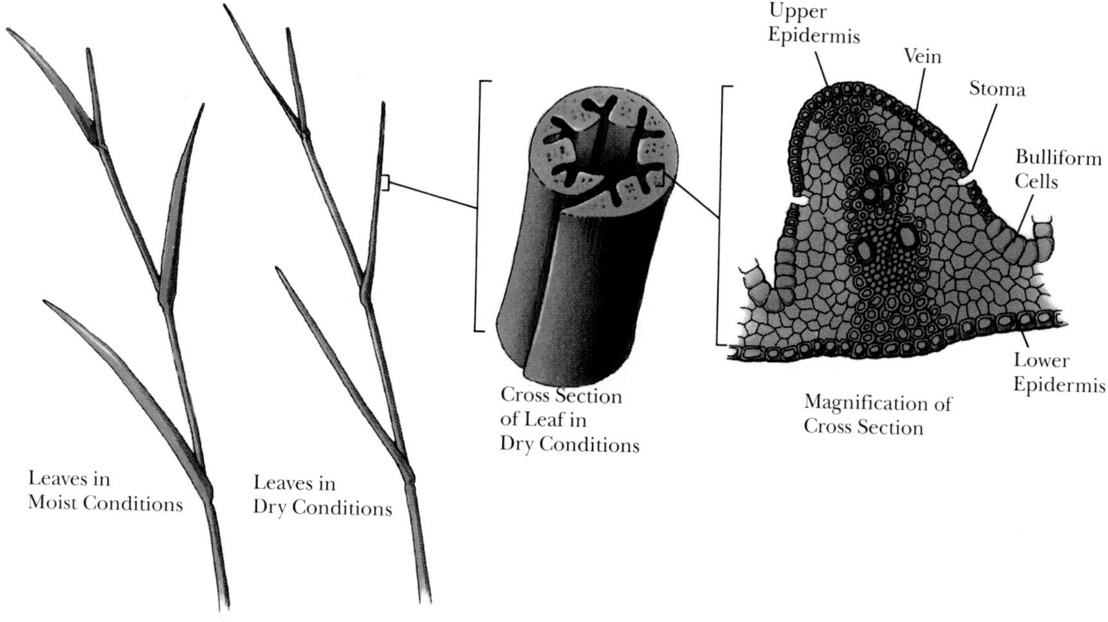

Figure 3-9

Many xerophytic grasses have leaf structures similar to that of western wheat grass, shown here. The stomata are located on the top of the leaf in deep channels. Presumably the air in these channels is less apt to be swept away by wind and, accordingly, is more likely to maintain a high humidity and consequently lower transpiration water loss. The lower epidermis is covered with thick-walled cells through which little water can escape. Under moist conditions the leaf is flat like any grass blade in a lawn. With drought, however, thin-walled cells *(bulliform cells)* on the upper surface of the leaf lose water rapidly, become limp, and allow the leaf to roll up. When this happens the stomata are totally enclosed within the inrolled leaf and the only surface of the leaf that is exposed is the heavily cutinized lower epidermis. The thick-walled cells surrounding the vein are like I-beams, preventing the leaves from drooping and bending the water-conducting tubes. Notice also that the amount of water-conducting tissue is large compared with that in the mesophytic leaf, Figure 3-8. (Based in part on Weaver and Clements 1938, p. 450.)

Desert ephemerals are annuals that look much the same as their relatives living in moister regions. Their adaptation to desert life is in their life history rather than in their structure. When enough rain falls, the seeds germinate, the plant grows, flowers, sets seed, and dies, all in a short period when conditions are fairly moist. The new seeds then lie in the ground until enough rain falls again; this may occur the next year or not for several years.

Another category of xerophytes is the **phreatophytes** (from a Greek word meaning well). These plants grow in the deserts but have roots that go down far enough to tap the water table. Mesquite is a phreatophyte that may have a root system reaching down 175 feet.

Many xerophytes are C_4 plants, which do well in arid habitats in part because of an ability that may seem unrelated to drought. They are able to use carbon dioxide (for photosynthesis) at very low concentrations. This is an advantage to a xerophyte because it means that it can keep making food when its stomata are nearly closed, and even afterward as long as some small amount of carbon dioxide remains in the air spaces. C_3 plants, lacking such ability, have to keep their stomata open to continue photosynthesizing. There are, nevertheless, C_3 xerophytes; they do not have this big advantage, but they deal with the stresses of aridity in other ways (Hanson and Nelsen 1980).

Water conservation is a serious matter for xerophytes. As the foregoing discussion may suggest, however, it is not their central problem. That, as for every organism, is success in a community of competitors and predators. The Russian plant physiologist N. A. Maximov (1931) noted that xerophytes tend to have larger bundles of water-conducting tissue in their stems and leaves than do mesophytes. Clearly, this is no adaptation for conserving water; in fact, it allows faster water loss. Consider a xerophyte and a mesophyte growing near one another in the desert and drawing on similar amounts of soil moisture. In the heat of the day the mesophyte wilts because it cannot transport water as fast as it is lost. The wilted plant can no longer carry on photosynthesis, and eventually it dies without having thoroughly depleted the soil moisture around it. The xerophyte remains unwilted and carries on photosynthesis until the moisture around its roots is gone; only then are the water-conserving mechanisms used. In the meantime it has stored food and possibly produced flowers and seeds.

We can imagine the perfect water conserver: possibly, it would be ball shaped, with a thick waxy covering. It survives for a long time—until perhaps a mule deer comes along and eats it. But meanwhile some less conservative plant will have reproduced and populated the desert.

Hydrophytes and Hydroperiod

For hydrophytes, the problem is not too much water but too little oxygen or carbon dioxide, because the concentration of these vital gases tends to be lower in water than in air. Leaves of hydrophytes that grow submerged or floating tend to have big internal open spaces (lacunae) where the carbon dioxide given off when the cells respire and the oxygen given off in photosynthesis can accumulate and be recycled (Figure 3-10). In floating-leaved and emergent hydrophytes, such spaces, continued in stems and roots, also allow for the transport of air from the leaves to the underground or submerged parts. Various sedges, grasses, and rushes have such structures as do some wetland trees including slash pine, cabbage palm, and cypress (Brown et al. 1990).

Some sites are perpetually wet or dry but many others are wet for a part of the year. The length of time that the soil is saturated is called the **hydroperiod.** Here we are referring to a portion of a year rather than a portion of a day as in the terms photoperiod and thermoperiod.

On sites where the hydroperiod is short—days or weeks—many of the same species that live on unflooded sites occur. Relatively few plant species do well when the soil is waterlogged. These are ones that can tolerate the lack of soil oxygen and also the high concentrations of dissolved iron and manganese that may develop. As hydroperiods lengthen, such species tend to enter the community and as hydroperiods lengthen even more, the flora tends to be reduced to such species.

Some wetlands plants have xeromorphic features, that is, structural features often associated with xerophytes. This anomaly is at least partly dependent on hydroperiod. Plants growing in waterlogged soil have shallow roots. During a severe drought season, the water table may drop so low that only species possessing features that allow stringent water conservation do well (Lugo et al. 1989).

Halophytes

High concentrations of salts (sodium chloride and others) in soils make it difficult for ordinary plants to obtain sufficient water. **Saline soils,** which include the salt flats of deserts and salt marshes along the sea coasts, are occupied by plants called **halophytes.** Some halophytes have structural features found in xerophytes: for example, several are succulents. Examples are samphire and sea blite. An important early insight in ecology, by A. F. W. Schimper, a German plant geographer, was that saline areas (and certain other habitats, such as bogs) were "physiologically dry," even though physically wet. That is, water balance was a problem, not because of a lack of water but because something—high osmotic concentration in this case—prevented the plant from absorbing it.

Although the average mesophyte has trouble extracting water from saline soil, halophytes do not. They manage by increasing their own osmotic concentration. This generally involves allowing increased salt concentration in their cells but may also involve other solutes, such as malate.

Even though halophytes in general tolerate higher salt concentrations in their cell sap than nonhalophytes, they may also have ways of getting rid of the excess. Several, such as mangroves (Chapter 18) and tamarisk, have glands that excrete salt to the surface of the leaves. Halophytes also sidestep high salt concentrations in other ways. They tend to make most of their growth after heavy rains when the salt concentration is lowest, and this is also when germination ordinarily occurs (Ungar 1978). Although all halophytes germinate better in fresh

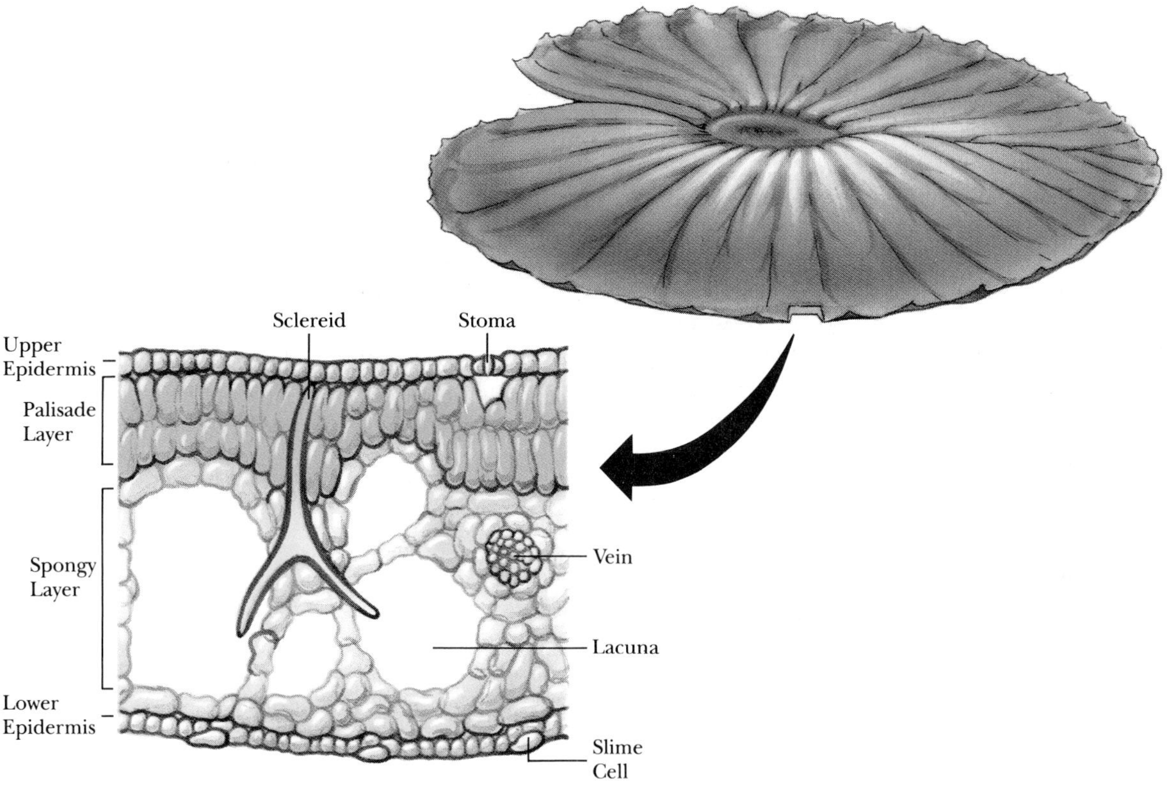

Figure 3-10

A hydrophyte leaf. This is a cross section through a small part of a water lily leaf, which floats on the water surface. The stomata are in the upper surface, the only part in contact with air. The large spaces in the spongy layer, called *lacunae,* do not open through stomata to the outside. They provide buoyancy; they also allow carbon dioxide to accumulate at night to be used by day, and oxygen to accumulate in the daytime for nighttime use. The upside-down Y is a thick-walled structure called a *sclereid.* It gives support to the leaf without giving it side-to-side rigidity, allowing it to ripple with wave movements. There is little water-conducting tissue. (Based in part on Weaver and Clements 1938, p. 431.)

than salt water, pregermination exposure to salt water increases the percentage of seeds germinating in some species (Woodell 1985). Many halophytes will grow perfectly well on nonsaline soils if given a chance, but some, such as most mangroves, are obligate halophytes.

Water Balance in Animals

For animals, water gain and loss are somewhat more complicated than in plants. The usual sources of gain are drinking water, water in food, and metabolic water (water molecules produced in the breakdown of foods). Loss is through urine, feces, and the water evaporated from the skin and lungs. Two kinds of adaptations are prominent in animals living in arid regions. One is lowering water loss as much as possible. This is achieved by reabsorbing water in the intestine to produce feces as nearly dry as possible and by using only the minimum water necessary to excrete the nitrogenous wastes of the urine. Cutting water loss from the lungs and skin is a dilemma for desert vertebrates similar to reducing transpiration loss for plants. Water loss from the lungs is an inevitable consequence of breathing; this water loss, along with that from the skin, also keeps the animal from becoming too hot. About the best that can be done is for animals to become nocturnal and, if they are small enough, to stay in burrows during the day. Inside the burrow it is cooler and the humidity becomes high enough that there is less evaporation from the lungs.

The second means of adapting to life in arid regions is the ability, when the animal does suffer a temporarily unfavorable water balance, to tolerate the resulting dehydration and elevated body temperature (Schmidt-Nielsen et al. 1957). Camels can readily go 8 days without drinking water and can tolerate water losses of 25% or even 40% of their body weight (Gauthier-Pilters and Dagg 1981). Humans can tolerate a loss up to about 20%, but by the time they have lost about 10%, they are physically and mentally unable to take care of themselves.

For aquatic animals the corresponding problem is one of maintaining **osmotic concentration,** keeping a proper salt balance (Prosser 1973). For marine invertebrates there is rarely a problem because most have a salt concentration similar to that of seawater (they are said to be **isotonic** with seawater). Fresh-water organisms, whether invertebrates, fish, or amphibians, are **hypertonic** to the water in which they live (they have a higher salt concentration) and, consequently, must have ways of actively absorbing salts and also of recovering salts from the urine before it leaves the body.

Most marine vertebrates are **hypotonic** (a lower salt concentration) to seawater and thus must be able to excrete a concentrated urine to get rid of excess salt that is taken in with the seawater or in the food they eat. Seawater contains 3% salt. Whales can excrete urine more concentrated than this, but the best humans can do is 2.2%. Marine birds and reptiles have an additional organ that gets rid of excess salt; this is a pair of nasal glands that lie on the top of the head just above the eyes and open into the nasal cavity. These glands secrete sodium chloride (table salt) in concentrations up to about 5%.

The interaction of moisture and temperature is sometimes important in the life of an organism. A graphical method of studying such interaction, the **climograph,** is shown in Figure 3-11.

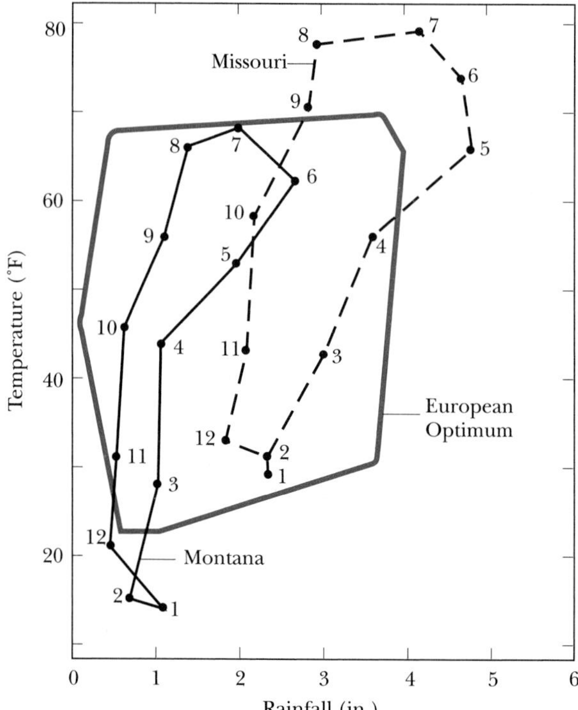

Figure 3-11

One method of considering temperature and rainfall at the same time is in a graph such as this one, called a *climograph.* It may be used in a purely descriptive way just to see the difference, for example, between the climate of rain forest and desert. It may also be used as a tool for prediction. In this climograph the climate of the best European range of the gray (Hungarian) partridge is compared with that of Havre, Montana, and Columbia, Missouri. The partridge was introduced successfully in Montana but failed in Missouri, implying that hot (or hot and wet) summers are limiting but that low winter temperatures are not. Future introductions (if justified) should not be in areas having a climatic regime such as that found in Columbia, Missouri. (From E. P. Odum, *Fundamentals of Ecology,* Philadelphia, W. B. Saunders, 1971, p. 125.)

Light

Light is a kind of radiation, specifically the visible wavelengths of electromagnetic radiation. This portion of the spectrum runs from violet (short wavelength) to red (long wavelength) (Figure 3-12). Wavelengths just shorter than violet are ultraviolet radiation, which is visible to insects and fish and at least some birds but not to us, and X-rays. Wavelengths just longer than red are infrared and microwave radiation. Ecologically important light is generally sunlight, or solar radiation. Solar radiation contains wavelengths other than visible light. About one half the energy in solar radiation reach-

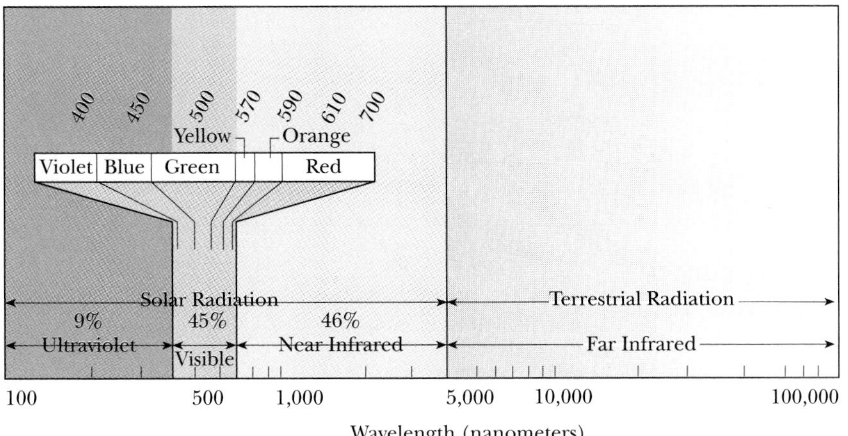

Figure 3-12

Electromagnetic radiation ranges from the very short, high energy cosmic rays to the very long, low energy radio waves. Most of the energy of sunlight is in the range between 100 and 4000 nanometers. "Terrestrial radiation" is the longer wave radiation, in effect, heat rays, given off by the earth, including its vegetation (and animals), water, and air. (Diagram adapted from Halverson & Smith 1979.)

ing the earth's surface is in the visible range, and about one half is in the near infrared (the infrared wavelengths near the visible spectrum). Because of the ozone layer of the atmosphere, only a little ultraviolet radiation reaches the earth's surface.

Light, or solar radiation in general, has many effects of ecological importance ranging from the role of light (mainly red and blue) in photosynthesis to the role of ultraviolet radiation in vitamin D production in animals. At this point we will deal with only two ecological aspects of light, shade tolerance and photoperiodism.

Shade Tolerance

Shade tolerance is the ability of a plant to survive and grow in the shade. We think of shade as meaning lowered light intensity, which it does. The canopy of a beech-maple forest may screen out over 99% of the sunlight hitting it. In the middle of a summer day, the light intensity outside the forest may be over 10,000 footcandles (1×10^{10} lux) and inside the forest less than 100 footcandles (1,076 lux). Shade also means other things, however, including reduced wind, high humidity, and root competition. Still, the ability to stay alive and grow in low light levels is probably the one most important feature of shade tolerance.

Foresters classify trees in five categories of shade tolerance, from very intolerant to very tolerant. The subjective rankings given in Table 3-2, based on a survey of foresters, are widely used, but evaluation of shade tolerance is complex. For example, tolerance may change with age. Seedlings are usually more shade tolerant than older trees. In an experiment spanning several years, the Canadian forester K. T. Logan grew several species of trees from seed under different levels of shade. He found that no species, even those considered most shade tolerant, made their best growth at light intensities below 25% full sunlight. Even sugar maple, American beech, and eastern hemlock had peak growth rates at about 45% full sunlight.

In general, the most shade-tolerant trees had low growth rates, but their growth was not reduced as much by low light levels as less tolerant species. This was especially true for root growth. Root weight (oven-dry) of three-year-old seedlings of the intolerant paper birch averaged 13.7 g for plants grown in full sunlight and only 3.5 g in 13% full sunlight (Logan 1965, 1973). This is a 75% reduction. The very tolerant sugar maple, however, had root weights of 7.8 g in full light and 6.8 g in 13% light. The reduction is only 13%; furthermore, the root mass produced by the maple in this deep shade is almost twice that of paper birch.

Knowledge of the light requirements of plants is useful in managing forest lands and also simply in gardening. A shaded garden is of very little use unless one plans to grow native forest wildflowers or their counterparts from elsewhere in the world. Plants from which you can obtain an edible crop are almost all shade intolerant.

Table 3-2
Shade Tolerance of American Forest Trees

Some trees are difficult to place in one category with assurance (some seem to behave differently in some regions of the country, for example).

Tree Category	*Tolerance Class*									
	Very Intolerant		*Intolerant*		*Intermediate*		*Tolerant*		*Very Tolerant*	
Eastern Conifers	Tamarack Jack pine	Longleaf pine	Eastern red cedar Red pine Pitch pine	Shortleaf pine Loblolly pine Virginia pine	Eastern white pine	Bald cypress	Red spruce Black spruce	White spruce Northern white cedar	Balsam fir	Eastern hemlock
Eastern Hardwoods	Most willows Quaking aspen Bigtooth aspen	Cottonwoods Black locust Osage orange	Black walnut Butternut Pecan Hickories Paper birch Yellow poplar Sassafras	Sweetgum Sycamore Black cherry Honey locust Kentucky coffeetree Catalpas	Yellow birch American chestnut White oak Red oak Black oak American elm	Rock elm Hackberry White ash Green ash Black ash	Red maple Silver maple Box elder Basswood	Tupelos Persimmon Buckeyes	Eastern hophornbeam American hornbeam American beech	American holly Sugar maple Flowering dogwood
Western Conifers	Whitebark pine Foxtail pine Bristlecone pine	Digger pine Western larch Alpine larch	Limber pine Pinyon Ponderosa pine Jeffrey pine Lodgepole pine Coulter pine	Knobcone pine Bishop pine Big-cone spruce Noble fir Junipers	Western white pine Sugar pine Monterey pine Blue spruce	Douglas fir Big-cone spruce Red fir Giant sequoia	Sitka spruce Engelmann spruce Mountain hemlock Pacific silver fir Grand fir White fir	Redwood Incense cedar Port Orford white cedar Alaska yellow cedar	Western hemlock Alpine fir Western red cedar	Pacific yew California torreya
Western Hardwoods	Quaking aspen	Cottonwoods	Red alder		Golden chinquapin Oregon ash	California white oak Oregon white oak	Tanoak Canyon live oak Bigleaf maple	Madrone California laurel		Vine maple

Increasing shade tolerance ⟶

Based on Baker, F. S., "A revised tolerance table," *J. Forestry,* 47:180–181, 1949.

Photoperiodism

Photoperiod refers to the length of the light and dark portions of the 24-hour day. Because of the way in which the earth's axis is tilted, the length of day and night changes seasonally everywhere except at the equator (Figure 3-13). In the northern hemisphere, the longest day occurs about June 21 and is called the summer solstice; the shortest day is December 21, the winter solstice. The spring and autumn days when day and night are each 12 hours occur on March 20 and September 23, the vernal and autumnal equinoxes. Where I live, 42° north of the equator, the longest day is about 15 hours long and the shortest about 9. Further south the difference diminishes, and further north it increases.

For most of the earth, the change in day length is the most accurate information available to an organism on the advance of the seasons. For many organisms, day length, or photoperiod, is an important factor in the scheduling of their life history events. Most birds of the temperate regions begin to come into breeding condition not because of increased temperature or a better food supply but because of a lengthening photoperiod. At some point, temperature or food supply or some other factors may become important in determining just when egg-laying actually begins.

Like most other animals, humans are probably affected by photoperiod (Reiter 1991), but almost no information other than anecdotes are available about the role of photoperiod in human societies without electric lights. One apparent effect in today's society is the effect of short days on mood (Hyman 1990). Winter depression, or seasonal affective disorder (SAD), may be triggered by the short days of winter. Light treatments, such as exposure to bright light early in the morning, have shown promising results.

Among plants the timing of many events, including flowering, may be based on photoperiod. **Short-day plants** require days shorter than some maximum length to come into flower; if days stay longer than this, the plants will continue growing vegetatively and will not produce flowers. Examples of short-day plants are lespedeza, chrysanthemum, soybean, poinsettia, and violet. The critical day length varies from one kind of plant to another; usually it is somewhere between 11 and 14 hours. **Long-day plants** need day lengths longer than some critical minimum to come into flower. Here again, the shortest day that will stimulate a long-day plant to prepare for flowering varies between species but is usually 10 to 13 hours. Some long-day plants are red clover, evening primrose, dill, spinach, timothy, and wheat grass. There are also day-neutral plants

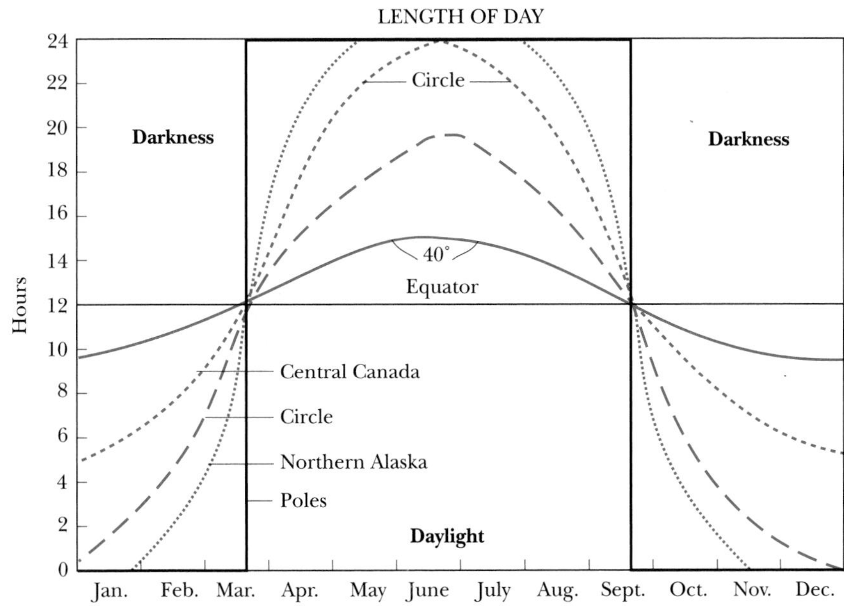

Figure 3-13

Photoperiod through the year at various latitudes. At 40° north, approximately the latitude of Indianapolis and Denver, day length is slightly below 10 hours on January 1 and slightly less than 15 hours at the summer solstice toward the end of June. (From A. M. Woodbury, *Principles of General Ecology*, New York, Blakiston, 1954, p. 122.)

that will flower under a variety of photoperiods if other conditions are favorable, such as buckwheat, cucumber, impatiens, and corn.

If you think at this point that long-day plants flower on photoperiods longer than 12 hours and short-day plants flower on photoperiods less than that, you have not read the definitions carefully enough. It should be clear that, depending on the species, both long- and short-day plants may be induced to flower by photoperiods of 12 hours, or 11 hours, or 13 hours (among others). This points up the fact that photoperiod is rarely important to plants and animals as an ultimate factor. Rather, photoperiod is used as a timing device (that is, as a proximate factor) to schedule an activity at the appropriate season.

Most biologists still live in a fairly small part of the globe, the forested and grassland parts of the temperate regions, where photoperiod change is pronounced and the main seasonal change is winter to summer. In other parts of the earth, such as the tropics and some deserts, photoperiod may change little or a change from dry to wet season may be the important seasonal change. For these regions natural selection has favored factors other than (or in addition to) photoperiod as the environmental cue for when to breed or flower (Figure 3-14). In some desert birds, courtship and mating may begin within hours after a substantial rain. Here, something about the rain itself seems to be the trigger. In other birds, there is evidence that chemicals that inhibit reproduction become concentrated in the plants eaten by these birds during the dry season. With the coming of the rains, rapid plant growth evidently lowers the concentrations of these chemicals, reproduction is no longer inhibited, and the birds begin to breed (Leopold et al. 1976).

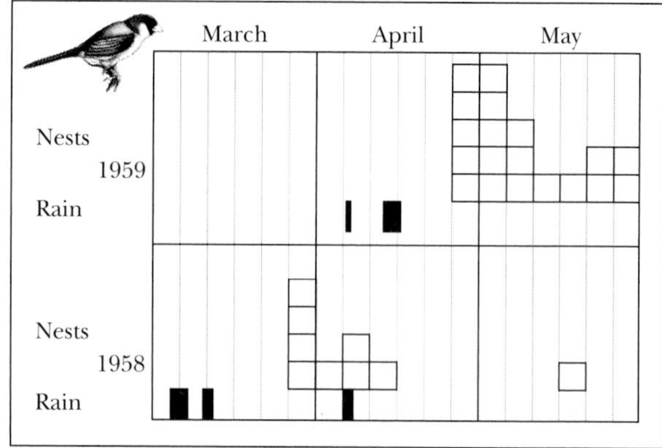

Figure 3-14

Although photoperiod may be important in breeding of the Abert's towhee, the immediate cue for nesting seems to be a factor associated with heavy spring rains. Nesting begins ten days to two weeks after a heavy rain, as early as February or as late as April. The diagram indicates nesting for a mesquite woodland area near Tucson, Arizona. The black bars represent rains of 2.5 cm or more; the boxes represent new nests started at five-day intervals. In 1959 nesting was late, following the only period of heavy rain of the spring; in 1958 nesting was early, following heavy rains in early March. (Adapted from Marshall 1963, p. 620.)

Circadian and Other Rhythms

The first biologists working with photoperiod thought that it was the length of the light period that was important in producing a response. Later experiments showed that long-day organisms kept on a short-day photoperiod would respond as though they were really getting long days if the nights were interrupted even with just a few minutes of light. From this, it was concluded that night length rather than day length was important. Further research, using photoperiods that could never occur in nature, such as 6 hours of light and 36 hours of darkness, have shown that the explanation is more complicated. Organisms have a daily rhythm of susceptibility to light stimulation. During part of the 24-hour day they get no stimulatory effects from light; during the rest, light is stimulatory. It is the light falling in this phase of the cycle that produces the effect, whether the light is a continuation of a long day or merely some brief period during the night (Bünning 1967).

Many recurring biological events are based on underlying rhythms. Included are a host of behavioral and physiological phenomena in many kinds of organisms. Included are the activity period, as in nocturnal versus diurnal organisms, sleep movements of plants, hormone secretion, and body temperature of homoiotherms (Saunders 1977). Daily rhythms are referred to as **circadian rhythms.** The primary features of circadian rhythms are (1) they persist when organisms are shielded, insofar as is possible, from natural environmental periodicities such as the daily cycles of light and temperature, and (2) their period when so shielded tends to be about 24 hours but to drift to slightly longer or slightly shorter.

Circadian rhythms seem to serve the function of a biological clock; they measure the passage of time, so that events occur when they need to. For most organisms under natural conditions, the biological clock is reset every day by environmental periodicities, in particular sunlight. The term generally used for such an environmental cue is *zeitgeber,* German for "time giver."

For many birds, photoperiod changes seem to be the proximate factor for coming into migratory condition. For example, many birds that winter in the southern United States seem to come into migratory condition with the lengthening days of late winter. Birds that winter close to the equator, however, have only slight photoperiod changes on which to cue, and there is evidence suggesting that some of these birds have a **circannual rhythm** (about a year) in migration readiness (Hagan et al. 1991). There is much to be learned about such birds; it may be that events experienced in summer in the temperate zone set their calendar, much as sunrise sets the biological clock.

Soil

Soil is the loose surface material of land in which plants grow. It cannot be placed in the same category as temperature or light as a simple "factor" affecting organisms, since soil is itself a complex system. It is composed of fragments of the parent mineral material, organic matter in various stages of breakdown, soil water and the minerals and organic compounds dissolved in it, soil gases in the spaces not containing water, and living organisms.

Soil plays obvious roles in plant growth. From it come the water that plants use in transpiration and photosynthesis, the calcium, nitrate, phosphate, and other mineral nutrients that the plant requires in the manufacture of various organic compounds, and the oxygen that the cells of the roots need for respiration. But the soil is a great deal more than this. When a plant or any organism dies, the process of decay occurs in the soil through the activities of soil organisms, especially bacteria and fungi. **Humus,** finely ground organic matter mixed with the mineral part of the soil, is produced, and eventually the minerals originally taken in by the plant roots are returned to the soil. The soil is home to many types of animals, from moles to salamanders to earthworms to beetles to protozoans. The number of individuals of some of the smaller invertebrates is incredible to anyone who has never used the special techniques for studying soil organisms. In a patch of soil 1 m on each side, or about 1 yard square, there may be 10,000 small insects called springtails, 100,000 small relatives of the spider called mites, and 1,000,000 thin white roundworms, or nematodes.

Soil Structure

The formation of soil, whether from solid rock or from mineral material deposited by a glacier, wind, or water, is a complex process. **Mechanical weathering,** such as the cracking of big rocks into little ones by the wedging action of water freezing in a crack, is important. **Chemical weathering** is also important, especially the leaching downward of alkaline materials such as calcium salts by rainwater containing dissolved carbon dioxide. Also important are the activities of organisms, such as decomposition and the aggregation of soil particles produced by their passage through an earthworm's gut. The biological aspects of soil formation are described in Chapter 12.

Because most soil-forming processes tend to act from the top down, soil develops a vertical structure referred to as the **soil profile** (Table 3-3); Figure 3-15). Soil scientists recognize three main layers, or **horizons.** The uppermost layer is topsoil, or the **A horizon,** where most of the plant roots are located. Dead organic matter is added to the top as **litter** and, as it is partially broken up, mixed as humus with the mineral soil below. Organic material is largely absent from the **B horizon,** and the mineral parent material is less thoroughly weathered. Materials leached from the A horizon may be deposited here. Below this is the **C horizon,** consisting of more or less unaltered parent material such as glacial drift or bedrock.

What the soil of a particular area is like, the end product of soil formation, depends on many things. Three of the most important are the nature of the parent material, climate, and vegetation (or, better yet, the whole community including animals and microorganisms). For example, where the parent material is a rock called serpentine, the soil that develops is deficient in calcium and usually high in magnesium and nickel. On such areas in the southeastern United States, Sweden, and elsewhere grow a collection of species very different from those on adjacent areas formed from different types of rock (Proctor and Woodell 1975).

Soil development is strongly influenced by rainfall, evaporation, and temperature, so that the arctic, tropic, temperate, and arid regions all tend to develop different types of soil (Table 3-4). In the northern Midwest, however, coniferous forest, deciduous forest, and grassland grow within a few miles of one another, and the soils, all derived from the same basic parent material, may be quite different owing to the differing effects of the vegetation. Under the coniferous forest may be a strongly acid soil with a heavy layer of undecomposed litter and a hardpan of leached clay in the B horizon. Under the deciduous forest the litter is thinner and grades into the mineral soil below; the soil is less acid because basswoods, maples, and dogwoods absorb calcium and other minerals from the subsoil, incorporate them into their leaves and stems, and return them to the soil surface each autumn. Under the grassland the soil may be still less acid, possibly almost neutral, and still more fertile because the nutrient pumping action is even more pronounced. The grass roots penetrate deeply, the plants use large amounts of calcium and other nutrients, and each year not only leaves but the entire aboveground part of the plant dies back, allowing decomposition to free the nutrients at or near the soil surface. The grassland soil is dark through the addition of this organic material plus that added to the soil at various depths by the death each year of many fine roots.

Table 3-3
Major Soil Horizons and Their Divisions

0 (or A_0) Forest floor—the organic layer on top of the mineral soil—litter and humus.

(A) Mineral soil in which organic matter is accumulating from above and from which clay, iron, or aluminum are being leached.

A_1—zone dark from humus

A_2—zone light (ash-white in podzols) because no humus and leached.

(B) Zone of deposition of clay, iron, or aluminum. Often blocky or columnar structure. In many soils, reddish from iron oxides.

(C) Little weathered. In areas where precipitation/evapotranspiration ratio <1, calcium or magnesium carbonate layer (ca) usually present.

Soil Texture and Fertility

Soil texture is based on the sizes of mineral particles making up the soil, classified as **gravel, sand, silt,** and **clay** (Figure 3-16). Soils made up mainly of small particles are called heavy soils. Water soaks into heavy soils slowly but is retained well; such soils have the potential of being very fertile. **Light** soils, such as sandy soils, are well aerated and allow free movement of roots and water but are relatively infertile.

The difference in potential fertility between light and heavy soils results from the way minerals are retained. Such minerals as calcium and magnesium (specifically, most elements that form positive ions, or **cations,** when dissolved) are stored on the surface of particles. Anions, by contrast, are dissolved in soil water.

Plant roots remove cations and replace them with hydrogen ions. The potential fertility of a soil depends primarily on **cation exchange capacity,** a measure of the number of sites per unit of soil (usually 100 g) on which hydrogen can be exchanged for a mineral cation. Clay particles are small and provide more surface area in a given weight or volume compared with the same amount of sand, pebbles, or boulders. Clay soil may have a cation exchange capacity from twice to twenty or more times that of sand.

A soil with a high cation exchange capacity and, consequently, high potential fertility may be actually infertile if most of the sites on the particles are filled with hydrogen ions. Such a soil will be acid and will have little calcium to supply to the growing plant. **Percentage base saturation,** the percentage of the exchange capacity satisfied by calcium, magnesium, and similar elements, measures this aspect

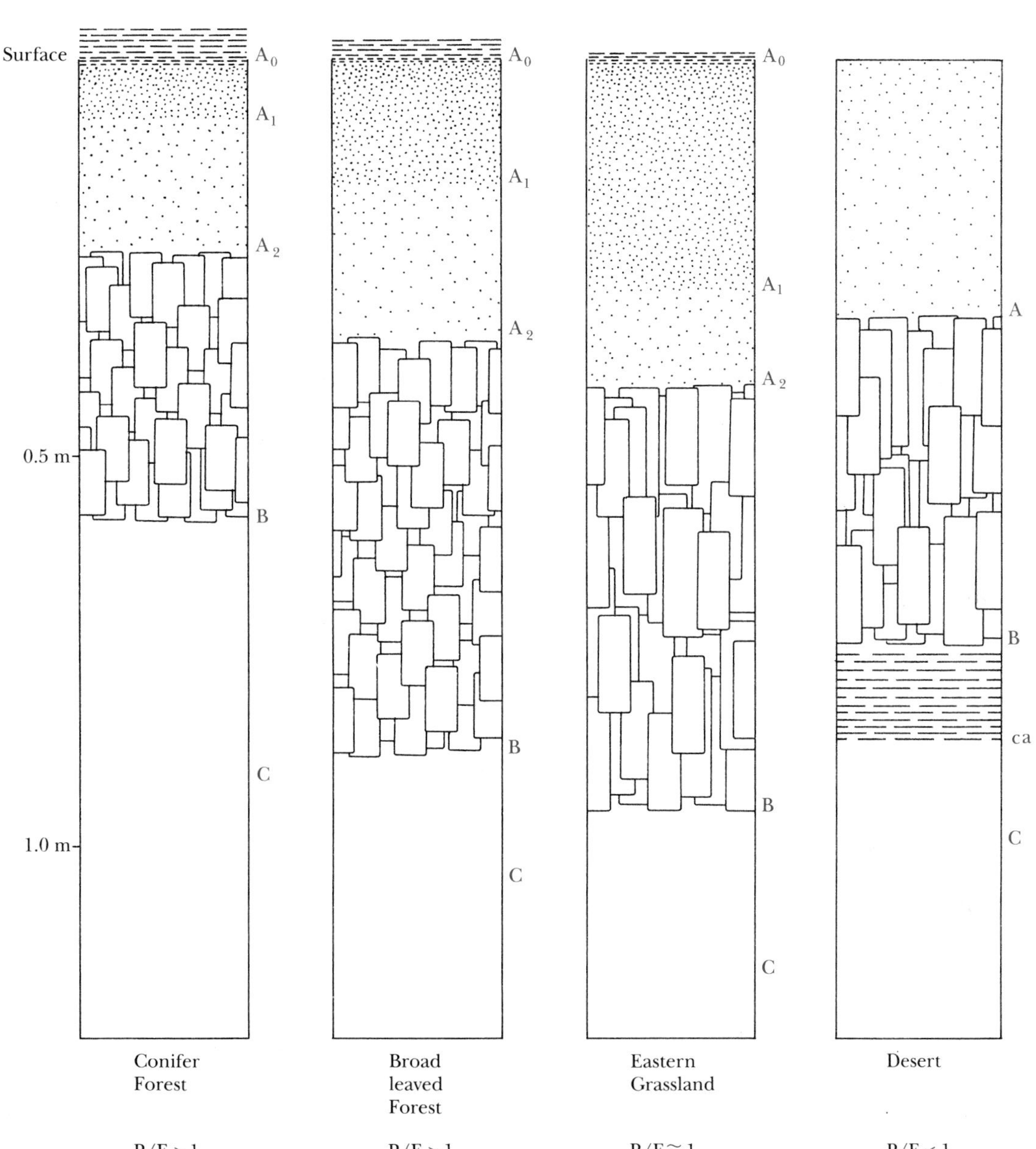

Figure 3-15

Diagrams of four types of soil profiles. Heavy litter with an abrupt transition to the A_1 horizon and the ashy-white A_2 horizon are features of the spodosol. The mollisol has a deep, dark top soil as the result of large amounts of organic matter, especially from roots. The horizons are less distinct in these grassland soils than in forest soils. In the aridosol there is little organic matter in the top soil and a calcium carbonate layer (ca) is present in the C horizon. Much of the calcium has been wholly leached out of the forest soils. P/E is precipitation:evaporation ratio.

Table 3-4
Types of Climatic-Vegetational Soil Development

The processes described in this table give rise to *zonal* soils which occupy much of the area of a region having a certain climate and vegetation. Various *intrazonal* and *azonal* soils may occur in all regions—for example, serpentine soils, hydromorphic soils (such as in bogs), halomorphic soils (salty soils of arid, poorly drained areas), lithosols (bedrock outcrops), etc.

Type of Soil Development	*Climate*	*Vegetation*	*Important Processes*	*Kinds of Soils*
Gleization	Cold, little rainfall	Tundra	In summer, soils remain wet due to poor drainage caused by permanently frozen lower layers *(permafrost).* In winter, freezing at top compresses middle layers *(glei)* which are sticky and blue-gray because of iron in the reduced state. Glei is forced up through cracks and thoroughly mixed. Freezing and thawing sort out rocks on the surface.	Tundra
Podzolization	Cool, fairly moist	Coniferous or deciduous forest	Seen in extreme form only under certain types of coniferous forest. Percolation of acid water downward leaches out carbonates, A horizon becoming acid. If extremely acid, clays may be leached and deposited in the B horizon and form a hardpan. Decomposition is slow under acid conditions, and there is often a sharp break between the humus layer and the upper mineral layer, which may be sandy and ashy gray.	Podzols (coniferous forests) Gray-brown podzols (deciduous forests; podzolization less extreme so that soils are less acid and more fertile) Red-yellow podzols (southern forests; transitional to lateritic soils)
Laterization	Warm, moist	Tropical forest	Clay minerals decompose rapidly and release bases, keeping soil from becoming highly acidic. Humus decomposes rapidly and does not accumulate. Silica fraction is leached, and oxides of iron, aluminum, and manganese remain, giving solid reddish or yellowish color.	Lateritic soils
Calcification	Warm to cool, not moist	Grassland, savanna, desert	Leaching not sufficient to carry away calcium carbonates that may accumulate as a hardpan; soil remains neutral to alkaline, and clay is not leached. Death of roots and tops adds organic matter and nutrients. Darkness is related to amount of organic matter and thus to rainfall. In desert, wind erosion may remove finer particles.	Prairie soils (eastern grasslands where rainfall is sufficient to leach calcium carbonate) Chernozem soils (humid grassland Chestnut and brown soils (mixed and short grass prairie) Gray desert soils (sagebrush) Red desert soils (shrubs and cacti)

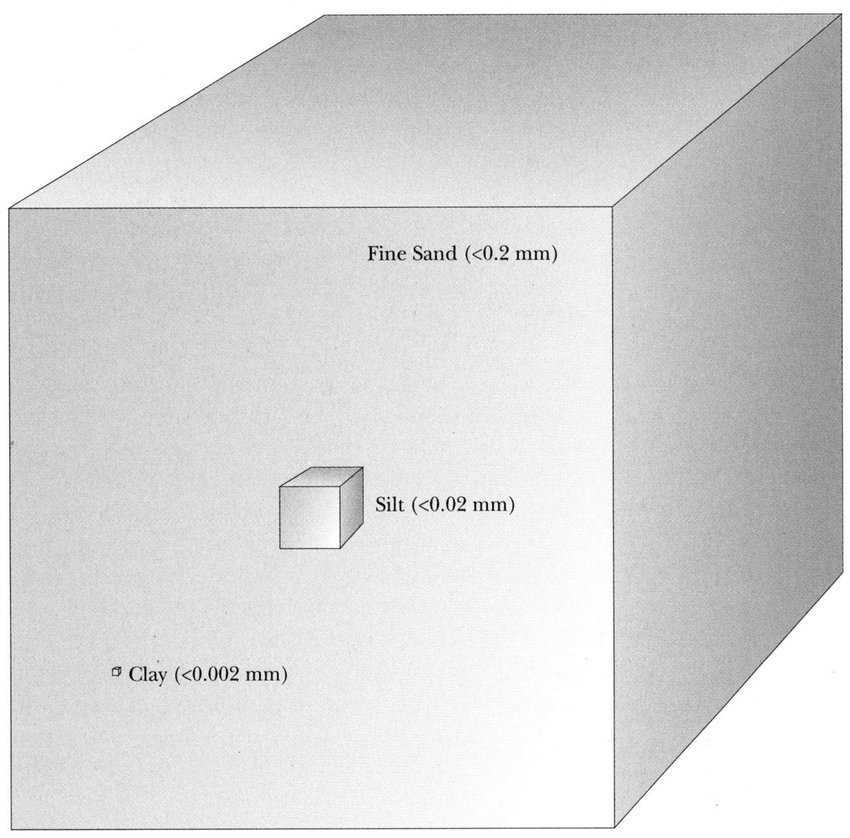

Figure 3-16

This shows the relative sizes of the largest clay, silt, and fine sand particles, each magnified 500 times. At the same scale, fine gravel (2 mm) would be a cube a little more than a yard on a side.

of fertility. If percentage base saturation is 60%, then 60% of the sites are filled by basic ions and 40% by hydrogen ions.

Soil Water

Water in soil can be put in three categories: gravitational water, capillary water, and hygroscopic water. When water is added to soil by rain or irrigation, the water in the larger pore spaces is only temporarily available. This **gravitational water** soon drains below the root level of most plants but is used as it goes by. The amount of water left in soil one to three days after a thorough soaking is referred to as the **field capacity** of the soil. It consists of **capillary water** plus **hygroscopic water.** The former is the water in smaller pores, ones small enough that water is retained by capillary action against the pull of gravity. This is the main source of water in the soil for plants. Hygroscopic water is adsorbed on the surface of soil particles and is virtually unavailable for plant use.

Obviously the actual amount of water that soil contains depends on time since the last rain, usage of water by plants, the proximity of the soil to the water table, and other factors. Other things being equal, availability of soil water to plants depends mainly on soil texture. Texture is the main factor determining **field capacity** and also another important trait, **permanent wilting percentage** (Briggs and Shantz 1912). These are still useful concepts, although a student of soil physics would describe plant-water relationships differently (Slatyer 1967, Kramer 1969).

The permanent wilting percentage is the amount of water in a soil when plants growing in it are irreversibly wilted (taken to be when giving them a saturated atmosphere overnight does not revive them). Obviously this depends partly on the plant, but the force with which soil holds water increases sharply over a fairly narrow range where soil moisture is reduced to hygroscopic water plus water in the smallest pores. This soil moisture level depends strongly on soil texture because the surface area in a given volume of soil increases as the percentage of fine particles increases. Field capacity also goes up in heavier soils because soils with small particles have more pores and also the proportion of them small enough to hold water by capillary action goes up.

The relationship between these two traits and soil texture is diagrammed in Figure 3-17. As expected, clays have the greatest water-holding capacity, but they also have a high permanent wilting percentage. Given adequate rainfall and drainage, the soils of a region with the biggest distance between field capacity and permanent wilting percentage will have the largest amount of **available water** (that is, available for plant growth). These are clays and the heavier loams.

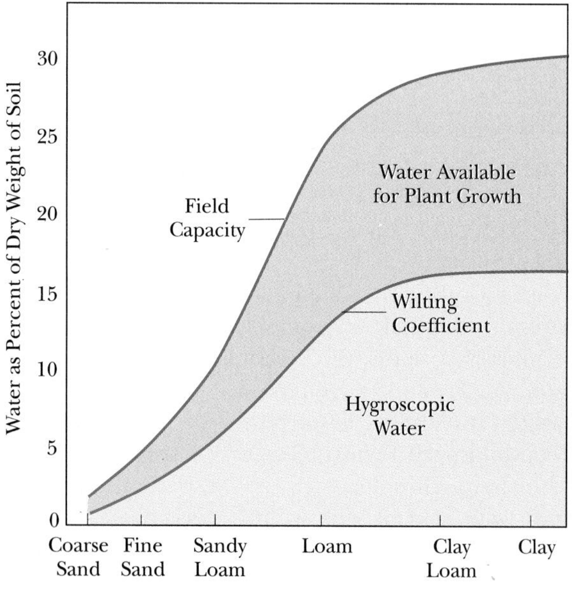

Figure 3-17

Generalized relationship between water availability and soil texture. The distance between the two curves at any point is approximately the amount of water available for plant growth. For 100 grams of soil (oven-dry), the values on the Y axis would be equivalent to grams. Based on values given in Briggs and Shantz 1912 and Daubenmire 1959.

Loams and Humus

The soil structure that is generally best for plant growth contains both large and small particles, combining the desirable features of good water retention, drainage and aeration, easy root penetration, and potentially high fertility. A soil that has both large and small particles well represented is called **loam,** and loams are almost always productive.

Even soils made up mainly of sand or clay can be given a favorable structure by adding organic matter. Knowledge of the importance of humus in soils is not new to ecology. Long ago the pioneer American ecologists John Weaver and Frederic E. Clements wrote: "... its physical effects are so marked that when the organic matter present in a soil is very high the distinctions between sands, loams, and clays are practically obliterated." In clays, the particles may be bound together, improving aeration and water movement, by organic compounds released in the breakdown of humus (and also by gums produced by bacteria and blue-green algae and by the activities of fungi and earthworms). Humus has an even higher capacity for holding water and nutrients than does clay; consequently, deficiencies of sandy soil can be largely remedied by the addition of organic matter. Adding peat moss to sand in a mixture of one fifth peat, four fifths sand more than quadruples the ability of the sand to retain available water (Baver 1940).

Soil Names

The names of soil groups in Table 3-4, such as podzol and chernozem, are being replaced by a new system (U.S.D.A. 1960) described in detail by the U.S.D.A. Soil Survey in 1975. The U.S.D.A. Comprehensive System is described as more empirical than the old one, being based on observable soil traits independent of assumptions about vegetational-climatic soil-forming processes. The new system is arranged in a hierarchy, rather like the kingdom, phylum, class, etc., of biological nomenclature. The most inclusive category is the **order,** of which there are 11 (Soil Survey Staff 1990). The smallest category (more or less analogous to the species of biology) is the **soil series.**

Table 3-5
Soil Orders, With Brief Descriptions and Examples

The examples relate the orders to older soil names but do not necessarily include the complete range of soil groups in the order.

Order	*Description*	*Examples*
Alfisols	Gray to brown surface horizons; medium to high base supply; subsurface horizons of clay accumulation	Gray-brown podzols
Andisols	Volcanic soils, generally deep with low bulk density and amorphic iron– and aluminum–containing materials	Soils developing in ash, pumice, cinders
Aridisols	Low in organic matter; dry more than half the year; light in color	Desert soils
Entisols	Immature soils; little or no horizon development	Azonal soils such as sand dunes and thin, rocky soils
Histosols	Organic soils	Peat and muck
Inceptisols	Youngish soils, usually wet, with weakly developed profiles	Tundra and alluvial soils
Mollisols	Nearly black, organic-rich surface horizons; high base supply	Prairie soils, chernozems
Oxisols	Deeply weathered soils; silica leached	Lateritic soils
Spodisols	Subsurface accumulation of organic matter and aluminum and iron oxide; acid	Podzols
Ultisols	Weathered forest soils with horizons of accumulation and low base supply	Red-yellow podzols, red-brown laterites
Vertisols	Soils with high content of clays that swell when wet; wide, deep cracks in dry season	Grumusols

There are about 7000 of these in the United States. They have familiar names like "Plainfield" or "Swift Creek," which may be attached to textural classes to produce Plainfield sandy loam and similar combinations.

Names of the 11 orders, all of which end in "sol," and which we must now learn, are given in Table 3-5. Names of the intervening categories, such as **great groups** and **families,** sound to the uninitiated like visitors from an Ursula LeGuin novel. "Typic Hapludalf" is an example.

Fire

Many early ecologists in this country regarded fire as a destructive, unnatural phenomenon, associated with humans. Until recently the U.S. Forest Service shared and spread the same view; Smokey the Bear still represents the Forest Service to most of us. Fire certainly is destructive, but most ecologists and foresters now believe that it is a natural part of the environment and that it is a useful, even necessary, tool in managing many ecosystems (Cooper 1961).

It is true that humans start many fires, but the overall result of our activities has probably been a reduction in the amount of burning of vegetation. When fires start, people put them out, and our land-use patterns have produced frequent firebreaks in such forms as roads, plowed fields, and asphalt parking lots. Native Americans and other indigenous peoples were responsible for burning a great deal of the landscape. This seems to have been partially accidental in that they did not take fires very seriously and so took few pains to prevent them from spreading. There is evidence, however, that Indians deliberately set fires to drive game, and probably they knew that fires would favor certain useful plants or vegetation types (Barrett 1980). Fires started by lightning are so much a natural occurrence in many geographical areas that they should be considered an aspect of climate.

Figure 3-18

Dense stands of lodgepole pine accumulate flammable debris from needle fall and the death of trees from shading and disease; these stands are susceptible to crown fires. On the left is a lodgepole pine stand in Yellowstone National Park in the summer of 1988. On the right is a similar stand that was burnt during that summer. This site is likely to regenerate a dense stand of young lodgepoles from the seeds released from serotinous cones. (Photos by the author.)

Types of Fires

There are three types of fires. **Surface** fires sweep rapidly over the ground, consuming litter and herbs, often killing the aboveground stems of shrubs and scorching tree bases. The temperature at the surface of the ground may be high, 90 to 120°C or even higher, but a few centimeters into the soil, it may not rise at all. Surface fires at intervals of years to tens of years were a natural occurrence in many kinds of vegetation in prehistoric times.

Ground fires may occur where there is a thick accumulation of litter. A ground fire is a flameless, subterranean fire that may burn slowly for long periods. It is hot and slow moving and kills most of the plants rooted in the burning material. Even bogs can burn in this way because the heat from the ground fire dries out the peat ahead of it.

Crown fires occur in dense woody vegetation, usually coniferous. Fire spreads rapidly through the canopy, killing the trees and most vegetation from the ground up. Usually crown fires occur when surface fires are too infrequent to prevent the buildup

of flammable material below the canopy (Figure 3-18).

Effects of Fire on Plants

We can look at direct and indirect effects of fire on plants and animals (Daubenmire 1968). By indirect effects, we are usually referring to responses to fire-produced changes in the physical environment, dealt with in the next section. Many of the direct effects of surface fires on plants are obvious. Most herbs and shrubs are killed above ground but sprout from underground parts.

Whether or not fire kills a tree is related to characteristics of the tree and the fire. In general, cambium (the layer of actively dividing tissue in the outer part of the trunk just inside the phloem) is killed by temperatures of 64°C or above (Rouse 1986). If temperatures are that high all around the trunk, the above-ground portion of the tree dies. Mainly for this reason, mortality tends to be higher for trees of small diameter than for large trees. The often-thinner bark of young trees is another factor.

Species-specific differences in bark thickness are also important. Thick-barked trees, such as bur oak, tamarack, and ponderosa and longleaf pines, will usually survive a moderate surface fire, whereas a fire of the same intensity may kill thin-barked species, such as black oak, white cedar, or lodgepole pine. Some trees die a year or more after a fire, so total tree mortality from surface fires may not be immediately evident.

Many trees, if top-killed, have the ability to send up sprouts from their roots, allowing them to sur-

Figure 3-19

The top photo shows an oak forest a few days after a spring surface fire. The bottom photo shows the same forest two years later in an area where most of the trees were killed. Heavy root sprouting by oaks, sassafras, and wild black cherry had produced this sunny, brushy landscape. A dense growth of grasses filled in the spaces between the clumps of woods sprouts. (Photos by the author in Allegan County, Michigan.)

vive fires in the same way as perennial herbs (Figure 3-19). Examples are black oak, black cherry, beech, basswood, and trembling aspen. Often the roots send up several sprouts, so that a forest with trees having multiple trunks is a good indicator of past fire.

To summarize, recurrent fires favor herbaceous plants, which can regrow quickly, over woody plants. Fires favor thick-barked and root-sprouting woody plants over less fire-adapted ones and, as discussed in the next section, shade-intolerant over shade-tolerant species.

Some species of plants are virtually dependent on fire to reproduce successfully. Probably the best-known examples are the pines having cones that stay on the tree and retain the seeds for many years unless the cones are heated or the tree is killed (**serotinous** cones). Lodgepole, pond, and jack pine are examples (Figure 3-20). Fire generally causes the prompt shedding of the seeds with the resulting seedlings growing up to produce an even-aged stand. For certain plants, such as the sandhill laurel, some prairie plants, and some chaparral plants (Keeley 1987), exposing the seeds briefly to high temperature breaks dormancy, allowing them to germinate promptly in the aftermath of a fire. Features such as these would be advantageous only in environments in which fires are reasonably frequent; they are evidently adaptations to fire.

Still other species seem to need the stimulus of fire to flower and produce seeds. An example is wiregrass in the longleaf pine ecosystem of the Southeast. Vigorous growth of this species is maintained by annual or biennial fires at any season but under a regime of winter fires—often used for wildlife and timber management purposes—very little flowering occurs. Fire during the spring, the usual fire season under natural conditions, results in profuse flowering (Platt et al. 1988) (Figure 3-21). Several other species characteristic of longleaf pine savannas flower much more heavily after fire (Abrahamson and Hartnett 1990).

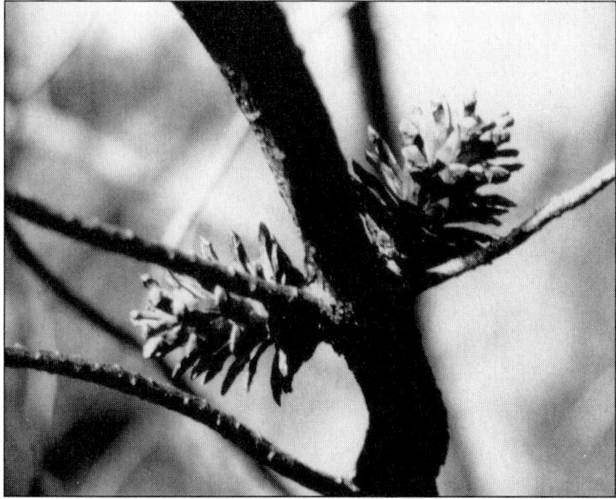

Figure 3-20

These serotinous cones of jack pine are open and have shed their seeds following a fire. (Photo by the author.)

Figure 3-21

The season of fire is important for reproduction by wire grass in the longleaf pine ecosystem. Little or no flowering typically occurs following winter fires. Here wire grass is fruiting copiously in the autumn following a growing-season burn. (Photo in southern Georgia by the author.)

Figure 3-22

These two photographs compare the forest floor in oak forest before (left) and after (right) a surface fire. About a week after the fire, the charred base of black-oats grass and the great reduction in litter are noticeable. Less obvious is the fact that the general color of the floor is black, rather than the brown of dead oak leaves. Photos in Allegan County, Michigan, by the author.

Effects of Fire on the Physical Environment

Effects on the physical environment vary, depending on the ecosystem and the season and severity of the fire. Here we examine them under three categories.

1. Removal of plant cover. Light levels are increased through the loss of plant cover. This may allow the entry or improved growth of surviving shade-intolerant species. The elimination of dead biomass in grassland (both standing dead stalks and litter) leads to much higher biomass production as the result of many more shoots, or tillers. This, rather than greater nutrient availability, is evidently the main cause of the increased productivity that results from fire in grassland (Svejcar 1990).

   Removal of plant cover reduces whatever moderating effects the plants in the specific habitat had on wind and temperature extremes. The loss of a large volume of foliage leading to less evapotranspiration may temporarily produce a higher water table.
2. Physical effects of litter removal by fire. The burning of litter can vary from complete to nil, even in the same burn, depending on fire intensity. Wherever the surface is blackened, the soil tends to heat up faster in the day and in the spring and to reach higher temperatures. This is especially pronounced if most of the insulating litter was consumed (Figure 3-22). The absence of both plant and litter cover favors greater temperature extremes, but overall a warmer soil results because of the greater amount of sunlight reaching the surface.

   A higher water table, mentioned above, is not a universal outcome of fire. In permafrost regions, the greater heating of the soil lowers the level of the permafrost, encouraging drainage (Van Cleve et al. 1983). Melting of the upper layer of permafrost also provides a greater volume for plant root growth. In grassland, there tends to be more input from precipitation on burnt areas, from lowered interception of rainfall, but also more evaporation from the soil, as the result of the lack of litter (Seastedt and Ramundo 1990).

Certain seeds germinate best, or only, on bare mineral soil and so are virtually dependent on fire for entry into many situations. The aspens are an example (DeByle and Winokur 1985).

3. Effects of litter burning on minerals. Carbon is, of course, lost from a burned site, mostly as carbon dioxide from the burning of living and dead biomass. "Literally hundreds" of other carbon-containing compounds are also given off (Sandberg et al. 1979), though what ecological effects these emissions may have, if any, are unknown. Nitrogen is also volatilized and lost (Seastedt and Ramundo 1990), but the effect of burning on nitrate availability seems complex. Sometimes the immediate effect may be a reduction in available nitrogen and consequent flourishing of plants with low nitrate requirements; however, the growth of nitrifying bacteria may be promoted, followed promptly by the invasion or increase in plants with high nitrogen requirements, such as fireweed.

   Most minerals are left as ash on the soil surface (2900 kg/ha, or 290 g/m^2 in the Entiat fire in conifer forest on the east slope of the Cascades in central Washington, Grier 1975). Included are calcium, magnesium, potassium, phosphorus, and sodium, among other elements. Rainfall carries these into the soil, though there may be some loss by wind or runoff. In the soil, the cations tend to replace hydrogen ions, improving the percent base saturation and raising the pH of acid soils. Minerals are rapidly taken up in the regrowth of vegetation, and in some ecosystems productivity is increased as the result of the increase in available nutrients, at least for phosphorus.

Effects of Fire on Animals

The responses of animals to fire have not been as well studied as those of plants. Some highly motile animals, such as birds and some mammals and insects, tend to vacate the area ahead of a fire and return if the burnt area remains suitable habitat (Gillon 1971). Small mammals, such as meadow voles, may hunker down and survive a fire but then desert and only return weeks or months later (Vacanti and Geluso 1985). Animals that burrow or live under logs often escape injury.

Nevertheless, there is mortality among a variety of animal groups, either from heat or asphyxiation (Kozlowski and Algren 1974, Kaufman et al. 1990). The Yellowstone fires of 1988 caused the death (from smoke inhalation) of about 250 elk out of the summer herd of about 31,000 (Singer et al., 1989).

Many species of predators and scavengers are known to gather at the edges of fires, eating animals that are flushed out or killed by the fire (Komarek 1969). Winter fires in the Georgia pinelands attract several species of small birds, most of which feed downwind of the smoke and flame, often on the ground, evidently eating insects fleeing the burn. Pine warblers, eastern bluebirds, and chipping sparrows were the most abundant species at one such fire (Hopkins 1975). Kites and other raptors are attracted to fires in African savanna where they feast on the large grasshoppers moving ahead of the flames (Gillon 1971).

The kites are apparently attracted from long distances by the plume of smoke (Figure 3-23); however, behavioral, physiological, or morphological features that are clearly adaptations to fire seem rare in animals (Meredith et al. 1984). Perhaps the best example of truly fire-adapted species are the "fire beetles" in the family Buprestidae, whose wood-boring larvae develop in fire-damaged trees. The adults have infrared sensors that are said to lead them to fires in conifer forest from a distance of several kilometers (Evans 1966). Mating occurs while the forest is still smouldering, and the female lays eggs under the bark of scorched trees.

Dead and injured conifers are also invaded almost immediately by other beetles, especially bark beetles in the genus *Dendroctonus,* and later by still others. The developing larvae are fed upon by in-

Figure 3-23

The smoke from the North Fork fire in Yellowstone National Park on 14 August 1988 can be seen from miles away. (Photo by the author.)

creased populations of certain birds, especially woodpeckers. It seems reasonable that the black plumage of the black-backed woodpecker, a North American species that reaches high densities only in burnt-over conifer forest (Evers 1991), is an adaptation for camouflage on the burnt trunks (Figure 3-24).

In the primeval landscape of eastern North America as well as many other parts of the world, a large percentage of the less-wooded part of the landscape was the product of fire. In Michigan, for example, most grassland, open wetland, thicket, and savanna resulted from fire (Brewer 1991); in the Southeast, the pinelands were fire dependent. The species of animals characteristic of these habitats, consequently, were almost completely dependent on fire for their existence in these landscapes. Among bird species in this category in Michigan are eastern meadowlark, brown thrasher, eastern bluebird, song sparrow, chipping sparrow, and rufous-sided towhee.

Although European-style settlement reduced fire frequency, the overall result was not, of course, a decline in the numbers of grassland and forest-edge animals. Instead, human activity produced many types of vegetation that mimicked the physiognomy of natural prairie, savanna, and brush, so most of these species greatly increased.

The Kirtland's warbler, a species of fire-dependent jackpine habitats, probably increased for a time after the slash fires in northern Michigan that followed cutting of the pine forests. In today's landscape this now-endangered species is dependent on management of the jackpine plains that employs fire.

Fire Frequency

For a fire to occur, two things are needed—an ignition source (in prehistoric times, almost always either lightning or Indians) and fuel. The fuel has to be sufficient in quantity and dry enough to burn. Fire frequency under natural conditions, accordingly, tends to be a function of climate (Martin 1983). Climate has a direct effect by determining the occurrence of drought periods during which litter and vegetation dry up enough to burn and also by determining the frequency of electrical storms, especially ones with dry lightning (no accompanying rain). Climate also affects fire frequency by influencing what kind of vegetation will grow, how much biomass is produced, and how fast litter decays. These factors determine the kind and amount of fuel available.

In hot, dry regions, fires are infrequent (Figure 3-25) because not much litter is produced, and it tends to dry up and blow away. In warm, wet regions, litter buildup is scant because of rapid decay. In cold climates, annual biomass production is low, humidity tends to stay high so the litter rarely dries out, and dry lightning is rare. In certain intermediate climates, fuel buildup, drought, and lightning combine to make fire likely on a scale of one to a few years.

In these habitats, such as mesic prairie, ponderosa pinelands in the West, and longleaf pinelands in the Southeast, any one fire produces only slight alterations in the community. Not much litter accumulates from one fire to the next, so fires tend not to be intense. The characteristic plants are fire

Figure 3-24

The black-backed woodpecker nests at high densities after fires in conifer forest. (This photo was taken in Oscoda Co., Michigan, by L. H. Walkinshaw.)

adapted; nonfire-resistant species are killed before they make much headway.

As fire frequency decreases, the effect of fire on community structure becomes more severe. With a long absence of fire, fire-intolerant plants may invade and even come to dominate the landscape. Litter and other flammable debris accumulate. When a fire eventually comes, it burns hot, killing most plants, including trees. The site may be set back to a grass- or fern-dominated stage, followed by shrubs, and later, shade-intolerant trees such as white pine in the East or Douglas fir in the West.

We need to recognize that other factors besides regional climate also affect fire frequency. In the northern Midwest at the time of European settlement, tallgrass prairie that burned almost every year, oak savanna with fires every few years, and beech-maple forest with decades between fires might all grow within an area of a few square kilometers. Much of the difference was related to topography. Prairie tended to be located on flat areas conducive to the spread of fire. Beech-maple forest tended to be located on moist sites, especially ones downwind of lakes, rivers, and wetlands—and thereby protected from the sweep of fire.

Another factor related to fire frequency is the flammability of foliage and litter. Robert W. Mutch (1970) proposed that evolution of plants in fire climates may proceed along two lines. (1) As already suggested, plants of habitats where fire is frequent may evolve features that allow them to resist fire or to recover rapidly, thereby gaining an advantage over species that fail to evolve such mechanisms. (2) Any genetic tendencies toward producing a more

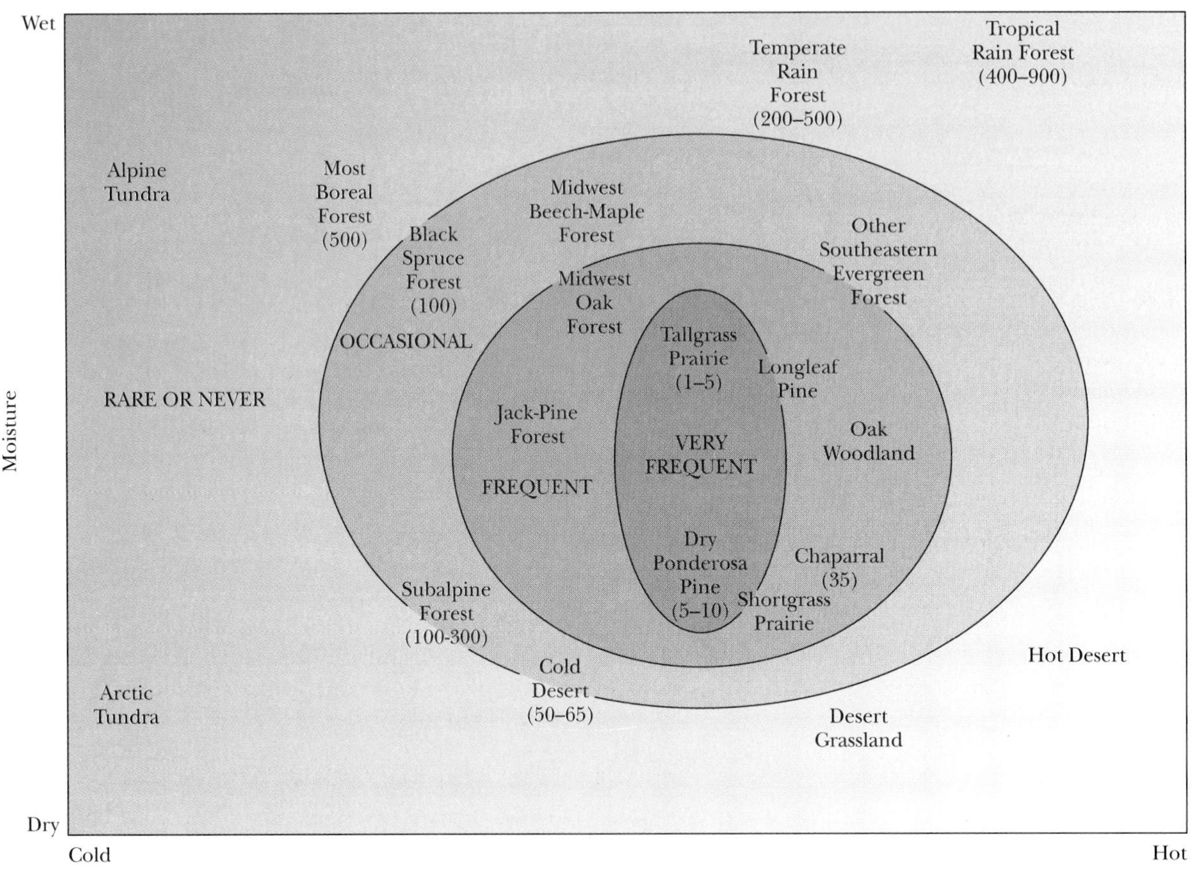

Figure 3-25

Generalized relationship between fire frequency and climate. In the middle of the graph, fires are frequent and tend to maintain the status quo. At the fringes of the figure, fires are rare and tend to alter the community drastically. Placement of vegetation types is tentative. Some estimates of the average number of years between fires are given (in parentheses).

flammable foliage among species that had achieved such fire adaptations would put their nonfire-resistant competitors at a further disadvantage. That is, highly flammable litter and foliage would tend to facilitate the spread of fire in these communities eliminating the nonfire-resistant species more quickly and thoroughly. At the community level, such an evolutionary tendency would lead to still more frequent fires in vegetation already characterized by high fire frequency.

The evolutionary aspect of this hypothesis is difficult to test, but it is a fact that many species of fire-dependent vegetation have highly flammable foliage. The eucalyptus trees of Australia, with their high oil content, are striking examples (Mutch 1970).

In the case of a community like prairie, it is easy to see fire as a factor in the normal functioning of the community. Stands of tallgrass prairie are rapidly invaded by woody seedlings (Kucera 1960) which, if they persist, turn the site into brush and eventually forest so that the perpetuation of the community depends on near-annual fire (or some other factor that removes the woody invasion).

Ecologists tend to see the chaparral of the foothills of southern California as a vegetation type that, under natural conditions, exists in a fire-initiated cycle. Areas of old shrubby growth build up a heavy, highly flammable accumulation of leaves, twigs, bark, and acorns. Fire removes the litter and dead stems and kills most of the above-ground shrubs, allowing grasses and other herbs to flourish, but the

Table 3-6
Fire Adaptations Characteristic of 42 Important Trees and Shrubs of the Arizona Chaparral

Trait	*Percent Showing*	*Examples*
Sprouts from root crown or roots	69	Gambel oak, birchleaf mountain mahogany
Seed cones opened	2	Arizona cypress
Thick bark	2	Ponderosa pine
Germination stimulated	7	Desert ceanothus, pointleaf manzanita
Slight or none evident	19	Utah juniper, Parry agave

Based on Knipe, O. D., Pase, C. P., and Carmichael, R. S., "Plants of the Arizona chaparral," U.S. Forest Service Gen. Tech. Rep. RM-64, pp. 51–52, 1979.

banishment of the shrubs is only temporary (Table 3-6). Various effects of their regrowth, such as shading and perhaps allelopathic chemicals, cause the herbs to decline. The shrub litter builds up again, and in 30 years or so another fire is likely.

In chaparral, compared with prairie, the fire interval is longer and the changes wrought by fire are greater, but viewing the whole cyclic system as one community maintained by periodic fire makes sense to most students of vegetation. When the interval is still longer and the fire is a catastrophic occurrence for the dominant vegetation of the site, many persons would consider the fire as destroying the existing community and initiating a new sequence of communities on the site (secondary succession; see Chapter 13). This was the view of most early ecologists (Weaver and Clements 1938) and some still employ a successional framework and terminology (e.g., Rego et al. 1991).

There is nothing wrong with this point of view, but we need to recognize that even in these situations, fire may be a natural, expectable event at the landscape level. For example, the mosaic of black spruce and aspen-birch forest characteristic of much of the taiga depends on periodic fires or other catastrophies periodically killing the spruce, allowing the hardwoods to flourish briefly. Although fire is rare in the beech-maple forests of the Midwest, there is evidence that much of the herb diversity that our pioneer ecologists considered characteristic of these mature forests was the result of past fires and that in the absence of fire, the number of herb species steadily drops (Brewer 1980).

Fire as a Management Tool

The scientific use of fire as a tool for ecological management originated in the southeastern United States through the work of such men as the Yale forester H. H. Chapman and the wildlife biologist Herbert Stoddard (1936). Longleaf pine is a valuable timber tree, and the grasses and legumes that grow between the wide-spaced trees support good populations of the northern bobwhite, a favorite game bird. Controlled burning to produce light surface fires eliminates or suppresses the oaks and other hardwoods that would eventually dominate the area and also favors the grasses and legumes.

Longleaf pine has a life history adapted to recurrent fires (Figure 3-26). Early growth produces a short stem surrounded by dense needles; the young plant looks much like a clump of grass and is called the *grass stage.* The plant shows little growth in height for three to seven years; at this stage it is resistant to damage from fire because the terminal bud is buried in the middle of the dense clump of long, green needles, which insulate the bud and burn very poorly. Although it doesn't get taller, the plant produces a deep, extensive root system, allowing it to grow rapidly at the end of the grass stage. It may grow 4 to 6 feet a year for 2 or 3 years, carrying the terminal bud up out of the reach of surface fires. It is during this period of rapid growth that the plant is most vulnerable to fire. Soon afterward the bark thickens enough to make the large sapling again resistant to fire.

Ecologists, foresters, and wildlife ecologists of the Southeast have begun recently to direct their attention to managing longleaf pine land as an ecosystem, rather than merely as a site for timber (or even timber plus game) production (Landers et al. 1989). The management goal is ecosystem integrity, including birds such as the red-cockaded woodpecker and the characteristic understory herbs. In this attempt, the Southeast has been far ahead of most parts of the country.

Periodic burning is necessary to maintain prairie vegetation in the Midwest (Henderson 1982). This is true for patches of relict prairie as well as restored and reconstructed prairies—that is, areas which have been seeded or planted to prairie

(A) (B)

(C)

Figure 3-26

Some aspects of adaptation to fire in longleaf pine. In the grass stage (A), longleaf pine is resistant to surface fires because the bud is protected by a dense cluster of green, very long needles. This is in contrast to the seedling stage of most pines, such as slash pine of the same region, in which the terminal bud is much more exposed (B). The grass stage may last three to seven years, during which the seedling grows little in height but instead puts its energy into food reserves and an extensive root system. These prepare the plant for rapid growth, with the tree shooting up eight to eighteen feet over a period of two to three years (C). This is the period of greatest vulnerability to fire but, by the end of it, the terminal bud is out of reach of surface fires and the bark has thickened enough to protect the trunk. Photographs are of longleaf pine flatwoods in Alachua County, Florida. Older longleaf pines are visible in the background. (Photos by the author.)

(Schramm 1978). Burning every one to three years tends to kill woody plants and weeds and also to increase the growth (Hadley and Kieckhefer 1963) and flowering (Petersen 1983) of many of the native prairie species (Table 3-7).

Fire has also been recognized as a necessary tool in regenerating the even-aged stands of jack pine that the endangered Kirtland's warbler needs as nesting habitat. In the early 1950s the U.S. Forest Service and the Michigan Department of Natural Resources began a program of controlled burns on about 135,000 acres of federal and state land in northern Michigan (Anon. 1984). This management for the warblers seems not to conflict with commercial forestry designs on the land because the warblers discontinue breeding in the jack pines

Table 3-7
Effects of Fire on Growth and Flowering Stalk Production in Big Bluestem in a Restored Illinois Prairie

The last fire was in May of the year in which the measurements were taken. Net production was approximated by ovendry weights of samples clipped at the end of the growing season, in October.

Years Since Last Fire	*Net Production of New Shoots (Ovendry, gm/m^2)*	*Number of Flowering Stalks per m^2*
19	362	28
3	359	30
1	591	53
0	1360	130

Data from Hadley, E. B., and Kieckhefer, B. J., "Productivity of two prairie grasses in relation to fire frequency," *Ecology*, 44: 389–395, 1963.

before the pines reach the size at which they are cut for pulp. Consequently, the bitter battles between persons wanting to save the birds and ones wanting to make money off the timber, such as have been fought over the red-cockaded woodpecker in the southeastern United States and the spotted owl in the Northwest, have so far not developed.

All went well until May 5, 1980. A fire was set on a 210-acre (85 hectare) clearcut near Mack Lake by U.S. Forest Service personnel on a day that was probably too dry and windy for a safe burn. It broke out of control about noon and by 7 P.M. had burnt 25,000 acres (10,000 hectare) of jack pines and damaged 44 homes and summer cottages.

This was an unfortunate occurrence, but no one who knows anything about ecology would believe that a person could live amongst jack pines without eventually being involved in a fire. The jack pine plains are a fire-dependent ecosystem, and even with the best human efforts to prevent fire, one will eventually come. The same lack of ecological knowledge leads to home building in the California chaparral. The scenes on the evening news of persons losing their $800,000 houses in a chaparral fire bring tears to the eyes of us all, to be sure, but fires there also are predictable events. Delaying fires in either chaparral or jack pine simply means more litter buildup and a hotter fire when it comes.

In the 1970s, the federal government adopted a "natural burn" policy, called by the media the "let-burn" policy, for many of the national parks and the wilderness areas in national forests. This was part of an attempt to restore natural processes to these communities. Before this policy was initiated, fires were suppressed. Roughly, the let-burn policy consists of letting fires that start from lightning burn as long as they don't threaten property. Property-threatening fires, as well as those started by humans accidentally or deliberately, are fought.

The let-burn policy came to national attention in the long, hot summer of 1988. A series of fires began in Yellowstone National Park on June 23 and eventually burned a sizable fraction of the park. The summer of 1988 was a period of drought in many locations and it was also a time of increasing awareness of the threat of global warming from the greenhouse effect. The drought caused the Yellowstone fires in the sense that normal rainfall in July and August would probably have limited the area burned to a few thousand acres. The anxiety of the public about our climatic future focused attention on the fires as a possible portent of that future.

Although television news programs talked of a million acres of Yellowstone being "destroyed" that figure was arrived at by drawing lines around all the areas where fire occurred. Within those perimeters were large areas that the fire skipped. Something less than half a million acres actually burned, and areas of severe burn probably amounted to about 1% of the park's 2.2 million acres.

One indication of the importance of fire in Yellowstone was the acreage of lodgepole pine, which occupied about three fourths of the total area (Cundall and Lystrup 1987). Although this fire-dependent community must have been the result of past fires, recent history gave little direct evidence of the importance of fire. Since the implementation of the let-burn policy in 1972, only about 14,000 ha (about 34,000 acres) had burned (Varley and Schullery 1991). The last big fire occurred nearly 50 years earlier, in 1931, and had burned only about 8000 ha; the total area burned from 1885 to 1987 was 80,000 ha (Mathews 1988).

The great buildup of flammable material—standing and fallen trees—and the long summer drought that reduced the moisture content of the dead wood to below that of kiln-dried lumber set the stage for the Yellowstone fires. It is possible that the fire-suppression policy of past decades exacerbated the fuel buildup (Romme and Despain

1989), and it is true the drought was one of unusual severity; nevertheless, it would be a mistake to think that the Yellowstone fires were a freakish occurrence. Instead, when we consider the dominance of the landscape by lodgepole pine in the context of the near incombustibility of the vegetation as shown in the years 1972 to 1987, it looks as though infrequent, catastrophic fires in extreme drought years is the way in which this vegetation regenerates (Romme and Despain 1989).

The public, fanned by elements of the media that spoke of the destruction of Yellowstone, reacted unfavorably to the let-burn policy. Noisiest were the owners of businesses in the surrounding area and the politicians. In fact, the let-burn policy probably had no influence on the outcome of the Yellowstone fires. After first enthusiastically embracing the prospect of finally seeing how fire was supposed to work in Yellowstone, the park administration grew timid and on July 21 (after about 7000 ha had burned) suspended the policy and fought all fires thereafter (Hackett 1989 (Figure 3-27). In the tinderbox conditions of the summer of 1988, it made little difference. If every fire that started in early summer could have been put out, the total acreage burned would probably have been trivially different.

Figure 3-27

Not long into the summer of 1988, the policy of letting natural fires run their course was abandoned and all the resources of the Park and Forest Services were directed toward fighting the fires. This photo of a helicopter with a large water bucket just filled in Lake Yellowstone was taken in August 1988 by the author.

There are three general approaches to fire as a management tool: fire suppression, prescribed (management-set) burning, and the let-burn approach. Fire suppression is expensive and is nearly always for timber management rather than ecological management. That is, taxpayer dollars are spent to protect trees that will be later cut for profit by lumber companies. It is still unclear whether some preserves set aside to preserve vegetation types with very low natural fire frequencies ought to be protected from all fire.

We have already described examples of prescribed or controlled burning in the management of prairie and jack pine and longleaf pine ecosystems. Some have argued that this approach would also be better in the national parks and wilderness areas of the national forests. In the unnatural world we have created, the argument goes, Mother Nature's hands may be tied. In the case of Yellowstone, she waited too long; prescribed burns in the lodgepole forest over the past 20 years would have been better. The enhanced diversity now being seen in Yellowstone would have begun earlier and the conflagration of 1988 would have been avoided.

Prescribed burning is an adequate answer in many circumstances. As the end-product of sound research, a controlled fire in a fire-dependent ecosystem is better than no fire at all. A major lesson of the last couple of decades, however, is the significance of rare events in the normal functioning of populations and ecosystems. We talked earlier of controlled burns in the management of Kirtland's warbler and the one—the Mack Lake fire—that escaped to become the only significant wildlife in recent Kirtland's warbler history. Today, the bulk of the warblers are on the Mack Lake burn (Probst 1991). For the warbler, and probably for other species, the timid controlled burns are often too small, too cool, or otherwise incapable of producing optimal habitat.

For Yellowstone, a hybrid fire management plan has been devised (Yellowstone National Park 1992). It includes three zones: (1) A *suppression zone* that consists of about ten large spots around the developed areas of the park. (2) A *prescribed natural fire* zone covering most of the park where natural fires "will be allowed to burn under predetermined prescriptions to perpetuate natural processes."

Natural fire that "exceeds prescription will be declared a wildfire" and suppressed. (3) A *conditional zone* that is basically a strip 2.4 km wide just within the park borders. This strip will intercept fires that start off the park, as did several of those of 1988. Natural fires in this strip will be scrutinized closely; they may be allowed to burn but "under more conservative prescriptions" than in zone 2. In all three zones, controlled burns may be used for hazard fuel reduction, though the low combustibility at most times makes this approach of dubious utility (Schullery and Despain 1989).

Whether this is a plan that will allow ecological reality to coexist in Yellowstone with sagebrush politics remains to be seen.

Pollution

Not long ago the word "pollution" generally referred to adding filth or poison to water or air, but it has recently, and quite sensibly, been given a broader meaning. A good formal definition is that **pollution** is the unfavorable modification of the environment by human activities.

Pollution is not new. Air pollution caused by burning coal was recognized as a human health problem by the mid-1600s, but pollution of various kinds was producing unrecognized problems long before that. Lead poisoning occurred among the upper classes in Ancient Rome as the result of lead-lined cookware and lead water pipes (our word "plumbing" comes from the Latin word for lead, *plumbum*). Even the scientific study of pollution goes back many years. Stephen A. Forbes and R. E. Richardson (Forbes and Richardson 1919) of the Illinois Natural History Survey studied the effects of sewage pollution on the Illinois River before 1920 (Figure 3-28), and R. C. Osburn reported serious pollution of Lake Erie waters around the larger ports at the 1916 meeting of the Ecological Society of America.

Pollution affects the individual organism di-

Figure 3-28

The Chicago Sanitary Canal. Although there were intervals during the post-glacial history of Lake Michigan when it drained south through the DesPlaines River, in modern times there was no connection between the two. Then in 1899 the Sanitary District of Chicago dug a canal into which it could dump its sewage. From the DesPlaines the sewage was carried to the Illinois River and thence to the Mississippi. By 1915, the Illinois River was badly polluted more than 120 miles downstream and by 1920 the heavily polluted section had extended another 80 miles, well south of Peoria. (Photo by Joseph G. Engemann.)

rectly by weakening or killing it, but it also has effects at the population and ecosystem levels. One species may increase in number because pollution has harmed a species that competes with or preys upon it. Through such effects the community may change, often becoming less diverse.

Pollution will be discussed in more detail in Chapter 20. It is mentioned here to make the point that today pollution of one kind or another is an important factor of the habitat, often on a par with those discussed in the preceding pages. Although degrees of pollution exist, it is no exaggeration to say that there are no unpolluted habitats left. It is not a question of whether pollution affects the particular organisms or community we are studying but, instead, whether we are aware of it.

Population Ecology: *Growth and Density*

OUTLINE

- Birth and Death
- Population Growth
- Population Density and Population Regulation
- Growth and Regulation of Human Populations

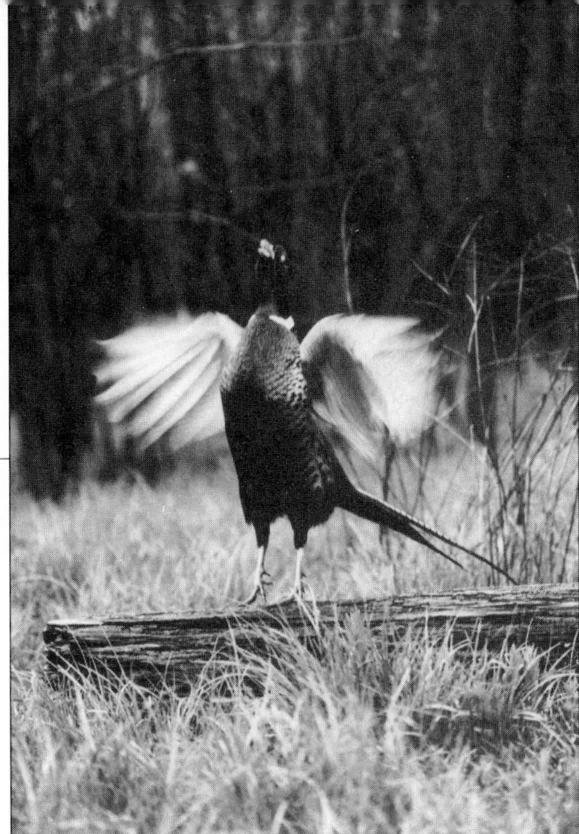

Ring-necked pheasants have a high reproductive rate. *(Sharon Cummings/Dembinsky Photo Associates)*

Populations have traits of their own that differ from those of the individuals composing the populations. An individual is born once and dies once, but a population continues, perhaps changing in size depending on the birth and death rates of the populations. An individual is male or female, old or young, but a population has a sex ratio and an age structure. This chapter and the next examine some of the ecological traits unique to the population level of integration.

Birth and Death

Birth Rate and Death Rate

New individuals can be added to populations in two ways, by birth and by immigration (Figure 4-1). Individuals can leave populations in two ways—death and emigration. Generally births and deaths are expressed as rates, that is, as numbers in a given time. If there are 120 births in a population during a year,

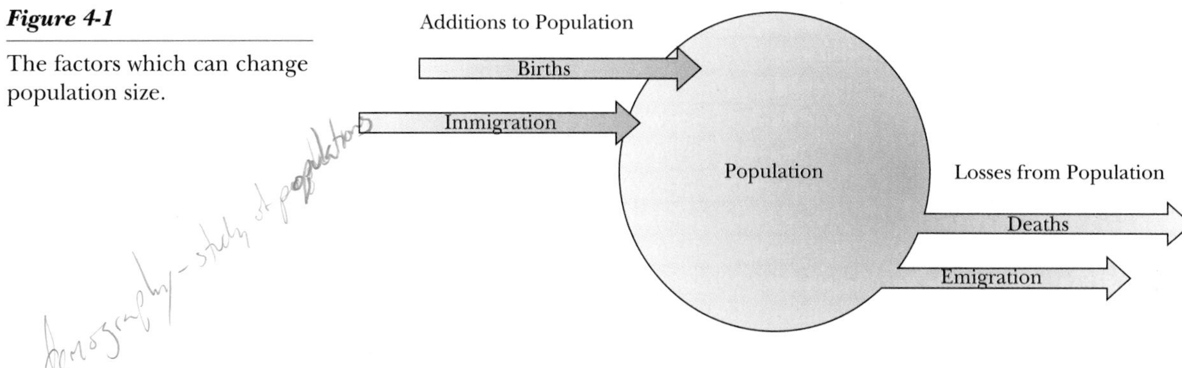

Figure 4-1

The factors which can change population size.

the birth rate **(natality rate)** is 120 per year, or 10 per month. We can express the death rate **(mortality rate)** in the same way. Often rates like this that deal with the population as a whole, rather than specific segments of it, such as a particular age class, are called **crude birth** and **deaths rates.**

If we ignore immigration and emigration—which is easy enough to do on paper but generally impossible in nature—then changes in population size depend on the balance between birth and death rates. If the two are equal, then the population size is stable. If birth rate exceeds death rate, the population grows; and if death rate exceeds birth rate, the population declines.

When we deal with human populations, crude birth rate and crude death rate are usually given per thousand persons in the pouplation. If there were 250 children born to a population of 10,000 during a year, then the yearly crude birth rate would be 25. Crude death rate is calculated in the same way; if 150 persons died in the same year, then the crude death rate would be 15 per year.

Such a population is growing because birth rate exceeds death rate; determining how fast it is growing is simple enough. There were 10 more births than deaths per thousand persons (a crude birth rate of 25 per thousand minus a crude death rate of 15 per thousand). Expressed as a percentage (per hundred rather than per thousand) this is an increase rate of 1% per year. There were 10,000 in the population last year; there were 150 deaths and 250 births, for a net increase of 100. This 100 is the 1% increase over the original 10,000. Birth rate can also be expressed per head (per capita), which in this case is 0.01 per year (100/10,000).

Life Tables and Longevity

A **life table** summarizes the statistics of death and survival of a population by age. A made-up life table, just to illustrate the terms and calculations used, is shown in Table 4-1 for an odd bird, the McKinley murre population on Mount Deevey. Ages are in years, but for other kinds of organisms they could be days, months, or hours. The age at the beginning of the interval is represented by x; l_x is survivorship, the number of individuals alive at the beginning of the interval; d_x is mortality, the number dying within the interval; q_x is mortality rate, the number dying within the interval divided by the number

Table 4-1
Life Table for McKinley Murre Population

Age x	*Survivorship* l_x	*Mortality* d_x	*Mortality Rate* q_x	*Life Expectation* e_x
0	100	55	0.55	1.15
1	45	30	0.67	0.94
2	15	10	0.67	0.83
3	5	5	1.00	0.50
4	0	—	—	—

alive at the beginning of the interval; and e_x is life expectation, the average time left to an individual at the beginning of the interval.

In this table we start out with 100 individuals at the time they are born (age 0). Such a group, all born at the same time, is called a **cohort;** all the children (or murres) born in the United States in June 1971 would be a cohort. All are alive at the beginning of the interval, but in the table over one half (55) died during the first interval. For the murres in their first year, then, mortality rate is 55% (55/100 × 100). A murre at birth can expect to live, on the average, just over one year. Since 55 animals died in the first year, 45 survived to begin the second. During this period, from age one to two, 30 died. Mortality rate in the second year, in this case, is slightly higher than for the first year, 67% (30/45 × 100). The murre aged one year has, on the average, slightly less than one year of life remaining.

About the only part of such a table that is not easy to understand is the calculation of **life expectation.** A method of calculating expectation of further life is given in Brewer and McCann (1982). The concept itself is simple enough; it is the average additional time still left to individuals alive at a particular age. The life expectation for a male in the United States as shown in Table 4-2 is 71.5 years when he is born. When he is one year old, his life expectation is not 70.5 years but 71.3, virtually the same as at birth. This is because the death rate for children under one year old is high, higher than at any other time until after the age of 40 years. Similarly, the life expectation of a 71-year-old man is not half a year. Most who make it to age 71 live a while longer—about 11 years longer, on the average.

Life expectation at age 0 is the same thing as **mean natural longevity.** Another aspect of longevity is **physiological longevity,** the age reached by individuals dying of old age—individuals living under conditions where such causes of death as predation, accident, poor nutrition, and infectious disease are not factors. Although this may seem to be an ill-defined concept, it is nevertheless useful. There are upper limits to the ages of most kinds of organisms, a time of life when senility takes over no matter how good the conditions of life are. There are no 2000-year-old men, or even 200-year-old ones, just as there are no 100-year-old horses or 50-year-old dogs.

Types of Life Tables

Life tables may be constructed in several ways, using different sorts of information; however, there are two basic types: **age specific** and **time specific.** Age-specific tables are straightforward; the data can be obtained by keeping track of the ages at death of a cohort. The columns we have are age (x) and deaths (d_x); using these we calculate the rest, as described in the preceding section.

Time-specific tables are less straightforward. Information from a single time period is used to estimate one of the columns of the table. For example, we may be able to census the population and, thereby, obtain (with a little more arithmetic) an estimate of the survivorship (l_x) column. From that we can calculate the other columns. Or we may be able to estimate death rates (q_x) for the current year and use that information to calculate the other columns. This second approach is easy for human populations (Table 4-2), but for most others it is hard to get the relevant data.

A confusing variety of synonyms have been used for both age-specific and time-specific tables. Anyone delving into the literature on the subject will have to learn them, but they are otherwise unnecessary. One additional term may be useful, however. **Composite** life tables combine data from several cohorts or several time periods (generally done to get a sample of satisfactory size). To illustrate, suppose we spend a lot of time hiking around the farm, and every time we come across a deer skull, we pick it up and figure out its age (by tooth wear). We believe that these are the remains of deer that have died of old age or starvation or were killed by predators. After several years, we have data that can be used to estimate the mortality (d_x) column for an age-specific table; however, the cohort is composite, composed of individuals that actually began their life in several different years. Composite time-specific life tables may also be constructed.

Age-specific and time-specific life tables will be identical for a population only if conditions are not changing with time or, specifically, only if there are no important differences between years in birth rate and death rate (differences between ages are,

Table 4-2
Life Tables for Humans in the United States, as of 1988

Women have considerably better survivorship so that the mean expectation of further life at birth is 7 years longer for females than for males. Notice, though, how a life table such as this one is constructed. It does not really start out with 100,000 children born in 1988; we could not get their vital statistics until well into the twenty-first century. Instead, it uses current death rates for each age group. The mortality rate for young children is for persons born in the 1980s, but the mortality rate for the oldest age categories is for persons born around 1900. The other columns are all calculated from the mortality rate column. Therefore we cannot really say that a girl born now will live longer, on the average, than a boy. It may be so, but it also may be that social conditions will change enough that women will suffer increasingly from the stress-related diseases such as heart disease and high blood pressure that now carry off men at early ages. Life tables cannot predict the future or, rather, they predict the future only when conditions do not change.

Age x (Years)	*Mortality Rate* q_x	*Survivorship* l_x	*Mortality* d_x	*Life Expectation* e_x
Male				
0–1	.0110	100,000	1,104	71.5
1–5	.0022	98,896	220	71.3
5–10	.0014	98,676	138	67.5
10–15	.0017	98,538	165	62.5
15–20	.0062	98,373	615	57.6
20–25	.0087	97,758	854	53.0
25–30	.0089	96,904	865	48.4
30–35	.0107	96,039	1,024	43.8
35–40	.0134	95,015	1,276	39.3
40–45	.0170	93,739	1,592	34.8
45–50	.0245	92,147	2,261	30.3
50–55	.0385	89,886	3,457	26.0
55–60	.0610	86,429	5,273	22.0
60–65	.0941	81,156	7,639	18.2
65–70	.1360	73,517	9,996	14.9
70–75	.2022	63,521	12,842	11.8
75–80	.2931	50,679	14,852	9.1
80–85	.4239	35,827	15,189	6.9
85 and over	1.0000	20,638	20,638	5.1

of course, okay; they are what life tables are about). Neither kind of table is better than the other; they tell us different things.

Survivorship Curves

Plotting the l_x (survivorship) column of a life table against the x (age) column gives a **survivorship curve**. This is convenient for use as a visual aid to detect changes in survivorship (and mortality) by period of life.

It is conventional and, to a degree, useful to recognize three general categories of curves, as outlined by E. S. Deevey (1947) in an early ecological treatment of survivorship curves. These are called **Types I, II,** and **III,** or, better, **convex, diagonal,** and **concave.** Note that these are the shapes (Figure 4-2) when the logarithm of the number surviving is plotted; in an arithmetic plot, the Type II curve is not diagonal but concave, though less so than the Type III curve.

A convex curve indicates high survivorship up

Table 4-2
(continued)

Age x (Years)	*Mortality Rate q_x*	*Survivorship l_x*	*Mortality d_x*	*Life Expectation e_x*
Female				
0–1	.0089	100,000	890	78.3
1–5	.0018	99,110	175	78.0
5–10	.0010	98,935	101	74.2
10–15	.0010	98,834	100	69.2
15–20	.0024	98,734	240	64.3
20–25	.0028	98,494	272	59.4
25–30	.0033	98,222	321	54.6
30–35	.0041	97,901	405	49.8
35–40	.0058	97,496	567	45.0
40–45	.0084	96,929	817	40.2
45–50	.0135	96,112	1,293	35.5
50–55	.0219	94,819	2,077	31.0
55–60	.0347	92,742	3,217	26.6
60–65	.0537	89,525	4,810	22.5
65–70	.0793	84,715	6,716	18.6
70–75	.1210	77,999	9,435	15.0
75–80	.1843	68,564	12,640	11.7
80–85	.2981	55,924	16,671	8.7
85 and over	1.0000	39,253	39,253	6.3

From U.S. Dept. of Health and Human Services. *Vital Statistics of the United States 1988,* 1991, vol. 2, part A, Table 6-1, Hyattsville, Md, Public Health Service.

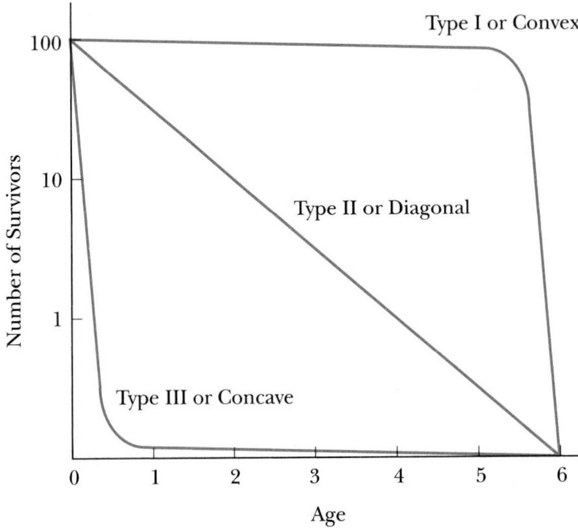

Figure 4-2

Types of survivorship curves. Note that the y axis is logarithmic.

to a particular age when most of the population dies. This could happen if environmental factors were unimportant and most of the organisms lived out their full physiological longevity. The abruptness of the drop in survivorship would depend on how variable the population was in genetic factors affecting length of life.

A diagonal curve indicates a constant probability of dying; that is, the same percentage of the population is lost each time period. Here the environment is very important indeed since it lops off the same proportion whether the animals are young or old.

A concave curve indicates high juvenile mortality and then low mortality as the survivors slowly succumb to environmental factors and, eventually, senility. The high juvenile mortality would result from such factors as inexperience in foraging and avoiding predators and lack of immunity to disease.

Probably no organism for which adequate birth-to-death information is available actually fits any one of these idealized curves. The most frequent real pattern seems to be one in which there is a juvenile segment of high mortality, followed by an adult segment of low, nearly constant, mortality, and a final senile segment in which mortality again rises. Such a pattern is shown in Figure 4-3.

Apparent deviations from this pattern often are the result of omitting part of the organism's life span. For example, annual plants, with their preprogrammed senescence, often show a convex curve if we start with seedlings. Doing this, however, omits a long stretch of their life, as seeds, when mortality is higher (Figure 4-4). Similarly, birds are often put forward as cases of diagonal curves, but they seem so only because survivorship curves are usually based on recoveries of banded birds and, for methodological reasons, typically start when the birds are about 6 months old. If juvenile mortality is plotted, birds show high early mortality. It appears, however, that survivorship curves for wild birds rarely show the final, senile segment. Instead, their life span in nature is almost always far below that which they can reach when kept in comfort in the aviary, so they rarely attain the age at which senility sets in. Seeing the very brief average age reached by birds, as shown by banding data, brings home to us the shortfall for most individual animals between the possible and the actual (Table 4-3).

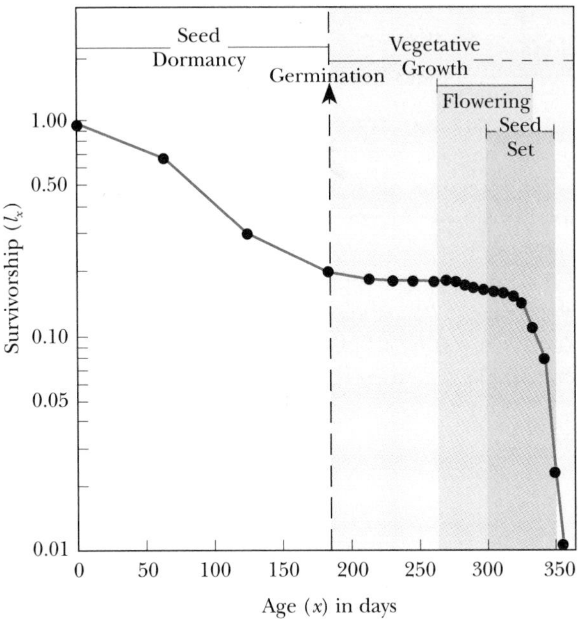

Figure 4-4

A survivorship curve for Drummond's phlox that began at germination (the vertical broken line) would be convex. The actual curve including the high seed mortality is more complex. Drummond's phlox is an annual that germinates in the winter, makes slow growth until early spring, then grows and flowers through May, when plants begin to die. (From W. J. Leverich and D. A. Levin, "Age specific survivorship and reproduction in Phlox drummondii." *Am. Nat.*, vol. 113, p. 885. Copyright © University of Chicago Press. Reprinted with permission.)

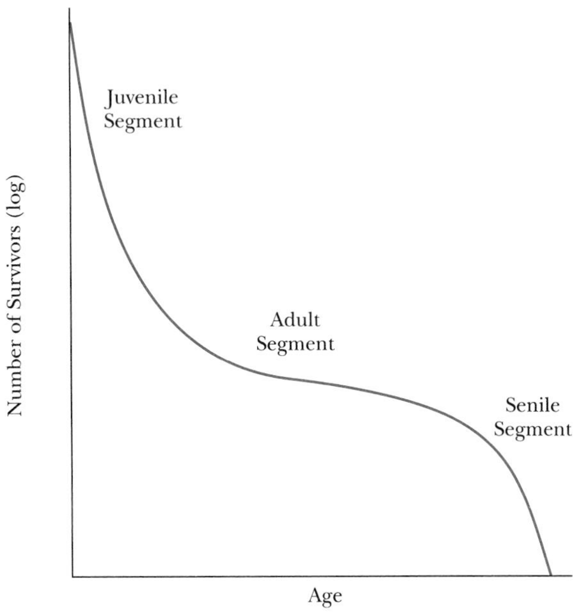

Figure 4-3

A frequent type of survivorship curve with high early mortality, lower, nearly constant mortality in adult life, and higher mortality again among the elderly.

Population Growth

Exponential Population Growth

When a very small population is living in a very large area of favorable habitat, population growth rate tends to depend on two factors: the size of the population and the capacity of the population to in-

Table 4-3
A Comparison of Estimates of Mean Natural Longevity and Potential Longevity in Various Birds

Mean natural longevity is calculated from banding data. Potential longevity is the oldest bird reported from banding (potential natural longevity) or in zoos (the latter, probably a fair estimate of physiological longevity, is indicated by †). Mean natural longevity is from data in Farner, D. S., "Birdbanding in the study of population dynamics," *Recent Studies in Avian Biology*, Wolfson, A. (ed.), University of Illinois Press. Potential longevity is from Table 18-1 in Welty, J. C., *The Life of Birds*, Philadelphia, Saunders College Publishing, 1982, except as indicated.

Species	*Mean Natural Longevity**	*Potential Longevity*
Mallard	1.5 years	29 years
Herring gull	6.2	41†
American robin	1.4	11
European robin	0.9	11
European starling	1.4	20
Northern cardinal	1.6	22†

* As of the first August 1, September 1, or January 1 following banding as a young bird.

† From Flower, S. S., "Further notes on the duration of life in animals. IV. Birds," *Proceedings of the Zoological Society of London*, ser. A, 108:195–235, 1938.

crease (in other words, the **biotic potential**). A situation of this sort occurs for aquatic organisms when a new lake is formed, whether by glaciers or the Corps of Engineers. For deer it may occur when they somehow manage to reach an island where they have never lived before. In the laboratory we can readily create such a situation by mixing up a bottle of culture medium and adding a microorganism to it. Let us examine such a case. For the sake of simplicity, we will use an organism that splits into two and assume that under thoroughly favorable conditions this occurs every 4 hours.

We will start at time zero with one such animal. Four hours later it will divide, and we'll have two. By the end of 8 hours, these two will divide, and there will be four swimming in the culture medium. Four hours later, after half a day, there will be eight, and after another 12 hours, at the end of the first day, there will be 64 organisms from the original one. Four hours into the second day there will be 128 organisms, and by the end of the day there will be 4096. By the end of the third day there will be 262,144 organisms. By the end of the fourth day there will be 16,777,216, which is not so many, but by the end of the fifth day there will be 1,073,741,824.

This type of growth, in which the curve of numbers versus time becomes steeper and steeper, is called **exponential growth** (Figure 4-5). The growth depends on the biotic potential, but this does not change; a division occurs every 4 hours all through the five days. Growth also depends on the size of the population; this does change, continually growing larger. As a consequence, the growth rate of the population steadily increases from a slow rate, when the population is low (an increase of only 63 animals in the first day) to a fast rate when the population is high (an increase of over 1 billion animals the fifth day).

Although there are many examples of exponential growth in laboratory studies, good data from the field are scarce because of the large

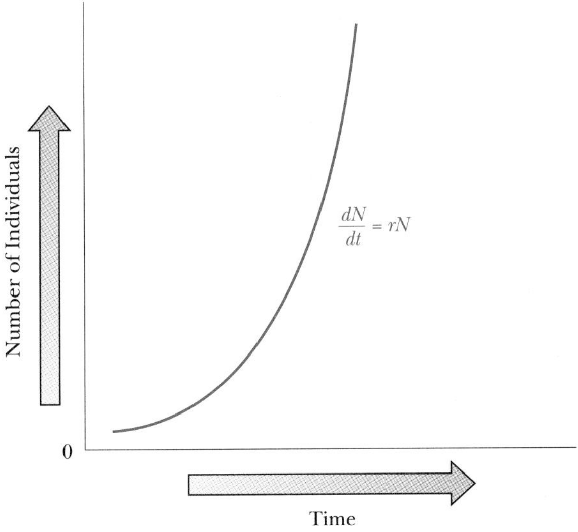

Figure 4-5

An exponential growth curve. In exponential population growth, growth rate depends on the biotic potential, which is a constant, and the number of reproducing individuals, which is steadily increasing. Both growth rate and the total population size (plotted here) rise at a steadily increasing rate.

amount of hard work usually required for accurate censusing. One widely cited example is the increase of a ring-necked pheasant population introduced on Protection Island off the coast of Washington (Einarsen 1945). A starter population of 8 birds increased to 1898 over six breeding seasons. Here we illustrate exponential growth with the study of a herd of tule elk introduced onto Grizzly Island northeast of San Francisco, California (Gogan and Barrett 1988). From an initial release of 8 animals in mid-1977 developed a population of 150 by 1986 (Figure 4-6).

For a stable population the rate of population increase is zero, because birth rate is equal to death rate. If a population grows at a fairly constant rate, whether it is 1%, 5%, or one half of 1%, population size will increase exponentially. This is exactly the same situation discussed earlier in more general terms. The only difference is that the rate of increase may be slower or faster than would be predicted on the basis of the biotic potential of the organism. Growth faster than predicted from the biotic potential can occur if the population does not have a stable age distribution, as discussed in Chapter 5; however, a stable age distribution is quickly established if growth persists at a constant rate.

One way to understand the consequences of exponential growth resulting from even low percentage rates of increase is to consider the relationship between the yearly rate of increase and the number of years it will take the population to double in size,

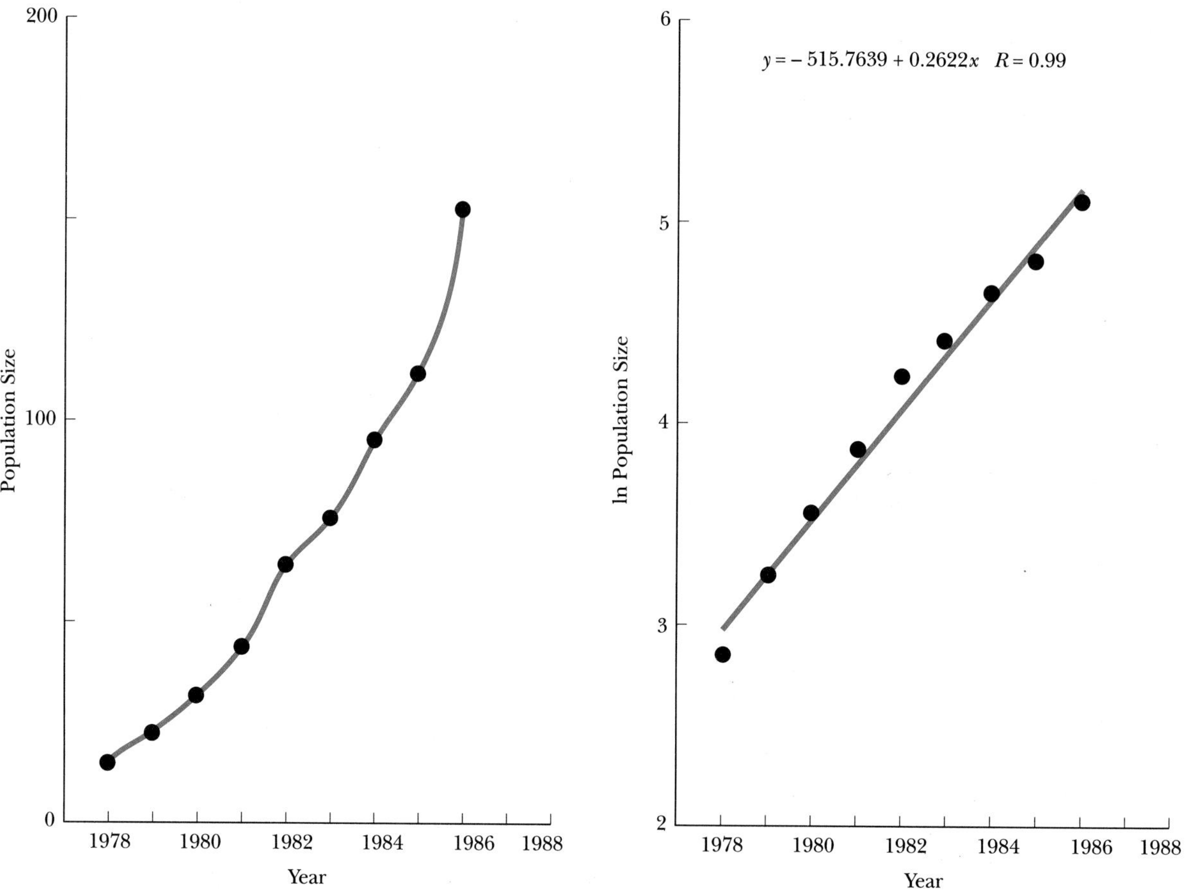

Figure 4-6

Exponential population growth by tule elk introduced on Grizzly Island, California. The left-hand graph plots population size arithmetically starting with the 17 animals present in 1977; the graph on the right shows the same data in a semi-log plot. (Data from Gogan and Barrett 1988.)

assuming that the yearly rate continues. Some of these figures are given below:

If the Yearly Percentage Increase is	*The Years Required for the Population to Double Will be*
0.5	139
1.0	69
1.5	46
2.0	35
2.5	28
3.0	23

As an approximation, good enough for everyday purposes, doubling time can be calculated by dividing the percentage increase into 70. The mathematically inclined may wish to derive this "rule of 70" from the exponential growth equation, $N_t = N_0e^{rt}$.

Biotic Potential

The **biotic potential** is the capacity of a population for increase. It is often represented by the symbol r (little "r"). Another name for biotic potential is **intrinsic rate of natural increase,** a term coined by the insurance actuary A. J. Lotka (1925), a pioneer in population theory. Any thorough discussion of biotic potential quickly becomes mathematical; at this point we will say only that it is a population increase rate and is determined by both birth and death rates. More specifically, it is the population growth rate per head (per individual member of the population) when the population is uncrowded and has a particular type of age distribution (the stable age distribution). Note that the population will actually grow at this rate only when it is uncrowded; much of the time the biotic potential is, in fact, potential and not actual. Nevertheless, if two populations, one with a high r and one with a low r, each suffers some catastrophe that drastically lowers their numbers, the species with the high r will be able to rebound more rapidly.

Some of the life history traits that influence whether the biotic potential of a population is high or low are the following:

1. Number of offspring per breeding season (for example, number of young in a litter and number of litters per year)
2. Survival up to and through reproductive age
3. Age at which reproduction begins
4. Length of the reproductive age of the organism (at one extreme are organisms such as salmon that breed only once; many other types of organisms breed every year until death or old age)

Obviously if an organism has a large litter, has a high survival rate, breeds at an early age, and reproduces repeatedly, it will have a higher r than one that has small litters, has low survival, does not mature until older, and breeds only once. However, it is not easy to see which of these factors tend to be relatively more important without some mathematics. Which of these traits, if altered, will most affect biotic potential? LaMont C. Cole (1954) showed that age at first breeding is exceedingly influential in determining whether biotic potential is high or low (Figure 4-7). This may be surprising at

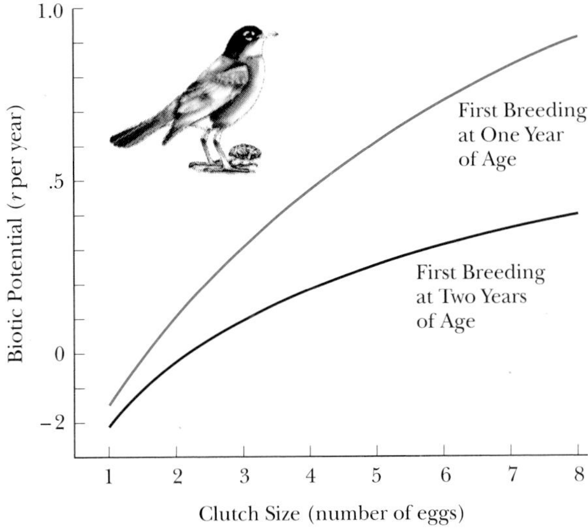

Figure 4-7

The influence of number of eggs and age of first breeding on biotic potential, or intrinsic rate of natural increase, in birds. Biotic potential is plotted against clutch size (number of eggs laid for one nesting). Calculations are based on life-history data intended to represent a small- or medium-sized open-nesting bird such as a song sparrow or an American robin. Increasing clutch size increases r; delaying maturity decreases r. A bird which does not breed until it is two years old but then lays eight eggs every year has a lower r than one that lays only four eggs per year but begins breeding when it is one year old. (Adapted from Brewer & Swander 1977, p. 218.)

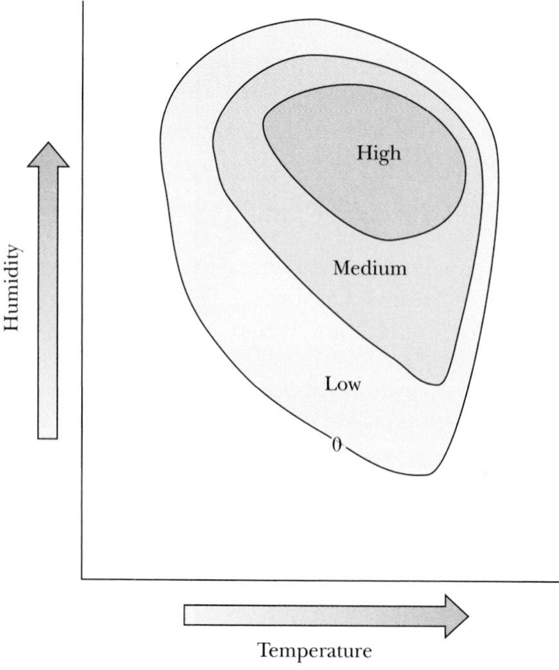

Figure 4-8

Hypothetical relationship of the intrinsic rate of natural increase (for example, *r* per week) of an organism to temperature and moisture. The figure is similar to a topographic map that shows elevation by contour lines. Here *r* rather than elevation is plotted and the coordinates are temperature and moisture instead of latitude and longitude. The peak *r* is at a medium temperature and a fairly high humidity. Moving away from the peak, *r* decreases, rapidly at higher temperatures and humidities, more slowly at lower temperatures and humidities, until it equals zero. For temperature-humidity combinations beyond this line, the population could not grow but only decline. Habitats having these combinations of temperature and moisture would not be permanently habitable.

first, but a little reflection will show the logic of it. Imagine two species that are similar except that one begins to breed at age one year and the other at age three years. At the end of the second year, the first species will have daughters who will be breeding and producing daughters who will, themselves, be breeding and producing more daughters in the third year, when the second species is just beginning to breed.

The loss of individuals through mortality makes the first reproduction especially important in determining *r*. Consequently, a long-lived organism that breeds repeatedly does not have an intrinsic rate of natural increase much higher than an otherwise similar one that breeds only once and then dies. These topics are discussed further under the evolution of life history traits in Chapter 5.

In suboptimal environments, even species with life history traits favoring rapid, prolific reproduction will have a low *r* because of abiotic factors that lower birth rate, raise death rate, or slow development (Figure 4-8). In fact, looking at the geographical range of a species in terms of population ecology, the range boundaries are approximately the line at which *r* is zero. Outside the line, populations of the species, if they somehow get a start, will decline, while inside the line, the populations will have a positive capacity for growth (Figure 4-9).

The ranges of the intrinsic rate of natural increase for various kinds of organisms are shown in Table 4-4. Because the exact determination of biotic potential is tedious, time-consuming, or difficult for most kinds of organisms, the values are based on a few direct studies plus several estimates (such as Robinson and Redford 1986). These ranges include the bulk of the species in each group, but there are exceptions. For example, the

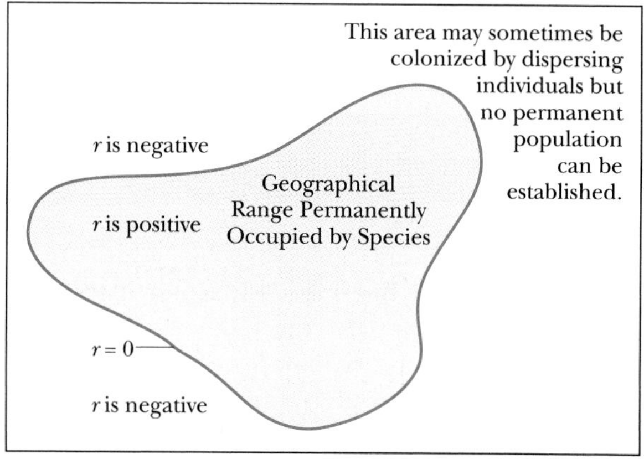

Figure 4-9

The geographical range of a species can be thought of as the region within which the intrinsic rate of natural increase (*r*) of its populations is greater than zero. Although this is generally true, it ignores a few complications. The geographical range of long-lived plants may, for example, include areas in which, because of climatic changes, they can no longer reproduce but where they can continue to survive very well.

Table 4-4
Intrinsic Rate of Natural Increase

Kind of Organism	*Approximate Biotic Potential, r (Per Year)*
Large mammals	0.02–0.5
Birds	0.05–1.5
Small mammals	0.3–8
Larger invertebrates	10–30
Insects	4–50
Small invertebrates (including large protozoans	30–800
Protozoa and unicellular algae	600–2000
Bacteria	3000–20,000

17-year cicada is older when it reaches breeding age than elephants, rhinoceroses, or humans; it has a biotic potential somewhere around 0.4 per year, far below that of most insects.

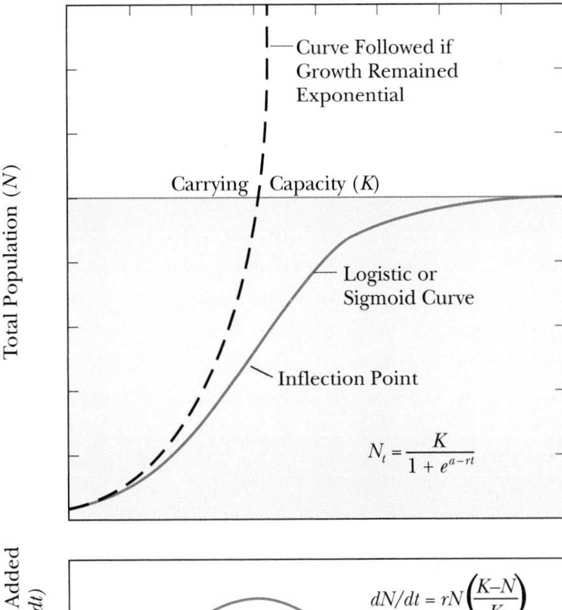

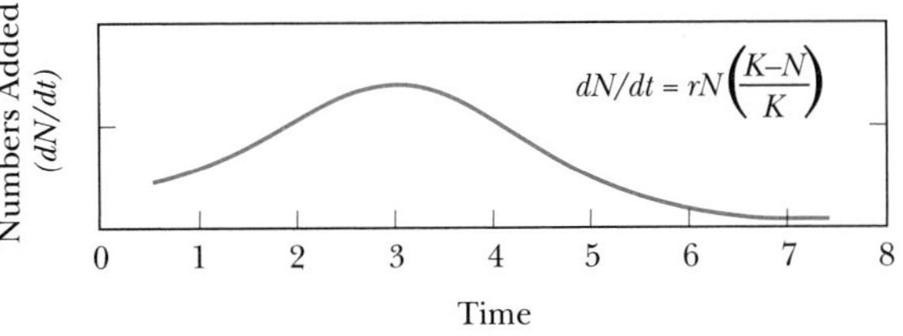

Figure 4-10

Logistic growth curve. In logistic population growth, the total size of the population grows in an S shape, first growing slowly, then faster, then more slowly, levelling off near the carrying capacity. Growth rate (plotted in the lower graph) increases, reaches a peak in the middle of the curve of numbers (the point of inflection, where the curve stops bending to the left and starts bending to the right), and then decreases toward zero as population numbers approach the carrying capacity.

Logistic Population Growth

Calculations demonstrating exponential population growth always seem faintly ridiculous because the numbers of most organisms we are acquainted with usually stay about the same from year to year (the human population is a temporary exception). If we have a pair of house wrens nesting in our yard this year, we know that the likelihood is small that we will have 4 pairs next year, 16 the next year, and so on. At some point in the growth of a population, then, growth must slow down and tend toward a zero growth rate.

When the growth curve of many populations is drawn—when population size is plotted against time—the curve that results looks like a flat S and is called a **sigmoid growth curve** (Figure 4-10). In this curve, population growth is exponential or approximately so at the beginning; the growth rate starts out slowly and then gets faster and faster. Then, when the population is medium sized, the growth rate begins to slow until it finally reaches zero when births balance deaths. The simplest model for describing this type of growth is the **logistic** equation, introduced to ecology by Raymond Pearl and L. J. Reed in 1920. What it says, in effect, is that the growth rate of the population is determined by the biotic potential and the size of the population as modified by the **environmental resistance,** or, in other words, by all the various effects of crowding. These effects may include lowered reproduction because of the mother's poor nutrition, high death rates because of predators, and increased emigration, among other things. Environmental resistance increases as population size gets closer to the **carrying capacity** (usually represented by *K*), which is the population size that the area has the resources to support.

Not all populations show a growth curve such as the one just described. Population growth very frequently seems to overshoot the carrying capacity and then drop back rather sharply, so that the first part of the curve looks something like a J (Figure 4-11). What causes such a curve? One likely reason is that there is a lag between the time at which the population attains a certain size and the time at which the unfavorable effects of that level of crowding are felt; in the meantime, the population continues to grow. What could cause such a time lag? If predators are an important aspect of environmental resistance, they might take a while to reproduce or to immigrate into the area of high population. Until their numbers increase, population growth continues. Or suppose that the nutrient supply in a pond is sufficient for 5 million individuals of some organism that reproduces by splitting in two. Population growth continues until there are 4 million, at which point the environment is still relatively uncrowded. One more division of each individual would raise the number to 8 million, well above the carrying capacity. A die-off to or possibly below the original carrying capacity would follow. Or suppose that the unfavorable effects of high populations include a slow decline in soil fertility and a slow buildup of toxic materials produced by the organisms, effects that are not serious until they reach a certain level, a threshold, at which point

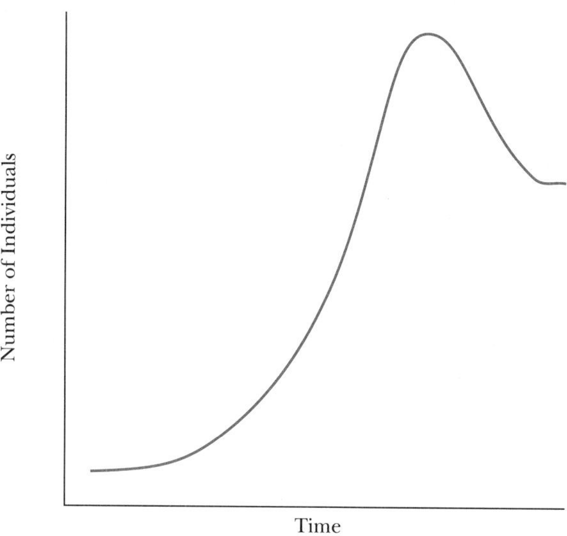

Figure 4-11

Often population numbers seem to overshoot the carrying capacity and then drop back to (or below) it.

they become very serious. If you think that this last example sounds as though it might have something to do with human populations, you are not alone.

The Logistic Curve as a Model

Remember that the logistic equation is just one mathematical way of representing population growth. It is a **model** of what the population may be doing. A model is defined as a simplified representation of a real system. In a sense, the "simplified" goes without saying, because the only non-simplified representation of a population of *Paramecium* would be a population of *Paramecium.*

We have already made use of models without being very explicit about it. For example, Figure 3-18 is a model of the relationship between fire frequency and climate, and Figure 3-1 is a model of heat transfer between organism and environment. All the ecological principles we have stated or will state in the rest of the book are models; they are our versions of how a part of the world works.

In ecology there tends to be a trade-off between how well a model represents a broad range of systems and how precisely it will predict the behavior of any one system. The logistic, for example, is too oversimplified to to used, unadorned, to calculate what the population of grouse in Portage County, Wisconsin, will be next year. Some of the ways in which it may be unrealistic include the following:

1. It assumes a straight-line relationship between density and the unfavorable effects of crowding, which is probably sometimes the case and sometimes not. There are indications for some organisms that at very low densities undercrowding is also unfavorable (the Allee Effect; see following section).
2. The simple logistic does not allow for time lags; everything is instantaneous. This is clearly not true in nature where predators may respond to a crowded prey population by immigrating, a process that may take days or weeks.
3. The logistic equation assumes that every individual in the population is the same. This may be valid for bacteria, but in a population of mammals there will be young and possibly old animals who will not be reproducing at all,

and females of different ages who may have different-sized litters.

If the logistic has these and other faults, why use it? Because (in addition to its being mathematically easy), it is still an adequate representation of population growth for many purposes. It represents, in a general way, how growth rate in many populations responds to density.

More complex models could be devised to represent population growth. For example, we could make the logistic more realistic by including time lags (as in the next section), by breaking *r* down into birth rate and death rate, or by including immigration and emigration. We could dispense with the logistic entirely and put all these variables into a series of equations treating the different age classes separately and solve the whole thing on a computer. The Leslie matrix (Leslie 1945) is one technique for doing this; it is discussed further in a later section of this chapter.

From such more complex models we might be able to ask questions for which we could get numbers as answers. For many practical purposes, such as managing grouse populations, this is valuable. For the purposes of understanding the world, however, mathematical models are not necessarily superior to graphical ones or even ones stated in words. It is sometimes easier to test a model, to see how well it represents the universe, if the answer is a number, but not always; it depends a great deal on how ingenious we are in making predictions and devising tests.

Model making is not some esoteric pursuit restricted to science. We all construct models in our efforts to deal with everyday life. Although we would probably not dignify it with the name, working out a plan for freeing cars stuck in snowdrifts is model making. Most of us would model the behavior of politicians by assuming that in choosing between alternatives they take the course that will yield the most votes.

We need models; our brains do not equip us for keeping whole systems continually in mind. At the same time, we need to remind ourselves frequently that a population is not a logistic curve. It is a bunch of animals, male and female, young and old, in good health and bad, with differences in genes and experiences, running around, escaping enemies, mating, and giving birth. If we are not to mislead ourselves, we must be vigilant in remembering that the logistic or any other scientific model is simply a guide to the system. Confusing model and system, as Alan Watts has said, is like eating the menu instead of ordering a meal.

The Allee Effect

To W. C. Allee, a Quaker, V. E. Shelford's first Ph.D. student, and a long-time faculty member at the University of Chicago, the emphasis in ecology on competition and overcrowding seemed one sided. Cooperation is important for some organisms and, consequently, **undercrowding** could also be detrimental (Allee 1958). This view that the optimal density of certain organisms is above the minimal density (one individual or pair) is the **Allee principle** or **Allee effect.**

The most comprehensive way of looking at the Allee effect is in connection with population growth rate. If population growth is strictly logistic, there will be a straight-line increase in per head growth rate as density decreases. Where the Allee effect operates, there is, instead, a decrease at some low density (Figure 4-12). Not many tests of this idea have been made. It seems reasonable that it will hold for some, if not all, highly social species, but how widely it exists in more solitary organisms is not clear. One of the few studies that seems directly applicable to the question shows nothing that could be interpreted as the Allee effect—but it does not show a straight-line relationship either (Figure 4-13).

Loosely, any optimal functioning—faster body growth, increased reproduction, or longer life—at an intermediate rather than minimal density has been regarded as the Allee effect. Instances of a drop in reproductive rate at very low densities seem fairly numerous and include a wide range of organisms. Presumably some females going unmated because they simply were not found by males or because of an unbalanced sex ratio accounts for much of this.

The classic example of a population that may have gone extinct for such reasons is the heath hen, the eastern form of the greater prairie chicken (Figure 4-14). Although the bird was originally abundant in the scrubby oak plains of the Eastern seaboard, overhunting had, by 1870, restricted it to

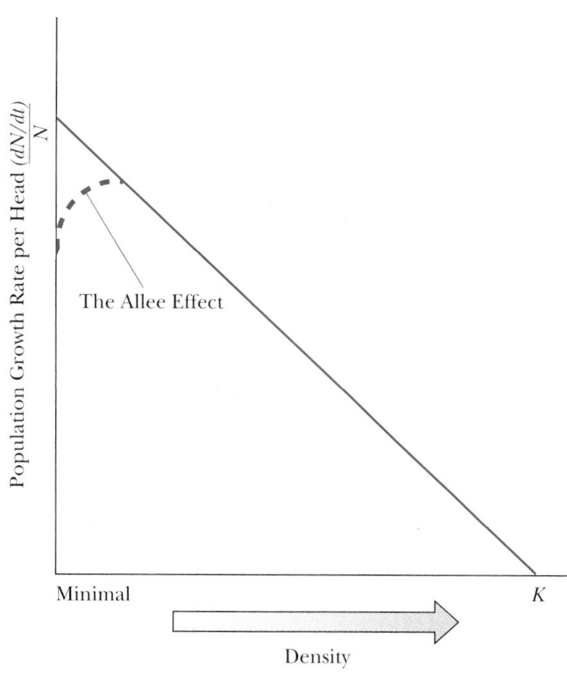

Figure 4-12

The straight line shows the relationship between population growth rate per head and density expected if population growth rate is strictly logistic. The dotted line shows how the relationship should change at low densities if the Allee effect holds.

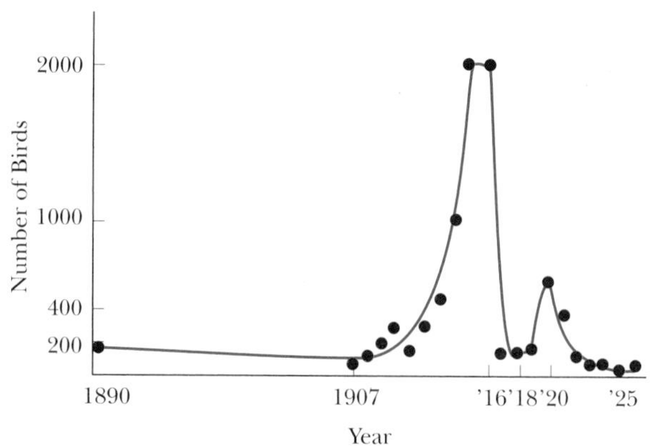

Figure 4-14

Population size of the heath hen on Martha's Vineyard from 1890 to 1928. From Allee et al. 1949 (after Gross).

Martha's Vineyard, a large island off the Massachusetts coast. Its population there fluctuated from 50 to 200 birds for several decades.

In 1908, the state of Massachusetts and private groups set up a 1600-acre (650 ha) sanctuary where the birds were protected by wardens. The population grew rapidly to about 2000 birds. Then an unlikely combination of catastrophes occurred (Gross 1928). A fire broke out during a gale on May 12,

Figure 4-13

Relationship between population growth rate per head and density found in water-flea *(Daphnia magna)* cultures. Per head growth rate continued to increase through the lowest densities investigated. (From "Population dynamics in *Daphnia magna* and a new model for population growth" by F. E. Smith, Ecology, 1963, 44, 651-663. Copyright © 1963 by the Ecological Society of America. Reprinted by permission.)

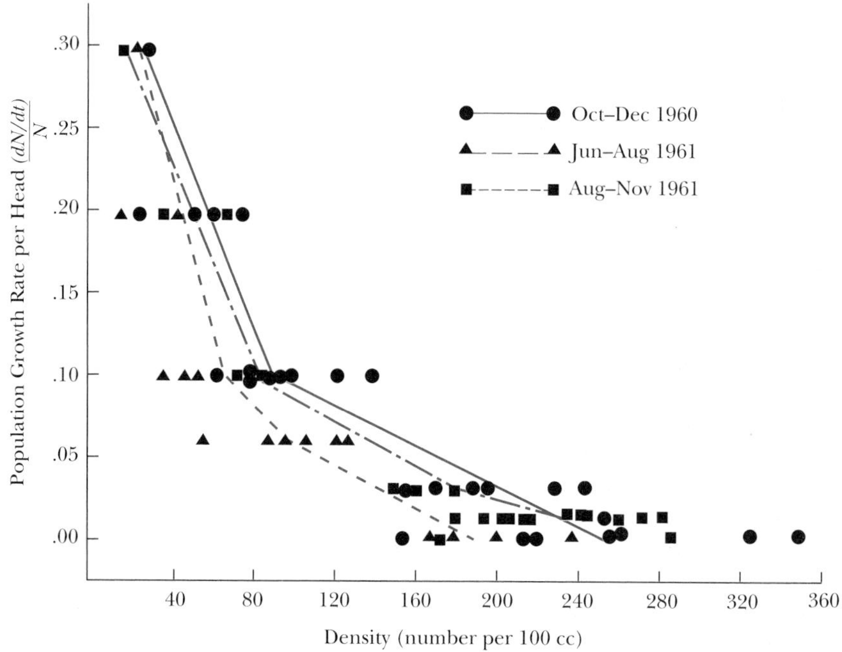

1916, and swept over the interior of the island, destroying nests and eggs and killing many females on their nests. The following winter was hard and was made worse by predation from an unusually heavy concentration of goshawks. By the next breeding season, the population had been reduced to fewer than 150 birds. After a slight increase around 1920, the population entered its final decline.

By 1928, only a single bird, a male, was known to exist (Figure 4-15). He lived on for a few years and was last seen alive on February 9, 1932. It is possible that the small population following the fire consisted mostly of males and so the potential growth rate was lower than the total population figures would suggest. There were other possible factors in the final decline, including the spread of the disease blackhead from domestic poultry on the island.

A currently endangered species of bird, the Kirtland's warbler, may also be suffering from its own rarity. The species nests only in a small section of central Michigan and winters in the Bahamas. Since 1971, its breeding population has rarely been above 500. With numbers so low, it is likely that an individual that disperses more than a few miles from the central breeding range will be lost as a breeder (Mayfield 1983). The chances are small that any other bird, let alone one of the right sex, would happen to disperse to the same spot. Occasional unmated birds have been found in Ontario, Wisconsin, and far northern Michigan.

For plants, pollination failures would be an analogous result of undercrowding.

Mathematical Treatments of Population Growth Skip →

Students sometimes get the idea (from Descartes, perhaps) that we inhabit a mathematical universe. Such a notion would be discouraging to those of us who are not very good with numbers, if it were true; but it is not. The universe exists, and mathematics is useful in describing and analyzing it, as are pictures (including graphs) and words. If God were a mathematician (so to speak), both pi and *e* would be 3.0000, rather than the irrational numbers they

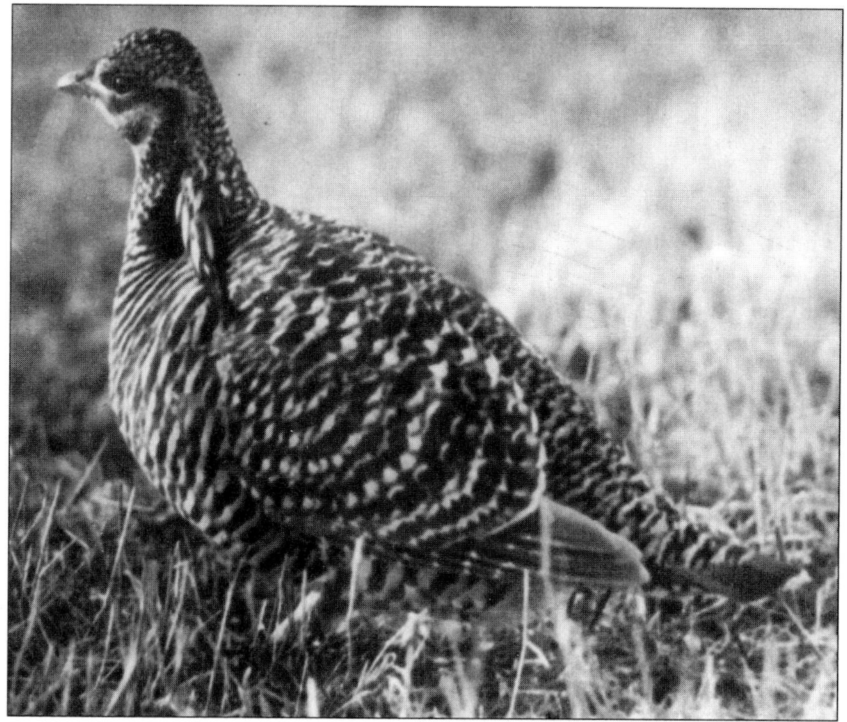

Figure 4-15

The last heath hen. By December 1928, only this single member of the eastern form of the greater prairie chicken still survived. In the spring it appeared daily at the traditional booming ground between Edgartown and West Tisbury on Martha's Vineyard. It was photographed there March 31, 1930 by Alfred O. Gross.

are. Nevertheless, mathematics is such a useful tool that we will, from time to time, include some of the more important mathematical approaches to ecological principles.

Exponential Growth

The formula by which exponential population growth is generally represented is

$$dN/dt = rN$$

This is a differential equation in which dN/dt is the population growth rate. It refers to the change in numbers (dN) per time interval (dt) when this interval is very small, so that we are taking an instantaneous reading of rate of growth. (A speedometer gives us such an instantaneous reading of our rate of travel in a car, but regrettably there is no "growthometer" to attach to populations.) Biotic potential, r, is the increase in numbers of individuals per time period per head (that is, per individual). It combines birth rate and death rate, being equal to the instantaneous birth rate minus the instantaneous death rate. N is the number of individuals in the population.

Growth rate is higher in a population with a high r compared with one with a low r. For example, if two populations each have a population size of 100, and r for one is 0.5 and for the other 0.1, then dN/dt for the first population will be 100×0.5, or 50, and for the second it will be 100×0.1, or 10.

In this formula for exponential growth the growth rate also depends on N. When N is small, the growth rate is slow, and when N is large, the growth rate is rapid. Thus, if r is 0.5, and population size is 10, then dN/dt will be 5. If population size is 1000, then dN/dt will be 500.

The formula above gives **growth rate** in an exponentially growing population. If we wish to look instead at the **population size** at various times during exponential population growth, an equivalent expression is the integral equation

$$N_t = N_0 e^{rt}$$

Here N_t is number at time t; N_0 is number at time 0, the beginning of the period being studied; e is the base of natural logarithms (about 2.718); r is biotic potential; and t is the time period being studied. Calculators with e^x keys make this formula easy to use directly. Formerly it was often easier to take the natural logarithm of each side of the equation and use the resulting formula:

$$\ln N_t = \ln N_0 + rt$$

As a numerical example of calculating population size during exponential growth, suppose that a population of 10 duckweeds grows for 4 days and that r is 0.20 per day. Then

$$N_4 = 10 \times e^{(0.2 \times 4)} = 10 \times 2.22$$

Population size at the end of 4 days is 22 or 23 duckweeds.

Logistic Growth

Population growth rate in the logistic equation is given by

$$dN/dt = rN\left(\frac{K - N}{K}\right)$$

This is the same as the formula for exponential population growth except that the term $(K - N) / K$ has been added. K is known as the carrying capacity, and the whole expression $(K - N) / K$ is a measure of environmental resistance or the effect of crowding. When N is small, $(K - N) / K$ is close to 1, so that biotic potential is nearly completely realized and population growth can be rapid. Suppose that $r = 1$ and $K = 50$. Then, when $N = 5$,

$$dN/dt = 1 \times 5 \times \frac{(50 - 5)}{50}, \text{ or } 5 \times 0.9$$

and dN/dt is 4.5.

When N is large, growth is slow. If $N = 45$, then

$$dN/dt = 1 \times 5 \times \frac{(50 - 45)}{50}, \text{ or } 5 \times 0.1$$

dN/dt in this case is 0.5.

$(K - N) / K$ is a measure of environmental resistance but is an inverse measure; when environmental resistance is low, the numerical value of $(K - N) / K$ approaches 1, and when environmental resistance is high, the value of $(K - N) / K$ approaches 0.

We have just discussed growth rate during logistic growth (the bottom graph in Figure 4-10). To determine population size at any particular time (the top graph in Figure 4-10) we would use the equivalent integral equation

$$N_t = \frac{K}{1 + e^{a-rt}}$$

The terms are all the same as before except for a. Basically a specifies how close the population is to the carrying capacity when you start out. (More details on using the logistic equation, including how to calculate a, are given in Brewer and McCann's *Laboratory and Field Manual of Ecology*, [pp. 92–99].)

Adding Time Lags to the Logistic Equation

The logistic equation assumes that the effects of population size, such as crowding, are instantaneous. Let's explicitly add time notations:

$$dN/dt = rN_t[(K - N_t)/K]$$

The time, t, in this equation is right now. Population growth rate is dependent on the size of the population at this instant.

We have already said that it would be surprising if all of the effects of an increased population size were instantaneous. In the first part of the equation, for example, new individuals do not necessarily raise the rate of population growth immediately; in most kinds of organisms they have to reach reproductive age, so that there is a certain period between the birth of a new individual and the time when it will be contributing to reproduction.

In the environmental resistance part of the equation, there will probably also be time lags. As mentioned, it may take a while for high populations of insects to attract predators or for the predators to reproduce. A population of deer that has grown too large to live indefinitely on the available food supply may still be able to live for weeks or months before the vegetation is damaged by overbrowsing. Some plants, as they are grazed on, apparently begin to produce chemicals that will be harmful to the animals eating the next set of leaves, which, in the case of some insects, might be the next generation.

The logistic could be improved, at least as far as using it in our thinking about populations, by adding some terms for time lags:

$$dN/dt = rN_{t-c}[(K - N_{t-w})/K]$$

where w and c are time lags.

In the first part of the new equation, population growth rate at time t will be influenced not by the population size right now (N_t) but by the population size at some earlier time (N_{t-c}). In the second part of the equation, we are saying that the depressing effects of crowding are not being felt immediately, proportional to the whole population size now (N_t), but to the size of the population at some earlier time (N_{t-w}).

This is clearer with numerical examples. Suppose a population in which $r = 0.6$, $K = 500$, and $N_t = 450$. In the ordinary logistic

$$\begin{aligned} dN/dt &= 0.6 \times 450[(500 - 450)/500] \\ &= 27\text{, and so the new population size is } 477 \end{aligned}$$

Now suppose that only time lag c is important and that it is of a size such that the effective population is only 400, not 450. We would have

$$\begin{aligned} dN/dt &= 0.6 \times 400[(500 - 450)/500] \\ &= 24\text{, and so the new population size is } 474 \end{aligned}$$

The effect of this time lag, by itself, is to slow population growth compared with the ordinary logistic.

Now suppose that only time lag w is important, so that at the time the population is 450, the unfavorable effects of crowding are being felt as though the population were still only 400. We will have

$$\begin{aligned} dN/dt &= 0.6 \times 450\ [(500 - 400)/500)] \\ &= 54\text{, and so the new population size is } 504 \end{aligned}$$

The effect of this lag affecting environmental resistance is to increase population growth rate over the ordinary logistic. Not only that, we also see a mathematical way in which a population can overshoot carrying capacity.

Time lag c produces no qualitative changes in the way the population growth curve looks compared with the ordinary logistic. The time lag in the environmental resistance term (w) does. The size and nature of the effect depend on the length of the time lag, that is, how long before the effects of overcrowding are felt. In comparing different species or populations it also depends on the size of r. Small time lags will cause the population to come to rest on K by damped oscillations after overshooting and undershooting slightly (Figure 4-16). With large time lags, or, specifically, when the product of r times $w >$ about 1.6, the oscillations are not damped. Instead population size cycles from above K to below and back, apparently stably (Cunning-

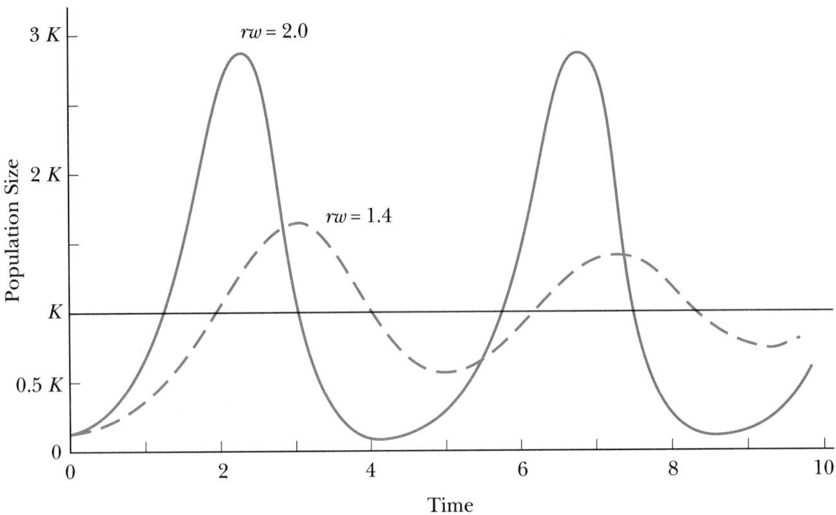

Figure 4-16

Population growth with a time lag (w) in the environmental resistance term of the logistic. With no lag, the population would approach K by a smooth sigmoid curve. The y-axis is calibrated in multiples of carrying capacity (K). Two situations are graphed, one in which $r \times w$ is below 1.6, so that K is reached by damped oscillations, and one where $r \times w$ is above 1.6, producing cycles of population size above and below K. (Adapted from Cunningham 1954.)

ham 1954). Of course, any cyclic population becomes susceptible to extinction from chance causes, such as weather extremes, during the time it is near zero.

G. E. Hutchinson, who first discussed time lags in connection with the logistic (in 1948), noted that such lags could result in population fluctuations even in stable environments. In large ungulates, there is evidence that fluctuations in numbers are associated with time lags in growth and survival of young animals. In white-tailed deer fluctuations in southeastern Ontario between 1953 and 1986, the time lag seemed to be associated with the food supply (Fryxell et al. 1991). Perhaps the most likely explanation is that the condition of the vegetation lags behind population size; for example, a high population may overeat the vegetation to a degree that, despite a drop in numbers, satisfactory survival of young deer is delayed a few years until the vegetation recovers.

Ways of Looking at r

Per capita growth rate is usually represented in equations by the symbol r. There are two ways of looking at r. We can consider it a constant as in the logistic equation

$$dN/dt = rN([K - N]/K)$$

So conceived, it is a population parameter that expresses the intrinsic capacity of the population for increasing. It is the biotic potential; however, the potential is realized only during times when the population is very low relative to K, the carrying capacity. At other times the actual per capita increase rate is below the value of r.

A simpler equation for population growth is

$$dN/dt = rN$$

If r is constant this equation is an expression for exponential growth. If, however, r is a variable, curves of other shapes may be obtained and, in particular, if r varies in the course of population growth from a high value to zero, we obtain a sigmoid curve like the logistic.

In this book r is used as a constant, the intrinsic rate of natural increase. If r were to take different values, we could designate the value of r as used here, r_m. The value that this constant takes is a function both of the life history traits of the population and the effects of the environment; consequently, $r(r_m)$ for a given species will vary from one habitat to another.

Intrinsic Rate of Natural Increase and Net Reproductive Rate

During the period when growth is exponential, that is, when this relationship holds:

$$N_t = N_0 e^{rt}$$

we can solve for r, the intrinsic rate of natural increase, using the formula

$$r = \frac{\ln (N_t/N_0)}{t}$$

(This formula is obtained from the preceding one by dividing through by N_0, taking the natural log of both sides of the equation, and then dividing by t; see Brewer and McCann [1982:94].)

For example, the pheasants on Protection Island mentioned earlier increased from 40 to 426 between the first and third years (that is, in two years). If it is assumed that population growth in this period was exponential and that a stable age distribution had developed, then

$$r = \frac{\ln (426/40)}{2}$$
$$= 1.18$$

It is rare to have good field data on a population that we can be sure is growing exponentially, but we can keep animals in the laboratory and do the same calculations. Of course, to keep even 426 pheasants might tax our facilities, especially since we have to make sure they don't feel crowded. P. H. Leslie and R. M. Ranson (1940) devised a way to make the same calculations without accumulating vast numbers of animals. We start with a small, manageable number of pheasants and record the production of young through their lifetimes. In practice, we will just keep track of females and assume that the male segment of the population will follow along. We only need to follow the females through the end of their reproductive life. What they do after they have quit reproducing has no effect on r, at least for these calculations. We don't need to keep all the little pheasant chicks; we just tally how many females are produced.

At the end of six years, let's say, we have sold quite a few young pheasants to other pheasant breeders, and our last pheasant has quit laying eggs. We put the data we gathered in a table (Table 4-5). The first two columns are part of a life table for our pheasants. Age is given by x, and l_x is survivorship. After one half year, 9 of the 10 birds were still alive; 2 died in the next year, 1 in the next, and six and one half years after we started, 1 bird was still alive.

The m_x column is a maternity column. It records the number of live births per female. In this case, it is the number of (female) chicks hatched per number of hens alive at that age. In the first year a total of 27 (female) chicks were hatched. Because there were 9 hens (some of which may have laid no eggs the first year), the average number of chicks per hen was $(27/9) = 3.0$.

To calculate r, we'll use the same equation as before but make one change, changing t to T:

$$r = \frac{\ln (N_T/N_0)}{T}$$

We will define T as generation length. A generation is the length of time from the birth of parents to the birth of their offspring. The mothers in this case were all born at the same time because we arranged it that way. The daughters were born over a period of six years. Obviously, deciding what to use as our figure for generation length will require some arithmetic, which we will get to shortly.

Because our table is set up on a per capita basis, N_0 is 1. What do we use for N_T? N_T will be the number of daughters born to the average female. We get this simply by adding up the $l_x m_x$ column. Note

Table 4-5

***Fabricated Data About Ring-Necked Pheasants Reared in the Laboratory for Calculating* R_0 *and* r**

x	l_x	m_x	$l_x m_x$	$l_x m_x x$
0.5	0.9	3.0	2.70	1.35
1.5	0.7	6.0	4.20	6.30
2.5	0.6	6.0	3.60	9.00
3.5	0.5	6.0	3.00	10.50
4.5	0.4	5.0	2.00	9.00
5.5	0.2	4.0	0.80	4.40
6.5	0.1	0.0	0.00	0.00
Totals (Σ)			16.30	40.55

that any females that died in their first few months left no offspring. Any that lived six or more years may well have left 40 daughters apiece. But, on the average, each hen we started with left 16.3 daughters apiece.

The ratio N_T/N_0 has been represented by R_0 and called **net reproductive rate.** It is an easier idea for many people to think about than r, because it tells directly the capacity of the population to multiply in one generation. For example, if R_0 is 2, then that population has the ability to double in number in one generation. In using R_0 in this way, however, it is necessary to remember that in the real world, events occur in real time. If bears and bugs both had an R_0 of 2, bugs might double in population several times a year and bears might double every decade.

We can rewrite our earlier equation as

$$r = \frac{\ln R_0}{T}$$

Now, how do we calculate T? We want the average interval between the birth of the mothers (which is time 0) and the birth of their daughters. This is calculated (Dublin and Lotka 1925) as

$$T = \frac{\Sigma l_x m_x x}{\Sigma l_x m_x}$$

T is generation length in whatever time units x is in. In our example, it is (40.55/16.30) = 2.49. r, then, is (ln 16.30)/2.49 = 2.79/2.49 = 1.12.

The equation

$$r = \frac{\ln R_0}{T}$$

is useful conceptually because it shows formally some of the things we talked about before. Anything that extends generation length (such as deferred maturity) will lower r (Figure 4-17). Anything that increases net reproductive rate (such as a bigger clutch) will raise r. Increasing clutch size for young animals, but not older ones, will raise net reproductive rate *and* lower generation length, and so that will raise r a lot. Increasing clutch size for older animals while keeping clutch size the same for younger ones will raise net reproductive rate but also raise generation length; it will change r little. Table 4-6 begins with the values used in Table 4-5 and alters them to illustrate the effects of a few such

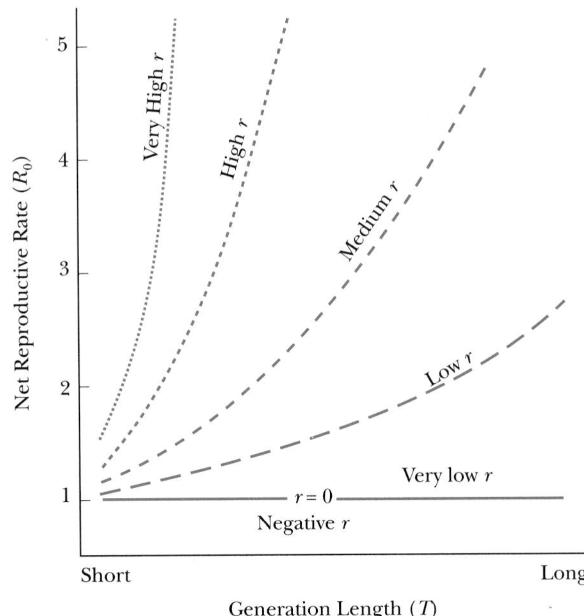

Figure 4-17

The relationship between the intrinsic rate of natural increase (r), net reproductive rate (R_0) and generation length (T).

life-history changes. The evolution of life-history traits is discussed in a later section of this chapter.

For organisms with overlapping generations, the formula $r = (\ln R_0)/T$ is approximate (Leslie 1966). In practice, we would use the data in Table 4-5 and calculate r using the Euler (pronounced "Oiler")-Lotka equation. This is tedious and not particularly informative in understanding what r means. The Euler-Lotka equation and how to do the calculations are shown in Brewer and McCann (1982: 109–110).

Adding the Allee Effect to the Logistic

As already mentioned, the basic logistic model assumes that per-head rate of increase is greatest at the lowest population density, ignoring any unfavorable effects of undercrowding. Such unfavorable effects may exist either by reduced reproductive rate or higher mortality rate. If we consider that r at minimal population density may be less than r_m and represent it as r_0, we can write a new equation for logistic population growth. We might write the ordinary logistic as

Table 4-6
Effects on R_0, T, and r of altering various life-history features. Only the changes stated are made.

Life-History Features	*Net Reproductive Rate (R_0)*	*Generation Length (T)*	*Intrinsic Rate of Natural Increase (r)*
Original values (Table 4-5).	16.30	2.49	1.12
Clutch size increased by 1 egg for all ages.	19.70	2.48	1.20
Lifespan shortened: All birds alive at age 4 die before age 5.	15.50	2.33	1.17
Maturity deferred to age 2 (m_x for 0.5 and 1.5 age classes = 0).	9.40	3.50	0.64

$$dN/dt = r_0(\times \text{ negative effect of crowding}) \times N$$

Here r_0 and r_m are equivalent and the negative effect of crowding is $[K - N/K]$, so this is exactly the expression already used for the logistic.

The general form of a logistic taking the Allee effect into account could be something like

$$dN/dt = r_0 \times (\text{positive effect of crowding up to an optimal density} - \text{negative effect of crowding}) \times N$$

Versions of the logistic taking this approach tend to be oversimplified or too cumbersome to delve further into here (Jacobs 1984); however, the more realistic ones can be useful in computer modeling of small populations for conservation purposes.

Projection Matrices

Another approach to studying population growth dispenses with the assumptions of the logistic and uses matrices involving age classes, reproductive rates, and survival rates to project population size at a later time. In the Leslie matrix (Leslie 1945), a square matrix containing reproduction and mortality rates by age class is multiplied by a column vector containing numbers of organisms in each age class. The biological basis for such a matrix is shown in Figure 4-18. Reproduction by all age

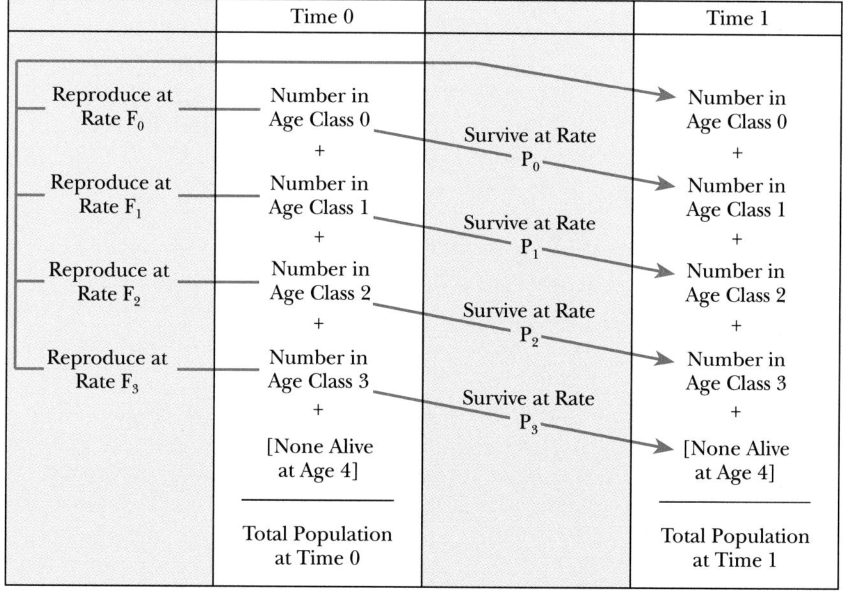

Figure 4-18

Biological basis for the Leslie matrix. Adapted from Brewer and McCann 1982.

classes in the first time period gives rise to age class 0 of the next time period. Meanwhile, the members of each age class die at a rate that determines the number alive in the next higher age class at the next time period; in other words, the surviving members of age class 1 at time 0 form age class 2 at time 1.

The usefulness of this approach when we wish to know something about age structure is clear. One obvious example is humans. A population in which the older age classes were growing and the younger were shrinking would require different social planning than a population of the same size but with an increasing number of young members. Formulas and details of calculation are given in *Laboratory and Field Manual of Ecology* (Brewer and McCann 1982: 100–104). It is worth noting that the rates designated by F in Figure 4-18 include not only reproduction but also the survival of the young from birth to time 1 (Michod and Anderson 1980).

Population Density and Population Regulation

Density-Dependent Factors

It is often convenient to discuss population size in terms of **density,** numbers per unit of space. Five hundred rabbits in a square mile is very different from 500 rabbits in a square block, and if we express the population size as rabbits per hectare we can make the comparison much more readily (500 rabbits per square mile would be about 0.5/ha; 500 per square block would be about 100/ha).

Many organisms show a similar density from one year to the next, measured from June to June or from October to October. By this we do not mean that their numbers are exactly the same every year; nature is more variable than that. The population does not, however, grow exponentially year after year, nor does it decline to extinction (Figure 4-19). And for some kinds of organisms the numbers from one year to the next may be very close. Over a period of 18 years the number of red-eyed vireos breeding in a 65-acre beech–sugar maple forest near Cleveland, Ohio, studied by A. B. Williams (1950), never went below 18 nor exceeded 36.

This type of stability leads us to speak of such populations as being regulated. Note that we are speaking here of local populations. Regional and whole species populations are limited but are rarely, if ever, regulated in the sense discussed below for local populations. Limitation of the total numbers of a species is, in the broadest sense, based on the smallest amount of favorable habitat available during the year.

For local populations, however, we can readily see how regulatory factors can work. Information about density can be transmitted and received. It is evolutionarily advantageous to animals to be able to recognize crowding and avoid it, if possible. For example, the regulation of local populations of birds probably occurs by individuals giving signals, such as territorial songs, which allow one another to assess the degree of crowding. As areas become crowded, newly arrived individuals tend to go elsewhere. These ideas are developed more fully in a later section.

Population regulation occurs by the action of factors that vary in their effect according to population density—**density-dependent factors** in the terminology of the economic entomologist J. S. Smith (1935). As long as birth rate exceeds death rate, a population will grow (and if death rate exceeds birth rate, it will decline). Over the whole span of the graph in Figure 4-20, population growth continues. From A to B, the per head population growth rate is the same. From B to C the rate drops, but even at C, population growth continues.

Only if the lines cross, marking the point at which death rate equals birth rate, will population growth stop. More realistically, population growth stops when birth rate (b) + immigration rate (i) equals death rate (d) + emigration rate (e). This happens only if one or more of the rates vary with density, as shown in Figure 4-21. Where the lines cross is equilibrium population size, or carrying capacity (K). Table 4-7 shows how density-dependent variation in mortality rate has the effect of regulating density.

Density-dependent factors vary in such a way as to lower population size when the population begins to become too large and to allow population size to increase when it becomes small (Figure 4-22). An approximately analogous system is the control of room temperature by an air conditioner. As

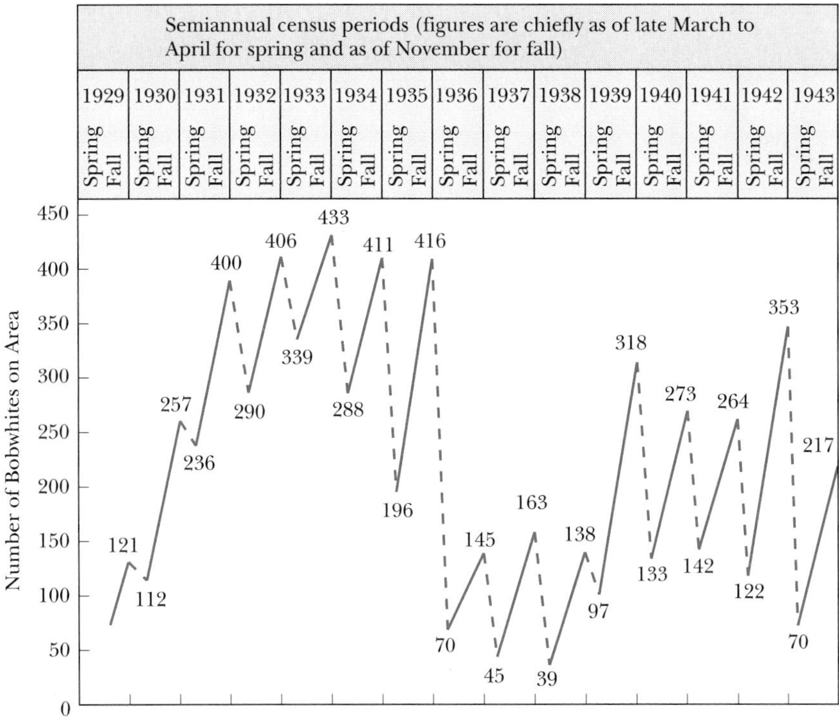

Figure 4-19

Population sizes of the northern bobwhite on an 1800-hectare (4500-acre) area in Wisconsin over a 14-year period (1929–1943). (From "Some contributions of a 15-year local study of the northern bobwhite to a knowledge of population phenomena" by P. L. Errington, *Ecological Monographs,* 1945, 15: 1-34. Copyright © by the Ecological Society of America. Reprinted by permission.)

the temperature increases, the thermostat turns on the air conditioner to cool the room. A little later, as the temperature drops, the air conditioner shuts off. When the room temperature rises again, this series of events is repeated. This general type of control is called **negative feedback.** In the analogy room temperature is, of course, population size or density and the air conditioner is the density-dependent factor.

As density rises in a population, therefore, various factors come into play that raise the death rate, lower the birth rate, or raise the emigration rate. When density falls, density-dependent factors allow the birth rate to rise, the death rate to fall, or the immigration rate to rise.

In talking of population regulation we are dealing with two closely related but slightly different questions. One is, What limits population size, or what sets the carrying capacity? The carrying capacity is approximately equivalent to the thermostat setting in our air conditioner analogy. The second question is, What are the actual factors that change in a density-dependent fashion (the factors of environmental resistance)?

At this point, it is desirable to pause and look again at Figure 4-19, Paul Errington's data showing northern bobwhite populations on a good-sized study area in Wisconsin. Frequent inspection of numbers from real populations is desirable to keep one's ideas grounded in reality. This graph shows that, despite our discussion of populations being regulated, the breeding (spring) population of

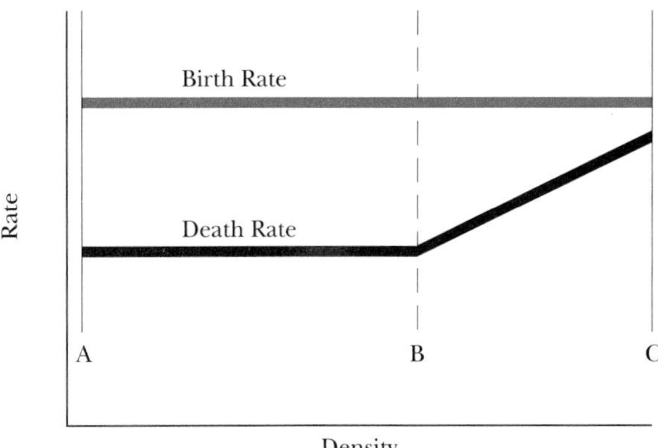

Figure 4-20

Relation of birth rate and death rate to density. At any given point, population growth rate is equal to the distance between the two lines.

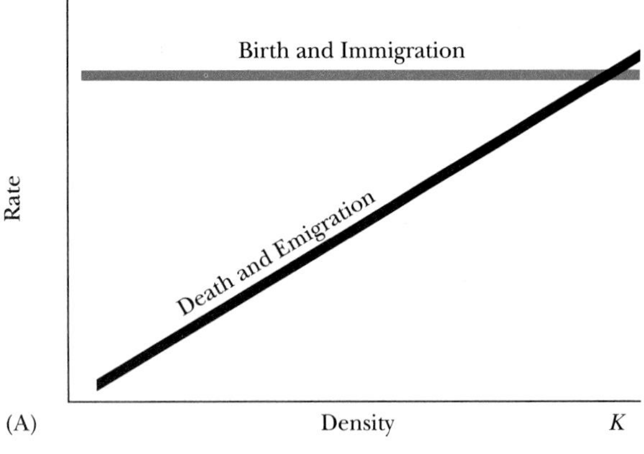

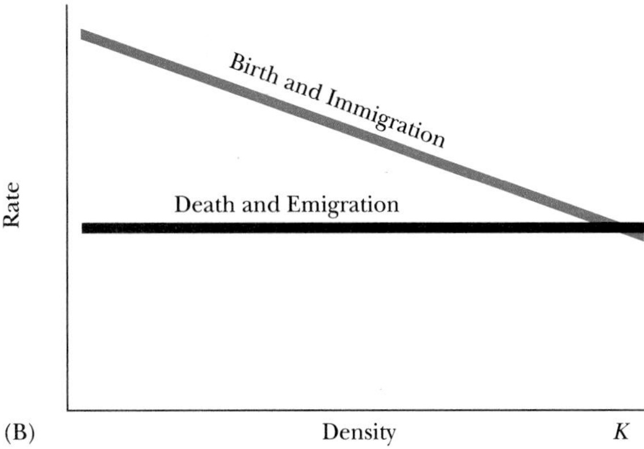

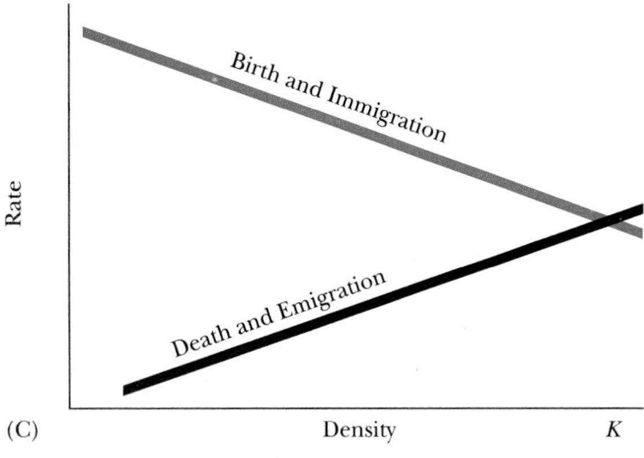

Figure 4-21

Equilibrium density, or carrying capacity, may be set by density dependence of death and/or emigration (as in graph A), birth and/or immigration (as in graph B), or by density dependence in both increase and decrease factors (as in graph C). (Adapted from Enright 1976.)

Table 4-7
Comparison of Density-Dependent and Density-Independent Mortality on Population Regulation

This table compares the effect on population size of a mortality factor that increases the proportion of the population killed as the population size (density) increases with a factor that kills a constant proportion no matter what the population size. In one sense, the second kind of factor is also density dependent, because the number killed goes up with increased density.

Population Size at Beginning	*Number Killed*	*Mortality Rate (Percent)*	*Population Size at End*
	Density-Dependent Mortality		
100	0	0	100
150	50	33	100
200	100	50	100
250	150	60	100
300	200	67	100
Density-Independent (Density-Proportional) Mortality			
100	50	50	50
150	75	50	75
200	100	50	100
250	125	50	125
300	150	50	150

In the first case, the population size is regulated at 100 by the mortality factor killing an increased percentage with higher densities. The factor that kills a constant percentage does not have a regulatory effect. A second kind of density-independent factor, in which a constant number is killed no matter what the population density, could exist. It would not have a regulatory tendency either.

bobwhites varied from below 50 to nearly 250 over a five-year period.

A second fact sometimes forgotten in looking at smooth theoretical curves is that many populations live in strongly seasonal environments. These populations often show fluctuations like those in Figure 4-19, resulting from the existence of a season of increase, when deaths are much overbalanced by births, and a season of decrease, when there are few or no births but deaths continue or increase.

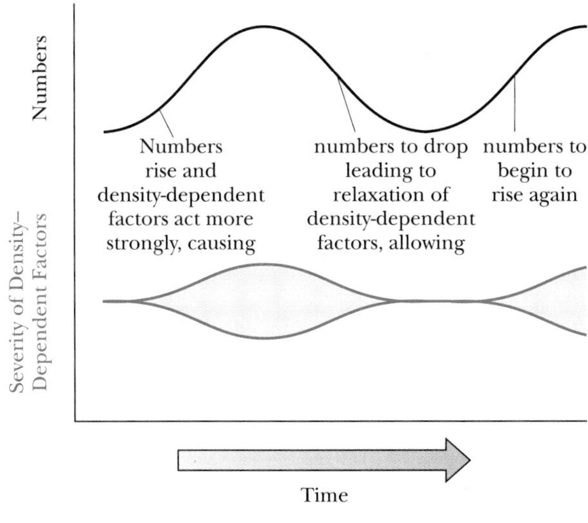

Figure 4-22

Basic features of population regulation by density-dependent factors.

Habitat Distribution

In Chapter 2, we discussed habitat selection as a part of the behavior of individuals; however, habitat selection also involves populations. Specifically, how crowded a habitat is may influence an individual's decision on whether to settle. Also, one of the results of habitat selection in a landscape is the population-level trait of relative abundance among the habitats—In which habitats is the species abundant, in which common, and in which rare?

Some animals occur over a very narrow range of habitat. The Kirtland's warbler of the jack pine savannas and the Swainson's warbler of the canebrakes are such species and are now rare in part because their habitats are rare. Other animals with scarcely wider habitat distributions may be common because their habitat is common; the horned lark, which occurs only where there is bare ground and short vegetation, is common in much of North America because it is able to inhabit the vast hectarages of rowcrops.

Many species accept a variety of habitats; for these we may wonder how habitat selection works in sorting individuals among them. One possibility developed by Stephen Fretwell and H. L. Lucas, Jr. (1970, Fretwell 1972) is that a range of basic, or intrinsic, habitat suitabilities exists, from most preferred to unacceptable, and that animals initially choose the most preferred sites. As these fill up, they become less suitable because of the unfavorable effects of crowding. Later-arriving individuals go to habitats that have lower basic suitabilities but that are uncrowded.

Fretwell and Lucas put these ideas into a graphical model that is shown in Figure 4-23. They termed the result the **ideal free distribution.** If populations do behave something like this, it is presumably because the animals that include crowding in assessing overall habitat suitability do better than ones that fail to take crowding into account. One prediction that follows is that the fitness of birds should be the same in all the habitats the species occupies in a given landscape.

Several studies have compared field data with this model. Fitness is hard to measure, but various measures of success can be looked at as more or less satisfactory stand-ins. For a species of shorebird that utilizes different foraging habitats on its wintering grounds, for example, we might keep track of feeding success—biomass eaten per day, for example. If its habitat distribution follows the ideal free distribution, we will expect food per individual per day to be the same on sand beaches, mud beaches, salt flats, etc.

Such studies have sometimes found that success is the same in different habitats (e.g., Lemel 1989) and sometimes have not (e.g., Messier et al. 1990). The importance of the latter group is not so much that they disprove the concept of the ideal free distribution but rather as starting points for investigating how it is inadequate. One obvious shortcoming for birds and probably most vertebrates is that it ignores the influence of philopatry, or site tenacity. Not all the individuals of a species reassort themselves every breeding season according to habitat suitabilities; rather, most older birds tend to stay where they were last year.

Another set of complications is connected with the Allee effect and sociality in general. If basic suitabilities are not highest at zero density, such features as the time when different habitats ought to be occupied and comparative densities and success rates are hard to predict—but interesting to think about.

Figure 4-23

The "ideal free" model of habitat distribution. Based on S. D. Fretwell (1972).

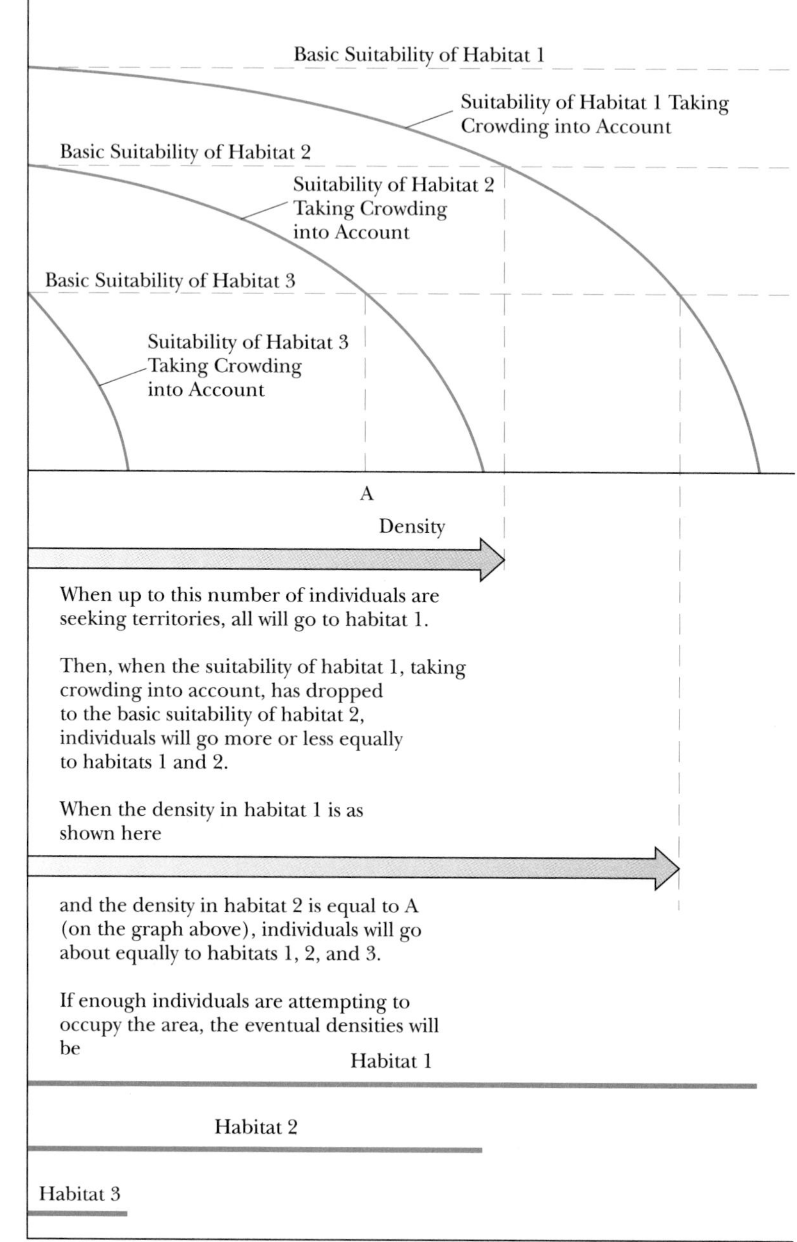

Carrying Capacity

Anything that influences birth rate or death rate can influence carrying capacity. The effect on birth or death rate may vary with density, but it does not have to. Figure 4-24 suggests how a benign environment, in which density-independent mortality is low, can have a higher carrying capacity than a harsher environment, even though density-dependent mortality and emigration are (in this model) unchanged.

From this we can see how factors such as soil and climate may affect carrying capacity even though their effects on mortality rate may not change with density, at least over a considerable range of density. Alternatively, factors that raise the

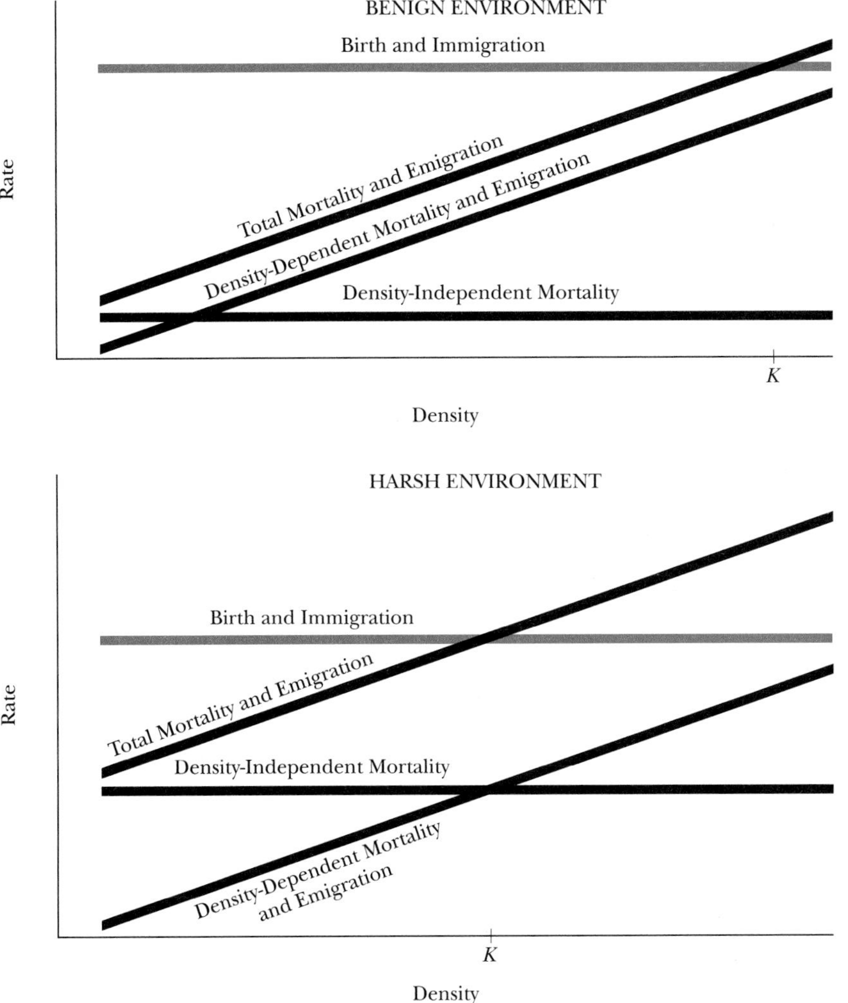

Figure 4-24

Carrying capacity may differ between two environments because of differences in density-independent mortality. (Adapted from Enright 1976.)

general birth rate will raise K, if nothing else is affected. If a benign environment means bigger litters or a longer breeding season, for example, the birth rate line would be raised. In the top graph in Figure 4-24, raising the birth rate would shift K off the graph to the right. Realistically, of course, other things will be affected if birth rate is higher, and the specifics of mortality and emigration will determine whether carrying capacity actually does increase.

Usually the carrying capacity seems to be determined by one or another of the following: climate, nutrition, or favorable space. An example of the effect of climate is given in Figure 4-25, which shows equilibrium densities of the chestnut grain beetle at different temperature-humidity combinations. Nutrition includes food (Table 4-8) and also various other kinds of nutrients, such as phosphate for lake algae.

Favorable space covers a multitude of virtues, such as nest sites, escape cover, and suitable soil texture for plants. If we find that by putting up more and more birdhouses we steadily increase the number of a bird species, we can conclude that nest sites had been the limiting factor for population size. If eventually, despite our putting up more nest boxes, the population size does not rise anymore, we can conclude that some other factor is now limiting, and we would need to devise other experiments to find it.

Other factors that have been shown to limit population size include competition, predation, parasitism, and disease. Examples are given under discussions of these factors in later chapters. We will

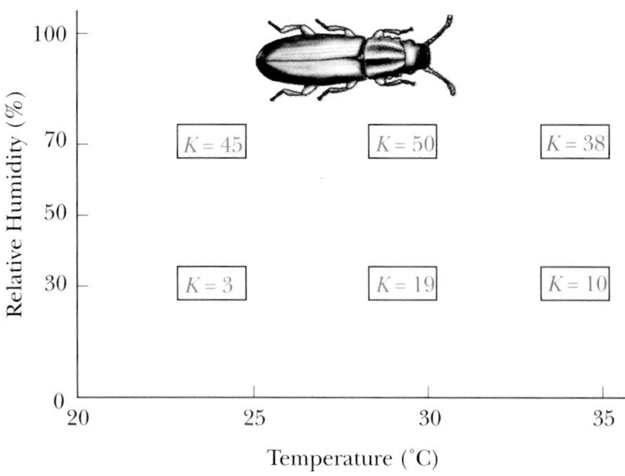

Figure 4-25

Equilibrium densities of the chestnut grain beetle at different temperature-humidity combinations. Cultures consisted of 8 grams of whole wheat flour. Densities are given as total population (adults + pupae + larvae) per gram. (Based on T. Park 1954.)

settle for one example here. When grain beetle populations, ordinarily infected with a protozoan parasite, the barbershop coccidian, were cultured without it, *K* for the chestnut grain beetle was more than tripled, from a little more than 10/g to about

Table 4-8
The Influence of Food on Carrying Capacity*

The work of G. F. Gause on *Paramecium* is one of the classic studies of competition, but we mention it here for another purpose. Gause cultured the two species separately in a water and salts solution and fed them daily with bacteria. The amount of food given was either one loopful or half a loop. The carrying capacity with one-half loop and one loop food supplies for two species are shown in the table.

Species	*Food Supply*	
	Half Loop	*One Loop*
Paramecium caudatum	64	137
P. aurelia	105	195

Thus, doubling the food supply while maintaining other conditions constant approximately doubled populations for both species.

* Data from Gause, G. F., *The Struggle for Existence,* Baltimore, Williams & Wilkins, 1934, p. 104.

40 (Park 1948). This did not, however, happen with the very similar confusing grain beetle, indicating that parasitism was not important in setting its carrying capacity.

Population Regulation

The factors that act in a density-dependent way to regulate population size can be classified as **extrinsic** and **intrinsic.** Intrinsic factors are the population's own response to density; extrinsic factors involve interaction with the rest of the community. The main extrinsic factors are predation, parasitism, disease, and interspecific competition. It is hard to produce a completely satisfactory classification of intrinsic factors because they form a tightly interrelated group of processes; for purposes of discussion we will say they include intraspecific competition, immigration and emigration, and physiological and behavioral changes affecting reproduction and survival.

All the extrinsic factors can act in a density-dependent way, and probably all of them, at some time for some populations, are important in regulation of population size. Clearly disease can spread more rapidly in a dense population because of the greater number of contacts between individuals. A dense population may be more vulnerable to predation because predators will be drawn to the area of concentrated food and because some members of the prey population will be living in poor cover.

On the whole, extrinsic factors are probably a little more chancy than intrinsic factors. For example, the predators that could be taking the surplus animals may be concentrated in some other area where the prey population has reached a still higher density. Teleologically speaking, there may not be a predator around when you need one. Intrinsic factors, on the other hand, are part of the evolved behavior of the organism, keyed to the fitness of the members of the population.

It seems likely that disturbance of the ecosystem will have greater effects on species whose population regulation has had a strong extrinsic component. Probably large ungulate populations in North America evolved in communities in which predation was an important force in both limiting and regulating their densities. In this context, the failure to evolve intrinsic behavioral responses for avoiding crowding was rarely penalized (Pimlott 1967). With the large predators exterminated and the ecosystems disrupted in other ways, deer and other ungulate populations frequently grow to densities where they overwhelm their winter food supply and have their numbers regulated by starvation.

For many kinds of organisms intraspecific competition is one of the most generally important density-dependent factors. The word "competition" comes from roots meaning "to seek together" and is usually defined as a combined demand in excess of the immediate supply. Competition may be intraspecific, in which individuals within one species compete, or interspecific, in which individuals of two or more species compete. Interspecific competition, which may sometimes be important as an extrinsic density-dependent factor, has another important ecological role to be discussed later in Chapter 8. For the moment we will concentrate on intraspecific competition.

Intraspecific Competition

Competition may be simply a matter of each individual taking as much as it can get of some resource, called **exploitation** (Park 1954). A thicket of young trees compete in this way; for a time there is enough sunlight or water for all of them to survive, although they do not grow as fast as they would if they were less crowded. Later some begin to lose out; they die, and the density drops.

Pigs at a trough would be an example of exploitation if each had a place and ate as much as it could. On the other hand, if some pigs—smaller or less aggressive—couldn't get a place at the trough until more dominant individuals had eaten, the focus of competition would have shifted from the prime resource, food, to space at the trough or access to the food. This type of competition involving interaction among the competing individuals is called **interference.** Two nearly synonymous terms (Nicholson 1955) are **scramble** (exploitation) and **contest** (interference).

Two bruchid beetles that develop within bean seeds are good examples of contest and scramble competition (Toquenaga and Fuji 1990). Eggs are laid on the outside of the bean; the larvae burrow into it and feed. In one species, the retentive bean beetle, only a single adult emerged from each bean, as a result of fighting among the older larvae. This was true no matter whether the initial larval number was one or six. The other species, the kidney bean weevil, showed no interference behavior. As many as six or seven adults might emerge from a large bean, starting with initial densities of seven to nine. At higher beginning densities, numbers of emerging adults began to drop. At initial densities of 14 or more, many beans produced no successfully emerging adults. A diagram showing in a very simplified fashion the usual outcome of contest compared with scramble competition is given in Figure 4-26.

Among animals, dominance hierarchies and territoriality (Figure 4-27) are good examples of interference. An example among plants is the release of inhibitory chemicals, discussed in later chapters.

In many different kinds of organisms, an individual defends an area, called a **territory,** against

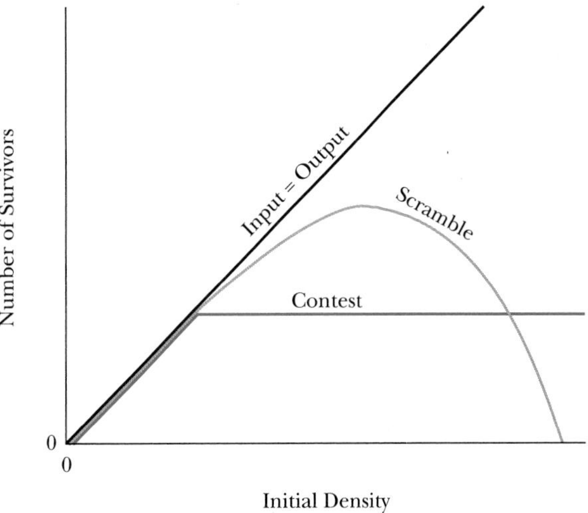

Figure 4-26

A generalized view of the differing outcomes of scramble and contest competition. Output (survivors) increases with increased input (initial density) in both types, up to a point. Then, when a relatively inflexible carrying capacity is filled in contest competition, output levels off and remains the same, no matter how many competing individuals there were to start with. In scramble competition, output in absolute numbers continues to climb with increased input (though percent survival drops). Past some point, output begins to drop and at very high initial densities, survivorship may go to zero, with none of the competitors obtaining enough of the resource to survive.

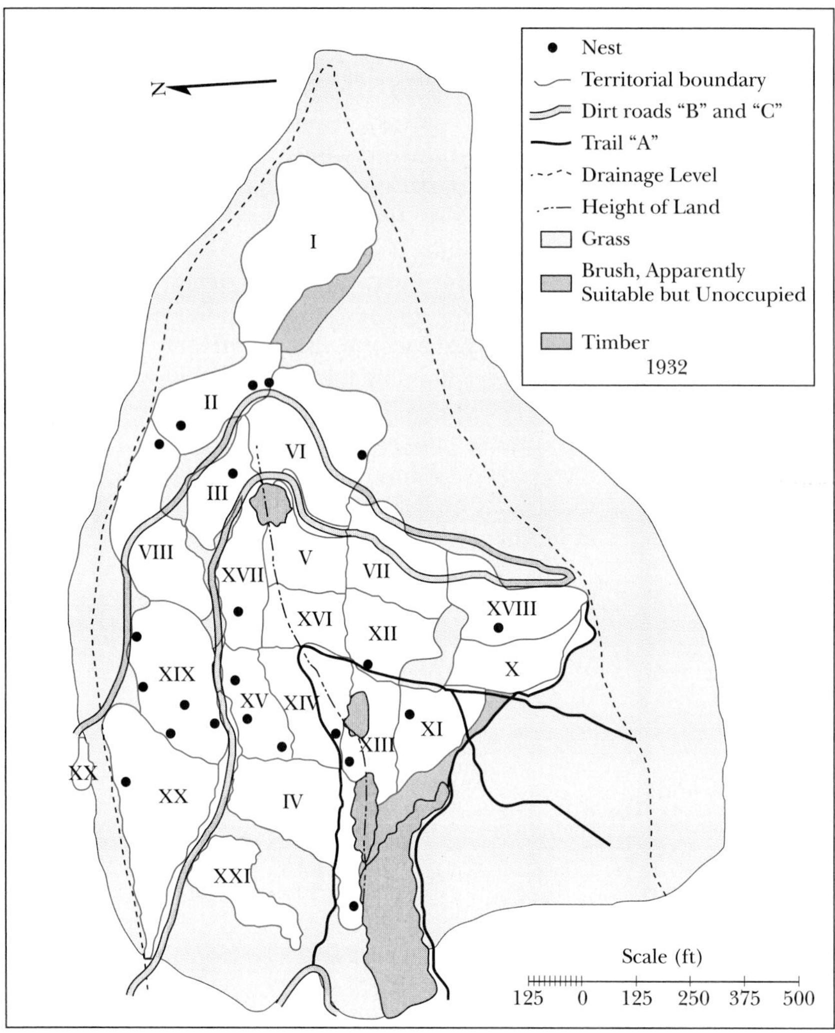

Figure 4-27

One of the first thorough studies of territoriality was by Mary Erickson on a small strange bird called the wren-tit, the most characteristic bird of California chaparral. The map shows a chaparral-filled "gulch" near Berkeley, California, fully occupied by wren-tit territories an acre or two in size. (From Erickson 1938, p. 255. Published 1938 by The Regents of the University of California; reprinted by permission of the University of California Press.)

other members of the same species. Much of the singing of birds advertises the fact that the male is in possession of the piece of land. There are many different kinds of territories, but if we take the simplest case, a territory on which a pair mate, nest, and find food for themselves and their young, it is clear that, for a particular area, density of that species can be regulated fairly precisely. If there is room for about five pairs of scarlet tanagers in a 40-acre oak forest, there is likely to be about five pairs there, no matter how many are seeking to occupy the space. If numbers are high, there will be some chasing, a lot of singing, and perhaps a little actual fighting early in the spring. Within a few days the situation will settle down, with about five pairs occupying territories. What of the other birds, the surplus? Several possible fates await them. They may go elsewhere, to areas that may be physically less suitable but that may be more favorable to them because they are uncrowded (Fretwell 1972). When all the habitats that are even marginally acceptable are full, some birds may not be able to establish territories at all but may simply form a floating, nonbreeding population (Hensley and Cope 1951).

Territoriality is social behavior. It involves communication and reaction among members of a population. The other intrinsic factors regulating population size are also socially based. Many organisms maintain no territories or perhaps only very small ones, and consequently several individuals may have overlapping home ranges; the number of individuals whose home ranges do overlap will tend to increase as density increases.

In some laboratory situations "social pressure" increases as density goes up and has some far-reaching physiological and behavioral effects (Calhoun 1963, Christian 1963a,b). Social pressure includes factors as simple as the frequent disturbance that occurs when things are crowded or as complicated as competition for rank in the social order (that is, which animals are to be dominant). Physiological pathways may involve endocrine glands such as the anterior pituitary and the adrenal cortex, but we will leave these details to physiologists and concentrate on effects that could lower population size. A partial list of these includes lowered resistance to disease, failure of females to build proper nests or to nurse young, and even failure of males to produce sperm or females to come into heat. It is certain that these effects befall populations when they get high enough, as in enclosed laboratory situations. What is not clear is how far along this road natural populations go. Fairly often social pressure probably leads to emigration before some of the more dramatic results occur.

Note that many types of interference occur even though no resource is obviously in short supply. The justification for regarding these cases as competition rests on two assumptions. One involves an evolutionary judgment. We are usually assuming that competition directly for a resource has occurred in the past and that some interference mechanism has evolved in response (MacArthur 1972). For example: given two genotypes of a plant one of which simply grew and took what soil moisture it could get and one of which produced chemicals that inhibited other plants from growing near it and thus protected a water supply, the second genotype came to prevail. A subsidiary assumption is that, in the absence of the interaction, a resource shortage would develop.

Red Grouse Populations

The red grouse lives on heather moors in Britain, which are treeless areas largely covered with low evergreen shrubs in the heath family (Lack 1966). The bird is generally considered a subspecies of the willow ptarmigan that occurs in tundra across Eurasia and North America; the red grouse differs by lacking the white winter plumage seen elsewhere. The main food of this territorial and monogamous bird is heather shoots which are not very nutritious, being especially low in protein. Most of the moorland is grazed by sheep and burned at intervals. The red grouse is the most important game bird in Britain and so there are strong financial reasons for landowners and the British government to be concerned about grouse populations.

The red grouse in Scotland has been the subject of one of the longest-running population studies of any species, spanning (up to now) more than 30 years. The studies, combining field and laboratory work, have produced a great deal of information concerning changes in grouse numbers. The general picture fits well with the model described in preceding sections of a carrying capacity set mostly by food and favorable space and population regulation in the vicinity of carrying capacity by density-dependent factors such as territoriality (Jenkins et al. 1963, Watson et al. 1984).

Numbers of territorial grouse on a particular moor tend to increase over a period of three to four years and then decline over a like period. When populations are low, aggressiveness is low, and young birds have a relatively easy time of establishing themselves. When populations are high, aggressiveness is high, and few young birds can manage to establish territories. Nonterritorial birds are vulnerable to predation and starvation, and during the period from fall through spring many die. When birds are experimentally removed from territories, previously nonterritorial birds move in and survive as well as the other territorial birds (Watson and Jenkins 1968).

This cyclic sequence of changes—an increase in numbers followed by a decrease and correlated trends in aggressiveness—tends to occur whether the general population level is high or low. The general population level seems to depend on habitat quality. The grouse prefer young heather shoots, high in protein and phosphorus, for food, but they also need older, taller heather for nest and escape cover. Populations can be increased (mostly as result of acceptance by the birds of smaller territories) by fertilizing the moors or by burning strips through the heather (Watson et al. 1984).

Although these general results are clear, much else is not, despite the concentrated work on the species. The proximate cause of the variations in aggressiveness is probably hormonal, but the rest of the linkage with population size is uncertain. One recent suggestion is that during the increase phase, most of the new males being recruited into the breeding population are kin of the resident males, so that aggression is low and smaller territories are defended (Moss and Watson 1991). At high populations, production of young declines from crowding, and many of the young males now seeking territories are strangers, which provoke greater aggression.

The role of density-dependent factors other than the territoriality-emigration complex is also uncertain. The idea that parasitism by a nematode, the threadworm, causes the periodic drops in numbers goes back to the earliest scientific studies of the grouse (Lovatt 1911) and has recently been revived (Hudson 1986). It is possible that different factors are important in different localities. It is also possible that the relationship between numbers and environment—even in this simple system—is inherently complex and the roles of territoriality, parasitism, food, and perhaps other factors such as predation vary from time to time.

Intercompensation

Paul Errington (1967) emphasized the interacting nature of factors affecting population size by calling them **intercompensatory.** Weather calamities may kill large numbers of organisms and, particularly at the edge of the geographical range of a species, may even destroy a local population. However, the effect is usually only a temporary (sometimes a very temporary) reduction of numbers. The individuals that survive have the best of habitat available to them, competition for food is low, and consequently they thrive and multiply. This is simply a restatement of how density-dependent population regulation works, but it is worth emphasizing.

Two populations of feral horses live on Assateague Island, a barrier island off the coast of Maryland and Virginia. The horses occur in two separate herds, one in Assateague Island National Seashore and one in Chincoteague National Wildlife Refuse (Kirkpatrick and Turner 1991). The former herd is unmanaged; the latter is maintained at a near-constant number by annually removing about 80% of the foals. This mimics a substantial amount of mortality on the young. Intercompensation is shown by substantially increased fecundity in the managed population. October pregnancy rate in this population was 67% compared with 35% in the unmanaged population.

Hunters often believe that getting rid of the predators of some game species will mean more of that species available for them to shoot. This is sometimes true when predation is acting as a limiting factor. But it is often not the case, because saving the animals from predators results in the operation of other intercompensatory factors such as increased starvation or lowered reproductive success.

Nonequilibrium Populations

In the preceding sections, we have been talking mostly about equilibrium population densities. The idea of an equilibrium density may not apply to some populations. Probably most populations are, in fact, not at equilibirum. At any given time, they are growing toward a summer *K*, or declining toward a winter *K*, or recovering from some unusual event.

Some ecologists believe that it is virtually pointless even to discuss equilibrium population densities. The processes just mentioned, they believe, dominate the populations they study rather than any equilibrium tendencies.

For many years, J. Davidson and H. G. Andrewartha (1948a,b) studied small insects called thrips in an Australian rose garden. The yearly trend in numbers tended to look like Figure 4-28. Populations were low during the Australian winter, grew rapidly with the onset of spring, and then crashed in late summer with the arrival of dry weather.

Davidson and Andrewartha suggested that each spring the thrips began growing toward some hypothetical *K* but never reached it because the length of the favorable period (the period when $b + i > d + e$) was too short. Each year the population got more or less close, depending on how early favorable weather arrived and how long it lasted (Figure 4-29). The population did not go extinct in the winter because the bad weather did not last long enough; however, if it had, the garden would have eventually been repopulated from some other site where the insects had successfully overwintered.

Many ecologists believe that the patterns just described characterize populations of certain organisms, such as many insects, whereas equilibrium

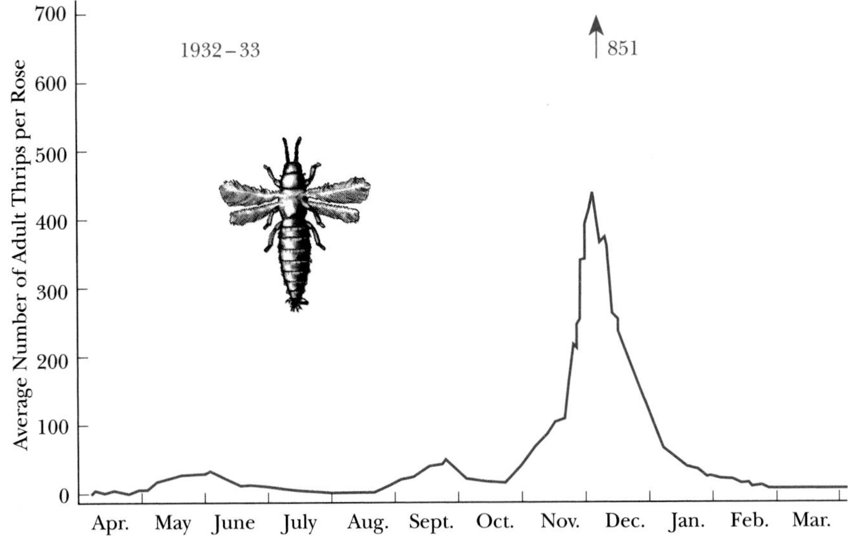

Figure 4-28

The density of thrips in an Australian rose garden during one year, April 1932–March 1933. Thrips are little insects with fringed wings. Each year in the Australian spring, with rising temperatures and favorable moisture from the winter rains, numbers rose exponentially. Each summer—a little after Thanksgiving in the United States—the soil dried out, the flowers the thrips feed and breed on died, and the number of thrips dropped precipitously. (From Davidson & Andrewartha 1948a.)

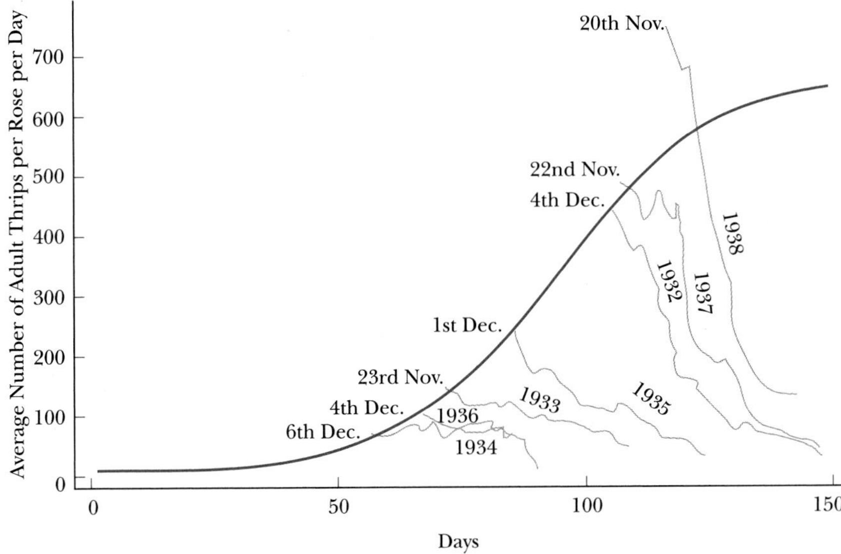

Figure 4-29

The peak population reached by the thrips populations discussed in the preceding figure depended on the length of time birth rate exceeded death rate and this varied with each year's weather. Accordingly, peak numbers varied from year to year, from about a hundred to about a thousand per rose. One way of looking at the data is to consider that each year the population set out on a curve of logistic increase which was never completed owing to the onset of unfavorable weather. The 1938 population went above the hypothetical curve, perhaps because the carrying capacity was unusually high that year. (From Davidson & Andrewartha 1948b.)

tendencies characterize others, such as most birds and mammals. Even if we believe that bird populations in prime habitat are closely regulated, however, populations in marginal habitats may not be. Their size may be little more than an accidental result of events in more nearly optimal habitats (Kluyver and Tinbergen 1953, Brewer 1963).

Sources, Sinks, and Metapopulations

Many populations exist as a set of subpopulations that are more or less isolated but that have some exchange of individuals (and genes) by way of dispersal. This view of how populations are arranged was discussed by H. G. Andrewatha and L. C. Birch (1954) and Richard Levins (1970), among others. Levins used the term **metapopulation** to refer to such a subdivided population.

Success is likely to vary among subpopulations. Some sites, though triggering a positive response in habitat selection, are unfavorable enough that reproduction is below replacement. Such sites can remain occupied only by immigration from outside. Examples include Caspian tern populations in the U.S. Great Lakes (Ludwig 1979) and montane populations of white-crowned sparrows (King and Mewaldt 1987). The term **sink** is used to refer to habitats where population increase does not balance population loss. Habitats producing surplus individuals available for dispersal are **sources.**

Because of the fragmented nature of our natural landscape, these concepts of metapopulation, sources, and sinks are particularly appropriate as a framework for studying many of our threatened and endangered species. We will meet them again in Chapter 19 on conservation biology.

Growth and Regulation of Human Populations

Throughout much of human history population size stayed steady or increased slowly as new lands were discovered and colonized. Fairly high birth rates were approximately balanced by fairly high death rates. Ten thousand years ago, as the last Ice Age ended, it has been estimated that there were no more than 5 million humans. With the rise in agriculture and the shift away from the hunter-gatherer way of life a cycle of population growth was begun. World population may have first reached 100 million shortly before the time of Christ, or somewhere between 1 and 2 million years after humans first appeared on earth. The world population had reached 500 million about A.D. 1650, when Rembrandt was painting, Leeuwenhoek was inventing the microscope, and Louis XIV was doing whatever kings of France did in those days.

In the following years world population growth was accelerated by events associated with industrialization, urbanization, and, particularly, the exploitation of sparsely occupied lands in North and South America and Australia. By 1850, about the time Henry Wadsworth Longfellow was writing *The Song of Hiawatha,* Abraham Lincoln was practicing law in Springfield, and Charles Darwin was working on barnacles, the world population had reached 1 billion. The next billion took 80 years, to 1930. The third billion took 30 years, to 1960, the fourth about 15 years, and the fifth billion 12 years. By 1995, we will be three fourths of the way to 6 billion people.

In many hunter-gatherer societies a reasonably close regulation of local population size was achieved, with both physiological and behavioral factors probably involved. Prolonged nursing of young suppressed ovulation, tending to keep women infertile for three or four years after a birth. Postpartum sexual abstinence, induced abortion, geronticide, and infanticide were all practiced. Infanticide, apparently, was particularly widespread. Expulsion and emigration also tended to maintain local population size below the carrying capacity of the band's territory.

In regions suitable for the growth of crops, agriculture raised the carrying capacity of the land. Once a sedentary village life was adopted, transporting infants was not a problem, so there was no longer that penalty for shortening the intervals between births. In fact, most penalties for overreproduction were reduced for the individual parents because child care was spread over extended families. Along with the fact that large families are often an economic asset in agricultural societies, these features seem to explain the jump in population growth after the agricultural revolution. It would be a mistake, however, to think that no preindustrial agricultural societies regulated their population size. Most did, by delaying the age of marriages and, especially, by restricting the inheritance of land to the eldest son.

In the 18th and 19th centuries, as industrialization occurred in Western Europe, there occurred a series of events that sociologists have called the **demographic transition** (Coale 1974). What happened was approximately this: There was a decline in death rates, initially through better sanitation and attention to public health. (One of the first public health laws in England made it a capital offense for anyone restricted to his house with the plague to break quarantine.) Later, in the 19th century, improvements in medical practice also became important. With the decline in death rates, there was a burst of population growth. England's population increased from not much more than 10 million in 1815 to about 29 million in 1890, while, at the same time, about 11 million emigrants left for the United States and elsewhere.

Birth rates also dropped but later and more slowly, and rarely low enough to balance death rates. A typical pattern was for the death rate to drop from 40 to 50 per thousand to about 10 per thousand while the birth rate dropped only to about 15 per thousand. The University of Oregon anthropologist D. E. Dumond (1975) has suggested that birth rates dropped because, in an industrial society, many of the responsibilities of child care were returned to the nuclear family. The parents again bore the cost of rearing children at the same time that the rise of compulsory schooling decreased their economic value (as wage earners).

This pattern of the demographic transition was repeated in many countries as they industrialized. What of the unindustrialized nations? Until shortly before World War II they continued the primitive pattern of high birth and death rates. Sanitation and medicine were exported to these countries beginning about that time and at an accelerated rate after World War II, with a consequent decline of death rates. Birth rates have not followed suit, so that population growth is extremely rapid in the developing countries. From 1850 to 1920 the increase in the world's population was about equally divided between industrialized and developing parts of the world. Now, however, about 85% of the world's population increase comes from the nonindustrialized nations.

Following are some figures for percentage population growth in 1990 for various countries from the United Nations Population Fund.

Country	*% Growth Per Year*
Brazil	1.9
Mexico	2.0
India	2.1
Kenya	3.7
Nigeria	3.2
Japan	0.4
United Kingdom	0.2
Poland	0.5

It is worth looking at recent figures for the United States separately. In 1960, birth rate was still at the high post-World War II levels, and the percentage growth rate was about 1.4. Sharp declines in birth rate and generally lower death rates resulted in a 1965 rate of 1.0%, and a 1970 rate of 0.9% (Figure 4-30). The years 1973–1976 had the

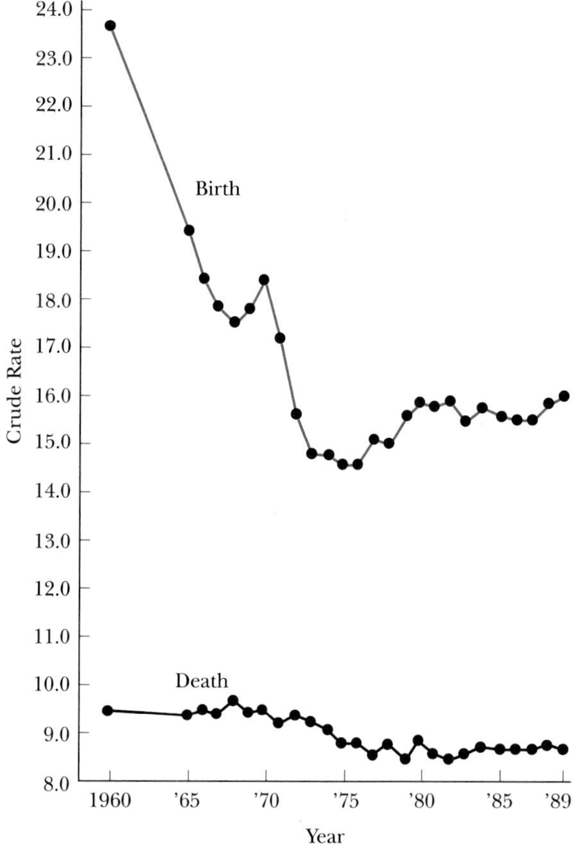

Figure 4-30

Crude birth and death rates from the U.S. population between 1960 and 1989. Data from U.S. Dept. of Commerce, 1990. (Statistical abstract of the United States 1989. Bureau of the Census, Washington DC.)

lowest U.S. growth rates yet seen, 0.55–0.58%. Birth rate turned up in 1977, and death rate, which had dropped from 1968 to 1977, is no longer declining. As a consequence, population growth rate has been around 0.7% the past few years. Given the current age structure, these levels of birth and death would allow population stabilization at some point 40 or so years in the future. (Remember that, owing to the large number of young persons in the population, birth rate will remain higher than death rate and the population will continue to grow for some years, even though family size is at or below replacement level.)

However, there is another complication here. Immigration is as much a reality in human populations as any others. Nobody knows how many illegal immigrants arrive yearly in the United States, but 1.1 million were *caught* along the U.S.–Mexican border in 1990. If we take a probably conservative guess that 1.5 million illegal immigrants successfully crossed our borders somewhere and add that to the legal immigrants (currently about 750,000), we can see that the current contribution of immigration to U.S. population growth may be about 2,250,000 per year. U.S. population growth as the result of births and deaths of citizens and other

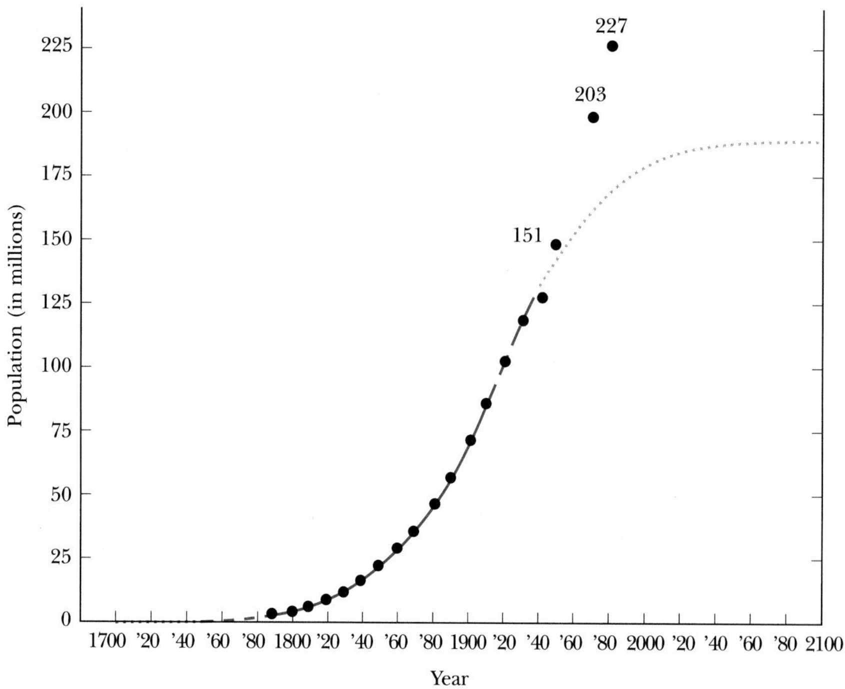

Figure 4-31

A population projection that went wrong. Just before the 1920 census Pearl and Reed found that population growth in the U.S. from 1790 to 1910 seemed to have taken the form of a logistic curve which would tend to level off below 198 million somewhere after the year 2100. The 1920 census figures fit fairly well, as did the 1930 figures. The 1940 figures, however, definitely did not fit and Pearl and his coworkers wondered if the curve needed revision. They revised the curve *downward,* predicting an upper limit of around 184 million. We know now, of course, that the slowing in growth resulting from the drop in birth rate during the 1930's depression was temporary. As shown by the points added to the curve, the 1950 data was close to their projections but by 1970 the U.S. population was already far greater than their first projected limit of around 198 million. None of this is a reflection on Pearl or his colleagues, who clearly stated the limitations of any population projection. (Adapted from Pearl et al. 1940, p. 487.)

residents is at the rate of 0.7% on a population base of about 250,000,000, or about 1,750,000. In other words, the biggest factor in the increase of the U.S. population is now probably immigration. The actual growth rate may be closer to 1.7% than 0.7%, or, in other words, at least as fast as in the baby boom years just after World War II. The actual time to double to 500 million may be closer to 40 years than 100.

Any precise prediction that we make about the future population of the United States or the world is likely to be wrong. Such predictions are merely statements of what will happen if growth rates stay the same, but history has shown that growth rates are changeable (Figure 4-31). The world population (we need not worry about immigration and emigration here) has been growing at 1.7% per year, which would give a doubling of the current more than 5 billion to nearly 11 billion in about 40 years. We may hope that this prediction will be wrong, on the high side, but one imprecise prediction that we can make with confidence is that the world population is going to get larger before it gets smaller.

***Thomas Robert Malthus** (English, 1766–1834)*

Because of his Essay on the *Principle of Population,* Malthus is considered a pioneer in demography and, thus, in population ecology. Besides that, his comments on the tendency of populations to grow at an exponential rate influenced Darwin's thinking on natural selection.

Malthus, called Robert or Bob, was born at the Rookery near Dorking and educated at Cambridge. He spent several years as curate of Okewood Chapel in a pleasantly rural but bitterly poor area of Surrey. In 1804, at the age of 38 years, he married and a year later was appointed Professor of History and Political Economy at the East India Company's College in Hertfordshire. He had two surviving children, a zero-population-growth family, as G. E. Hutchinson has noted.

The first edition of Malthus' essay was published in 1798, a much-enlarged second edition in 1803, and later revisions up to 1826. The context of the essay was the discussion on whether human society is perfectible. Malthus thought not because of the tendency of human populations to increase beyond the level of ample food (and adequate wages). This led to misery and vice; it stemmed from the tendency of populations to grow exponentially while the food supply grew, at best, at an arithmetic rate.

By misery, Malthus meant hunger, disease, inadequate shelter, and the other living conditions associated with poverty. What he meant by vice is not so obvious to a 20th century American audience. He was suggesting that a young man, seeing that he could not adequately support a family, would not marry but instead would visit prostitutes, masturbate, or engage in homosexuality. Probably Malthus also had contraception and abortion in mind.

Malthus received a good deal of contemporary criticism, but according to his biographer in the 11th edition of the Encyclopaedia Britannica, "he bore popular abuse and misrepresentation without the slightest murmur or sourness of temper. In all his private relations he was not only without reproach but distinguished for the beauty of his character."

The idea that agricultural production could increase at no faster than an arithmetic rate has, in a strict sense, been disproved. Nevertheless, few who have thought seriously about populations and resources disagree with Malthus's main conclusions: that populations tend to increase past the point where food and other resources are ample and that, however much we may yet increase food supply, the time will come when no further increase occurs. As Kenneth Boulding commented, anybody who believes exponential growth can go on forever in a finite world is either a madman or an economist.

Population Ecology: *Organization and Evolution*

The eastern mountain gorilla is a social member of the primates usually living in groups of 8–12. *(Animals Animals © 1994 Bruce Davidson)*

OUTLINE

- Organization
- Evolution
- The Evolution of Behavior
- The Ecological Theater and the Evolutionary Play

The preceding chapter dealt with various aspects of population growth and size. Here we look at how populations distribute themselves in space and also at the internal organization of communities, including age structure and mating systems. The differences between solitary and social organisms are explored. The population is the level at which evolution occurs, so natural selection is considered. Special attention is focused on the evolution of life history traits and social behavior.

Organization

Spacing

Spacing refers to the positions of members of a population relative to their neighbors. There are three basic kinds of spacing, **random, clumped,** and **even** (Figure 5-1). Either nonrandom type of spacing is referred to as **pattern.**

To illustrate the three types, visualize a long lunch counter with 50 stools with ten persons on them. If there is no pattern to their spacing, then

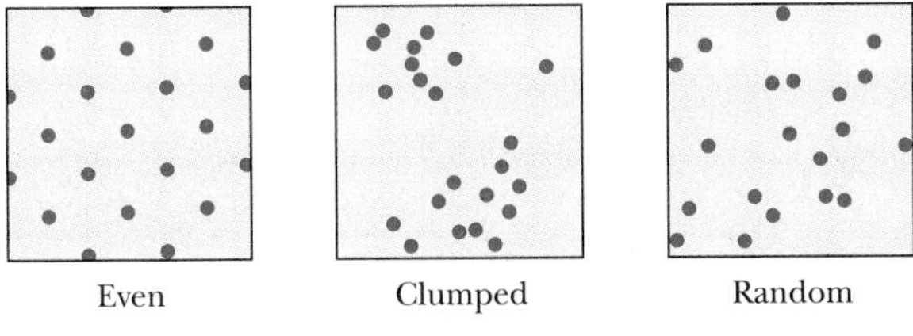

Figure 5-1

The three general types of spacing.

we say that the distribution is random. Specifically, it is random if every stool has an equal and independent likelihood of being occupied. This may well be true if each person came in separately and sat down where it was convenient to do so. But if the ten persons came in as two groups, perhaps two families, then each group would probably sit on stools next to one another. This distribution, in which individuals tend to occur in bunches or clumps with empty spots between, is a clumped (also called **aggregated**) distribution. A flock of birds and a bed of flowers are other examples of clumped distributions.

Suppose now that we start out with 10 persons in a random distribution and 15 more people come in until there are 25 persons on 50 stools. Although there are friendly types who will sit down next to you and start a conversation, most people tend to leave some space between them and their neighbors. Consequently, the pattern at the lunch counter would probably look like this: person, stool, person, stool, person, etc. This is an even distribution, in which individuals are spaced more regularly than in a random distribution. Other examples of even distributions are the trees in an orchard or the honeybee larvae in a comb. In the honeycomb we see the most extreme case of even spacing, in which each individual is equidistant from six others (Figure 5-2).

It is reasonable that individuals should be randomly spaced unless something is happening to warp the spacing toward being uniform or aggregated. Actually random spacing is not very common in nature, nor is even spacing. By far the most common pattern is clumping.

Several factors may operate in different populations to produce clumping. First, variation in the spatial distribution of environmental factors, such as moisture, food, or cover, is often patchy. If a wildflower grows only in the moister parts of the forest, it is apt to occur in patches corresponding to the low spots. Second, dispersal patterns could cause clumping. For example, many plants reproduce vegetatively by spreading underground stems or by some similar method. We may find well-defined, nearly circular clumps of May apple, for example, the result of vegetative spread by the single individual that started years before from seed. Third, for many animals, social behavior is responsible—that is, the animals associate with other animals like themselves (Figure 5-3).

Factors tending to produce an even distribution are mainly competitive interactions between individuals. The territoriality of birds causes them to be evenly spaced (this is, of course, no less a social response than flocking). The trees making up a forest canopy may be on the even side of randomness as the result of competition for sunlight (Table 5-1).

One final point about spacing is that a population may show one pattern at one scale and another at a smaller scale. Herds of wildebeest in African grasslands are clearly aggregated, but if our scale is not one of several square miles but a few hundred square feet, the picture changes. If we

Figure 5-2

Even spacing in the cells of a honey bee hive. Grubs (larvae) are in some of the center cells; the capped cells contain pupae. (Photo by the author.)

Figure 5-3

Alaska fur seals are a social species. More than that, they are a contact species, individuals showing little or no aversion to bodily contact and not maintaining an individual distance. (Photo on St. Paul Island, Alaska, courtesy U.S. Fish and Wildlife Service.)

look just at spacing within the herd, we see even spacing because of the tendency of the animals to maintain a certain minimum **personal distance** between themselves and their neighbors (Hall 1969) (Figure 5-4). Infringement of this circle of personal space by another animal will cause the first animal to shift away or, in the appropriate circumstances, to fight. The circumstances in which close contact between two individuals is tolerated or welcomed vary between species, but two fairly widespread cases are mating and care of dependent young.

Sociality

Animals grade from virtually solitary, like chipmunks and hummingbirds, to highly social, like gulls and chimpanzees. A **society** is defined as a group of animals of the same species organized in a cooperative manner. The organization is based on reciprocal communication, which can be visual, auditory, or chemical.

Even solitary animals usually show some social behavior, such as territoriality or the interactions of a mother and her brood. In discussing sociality, though, we are primarily concerned with animals that live in large groups. From an ecological viewpoint, the disadvantages of crowding are so obvious that the existence of group living seems to require special explanation.

All the disadvantages of crowding are put in the $(K-N)/K$ term of the logistic. We have discussed several of the disadvantages—attracting predators, transmission of disease, social stress, overeating the food supply, and other unfavorable modifications of the habitat. What are the benefits of living together in colonies, bands, troops, or flocks that for some species at some time counterbalance the

Table 5-1
Density and Spacing in a Sand Pine Scrub Community

The sand pine scrub is a unique chaparral-like community in Florida. It is maintained by fire, and the dominant species, sand pine, has serotinous cones like those of jack pine and logdepole pine. After a fire, a dense even-aged stand of small trees grows up. Their spacing is random or slightly clumped; with time, density declines and the spacing becomes even.

Time Since Burnt (Years)	*Density (per m^2)*	*Spacing*
12	2.6	Random
20	0.5	Random
25	0.2	Even
51	0.08	Even
66	0.04	Even

The tendency toward even spacing is caused mainly by the death of trees from competition with their neighbors. Deaths could keep the spacing random if they occurred randomly, or they could produce a clumped distribution if the deaths occurred in patches. In fact, the deaths produce an even distribution, showing that crowded trees are more likely to die than uncrowded ones. This is shown below, where data from the 51-year-old stand compare the spacing of all trunks (alive plus still-standing dead) with the live trunks.

Trunks Measured	*Density (per m^2)*	*Spacing*
Alive plus dead	0.16	Random
Alive only	0.08	Even

From Laessle, A. M., "Spacing and Competition in Natural Stands of Sand Pine," *Ecology*, 46:65–72, 1965.

costs? These fall into three main categories: protection from predators, improved foraging, and favorable modification of the environment.

Predator Protection

Groups of animals living, feeding, or traveling together could, under some circumstances, improve their ability to escape predation in various ways. If several birds or mammals are feeding together, the likelihood that at least one will be looking the right way to see a predator approaching is higher than for a single animal by itself. If awareness of an approaching predator is communicated to the rest of the flock or herd and they are able to escape its attack, such **mutual vigilance** would be adaptive. Several studies have shown that groups of animals spot predators quicker than single animals (Powell 1974, LaGory 1987).

Suppose a predator does attack a group. How might being in a group be beneficial? **Confusion** of the predator is one possibility; a covey of bobwhite exploding out of a roost may startle a fox or a hawk so much that it is unable to single out one of the birds to pursue. For some animals actual **physical defense** may be possible against some predators. Some ant and termite species have soldier castes whose role is defense of the colony. They bite or stab intruders or spray them with acid or glue. When muskoxen are attacked by wolves, the females form a circle with calves inside while the adult males stay outside the circle to harass the wolves. **Mobbing** of a screech owl by chickadees, of a great horned owl by crows, or of a leopard by baboons is in this category.

The sorts of behavior discussed so far not only help the individuals in the group, but also tend to immunize the group itself against predators. The last category under predator defense—**geometrical effects**—is, however, purely selfish. The "selfish herd" idea (Williams 1964, Hamilton 1971) is that each individual uses the other individuals as cover, hiding behind them much as it might hide behind trees (Figure 5-5). On the average the individuals in the middle of the flock or the colony (for nesting birds, for example) ought to be safest and those on the outer rim ought to be most vulnerable. Several studies, such as the one shown in Figure 5-6, and many casual observations suggest that this is often the case.

If these various kinds of group behavior do help in predator defense, we should be able to detect it, and we can. A British ornithologist (Kenward 1978) used a goshawk trained for falconry to attack woodpigeons (Figure 5-7). He found that attacks on solitary woodpigeons were successful in a high percentage of the trials, while very large flocks were practically invulnerable to goshawk attack.

Increased Foraging Efficiency

Under some circumstances, groups can obtain food more efficiently than solitary individuals. The other

Figure 5-4

Although barn swallows may travel in flocks, when they perch on a wire their spacing tends toward evenness, since they leave at a minimum about one bird width between each other. (Photo by R. Bruce Moffett.)

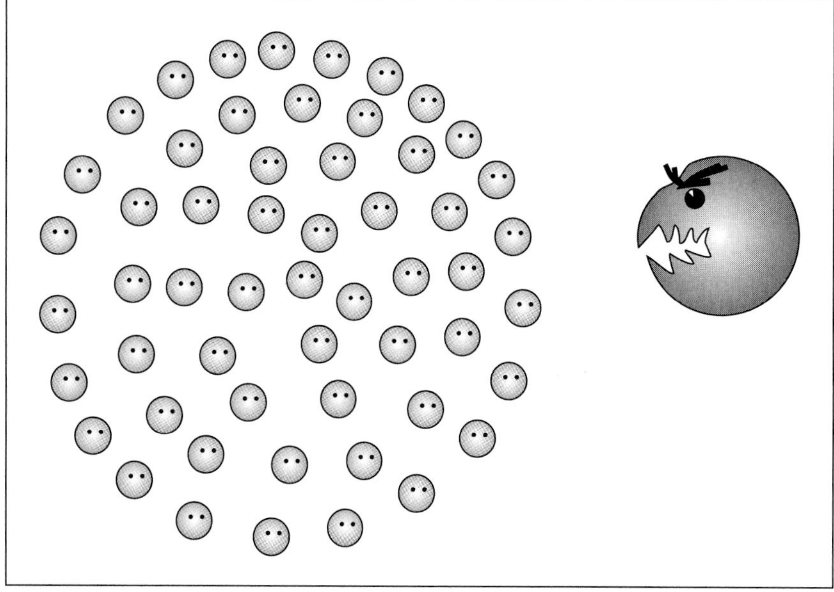

Figure 5-5

The selfish herd. Each individual seeks to hide among the fellow members of the herd.

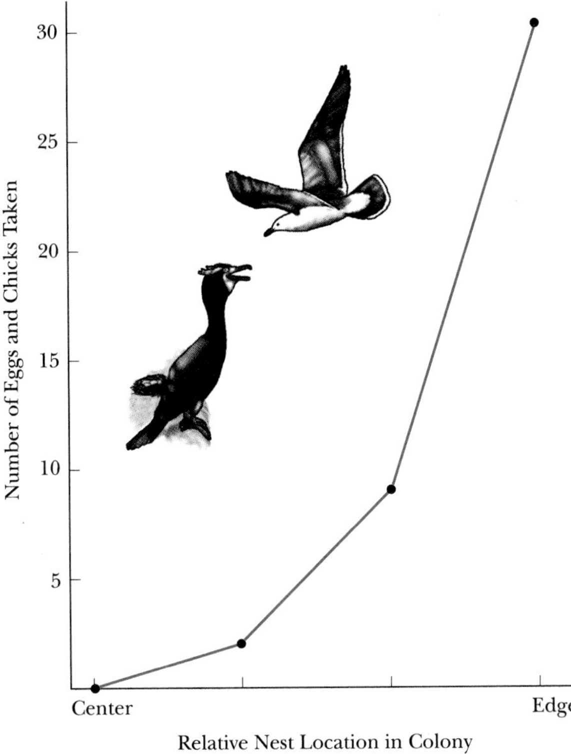

Figure 5-6

Predation on double-crested cormorant nests in relation to location in the colony. Only nests on relatively level ground are included. Few predatory attempts were made on centrally located nests and they were unsuccessful. Nests on the periphery were often successfully attacked. The main predators were crows and gulls. (Data from Siegel-Causey & G. L. Hunt 1981.)

side of the coin of mutual vigilance for predator defense is that each individual in a flock can get by with **reduced vigilant time,** allowing more time for foraging (Pulliam 1973) (Figure 5-8). Ostriches raise their heads at intervals while they are feeding. Only when their heads are up are they able to see if predators are approaching over the savanna. One researcher found that they raised their heads three or four times a minute while feeding alone, but less than twice a minute in groups of three or four (Bertram 1980). So, the number of head raises per bird was reduced, whereas the total for the group was as high as or higher than for a solitary bird. A negative relationship between the time each animal spends scanning for predators and group size has now been demonstrated for a variety of birds and mammals (Elgar 1989).

Each individual in a group moving together tends to flush possible prey items. These may go off in a direction or manner that makes them difficult to catch for the individual that kicked them up, but vulnerable to another member of the group. This avenue of increased foraging efficiency in a group has been called the **beater effect,** named for the peasants or slaves who, in the old days, drove game past shooting stations for their masters.

Another way that an individual in a group may fare better in foraging than it can by itself is by **social facilitation;** that is, it can imitate or learn from the actions of its flockmates. A chickadee may learn from watching flockmates that it is apt to find a larva if it pecks open a rolled leaf. As simple a matter as noting that another bird has just covered a particular tree branch can improve efficiency by saving time.

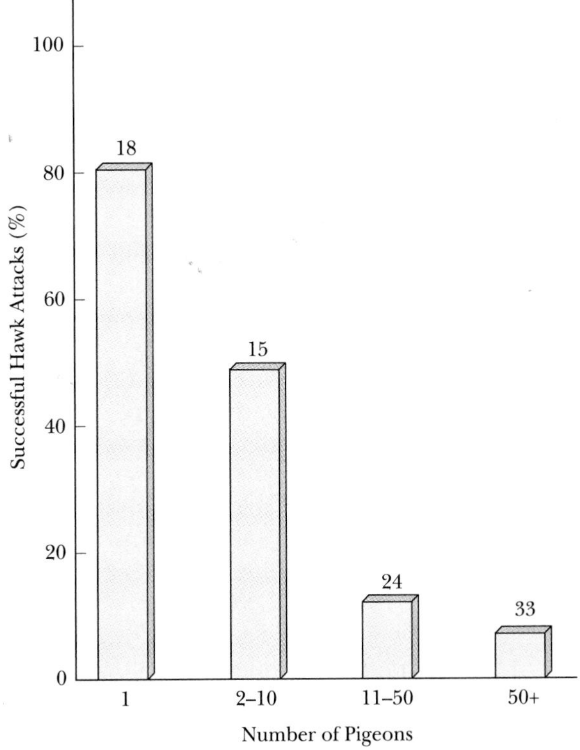

Figure 5-7

The percentage of goshawk attacks that were successful on different-sized flocks of woodpigeons. The number of attacks observed is given above each bar. (From Kenward 1978.)

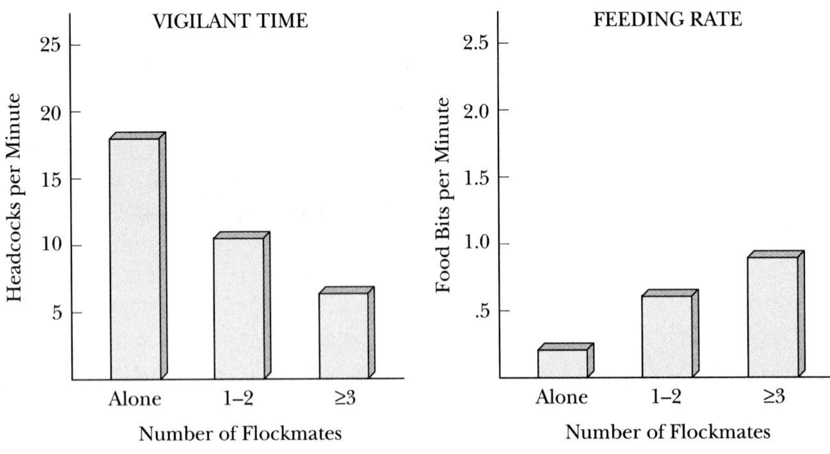

Figure 5-8

Some of the advantages of group living apply interspecifically. Downy Woodpeckers, by utilizing the calls of other birds they travel with in feeding flocks (such as black-capped chickadees) can reduce the time they need to be vigilant compared with when they are alone. Vigilance time is measured here by "head cocks," that is, peering from side to side. By reducing head cocks, they have more time to use for feeding. (From Figures 2 and 3, Sullivan 1984.)

Colonies may be **information transfer centers,** at which information about where food or other resources can be found is exchanged (Ward and Zahavi 1973). This is clearly the case with certain social insects, such as honeybees. The "waggle dance," described in 1945 by the German entomologist Karl von Frisch (1967), functions in this capacity. A worker bee that has found a good source of nectar does a figure-8 dance in which the direction of the crossbar tells the direction of the food from the hive and the length of the crossbar tells the distance.

It would be unusual if colonies of other kinds of animals did not show similar information exchanges, but the subject has not yet been well studied. Is there a transfer of foraging information within large breeding colonies of gulls or pelicans or within large winter roosts of starlings or grackles? In groups where many of the individuals are related, like honeybee colonies, we might expect active communication, but even in unrelated groupings it ought to be advantageous for hungry individuals to follow any unusually sleek or contented-looking neighbors. Such seems to be the case in cliff swallow colonies (Brown 1986). In Norway rat aggregations, not only do unsuccessful foragers learn from successful ones but also information seems to be transferred between successful individuals about the availability of different food types (Galef 1991).

Finally, some social animals can use group effort to **overwhelm prey** larger than they could successfully attack singly. There seem to be few, if any, birds where this is important, but soldier ants can catch large invertebrates and even lizards. Several of the large carnivorous mammals, such as lions, wolves, and wild dogs, hunt socially. The hyena is a good example. Although the hyena used to be thought of as strictly a scavenger, much of its food comes from predation. Hyenas cooperate in preying on wildebeest calves. A wildebeest can keep a single hyena at bay, but with one or two hyenas harassing the mother and another grabbing the calf, the attack is usually successful.

A group (of lions, for example) has a greater ability to **protect the kill** from scavengers, such as hyenas and vultures. At the same time, the chances of an individual scavenger stealing a bite are increased when it is in a group; while the lions are chasing one vulture, another can tear off a piece and flap away.

Modifying Environment

Many of the effects of organisms on their physical environment (their **reactions;** see Chapter 12) become harmful as density increases, but some social organisms are able to modify their environment advantageously. Humans are a social species that produces both favorable and unfavorable reactions in spectacular numbers and extent. There are many fairly modest examples of favorable habitat modification, such as bats clustering to conserve energy on a roost, but the classic example is thermoregulation by honeybee colonies (Seeley and Heinrich 1981). A honeybee hive is maintained within a temperature range of less than a degree on either side of 35°C in the summer and in a slightly greater range (±5°C) near 25°C in the winter. The details are fairly complicated, but basically the bees warm the hive in the winter with body heat from exercising (with honey as the fuel) and cool it in summer by carrying water (whose evaporation removes heat) and by sitting at the entrance beating their wings like exhaust fans.

Other Benefits of Sociality

When we look at social species, we can see other benefits that seem to derive from social living. For example, **allogrooming** may occur, in which troopmates pick off lice or other external parasites, aiding hygiene. The formation of mating swarms may promote outcrossing. These are real benefits, but it is important to remember that they, as well as some of the ones we have already mentioned, may be benefits that depend on the existence of sociality to arise. Some may even be unnecessary in less social species; it is possible that neither lice nor inbreeding is a serious problem for territorial animals. Likewise, some of the elaborate antipredator systems we see may have evolved because of the increased predator attention that groups of prey bring.

From the list of benefits from group living, it would be easy to conclude that sociality is the only way to go. It is worth re-emphasizing that group living also has costs; in some ecological circumstances the balance favors group living, but in many others it does not.

Territorial spacing seems to depend on the dispersion of resources, especially food and nesting habitat. If food conjoined with adequate nest sites is reasonably evenly spaced, classic Type A territories (in which most of the activities of the animals occur) are the rule. If, on the other hand, the food supply is unmonopolizable, colonial nesting tends to evolve (Brown 1964). This could be the case if the food always occurred in the absence of suitable nest sites (as for birds that feed on organisms of the open ocean) or if the food supply was patchy and unpredictable in time, so that establishment of a territory offered no assurance of a food supply lasting through a breeding season.

Even for animals in which coloniality is highly developed, group living has its costs. Studies of several species of colonial seabirds have shown poorer reproductive performance—lower clutches, fewer young produced per nest, and slower growth—at very high densities (Hunt et al. 1986). Figure 5-9 illustrates this for the black-legged kittiwake, a cliff-nesting gull of the Arctic seas.

Like other traits, social behavior in many organisms may show flexibility involving adaptive changes related to changed circumstances. Arctic ground squirrels studied in Alaskan tundra by E. A. Carl (1971) live through the spring and summer in dense colonies in which a male, a female, and the young produced inhabit a burrow system. Unmated males inhabit smaller burrows. During this time, the prime predator is the red fox, which hunts from ambush. Its success is reduced by the high level of vigilance in the colony coupled with alarm calls, which cause the other ground squirrels to become alert or to retreat to their burrows. In the burrows, the squirrels are virtually secure because the ground is still frozen not far below the surface, mak-

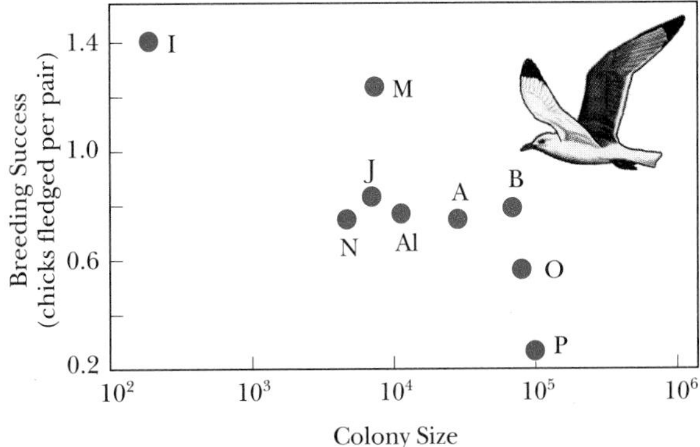

Figure 5-9

Relationship of reproductive performance with population size in the black-legged kittiwake. Letters refer to colonies listed in Hunt et al. 1986 from which this figure is drawn.

ing excavation of the burrows impossible for the foxes or other predators. In late summer, heavy rains lower the frost line below the bottom of the burrows. By this time, however, the foxes have switched to other food sources that have become readily available, and the primary predator is the grizzly bear, which preys on ground squirrels by excavating the burrow. The ground squirrels now become individually territorial, and the population spreads out. Each squirrel, whether old male, old female, or young of the year, occupies a separate burrow system, in which it will later hibernate. One effect of this shift is that the best a grizzly bear can get for its work in excavating a burrow is a single ground squirrel.

Social Organisms

Social species occur in many groups of organisms—there are colonial bacteria—but the most complex societies are found in three groups of animals. First are the colonial invertebrates, such as corals and bryozoans. In some of these the integration is so complete that anyone who had not taken zoology would consider them individuals rather than colonies. Some of the best examples are in a group of coelenterates called siphonophores (the Portugese man-of-war is one); these look and function much like an individual jellyfish, or medusa, but each is a colony consisting of nutritive, tactile, defensive, and reproductive members (zooids).

The second group showing a high level of social organization comprises the insects. There are many insects about as social as the average bird or mammal species, but in three groups "eusocial" species occur. **Eusocial,** or truly social, insects are defined by three traits (Wilson 1975):

1. Rearing of young is cooperative.
2. The colony is permanent enough that offspring remain cooperatively associated with parents.
3. There is a division of labor among individuals, in particular, a separation into sterile and reproductive castes.

Eusocial insects consist of the termites, the ants, and some bees and wasps. Colony structure and life history differ between each of these groups, and details differ between each separate species. As one example, a generalized ant society functions something like this: The colony, most often located in the ground, runs from a few hundred to hundreds of thousands of individuals. A typical array of castes is

Queen	Reproductive (female)
Major worker (= soldier)	Sterile (female)
Minor worker	Sterile (female)
Male	Reproductive (male)

A colony is started by a new queen that has been fertilized during a mating flight. In a cell that she constructs, she sheds her wings and begins laying eggs. The first larvae to hatch from these eggs are fed by the queen on saliva containing nutrients derived from her resorbed wing muscles. These first larvae develop into minor workers that begin the serious work of the colony—enlarging the structure, foraging for food, etc. The queen settles into her reproductive role in earnest. Major workers may be produced that serve defensive functions such as guarding the entrance or repelling threats to the colony. The minor workers tend the queen; take care of eggs, larvae, and pupae; and hunt for food to bring to the colony.

All of the reproduction in the colony is done by the queen. Some of the eggs she lays are fertilized (with sperm deposited by a male on her mating flight), and some are not. The fertilized eggs produce females—the workers—while unfertilized eggs produce males. This type of sex determination is called **haplodiploidy** (the males, resulting from unfertilized eggs, have a haploid chromosome number, and the females, from fertilized eggs, have a diploid number).

Basically, then, an ant colony is a bunch of sisters, the workers, taking care of their mother, who continues to produce more offspring that are added to the colony. New queens are produced occasionally (whether a fertilized egg produces a major worker, a minor worker, or a queen depends on the nutrition it receives) to go off on mating flights or to take over if the old queen dies. Males are also produced occasionally that go off and mate with a newly emerged queen from another colony. "Try to mate" is perhaps a better way to put it; many more males than queens are produced, and most males never manage to copulate with a queen.

The sterile castes are a particularly interesting

feature of social insects. They have always been difficult to explain evolutionarily. We generally think of traits as being adaptive according to their effect on individual fitness, that is, on the number of an individual's offspring reaching the next generation. How, then, could sterility, which means leaving *no* offspring, evolve? We will return to this question a little later in the section on sociobiology.

The third group in which sociality is especially well developed is the mammals. By no means are all mammals social, but highly social species have evolved in virtually every order. Just how the societies are organized differs among kangaroos, prairie dogs, dolphins, antelope, elephants, lions, wolves, and all the other varied group-living mammals. Many of the societies are exceedingly complex, but few, if any, are eusocial in the insect sense. One close approach may be human societies, in which some have suggested that homosexuals amount to a sterile caste.

Another mammal considered to be eusocial is the naked mole rat. This is a hairless, nearly sightless, burrowing rodent of arid East Africa. Studies suggest that its colonies of 40 or more individuals consist of several generations that join in gathering food and other tasks. There is a separation into castes, though not very well defined ones. Only a single female breeds. Most of the colony's females probably will never reproduce, but if the breeding female dies, one will replace her. Caring for young is cooperative; however, only the breeding female nurses. These traits perhaps qualify the naked mole rat to be called eusocial (Jarvis 1981, Sherman et al. 1991).

Sociality in one order of mammals, the Primates, is of special interest because of our own membership in the order. Primates are surprisingly varied socially. The forest-dwelling orangutan is practically solitary, individuals meeting to mate and then separating. Most primates are, however, highly social. Among well-studied species are the small *Hamadryas* baboon (Kummer 1968), the large, gentle mountain gorilla (Schaller 1963), and our close relative, the chimpanzee (Goodall 1968).

Mating Systems

There are three basic mating systems in a population, **monogamy, polygamy,** and **promiscuity.** In monogamy a persistent pair bond is formed between one male and one female. Polygamous unions involve a bond between a single member of one sex and several of the other sex. The most common type of polygamy, at least in birds and mammals, is **polygyny,** in which one male has a harem of several females. **Polyandry,** in which one female has several male consorts, is rather rare. In **promiscuity** the individuals meet for copulation but no continuing relationship is established.

Monogamy lasting through a brood or a breeding season is the usual situation with birds. Longer unions lasting several years or up to life are known for many hawks, swans, geese, crows, and chickadees. Among mammals, many of the terrestrial carnivores are monogamous. Polygynous animals include a good many birds such as the bobolink, most pheasants, and the ostrich; polygynous mammals include many of the primates, such as baboons (however, gibbons are monogamous and chimpanzees promiscuous) and some of the large grazing animals such as deer. Polyandry is known for a few birds and is usually accompanied by a reversal of sexual roles, the male incubating the eggs and caring for the young. Promiscuity is the usual condition among smaller mammals and also occurs in some birds, notably **lek** species. These are birds in which the males gather in an area (referred to as a lek or an **arena**) and go through various displays. Females visit the lek for copulation and go elsewhere in the vicinity for nesting. Grouse, such as prairie chickens and sharp-tailed grouse, are examples.

There are often clear ecological and evolutionary reasons for the existence of various mating systems. Female mammals, unlike female birds, are not tied to a nest and eggs that require incubating and defense from predators. Furthermore, they produce milk as a food supply for the young once they are born. For these reasons, the continued presence of a male is not a necessity in mammals. In fact, by eating food that the female could turn into milk and by attracting the attention of predators, the male may be a liability after impregnation has occurred. Hence, promiscuity has evolved as a mating system among many mammals. For most birds care of the nest and young by both parents is advantageous and most species are monogamous. The monogamous mammals tend to be the hunting mammals, in which the male is useful for obtaining meat for the female and young.

In birds and perhaps in some other animals, we

can sometimes relate the occurrence of polygyny to habitat variability. Figure 5-10 shows a graphical model developed by G. H. Orians (1969) illustrating this. The model assumes, among other things, that fitness of the female is related to territory quality and that, with equal territory quality, females in monogamous relationships (one female per territory) will be fitter than those in bigamous ones (two females per territory). If there is a wide range of territory quality, or if female fitness is a steep function of quality, then a newly arrived female would be better off being the second mate on a good territory than the only mate on a poor one.

The distance, P.T. (which stands for "polygyny threshold"), shown on the bottom graph of Figure 5-10 illustrates the point. If the range of territory quality is greater than this, some females would be fitter in bigamous matings. In such habitats we would expect polygyny to arise. If, however, the range of territory quality is low (or the dependence of female fitness on territory quality not very strong), the polygyny threshold would not be exceeded and tendencies toward polygyny would be selected against (top of Figure 5-10).

The model was developed in connection with the observation that most polygynous birds in North America are characteristic of marsh or grassland (Verner and Willson 1976). Examples are red-winged blackbirds and dickcissels. These two habitats are certainly temporally variable, although perhaps not more variable spatially than some other vegetation types having few or no polygynous species. Tests of the polygyny threshold model have generally supported it (for example, Petit 1991), but there is clearly much yet to be learned about the ecological context of polygyny (Searcy and Yasukawa 1989, Bensch and Hasselquist 1991).

Cuckoldry, Pro and Anti

Several years ago, a team of wildlife biologists published a short paper reporting a discovery whose full

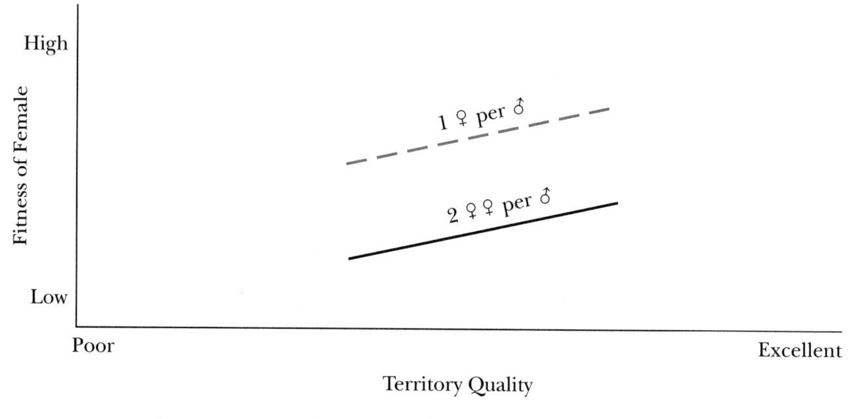

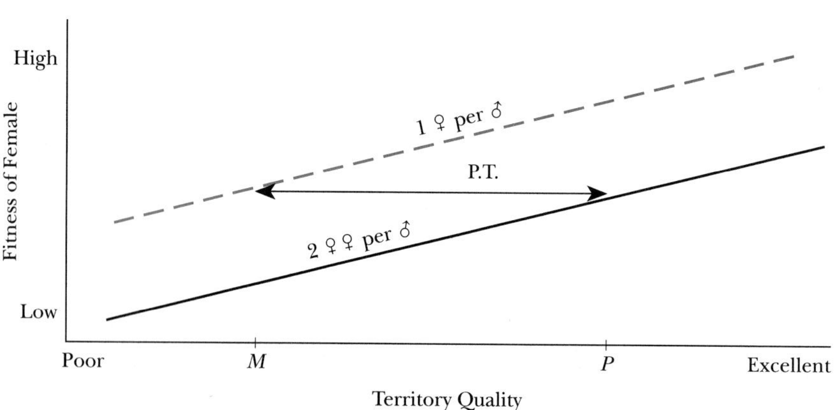

Figure 5-10

The "polygyny threshold" model. If the habitat is spatially variable, some territories may be so poor that a female could do better as a second mate on good territories that mated monogamously on a poor one (Orians 1969). If the only unmated male available had a territory of quality *M* or below (bottom graph), a female would be better off becoming the second mate on any territory with quality higher than *P*.

significance has only recently begun to be appreciated (Bray et al. 1975). They were interested in the feasibility of sterilization as a technique for reducing blackbird populations, the objective of which would be to reduce agricultural damage that large concentrations are claimed to produce. Toward this end, the team vasectomized some red-winged blackbirds. Their surprising discovery was that a substantial percentage of the eggs produced by the females mated to the sterile males were fertile.

Since then, techniques for studying relatedness among individual animals using electrophoresis and DNA fingerprinting have come into wide use. They have shown that a considerable fraction of the nests of many species have one or more young that are the result of matings of the female with males other than the owner of the territory where she resides. This has been the outcome in both polygynous species, such as the red-winged blackbird (Gibbs et al. 1990), and monogamous species, such as the indigo bunting (Payne and Payne 1989). In some species, however, the percent of young resulting from extra-pair copulations is low; only 15% of pied flycatcher broods had any young that did not belong to the male territory holder (Lifjeld, et al. 1991).

Clearly, both monogamy and polygamy are complicated states. From the standpoint of the male, fertilizing as many females as possible will be selectively advantageous. That is, males that fertilize many females will contribute more offspring and genes to the next generation than ones that fertilize one or a few, other things being equal. On the other hand, males that contribute to rearing young, by gathering food and feeding the young birds, for example, will be wasting time and energy helping genes that are not their own if their female mates with other males. There is, consequently, a great deal of behavior shown by males toward females that has been interpreted as **anticuckoldry** devices (a cuckold is a man whose wife has committed adultery). One prominent type is mate-guarding behavior which occurs very widely, from mammals (such as humans) to birds to insects. One of the best-studied examples is the yellow dung fly (Parker 1974).

All this is from the viewpoint of the male. Let's consider the interests of the female. Although more complicated considerations may also apply (for example, Travis et al. 1990), a female with a sterile mate in a strictly monogamous union obviously suffers a great drop in fitness. Perhaps the problem is most serious for birds, where the female, at heavy energetic cost, produces a clutch of eggs and then spends time and energy incubating for days or weeks. If the eggs are infertile, she may continue to sit until she has lost her chance for reproduction that season. If sterile or nearly sterile males occur in the population, it is obviously to the advantage of the females to hedge their bets and copulate with other males.

The overall interest of the pair is to produce a brood that yields as many offspring as possible surviving to the next season. If all the offspring are the progeny of the resident male and female, their interests coincide; if not, their interests diverge considerably. One of the lessons of behavioral evolution is the importance of these conflicts of interest—between the sexes and between parent and offspring.

Sex Ratio

Sex ratio refers to the proportions of the two sexes (among bisexual, or dioecious, organisms). It is most often expressed as the number of males/number of females. A population of 110 males and 100 females would have a sex ratio of 1.1. Sex ratios are of interest for both theoretical and practical reasons. A wildlife manager, for example, would like to know whether a herd of 1000 deer consisted of 800 does and 200 bucks, which might be considered a desirable ratio, or 950 does and 50 bucks, which might lead to overpopulation and sportsman unhappiness (Leopold 1933). A conservation biologist might need to know if events had conspired to skew the sex ratio in an endangered species of songbird (Zimmerman 1991).

We can measure sex ratios at various points in the life history of a species. One set of terms (Robinson and Bolen 1989) used to express this is the series

1. Primary sex ratio, at conception
2. Secondary sex ratio, at birth or hatching
3. Tertiary sex ratio, of juveniles
4. Quarternary sex ratio, of adults

These ratios tend to differ because of mortality and other phenomena that affect one sex more than the other. In humans in the United States, secondary sex ratio is about 1.05, the ratio at 15 years of age about 1.03, and the ratio between 20 and 45 is 1.01, dropping to about 0.95 (Weller and Bouvier 1981).

The first three of these terms tend to refer to sex ratio in a cohort of animals, but the last term is a population sex ratio, based on the adult members of all the cohorts hatched or born at various times in the past. It is influenced by sexual differences in rates of maturation, mortality, and the balance between immigration and emigration. Actually, the interesting figure is the ratio between sexually mature males and sexually mature females. This sex ratio in many turtle populations is male biased because males mature several years earlier than females (Gibbons 1990). If mortality rates are the same in the two sexes, the difference in maturation rates means that there are more sexually active males in a population at any one time than females. If females have a higher mortality rate (from being killed by skunks while digging a nest, for example), the proportion of males would be still higher.

In birds, one could make a case that sex ratios should be female biased because of the vulnerability of males to predation because they are more conspicuous from singing, bright colors, and larger size. Alternatively, one could argue that predation should be higher on females because of their generally greater role in care of the eggs and young. Available data seem to suggest that most bird populations have more males than females (Welty and Batista 1988).

Most organisms that have two sexes and do not practice close inbreeding have a sex ratio that is close to 1.0 at the time parental care ceases and their own reproduction starts. This is true even in polygamous species, even though an engineer or a social planner might design things differently. After all, some of those males are just going to waste, and a population that produced mostly females and just enough males to fertilize them all would be more efficient.

The evolutionary basis for the 1.0 ratio was first stated by R. A. Fisher (1930). We need to consider at least three generations to understand the basics of the argument. Suppose first that genetic factors exist that predispose a female to produce mostly males, mostly females, or half and half. Let us start with the situation in which females producing mostly daughters predominate and the population accordingly has a sex ratio of, say, 0.1. All these daughters have to be fertilized (or if they're not we can forget about them for the future). Most of them will be fertilized by sons produced by the females in the mostly male and 50:50 categories. Accordingly, in the next generation, the genes from these two groups will increase and the genes favoring producing an excess of daughters will decline.

Suppose that the process continues until the population is at an equal sex ratio (1.0) and that virtually all individuals are either the 50:50 or the mostly male-producing genotypes. Although the daughters of this generation will be fertilized by males from both categories, nearly all daughters are from the 50:50 type. So nearly all the zygotes will be at least heterozygous for the 50:50 type. Accordingly, the mostly male-producing genotypes will decline.

We could follow the same reasoning from the opposite direction, starting with a population in which males predominated. Only the 50:50 ratio is stable. One further consideration is that if parental investment (Trivers 1972) in the two sexes of offspring is different, something other than a 1.0 ratio of individuals may be expected (Fisher 1930). In other words, if sons cost slightly more for a pair to produce, natural selection should result in a situation in which pairs produce slightly fewer of them. These predictions are for the time at which parental caregiving ends.

Two icthyologists tested Fisher's sex ratio principle using a small marine fish, the Atlantic silverside (Conover and Van Voorhees 1991). In this species, sex determination is by the joint effects of temperature (lower temperatures tending to produce females) and sex-determining genes. Wherever the species occurs the natural sex ratio is about 1.0. By using species from different geographical areas and different rearing temperatures, the investigators were able to set up five populations with varying sex ratios. For example, by putting South Carolina fish at 28°C, they produced a population that was initially 80% males. After eight generations, the ratio was almost 1.0, and the other populations showed the same outcome: they tended toward a 50:50 ratio.

Age Structure

The age structure of a population refers to the proportions of individuals of various ages. Age may be expressed in days, months, or years, or the categories prereproductive, reproductive, and postreproductive may be used.

For most kinds of animals, the age of an individual is important in specifying its role in the population. Knowing age we have a good idea of an animal's life expectation and reproductive rate; we can even make reasonable guesses about such things as its social relationships and energetic requirements.

A population growing (or declining) at a constant per head rate comes to have a characteristic age structure called the **stable age distribution.** If the growth rate is zero, that is, if the population is not changing in size, the stable age distribution is called the **stationary age distribution.** It is calculated from the l_x column of the life table and tends to have more older and fewer younger individuals than in a growing population. Figure 5-11 graphs age structure in an exponentially growing population and in one stable in size. Any real population usually has an age structure different from either of these because of various events in its recent past (Figure 5-12).

Age structure is important in population growth. We can visualize two sets of sailors who have mutinied, cast their captain adrift, and set sail with a bevy of South Seas maidens for an uninhabited island to found a new colony. In one case imagine the sailors and maidens as all being about 19 years old and in the other about 55 years old. The potentiality for exponential growth is high in the first case (higher, in fact, than would be predicted from the biotic potential which assumes the stable age distribution) and almost nil in the second. The presence of a high proportion of persons who have recently entered reproductive age is responsible for the fact that, even if the average number of children per family moved to the replacement figure of just over two and stayed there, the population of the United States would continue to grow for a few decades.

The average age of an individual in the United States today is nearly 33 years. If the population stabilized, with death rates remaining about the same as now, the average age would go up to about 38 years halfway through the 21st century. This is an inevitable consequence of population stabilization—inevitable unless stabilization is achieved by raising death rates rather than by lowering birth rates, in which case a young population could be maintained. The average age has moved up fairly steadily through the years; in 1800 it was about 16,

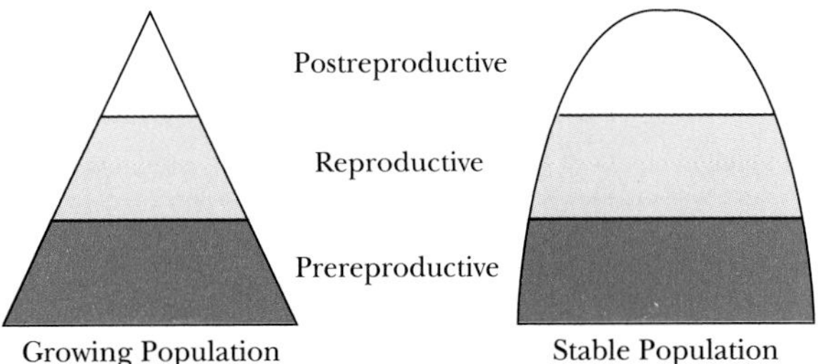

Figure 5-11

Diagrammatic representations of age structure in a growing population and a stable population.

in 1900 about 23. With stabilization, the proportions of various ages will shift. There will be slightly more persons over age 50 years than under age 20 years, whereas now the youngsters outnumber the oldsters 3 to 2.

The many possible economic and sociological consequences of this change in age structure are fascinating but too far afield. It is evident that a greater proportion of our resources will have to be used for helping the elderly and a smaller proportion used for education and other child care services. However, the total economic burden should be no greater because the dependency ratio (the number of nonworkers compared with the number of workers) will not be greatly different. In Table 5-2, for example, it can be seen that 38% of the present population in the United States is either under the age of 17 or over the age of 65 years. With a stationary population in A.D. 2010, the total of these two groups will constitute 36% of the population.

Age Structure in Plants

At the beginning of this section, we said that the age of an individual animal tends to specify its role in the population. This is also true of some kinds of plants. We can study them from seed to death with survivorship curves and other methods devised for animals. Other kinds of plants (and many poikilothermic animals) fit this mold less well. A ragweed seed, when it germinates after a Midwest woodlot is cut, may be 25 or 30 years old. How do we treat this in a life table? We could consider the newly germinated seed as age zero, but if we do, we have missed the mortality that occurred between seed production and eventual germination.

Then, too, what do we do about trees that ger-

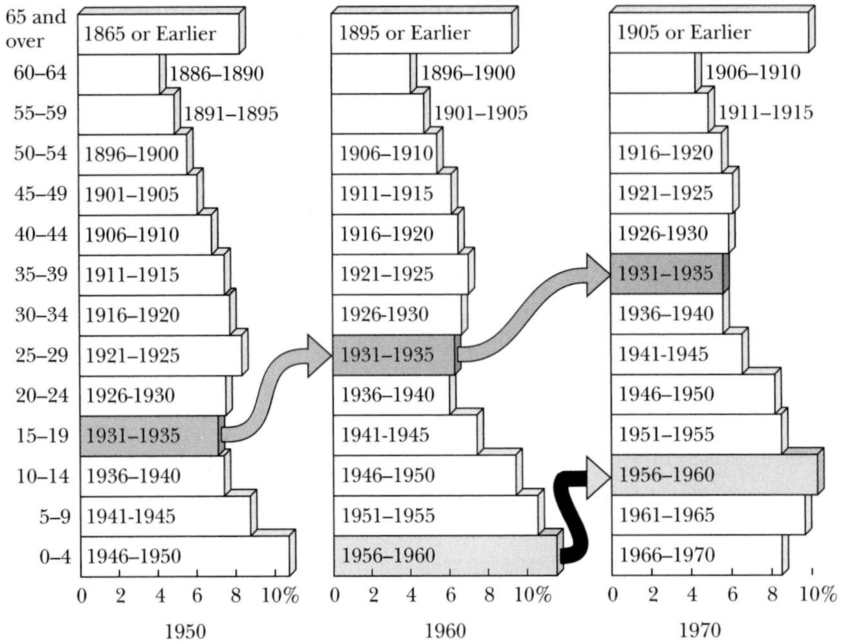

Figure 5-12

Age distribution in a real population, in this case the U.S. human population in 1950, 1960, and 1970. The 1950 population shows a small 15–19-year age class as a result of the low birth rates of the Depression years (1931–35). That same small age class is 25–29 years old in 1960 and 35–39 years old in 1970. The baby-boom babies were produced during the high fertility years of the late 1950s and early 1960s. Accordingly, there is a large 0–4-year age class in 1960 which then occurs as a 10–14 year age class in 1970. The large 65-and-older age class is the result of combining several 5-year classes; however, the baby boomers will reach retirement age starting about 2020 and then it will really jump. (Based on Historical Statistics of the United States.)

Table 5-2
Approximate Age Distribution in the United States at the Present Time and in 2010*

Age Class (Years)	*Today (%)*	*2010 (%)*
0–17	25.7	22.2
18–44	43.1	36.1
45–64	18.7	27.8
65 and over	12.5	13.9

* Assuming 1.8 births per woman, 500,000 immigration per year.

Based on Tables 12 and 18 of the *Statistical Abstract of the United States,* 111th ed., Bureau of the Census, 1991.

minate in a forest and hang on, growing very slowly (**suppressed** is the foresters' term) until the canopy is opened. In such species as hemlock and sugar maple two trees, one sapling sized and one in the canopy, can be about the same age. The difference is that the second germinated in a spot where the canopy was opened soon afterward, while the first has had to grow year after year in dense shade.

For these and other reasons, age structure may not be directly usable for some purposes. Size, state (Rabotnov 1969), or stage may be the appropriate base for graphs and calculations. Even in these cases, however, the additional knowledge of age allows interpretations not possible based on state or stage alone (Ballegaard and Warncke 1985).

Evolution

Natural Selection

Organisms evolve. An ecological problem for an organism today may not be a problem a few generations from now because the organism has the potential to change, through natural selection, so that it becomes adapted to the situation. Evolution, then, is the adaptive modification of organisms with time, or, as Charles Darwin put it, descent with modification. Scientists have seen evolution occurring, in the development of resistance to antibiotics in bacteria and in the now-familiar story of the development of industrial melanism in the peppered moth (Kettlewell 1973). And fossils are a record, more or less complete depending on the kind of organism, of evolution in the past.

The mechanism by which evolution occurs is primarily natural selection, operating almost exactly as Charles Darwin described it in 1859. The main supplement has been an understanding of the genetic basis for selection: Genetics as a field of study did not begin until 1900 with the rediscovery of Mendel's work.

Individuals, Darwin said, vary. Those whose variations suit them best to the existing environment tend to leave behind more descendants in the next generation, because they survive longer and reproduce more. Consequently, any of the favorable variations that are hereditary will tend to accumulate in succeeding generations; in other words, an increasingly large percentage of individuals of the population will possess these traits. The percentage of individuals possessing the unfavorable traits will decline because in each generation they die early or have few offspring. Evolution, then, in its simplest sense, is a change in the proportions of individuals in a population possessing some given hereditary factor, or gene. Note that evolution is a process of **populations.** Individuals do not evolve; they survive a longer or shorter period and leave a larger or smaller number of descendants.

The ultimate source of the genetic variability important in evolutionary processes is presumably mutation. However, at any one time, natural selection works mainly on the variability already present in the population. It has been discovered in the last couple of decades that most populations have a lot more variability than previously believed.

Most mutations that have been studied are disadvantageous in some way. At first this may seem puzzling. How can mutation be important in adaptation if mutations are harmful? The explanation seems to be this: Most species have been around for hundreds or thousands of generations, so that most mutations that are likely to occur *have* occurred. The favorable ones have already been incorporated as one of the normal conditions of the organism; the harmful ones tend to be continually eliminated. It is these harmful mutations, as they crop up again, that we detect as new.

This situation prevails as long as the environment of the organism does not change materially. Various insect populations possess mutations that, as one of their effects, make the insects resistant to DDT. In a pre-DDT environment this effect was rarely of any use to the insect and was usually coupled with some other effect that lowered the fitness of the insects possessing the gene. In the DDT era these genes became extremely advantageous. A good spraying of DDT would all but wipe out most of the insect population, leaving the environment uncrowded for the few resistant individuals. These would prosper to such a degree that only a few years would be enough for a population to evolve from a situation in which nearly every individual was killed by DDT to one in which practically none was (Figure 5-13).

Some have suggested human evolution as a solution to our environmental problems. If some pesticide is injurious to us today, we will evolve immunity to it. If the genes for such immunity occur in the human gene pool or if mutation should happen to produce them, then this is a possibility. Few biologists would regard it as a feasible or humane solution, though, because they know that to produce a largely immune population in a few generations would require the death (or at least the nonreproduction) of nearly every person who was not immune; anything less drastic would take many generations.

The Hardy-Weinberg Law

In two papers that are the beginning of population genetics, G. H. Hardy (1908) and W. Weinberg (1908) pointed out that in diploid, sexually reproducing organisms phenotypes, genotypes, and genes all tend to come to equilibrium in popula-

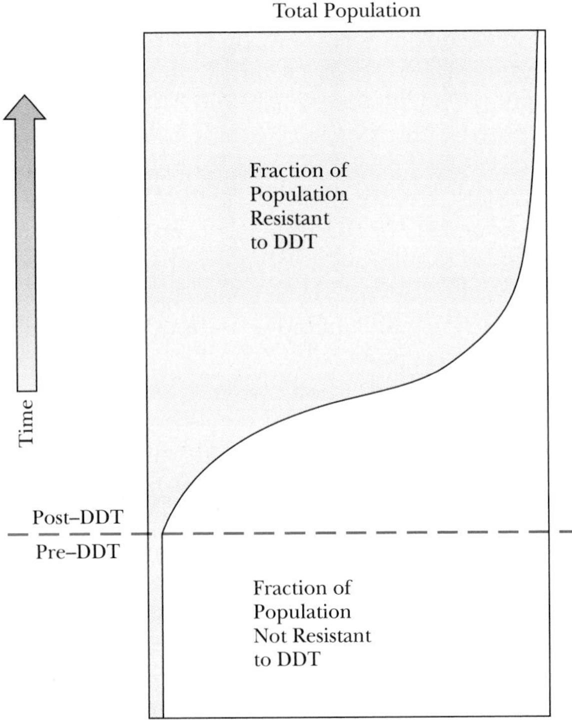

Figure 5-13

Evolution of resistance to DDT in an insect population. In a pre-DDT environment some small fraction of the population contains genes giving some measure of resistance to the insecticide. When applications of DDT kill most of the nonresistant individuals, the environment becomes uncrowded for the resistant ones. These reproduce heavily and survive well so that, very rapidly, the population changes to one in which a large proportion possesses genes for DDT resistance.

tions. The equilibrium frequencies are given by the expression

$$1 = p^2 + 2pq + q^2$$

where p is the frequency of one allele and q the frequency of the other. (For simplicity, we assume that there are only two alternative forms of the gene, but the result is the same if more than two are involved.)

For example, suppose that some phenotypic trait, such as pale eyes, is seen only in homozygous recessives, which we will represent as *ee*. The other two genotypes will be the homozygous dominant condition, *EE*, and the heterozygote, *Ee*. Observation shows that only 6% of the population displays pale eyes. In other words, $q^2 = 0.06$, and q (the frequency of gene *e*) is $\sqrt{0.06} = 0.24$. Because $p + q = 1.0$, p (the frequency of gene *E*) is 0.76.

We now know the frequencies of the two genes in the population. Knowing these values, we can readily calculate the frequencies of the three genotypes. *EE* is p^2, or $0.76^2 = 0.58$; *Ee* is $2pq$, or $2(0.76 \times 0.24) = 0.36$; and *ee*, or q^2, as we already know, is 0.06. If dark eyes are dominant, then the phenotypic ratio is 6% of the population with pale eyes and 94% with dark eyes.

Numbers like these can be useful, and the relationship itself provides some perceptions we might not otherwise have. For example, with only 6% of the population displaying pale eyes, we may be surprised to learn that 36% of the population are carriers of the trait. But the most important role of the Hardy–Weinberg law is that it tells us that gene frequencies in a population stay the same with time as long as certain conditions are met. This gives us a starting point for knowing what factors will change gene frequencies and by how much.

Specifically, the Hardy–Weinberg equilibrium is maintained only if none of the following are occurring:

Nonrandom mating. Tendencies for like to mate with like or for self-fertilization increase the proportion of homozygous individuals. Outbreeding will increase heterozygosity.

Random genetic drift. This is a change in gene frequencies that occurs in small populations as the result of sampling errors. Visualize a population of mice inhabiting a small island. Let us say there are ten mice now and will be ten mice a generation later. Some alleles, purely by chance, may fail to make it from one generation to the next. The only mouse with a gene for darker coat color, for example, may fall in the lake and drown, not because it is clumsier than other mice but by accident. Or its young may all be killed by a freak storm. Or suppose that there are four females and six males, and this mouse is one of the males that fails to fertilize any eggs. The point is that in large populations the likelihood of chance events affecting gene frequencies in the next generation is slight, but in small populations it is high. Genetic drift removes genes from the population but cannot add them, and so the general result is loss in genetic variability.

Immigration and emigration. The movement of individuals, and thereby alleles, into and out of a population will have effects on gene frequencies that can be predicted only if we know more about the rate and nature of the migration.

Mutation. Mutation will alter the Hardy–Weinberg equilibrium values except in the unlikely case that the reverse mutation occurs at the same rate.

Natural selection. The basis of natural selection is the change in gene frequencies from one generation to the next. This occurs by differences in contribution to the next generation of different phenotypes and, thereby, genotypes. The Hardy–Weinberg law allows us to put natural selection on a quantitative genetic basis.

Genetic Fitness

The measure of an individual's or genotype's contribution to the next generation is **fitness.** Suppose we manage the difficult feat of determining the number of descendants that individuals of three genotypes, B_1B_1, B_1B_2, and B_2B_2, leave in the next generation. They are, let us say,

B_1B_1	0.6
B_1B_2	1.0
B_2B_2	1.5

These are **absolute fitnesses.** Conventional population genetics has been more interested in **relative fitness**—how do the different genotypes stack up against the best one? Dividing each by 1.5, accordingly, we get relative fitnesses of

B_1B_1	0.40
B_1B_2	0.67
B_2B_2	1.0

The B_2B_2 genotype, therefore, has high fitness.

Although natural selection acts on phenotypes, the Hardy–Weinberg formula lets us look directly at gene frequencies in the population without necessarily worrying about what bodies they are in. We can refer to the sum of all the genes of a population as the **gene pool.** We have talked about evolution as changes in traits of organisms, but genetically, it is a change in gene frequencies in the gene pool from one time to a later time.

The strength of selection against one of the less favored genotypes is expressed in the coefficient of selection, s. This is simply 1 − fitness. For the genotype B_1B_1 in the madeup example used earlier, the coefficient of selection is 1 − 0.4 and, consequently, $s = 0.6$. (In practice, one is more apt to estimate s and calculate fitness from it.)

We can plug the values for s into the Hardy–Weinberg formula and calculate how much change in genotypes will occur in the next generation (given as Δq). As an approximation (Falconer 1989), good enough for work in the private sector,

$$\Delta q = \tfrac{1}{2}sq(1 - q)$$

Δq is the change in gene frequency for the gene with frequency q. This approximation is useful when either s or q is small and assumes no dominance.*

Let us concoct a situation of no dominance, an s of 0.6, and a q (frequency) of gene B_1 of 0.1. The change in gene frequency in the next generation will be negative because the gene is being selected against. We will calculate the change as

$$\tfrac{1}{2}(0.6 \times 0.1)(1 - 0.1) = -0.027$$

Consequently, the frequency of this gene (q) drops by 0.027 to 0.073 (1 − 0.027). The frequency (p) of the other gene B_2 rises from 0.9 to 0.927.

Note that the effectiveness of selection, that is, the size of Δq, depends on both s and q (Figure 5-14). The change in gene frequencies is fastest when both alleles are at intermediate frequencies. Selection against the less fit allele goes slowly when it is rare and especially so if it is recessive. Likewise, selection for a newly arisen more fit allele would go slowly at first and fastest when it had reached an intermediate frequency.

When selection is occurring, the individuals that die or fail to reproduce are said to have suffered **genetic death,** and this proportion of the population is termed the **genetic load** of the population. The genetic load will not necessarily result in

* The equivalent approximation for complete dominance is $\Delta q = \pm sq^2(1 - q)$.

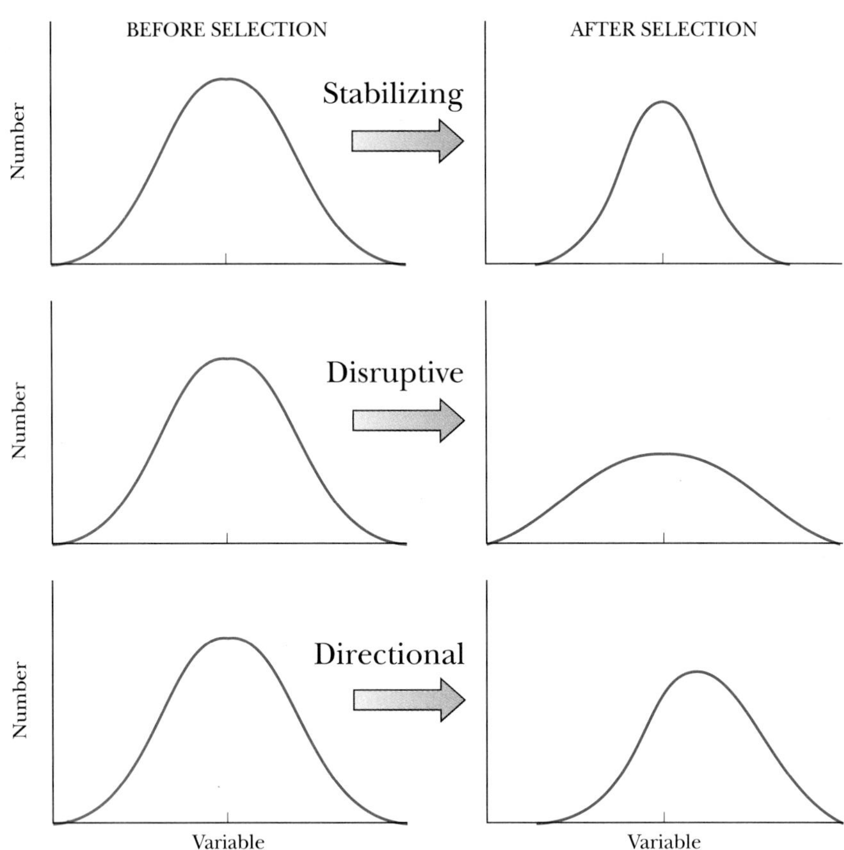

Figure 5-15

Types of selection. These diagrams suggest the outcome of stabilizing, disruptive, and directional selection on some trait (variable)—for example, beak length. Numbers of individuals showing each value are plotted before and after selection. Disruptive selection, if extreme, can produce a saddle-shaped curve or even two disjunct curves.

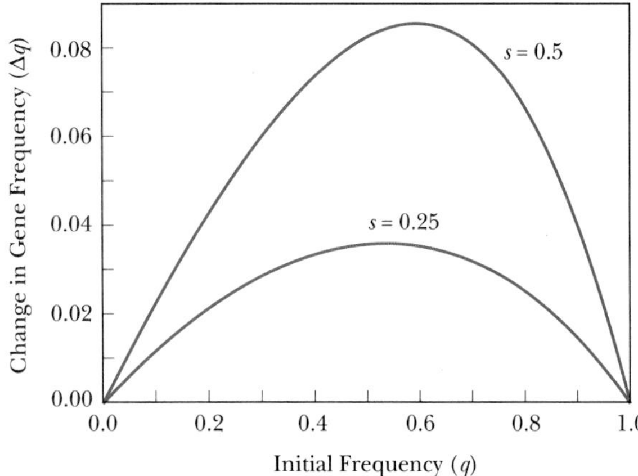

Figure 5-14

Change in gene frequency (Δq) resulting from one generation of selection against a gene (no dominance) with various initial frequencies (q) from 0 to 1 and with two selection intensities, s = 0.25 and 0.5. Calculated from the formula $-(\frac{1}{2} sq\,[1 - q])/(1 - sq)$ (Falconer 1989).

any lowering of density because most populations produce a surplus of offspring every year. The deaths as a result of selection can occur from a variety of immediate causes, such as predation and starvation; they are more a matter of, Who? rather than, How many?

Types of Selection

After each period of reproduction, the range of phenotypes in a population tends to broaden as a result of genetic recombination. Natural selection can work on these phenotypes and the underlying genes in three general ways: stabilizing selection, disruptive selection, and directional selection (Figure 5-15).

In **stabilizing selection,** the extremes of the distribution don't survive as long or reproduce as well as the intermediate values; intermediate individuals have higher fitnesses. The overall result is to keep the population about the way it has been. For example, young and old threespine sticklebacks, small fish of western North American lakes, have the same average number of lateral bony plates, but the variance is high in the young fish and low in the older (Hagen and Gilbertson 1973). The selection is due to predators taking more fish with five, six, eight, or nine plates, but why the seven-plated fish escape predation more often is not at all clear (Moodie et al. 1973). In stable environments, stabilizing selection must be a common mode.

In disruptive selection, the ends of the curve do better than the middle. One way that this could happen would be if predators tend mostly to take the common forms, perhaps because the predators are less likely to recognize the rarer forms as prey. Such selection, if strong enough, could tend to divide the population in two; at lesser levels, it may simply maintain the curve for the trait as lower and flatter than it would be under stabilizing selection.

Usually when we talk of natural selection, we are thinking of directional succession. Here, the individuals at one end of the curve are fitter than individuals at the other end, and so the average value of the trait tends to move in a particular direction.

Suppose that some trait of an organism is not perfectly adapted to the environment. Suppose, for example, that a bunch of insects are carried by a storm to an oceanic island. Although the wings are what got the insects to the island, they are now a disadvantage because they predispose the insects to being blown out over the ocean where they generally drown rather than reaching some other land. Perhaps the fittest condition would be winglessness. How quickly might this evolve?

To begin with, if every individual has the same genes for wings, there will be no selection since there is no variation for selection to work on. It makes general sense that the more variability there is, the faster the response is likely to be. R. A. Fisher's (1930) fundamental theorem of natural selection states that the rate of increase in fitness of a population is equal to its genetic variance in fitness at that time. There is, however, both more and less to this statement than meets the eye (see, for example, Falconer 1989 and Frank and Slatkin 1992), and we will leave the matter at the intuitive level that the rate of natural selection will be proportional to genetic variability.

It is worth remembering that the phenotype of an organism consists of many traits, many of which have a complex rather than a simple genetic basis. It is no settled question whether selection tends to produce traits with individually optimal values or an optimal organism in which some or many of the traits are compromises (Levins and Lewontin 1985, Travis 1989).

We may also wonder whether selective deaths are daily events or whether there are benign weeks, months, or even years when the culling effect of selection is slight. It is a question that Darwin mentioned in the marvellous last paragraph of Chapter 3 of *On the Origin of Species.* "Each [organism] at some period of its life, during some season of the year, during each generation or at intervals, has to struggle for life and to suffer great destruction. When we reflect on this struggle, we may console ourselves with the full belief, that the war of nature is not incessant, that no fear is felt, that death is generally prompt, and that the vigorous, the healthy, and the happy survive and multiply."

Whether selection occurs only in rare crunches, once a generation perhaps, or more frequently probably varies with organism and situation; nevertheless, currently accumulating evidence suggests that selective events tend to be more, rather than less, frequent (for example, Martin 1987, Lindén et al. 1992).

The Evolution of Life History Traits

r *and* K *Selection*

Different traits are advantageous in different situations—wet versus dry, cold versus hot. It occurred to Robert MacArthur and E. O. Wilson (1967) that different traits are also advantageous in crowded versus uncrowded conditions. In stable habitats, populations will spend most of their time at or near K, the carrying capacity (Figure 5-16). Examples of stable habitats might be climax forests, caves, and the ocean depths. MacArthur and Wilson called such environments K selecting. Species that range over large areas also tend to live under stable conditions, and, thus, to be K selected, even though individual portions of the landscape over which they roam may fluctuate.

Populations would rarely be at K in unstable—temporary or highly variable—habitats. When conditions were very favorable, they would be growing toward *K*. The biotic potential (r) would be fully realized, or close to it, and so MacArthur and Wilson called such conditions r selecting. For many species in such environments, however, this is just one side of the coin. When conditions shift to being unfavorable, or, in other words, when *K* drops, they may be overcrowded at the same population size (Brewer and Swander 1977). So r selection is a more complex phenomenon than *K* selection. Some examples of r-selecting environments might be pioneer communities, river sandbars, openings in forest, temporary ponds, carrion, and small marshes.

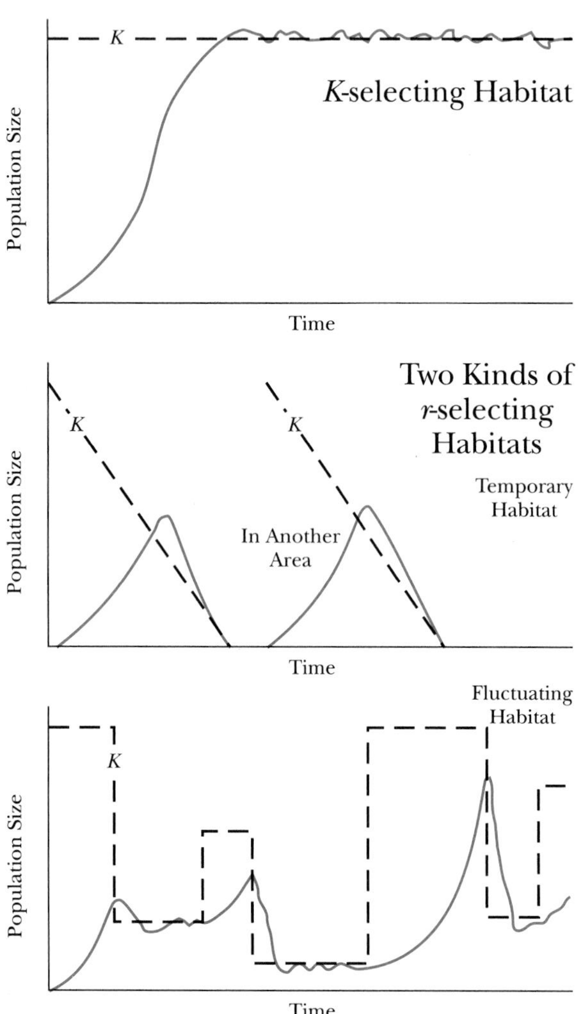

Figure 5-16

K-selecting and r-selecting habitats. In stable habitats, where *K* varies little, organisms tend to live under relatively crowded conditions in which their biotic potential is not fully expressed. Organisms of r-selecting habitats (fluctuating or temporary) are more likely to encounter uncrowded conditions at some time.

Under *K*-selecting conditions, there is a premium on getting a piece of land and holding onto it as long as possible. Intraspecific competition for resources such as space, water, sunlight, or nutrients is likely to be an important factor in shaping the population. Under r-selecting conditions, getting a piece of land is rarely a problem, though getting *to* it may be, and it may be pointless to hold onto it for very long because it may soon become unsuitable. Intraspecific competition for resources is variably important in shaping these populations.

Selection of Low Reproductive Rate. What traits would be generally advantageous in these two contrasting situations? The most obvious is reproductive rate. Populations characteristic of r-selecting environments often have an opportunity to grow (Figure 5-16), but populations of *K*-selecting environments don't. Consequently, genotypes that raise reproductive rate in ways such as we discussed in Chapter 4 will tend to be selected in r-selecting habitats.

Will, contrariwise, low reproductive rates evolve in *K*-selecting habitats? Evidently so; but note that this goes against the traditional view of natural selection in which biotic potential (r) is maximized. In other words, we ordinarily think that if any population has two genotypes differing in reproductive rate, the one that reproduces faster will increase relative to the slower one. It must generally be true that genotypes favoring fewer eggs, later puberty, and the like lose out in natural selection. However, there is indirect evidence that evolution sometimes produces lower reproductive rates. For example, certain very large birds, such as the California condor and various seabirds, lay only one egg per year (or sometimes only one every other year) and require several years to become sexually mature. Unless we believe that the earliest birds had a single-egg clutch and deferred maturity, and that the condors, almost alone among birds, retained the combination, then we must conclude that they evolved from birds having a higher reproductive rate.

How could lower reproductive rates evolve? If one bird produces six eggs a year and another produces two, won't the genotype for six come to predominate in a population? Two different approaches have been taken to explain the evolution of low reproductive rates. V. C. Wynne–Edwards suggested that group selection was at work (1962). He reasoned this way: Suppose we have two populations of the same species that differ only in that one has a genetic factor that keys its number to *K*, whereas the other reproduces maximally. There is a much greater chance, Wynne–Edwards said, that a year will come in which the high-*r* population, through imprudent overreproduction, will overwhelm its resources and die out. Into that ecological vacuum would step dispersers from a low-*r* population. In this way, primarily by extinction of local populations, would prudential restraint on breeding rate evolve.

Most recent students have thought that low reproductive rates could evolve by individual selection—by *K* selection—as follows: To begin with, it is not production of young that determines evolutionary fitness, but recruitment of young into the next generation. *K* selection works as shown in Figure 5-17, in which the slower-breeding genotype is increasing.

The crux of *K* selection is better recruitment rate of the genotype with the lower *r*. This could be achieved by the low-*r* organisms using energy, resources, or time, which they would otherwise use in producing more young, for instead improving the survival of the few young they do produce. They could, for example, produce a few seeds with much stored food instead of many with a little. They could lavish parental care on the few babies they give birth to rather than producing a lot and letting them fend for themselves. They could grow armor for predator protection.

It is worth reflecting that *K* selection is not inevitable. If the slow-breeding allele in our example (Figure 5-17) never arose, the high-*r* genotypes might do perfectly well competing among themselves. The population would persist in the habitat, very possibly at the same density but with a higher birth rate and higher death and emigration rates.

Most ecologists are satisfied that the way just described—trade-offs favoring low reproductive rates in certain situations—is the usual route of *K* selection, but this does not mean that group selection never plays a role. In a later section, we will explore the idea a little further.

Characteristics of* r- *and* K-*Selected Organisms. Figure 5-18 presents in simplified fashion some of the major pathways involved in *r* and *K* selection. An important conclusion to be drawn concerning the two processes is that once a species is started along one road or the other, the process tends to be self-reinforcing (Horn 1978). A fluctuating environment leads to higher reproductive rates, which lead to smaller offspring, leading generally to smaller adults, leading to a greater susceptibility to death from environmental fluctuations, leading back to an enhanced tendency for populations to fluctuate.

Table 5-3 summarizes some traits that tend to be opposed in *r*-versus-*K* selection. The last entry deserves some additional comment. Clearly, species of most temporary or fluctuating habitats need good dispersal ability. If we did not mow our lawns, dandelions would be shaded out by taller herbs and trees in a few years; their windblown seeds are probably adapted to traveling far to other lawnlike places. But low dispersal ability is not advantageous to*K*-selected species either (Hamilton and May 1977). It is true that, for adults, the spot where they live now is likely still to be present and suitable next year. But the young, which do most of the dispersing anyway, are going to need to find a spot, and it does neither adult nor young any good to be competing among themselves.

If young have to compete for a spot—and by and large in *K*-selected species, they will—they ought to compete with somebody other than their parents. This is partly because it is bad for your genetic fitness to have your offspring competing with you, but also because if the offspring move, they have at least some chance of competing with another young individual, which they may beat, rather than with an adult. So, *K*-selected species cannot dispense with dispersal. Just how dispersal patterns differ betwen *r*- and *K*-selected organisms is a topic worthy of further research.

Many of the predictions of *r* and *K* selection are borne out in a comparison of life history traits of birds of the eastern deciduous forest region (Brewer and Swander 1977). Presumably the most stable community is climax (mesic) forest, and the most variable communities are grassland and

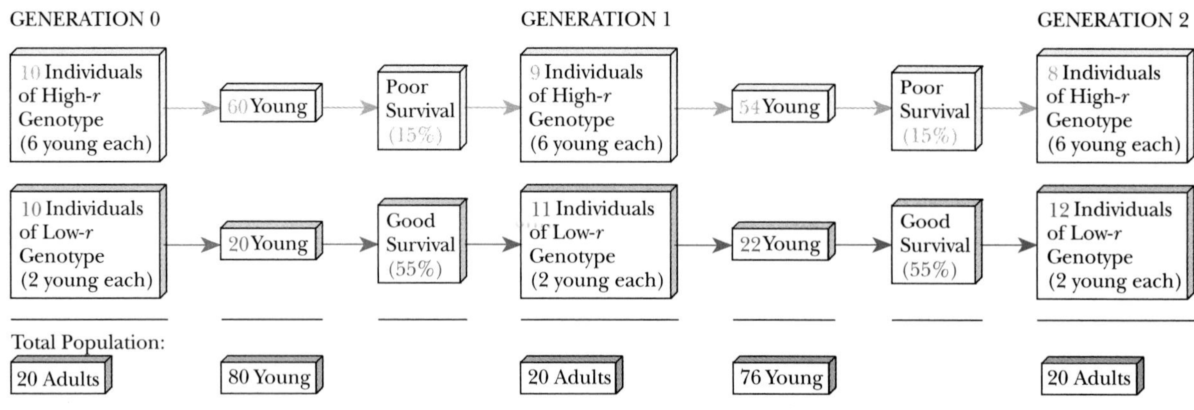

Figure 5-17

K selection. In this hypothetical population there are slow-breeding and fast-breeding genotypes. The slow-breeding genotype uses time and energy saved from lowering reproductive rate to improve the survival of its young (they are, perhaps, larger and more competitive). In the situation as diagrammed, more offspring from the slow-breeding than of the fast-breeding genotype are recruited into the next generation, so that the population evolves toward a lowered reproductive rate.

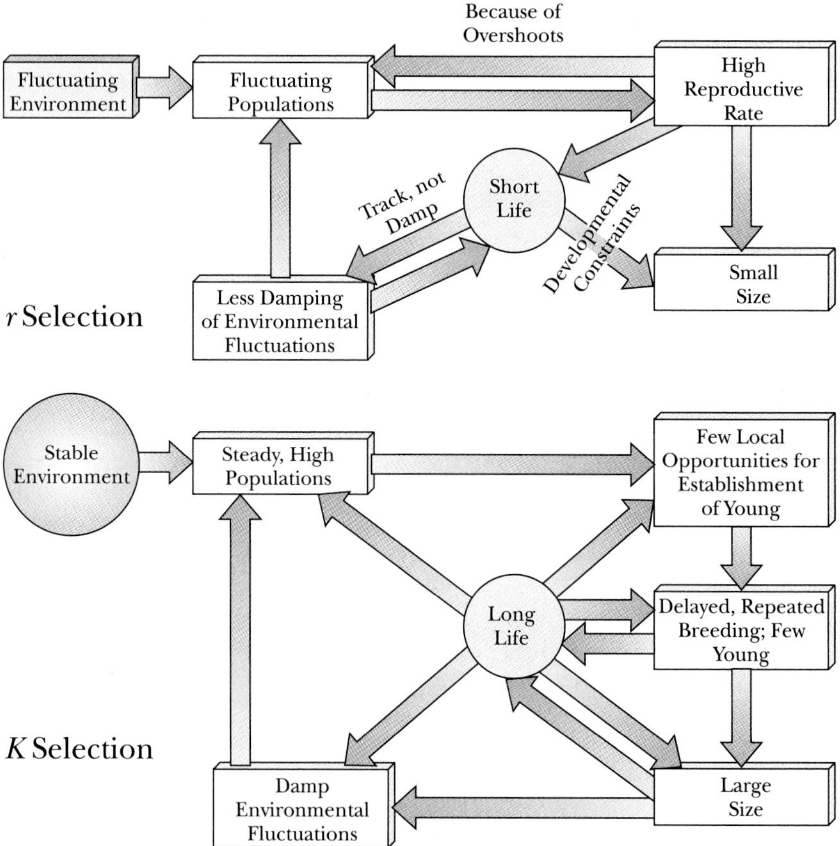

Figure 5-18

Some important pathways in *r* selection, above, and *K* selection, below. (Adapted from Horn 1978. Used by permission of Blackwell Scientific Publications Ltd.)

Table 5-3
Opposing Trends in* r*-versus* K*-Selected Species

Parenthetical comments are brief suggestions about the selective basis of the trait.

Traits Favored by r *Selection*	*Traits Favored by* K *Selection*
Early maturity (increases *r*)	Late maturity (avoids risks of breeding)
Many young (increases *r* and dispersal)	Few young (allows diversion of energy to other things)
Small young (so can make many)	Large young (increases their chances of survival)
Short life (conditions may be unfavorable soon)	Long life (good spots don't open up often, so it is advantageous to hang onto one)
Annual	Perennial
Less parental care (no time or energy)	More parental care (increases young's chances of survival)
Less competitive ability (conditions often uncrowded)	More competitive ability (conditions nearly always crowded)
Very good dispersal ability (to get to favorable patches from unfavorable ones)	Good dispersal ability (to avoid competing or breeding with relatives)

marsh. Overall clutch size for open-nesting passerine birds in the region centers on four but mesic deciduous forest species more often lay only three eggs (Table 5-4). In grassland-marsh, clutches of five or more eggs are as common as those of four. Also, most forest birds lay only one clutch per year, whereas many grassland-marsh birds regularly lay two or more (Table 5-5).

Annuals and Perennials. Some organisms are **annuals,** and some are **perennials;** that is, some grow up and reproduce in a year and then die, whereas others live long and reproduce often. Two not-very-melodious terms, **semelparous** (meaning breeding once) and **iteroparous** (meaning breeding again), have been used to mean about the same thing (Cole 1954). The distinction is that semelparous organisms are not necessarily annuals. Such plants as yuccas and bamboos and such animals as 17-year cicadas and certain salmon live for years before they finally breed and die.

Our first thought might be that a perennial ought to have a much higher *r* than an otherwise similar annual. Further reflection might make us uncertain; being a perennial would increase net reproductive rate (R_0), certainly, but it would also increase generation length (T). Without more information, it is difficult to estimate the overall effect on the expression

$$r = (\ln R_0)/T$$

Table 5-4
Percentage of Species With Different Clutch Sizes Among Open-Nesting Passerine Birds in the Eastern Deciduous Forest Region

		Number of Eggs		
Community	*N**	*1–3*	*4*	*5 or More*
Mesic forest	7	57%	29%	14%
Other forest	8	0	87.5	12.5
Grassland-marsh	12	8	46	46
Forest edge	20	15	70	15

* Number of species.

Data from Brewer, R., and Swander, L., "Life history factors affecting the intrinsic rate of natural increase of birds of the temperate deciduous forest biome." *Wilson Bulletin,* 81:211–227, 1977.

In his important paper on the role of life history factors in setting *r*, L. C. Cole (1954) concluded that the maximum increase in *r* that a switch from semelparity to iteroparity could achieve is exactly the same as would be achieved if the organism were to remain an annual but produce one more offspring in its single reproduction.

The truth of the proposition under a certain set of conditions is easy to demonstrate, as seen in Figure 5-19. The originally annual organism (at the top) is represented in generation 2 by four individ-

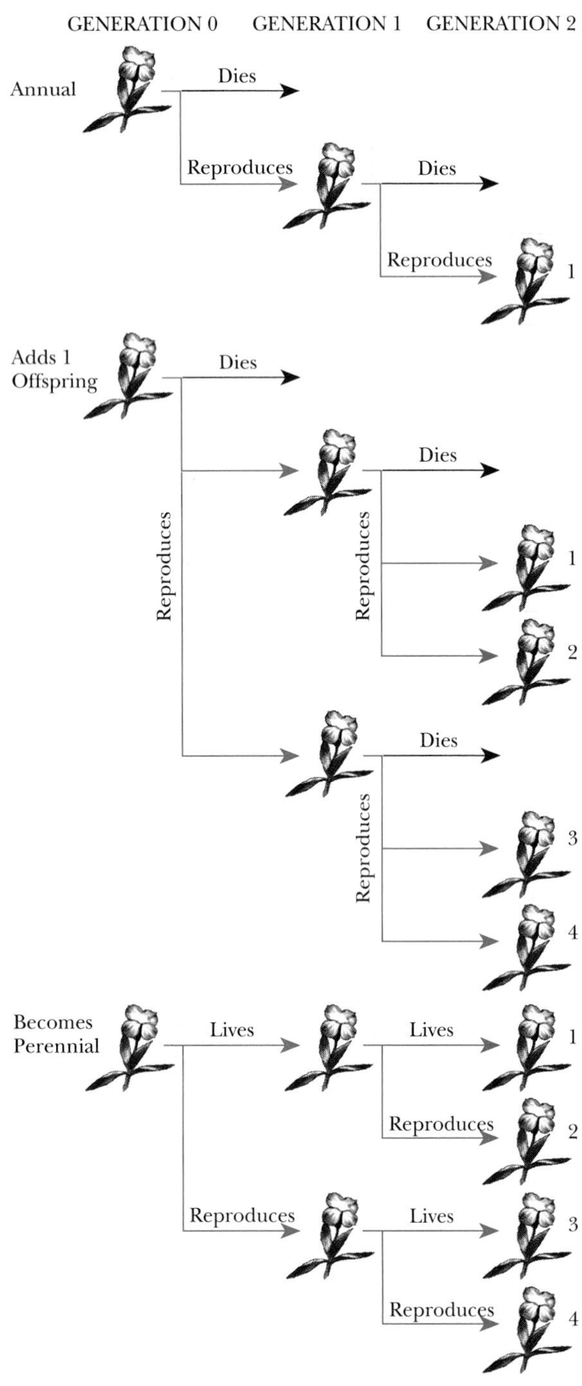

Figure 5-19

Under certain circumstances, a change from an annual to a perennial habit adds no more to biotic potential than a change that adds 1 more seed, egg, or baby.

Table 5-5
Percentage of Bird Species That are Single Brooded and Multiple Brooded in Communities of the Eastern Deciduous Forest Region

		Number of Broods per Breeding Season	
Community	*N**	*1, Sometimes 2*	*Regularly 2 or More*
Mesic forest	11	82%	18%
Other forest	10	90	10
Grassland-marsh	16	50	50
Forest edge	23	43	57

* Number of species.

Data from Brewer, R., and Swander, L., "Life history factors affecting the intrinsic rate of natural increase of birds of the temperate deciduous forest biome," *Wilson Bulletin,* 81:211–227, 1977.

uals whether it becomes perennial or adds one offspring to its first reproduction.

Can this conclusion be correct? Can it really be true that a frog with a life span of a year would increase its biotic potential as much by laying 201 eggs, rather than 200, as it would by living to an old age and laying 200 every year? The answer is, only under certain circumstances.

Suppose a population has an annual and a perennial genotype equally represented. Under what circumstances will this representation be perpetuated, and under what circumstances will one genotype or the other increase? For the sake of simplicity, let's deal with plants, although the same arguments should apply to animals. The numbers per annual individual that will be around next year just prior to flowering (Charnov and Schaffer 1973) will be

$$\left(\text{Number of annual's offspring} \times \text{their survival rate}\right)$$

The numbers per perennial individual will also be given by this expression; however, because it is a perennial, there will also be older individuals still living. Consequently, the total representation (per individual) just before flowering will be

(Number of perennial's offspring × their survival rate) + (survival rate of adults)

Clearly, then, for the two genotypes to remain at the same level in the population

$$\left(\begin{matrix}\text{Number of}\\ \text{annual offspring}\end{matrix} \times \begin{matrix}\text{annual offspring}\\ \text{survival rate}\end{matrix}\right) = \left(\begin{matrix}\text{number of}\\ \text{perennial offspring}\end{matrix} \times \begin{matrix}\text{perennial offspring}\\ \text{survival rate}\end{matrix}\right) + (\text{survival rate of perennial adult})$$

If this is an inequality, then one genotype or the other will be increasing.

Let's insert some numbers to see the behavior of this model. Suppose that

Number of annual offspring = 1.0
Number of perennial offspring = 1.0
Annual offspring survival rate = 1.0
Perennial offspring survival rate = 1.0
Survival rate of perennial adult = 1.0

Then

$$10 \times 1 < (10 \times 1) + 1$$
$$10 < 11$$

and the perennial genotype should be favored. However, if we simply add 1 to the annual offspring number, then

$$11 \times 1 = (10 \times 1) + 1$$
$$11 = 11$$

which is Cole's result that we obtained earlier in Figure 5-19.

These figures are, however, not realistic. Suppose we alter the survival figures to 0.5 for both annual and perennial young. The 1.0 survival rate for the adult perennial is not realistic either, but a large tree often lives a hundred years after reaching reproductive age, giving an annual survival rate of 0.99, which is pretty close to 1.0. With the altered survival figures for the offspring, we have

$$10 \times 0.5 < (10 \times 0.5) + 1$$
$$5 < 6$$

Now adding 1 to the annual's offspring production no longer makes up the difference, though it comes close:

$$11 \times 0.5 < (10 \times 0.5) + 1$$
$$5.5 < 6$$

Realistically, though, the perennial is unlikely to produce as many seeds as the annual, because for the adults to survive, energy that could otherwise be turned into seeds will be stored in tubers or some other organ. This particular trade-off was noted by Aristotle (as Haukijoa and Hakala [1979] pointed out): "[Annuals] consume all their nutriments to make seed, their kind being too prolific." And there is a great deal of data showing that semelparous plants devote more energy to reproduction—usually two or three times as much—than iteroparous relatives (Young 1990). So let's change seed production to 20 for the annual.

$$20 \times 0.5 > (10 \times 0.5) + 1$$
$$10 > 6$$

The annual genotype is clearly favored.

Under just what reasonable circumstances, then, would it be advantageous to be perennial? The remaining feature that can vary between the two genotypes is survival of the offspring. This feature can vary enormously in both genotypes. Consider perennials. Although they produce few seeds and these are generally well supplied with food, there may be few spots for them to grow because adults are already occupying the space.

Annuals produce many seeds that grow into seedlings without competition with adults of their own species but may encounter many other vicissitudes. If we say that offspring survival in perennial and annual is equal, we can rewrite the original model, which was

$$\left(\begin{matrix}\text{Number of}\\ \text{annual offspring}\end{matrix} \times \begin{matrix}\text{offspring}\\ \text{survival rate}\end{matrix}\right) = \left(\begin{matrix}\text{number of}\\ \text{perennial offspring}\end{matrix} \times \begin{matrix}\text{perennial offspring}\\ \text{survival rate}\end{matrix}\right) + (\text{survival rate of perennial adult})$$

(after dividing by offspring survival rate) as

$$\text{Number of annual offspring} = \begin{matrix}\text{Number of}\\ \text{perennial offspring}\end{matrix} + \frac{\text{survival rate of perennial adult}}{\text{survival rate of offspring}}$$

Because the number of offspring will generally be greater in the annual, the perennial habit will be favored only when this advantage is offset by considerably lower adult compared with juvenile mortality. Using 20 annual offspring, 10 perennial offspring, and 100% adult survival, juvenile survival has to drop below 10% before the perennial habit is favored:

$$20 = 10 + (1/0.1)$$
$$20 = 20$$

How might a switch from one condition to the other come about? Suppose the climate in which a perennial herb lives becomes steadily more seasonal with, say, colder and drier winters. The percentage of adults surviving the winter may drop, lowering the ratio

Adult survival rate/juvenile survival rate

The point may come where individual plants that put all their energy into producing seeds and none into surviving to the next year will be favored.

How, then, are we to rate annuals versus perennials in terms of r and K selection? The answer is clearly more complicated than for more versus fewer offspring or earlier versus later breeding. The crux, however, is high adult survival combined with low juvenile survival (Charnov and Schaffer 1973, Young 1990), a situation that is favored by a stable environment. So, K selection will tend to produce perennials, and r selection, annuals.

This is clearly not the whole story, however. One significant additional factor not included in the model is great year-to-year variation in juvenile survival (Murphy 1968). In terms of human clichés, annuals put all their eggs in one basket whereas perennials spread the risk; they hedge their bets. Circumstances in which complete reproductive failure can occur occasionally are circumstances in which only perennials (or annuals with a seed bank) can be permanent inhabitants. But this situation, if based on environmental fluctuation, is one we would ordinarily consider r selecting.

Salmon and Century Plants: Long-Lived Semelparous Organisms. Semelparous but nonannual organisms—ones that live for years, reproduce once, and then die—seem to have the worst of both worlds. Because of delayed maturity, they combine a low r with the annual's failure to spread its reproductive risk over a series of years. There are, nevertheless, several spectacular examples of this life style, including bamboos and century plants, squid, cicadas, and Pacific salmon.

It may be that no single evolutionary explanation will suffice to explain this life history pattern (Young and Augspurger 1991). Contrasting it with an annual life history is, for many of these species, asking the wrong question because their relatives, though iteroparous, also have delayed maturity. The right question is the same as that asked earlier for annuals and perennials, Why reproduce more than once? Some semelparous organisms clearly live in circumstances where conditions suitable for reproduction come infrequently and mortality is high between reproductive episodes (Young 1990).

For some long-lived semelparous organisms, additional considerations may be involved, one being escape (in time) from predators. For example, the herbivores that have been making a living off a grove of one of these semelparous species may perish or emigrate following the old plants' death, allowing better survival of the young plants.

Pioneer Versus Climax Species. A good illustration of r-selected versus K-selected species is seen in a comparison of pioneer and climax plant species. In the earliest stage of old-field succession, which lasts only about a year in the midwestern United States, plants are widely spaced, suggesting that the area is not crowded. They are all herbs and mostly annuals that produce a large number of seeds, and many of the plants are wind or animal dispersed.

The dominant plants of the climax forest are not herbs but trees. They are large, live for tens or hundreds of years, and reproduce repeatedly. Nearly all herbs that are a part of the climax forest are also long lived. Most of the climax forest herbs produce relatively few seeds, but they reproduce readily by rhizomes and other vegetative means, so that when they reach a spot, they are able to spread over and dominate it.

Several of the most important pioneer plant species do not show any obvious means of rapid, distant dispersal. At first this is puzzling, but further study shows that these species have an alternative way of attaining the same effect. Their seeds are very long lived; annual ragweed seeds may live for 40 years, evening primrose seeds for 80 years, and moth mullein for 90 years. In the course of time, with an occasional ragweed plant growing here or there and producing seeds, seeds become widespread in the soil. When an area is disturbed—a forest cut over, for example—ragweed does not need to disperse to the area. It is already there in the form of seeds that now find favorable condi-

Table 5-6
Seed Bank of a 5-Year-Old Clear-Cut and a 95-Year-Old Northern Hardwood Stand in the White Mountains, New Hampshire*

	Number of Seeds per m²	
Species	*Young Clear-Cut*	*Hardwood Forest*
Herbs	407	13
Raspberry	1016	68
Blackberry	31	21
Red-berried elder	42	3
Other shrubs	14	15
Large-toothed aspen	13	8
Yellow birch	181	961
Paper birch	87	445
Pin cherry	2	52
Other trees	4	26

If 95-year-old hardwood forest is clear-cut, what will the vegetation be like after 5 years? It will have a lot of raspberry and blackberry (though these may already be declining). The tree composition of the 5-year-old clear-cut may look like the table below.† The paper birch, pin cherry, and perhaps, yellow birch are from the seed bank. The aspen is mostly from root sprouts but also from seed dispersal. The other species are largely from "advance reproduction," that is, young trees that were present before cutting.

Species	*Trees ≥ 1 ft High (per m²)*
Beech	1.6
Sugar maple	2.4
Large-toothed and trembling aspen	0.2
Yellow birch	0.7
Paper birch	0.8
Pin cherry	2.0
Other trees	0.5

* From Graber, R. E., and Thompson, D. F., "Seeds in the organic layers and soil of four-beech-birch-maple stands," *U.S. Forest Service Research Paper,* NE-40j1:1–8, 1978.

† From Marquis, D. A., "Clearcutting in northern hardwoods: Results after 30 years," *U.S. Forest Service Research Paper,* NE-85:1–13, 1967.

tions for germination. The viable seeds contained in the soil and litter form the **seed bank** (Table 5-6). This pattern of pioneering has been termed the **buried seed strategy** (Marks 1974).

Soil seed banks are one way of being there first. Another is canopy seed banks (serotiny). This strategy of storing seeds in the canopy is nearly confined to woody plants of fire-prone regions. Seeds are retained in fruits on the plants and released slowly or not at all until a fire or death of the plant. Few seedlings from any interfire dispersal survive a fire, and the post-fire generation consists of individuals that grow from the seeds of the serotinous fruits (Lamont et al. 1991).

Worldwide, 40 genera of plants show canopy seed storage. The habit is best represented in Australia and Africa. In North America and Eurasia, canopy seed storage seems restricted to conifers, the best-known examples being certain of the pines.

These two alternative pioneer patterns—getting there fast and being there first—are not restricted to plants. Temporary pools on rock outcrops in Africa are habitats for fly larvae. The pools hold water for a few hours to a few weeks. The pool-colonizing midge is the typical pioneer or fugitive species. Females find new pools rapidly and lay eggs that give rise to adults, ready to fly off and colonize other new pools, in two or three weeks (McLachlan and Cantrell 1980).

Another species, Vanderplank's midge, may take two weeks to reach a new pool (of course by chance, it may colonize some new pools immediately), but it has the ability to withstand the pool drying up completely. Such drying kills the larvae of the pool-colonizing midge. Vanderplank's midge may go through several cycles of rain and drying before it finally emerges as an adult. Each time the pool is refilled, the larvae rehydrate and become active again, and each time, they are there occupying the pool and discouraging (apparently by chemical means) the invasion or growth of other species. By the time these larvae mature, newer eggs of the same species may have been laid in the pool.

Both of these species occupy the same temporary habitat, rain pools, but the pool-colonizing midge tends to dominate larger pools that, once filled, last three or four weeks. Vanderplank's midge tends to occupy pools that hold water a shorter length of time.

Beyond the r *and* K *Selection Model*

The idea of *r* and *K* selection is a powerful way of looking at the world, but it is, nevertheless, an over-

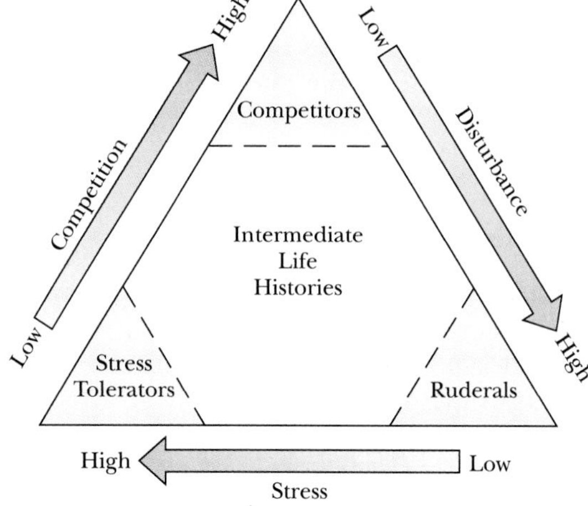

Figure 5-20

Expected plant life histories based on a model in which stress, disturbance, and competition are the important selective factors. (Based on Grimes 1979.)

COMPETITORS	RUDERALS	STRESS TOLERATORS
•Herbs, shrubs, or trees •Large, with a fast potential growth rate •Short to long life span •Reproduction at a relatively early age •Small proportion of production to seeds •Seed bank sometimes; vegetative spread often important	•Herbs, usually annual •High potential growth rate •Reproduction at an early age •Large proportion of production to seeds •Seed bank and/or highly vagile seeds	•Lichens, herbs, shrubs, or trees •Usually evergreen •Potential growth rate slow •Reproduction at a relatively late age •Small proportion of production to seeds •Vegetative spread important

simplification. Adaptation to fluctuating environments is particularly complex because fluctuations come in all types, sizes, and frequencies (Stearns 1976: Table 6).

The English plant ecologist J. P. Grime (1979) has suggested that stress forms another dimension to which adaptation in life history is important. Stress consists of external factors that limit productivity, such as low temperature, aridity, low mineral nutrient levels, and for understory plants, heavy shade. In this view, then, habitats that are cold, dry, or otherwise deficient tend to select for **stress tolerators,** plants that grow slowly, live a long time, and reproduce only when they can afford to. Habitats that suffer frequent disturbance select for **ruderals** (botanical jargon for weeds). These are the classical *r*-selected fast-growing, heavy-reproducing annuals. Habitats that are neither particularly stressful nor frequently disturbed are occupied by **competitors,** large-sized, relatively long-lived plants that occupy a lot of space both above and below ground. These are *K*-selected species in the *r-K* classification but, clearly, so also are the stress tolerators. Habitats that are intermediate as to disturbance or stress would select for intermediate life histories.

Although developed for plants, the classification can also be applied with some justification to animals. Many desert animals show stress tolerator life histories involving such features as long lives, inactivity during the unfavorable parts of the year (aestivation), and intermittent reproduction.

A graphical view of the stress-disturbance-competition model is given in Figure 5-20. The triangular representation is valuable because it emphasizes the idea that differing life history strategies generally require trade-offs. A one-year-old bird can put its energy either into attempting to reproduce or into keeping itself alive until it is two years old. Predation may lead to the evolution in the prey of armor or speed, but not both in the same species.

For a tree to be long lived, it needs thick bark, defensive chemicals, and dense wood to withstand fire, insects and fungi, and wind. Achieving these things, however, generally means sacrificing fast growth in height and early reproduction. These general predictions were fulfilled in a test involving North American hardwoods (Loehle 1987). For example, the slow-growing hardwoods, such as hickories and oaks, tend to live just under 200 years as a typical value, whereas the fast-growing ones, such as box elder and cottonwood, tend to die at well below 100 years.

As logical as all this is, trade-offs are sometimes hard to find in specific cases (Mueller et al. 1991). In the analysis of North American trees (Loehle 1987), significant trade-offs were scarce among the conifers. One reason for the failure to detect trade-offs is that one big modification may be achieved, not as the result of one big loss, but as the result of small subtractions from a number of traits. Another reason is that trade-offs need not be one-to-one (Loehle 1988). Energy, time, or other resources used to achieve one advantage may serve multiple purposes. Spines on cactus protect the plant from herbivore damage, but they also shade it, reducing heat stress. A bug gaining the capacity to eat a chemically protected plant may lose some digestive efficiency but it may gain protection from predators.

Grime's triangle model is oversimplified (Loehle 1988). In the last analysis, each species has a unique combination of traits—life history and others—that fit it to its particular habitat and to the particular position it occupies in that habitat. Most species show a combination of traits, some that we might call *r* or *K* selected, some that we might call stress tolerating or competitive, and some that we cannot readily categorize.

These models are most useful when comparing related organisms. Insects of climax forests may be *K* selected compared with insects of crop fields, but most will be *r* selected compared with forest birds. Partly, this is another way of saying that habitats can be unstable or stressful for some organisms but stable or nonstressful for others. This is a feature of the interaction of the organism with the habitat.

As an example, cleavers is an annual herb that thrives in climax deciduous forests. It does so, in part, by germinating in the fall when other herbs have died and growing whenever it can in the fall and winter so that it is ready to make rapid growth in spring as soon as conditions become favorable. It produces copious seeds that are spread far and wide by being carried on the fur of mammals. Despite being a member of the climax plant community, it is as *r* selected as you could wish, and, indeed, it occupies a temporary habitat—bare sunny ground in deciduous forest.

We need also to remember that most species show some degree of phenotypic plasticity in life history traits, just as in morphology and physiology. Bay-breasted warblers, characteristic birds of the boreal coniferous forest, lay bigger clutches (mostly six eggs instead of mostly five) in years of spruce budworm outbreaks than in nonbudworm years (MacArthur 1958). A small annual knotweed of the higher elevations of the Cascade Range in Oregon devoted nearly 60% of its biomass to seeds in harsh environments and only 40% in milder ones (Hickman 1975).

We might suppose that plasticity itself is a trait that is advantageous in unpredictable environments. There may be something to this, but it is not only *r*-selected species that show such flexibility. Various colonial birds that ordinarily defer breeding until they are several years old breed earlier when densities are low, and the eagle breeds at a younger age than usual when many older birds are shot. Lack (1968: 298) gives other examples of the same phenomenon.

Brewer and Swander (1977) found little evidence that birds of mesic forest, with their small clutches (Table 5-4), had extra energy to divert to activities useful in stable, competitive environments. Their eggs, for example, were not consistently larger (relative to body size) than those of open-country relatives. It is possible that a three-egg clutch is simply the biggest that they can raise in mesic forest. Similarly, it has yet to be shown that the single brood of forest birds (Table 5-5) reflects *K* selection rather than a more limited time over which food suitable for raising young is available relative to other habitats.

Semelparity, although clearly genetically programmed in some instances, has been shown to be a phenotypic response to environment in others. A northern leech that is semelparous in all known field populations, breeding at one or two years of age, was raised in the laboratory. In this environment, many individuals, especially larger and well-fed ones, survived one breeding period and re-entered reproductive condition (Baird et al. 1986).

It is possible, then, that some of the life history traits we have looked at in preceding sections are not adaptations to stability, or stress, or anything else. They may simply be the best that the organism is capable of in the prevailing circumstances (Järvinen 1986, Wootton et al. 1991).

Practical Aspects. Perhaps the ideas developed in preceding sections about the evolution of life history traits seem abstract and remote from any practical applications. This is far from the truth. Consider the question of endangered or threatened species. Suppose that something (such as a pesticide) makes some proportion of the population sterile or, at least, prevents successful reproduction. The results are very different in *r*-selected versus *K*-selected species. Table 5-7 compares an *r*-selected species (breeds at age one year, has a clutch of 5.6 eggs, and a mean annual mortality rate of 73.8%) with a *K*-selected species (breeds at age six years, 1.3 eggs, 50% mortality first year, 12.5% each succeeding year). The particular species used as models are the American robin and the bald eagle. If 90% of the population fails to reproduce, population size in the two species decreases, as shown in Table 5-7.

For the *r*-selected species the decrease in numbers will be detected quickly, almost certainly by the second year, and the potential exists for doing something about it. In the *K*-selected species, however, almost no successful reproduction has gone on for four years and the population is not yet halved. Consequently, *K*-selected species require much closer monitoring of their numbers than do *r*-selected species if we are to detect problems early enough to deal with them. Otherwise the *K*-selected population may become so top-heavy with old individuals that prospects for recovery are poor before we even realize a decline in population size is occurring.

On a more general level, it is primarily *r*-selected species that make suitable game animals. *K*-selected species rarely have the biotic potential to replace animals lost to hunting. Possibly the best example is the passenger pigeon, a forest species that laid one one-egg clutch per year. It was still fantastically abundant in the early 1800s, but by 1900 it was extinct in the wild.

Table 5-7
Population Decrease in* r- *and* K*-Selected Species

Year	*r-Selected Species*	*K-Selected Species*
0	1,000,000	1,000,000
1	270,526	801,984
2	73,185	702,976
3	19,799	616,345
4	6,669	540,544

From Young, Howard, "A consideration of insecticide effects on hypothetical avian populations," *Ecology*, 49:993, 1968.

The Evolution of Behavior

Heritability

Beginning biology courses encourage us to think of the genetic basis of eye color in humans or fruit flies and seed types in peas and corn. The courses often say little about the genetic basis of behavior, but most biology students would expect behavioral phenotypes, like morphological and physiological phenotypes, to be the result of the interaction between genotype and environment. Where studied, this has been true. Among birds, for example, a genetic component has been shown for such aspects of behavior as songs and calls, courtship displays, territoriality, and migratory behavior (Buckley 1987, Boag and van Noordwijk 1987).

This section tries to provide a framework for thinking about the genetic basis for behavior. Because the behavior of humans has intrinsic interest, we use as our example the Minnesota Study of Twins Reared Apart (Bouchard et al. 1990a,b). Another way of saying that phenotype depends on genotype and environment is to say that the total phenotypic variance (Vt) within a population is equal to the variance due to genetic differences (Vg) plus the variance due to environmental or experiential factors (including learning) (Ve), thus:

$$Vt = Vg + Ve$$

A useful term in quantitative genetics is **heritability,** defined as Vg/Vt. In words, heritability is the proportion of total phenotypic variance that can be attributed to genetic variance (Falconer 1989). A point worth making at the outset, and one to which

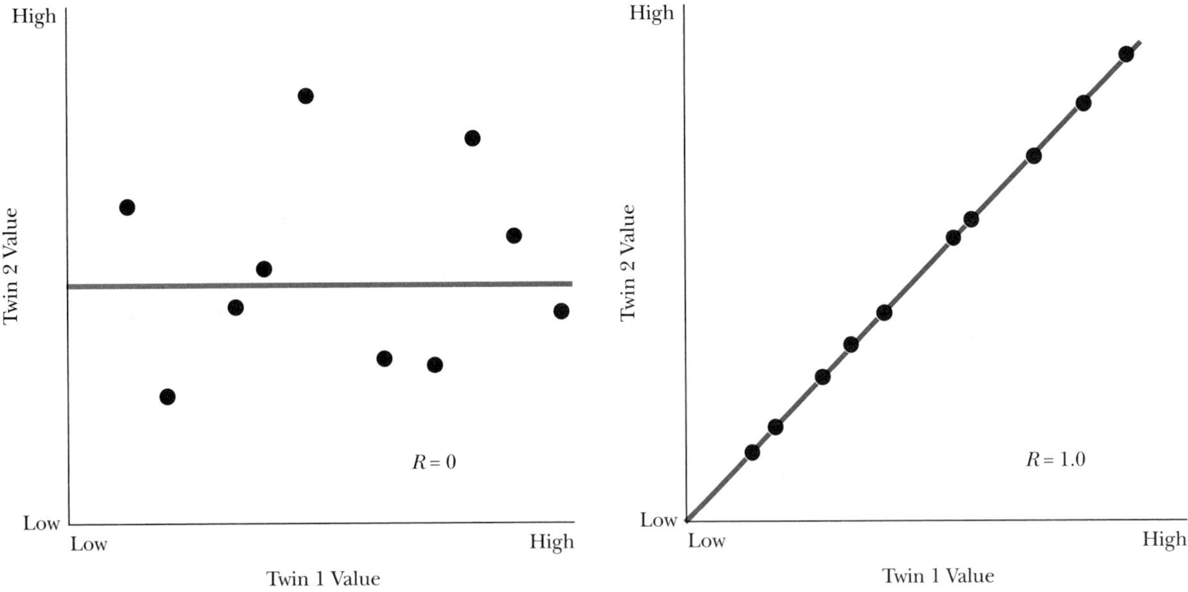

Figure 5-21

These graphs compare two possibilities for pairs of monozygous (identical twins) reared apart. If genes have nothing to do with the trait, the value of some trait shown by twin 2 may bear no relationship to the value of the trait shown by twin 1 ($R = 0$). If, on the other hand, the trait is wholly specified by genotype, then, the trait in one twin will have the same value as the trait in the other ($R = 1$).

we will return, is that measured heritability is affected by the particular range of environments available.

Monozygous (identical) twins are essentially identical genetically. If they are reared together, similarities in their behavior could be due to genetics or to their shared environment—the fact that they were treated alike, saw the same TV programs, ate the same food. If, however, they are reared apart, they could be much more different from one another, depending on how important the contribution of the post-womb environment is (note that any variance from differences in, say, position in the uterus would be included in the genetic variance).

What was studied was the coefficient of correlation (R) of twin compared with twin, illustrated in Figure 5-21. If heredity is unimportant, the value of some trait, IQ for example, in twin 2 should not be correlated with IQ in twin 1. If, on the other hand, there is a genetic component to IQ, then the IQ of one twin should correlate with the IQ of the other. Correlation coefficients can run from 0 to 1, being zero if the trait in twin 1 is random with respect to the trait in twin 2, and 1.0 if they are perfectly correlated (that is, all the values fall on one line). If statistically significant correlation exists, the spread of points around the line (that is, the additional variation not explained by genetics) can result from environmental factors or from measurement errors—for example, one twin not feeling very well on the day the IQ test was given.

Table 5-8 provides a sampling of results from the Minnesota twins study. A large value of R in the MZA (monozygotic twins reared apart) indicates high heritability for a trait. A large difference in R between MZA and MZT (monozygotic twins reared together) suggests a strong environmental or experiential component.

For a morphological trait, fingerprint ridge count, differing environments produced little or no difference; virtually all the variance is the result of genetic factors (Table 5-8). Weight is to some degree morphological and physiological and to some degree behavioral. The R value for MZA suggests that 73% of the total variance we see in adult weights has some sort of hereditary basis (see Bou-

Table 5-8
Correlations for Monozygotic Twins Reared Apart (MZA) and Monozygotic Twins Reared Together (MZT) for Different Variables

Sample size is number of pairs of twins. Based on Bouchard et al. (1990), who provide citations to the specific tests used.

	MZA		*MZT*	
Variable	*R*	*Sample Size*	*R*	*Sample Size*
Fingerprint ridge count	0.97	54	0.96	274
Weight	0.73	56	0.83	274
Systolic blood pressure	0.64	56	0.70	34
IQ	0.69	48	0.88	40
Religiosity	0.49	31	0.51	458
Occupational interest scale	0.43	40	0.49	376

chard et al. 1990b). This leaves, of course, an important 27% based on other factors.

IQ as measured by the Wechsler Adult Intelligence Scale showed a heritability of .69. Clearly this does not mean that intelligence is set at birth because this leaves 31% of the total observed variation to be explained by differences in diet, schools, how hard the individual works, etc.

The Minnesota twins study tells us something about the relationship between genetic and environmental variation in IQ and other traits in a particular subset of environments—the middle class of a 20th century industrialized society. It is likely that if some of these twins had been reared in extreme environments—the slums of a Third World country or wealthy yet intellectual homes—the environmental component of variability might be greater.

These data do not necessarily tell us what the effects of greatly improved environments might be on "low IQ" versus "high IQ" genotypes. This is the same general question we addressed in Chapter 2 in our first consideration of reaction norms. Figure 5-22 looks at the situation graphically. The shaded area in the middle is supposed to represent the range of environments included in the Minnesota twins study. We can imagine, as is shown in the top graph of Figure 5-22, that an improved environment helps genotype B greatly so that its IQ phenotype rises substantially. Genotype A, though higher than B in the shaded range of environment, shows little improvement and eventually drops below genotype B. We can also imagine, as in the lower graphs, that both genotypes are improved somewhat, or that low IQ genotypes are little improved whereas high IQ genotypes are greatly improved (Figure 5-22).

Although we have dealt here with the interplay between heredity and environment in a few aspects of human behavior, the same considerations are involved in the hunting behavior of coyotes, the song of birds, or the pollination behavior of bumblebees.

Sociobiology

E. O. Wilson, the Harvard population biologist, defined **sociobiology** in 1975 as the study of the biological basis of social behavior. In practice, "biological" has meant "evolutionary," and so sociobiology has concerned itself with the evolution of such phenomena as flocking and schooling, territoriality, parental care, and the complex behavior of insect and vertebrate societies.

As we have already shown, many kinds of social behavior are of direct benefit to individuals or their offspring; explaining their evolution poses no special problem. For example, cooperating with other individuals in foraging can help the individual that joins in. It gets more food than if it went it alone; its fitness is raised.

To a biologist, the contribution of sociobiology

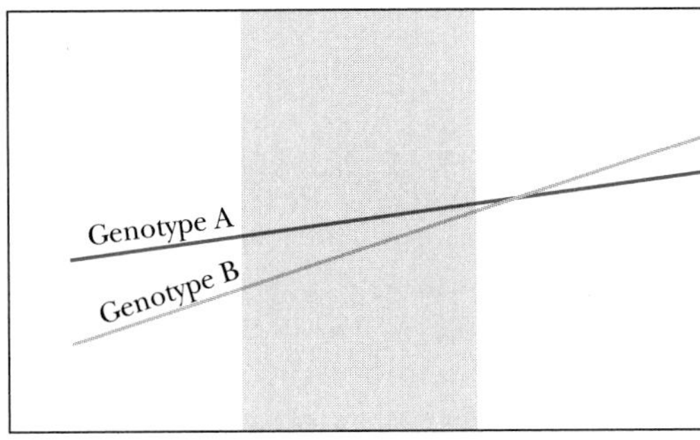

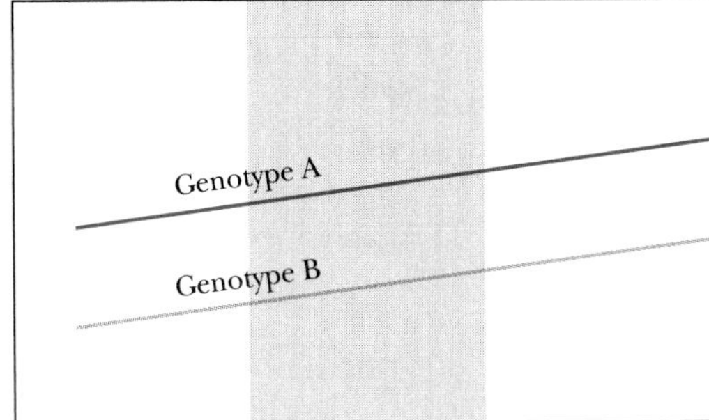

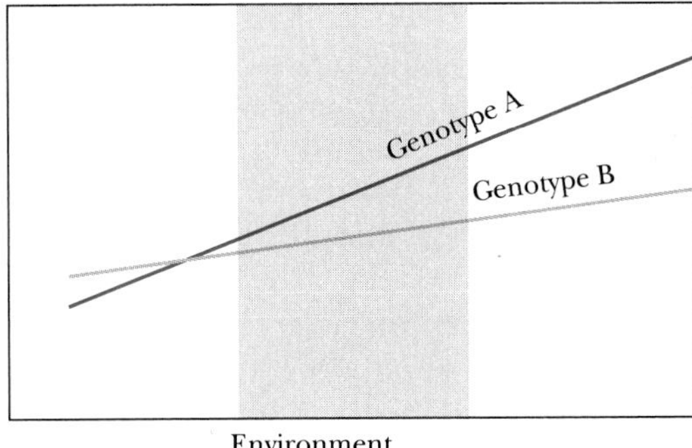

Figure 5-22

Three possible arrangments for IQ reaction norms for two genotypes. In the top graph, genotype A has higher IQ phenotypes in the medium range of environments but drops below genotype B in better environments. In the middle graph, both genotypes show increased IQ with increasing environment quality. In the lower graph, genotype A lies above genotype B in medium and good environment but falls below it in environments at the left of the scale.

has been in providing credible hypotheses for the evolution of certain kinds of social behavior that were hard to explain in terms of individual selection. Prime examples are (1) the evolution of less than maximal reproduction (*K* selection, already discussed), (2) altruism, and (3) the evolution of sterile castes in the eusocial insects.

Altruism

Altruism is doing good for others. Sharing food, warning others of danger, adopting orphans, defending one's country or colony—these are altruistic acts. There is always a cost to altruism, though it may be slight. At the least, the altruist spends time that could have been used for other purposes. Sociobiologists have formulated altruism in genetic terms: Altruistic behavior raises the fitness of another individual while lowering one's own fitness.

To reiterate, the evolution of such behavior has been difficult to explain in terms of individual selection. Suppose there are two alleles in a population of animals, a and a'. a favors the possessor going to save another member of a species, which is, let us say, in the clutches of a predator; a' favors not getting involved. In most circumstances, the possessor of a' is going to live longer and, therefore, reproduce more than possessors of a. Allele a, favoring altruistic behavior, will decline in the population.

Nevertheless, seemingly altruistic behavior occurs in humans and many other animals. How has it evolved? Research in the past 20 years has favored three explanations.

Selfish Behavior. Probably the explanation most frequently invoked has been that the seemingly altruistic act really benefits the actor. It is, in other words, selfish behavior, although others in the group may also benefit. In ordinary speech, "selfishness" has unpleasant connotations, but those connotations should not be attached here. All that is meant is that the behavior benefits the individual doing it; it could be as noncontroversial as coming in out of the rain.

Take warning calls. A hawk approaches a chickadee flock and a chickadee gives an alarm note. Isn't the chickadee putting itself at risk to save its flockmates? Possibly not. Ingenious biologists have thought up all sorts of individual advantages for such behavior. Suppose that the hawk is unlikely to stop unless vulnerable prey are visible, then the warning note helps the one who gives it by getting its flockmates out of sight. Suppose that the hawk is likely to move on quickly unless it gets reinforcement in the form of food. By giving a warning note, the bird has helped itself by decreasing the chances of predation on its flockmates.

Another suggestion, among many others, has been that the call itself may be less a warning to flockmates than a signal to the predator that the bird giving the call has spotted it and is, thus, a poor candidate for an attack (Dawkins 1976). Although the idea is appealingly ingenious, it is probably wrong, at least for the great tit, a European chickadee, and its hawk predator. The chickadee's alarm notes are high pitched, above the frequencies at which the hawks hear well but in the range of greatest sensitivity for chickadees. Consequently, the calls are audible to flockmates but not to approaching hawks (Klump et al. 1986).

Similar explanations have been suggested for such mammalian behavior as **stotting,** the strange bounding movements shown by gazelles. It could be a signal that the stotter is strong and healthy and likely to be hard for a predator to run down. **Tail flagging**—raising the tail to show its white underside—of white-tailed deer has also been suggested to be a signal to the predator that it has been discovered. Although there is probably much still to learn about such behavior, evidence seems currently to support the view that the signals promote group movements that would discourage predation. One reason to believe this is that the behavior in both gazelles and deer occurs in quite young animals (Smith 1991) that are not particularly adept in eluding predators but that would be helped by the presence of adults. This qualifies as selfish behavior; however, it may also be significant (as we see in the next section) that tail flagging in adult deer is frequent among groups of does, which are typically mother, daughters, and sisters, and infrequent among buck groups, composed usually of unrelated individuals.

Raising orphans has been suggested as an especially altruistic act. The behavior remains a puzzle in many ways (see Shy 1982, for example) but a hypothesis that may cover some cases is that the adopters are young animals practicing being parents. In species with deferred maturity, the adopters will have avoided much of the cost of attempting an early breeding but will have gained experience that may make their first attempt with young carrying their own genes more likely to be successful. It is not unlike helping your neighbor with his drywall, expecting that when the time comes to build your own house you'll know how to do it right.

Kin Selection. The concept of kin selection is probably the one biggest achievement of sociobiology. In a 1963 paper that can be thought of as the beginning of sociobiology as a field of study, the English population biologist W. D. Hamilton proposed a mechanism for the evolution of altruistic behavior even in cases where individual fitness was lowered. The mechanism he proposed is kin selection, although this term was first applied a year later (Maynard Smith 1964).

Remember that evolution can be thought of purely in terms of genes, as a change in the proportions of alleles in the gene pool from time 1 to time 2 (Figure 5-23). How does a gene increase from one time to another? The usual answer is by one genotype leaving behind more offspring than another genotype. Hamilton pointed out that, as far as the gene is concerned, it makes no difference whether it gets from time 1 to time 2 in offspring or in brothers, sisters, cousins, or nephews. In other words, a gene that you possess can increase in the next generation by your leaving a lot of offspring or by your making sure a lot of other relatives (some fraction of whom possess the gene) get there.

Let's look again at the situation we examined

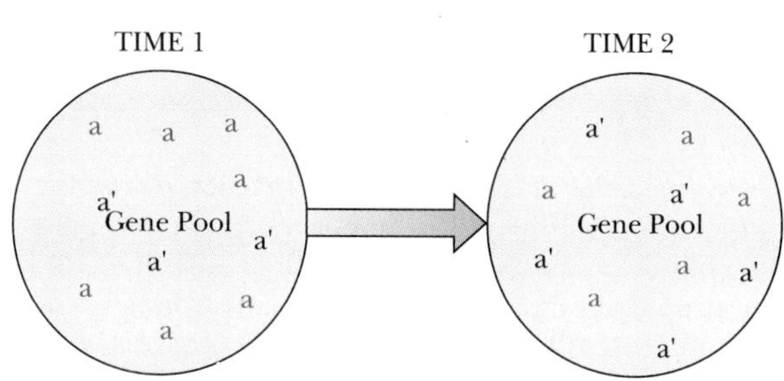

Figure 5-23

Evolution consists of genetic modification with time. We can think of it as change in the percentages of individuals displaying certain genetically based traits or as a change in the percentages of individuals with certain genotypes. Another way of looking at it ignores the individual organisms and simply looks at the percentages of different genes in the population. In this diagram, the gene pool at time 2 has a higher percentage of gene a than at time 1.

earlier, but this time suppose that the individual needing help is a relative. If an individual ignores a relative in trouble, about to be eaten by a predator, the percentage of the selfish individual's genes that will survive through the incident is 100%, or 1.00.

Under what circumstances would a higher percentage of the same genes survive? What if the relative is a brother? The brother shares 50% of the individual's genes, on the average. Suppose the individual goes to the help of his brother knowing he himself will die but his brother will survive? Does this make evolutionary sense? Obviously not, because only 50% of his genes—carried by his brother—survive instead of 100%. Altruism is not going to evolve in such circumstances.

Hamilton specified the condition in which hereditary factors favoring altruism will be selected for. If we let

r = the percentage of genes shared by two individuals by common descent (this little r is not the intrinsic rate of natural increase but instead a genetic measure called the coefficient of relatedness)
B = gain in fitness by the relative
C = loss in fitness by the altruist

then an altruist allele is favored when

$$(B/C)r > 1.0$$

In words, this says that altruism will be evolutionarily favored when an altruistic act in the long run saves more than 100% of an individual's genes—though they are in somebody else's body.

Let's take an example, using for convenience's sake, humans. Suppose someone is drowning. By jumping in to save that person, we risk drowning ourselves one time in ten, but five times in ten we will rescue the drowner, and so

$$B = 0.5$$
$$C = 0.1$$

r will depend on the relationship. Suppose that the person drowning is a brother or sister, a sibling. Then r is 0.5; the drowner and we share one half our genes. Then $(B/C)r = (0.5/0.1)0.5 = 2.5$. $2.5 > 1.0$, and so under this set of circumstances, a tendency to jump in and save siblings would be selected for.

B and C may in practice be difficult to measure. If we know the relationship between the two individuals, r (as a probability) is easy to specify. As long as inbreeding is not a consideration, relatedness can be reasoned out fairly easily; however, most of the relationships are given in Table 5-9.

Let's look at one further example. Suppose that the altruist's chances of drowning are 1 in 15 and that a cousin is the person going down for the third time. Will selection favor saving cousins in these circumstances?

$$B = \text{(again) } 0.5$$
$$C = 1/15 = 0.066$$
$$r = 0.125$$
$$(0.5/0.066)0.125 = 0.95 < 1$$

so saving a cousin, even though the risk is slight (1 chance in 15 of drowning) will not be selected for, when B and C are as given.

The idea of kin selection is supposed to have been stated very neatly in a (possibly fictitious) conversation many years ago between J. B. S. Haldane,

Table 5-9
The Fraction of Shared Genes (Coefficient of Relatedness, r) Between Different Relatives

The values are for diploid, biparental, noninbred organisms. Figures for parent-child relationship are virtually exact; the rest are averages. The values result from the fact that in gamete formation half the chromosomes from a parent's cells (one member of each chromosome pair) end up in each egg or sperm. Consequently, each child gets half of each parent's chromosomes (and, thereby, its genes). Following this process through family trees gives the values summarized here.

Relationship to Individual A	*Fraction of Genes Shared With Individual A*
Parent	0.5
Child	0.5
Brother or sister (sibling)	0.5
Grandparent	0.25
Grandchild	0.25
Half sibling	0.25
Uncle, aunt	0.25
Niece, nephew	0.25
Cousin	0.125

an English biologist of diverse interests, and some unknown person. "Haldane," the person said, "what's your opinion of altruism?"

"I would be prepared," Haldane said, "to lay down my life for three brothers or nine cousins."

Let's pause at this point to say that we are not implying, even when talking about humans, that cognition has to be involved in these decisions. We are saying that if "altruist alleles" are present, the outcomes we have shown are the inevitable result of natural selection. It makes no difference whether the individuals know what the costs and benefits are or even what altruism is.

It is, however, necessary that individuals recognize their relatives or, at least, behave differently toward them. Obviously this occurs in humans, and it is now clear that many other mammals, birds, bees, wasps, tadpoles, and doubtless many other organisms are able to tell kin from nonkin and often to distinguish degrees of relatedness (Hepper 1991). In modern U.S. society, we may not know all our more remote relatives, but in most societies, people do. Every anthropological description of a tribe or a society contains a long chapter on kinship describing the meticulous detail with which most societies keep track of genealogy. The Australian aborigines, for whom "relationship is the anatomy and physiology of society" (Elkin 1964) are typical, not unusual. To an outsider, a society's emphasis on relationships can be boring, but it is now understandable in terms of kin selection theory.

Kin selection seems to be a promising explanation for many kinds of altruistic behavior. In purely genetic terms, of course, the altruism has been taken out of such acts because the altruist is helping replicas of his own genes that happen to be in a relative's body rather than his own or that of his offspring. But in terms of individual organisms, the altruism remains.

Note that even if we can come up with an individual advantage for some behavior, kin selection may still be involved. The total increase in fitness would be the sum of the two components (Wade 1980).

Eusociality. Kin selection is probably the best explanation available for the evolution of sterile castes in ants, bees, wasps, and termites. Most of the members of these colonies show what might be regarded as the ultimate in altruism, giving up individual fitness entirely and devoting their efforts to the good of the colony. But remember that these colonies are, in fact, families, consisting mostly of a queen and her offspring.

Let's examine the genetics of the situation from the standpoint of an individual ant, bee, or wasp worker. How can she best help transfer genes she possesses to the next generation? She could breed, and produce offspring (this may not be a real alternative now, but if eusociality has evolved, it was at some point in the evolutionary history of the organism), or she can take care of her sisters. Because of haplodiploidy, the female hymenopteran may well do better by taking care of her sisters. She shares 75% of her genes with each sister, on the average, whereas any offspring she produced would only have 50% of her genes. These genetic relationships are portrayed in Figure 5-24.

Because of the way haplodiploidy inflates r for sisters, it probably predisposes the Hymenoptera to evolve eusociality when ecological conditions are favorable. It is not, however, necessary for eusociality; termites do not show haplodiploidy.

In understanding the evolution of eusociality, it may help to look at a system that seems less outlandish to vertebrates like us. Tropical wrens studied by K. Rabenold (1984) show the phenomenon Alexander Skutch (1935) called "helpers at the nest." The pattern is now known for many tropical bird species. Probably the best known North American example is the Florida scrub jay (Woolfenden 1975, Woolfenden and Fitzpatrick 1984). In the wrens, young birds tend to remain in the territory where they hatched (later, females shift to a nearby territory).

In this social grouping, the birds help feed younger siblings and defend the nest from predators. In the social group occupying a territory, only one female lays eggs and only one male fertilizes them. When either of these dies, his or her place is taken by one of the helpers.

Rabenold found that twosomes almost never managed to raise any young. A male and a female with one helper did little, if any, better. A pair with two helpers raised four times as many young as twosomes and trios, and groups of six to eight raised twice as many young as the quartets (Figure 5-25). The big factor in improved nesting success seemed to be better predator defense. Predation is an important mortality factor in the tropics.

In this case, the reasons for helping siblings

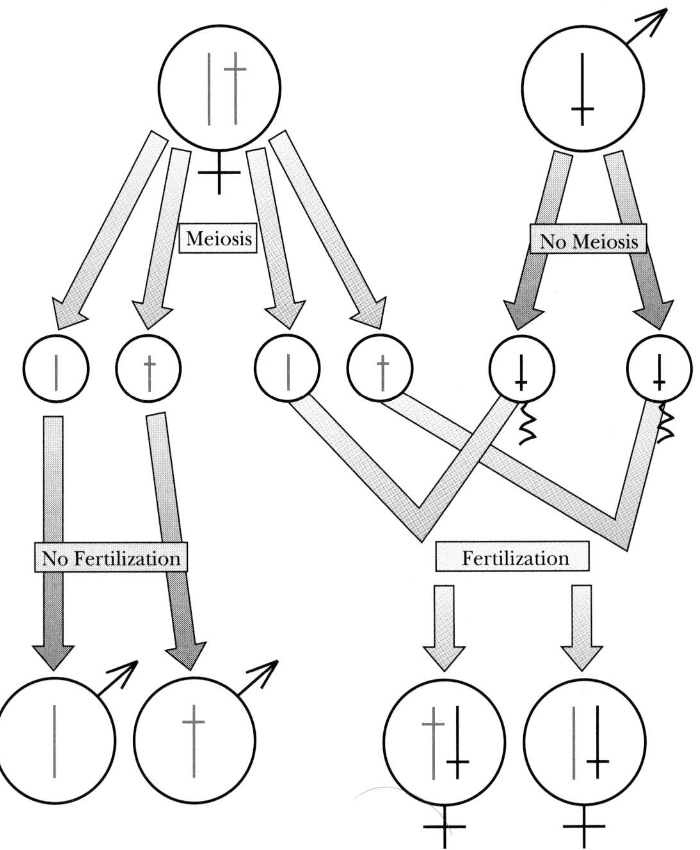

Figure 5-24

Diagrammatic representation of gamete formation, fertilization, and development in Hymenoptera, yielding diploid females and haploid males (no fertilization). This haplodiploid mechanism of sex determination results in asymmetrical coefficients of relatedness, sisters sharing 75% of their genes with one another but having only 50% of their genes in common with their mother or daughter.

Coefficients of Relatedness

Mother to daughters—50%
Mothers to sons—50%

Fathers to daughters—100%
Fathers to"sons"—0%

Daughters to sisters—75%
Daughters to brothers—25%

Daughters to mother—50%
Daughters to father—50%

Sons to sisters—25%
Sons to brothers—50%
Sons to mother—100%
Sons to"father"—0%

may be clearer to us. Individual wrens can and do strike off on their own to form pairs, but they are wasting their time and energy and exposing themselves to predation for virtually nothing, genetically. They got essentially no replicas of their genes into the next generation. Helpers, on the other hand, are contributing to the success of their brothers and sisters, who share 50% of their genes. Genes favoring helping are being passed along.

Kin selection is not the only factor, however. Helping for a time in the wrens and most other vertebrates with cooperative breeding is the only path to successful reproduction (Stacey and Koenig 1990). This can be because predation or other factors make only cooperative efforts successful (Stacey and Ligon 1991) or because all suitable habitat is already occupied (Fitzpatrick and Woolfenden 1986). Accordingly, helping can also increase individual fitness. But this increase often stays potential. Most helpers in cooperatively breeding birds and mammals live out their lives as helpers, never becoming the breeding pair on the ter-

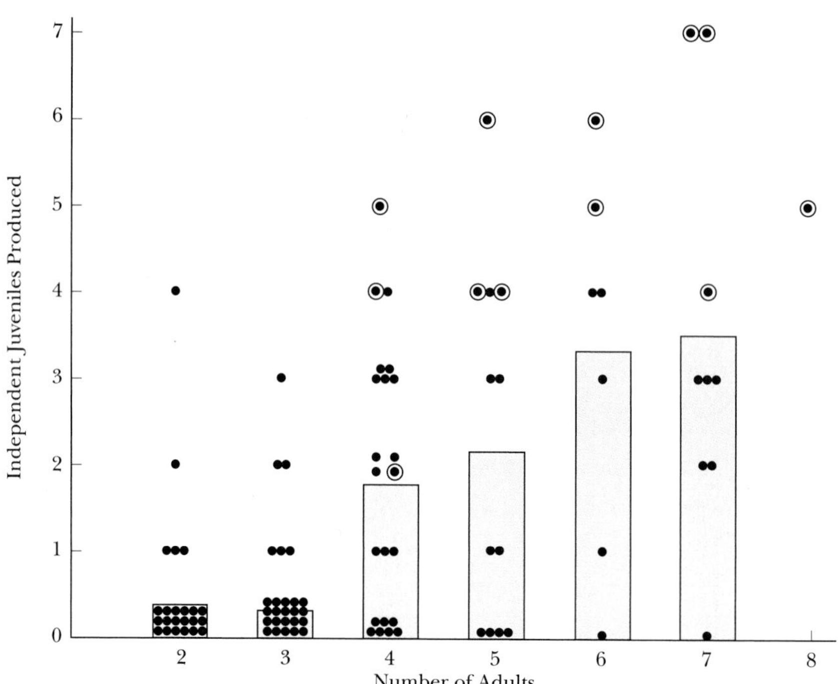

Figure 5-25

Annual production of independent juvenile stripe-backed wrens on a Venezuelan study area by group size. Dots represent individual nesting attempts; circled dots indicate two broods in a single season. Most nesting attempts for twosomes and trios produced no young birds. (From "Cooperative enhancement of reproductive success in tropical wren societies" by K. A. Rabenold, *Ecology*, 1984, *65*:871–885. Copyright Ecological Society of America. Reprinted by permission.)

ritory. It is a short step from this situation to eusociality.

Reciprocal Altruism. Robert Trivers (1971) suggested another mode for the evolution of altruism that does not require that the interaction involve kin. How widely important **reciprocal altruism,** or **reciprocity,** is, outside human societies, is still unclear. The basic idea is that one individual ought to help another if there is a good chance that the altruist (or its offspring) will receive help from the helpee later on.

In a way, reciprocal altruism provides a basis for the golden rule of "Do unto others as you would have them do unto you," although the old saying, "You scratch my back and I'll scratch yours," catches the spirit a little more precisely. The Latin *quid pro quo* is another statement of the idea. We are, of course, participating in quid pro quo when we give a shopkeeper money for an ice cream cone, but the quid need not be that immediate. We may donate blood to the Red Cross with an only partially conscious expectation that, should we or our relatives need a transfusion some time in the future, blood will be available.

The evolution of reciprocity seems to have several fairly stringent requirements. It depends first on relatively stable associations; otherwise the helpee may move and not be around to return the favor it received. Under the right set of circumstances, highly stable associations might be enough to promote tendencies toward reciprocal altruism, but the system seems to invite cheaters—individuals that receive help but don't return it. Ordinarily, then, the evolution and maintenance of reciprocal altruism is thought to require the additional conditions of (2) individual recognition among the organisms, (3) a good memory, and (4) penalties for the cheater, such as being cut off from future altruistic acts.

The conditions may be possible in various vertebrate societies and certainly can be met in human society. The variety of expressions for reciprocity that we already listed is one indication of the probable importance of the relationship in human affairs. Garret Hardin (1977) suggested that "The process of growing up is in part a process of gradually becoming aware of the quid pro quos of the world, its obvious ones first, the less obvious ones later." Of course, Hardin noted, some people grow up faster than others.

Human Sociobiology

The last chapter of Wilson's book *Sociobiology* deals "in the free spirit of natural history, as though we were zoologists from another planet completing a catalog of social species," with human social behavior and its evolution. Some scientists have found this approach valuable. To them, human behavior, like that of other organisms, is a product of both genetic predispositions and experience, with a strong ecological component in the eventual expression of the behavior (Lopreato 1984). Understanding the hereditary contribution, understanding, that is, "human nature," seems to some to have practical importance. It might help us understand what human societies safely could do and should not do in the future, under pressure for social change coming from increasing population size and technological innovations.

Opposition to such views has come, expectably, from some sociologists, psychologists, and anthropologists who argue that human culture cannot be studied as biology. Most of these tend to think that the genetic influence on human social behavior is so slight that it can be virtually ignored. Some other social scientists have found that sociobiology has provided a theoretical framework previously lacking in their field.

Opposition has also come from certain scientists who argue that any hereditary basis of human behavior should not be studied because such study might be used as justification for repressive political policies, such as racism and sexism.†

Because of its willingness (some would say, eagerness) to include humans in its deliberations, sociobiology has become a controversial field of study. As a result of the controversy, or rather because of the political nature of the controversy, few people today readily admit to being sociobiologists. Behavioral ecologist is now an approximate synonym.

Group Selection

In the past, biologists—let alone nonbiologists who have tried to talk about evolution—have sometimes been uncritical in their thinking about the evolution of features that are advantageous to the population or the community. They have talked about certain kinds of behavior, such as lemmings marching into the sea, as being "for the good of the species." In earlier sections, we emphasized that evolution is largely a matter of whether an individual leaves more or fewer descendants in the next generation relative to others of its species. The casual formulations of past years that require an organism to act against its own interests to favor its group are probably wrong.

In most cases of behavior "for the good of the species," advantages to the individuals can readily be discovered. Lemming emigration occurs from areas of overpopulation. Doubtless many of the emigrants perish, but some may find suitable, uncrowded habitat. It might well be that a higher percentage would die if they all stayed put in an overpopulated range.

The case for individual selection was made clearly in a small book *Adaptation and Natural Selection* by G. C. Williams (1966). As an antidote to the fuzzy thinking about evolution that was then current, the case was a good one, but it is not necessarily complete. We have already suggested that kin selection, in which an individual's alleles reach the next generation via other relatives than that individual's offspring, may be necessary to explain many kinds of altruistic behavior.

Williams' book was, in part, provoked by the writings of V. C. Wynne–Edwards (1962), mentioned earlier in the section on *K* selection. Wynne–Edwards came down squarely on the side of group selection. He did so, not fuzzily, but with a clear formulation. Wynne–Edwards was particularly concerned with the regulation of population size.

To most people's way of thinking, the ability of a population to regulate its size below the limit set by resources is of obvious benefit to the population. With this trait, it consists of healthy, well-fed animals living in an uncrowded environment. Without it, the population may consist of crowded, sickly, half-starved animals. But for the population to limit its size below carrying capacity, some sort of restraint on the part of the animals is required. They may have to breed at a lower than maximal rate by delaying maturity or having small litters or even by giving up breeding altogether in years when density has edged up. Some animals may have to leave a favorable habitat and find some less suitable place

† This latter viewpoint is set forth in "Sociobiology—another biological determinism," *BioScience*, 26:182, 184–186, 1976.

or, perhaps, wander about and die of predation. How could such kinds of behavior arise?

Wynne–Edwards said they could evolve like this: Visualize two populations, one that limits its population size as a result of some genetically based behavior and one that doesn't. The one that practices population limitation may do it, let us say, by having fewer young when food is scarce.

In an average year, both populations survive well enough. But what happens in a hard year? In the population that has not limited its size, no individual gets as much food as it needs and none survives. The other population comes through basically intact. These individuals reproduce and send out offspring that, along with dispersers from other such prudent groups, repopulate the range left vacant by the other population. In the future, the new population will also consist of individuals that limit their population by social means. In Wynne–Edwards' (1962) view, the lemming emigrations really are for the good of the species.

Such events occur; local populations go extinct and are replaced by dispersers from other populations. But most students of the subject have doubted that the process can counteract opposite evolutionary trends based on individual selection. If we have, for example, a local population of 19 individuals that decrease their reproductive rate as density goes up and 1 individual that doesn't, this cheater will reproduce freely, causing the others to slow their reproduction. Eventually, the local population will consist mostly of the cheater, fast-reproducing genotype. "Selection among populations," Williams (1966) wrote, "cannot cause evolution to go in one direction, when each of the populations is evolving in the opposite direction."

There the matter stood for a few years—and still stands in the opinion of many. If group selection occurs at all, it occurs only in the situation in which the group consists of relatively close kin (kin selection). Group selection, as such, became as unpopular as a ZPG button at the Right-to-Life potluck.

More recently, several evolutionary ecologists have produced credible models and experimental evidence (Wade 1978, 1982) showing that group selection is a feasible enterprise. Group selection much as Wynne–Edwards envisaged it remains a possibility in some circumstances; however, the most interesting new models (Wilson 1980, 1983b) emphasize differences in productivity or reproduction of local groups, rather than extinction. The argument for **intrademic group selection** consists of these parts:

1. Populations tend to be *recurrently* subdivided into local groups.
2. These groups tend to vary in their genetic makeup through chance and other factors.
3. Some genetically based traits, even if individually disadvantageous, may be of benefit to the local group.
4. Local groups having a higher percentage of individuals with these traits will produce many or highly successful young even though the individuals possessing the traits have a lower fitness than the other members within the group.
5. If the differences in productivity among the groups are large relative to any individual disadvantage, the genetic factors for group-beneficial behavior will increase in the population in the next generation.

An example of the process is illustrated in Figure 5-26 where gene A′, which is group beneficial, increases from 50% to about 51% in a generation, even though it declines within each of the local groups.

The group-beneficial behavior we are talking about can include the familiar examples of altruism, such as sharing food and cooperating in the rearing of young, but also some other sorts of behavior that previously have been hard to explain using either individual or kin selection. In winter dominance hierarchies it might be altruistic for subordinate birds to pack up and head out as food grew tight. For the dominant two or three animals to drive the subordinate animals out is hardly altruistic in the usual sense, but it is, nevertheless, of benefit to the remaining members of the group.

Although the model has been stated in terms of differences in productivity among groups, this need not involve immediate reproduction and multiplication. For example, the model also fits situations such as winter flocks from which the following year's breeding population is to be drawn. If the over-winter survival of members composing these flocks varies greatly depending on genetically based behavior, group selection such as we have described could be occurring at this stage of the life history.

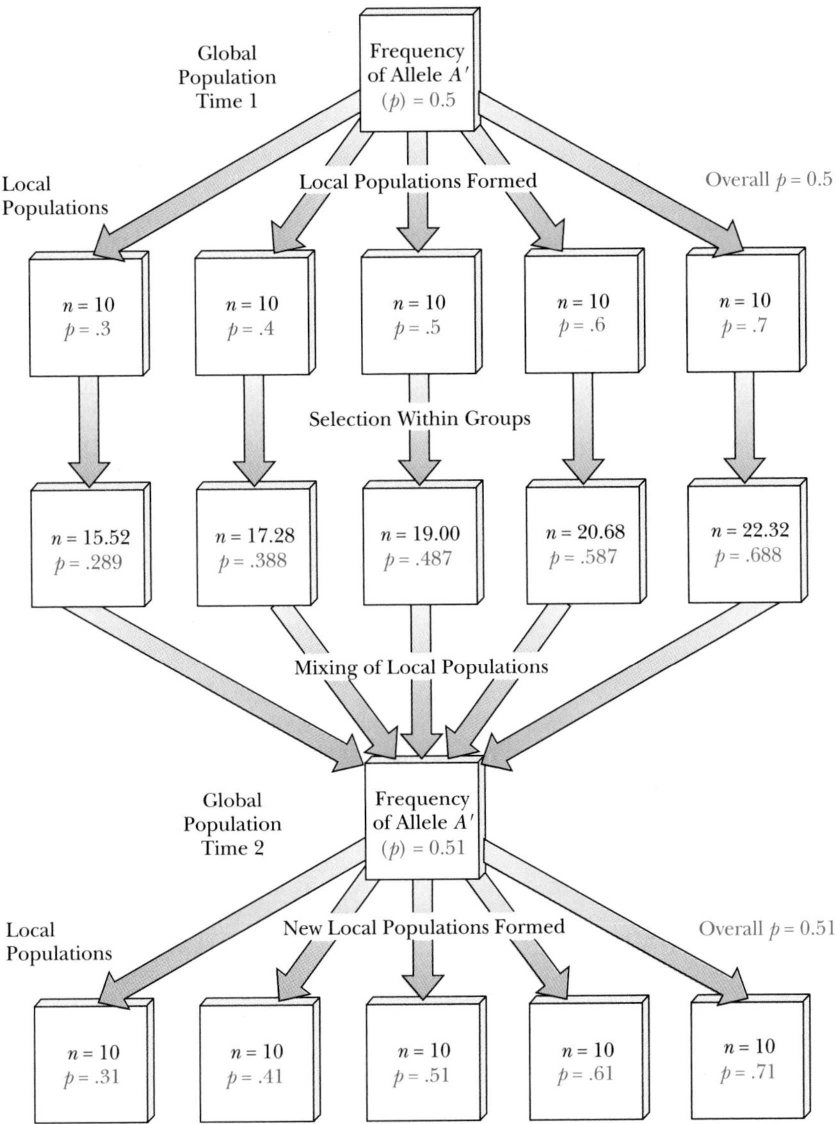

Figure 5-26

Group Selection. In this example, the fitness of individuals carrying the allele A' is reduced but their presence in the group causes it to prosper. Accordingly, the frequency of the allele (p) increases in the population as a whole, even though it declines within each local population. n is size of the local population. (Additional details are given in D. S. Wilson 1983, from which this diagram is adapted.)

There is no question that group selection can occur. It is probably of widespread occurrence in furthering the spread of features, such as alarm calls and emigration from overcrowded habitats, that are also favored by individual selection. There has been a tendency to consider that group selection has been disproved if a plausible case can be made that a trait has individual benefits. This is wrong. Showing that behavior is individually advantageous shows that it is individually advantageous; group selection could play an additional role in its evolution.

The question that remains is, How frequently does group selection further the evolution of features that are individually neutral or disadvantageous? It is still possible that one of the assumptions stated or implied in the five points listed above will prove to be so rarely met in nature that group selection will have to be put back on the shelf. Perhaps the most crucial point is whether there is regularly a period in the organism's yearly cycle in which the individual disadvantages of the feature persist but the group-beneficial aspects disappear.

At present the necessary conditions all seem

reasonable for many different kinds of organisms. It is possible that group selection of this form is a powerful way by which behavior that is slightly disadvantageous individually but very beneficial to the group has been generated.

The Ecological Theater and the Evolutionary Play

Evolution through natural selection is a process of populations; however, the process goes on in the context of the ecosystem. This idea is captured neatly in the title of a book by G. E. Hutchinson (1965), *The Ecological Theater and the Evolutionary Play* (although the phrase "ecological theater" was used earlier in the same sense by John Steinbeck).

In some cases two or more species interact so closely that evolutionary changes in one tend to be followed by evolutionary changes in the other, so that they form an evolving system. This evolving together was termed **coevolution** by P. H. Ehrlich and P. R. Raven (1965). Coevolution probably occurs among many pairs (or larger numbers) of species linked by coactions. Prey populations, under the selective pressure of their predators, may tend to evolve toward running faster or hiding better. Predator populations, under the selective pressure of starvation, may become more stealthy or develop keener eyes or noses.

Coevolution can be involved in mutualistic associations; in fact, no other explanation seems possible for obligatory mutualistic associations. Reproduction of the yucca plant is dependent on the yucca moth, its only pollinator, but reproduction of the yucca moth is dependent on the yucca plant, in whose flowers its eggs develop. The same is true of the fig and the fig wasp. At some time the relationship between these plants and insects must have been looser; the tight interaction in which one cannot exist without the other must have developed stepwise in coevolution.

Coevolution involves changes in both species. Adaptation of organisms to the biotic features of their environment occurs, just as to abiotic features, but such changes are not necessarily coevolutionary. A moth may evolve to resemble tree bark, but as concerns moth and tree, this is not coevolution. A moth may evolve to resemble another species of moth, but unless the second species also changes, this likewise is not coevolution.

Selective pressure in the community context may come not from a single species but from several, and, similarly, several species may undergo evolutionary changes. The shade that has helped to shape the life histories of forest herbs comes from many species of trees, and many species of herbs have come to show the spring ephemeral schedule.

Coevolving systems, also, can involve several species. Herbivory by many species of insects may be the selective agent that gives the advantage to the genotypes of a plant species that produce a toxic chemical (Marquis 1991) and it may be that more than one of the insect species evolves detoxification or other mechanisms that allow them to continue feeding, and exerting selective pressures, on the plant.

It is well to remember that fine tuning of coevolutionary systems becomes less likely the greater the differences in generation length among the coactors. Other things being equal, rapidity of evolutionary change depends on generation length, so insects can adapt faster than vertebrates, and vertebrates faster than trees. Frugivorous birds, for example, may change rapidly in response to a change in a character of a fruit tree; however, any evolutionary response of the trees will take much longer. This does not mean that the features of such a system—for example, fruit color and sugar content, bird life history and digestive apparatus—cannot be the result of coevolution. It does mean that the cast of the evolutionary play may have changed: The plant traits that select for certain frugivore traits may be the result of selective pressures from the "evolutionary predecessors" (Herrera 1985) of the current frugivores—different species or even species of different families, orders, or classes.

Many aspects of community organization seem to be based on coevolution leading to the reduction or avoidance of competition. A species extends its range and enters a new community. Evolutionary adjustments occur in it or its competitors or both, allowing coexistence by dividing up, or **partitioning,** the shared resources. One species comes to feed on larger items and one on smaller, for example. Or they may divide the community up spatially, one feeding in the shrub stratum and one in the trees. Temporal partitioning may occur with two grass-

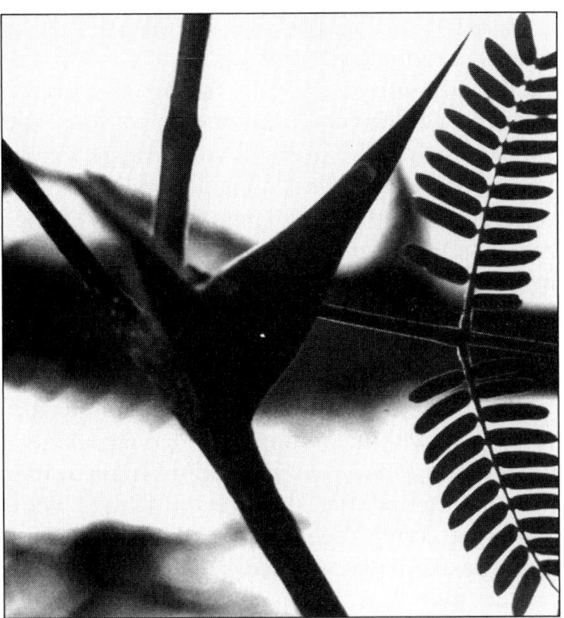

Figure 5-27

A thorn of a swollen-thorn acacia, showing an entrance hole made by ants. (Photo by the author.)

Figure 5-28

Beltian bodies are the pale-colored projections at the tips of the leaflets. They are high in protein and oil. (Photo by the author.)

land herbs diverging so that one grows and flowers in spring and one in summer.

Most communities seem to be the product of this and other sorts of evolutionary and coevolutionary change occurring repeatedly over the course of geological history. The deciduous forest of eastern North America, with its stratification and seasonal changes, its plants adapted to shadier and sunnier, wetter and drier sites, and its mycorrhizae and root parasites, is one such case. The roles of competition and coevolution in community structure are explored more fully in a later section.

The following sections discuss a mutualistic system and a predator-prey system in which coevolution seems to have been involved.

The Ant–Acacia System

In Central America certain kinds of acacia trees have thorns with very large bases. Members of an ant colony live within these thorns, which they hollow out. There are related types of acacias without swollen thorns that do not harbor ant colonies and related types of ants that do not live on acacias. Both the acacia ants and the swollen-thorn acacias have traits missing in their relatives that would seem to be of no benefit to them except in the context of their association.

The ants live and raise their young in the enlarged thorns (Figure 5-27). The adult ants feed on the nectar produced by the acacia, which is produced in nectaries located on the leaves, rather than in the flowers. Such extrafloral nectaries are present on nonswollen-thorn acacias but are smaller than those on swollen-thorn plants. The larvae of the ants are fed material from specially modified leaf tips called Beltian bodies (Figure 5-28), which do not occur on nonswollen-thorn species.

The advantages to the ants of this coaction are obvious, but what advantages does the plant derive? It is protected from most kinds of herbivores, whether insect or mammal. Insects that land on the plant are attacked by the ants and killed or driven off. Likewise, a mammal that brushes against the plant is bitten and stung. The tropical ecologist D. H. Janzen (1966), who worked out many of the details of this case of mutualism, showed that these activities by the ants do, in fact, benefit the plant. He prevented ants from colonizing new sprouts and found that these sprouts grew slower, produced fewer leaves, and had a lower rate of survival than similar sprouts that the ants had colonized.

Unlike related species, which are active only in the daytime, the acacia ants patrol 24 hours a day, thus protecting the plant from both diurnal and nocturnal herbivores. The plants may gain another benefit from the ants' activities. The ants not only attack animals but also maul any other plants that touch the acacia foliage or that grow up below the acacia. Acacia is intolerant to shade and susceptible to fire, which is frequent in the area it inhabits. The clearing activities of the ants may tend to keep the acacia plant from being shaded out and may keep the vicinity clear of flammable material. In general, swollen-thorn acacias cannot survive to reproductive maturity without the patrolling activities of an ant colony (Janzen 1969a).

The anti–acacia interaction just described is obligate; this particular ant and acacia do not occur separately. There are, however, species of acacias and related species of ants in which the relationship is not as close; they often occur together but also may be found not associated with one another. Coevolution seems to be the most likely explanation for the obligatory relationship. The association must have arisen step by step through small changes that had the effect of fitting the ants and the acacia into a closer, more efficient system.

Bat-Moth Coevolution

Everyone knows that some kinds of bats use **sonar** to catch prey and to maneuver between obstacles in the dark. That is, they send out sound pulses and use information provided by the echoes. Bat signals are mostly ultrasonic, above 20 kilohertz (kHz). These high frequencies give better resolution than the lower ones that oilbirds, for example, use. Using their sonar, bats can fly, not just between tree trunks but through a maze of piano wires, as the first modern study of echolocation showed (Griffin 1953). They can also detect fruit fly-sized insects. Physics being what it is, high frequencies also have a disadvantage. They attenuate rapidly; the intensity drops rapidly with distance.

As a consequence, most bat vocalizations are very loud. The little brown bat, a common species over much of North America, produces sounds equivalent to 109 decibels, about the same as an unmuffled motorcycle. It is just as well for us that bats use ultrasound and not frequencies we can hear. A part of the echolocation signal of some bats is low enough to be in the audible range for the small percentage of the human population with very good hearing in the higher frequencies. When these people complain of bat noise, they tend to be met with skepticism. Of course, bats also make nonecholocation sounds that would be audible to any of us; they have a variety of squeaks by which they communicate with one another.

The ultrasonic signals of bats are produced in the larynx, but by vibrations of laryngeal membranes rather than by the vocal cords. In many bats, the sounds are emitted through the mouth. These bats tend to catch their prey in a kind of baseball mitt or jai-alai basket formed by the skin that runs between their legs and includes their tail. From here the insect is transferred to the mouth. Other bats, however, emit sounds through their nose; accordingly, their noses have an odd structure rather like the front grill of a BMW, and their names—horseshoe-nosed bats—reflect the peculiarity. Evidently the advantage to this arrangement is freeing the mouth to grab the prey.

Typically, bats when searching for prey fly along giving fewer than ten notes per second. As they close on an insect, pulses go up to 200 per second. The process of catching a prey—detection, tracking, fixation, and capture—may take 300–500 milliseconds.

Bats hear sounds up to about 40 kHz; above this they are generally deaf except for the frequencies they use; these vary from species to species. The horseshoe-nosed bat hears well in the range centered on 83 kHz but is essentially deaf below 81 and above 86 (Remmert 1980).

Different bats use different methods of echolocation (Hill and Smith 1984). The horseshoe-nosed bat, when searching for insects, emits a signal at 83.4 kHz. From the echoes it receives, it can determine the distance of prey, prey speed and direction of flight, and something of its shape, texture, and actions. It can determine distance to prey by the time required for the signal to bounce back. Sound travels approximately 330 m per second, varying slightly with temperature, moisture content, and atmospheric pressure. If we yell at a cliff and the echo comes back in 2 seconds, we know that it

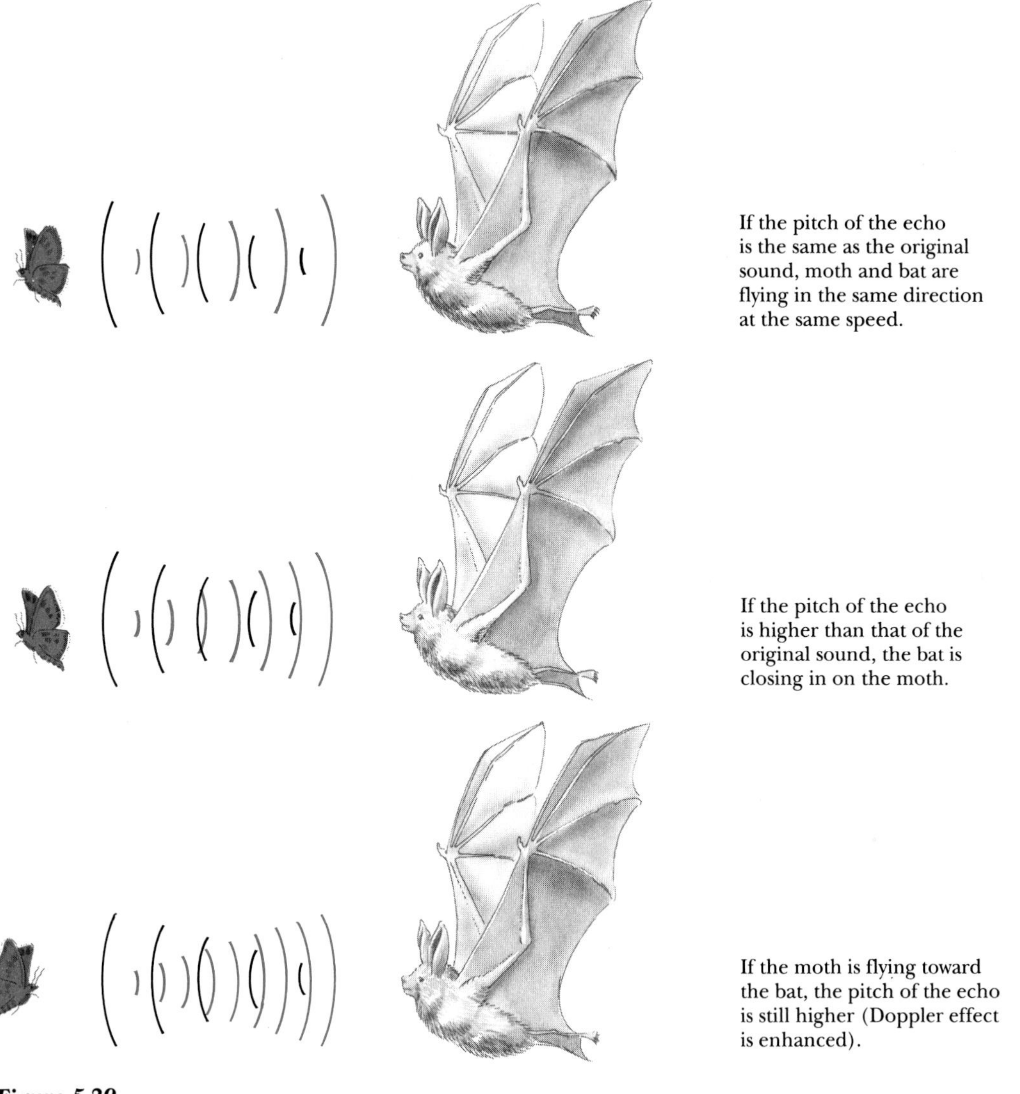

If the pitch of the echo is the same as the original sound, moth and bat are flying in the same direction at the same speed.
If the pitch of the echo is higher than that of the original sound, the bat is closing in on the moth.
If the moth is flying toward the bat, the pitch of the echo is still higher (Doppler effect is enhanced).

Figure 5-29

Bats can detect the movements of insect prey relative to themselves by use of the Doppler effect.

is about 330 m away (a second for the sound to get there, and a second for the echo to get back). The bat's analysis is similar, except that it is working in milliseconds.

The echoes coming from the insect show the **Doppler effect.** If a sound source is moving toward us, the sounds will have a higher pitch; if it is moving away, the sounds will seemingly be lower pitched. This is what causes the rise and then drop of a locomotive whistle as it zooms past us at a railroad crossing.

If the echoes coming back from the insect are at 83.4 kHz, they tell the horseshoe-nosed bat that the insect is flying away from it at the same speed that the bat is traveling (Figure 5-29). If the sound comes back lower than 83.4 kHz, the insect is moving away faster than the bat; and if the sound comes back higher, as will usually be the case, it means the bat is closing on the insect. The bat now lowers the frequency of its signal, so that the echoes have a frequency of 83.4 kHz.

Perhaps this is a simple way of calculating speed

of closure. Another function may be to provide a constant carrier frequency to allow interpretation of frequency modulations produced by the prey (Schnitzler 1978). Wingbeat and other features give different frequency modulation patterns that help the bat tell insects from leaves, one kind of moth from the other, etc.

All this is a remarkable set of adaptations for nocturnal predation on insects. Where, though, does coevolution come in? The insects have not remained evolutionarily static. What changes, if they were possible genetically, might help nocturnal insects, such as moths, avoid being taken by bats?

1. Moths could become diurnal. Diurnal moths exist, though it is unclear whether escape from bat predation was an important factor in their evolution. A switch to diurnal activity exposes moths to diurnal predators, particularly birds. The clear-winged moths (Aegeriidae), a diurnal group, are wasp mimics.
2. Moths could become flightless. This drastic step has occurred in a few species but seems generally connected with habitats in which long-distance movement would be disadvantageous. some bats do feed on insects on the ground or on foliage.
3. Moths could become poisonous. This is useful only if combined with a warning. Warning coloration, usual in day-active poisonous animals, would not be helpful against nocturnal predators. Tiger moths (Arctiidae), which are distasteful, emit ultrasonic clicks that cause bats to veer off (Dunning and Roeder 1965). It is possible that these are a sound version of warning coloration (Dunning 1968). As the name suggests, tiger moths are also warningly colored. There is no paradox in this; the moths' main predators when they are active at night are bats, but while the moths are resting in the daytime, they may be found by foraging birds.
4. Moths could jam the bat's sonar by producing loud noises containing the bat's frequencies (Fullard et al. 1979); however, bats seem to have little trouble picking out their echoes from background noise. The insect would have to put a large amount of energy into its call to make it loud enough to drown out the bat's signal. The jamming noise itself might be a useful target for the bat.

   An alternative explanation for the tiger moth clicks is that they somehow confuse bats, rather than warn them (Fullard et al. 1979). It is possible that the clicks are interpreted by the bat as another insect or some other obstacle very close to it. That is, if the clicks are interpreted as echoes, they would indicate that the object was only half as far away as the actual distance. Another possibility might be that the clicks are interpreted as another bat nearby.
5. Moths could become anechoic. Some surfaces echo more than others; a room with hard, flat walls is full of echoes, whereas one with draperies and wall hangings is much quieter. One possible evolutionary pathway, then, might be to reduce the echo as much as possible, which should make the moth harder to pick up at a distance. It seems possible that some of the furriness of moth bodies might be adaptations serving this function.
6. Moths could use the bat's sonar as a signal to take evasive action. This involves the evolution of organs of hearing—ears, in other words. Around a dozen families of moths do have such organs, though many of the moths do not themselves produce sounds. The organs were a puzzle to entomologists for many years; the insects seemed to have no need for them, and also they seemed to be deaf. The ears, in fact, do not respond to sounds in the audible range but instead to ultrasonic frequencies (Roeder and Treat 1961).

It is hard to avoid the conclusion that moth ears have evolved as antibat devices. Especially instructive are the noctuid moths, which, as adults, have hearing organs tuned to ultrasonic frequencies. The larvae also have hearing organs, but these detect frequencies below 1 kHz. One of the main predators of noctuid larvae is the polistes wasp which paralyzes them and takes them to its nest as provisions for its brood. The wing hum of the polistes wasp is in the range of frequencies to which noctuid larvae are sensitive.

The evasive action taken by moths varies. In general, if the bat call is detected at a distance, the moth flies directly away from the sound. The bat cannot detect the moth further than a few meters away, whereas the moth can hear the bat at several times this distance. If the bat is already close when the moth first detects it, however, the moth either folds its wings and plummets to the ground or else flies to the ground in an erratic spiral (Roeder and Treat 1961). Tiger moths click in response to bat calls but also take evasive flight (Dunning 1968).

As the German ecologist Hermann Remmert (1980) points out, you can readily observe evasive behavior if you see moths around a streetlight or porch light. Shake a bunch of keys at them—ultrasonic as well as audible tones are produced—and many of them will go looping and fluttering to the ground. If you feel the need for a control, you can find another streetlight and simply shake your fist at it.

The evasive behavior described is effective. One set of observations showed that bats were successful in 48% of their attacks on moths that took no evasive action but in only 7% on ones that did show evasive action (Roeder and Treat 1961).

Several other groups of nocturnal insects are now known to have developed similar defenses. Included are praying mantises, which go into a spiral power dive when they pick up the bat's sonar (Yager and May 1990), and crickets, which respond with the flying equivalent of tripping themselves, causing zigzag flight (May 1991).

It is likely that some bat features represent coevolutionary countermoves to moth coevolutionary changes. There are, for example, "whispering bats" that produce much softer signals than do most bats. This limits the distance at which the bat can pick up a potential prey, but it also limits the distance at which the insect can detect the presence of the bat. Because sound intensity drops as the square of the distance from the source, the loss of distance is proportionally greater for the insect.

Extinction

In the long run we are all extinct. The extinction of a species may be merely technical; a line of evolutionary descent may be modified in the course of time so that species "y" disappears, having become species "z." But most species of the past have come eventually to a definite end of the line; the last individual dies, the gene pool is gone. As the Cretaceous Period of geological time drew to a close, one species of dinosaur after another died out, until none was left. Two thirds of the families of the marine relatives of crustaceans called trilobites disappeared near the end of the Cambrian Period; other species lived on for another 250 million years, but none saw the end of the Permian.

To say that extinction occurs when death rate remains enough greater than birth rate that the population declines to zero may seem to say very little. It does, however, focus attention on extinction as a process of populations, different from other changes in population size only in its finality.

The basic cause of extinction appears to be a change in conditions causing increased mortality or decreased natality, for which the population cannot compensate. The change may be a new climate, a new predator, or a new competitor. If the species can adjust behaviorally or physiologically to the change, or if it can adapt genetically, then it can survive.

The "specialist" species, those with a narrow range of tolerance or adapted to a single food, are more prone to extinction than are the "generalist" species. This is because a change in conditions is more likely to move outside a narrow tolerance range than a broad one. Also, the evolutionary remaking of extreme specializations (such as saber teeth) is probably a slower process than remaking modest ones. Specialization is, of course, no crime as long as conditions do not change. As we shall see, competition theory suggests that ordinarily species must specialize, relative to one another, to avoid extinction from competitive exclusion.

Some specializations associated with extinctions have included restriction to a single habitat, restriction to a single food item, and adaptation to stable conditions. The latter often involves a reduction in biotic potential and an increase in body size (see the discussion of *r* and *K* selection earlier in this chapter).

Extinction, then, no less than evolution, is a natural event. Our role in extinction has been to speed up its rate through our remarkable effectiveness in causing change. Some of the most important changes by which humans have been involved in extinctions have been habitat changes, including the reduction and elimination of natural ecosystems, overhunting, and the introduction of predators and competitors.

Loss of habitat was probably a prime cause of the extinction of the heath hen and the ivory-billed woodpecker in the United States. Overhunting was probably an important agent in the disappearance of the passenger pigeon and the great auk. Many species of vertebrates known to have become extinct in modern times have been island forms—the moas, the Sandwich rail of Hawaii and several other flightless rails on other islands, the Oahu thrush, the Laysan apapane, the Tristan coot. Although habitat changes and overhunting were involved in the extinction of some island forms, most of them seem to have perished from predation or competition (through either exploitation or interference) by animals introduced by man. Some introductions were deliberate; in earlier times, sailors would release goats on islands so that there would be a source of meat and milk on their next voyage. Others have been unintentional, as in the spread of the Norway and black rat to islands throughout the world. Other animals introduced have been mongooses, cats, rabbits, and the familiar list of game mammals and birds.

The frequency of extinctions on islands suggests that local extinctions of populations occur more often than has been generally realized. If a population becomes extinct in a nonisolated habitat, it may be reestablished by dispersal from surrounding areas so rapidly that we have no inkling that an extinction occurred. Recolonization is slower on islands, so that we detect the extinctions. For the many species that occur only on a single island (generally through the colonization of an island by mainland stock and subsequent evolution there), no difference exists between local extinction and extinction of the species.

Extinctions brought on by human activities are occurring throughout the world. The potential for loss is greatest in the tropics where perhaps half of all plant and animal species live. The incredibly rapid destruction of tropical forests (described in Chapter 16) will probably cause the extinction of thousands of species within the next few years. One estimate is that only about 15% of the species of tropical forests have been described and named up to this time. Without drastic conservation measures, it seems sure that many of these uncatalogued species will be extinguished without our ever having the chance to see what insights they may give us about evolution or other aspects of science or what benefits they might have for us in the way of beauty or utility. It is a loss like the destruction of the libraries of Alexandria but immeasurably more calamitous.

There is more about the process of extinction in Chapter 19, Conservation Biology.

The Population—Community Ecology Interface: Herbivory and Predation

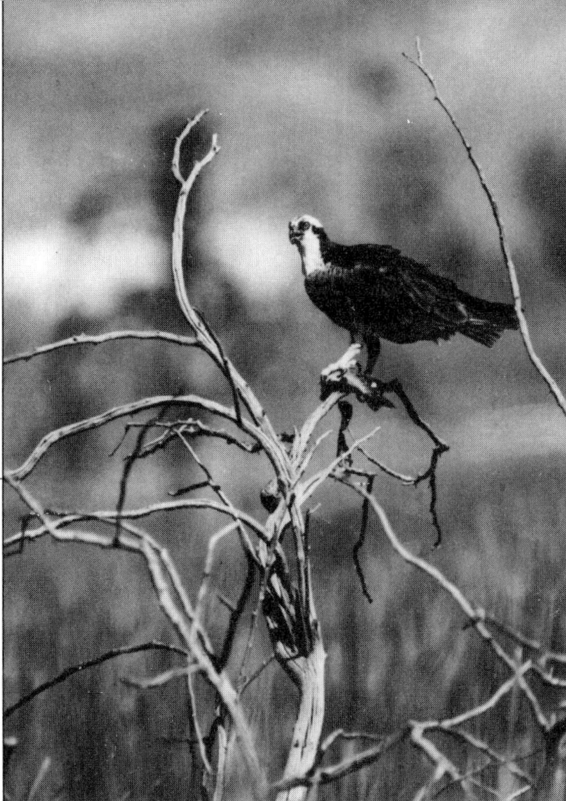

The osprey is a predator that specializes on fish that it catches by plunging into lakes or rivers. *(Dominique Braud/Dembinsky Photo Associates)*

OUTLINE

- Types of Species Interactions
- Trophic Interactions
- Optimal Foraging
- Herbivory
- Frugivory
- Seed Predation
- Predation
- Mimicry and Other Kinds of Advantageous Resemblance
- Biological Control
- Cannibalism, Siblicide, and Intraguild Predation

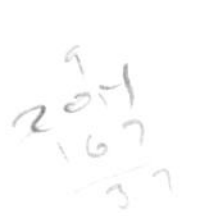

Types of Species Interactions

Species interactions, or coactions, are at the heart of ecology. The focus here and in the next chapters is on describing coactions and evaluating their effects on the interacting species. The relationship of one species to another is the interface between population ecology and community ecology. Although some community-level effects are touched on, general discussions are reserved for later chapters on community ecology.

The classification of coactions in current use was formalized in a 1949 paper by philosopher E. F. Haskell. Its basis is whether the effects of the coaction are helpful, harmful, or neutral for any two species. The classification uses a gain-loss matrix such as given in Table 6-1. Predation, for example, is the interaction in which the strong (usually meaning bigger) member is helped and the weak member is hurt in the interaction.

The classification has been a useful step in thinking about interactions (Malcolm 1966), but it and various proposed modifications have serious deficiencies. Applications of Haskell's classification are rarely clear about whether the effects are on the interacting individuals or on the populations and, if the latter, whether the effects to be considered are those on population growth rate, eventual population size, or evolutionary modification.

A second problem is that some of the cubbyholes contain several different interactions. The competition, parasitism, predation, commensalism, and mutualism cells of the matrix all contain a variety of phenomena. Parasitism, for example, is the coaction in which the weak member benefits and the strong member loses. Even persons believing that this kind of classification represents a fundamental way of looking at the world might wish to discuss separately such varied "parasites" as bacterial pathogens, tapeworms, acorn weevils, foliage-eating caterpillars, aphids, and yellow-bellied sapsuckers.

Some cells of the matrix, on the other hand, are all but empty. "Allolimy," in which the strong species is harmed and the weak species unaffected, seems fairly unimportant in the overall scheme of things.

Perhaps the most important statement we can make about classifications of coactions is that the relationship between any two species can pop from one cubbyhole to another as the result of disconcertingly small changes in circumstances (Figure 6-1). Microorganisms commensal in humans can become pathogens as the result of a round of antibiotics. Pea crabs that live in the mantle cavity of mussels are parasites in low-nutrient environments but cause little, if any, decrease in mussel growth in high-nutrient environments (Bierbaum and Ferson 1986). The interaction between bark beetles and trees wobbles back and forth between parasitism and saprobism. Added to such complications is a strong evolutionary pull toward the development of mutualism from a variety of starting points.

So let's not chain our thinking too closely to the Haskell gain-loss matrix. Here it is used as a starting point for the descriptive classification given below. Gain and loss are evaluated by the question, Who does what to whom? on the individual level not, What is the eventual effect on whom's population?

The following interactions are discussed:

Trophic interactions (in which one species uses another as food)
- Herbivory
- Grazing and browsing
- Frugivory
- Seed predation
- Other kinds of herbivory
- Predation
- Parasitism and disease
- Saprobism

Table 6-1
The Gain-Loss Matrix Conventionally Used for Classifying Coactions

Terminology is modified slightly from Haskell (1949). The terms in quotation marks are rarely used.

Effect on "Strong" Species	*Effect on "Weak" Species*: –	*0*	*+*
–	Competition	Amensalism, "Allolimy"	Parasitism
0	Amensalism	Neutralism	Commensalism
+	Predation	Commensalism, "Allotrophy"	Mutualism, Protocooperation

Figure 6-1

The boundaries between commensalism, mutualism, and parasitism are not always clearcut. (Reprinted by permission: Tribune Media Services.)

- Nontrophic interactions
  - Commensalism
  - Competition
  - Amensalism and allelopathy
  - Neutralism
  - Mutualism (many types include a trophic element)
  - Mycorrhizae
  - Pollination

Trophic Interactions

The use of one organism as food by another is a complex subject. First, the activity is the basis of the transfer of energy through the community (see Chapter 11). Second, it is a subject that must consider the diverse kinds of hunting and foraging techniques used by animals to find and catch food, and the diverse ways by which plants and animals avoid being eaten. Third is the numerical effect, the influence on population size of both the eater and the eaten as a result of the interaction. Fourth is the evolutionary effect, the selective effect that predation has on the prey and that food shortage has on the predator.

Ways of feeding are exceedingly diverse, ranging from clams that filter water for microscopic organisms and bits of organic matter to wolves that catch and eat moose, from bison that crop off grass to plant lice that suck sap out of leaves, from bird lice that eat feathers to blood flukes, liver flukes, or intestinal flukes that soak up nutrients from body fluid. There seems to be a fairly basic subdivision between the organisms that routinely kill another individual organism in their feeding and those that get their food without necessarily killing the individual supplying it. The first category includes predatory animals such as insect-eating birds and rat-eating snakes, but also seed-eaters such as many kinds of birds, small mammals, and insects. The second category includes grazing and browsing plant-eaters and external and internal parasites.

If we simply take a particular animal as it is, with sharp teeth or blunt, long legs or short, a digestive system built for grass or one for flesh, then its diet probably depends mainly on two factors: the availability of different food items and the animal's food preferences. In starting at the present and ignoring how the horse came to be fitted for eating grass or the Everglade kite for eating a single kind of snail we are ignoring a great deal of ecology. The relationship between the eater and the eaten is a strong evolutionary force for both, and we will return to the topic a little later.

In the summer, red foxes may eat meadow mice; in the winter, when snow is deep, they may eat rotting apples because that is what is available to them. Availability depends on abundance but not strictly on abundance. Meadow mice may not be much less abundant in winter than in summer but they are much less available, much less easily found, when the snow is deep. Food items may be unavailable for other reasons; they may be camouflaged or have some other kind of protective coloration or shape, or they may have protective devices such as thorns, spines, or prickles. It is no accident that in an overgrazed pasture everything may be cropped off low except the thistles. The protective devices may consist of unpleasant or poisonous chemicals; most people know that toads and monarch butterflies are poisonous, but there are many other animals and plants that produce poisonous or repellent substances. Millipedes spray cyanide; the bombardier beetle through a most remarkable system shoots out noxious, boiling hot chemicals. Nicotine, caffeine, pyrethrin, and rotenone are examples of chemicals that seem to protect the plants producing them from grazers. Organic gardeners believe that interplanting species with poisonous or repellent qualities, such as marigolds, chrysanthemums, or garlic, with more palatable crops cuts down on insect damage to the crops. This seems reasonable, but it is not a topic on which the agricultural colleges have done much research.

Probably no protective device is perfect, at least for very long. The evolutionary development of a protective poison by a plant presents an enormous opportunity for a herbivorous species if it can somehow evolve immunity; that is why there are tobacco worms. But for many herbivorous insects tobacco plants are unavailable even where there is a field full of them.

Within the range of items that animals are able to catch and use as food, some are usually preferred. Red foxes do not eat many apples if they can get white-footed mice, and they do not eat many white-footed mice if they can get meadow mice (Scott and Klimstra 1955). When food is scarce, an animal may take what it can get, but when food is abundant, the animal tends to concentrate on its preferred food.

Starvation is a powerful selective force for the predator to improve its hunting, and being eaten is a similar selective force favoring better methods of escape by the prey. Predator and prey are in a kind of evolutionary race. If one wins the race, then we no longer see them as a predator-prey system. If the prey become so adept at escape that they are no longer taken, then the predator must turn to another food (or become extinct). If the predator becomes so efficient a hunter that it exterminates not only the old and sick but the young and healthy, then the prey is gone and so is the predator, again unless it can turn to other foods. Thus the predator-prey systems that we see are those that work, those in which the predator does not overeat its prey, nor the prey starve its predator.

The occasional situation in which predator, parasite, or disease does virtually exterminate a prey or host is usually one in which the species are newly exposed to one another; they have not yet started an evolutionary race. Familiar examples are the almost complete annihilation of the American chestnut by the chestnut blight introduced from Asia around the turn of the century, the virtual elimination of American and slippery elm in much of the eastern United States by the Dutch elm disease brought to this country from Europe in the late 1920s, and the near extinction of the lake trout in the Great Lakes once the sea lamprey was allowed entrance (by the construction of the Welland Canal around Niagara Falls).

These occurrences, along with the success stories in biological control, make it clear that the numbers of an organism can be limited by predation—that is, predation can set the carrying capacity. Hunters have never had any doubts about this.

Hawks eat quail; consequently, a hawk killed means more quail to be hunted. Game biologists and ecologists in general are less certain that predation is a limiting factor for many species. Paul Errington's research on bobwhite (and muskrats) convinced him that up to a certain density these animals were almost immune to predation, except as an occasional accident. This density was set by food and cover and by the animal's perception of what was crowded and what was not. Surplus birds were forced from favorable habitat. Errington (1962) wrote:

> *Although there can be fighting (including fighting between social groups that may not be dissimilar to human warfare), the limiting factor of social intolerance need not always take the form of overt antagonism or fighting. Some of the most significant intolerance can have such benign bird-between-bird manifestations as frictionless avoidance or withdrawals on the part of the individuals or groups that recognize their own superfluity in places where they do not belong.*
>
> *Bobwhite equivalents of displaced persons wander in strange places or try to live in uninhabitable areas. In their wanderings, they tend to be harassed by and vulnerable to predatory attacks. Not only may they be vulnerable to such formidable predators as great horned owls and dashing blue-darter hawks and agile and clever foxes but also to rather weak and clumsy predators having no special aptitudes for preying upon grown bobwhites unless something goes wrong. To a considerable extent, it may not seem to make much difference what kills the birds that are trying to live under highly adverse, if not hopeless conditions.*

Optimal Foraging

Animals have to eat, and those that can find and catch food efficiently should do better than those that can't. This will obviously be true for situations where food is in short supply, but it is more generally true because animals that use less time in foraging can use more time for taking care of young, defending territories, etc.

As a consequence, animals should tend to forage optimally. That is, they ought to consume the things that give them the best return for what they spend, which is mostly time and energy. If they behave in this way, they become counterparts of the rational man of the economists, that mythical creature who knows the price of all things and always chooses the best buy. Many things may conspire to prevent a human from being a perfectly—or even a halfway—rational consumer. However, one big thing has a strong tendency to make animals forage optimally, and that is natural selection.

What might be, for example, the optimal size prey for an animal to eat? Cost goes up with increasing size of the food item (Figure 6-2) because, we will say, bigger ones are scarce and, consequently, require more search time or, perhaps, they put up more of a fight, requiring more energy to subdue. Energy intake, or benefit, increases at first because larger items have more energy than smaller ones, but then decreases because the big food items are scarce, have more bone, etc. Where the curves are the greatest distance apart, energy intake per item

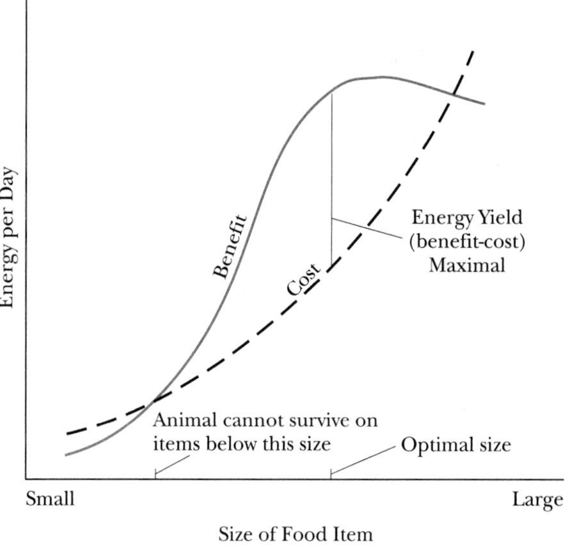

Figure 6-2

An example of determining optimal diet. If we can determine assimilated energy per day by size of food item and the energetic cost of finding, catching, and devouring the different sizes, we can determine where the yield of energy is greatest. This would represent the optimal size.

Table 6-2
Characteristics of Different Kinds of Flowers in the Enclosure Used for Bumblebee Foraging

The last column, Potential Benefit, is obtained by multiplying column 1 (sugar content per flower) by column 2 (number of flowers handled per minute).

Species	*Sugar per Flower (mg)*	*Handling Time (Flowers per Minute)*	*Potential Benefit (mg Sugar per Minute)*
Touch-me-not	2.8	10.7	30
Turtlehead	3.3	2.8	9
Red clover	0.05	44.0	2
Others	—	—	0.01 or less

Data from Heinrich, B., " 'Majoring' and 'minoring' by foraging bumblebees, *Bombus vagans:* An experimental analysis," Ecology, 60:245–255, 1979.

is maximized, and in this oversimplified example, foraging behavior would be optimized if the animal concentrated on this size of prey.

Do animals, in fact, tend to select optimal diets? In many studies that have been done in the last few years, they show more or less close approaches. Bernd Heinrich (1979) studied bumblebee foraging for nectar. In one experiment, he placed individual bumblebees in an enclosure that held several different kinds of flowers. He measured the amount of sugar per flower and the handling time (the time required to extract the nectar) and calculated the potential benefit of visiting a single flower of each species expressed as milligrams of sugar per minute (Table 6-2). On early visits, most bumblebees tended to try flowers of several different species, but after the fifth visit to the field (that is, after Heinrich had put each bee into the enclosure five separate times), most bees visited only the most profitable species, the touch-me-not (Figure 6-3). In doing this, they were selecting flower species optimally.

Robert MacArthur (1972) broke foraging down into a series of steps consisting approximately of the following:

1. Deciding where to look for food
2. Looking (searching)
3. Deciding whether to pursue a food item once one is spotted
4. Pursuing
5. Capturing and handling

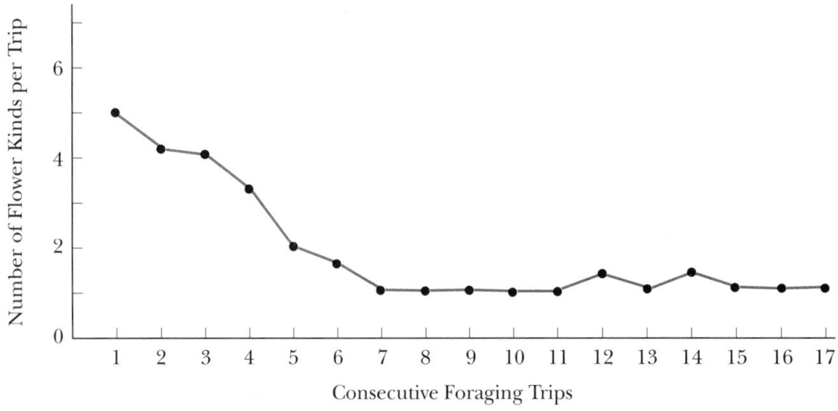

Figure 6-3

Average number of different flower species visited by bumblebees placed by themselves in an enclosure in a meadow. In their first experience in the enclosure, they tended to visit several species; on later trips, they spent their time on fewer kinds. (From " 'Majoring' and 'minoring' by foraging bumblebees, *Bombus vagans:* An experimental analysis" by B. Heinrich, *Ecology,* 1979, 60:245–255. Copyright © by Ecological Society of America. Reprinted by permission.)

The two decisions, 1 and 3, MacArthur approached through the use of economic models. In the following discussion, the language used sometimes suggests that animals are making conscious choices. This is doubtlessly untrue in most cases; they are responding more or less automatically, taking big seeds instead of little ones, leaving an environment where they are going hungry, etc.

Where to Look

The question the consumer or predator has to answer is, Do I look here or do I look someplace else? By someplace else, we mean another patch of suitable habitat. When should an animal give up on one patch and go to another? It will be foraging optimally if it moves to a new patch when

Time to find an average item in a new patch + travel time between patches < time to find an average item in the old patch

We might use a similar model to decide whether to stay in a slow line at the bank or switch to another one. Will we get done faster by staying in our line or by switching to what seems to be a faster one if we include the time lost by going back to the end of the line?

The same ideas are shown graphically in Figure 6-4. The rate of energy gained from a patch is rapid soon after arrival (*A* in the graph) but then declines because of such factors as prey depletion by the predator or through the prey getting scared and hiding or leaving. A line tangent to the gain curve gives the rate of energy intake at any one time. The highest rate, taking travel time into account, is given by the tangent passing through time zero (that is, the time of leaving the old patch). The time corresponding to this rate (*R* on the graph) is the optimal patch residence time. If the animal leaves earlier, it will have a lower net rate of energy gain because it will be wasting time traveling. If it leaves later, it will be spending too much time in depleted patches that have low rates of return.

If the patches are all alike, the animal following such a rule would spend the same length of time in each. If the patches vary in quality (as in Figure 6-5), the animal should leave a patch when its rate of return is the same as that at the optimal residence time in an average patch. This rate is given by lines tangent to the curves for rich and poor patches but parallel to the line for the average patch. If the animal were, instead, to follow the same rules as when the patches are all alike, it would stay too long in the poor patches and not long enough in the rich patches (top of Figure 6-5).

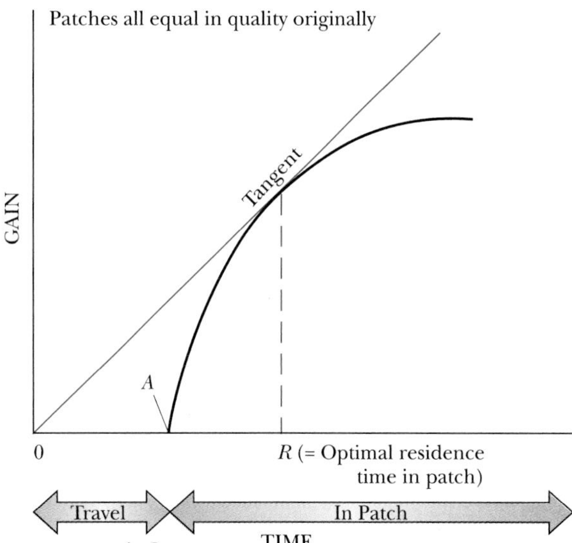

Figure 6-4

Optimizing patch use in foraging (based on Parker & Stuart 1976). The rate of energy gained from a patch is rapid soon after arrival (*A*) but then declines. A line tangent to the gain curve gives the rate at any one time. The highest rate, taking travel time into account, is given by the tangent passing through zero (when the animal leaves the old patch). Past this point, the energy gain per unit time drops. If the animal is to be foraging optimally, it should leave at this time.

One prediction from this model is that an animal will stay a longer time in a rich patch and a shorter time in a poor one—an idea that we might have reached more simply by other means. Another prediction not quite so obvious is that prey abundance (or, at least, the rate at which the animal is taking food) ought to be the same when the animal leaves a patch whether the patch started out as rich, poor, or average. In economics, one would speak of similar situations (buying raw materials, for example) as having the same "marginal" cost, so Charnov (1976) termed this model of patch utilization the **marginal value theorem**.

Clearly, this model would need to be made a

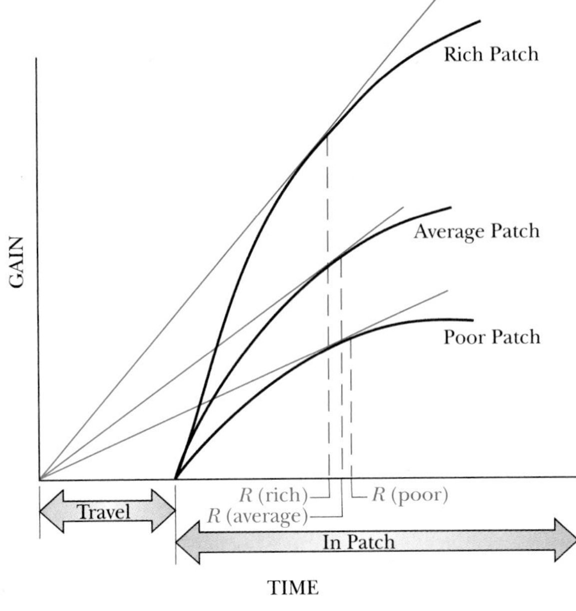

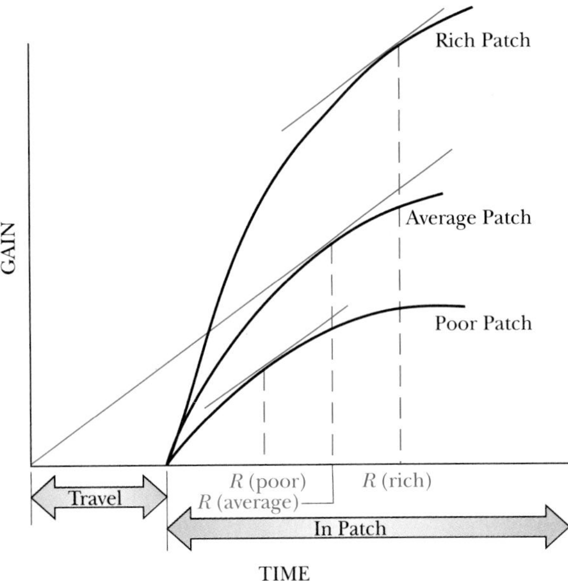

Figure 6-5

If patches vary in quality, the animal will no longer be foraging optimally if it treats each patch the same (top graph). Rather, it should leave the poor patches quicker and stay longer in the rich patches. Specifically, the optimal residence times (R) are given by the tangents parallel to the average patch tangent (bottom graph). In similar fashion, management of a manufacturing company when they need to close factories will shut down the less efficient ones before the average ones and average ones before they shut down the most efficient ones.

little more complicated for real-life applications. For example, if the average item in each patch is not the same size, we will have to take differences in energy content into account. But this will do for a start.

What sort of information will organisms need to make such decisions? Possibly the simplest approach would be to have a **giving-up time**, related to travel time and the time needed to find a food item in an average patch. Once the animal went that long without finding a new food item, it would take off.

Of course, having specific knowledge about one or more new patches would be even better information. Bumblebees tend to concentrate on one productive flower species at any one time—their "major"—but also spend a smaller amount of time at one or more other species—their "minors" (Heinrich 1979). In experimental setups, animals such as European starlings (Inglis and Ferguson 1986) and Mongolian gerbils (Forkman 1991) will pass up freely available food to visit and sample other food sources. In one sense, such behavior is a deviation from optimal foraging because the animals are spending time traveling to and foraging in suboptimal patches. But the behavior will ordinarily improve efficiency over the longer term because of the animal's awareness of its environment and consequent ability to avoid later food shortage or starvation.

Animals need not use just time since last capture as their measure of whether to leave a patch or not. They could, for example, use a moving average of the last three or four captures or any of a variety of more sophisticated indices.

Whether to Pursue

When an animal spots a prey item, it needs to respond either by attempting to take it or by ignoring it and continuing the search. MacArthur suggested that an animal should pursue a given item if

$$\text{Pursuit time for this item} < \text{average pursuit time} + \text{average search time}$$

This is assuming that the item is average in energy content and handling time. A very fat mouse might be worth chasing even if the pursuit time was going to be high. A heavily armored bug, requiring time-consuming shucking, ought to be ignored.

These economic models provide a general explanation for a variety of well-known natural history observations. Suppose, for example, that search time is always large compared with pursuit time. Suppose we are dealing, for example, with warblers gleaning insect larvae off leaves or woodcocks probing for earthworms in the mud. When these predators find their prey, pursuit time is virtually nil, and so it will almost always be the case that

$$\begin{matrix}\text{Pursuit time} \\ \text{for this item}\end{matrix} < 5 \begin{matrix}\text{average} \\ \text{pursuit} \\ \text{time}\end{matrix} + \begin{matrix}\text{average} \\ \text{search} \\ \text{time}\end{matrix}$$

These organisms should, accordingly, always pursue. They should eat every prey item they find or, in other words, be nonselective feeders.

If, on the other hand, most prey are easy to find but hard to catch, the predator should be selective. An example might be a lion which might rarely be out of sight of herds of grazing mammals that are suitable meals but are difficult to catch or subdue. Here the predator should be selective, choosing old or sick animals or ones that can be taken by surprise.

Similar reasoning applies to productive versus unproductive environments. In a productive environment, average search time will be lowered, and so the same predator should be more selective in a productive environment (or a year of plenty) than in an unproductive environment (or a year of scarcity). This is the basis for the frequently made natural history observation that predators may make a greater impact on their preferred foods during times of plenty than times of scarcity.

Note that on the basis of this model, rarity is no protection against predation, because pursuit time is not related to rarity. Search time for that one species would be, of course, but only *average* search time is important in deciding whether to take an item or not. Doubtless, this model like the last is an oversimplification, however. One way in which foraging is more complicated than this is that many animals probably form **search images** (Tinbergen 1960). A predator learns to look for a specific kind of food because it is common and full of energy; the result is that the predator finds this item more efficiently and other items less so. If search images are formed, rarity could offer protection because the predator does not immediately recognize the rarer item as suitable prey.

Central-Place Foraging

In many situations an individual is not free to flit hither and yon within its home range. It may instead need to return at intervals to a nest or some other more-or-less central place. If so, the models just looked at are no longer adequate. Other simple models have been developed to deal with this situation (Stephens and Krebs 1986). In general, these predict that to maximize the rate of energy supplied to the central place, the forager should be more selective in distant patches. That is, there is no point in a kingfisher flying 2 miles downstream and settling for a baby minnow that it could have caught 10 yards (9.1 m) from the nest. It should stay a little longer, catch something a little bigger, and make the trip worthwhile.

Beavers cut down aspen and cottonwoods, trim off the branches, and haul them back to the lodge for food. Several studies have shown that beavers tend to conform to the predictions of the central place models. In a study on a Utah stream, for example, beavers cut branches with a larger mean diameter from trees far from the stream and branches with a smaller mean diameter from trees near the stream (McGinley and Whitham 1985).

What Do Tests of Optimal Foraging Tell Us?

Suppose that we compare the foraging of some animal to one of the models and find that it does not fit particularly well. Does that tell us that the population is doing it wrong? Probably not, though we should remember that individuals vary in ability; young animals, for example, need practice before they can forage as efficiently as adults (Wunderle 1991). But suppose that our sample is good. At their basis, optimal foraging models simply say that foraging, like other activities, is molded by natural selection. If we find a mismatch between what the animal is doing and the model, it is generally a signal that the simple model of search, pursuit, and energy maximization has neglected something important to the animal's fitness. We need to learn more about the natural history of the species. One likely problem area is the failure of energy as a stand-in for fitness.

1. Every animal has specific nutrient requirements that have to be satisfied along with getting sufficient calories. For example, the food of many herbivores is high in potassium and low in sodium. Excretion of the heavy potassium load also results in sodium loss with the result that many herbivores have to resort to special techniques to replenish their sodium supply. Porcupines may deviate from optimal foraging in energetic terms by leaving their territories to travel to lakes where they swim out to eat water lilies and other aquatic plants high in sodium (Roze 1989).
2. Some prey that would otherwise be highly satisfactory are not eaten because they are unpalatable, poisonous, hard to find, hard to reach, not available at the time of day when the predator is active, unrecognized as food, or for other reasons.
3. Fitness will not be maximized for an animal that eats the most energy-laden food if, in the process, it exposes itself unduly to predation. Many studies have shown that predator avoidance often causes animals to avoid food-rich but dangerous habitats. Male moose and solitary cows tend to occur in the areas of Isle Royale National Park where the food supply is best; cows with calves, however, retreat to islands where the the abundance of preferred foods is lower but where wolves are nearly absent (Edwards 1984).

   Gray squirrels foraging in the open are more apt to carry larger food items back to the relative safety of a tree (Lima et al. 1985). This adds to travel time but reducing risk of predation during the time required to eat the larger item evidently makes it worthwhile. Rumination may be a more elegant solution to the problem of combining predator escape with food intake. Cows, giraffes, camels, and other ruminants clip off grass and swallow it unchewed into their rumen. Then they trot off to a safe place, regurgitate the grass, and peacefully chew their cuds.

   Predation enters the equation for central-place foragers also. Birds feeding young need to supply them with calories, but they may also need to protect the nest (Martindale 1982). Following rules for maximal delivery of calories to the nest could cause the adult birds to stay away so long that the nest was left vulnerable to predators and nest usurpers. Optimal foraging rules in such cases would be a compromise between the two conflicting imperatives of the most food and the best protection.
4. Even if energy makes a good currency for measuring fitness, is the organism fittest if it maximizes energy intake or if it minimizes time to acquire some adequate amount of energy? The answer probably differs from time to time and organism to organism. Still more subtle or complicated ways of optimizing can be imagined. For example, when would it be fittest not to maximize energy but to minimize the risk of going hungry? That is, what circumstances would lead to behavior where a low but certain energy supply (like buying government bonds) is chosen over a higher, but chancier, supply (corresponding to speculative stocks)?

Finally, it is well to remember that evolution does not yield perfection; some of the reasons are discussed in the next section.

Optimization

If we took all the problems listed above into account, would we find that foraging was optimal? We would probably find that it was pretty close, but there really is no reason to expect evolution to optimize anything. First, evolution works with what it has. There may be better ways of foraging for some animals that simply are not available genetically. Second, natural selection is not so much survival of the fittest as loss of the least fit. This means that it leaves not only the most fit but also the fairly fit, semi-fit, etc., depending on how benign the environment is in a given generation. This leads us to the third point, that evolution is an ongoing process and that the environment itself is in constant change. Adaptation, accordingly, is a variable approaching a variable (to reuse H. C. Cowles' description of succession). At any one time, foraging could conceivably be a long way from optimal—but getting better.

Herbivory

Herbivory is plant eating. Herbivores are numerous among insects; about one half the insect species in temperate regions are herbivorous (Table 6-3). There are also many herbivorous mammals, ranging in size from meadow mice to elephants. Few birds other than geese and swans subsist on foliage, but a good number eat seeds, fruit, or nectar. Among other animal groups only the mollusks, turtles, and, in water, the algae-eating invertebrates such as rotifers and copepods contain many herbivorous species.

Types of Herbivores

Grazers, such as bison and grasshoppers, eat herbaceous plants ("graze" and "grass" have the same Anglo-Saxon root). **Browsers**, such as deer and rabbits, eat leaves and twigs of woody plants (Figure 6-6). The same terms are sometimes used to describe aquatic animals, but inconsistently. Animals eating unicellular algae are often said to be grazing, but because they are consuming individual creatures, what they are doing is akin to predation.

Table 6-3
Food Habits of Insects

An entomologist, H. B. Weiss, compiled the food habits of the insect species of various regions. Temperate regions of the United States seemed to converge on the ratio given below.

Food Habit	*Percentage of Fauna*
Herbivorous	52
Saprophagous	19
Predaceous	18
Parasitic	11

From Weiss, H. B., based on page 298 of Frost, S. W., *Insect Life and Insect Natural History*, Dover Publications, 2nd rev. ed., p. 526.

The distinction between grazing and browsing is handy, but it just scratches the surface of the way animals make use of plants as food. Some other modes of feeding include the following:

Eating fruits, seeds, nectar, and pollen, discussed in later sections (Frugivory, Seed Predation, and Pollination).

Leaf mining. Larvae of several species of small beetles, flies, sawflies, and especially moths eat away the tissue between the upper and lower epidermis of leaves. More than 50 species of leaf miners attack oak leaves, for example (Frost 1959).

Boring. A large number of beetles, moths, and flies obtain their nourishment by boring in plant tissues, including roots, stems, trunks, buds, seeds, and fruits.

Root eating. Some invertebrates such as nematodes and cicada larvae concentrate on the below-ground parts of plants. Such burrowing mammals as pocket gophers and the naked mole rat of Africa also eat roots, bulbs, and tubers. Pigs do the same but by rooting them out rather than burrowing.

Sap sucking. Many insects in the orders Hemiptera and Homoptera are adapted to living on plant juices. Well-known examples include aphids, spittlebugs, and leafhoppers. Only a few vertebrates have adopted sap as a diet. The yellow-bellied and other sapsucker species lap sap from holes they make in tree trunks, also eating the insects attracted to the sap. At least one mammal does much the same thing, and it is a relative of ours: Marmosets, South American monkeys, use specialized incisors that point forward to gouge holes in tree bark and then lick up the gummy sap (Izawa 1975).

Gall formation. Certain insects (mostly small flies and wasps) develop within structures called galls. These are produced by exaggerated growth of the plant, resulting from the activities of the animal. The developing animal lives within the gall and feeds on the abundant tissue being produced. An example is the round, ball-like gall seen on goldenrod stems (Figure 6-7) and caused by a fly.

Figure 6-6

A "browse-line" on red cedar produced by white-tailed deer in a Michigan old field. In winters of food shortage, all the branches on preferred species may be eaten as high as the deer can reach, often permanently deforming the plant in this way.

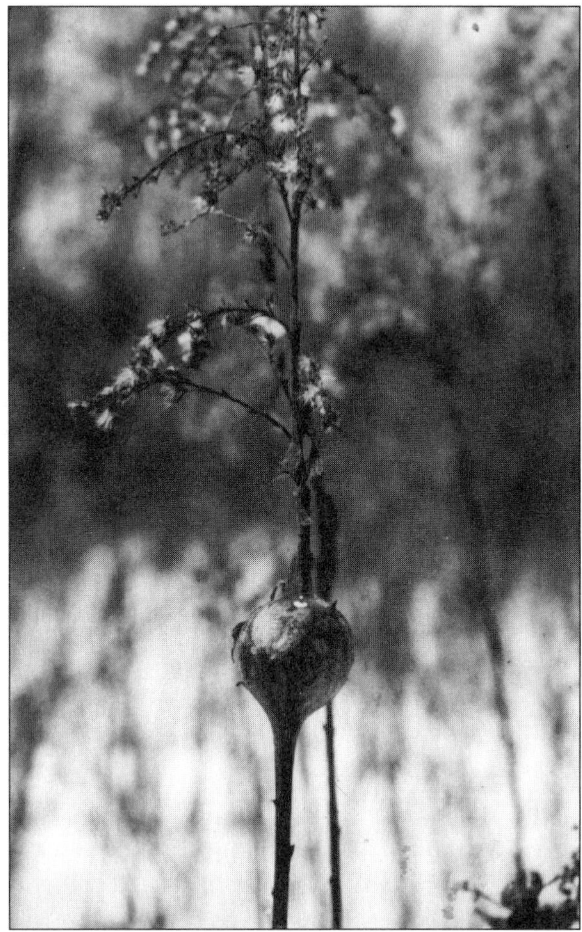

Figure 6-7

A ball gall on Canada goldenrod. The larva of the fly responsible overwinters in the gall, pupates there in the spring, and the adult emerges in May. After mating, the female lays eggs in the terminal bud of a new goldenrod stem.

So, herbivores eat plants. Something like a thousand insect species are known to attack oak trees (Frost 1959), and to these we need to add the vertebrates such as mice and rabbits that clip off oak seedlings and squirrels, woodpeckers, and jays that eat the acorns. The herbivores divide the plant bodies up along lines we have just suggested (Figure 6-8).

Plant Defenses Against Herbivory

At first glance, it may seem that for most herbivores, plants are free for the picking. There are some obvious protective devices such as the thorns and prickles that discourage browsers from eating certain savanna and forest-border trees and shrubs. But the world is green (Hairston et al. 1960); isn't food unlimited for many plant-eaters?

It is now clear that plant defenses are numerous and subtle. Most defenses can be included in three categories: (1) morphological, (2) chemical, and (3) associational.

Besides obvious **morphological defenses**, such as spines and stinging hairs, there are a variety of less obvious ones. The heavily calcification of certain algae protects them from many aquatic grazers (Duffy and Hay 1990). Janzen (1969a) lists a variety

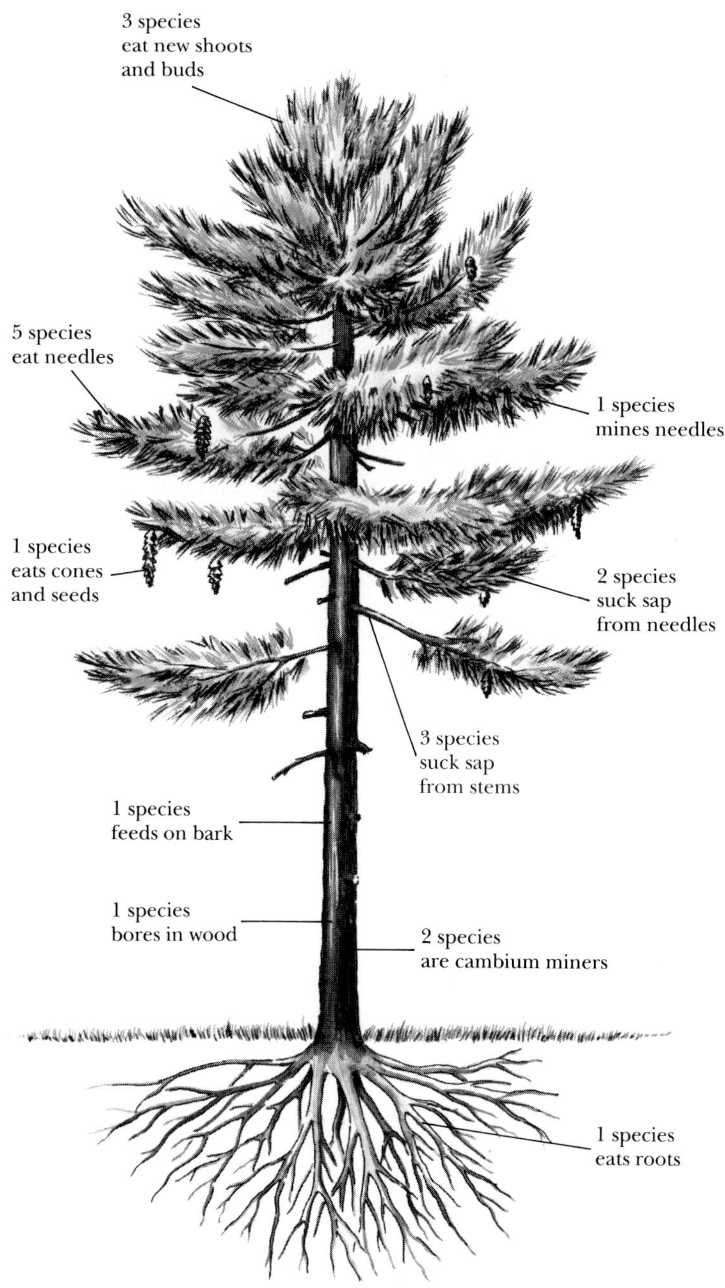

Figure 6-8

Twenty species of insects that specialize or are common on eastern white pine divide the tree up as shown in this diagram. (Based on L. F. Wilson 1977, p. 218.)

of features of legumes that help to discourage seed predation by bruchid beetles.

The complete picture of **chemical defense** by plants is complex (Kogan 1986, Iason and Palo 1991). Many chemical defenses, nevertheless, fall into two general categories.

The plant may accumulate compounds that make tissues hard to eat, difficult to digest, or unpalatable. Tannins are an example. It appears that one group of tannins (hydrolyzable tannins) serves a protective function against many insects by inactivating digestive enzymes (Feeny 1970). Condensed tannins, on the other hand, seem to defend plants against microbial and fungal attack (Zucker 1983). This results in poor digestion of leaves high in condensed tannins by ruminants (which depend on microbial fermentation of food in their digestive tract). In African savanna, ruminants tend to avoid leaves with more than 5% condensed tannins (Cooper and Owen-Smith 1985).

On the other hand, the plant may produce chemicals that are strongly aversive or even toxic. Examples include the mustard oils (glucosinolates) found in several families but most notably in the mustards, alkaloids in the nightshade family, and cardiac glycosides (cardenolides) in the milkweeds and dogbanes.

In general, this type of defensive chemical tends to occur in plants of disturbed areas and early successional stages. Chemicals of the first type, such as the tannins, are more frequent and the fraction of the plant leaf devoted to defensive compounds is larger in longer-lived plants and more stable communities. Several hypotheses have been proposed to explain such trends (Coley et al. 1985); whether a general explanation exists is unclear.

Protective chemicals such as these are generally referred to as **secondary compounds**, implying that they are not essential components of the plant's basic cellular metabolism but by-products instead. Although some biologists regard any connection between secondary compounds and protection from herbivory as accidental, most suppose that natural selection would lead to the spread of genes favoring production of chemicals that conferred protection against the plant's enemies.

The realization that the production of defensive chemicals in plants can be induced by browsing is fairly new (Rhoades 1979). The plant may, for example, begin to turn sugars into tannins rather than translocating them to storage. Chemical responses occur in damaged leaves but also in undamaged ones on the same plant. Most such studies have been on terrestrial vegetation, especially trees (Khan and Harborne 1991), but the same pattern has been observed for the marine brown alga Fucus; clipped plants increased polyphenols both in the damaged branches and adjacent undamaged ones (Van Alstyne 1988).

In large doses, protective substances may kill herbivores. In practice, the main effects may be that odor or taste repel the animal or deter feeding. Some protective substances seem to have subtler effects than simple toxicity. Many plants are now known to produce various animal hormones or mimics of them. Potentially, these could reduce or halt feeding on the plant by causing the animal to go into diapause, metamorphose prematurely, or produce fewer young. Examples are the ecdysone (molting hormone) present in many ferns, among other plants, and estrogen-like substances in many legumes. How prevalent such effects are under natural conditions is still uncertain.

Plant defenses are both a barrier and an opportunity for animals: One man's meat is another man's poison, and vice versa. Any organism that can evolve the ability to tolerate or otherwise circumvent an otherwise widely repellent or toxic compound potentially gains several advantages:

1. A food source for which there are few or no competitors becomes available. Most insects that feed on oak leaves do so in the spring when tannin and lignin concentrations are low. A few species, however, are able to thrive on the full-grown leaves of late summer (Figure 6-9). These species produce highly alkaline conditions in a midgut, a condition that maintains the functioning of protein-digesting enzymes despite high tannin concentrations (Lawson et al. 1982).

   As another example, the seeds of a leguminous vine, *Dioclea megacarpa,* occurring in tropical deciduous forest are protected (as are some other legumes) by a nonprotein amino acid, canavanine. Canavanine is toxic to insects and mammals because it mimics arginine and is incorporated into proteins that then fail to perform necessary physiological functions. A bruchid beetle, *Caryedes brasiliensis,* studied by G. A. Rosenthal (1983) is virtually the only insect attacking the seeds of this plant. It is able to do so because it has evolved the ability to detoxify canavanine (its transfer RNA is not fooled) and breakdown products. The weevil is even able to make use of the canavanine as a nitrogen source for building new amino acids because it, unlike most insects, can convert canavanine (and other amino acids) to ammonia.
2. The defensive features of the plant may, in one way or another, help to ward off the her-

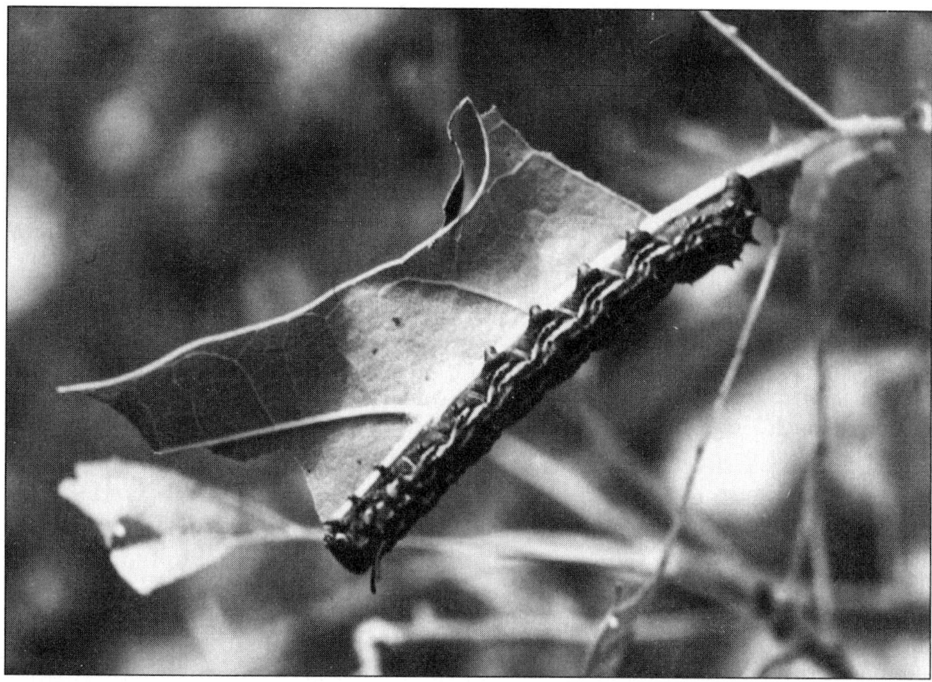

Figure 6-9

The orange-striped oak worm feeds on oak foliage in the late summer.

bivore's predators. It seems to be generally true that, except for venoms, poisonous insects do not manufacture their own poisons. Rather they get them from the plants on which they feed (Figure 6-10). Much the same is true of marine herbivores (Duffy and Hay 1990). Sea hares—big shell-less marine snails—eat seaweeds, such as red, brown, and blue-green algae. Many of these contain secondary compounds, which the sea hares concentrate in their digestive glands. A particular sea hare that preferentially feeds on one specific blue-green alga (cyanophyte) was shown to be avoided by most carnivorous fishes in the waters off Guam (Paul and Pennings 1991). Food flavored with extracts of the seaweed or the sea hare were also avoided by fish. Several deterrent compounds could be involved, but two malyngamides were identified in relatively large quantities.

3. The herbivore may be able to use the defensive chemicals as easy signposts to its food supply. As an example, cabbage butterflies are attracted to sinigrin, a mustard oil found in cabbage, and will lay their eggs on filter paper if it has been soaked in a sinigrin solution.

Figure 6-10

This monarch butterfly larva, feeding on common milkweed leaves, accumulates a cardiac glycoside that does not harm it but is unpleasant or toxic to its potential predators.

Associational defense means the protection gained by a plant by living in association with another species. It may be the simple matter of unpalatable plants happening to form refuge zones for palatable ones growing nearby; some aquatic examples are given by Duffy and Hay (1990). Much more complicated situations are possible, as in the central American tree, big cassia (Janzen 1971). Seeds in pods that remain on the tree are almost all destroyed by various beetles; however, the sweet pulp of the pods is attractive to herbivorous mammals. Where herbivorous mammals are available (that is, in less disturbed habitats), they eat the pods or carry them away, and these seeds removed from the vicinity of the parent plant usually survive.

One of the most often cited cases of associational defense is the attraction of ants by certain species of plants, usually by having extrafloral nectaries—that is, nectar-producing organs somewhere on the plant other than the flowers (Schupp and Feener 1991). Bracken fern and partidge pea are examples.

This ant-plant coaction is a type of mutualism in which the ant receives energy (and sometimes other perquisites), and provides defense against herbivores and seed predators by attacking invading herbivores, especially insects, or by removing insect eggs laid on the plant. Several studies have shown that the ants are effective in cutting the loss of plant tissue to herbivores, but an increase in fitness for the plant has not always been demonstrated (Kelly 1986).

Providing sugar water for ants costs energy. So too do most other forms of plant defense, such as growing thorns and making poisons. They involve using carbohydrates for something other than growing larger and making fruits and seeds. A part of the breeding process for cultivated crops is developing strains that channel as much energy as possible into biomass directly usable by humans. Often this has meant breeding out plant defenses, such as spines, toughness, hairiness, or nasty chemicals, either purposely or as a by-product of selecting for some other trait (Cox and Atkins 1979, Davis et al. 1990). The main human response to this problem has been pesticides, but there has also developed a large industry attempting to breed plants resistant to insect herbivory (Maxwell and Jennings 1980). The latest technique of this particular technological fix is gene splicing, such as inserting *Bacillus thuringiensis* (Bt) genes into cotton, tomato, and potato (Moffat 1991).

Effects of Herbivory on Plant Distribution and Abundance

The role that selective herbivory plays in determining the species composition of vegetation is substantial but poorly known. The role may be difficult or impossible to appreciate simply by observation, no matter how careful. A justifiably famous example involves common St.-John's-wort, called klamath weed in the West. This poisonous plant, naturalized from Europe, became a problem on rangelands, from Washington and Montana south to California. (The species occurs only at low densities in grassland unless the area is overgrazed; also cattle will not eat it except under starvation conditions. Its achievement of pest status is a measure of how scarce good husbandry has been on these grazing lands.) In the mid-1940s biological control efforts were begun using two species of beetle imported from Europe. One of these was successful in nearly eliminating the plant from much of the problem area. Repeated defoliation causing the plant to exhaust its food reserves was the apparent mechanism.

Neither the St.-John's-wort nor the beetle is now common in the rangelands. Furthermore, the plant is now confined mainly to shadier sites, marginal in terms of its former distribution, but sites that the beetle tends to avoid on its egg-laying flights (Huffaker 1957). Without knowledge of the history of this system or without experimentation, we would be unlikely to conclude that herbivory was limiting the plant's abundance in rangeland, and we would probably misinterpret its habitat preference.

Many studies have focused on the effects of herbivory through the use of **exclosures**—plots from which various herbivores are excluded. The effects of protecting the vegetation range from dramatic to nil. In certain areas where deer are abundant, their browsing can alter the composition of the forest that regenerates after clearcutting or even prevent woody regeneration, maintaining herbaceous openings in the forest (Figure 6-11).

Exclosure studies in clearcuts in the Allegheny plateau region show the selective impact of brows-

Figure 6-11

Deer browsing has prevented tree regeneration in this clearcut in northern hardwood forest, as shown by the satisfactory regeneration within the exclosure on the right. (Photo from the Allegheny Plateau of northwestern Pennsylvania, courtesy of the U.S. Forest Service.)

ing (Marquis 1974). Although most woody species show reduced densities outside exclosures, the effect is especially strong on pin cherry, sugar maple, and beech (Table 6-4). Species less favored by the deer, such as black cherry, birch, and striped maple, show proportional increases. How often deer reached densities high enough to produce similar effects under primeval conditions when large predators were present is not known.

An example in which one plant is excluded from a habitat by herbivores attracted to another plant was provided by M. A. Parker and R. B. Root (1981). Two composites, one a shrub known as broom snakeweed and the other a biennial herb, are rarely found together as mature plants. Seedlings of the biennial are, however, sometimes found in the arid overgrazed grasslands where the shrub is common.

Plants of the biennial were transplanted into the vicinity of individual broom snakeweed shrubs. Some of the herbs were protected by exclosures, and others were surrounded by dummy cages that did not keep insects out. A grasshopper feeds heavily on broom snakeweed but ordinarily does not kill it. Of the transplanted biennials protected against the grasshopper, 70% survived to flower, whereas none (of 12) lived as long as 2 weeks in the dummy cages open to the grasshopper.

It is easy enough to classify the effect of the grasshopper on the biennial composite as herbivory. If we wish to give a name to the effect of broom snakeweed on the biennial, probably amensalism is the best choice. Amensalism is discussed at greater length in Chapter 8. Robert Holt (1977) called such a relationship "apparent competition," because the more abundant plant is seemingly excluding the other. Whatever we call it, we need to be alert to the possibility of such interactions. As Parker and Root note, the extinction of small colonist populations by the resident herbivores of a community

Table 6-4
Average Number of Stems over 5 Feet (1.5 m) Tall in 5- to 6-Year-Old Clearcuts Inside and Outside of Fenced Areas

	Protected From Deer		*Unprotected*	
Tree Species	*No.*	*(%)*	*No.*	*(%)*
Black cherry	2.2	(30.5)	1.8	(39.1)
Sugar maple	0.3	(4.2)	0	(0)
Red maple	0.4	(5.6)	0.3	(6.5)
White ash	0.4	(5.6)	0.1	(2.2)
Beech	1.0	(13.9)	0.3	(6.5)
Birch	1.0	(13.9)	0.8	(17.4)
Aspen	0.2	(2.8)	0.2	(4.3)
Pin cherry	1.2	(16.7)	0.1	(2.2)
Striped maple	0.2	(2.8)	0.7	(15.2)
Others	0.3	(4.2)	0.3	(6.5)
Total	7.2		4.6	

Based on Marquis, D. A., "The impact of deer browsing on Allegheny hardwood regeneration," USDA For. Res. Paper NE-308, 1974.

may be a pervasive factor in determining the species composition of communities.

The conventional wisdom of ecology, dating from Darwin, is that herbivory and predation increase species diversity by preventing competitive exclusion (Chapter 8) from going to completion. That is, herbivores keep down the more dominant plant species preventing them from crowding out the weaker ones and, similarly, carnivores keep the dominant herbivore species from forcing their competitors to extinction. This must often be the rule but it is not a universal outcome. My classes and I have found that mowed old field vegetation in southern Michigan has about two species per square meter fewer than adjacent unmowed old field. Frequent, close mowing or grazing seems to be acting as a stress to which only a small number of plants are well adapted. In both the deer and the grasshopper examples mentioned, the immediate effect of grazing was reduced diversity though we do not know what other, indirect effects on other species there may have been.

Frugivory

Frugivory is fruit eating. In this context, "fruit" means fleshy fruit, such as cherries, blackberries, figs, and bananas. Most fruit-eaters are birds, mammals, or insects; bacteria and fungi also invade fruits and make use of them as foods.

Some organisms may gain their energy from the flesh of fruits without dispersing them; in fact, some may eat the flesh and crack the pit to get the kernel, becoming seed predators. Parrots, pigeons, and rodents, among other animals, may do this. The relationship between fruit and animal in many cases, however, is mutualistic. The animal gains calories and moisture from the flesh, and the plant gains dispersal when the bird or mammal regurgitates or voids the seeds.

Frugivory as a year-round way of life occurs only in the tropics, where 50 to 90% of the trees and shrubs produce fleshy fruits (Fleming 1979). Among the typical frugivores of the New World tropics are such birds as toucans, cotingas, and manakins and such mammals as phyllostomatid bats and platyrrhine monkeys (Snow 1981, Fleming 1979). Old World frugivores include musophagids, colies and barbets, megachiropteran bats, and many primates.

Fruit is an important summer and autumn resource for some animals in temperate regions as well as a much less widely used winter resource. Examples of North American animals that are heavy fruit-eaters include catbirds and other mimic thrushes, cedar waxwings, robins and other thrushes, raccoons, bears, and foxes.

Temperate shrubs and trees that bear fleshy fruits tend to fruit in late summer and autumn. This is a time when birds are most abundant (young are out of the nest) and most on the move (territories break down and, later, fall migration is occurring). Both matters aid seed dispersal. An additional evolutionary factor for most plants—a constraint—is that only a short period—after growth and flowering but before winter—is available for fruit ripening.

The period of fruiting of any one species of tree or shrub in the tropics is also short, but, over the entire flora, some species is almost always in fruit. It is likely that the frugivory interaction was involved in the evolution of flowering-fruiting periods (Snow 1976). If we hypothesize a situation in which all the trees of a region bear fruits at the same time, it is clear, first, that year-round frugivory, and thus close adaptation to fruit as a diet, is not possible. Second, it is clear that some fruit would be likely to go uneaten and, accordingly, seeds undispersed. Species that fruited earlier or later than the rest would have the frugivores to themselves. The end result of such a process would be that different species fruit at different times throughout the year. The same effect could be achieved by each species spreading out its fruiting period, some individuals bearing fruits now, some earlier and some later, but this might be disadvantageous from the standpoint of efficient pollination. If only one tree per hectare of a given species came into flower every month, its likelihood of being pollinated might be low.

When fruit is available year-round, the opportunity for close adaptation of frugivores to fruit—and fruit to frugivores—becomes possible. Some features of plants that are probably adaptations (in tropical or temperate regions or both) to the vertebrates doing the dispersing include flower and fruit placement, fruit size and color, and pulp composition (Fleming 1991). Some features of birds associated with fruit eating that are probably evolutionary consequences of that interaction include the following:

1. Social feeding parties occur even in the breeding season. With different trees scattered here and there bearing fruits at different times, the maintenance of a feeding territory is of little utility to a frugivore. By contrast, a pair of insect-eaters, feeding on a resource spread more evenly, can defend an area big enough to provide a continuous supply of insects.
2. Polygyny and promiscuity are frequent mating types. The time needed for foraging is not very great for frugivores. Although individual fruiting trees may be scattered, each one tends to bear large quantities. Search and handling times for fruit tend to be low, and pursuit time, nil. As a result, a single parent can usually do an adequate job of raising young. This is ordinarily the female. In many species, males gather in communal display areas, called **leks,** to which the females come for mating. The dominant males at leks fertilize most of the females. Lek species are far more common among frugivores than elsewhere. Bizarre plumages, presumably the result of sexual selection, are characteristic of lek species. Birds of paradise are probably the most famous example.
3. Frugivorous birds in which the nestlings are fed fruit (total frugivores) tend to nest high in trees (or in cavities). Pits or seeds regurgitated by the young form a pile under low nests, inviting predation. They scatter more when released from a high nest. Height in itself offers protection against terrestrial predators.

Most frugivorous birds, however, feed their young primarily on insects instead of fruit. The main selective force counteracting the evolution of total frugivory seems to be predation. Young fed exclusively on fruit develop more slowly than young fed on protein-rich insects; the time to reach adult weight may be over 70% longer (Morton 1973). This means that the very vulnerable nestling stage is prolonged. The tree species involved in the interaction with total frugivores produce different fruits with a different collection of traits than those that are eaten by the bird species that are frugivorous only as adults. Fruits eaten by total frugivores are high in protein and fat, dryish, dull colored, and not very tasty. Figs are an example. Fruits that are fed on by the wide variety of less specialized frugivores are bright colored, sweet, and juicy, like cherries.

Seed Predation

The use of seeds as food by animals is more similar functionally to predation than it is to grazing and browsing. Predation of a seed removes an individual from the population (of seeds) until seeds are again produced; it is not like eating a leaf where the plant is left otherwise intact and, in fact, may be stimulated to produce more leaves. We are talking of the actual destruction of the embryo plant inside the seed kernel and not just the eating of fruit by an animal. Fruit eating may simply result in dispersal if the seeds pass unharmed through the animal's gut or are regurgitated. Passing through the digestive tract of a bird or mammal even improves germination of some plants.

By contrast, an acorn that a squirrel eats has been depredated; it is not going to become a little oak. Most seed predators are insects (especially moths, flies, beetles, and wasps), birds (such as jays and woodpeckers), and mammals (such as mice and squirrels).

One of the interesting things about the size of seed or fruit crops is that like animal numbers, they can be about the same from year to year, or they can vary in either an irregular or relatively regular fashion. Figure 6-12 shows a 12-year record of seed production by white oak. In three years, seed production was very high; in six, it was very low, and in three, it was intermediate.

Such observations have been explained in two ways. The first hypothesis relates the pattern primarily to the tree's response to climate. It is assumed that good years are ones with a favorable combination of physical factors; for example, plenty of food-making capacity is available, and the conditions for both flowering and getting polli-

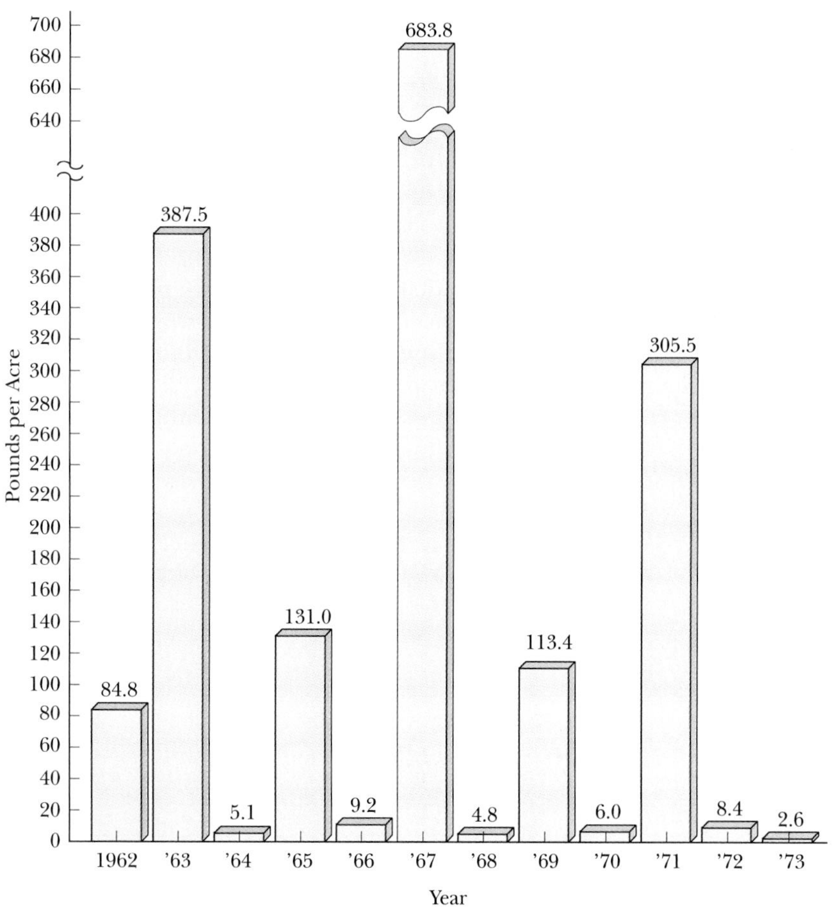

Figure 6-12

White oak acorn production at Bent Creek Experimental Forest in western North Carolina. The biggest year was nearly 300 times the poorest year. Data from D. E. Beck. 1977. Twelve-year acorn yield in southern Appalachian oaks. (USDA Forest Service Research Note SE-244: 1–8.)

nated are favorable. The fact that many trees, especially those with large fruits, such as apples or acorns, rarely have two good years in a row is explained by one further constraint: Several studies have shown that a heavy fruiting year uses a great deal of energy. Most trees must devote their energy to growth and maintenance and to building up future reserves during one or more years following a bumper crop.

These ideas are probably correct but may neglect an important evolutionary cause. The Swedish ecologist G. Svärdson (1957) and, later, Daniel Janzen (1971, 1975b) developed the idea that bumper crops are adaptive by overwhelming, or satiating, the predator population. If a tree produced a medium number of fruits every year, a stable population of seed predators could be supported. By concentrating production in an occasional year, the seed predators that built up during a food year will die or, at least, move on (Silvertown 1980). In a Swedish beech forest, birds, mammals, and a seed-eating moth had removed 50% of the beech seeds by November in a year of heavy seed production but only 9% by the same time in a year of light seed production (Nilsson and Wästljung 1987).

For this approach to predator defense to work well, individual trees have to be synchronized in their fruiting; their good and bad years ought to correspond. We would expect this to be true anyway in a general way. If the weather has been favorable for reproduction by a species, many of its members ought to respond. But there should be an enhancement of this tendency if predator defense is the function of irregularity in crop size.

There ought to be a cue or set of cues that most members of the species use to switch reproduction on or off. Nothing magical is needed for such synchronization to evolve (Figure 6-13). The individual tree that is out of step, producing plenty of seeds in a year when few others do, will contribute to the survival of the mice and weevils but not to its own future generations.

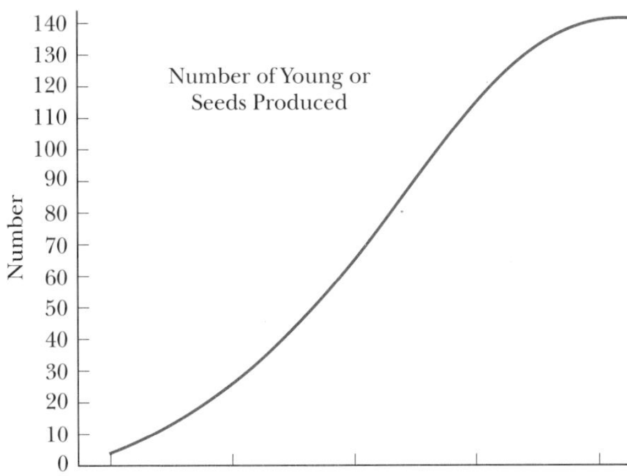

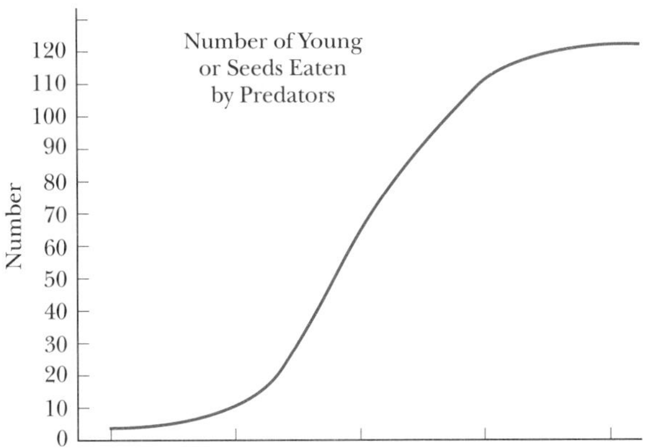

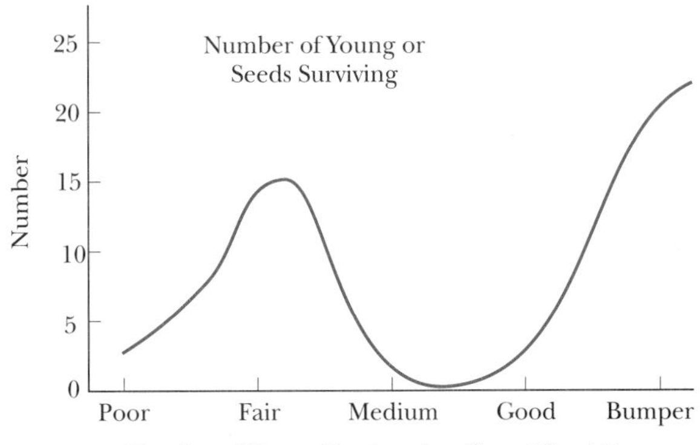

Figure 6-13

If predation on young animals (or seeds) is related to their numbers as shown in the middle graph (opposite), the actual number of survivors could be highest in very good years and second highest in fairly bad years (bottom graph). This could lead to a responsiveness to environmental influences that would contribute to a cyclic population pattern.

Predation

Types of Predators

Predators, organisms that kill and eat animals, range in size from microscopic protozoans to 400-

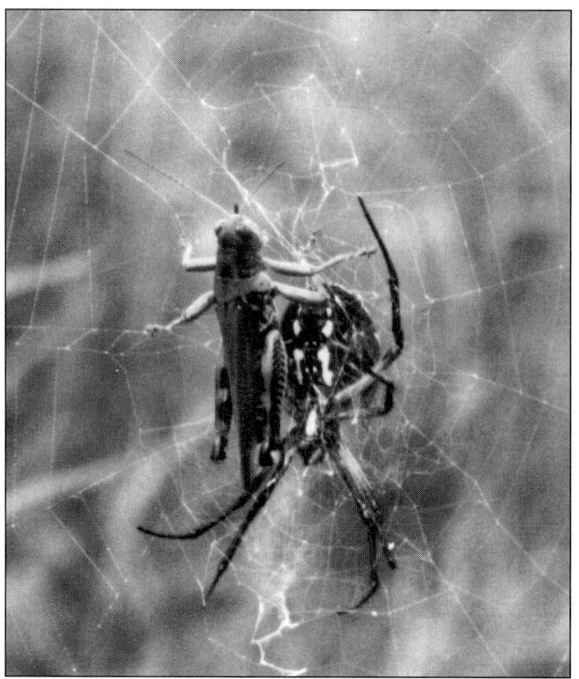

Figure 6-14

An example of a sit-and-wait predator, the garden spider, and guest. (Photo by the author.)

pound (180 kilograms) lions to killer whales weighing close to 5 tons (4500 kg). The way they deal with their prey is also diverse, from anteaters that lick up ants with a sticky tongue, to spiders that suck the vital fluids out of insects, to great horned owls that grasp rabbits in their talons and tear them up and eat them. Even plants can be predators; examples are the Venus flytrap and bladderworts (Schnell 1976).

One way of categorizing predators that cuts across size and feeding method is to divide them into those that sit and wait for prey to come and those that actively seek prey (Huey and Pianka 1981). Sit-and-wait predators may work from concealment, in which case, they would be hunting from ambush. Examples of ambush predators include mantids that look like flowers and feed on the insects that come for nectar, and anacondas, lying in wait beside a jungle stream for unsuspecting animals. "Ambush" is hardly the correct word, however, for the attack of many kinds of sit-and-wait predators such as kingbirds that perch in plain sight between flycatching sallies. Other examples of sit-and-wait foragers include most spiders (Figure 6-14) and owls, ant-lions, and snapping turtles.

Predators that move about actively searching for prey to snap up are more numerous. Included are northern harriers that criss-cross a field looking for mice, warblers that flit through the trees gleaning caterpillars, and dragonflies that cruise the airspace over ponds for midges and mosquitos.

In general, sit-and-wait is an energy-efficient strategy. The ant-lion does not get many calories (or even joules) in a day, but it does not use many. The dragonfly darts about, snapping up mosquitos by the hundreds but also using much energy.

We can see several other features that tend to be correlated with these two foraging types (Huey and Pianka 1981). Sit-and-wait predators depend on active prey, whereas actively moving predators can exploit relatively sedentary types. Sit-and-wait predators often take large prey, relative to their own body size (O'Brien et al. 1990). They are themselves probably vulnerable to a smaller number of predators than are actively moving foragers.

It is unclear, nevertheless, whether these two categories represent some fundamental way of looking at predation or whether they are just ends of a (possibly multidimensional) continuum. Certainly, a large proportion of all foraging animals do some moving and stopping, although the fraction of time spent in each can vary greatly. This kind of foraging has been termed saltatory by one set of authors (O'Brien et al. 1990). Although "saltatory" means "hopping," they were not referring to foraging by kangaroos but instead to the very common situation in which an animal moves a while, stops briefly or for a long time in a new spot to search for prey, and then moves again.

The Lotka–Volterra Predator-Prey Model

The Lotka–Volterra predator-prey model (Lotka 1925, Volterra 1931) is simple; that is the problem. It assumes that the numbers of a predator depend on the prey population, acting on the predator birth rate. That is, if the mouse population is high, the foxes will eat a lot of mice and have a lot of babies. If the mouse population is zero, the foxes

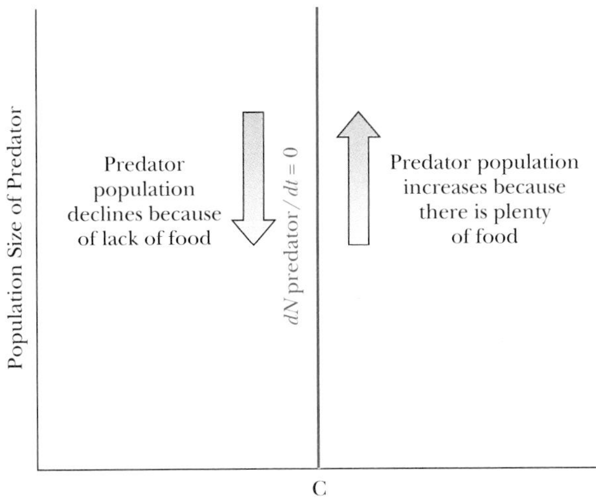

Figure 6-15

Trends in predator populations according to the Lotka-Volterra model.

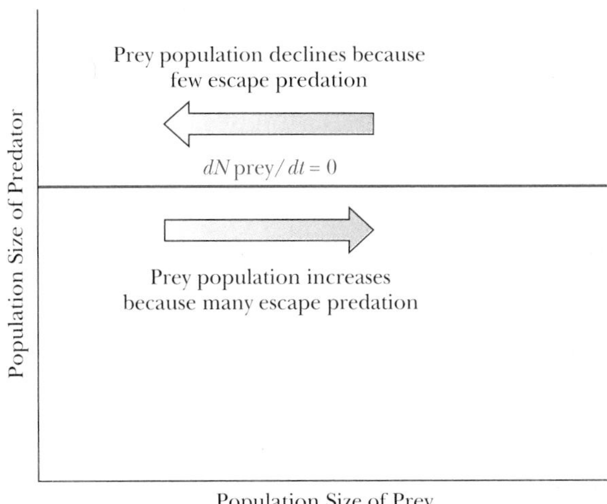

Figure 6-16

Trends in prey populations according to the Lotka-Volterra model.

will not reproduce at all. The numbers of a prey species depend on the predator population, acting on prey death rate. In other words, if the fox population is low, most mouse babies will grow up and the old mice will also live a long time. So prey are predator limited, and predators are food limited.

It is not particularly edifying to go into the equations of the Lotka–Volterra predation model, but they lead to a graph for the predator that looks like Figure 6-15. A population of predators at a point on line C will neither grow nor decline ($dN_{predator}/dt = 0$).

We can produce a similar graph for the prey population (Figure 6-16). Note that this graph is read sideways; that is, the mouse population size is on the bottom and numbrs go up to the right. A population of prey at a point on line D will neither grow nor decline ($dN_{prey}/dt = 0$).

We can now combine these two diagrams to examine the behavior of the system (Figure 6-17). At point E, $dN_{predator}/dt = 0$ and $dN_{prey}/dt = 0$. Neither population is changing in size; the system is at equilibrium. In any real system, chance events would come along to bump one population or the other away from this point, and the behavior of the system is not such as to restore that pair of densities.

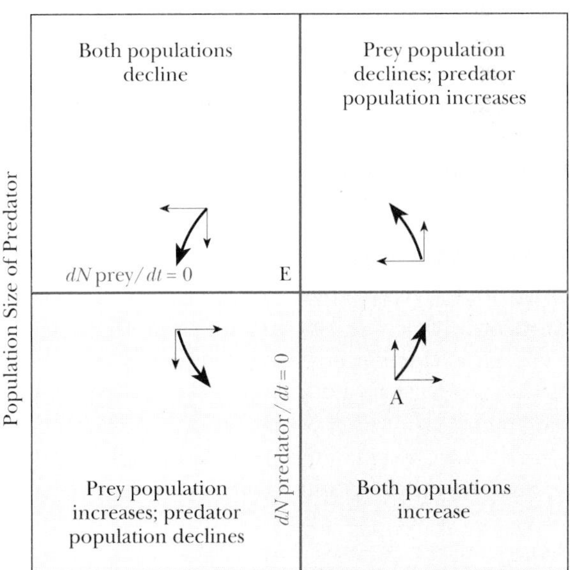

Figure 6-17

Combining Figures 6-15 and 6-16 gives the behavior of the predator-prey system according to Lotka and Volterra.

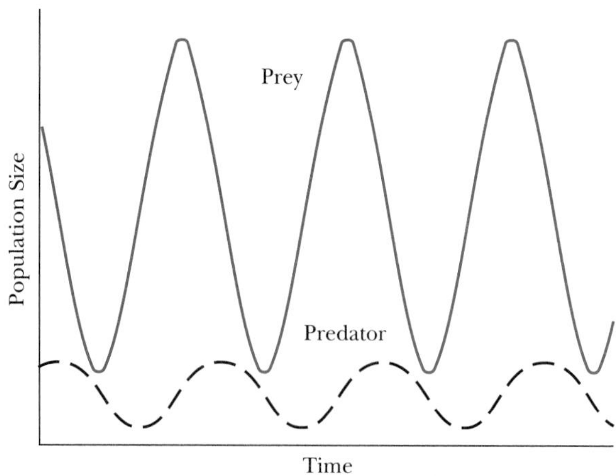

Figure 6-18

Plotting numbers of predator and of prey (as in Figure 6-17) against time gives predator-prey cycles. In this example, the predator averages about one-sixth as abundant as the prey.

Rather, the tendencies of the system are as shown by the dark arrows. To figure out the details would require some further refinement, but clearly the general behavior is circular. With two population sizes as shown by point A, both populations would increase, moving the system into the upper-right quadrant. In the upper quadrant, the changes would be toward predators continuing to increase while prey declined. This would move the system into the upper- left quadrant, etc. The system would eventually return to A.

This was the interesting result of the Lotka–Volterra model: If we pulled out the numbers of predators and prey and plotted them against time, we saw cyclic fluctuations as in Figure 6-18. These seemed to correspond to events biologists were seeing in certain natural populations.

Michael Rosenzweig and Robert MacArthur (1963) used the graph from the Lotka–Volterra model as a starting point for a family of slightly more realistic models. They added a carrying capacity for both predator and prey, because it seems likely that something will set an upper limit for prey numbers even if there are no predators to eat them—their own food supply or social stress, perhaps. Similarly, it seems likely that even with enormous prey numbers, the incompressibility of territories beyond some limit or some other factor will set a *K* for predators. The model also included an Allee effect for the prey, with a lowered ability to grow at very low populations.

Depending on the particular form of the curves, stable oscillations like the original Lotka–Volterra model could be produced. For situations in which the predator is efficient at taking prey even when prey density is low, the result was diverging oscillations, leading to the extinction of one or both species. Most laboratory predator-prey systems give this outcome.

Forms of the model in which the predator cannot increase until the prey is at high densities result in damped oscillations or go directly to a noncyclic equilibrium. Recently, considerable attention has been given to ratio-dependent models that relate predation levels to the ratio between predator and prey numbers (Berryman 1992).

In a famous laboratory study of predation, G. F. Gause (1934) found that when the populations of the protozoans *Paramecium caudatum* (prey) and *Didinium nasutum* (predator) were placed in a homogenous culture medium, the predator ate up all the prey and then followed it to extinction. If, however, the culture medium had sediment in the bottom, prey in the sediment were immune to attack because the *Didinium* would not enter the sediment. The wildlife biologist Paul Errington believed, based on his studies of bobwhite and muskrat, that the latter situation was frequent. Below some threshold density, the prey was virtually secure; it was only at higher densities, above the threshold of vulnerability, that other-than-accidental predation became important.

Rosenzweig and MacArthur (1963) also attempted to diagram this situation (Figure 6-19). Within the shaded area, prey numbers can increase. (To the left, at very low densities, the Allee effect kicks in, and prey numbers decline.) The upward projecting strip represents the refuge, the range of prey densities in which the prey is nearly invulnerable to predation and can increase no matter what the predator density. The general tendency of such a system is to produce stable cycles.

Fluctuations in Population Size

Big changes in animal numbers, especially cyclic ones, have fascinated people throughout history.

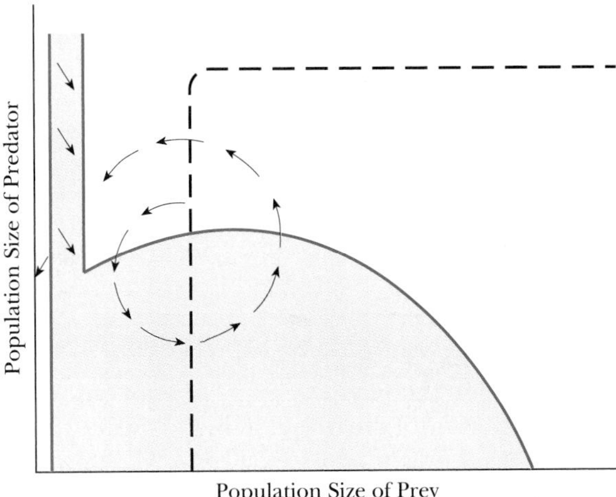

Figure 6-19

In this version of the Rosenzweig-MacArthur model (1963), which may correspond fairly closely with some natural situations, a refuge exists in which prey numbers can grow even though predators are abundant.

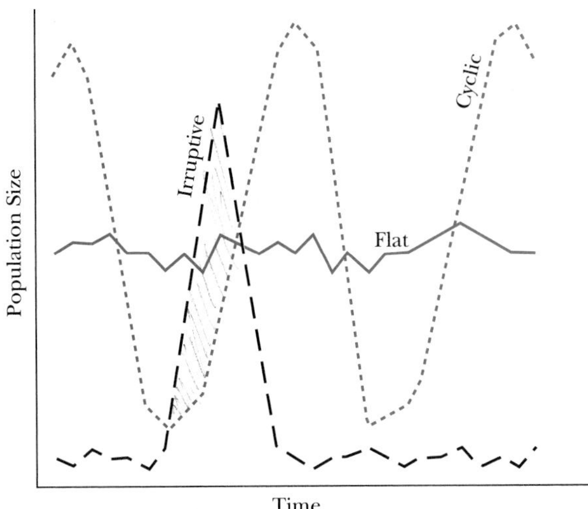

Figure 6-20

Three types of population curves.

The Bible records locust plagues—destructively high densities of the migratory locust—in Egypt. Robert Browning's poem *The Pied Piper of Hamelin* describes a rodent plague in a German city in the 14th century:

> *Rats!*
> *They fought the dogs and killed the cats,*
> *And bit the babies in the cradles. . . .*

Charles Elton (1942) suggested that a mouse plague influenced the course of World War I through its effect on the German food supply. In 1917 to 1918, high densities of field mice, plus five other rodents, ravaged hay fields and pastures that would have supported livestock. They ate seeds and seedlings out of the fields and made heavy inroads into such crops as potatoes, sugar beets, and garden produce like carrots and cauliflower.

Aldo Leopold (1933) in his book *Game Management* recognized three types of population curves—**flat, irruptive,** and **cyclic** (Figure 6-20). Both the irruptive and the cyclic types show large changes in numbers; they differ in that the peaks in the cyclic populations tend to recur at fairly regular intervals. About the same time Lotka and Volterra were writing, cyclic populations were being studied in nature by Leopold and also by Charles Elton and V. E. Shelford. At first glance, the diagrams resulting from the Lotka–Volterra model (Figure 6-18) look appealingly like some of the actual data from nature, such as the plot of varying hare (prey) and lynx (predator) numbers in Canada (Figure 6-21).

The idea that we are seeing Lotka–Volterra cycles in the lynx-hare fluctuations can be readily disproved. The cycles of the two are sometimes in phase rather than with the predator peak one-fourth cycle after the prey peak as should be the case (Figure 6-21). This is suspicious. Besides that, snowshoe hares show a similar cyclic pattern on Anticosti Island where there are no lynx (Keith 1963).

It seems likely that Lotka–Volterra cycles do not exist off paper. The model is just too simplified to fit anything biological. As we mentioned, Gause (1934) attempted to set up a cyclic predator-prey system using *Paramecium caudatum* and *Didinium nasutum.* He was unable to do so. What generally happened was that *Didinium,* a voracious predator, ate up all the *Paramecium* and then died. When Gause provided a refuge for the *Paramecium* in the form of sediment in which those at the bottom of the test tube were safe from attack, the *Didinium* population died out after demolishing the bulk of the *Paramecium* population. The *Paramecium* population then increased rapidly from the individuals that had survived in the sediment.

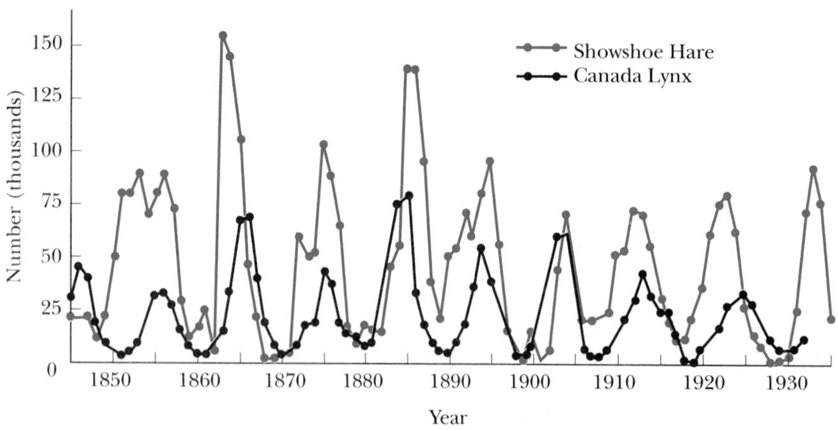

Figure 6-21

Population numbers of varying (snowshoe) hare and Canada lynx in northern Canada based on numbers of pelts handled by the Hudson Bay Company. (Adapted from Kendeigh 1974; after MacLulich.)

Gause obtained cyclic fluctuations only when he began to add one *Paramecium* and one *Didinium* every three days to the culture. Extinctions continued to occur, but now the immigrants were able to set off new periods of growth for whichever species was extinct.

Gause's study and others since, such as Huffaker's (1958) complicated experiments using predator and prey species of mites, lead us to this conclusion: For predator and prey to coexist, enough diversity has to exist in the system for the prey, when scarce, to have a refuge from the predator and, also, for the predator to avoid starving when the prey are no longer available. The second would be possible if (1) spatial diversity existed so that the predators could move away to some place where prey populations were still high or (2) there were alternative prey species to which predators could turn.

But these conditions violate the basic Lotka–Volterra model. If there is a refuge for prey their numbers, to that degree, are not controlled by predation. If the predators eat alternative food, predator numbers are not controlled by the original prey species.

Cyclic Populations

Cyclic populations are ones in which the peaks and troughs recur at fairly regular intervals. A few biologists (notably Cole 1951) have doubted that cycles in animal numbers exist. Partly, this results from the influence of the physical scientists, who define cycle for their own purposes as having a constant amplitude (height) and phase (length), like the sine waves of alternating current. Few, if any, ecological phenomena are that regular. However, the word "cycle" from a Greek root meaning "circle" or "wheel," has developed many definitions over the centuries, mostly figurative. The ecological usage meaning recurrent changes in population size, with a regularity greater than expected by chance, is perfectly reasonable (Butler 1953).

The two most widely reported cycle lengths (Table 6-5) are three to four years and nine to ten years. There is much yet to be learned about cyclic population changes, but, descriptively, we can say several things:

1. Cycles, in general, are characteristic of high-latitude ecosystems, especially tundra and boreal forest. Species, such as the muskrat, that show cyclic changes in the far north, may show little or no evidence of cycles in temperate regions (Hansson and Henttonen 1985).
2. The three- to four-year cycle is more characteristic of tundra, by habitat, and of Eurasia, by geographical location.
3. Where three- to four-year cycles occur in North America, they occur in tundra.

Table 6-5
Examples of Birds and Mammals That Often Show Cyclic Changes in Population Size

Species	*Habitat*	*Location*
	3- to 4-year cycle	
Snowy owl	Tundra	North America
Willow ptarmigan	Tundra	Northern Europe
European lemming	Tundra	Europe
Collared lemming	Tundra	North America
European meadow vole	Grassland	Europe
Arctic fox	Tundra	Greenland
	9- to 10-year cycle	
Ruffed grouse	Mixed and boreal forest	Northern North America
Willow ptarmigan	Tundra	North America
Snowshoe hare	Boreal forest	North America
Muskrat	Marshes	Northern North America
Canada lynx	Boreal forest	North America

4. Cycles of nine to ten years are more characteristic of boreal forest animals but are rarely or never seen in Eurasia.

What causes population cycles? Why do the populations go way up (the ratio of peak to trough in the Hudson Bay pelts was about 240:1 for the snowshoe hare and about 70:1 for the lynx, according to Keith [1963]) and then drop back? There are four basic hypotheses:

1. The fluctuations are somehow inherent in the predator-prey system. Although the Lotka–Volterra model is oversimplified, more realistic models of the situation in which the vital statistics of predator and prey are closely linked can also have stable cycles as the outcome (Henttonenen et al. 1987).
2. The fluctuations are the result of time lags in the unfavorable effects of crowding expressing themselves. Such fluctuations would be inherent in the relation of the organism with the environment but not through predator-prey interactions.

Most of the hypotheses in this category have not addressed the idea of a time lag explicitly but instead have focused on the agent that lowers population size after very high levels have been reached. This truncated, and to the uninitiated, rather cryptic approach goes back to the earliest considerations of the Hudson Bay data. Ernest Thompson Seton (1911), the great American naturalist of the early 20th century, was the first person to use these data for ecological purposes. He claimed, "To explain the variations we must seek *not* the reason for the increase—that is normal—but for the destructive agency that *ended the increase.*" In other words, populations build up based on their biotic potential, overshoot the long-term carrying capacity, and crash.

For lemmings and voles, various investigators have suggested the crash comes because the animals have overeaten their food supply or at least have caused food quality to deteriorate (Pitelka 1964, Schultz 1969), because of endocrine and behavioral changes resulting from stress (Christian and Davis 1964, Andrews 1968), or because of genetic changes that favor fighters not lovers at high densities and the reverse at low (Chitty 1960, 1967).

A time lag associated with the buildup of inducible defensive chemicals in plants has been suggested as a cause of cycles (Haukioja 1983). In this idea, increasing grazing by lemmings or voles causes an increase in the chemical defenses of the vegetation, leading to a decline of the animals. However, the increase in plant chemical defense continues past the time numbers have turned down, dropping the numbers to still lower levels. With a great decline in grazing, the plant chemical defense level eventually drops. Lindroth and Batzli (1986) found that grazing by prairie voles induced the production of phenolics in alfalfa in an Illinois old field, but the high level of defensive chemicals did not carry over to the next year and did not seem to be involved in the cyclic changes in prairie vole numbers.

It is worth repeating a point made in Chapter 4, that the operation of many of the population-regulatory density-dependent factors are likely to have at least some lag associated with their effect. That is, a population has to get just a little too high before it starts to go back down and a little too low before it goes back up. If that is how population regulation works, the question of why more populations are not cyclic may be at least as interesting as why some are.

3. The fluctuations are the result of an outside periodism in the environment. For example, V. E. Shelford (1951) suspected that the ten-year animal population cycle was based on ultraviolet fluctuations that could be measured by the sunspot cycle. Relatively few persons still credit this particular idea. The sunspot cycle is actually more like 11 than 10 years. The sunspot cycle and the lynx cycle coincided beautifully from about 1825 to 1845 but were almost exactly out of phase from about 1850 to 1880.

   Outside influences other than solar radiation could be important. Recently, some investigators have correlated animal numbers with changes associated with El Niño. The food supply, for its own reasons, could fluctuate, and the animals eating it could simply track the food supply. It is likely that many *predator* cycles are of this nature. The predators increase because of the prey increase but may have little influence in the eventual reduction of prey numbers.

4. The preceding categories of hypotheses assume that the cycles are evolutionarily passive. Another hypothesis suggests that the cycles of prey are evolutionarily advantageous in the following way:

   Assume that at low population density, a lemming population will attract almost no predators, so that the few individuals that are there will get along well. Assume that medium-to-large populations will attract heavy predation, but that when populations are very large, the proportion of individuals actually taken by predators drops. Assume, in short, the pattern shown in Figure 6-13. If all this is true, it is to the advantage of an animal to produce a lot of offspring at the same time his or her neighbors are doing the same thing and, likewise, to produce only a few when they are producing only a few. Animals that do not conform to this pattern are likely to produce young that serve only to tide the predators over a lean time.

   This could be an ultimate, or evolutionary, explanation for the existence of highs and lows, rather than a flat population. Reproduction is concentrated in such a way that the predators are overwhelmed, or satiated, and can't keep up. How could this pattern be produced proximately? Perhaps by having a precise and stereotyped response to the physical environment (such as Mullen [1969] found for lemmings). When the weather was such as to be a good year, it would be advantageous to breed, breed, breed. In any other kind of year—medium-good, average, so-so, bad—keep reproduction low.

   Different species could evolve to respond similarly to similar climatic cues if the same predators were involved. If one species of mouse were so common as to overwhelm the foxes (in the sense that the foxes were eating a smaller percentage than at a lower density), then if another mouse species could be common at the same time, it would reap some of the advantages too. This would be true even if it were a rare species that, by itself, never could be abundant enough to satiate the predator. There is a suggestion of this in Figure 6-22. The gray-sided vole contributes the largest part of the peak in most of the highs, but the years of its highs are also highs for several of the other species. (A similar response to fluctuations in the physical environment could produce the same pattern.)

   Although this idea was foreshadowed by earlier authors, most of its recent development has been in connection with seed predation, discussed in an earlier section.

Chaos

One more reason why populations may fluctuate should be mentioned. They could be behaving chaotically. Simple population models can produce seemingly random fluctuations in population size if they are nonlinear—if they have some term with

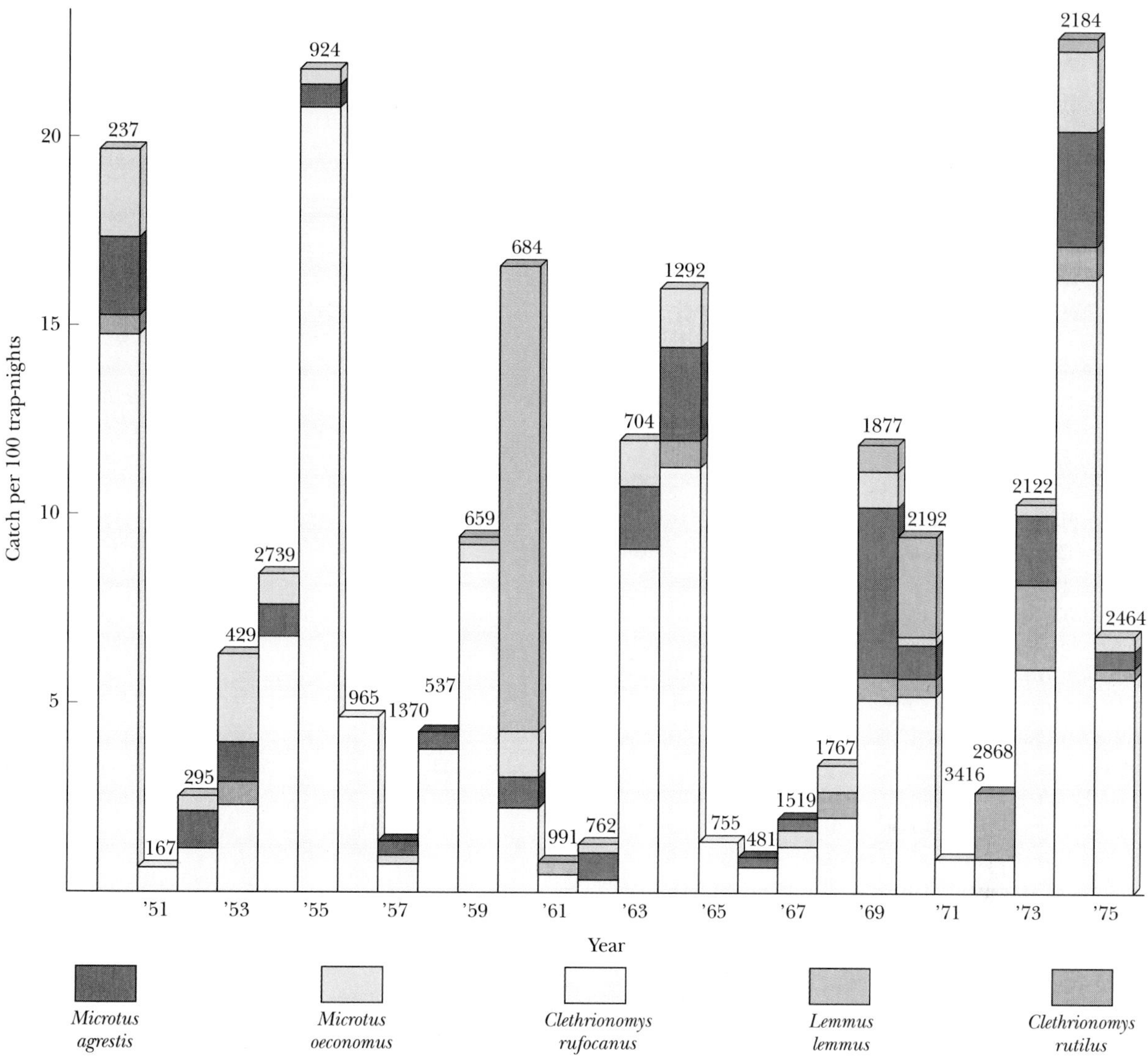

Figure 6-22

Numbers of small mammals trapped in northern Finland from 1950 to 1976. Most of the change in small mammal populations is the result of fluctuations in the numbers of *Clethrionomys rufocanus.* There is, however, a tendency for other species also to be common during the *Clethrionomy* peaks. Numbers at the top of the bars are trap-nights. (From Lahti et al. 1976.)

an exponent. This is what is meant by *chaos*—the tendency of simple, deterministic systems under some circumstances to exhibit complicated and effectively unpredictable dynamics (May 1989).

Robert May (1974, 1976)—and many others since—noted that the simplest possible expressions for density-limited population growth of annual species can have remarkably complex dynamics. These organisms, such as insects in which the adults are killed off every winter, do not have overlapping generations. To model the growth of such populations we would not use the differential logistic equation of Chapter 4 and most other places in this book, which is appropriate for continuous growth.

We would instead use an analogous **difference equation.** One difference equation that has been used to represent logistic growth is

$$N_{t+1} = \lambda N_t(1 - N_t)$$

Here N_t is the population at time t, N_{t+1} is the population the next year, and λ is the average number of offspring produced per individual when conditions are uncrowded. Note that, since one year is also one generation for annuals, λ is also basic (intrinsic) reproductive rate, R_0. To connect this equation with the logistic as given before, note that everything is scaled to K. The values for N go from 0 to 1.

May found, and you can readily see on the graphing calculator that you bought for algebra class, that for values of λ from 1 to about 2 the population reaches a stable level (after a few overshoots and undershoots) and stays there (Figure 6-23). As reproductive rate is increased, from $\lambda = 2$ to about 2.5, stable cycles of "period 2" are produced; the population goes to a high value one year, drops to a low value the next, goes back up to the same high value the next, drops to the same low value the next, and so on.

This stable cycling is not an unexpected result; the difference equation has a built-in one-year time lag. The cycles we see here are the same as those resulting from time lags as discussed in Chapter 4 (see Cunningham 1954) and in point 2 of the preceding section.

What happens at still higher values of λ is new. There is a range of values of λ yielding cycles of period 4 (for example, populations of 0.8, 0.3, 0.5, 0.4, 0.8, 0.3, 0.5, 0.4, etc.), where the same population size recurs every fourth year. There may be another bifurcation, to period 8, but then as reproductive rate is increased still more, the realm of chaos is reached. At these higher values of λ, population changes follow no pattern, going up, down, or sideways from year to year.

Let us be clear that we are speaking here of the intrinsic behavior of this population model. At these values of λ, it behaves in this way, just as at somewhat lower levels it behaves so as to yield a stable limit cycle and at much lower values, a single stable population level. Even though the behavior looks random, we are specifically not tying this randomness to random variation in the demography of the population or its environment. Demographic and environmental randomness, or stochasticity, are also likely to be present, and one problem in deciding the real-life importance of chaos is in separating regulation + stochasticity from chaos + stochasticity. (Stochasticity is discussed more thoroughly in Chapter 19).

Although the presence of built-in time lags in certain types of populations (and in difference equations) may predispose them to show chaotic behavior, more complicated systems of interacting species with overlapping populations can also potentially do so (Schaffer 1985). Examples could include predators, such as the lynx (Schaffer 1984), and prey, such as lemmings (Schaffer 1988).

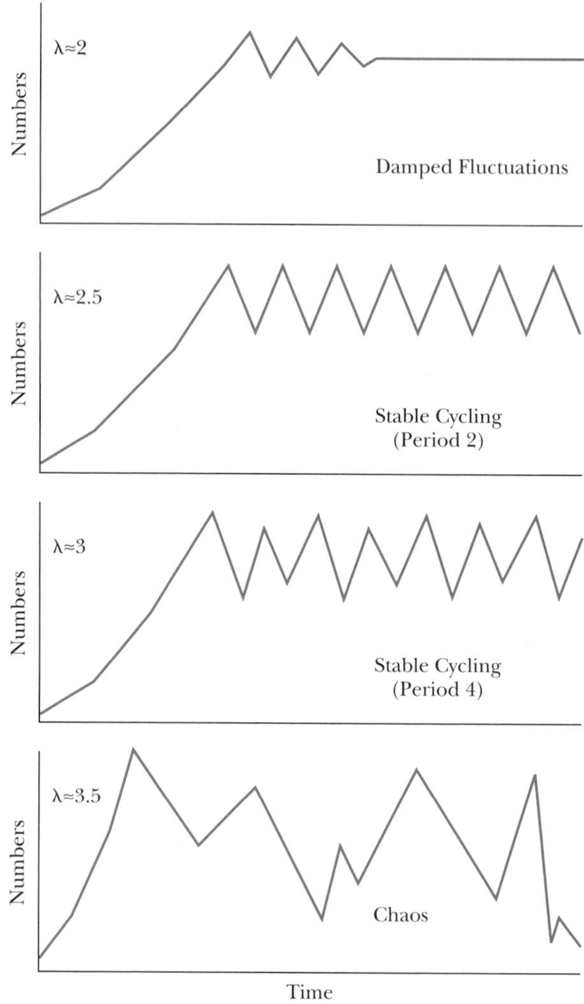

Figure 6-23

Generalized results of population growth for annual organisms modeled by difference equations for various values of λ.

Just how important chaos is in the observable fluctuations of animal numbers is still not clear. Although chaos in some biological and physical systems seems well demonstrated (Gleick 1987), several studies suggest that relatively few field or even laboratory populations show chaotic dynamics (Thomas et al. 1980). This is evidently because most species, for some evolutionary cause, have values of λ below the range at which chaotic dynamics take over or otherwise fail to conform to the models leading to chaos.

If it is true that chaos is rarely important in population fluctuations, it is uncertain whether this is good news or bad news. Schaffer (1985) suggested that if animals have dynamics that are inherently stable, the fact that violent fluctuations occur must mean that environmental stochasticity is what we will have to study. "Viewed in this light," Schaffer wrote, "low-dimensional chaos may be the last hope for a deterministic theory of ecology."

Functional and Numerical Responses and the Control of Prey Numbers by Predators

Predators can regulate the population size of their prey only if they can eat (or at least kill) more of the prey as the prey population size (density) increases. More precisely, the predators must eat a higher percentage of the prey population as the prey density increases because only then will predation be acting as a density-dependent factor on the prey.

A population of predators can eat more prey in two ways: (1) each individual predator can eat more; and (2) the number of predators can increase. The second way would occur by either immigration or reproduction. The change in feeding rate with prey density has been called the **functional response** (by the English ecologist M. E. Solomon, 1949) and the increase in predator numbers termed the **numerical response.**

C. S. Holling (1959) used Solomon's terms in a study of predation in Canada by small mammals on an insect pest, the European pine sawfly. He found that the functional response was an increase in number of prey eaten with increased prey density, but that the increase leveled off. A functional response curve might look something like Figure 6-24. As prey density increases, each predator eats more of the prey; however, at very high densities the number eaten only increases a little. The leveling off occurs for various reasons; digging up, running down, peeling, and pitting—or otherwise capturing and preparing the food—take a certain irreducible amount of time. Also, animals get full, or, in the technical term, satiated.

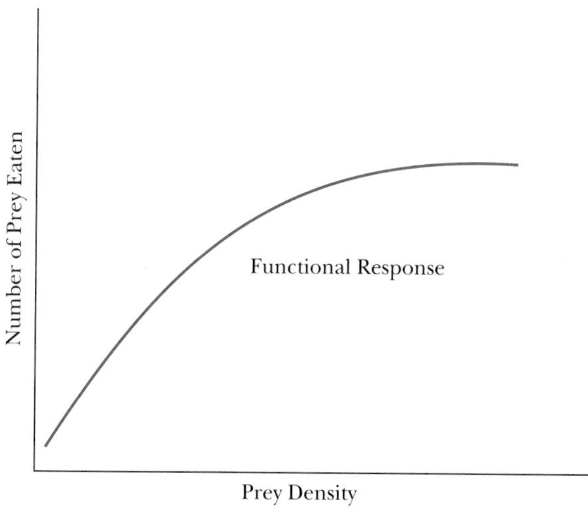

Figure 6-24

Functional response curve. This and the next two figures are adapted from Holling 1959. (Canadian Entomologist 91: 293–320.) Note that the y-axis is number of prey eaten.

The numerical response may take a variety of forms, depending on whether an increase in prey density causes an increase in immigration, reproduction, or both or neither. In Holling's study one species of shrew showed no increase in numbers, but another species of shrew and a species of mouse did increase. These two kinds of numerical response curves, then, looked like Figure 6-25.

The combined response of any one species of predator would be the product of the functional and numerical responses. That is, if at low densities 10 predators were each eating 10 prey per day, then 100 prey per day would be consumed. At high densities, if prey per predator increased to 20 and predator numbers also doubled, to 20, then 400 prey per day would be eaten.

Holling calculated the combined effect, functional and numerical, for all three species of small mammals preying on the pine sawfly and prepared another graph. The graph plotted the percentage of the prey population eaten by all three species

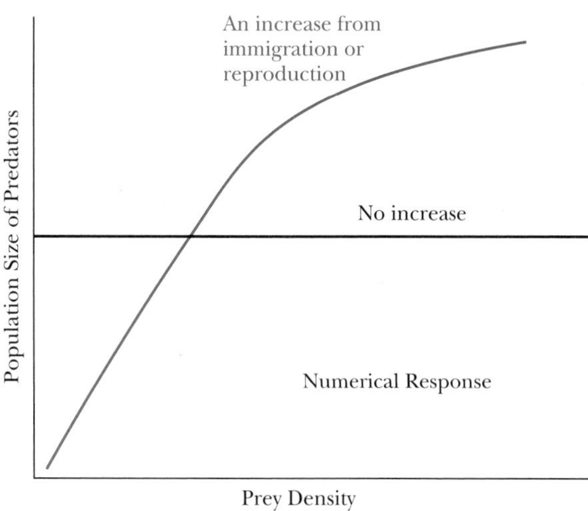

Figure 6-25

Numerical response curves.

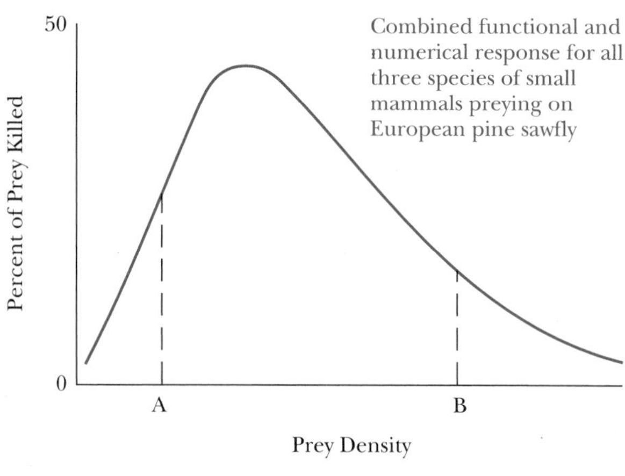

Figure 6-26

Combined functional and numerical response. Note that the *y*-axis is percent of prey population that is killed.

against prey density. The curve (simplified a bit) looked like Figure 6-26.

We may note that, because the percentage of prey killed did increase over a wide range of densities, the small mammals of this ecosystem are acting in a density-dependent way and can regulate the population size of the species of insect. This could be true even for prey densities that are high enough to have gone past the peak of the curve (where the percentage killed is declining) as long as the percentage of insects killed is no less than the percentage added by insect reproduction. If, however, the insect is able to increase explosively from very low to very high densities, perhaps from A to B on the curve, then it can escape regulation by these predators.

Mimicry and Other Kinds of Advantageous Resemblance

Protective devices of prey are various including armor, speed, and evasive movements. One more device that is particularly fascinating is mimicry and its relatives, such as cryptic coloration. **Mimicry** is the resemblance of one organism to a not very closely related organism living in the same area. It was a puzzling phenomenon until the Darwin–Wallace theory of natural selection was put forth; then it, along with many other problems, became immediately comprehensible. H. W. Bates was the first (in 1862) to discuss mimicry specifically in terms of natural selection. He described the case in which some distasteful butterflies, not eaten by birds, were mimicked by another perfectly wholesome group of butterflies. He pointed out how a slight resemblance of some individuals of an edible form to a poisonous or otherwise distasteful form would lead to those individuals being passed over by predators, causing this kind of favorable variation to accumulate in later generations.

The explanation was greeted enthusiastically, and in the next few years more cases of mimicry were described and similarly explained. Not much experimental work accompanied these descriptions, and some persons grew disenchanted with the protective resemblance—the delusive similarity—explanation of mimicry. The naturalist and explorer William Beebe recounted an incident that occurred early in this century when the great entomologist William Morton Wheeler, on his first trip to the tropics, visited Beebe in British Guiana. They were chatting, and Wheeler told him, "Beebe, in regard to mimicry, be sure to hold back; don't accept any new instances without complete evidence."

Just then Wheeler's son, who had been exploring the area, brought in a tobacco can of half-dead insects and poured them out for his father to look at. Wheeler picked one up, looked at it carefully, and then abruptly threw it down. Then he looked at it again. "Beebe," he said, "believe anything you damned please about mimicry."

"He had," Beebe (1945) wrote, "picked up a dead, yellow-banded, small-waisted wasp, which suddenly came to life and tried to sting him. On more careful scrutiny the insect proved to be a perfectly good moth, which in shape, proportions, antennae, pattern, color, and movement was an amazingly exact imitation of a wasp."

Since that time, much experimental work has been done on mimicry, and it has mainly supported Wheeler's second opinion; that is, it has supported the protective function of mimicry and its evolution through natural selection.

Mimicry in which an edible mimic resembles a distasteful or poisonous model is known as **Batesian mimicry.** The example familiar to almost everyone is that of the monarch and viceroy butterflies. This system has been well studied by Lincoln Brower and associates (examples: Brower 1958, 1960; Brower and Brower 1964, 1969; Glendinning and Brower 1990). In this case, the viceroy is the more-or-less edible form that closely resembles the poisonous monarch. The poisons include cardiac glycosides (cardenolides) that are distasteful and produce nausea and vomiting in birds within a few minutes after being eaten or, in other words, fast enough for the birds to associate the effects with the food item. The poisons are manufactured by the plants (milkweeds) on which the monarch feeds rather than by the monarchs themselves.

There are many good examples of Batesian mimicry. There are harmless snakes banded like poisonous ones (Greene and McDiarmid 1981). In addition to moths there are flies and beetles that resemble various types of bees and wasps. Mimicry in plants has been described less frequently, but there are weeds that look like crop plants and, hence, do not get weeded (Wickler 1968).

Müllerian mimicry, first described by F. Müller in 1879, occurs when several poisonous or unpalatable species resemble one another. The success of Batesian mimicry depends on the mimic staying fairly rare as compared with the model. Otherwise, predators may tend to associate the pattern with edibility rather than inedibility. In Müllerian mimicry this problem does not exist; a predator eating either (or any, because often more than two species are involved) of the Müllerian mimics will find it distasteful. Julian Huxley compared Batesian mimicry to an unscrupulous company copying the advertisement of a successful firm and Müllerian mimicry to several companies adopting the same advertisement to save money.

By and large, the avoidance of distasteful or dangerous organisms involved in mimetic relationships is learned. Learning is fastest when the ratio of model to mimic is high (Brower 1960) because reinforcement of mistakes is least then. Although the early view was that model and mimic needed to occur together, it is now clear that birds (possibly not some lower vertebrates or invertebrates) have long enough memories that this is unnecessary. In fact, the ideal situation for the mimic is to be totally separated in time from the model.

Fly mimics (harmless and palatable) of stinging bees and wasps have nearly achieved this situation in northern Michigan (Waldbauer and LaBerge 1985). They fly about in the spring and early summer before the bees and wasps are active. At this time, their potential bird predators are all experienced individuals, a year or more old. By the time the young birds leave their nests, the flies are nearly gone (Figure 6-27). The young birds blunder around, encountering the bees and wasps and getting stung and rarely encountering the mild-mannered mimics.

In most of the examples of mimicry mentioned so far, the models are potentially harmful to the predators, but the term can be logically used in other situations. Given the judgments that an animal must make in foraging, many ways of misleading it can be advantageous. Slow, vulnerable animals might get a favorable verdict in the predator's decision to pursue or not if they resemble speedy or elusive ones (Gibson 1974). It is conceivable that animals requiring a large amount of handling time per calorie yielded—animals to be avoided not because they are dangerous but because they are uninteresting—might serve as models for more succulent species.

There are several other kinds of protective resemblance similar to mimicry (Malcolm 1990). Cryptic coloration hides the organism. Desert animals tend to be sand colored, grassland birds have brownish, streaked backs, and moths that live on tree trunks are bark colored. There are caterpillars

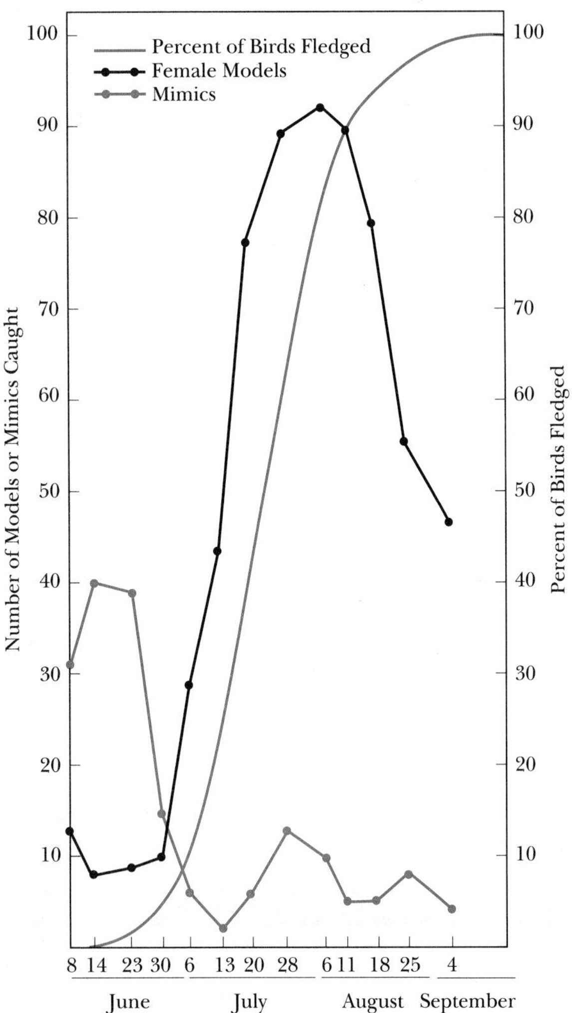

Figure 6-27

Seasonal occurrence of stinging bees and wasps, the flies that mimic them, and young birds. The mimics occur early in the season when the only birds around are ones that are at least a year old. By the time young birds leave the nest (fledge) and begin to feed on their own, the flies have become scarce and the fledglings encounter mostly the hymenopteran models. (Adapted from Waldbauer & LaBerge 1985.)

shaped and colored like leaves (Figure 6-28), twigs (Figure 6-29), and bird droppings (Owen 1980). Although most examples of camouflage are animals, at least a few plants are involved. The best-known are the African desert "pebble plants," succulents that look like small rocks.

Many poisonous, stinging, or otherwise-to-be-avoided animals are brightly colored or strikingly marked (Figure 6-30). They are said to show **warning coloration.** ("Aposematic coloration," a term used by E. B. Poulton in the late 1800s, is a synonym.) Examples of warning coloration are the black and white striping of the skunk and the black and orange striping of stinging insects. Many warningly colored animals are red, such as the red milkweed beetle and the monarch butterfly, both of which are poisonous by virtue of feeding on milkweed. The poison arrow frogs of the South American rainforests are an incredibly flashy black and red. Animals may have warning signals other than color, such as the rattle of the rattlesnake.

Many of the warningly colored organisms are the "successful advertisers" imitated by Batesian mimics. The evolutionary sequence, oversimplified somewhat, has involved, first, the greater survival of individuals having some protective feature, such as a distasteful or dangerous chemical (Malcolm 1990). Any individuals who combined the basic protection with a color or pattern that predators could note and avoid would survive better than those without the pattern—or if they didn't survive the first encounter, the chances of their similarly colored kin would be improved (Malcolm 1986). With the establishment of warning coloration, the stage is now set for the development of Batesian mimicry, as already described.

In **aggressive resemblance** (or **aggressive mimicry**), a predator or parasite looks like its prey or host. In common language, it is a wolf in sheep's clothing. The term has also been used if the predator is merely camouflaged (to look like part of a flower, for instance) or looks like some harmless species of the community. There are parasitic flies whose larvae feed on bee larvae; the adult flies resemble the bees that they parasitize and consequently are able to enter the hive to lay eggs without being attacked.

Another example of aggressive mimicry is seen in lightning bugs, or fireflies (which are really beetles). The flashings of lightning bugs are signals between the sexes. The males fly about on a summer night and produce flashes of light that are different for each species (depending on the species they may be bright or dim, long or short, fast or slow, and the lights may be slightly different colors). The females sit on plants and if they are ready to mate respond to the male by a flash that occurs at a char-

Figure 6-28

This caterpillar (the saddled prominent) looks like the eaten edge of the basswood leaves on which it had fed. (Photo by the author.)

Figure 6-29

The shape, color, size, and posture of this geometrid larva make it look much like a twig. (Photo by the author.)

Figure 6-30

Apparent warning coloration in a caterpillar. This gregarious caterpillar (*xanthopastis timais*) is boldly marked with black and white stripes and has a red head. It devours patches of swamp lily in wet forest in Florida. Fecal pellets, or frass, are noticeable on the leaf blades. (Photo by the author.)

acteristic interval (say, 5 seconds or 2.2 seconds) after the male's flash. The female continues to flash in response to the male's flashes as he homes in for mating. Large females of a different group of lightning bugs mimic the intervals of the females of the first group; the males are attracted to these females, but when they arrive, the males become not a mate but a meal (Lloyd 1965).

In one further twist in what is evidently a complex system, the males of the species to which the female mimics belong may, themselves, mimic the flash pattern of the intended prey. They change to their own, nonmimetic mating flashes only when they had landed near the female. These males, Lloyd (1980) concluded, may be using their mimicry to locate and seduce their own hunting females.

The foliage of tropical American passionflowers (*Passiflora* spp.) is fed upon by the larvae of *Heliconius* butterflies. The female butterflies avoid laying eggs on a plant that already has eggs on it, to reduce competition, presumably, and also because the *Heliconius* larvae eat smaller larvae and eggs. The leaves of certain species of passionflowers have small yellow bumps resembling *Heliconius* eggs; these also inhibit egg-laying by the butterflies (Williams and Gilbert 1981). Here, then, is a case of mimicry in which the prey (the plant) looks like its predator.

Biological Control

We are emerging from a period in which an attempt was made to control pests almost solely through the use of chemical pesticides. **Biological control** generally refers to the use of predators or parasites of a pest to reduce its numbers to a point where it is no longer an economic problem. There have been several spectacular successes in biological control and a great many failures.

Although predaceous ants were used against citrus pests as early as the third century in China (Itô 1980), most applications have been in the past hundred years. The earliest success in the United States was the control of the cottony-cushion scale in California, a small insect native to Australia that attacks citrus trees. Late in the 1800s, the vedalia beetle was imported from Australia and preyed on the scale insects so effectively that they were virtually eliminated from the citrus groves within two years. Since then, some 500 organisms have been imported into the United States and released for biological control purposes. About 100 became established, and about 20 were successful in controlling the pest. Another success, already mentioned, was the control of klamath weed on western U.S. grazing land.

For many pests the predators or parasites that have been tried have evidently not been able to limit the pests' abundance. It may well be true that for some pests no organisms will have this ability, and this type of biological control will fail.

Many pests are introduced species, freed from the ecological context in which they evolved. This is also the status of the control organisms. So far only a few organisms imported into this country for biological control are known to have become important pests themselves. One is the house sparrow, introduced—incredible as it now seems—to control insect pests of shade trees (Leahy 1984) but which was much more efficient at lowering numbers of various native birds, such as the cliff swallow (Brewer 1991). Other locales have suffered still more serious consequences; one example is the depredations on island bird life by the mongooses introduced with the aim of controlling poisonous snakes or rats (Oliver 1963).

Various other pests are native species, but for these also, the situation is usually one of ecological disruption (Dreistadt et al. 1990). The Colorado potato beetle is a native species that once lived a low-profile existence on buffalo bur, a spiny annual of disturbed patches in western grassland. Then humans began to produce huge disturbed patches called potato fields, where nothing grew except a buffalo bur relative, the potato. The Colorado potato beetle spread through potato fields from the east side of the Rockies to the Atlantic coast in 15 years, between 1859 and 1874 (Palmer and Fowler 1975). The path for biological control of native pests is to find and release predators and parasites of related species that occur in different geographical regions.

One of the most-publicized biological agents of the past few years is the soil bacterium Bt. When sprayed on tree foliage, a mixture of its spores and the toxic protein crystals it produces is relatively effective in protecting trees against the gypsy moth, spruce budworm (Grimble and Lewis 1985), and many other insect pests. Various other bacteria, fungi, and viruses are being studied for similar uses (Moffat 1991).

Note that Bt is not a biological control agent in the classical sense of the term (DeBach 1964, Garcia et al. 1988). Biological control is the establishment of natural enemies of a pest; after that, pest and enemies persist in a biological relationship requiring little or no further intervention by humans. Companies in the agriculture business, for obvious economic reasons, have little interest in biological control in this sense. They are much more interested in agents like Bt that require repeated application and that should be thought of as pesticides, though biological ones.

Other biological lines of defense against pests that are favored by companies in the business include development and use of resistant crop varieties, including genetically engineered ones, and the use of pheromones, chemicals used as signals between members of a species (see the section on Chemical Ecology in Chapter 10), which can function to attract a single species of pest to traps. These methods are often more environmentally attractive than chemical pesticides but, like them, require continuing expenditures of money.

Biological methods not so interesting to the agricultural and chemical companies include increasing crop diversity in space and time by planting smaller fields and returning to crop rotation; renewed attention to phenology, as in delaying the planting winter wheat until after emergence of the Hessian fly; and repelling pests by companion plantings. This use of allelochemicals is already practiced by organic gardeners who interplant marigolds and garlic as protection for other crops but potentially has applications on a much broader scale.

Integrated pest management (IPM) is a combined approach to pest control that, in theory, includes biological control in the strict sense, additional methods, such as some of those described in the preceding two paragraphs, and pesticides (Chapter 20). One useful advance of IPM has been the inclusion of pest population monitoring as a guide to the pesticide application, rather than routine use without regard to pest numbers.

Cannibalism, Siblicide, and Intraguild Predation

Cannibalism is not a species interaction; it is predation but is intraspecific, not interspecific. The killing of one individual by another member of the same species is not common; eating the dead individual is even less so. Occasionally, an adult animal kills another in some bout of interference competition, but that occurs as a rare accident. Access to a food item or a mate is important, but rarely is it worth a fight to the death.

More frequent, but still restricted to a few species, is one nest or littermate killing another. Here, too, competition is involved. In the battle for food brought to the nest, the smaller, weaker young is sometimes injured or killed. This **siblicide** or **cainism** may (but doesn't necessarily) occur in species (1) that have the physical weapons to kill, (2) in which the young are different sizes (in birds, as the result of the eggs hatching over a period of days, instead of all at once), and (3) that live in situations where the food supply fluctuates (O'Connor 1978, Forbes and Ydenberg 1992). When there is ample food over the whole nestling period, all of the young may survive in some species. In others, such as certain of the large, long-lived raptors that lay two eggs, one nestling seems invariably to kill the other (Gargett 1978, Simmons 1989). Besides raptors, herons (Mock 1984) and seabirds are groups in which siblicide is relatively frequent. Occasionally, but not usually, the victim may be eaten by its murderer (in the barn owl, for example, Baudvin 1979).

Cannibalism seems to be a regular occurrence in some populations of amphibians that develop in temporary ponds. Rapid growth is highly advantageous, to complete metamorphosis before the pond dries up. Here, as in the case of nestling birds, the cannibalism involves young and, potentially, kin.

The rarity of cannibalism seems puzzling. As a group of Arizona State researchers (Pfennig et al. 1991) studying the tiger salamander wrote, "More than any other intraspecific interaction, cannibalism consistently generates the greatest fitness differential between interactants." That is, not only is the potential competitor removed, the murderer gets a good meal out of it. The rarity of cannibalism seems related to three problems: (1) The process of killing another member of the same species can be physically impossible, excessively time-consuming, or dangerous. For example, a fox, or a fox squirrel, attempting to kill another member of the species might have to spend a lot of time and energy and might be pretty well chewed up at the end. (2) Killing kin involves getting rid of shared genes, lowering inclusive fitness. (3) The predator may catch something—viruses, bacteria, worms—from the prey. Because the two belong to the same species, the disease organisms are likely to infect and thrive in the new host individual. Cannibal tiger salamanders are more likely to die of disease than noncannibals (Pfennig et al. 1991).

It is worth observing that problem 1 is reduced if there is a big difference in size and strength, as in adults preying on young (or eggs). Likewise, problem 3 is apt to be less when the victims are young, because their parasite load is likely to be lower. Consequently, cannibalism, when it occurs, will generally involve young animals as victims.

Cannibalism occurs in humans. Sometimes it involves individual psychopathology and so is relatively uninteresting except to the media, but cannibalism has occurred episodically in most human societies in times and places of calorie or protein shortage (Harris 1977). The risk of disease may also be part of an evolutionary explanation for the general taboo against cannibalism in human populations.

Probably the best-known example of a human disease related to the practice is kuru. This fatal viral disease affecting the central nervous system (''kuru'' means ''trembling'') afflicted inhabitants of central New Guinea. This virus was evidently spread by the ritual eating of brains of infected persons (Gajdusek 1977) and with the cessation of that practice, the disease began to disappear.

If one animal kills and eats a member of a competing species, it profits in the same way as it would by cannibalism but avoids problem 2—killing kin—and at least reduces the risk of infection—problem 3. Such ''intraguild predators'' (Polis et al. 1989) are, however, still faced with problem 1. This is probably the reason why intraguild predation is mostly restricted to adults preying on the young of competitors and large animals preying on small animals with which they share foods. Grain beetle larvae and adults eating eggs (see Chapter 8) and big spiders eating little spiders (Jackson 1992) are examples of the two categories.

The Population—Community Ecology Interface: *Parasitism, Commensalism, and Saprobism*

White-backed vultures function as commensals and saprobes at this lion kill in Kenya. *(Fritz Polking/Dembinsky Photo Associates.)*

OUTLINE

- Parasitism
- Health and Disease
- Commensalism
- Saprobism

The preceding chapter dealt mostly with interactions involving organisms that live separately but come together in a trophic interaction—one in which energy passes from one species population to the other. In this chapter the associations usually involve organisms that live intimately together—usually by one living in or on the other member of the coaction. Symbiosis is a term applicable to any such close association, including certain types of mutualism discussed in Chapter 9. Food is the basis of the coaction in parasitism and saprobism and may be indirectly involved in commensalism.

Parasitism

The Prevalence of Parasites

A parasite gets its food from another organism, but that is true of most of us. To be what is ordinarily meant by the word ''parasite,'' it must also live in close association—in or on—its host and should not kill its host, at least not right away.

Those of us living in the temperate regions in the 20th century may get the idea that parasitism is not a very common way of life. If we knew the medical histories of all our acquaintances we could perhaps put together a list of half a dozen species (ignoring viruses and bacteria) that have infested them sometime in their lives. Included might be swimmer's itch (caused by a trematode worm), head lice (an insect), pinworms (a nematode worm), chigger bites (a mite), and, depending on how old we are and how diverse our friends, possibly a few others, such as hookworm or crab lice. In fact, this virtually parasite-free existence is totally atypical for the human species for most of our history and for any wild animal—or plant—today. There are enormous numbers of parasitic species, and it is a rare individual organism that does not harbor several kinds. Figure 7-1 illustrates this and also shows that different species differ in the numbers of species and individuals of parasites they typically harbor.

Fish, amphibians, and reptiles have fewer parasites than birds and mammals (Kennedy et al. 1986, Aho 1990). The difference seems to be between homoiotherms and poikilotherms, but just what features of homoiotherms are important is not clear. Parasitologists favor the idea that the greater diversity is the result of more opportunity for infection, based on homoiotherm's moving about more and eating more often. Also worth considering are

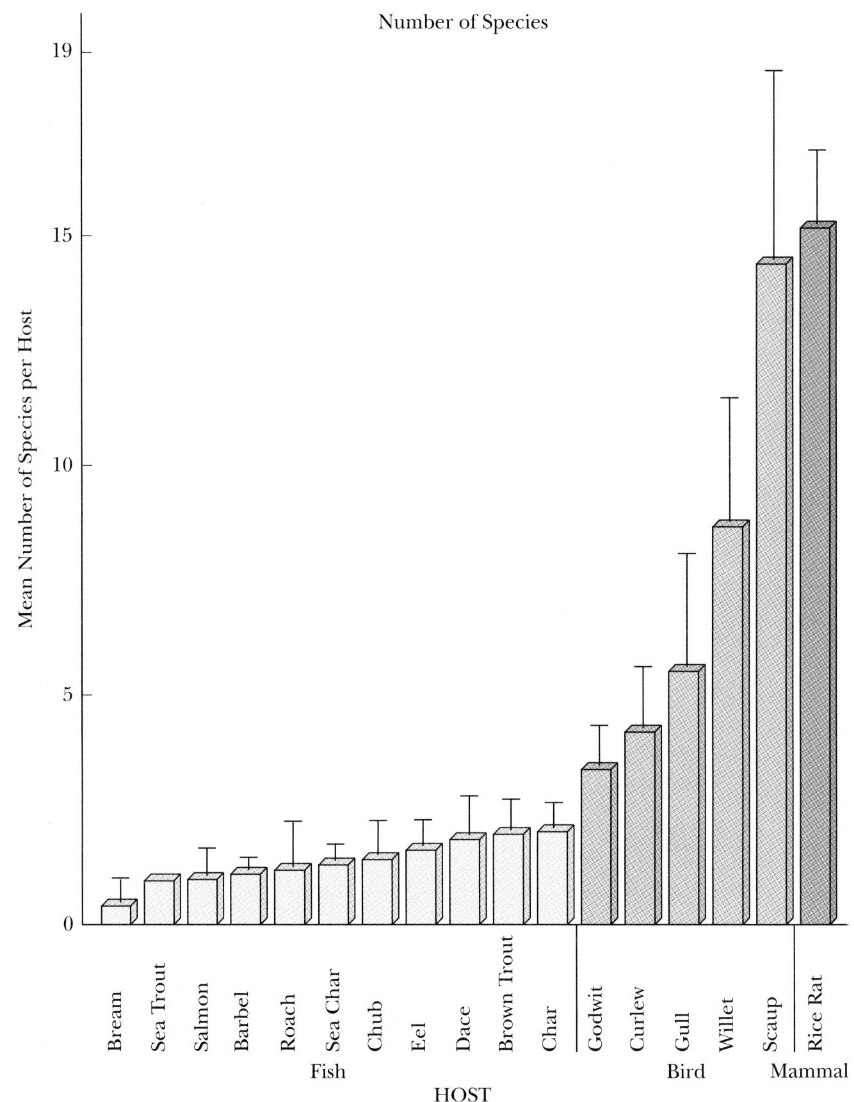

Figure 7-1

Helminth (worm) parasites of the intestine of fish, birds, and mammals. Fish tend to have between 2 and 100 individual parasites of 1 or 2 species. Birds usually have 50–20,000 individuals of 3–14 species. Mammals are like birds. Lines on the species graph are standard deviations. (From Kennedy et al. 1986.)

that the physical and chemical habitat conditions provided by the homoiotherm are more stable and the energy base is likely to be greater, at least for digestive tract parasites.

Parasitologists claim that there are more parasites, both species and individuals, than there are free-living organisms. Their argument is that all free-living forms have several parasite species, usually including one or more specific to them. Therefore, the number of parasitic species is some multiple of the number of free-living species.

Types of Parasites

Parasitism as a mode of life has obviously evolved repeatedly; few kinds of organisms are without their parasitic representatives. Some of the groups where parasitism is especially prevalent include the vi-

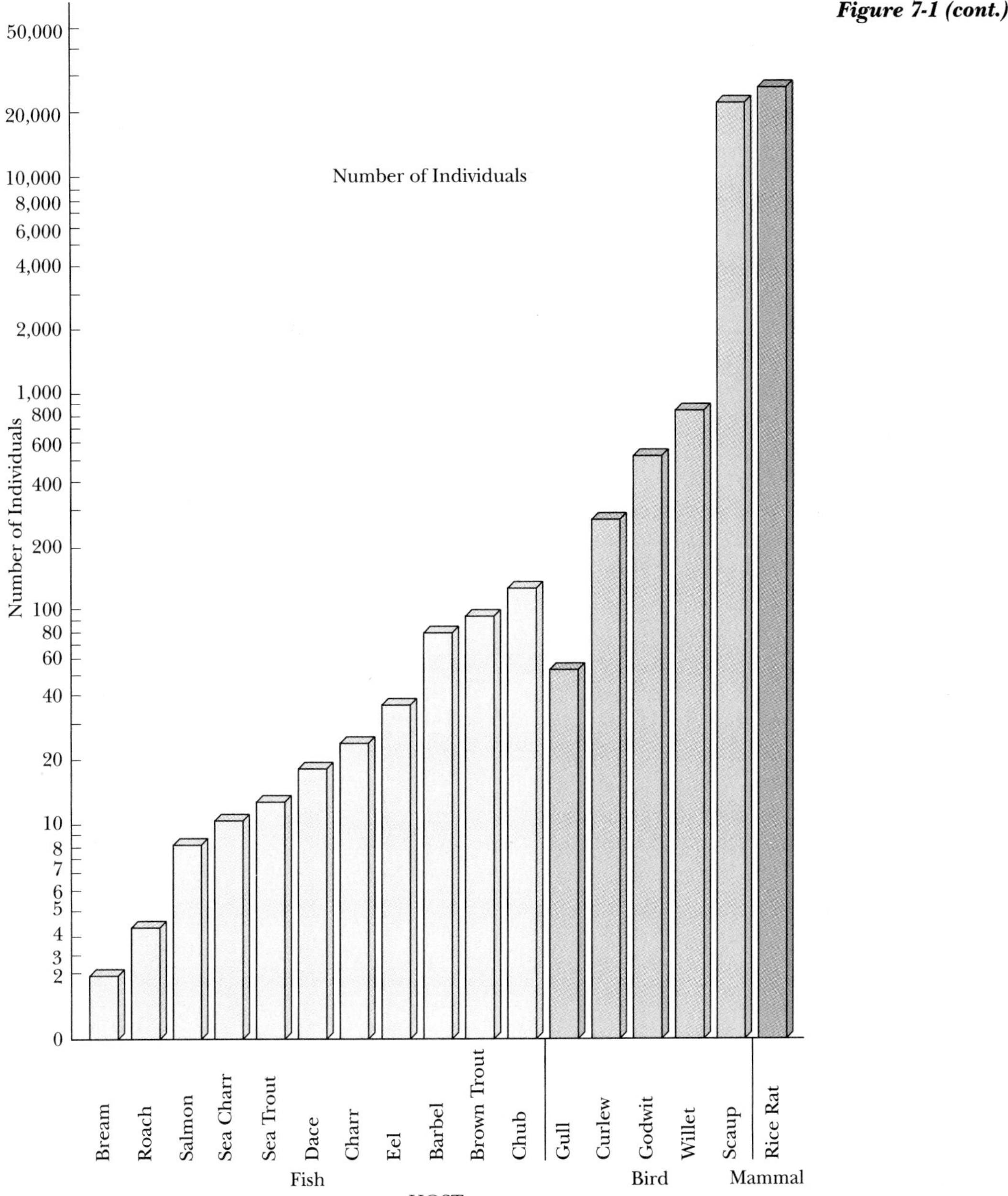

Figure 7-1 (cont.)

ruses, bacteria, fungi, protozoans, flatworms, roundworms, spiny-headed worms, some groups of crustaceans such as copepods (fish lice), and some groups of insects (such as lice and fleas).

Only a small percentage of algae and higher plants are parasites, but in total numbers there are a good many. The dodders, chlorophyll-less and reduced to long, stringlike bodies, twine around other plants and absorb nutrients from them by means of outgrowths called **haustoria** that penetrate their tissues. Mistletoe, though it is green and makes some of its own food, is obviously a parasite growing, as it does, on the branches of trees. Some other green plants are less easy to recognize as parasites because their haustoria are below ground, attached to the roots of other plants. Widespread plants that are examples of this kind of parasite include the false foxgloves, bastard toadflax, and louseworts. All of these are capable of parasitizing a variety of trees (Musselman and Mann 1978). Beechdrops is a similar, but chlorophyll-less, parasite occurring only on beech roots.

One of the most remarkable parasitic plants is rafflesia, which grows entirely within the roots of tropical possum-grape vines until the time for reproduction arrives. Then it sends up a stem that produces a cabbage-like bud on the surface of the ground. After several months of development, the bud opens into an enormous, reddish, mottled flower, the world's largest, up to a meter across. Rafflesia grows in tropical rain forest in Sumatra and Borneo, but most of its original habitat has been turned into rice paddies and it is now rare (Sandved and Emsley 1979).

The only vertebrates that have become parasites are among the fish, and there are only a few of them. The lampreys attach themselves to other fish, rasp an opening, and suck body fluids. Small South American catfish live in the gill cavities of large catfish and suck their blood. In the ceratioid anglerfish, which lives in the middle depths of the ocean, the male is parasitic on the female. Early in life the small male attaches by means of its mouth to the body of the female and obtains nourishment by means of a placenta-like fusion of their skin.

Parasites can be divided into **ectoparasites** and **endoparasites.** Ectoparasites live on their hosts; examples are fleas and ticks. Endoparasites live in their hosts; examples are viruses and blood flukes.

Parasites may also be categorized as **microparasites** and **macroparasites** (May 1983). Microparasites, which include disease organisms—viruses, rickettsiae, bacteria, fungi, and protozoa—increase in numbers within their vertebrate host by reproduction. Macroparasites are the various parasitic worms and arthropods. They generally do not multiply within the body of the vertebrate host but produce eggs or larvae that pass out of it. The response of the host to microparasites is usually either death or recovery with the development of a relatively long-lasting immunity. The response to macroparasites is more variable depending, among other things, on parasite load. A dog with a flea may be healthy and carefree; a dog with a thousand fleas may be a very sick animal.

Parasitoids are insects (flies or wasps) that live as parasites within the egg, larva, or pupa of another insect, consuming its tissues (Figure 7-2). Eventu-

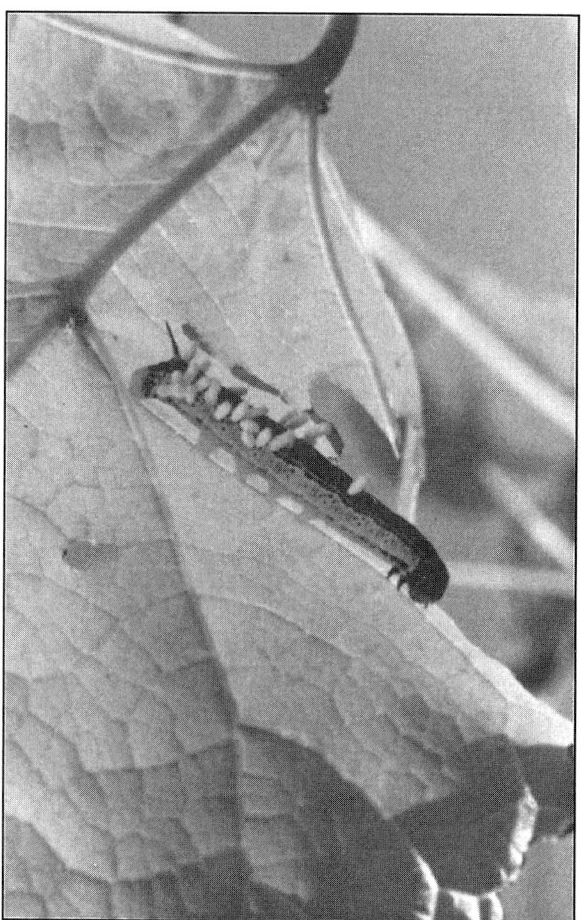

Figure 7-2

The white capsules on this catalpa sphinx larva are cocoons of a wasp that is a parasitoid of the moth larva.

ally they emerge and the host dies. Thus, parasitoids combine features of both parasitism and predation. One familiar example of the relationship is seen in the tomato worm, the larva of the sphinx moth. A braconid moth is a parasitoid of tomato worms, spending its larval life within the worm, and then pupating in small egglike cocoons on its back. Some parasitoids are absurdly small; there are parasitoids of scale insects and cockroach eggs and even parasitoids of small parasitoids.

The lines between parasitism and other coactions, such as commensalism and mutualism, are not always clear. Several coactions described earlier under herbivory could just as well be considered parasitism on plants. Included are leaf mining and gall making, both of which could be considered forms of endoparasitism. Sap sucking by aphids and spittlebugs is analogous to the blood sucking—usually considered parasitism—of fleas, ticks, and mosquitos. Mosquitoes, in fact, also suck plant juices; the female sucks blood as a protein source during egg production.

Two aspects of the ecology of parasitism can be recognized. One is the ecology of the parasites—their individual, population, and community ecology with the host as habitat. The second is the ecology of the interaction—the numerical and evolutionary relations between parasite and host. The study of both aspects is still in its infancy.

The Ecology of Parasites

If we regard the host simply as the parasite's habitat, we can study the ecology of parasites like any other organism. They disperse, reproduce, have interspecific relationships with other organisms besides the host, form communities, transfer energy, and participate in the other ecological activities better known for free-living forms.

Most of the special features of the parasite life history are related to (1) the security and stability of the parasite's life once in a host and (2) the hazards and uncertainty of the transfer. One of the earliest observations about parasites is that they are degenerate, by which is meant that certain structures such as sense organs, legs, and digestive tracts are reduced or missing. Their reproductive organs and organs for keeping them fastened to where they need to be in the host are not degenerate, however, but rather the opposite.

In fact, the set of characteristics shown by parasites are adaptations for life in a highly limited, stable, but temporary, environment. Perhaps the extreme of this trend is seen in tapeworms. They live in the vertebrate intestine, surrounded by food, at a near constant pH, protected from temperature changes, and they consist of a head supplied with hooks and suckers but no mouth or eyes, and a series of identical segments occupied primarily by gonads.

Everyone knows the high reproductive rate of parasites, and everyone is right. A female roundworm of the species studied in freshman zoology is said to produce 200,000 eggs a day for a year, an adult beef tapeworm produces 100,000 eggs a day, and other parasites have comparable abilities.

High rates of egg production are necessary (if the species continues to exist) because of the limited life span of its habitat. Hairston (1965) used a life table approach to study schistosomiasis, caused by small flukes that live in the blood vessels of humans (Figure 7-3). At least 300 million people and possibly many more are infected with schistosomes, mostly in Africa, South America, and southeast Asia.

Although a female schistosome (in a human's blood vessels) may produce tens of thousands of eggs in her lifetime and although one miracidium entering a snail may give rise to a few thousand cercariae, the probabilities of infection (of snail by egg and of man by swimming cercaria) are very small. About 2 of every 100 eggs produced by the adult flukes in a human were effective in infecting snails, and somewhere between 1 in 1,000 and 1 in 1 million cercariae produced in snails successfully infected a human.

Within the host, each parasite occupies a specific site. Much of this is due to habitat selection on the part of the parasite, as several experiments have shown. Interspecific competition also occurs, and in some cases, results in the exclusion of a species from a site that it would otherwise occupy. For example, a tapeworm and a spiny-headed worm both prefer the upper part of the intestine of rats. When they occur together, the spiny-headed worm is able to occupy the preferred site and the tapeworm is displaced posteriorly (Holmes 1961).

Some other studies have failed to show effects of interspecific competition. A fluke that is a parasite on the gills of topminnows, for example, showed no restriction of its distribution—no narrowing of its ecological niche—when other gill par-

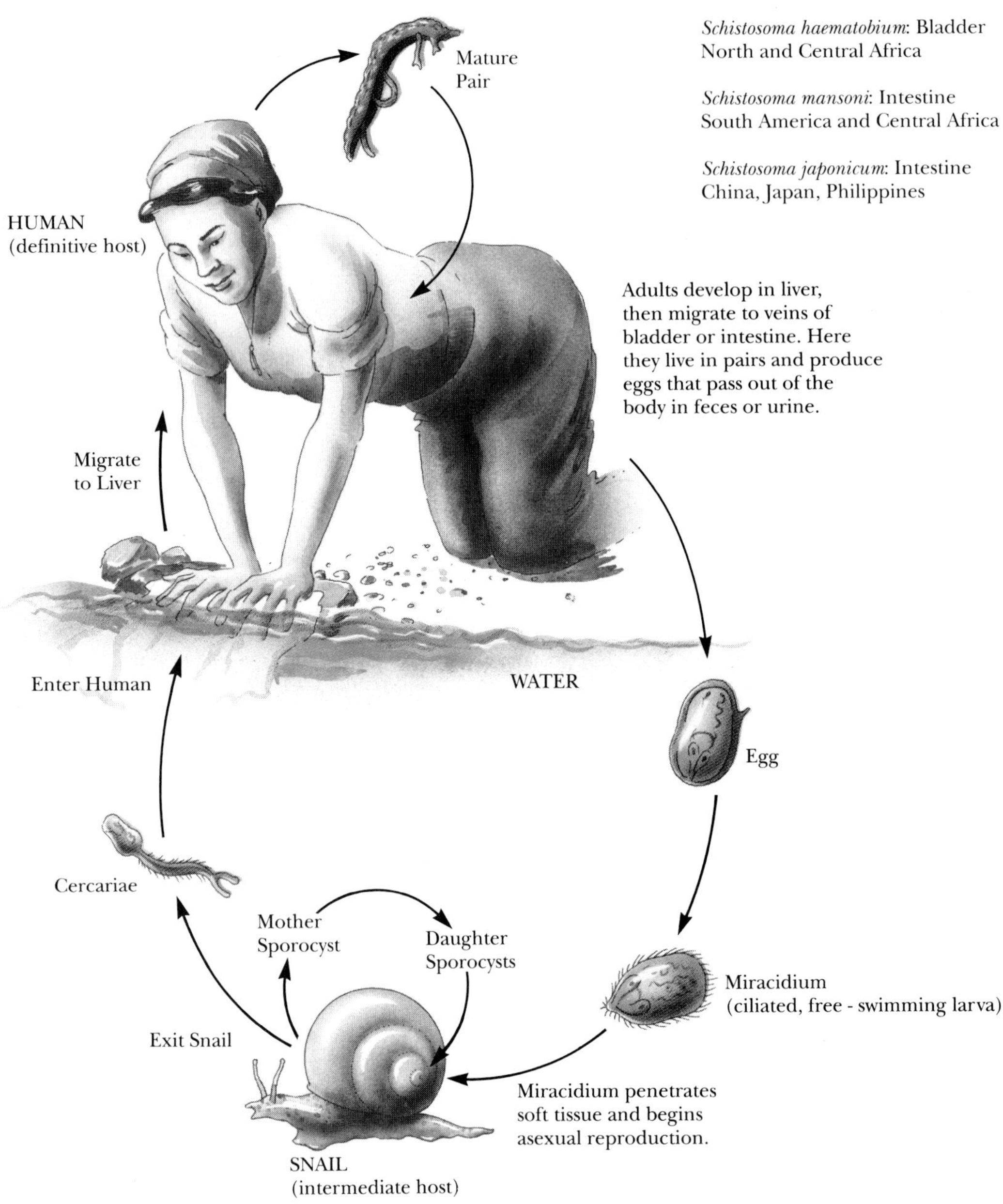

Figure 7-3

Generalized life cycle of a schistosome.

asites were present (Janovy et al. 1991). Possibly the role of interspecific competition differs between the species-rich homoiotherm parasite communities and the sparse poikilotherm ones, but the relevant studies are few.

It is known that the species of parasites likely to be found in a host may vary with its age, a phenomenon possibly similar to succession, but this, like most aspects of population and community ecology of parasites, needs more study.

Rates of Increase

Usually students of parasite populations have used net reproductive rate, R_0, as the measure of the potential of parasite populations for growth. Net reproductive rate in the absence of crowding is called basic, or intrinsic, reproductive rate. "Basic" reproductive rate seems to be the prevailing usage but "intrinsic" would be more satisfactory, echoing the usage for the intrinsic rate of natural increase. (Recall that the intrinsic rate of natural increase r, is $[lnR_0/T]$, where T is generation length.)

For parasites, population growth terms have been given slightly specialized definitions. For microparasites—viruses, bacteria, etc.—intrinsic reproductive rate is defined as the average number of secondary infections produced when one infected individual is introduced into a host population composed wholly of susceptible individuals.

For macroparasites, such as worms, net reproductive rate is the average number of daughters reaching sexual maturity produced by a mature female over her lifetime. If the parasite population is suffering no detrimental effects from crowding, this will be the intrinsic reproductive rate. This definition of net reproductive rate for macroparasites is basically the same as the general definition given in Chapter 4. Bear in mind, however, that the production of a sexually mature parasitic worm may involve a long time period that includes one or more episodes of asexual reproduction. A few examples of the intrinsic reproductive rate for human parasites are given in Table 7-1.

For directly transmitted diseases newly introduced to a population of hosts, basic reproductive rate (Fine 1982) is approximately given by

$$R_0 = bND$$

where R_0 is basic reproductive rate, b is the rate of effective transmission between individuals, N is population size, and D is the average duration of infectiousness.

Just as in our earlier discussion of r and R_0, the basic (intrinsic) rate will be fully expressed only when the population (of parasites in this case) is low, such as when a parasite is introduced into a population of hosts where it has not previously existed. In a situation where the parasite population is stable, net reproductive rate will be 1, as each parasite in this generation replaces itself in the next. This will be true whether the basic rate is 1.1 or 10. In the schistosomiasis study mentioned in the preceding section, the best estimate of net reproductive rate for *Schistosoma japonicum* in one area of the Phillippines was 0.6. Inasmuch as schistosomiasis is still with us, the discrepancy was probably the result of difficulties involved in measuring reproduction and mortality rates in this complicated life cycle.

Table 7-1
Intrinsic (Basic) Reproductive Rate (R_0) for Human Parasites

Parasite	R_0	*Source*
Viruses		
Whooping cough	13–17	May 1983
Measles	12–13	May 1983
Poliomyelitis	6	May 1983
Smallpox	3–5	May 1983
HIV (AIDS)	5–10*	Calculated from May and Anderson 1987
Protozoans		
Malaria (*Plasmodium falciparum*)	70–80	Anderson 1982
Helminths		
Hookworm (*Necator americanus*)	2–3	Anderson 1982
Roundworm (*Ascaris lumbricoides*)	4–5	Anderson 1982
Schistosomiasis (*Schistosoma mansoni*)	1–2	Anderson 1982

* For homosexual spread in the early 1980s.

If basic reproductive rate is below 1, the new disease will be unable to sustain itself and will die out. Note that, from the equation above, R_0 depends in part on population size of the host. Consequently, a threshold population size will exist, above which a directly transmitted disease can be indefinitely sustained and below which, if newly introduced (by an infected immigrant), it will eventually go extinct.

Parasite-Host Interactions

Epidemics

One response in vertebrates to infectious disease (or sometimes merely exposure without the development of recognizable symptoms) is the development of immunity. The disease organisms may serve as antigens that, through a complicated series of steps, result in the body constructing defenses specific for the particular pathogen. (Parasites that can be connected with a particular set of symptoms are usually called "pathogens.") Involved is a spectrum of defenses that includes (1) production of antibodies—proteins in blood and lymph that destroy or prevent the spread of pathogens and neutralize their toxins and (2) cell-mediated immunity, in which lymphocytes destroy target cells, such as those containing pathogens. Because of the absence of specific immunities, children are more susceptible to many infectious diseases than adults, and adults are more susceptible to "new" diseases, perhaps imported from another locality.

One of the major differences between most microparasites and most macroparasites is that immunity to macroparasites is relatively weak or infrequent. Why hosts fail to develop immunity to these parasites is an important field of research in parasitology. One proximate reason that has been discovered for schistosomiasis is a kind of mimicry or camouflage at the molecular level. The adult schistosome becomes coated with protein molecules of the host, so that the host's immune system is unable to tell the parasite from the background (Kolata 1985).

An **epidemic** is the spread of a communicable pathogen through a population. Its occurrence depends on the presence of a threshold number of individuals lacking immunity **(susceptibles),** and its end depends on the decline of susceptibles (by immunity, death, or quarantine) to below the threshold. Some epidemics, such as plague, end with the population almost totally immune or dead and the pathogen locally extinct. For many disease organisms, however, epidemics recur by the pathogen persisting until the number of susceptibles rises above the threshold density (usually through births). Poliomyelitis and many childhood diseases showed such patterns before the widespread use of vaccines. Recurrent epidemics show many of the same features as predator-prey cycles.

A simple model for a directly transmitted microparasite is shown in Figure 7-4. Four compartments are shown, susceptibles, infected, immune, and dead (for animals dying in the interval shown). Two times are shown, 0 and 1, with the interval being long enough for infected individuals either to recover or die. Susceptible and immune individuals die at rate d; infected individuals die at rate $d + \alpha$, with α being disease-induced mortality (*virulence* is the epidemiologist's term). In general, births from all categories of living individuals give rise to susceptibles, though transmission from parent to unborn offspring can occur in certain diseases. Loss of immunity turns immunes back into susceptibles at rate λ.

To repeat, if net reproductive rate of such a parasite is 1, it will be perpetuated in a population. If below 1, it will tend to go extinct. The factors that influence whether the disease is maintained in the population are given below (Yorke et al. 1979). The first five all affect the population size of susceptibles.

1. Size of population. A disease will die out in a population below the threshold number.
2. Rate of population turnover. Individuals lost from the population by death include some with immunity; births give rise to susceptibles (though newborns often have a temporary passive immunity to diseases for which the mother has antibodies).
3. Density of population. Contacts between susceptibles and infectious individuals tend to be more frequent in denser populations.
4. Persistence of immunity. The rate at which immunity is lost varies among pathogens and individual hosts. The longer the immunity, the more likely the disease is to be eliminated from the population.
5. Generation period. This is the average interval between acquisition and transmission of infection. Although a short generation period favors rapid spread of the disease, the disease may spread throughout the population so quickly that host reproduction has no time to produce a new crop of susceptibles.
6. Transmissibility or infectiousness. How easy is it to catch the disease? Diseases with low transmissibility will be more likely to be eliminated from a population.
7. Duration of infectiousness. Other things being equal, diseases with short periods of infectiousness will be more likely to be eliminated.

These factors interact. For example, a larger population is required to maintain a disease that is

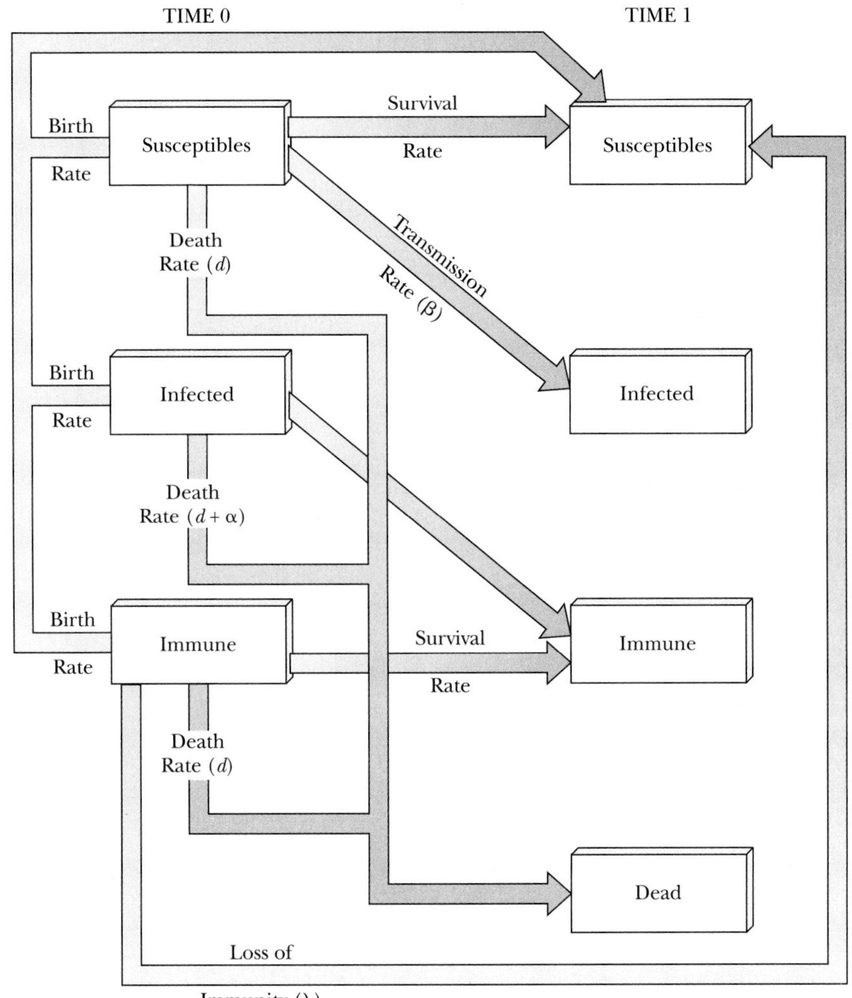

Figure 7-4

A model of population structure for a directly transmitted microparasite. The interval between time 0 and time 1 is long enough for infected individuals to recover or die.

not very infectious or that has a short duration of infectiousness.

Threshold Size and Human Diseases

It is likely that hunter-gatherer societies had various macroparasites and microparasites, specifically forms with traits that allowed them to persist despite the small size of the human bands. Examples would include species that are maintained in other organisms (such as the arboviruses that cause encephalitis and have a reservoir in mosquitos or other arthropods), diseases that are capable of being carried asymptomatically by some individuals (such as hepatitis or typhoid), and disease organisms that can survive in the soil (such as tetanus).

However, the microparasitic diseases that caused heavy mortality in historic times, such as cholera, smallpox, measles, and other of the children's diseases have high threshold populations. Precise measurement of threshold size has been accomplished for relatively few diseases. For measles, it looks to be on the order of 300,000 to 500,000 (Yorke et al. 1979). Cities with a population above this level have at least one measles case every month (or did in the premeasles vaccine era). For cities of 300,000 people, months with no measles cases—indicating local extinction of the pathogen—show up (Figure 7-5). These diseases have not always been a part of the human condition; they became important only with agriculture and urbanization.

It is worth reconstructing the situation that de-

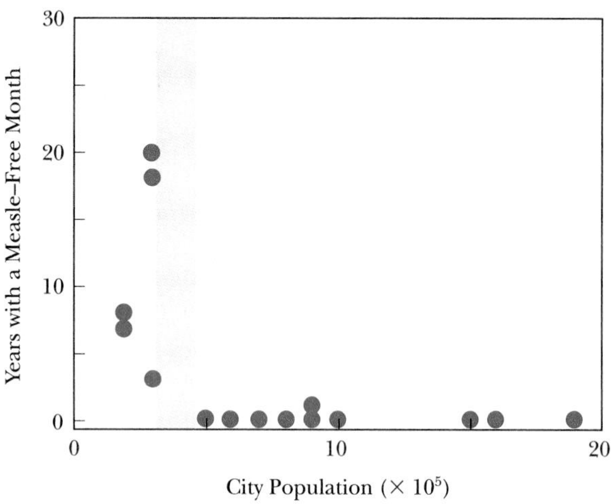

Figure 7-5

The threshold population, or critical community size, needed to maintain measles is suggested here by plotting number of years (between 1921 and 1940) when no cases of measles were reported in at least one month versus population size of U.S. and Canadian cities. Omitted from the graph at the right end are Chicago (3,400,000) and New York City (7,500,000). Like other cities with populations above 500,000, these had no months without reported measles cases. (Based on data in Yorke et al. 1979.)

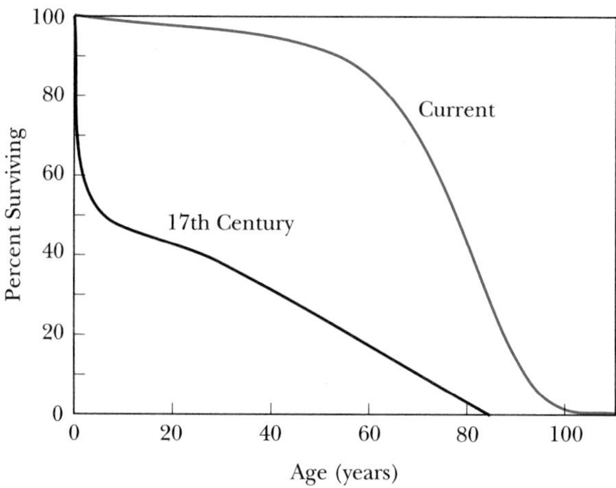

Figure 7-6

Survivorship curves for a 17th-century European city and for the United States in 1988. Data from Halley 1694 and National Center for Health Statistics, Vital Statistics of the United States, 1988, vol. 2, mortality, part A. Washington, Public Health Service.

veloped with the growth of large cities. Microparasites that previously tended to go extinct whenever they arose were now perpetuated in the human population. Although they often killed children, surviving individuals were immune, and so, in time, these became "childhood diseases." Such diseases evolved in the scattered urban areas of China, the Middle East, and Europe. When regular trade began among those areas, diseases were exchanged with temporarily catastrophic consequences. They killed off not just children, who were considered easily replaced, but also adults—skilled workers, soldiers, bureaucrats, parents. Smallpox, for example, seems to have arrived from the Near East and to have been the cause of great mortality in the Mediterranean area in the second and third centuries AD (McNeill 1976).

When explorers, soldiers, or traders from the crowded countries of the Old World visited smaller, isolated communities, they took their diseases with them. This is what happened in the New World, where during the period from 1500 to 1700 smallpox, measles, and other diseases endemic in Europe decimated the North American Indians, the Aztecs, and the Indians of Peru (McNeill 1976).

The survivorship curves shown in Figure 7-6 give an idea of the impact of childhood diseases in a European city in the 17th century. Specifically, survivorship is plotted for the city of Breslau (now Wrcolaw, Poland) for the years 1687 to 1691 in comparison to the current (1980 census) survivorship curve for the United States. The Breslau data from a 1694 paper by Edmund Halley (of Halley's comet fame) provide perhaps the first satisfactory life table (Hutchinson 1978). The full title is "An Estimate of the Degrees of the Mortality of Mankind, drawn from curious Tables of the Births and Funerals at the City of Breslaw: with an Attempt to ascertain the Price of Annuities upon Lives." The drop in childhood deaths, from a situation in which one half of every cohort died by age 8 years to the current situation in which one half of the population survives nearly to age 80 years has meant a revolutionary change in human social structure.

AIDS

Acquired Immunodeficiency Syndrome (AIDS) is a new human disease. Although it can probably be traced back to the 1950s (Grmek 1990), the first glimmerings that a new epidemic was beginning appeared in the 1970s, and the disease was first characterized in 1981. **Human Immundeficiency Virus** (HIV) is a causative agent, though the actual development of AIDS after HIV infection is a long process. The role of nutrition and other aspects of health and of other diseases as cofactors in AIDS is still being debated. In the United States and western Europe, AIDS appeared first as a disease of homosexual males, but it apparently originated in Africa where most transmission is evidently heterosexual.

The virus is transmitted via bodily fluids, especially semen and blood. Transmission is usually by sexual contact, though it also happens through transfusions of infected blood and by intravenous drug users sharing needles.

AIDS results from the sabotage by HIV of the host's immune system. The incubation period (the time from HIV infection to AIDS symptoms) tends to be long. Generally the starting point is taken as the first positive test for HIV antibodies, which comes a few weeks or months after infection. The median time to the development of AIDS symptoms is about ten years, although the progression sometimes occurs in less than two (Brookmeyer 1991). Because of the long incubation period, the percentage of persons infected by HIV that develops AIDS is not yet known but it is clearly high. HIV targets CD4 lymphocytes, and it is the decline of these cells that seems to be most correlated with the development of AIDS.

The disease is characterized by the development of cancers and frequent infections. These include types of pneumonia and meningitis caused by organisms that are rarely pathogenic in persons with intact immune systems. AIDS sufferers also have high rates of tuberculosis and other non-opportunistic infections. Infection or cancer is usually the immediate cause of death. The percentage of those who develop AIDS that die as a result of it is unknown but high, currently probably close to 100%.

The population characteristics of sexually transmitted diseases differ in several respects from those transmitted by other means (Anderson 1982). Although AIDS shows some unique features, it is like many other sexually transmitted diseases in several ways. The duration of infectiousness is long, asymptomatic carriers are often important in spread of the pathogen, and acquired immunity to the pathogen does not occur or is of negligible importance.

Another important feature is that basic reproductive rate as well as the persistence of the disease in a population do not depend on host population size or density. This is for the obvious reason that transmission is based not on the average number of susceptibles an infected individual stands next to on the subway or shakes hands with—contacts that increase with density. Rather, transmission depends on the number of new sexual partners per unit time, contacts that depend less on density levels than on promiscuity levels.

The equation giving R_0 for a disease such as AIDS is slightly different from that given earlier for directly transmitted diseases in general:

$$R_0 = \beta c D$$

Here R_0 is the basic reproductive rate, β is the average probability that infection is transmitted from an infected individual to a susceptible partner (per partner contact), c is a measure of the average rate at which new sexual partners are acquired, and D is the average duration of infectiousness (May and Anderson 1987). None of these values for HIV is yet well known individually.

Certain features are, however, evident based on the current limited knowledge about HIV taken in conjunction with existing ecological theory on disease. It is clear, for both homosexual and heterosexual transmission, that changes in behavior can lower β and c. β, the probability of transmission, can be lowered by the use of condoms, for example, and c, the rate of acquisition of new sexual partners, can be reduced by monogamy and, of course, abstinence. So both of the popularly prescribed approaches to reducing the AIDS epidemic are sound.

The rate of acquisition of new sexual partners in a population tends to be a highly skewed distribution, with the peak at a low value followed by a long tail consisting primarily of promiscuous individuals and prostitutes who may have 5, 10, or 100 new partners in a given time period. c and, thus R_0, are heavily influenced by this latter group. It seems likely that such sexually transmitted diseases as gonorrhea have an R_0 below 1 in most of the population and would die out in the absence of this group (Anderson 1982).

Probably both β and *c* are currently much lower in the homosexual community than they were in the early days of the epidemic; consequently, R_0 based on current behavior would be well below the estimate of 5 to 10 given for the period of the late 1970s and early 1980s (Table 7–1).

Probably, also, R_0 for heterosexual transmission is lower than for homosexual transmission. Whether it is below 1, making HIV infection not self-sustaining in a strictly heterosexual setting, will require more data to answer. This answer also depends on values of β and *c*. These values probably are different for men and women. The value of *c* will also vary depending on the prevalence of monogamy and on the size of the highly promiscuous portion of the population.

Currently the World Health Organization believes that about 10 million persons worldwide are infected with HIV (Palca 1991). Prevalence varies geographically, between the sexes, and among various subpopulations. Overall, the highest rate of infection is in sub-Saharan Africa (excluding South Africa), where the rate is about 1 person in 40, in both men and women (Palca 1991). Rates are currently very low in China, about 1 in 4000 men and 1 in 20,000 women. In the United States, current rates are about 1 in 75 men and 1 in 700 women.

The peak period in the United States for infections was between 1984 and 1986 when there were probably about 150,000 to 170,000 new infections per year (Brookmeyer 1991). A fairly steep decrease in new homosexual infections began about 1984 as a result of the behavioral changes already mentioned. This does not mean that the numbers of infected persons declined; they have continued to grow, but at a lower rate. There has been a slight decline in the rate of infection among intravenous drug users since 1986. Heterosexual transmission in the United States was very low until 1982, increased into the late 1980s, and may now be about stable with approximately 10,000 new cases per year.

Massive amounts of money have been spent studying cellular and molecular aspects of AIDS, and this will continue. Rapid strides have been made in understanding the functioning of HIV within the body and in developing therapies for treating infected individuals. Currently, it looks as though treatment will come to involve a combination of new drugs soon to appear on the market that will delay the onset of AIDS. The cost for treating one AIDS victim over a lifetime is approximately $120,000. This will inevitably increase; the only questions are how much and how fast.

Nearly everyone would agree that the most desirable development would be a preventive vaccine, so that we could all be immunized against HIV and forget about it, as is the case with polio and measles. Such a vaccine probably cannot be produced, primarily because retroviruses, such as HIV, show enormous genetic diversity (because of uncorrected errors in replication). However, the possibility is not yet absolutely foreclosed (Cherfas 1989, Wolinsky et al. 1992). A vaccine to boost the immune response and to give to persons soon after infection (Salk 1987) could be part of the combination of treatments described in the preceding paragraph.

As important as this cellular, molecular, and medical research is, more ecological research on basic population parameters is needed. We need to know what the public health burden is likely to be at various times in the future. For how many people, of what ages and sexes, is AIDS-related health care going to have to be provided? What alterations of mores and what limitations on personal liberties are strongly indicated to limit spread of the disease, and which are irrelevant?

For an individual who has AIDS, the danger is very great. The magnitude of the danger to the general population of the United States and most other countries may be no greater than that from many other diseases and conditions. By allowing us to assess the magnitude of the danger, basic epidemiological information can help decide where limited government dollars should be spent. Would some of the money currently being spent on AIDS save more lives if it were spent on research on hepatitis B or on stroke, or not on medical research at all but perhaps on improving safety in the use of automobiles or pesticides?

Effects of Parasitism on Host Numbers and Distribution

The effects of parasites on host numbers may range from unnoticeable to severe. Ample evidence exists that some parasites limit and regulate host numbers and thereby affect their distribution and the structure of the community (Minchella and Scott 1991). At one extreme, parasites presumably can drive the host to extinction. If the parasite is specific to that host, its extinction would then follow. Another possible result might be that when the hosts became very rare, a few would survive unparasitized and the parasite, instead, would become extinct. Thoroughly studied cases of the extinction of a species by a parasite are scarce (but see the comments on Hawaiian land birds below); however, there are several cases of virtual extinction.

The American chestnut once was an important element in the deciduous forests of eastern North America; most of the southern Appalachians, from Pennsylvania to Georgia (Chapter 15), were occupied by oak-chestnut forest. The chestnut blight, a fungus, was introduced to this country from China about 1904. Its effect on Chinese chestnut trees was not serious, but it was lethal to the American species. Within 50 years, chestnut was eliminated as a significant constituent of the forests of this country. Chestnut trees do still exist. A few trees planted far from the original range of the chestnut are, for the moment, disease free; occasional trees whose aboveground parts were killed root-sprout; and some trees hang on even though infected.

Parasites can influence the geographical or habitat distribution of their hosts. Evidence is available that suggests some Hawaiian birds have been limited to higher elevations—and some may have been eliminated entirely—by one or more bird diseases. When Captain Cook arrived at the Hawaiian Islands in 1778, they possessed a unique land bird fauna distributed among six families of perching birds. By the late 1800s, great reductions of many of these species were noted. Among the honeycreepers, a family confined to the Hawaiian Islands, eight of the original species are now extinct and the remaining species do not occur below 600–700 m (Warner 1968).

Although destruction of the native vegetation and competition with introduced species of birds may also have been factors in some of the extinctions, there is persuasive evidence of the role of diseases, especially bird pox and avian malaria. These diseases were present on the islands in migratory bird species but were no threat to the resident birds because of the absence of insects to spread the pathogens (bird pox is caused by a virus, malaria by a protozoan). There were, for example, no mosquitos; Hawaii was an island paradise, indeed.

In 1826, the night-flying mosquito was introduced. A party from the ship *Wellington*, ashore on Maui to fill the ship's water tanks, emptied the dregs full of mosquito larvae into a stream. They had last filled the tanks on the west coast of Mexico, and so the population introduced was a tropical one. Significant populations of the night-flying mosquito do not occur in Hawaii at the cooler temperatures above 600–700 m.

The honeycreepers that have become extinct were those that inhabited ranges below that elevation or that made periodic migrations to the lowlands. The species that still persist no longer occur below these elevations, even in the remnants of native vegetation that exist. The populations at higher elevations show little immunity, and if they are caged and brought to lower elevations, they quickly contract the diseases. One small island, Laysan, has one species of honeycreeper, the Laysan finch, and no mosquitos. When Laysan finches were brought to Honolulu and put in cages that allowed mosquitos to enter, all of 13 birds died with symptoms of malaria in a little more than two weeks. None of the 13 birds kept in mosquito-proof cages died.

Parasites, like predators and herbivores, by keeping potentially competing species below their carrying capacities, can prevent competitive exclusion and, thereby, increase species diversity. But, also, like predators and herbivores they can lead to the local extinction of species and thereby reduce diversity.

One example of the latter situation is the infection of moose populations by the brain worm, a roundworm, that occurs in the brain and spinal cord. It produces odd behavior—aimless wandering, a lack of fear of humans, and if poachers or accidents don't intervene first, death. The usual definitive host of the parasite is the white-tailed deer (Figure 7-7), which seems not to be seriously harmed by the infection.

Logging and other human modification of the vegetation in Canada and the northern United States tended to increase habitat favorable for deer through much of the early and middle part of this century. Deer spread and increased in these regions, bringing with them the brain worm. Declines in the moose population followed (Karns

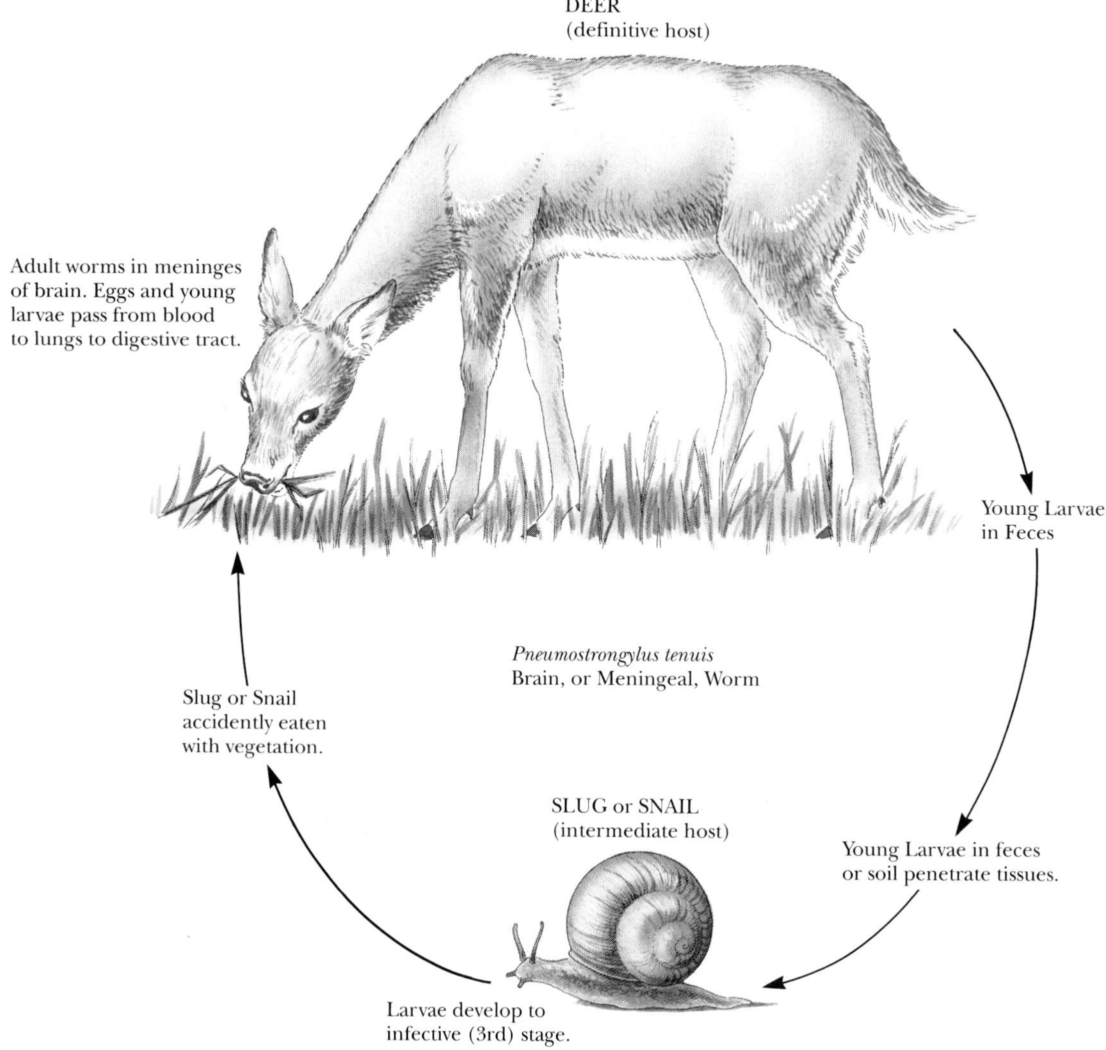
DEER
(definitive host)
Adult worms in meninges of brain. Eggs and young larvae pass from blood to lungs to digestive tract.
Young Larvae in Feces
Pneumostrongylus tenuis
Brain, or Meningeal, Worm
Slug or Snail accidently eaten with vegetation.
SLUG or SNAIL
(intermediate host)
Young Larvae in feces or soil penetrate tissues.
Larvae develop to infective (3rd) stage.

Figure 7-7

A simplified diagram of the life cycle of the brain worm of white-tailed deer. Sexual reproduction occurs in the definitive host resulting in the production of many thousands of eggs. Development but, in this case, no asexual reproduction, occurs in the intermediate host, one of various terrestrial snails or slugs.

1967, Gilbert 1974). We are often tempted to hunt for a competitive exclusion explanation for such replacements. Probably competition is frequently involved, but a shared parasite may be the immediate cause, as it apparently is in this case.

In the past few decades, the forest-edge conditions favored by deer have declined and moose populations have returned in various areas. Reintroduction of moose into areas from which they had disappeared, such as northern Michigan, has been hindered by high rates of parasitism by brain worm. Note that this case is exactly parallel to the situation described earlier in which the biennial composite was excluded, through the action of a shared herbivore, from habitats where broom snakeweed was common.

Certain habitat features, if present, seem to allow deer and moose to coexist. If the region is largely conifer forest, so that deer populations are small and localized, as in various regions in prehistoric times (Telfer 1967), the moose population seems not to be seriously affected. If there are substantial differences in snow depth, deer tend to be restricted to the areas of least snow cover, and transmission of the disease is lowered (Figure 7-8). In Fundy National Park, deer do not winter at higher elevations, evidently because of deeper snow there. Moose survive in these areas, but any that descend to the low country die before the end of the winter (Kelsall and Prescott 1971).

The extinction of one species in a shared range is more likely if one of the two species is less susceptible to infection, has a low death rate from the disease, or has a higher reproductive rate (Holt and Pickering 1985). Coexistence can be achieved by various means. Even if the disease is the limiting factor for population size in both species, coexistence can occur if within-species disease transmission is stronger than between-species transmission. This is probably the case when deer and moose contacts are reduced by sharp vegetational boundaries or snow depth.

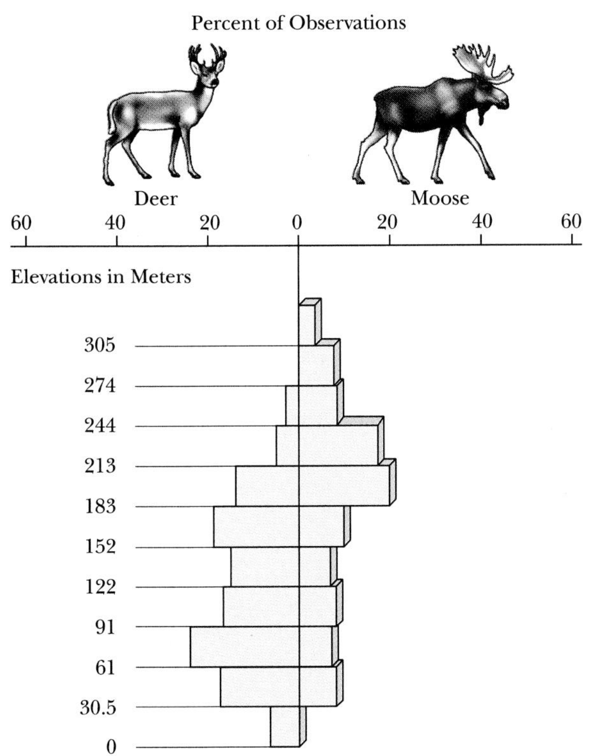

Figure 7-8

Deer and moose distribution by altitude in Nova Scotia in February 1964. Frequency of observation (of animals or tracks) as a percent of total is plotted. (Adapted from Telfer 1967. Copyright the Wildlife Society.)

The effects of parasitism on the dynamics of populations and communities can be exceedingly subtle and difficult to detect simply by observation. A series of experiments on the Sierra treehole mosquito in its natural habitat and in laboratory microcosms investigated the effects of parasitism by a ciliate, Clark's lambkin (Washburn et al. 1991). Under natural circumstances, mosquito eggs hatch as treeholes fill with water at the beginning of the rainy season. Some young larvae are infected by Clark's lambkin, which also can pass the dry season in an inactive form. Not all treeholes in a given season have the ciliate parasite. Infected mosquito larvae die in three to four weeks; however, the ciliates multiply inside the larvae, escape, and parasitize other larvae.

Experiments used low and high densities of mosquitos (relative to the food supply of treehole detritus or, in the laboratory, Purina rat chow) with and without the parasite. In the uncrowded microcosms, unparasitized populations of the mosquito survived the 25 to 30 weeks to emergence as adults better than parasitized ones, an unsurprising result. The figures are that 64 of an original cohort of 100 survived in the unparasitized cultures compared with 48 in the parasitized ones.

In the high density microcosms, despite the fact that they began with five times as many larvae, fewer adults were produced than in the low density ones—only 25 or 30 adults out of the original 500. This outcome—lower production in absolute terms at very high densities—is a frequent outcome of intraspecific competition of the scramble type (see Intraspecific Competition, Chapter 4).

In these high density, or crowded, cultures, there was no difference in survival between parasitized and nonparasitized populations. To use another term introduced in an earlier chapter, **intercompensation** was occurring in the high density situation: Parasitic mortality was being substituted for, rather than added to, the other mortality resulting from the crowding.

There was one more important result: The

Table 7-2
Effects of a Parasite, the Barbershop Coccidian, on the Outcome of Competition Between Two Species of Grain Beetles

Outcome	*Parasite Present (74 Trials)*	*Parasite Absent (18 Trials)*
Chestnut grain beetle wins	11%	67%
Confusing grain beetle wins	89%	33%

Data from Park, T., "Experimental studies of interspecific competition. I. Competition between populations of the flour beetles *Tribolium confusum* Duval and *Tribolium castaneum* Herbst,"*Ecol. Monogr.*, 18:265–308, 1948.

adults emerging from the crowded parasitized tree-holes were larger than ones from the crowded non-parasitized holes, and about the same size as ones emerging from the uncrowded holes. Lifetime reproduction in these mosquitos depends strongly on body size, so at high densities, parasitism probably raised genetic fitness.

Another example of the subtle effects of parasitism, this time on the outcome of interspecific competition is seen in some of the early work by Thomas Park (1948) on grain beetle competition. He found that the chestnut grain beetle usually lost out to the confusing grain beetle under moist and relatively warm conditions (29.5°C, 70% relative humidity). Then he discovered, as we have noted earlier, that the cultures were infected with a protozoan parasite, the barbershop coccidian. When he was able to rear parasite-free cultures, the outcome of the competition (for this set of physical factors) was reversed. The chestnut grain beetle, whose equilibrium population size was much more strongly depressed by the parasite, now usually persisted, and the confusing grain beetle became extinct (Table 7-2).

Evolution Within the Parasite-Host System

Subtle effects of parasite on host, such as seen in the effects of the barbershop coccidian on the confusing grain beetle, are common; severe, debilitating effects and death are not. We see them mostly in one of two situations: (1) a relatively recent contact such as those described for the American chestnut and the Hawaiian honeycreepers and (2) cases where the circumstances of the interaction have been altered by environmental changes.

A frequent evolutionary outcome of a parasite-host interaction seems to be the development of greater resistance by the host or lowered virulence by the parasite or both. There are probably several ways in which these can evolve, but a major feature of the coevolution (Chapter 5) is probably that hosts with the least degree of resistance aren't able to reproduce as much—or at all—compared with hosts with some resistance. Consequently, the resistant genotypes increase.

The resistance may be of several kinds. One very important category is immunological, whereby cells or antibodies kill off parasites. But more obvious and straightforward defenses can also be involved. Evidently, one reason for the greater susceptibility of honeycreepers to mosquito-borne diseases is the honeycreeper's sleeping posture. Unlike some of the introduced species that survive in the lowlands of Hawaii, the honeycreepers sleep with their legs and the unfeathered areas around the beak and eyes—the parts of the body that are most vulnerable to mosquito bites—exposed, instead of tucked under feathers.

A remarkable example of what seems to be anti-parasite behavior has been described for certain birds that reuse nests, such as tree cavities and nest boxes. Populations of fleas, lice, and mites may build up in these nests to levels harmful to the young. Several species of such birds are known to add fresh foliage from aromatic plants, such as wild carrot, to the nest at intervals, and this has been demonstrated to reduce mite populations (Clark and Mason 1988). Birds showing the behavior of adding fresh plant material to the nest through the nesting cycle include the European starling (Clark and Mason 1988), purple martin (Johnston and Hardy 1962), and various hawks (such as the goshawk and red-tailed hawk, in Brewer 1991). The starling evidently selects the plant material on the basis of smell (Clark and Smeraski 1990).

The evolution of lowered virulence in the parasite also comes about because of the lengthened life span of the hosts. In this case, though, the importance of the host's living longer is in allowing greater reproduction by the less virulent parasites. Parasites that kill their hosts in a hurry will have a reduced opportunity to reproduce and to spread to other hosts.

So the general picture of an evolutionarily old host-parasite relationship is one in which the parasite makes a good living without crimping the life style of the host very seriously. Even so, the relationship has its costs (Walkey and Meakins 1970, Kennedy 1972). A study of tapeworm infection in white-footed mice found that they digested their lab ration food less efficiently than uninfected mice (Munger and Karasov 1989). The drop was slight, however, about 2%, and the mice lost no weight, nor even showed any obvious signs of compensating for the loss. It is possible that tapeworms in young animals or in pregnant or lactating females might be more harmful. There is much yet to learn about energy flow in host-parasite systems.

Even relationships in which parasite and host are well adapted to one another can show harm to the host in special circumstances. The populations of most parasites are rarely very high in a particular individual host. This is especially true of the macroparasites that (unlike the bacteria, viruses, and protozoans) do not reproduce within the host. A host may do very well indeed with a single tapeworm, even one 6 feet long, attached to its gut. But if it is infested with 20, it may die or be weakened for other disease or predators. The way such heavy infestations arise on a populationwide scale is from some ecological dislocation that makes the likelihood of transmission very high. For example, much of the damage to agriculture from plant pests results from the effects of monocultures and year-after-year cultivation of the same spot with the same crop, allowing parasites to build up and to spread freely.

We see the same problems of ecological dislocations with humans. We have already talked about the role of high densities and large populations in the transmission and perpetuation of human parasites. Higher death rates from disease in developing countries are often the result of greater susceptibility because of malnutrition (Pereira 1982), the root cause of which is usually overpopulation.

Other ecological dislocations besides crowding can be important. Most of us are aware that humans carry around small populations of potentially harmful bacteria that produce disease (that is, multiply to problem proportions) only after some sort of interference with the body's normal operation; staphylococcus infection of the digestive tract after antibiotic treatment is an example. Microbes are implicated in respiratory diseases such as emphysema but, again, only when the ecology of the respiratory tract is changed by smoking or air pollution.

When Should Parasites Harm Their Hosts?

There are exceptions to the generalization that well-adapted parasites should not harm their hosts. Many parasitic worms have a life history pattern similar to Figure 7-7. The transfer of parasites from intermediate host (where asexual reproduction may occur) to definitive host (where sexual reproduction of the parasite takes place) occurs through the definitive host using the intermediate host as food (Figure 7-7). In this case, it might be advantageous for the parasite, after achieving some level of development, to sicken the intermediate host enough to make it more likely to fall prey to the definitive host (Kennedy 1975). Studies have suggested that this is sometimes the case. Larvae (metacercariae) of certain kinds of flukes live in the eyes of fish, causing partial blindness and presumably increasing the vulnerability of the fish to predation by birds and mammals.

A still more elegant evolutionary solution to the problem would be to make the intermediate host more vulnerable to predation without weakening it (Moore 1984). This also happens. Larvae (sporocysts) of another fluke migrate to the tentacles of the snails that are their intermediate hosts. The sporocysts are brightly colored, and they pulsate, advertising the infected snail to its bird predators. Where I live, businesses would be violating a city ordinance if they used advertising signs as garish as that.

Several similar cases are known in the Acanthocephala, the spiny-headed worms. One of these parasitizes the digestive tracts of mallards and muskrats. Eggs in the bird or mammal feces are eaten by amphipods, small aquatic crustaceans. Ordinarily amphipods avoid light, but when the developing spiny-headed worms reach the stage (cysticanth) in which they can infect their definitive hosts, the phototaxic response reverses: The amphipods move toward light. This causes them to stay near the surface of the water where they are easy prey for mallards and muskrats. It would be interesting, indeed, to know if the slugs or snails carrying brain worms (Figure 7-7) show alterations of behavior that make them more prone to "accidental" ingestion by deer when the brain worm larvae have reached the infective stage.

Health and Disease

Health and disease are ecological terms that describe relationships between an organism and various features of its environment. Most of us learn the germ theory of disease in a regrettably oversimplified version: A particular pathogen enters the tissues of a host, multiplies, and produces a particular set of symptoms. This is true as far as it goes, but as René Dubos pointed out in "Second Thoughts on the Germ Theory" (1955), disease is a coaction, an interaction between pathogen and host. It is not the only interaction possible between these organisms—commensalism, neutralism, and even mutualism are other possibilities—and whether or not disease develops, as well as what form it takes, depends on the relationship of both organisms to their environments.

"This ecological concept," Dubos wrote, "is not merely an intellectual game; it is essential to a proper formulation of the problem of microbial diseases and even to their control." These ideas are discussed further in the section "The Role of Environment in Health and Disease" (Chapter 21).

Commensalism

A **commensal** lives on or around individuals of some other species (which may be called **hosts**) and derives benefit from the association. The distinction between commensalism, in which the host suffers no negative effects, and parasitism, in which it does, is vague and fluctuating. Likewise, the point at which commensalism comes to be mutualism is clearer in definition than in practice (Figure 6-1).

P. J. Van Beneden (1876), in coining the word, wrote, "The commensal is simply a companion at the table." Good examples of commensals in this sense are the scavengers, such as vultures, that live on the scraps from the kills of large carnivores, such as lions. A favorite textbook example of a commensal is the remora, a fish with a suction cup on the top of its head by which it attaches itself to a shark. It thus travels with the shark and eats the leftovers from the big fish's meals. The coliform bacteria in the intestine of vertebrates are ordinarily commensals; they subsist on the food residues passing through.

The term "commensalism" is now used in a broader sense to refer to coactions in which the gain is something other than direct access to food provided by the host. Usually the gain is some combination of transportation, support, or shelter. The use of prairie dog burrows as nest sites for burrowing owls and the use of old bird's nests by deer mice as sites for their own nests are examples of commensalism. Turning to plants, simple **epiphytism**, in which one plant uses another for support, is considered commensalism. Spanish moss, which festoons live oaks and cypresses in the Gulf states, seems to be in this category.

Dispersal by animals is often commensalistic. Although some of the relationships may be mutualistic, such as frugivory, the individual heron that carries seeds of aquatic plants in the mud on its feet or the swallow that spreads algae on its plumage probably receives no immediate gain from the event.

Commensalism is rampant in aquatic habitats, especially marine ones. Burrows, such as those of polychaete worms, are full of crustaceans and other guests. Organisms encrust other organisms: Barnacles cover whales. One clam that washed up on a Florida beach had more than 100 animals of 25 species attached to it (Perry 1936). "Sponges," according to A. S. Pearse (1939), "are often veritable living hotels." One loggerhead sponge he examined at Dry Tortugas sheltered over 17,000 animals of 10 species.

Many kinds of commensalism do not involve highly specific relationships. Who killed a wildebeest may make very little difference to a griffon vulture, just as long as the wildebeest is dead. Spanish moss grows on many species of trees, though some are unsuitable because of the nature of their bark. In other cases, such as the invertebrates that live in the nests of social insects, the relationship is more specific. The water in North American pitcher plant leaves serves as the pools in which the larvae of the mosquito *Wyomeia smithii* mature, in a close commensal relationship.

Humans are surrounded by commensals. They range in size from the coliform bacteria in your gut to the white storks nesting on your chimney (if you live in western Europe). In between in size are the mites that inhabit the oil glands around your nose (commensals rather than parasites because they live on the sebaceous secretions of the glands) and the cockroaches and house mice that, through history, have shared human dwellings and leftovers. Every organism with the word "house" in its name is a human commensal.

Phoresy

Transport of one animal by another has been termed **phoresy.** Although members of several taxonomic groups, such as insects and mollusks, use other animals to carry them, phoresy is probably best developed among the mites, small eight-legged creatures related to spiders. Phoretic relationships span the range from predation and parasitism through commensalism and mutualism (Wilson 1980). Some mites seem to use insects (or other animals; some mites move from flower to flower on hummingbird beaks [Colwell 1982]) simply as a way to get where they need to go.

Other mites do things of probable benefit to the insect that transports them. The burying beetle lays eggs on the body of a mouse that it buries as a food supply for its larvae. Mites that it carried to the site also live on the corpse, but the first thing they do after hopping off the beetle is to run over the dead mammal eating blowfly eggs. In so doing they (perhaps) help prevent the development of the maggots that are the main competitors of the burying beetles (Springett 1968, Wilson 1983).

Some mites are predaceous on the insects that carry them. This is true also of larvae of certain blister beetles (Meloidae) that lurk on flowers, waiting for bees. They hitch a ride back to the bee nest and then eat the bee's larvae. Many species of mites live on ants, from whom they beg food in relationships that seem basically parasitic (Burton 1969).

Saprobism

Saprobes are organisms that gain their energy from dead or dying tissue. They include the scavengers and decay organisms. One member of the interaction, being dead or dying, may not take a very active role in the coaction, but this is also true for the less personal kinds of commensalism. Some cases of saprobism may involve specific relationshps between saprobe and host, but many of these relationships are very general. The taxonomic origin of detritus may make little difference to most detritivores; on the other hand, they may be specialized as to kinds of chemical compounds they utilize.

Freshly dead material available for saprobes includes the following categories:

- Dead bodies of animals (carrion)
- Feces (dung) and excreted organic compounds
- Standing dead trees, logs, stumps, and tree roots
- Dead leaves and stems (litter)
- Small woody parts
- Overripe fruit

After saprobes and physical agents have fragmented these materials, they form detritus. Eventually, through continued decomposition, the materials are mineralized to simple inorganic compounds. The roles of saprobes in energy flow and nutrient cycling in ecosystems are discussed in later chapters. Here we outline some of the important features of saprobism in carrion, dung, and logs.

Carrion

Dead bodies are a rich nutritional resource; nothing else (except live bodies) combines readily digestible calories with such a concentration of nitrogen (protein). When a fair-sized animal dies, a race begins among the scavengers. It is rare for birds and mammals to beat the insects to the carcass, but they may still, by the rapidity of their consumption, win the race. The microorganisms, bacteria and fungi, are also contestants and generally obtain a share of the prize.

Because humans tend to be repulsed by the

smell—or even the idea—of decaying flesh,* there are few detailed investigations showing the relative importance of various carrion feeders. Charles Elton (1966) cited a Russian study of dead ground squirrels in steppe that found that about two thirds were carried away by vertebrate scavengers. The rest were decomposed in place by insects and microorganisms.

Vertebrates

The best-known vertebrate carrion-feeders include the condors, vultures, hyenas, and jackals, but many other animals also make use of carrion. Gulls have always been important scavengers along the beaches. More recently they have undergone a population explosion as they have taken over the role of scavenger of the garbage dump and shopping center parking lot. Several members of the crow family eat carrion; in fact, one European species is called the carrion crow.

The purest of the scavenging vertebrates are the vultures and condors; otherwise the line between scavenger and predator is rarely distinct (Houston 1974). Many predators, such as lions and bears, readily accept carrion. Certain storks and eagles, although basically carrion-eaters, are not averse to speeding matters up when they encounter injured or otherwise vulnerable prey. For many species, switching from predation to scavenging occurs predictably based on prey and carrion availabilities.

In areas with large herbivores and a rich scavenger fauna, notably east Africa, there may be some specialization by different species on different parts of the corpse, although the large, tender muscle masses around the shoulders and hips are prime fare for many. Some, nevertheless, are able to make good use of skin, tendon, and bone. Hyenas, for example, have jaws powerful enough to crush all but the heaviest bones to extract the marrow. One bird, the lammergeyer, or bearded vulture, also specializes on bone marrow. It is a large bird, nearly 4 feet from beak to tail. After soaring to a great height, it drops a bone, shattering it on the rocks below, and then descends to pick out the marrow.

The most obvious anatomical adaptation of scavenging birds is a bare head and neck, allowing them to probe the depths of large carcasses. Feathers matted with blood or soaked with fat would be a detriment rather than an asset, and birds cannot preen the feathers on their head. Most avian scavengers also have keen eyesight, to pick out dead bodies while soaring high and to keep track of other vultures in the sky. One bird beginning a descent will quickly draw others. It is said that hyenas, jackals, and lions also watch vultures for the same reason.

Curiously, few birds use smell as a cue to the presence of carrion. Perhaps in most circumstances if they wait until the scent is obvious, they will be too late for a place at the table. There is at least one possible exception. The turkey vulture is attracted to carcasses it cannot see and has been shown to be able to detect chemicals released by decaying flesh (Stager 1964, Smith and Paselk 1986).

Insects

Most of the insects that lay eggs in carrion or that feed on it as adults belong to the beetles and the flies. Widespread examples are fleshflies and blowflies and carrion and burying beetles.

Flies may arrive at a corpse with incredible swiftness, so much so that it seems possible that they visit on speculation any animal they notice to stop moving. Volatile chemicals produced by decay are also important slightly later. It is an interesting sidelight that certain flowers smell like carrion—skunk cabbage and pitcher plant, for example—and are pollinated by flies. Stinkhorn mushrooms also smell like carrion (or worse) and have their spores spread by blowflies and other carrion- and dung-feeders.

A succession, a microsere, occurs on carcasses, from the first blowflies to the last skin beetles, and is described in detail in Chapter 13. If burying beetles (*Necrophorus* spp.) find a carcass soon enough, however, this sequence of events is likely to be canceled.

Burying beetles are handsome black and orange beetles that are about 2 centimeters long. In a remarkable feat, they remove soil from underneath a small animal and put it on top so that the

* Daniel Janzen (1977) suggested that the putrefaction of meat and fruit caused by microorganisms is an example of interference competition. The selective advantage to the microorganism is the discouraging of consumption by animals. The idea is an example of rampant adaptationism and, hence, very possibly correct. The variation in tolerance to the smell of decay among different animals and, to a degree, among human cultures is interesting.

carcass seems to sink into the ground (Milne and Milne 1976). In soft soil, the process can be swift, although usually about a day is required. Once the body has reached a suitable depth, generally 3 cm or more below the surface, the beetles shave it or skin it and roll it neatly into a ball. The shaving serves to remove the eggs of blowflies that are the beetles' main competitors. As already mentioned, mites carried by the beetles also search out and destroy blowfly eggs, in an apparent case of mutualism.

Although several beetles may cooperate in burying a corpse, eventually all but one pair leave or are driven away. The female lays eggs in a short tunnel off the main burial chamber. The young larvae are fed regurgitated fluid by the adults, which feed on the carcass, and later the larvae also use the carcass directly as their food source. Although the beetles rarely bury anything larger than a mole, they will feed on much larger carcasses—deer, for example—when they are not raising a brood.

Dung

Every animal egests wastes, and there has evolved a large group of organisms specializing on this habitat—or this set of habitats, inasmuch as dung varies from frass (insect fecal pellets) to bird dropping to cow pats. Only for the last, the dung of large herbivorous mammals, is there much ecological information.

As with carrion, flies and beetles are the most important insect groups utilizing dung. Most of the specialized dung species reach newly deposited feces within a half hour, in response to chemical stimuli (Figure 7-9). One possible exception is the first arrival, the hornfly or buffalo fly *(Haematobia irritans).* These, which resemble houseflies reduced by one half, follow cattle or other large herbivores, from which they suck blood. As soon as the cow deposits feces, the hornflies drop down to lay eggs and then immediately rejoin their host (Mohr 1943).

Dung consists of undigested food residues, water and other fluids including secretions and excretions from the digestive tract, sloughed-off epithelium, bacteria, yeasts, and molds. The larvae of the numerous fly species feed on the fluid or the microorganisms in it. The outside of the dung dries quickly, so the fly larvae are confined to the interior. To a considerable degree, the flies are dependent on the activities of certain small scarab beetles *(Aphodius).* These and their larvae burrow in the

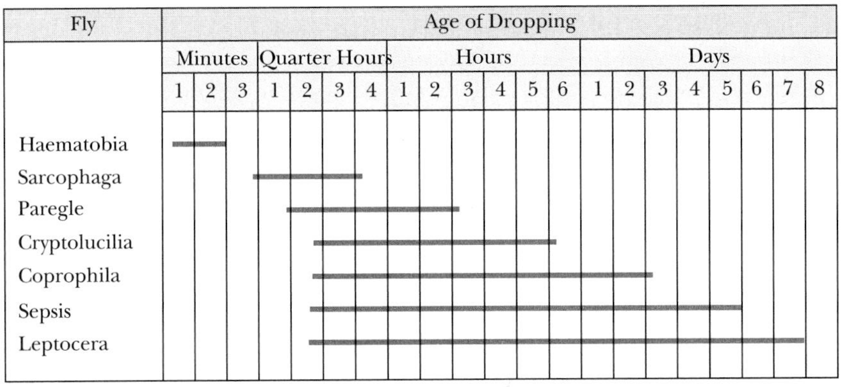

Fly	Age of Dropping																				
	Minutes			Quarter Hours				Hours						Days							
	1	2	3	1	2	3	4	1	2	3	4	5	6	1	2	3	4	5	6	7	8
Haematobia	—	—																			
Sarcophaga			—	—	—	—	—														
Paregle				—	—	—	—	—	—	—											
Cryptolucilia					—	—	—	—	—	—	—	—	—								
Coprophila					—	—	—	—	—	—	—	—	—	—	—	—					
Sepsis					—	—	—	—	—	—	—	—	—	—	—	—	—	—			
Leptocera					—	—	—	—	—	—	—	—	—	—	—	—	—	—	—	—	

Figure 7-9

Here is shown the succession of adult flies on cow dung. The flies, whose generic names are shown, are egg-laying or feeding or both. Their occurrence is based primarily on the moistness of the surface which may initially be too wet for some species and eventually becomes too dry for all of them. Note the quasi-log scale, related to the fact that the initial changes in moisture and, consequently, in the kinds of ovipositing flies, occur rapidly. (From "Cattle droppings as ecological units," by C. O. Mohr, *Ecological Monographs,* 1943, 13:275–298. Copyright © 1943 by Ecological Society of America. Reprinted by permission.)

cow pat, providing aeration for the otherwise anaerobic mass (Elton 1966). These beetles are saprobes, but several others, both adults that visit the dung plus larvae that develop in it, are predaceous, preying on the fly and scarab larvae.

Bacteria, nematodes, and a variety of mites also inhabit dung. Some species of mites subsist directly on the fecal material; others eat fungi, fly larvae, or nematodes (Stamatiadis and Dindal 1990).

As the cow pat dries up, it is invaded by more generalized soil organisms and eventually is scattered to a degree that is no longer identifiable. However, the site is often marked by heavy growth of nitrogen-rich grass which is relatively unpalatable to cattle.

Tumblebugs and Other Buriers

One group of dung beetles plays a role that is equivalent to that of the burying beetles with carrion. These beetles are a variety of species of scarabs of different sizes and habitats that bury dung. They have two general approaches. Some bury in much the same manner as the burying beetles, by excavating beneath the dung. Others, the tumblebugs, shape the dung into balls which they roll away before burying. In either case, eggs are laid in the dung and the developing larvae feed on it.

As with carrion, the microsere that would otherwise develop is aborted by the burying and related activities of these beetles. Presumably this removal from competitors, other beetles and flies, is the advantage of burial. The larger tumblebugs can remove most of a cow pat in a day, leaving any fly larvae that survive the process to perish in the dry crust that remains. Another advantage, and probably the main one for actually removing the dung from the site, is that the beetles get their eggs and larvae away from possible predation by birds and mammals drawn to the activity at the cow pat.

Someone without experience around a farm—or a zoo—may not appreciate the amount of fecal material produced by large herbivores. A cow produces a dozen pats a day, amounting in weight to about 50 pounds (23 kilograms) and in ground area covered to nearly a square meter. In communities such as grassland and savanna where large herbivores are numerous, dung is an important microhabitat and a significant influence on the soil and vegetation. Everywhere, of course, it and its associated organisms form an important link in biogeochemical cycling.

In Australia, the only large native herbivores are kangaroos. No native beetles were adapted to process the large pats such as are produced by domestic cattle. When cattle were introduced, ranges tended to deteriorate owing to the persistence of the pats and the rank growth of grass surrounding them. At the suggestion of a Hungarian agricultural scientist, G. Bornemissza, African tumblebugs were imported a few years ago (Hughes 1975). Several species have spread, one explosively. So far they have done what they were expected to do and have had basically beneficial effects. It is still too early to tell whether their activities will markedly increase carrying capacity of the cattle range.

Dead Trees

A few insects live or reproduce only in dying or newly dead trees. Notable examples are the bark and ambrosia beetles (Scolytidae) and the horntail wasps (Siricidae). The beetles in particular may have a narrow range of trees that they will accept. There is, for example, a species of ambrosia beetle that specializes on birch and another that specializes on aspen.

As with some of the carrion and dung insects, the speed with which these insects find their hosts is remarkable. Horntail wasps are said to "settle on freshly felled trees, sometimes before the woodsmen have finished cutting them into logs, and on fire-killed trees before the fire is out" (Furniss and Carolin 1977). For the ambrosia beetles, the attractant may be ethanol produced in the beginning of anaerobic decay of the tree sap. Some bark beetles have been shown to be attracted to chemicals that leak from the tree when it is wounded—terpenes in the case of the Douglas fir beetle. Others seem simply to fly around trying every tree. Once a bark beetle has located a suitable tree, by whatever means, it releases a pheromone that quickly draws others.

Many types of fungi are important in wood decay. Some are introduced by the bark and ambrosia beetles, carried in special chambers (mycangia) in the thorax. More than one kind of fungus may be carried. Some of the beetles cultivate fungi in their tunnels on which they or their larvae feed. Other kinds of fungi spread through the tree, hastening death or contributing to decay.

The course of breakdown and decay differs in detail according to whether the trunk is large or small, standing or fallen, in sun or in shade. The general trend of the microsere, however, is for the early invaders to loosen the bark and soften portions of the wood, providing situations that many other organisms can utilize (Shelford 1913). These later organisms tend to be unspecialized as to species of tree. Several of the organisms characteristic of this stage are flat, such as fire-colored beetle larvae (Pyrochroididae), isopods, and centipedes.

If the bark remains intact, this stage may persist for years as the decay of wood continues within. The surface of the trunk beneath the bark is a mass of fungal hyphae which also increasingly invade the wood itself. Within the decaying wood may be larval and adult beetles of several families (the bessbugs, Passalidae, are an example), wasps, flies, roaches, spiders, and the soil-related fauna such as springtails, mites, millipedes, and snails. Some of these eat the wood, some eat the fungi, and others are predators or parasites of the first two.

Logs on the forest floor also provide homes or hiding places for several vertebrates, especially snakes and salamanders. Standing dead timber (called "snags" by foresters) is necessary for most hole-nesting birds—about 85 species in North America (Scott et al. 1977). Included are those that excavate their own cavities, such as woodpeckers and chickadees, and those that use old woodpecker cavities or natural hollows, including bluebirds, house wrens, prothonotary warblers, screech owls, and wood ducks. Among mammals, flying squirrels and certain bats are strongly dependent on large dead or dying trees for homesites, and many other species regularly use tree cavities. White-footed mice, gray squirrels (Figure 7-10), and raccoons are examples. None of these vertebrates is a saprobe, though some of them contribute to the breakdown of the dead tree. They occur because the logs or snags provide suitable microhabitat; the relationship in Haskell's terminology is commensalistic.

Conventional foresters detest "overmature" trees—that is, ones that are large and old, with dead heartwood and likely to die in the next decade or two. Their reasoning is that trees are for people: By

Figure 7-10

A gray squirrel with a tree cavity nest. (Photo by George and Betty Walker.)

letting a tree die, we are wasting it. But as we have seen, the forests produced by this line of reasoning are impoverished (Davis 1983, Albrecht 1991). When we do without standing dead trees and logs on the ground, we also do without many kinds of mushrooms, beetles, red-backed salamanders, king snakes, and pileated woodpeckers.

Litter

A large part of the organic production of most communities ends up each year as dead leaves and stems. This tends to be distributed fairly uniformly over the soil, forming the background on which are superimposed the carrion, dung, and logs that we have discussed here. The general process of breakdown and decomposition of litter is discussed in the section on soil formation (Chapter 12).

Aquatic Habitats

Dead organic matter is a major source of energy for aquatic food chains (Chapter 11). Bodies, leaves, feces, etc., are rapidly converted by biotic or abiotic means to **particulate organic matter** (POM) or **dissolved organic matter** (DOM), which passes through a 0.5-micrometer filter. Aquatic habitats have few, if any, animals specialized for carrion, dung, or logs. On the other hand, a high percentage of the fauna are detritus-feeders, gaining their nourishment either as saprobes from the dead organic matter or from the saprobic bacteria.

The Population-Community Ecology Interface: *Competition and Related Coactions*

Black-capped chickadees (shown here) replace Carolina chickadees in the northern United States and at high elevations in the Appalachians. *(George E. Stewart/Dembinsky Photo Associates)*

OUTLINE

- Interspecific Competition
- Amensalism
- Allelopathy
- Neutralism

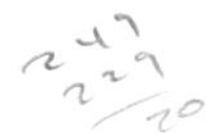

This chapter continues the exploration of species interactions. In it are discussed interspecific competition and two related coactions, neutralism and the neglected topic of amensalism.

Interspecific Competition

Outcomes of Competition

Here we are concerned with the interaction of two or more species that share a resource. A **resource** by dictionary definition is something that can be drawn on for help or support. Tree seedlings in the forest understory might share the resources of soil moisture, nutrients, and sunlight. Two bird species might share a food supply of small insects. In discussing intraspecific competition, we defined competition as a combined demand in excess of the supply. Whenever the combined demands of, say, the insectivorous birds exceed the supply of insects, whether by the birds being common or the insects scarce, then the two species would be competing.

What are the possible outcomes of the situation in which species share a resource?

1. Competition may be severe enough to result in the elimination of one or the other of the two species. This is **competitive exclusion,** discussed in the next section. It is the outcome predicted by cases A, B, and C of the Lotka–Volterra model (see Figures 8-7 to 8-12).
2. The shared resource may rarely or never be in short supply, so the two species rarely or never compete. There are three obvious ways in which this can happen.

   A. Some resources may be nondepletable or nonmonopolizable, such as oxygen for most terrestrial organisms or tidal energy for oysters and sea urchins. Some authors have wished to define "resource" more narrowly than the commonsense dictionary meaning. One example is "any substance or factor which can lead to increased growth rates as its availability is increased, and which is consumed" (Tilman 1982). Such definitions come close to defining a resource as "that which is competed for," and, hence, run the risk of circularity in defining competition.

   B. Predators (or herbivores or parasites) may keep densities of potentially competing species below the level at which they would have serious competitive effects on each other. The idea goes back at least to Darwin (1859), who wrote, "If turf which has long been mowed, and the case would be the same with turf closely browsed by quadripeds, be let to grow, the more vigorous plants gradually kill the less vigorous, though fully grown plants; thus out of twenty species growing on a little plot of mown turf (three feet by four) nine species perished from the other species being allowed to grow up freely."

   The Japanese ecologist Syunro Utida (1953) showed that when two species of bean weevils were cultured together in the laboratory, one, the Chinese bean weevil, was eliminated by the four-spotted bean weevil. However, when a parasitoid of both, the otherwise-spotted wasp, was added to the system, the period over which the two species coexisted was extended.

   A field experiment that showed a similar effect involved the rocky shore fauna along the Pacific coast. Here, experimental removal of a starfish predator resulted in domination of the rock surfaces by a few species of bivalves and barnacles and the elimination of several competing species (Paine 1966).

   C. The third situation is that in which seasonal change prevents numbers from reaching a level at which competition would begin. Many ecologists studying insects, especially temperate-zone herbivores, claim that this is the situation they usually encounter. With favorable weather and plant growth in the spring, insect numbers take off on an exponential rise (see Figure 4-28). But before they reach the point at which sustained competition for food or breeding sites would occur, unfavorable weather or similar factors intervene and their numbers plummet (Figure 4-29).
3. Competition between the two species may be mild enough that they can coexist. Because each is depleting a resource the other can use, each should be less common than if it occurred alone. This corresponds to case D (Figure 8-12) of the Lotka–Volterra model of interspecific competition. In nature, we may see it in such cases as the drop in number of cavity-nesting native birds, such as bluebirds, following the introduction to North America of starlings and house sparrows.
4. Competition may initially be severe enough that the eventual outcome would be competitive exclusion, as in point 1 above; however, behavioral or evolutionary changes have the effect of reducing competition to the level described in point 3.

Coexistence is discussed in a later section of this chapter. Community-level effects of mechanisms reducing competition are discussed in connection with community organization in Chapter 10.

Ecological Niches and Guilds

Both of these concepts are properly part of community ecology but were developed in connection with interspecific competition and so are mentioned here. The word "niche" (pronounced "nitch") was first used about the same time but in slightly different ways by the English animal ecologist Charles Elton (1927) and the University of California naturalist Joseph Grinnell (1917). The term seems valuable as a shorthand expression indicating the entire role played by a species in an ecosystem. "Tree trunks" might specify in a general way the foraging dimension of a bird's niche, "insect eggs" the food dimension, and "natural cavities" the nest dimension. Eugene Odum, in his classic textbook (1953), took pains to distinguish *"niche"* from *"habitat."* Habitat, he wrote, corresponds to the organism's address, niche, to its profession.

Both Elton and Grinnell believed and stated that (to use Elton's words) "as a result of competition two similar species scarcely ever occupy similar niches." G. E. Hutchinson of Yale University put forth (in 1957) a formal view of the ecological niche that was extremely influential in the mid-20th century. This is developed in Chapter 10. Suffice it to say that he recognized a "fundamental niche" that a species could occupy and a "realized niche" that it did occupy. The latter might be restricted compared with the former because of interspecific competition.

A **guild** was defined as "a group of species that exploit the same class of environmental resources in a similar way;" that is, they "overlap significantly in their niche requirements" (Root 1967). An example of a guild might be the birds that glean foliage (for insects) in deciduous forest. Most usage of the term has been in the context of defining potential competitors. Like "niche," it is useful primarily as shorthand for all the species sharing a resource rather than as a rigorous formulation.

A widely cited experimental study by James H. Brown and coworkers has been examining the seed consumers of the southwestern U.S. desert. Although rodents, ants, and birds all gather and consume seeds (mostly of annual herbs), the study has focused on rodents and ants. The primary method of attack has been to use 0.1 ha (quarter-acre) enclosures, from which ants or rodents or both were removed (Davidson et al. 1980).

Some persons have considered that all the animals mentioned constitute one seed-foraging guild. However, the ants and rodents differ in the microhabitats they search for seeds and the seed sizes they prefer (ants small, rodents large), and the seeds differ in size, chemistry, and the speed with which they become buried (the fruits in one large-seeded genus screw themselves into the ground). Based on these considerations, other persons have suggested that some larger number of guilds, up to six, are represented (Simberlof and Dayan 1991).

One conclusion from the study was that, in the two enclosures from which rodents were removed, ant colonies (of the predominant harvester ant genus *Pheidole*) increased (Brown et al. 1979). This suggests that competition with rodents for seeds may have been limiting ant populations. Whether or not we consider ants and rodents as belonging to the same guild, demonstration that such dissimilar organisms may compete is interesting.

It is also possible that rodents increased in the two enclosures from which ants were removed, which might be still more interesting. The results from this reciprocal side of the removal experiment are, however, less persuasive (Galindo 1986, Brown and Davidson 1986).

Competitive Exclusion

One possible outcome of competition is for one of the two species to become extinct. There is, in fact, a generalization to this effect known as **Gause's rule,** after the Russian ecologist G. F. Gause (1934), or the **competitive exclusion principle,** a name proposed by Garrett Hardin (1960). One form of Gause's rule makes use of the concept of the ecological niche: Two species cannot occupy the same ecological niche. Incorporation of the niche concept is, however, unnecessary.

The baldest statement of Gause's rule is that complete competitors cannot coexist. This sounds reasonable; if we are willing to imagine two species that eat exactly the same thing, it seems likely that one of them will prove a little more efficient at getting the food or surviving starvation—or at something—that will give it the advantage and allow it to be around after the other is gone.

Competitive Replacement

The complete replacement of one species by another, similar, species over a considerable geographical area has occasionally been observed (Diamond and Case 1986). Although we may favor competitive exclusion as an explanation for many such episodes, they are rarely documented well enough for firm conclusions.

Many of the most attractive possibilities come from times or places where the details of the process seem likely to remain obscure. What are we to make, for example, of the replacement of the broad-legged by the long-legged crawfish (Gause 1934) in the lakes of the Russia in the early 1920s? The moth skink, a kind of lizard, disappeared from the Hawaiian Islands, where it had once been fairly common, soon after the introduction (around 1917) of the metallic skink (Oliver and Shaw 1953), but little is known of the episode other than that it occurred and that the species are related.

A few cases exist in which replacement through competitive exclusion seems strongly indicated. The Seychelles kestrel, a small falcon, has disappeared from most of the islands in the Seychelles group, following the introduction of the barn owl from South Africa. Both are hole-nesters; the likelihood is that the kestrel has been eliminated by the aggressive newcomer as a result of nest-site competition. The hole-nesting Seychelles owl has also declined drastically (Fisher et al. 1969).

The economic entomologists P. DeBach and R. A. Sundby (1963) have given us one of the better documented cases in which good field coverage was bolstered by laboratory experiments. The California red scale, a small insect in the order Homoptera, is a pest in the citrus groves of southern California. It is parasitized by small—very small—wasps in the genus *Aphytis*. The female wasp lays an egg beneath the scale of the scale insect. The egg develops into a larva that sucks the body contents of the scale insect, causing its eventual death.

In the early 1940s, one species of this wasp, the golden-naveled scalesucker, occurred pretty much throughout the southern California citrus areas infested with red scale (Figure 8-1). It had probably been introduced to the region near the beginning of the 20th century, perhaps from the Mediterranean region. The scalesucker was most common in the coastal areas where the climate was mild and was, in fact, limiting the scale insects to low population densities in these areas.

Nevertheless, another, very similar, member of the same genus, the Lingnan scalesucker was brought to this country from China, the idea being that it might give still better control. It was deliberately released at various sites throughout the range of the golden-naveled scalesucker. Replacement occurred fairly rapidly. At a given site, the population of the Lingnan scalesucker usually exceeded that of the golden-naveled scalesucker within a year after release. Within two years, complete replacement had occurred on about one third of the plots, and within four years, the Lingnan scalesucker made up 95% of the parasite population on all the release sites studied.

Although spread of the Lingnan scalesucker was not very rapid—about ten tree rows per year—in 11 years it had displaced the golden-naveled scalesucker from nearly all of its former range, an area of more than 10,000 square kilometers (Figures 8-2 and 8-3). Although the Lingnan scalesucker was a good competitor, it gave poorer control of red scale than the golden-naveled scalesucker, so later a third species, from India and Pakistan, was introduced. It, in turn, replaced the Lingnan scalesucker but only from interior citrus areas which have more strongly seasonal climates.

Laboratory experiments showed that in mixed cultures of scalesuckers, competitive exclusion always occurred (within eight generations). Survival or extinction in the cultures was based entirely on reproductive success, with the species producing the greatest number of offspring that reached maturity under a particular experimental regime being the invariable winner.

It may be no coincidence that this excellent example of competitive replacement involves introduced species in a highly simplified community. Undisturbed species-rich natural communities are highly resistant to invasion by new species (Weaver and Flory 1934, Elton 1958). Undisturbed prairie in the midwestern United States includes only a handful of nonnative plant species and undisturbed forest has still fewer. Evidently the same is true at the level of the soil microfungi. According to Martha Christensen (1989), "Spores of alien species are apt to germinate and die or be consumed long before gaining entry into the adapted and interactive community of soil fungi."

Competition as a Factor in Geographical and Habitat Distribution

The inability of a species to invade a geographical area because of the presence of a competitor is the other side of the coin of competitive displacement.

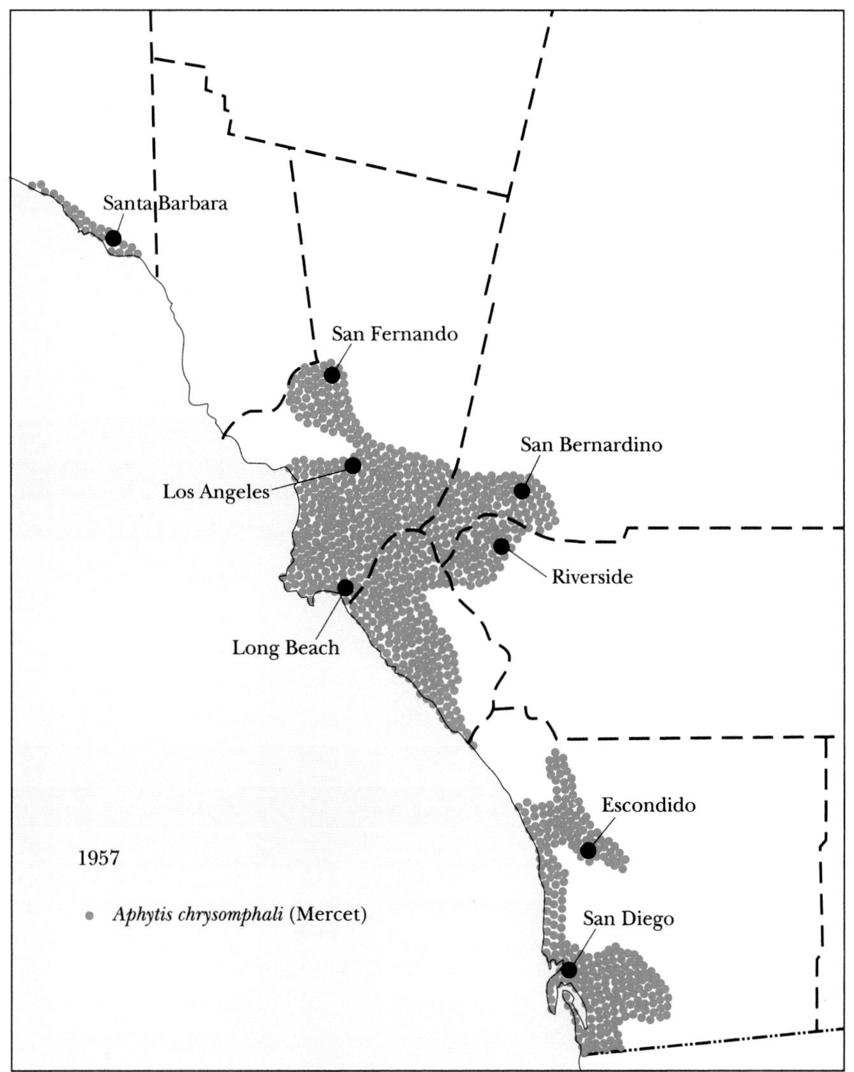

Figure 8-1

Distribution of the golden-naveled scalesucker in the citrus-growing region of southern California just prior to the introduction of the Lingnan scalesucker *(Aphytis chrysomphali)*. (Adapted from DeBach & Sandby 1962.)

Competition has often been suggested as a factor in setting the limit to geographical ranges but, again, convincing cases are scarce. One example is the distribution of black-capped and Carolina chickadees in central Illinois (Brewer 1963). The presence of black-capped chickadees in the forests along stream systems (most of the land between streams is farmland unsuitable for chickadees) seems to prevent the Carolina chickadee, a similar southern species, from nesting there. Only black-capped chickadees occur along the Kaskaskia River north of 39° north latitude (Figure 8-4). On parallel stream systems only about 20 miles eastward, black-capped chickadees do not occur, and here Carolina chickadees extend another 70 miles north.

A similar situation exists, altitudinally, in the Great Smoky Mountains. There breeding populations of black-capped chickadees inhabit higher elevations on the taller mountains, and Carolina chickadees occur only up to about 3500 feet. Lower mountains lack black-capped chickadees; on these, Carolina chickadees breed at higher elevations, up to 4000 feet (Tanner 1952). It is hard to avoid the conclusion that the presence of black-capped chickadees is what keeps Carolina chickadees out in both situations.

Much more common are cases in which one species is excluded from a habitat. Where the willow tit and the marsh tit—European relatives of the chickadees—are **sympatric** (occur in the same geo-

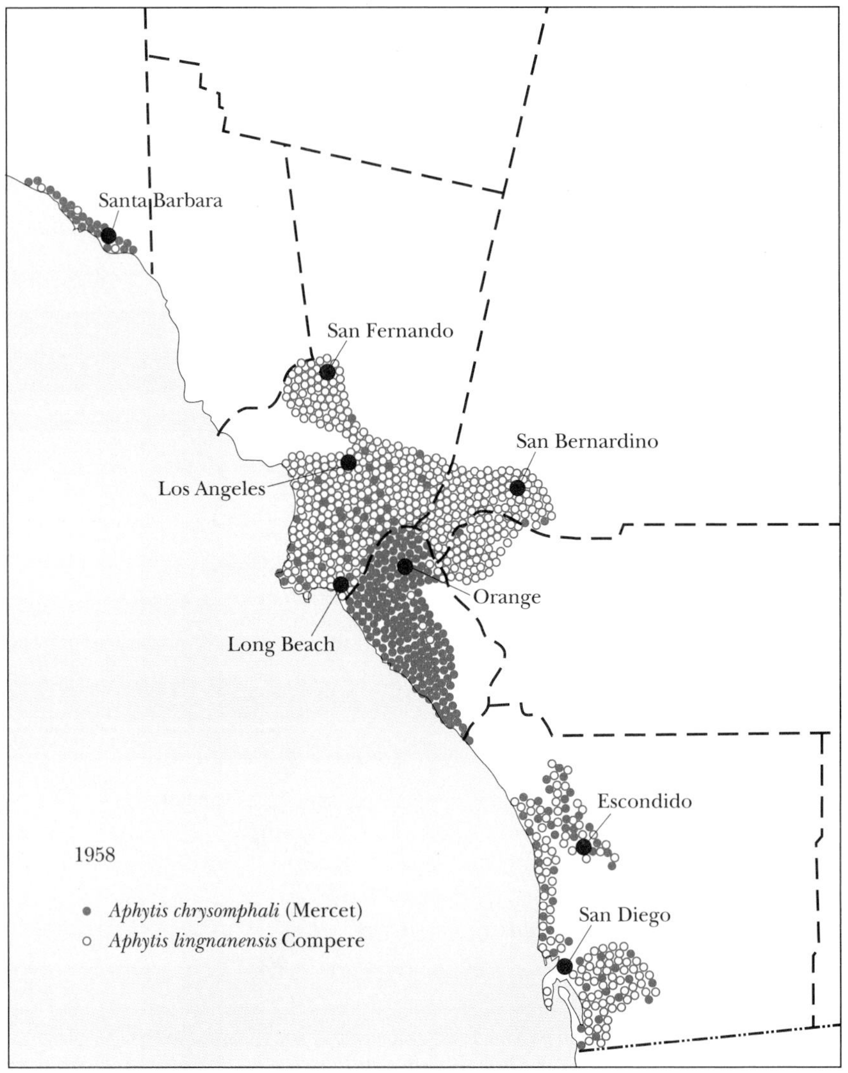

Figure 8-2

Ten years after the Lingnan scalesucker *(Aphytis lingnanensis)* was introduced, it had spread widely. (Adapted from DeBach & Sandby 1962.)

graphical area), the willow tit lives in coniferous and mixed forest and the marsh tit in broad-leaved forest (Snow 1954). In at least some localities where one species is absent, the other occupies both coniferous and broad-leaved woods. This suggests that the habitat distribution in areas of sympatry is the result of competition. Such a case may be considered competitive exclusion (from a habitat); however, it may just as logically be considered coexistence (within a geographical area).

Table 8-1 compares tit populations in Uppland, on the mainland of Sweden, where both willow and marsh tits occur, and on the adjacent Åland Islands, where the marsh tit is absent. In the absence of marsh tits, willow tits are much more common in oak-ash (broad-leaved) forest. Two other conifer forest titmice, the crested and coal tits, show no similar shifts. For them broad-leaved forest is evidently unsatisfactory as a habitat in some other way.

Another apparent example is the limitation of the least chipmunk to open habitats in the western United States by the more aggressive yellow pine chipmunk. In geographical areas where other chipmunks are absent, the least chipmunk is said to occupy forest habitats (Sheppard 1965, cited in Meredith 1977). Perhaps aggression sets the habitat boundaries between several of the western chipmunks (Heller 1971), but here, as in most such

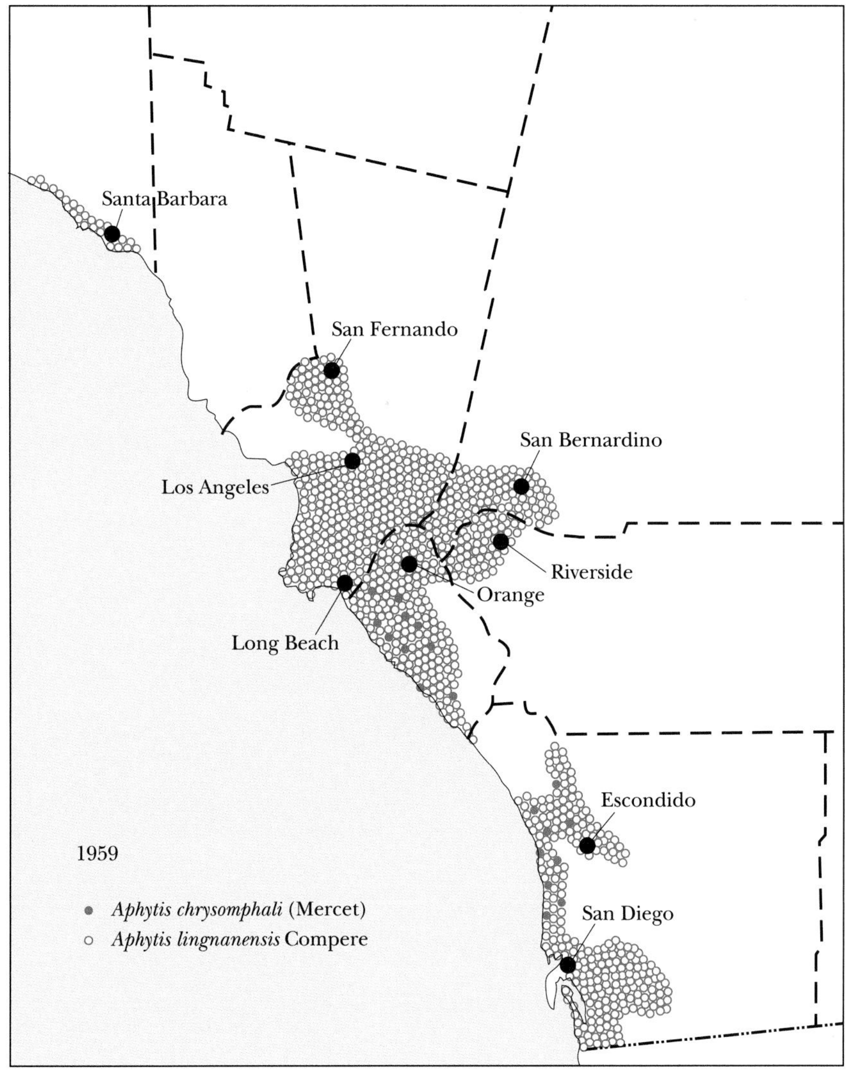

Figure 8-3

Eleven years after its introduction, the Lingnan scalesucker had replaced the golden-naveled scalesucker through much of its former range. (Adapted from DeBach & Sandby 1962.)

cases, the situation needs more study. It is possible that habitat selection, not current competition, is the proximate cause for many of the observed distributional patterns—as is presumably the case in the failure of coal and crested tits to invade broad-leaved woods even in the absence of marsh tits. Of course, one may suppose that the evolution of such habitat selection is a frequent evolutionary outcome of competition.

Apparent Competition

Several cases in which one species has been replaced by a related species have turned out on further study not to be the result of competition or, at least, to be more complicated than they seemed. The gray squirrel of North America was introduced in Britain around 1900. By 1930 it became evident that the native red squirrel was disappearing from sections occupied by the invader. The replacement has continued, the gray squirrel spreading to large areas of England and Wales and parts of Scotland and Ireland. This seems to be a good candidate for competitive replacement, but when the details are looked at, it becomes much less certain.

In many localities, the red squirrel disappeared before the first gray squirrel was recorded—in some cases, 16 years before (Reynolds 1985). For these

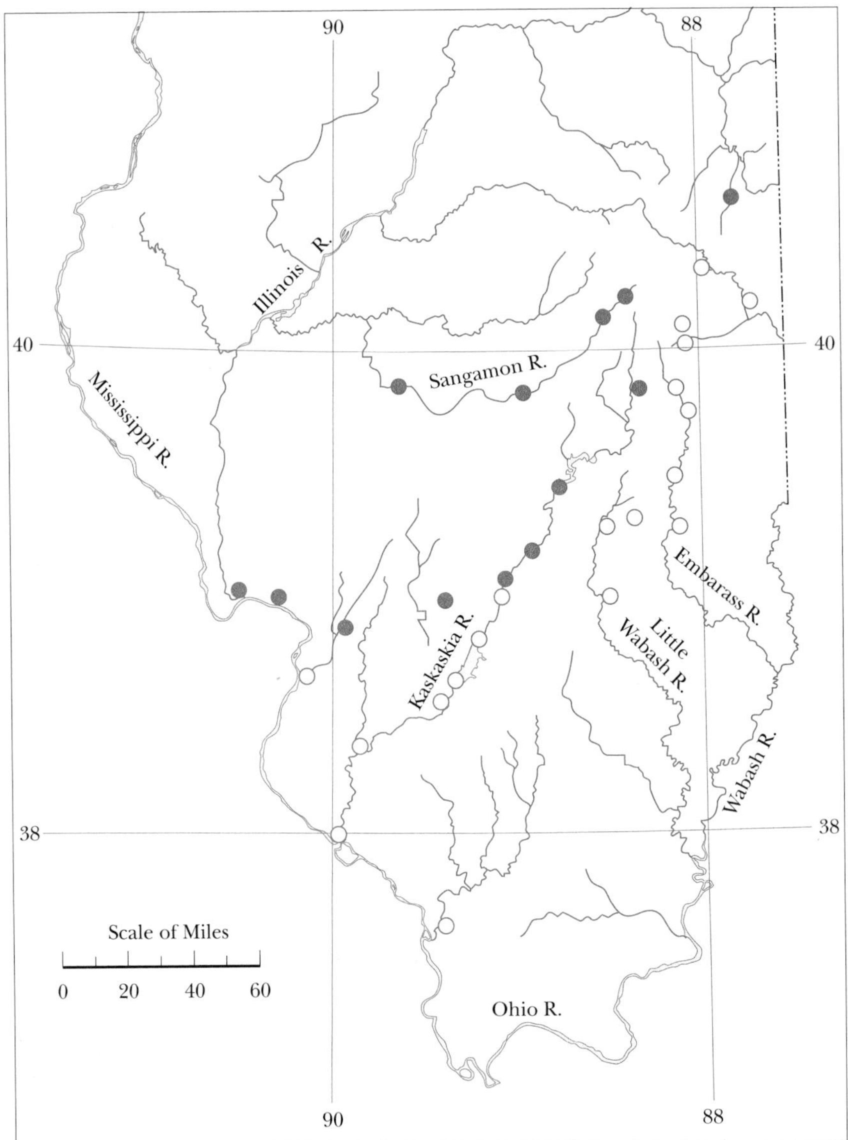

Figure 8-4

On the Embarass and Little Wabash Rivers, black-capped chickadees *(blue dots)* do not occur and Carolina chickadees *(circles)* extend all the way north, above 40°N latitude. On the Kaskaskia River, only about 20 miles west, black-capped chickadees occupy the northern section of the stream and Carolina chickadees extend north only to about 39°N latitude. (From Brewer 1963*a*, p. 11.)

localities, competitive exclusion of red by gray squirrels seems less likely than three other possible explanations.

1. Red squirrels have declined from other causes (climatic change, for example), and, purely as a coincidence, gray squirrels happened to arrive on the scene.
2. Red squirrels declined from other causes that also favored the successful invasion of gray squirrels.
3. Red squirrels declined from other causes, and their disappearance allowed gray squirrels to move into areas that had always been suitable but from which they had earlier been excluded by competition.

The term "apparent competition" has been used by one author (Holt 1977) to refer not to just any situation that looks like competitive exclusion but isn't, but specifically to the situation in which one member of a pair of potential competitors is driven to extinction by the action of a shared predator, herbivore, parasite, or disease. The more productive prey (or host) allows the predator to maintain a high enough density to lead to this outcome.

Table 8-1
Winter Titmouse Populations (Birds per Kilometer) on Transects in Two Areas of Sweden

	Forest Type		
Tit Species	*Conifer Forest*	*Birch-aspen*	*Oak-ash*
Mainland Sweden (willow and marsh tit both present)			
Willow tit	**2.8**	**1.2**	**0.4**
Marsh tit	**0.3**	**1.1**	**3.5**
Crested tit	2.3	—	—
Coal tit	2.0	0.1	0.1
Great tit	1.3	2.9	4.0
Blue tit	<0.1	2.3	3.8
Åland Islands (marsh tit absent)			
Willow tit	**2.6**	**1.2**	**1.8**
Crested tit	0.7	—	—
Coal tit	1.9	—	—
Great tit	0.8	1.6	2.7
Blue tit	0.2	2.2	3.2

From Alatalo, R. V., Gustafsson, L., Lundberg, A., and Ulfstrand, S., "Habitat shift of the willow tit *Parus montanus* in the absence of the marsh tit *Parus palustris,*" *Ornit. Scandinavica,* 16:121–128, 1985.

Examples of what seem to be this situation have been given in Chapters 6 (broom snakeweed-biennial) and 7 (moose-deer).

One suggested example of apparent competition was called "The Case of the Missing Hares" (Holt 1977). The general situation is this: Arctic hares are native to Newfoundland. The smaller snowshoe hare was introduced around 1870. Currently, the Arctic hare occupies coastal and mountain barrens, the snowshoe hare forested areas (Fitzgerald and Keith 1989). A species related to the Arctic hare occupies both forest and tundra in Eurasia, and some authors have suggested that this was true of the Arctic hare prior to the introduction of the snowshoe hare.

Early authors attributed the Arctic hare's restriction to open areas to competition with the snowshoe hare, either by interference or exploitation (Table 8-2). Arthur Bergerud (1967) was the first to propose that a shared predator was responsible. He suggested that the arrival and growth of snowshoe hare populations allowed lynxes—predominantly forest dwelling but given to widespread wandering during snowshoe hare lows—to increase so much that they kept the Arctic hares out of forests and adjacent areas.

Lloyd Keith and colleagues at the University of Wisconsin have clarified habitat relations and coactions of the two hares. A series of laboratory and field experiments showed that Arctic hares quickly died of starvation when forced to spend the winter in forested habitats, whether or not snowshoe hares were present (Barta et al. 1989). Snowshoe hares forced to live in treeless areas were decimated by predation, especially by great horned owls and northern goshawks. Interspecific encounters between the two hares were rare in nature, but the Arctic hare, not the smaller snowshoe hare, dominated (Fitzgerald and Keith 1990).

Table 8-2
Three Explanations for the Habitat Distribution of the Arctic and Snowshoe Hares in Newfoundland. The Last Is Probably Correct.

	Barrens	*Forests*
Species present	Arctic hare	Snowshoe hare (Introduced)
Competition explanation		The snowshoe hare is dominant and excludes the Arctic hare from this, the snowshoe hare's preferred habitat.
Apparent competition explanation		Introduction of snowshoe hare raised lynx populations enough to allow this forest-dwelling predator to exterminate Arctic hares in this habitat.
Habitat factors explanation	Snowshoe hares in this habitat killed off by great horned owls and northern goshawks.	Arctic hares starve in this habitat, whether or not snowshoe hares are present.

Consequently, lack of suitable winter food seems to keep Arctic hares out of forest, and high predation rates keep snowshoe hares out of tundra. Competition is unlikely to be important, and the importance of predation is related to the vulnerability of the snowshoe hare to predation by birds that rarely eat Arctic hares. One conclusion to be drawn is that not all situations that appear to be competition, are. The second conclusion is that not all situations that appear to be apparent competition, are either.

Competitive Exclusion in the Laboratory

Much work has been done in which laboratory ecologists have made complete competitors out of two species by confining them in cans or vials with a very simple environment and a single kind of food. Such work has been performed with protozoa and grain beetles, among other animals. Almost without exception, one species becomes extinct. These studies are valuable for showing that competitive exclusion really does occur. Two very similar species, each of which will live alone quite satisfactorily in a certain simple environment, may not be able to live there together.

In a series of studies spanning about 40 years, Thomas Park and his colleagues at the University of Chicago thoroughly investigated the population growth and competitive relations of two small grain beetles, referred to here as the chestnut (-colored) grain beetle and the confusing grain beetle (confusingly like the chestnut grain beetle). Extinction of one species was the invariable result of many trials with mixed cultures of the two species, but the species that became extinct varied according to environmental conditions (Park 1954).

In a cool dry environment, the confusing grain beetle always survived, but in a hot moist environment the chestnut grain beetle always survived (Figure 8-5). Under intermediate conditions the survivor was sometimes one species and sometimes the other, showing the importance of chance events whenever biological systems are considered, even under controlled laboratory conditions. Apparently cannibalism of eggs by larvae and adults, a form of interference competition, is an important factor in limiting population size intraspecifically and also in the process of competitive exclusion.

Note that the time required for extinction was long, more than two years, even for these strongly similar species in this highly simplified environment. Generation length for the confusing flour beetle under the conditions shown in Figure 8-5 is around 60 days (Chapman 1931), so extinction took more than ten generations.

Coexistence of Competing Species

The competitive exclusion principle may be satisfying in its simplicity, but it is not otherwise very satisfactory. Complete competitors must be rather rare, after all; the potentialities of the natural world are so vast that it is hard to imagine that the situation in which two species are *exactly* the same in their diet or nest sites is very common. The realistic question, then, is to ask just how strong competition has to be before coexistence is impossible.

Unfortunately, the general answer is simple when expressed mathematically (as it was by A. J. Lotka and Vito Volterra in the mid-1920s) but rather difficult to state briefly and clearly in words. Roughly the answer is this: Two competing species will be able to coexist if the effects of crowding are more severe intraspecifically than interspecifically. Where this is not the case, one of the two will be eliminated. For many cases, this is equivalent to saying that coexistence is possible where intraspecific competition is more important than interspecific competition. This is achieved principally by small differences in the use of resources that are limiting to the populations at one time or another. Two kinds of birds may forage for insects in slightly different places, say, one on tree trunks and the other on branches. The trunk-feeding bird, when it is common, will deplete its own food supply while leaving much of the food supply of the branch-feeding species untouched. The same is true in reverse for the branch-feeding species.

The differences in the use of resources sufficient to permit coexistence may be large, such as the restriction to different habitats in the case of the willow and marsh tits, or they may be very subtle. In a later section we will describe a situation in which two species of bees live off pollen from the same plant but manage to coexist by one concentrating on clumps of the plant and one concentrating on dispersed individuals of the plant.

Probably many of the differences that act now to limit interspecific competition and thereby to allow coexistence have evolved with the unfavorable

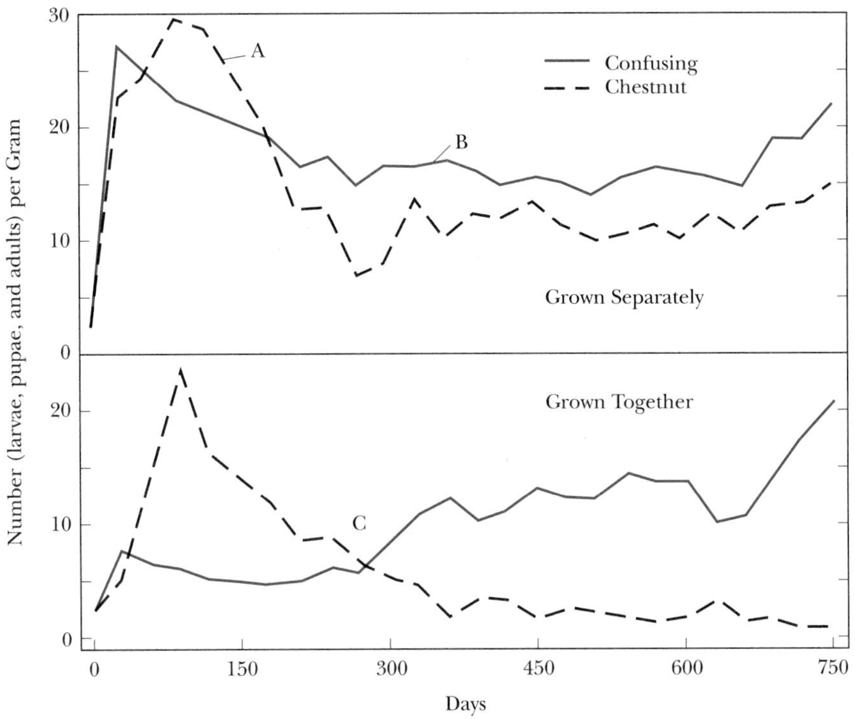

Figure 8-5

Competitive exclusion in grain beetles. In *A*, the chestnut grain beetle was grown separately in a culture consisting of 8 grams of flour and yeast, with the temperature maintained at 29.5°C (about 85°F) and relative humidity 60 to 70%. In *B*, the confusing grain beetle was grown separately under exactly the same conditions. In *C*, the two species were grown together in the same container. It can be seen that the chestnut grain beetle in *A*, living by itself, persists indefinitely (the graph covers more than two years). The same is true of the confusing grain beetle in *B*. But when the two are grown together, *C*, the chestnut grain beetle eventually becomes extinct. (Adapted from Andrewartha & Birch 1954, p. 428; after Park.)

effects of interspecific competition as the main selective force. As a simple example, suppose we have two species that overlap widely in the size of food they eat, and suppose also that the size of the food taken is at least partly based on hereditary factors (for example, the bird that takes slightly more small items has a slightly smaller bill). Interspecific competition will lead to poorer survival and reproduction among the individuals feeding on the medium-sized food items. Accordingly, natural selection will favor the individuals feeding on smaller items in the small-beaked species and individuals feeding on larger items in the large-beaked species (Figure 8-6). If one species is more efficient, then the effects on survival and reproduction will be greater on the other species. In this way the species diverge slightly and intraspecific competition becomes more important than interspecific competition; an accommodation is reached, and the community can support both of them. This sort of thing, discussed in more detail in the next chapter, has probably happened repeatedly in the history of all communities. Competitive exclusion is what happens when, for some reason, competing species do not find the means for cutting down interspecific competition.

The example just described deals with evolutionary modification, but the accommodation does not have to be based on genetic change. Among vertebrates, at least, it is often the result of experience. One kind of squirrel that is attacked by a larger or more aggressive species every time it infringes on the other's preferred habitat may learn

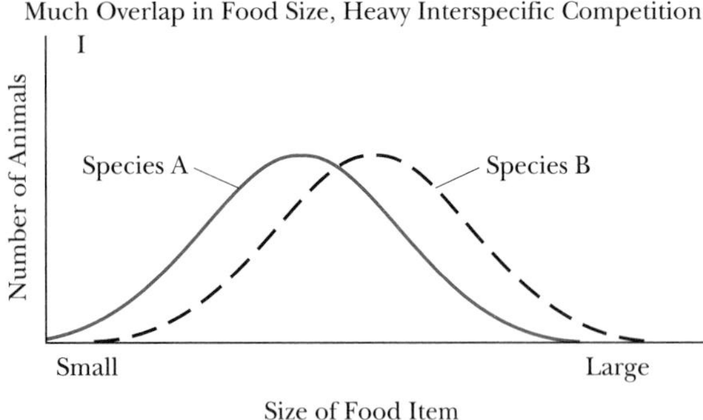

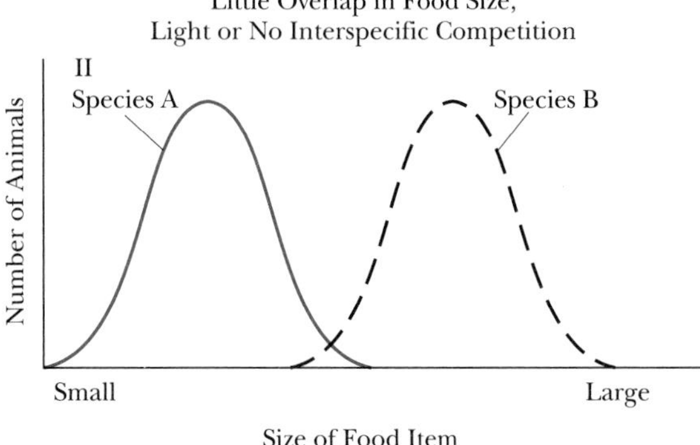

Figure 8-6

Natural selection reducing competition by reducing overlap in size of food items eaten; before *(I)* and after *(II)*. The animals of species *A* eating small items and those of *B* eating large ones are more successful than individuals of either species eating medium-sized food; they have more offspring and live longer. Consequently, in the course of time, the "small" genotypes of *A* and the "large" genotypes of *B* increase and the diets diverge.

early in life to stay out, with only an occasional reminder needed (see, for example, Meredith 1977).

Mathematical Representation of Interspecific Competition

The Lotka–Volterra Model

In an earlier section on the logistic curve, we represented the growth of one population by

$$\frac{dN}{dt} = rN\left(\frac{K - N}{K}\right)$$

where N is population size, r is the intrinsic rate of natural increase, and K is carrying capacity. If two species are competing—if one is using some of the resources of the other—then we may rewrite the equations for the two species as follows.

For species 1:

$$\frac{dN_1}{dt} = r_1N_1\left(\frac{K_1 - N_1 - \alpha N_2}{K_1}\right)$$

For species 2:

$$\frac{dN_2}{dt} = r_2N_2\left(\frac{K_2 - N_2 - \beta N_1}{K_2}\right)$$

The only change is that for each species the term αN_2 or βN_1 has been added. The coefficients α and β are competition factors indicating, respectively, the effect of species 2 on 1 and the effect of 1 on 2. Note that the addition of each new individual of a species has an inhibitory effect on its further population growth. The inhibitory effect of one more individual of species 1 on itself is $1/K_1$; the inhibitory effect of one more individual of species 2 on itself is $1/K_2$. α and β are coefficients that express the inhibitory effects of each species in terms of the number of individuals of the species with which it is competing. The inhibitory effect of one new individual of species 2 on the growth of species 1 is α/K_1. For example, if two species are each competing for grass and species 2 eats three times as much as species 1, then α is 3. In other words, the inhibitory effect of species 1 on itself is $1/K_1$, but the inhibitory effect of species 2 on species 1 (that is, α/K_1) is $3/K_1$.

Any one species will stop growing when its carrying capacity has been reached by the *combination* of its own numbers plus the individuals of the other species (multiplied by the appropriate competition coefficient). That is, species 1 stops growing when

$$N_1 + \alpha N_2 = K_1$$

and species 2 stops growing when

$$N_2 + \beta N_1 = K_2$$

Notice that the two-species system will be at equilibrium only when the point at which growth stops coincides for the two species, that is, when $dN_1/dt = dN_2/dt = 0$. Otherwise one species will keep growing, meaning that the other species will have to decline.

We can understand these ideas more easily using graphs. In Figure 8-7 the straight line consists of all the mixes of species 1 and species 2 that add

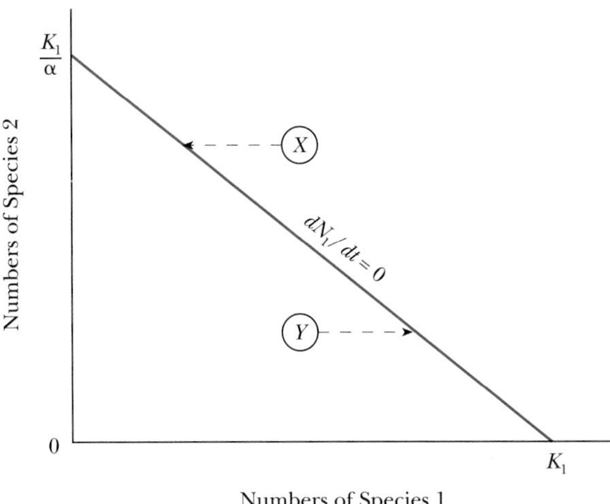

Figure 8-7

ZPG (zero population growth) line for species 1. With any mixture of numbers of species 1 and species 2 that lies on this line, species 1 will cease population growth.

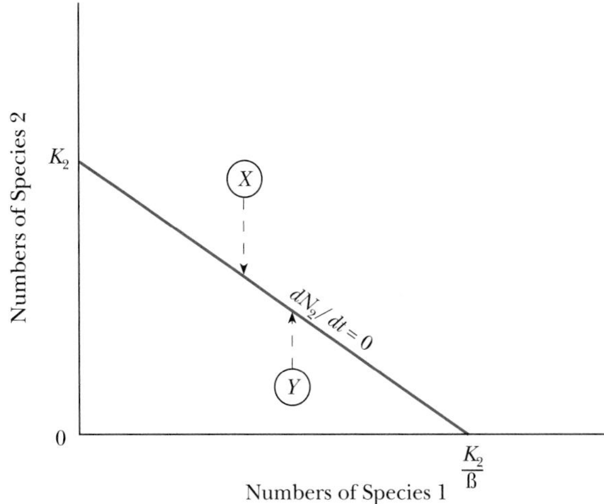

Figure 8-8

ZPG line for species 2. With any mixture of numbers of species 1 and species 2 that lies on the line, species 2 will cease population growth.

up to K_1. When N_2 is 0, then N_1 is equal to K_1. When N_1 is 0, then there is a number of N_2 equal to K_1/α. The line represents all the combinations of numbers at which species 1 will stop growing. If there is a mix of N_1 and N_2 such as at *X*, then N_1 must decrease (as shown by the arrow). Similarly, if there is a mix such as *Y*, then N_1 will increase (as shown by the arrow).

We can prepare another graph like the previous one but this time for species 2. It will look like Figure 8-8. With a mix like that at *X*, species 2 must decline. With a mix such as that at *Y*, species 2 can increase.

We can now put the lines for the two species together on the same graph, as in Figure 8-9. We can determine how the densities of the two species will change from any point by drawing arrows for each species, as we did before, and then drawing the arrow in between that will be the resulting vector (just as in physics). Consequently, we can see that anywhere below line $K_2 - K_2/\beta$ (in other words, in the gray area), both species will be able to increase their numbers. For example, if point A represents 20 individuals of species 2 and 30 individuals of species 1, they might be able to grow to 40 of species 2 and 50 of species 1. Once the mix of species forms a point lying between the two lines (in the blue area), however, species 1 can continue to increase but species 2 must decline. Note that

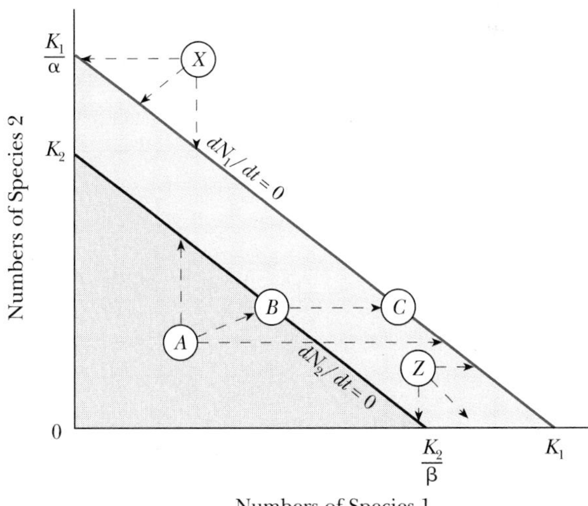

Figure 8-9

Combining Figures 8-7 and 8-8 yields this diagram showing the behavior of the two species system. In this case, the ZPG line for species 1 lies outside that for species 2, indicating that species 1 can increase after species 1 has lost the power to increase. The eventual result is the extinction of species 2.

the combination of the two densities could move into this zone without any growth in numbers of species 2. If species 2 stays the same but species 1 grows in numbers, this will slide the point horizontally across the line (as in going from B to C), whereupon species 2 will have to begin to decline in numbers. In the clear zone above the line $K_1/\alpha - K_1$, species 1 as well as species 2 have to decline.

If other arrows are drawn on the graph, they will always end up in the same place. This is where $N_1 = K_1$ and $N_2 = 0$. In other words, the extinction of species 2 is the invariable result.

Three other relationships are possible in a graph of this sort (as long as we use straight lines). The second case is the opposite of the one just examined and is shown in Figure 8-10. Here, because species 1 cannot increase past the mix of densities represented by the blue area, and species 2 can increase in this area and also in the gray area, species 2 is the invariable winner, and species 1 becomes extinct.

In the third and fourth cases, the lines cross. Figure 8-11 is the third case. Here there are three possible equilibria. One of these is unstable; this is the one in the middle of the graph in which the two species coexist. Any event that shifts densities into the triangles will lead to the extinction of one species or the other. Which species wins and which becomes extinct, however, differs according to which triangle the mix of species enters. Consequently, if we begin with initial densities as in *X*, species 2 will win, but if we begin with initial densities as in *Y*, species 1 will win.

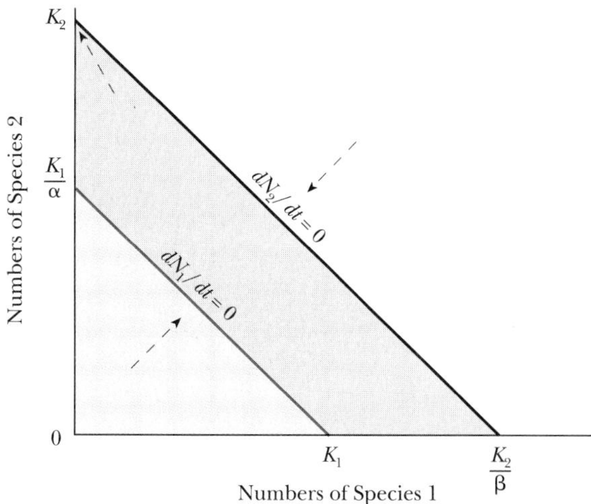

Figure 8-10

The opposite of Figure 8-9. When the ZPG line for species 2 lies outside that for species 1, the extinction of species 1 is the result.

The fourth case, given in Figure 8-12, is the only one that predicts coexistence. Here each species, as it becomes abundant, loses the capacity to increase before the other does. The arrows converge on a point representing some number above zero for both species, and neither becomes extinct.

We can summarize these various outcomes as follows:

Species 1	*Species 2*	*Situation*	*Outcome*
$K_1 > \frac{K_2}{\beta}$	$K_2 < \frac{K_1}{\alpha}$	Species 1 inhibits the further increase of species 2 while it can still increase itself	Species 1 wins (species 2 becomes extinct) (Figure 8-9)
$K_1 < \frac{K_2}{\beta}$	$K_2 > \frac{K_1}{\alpha}$	Species 2 inhibits the further increase of species 1 while it can still increase itself	Species 2 wins (species 1 becomes extinct) (Figure 8-10)
$K_1 > \frac{K_2}{\beta}$	$K_2 > \frac{K_1}{\alpha}$	Each species, when abundant, inhibits the increase of the other species while still able to increase itself	One species or the other wins, depending on initial numbers (Figure 8-11)
$K_1 < \frac{K_2}{\beta}$	$K_2 < \frac{K_1}{\alpha}$	Each species, when abundant, inhibits its own further increase more than it inhibits the further increase of the other species	The two species coexist (Figure 8-12)

Although the outcome of interspecific competition (in this model) depends on the inhibitory ef-

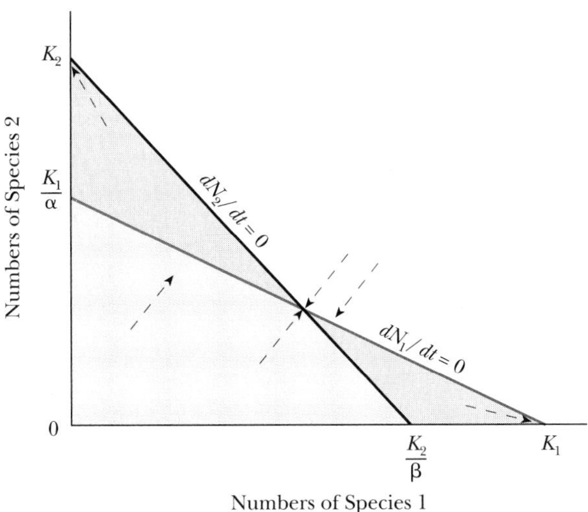

Figure 8-11

Here the two lines cross in a way that leads to extinction of one species or the other, depending on which of the darker triangles the mix of populations enters.

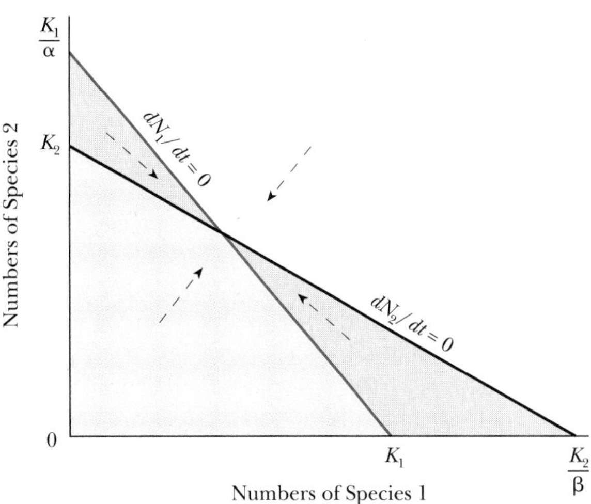

Figure 8-12

Here the two lines cross in such a way that the behavior of the system is toward coexistence of the two species with a mixture of the two populations as given by the intersection.

fect of each species on the other, it also depends on the relative sizes of K. One species may have the advantage if we look only at the competition coefficients but may nevertheless lose if its K is very low relative to that of the other species. Suppose that species 1 has a competition coefficient (β) of 1.2 and species 2 a competition coefficient (α) of 0.8. This indicates that one individual of species 1 outcompetes one individual of species 2. But suppose that K for species 1 is 20 and K for species 2 is 80. The graph would look like Figure 8-13, and the outcome is clearly the extinction of species 1.

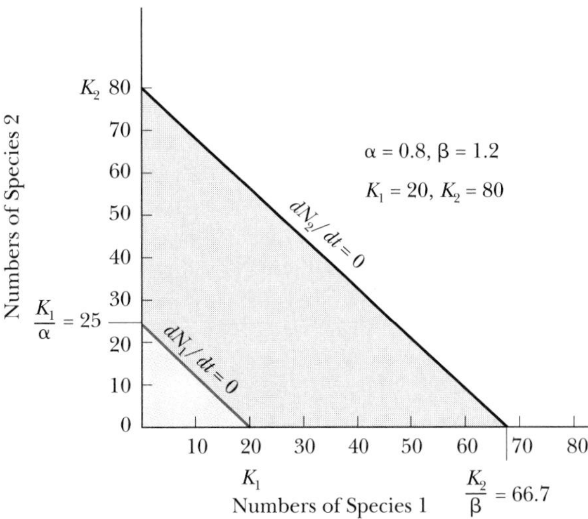

Figure 8-13

The outcome of a competitive situation depends in the Lotka-Volterra model on both the competition coefficients (α and β) and on carrying capacities (K). Here a species with a lower competition coefficient but a higher K is the winner.

Adding Species to the Lotka–Volterra Model

Mathematically, the Lotka–Volterra model can readily be extended to three or more species; we simply put additional terms in the numerators of the equations corresponding to the competition coefficient and population size of each additional species. For example, in a three-species system, we can write for species 1

$$\frac{dN_1}{dt} = r_1 N_1 \left(\frac{K_1 - N_1 - [\alpha N_2 + \gamma N_3]}{K_1} \right)$$

where γ is the competition coefficient equating species 3 to species 1. (More often when systems of three or more species are represented, all the competition coefficients are represented by α with appropriate subscripts. For example, α_{13} is the effect on the future growth of species 1 of one individual of species 3.)

Once we enter this realm, however, we have another question to address: Is the combined effect of competition from two or more species simply predictable from their separate effects? Suppose we measure K_1,K_2, and K_3, and α, β, and γ. Is the combined competitive effect of species 2 and species 3 on species 1 equal to $(\alpha N_2 + \gamma N_3)$, as the equation above states? Or is the combined effect something more or less than this? If the competition effects are simply additive, we can readily predict the outcome of the three-species system before we put it together. If they are not, we may not be able to.

Biologically, we can think of various reasons why the combined effects of two (or more) competing species might not be the same as a summation of their separate effects. In perhaps the simplest case, species 2 and 3 may each take a different end of species 1's resource axis, so that species 1 is squeezed out of a three-species system, although it survives with either one of the competitors (Figure 8-14). The term "diffuse competition" has been applied to this concept of the combined competitive effects of several species (MacArthur 1972).

Alternatively, it may be that part of the competitive effects of species 2 and 3 are dissipated against one another, and so their combined effect on species 1 is lessened. This might result in the survival of species 1 when, in pairwise trials, it would go extinct.

More complicated interactions among three or more competing species can be imagined. Given the diversity of the living world, some probably exist, but it is unclear how widely important they are. Only a few experiments have been directed to the question (such as VanderMeer 1969, Neill 1974, Richmond et al. 1975, and Case and Bender 1981). The general impression from such studies is that the effects of competition from two or more species are not simply additive but that the interaction effects may not very often change the outcome (survival or extinction) predictable from pairwise comparisons. The subject is, however, far from settled.

To the degree that these "higher-order interactions" exist, the Lotka–Volterra model and the resulting arithmetic would have to be further complicated (see VanderMeer 1969). Of course, the Lotka–Volterra model is based on the logistic equation, and this has its own flaws. For these and similar reasons, some other ways of modeling competition have been devised (for example, Gilpin and Ayala 1976 and Schoener 1976); nevertheless, the Lotka–Volterra model has been the cornerstone of most thinking about competition up to now.

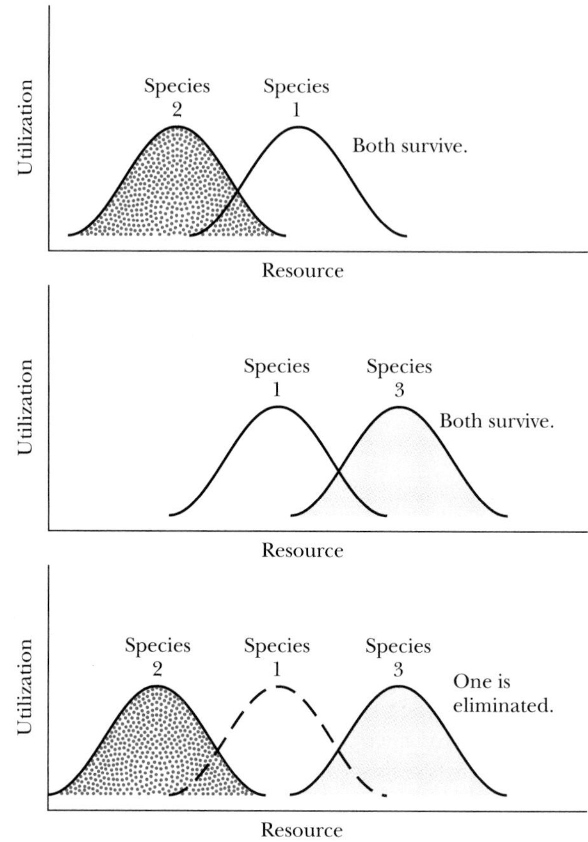

Figure 8-14

Coexistence with two competing species may be impossible (here, for species 1), depending on the patterns of resource usage.

Can Two Species Live on One Resource?

Whether there can be more species than resources depends entirely on what a resource is. Garbage is garbage until someone makes a profit producing methane from it, at which point it becomes a resource. Below is described a study in tropical dry forests in Costa Rica (Johnson and Hubbell 1975) in which two species of stingless bees were able to coexist on the pollen of a single species of flowering shrub.

The researchers observed that one species foraged in groups and gathered pollen from large clumps of the flowering shrub. The other foraged singly on scattered individual shrubs (plus small clumps). The researchers constructed the diagram that economists use in the analysis of cost-benefit ratios. They assumed that feeding rate would go up and tend to level off, as plotted against resource density. (This is the same curve called a functional response curve in the earlier section on predation.) This, then, is the benefit derived from greater densities.

As resource density increased, it was assumed that the costs of foraging would decrease. The costs are the amount of energy required to find the food source and to make use of it (including any energy used to defend it). Some measure of energy, such as kilocalories, would be appropriate for expressing both the benefits (for example, kilocalories obtained per day of foraging) and costs (kilocalories used per day). A graph combining the two curves would look like Figure 8-15. D is the break-even point. At any density below D the species could not make a living off the resource; it would have to spend more energy than it obtained. At any density above D it could make a living.

Plotting both species (A and B) on the same graph would look like Figure 8-16.

The species foraging on clumped resources (the high-density specialist) has the potential for increasing its feeding more with higher density for several possible reasons. Once in a large clump, it would only have to spend a little time in traveling from plant to plant. Also, because it relies on scouts and then forages in groups at the site found by the scouts, less time may be spent per individual in locating clumps. This species also has higher energy costs, however, especially at low densities, because it must find and travel to clumps which at low densities will be far distant rather than making use of nearby individual plants. Also, for the model to work, the high-density specialist must defend clumps against the low-density specialist, and this takes energy. Such defense was observed.

The situation then is that the high-density specialist uses clumps of the resources that it defends against the low-density specialist. The low-density specialist uses the isolated plants that the high-density specialists cannot make a living on. The low-density specialist could, of course, make profitable use of the clumps if it were not prevented from doing so by the other species.

The question that this section asks, then, is nearly meaningless. A resource, such as food, can

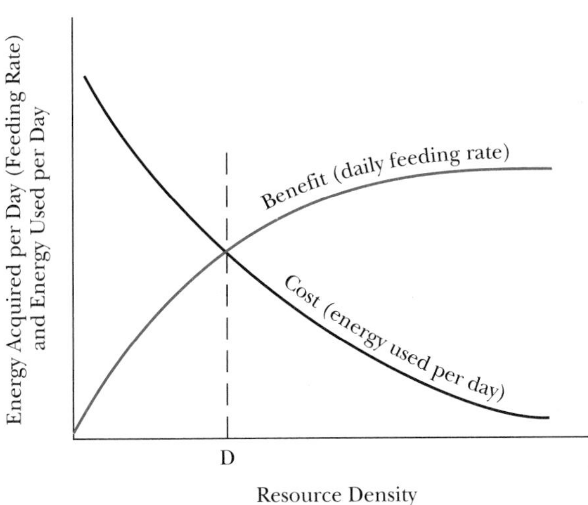

Figure 8-15

Cost-benefit diagram for foraging bee.

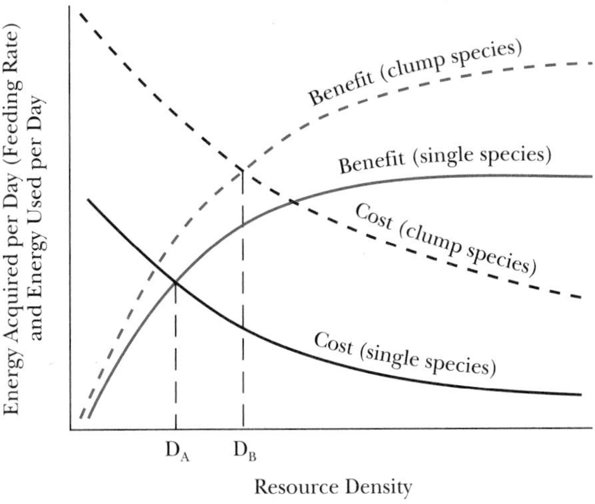

Figure 8-16

Cost-benefit curves for bee species specializing on clumps of plants or on isolated individuals of the same plants.

be subdivided in many ingenious ways, by both evolving organisms and theoretical ecologists. For one herbivore species to concentrate on one part of a plant and another species on another part of the same plant is biologically commonplace (for example, Figure 6-8). This example for the stingless bees in which the resource is subdivided on the basis of density is less obvious.

A similar case in which the resource varies in time, as for example a food supply that changes seasonally, might also allow two species to coexist (Stewart and Levin 1973). In this case, one species should be an especially efficient forager so that it does better relative to the other species when the food supply is low. The other should be better at reproducing rapidly when the food supply is high. It might even be that the oscillations in food supply could be the result of the activities of the organisms (DeJong 1976); the extravagant, fast-reproducing species overeats the food to a degree that its population crashes. The efficient, low-reproducing species now has the advantage, but its low rate of utilization allows the food supply to recover to the point where the other species can again increase.

In the real world, nevertheless, limiting resource axes are not infinitely divisible. Food cannot be subdivided past the point where populations of each species can obtain enough energy to be numerically viable. This suggests that there is some limit on how alike species can be in their use of limiting resources and still coexist—that there is a "limiting similarity" (MacArthur and Levins 1967). A large, repetitious, and generally unheuristic literature has grown up on this subject. The most important generalization to emerge seems to be that limiting similarity ought to vary among ecosystems (Abrams 1983). Some, such as high-energy systems, should allow species to be more alike than others. These ideas are explored further in the discussion of species diversity in Chapter 10.

Amensalism

Haskell (1949) termed the 0/− interaction, in which one species is harmed and the other unaffected, amensalism. Amensalism is an interaction in which one organism harms others as a by-product of its activities, as in the African proverb, "When elephants fight, it is the grass that suffers." Humans are involved in amensal relations with many organisms, such as the lichens that die of air pollution and bees that die when the health department sprays for mosquitos.

Bird colonies produce guano that affects the vegetation below them. This occurs in colonies of great blue herons and great egrets, but little seems to be known about the effects, other than that understory plants including seedlings of nest trees may be killed (for example, Weseloh and Brown 1971). Better known are the effects on intertidal plants affected by seabird guano (Figure 8-17). A study on rock faces on the Pacific coast of Washington found, among other effects, that the seaweed *Fucus* covered around 25 to 40% of the rock surface at sites with no colonies above but was almost absent from sites subjected to feces of guillemots, cormorants, and gulls nesting on the cliffs above (Wootton 1991).

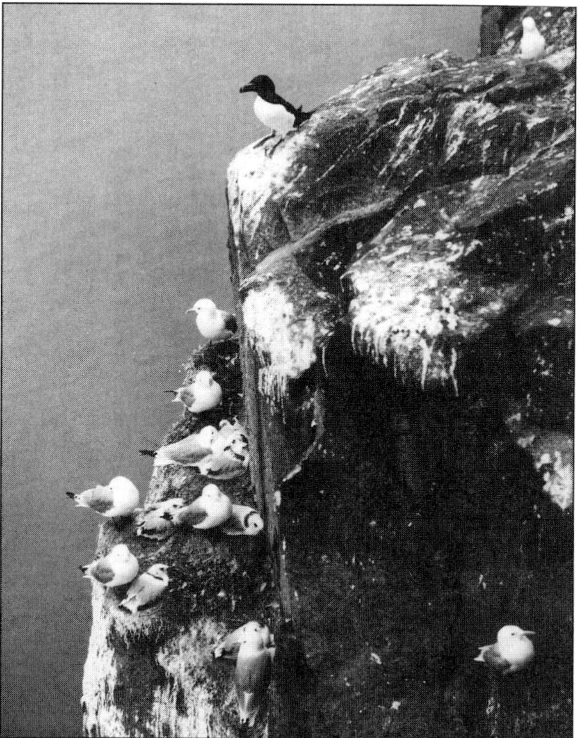

Figure 8-17

The guano from bird colonies can have harmful (amensalistic) effects on algae growing on the rocks below. Here the guano is being produced by kittiwakes nesting at Craigelieth, Scotland. The lone bird above is a razorbill. (Roger Wilmshurst/Dembinsky Photo Associates.)

We could hardly wish for a more perfect example of amensalism than this. Some algae on the same rocks showed better growth in areas of guano deposition, a coaction between bird and plant that in the Haskellian system would be a form of commensalism.

The Haskell classification requires that competition be a −/− interaction, that is both species have to suffer. However, many cases of competition are asymmetrical, with the effects on one species being less severe. These are generally situations in which one species can, in some way, exclude the second from the optimal range of resources. Where the two species occur together geographically, the second occurs only in low densities in preferred habitats or is relegated entirely to marginal habitats. If removal experiments are done, the subordinate species increases greatly in numbers when the dominant species is removed, but the dominant species changes little or undetectably when the subordinate species is removed.

Most ecologists have preferred to retain such situations under the name "competition." The assumption is that, in the absence of interference by the dominant species, the subordinate species would be competing with it for resources, and a mutual reduction in numbers would be detectible. As a general pattern, the asymmetrical situation would be an evolutionary derivative of an earlier symmetrical interaction. Probably this view is often correct; however, there are now several cases known in which one species, simply as a by-product of its existence, harms another, conceivably competitive species. The case of the deer, the moose, and the brainworm (Chapter 7) is one such. It is possible that the term "amensalism" should be more widely applied and, further, that not all cases of interference should be equated with competition.

Allelopathy

Other organisms besides humans release chemicals into the environment with unfavorable effects on other species. The name now generally used for this interaction, **allelopathy**, dates only from a 1937 book by the Austrian botanist Hans Molisch, but the effect has been known for a long time. Most of the early knowledge involved plants of direct importance to humans. Pliny's *Natural History,* written in the first century A.D., comments that radishes and laurels "hurt and offend" grapevines. Walnut trees, Pliny said, cause headaches in humans and injury to anything planted in their vicinity. The inhibitory effect of a related species, the black walnut, was one of the first cases of allelopathy to be studied scientifically (Harborne 1982). A series of studies by several botanists culminated in the identification in 1928 of the compound juglone as an important agent of the inhibition. Juglone, a quinone, is the substance that makes your fingers brown if you handle walnuts.

Allelopathy in walnuts could be playing any— or several—of a number of roles. Juglone could inhibit insect consumption of the foliage, for example, or it could prevent squirrels from taking the fruits while they are green (it disappears from the nuts as they ripen). Preventing other plants from growing beneath the tree could, in this case, be simply a side effect, and of no real benefit to the plant. On the other hand, if more shade-tolerant trees are inhibited, the effects could be viewed as a part of interference competition.

Allelochemicals may be released into the environment directly from a plant, by the release of volatile compounds from the leaves, through root exudates, or by leaching from the leaves or litter by rain, fog, or dew (Rice 1974). Or they may be released indirectly as the result of decay organisms acting on litter or dead roots.

Most allelopathic chemicals so far recognized are terpenoids or phenolics, but this may merely reflect incomplete knowledge; many other kinds of compounds could potentially be involved (Putnam 1983). Although the specific physiological mechanisms are probably diverse, reduced nutrient uptake, inhibition of respiration, and interference with growth hormone metabolism are probably frequent effects. As a result of these and other physiological effects, germination may be inhibited, growth may be reduced, and survival may be lowered.

These are autecological effects. At the population level, one plant prevents others from growing near it. These may be individuals of other species, but fairly often plants produce materials that inhibit germination of other members of their own species; however, there exists at least one case of self-stimulation (Newman and Rovira 1975)—a botanical analog of sociality. It seems likely that in some of these cases competition for mineral nutrients or water would be thereby reduced, and possibly some other benefits might be conferred.

One of the most thoroughly studied examples of allelopathy involves patterns in the distribution of plants in the California chaparral. Grasses and forbs, mainly annuals, are absent under some shrubs such as chamise, and are absent even from the vicinity of others. Around sagebrush there is typically a bare zone often more than 1 m wide.

In a series of studies that spanned more than a quarter century, C. H. Muller and his coworkers at the University of California showed that the shrubs contained chemicals inhibitory to germination and growth of several of the important herbs of the chaparral. For sagebrush, there are terpenes that are released into the air (Muller 1965). For chamise, the allelochemicals are water-soluble phenols and phenolic acids that are leached from foliage by the winter rains (McPherson and Muller 1965). Herb seeds planted in soil from beneath the shrubs (and from the bare zone beyond, in the case of sagebrush) grew poorly. Germination and early growth were inhibited by sagebrush leaves simply placed in the same chamber, so that the only contact was through the air.

Under normal circumstances, a stand of chaparral burns every 20 to 35 years. The fires destroy the aboveground portion of the shrubs, temporarily removing them as a source of allelochemicals and also perhaps detoxifying soil residues. In the year following a fire, abundant germination of grasses and forbs (and shrubs) occurs. Much of this germination is evidently from seedbanks. In subsequent years, as the shrubs that have resprouted or germinated grow and spread, the herbs are reduced to smaller and smaller areas. If fire recurrence is delayed for many years, herbs may be all but pinched out of large areas.

The current trend is to downplay the importance of allelopathy in keeping herbs out of unburnt chaparral (Keeley and Keeley 1989). It is suggested that a high proportion of the seeds of several of the typical herbs germinate only after a fire cue. For several, the cue seems not to be heat but instead the presence of charred wood. Allelopathy probably plays a role but mainly in reducing success or survival of various herbs. Also perhaps involved are herbivory by small mammals (Bartholomew 1970) and unfavorable abiotic factors such as light or soil fertility.

Sponges contain a variety of biotoxic chemicals that, extracted and tested in the laboratory, have proved to affect everything from bacteria to fish. The possibility has been raised that some of these chemicals are released into the water and inhibit the settling and growth of other, potentially competitive invertebrates. A study at ocean sites in Florida and Belize compared the rates at which invertebrate larvae were recruited to tiles with real sponges, synthetic sponges, and no sponges (Bingham and Young 1991). It was found that the spongeless tiles tended to recruit more larvae than tiles with sponges; however, few species of invertebrates, if any, were more inhibited by real sponges than plastic ones. Because the synthetic sponges were not producing allelochemicals, it seems likely that the lower recruitment on the sponged tiles was the result of effects on water flow, such as increased turbulence.

Allelopathy has been suggested as the cause of a variety of patterns in certain seres and in the composition of climax communities. How many of these suggestions will prove true is unclear, but we can be sure that the role of allelopathy, whether large or small, will be intertwined with other interactions such as shading, nutrient depletion, competition for water, and selective herbivory.

Evaluating the potential of allelopathy for agriculture has been remarkably slow. It seems possible that allelochemicals could be used as herbicides that would be less harmful in the environment than most current herbicides. Is it likely that weed infestations can be reduced in fields by the right kind of crop rotation (sorghum residues, for example, contain potent inhibitors of many weed species)? Could companion plantings keep down weeds or insect pests? These are attractive possibilities to the ecologist and to the aware citizen but are much less attractive to the chemical companies that finance most agricultural research.

Neutralism

Neutralism occurs when neither of two species has any effect on the other. At the level of the population and considering a considerable length of time, it may be that every species in an ecosystem has effects on every other one, in the manner of cat lovers influencing seed set in red clover. But our focus here is on immediate individual-individual effects. At this level neutralism is probably a common interaction—or lack of interaction—among species that have no trophic interactions and do not share a limiting resource. A falcon ordinarily has no interaction with a fern, except very indirectly, and it may be that a falcon and a fish hawk are all but neutral in their effects on one another.

An important evolutionary question is whether neutralism tends to evolve from other coactions, especially competition. Many ecologists believe that natural selection based on competition causes competing species to diverge and, thus, to divide up formerly shared resources. Competition is thereby reduced. Looked at from the standpoint of species interaction, this implies that neutralism can evolve from competition (Rosenzweig 1981).

The Population–Community Ecology Interface: Mutualism

The cleaning symbiosis, a case of mutualism. The cleaner fish is seen at work just behind the eye of the tiger grouper. *(Animals Animals/© 1994 W. Gregory Brown)*

OUTLINE

- Symbiotic Mutualism
- Nonsymbiotic Mutualism
- Mycorrhizae
- Pollination
- The Evolution of Mutualism
- The World's Greatest Symbiosis: Evolution of the Eucaryote Cell

Mutualism is the coaction in which both partners benefit. A useful distinction is between symbiotic and nonsymbiotic mutualism. **Symbiosis** is a term coined long ago from roots meaning "living together." Logically, any close, continued physical association is symbiosis including many cases of parasitism and commensalism, as well as mutualism.

Symbiotic Mutualism

Probably the best-known example of mutualism is the symbiotic union of alga and fungus, forming lichens (Figure 9-1). In lichens the association is so intimate that biologists, for most purposes, regard the union, rather than the separate alga and fungus, as the unit of study. Lichens are able to reproduce the union by producing bodies (soredia) that consist of a small amount of fungus tissue enclosing a few algal cells.

The main benefit of the relationship to the fungus is believed to be food, that is, carbohydrate pro-

Figure 9-1

Lichens—the symbiotic union of alga and fungus—are able to inhabit sites so extreme as to be intolerable to fungi or algae separately or to most other plants. Here lichens flourish on a bare rock ledge in southern Illinois. (Photo by the author.)

duced in photosynthesis by the alga. Important benefits to the alga are water and protection from desiccation. The lichen is able to flourish in harsh environments, such as bare, dry rocks and arctic tundra, where neither algae nor fungi do well alone.

Lichens are probably the premier example of mutual aid from the standpoint of popular recognition, but there are alga-animal unions that are very similar. Algae occur within the animal bodies, supplying food and perhaps oxygen to their animal associates, and receiving shelter, carbon dioxide, and nitrogen.

In one striking and well-studied example, a marine flatworm, the Roscoff twister, and a unicellular alga form a mutualistic union (Muscatine et al. 1974). The flatworm is free of algae when it hatches but soon becomes infected. These algae may be of several species, but when the right one comes along, the others are eliminated. When the algae enter the worm, they lose their flagellae and eyespots and live within the flatworm cells. In this respect the union is closer than the association between alga and fungus in lichens.

The now-green flatworms soon stop feeding and live from that time forward on food provided by the alga. The joint organism, which lives in the intertidal zone of the ocean shore is functionally a seaweed but one with unusual abilities to move around and burrow down into the sand. Curiously, the food transferred in this case is not carbohydrate, although there are similar unions involving other invertebrates, such as hydra, in which this is the case. Rather, the food that the flatworm gets from its algal symbionts is amino acids, mostly glutamine, as shown in Figure 9-2.

Other relatively well-known examples of symbiotic mutualism include the digestion of cellulose by microorganisms living in the gut of many kinds of herbivores from termites to hippopotamuses; nitrogen fixation by bacteria in the nodules of legumes (Chapter 12); chemosynthetic organisms living within the worms and clams of marine hydrothermal vent ecosystems (Chapter 11); and mycorrhizae, discussed later in this chapter.

Some not so well-known examples are the luminescent bacteria that live in and provide the light for the light organs of certain kinds of marine fishes (McFall–Ngai 1991) and the ascomycete fungi that live in grasses and sedges and produce toxins that have detrimental affects on grazers (Kendrick 1991).

Mutualism is apparently the basis for **domatia** (singular, domatium), little shelters produced on the bottom of the leaves of many trees and shrubs (O'Dowd and Willson 1991). Predaceous and fungivorous mites live in these structures. Details of the interaction are yet to be studied, but it is plausible that the mites raise the fitness of the host plants by removing fungi and small herbivores, including scale insects and herbivorous mites (Pemberton and Turner 1989).

Nonsymbiotic Mutualism

Pollination, a widespread type of nonsymbiotic mutualism, is discussed later in this section. "Cleaning" of coral reef fish by other organisms is discussed in Chapter 18. Some other cases of nonsymbiotic mutualism involve dispersal (squirrels store nuts but do not retrieve some of them which, planted in a good seedbed, produce more

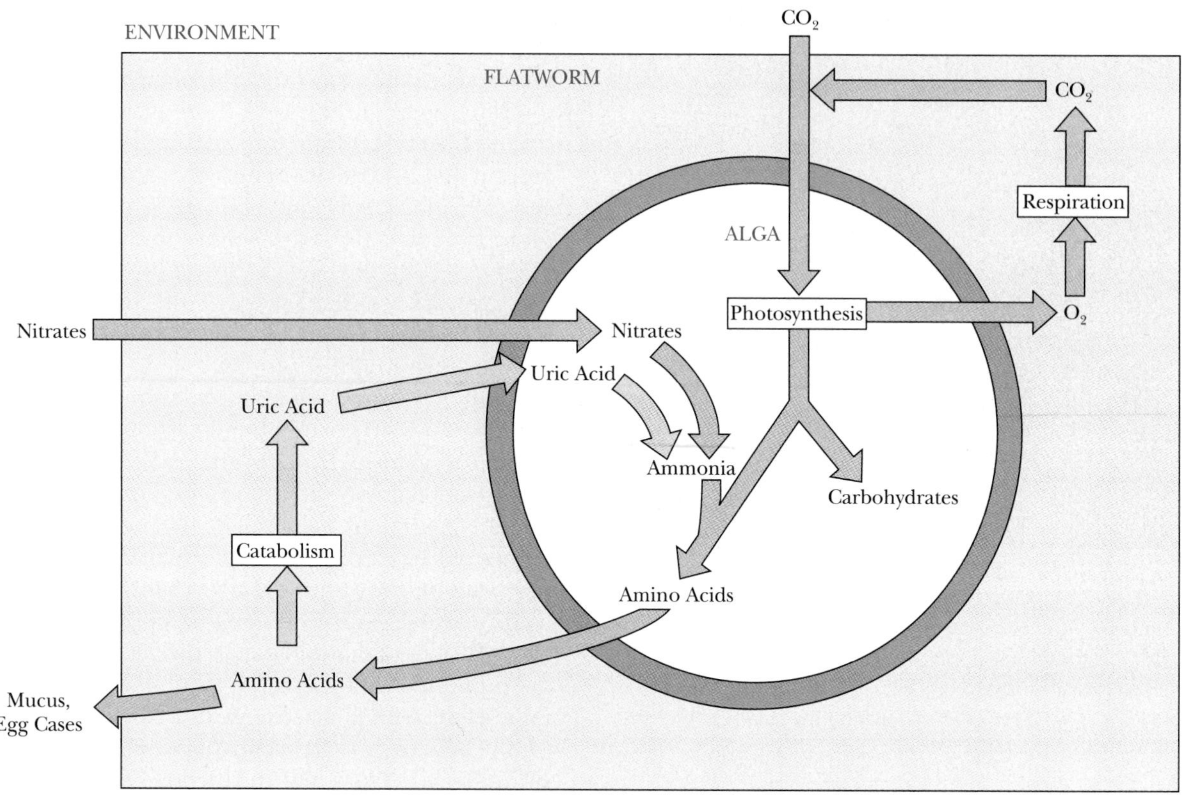

Figure 9-2

Some major metabolic interrelations between the Roscoff twister and its intracellular algal symbiont. Nitrogen recycling seems to be an important feature of the symbiosis for the alga, and the main foods that the flatworm gets from the alga are amino acids. (Simplified from D. C. Smith 1979.)

nut trees) and protection (ants protect the aphids producing honeydew, a sugary secretion, upon which the ants feed).

Many seemingly obvious cases of parasitism or predation show mutualistic aspects when closely examined. Grazing by bison and other herbivores seems clearly to be a +/− interaction with the grass getting the minus. Range managers, however, have known for a long time that a square meter of grazed grassland may produce more total biomass in a year than one left undisturbed (Jameson 1963, Paige and Witham 1987). Part of the reason seems to be faster cycling of nutrients. Grazers deposit grass fragments, feces, and urine in the grassland, allowing a faster return of nutrients to the soil than if they remained tied up in grass foliage until the end of the growing season (Owen 1980). In some cases it may be merely that plants, in response to grazing, overcompensate for the lost tissue.

Similar statements can be made about certain cases of herbivory in aquatic habitats. Filamentous algae of coral reefs have higher productivity when grazed, probably because of fertilization by herbivore excretions and the reduction of self-shading (Duffy and Hay 1990). The fresh-water colonial green alga Schroeter's roundie, a member of the phytoplankton, is eaten by zooplankton, including members of the genus *Daphnia;* however, the fact that it is eaten does not mean that it is consumed (Porter 1976). Many cells (more than 90%) pass through the gut unharmed and have their growth enhanced by the added nutrients (from other, more easily digested algae and gut secretions of the *Daphnia*) gained in the passage. Schroeter's roun-

Figure 9-3

Aphids are small, soft-bodied insects that suck sap and discharge a sugary fluid, honeydew, from their anus. (Photo by Carol Snow.)

die does better in areas with daphnia than without. One author preferred to call this relationship algal parasitism rather than zooplankton herbivory (Belsky 1986), but, as we have said several times, few coactions fit neatly into the cubbyholes that humans construct.

A similar relationship was found between a fish, the blue tilapia, and another green alga, the copperhead desmid (McDonald 1985). The fish grew and the algae population increased with grazing.

Aphids (Figure 9-3) produce vast quantities of honeydew from the plant sap that they suck, much more than they themselves use and generally even more than the tending ants use. This falls to the soil at the base of the plant where it is utilized by still other organisms. Owen and Wiegert (1976) suggested that one such group of organisms is heterotrophic nitrogen fixers. The increase in soil fertility resulting from the nitrogen fixation may well outweigh the loss of photosynthate. The profit could be increased vegetative production, increased seed production, or both—or, of course, neither, because there is much yet to learn about such relationships (Crawley 1987).

Mycorrhizae

Nearly everyone recognizes that lichens are a symbiotic union composed of two different organisms; almost no one realizes that the same statement is true of most vascular plants—ferns, herbs, trees—growing in nature. One of the most widely important mutualistic associations is the union of vascular plant roots with fungi to form mycorrhizae (meaning ''fungus-root''). It is almost impossible to overstate the importance of this coaction in plant growth. One author (Wilhelm 1966) commented that plants in nature don't, strictly speaking, have roots; they have mycorrhizae. Many functions that we regard as being done by roots are actually performed by the plant-fungus union and are done poorly or not at all if the fungus is absent.

Although several types of mycorrhizae have been recognized (Brundett 1991), two are particularly important. In **endomycorrhizae**, strands of the fungal body (hyphae) penetrate the cells of the roots. Vesicular-arbuscular (V-A) mycorrhizae is an approximate synonym. In **ectomycorrhizae**, they do not, but, instead, they form a sheath or mantle around the roots and also grow between the individual cells in the outer layers of the roots, replacing the cells' middle lamellae.

Endomycorrhizae are found on—or as a part of—many terrestrial plant species (except for some bog plants, aquatics do not have mycorrhizae). Most agricultural crops are endomycorrhizal. Ectomycorrhizae occur on many tree species, including pine, spruce, hemlock, aspen, hickory, oak, beech, and eucalyptus. A few groups of trees are endomycorrhizal (maple, elm, ash, sweet gum, and sycamore, among others). Some plants, including heaths, legumes, and roses, may possess both endomycorrhizae and ectomycorrhizae.

For many endomycorrhizae, the fungi involved in the symbiosis are Phycomycetes (Zygomycetes)—evidently a fairly small number of very widespread fungal species (Powell and Bagyaraj 1984). A large number (more than 2000) of Basidiomycetes are the fungal element in ectomycorrhizae. For either type, few obligate relationships between one species of fungus and one species of vascular plant are known. In fact, a single tree may have mycorrhizae incorporating several fungal species. The Basidiomycetes involved in ectomycorrhizae include many of the familiar mushrooms and puffballs of forest and savanna; however, not all mushrooms are mycorrhizal; many species are saprophytes (decay organisms) or parasites.

Although the relationships may be loose as far as requirements for a particular species are concerned, the mycorrhizal relationship itself is obligatory for many plants and fungi. Under natural conditions, many plants simply do not survive without a mycorrhizal fungus, even though this can be any one of many species. A nursery experiment comparing seedlings of sweet gum with and without an appropriate fungus found that uninfected seedlings ceased growth after the primary leaf stage. The result was that after 24 weeks, dry weight of the mycorrhizal plants was over 5000% greater than that of nonmycorrhizal plants (Bryan and Kormanik 1977). Similarly, many fungi will not fruit (that is, produce the aboveground, spore-producing mushrooms or puffballs) without one of their appropriate hosts. All the V-A fungi have so far proved impossible to grow in culture (Hepper 1984).

One obvious benefit that vascular plants receive from the symbiosis is greatly improved mineral uptake. Good mineral nutrition, in turn, improves the plant's ability to withstand water stress. Infection with the mycorrhizal fungi also seems to protect plant roots against infection by parasitic fungi.

Evidently the main reason for improved nutrient uptake by mycorrhizal plants is the greatly increased soil volume from which the plant can draw because of its network of fungal hyphae extending to several meters. Mycorrhizae are particularly effective in improving phosphate uptake. In fact, little mycorrhizal development occurs in soils with high phosphate levels. In experiments where plenty of phosphate is provided, infected plants do little, if any, better than nonmycorrhizal ones (Kormanik, Bryan, and Schultz 1977).

The main benefit that fungi derive from the mycorrhizal association is, depending on how you want to put it, food, carbon, or energy. Having no means of foodmaking, they instead use some of the vascular plant's photosynthetic production. Just how high a tax the fungi exact has not been widely studied, but two estimates suggest they take from 15 to 25% of the trees' photosynthate (Vogt et al. 1982, Odum and Biever 1984).

How often does the absence of suitable fungi keep plants out of a given habitat or region? The fungi involved in endomycorrhizal associations are widespread, so probably they rarely limit a plant's distribution. The basidiomycetes required by many trees are much less widespread, so it is possible that treeless areas such as desert, steppe, and alpine tundra may tend to remain so because of the lack of appropriate fungi. This idea has been one of the regularly mentioned but scarcely tested hypotheses for the treelessness of the North American prairie. There is good evidence that certain new soils, such as strip-mine spoil banks, have few mycorrhizal fungi: A considerable amount of forestry research in the past 20 years has been directed toward improving reforestation of strip-mine spoils by using mycorrhizal seedlings.

The appropriate fungi may be absent from a

site because they haven't managed to get there or because physical conditions are unsuitable for them; however, they may also be excluded by biotic means. Extensive areas in the British Isles are dominated by the shrub known as heather. One reason why the heathlands are resistant to invasion by birch or other trees, which would then shade out the shrubs, is that the mycorrhizal fungi required by the trees are absent. The fungi are absent, it appears, because of allelopathy; the mycorrhizal roots of heather produce allelochemicals that are toxic to the ectomycorrhizal fungi needed by the trees but not to their own fungi (Robinson 1972).

The potential range of interactions based on mycorrhizal connections is vast. It is known that the fungi of orchids are saprophytic or parasitic basidiomycetes. The orchids, whether or not they themselves have chlorophyll, share in the energy the fungi obtain elsewhere. Chlorophyll-less members of the heath family, such as Indian pipe and pinesap, were once thought to be saprophytes. Instead, they are now known to be involved in a mycorrhizal union with a fungus and a green plant (Björkman 1960). The fungi involved are ectomycorrhizal forms, such as the boletes, and the green plants include trees such as oak, hemlock, and pine. The fungus-tree association is mutualistic. The pinesap is a parasite once removed, or **epiparasite,** on the tree; however, the relationship may be still more complicated. There is evidence that the pinesap may stimulate fungal growth that could benefit the fungus and, indirectly, the tree (Castellano and Trapp 1985).

It is worth pausing and trying to visualize what the below-ground part of a forest ecosystem is like. Above ground we can see the individual trees standing here and there and, beneath them, shrubs and herbs. Below ground, the distinction between individual plants is much less clear. From the mycorrhizal roots, fungal hyphae radiate considerable distances and fuse with the hyphae from other plants, so that, below ground, plants tend to be interconnected by a vast, fine fungal network.

Something of the potential importance of these interconnections is indicated by a study by two forest ecologists (Woods and Brock 1964). They introduced radioactive phosphorus (^{32}P) into recently cut stumps of red maple in a mixed hardwood forest in North Carolina. (They used modeling clay to make a small dam around the top of the stump, filled the reservoir with water containing the radioactive phosphorus, and covered the whole thing with a plastic bag.) After 8 days, 72% of the trees within 2.5 m of the treated stumps showed detectable levels of radiophosphorus; 43% of the trees they checked within 8 m had received the labeled phosphorus. Besides other red maples, 18 species were involved, including oaks, hickories, hollies, red cedar, gums, tulip tree, mulberry, and Virginia creeper. Most avenues by which the mineral could be transferred from the maples to other trees, other than mycorrhizal unions, were ruled out. Root grafts could have played some role, although they are usually between individuals of the same species. Whatever the relative roles of root grafts and mycorrhizae, the soil of a forest clearly is a place of interconnections.

It seems likely that ecologically significant exchanges of carbohydrates, minerals, and other compounds occur among the plants of forest ecosystems via mycorrhizal unions. The results of their radiophosphorus studies led Woods and Brock to take an organismic point of view. Perhaps, they suggested, we should "regard the root mass of a forest ecosystem as a single functional unit, rather than a number of interwoven but independent entities." The possibilities for cooperative interactions, such as sharing carbon during bad years or synchronizing fruiting, are immense. So also are the possibilities for manipulative interactions by individual trees, such as taking more carbon than one gives or sending misinformation rather than the right stuff.

Pollination

Pollination is an essential part of sexual reproduction in seed plants. For flowering plants, we can define it as the transport of pollen from anther to stigma. For the plant, pollination is dispersal of its microspores. For animals, pollen and nectar are food sources. Frequently, the two processes of dispersal and herbivory have become evolutionarily locked into a mutualistic coaction (Faegri and Van der Pijl 1979).

Not all pollination is accomplished by animals, however. Wind pollination is the prevalent mode in conifers (Figure 9-4), some broad-leaved trees, and most grasses, sedges, and rushes. Wind-pollinated

Figure 9-4

A shower of pollen released from staminate cones of a jack-pine.

flowers tend to be small and greenish with no nectar or scent. Flowering often occurs before the vegetation leafs out (such as oaks and aspen), or, otherwise, the flowers are produced above the general level of vegetation. In Michigan prairies, the general vegetation level in mid-May is about 18 cm high. The insect-pollinated flowers are borne at this level, but the wind-pollinated meadow rue plants tower above everything else, up to 55 cm tall (Brewer 1985).

Only a small percentage of wind-dispersed pollen grains end up on the stigma of the right species; accordingly, successful wind-pollinated species produce copious pollen. Walking through a pine forest during the pollination period raises clouds of yellow dust—pollen—from the ground, and pollen covers the lakes, caught for a time in the surface film. The pollen grains of wind-pollinated plants are usually smooth and dry. Individual plants tend to be either male or female, specializing either in producing the vast quantities of pollen or receiving it. **Dioecious,** meaning two houses, is the technical term for the condition.

Most flowering plants are pollinated by animals, usually insects but occasionally birds or bats. In general these plants have traits opposite to those listed for wind pollination. The flowers, for instance, are usually conspicuous in some way and produce pollen grains that stick to another and to the pollinating animal. It seems clear that many features of plant and pollinator are the result of long coevolutionary sequences that began with the first flowering plants in the Mesozoic era (insects had been around since the late Paleozoic).

Although a great deal was known about pollination biology by 1900, the past 25 years have seen concentrated study on ecological and evolutionary details. Here we can outline only the main features of the coaction.

Table 9-1 summarizes some of the flower traits that tend to be associated with the important pollinator groups. These features tend to fit the flowers to various morphological, physiological, and behavioral traits of the pollinators. Bumblebee flowers, for example, are large and closed (Figure 9-5). Few insects other than bees are strong enough to open them, and smaller bees, even if they open them, cannot reach the nectar. Butterfly flowers, pollinated by diurnal insects with color vision, are red and open in the daytime. Moth flowers are white and open at night. Bat flowers are also white and open at night, but they tend to be short and large mouthed rather than the long slender tubes of moth flowers (Figure 9-6).

It is rarely possible to say which traits came first, those of the animal or the flower. In any case, the situation that we see today is ordinarily the result of a long series of evolutionary changes on the part of both pollinator and plant. There are obvious advantages to both individuals and populations in the usual pollination system. The animal gets calories in pollen and protein in nectar. The plant gets the opportunity to fertilize many other plants. The animal population participates in the reproduction of one of its food sources. The plant population is assisted in outcrossing, helping to maintain genetic diversity.

The Exploitation of Mutualisms

The existence of a coevolved mutualism provides an opportunity for exploitation by other species. Thus, we find some animals short-circuiting the flower's pollination apparatus, stealing nectar without helping to accomplish the spread of the plant's pollen. Birds, for example, may peck a hole in the side of a flower and remove the nectar through it. Ants have a great appetite for sugar but are poor pollinators, being small and smooth. Most

Table 9-1
Features Associated With Various Pollinators

		Flower Features				
Pollinator	*Examples of Flowers*	*Shape*	*Color*	*Scent*	*Nectar*	*Actions of Animal*
Beetles	Magnolias, umbellifers	Large and shallow; small and grouped	Dull, greenish or white	Strong, fruity	Sometimes	Walk around and feed on blossoms
Carrion flies	Skunk cabbage, purple trillium, carrion flower	Often deep	Dark reddish brown	Decaying flesh	None	Go into flower; may be trapped for a time
Bees	Clovers, yellow violets, mints, fringed gentian, some orchids	Closed, deep; bilaterally symmetrical, with a landing platform	Blue or yellow	Sweet	Moderate	Open flower and lean in or stick tongue in
Butterflies	Some milkweeds, primroses, orchids	Erect tubes, with rim, fixed anthers	Vivid, reddish	Weak, agreeable	Ample	Light on flower and stick proboscis in
Moths	Morning glory, phlox, jimson weed	Horizontal or hanging tubes without rims; free anthers; open at night	Usually white	Strong, sweet, at night	Copious	Hover in front and stick proboscis in
Birds	Trumpet creeper, columbine	Deep, wide tube	Vivid, often scarlet	None	Copious	Hover in front and stick beak or tongue in
Bats	Mimosas, agaves, saguaro, sausage tree	Large-mouthed single or small and grouped, may be on tree trunks; open at night	Usually dull whitish	Strong, stale, at night	Copious	Land and stick snout and tongue in

Fiugre 9-5

The large, closed flowers of wild indigo are pollinated by bumblebees.

Figure 9-6

The Sonoran desert cactus saguaro is bat pollinated. (Photo by George and Betty Walker.)

flowers probably lose some nectar to them, even though many have some sort of ant defense. Examples are sticky areas on the stems and nectaries that are buried in hairs longer than an ant's tongue.

Most of the 700 species of figs are pollinated by a different species of wasp, each of which can reproduce only in the flowers of that particular fig. Female fig wasps pollinate the flowers within a fig inflorescence and die. Their daughters develop within the flowers, feeding on developing seeds. After mating with the wingless males, the daughters emerge from the fig fruit with pollen from the newly matured anthers and fly off in search of another tree that has flowers at the appropriate stage for pollination.

The existence of this complex pollination system is well known. Less well known is the fact that each species of fig supports, besides its pollinator, several other species of small wasps that have similar life histories but do not pollinate the fig (Bronstein 1991). These are parasites on the figs, although their impact may be slight. Some species kill developing pollinator larvae to take over their seeds and thus are interference competitors with the pollinators.

One group of authors (Addicott et al. 1990) called the nonpollinating wasps **aprovechados,** a Spanish word meaning approximately "those who help themselves." The term could very well be applied more widely to other organisms that butt in on mutualisms. The examples of aprovechados mentioned so far involve pollination, but other mutualistic relationships can also be taken advantage of. A good example is the fish that masquerade as cleaners, taking advantage of the cleaning symbiosis described in the section on coral reefs (Chapter 18).

In the cases discussed so far, it has been the animal that has failed to deliver, but plants can default also. The flowers pollinated by carrion flies typically have no pollen or nectar for the insect. The insect visits the stinking, meat-colored flowers expecting to find a suitable place to lay eggs. Instead, it is dusted with pollen and sent on its way. This is not mutualism; it is deceit.

The case just described is carrion mimicry. Other sorts of mimicry may occur, of which the most striking is the similarity between certain orchid flowers and wasps or bees. The mimicry is perhaps most striking in the genus *Ophrys* (Kullenberg 1961). The lip, or labellum, of these flowers looks very much like the topside of a female insect. The resemblance is obvious and is even embedded in the scientific names of the various species: *Ophrys apifera,* for example, means the ophrys with a bee on it.

In size, shape, color, and texture, the lip of the flower is a very good mimic, good enough to fool the male wasp. He attempts to copulate with the flower, finds that he gets only so far, flies to another flower, and gets fooled again, but, in so doing, spreads pollen sacs from the first plant to the second. Not only visual and tactile mimicry, but also chemical mimicry, is involved. The flowers smell like the female wasps or bees. One botanist recounted having his car chased by wasps when he was carrying a bunch of *Ophrys* orchids in the backseat on the floor (Stoutamire 1974).

The flowering of the orchid corresponds with the emergence of the male wasps or bees, which precedes the emergence of the females. Evidently only inexperienced males are fooled, but these may be in the majority even later in the season because of the small number of reproductive females that are produced in many species of Hymenoptera.

The Evolution of Mutualism

Mutualism is the hardest coaction to contain within the conventional loss/gain classification. Part of the problem is maintaining a consistent point of view. Here we have focused on harm and help to individuals, but many discussions switch back and forth from individual to population. By our definition, predation is clearly a $+/-$ coaction, in which the plus gets food and the minus gets eaten. There is, however, perfect justification for describing a predator-prey system as mutualistic at the population level. The predator gets food; the prey gets population management.

But there is another, more fundamental reason why mutualism is troublesome in the conventional classfication: Many kinds of coactions, however they start, tend to evolve mutualistic aspects.

For every species involved in an interaction that is harmful to it, there will be an evolutionary prize to any individual that can do either of two things: (1) **disaffiliate**—escape the interaction—or (2) **capitalize** on the interaction—put it to its own benefit. Genetic factors allowing either outcome should spread rapidly in a population.

Disaffiliation includes the evolution of traits that halt parasitism or predation. The immune system of vertebrates is a physiological example for parasitism. The trait can, however, be morphological or behavioral. If some individuals of the Hawaiian land birds that we talked about in Chapter 7 had genes that caused them to sleep with the bare parts of their heads and legs tucked under feathers, they might have survived to save the species from extinction. For predation, disaffiliation can be accomplished through the evolution of greater running speed, armor, cryptic coloration, or defensive chemicals.

Disaffiliation, if it is successful, may well be fatal for the other species in the interaction. But if the threat is great, so also is the opportunity. Any individuals of the predator or parasite species that can solve the new defenses will prosper extravagantly. Sometimes there are none like this, and the disaffiliation goes to completion; this particular parasite-host or predator-prey system then ceases to exist. Often, however, the outcome is a continuing arms race in which new defenses call forth new weapons. Coevolution of bats and their moth prey was discussed in Chapter 5.

Capitalization involves the host or prey turning a net loss into a net gain, or at least finding a way of getting some benefit from what is otherwise a dead loss. If an animal is parasitized, perhaps it may be able to make use of some of the nutrient-containing waste products of the parasites and, thereby, not have to waste time or energy obtaining the nutrients elsewhere. If a flower's pollen is going to be collected by insects for food, the flower may just as well arrange for the animals to spread it to other flowers.

Unlike disaffiliation, capitalization does not put an evolutionary premium on countermoves by the predator or parasite. To the contrary: If the host or prey can improve its health, raise its reproductive rate, or increase its density, then the parasite or predator will often be better off too. For example, if a host can live a long time despite the presence of the parasite, then the parasite will be able to produce more offspring. If a grass regrows after it is eaten, as we described in a previous section, then the grazer can revisit it to eat again.

To sum up, evolution of host and prey tends to break up parasitic or predatory interactions (disaffiliation) or to pull a profit from them (capitalization). Coevolution on the part of the predator or parasite tends to oppose disaffiliation but usually to favor capitalization. Thus, the overall trend is for the development of mutualistic coactions.

It is worth pausing long enough to say that in certain situations the interests of the predator or parasite would be harmed by certain trends in capitalization by prey or host and these, accordingly, would not be favored through coevolution. One example already mentioned is improvement of health or vigor of intermediate hosts of certain kinds of parasites.

It obscures rather than clarifies matters to conclude that most interactions are mutualism. Mutualism perhaps should be thought of as another dimension along which to classify coactions. Generally mutualistic coactions are mutualism plus something else. This was recognized long ago by the plant ecologist W. B. MacDougall (1931), who described lichens as "a case of double, or reciprocal, parasitism."

The World's Greatest Symbiosis: Evolution of the Eucaryote Cell

Eucaryotes are organisms whose cells contain a nucleus. They include animals, plants, fungi, everything except the Monera, which are the bacteria and blue-green algae (cyanobacteria). Compared with the moneran (procaryote) cell, the eucaryote cell is a complicated structure with several kinds of organelles. Included are a membrane-bound nucleus, mitochondria where respiration occurs, and, if it is a plant cell, plastids that capture solar energy and manufacture carbohydrates.

A considerable amount of evidence has accumulated that the eucaryote cell arose through a series of symbiotic mutualisms (Margulis 1981). The idea is an old one, stated independently by several ingenious scientists of the early 1900s. Development and testing of the idea awaited modern biochemical and electron microscopical techniques.

Two billion years ago—give or take a few million—a variety of metabolically diverse bacteria were in existence. There were photosynthesizers such as blue-green algae, aerobes that used oxygen produced originally by the photosynthesizers to respire their food, and other bacteria that lived in anaerobic habitats and still carried on a more primitive form of metabolism, fermentation.

The first step in the development of the eucaryote cell may have been the acquisition of mitochondria (Figure 9-7). This may have first involved an association between a wall-less fermenter and a rod-shaped aerobe. The aerobe may have started out as a parasite, benefiting by going to the source of the pyruvate that it used in its respiration. Through the process we termed "capitalization," mutualism could have developed in which the fermenter evolved the ability to make use of some of the energy released in the more efficient Krebs cycle practiced by the formerly free-living aerobe.

The double membrane of the mitochondria is considered to be a relic of the symbiotic past, the inner membrane being homologous to the cell membrane of aerobic bacteria, the outer membrane of host origin. Although the mitochondria retain some DNA, it is postulated that most genetic information has been transferred to the host's nucleus.

The development of the nuclear membrane is viewed not as the result of a symbiotic event but as

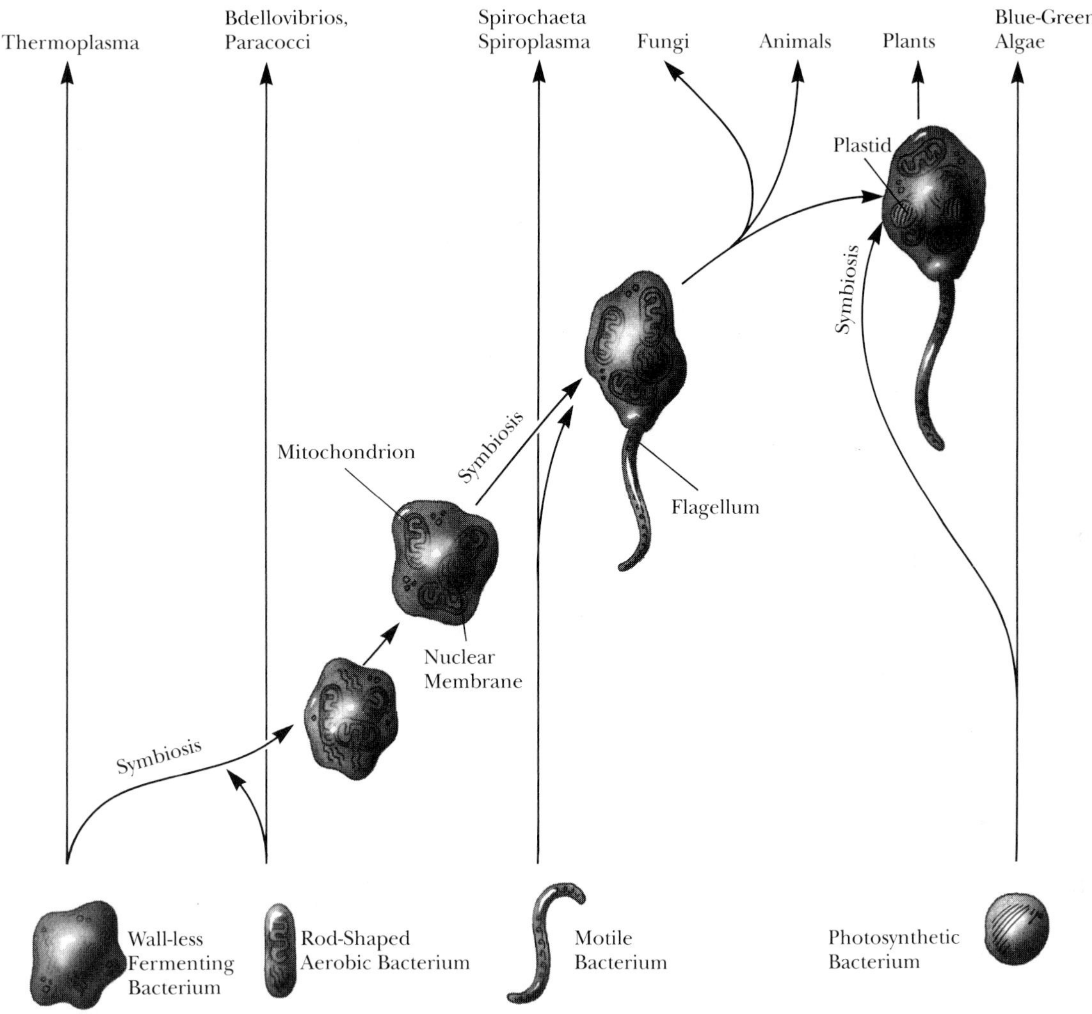

Figure 9-7

The evolutionary pathways leading to the major groups of organisms may have been something like this if symbiotic mutualisms were important steps in the development of the eucaryote cell. On the bottom line are bacterial types that became involved in the symbioses. On the top line are living groups that may have evolved along the various lines. (Based on Margulis 1981.)

a more standard evolutionary advance made advantageous by the increasing complexity and size of the evolving eucaryote cell. The development of mitosis, however, may have depended on elements derived from another symbiosis, this one with spirochaete-like or spirosplasma-like bacteria. Centrioles and the mitotic spindle that allow precise division of genetic material between daughter cells may have been derived from these symbionts.

Also derived from this association may have been a flagellum-like structure (Hinkle 1991). With this step, the cell would have achieved motility and the potential for all the other functions that flagella, cilia, and their derivatives perform in multicellular organisms as well as unicells.

By this point in the sequence, we have arrived at a primitive nucleated, mitochondria-containing flagellate. It is a good candidate to give rise, along one line, to protozoa and multicellular animals and, along another, to the slime molds and the multicellular fungi. With one symbiotic addition, the same basic stock serves as the progenitor of the algae and higher plants.

This next step is the addition of photosynthetic bacteria, perhaps by their entering the cell as parasites or commensals. Capitalization at this stage would lead to the host cell making use of food—organic molecules—that diffused from the photosynthesizer. This step is the beginning of the organelles that we now refer to as chloroplasts and other plastids.

A considerable amount of evolution occurred on the part of both the host cell and invader along each step of the path from invader to commensal or parasite to mutualist to organelle. Many of the details are uncertain and, given the fact that most of the evolution of the eucaryote cell occurred in the distant past, are likely to remain so. The general credibility of symbiotic mutualism as the basis for the process is enhanced by studies of current symbioses, such as the union between flatworm and algae described in an earlier section.

The broad biochemical functioning of the biosphere might not be much different if eucaryotes had never evolved. But the living world as we know it—the world of trees, grass, mushrooms, birds, and snakes, of algae, copepods, and fish—is a world of eucaryotes.

Community and Ecosystem Ecology: *Structure and Diversity*

The number of species on islands is related to their size and distance from the mainland. This is Bear Island just off the coast of Maine. *(Earth Scenes © 1994 Breck P. Kent)*

OUTLINE

- Community and Ecosystem
- Community Structure
- Ecological Niche
- What Produces Community Organization?
- The Importance of Competition
- The Integrated Versus the Individualistic Community
- Ecological Diversity

Community and Ecosystem

One of the first ecological observations to be recorded was that organisms live in **communities.** Certain species live together in an area characterized by certain environmental factors, and this combination of species tends to recur as the habitat recurs. If the organisms of a site are destroyed or removed, a similar community tends to be restored by successional processes (Chapter 13). Community-level traits include such things as species composition, diversity, stratification, and food chains.

The community is a system composed of species populations bound together by coactions. What each population does in the various coactions and what the effects are at the population level were extensively discussed in Chapters 6 to 9, but most community-level effects have been reserved for this and later chapters.

One of the more useful models of a community is that of a multidimensional network with each intersection a species and the links consisting of the coactions that join each pair of species. This is the image Darwin (1859) had in mind when he spoke of how the various plants and animals of a community, even ones that do not interact directly, "are bound together by a web of complex relationships." Figure 10-1 shows in a highly simplified fashion a small portion of the links in the vicinity of a single species.

The **ecosystem** consists of the community plus its habitat. A. G. Tansley (1935), in proposing the term, wrote, "The more fundamental conception [than biotic community] is, as it seems to me, the whole system, including not only the organism-complex, but also the whole complex of physical factors—the habitat factors in the widest sense." We could also recognize and study systems composed of an individual and its habitat or a population and its habitat, but this has only occasionally been done (as, for example, in studies relating individuals to their thermal environment, such as Porter et al. 1973).

In the ecosystem, community and habitat are bound together by action and reaction, the reciprocal effects of physical environment on organism and organism on physical environment. Actions of physical factors on organisms were discussed in Chapters 2 and 3; reactions are dealt with more fully in Chapter 12. Sometimes community ecology and ecosystem ecology are separated as forming different levels in a hierarchy. Energy flow and biogeochemical cycling then usually constitute the main topics of ecosystem ecology. It is true that the ecosystem represents a broader view than the community, but the level of biological integration is the same. More is gained, both conceptually and pedagogically, by an integrated approach.

Indirect Effects

The single most important ecological concept is that of the system. We have already mentioned Darwin's describing how organisms "remote in the scale of nature, are bound together by a web of complex relationships," and we have quoted Barry Commoner's first law of ecology: Everything is connected with everything else. The ramifying effects of the activities of each species were clear to the earliest ecologists. Clements and Shelford (1939), for example, pointed out that certain fish such as suckers may decrease trout populations by eliminating submerged aquatic vegetation (which supports invertebrates upon which trout feed).

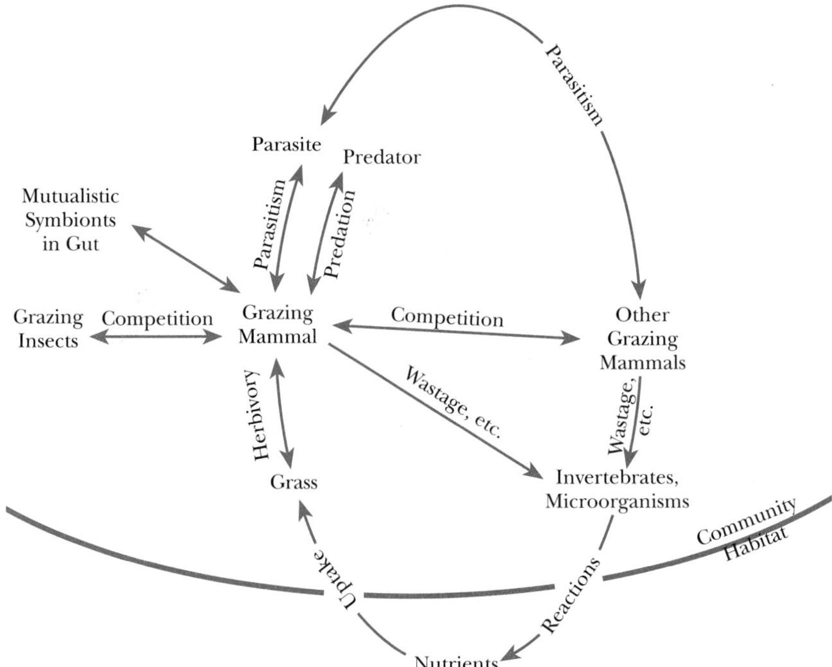

Figure 10-1

A small section of an ecosystem showing in highly simplified fashion a few of the interactions in which a species of grazing mammal is involved.

It is, nevertheless, easy to concentrate on the direct effects of one species on another and neglect effects further out in the web. **Indirect effects,** at their simplest, are how the interaction between two species is influenced by the presence of other species in the community (Miller and Kerfoot 1987). The influences that organisms exert on other organisms by altering the physical environment have sometimes been included under this term (Strauss 1991), but it is better to consider these effects under the heading of reactions (Figure 10-1) as is done in the next chapter.

Indirect effects can be positive or negative in their influence on abundance or growth rate, and they can occur within one trophic level or extend over two or more. Indirect effects are potentially even more varied than direct effects. In the preceding chapters on pairwise interactions, we broadened our vision in several cases to include indirect effects. Examples include the following:

Associational defense (Chapter 6). The situation in which a plant or herbivore gains protection by living on or next to a protected (poisonous or spiny) plant. Cases in which animals provide the associational defense are also known. Examples are birds nesting within colonies of gulls and terns, which provide protection by mobbing predators.

In a well-studied example, Norwegian ornithologist T. Slagsvold (1980) found that the bird community of spruce woodlands was substantially different depending on whether or not fieldfares were present. Fieldfares are colonially nesting thrushes of the taiga that actively defend the colony and surrounding areas against such nest predators as jays and crows. Observations and the experimental increase of fieldfares (by shooting crows) showed that several species, such as the brambling, tended to occur in the vicinity of fieldfares. Other species, such as the bullfinch, were less likely to live near fieldfare colonies than in similar areas lacking them.

The most important mechanism of increase was evidently active selection of areas defended by fieldfares. It is less clear whether decreases occurred by species such as the bullfinch avoiding fieldfare colonies or through competition with the increasers. Many aspects of community structure were substantially changed depending on the presence of the fieldfare. The increasers were northern-subalpine species and generally larger birds. The decreasers were species of less biomass and more southern affinities. Although the general tendency of associational defense is probably to increase species richness, in this particular case the bird community with the fieldfare was less diverse than the bird community lacking it.

Apparent competition (Chapters 6 and 8). The situation in which the presence of species A (the white-tailed deer, for example) supports a predator or parasite that eliminates or, at least, reduces the population of another member of the same trophic level as species A (the moose, for example). At the community level, this will generally lower diversity; however, it may generate more complex results.

Keystone predation (Chapter 6). The situation in which a predator (or herbivore) keeps the numbers of two prey (or plants) low enough that they coexist, rather than one suffering competitive exclusion. The experimental extension of bean weevil coexistence when a parasitoid was added to the system is another example (Chapter 8). The overall effect is to increase diversity.

Altered competitive balance (Chapter 8). Slightly more complex than the two preceding categories is the situation of the two grain beetles with and without the barbershop coccidian. The two beetles do, in fact, compete; which one wins depends on whether the parasite is present or absent (and on physical factors—moisture and temperature). The effect is to alter species composition without affecting diversity.

Vectors and alternative hosts of parasites (Chapter 7). The fact that parasite species A requires intermediate host B to infect definitive host C is a type of indirect effect that tends to escape mention simply because it is so obvious. Alternative hosts serving as reservoirs for a parasite that would otherwise disappear from host A and the community fit the definition of indirect effects also. A possible example is the swamp rat as an alternative host of schistosomiasis in Indonesia (Hairston 1965). In general, such effects require a relatively diverse community, but produce little or no change in diversity at the host level.

Diffuse competition (Chapter 8). The simplest effect on a two-species system of adding a third competitor is the increased likelihood that the middle species on the resource axis will become extinct (MacArthur 1972). The actual consequences are, nevertheless, potentially complicated and still poorly known.

Indirect benefit (Chapter 8). One possibility of adding a third competitor is that species A will affect species B in such a way (by lowering its numbers, for example) that B's harmful effects on species C are reduced. This is one form of indirect benefit produced by effects within a trophic level. Indirect benefit can also be mediated by effects on other trophic levels.

In removal experiments involving seed-eating ants and rodents of the desert, it was found, as already described in Chapter 8 that ants (of the genus *Pheidole*) increased in enclosures from which rodents were removed. The ant populations began to wane, however, after about a year and within two or three years were no higher than in control plots. The proposed mechanism is via effects on the annual plants that are the main food source of both groups of animals (Davidson et al. 1984).

Ants specialize on small-seeded species, rodents on large-seeded ones. Initially, removal of the rodents eliminated their occasional use of small seeds and perhaps gave ants the opportunity to eat some of the smaller seeds in the large seed category. Then, however, absence of the rodents permitted the more competitive large-seeded annuals to increase. The accompanying decrease of small-seeded plant species reduced the supply of the ants' prime food. The ants and rodents competed for seeds, it would seem, but also the rodents helped maintain the ants' food supply by their pressure on the seeds of the large-seed annuals. Note that, at the level of the plants, the rodents were acting as keystone predators.

The possibilities for indirect benefit are numerous; in fact, several of the categories already listed (as well as those listed below) involve indirect benefit, looked at from another direction.

Below we examine some cases of indirect effects not described in earlier chapters.

Habitat modification. This is an extremely important category of indirect effect. The example above of the suckers and trout mentioned by Clements and Shelford fits here. A recent study compared bird and mammal populations on grazed versus ungrazed grassland and shrub steppe in Idaho (Medin and Clary 1990). The habitat alteration in the case of cattle grazing was primarily reducing height and coverage of the herbaceous vegetation.

Grazing increased bird species diversity and also total bird biomass through the addition of three species of shorebirds (killdeer, willet, and long-billed curlew) to the grazed area. Red-winged blackbirds disappeared from the grazed plots and Brewer's blackbirds were added, again presumably because of the changed structure of the vegetation. The main effect of grazing on the small mammals was the loss of four of the rarer species.

Clearly, humans have produced enormous indirect effects in this category. One example was the widespread decline in the turkey vulture in the northern Midwest about 1870 (Brewer 1991). This evidently resulted, not from killing vultures, although that undoubtedly happened, but from a great decline in carrion availability. This came about because of the extinction or decline of the passenger pigeon, white-tailed deer, and many shorebirds and waterfowl. In the late 1880s, as cattle, horses, chickens, and other livestock increased, the turkey vulture returned to the region.

We briefly mentioned the direct amensalistic effects of seabird guano on the shoreline communities beneath. In intertidal areas, the guano kills the seaweed *Fucus;* as a result of the loss of this seaweed, desiccation rates are higher, and the barnacle *Balanus* also declines (Wootton 1971). In the splash zone, which is not regularly submerged, the guano contributes to heavy growth of a green alga and the decline of a lichen *(Verrucaria),* probably through shading by the alga.

Cascading trophic interactions. This is the idea that predatory effects at one trophic level affect community structure, such as amount of biomass, at trophic levels below the one affected directly. The classic case involves "grazing" food chains in lakes (Carpenter et al. 1985). Some lakes may lack piscivorous (fish-eating, first syllable pronounced like "pie") fish through winter kills, overfishing, pollution, or other reasons. If piscivorous fish are added, they may reduce the standing crop of planktivorous fish (fish that eat zooplankton). In the realm of indirect effects, the reduced stock of planktivorous fish may allow the zooplankton to maintain higher levels, in turn depressing the standing crop of phytoplankton (Figure 10-2).

There is no guarantee that such effects will cascade all down the food chain (Ginzburg and Akçakaya 1992). Nevertheless, such effects have been widely reported (Carpenter 1988). Mary E. Power (1990) described a case in which the standing crop of the green alga *Cladophora* was controlled by the presence or absence of larger fish in a California river. Larger fish ate smaller fish and predatory insects that limited numbers of midgefly larvae. The midge larvae ate *Cladophora.* Consequently, exclosures that kept the larger fish out resulted in abundant algal growth. As a further alteration of community structure, these algal clumps came to support epiphytic populations of diatoms and cyanobacteria (blue-green algae).

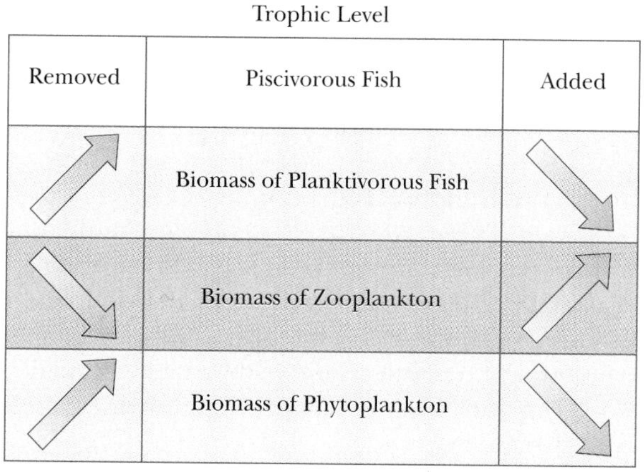

Trophic Level

Removed	Piscivorous Fish	Added
↗	Biomass of Planktivorous Fish	↘
↘	Biomass of Zooplankton	↗
↗	Biomass of Phytoplankton	↘

Figure 10-2

Cascading trophic interactions: removing or adding a higher trophic level may have these effects on the biomass of lower trophic levels.

This section does not include all possible categories of indirect effects, but it shows how varied and potentially ubiquitous they may be. How ubiquitous they actually are is an important question yet to be answered.

Community Structure

Communities show structure, or organization, in various ways. The major ways include the following:

Species composition. Each community has a characteristic group of species that is a subset of the total species pool of the region. Patterns of relative abundance exist among the species.

Physiognomic. A community has a characteristic physical structure (Chapter 15) including stratification and other spatial patterns.

Temporal. Communities have daily and seasonal cycles of activity.

Trophic. Communities have characteristic patterns of energy transfer involving food chains and trophic levels (Chapter 11).

Guilds and niches. We can view the community abstractly as a set of functional units.

In a given mesic hardwood forest or in all the mesic hardwoods of the Cumberland Plateau we can study these aspects and describe the community's structure. We then ask, How does the community get this way? Why, for example, is it made up of certain species and not others? Why are the majority of abundant herbs spring ephemerals?

Why do most of the shrubs ripen their fruit in early autumn?

This chapter describes some important aspects of community structure and considers how community structure is related to coactions, the life histories of the constituent organisms, and the physical environment.

Dominance

Organisms that exert control on the character of the community of which they are a part are called **dominants.** The control can be exerted through reaction or coaction. Trees are obvious dominants of forests that create, mainly through reaction, a microclimate suitable for a certain group of organisms and, accordingly, set the character of the community. Cattle are dominants of an overgrazed grassland; certain kinds of plants that are preferred by the animals or that cannot withstand being trampled have disappeared and been replaced by plants that are spiny, ill-tasting, or tough. Here the control is largely through coactions.

Dominance is not all or nothing, because virtually every organism of a community plays some role in determining its nature down to the most insignificant microbe. In fact some insignificant microbe of the forest may be just as important as the trees if the microbe is, for instance, essential for leaf decay or the conservation and cycling of phosphorus.

A **keystone** is the central wedge-shaped block in an arch that locks the other blocks together. If it is removed, the arch falls down. The term **keystone species** (stemming from Paine 1966) is sometimes used to refer to a species upon which several other species depend and whose removal would lead to their disappearance. Such species may exert their keystone role in a variety of ways. One example is the gopher tortoise of the Southeast, whose burrows provide habitat for various invertebrates, amphibians, and reptiles. The beaver, whose ponds provide homes for many organisms from pondweeds to black ducks, is another.

Chemical Ecology

A considerable amount of the organization of ecosystems is based on chemical interactions. The term **chemical ecology** refers to the study of the production and uptake or reception by organisms of chemical compounds having effects on other organisms (either intraspecifically or interspecifically). Chemical ecology is not basically different from any other kind of ecology, and we have dealt with several examples of chemical coactions in earlier chapters. It is worth a separate discussion at this time, because only recently has enough attention been given to chemical interactions for their importance to be appreciated.

Of course, when a lion eats an antelope's leg, it is eating chemicals, and when we breathe air, we are breathing oxygen and carbon dioxide produced by other organisms. Generally, chemical ecology is not used to include simple relationships of feeding and the movement of inorganic nutrients. R. H. Whittaker and P. P. Feeny (1971) set up a classification of interorganismic chemical effects that is used here (slightly modified). They have separated such effects into effects between different species (which they call allelochemic effects) and effects between individuals of the same species.

The second category includes the pheromones that serve as chemical messengers between members of a species. These may be compounds that waft in the air that attract male moths to female moths, or compounds in urine that mark the territorial boundaries of a wolf. In the form of royal jelly, they may determine whether a honeybee larva becomes a worker or a queen.

The fright reaction of fish is the result of a chemical released from the skin by injury (Barnett 1977). When other fish smell the chemical, they flee. This pheromone seems to function much the same as an alarm call does in birds.

It is often important for one animal to recognize another individually or, at least, as kin or as a nest-mate. This can be based on sight or sound, but in many species, pheromones seem to be involved (Gadagker 1985). One study showed that human mothers were able to distinguish their own infants from others by smell alone if they were allowed a brief time with their child soon after birth (Russell et al. 1983).

The allelochemicals have a variety of effects. There are repellents such as the distasteful and poisonous compounds in buttercup leaves or mustards and the noxious butyl mercaptan (butanethiol) sprayed by the skunk. Perhaps the most spectacular example of a defensive spray is the boiling hot pbenzoquinones ejected in a pulsed spray by the bombardier beetle from a nozzle on its posterior end (Dean et al. 1991).

These chemicals function for escape; there are, in addition, other kinds of escape substances. Some insects that live supported on the surface film of water release detergent-like substances to reduce the surface tension behind them, so that they are whisked rapidly forward and away from a predator.

There is the important suppressant group. Included are the materials produced by chaparral shrubs that inhibit herbaceous growth around them. Substances released by certain planktonic plants that slow the feeding rate of some herbivorous zooplankton are also suppressants. Included also are the antibiotics, produced mainly by various fungi, that humans have used so successfully for our own benefit. Presumably in the soil where these organisms live the antibiotic substance functions to inhibit the growth of competitors. In some cases the suppressant material eventually has inhibitory effects on the species producing it; it is then called an **autotoxin.**

Attractants may serve an aggressive function as in the case of attractants that lure insects to carnivorous plants, or a nonaggressive one, as in flower scents that bring insects to flowers. The attractant scent may benefit both the flower that is pollinated and the insect that obtains pollen or nectar. In this case the whole interaction is a case of mutualism. But when mosquitos use the lactic acid on our skin as a signal to bite, the attractant is only to their advantage.

The chemical signals emanating from an organism may be responded to by avoidance rather than attraction. Doubtless some poisonous plants and animals have odors that potential foragers can detect and avoid. From the other direction, predators may have their own distinctive odors that potential prey can use to advantage. The broad-headed skink, a good-sized lizard of the southern United States, can distinguish between the scent of king snakes, that eat lizards, and hognose snakes, that don't (Cooper 1990).

The same evolutionary and coevolutionary dances are being performed at the chemical level as at other levels. In a marvellously complex system, the female polyphemus moth (Figure 10-3) makes use of an airborne compound from growing red oak leaves as a signal that it is time to mate (Riddiford and Williams 1967). She then releases a pheromone received by the male polyphemus. The result is that larvae are produced at a time when the oak foliage is suitable for them to feed on.

Figure 10-3

In the presence of new red oak leaves, the female polyphemus moth releases a pheromone that attracts males. (Photo by George and Betty Walker.)

Often, three trophic levels (plant, herbivore, and predator or parasitoid) are involved in chemically mediated interactions. Plants of the mustard family produce mustard oils that are irritants to most animals, but some insects have evolved the ability to feed on the mustard plants and some even use the oils as cues for locating the plants. A wasp that parasitizes an aphid that feeds on the mustard uses the scent of mustard oil to find the aphids.

When corn seedlings are damaged by being fed on by larvae of a certain lepidopteran—but not just from general mechanical damage—the corn plants release volatile oils. A parasitoid of the caterpillars is attracted by the released chemicals. Release of the volatile oils can be induced experimentally by treating the corn leaves with oral secretions of the caterpillar (Turlings et al. 1990).

Because humans are so large and depend so strongly on sight, we have very little appreciation of the environment of plant and animal chemicals around us. We move through a world of chemical scent trails of ants, moth sexual attractants, poisonous plants, and the varied scents of dogs, cats, rabbits (and people), and we are almost totally unaware. Only when we smell flowers, use coffee, tobacco, or cinnamon (in doses that would kill a grazing insect), or perhaps take our dog to a new neighborhood and watch it explore with its nose instead of its eyes, does this world begin to infringe on our consciousness.

Species Composition

At any one time, a particular region has a pool of plant, animal, and microbe species. A given site with a given set of physical features, such as moisture conditions and soil type, has some subset of these species, and this subset tends to recur, but is not precisely repeated, on other sites with similar physical conditions. If we experimentally remove some segment of the community, such as the fish (Meffe and Sheldon 1990) or the birds (Hensley and Cope 1951), we are apt to find that the reassembled community on that site has about the same species in about the same proportions as before.

Few ecologists would disagree with these statements, but the next steps in elaborating these ideas would probably lose the consensus. Is the group of species simply those that found their way there and for whom the environment is tolerable? There are those who make this claim (Gleason 1926, James and Boeklen 1984), but to do so, "environment" must be defined so as to include everything external to the species. For example, beech-drops do not live in a forest unless there are beeches for them to live on, and cowbirds need at least one bird species with open nests and that will rear their young. These are coactions with community associates, but which coactions are important and how much? To what degree does current competition restrict some species from occupying sites that are otherwise tolerable? What are the roles of other coactions—predation, parasitism, mutualism? At any one time, what is the role of chance events? Which species got there first, for example?

We can also take one step further back and ask, What is the evolutionary origin of the habitat preferences, food habits, and other traits that bring organisms to certain sites and allow them to fit together in a community?

Just how similar are the communities on similar sites? When we speak of beech-maple forest, a certain constellation of species comes to mind—spicebush, red-berried elder, spring beauty, trout lily, etc. An experienced naturalist might readily think of dozens of species that are part of the beech-maple community, but he or she would not expect to find every species in every beech-maple stand.

Plant ecologists of the Zurich–Montpellier school (Braun-Blanquet 1932, Mueller-Dombois and Ellenberg 1974) developed a method designed to get at the question of community variation. They subjectively choose examples of a community and assemble an association table with species names along the left and stands across the top. The body of the table indicates the presence of each species by stand. From the table, they are able to identify species of high **constancy** (that is, ones that occur in a high percentage of stands). By reference to similar data from other communities, they can identify species of high **fidelity** (that is, ones that tend to be restricted to this particular association or community).

Table 10-1 is a simplified and fictitious association table for beech-maple forest. American beech and may-apple are species of high constancy, being present in every stand. Without more information, we cannot categorize species as to fidelity, but may-apple occurs in other forest types and so may have

Table 10-1
Simplified and Fictitious Stand Table for Beech-Maple Forest

X means the species is present in the stand.

	Stand Number				
Species	*1*	*2*	*3*	*4*	*5*
American beech	X	X	X	X	X
May apple	X	X	X	X	X
Beech drops	X		X	X	
Squirrel corn		X	X		X
Doll's eyes	X	X			X
New York fern		X	X	X	X
Wild leek			X	X	
Yellow trout lily	X		X	X	X
Smooth yellow violet	X				X

low fidelity to this association. The Zurich–Montpellier fidelity categories range through five degrees from exclusive species *(treue),* ones almost totally confined to the association, down to accidentals *(fremde).*

Subjectivity and statistical deficiencies are problems with the Zurich–Montpellier approach; it is, nevertheless, handy for descriptive purposes and focuses attention on important community attributes related to species composition.

Some of the questions asked in the preceding paragraphs are dealt with in later sections of this chapter.

Spatial Structure

Ecosystems generally have a noticeable vertical structure; that is, they have layers or strata. Many ecosystems show two broad trophic strata, an upper autotrophic and a lower heterotrophic. In a forest most food-making activities occur in the upper levels where the leaves are concentrated, and most consumption and decomposition occur on or beneath the forest floor. In a lake also, food production occurs mainly in the upper part of the water where light can penetrate. On the bottom are large numbers of animals and bacteria that live on the dead material that drifts down from above. The layers are not separate and distinct; herbs make food on the forest floor, and caterpillars and birds forage in the canopy.

Within or overlapping these trophic strata other structural strata may be distinguished. In many forests it is convenient to recognize the following layers, going downward: trees, shrubs, herbs, floor, and subterranean.

Some communities may lack some of these or may have other strata. In bog forest, for example, there are two herb strata, a low stratum of such plants as partridgeberry and goldthread and a high stratum of skunk cabbage leaves and ferns (Brewer 1966). Many temperate deciduous forests have a subcanopy tree stratum 5 to 10 m high, made up of such species as flowering dogwood, redbud, or blue beech. A possible physical basis for this stratum is discussed in Chapter 11 in the section on sunflecks. Tropical rain forest may have three tree strata, but herb and shrub layers are poorly developed (Chapter 17). A tree layer is absent from prairie, but underground layering is complex, based on depths reached by the roots of different species (Table 10-2).

Organisms may divide up the habitat in a more complex manner than by just occurring in layers. Three different species of nuthatches live together in the ponderosa pine forests of Colorado in the winter. One of these is the familiar white-breasted nuthatch, generally seen scrambling around a tree

Table 10-2
Underground Stratification in Prairie

The University of Nebraska student of the prairie J. E. Weaver specialized in the relationships of plants to the soil. He recognized three underground layers based on the depths to which roots penetrated.

Layer	*Depth (m)*	*Percent of Species*	*Examples*
Shallow	0.6	14	Blue grama, June grass
Medium	0.6–1.5	21	Needle grass, buffalo grass, many-flowered aster
Deep	1.5–6	65	Big bluestem, slough grass, compass plant

Based on Weaver, J. E., and Clements, F. E., *Plant Ecology,* 2nd ed., New York, McGraw-Hill, 1938, p. 315, and other publications by Weaver.

trunk probing for food in crevices in the bark. This is the way it also behaves in the Colorado pine forests. A second slightly smaller species, the red-breasted nuthatch, does most of its foraging on large branches of the trees, and a third, still smaller species, the pigmy nuthatch, obtains most of its food from small branches and clusters of pine needles. Such division of resources presumably reduces interspecific competition (Chapter 8), and is part of the differences between the ecological niches of the three species.

Synusia and Guild

A **synusia** (DuRietz 1930) is a subdivision of a plant community consisting of all the plants of the same life form. Often synusiae correspond to the layers of the community, as, for example, the canopy trees of a forest or the mosses of a bog; however, they may not form an actual layer. The larger epiphytes of a tropical rain forest might form a synusia, and the **epiphylls,** the algae and lichens that grow on rain forest leaves, might form another.

A **guild** (Root 1967) is a group of species that "exploit the same classes of environmental resource in a similar way," which generally means that they eat similar foods. The frugivores of a tropical rain forest or leafhoppers on sycamore might be considered guilds. The guild concept was discussed more fully in Chapter 8.

As a catchword for a group of species sharing a resource and, hence, possibly competing, the term has its uses. It has occasionally been applied to plants, sometimes where "synusia" would have been a better choice, but sometimes with logic as, for example, in describing the plants that depend on dirt mounds in prairie.

Temporal Structure

Day–Night Change

Most organisms have daily cycles; organisms that are active in the daylight hours and asleep or inactive at night are **diurnal,** and those that are active at night and inactive by day are **nocturnal.** Some organisms are most active at dawn or dusk or both; these are said to be **crepuscular** (Figure 10-4). Most crepuscular animals, such as the whippoorwill, show increased activity during times of bright moonlight (Mills 1986). On the other hand, many truly nocturnal animals, such as bats and moths, are **lunarphobic;** they become less active at such times.

These broad patterns of activity in the sense of being awake and running about are not the only 24-hour cycles that may be occurring in an organism, however. There may also be daily patterns of physiological activity such as the production of new cells or the secretion of a particular enzyme. Many of these cycles, whether of wakefulness, locomotion, or more subtle activities, will persist even dur-

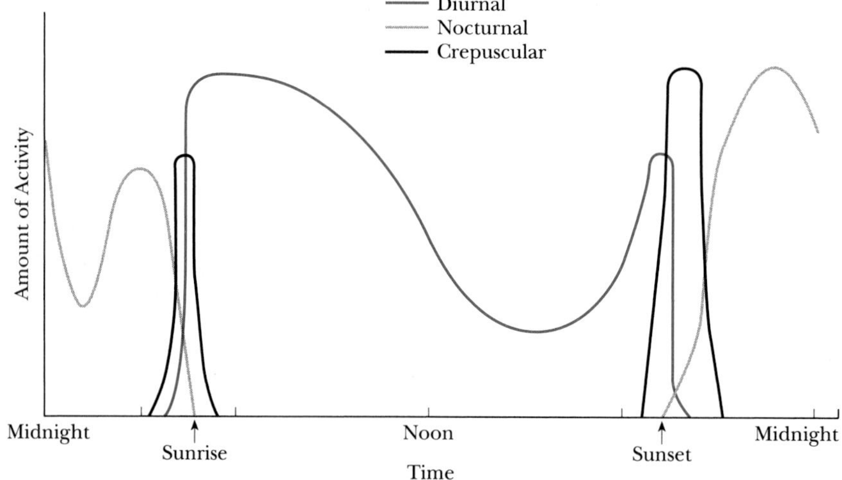

Figure 10-4

Generalized diagram of activity cycles of diurnal, nocturnal and crepuscular animals during the 24-hour day.

ing an experiment when the organism is placed in an environment with no alternation of light and dark, and with all other controllable factors such as temperature and humidity held constant. Daily cycles that continue under these constant conditions are called **circadian rhythms** (Pittendreigh 1960, 1981).

Humans are strongly diurnal creatures. Even those "night people" who sleep until noon and work or watch late movies until the early hours do not go walking through the forest at night; humans are just not built for it. Consequently few of us are aware of the day–night pattern of ecosystem organization.

Physical conditions at night are different—it is not just darker but cooler and more humid. The animals that are encountered at night are different. Except for owls and whippoorwills (and their relatives) birds are asleep. On the other hand mammals are abroad. This is, of course, true for bats; some of their adaptations to the night were described in detail in the section on coevolution (Chapter 5). It is also true for many other mammals. According to the Northwestern University ecologist O. Park (1940), one of the few students of nocturnalism, 60 or 70% of forest mammals are nocturnal. The types of invertebrates that are active are different too. The moist-skinned earthworms and slugs are out, lightning bugs may be flashing, if it is early summer, and crickets and katydids may be calling (by rubbing their forewings together) if it is late summer.

Plants are not photosynthesizing at night, but they are respiring and are translocating downward the food produced in the leaves during the day. Still other events are occurring; some plants show "sleep movements" in which the leaves fold up at night. Although most plants open their flowers either diurnally or throughout the entire 24-hour day, some species only open their flowers at night. Most of these flowers are moth pollinated and are white, which makes them more conspicuous in dim light.

In any case many nocturnal animals lack color vision. Nocturnal species of birds and mammals tend to have more rods and fewer cones in the retina of their eye as compared with diurnal species. Rods are stimulated by lower intensities of light so vision becomes more efficient in near darkness. Warning colors for animals likely to be preyed on at night are black and white, not red or orange. The skunk is an example (see Figure 10-5).

There are other features that nocturnal organisms have that seem to be adaptations to low light levels. If a light is shined on a raccoon or a whippoorwill, its eyes will glow. Eye shine is characteristic of many nocturnal species but few diurnal ones. It is light reflected from a layer in the back of the eye (the tapeta lucidum) and probably aids vision by, in effect, running the light through the retina twice, rather than just absorbing it at the back of the retina.

There is no single evolutionary answer as to why organisms are nocturnal or diurnal. For slugs and salamanders, it may be no more than a matter of high humidities at night being more favorable.

For many organisms, evading diurnal predators may be important. For example, a survey in England gathered 420 observations of bats flying in the daytime. Attempted or successful predation, mostly by hawks, was observed in 13 of these examples of daytime flight. The compiler (Speakman 1991) calculated that a bat could expect to be the object of a predatory attack once a day (once in 14.3 hours) if it flew in the daytime.

Avoiding a diurnal competitor might also be an important ultimate factor in nocturnality. This would most likely be of importance between organ-

Figure 10-5

Warningly colored nocturnal animals are black and white, like this striped skunk. (Alan G. Nelson/Dembinsky Photo Associates) isms in which interference in the use of resources was involved. If competition is solely by exploitation, it generally makes no difference to species A whether species B eats up their joint food supply at night or in the daytime; it is still gone. The British bats mentioned in the preceding paragraph were also occasionally harassed by birds, mainly swallows, that have a similar diet.

Seasonal Change

Most ecosystems show seasonal changes in structure, appearance, and function that are dependent on seasonal changes in the physical environment, especially temperature, precipitation, and photoperiod. As a community the tropical rain forest shows little in the way of seasonal changes. "All year round," wrote one student of tropical forests, P. W. Richards (1952), "the foliage is the same sombre green and in every month some species are in flower." But even in the tropical rain forest the individual species tend to be periodic in their activities, and for many activities the period may be approximately annual.

In Deciduous Forest Seasonal changes in temperate deciduous forest are probably as marked as anywhere on the earth. Six recognizable seasons, each blending into the next, and some of their major events are described in this section.

Early Spring (Prevernal Aspect) The buds of trees begin to enlarge; maple sap is running. The earliest spring flowers, such as harbinger-of-spring and skunk cabbage, bloom (Table 10-3). The earliest spring migrants, such as red-winged blackbirds and eastern phoebes, return. On sunny mornings, cardinals and tufted titmice sing. The big owls, barred and great horned, are nesting. Spring peepers begin to call in the marshes.

Late Spring (Vernal Aspect) The later spring flowers, such as white trillium and May apple, bloom, as does flowering dogwood. Trees leaf out, maples early, ashes and oaks late. This is the peak period of bird migration, but many birds are already beginning to nest. There is a local migration of insects from over-wintering sites in the forest floor to summer habitats. The spring overturn occurs in lakes, and fish run upstream for spawning.

Table 10-3
When Do the Spring Flowers Bloom?

Long records of the dates of seasonal events are scarce. Probably the longest record was kept by Thomas Mikesell. Around the small town of Wauseon in the northwestern corner of Ohio, he kept track of the dates of such events as flowering, leafing, and fruit ripening for many species of plants from 1883 to 1912. His average dates for the first blossom of some common spring flowers are given below.

Species	*Average Date of First Blossom*
Hepatica	April 13
Spring beauty	April 17
Bloodroot	April 18
Yellow trout lily	April 23
Pepper root	April 23
Early blue violet	April 25
Dutchman's breeches	April 28
Yellow violet	May 2
Wild blue phlox	May 2
Wake-robin	May 5
Jack-in-the-pulpit	May 14
May apple	May 16

From Smith, J. W., "Phenological dates and meteorological data recorded by Thomas Mikesell at Wauseon, Ohio," U.S. Department of Agriculture, *Monthly Weather Rev. Suppl.*, 2:23–93, 1915.

Summer (Aestival Aspect) This is the height of bird nesting as most species finish incubation and begin feeding the young (Figure 10-6). Consumption of tree leaves by caterpillars (and caterpillars by birds) peaks. Flies and mosquitos are numerous. Many forest flowers, the spring ephemerals such as Dutchman's breeches, trout lily, and spring beauty, die back entirely above ground and rest as corms or tubers until the next spring.

Late Summer (Serotinal Aspect) The roadsides are colorful with many open country plants blooming—black-eyed Susan, chicory, and turk's-cap lily. Buck-eye leaves die and fall. A few birds such as the goldfinch and the cedar waxwing are just beginning to nest, but family groups of many species are wandering about, and there is little territoriality and the singing that goes with it. Young spiders and toads

Figure 10-6

The nesting of birds is the hallmark of the aestival aspect. Shown here is a young yellow warbler in Kalamazoo County, Michigan. (Photo by John C. Stiner.)

are abundant. Cicadas emerge from the ground and become vocal.

Fall (Autumnal Aspect) The late open country flowers such as blazing stars and asters bloom; this is also a time of flowering in the marsh where New England aster, fringed gentian, and grass-of-Parnassus may be found in bloom. In the forest, nuts such as acorns and beechnuts ripen. Many of the herbs have ripe fruits of the beggar's-tick or stick-tight variety. Tree leaves turn color and fall, with willows, oaks, and tamaracks among the last. Fall bird migration is at its heaviest. There is a migration of forest insects down into the forest litter and soil, and a migration of open country insects into the litter and soil of the forest edge.

Winter (Hiemal Aspect) Most plants are dormant. Only winter and permanent resident birds are around; most of these gather in flocks. Also still active are most aquatic animals and some mammals such as mice, shrews, and foxes (Figure 10-7). Otherwise most animals are hibernating, such as many invertebrates and the woodchuck, or sleeping, such as the chipmunk.

Other Ecosystems In many tropical and subtropical ecosystems wet and dry conditions are more important in seasonality than warm and cold. This is true for the deciduous forests of the tropics, such as monsoon forests, and for many deserts (Table 10-4). In the more extreme deserts some rainy season events may not recur every year. In some years, possibly several in a row, not enough rain will fall to allow the growth of certain plants or to bring some animals out of the rest period (aestivation) in which they endure the drought.

Aquatic communities also show seasonal change. Even hot springs, which have constant temperature and salt concentration, show changes related to such factors as seasonal differences in sunlight. In the far north, light may be so dim and brief that algae die back because they use more food in respiration than is produced by photosynthesis. Farther south, the algae may not change much in biomass but may increase their chlorophyll content, evidently compensating for the decreased light. One of the best known sequences of seasonal change in an aquatic situation involves the thermal stratification of lakes (see the section on aquatic habitats in Chapter 18).

Phenology

Hopkins' Bioclimatic Law The scientific study of seasonal change is called **phenology,** perhaps the earliest branch of ecology. A Chinese calendar from the Hoang Ho Valley giving the dates of biological events is known from 700 B.C. (Shelford 1929). Primitive humans often planted crops according to the flowering times of wild plants. But despite its ancient origins, phenology has received relatively little attention in ecology. One early effort was by the entomologist A. D. Hopkins (1920). He gathered a vast amount of material on the timing of seasonal events to answer the question: How fast does the spring move north? Biologists indulge less in giving formal names to generalizations than chemists, but you may refer to Hopkins' bioclimatic law if you wish.

Hopkins stated that in the spring events occur about four days later for every degree of latitude northward, meaning that spring moves northward at about the rate of 17 miles a day. It means that, compared with Minneapolis-St. Paul, a given event of the spring occurs two weeks earlier in Des Moines, 26 days earlier in St. Louis, and almost a

Figure 10-7

A feature of late hiemal aspect or earliest prevernal is the opening of burrows and other potential den sites by red foxes. The fox pair prepare a grass-lined natal den in which the pups are born in the vernal aspect (March or April). (Photo by the author.)

month and a half earlier in Little Rock. Altitude also has its effect, with events occurring a day later for every 100 feet of elevation. In mountainous regions you can go from spring back to winter in a morning's hike from a valley up one of the peaks. Longitude is also significant, with events occurring one day later for every 1¼° of longitude eastward.

To summarize, events tend to be four days later for every 1° of latitude further north, every 400 feet (122 m) of altitude higher, and every 5° of longitude eastward, in the spring. The situation is reversed in autumn; however, much less information has been summarized for that season.

A study of the return of birds in the spring at three Missouri locations (Elder and Dierker 1981) found a tolerably close fit between actual arrival dates and predicted dates using Hopkins' law, but the fit was still closer if altitude was omitted (Table 10-5). With organisms as mobile as birds, a 100-m difference in elevation is evidently unimportant. Altitude is, however, strongly correlated with the flowering time of dogwood in the Great Smoky Mountains (Lieth and Radford 1971).

Phenology is affected by microclimate; spring flowers bloom earlier in warm microhabitats such as south slopes and protected coves (Gaddy et al. 1984). Early spring is different from late spring, with the timing of early events tending to be more variable (Leopold and Jones 1947). Every species (Figure 10-8) and locality have their own peculiarities. Despite these deviations, Hopkins' law is still a widely useful rule of thumb.

Ultimate and Proximate Factors. Hopkins' law presumably reflects ultimate factors important in the timing of life history events. As an empirical for-

Table 10-4
Seasonal Change in the Sonora Desert

In contrast to the six seasons in temperate deciduous forest, the Sonora desert in the vicinity of Tucson has four seasons.

Season	*Weather*	*Biological Events*
Winter wet (December–March)	Widespread rains totaling 5–8 cm; cool weather, last frost mid-March.	Vegetation shows signs of activity in January. Many plants, both winter perennials and winter annuals, bloom in February and have mature fruits in March and April. Bird nesting begins.
Dry foresummer (April–June)	Total precipitation about 2.5 cm; high temperatures.	Mesquite leafs out; however, many winter perennials lose their leaves during the course of the season and do not bear leaves again until the following January. Succulents—cacti (including saguaro), yuccas, agaves—bloom in profusion. A peak of bird nesting, including most of the desert species. Also heavy migration of birds which nest further north.
Humid midsummer (July–September)	Greatest precipitation of year in July and August, coming, however, in local heavy thunderstorms; temperatures very high—July average over 85°F (about 30°C).	Desert transformed by growth and flowering of vast number of summer annuals. Grasses and some succulents such as the great barrel cactus flower. Spadefoot toads breed. A second peak of bird nesting, including the rufous-winged sparrow and blue grosbeak. Heavy bird migration continuing into November.
Dry aftersummer (October–November)	Little rain; warm to cool. First frost mid-November.	The grasses and some other plants flowering in the humid midsummer ripen seeds. A nearly complete cessation of vegetative activity.

Based mainly on McDougall, W. B., *Plant Ecology*, Philadelphia, Lea & Febiger, pp. 235–236, 1941; and Marshall, J. T. 1963, "Rainy season nesting in Arizona," Proc. 13th Internat Ornith. Congr., Vol. 2:620–622.

Table 10-5
A Comparison of Spring Arrival Dates of Birds at Three Locations in Missouri

The expected differences between each pair of sites are calculated based on Hopkins' law.

	Expected Difference		
Comparison	*Using Latitude, Longitude, and Altitude*	*Using Only Latitude and Longitude*	*Observed Difference*
Hannibal versus St. Joseph	0.5 days earlier	2.8 days later	3.9 days later
St. Louis versus St. Joseph	4.6 days earlier	0.7 days earlier	0.1 days earlier
St. Louis versus Hannibal	4.0 days earlier	3.5 days earlier	3.9 days earlier

From Elder, W. H., and Dierker, W. W., "Do birds follow Hopkins' bioclimatic law?" *Inland Birdbanding*, 53:33–38, 1981.

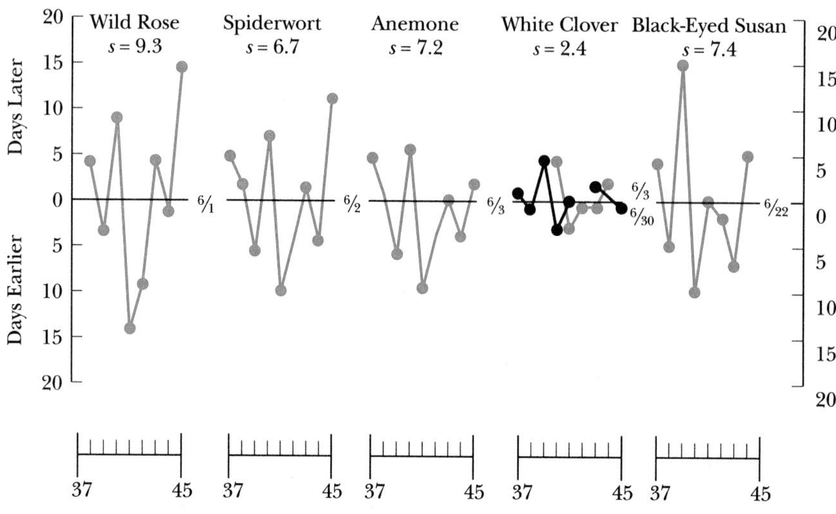

Figure 10-8

Shown here are the dates of first bloom of five plants that ordinarily flower in June. *S* is standard deviation. The straight line is the average date of first bloom and the circles are the actual date in a given year between 1937 and 1945. A considerable amount of variation from year to year is usual, presumably based mainly on the weather of the preceding weeks or months. An occasional species, however, shows much less variability. Such is the case with white clover. It is possible that photoperiod plays a greater role in determining flowering date in such species. (Adapted from Leopold & Jones 1947.)

mula, based on when events actually occurred in different localities, it integrates—or, at least, intermixes—whatever factors are important in the survival of the various species. Clearly some aspect of temperature is a major component. Isotherms move northward in the spring (and southward in the autumn) in a manner similar to Hopkins' law, and the altitudinal delay could scarcely be anything other than temperature based.

This is not to say that the ultimate factor may not be something that is dependent on temperature, rather than temperature itself. Many male birds seem to return to their breeding grounds as soon as they can expect to find food. Insect emergence, rather than temperature itself, presumably is the important factor for the birds. Starvation is a selective force opposing earlier arrival. Such factors as losing out on territories or females and the need to bring off young at the time when food supply is ample oppose a more leisurely return.

Natural selection seems to have scheduled the activity of the spring-flowering forest herbs to occupy the narrow space between the dual rocks of prevernal frost damage and lack of pollinators and the hard place of dense shade from the late-spring canopy.

The significance of imported seasonality has been underappreciated in the study of community organization. The presence of dispersers may be the prime evolutionary agent in the autumn scheduling of fleshy fruit ripening in temperate regions (Thompson and Willson 1979). If so, this schedule is based on an imported event, the arrival of vast numbers of migrant birds from forests to the north. When the flood of birds arrives in the tropics for winter, a strong seasonality is imparted to an ecosystem where there is little seasonality of physical factors. Streams, even if they are spring fed and have little seasonal change otherwise, show seasonal changes based on the import of organic matter in the form of leaf fall.

These are some of the ultimate factors involved in phenological events. The proximate factors, the environmental cues used in the scheduling, are diverse. They include temperature, photoperiod, moisture, and biotic interactions. For example, trees may require a cold treatment followed by warm temperatures to leaf out in the spring and may lose their leaves in the fall as the result of decreasing photoperiods.

Applications There are obvious practical functions of phenological data that, through most of the past few decades, have been appreciated more by the general public than by most applied scientists. Hopkins published maps of the average dates of emergence of the Hessian fly which, in those prepesticide days, were used as a guide for planting winter wheat. If plants are present, Hessian flies lay eggs on them, and the larvae, feeding on the plant sap, greatly reduce yield. However, the adult flies are short-lived, so that simply waiting until just after they emerge to plant prevents damage to the crop.

It is sometimes possible to tease out an important proximate factor or two and, thereby, predict when some event will occur. One can measure temperature and calculate when some pest insect will be at a stage susceptible to control measures, for example (McNeil and Stinner 1983). Usually a composite measure of temperature and time is most useful for predicting the scheduling of a life history event for poikilotherms. The number of hours or days (above a threshold temperature) are multiplied by the hourly or daily temperature to yield a "heat sum." For example, longleaf pine requires just over 1200° days (above a 50° base) to reach peak flowering (pollen production). This heat sum was accumulated by the first week of March in 1975 in southwest Alabama, but not until the end of March in the cooler spring of 1970 (Boyer 1978).

Still better, perhaps, is to use easily observed phenological events as ecological indicators of other, economically more important, events. Utilizing "the organism as the instrument" (Hopkins 1920) has advantages because organisms that respond to more factors than one are likely to measure and also automatically take interactions into account. Planting corn when oak leaves are the size of squirrel's ears is not a bad way to avoid frost while getting the crop started early enough for the corn to ripen. Keeping track of what herbs are in flower may be as good a way of predicting when grasshopper damage will start as feeding temperatures into a computer (Table 10-6). It is certainly more fun.

Table 10-6
Plant Flowering Associated With Grasshopper Life History in Shortgrass Prairie in Montana

Grasshopper Event	*Plant Events*
Grasshopper eggs hatch	When low larkspur is just beginning to bloom, white point locoweed is in peak bloom, and three-leaved milkvetch is in the last stage of bloom.
Grasshoppers are in their third instar (just prior to start of heavy grazing losses)	When plains prickly pear is just beginning to bloom, salsify is in peak bloom, and miner's candle is in the last stages of bloom.

Based on Hewitt, G. B., "Plant phenology as a guide in timing grasshopper control efforts on Montana rangeland," *J. Range Management,* 33:297–299, 1980.

Ecological Niche

In the chapter on interspecific competition, we defined niche as the role a species plays in an ecosystem—its profession. In discussing niches, we are viewing the community as a unit made up of the niches of the various species composing it. A proposed list of bird niches in mesic deciduous forest is given in Table 10-7.

In comparing different but similar communities we are sometimes struck by the similarities of niches. In deserts, for example, there seems to be a succulent niche filled in North America by cacti and in Africa by euphorbias. According to E. P. Odum (1971), the "bottom-dwelling carnivore" niche is filled in different coastal areas around

Table 10-7
Bird Niches in Climax Mesic Deciduous Forest

Two dozen niches are listed, mostly based on layer and type or size of food. "Large" animals eat, on the average, larger food items than "small" but also tend to be more versatile, taking some smaller food items.

Niche	*Species*
Large invertebrates on forest floor	Wood thrush
Small invertebrates on forest floor	Ovenbird (north), Kentucky warbler (south)
Understory gleaner	Hooded warbler

There seem to be more niches at the floor-understory level in southern forests. In addition to the Kentucky warbler and the hooded warbler (itself much more common in the South), the worm-eating warbler and Carolina wren are floor-understory species there.

Niche	*Species*
Generalized forager on shrub and low tree invertebrates	Black-capped chickadee (north), Carolina chickadee (south)
Generalized forager on medium and high invertebrates	Tufted titmouse
Large canopy gleaner	Scarlet tanager (north), summer tanager (south)
Medium canopy gleaner, leaves	Red-eyed vireo
Medium canopy gleaner, twigs	Yellow-throated vireo
Small canopy gleaner	Cerulean warbler
Very large woodpecker	Pileated woodpecker
Large woodpecker	Red-bellied woodpecker
Medium woodpecker	Hairy woodpecker
Small woodpecker	Downy woodpecker
Tree-trunk prober	White-breasted nuthatch
Small, low static flycatcher	Acadian flycatcher
Medium, high static flycatcher	Eastern wood pewee
Large, static flycatcher	Great crested flycatcher
Moving flycatcher, low	American redstart
Moving flycatcher, high	Blue-gray gnatcatcher
Large seeds and fruits	Northern cardinal
Small seeds and fruits	Indigo bunting

Both the cardinal and the indigo bunting tend to be species of openings in the forest. Because openings are natural, recurring from such causes as windthrow, both species are logically considered forest species. The scarcity of seed-eaters may be related both to relatively low production of seeds and to high populations of seed-eating mammals such as deer mice and chipmunks.

Niche	*Species*
Large omnivore	Blue jay
Very large omnivore	Common crow
Large owl	Barred owl (wetter forests)
	Great horned owl (drier forests)
Nectar feeder	Female ruby-throated hummingbird

Some other possible niches that seem to support species only occasionally or at low densities are large browser (ruffed grouse), small owl (saw-whet owl), large hawk (red-shouldered hawk), and medium hawk (Cooper's hawk).

North and Central America by the spiny lobster (tropical), king crab (upper West Coast), stone crab (Gulf Coast), and lobster (upper East Coast). Species filling similar niches in different regions are called **ecological equivalents** or **ecological counterparts.** The physical resemblance between the equivalent species may be only slight (as between kangaroos and bison, considered to be counterpart grazers in North America and Australia), but can at times be remarkably close. The yellow-throated longclaw, a bird of African grassland, bears a very striking resemblance to our meadowlark but is not closely related to it.

It would, however, be a mistake to think that every deciduous forest or desert or tropical rain forest has the same niches. The occasional examples of spectacularly similar counterpart species are at least balanced by cases of similar communities in which resources are divided up very differently.

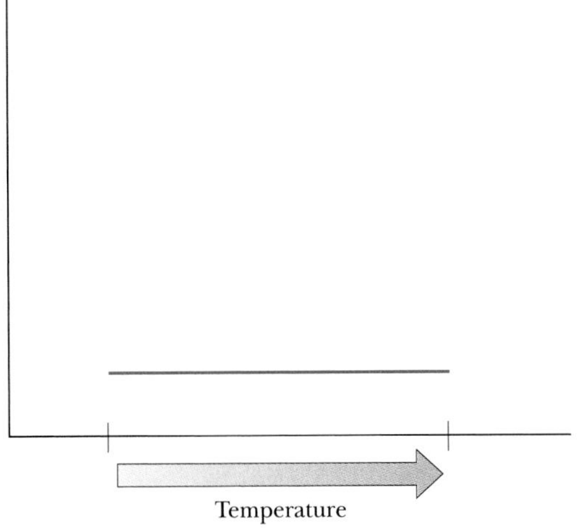

Figure 10-9

One dimension of an ecological niche.

The Niche as a Hypervolume

In this section is described the concept of the ecological niche as put forth by G. E. Hutchinson (1957) of Yale University, possibly the most influential ecologist of the mid-20th century in the United States. The general idea of the concept is given by the title of one of his essays, "The Niche: An Abstractly Inhabited Hypervolume." The intellectual atmosphere surrounding the development of the concept should be remembered—it was one in which the competitive exclusion principle was held as an important ecological rule. This concept of the niche was developed as a context for the competitive exclusion principle.

Suppose that we measure the range of some environmental variable (say, temperature) over which a particular species can live and reproduce (in effect, its range of tolerance) and we put this on a graph (Figure 10-9).

Suppose that we then do the same for another environmental variable (humidity, perhaps) and put this on the second axis of the graph (Figure 10-10). The space that is enclosed is a rectangle that represents two of the dimensions of the ecological niche of the species.

If we now erect a third axis for a third environmental variable (perhaps some nutrient), the space that is enclosed is now a volume in three dimensions (Figure 10-11).

If we now erect a fourth axis for a fourth environmental variable, the space enclosed is a hypervolume having four dimensions. Also, it gets hard to draw. We may include more and more dimensions, up to *n*, so that the niche as defined by Hutchinson has an infinite number of dimensions. This is the **fundamental niche** of the species. If the fundamental niches of two species overlap, then the

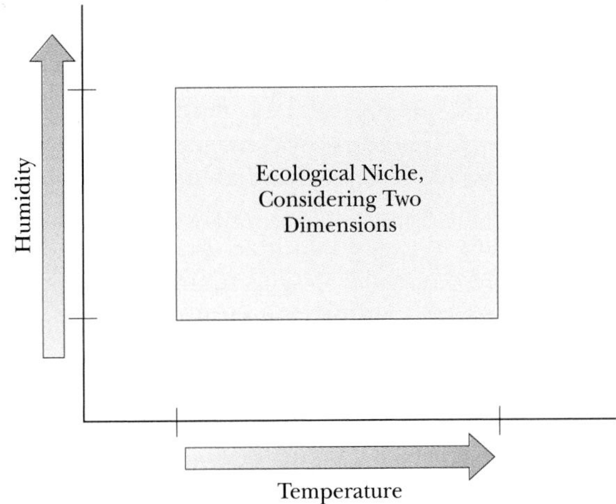

Figure 10-10

An ecological niche in two dimensions.

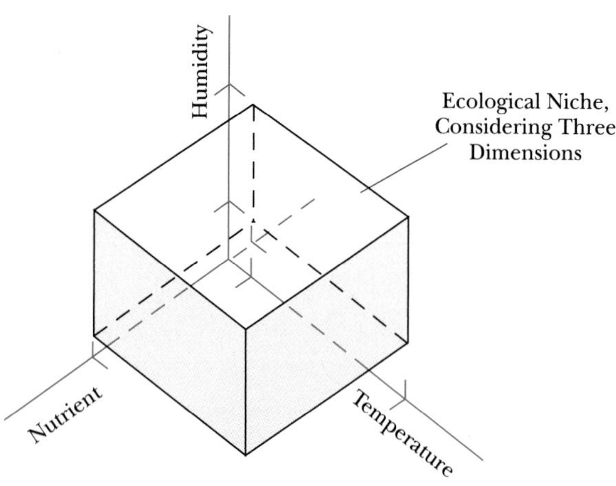

Figure 10-11

An ecological niche in three dimensions.

two species are competing. Note, however, that an apparent overlap in two, three, four, or ten dimensions may not mean any actual overlap; adding one more dimension may move the hypervolumes apart. (For an analogous situation, hold one hand a few inches in front of your eyes and one at arm's length. The two hands seem to overlap in two dimensions but you know that in three dimensions they are separated by a foot or so.) Two species, for example, do not overlap, or compete for food, even though they eat items of exactly the same size if they look for them in different places.

Hutchinson viewed the outcome of interspecific competition as either extinction or the development of differences allowing coexistence. Accordingly, he recognized a **realized niche,** a hypervolume bounded by the actual ranges, as found in the field, for each variable. To the degree that a species was excluded by competition from situations it was potentially able to use, its realized niche would be smaller than its fundamental niche.

Most well-integrated communities (such as a climax forest or a coral reef but possibly not a first-year weed field) would be made up of species with nonoverlapping niches. The defunct broken stick model of Robert MacArthur (1957), Hutchinson's best-known student, dates from this period. The community is represented by a stick and the species are segments formed by breaking the stick into lengths corrsponding to the abundance of each. It seems clear that during this period in the 1960s Hutchinson and MacArthur suspected that communities were made up of niches that fit together like the pieces of *n*-dimensional jigsaw puzzles (possibly resembling, in *n* dimensions, those little ball-shaped puzzles that sometimes come on key chains).

It now seems clear that some overlap can, and does, occur without extinction. Only if energy is included as a niche dimension is the broken stick model likely to be true for all species of all communities. Because a given quantum of energy is respired once and only once in its passage through the ecosystem, niches must be nonoverlapping in this sense.

If the niche concept has not proved useful in itself, then why not simply abandon it? First, the overall view of the community held by Hutchinson (and S. C. Kendeigh, among others) of largely complementary, noncompetitive use of resources seems basically, although not absolutely, right. Second, a given species does have a sort of abstract role in an ecosystem as defined by its relationships to the physical environment and other organisms. Ecological niche is a useful shorthand term for this concept. Such is the case for the work on niche breadth, which grew out of Hutchinson's concept of the niche but does not actually depend on it.

Partitioned Resources

Guilds of coexisting species generally are found to divide up the resources among them. A little earlier we mentioned how white-breasted, red-breasted, and pigmy nuthatches divide up the nuthatch foraging sites of the ponderosa pine forests. Three species of vultures are relatively common in southwestern Costa Rica. To study how the small-carrion resource is divided (to study, in other words, the differences in the ecological niches of these vultures), whole fish carcasses weighing 3 to 5 kg (around 7 to 10 pounds) were placed in three habitats (Lemon 1991). The habitats were primary (virgin or mature) tropical forest, second-growth forest, and open areas (gaps).

Turkey vultures tended to find the carcasses soon after the carcasses were put out, probably because of their well-developed sense of smell, and to use carrion in all three habitats. The king vulture, a considerably larger bird, also located the carcasses quickly but avoided the secondary forest with its dense understory. Black vultures did most of their feeding in open areas and also continued feeding on carcasses for a couple of days after the other species had quit.

Multivariate statistics have been widely used for studying such relationships (Carnes and Slade 1982). The general approach is to measure a variety of features of the abiotic and/or biotic features of the environment of a set of species—at the sites where they are found singing, for example, for birds—and then to use principal components analysis or discriminant function analysis to yield a smaller, abstract set of axes that define a space within which the species are arranged.

Figure 10-12 shows the locations (based on trapping) of three primarily herbivorous rodents in second-growth Tennessee forest relative to two discriminant function (DF) axes (Dueser and Shugart 1979). DF 1 is correlated strongly with the evergreenness of the overstory; low values indicate domination by broad-leaved trees, high values by pine. DF 2 seems to be a measure of "forestness." Low values indicate the open, nonbrushy understory characteristic of a more complete forest cover; high values indicate an area with denser shrubs under a more open canopy.

The upshot is that the golden mouse occupies vegetation that is structurally similar to that used by the white-footed mouse and the eastern chipmunk but occupies pine, not broad-leaved forest. The niches of the chipmunk and the white-footed mouse are similar, although the chipmunk is more accepting of a brushy understory. It should be realized that data were collected for only a limited number of vegetational and soil variables. If other niche variables, such as where food was sought (underground, surface, in trees) or the size of food items, had been added, different discriminant function axes would probably have resulted, and more aspects of the ecological relationships of these animals to one another might have become evident.

A similar study (Noon 1981) examined the habitat relationships of the five brownish thrushes of North America—the wood, Swainson's, gray-cheeked, and hermit thrushes, and the veery. All five species occur in the mountainous parts of New England. Here four of them live (DF 1) in rather similar forest considering canopy cover and height and the percent of conifers in the forest (Figure 10-13). The fifth species, the gray-cheeked thrush, occurs mostly at high elevations in stunted, open conifer forest. Along DF 2 (correlated with shrub density and ground cover), the other four species

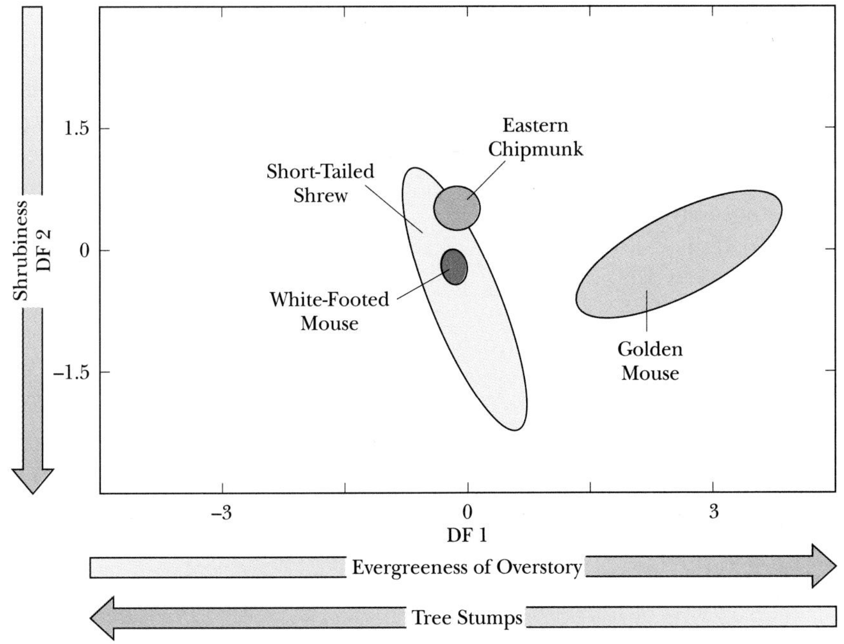

Figure 10-12

Habitat relations of four mammal species in Tennessee forests. Some habitat variables that were significantly correlated with the axes are shown by arrows. The ellipses represent the 95% confidence interval around the mean position of the species located with reference to discriminant function axes 1 and 2. Based on Dueser and Shugart 1979.

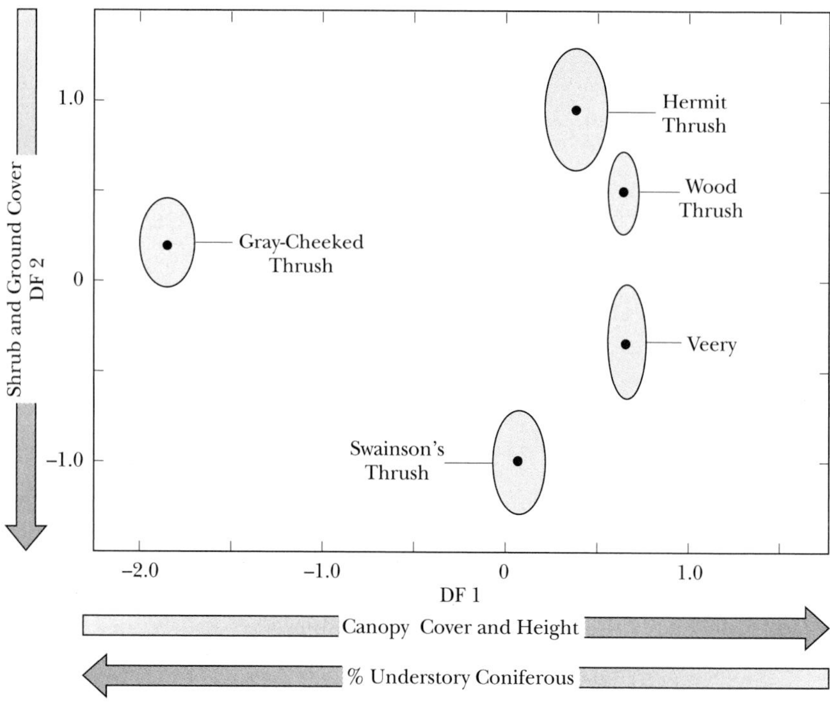

Figure 10-13

Habitat relations of thrushes in New England. Some habitat variables that were significantly correlated with the axes are shown by arrows. The ellipses represent 95% confidence intervals around discriminant function means. Based on Moon 1981.

show more separation than along DF 1. For example, the veery occurred on shrubbier sites than the wood thrush.

In the Great Smoky Mountains, 1400 km to the south, only the wood thrush and the veery occur. Here the wood thrush occupies about the same place on DF 1 as in New England, although it occupies a broader range along the axis. The veery, however, occupies a broader range and also shifts the center of its habitat occurrence far to the left along DF 1, in the direction of the habitat conditions used by the hermit and Swainson's thrushes in New England. Noon (1981) interpreted these results as "ecological release," that is, in New England, the presence of the other thrush species prevents the veery from making use of such sites.

What Produces Community Organization?

A spectrum of answers to the question of how community structure, or organization, is produced has been offered. Most current answers fall in or near three regions of the spectrum. These three answers are (1) coactions are unimportant in producing community structure; (2) some coactions are important, but interspecific competition is not, and (3) coactions, especially interspecific competition, are major determinants of community structure.

1. *Coactions are unimportant in producing community structure.* In its pure form, this view suggests that any given community consists of species that have evolved apart from one another and now occur together solely because their tolerance ranges and life histories allow them to do so. To any one species, all other species are either food or environment. H. A. Gleason (1926), who is as closely associated with the viewpoint as anyone, wrote: "Every species of plant is a law unto itself, the distribution of which in space depends upon its individual peculiarities of migration and environmental requirements. The vegetation of an area is merely the resultant of two factors, the fluctuating and fortuitous immigration of plants and an equally fluctuating and variable

environment. [A plant] association is not an organism, scarcely even a vegetational unit, but merely a coincidence."

The pure form of this view is not widely held because it is incredible in light of the clear importance in the structure of many communities of coactions, evolution in a community context, and coevolution, all discussed earlier. Two milder variants of this viewpoint are fairly commonly held. One variant holds that, except for a few spectacular and universally agreed on cases of obligate mutualism, parasitism, etc., coactions are unimportant. Because new cases, as they gain acceptance, can always be added to the list, this viewpoint cannot be disproved.

Another variant agrees that evolutionary processes in the past may have shaped the way species now fit together into communities but rejects the idea that this tells us anything useful for ecological purposes. This view has a long and honorable history, going back at least to H. C. Cowles and V. E. Shelford. It is a question, on the community level, of proximate and ultimate factors. Some ecologists may be willing to lump frugivores with gravity and wind as "dispersal agents." Others may find this viewpoint incomplete and consider the question of how the relationship came about an equally important aspect of ecology (Brewer 1967).

It is possible to believe that evolutionary factors are not just irrelevant but meaningless in answering current questions about the distribution and abundance of organisms. If one believes that most species now occur in assemblages different from those with which they spent earlier parts of their evolution, then what evolutionary factors shaped the diet of a species may be immaterial as well as unreconstructable.

Under this viewpoint, the features that allow organisms to fit together into communities today are not adaptations but preadaptations (or "exadaptations"), originally evolved in other circumstances. Perhaps the spring ephemerals of forests are spring ephemerals, not because they evolved under a light regime produced by forest, but because they evolved in some grassland habitat where spring growth was advantageous because of summer drought.

The current generation of palynologists provide support for this viewpoint by emphasizing the reassortment of species that occurred during the million or so years of Pleistocene. The pollen record suggests that large areas of the landscape were occupied by assemblages with no precise current counterparts (Chapter 14). Still, many pairs, trios, and quartets of species doubtless did continue to co-occur through most or all of the Pleistocene. Besides, many of the associations that we see in the current interglacial period probably began sometime during the 70 million years of the Tertiary, a long and stable time compared with the Pleistocene.

2. *Some coactions are important in producing community structure, but competition is not.* This merges with some of the less extreme forms of viewpoint 1, but competition is singled out as the coaction that is rare or nonexistent. Otherwise, a miscellaneous collection of viewpoints is included, some favoring predation (or herbivory, parasitism, or disease), some abiotic factors, and some the behavior of the individual species as the feature preventing numbers from reaching levels at which interspecific competition would become important.
3. *Coactions, especially interspecific competition, are major determinants of community structure.* Many proponents of viewpoint 2 agree that coactions and coevolutionary processes based on predator-prey, herbivore-plant, parasite-host, and mutualistic relations have been, and currently are, important in structuring communities. They diverge on whether the "horizontal" coaction of competition for some limiting resource is important. One of the more controversial topics in the 1980s was this dichotomy; accordingly, it is worth looking at it in some detail.

The Importance of Competition

Viewpoint 3 has a long history in ecology, beginning perhaps with Darwin (1859). Some critics seemed to see it as an invention of Robert MacArthur (1958, 1972), who creatively explored many of its ramifications; however, it is explicit in the writings of Forbes (Forbes and Richardson 1920), Clements et al. (1929), and Kendeigh (1945, 1947),

Robert H. MacArthur *(American, 1930–1972)*

Fame in the world of ecology came early to Robert MacArthur. So did death; he was 42 years old when he died. His Ph.D. thesis (MacArthur 1958) won a prestigious award from the Ecological Society of America. Between 1955 and 1974 over 50 articles and books appeared under his name, and nearly all set off reverberations in the research of other biologists. Some recurring MacArthurian themes were the role of competition in producing community structure, the ecological context of natural selection, the evolution of life histories, the basis of species diversity, island biogeography, and optimal foraging. It is no accident that these have also been some of the major topics of the whole field of ecology during the past 30 years.

MacArthur's boyhood was spent in Canada and Vermont. As an undergraduate, he majored in mathematics at Marlboro College in Vermont. After receiving a master's degree in mathematics at Brown University, he went to Yale. There the combination of an interest in natural history dating from childhood and the influence of G. E. Hutchinson caused him to switch from mathematics to ecology. His professional career remained in the Ivy League, first at the University of Pennsylvania and then Princeton.

MacArthur's contribution to ecology is partly his ideas but also his approach. He brought to ecology something of the methods and the excitement then current in molecular biology. He would produce relatively simple mathematical or graphical models of how he thought nature was constructed. From these he generated predictions that could be tested using field data. Philosophers of science have called this way of doing science the **hypothetico-deductive approach.** It was the way biologists were expanding knowledge of the structure and workings of the gene, but it was rare in ecology up to this time. In fact, any explicit hypothesis testing was rare in ecology in the 1950s, except in the physiological branch.

MacArthur's work has attracted many critics, more than any other ecologist since F. E. Clements. Much of the criticism has a curiously personal tone, more than seems justifiable on the basis of disagreements about the role of competition in community structure or how important equilibria are in nature. The underlying issue seems to be that of style—obviously a much more serious matter. One critic suggested that MacArthur had been so influential because his approach seemed easy and fun to do. The tone of the suggestion was not approving.

MacArthur believed, with Einstein, that imagination is more important than information. A bright idea about how nature is constructed and a clever way of testing it were what interested him. Although he evidently enjoyed field work—searching for patterns that would generate more ideas about the way nature was constructed—the relative drudgery of piling up data in the quantity necessary to satisfy the average reviewer of the average journal was unappealing.

Also unappealing was searching the previous literature. MacArthur's view of community organization is in the tradition of the American ecological mainstream, but it is uncertain whether he thought about this, or even realized it. Neither Clements nor Shelford are cited in the population biology textbook that MacArthur wrote with Joseph H. Connell.

These are flaws. A scientist should know, or learn, the literature and should be willing to put in the hard work to assemble the required data. But data uninformed by theory—which characterized much of the ecology of the 1950s and 1960s (and 1990s)—is sterile. If we have to choose between ideas with few data and data with few ideas, money on the former is more likely to pay off in advancing knowledge.

among others. MacArthur's sources were primarily his advisor, G. E. Hutchinson, and the writings of Gause (1934), Huxley (1942), and Lack (1944, 1947).

We introduced this view in Chapter 8 in discussing the coexistence of competing species; here we develop it more fully. Suppose that a community occupying some region of moderate habitat variability consists of a single species on each trophic level. Suppose now that a second species on one of these levels, let us say a second herbivore, appears. This may be by invasion from elsewhere, although for some kinds or organisms, the sympatric arisal of a new species may occur. Several possibilities now present themselves:

1. The new species may be adapted to a portion of the habitat axis—dry areas, perhaps—that the first species does not occupy, and so may be able to establish itself immediately, with no changes in either species. We may think that this event would be rare because, in the absence of other species, the first species would have evolved to be a generalist.
2. The two species may overlap in their habitat distribution. One of the two may be so much more efficient that the other suffers competitive exclusion before any changes that might permit coexistence can occur.
3. The two species may overlap in their habitat distribution, but one possesses features that allow it to dominate in certain situations. One, for example, does better in moist sites and, by its aggressiveness, restricts the other species, which could live in those sites, to medium and dry areas.
4. The two species may overlap in their habitat distributions, but because of differential reproduction or survival, genetic changes in one or both occur, with coexistence resulting. This was diagrammed in Figure 8-6. For example, suppose that both species eat both leaves and seeds but that one is slightly better at seed consumption and one at leaf consumption. The members of each species that are predisposed to feed on the wrong food (wrong now in the context of competition though not wrong previously) will be less successful than the ones that feed on the right food. This is because members of the other species will be depleting their food supply and will be doing it more efficiently than they are. Consequently, genes for feeding on the right foods will accumulate in the two species, and they will diverge ecologically.

We see, then, three ways in which competition could structure communities, corresponding to the last three points above. (1) Some species, even though they can disperse into the community, are absent because they are competitively excluded. (2) Current competition may restrict species ecologically—for example, to certain foods—causing them to be less abundant than they otherwise would be. (3) Prior competition may have caused species to diverge morphologically or behaviorally, reducing competition between them to a level at which coexistence is possible. Are we able to see these three points demonstrated in the way communities are put together?

Are There Regularities in Species Composition Based on Competitive Exclusion?

The Species/Genus Ratio

Darwin (1859) noted that "as the species of the same genus usually have much similarity in habits and constitution, and always in structure, the struggle will generally be more severe between them, if they come into competition with each other, than between the species of distinct genera." If competition is important in excluding species from communities, then, many ecologists have reasoned, we ought to find the different species of any one genus tending to be sorted out among the various communities of that region. The approach taken to test this has generally been to use the **species/genus (S/G) ratio,** comparing communities with the total species pool.

When this is done, the expectation is confirmed: The S/G ratio is lower in communities or small areas than for the whole regional fauna or flora (Elton 1946, Moreau 1966). Unfortunately, it turns out that any small subset of a regional fauna or flora, even one drawn quite at random, tends to have a lower S/G ratio (Williams 1964; Simberloff 1970). The reason is simple and statistical, not biological: In a small sample of the total species pool, a large genus will almost never be represented by all its species, whereas small genera often will be. Consequently, the average number of species per genus will be lower in samples than in the total pool. The effect is illustrated in the imaginary example in Table 10-8.

We cannot use a lower S/G ratio in communities as evidence for competition; it is just what would happen if communities were assembled randomly. The case of the S/G ratio points up the care with which tests must be devised for detecting the effects of competition when using biogeographical data. The "neutral model"—what we would expect in the absence of competition—needs to be clearly analyzed and stated. The same general problem has been encountered in other approaches using species occurrence data, such as the one described in the next section (Strong 1980).

Assembly Rules

In a 100-page paper, Jared Diamond (1975) put together a set of seven "assembly rules" for the bird communities on islands around New Guinea. As examples, the first two rules are:

> If one considers all the combinations that can be formed from a group of related species, only certain ones of these combinations exist in nature.
>
> Permissible combinations resist invaders that would transform them into forbidden combinations.

The conclusion was that the assembly rules, which had been more or less empirically derived from the species lists for the various islands, were the result of competition, tempered to a degree by colonizing ability.

Connor and Simberloff (1979) criticized Diamond's assembly rules as tautological and trivial, asserting that they were a summary of patterns that would, in any case, be expected if the species were distributed among islands by chance. A series of rebuttals and rejoinders (citations in Strong et al. 1984) and several similar discussions on the same general subject followed.

The main development from the papers on this general topic has been that it is hard to specify the

Table 10-8
An Example of the Species/Genus Ratio

At left is the regional pool of 16 species in 7 genera. Genera are indicated by capital letters. At right are 5 samples of 5 species apiece picked randomly from the 16 species. Species/genus (S/G) ratios are given; none of the 5 samples has an S/G ratio as high as the regional species pool.

Regional Pool	Samples
Genus A	
(1) Aa	*Sample 1*
(2) Ab	Aa, Ac, Fb, Ga, Gc
(3) Ac	S/G = 5/3
	= 1.7
Genus B	
(4) Ba	
Genus C	*Sample 2*
(5) Ca	Ca, Db, Dd, Fb, Gc
(6) Cb	S/G = 5/4
	= 1.25
Genus D	
(7) Da	*Sample 3*
(8) Db	Aa, Ab, Db, Fa, Gb
(9) Dc	S/G = 5/4
(10) Dd	= 1.25
Genus E	*Sample 4*
(11) Ea	Ba, Dc, Dd, Ea, Gb
	S/G = 5/4
	= 1.25
Genus F	
(12) Fa	
(13) Fb	*Sample 5*
	Ab, Ca, Da, Fa, Gc
	S/G = 5/5
	= 1.0
Genus G	
(14) Ga	
(15) Gb	
(16) Gc	
Overall S/G = 16/7 = 2.3	Average S/G = 1.3

biogeographical pattern to be expected if competition is unimportant (Colwell and Winkler 1984). Proponents of the view that competition is important hold that many features of the system (number of species per island, for example) have already been structured by competition. Accordingly, a null hypothesis (in this usage, a random or neutral model) that takes these features as givens is biased against detecting competition.

Is There Evidence That Current Competition Affects the Structure of Communities?

David Lack (1944) thought the answer to this question was no, although his reason was that he thought that species were already ecologically segregated through prior evolutionary processes. S. C. Kendeigh (1947) thought that both current and prior competition were important in the "segregation of species to particular niches." Most ecological thinking in the 1960s and 1970s tended to agree with this viewpoint, but the question has recently been critically re-evaluated.

Schoener (1983) and Connell (1983), in similar papers, evaluated field experiments on interspecific competition (laboratory and greenhouse studies have frequently demonstrated competitive exclusion). Such field experiments involve changing the numbers of one or more of the possible competitors. One then looks to see whether the community is affected. For example, a series of plots is set up; from one half of these, species A is removed, and the other one half are retained as controls. If, on the removal plots, species B increases in numbers, expands its diet to include the items formerly eaten by species A, or shows some similar effect reflected in community structure, a possible effect of interspecific competition on community structure has been demonstrated. The potential effects of other sorts of interactions must be included in the interpretation. We can envision, for example, situations in which the removal of an abundant species from a plot results in fewer visits from predators and a higher density of the subordinate species as a result of less predation. The removal of white-tailed deer from a region where they coexist with moose might result in a rise in moose populations, not primarily because of lessened competition but because of the disappearance from the area of brain worms.

When Schoener and Connell surveyed the field experiments, they found—unsurprisingly—that current competition is frequently important. Schoener reported that 90% of 164 studies showed some competition, and Connell, that 93% of 72 studies did so. One of the more striking results was that competition was frequently asymmetrical; for example, if species A was removed from some plots and species B from others, one species responded by increased numbers or an expansion of habitat or diet, but the other did not.

The removal study of granivorous ants and rodents in southwestern U.S. deserts (Davidson et al. 1980) described in Chapter 8 is one example of such a field experiment. Another involved insects that occur on the seaside daisy of the coastal bluffs and dunes of California (Karban 1986).

Two abundant insects on the daisy are the meadow spittlebug and the calendula plume moth. Both feed on the terminal shoot of the plant, and both tend to stay on a single plant to complete their development, so numbers are easy to manipulate.

The experiment involved populating a daisy plant with (1) five spittlebug nymphs, (2) one plume moth larva, (3) five spittlebug nymphs and one plume moth larva, or (4) neither spittlebugs nor plume moths. The outcome was that spittlebugs tended to disappear or die significantly faster on daisies with moths; however, the presence of spittlebugs had no significant effect on plume moth persistence (Table 10-9). Under the conditions of the experiment the plume moths tended not to last very long, evidently because of predation.

Table 10-9
Persistence of Spittlebugs and Plume Moths on Seaside Daisies, With and Without the Other Species

Species	*Alone*	*Other Species Present*
Moth	7.4 days	5.1 days
Spittlebug	23.4	16.9

Data from Karban, R., "Interspecific competition between folivorous insects on *Erigeron glaucus*," *Ecology*, 67:1063–1072, 1986.

The effect of the plume moths on the spittlebugs occurred through lowering production of new daisy leaves: The moths ate the terminal bud during their development, so the spittlebugs had no new terminal leaves for shelter or food. In nature, moths and spittlebugs co-occur on daisy plants less often than would be expected by chance, presumably because of higher mortality on, or avoidance of, moth-infested plants by the spittlebugs.

Is There Evidence That Evolution Based on Reduction of Competition Has Been Widespread?

Many authors have suggested that current competition is no more obvious than it is because prior competition has produced communities composed of species that now live harmoniously. An absence of current strong competition is not evidence against the importance of competition, but for it. Joseph Connell (1980) termed this result "the ghost of competition past," a phrase that captures the spirit of the idea.

A process of divergence would occur such as we have described and illustrated earlier (Figure 8-6). If this process is a frequent one, we ought to be able to find cases in which two species, when they occur separately (allopatric), are similar but differ when they occur together (sympatric). The changes could, of course, be in life history, behavior, or morphology. Morphological changes are easy to study and are less likely to be the result of current competition. The situation just described in which two sympatric populations are more different than when they occur allopatrically was termed **character displacement** by W. L. Brown, Jr., and E. O. Wilson (1956).

One of the best examples of character displacement involves the small and medium ground finches (*Geospiza fuliginosa* and *Geospiza fortis*) of the Galapagos studied by David Lack (1947). The medium ground finch occurs without the small ground finch on Daphne Island, and the small ground finch occurs without the medium ground finch on Crossman Island. In these cases of allopatry, the two species are almost identical in beak size (Figure 10-14 plots beak depth, the thickness from top to bottom). Where the two occur together, however, a considerable difference in beak size exists; the small ground finch has a smaller beak than it does on Crossman Island, and the medium ground finch has a larger beak than it does on Daphne Island. The underlying basis seems to be competition for food, especially seeds (Grant and Schluter 1984). Where either of the species occurs alone, it utilizes small and medium seeds; where they occur together, each specializes.

How many other clear-cut examples of morphological character displacement are known? The answer is, surprisingly few. Many examples that have been suggested have other possible explanations, such as geographical variation (Grant 1972b). Some of these may, of course, turn out to be cases of competitive character displacement when more complete information is available. There are, however, several reasons why character displacement may not be as frequent an outcome of competition as we might first expect.

1. If two species are similar, then one may simply be excluded from the community, with no opportunity for character displacement to occur. For example, *Geospiza conirostris* and *Geospiza scandens,* other ground finch relatives of the ones just discussed, have similar beaks and never occur together on the same island; perhaps coexistence long enough for natural selection to occur is not possible.
2. If moderate differences, developed allopatrically, already exist, these may be enough to allow coexistence without further change in beak size or other morphological features. Point 3 (below) may be important in this respect.
3. Phenotypic changes are probably often important. As the result of interspecific aggression and other experiences, animals learn to make ecological differences among themselves.
4. Some mathematical models suggest that character displacement is not an inevitable result of coevolution based on competition (Slatkin 1983 is an example). In fact, character convergence may occur. Suppose, for example, that the seeds in a given community show a normal curve of abundance when seed size is tallied. A newly invading species with a beak quite different from the optimal size for using these seeds might become more similar to the

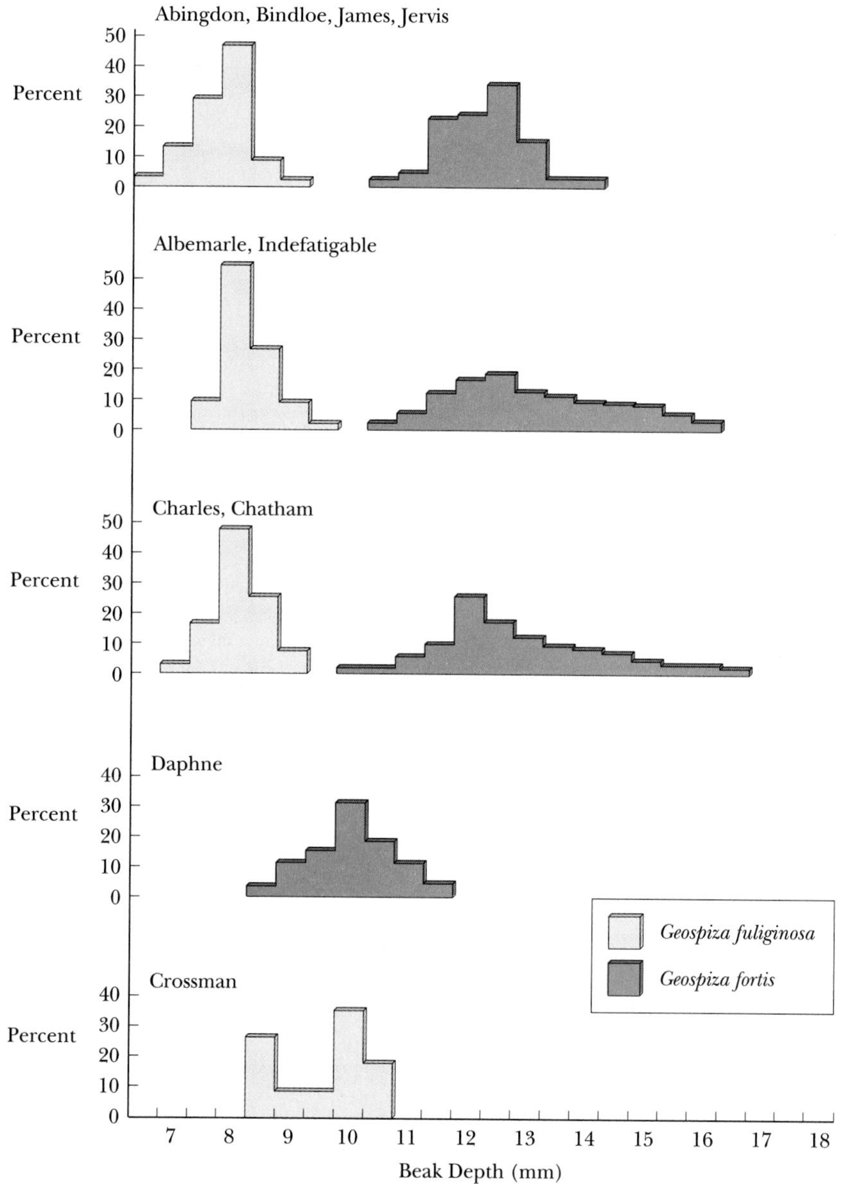

Figure 10-14

Beak sizes of two species of Galapagos finches are similar when only one is present on an island. On islands where both occur, *Geospiza fortis* has a thick beak and *G. fuliginosa* has a thin one. (Adapted from Darwin's Finches, by D. Lack, 1947, Cambridge University Press. Reprinted by permission.)

original species (Figure 10-15). Interspecific competition presumably would be originally very low and then increase up to some point where coexistence remained possible.

Examples of **character release** are, curiously, fairly frequent (Grant 1972b). Character release is the obverse of character displacement. Two species that generally occur sympatrically have a few allopatric populations. In these, whichever species is present shows a shift toward an intermediate condition.

Several recently described cases of character displacement and/or release seem fairly compelling (such as Davies 1987). One well-studied case involves the North American weasels (Dayan et al. 1989). Three species are involved, from large to small, the long-tailed weasel, short-tailed weasel (ermine), and least weasel. In various localities, two or all three species coexist. Sexual dimorphism exists

Figure 10-15

Species that share a resource might show character convergence rather than character displacement if they fit the conditions shown here.

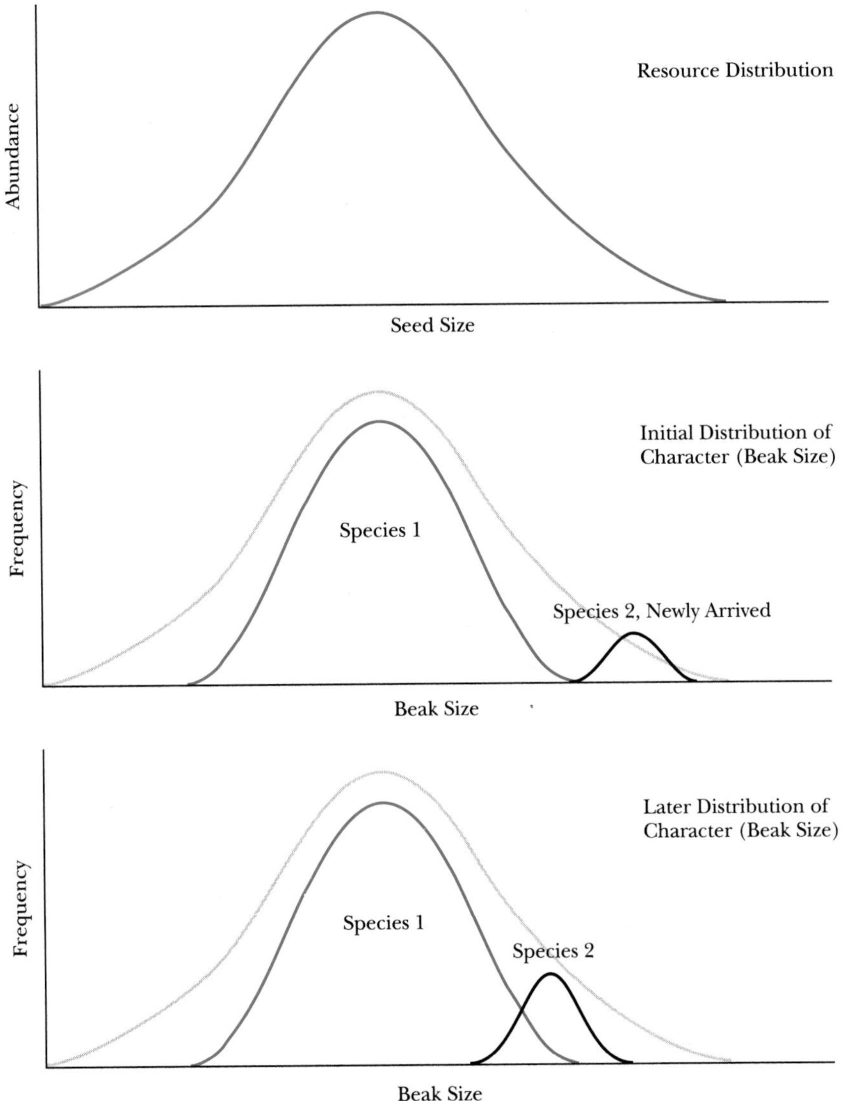

in all species, males being larger. Weasels kill their prey, mostly small mammals, by biting the neck, driving the canine teeth into the vertebral column, producing cervical dislocation or other trauma (Figure 10-16).

The maximum diameter of the upper canine tooth was measured on the assumption that this is correlated with the size of prey taken. Mean canine diameters for the different species-sex combinations tend to be spaced out in a way suggesting that they are dividing up the available food supply. In Minnesota, for example, average canine diameters (millimeters) running from the female least weasel through the male long-tailed weasel were 1.11, 1.27, 1.37, 1.77, 2.24, and 2.60.

The ratios between adjacent species are similar, on the order of 1.2 (Table 10-10). This recalls the idea of limiting similarity, that if two or more species utilize a resource in the same way, there is some limit to how alike they can be and still coexist. In this case, it looks as though canine teeth about 1.2× one's adjacent small competitor and about 0.8× one's adjacent larger competitor can do the job.

If weasels from different areas in North America are compared, the tooth sizes are similarly spaced with ratios around 1.2, although the actual

Figure 10-16

Weasels kill by driving the canine teeth into the vertebral column. Here a mink in Colorado is killing a redback vole. (Animals Animals © 1994 Stouffer Prod. Ltd.)

canine sizes of a given species differ by geographical region (Table 10-10).

It seems likely that the answer to the question of how communities are organized contains elements of most of the foregoing points of view. Neither coevolution based on prior competition nor the chance juxtaposition of preadapted species explains the way a forest or a grassland or a desert is constructed but they probably both play a role. So do current competition and so also do other coactions, such as predation and mutualism, both as they currently operate and as they have evolutionarily molded the responses of the various species. The interesting questions are whether forest, grassland, or desert differ in the relative roles of these processes. Similarly, how do successional or climax, tropical or temperate, aquatic or terrestrial communities differ? Do plants, vertebrates, and insects differ? Some beginnings have been made in answering such questions (e.g., Hairston et al. 1960, Schoener 1986), but there is far to go.

Table 10-10
Maximum Diameters (mm) of Upper Canine Teeth of Weasels in Vermont and Northern Alaska.

Locality	*Species and Sex*					
	Least		*Short-Tailed*		*Long-Tailed*	
	Female	*Male*	*Female*	*Male*	*Female*	*Male*
Vermont						
Measurements	(Absent)		1.34	1.62	1.92	2.32
Ratios				1.2	1.2	1.2
No. Alaska						
Measurements	1.23	1.49	1.76	2.26	(Absent)	
Ratios		1.2	1.2	1.3		

Figures are means. Data from Dayan et al. 1989.

The Integrated Versus the Individualistic Community

Tied to the discussion of how communities are structured is the dichotomy between the **integrated** (in its extreme form, the organismic) view of communities versus the **individualistic** view. The integrated view is that communities are highly organized by previous evolution and current coactions. The individualistic view, expounded by H. A. Gleason (1926) of the New York Botanical Garden and the Russian plant ecologist L. G. Ramensky (1924; see Whittaker 1975), emphasizes the independent behavior of the individual species. The integrated view generally leads to a classification of communities. Proponents believe that, because of the unity developed by community interactions, characteristic combinations of species tend to be repeated and can be given a name, such as beech-maple forest.

Usually accompanying the individualistic view is the continuum approach to describing vegetation. A leading proponent of this approach was J. T. Curtis (1959) in his studies of the vegetation of Wisconsin. Communities are supposed to vary continuously in space, with each point of the continuum being equally probable. That which might be called oak-hickory forest and beech-maple forest are simply regions on a continuum, and the combinations between these two are as likely to occur as any others (Figure 10-17). The continuum is more than one dimensional. Species might be arranged one way along a moisture gradient and another way along a pH gradient. Their actual distribution in nature would reflect their response to both, in addition to any other important factors.

There is evidence that continua do exist. Species tend to be distributed in nature according to their tolerances, and the range of one species differs from the range of most other species. It is not clear whether every point along the continuum is equally probable. It seems possible that coactions such as competition and predation, and reactions such as shading and soil formation, make some regions of a continuum improbable. For example, in a region where oak forest and maple forest both occur, oak-maple forests may be unusual (Figure 10-18). It seems possible that maple is able to invade an oak forest only very slowly, but when it does in-

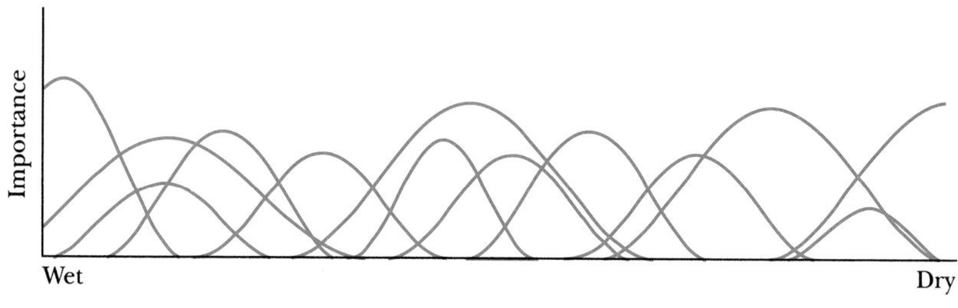

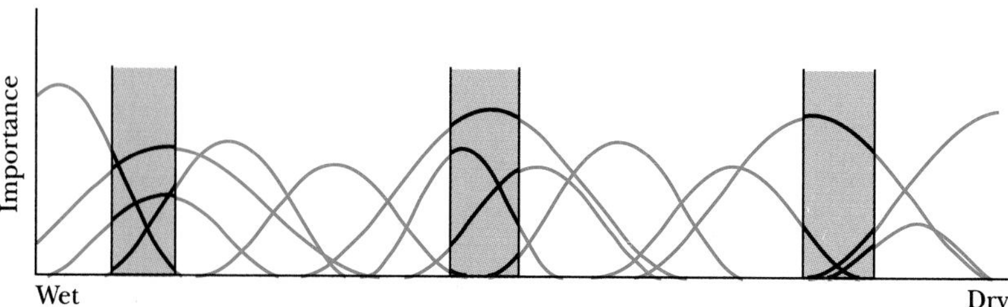

Figure 10-17

A diagrammatic representation of how species might be distributed individualistically, or in a continuum, but separate communities might exist because some parts of the continuum are improbable. The probability or improbability of various combinations would result from coactions and reactions of the dominants. The top graph is the continuum. The middle is a tally of actual stands of vegetation in a region. The bottom graph shows (blue screen) three recognizable communities having the species composition shown by the black segments of the curves.

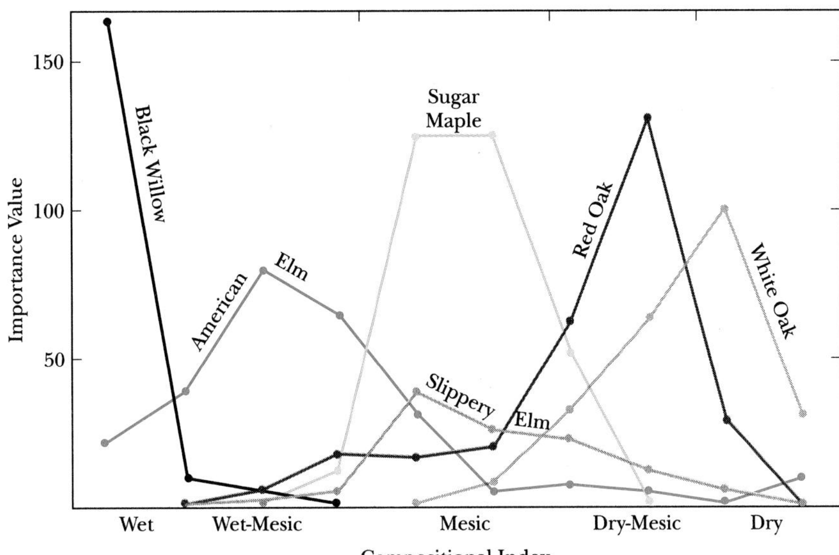

Figure 10-18

The distribution of some major tree species along a continuum of Wisconsin forests from wet to dry. The *importance values* plotted on the y-axis are figures that combine density, frequency, and size, thus measuring three aspects of the "importance" of a species in a forest stand. No two curves have exactly the same shape and location on the continuum, and this remains true when other species are added. (Adapted from Curtis 1959, p. 99; Copyright by the Regents of the University of Wisconsin.)

vade, succession to a maple-dominated forest occurs very rapidly. Consequently forests of intermediate composition are rare at any one time.

Also, not every species shows the single-peaked curve on a continuum diagram that would be expected if other species were unimportant in its occurrence. A striking example of this in Curtis' studies is the bimodal curve for basswood (Figure 10-19). It is hard to avoid the conclusion that something about the presence of sugar maple—its reduction of light intensity, competition for water, or allelopathy are possibilities—is the reason for the reduced importance of basswood in the most mesic forests. A long-term experiment eliminating sugar maple from the understory and eventually the canopy of some plots would be a good study—and a good trick, given the reproductive capacity of sugar maple.

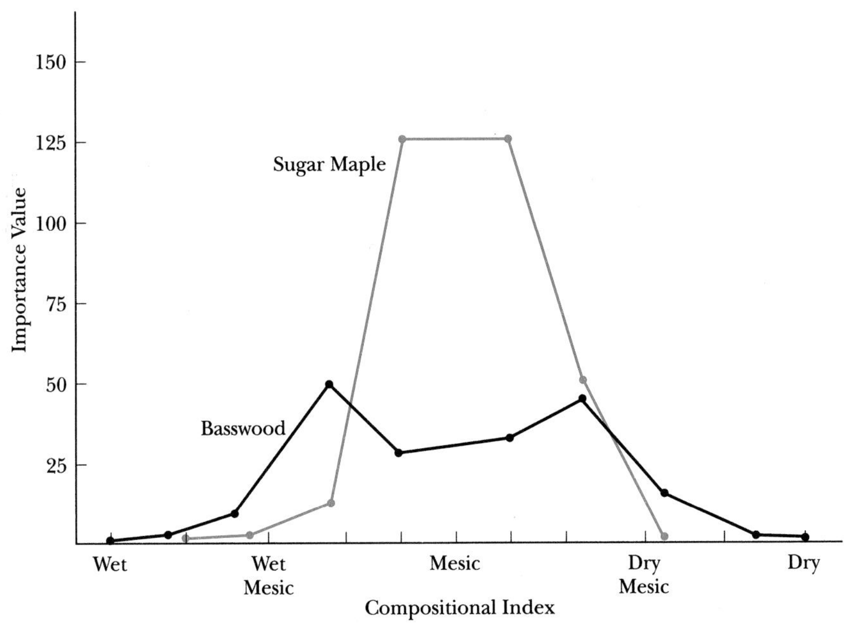

Figure 10-19

The curve of basswood importance in Wisconsin mesic forests is bimodal, with the trough corresponding to the greatest importance of sugar maple. (Based on Figures 10 and 11 in Curtis 1959.)

There is enough evidence on both sides of this question to indicate that the individualistic/continuum and integrated/classification views are not alternatives, one of which must be wrong, but rather complementary, equally valid ways of viewing communities.

Ecological Diversity

Ways of Being Diverse

A diverse community has a variety of species. Although there are other aspects to diversity, the simplest measure is the number of species present; a diverse community has more species than a less diverse one. This aspect of diversity is sometimes called **richness**.

If species number is based on a division of resources resulting from competition or by other means, two communities may differ in the number of species they possess for the following reasons:

1. One community may not be at equilibrium, considering the prevailing climate, regional pool of species available for colonization, and the dynamics of the community when undisturbed. A community may not have all the species that could live there; it is, in other words, **unsaturated** (Figure 10-20). Not all the potential niches are filled. Alternatively a community may be too diverse for its resources as a temporary stage following some change. It is **supersaturated** and some species will be lost in the relaxation to the equilibrium number. The rest of this section deals with equilibrium species number; we will return to the question of disturbance later.

Figure 10-20

The upper community has low diversity because not all niches are filled.

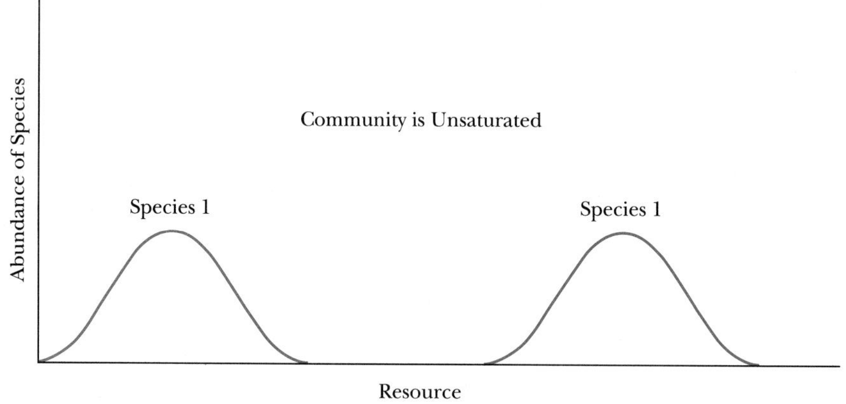

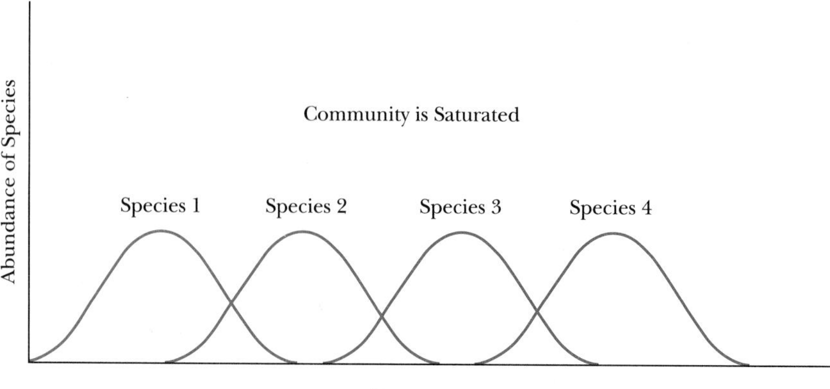

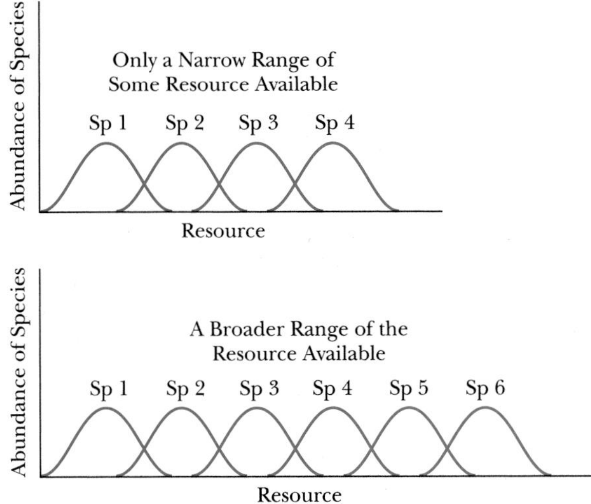

Figure 10-21

The upper community has low diversity because it has a narrow range of resources.

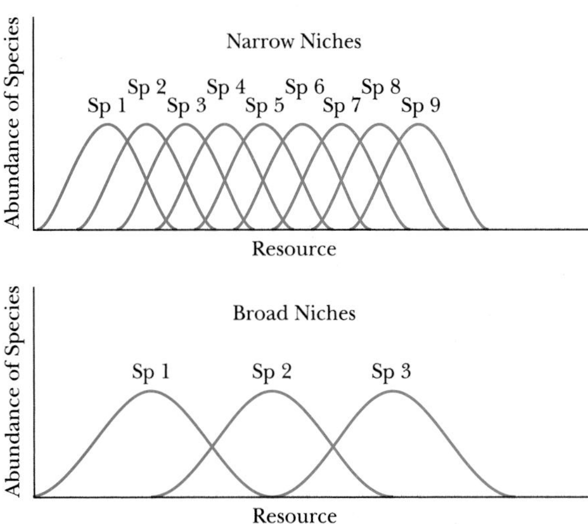

Figure 10-23

The upper community has high diversity because of much niche overlap.

2. Equilibrium species number may be higher in one community because it has a greater range of one or more resources. For some resource, then, one community may look like the first graph in Figure 10-21, and another, with a greater range of this resource and more species, may look like the bottom graph.

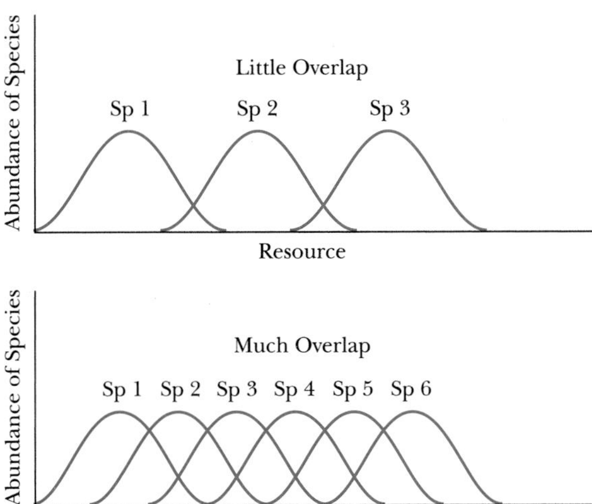

Figure 10-22

The lower community has high diversity because niches are narrow.

3. Species in one community may be more specialized (that is, adapted to a narrower range of food, soil, or other environmental factors) than in another. In other words, they may have narrower niches. A community with narrower niches (and more species) compared with one with broader niches (and fewer species) is shown in Figure 10-22. With narrower niches (and no increase in resources) each species will obviously be less abundant.
4. Species may share resources to a greater degree; that is, there may be more niche overlap (Figure 10-23). With increased overlap, interspecific competition would tend to increase as shown.

Relationships between diversity and stability are discussed in Chapter 13.

Factors Affecting Diversity

Many different influences have been suggested as affecting species diversity (in one or more of the ways just listed), and probably most of the suggestions are correct, at least for some community at some time. Some of these are the following.

Unique History

Each area has a history different from every other area, and past events may cause one area to be more diverse than another. The deciduous forests of eastern North America and western Europe are similar, but there are fewer tree species in Europe, probably owing to the different effects of the Ice Age. In Europe mountain ranges run east-west, and the forests that were forced southward by the glaciers and cold climate had only small areas of refuge between the glaciers to the north and the unfavorable climate and terrain of the southern mountains. Some of the forest trees became extinct. In North America the mountain ranges run north-south and do not form a barrier to southward movement. Here most species were able to find refuges somewhere to the south, survive, and invade northward again as the glacier receded. In this particular instance the European forests are presumably not saturated with species.

Another example has to do with the absence of snakes from Ireland. We can choose between two historical explanations for this lowered diversity. Legend has it that Saint Patrick expelled the snakes from Ireland, although the hagiologies tell us little of his motives or methods. If the story is factual, Patrick must have disapproved of the herpetofauna in general inasmuch as Ireland possesses only single species of lizards, frogs, toads, and salamanders and has no turtles.

An alternative hypothesis again has resort to the Pleistocene. At the height of the last glaciation, most reptiles and amphibians of Europe must have been restricted to the southern part of the continent or even what is now the Saharan region of Africa. With glacial retreat, the ranges expanded northward. At this time, with a great deal of water still tied up in glacial ice, there was a broad land connection between eastern Britain and the continent of Europe and between Ireland, the westernmost of the British Isles, and Britain. Some reptiles and amphibians crossed these connections before they were severed by rising sea levels (Street 1979). The connection between Ireland and Britain was cut off more than 7000 years ago when only a handful of the hardiest species had been able to get that far. The connection between Britain and the mainland was broken by the Straits of Dover somewhat less than 7000 years ago. A few more species had spread into Britain by that time, though its herpetofauna is also impoverished compared with mainland Europe.

Time

New communities arise when new habitats are created. Here we are not discussing a new habitat such as a new sandbar built by the Mississippi River but a habitat of a type not previously available for organisms. Dry land would have been such a habitat when oxygen began to build up in the atmosphere and the continents became habitable. Presumably new communities have few species, consisting of those with traits that, even though evolved in some other community, allow them to survive in the new environment. With time more species would be added through various processes, including the adaptation of already existing species to the new habitat and the evolution of new species.

Extreme Habitats

Habitats that are harsh tend to have relatively few species. By harsh we mean habitats that are extreme compared with most of the biosphere. For example, they may be extremely hot, cold, salty, acid, or polluted. In such habitats high populations of a few species are usually found. Presumably there are not more species because many kinds of organisms lack genetic capabilities for evolving the ability to tolerate a particular kind of extreme habitat.

Resource Diversity

If the resources for a particular trophic level or group of organisms are diverse, then that trophic level or group is likely to be diverse. The plant life of a physically diverse area—one with hills and valleys, rocky areas and good soil, wet spots and dry spots—will usually be more diverse than in a topographically uniform area such as a plain. There tend to be more bird species in vegetation with several layers such as a forest than in areas with little stratal diversity such as a grassland.

The area where two ecosystems come together, the transition zone between them, is called an **ecotone.** Species diversity of such areas tends to be high, and the main reason for this is clear. Where forest grades into grassland, the ecotone will have some characteristic forest species, some grassland species, and some additional species that require resources from both kinds of vegetation (food from the grassland and escape cover from the woody growth, for example). Members of the last group, characteristic of the deciduous forest-grassland ecotone, are called forest-edge species (Chapter 15).

Productivity

Generally more species can exist in areas of high productivity than low, the main reason being that in an unproductive area some features of an ecosystem—such as a certain kind of food—will be at too low a level to support a population of a species permanently. In a productive area the same feature is more abundant, becoming a resource that one or more species can use. The narrow niche becomes able to support a population of viable size.

Although this relationship between productivity and diversity seems typical of undisturbed ecosystems, the effect of increasing production artificially, such as by fertilizing a pond, is often just the reverse—species diversity is reduced. There may be several reasons for this, but probably the most important is a disturbance of the balance between competitive species.

Climatic Stability

Areas of stable or predictable climate tend to have more species than those of unpredictable climate. Many species with very low populations can survive in an area of stable climate, whereas chance weather events, such as a cold day in June or an unusually icy winter, may kill all the members (because there are not very many) of such species in areas of unpredictable climate. Also, organisms in a stable climate can have narrower tolerance ranges, which may allow narrower niches.

Predation

Predators or grazers may keep numbers of certain of their prey low enough that competitive exclusion does not occur even though the prey species show considerable niche overlap. Probably the best known example of predation keeping diversity high is the work of R. T. Paine (1966). In the rocky intertidal zone of the Olympic Peninsula in Washington the top carnivore is a starfish that preys on large invertebrates such as mussels and barnacles. When the starfish was experimentally removed, several species disappeared as a mussel that is a preferred food of the starfish came to dominate the area.

Disturbance

A disturbance is an interruption of the settled state. A fire, a flood, or the outbreak of a disease can be disturbances. At high levels of disturbance, the species list may be reduced to the species best able to tolerate the habitat conditions so produced. At low levels of disturbance, the most nearly climax species may come to dominate to a degree that most other species are lost. Consequently, the species list may be longest at some intermediate level of disturbance (Connell 1980) (Figure 10-24). For example, long-continued protection of mesic deciduous forest from fire and other disturbances may result in the decline or loss of various herbs that we ordinarily think of as typical mesic forest species (Brewer 1980). Some of the factors listed earlier, such as predation and grazing may be thought of

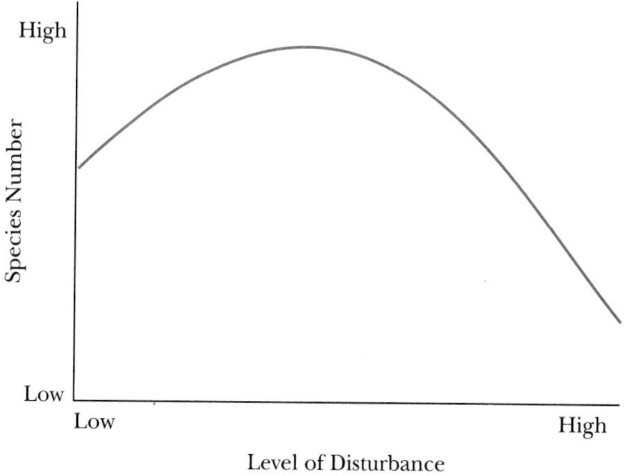

Figure 10-24

Where intermediate levels of disturbance increase species number, the relationship may look something like this.

as disturbance, and thus may have different effects at high and low levels.

Defining what constitutes a disturbance and what an intermediate level of it is (other than being a level that increases diversity) is not always easy. Diversity should increase with disturbance if the number of old species eliminated is fewer than the number of new species that enter as a result of the disturbance.

Three authors working in Canadian forests (Reader et al. 1991) compared three levels of selective cutting and three levels of trampling on understory plant species. They found that cutting and trampling removed some species and added others but the bottom line was that species number did not vary significantly across any of the treatments (Table 10-11).

Keddy and Reznicek (1986) concluded that year-to-year fluctuations in water level maintain plant species diversity along lake shores. The reasons for their conclusion were that

1. Low water levels kill certain species intolerant of drying that are replaced by other species emerging from the seed bank (which is very large in wetlands soils).
2. High water levels kill invading successional shrubs and trees and also kill certain highly dominant plants, such as cat-tails, allowing invasion of marsh plants from adjacent lake zones.

Table 10-11
Number of Understory Species (Mean ± 1 Standard Error) After No Disturbance, Intermediate Disturbance, and Heavy Disturbance in Southern Ontario Forest.

Disturbances were cutting canopy trees (0, 33, and 66% of basal area removed) and trampling (0, 50, and 100 passes). None of the differences was statistically significant at the 0.5% level. Data from Reader et al. 1991.

Type of Disturbance	*No Disturbance*	*Intermediate Disturbance*	*Heavy Disturbance*
Timber cutting	103 ± 3	112 ± 5	115 ± 3
Trampling	100 ± 2	96 ± 3	100 ± 5

The overall result is that a large number of species is maintained, most of them shifting back and forth in space depending on water level but others appearing only during years of low water.

Although this describes a common and important feature of wetlands diversity, water-level fluctuations may in some circumstances lower diversity. In a bog in a small, isolated basin, for example, low water levels may eliminate some plants, high water levels may eliminate others, and the bog may come to be dominated by only a few species.

Regional Diversity

The preceding sections have dealt mostly with diversity within one community on one site. We are answering such questions as, How many bird species do 20 ha of broad-leaved forest contain and why does the number differ from one locality to another? This local aspect of diversity was called **alpha diversity** by Robert H. Whittaker (1972), one of S. C. Kendeigh's numerous students.

Also interesting is regional diversity, or **gamma diversity.** How many bird species are there, for example, in the whole county of, say, 200,000 ha? Reasonable values for the midwestern United States might be 20 species of breeding birds on one forest plot (alpha diversity) and 145 in a county (gamma diversity).

A major reason for the higher value for gamma diversity is that any large area contains more than one habitat. As we move from 20 to 200,000 ha, we add streams, lakes, marshes, and other communities. If all the communities occupying these habitats had the same species, alpha and gamma diversity would be the same. But they do not and Whittaker designated species turnover between communities as **beta diversity.** Even if the whole region were occupied with one kind of forest, however, we would find more than 20 species of birds in the county. One reason is that in 200,000 ha we are bound to encounter some wetter forest and drier forest, openings resulting from wind throw, and various other minor variations in habitat that will add some new species of birds.

Even in the unimaginable case of no such added habitat diversity, we would, nevertheless, add new species when we added area because species distribution is not uniform. For example, some species are rare, perhaps occurring only in one 20-ha plot out of every 10, 20, or 100. So beta diversity is somewhat more complicated than species turnover between communities. Just moving around adds species; however, the big jumps tend to come at community boundaries.

In general, high beta and gamma diversities are positively correlated (for example, Cox and Ricklefs 1977). This corresponds with intuitive expectations and a certain amount of theory that if only a small pool of species is available, each species is likely to occupy more habitats than if it has competitors.

If we look at gamma diversity on a very broad scale, say the number of species in 20,000 square miles or 60,000 km^2, species number in temperate and arctic regions is highly correlated with some measure of solar energy input, such as potential or actual evapotranspiration (Currie and Paquin 1987, Currie 1991). Annual net production is strongly related to actual evapotranspiration (Rosenzweig 1968), and it is reasonable that energy flow can be partitioned into more populations (species) in productive than nonproductive areas. That a relationship between energy input and diversity should exist makes sense. Surprising, perhaps, is the small amount of variation left for other factors—such as history, habitat heterogeneity, and biotic interactions—to explain (Figure 10-25).

Diversity Indices

Species richness may be compared between two communities or areas by simply counting the number of species either in the whole community, if that is practical, or in suitable samples. However, there is at least one more aspect of species diversity that should sometimes be considered—relative abundance, or equitability. Imagine two communities each made up of two species. In one community there are 99 individuals of species A and one individual of species B, whereas in the second community there are 50 individuals of each species. The second community is more diverse in this sense: Almost every individual sample is predictable in the first community (it will be species A) but not in the second.

Three widely used measures of diversity include both richness and equitability in a single number, or index. The **sequential comparison index** (SCI)

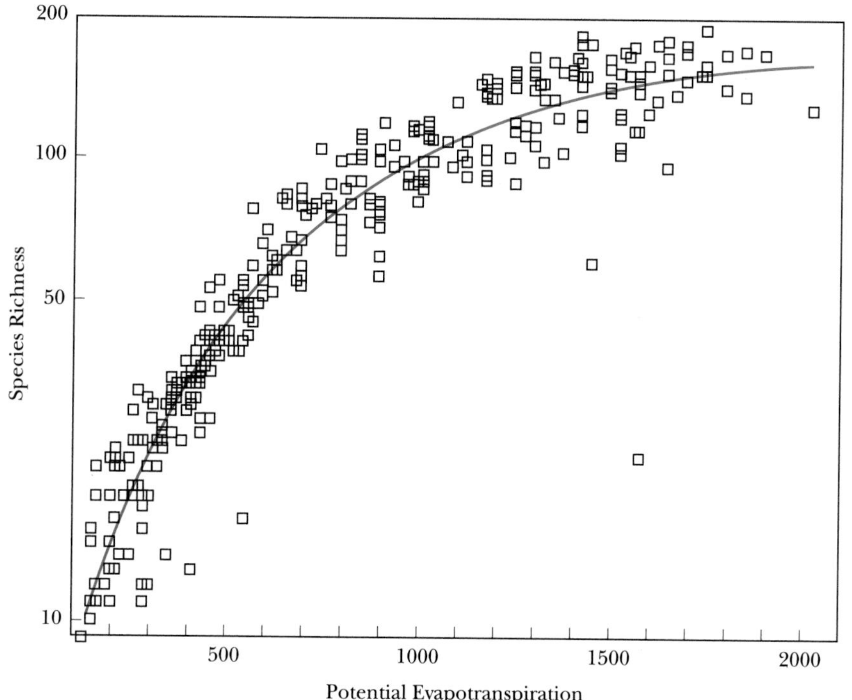

Figure 10-25

The relationship between number of non-volant terrestrial vertebrate species and potential evapotranspiration in North America. Species number is for quadrats about the size of West Virginia. "Non-volant terrestrial vertebrates" are mammals (except bats), reptiles, and amphibians. From Currie (1991).

(Table 10-12) measures the number of runs—a run is a series of individuals all of the same species—in relation to the total number of individuals in the sample (Cairns et al. 1968). If every individual in the sample is a different species, the SCI is 1; if all individuals belong to the same species, SCI is 1/number of individuals in the sample (Table 10-13). The main advantage of the SCI is that it is easily done in the field with little need for mathematics.

The **Shannon-Wiener index** (H′) is derived from information theory (Margalef 1958). It is a measure of the likelihood that the next individual will be the same species as the last. The **Simpson index** (C) does nearly the same thing; it measures the probability that both of two individuals drawn at random will belong to the same species (Simpson 1949). (The relationship between the two indices is described by Pielou [1974].)

Details of the calculation of these three indices are given in *Laboratory and Field Manual of Ecology* (Brewer and McCann 1982). These, and other, diversity indices have their individual strengths and weaknesses (Pielou 1974, MacArthur 1972). Simpson's index is straightforwardly based on probability theory, varies between 0 and 1 (which is approached asymptotically as sample size increases), and is easy to calculate. If a measure of diversity that includes equitability is needed, Simpson's index is as good as any.

A practical application of diversity indices has been in assessing stream pollution. Higher indices usually indicate little or no pollution. The index usually decreases with pollution because many species decline or disappear while a few species increase. For the Shannon-Wiener index ($\log_2$), $H' < 1$ usually indicates heavy pollution, and $H' > 3$ usually indicates very clean water.

Table 10-12
Formulas for Diversity Indices

Sequential comparison index (SCI)

$$SCI = \frac{\text{number of runs}}{\text{number of individuals}}$$

Shannon-Wiener index (H′)

$$H' = -\Sigma p_i \log p_i$$

Simpson index (*C*)*

$$C = 1 - \Sigma p_i^2$$

For both indices, p_i is the proportion of individuals of a given species; that is, number of individuals of species *i*/total number of individuals of all species.

* Note that the probability that both of two individuals from a sample will be of the same species *decreases* as diversity increases. Consequently, Simpson's index as originally formulated was simply Σp_i^2. It measures concentration or dominance and is inversely related to diversity. $(1 - \Sigma p_i^2)$ is one way to make it change in the right direction. MacArthur (1972) preferred the inverted Simpson's index $(1/\Sigma p_i^2)$ because it can be interpreted as measuring numbers of "equally probable species"; that is, it is the number of species that would give the calculated value of Σp_i^2, if they were all equally abundant.

Table 10-13
Upper and Lower Values for Three Diversity Indices With a Sample Size of 100

Diversity Index	*If Every Individual Is the Same Species*	*If Every Individual Is a Different Species*
Sequential comparison index (SCI)	0.01	1
Shannon-Wiener index (H′)	0	6.64*
Simpson index (C)	0	0.99

* Using $\log_2$.

What Determines the Number of Species on Islands?

Biologists have known for a long time that an island will contain fewer species than an area the same size on the nearest mainland. It was thought that islands had an impoverished flora or fauna because of the difficulties of dispersal over long stretches of water. This idea is valid for organisms such as frogs and salamanders, which are very poorly suited for long distance transport over seawater. But recent studies have shown that disseminules of most kinds of organisms arrive on islands at a high rate, although many of the species do not become established.

The island biologist Sherman Carlquist (1974) included as one of his "principles of dispersal and evolution" the statement that for species having the means of long-distance transport the eventual introduction to an island is more probable than its nonintroduction. The primary reason for the lower number of species on islands seems to be that they do not have an overflow from adjacent areas in the same way as a mainland area. There are other factors, however, including the fact that small islands may be too small to provide a large enough energy base to allow some of the top carnivores to exist.

R. H. MacArthur and E. O. Wilson (1967) developed a model to explain how the equilibrium number of species occupying an island is determined. They suggested that the number is the result of a balance between immigration and extinction rates (Figure 10-26). They theorized that immigration rate should decrease and extinction rate should increase as the number of species on an island increases.

Imagine an island having no species on it; immigration rate will be high because every species that lands as seed, spore, or adult will be a new one. Extinction rate will be low because there will be no species to become extinct. As species number increases, as it will if we started at zero, immigration rate will decline, since a larger and larger fraction of the possible species, or species pool, is already present. Extinction rate will increase because there are more species to become extinct and also because there is increased interspecific competition (and also increased predation and disease).

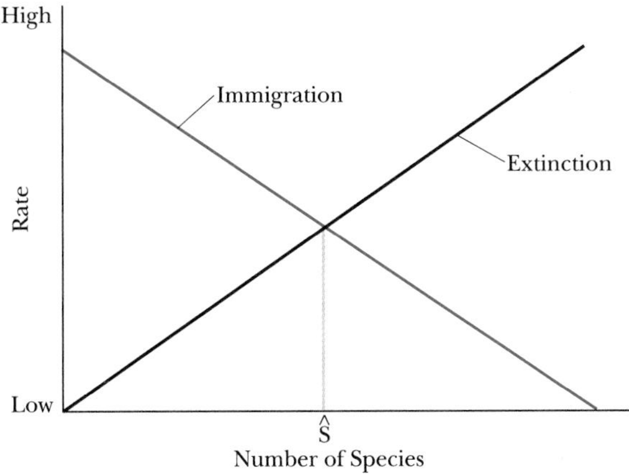

Figure 10-26

The basic MacArthur-Wilson biogeography model, showing extinction rate directly related and immigration rate inversely related to species number. Note that time is not on this graph.

To the degree that these sorts of coactions are important, the extinction line will curve upward (Figure 10-27). Similar interactions will also cause the immigration line to curve; for example, with an increased number of species present, the arrival of a seed may not lead to establishment because the places suitable for its growth are already occupied or because an animal eats it.

Also important in bending the immigration line is the fact that not all species are equally good immigrants. The first 20 species perhaps are the ones that are very good at traveling and get there in a hurry; the next 20 may be ones with lower dispersal abilities and may straggle in over a longer span of time. It is sometimes true that the early arrivals are *r*-selected species that tend to be pinched out as more *K*-selected species arrive (Diamond 1975), but in other cases, the early arrivals seem no more prone to extinction as species number increases than the later arrivals (Rey 1984).

The number of species at the point where migration rate and extinction rate are equal will be the equilibrium species number ($\hat{S}$, which is read

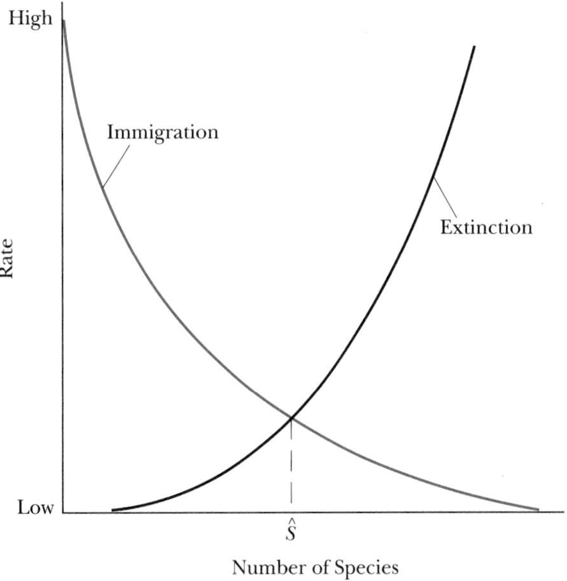

Figure 10-27

If, for the various biological reasons mentioned in the text, immigration and extinction are exponentially related to species number, the MacArthur-Wilson model might more realistically look like this, with curves rather than straight lines.

"S hat"). In some cases, we would not expect species number to be at equilibrium—if the island had just been formed, for example, or if a volcanic eruption had recently killed most of the plants and animals, or if it had recently shrunk (as by sea level rising).

If the model reasonably describes the way things work, we can make some predictions about the relative numbers of species we expect in different circumstances. Large islands should have more species than small islands, as shown in Figure 10-28. This is no surprise. The increase in number of species with area—the **species-area curve**—has been known since the early days of ecology (Gleason 1922, Braun-Blanquet 1932). Explanations such as increasing habitat diversity and the increased probability of including the rare species of the community have been the usual explanations. These may still be among the explanations in the MacArthur–Wilson model, but here the increased number is tied specifically to a lower extinction rate on larger islands.

Islands near a mainland as compared with islands further away should have a larger number of species, but in this case, it is the difference in immigration rate that is important (Figure 10-29). It should be higher for near islands. Note that here, too, we might reach the same prediction by other routes. We might suppose, for example, that many of the species that could live on far islands simply cannot get there in any reasonable length of time. Species numbers on far islands, thus, might be low because most are not at equilibrium.

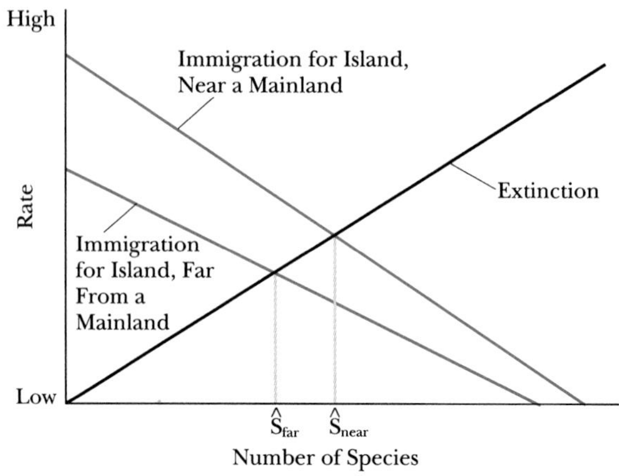

Figure 10-29

Near islands have a higher equilibrium species number than far ones because of a higher immigration rate.

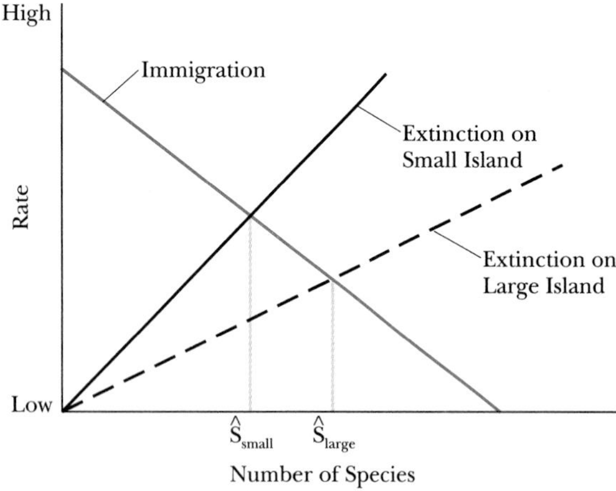

Figure 10-28

Large islands have a higher equilibrium species number than small ones because of a lower extinction rate.

Accordingly, many predictions that the MacArthur–Wilson model makes are not very valuable as tests because they are also the expected outcome of many other possible models. For example, species-area curves (Figure 10-30), which have been assembled in great quantity, are not evidence for or against the MacArthur–Wilson model. However, if a curve shows little variability, it may be evidence against nonequilibrium hypotheses.

Certain predictions, however, follow from the MacArthur–Wilson model and not, or less so, from others:

1. An equilibrium species number should exist. If we change the number of species, it should tend to be re-established. One of the first tests of this idea was by D. Simberloff and E. O. Wilson (1970). They censused the arthropods of several very small mangrove islands (10 to 20 m in diameter) in the Florida Keys and then hired an exterminator to fumigate them with methyl bromide. This killed the arthropods but left the mangroves intact. They then followed the recolonization process. Within 200 days species number had returned to about the same as that found in the "predefaunation" surveys.

A similar recent experiment used cord-

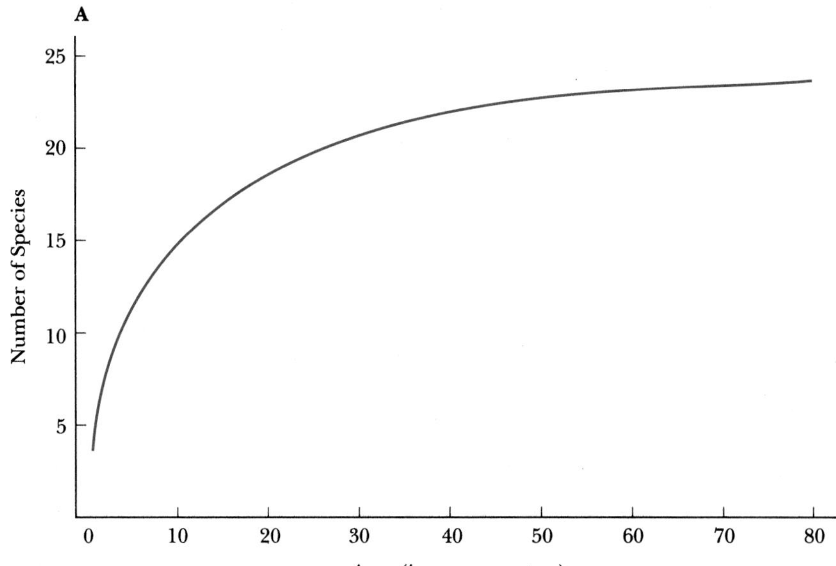

Figure 10-30

Two species-area curves. Graph *A* is an arithmetic plot of plant species vs. area in the aspen forests of northern Michigan. Increasing area to 240 m^2 increased species number to 27. (Data from Gleason 1922.) In graph *B*, both axes are logarithmic. Plotted are numbers of resident land and fresh-water bird species on islands of different sizes around New Guinea. The symbols others than dots are islands for which there is some reason to suppose species number is not at equilibrium. (Adapted from Diamond 1972.)

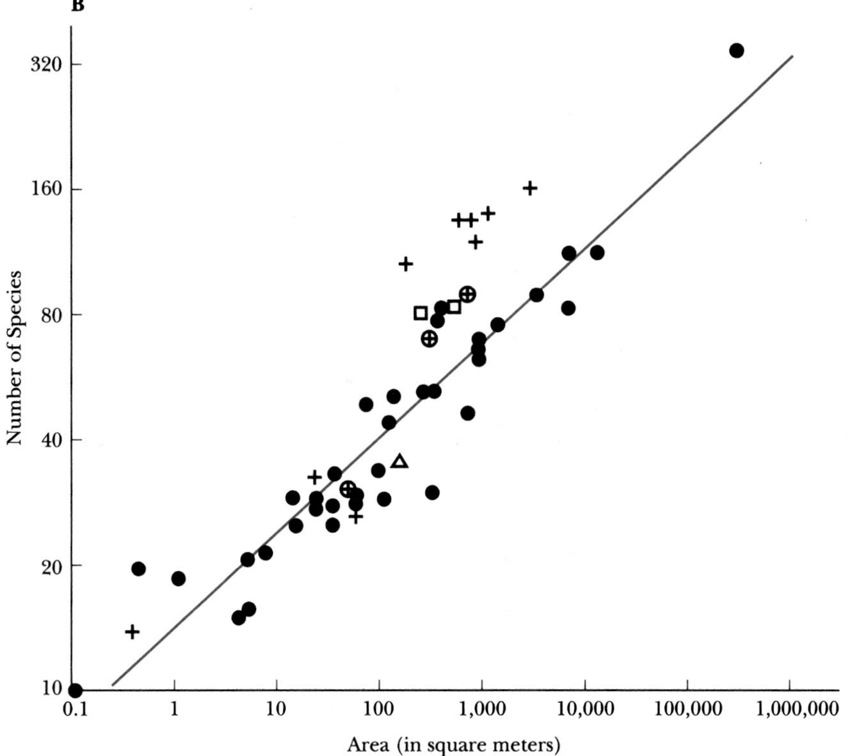

grass islands in Oyster Bay, northwest Florida (Figure 10-31). Here also, there was a quick (20 weeks or less) return to about the original species number (Rey 1984).

2. If the equilibrium number is the result of a balance between immigration and extinction rates, we expect that the same island checked at two different times will have about the same number of species but their identities need not be the same. This result was evident in the two experiments just cited but is also shown by other studies. The Channel Islands

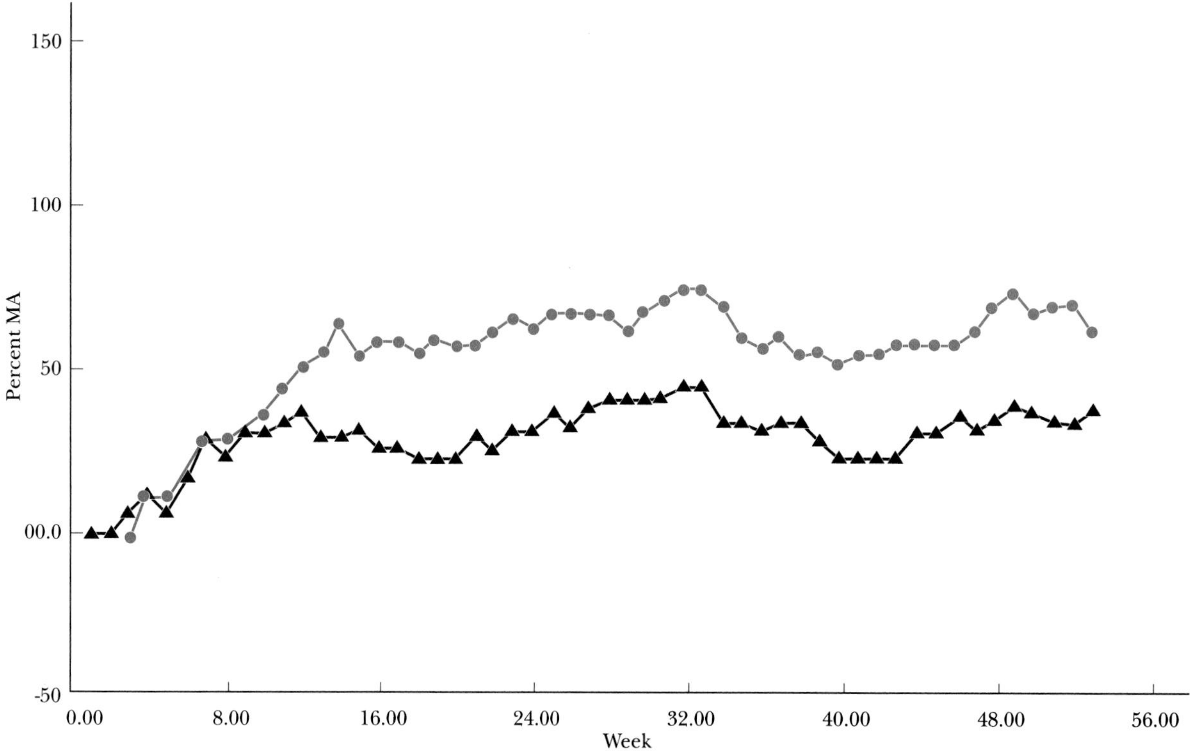

Figure 10-31

Numbers of arthropod species on cordgrass islands. The *y* axis (percent MA) is number of species on the islands expressed as a percentage of the number on the mainland. Immediately following fumigation, no arthropods were present. The number then rose rapidly to the vicinity of the "pre-defaunation" level, where it remained. The size relationship was also re-established: larger islands (*circles*) had more species than small ones (*triangles*). (From J. R. Rey, "Experimental tests of island biogeographic theory," in Donald R. Strong et al., eds., *Ecological Communities: Conceptual Issues and the Evidence.* Copyright © 1984 by Princeton University Press. Figure 7.2 reprinted with permission of Princeton University Press.)

off the California coast each had about the same number of bird species in 1968 as in a 1917 survey, but the identity of some 30% of the species was different (Diamond 1969).

3. Immigration rate ought to decrease and extinction rate ought to increase with species number. In the experiment by Rey (1984) extinction rate behaved in just this way, but the results for immigration rate were ambiguous.

In the same experiment, two other MacArthur–Wilson predictions, of a decrease in immigration rate with distance from the mainland and an accordingly lower species number on more remote islands, were not confirmed but, also, not strongly refuted. It is unclear whether such inconsistencies represent flaws in the tests or flaws in the model. Probably the model is incomplete, but it remains a useful way of thinking about species diversity relationships on islands.

The basic MacArthur–Wilson model deals with relationships in "ecological," not "evolutionary," time. If we ask what will happen if adaptation has time to work on the fauna of a group of islands (Wilson 1969), it seems likely that evolution favoring decreased competition, increased commensalism and mutualism, and the arisal of new parasites will raise the equilibrium species number.

Community and Ecosystem Ecology: *Energy*

OUTLINE

- Trophic Structure and the Food Chain
- Energy Flow
- Grazing Food Chains and Detritus Food Chains
- Biomass
- Productivity
- Ecological Efficiencies
- Food Chains and Food Webs
- Difficulties with Trophic Levels
- Chemolithoautotrophs
- Regulatory Processes and Energy Quality
- Energy in Agriculture
- Human Expropriation of the Earth's Primary Production

11

American robins when feeding on caterpillars are functioning as secondary consumers. *(Jack Dermid/U.S. Fish and Wildlife Service)*

The functioning of the community is based on energy transformations, just as is the functioning of the individual organism (Chapter 2) and, of course, the rest of the universe. In this chapter we look at the principles of energy flow through ecosystems in general and at some of the main features relating to the productivity of various natural and agricultural systems.

Trophic Structure and the Food Chain

The word "trophic" means "feeding." The trophic structure of communities is based on the **food chain,** the sequence of organisms in which one organism feeds on the one preceding it. A typical food chain might be oak leaf → caterpillar → scarlet tanager → Cooper's hawk. In most communities several or many food chains exist that have interconnections at different points, forming a **food web.** The food web for most communities is very complex, involving hundreds or thousands of kinds of

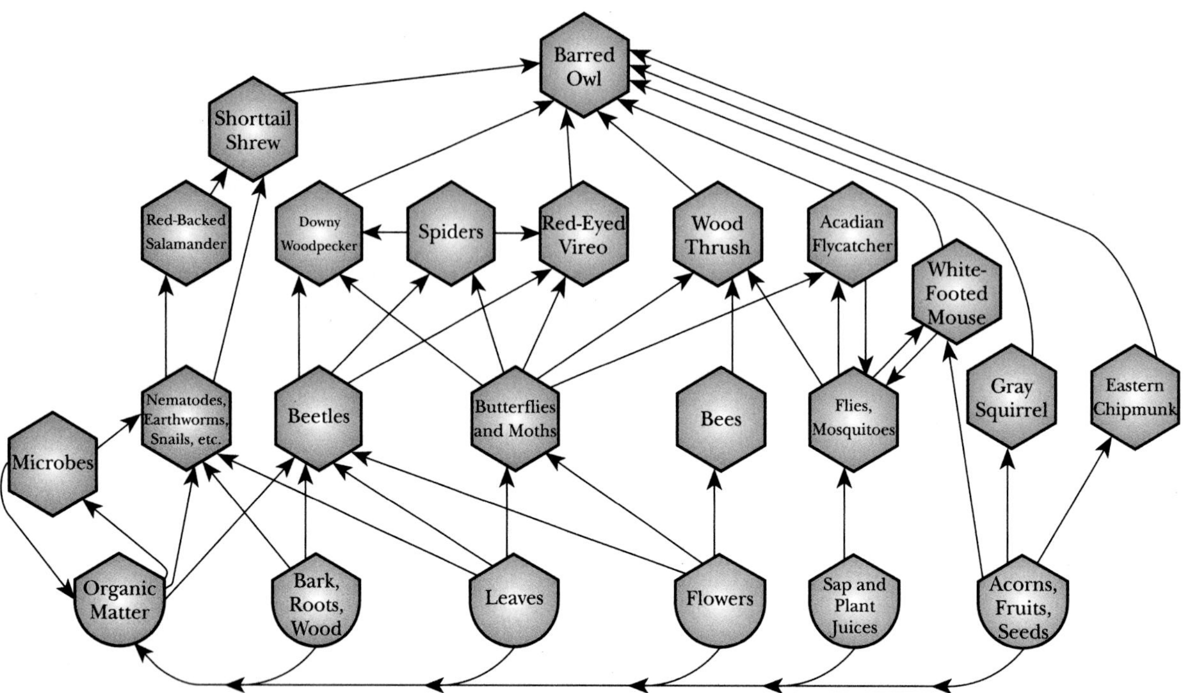

Figure 11-1

This food web for deciduous forest is greatly simplified as compared to the actual situation. Tens or hundreds of species have been lumped together in such categories as "beetles" and "nematodes, earthworms, snails, etc." Many species have been omitted, such as nuthatches, turtles, and fleas. Many connections are not shown. For example, mosquitoes bite other kinds of mammals in addition to mice, and other birds besides Acadian flycatchers eat mosquitoes. All of the animals provide food for the microbes. Despite these simplifications, the food web is still too complicated to comprehend easily. By grouping species into trophic categories, as in the next figure, the situation becomes somewhat clearer.

organisms (Figure 11-1). One useful simplification is to group organisms into categories known as **trophic levels,** based on their position in the food chain (Figure 11-2). The major categories are producers, consumers, and decomposers.

Producers

Producers (also called autotrophs) are organisms that can make food from simple inorganic materials. By food we mean complex organic compounds such as carbohydrates, fats, and proteins. Green plants are the producers with which most of us are familiar, and their food-making process is photosynthesis (chemosynthetic autotrophs are discussed in a later section). In photosynthesis plants use carbon dioxide, water, and some minerals, first to produce carbohydrates and later other organic materials, with oxygen being given off. Energy is as important as the materials involved. In the photosynthetic process the radiant energy of sunlight is converted to chemical energy and stored in the chemical bonds of the compounds made by the plants.

Consumers

These are organisms that obtain their food by consuming other organisms. If they consume plants, they are called herbivores, or **primary consumers.**

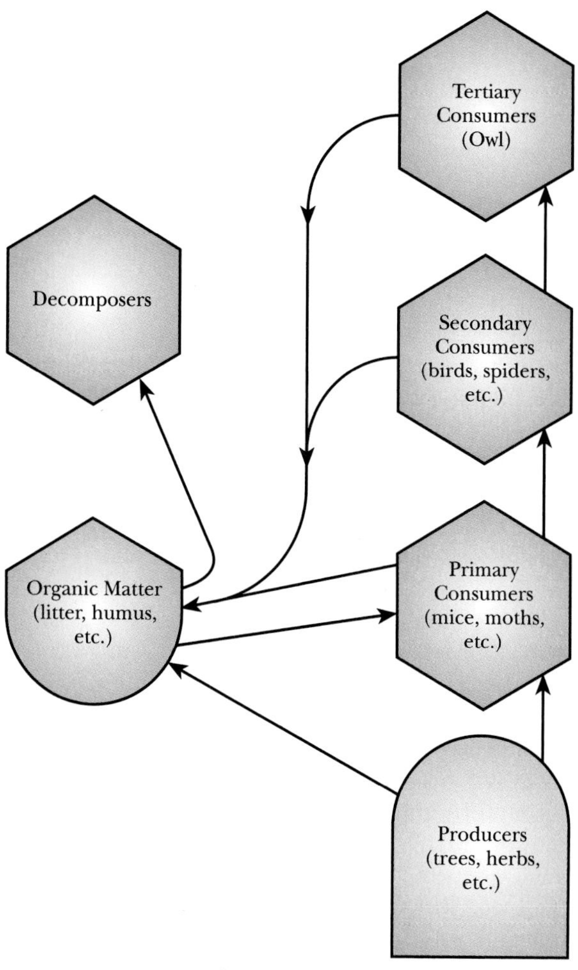

Figure 11-2

Here organisms of the deciduous forest are grouped into trophic levels based on the number of links that precede them in the food chain. All of the photosynthetic plants are in one trophic level, the producer level; all the organisms that eat mainly plants are in the primary consumer trophic level, etc.

If they obtain their food from green plants indirectly, by eating other animals, they are called carnivores, or **secondary** and **tertiary** consumers. All organisms that have to get at least some of their food prefabricated, are known as heterotrophs.

All animals are heterotrophs, as are fungi and many types of bacteria. We are most familiar with larger heterotrophs such as ourselves (or frogs or rats) that eat food, break it down partially in their digestive tracts, and absorb it mainly into the blood. The organic compounds from the bloodstream are absorbed by various cells in the body, where these compounds are used in two principal ways. They may be used as building blocks for other compounds and new cells, or they may be broken down to yield energy.

The latter process is respiration, in which organic compounds are combined with oxygen; the stored energy is released, and carbon dioxide, water, and some mineral wastes are formed. The energy is that used by the organism for all its work—such as repairing and constructing cells, moving around, courting, fighting, and catching more food. The energy used for these functions is then given off as heat by the organism into the environment. Most of the carbon dioxide, water, and minerals resulting from respiration are excreted in one form or another.

Respiration is a universal process; every organism, heterotroph or autotroph, including green plants, respires. The energy that a plant uses for its work, such as growing and flowering, comes from the respiration of foods previously produced by photosynthesis.

Decomposers

These include the organisms talked about earlier as saprobes—scavengers and the decay organisms including bacteria, actinomycetes, and fungi. They use dead plants and animals and excreta as their food source. The processes of digestion, respiration, and excretion are basically similar in all kinds of heterotrophs whether they are animals, bacteria, or fungi. Decay bacteria secrete their digestive enzymes into dead material outside their own bodies and absorb the food molecules rather than biting off a chunk and digesting it internally, but these are only details compared with the difference between autotrophy and heterotrophy.

Pyramids of Numbers and Biomass

If we go to the trouble to count the number of organisms in different trophic levels, we sometimes find that there are more plants than herbivores and more herbivores than carnivores, a pattern called the **pyramid of numbers.** A similar pattern, the **pyramid of biomass,** almost always results if dry weight is used instead of numbers. The pyramids of biomass and numbers are aspects of the structure of the community (Figure 11-3). The functional basis of the pattern is in the flow of energy in the ecosystem.

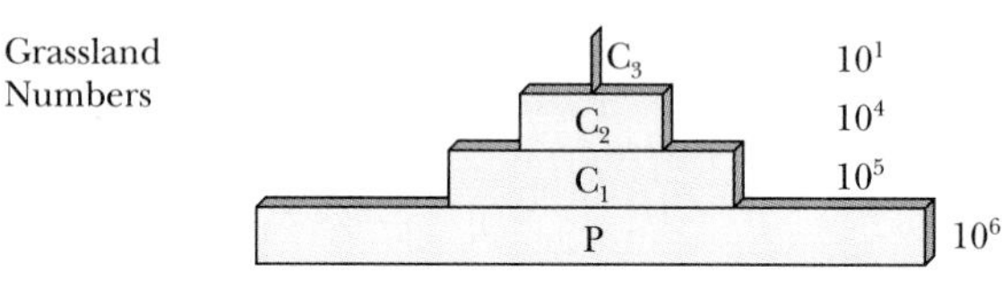

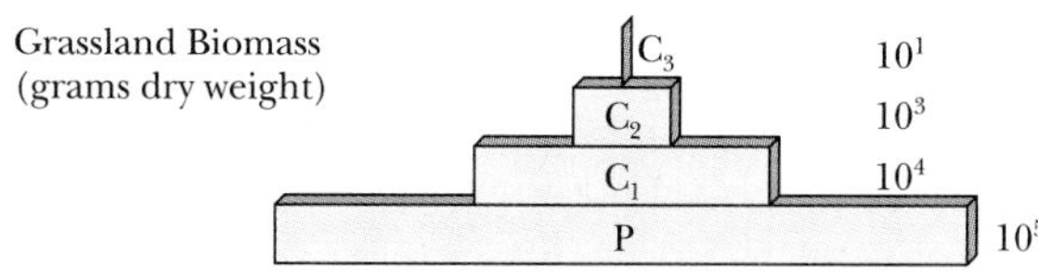

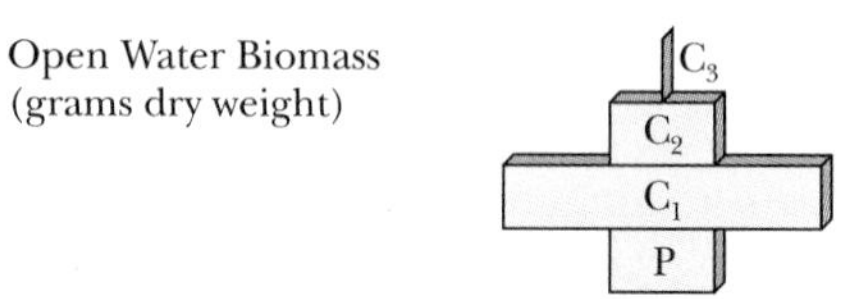

Figure 11-3

At the top are pyramids of numbers and of biomass (weight) such as might be expected in a thousand square meters of temperate grassland. C_1 = primary consumers, C_2 = secondary consumers, C_3 = tertiary consumers, and P = producers. There may be millions of grass plants, hundreds of thousands of grasshoppers, aphids, etc., thousands of carnivores such as spiders, and a few top carnivores like hawks or badgers. The pyramid in the middle represents the same situation but weight (specifically oven-dry weight) is used instead of numbers. The plants in a thousand square meters may weigh hundreds of thousands of grams, the primary consumers may weigh thousands of grams, etc. The third drawing, at the bottom, shows biomass by trophic level that does not form a pyramid. This sometimes occurs in open water of lakes or oceans where the producers are small and reproduce rapidly (single-celled algae, for example) and the consumers are large and long-lived (fish or large invertebrates).

Energy Flow

Solar Energy Input

The starting point for energy flow is sunlight. Of the solar radiation reaching the earth's atmosphere, about 30% is reflected into space, about 20% is absorbed by the atmosphere, and about 50% is absorbed as heat by ground, water, or vegetation. This does not seem to leave much for photosynthesis and, in fact, far less than 1% of the sunlight reaching the atmosphere is fixed in photosynthesis.

Although only a small fraction of the sun's energy actually enters the metabolism of the community, this is not to say that the energy absorbed as, or converted to, heat has no ecological effects. At the local level, heat from sunlight allows poikilotherms to warm up enough to become active and provides the energy for such processes as melting the ice in lakes in the spring and warming up the surface waters so that summer stratification occurs. On a global scale, sunlight powers the winds, the ocean currents, and the hydrological cycle.

Every point on the earth's surface is lighted for six months and is in the dark for six months. The only difference from place to place is the size of the packages, which range from six months of light alternating with six months of dark at the poles to a daily alternation of 12 hours of light and dark at the equator (Figure 3-13). This does not mean, however, that each spot on the earth receives the same amount of energy.

The main pattern of geographical variation in the amount of energy a spot receives is based on factors related to latitude. The sun's rays strike the earth's surface more obliquely at higher latitudes, spreading the energy out over a larger area. The total energy received per square meter at the equator is about 2.4 times that at the poles (Gates 1980).

Regional variations in solar energy input are produced by factors such as cloudiness. Overall, yearly energy input as sunlight, in kilocalories per square meter per year, varies from below 700,000 in the polar regions to above 2 million in tropical areas with few cloudy days (Figure 11-4).* These figures are for horizontal sites. A square meter of north-facing cliff may receive only one fourth as much energy as a square meter of horizontal surface (Gates 1980: Figure 6.26).

* For those who prefer SI units, recall that 1 kcal = 4.186 kJ.

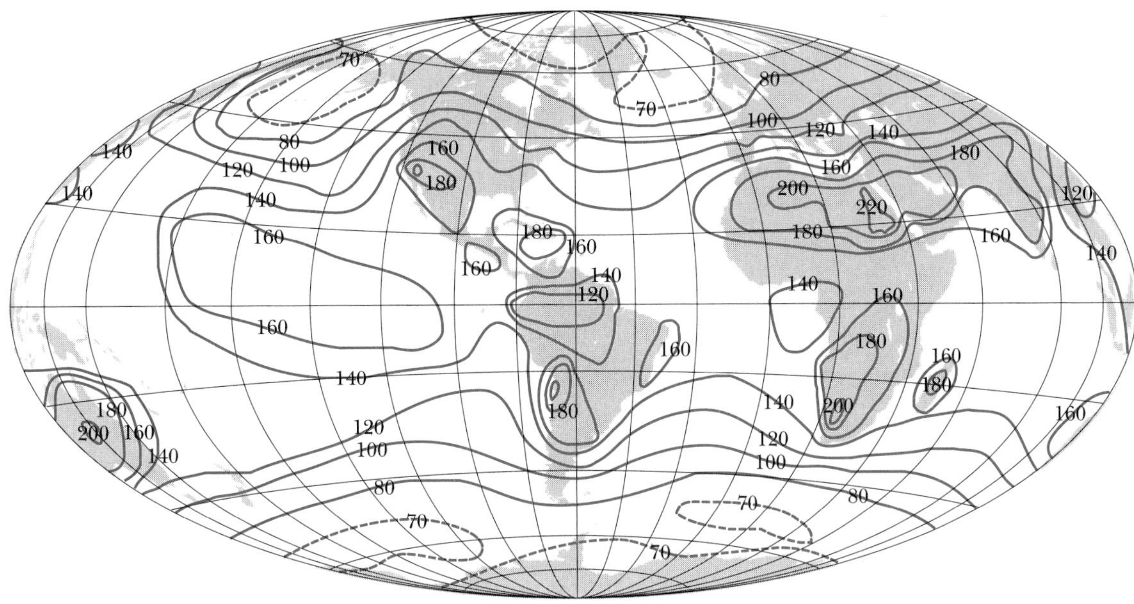

Figure 11-4

World distribution of solar energy. Figures are kilocalories per square centimeter per year. (From *Solar Energy 5*, H. E. Landsberg, "Solar radiation at the earth's surface," copyright 1961 Pergamon Journals, Ltd. Reprinted with permission.)

Energy Flow Within the Ecosystem

Let us trace the path of energy in the community (Figures 11-5 and 11-6). Of the energy stored in organic compounds in photosynthesis, the green plants themselves use some in respiration for their own growth and maintenance, with this energy being given off as heat to the surroundings. Primary consumers, or herbivores, eat plants. Some of the energy obtained in this way is stored in the growth

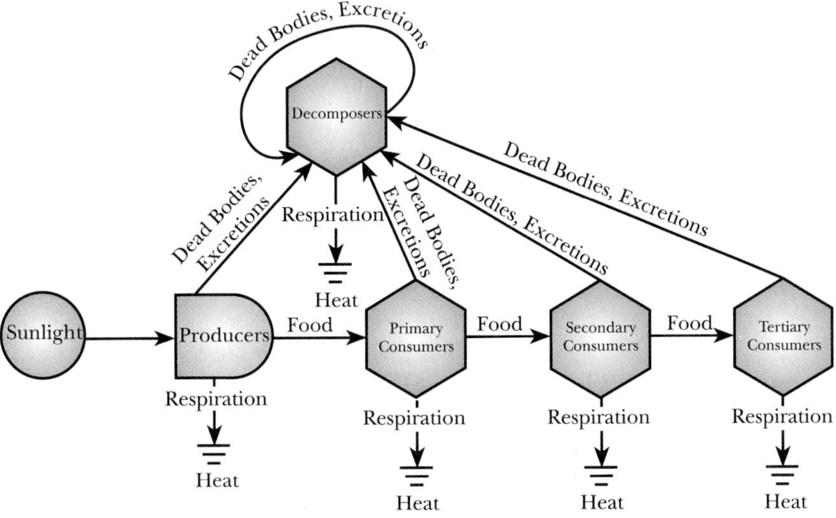

Figure 11-5

Energy flow in an ecosystem.

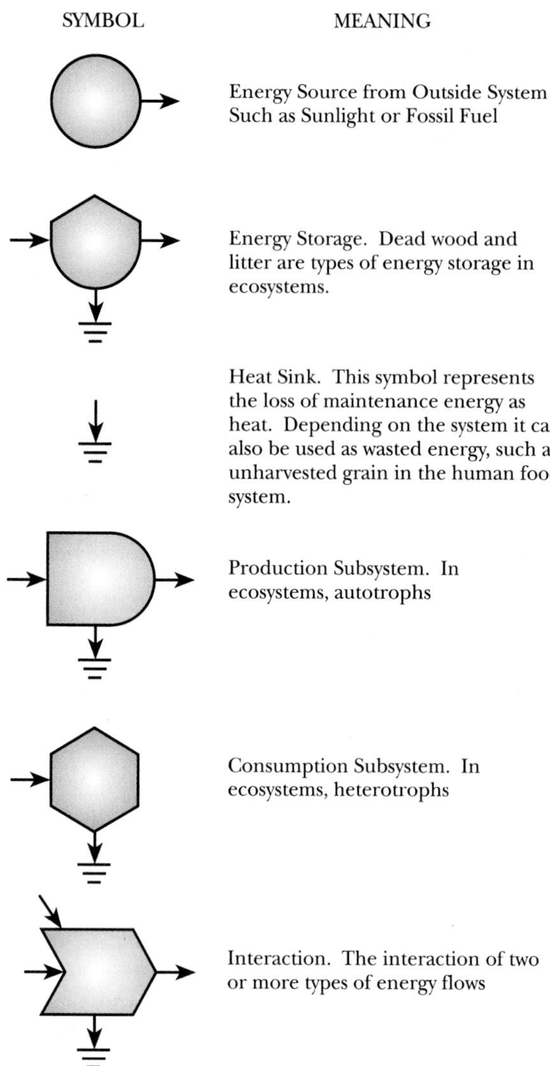
SYMBOL
MEANING
Energy Source from Outside System Such as Sunlight or Fossil Fuel
Energy Storage. Dead wood and litter are types of energy storage in ecosystems.
Heat Sink. This symbol represents the loss of maintenance energy as heat. Depending on the system it can also be used as wasted energy, such as unharvested grain in the human food system.
Production Subsystem. In ecosystems, autotrophs
Consumption Subsystem. In ecosystems, heterotrophs
Interaction. The interaction of two or more types of energy flows

Figure 11-6

The symbols used in most of the energy flow diagrams in this chapter are those employed by H. T. Odum (1971; also see Odum & Odum 1981). The diagrams are all understandable based just on the words and arrows but knowing the symbols allows quicker comprehension.

of new tissue (and, in reproduction, new individuals), but much is used for repair of tissues, locomotion, and other activities that herbivores must perform to maintain themselves. Like the respiratory energy of plants, this energy is eventually converted to heat. Note that the herbivores have much less energy available to them than the plants originally produced in photosynthesis. The energy that the plants used in respiration is gone; also, some parts of the plant die before being eaten by herbivores and become food for decomposers, rather than for herbivores such as grasshoppers or deer. A good part of a plant leaf or the rest of a plant body is indigestible to most herbivores. These indigestible remains still contain energy and, by forming the major parts of the feces of herbivores, become food for other decomposers.

The carnivores, or secondary consumers, obtain their energy from the herbivores in the same way that herbivores obtain theirs from plants. The carnivore uses some of the digested herbivore tissue for making new cells and tissue and for reproducing; the rest is respired to provide the energy for carrying on these activities. The energy available to the carnivore is, of course, much less than that taken in by the herbivore. Energy in materials that proved indigestible to the herbivore is gone, and so is the large amount of energy used by the herbivores in their own maintenance. Also, like the primary consumer, the secondary consumer is not completely efficient in harvesting the available food nor in digesting what it does harvest.

Tertiary consumers feed on secondary consumers, and quaternary consumers, if there are any in the ecosystem, feed on tertiary consumers. The foregoing processes occur at each of these levels (Figure 11-7).

It is worthwhile to look quantitatively at the amount of energy lost between one trophic level and the next in the flow of energy in the ecosystem. Let us say that an average figure for the energy in sunlight striking 1 square meter of ground in the United States is about 1.5 million kcal per year (the actual figure will vary a bit depending on such factors as latitude and cloudiness). Only a fraction of this is stored in photosynthesis; we can use 1% as a generous figure. Thus, about 15,000 kcal of the sunlight's energy are stored by plants in photosynthesis. The portion of this used by plants in their own maintenance varies. If 40% is used in plant respiration, then 60% of 15,000 is stored in new plant biomass. Consequently, the original 1.5 million kcal are now reduced to 9000 kcal potentially available to primary consumers.

In most ecosystems plant tissue is not harvested by grazers or browsers with high efficiency. Consumption by herbivores varies over a wide range but

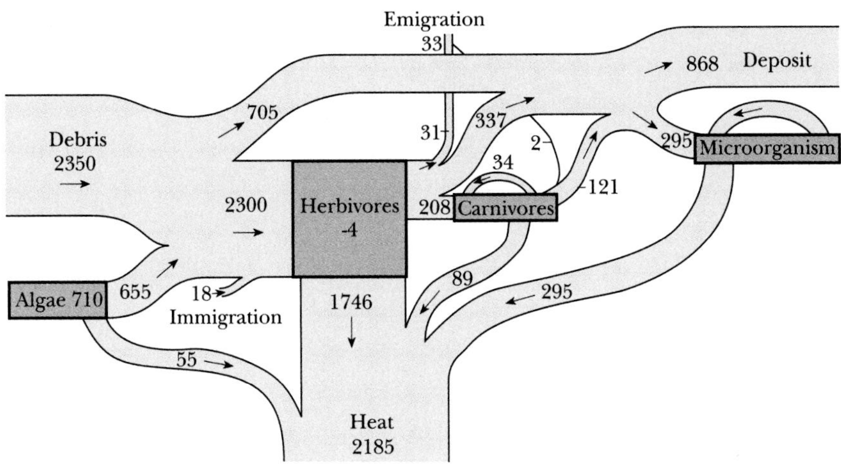

Figure 11-7

Energy flow for Root Spring, Concord, Massachusetts. Studying the energy flow through all the components of an ecosystem is an enormous task, and therefore most attempts involve many approximations. Many of the problems with precision were avoided in this study by dealing with a small, cold-water spring only 2 meters in diameter. The figures are kilocalories per square meter per year (figures in the herbivore and carnivore boxes are changes in standing crop). This ecosystem differs from the general example in the text in that import of energy in the form of organic matter, mainly leaves from surrounding trees, plays a large role. Not all of the energy entering is used, so that some goes into deposits on the bottom of the stream which will, in time, cause the spring to fill in. (From "Community metabolism in a temperate cold spring" by J. M. Teal, *Ecological Monographs,* 1957, *27,* 283–302. Copyright © 1957 by the Ecological Society of America.)

is usually less than 20%. The rest of the plant material dies and goes to decomposer food chains. If we take 20% as another generous value, then about 1800 kcal of plant material are eaten by herbivores, or primary consumers. Of this amount, very roughly 10% is stored as new herbivore tissue and is thus available to the secondary consumers, the first-order carnivores. The other 90% either is used in maintenance of the herbivores and lost as heat or goes to decomposers as feces and excretions. Accordingly, about 180 of the 1800 kcal of plant material eaten by herbivores become new biomass.

If we assume that secondary consumers are slightly more efficient at harvesting the new biomass available to them, taking 30% instead of 20%, then 54 of the 180 kcal available in herbivores are ingested by first-level carnivores. Using the same 10% efficiency figure that we used for the primary consumers for converting food to new biomass, these 54 kcal diminish to 5.4, which are stored as new carnivore protoplasm. This amount is the energy potentially available to tertiary consumers. If we assume that they catch 30% and convert 10% of that, as we did for secondary consumers, then they ingest food containing 1.6 kcal and are able to produce new biomass containing 0.16 kcal.

From this it should be plain why the pyramids of numbers and biomass occur. Less and less food energy is available at each level (Figure 11-8). If a herbivore can find as much food as it needs in 1 acre, a secondary consumer of the same size will need 10 acres, and a tertiary consumer will need 100. This is one main reason why hawks and trout are relatively rare and mice and bluegill sunfish are common. It is also the reason why two people per square mile can live on trout but 2000 must eat rice.

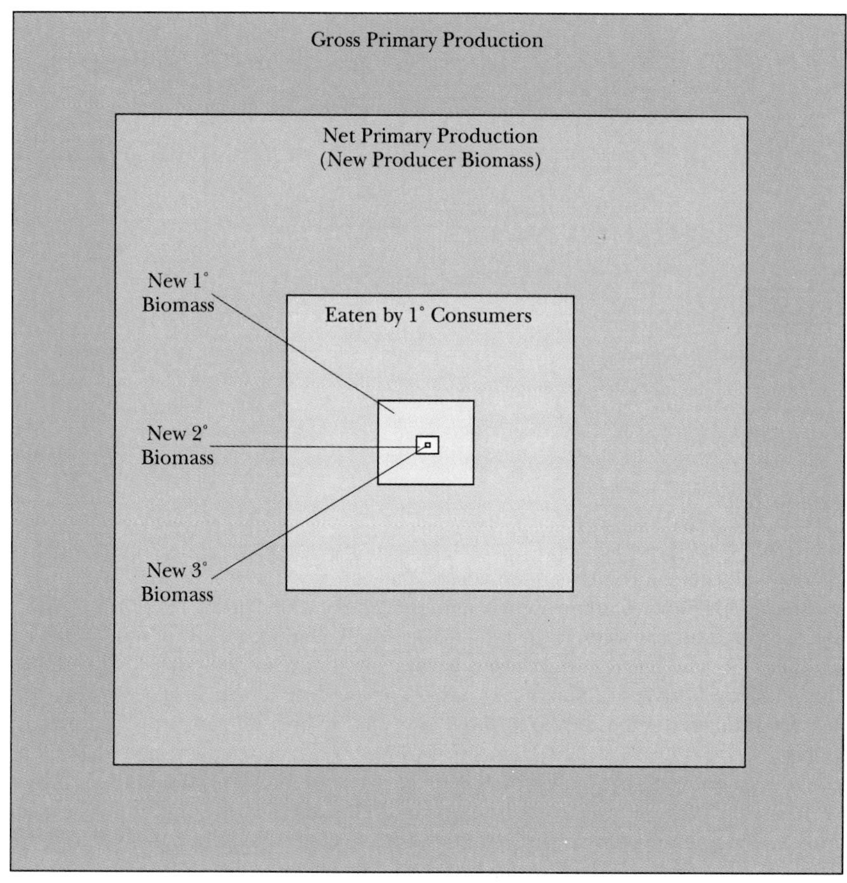

Figure 11-8

The area of each square represents the amount of energy at various stages of energy flow in an ecosystem. The large outer square is proportional to yearly gross production by the plants. The next square shows how much energy is left after plant respiration is subtracted. The three inner squares are new consumer biomass. In a sense, this is a view looking down on an energy flow pyramid.

Grazing Food Chains and Detritus Food Chains

Up to now, we have lumped under the term ''decomposer'' all the organisms that gain their energy from dead bodies or plants or animals, feces, and other wastes and excretions. In fact, dead organic matter, or detritus, is the starting point in most ecosystems for a series of food chains as complex as the grazing food chains that start with living plants (Darnell 1964). In most ecosystems, the total amount of energy flowing along these pathways is greater than that which travels the more conspicuous route from live plant to grazer to hawk or fox (Figure 11-9).

Bacteria, fungi, and actinomycetes grow on dead leaves, twigs, fruits, pollen, frass, and the other organic debris that fall to the ground (Dindal 1971). As shown in Figure 11-10, some animals also make direct use of this material. Earthworms, isopods, and slugs are examples. Other animals feed on the molds and other fungi and the bacteria. The actinomycetes, protected by antibiotics, seem to have relatively few consumers. As tertiary consumers, ground and rove beetles, centipedes, and spiders are found preying on some of the larger creatures and pseudoscorpions and predatory mites taking some of the smaller ones. Some of the larger invertebrates at all trophic levels may be eaten by ground-dwelling mammals and birds, such as short-tailed shrews, American robins, and ovenbirds.

Although detritus food chains are important in all ecosystems, they are outstandingly so in streams (Figure 11-11). A high percentage of the rich stream invertebrate fauna of caddis flies, mayflies, midges, stone flies, and isopods are detritus-feeders. In streams (as in small springs, Figure 11-7), much of the organic matter is imported (in the form of leaves, twigs, and logs) rather than being produced by the stream's own algae and macrophytes. In shaded headwater streams, the energy

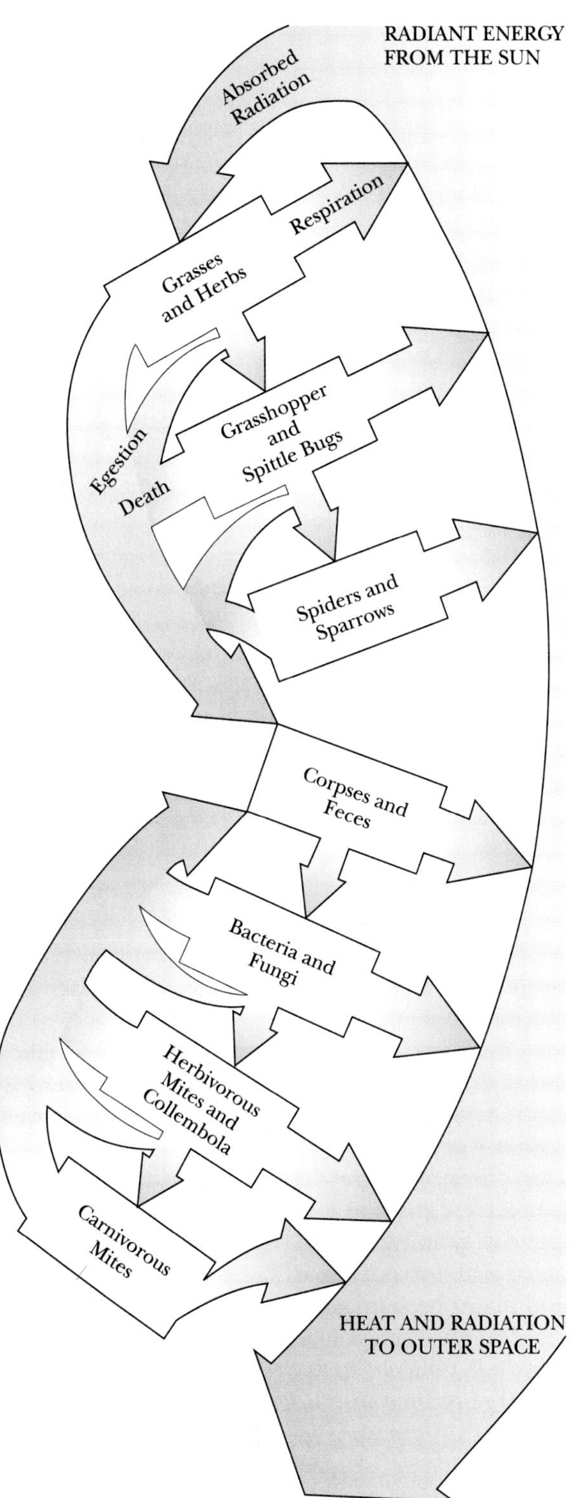

base of the stream's food web can be more than 99% imported detritus (Fisher and Likens 1977).

Biomass

In the previous sections we have been discussing energy flow. At any one time each trophic level contains some amount of energy stored as biomass, often referred to as the standing crop. The pyramid of biomass gives an indication of the amount of energy present at a particular time. To understand the relationship between energy flow and the energy present as standing crop we can compare energy flow with cash flow in and out of your checking account. Suppose that your account has $9.41 in it at the beginning of the year, and during the course of the year you put money in and take money out. If $1000 is deposited and $1000 taken out, then $9.41 still remains at the end of the year. The $9.41 is analogous to standing crop, and the cash flow of $1000 corresponds to energy flow.

Energy is deposited in the producer trophic level of an ecosystem by photosynthesis and in a consumer trophic level by assimilation (Figure 11-12). Energy is withdrawn from a given trophic level by the organisms' respiration and by consumption on that trophic level by the organisms of the next higher one. In some ecosystems import and export of biomass, discussed later, can be important.

If withdrawals equal deposits, biomass stays the same from year to year. Although there may be high and low points within the year, the trophic level is at a steady state. On the other hand, if you put

Figure 11-9

The importance of food chains that start with dead leaves, limbs, carrion, and "corpses and feces" in general is emphasized in this diagram of energy flow in an old field in Michigan. The food chains involving grasshoppers and sparrows are what we see when we look at a grassland but the food chains of bacteria and mites often process a greater proportion of the energy that the green plants have fixed. (From "The role of soil arthropods in the energetics of an old field community," by M. D. Engelmann, *Ecological Monographs,* 1961, *31,* 221–238. Copyright © 1961 by the Ecological Society of America.)

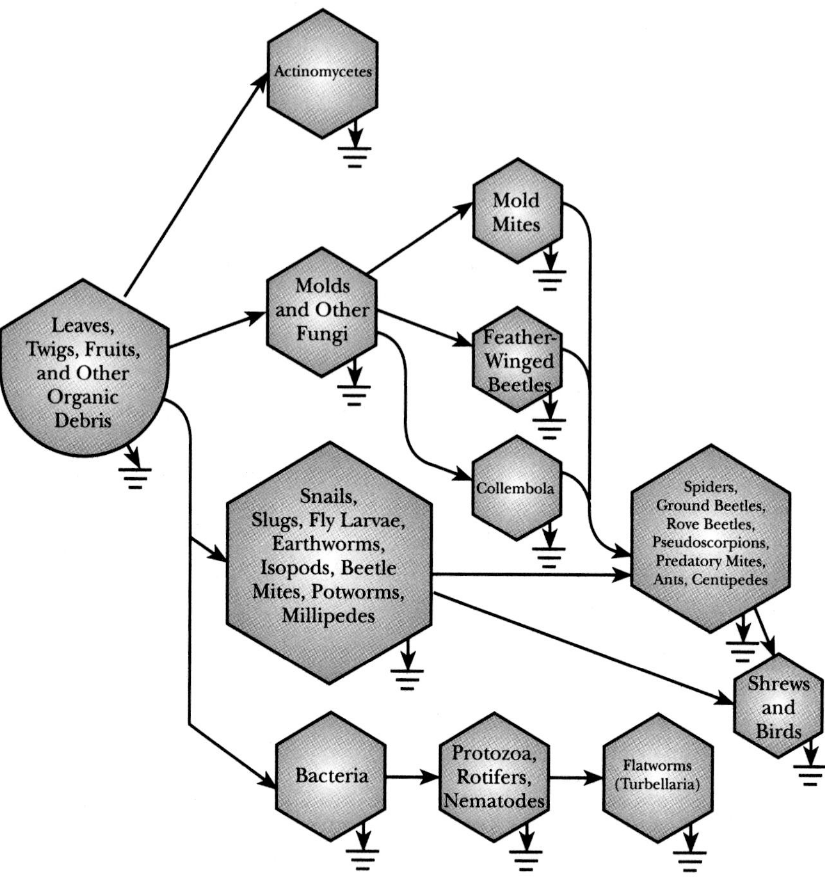

Figure 11-10

Energy flow in terrestrial detritus food chains. A small amount of production may occur within the soil by green algae or cyanophytes. Beetle mites are Oribatei and Cryptostomata, mold mites are Astigmata, and feather-winged beetles are Ptiliidae. Based mainly on Dindal 1971.

$1000 in your bank account during the year and take out only $900, your balance will be $109.41 at the end of the year. A positive balance for a trophic level implies that energy is accumulating somewhere—either as larger organisms (as in tree growth) or in more of them (population growth).

Productivity

Primary and Secondary Production

The total energy storage by autotrophs in an ecosystem is referred to as **gross primary production** (GPP). For most ecosystems, gross primary production is equal to total photosynthesis. The plants themselves use a considerable amount of this energy in their own respiration. The stored energy left after plant respiration is termed **net primary production** (NPP).

We are generally interested in production **rates,** that is, production per time unit. **Productivity** is the term used to refer to production rate. **Gross primary productivity,** for example, is the total photosynthesis in a specified time period. A year is a convenient interval to use in stating productivities, but the growing season could be used, and for intensive analysis of a particular ecosystem daily values might be calculated.

Energy storage at consumer levels of the ecosystem is referred to as **secondary production.** Secondary production is fundamentally different from primary production. In primary production, new organic matter is actually produced. Consumers, such as mice or cats, "produce" new mouse or cat tissue, but they do so by converting organic matter of the species preceding them in the food chain

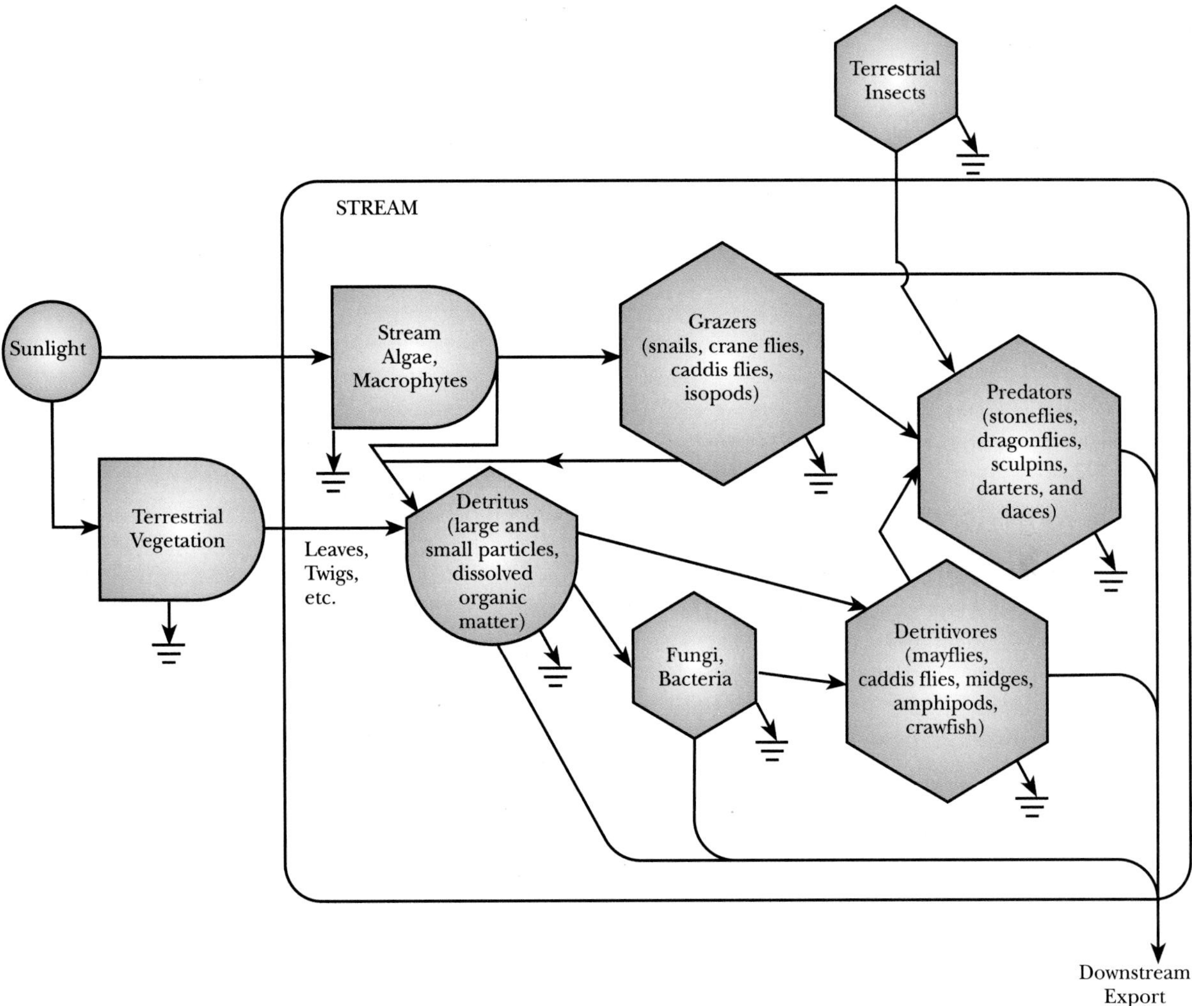

Figure 11-11

Simplified energy flow diagram of a stream ecosystem. The detritus food chain is especially important in stream ecosystems, where much of the energy flow is based on organic matter from terrestrial sources that is washed or blown in. The greatest amount of energy comes from terrestrial vegetation but also terrestrial animals, especially insects, provide food for many of the fish and predaceous invertebrates. (Based mainly on Minckley 1963 and Cummins 1974.)

(plants for mice, mice for cats) into organic matter stored in their own tissues. Of course, most of the energy in the tissues they eat does not end up stored in their tissues but is used in their respiration and given off as heat.

Usually, the terms "gross" and "net" are not applied to secondary production. Rather secondary production refers to new biomass at a consumer trophic level, that is, to energy storage after the consumers' respiration has already been subtracted. This is analogous to net production of plants. "Assimilation," or "assimilated energy," is the term usually applied to the consumer analog of gross primary production (Figure 2-10). That is,

Figure 11-12

Inputs to and outputs from biomass for a trophic level.

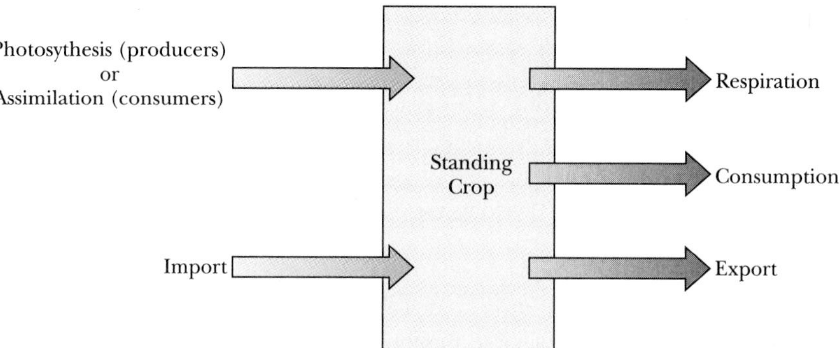

Assimilation = gross energy intake − egested and excretory energy

Annual net production is a convenient basis for comparing various ecosystems because it is the energy potentially available over a full seasonal cycle to the organisms of the trophic levels past the producers. It is also by far the easiest energy quantity to measure in most ecosystems, being roughly the new plant biomass added in a year. Problems of measurement do exist, including the following:

1. It is sometimes impractical to sample frequently enough to catch certain temporary plant parts. For example, some grassland goldenrods have large basal leaves that die back before flowering. Many techniques of measuring net production will miss this contribution (Brewer 1985).
2. Storage of new biomass in roots and other underground parts is hard to measure for perennial plants (Brewer and McCann 1982).
3. It may be difficult to correct for the removal of plant biomass by grazers. Because grazing can stimulate production, measurements in exclosures (that is, plots where grazers are fenced out) may underestimate production.
4. Use of primary production by such specialized primary consumers as parasites and mutualists that live symbiotically with the producer may be even more difficult to measure accurately.

Even so, determination of net primary production is straightforward compared with many features of ecosystem energetics and is the starting point for most energy flow studies. Gross production is usually calculated by adding in an estimate of producer respiration.

Energy units and weight (biomass) are two different ways of expressing productivity. Either may be used because the energy is stored in the organic compounds that compose the biomass. Dry weights are taken because the water content of different organisms varies greatly.

The energy content of a given weight of plant or animal material is calculated by burning the dry material in a calorimeter and determining the amount of heat given off. This has now been done for many kinds of plant and animal materials (Cummins and Wuycheck 1971). The amount of heat given off runs about 4 to 4½ kcal (16.7 to 18.8 kJ)/g of oven-dried plant matter; animal material is usually 5 to 5½ kcal (20.9 to 23.0 kJ)/g dry weight. Using these figures (which will be only approximate for any particular plant or animal), we can convert from weight to energy and vice versa. Some researchers, mainly limnologists, state productivity figures in grams of carbon (g C). Just under half the dry weight of organic matter is carbon, on the average, the rest being hydrogen, oxygen, nitrogen, and various minerals. Accordingly, 1 g of carbon ≈ 2.2 g oven-dry weight.

Energetic Steady States

In the section on biomass, we talked about steady states for trophic levels. We can apply the same con cept to whole ecosystems. If as much energy is deposited (photosynthesis) as is withdrawn (respiration at all the trophic levels), the ecosystem is at an energetic steady state. That is, when an ecosystem is at a steady state,

GPP = sum of the respiratory losses at all trophic levels.

To put it another way, a steady state exists when community production equals community respiration. Any community production in excess of community respiration is called **net community production** (NCP). In steady-state communities, NCP = 0.

Table 11-1 compares energy flow in an immature forest (a pine plantation in England) and a mature one (a rain forest in Puerto Rico). The rain forest is obviously much more productive, but the object of the comparison is to show that the rain forest is at a steady state but the pine plantation is not. In the rain forest, the sum of the energy used in respiration by the plants, animals, and bacteria equals the energy fixed in photosynthesis. In the pine plantation, more energy is being fixed in photosynthesis than is being used (NCP > 0).The NCP is accumulating as biomass in tree trunks, litter, humus, etc.

Another way of saying exactly the same thing is to say that for communities at a steady state $P/R = 1$, where P is production and R is respiration. If NCP > 0, then $P/R > 1$. Such communities, where production outweighs respiration, are referred to as autotrophic, by analogy with autotrophic organisms. Communities in which $P < R$ are heterotrophic communities.

Early successional communities, such as pine plantations (Table 11-1), have P/R ratios greater than 1; they are autotrophic. Streams are a prime example of natural heterotrophic communities. The regular import of organic matter from terrestrial ecosystems and its subsequent use by bacteria, fungi, and detritivores (Figure 11-11) give most streams a P/R ratio below 1. Only in the mid-reaches of river systems may production in the stream itself be high enough and import low enough to produce a P/R above 1 (Cummins 1977).

Import and export are generally important in aquatic ecosystems. Streams import a great deal of biomass, but they also export downstream both plant and animal biomass. Larvae of aquatic insects, for example, may drift downstream from one section of a stream to another. When insects become adults, they may leave the stream entirely, taking their biomass with them. In a classic early study of Silver Springs, Florida, H. T. Odum (1957) concluded that downstream export allowed the system to be at a steady state (Table 11-2). Without the export, organic matter would have been accumulating, causing the spring to fill in.

Logic, and some measurements, make us believe that climax communities are at an energetic steady state. If biomass were changing, then the community would be changing in other ways also. It seems likely, however, that the P/R ratio can sta-

Table 11-1
Energy Flow in Two Forests

	Immature Forest ($kcal/m^2/yr$)	*Mature Forest ($kcal/m^2/yr$)*
Total photosynthesis (GPP)	12,200	45,000
Plant respiration	4,700	32,000
New plant tissue (NPP)	7,500	13,000
Heterotrophic respiration*	4,600	13,000
NCP	2,900	≈0
P/R ratio	1.3	1

Modified from Odum, E. P., *Fundamentals of Ecology,* Philadelphia, W. B. Saunders, 1971, Table 3-5, p. 46.

* Includes herbivores, carnivores, bacteria, and other decomposers.

Table 11-2
Energy Balance Sheet for Silver Springs, Florida

Silver Springs is a large spring in central Florida coming up out of the limestone that underlies the region. Its flow is carried away by the Silver River. The energy import is in the form of bread that concessionaires operating glass-bottomed boats toss overboard to attract fish so that the tourists see them.

Energy Flow	*Amount (in $kcal/m^2$ yr)*
Energy inputs	
Photosynthesis	20,810
Import	486
Total	21,296
Energy outputs	
Plant respiration	11,977
Heterotroph respiration	6,819
Export	2,500
Total	21,296

Data from Odum, H. T., "Trophic structure and productivity of Silver Springs, Florida," *Ecol. Mon.,* 27:55–112, 1957.

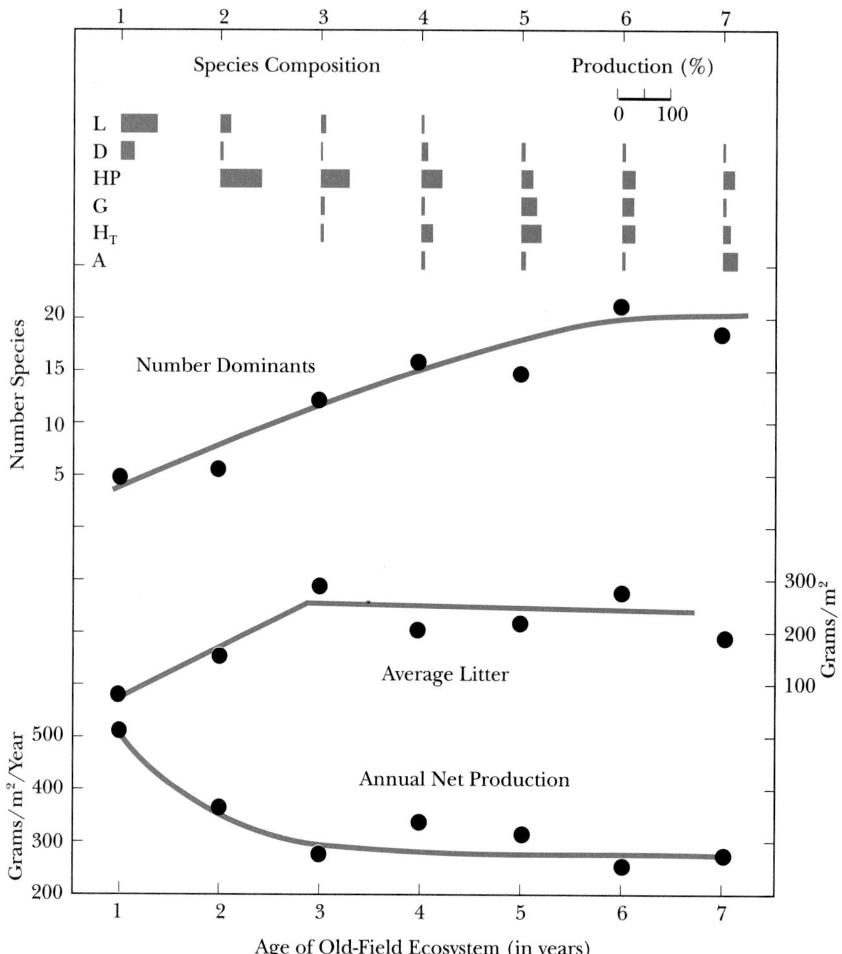

Figure 11-13

Although species composition was continuing to change through years 3–7 in Georgia old fields, net productivity and also the amount of stored energy had reached a temporary steady state. The plant species at the top are L = horseweed (*Conyza canadensis*), D = crabgrass (*Digitaria sanguinalis*), Hp = *Haplopappus divaricatus*, G = catfoot (*Gnaphalium obtusifolium*), Ht = camphorweed (*Heterotheca subaxillaris*), and A = broomsedge (*Andropogon virginicus*). (From "Organic production and turnover in old field succession" by E. P. Odum, *Ecology*, 1960, *41*, 34–49. Copyright © by the Ecological Society of America.)

bilize at 1 even though changes in composition continue to occur (Figure 11-13) (Odum 1960, Brewer et al. 1969). *P/R* ratios of various ecosystem types are diagrammed in Figure 11-14.

NPP in the Biosphere

Table 11-3 is an attempt (by R. H. Whittaker and G. E. Likens) to summarize information on the productivity of many different types of ecosystems. Some of the ecosystems could be subdivided further. For example, temperate grasslands vary greatly in productivity, with much of the variation associated with differences in moisture availability (Table 11-4). Using productivity figures from earth's ecosystems along with estimates of land area they occupy, it is possible to obtain global productivity totals. Such figures are given in the last section of this chapter.

Net yearly production ranges from zero in the driest deserts and other habitats too extreme to support plants to greater than 5000 g/m^2(Table 11-3). This latter value, upwards of 11 pounds of plant material added to 1 m^2 in one year, is a sizable amount of growth. Annual net production is probably between 500 and 2000 g/m^2 over much of the temperate part of the earth. On a pounds-per-acre basis, a familiar way of expressing agricultural production in English-speaking countries, this is about 5000 to 18,000 pounds per acre. These are not actually yields in the agricultural sense because they include all new protoplasm including stems, bark, roots, and pollen. Humans could obtain this

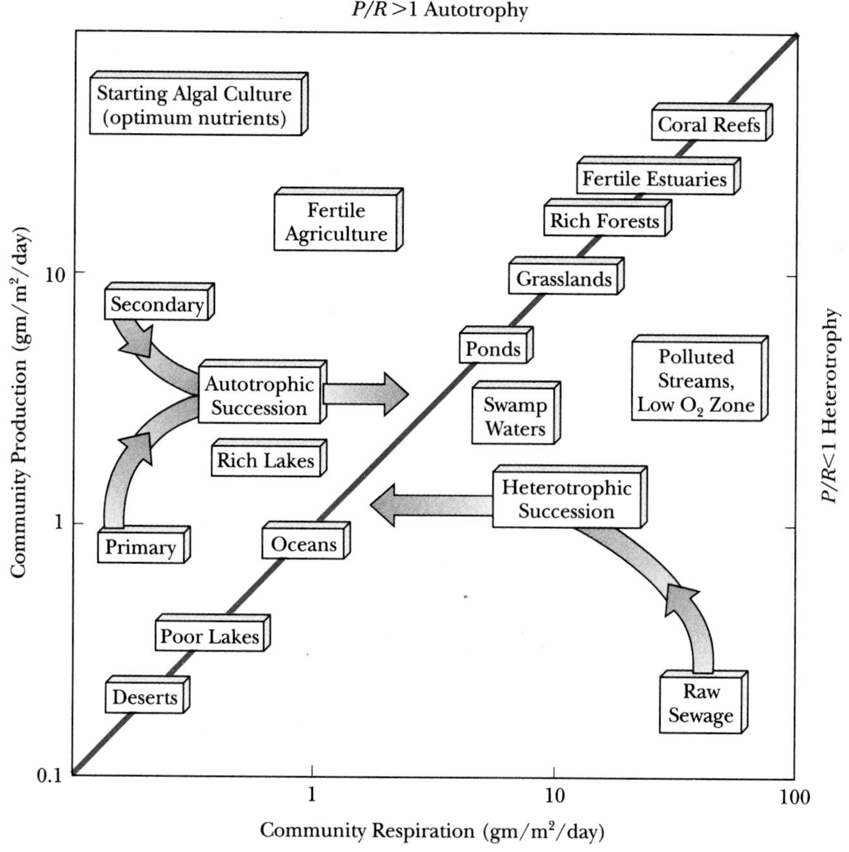

Figure 11-14

Community respiration plotted against community production for various ecosystems. Along the diagonal line, P = R, so the communities are at an energetic steady state. Heterotrophic communities, such as most streams, are in the lower right half, autotrophic communities, such as young forests are in the upper left. (Adapted from Odum 1983.)

amount of plant material as food only if they were able to eat all parts of the plant and even then only if they were able and willing to eliminate all other animals that might also consume parts of the plants.

What environmental factors contribute to high annual productivity? A good moisture supply, a long growing season, warm temperatures, and high fertility are all favorable. It is not surprising that marshes and estuaries where moisture is abundant and nutrients are supplied by runoff from the land have a high rate of production. Nor is it surprising that tropical ecosystems with a long growing season and high temperatures have a high annual production and that deserts with their low rainfall and high evaporation rate are unproductive. However, it is interesting to note that the productivity of the open ocean is only a little higher than that of deserts. The main reason appears to be a scarcity of chemical nutrients—phosphates, nitrates, or occasionally iron. The only really productive areas of the ocean are estuaries and coral reefs, where production may be as high as anywhere on earth. Areas of upwelling from the ocean floor (mainly along the western coasts of the continents) and other coastal areas where nutrients are carried out from rivers and estuaries have reasonably high productivity, similar to that of more terrestrial areas.

Relationships Between Productivity and Biomass

For the trees of a young oak-pine forest in New York, yearly net production was more than 10% of the biomass at that time (1060 compared with 9700 g/m^2; Table 11-5). In mature forests at or near a steady state, the ratio of production to biomass (*P/B* ratio) is much lower. The *P/B* ratio of the oak-pine forest was 0.11; for mature forests in the Great Smoky Mountains, whether hardwood or conifer, the *P/B* ratio was only 0.02 to 0.03 (Whittaker and Marks 1975).

Table 11-3
Estimates of NPP and Plant Biomass in Major Ecosystems

	Net Primary Productivity per Unit Area (g/m²/yr)		*Biomass per Unit Area (kg/m²)*	
Ecosystem Type	*Normal Range*	*Mean*	*Normal Range*	*Mean*
Tropical rain forest	1000–3500	2200	6–80	45
Tropical seasonal forest	1000–2500	1600	6–60	35
Temperate evergreen forest	600–2500	1300	6–200	35
Temperate deciduous forest	600–2500	1200	6–60	30
Boreal forest	400–2000	800	6–40	20
Woodland and shrubland	250–1200	700	2–20	6
Savanna	200–2000	900	0.2–15	4
Temperate grassland	200–1500	600	0.2–5	1.6
Tundra and alpine	10–400	140	0.1–3	0.6
Desert and semidesert scrub	10–250	90	0.1–4	0.7
Extreme desert, rock, sand, and ice	0–10	3	0–0.02	0.02
Cultivated land	100–3500	650	0.4–12	1
Swamp and marsh	800–3500	2000	3–50	15
Lake and stream	100–1500	250	0–0.1	0.02
Open ocean	2–400	125	0–0.005	0.003
Upwelling zones	400–1000	500	0.005–0.1	0.02
Continental shelf	200–600	360	0.001–0.04	0.01
Algal beds and reefs	500–4000	2500	0.04–4	2
Estuaries	200–3500	1500	0.01–6	1

Based on Whittaker, R. H., *Communities and Ecosystems,* 2nd ed., New York, Macmillan, 1975, p. 224. (After Whittaker and Likens in Lieth and Whittaker) Copyright 1975, The Macmillan Co.

Table 11-4
Productivity in Various North American Grasslands

Annual net aboveground production (grams per square meter, oven-dry weight) is measured.

Grassland Type	***Locality***	***Precipitation (mm/yr)***	***Productivity (g/m²)***	***Source***
Palouse prairie	Washington	150–175	82–114	Sims and Singh 1978
Desert grassland	Mexico	160–350	125–134	Sims and Singh 1978
	Idaho	372	207	Person 1965
Shortgrass prairie	Colorado	225–310	138–218	Sims and Singh 1978
	Texas	225–600	155–325	Sims and Singh 1978
Midgrass prairie	South Dakota	410–688	212–279	Sims and Singh 1978
	Kansas	475	363	Sims and Singh 1978
Tallgrass prairie	Oklahoma	650–940	290–416	Sims and Singh 1978
	Michigan	683	385–587	Brewer 1985
	Minnesota	696	807	Ovington et al. 1963
	Illinois	910	362–1536	Hadley and Kieckhefer 1963

Table 11-5
Biomass and Productivity Estimates for an Oak-Pine Forest in New York

This was a young forest with trees 40–45 years old and about 25 feet (7.5 m) tall.

	Annual Net Productivity		*Biomass*	
Trait	*g/m²/yr*	*% of Total*	*g/m²*	*% of Total*
Tree net production				
Stem wood and bark	174.9	16.5	4317	44.5
Branch wood and bark	247.0	23.3	1639	16.9
Leaves	350.9	33.1	408	4.2
Fruits and flowers	22.2	2.1	19	0.2
Roots	265.0	25.0	3317	34.2
Total tree production/biomass	1060	100.0	9700	100.0
Undergrowth production/biomass	134	—	460	—
Total NPP/biomass	1194	—	10,160	—

Based on data from Whittaker and Woodwell, "Structure, production, and diversity of the oak–pine forest of Brookhaven, New York," *J. Ecol.*, 57:155–174, 1969.

In these two situations, the *P/B* ratio is measuring different things. For immature forests, the ratio is a measure of biomass accumulation. When the forest is very young and wood and litter are being added rapidly, the *P/B* ratio is high. As the rate of biomass accumulation slows, the *P/B* ratio drops. Eventually, in the steady-state forest, the *P/B* ratio is a measure of population turnover rate.

Comparing steady-state communities, then, the *P/B* ratio is higher where generation length and life span are short, generally implying small body sizes, and low where generation length and life span are long and body sizes larger. Consequently, the *P/B* ratio is high in annual-dominated croplands (at a steady state by virtue of human intervention), fairly high in natural grassland and savanna, and low in forest (Table 11-3).

In aquatic communities where phytoplankton are the main producers, *P/B* ratios are several orders of magnitude greater. Fresh-water phytoplankton, for example, have a *P/B* ratio averaging 113 (Brylinsky 1980). In other words, yearly net primary productivity is 113 times the biomass that is present at any one time. In southerly lakes where the algae may go through more generations in the course of a year, the *P/B* ratio may run even higher.

We can make similar calculations for animals. In the size range between ants and antelopes, the *P/B* ratio runs from 0.1 to 10 (Table 11-6). Smaller

Table 11-6
Biomass and Productivity for Various Animals in African Tropical Savanna

Group	*Annual Productivity (P) (g/m²/yr)*	*Average Biomass (B) (g/m²)*	**P/B**
Primary consumers			
Rodents	0.07	0.02	2.8
Grasshoppers	0.40	0.06	7.1
Earthworms	7.49	3.60	2.1
Uganda cob*	0.16	2.18	0.1
Secondary consumers			
Birds	<0.01	0.01	0.4
Spiders	0.45	0.06	7.0
Driver ants	6.00	0.96	6.2
Large predators†	—	0.01–0.10	—

Based on Lamotte 1977, except as shown.

* From Buechner and Golley 1967.

† Includes lion, leopard, cheetah, spotted hyena, and wild dog in five African reserves. Based on Schaller, *The Serengeti Lion. A Study of Predator-Prey Relations*, Chicago, University of Chicago Press, 1972.

invertebrates, with more generations per year, may have higher ratios; zooplankton ratios are mostly between 10 and 15 (Brylinsky 1980). Associated with low P/B ratios are large size, fewer breeding periods per year (birds versus rodents, for example), and homoiothermy.

Ecological Efficiencies

Efficiencies are ratios of output to input. A large number of ratios can be calculated that trace the efficiency with which energy is transferred along the various steps of its movement through ecosystems. In the earlier section on energy flow, we used some generalized values. Here, we look at some of the more meaningful ratios in more detail.

Producer-Level Efficiencies

GPP/Sunlight

Algal cultures in dim light in the laboratory can approach the upper limit of about 20% set by the photosynthetic process (Bonner 1962), but both natural and cultivated ecosystems in the real world fall far below this. Record high field yields of corn biomass hit somewhere close to 25% of the potential set by the photosynthetic process for the growing season (see Bugbee and Monje 1992); however, few, if any, ecosystems have yearly efficiencies above 5%. E. P. Odum (1983, also see Gates 1980) suggests that an average efficiency for the biosphere is about 0.2%.

Why is sunlight converted to photosynthate so inefficiently? Several factors are important.

1. Only about 44% of the energy of sunlight is in wavelengths that are used in photosynthesis (Gates 1980:p. 135).
2. For several reasons, the efficiency with which leaves convert light to photosynthate decreases as light intensities increase (Good and Bell 1980), so the high efficiencies in dim light are irrelevant to the general productivity of most ecosystems.
3. Over much of the earth, some periods of the year are unsuitable for plant production. The growing season in most temperate and arctic regions depends ultimately on temperature. Elsewhere production may be curtailed during a dry season.
4. Even during the growing season, productivity on terrestrial sites may be limited by moisture, nutrients, or temperature; consequently, the available sunlight is not fully utilized.
5. On sites that are moist and otherwise favorable, plants will reach an upper limit of photosynthesis based on the shading of one leaf (or chloroplast) by another. For crop plants, 4 or 5 m of leaf surface over each meter of ground† seems to give maximum production; making the plants grow thicker than that does not add anything.
6. Productivity in aquatic habitats may be limited by nutrients or temperature. Also, because of the absorption and scattering of light in the water, only a small fraction of the light hitting a square meter of water surface is available for plants a few meters deep. Even in distilled water, less than half the visible light penetrates 20 m (Clarke 1954, also see Wetzel 1983: Table 5-2). In the ocean, the depth at which half the light is lost is usually 10 m or less; in an average lake, it is likely to be only 1 or 2 m.

NPP/GPP

The efficiency with which gross production is converted to net production depends on how much energy the plant populations of the ecosystem use in their own maintenance. It is not easy to measure autotroph respiration at the ecosystem level because of difficulties in detecting immediate utilization of photosynthate and also in separating plant respiration from the respiration of closely associated heterotrophs, such as parasites, commensals, and mutualists. Working with the data as they are, we find that autotrophic respiration lies generally between 20 and 75% of GPP, implying NPP/GPP efficiencies between 25 and 80%.

One fairly evident trend (Whittaker 1975) is a tendency for communities with a large amount of accumulated biomass (forests) to have higher values for respiration, presumably based on higher costs for maintaining a large amount of perennial

† This ratio, leaf area/ground area, is termed the **leaf-area index** (LAI).

nonphotosynthesizing structure (Kira and Shidei 1967). Communities where biomass is less than about four times annual NPP, such as grasslands, crops, and algal communities in water, tend to have lower values for respiration. As a very crude rule of thumb, the latter communities have efficiencies mostly 50% and above, whereas forests have efficiencies below 50%.

A second trend (Whittaker 1975, Saldarriaga and Luxmoore 1991) seems to be that communities in tropical regions tend to have higher values for respiration and, hence, lower efficiencies than communities in cooler regions. This may be related to higher respiration rates at the higher temperatures or to the greater energy requirements for maintaining leaves, etc., on a year-round basis, or both.

Harvesting Efficiency of Herbivores

The ratio gross energy intake of herbivores/NPP measures the efficiency with which the herbivores harvest new plant biomass. Material that is not harvested in this way goes sooner or later to detritus food chains. In a sense, the beetles, collembolans, fungi, and bacteria that make use of dead leaves or wood are also herbivores. So, likewise, are plant endoparasites such as nematodes, and mutualists, such as nitrogen-fixing bacteria. Most studies, however, have analyzed only the efficiency of grazing food chains under this category.

At times, harvest may be very high in any ecosystem. For example, trees may be defoliated by moth larvae, resulting in 30 to 40% of that year's net productivity being harvested (Figure 11-15). Ordinarily, however, the percentage of the forest leaves that is consumed is much smaller. If we stroll through a deciduous forest at the end of the growing season, a little before leaf fall, and look at the leaves, we see a few holes, a few edges eaten off, and a few areas in which the cells have been mined out. If we measure the total biomass consumed in these ways and add in a little for leaves wholly consumed and, thus, missing from our sample, the percentage consumption usually comes out below 10% and often below 5%.

In grassland and other ecosystems dominated by herbaceous plants, the percentage consumption also differs between average years and years of locust or rodent peaks. In average years on sites where large herbivores are absent, the percentage of aboveground NPP consumed seems similar to the figures for forest tree leaves—below 5% as an average figure (Scott et al. 1979, Pfeiffer and Wiegert 1981:Table 5.1). Where large grazers, such as zebras or bison, are present, the percentage may be much higher, 30% or more (Wiegert and Evans 1967). Below-ground consumption, on roots and rhizomes, has been studied on grassland sites and has

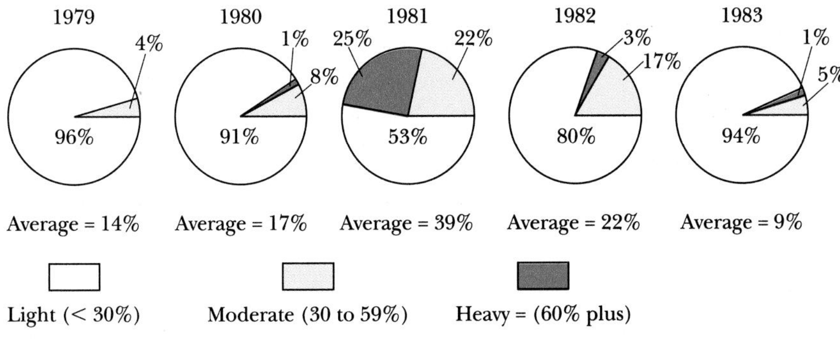

Figure 11-15

These pie diagrams show the percentages of forest plots in central Pennsylvania with various levels of defoliation over a five-year period. Gypsy moths had recently invaded the region and much of the defoliation was their work. In most years defoliation was light but in 1981 nearly half the plots showed 30% defoliation or higher. (Adapted from Gansner & Herrick 1985.)

been found to be higher than consumption above the ground. Below-ground consumption, mostly by nematodes, seems to average about 20% (Scott et al. 1979).

In aquatic ecosystems, where the plants are usually small and put little energy into hard-to-eat materials, herbivore consumption seems to run 20 to 40%.

Herbivores, in terrestrial ecosystems, at least, are often sloppy feeders, clipping off leaf fragments or whole leaves that they do not consume. This material, which can equal that actually eaten, goes more immediately to the decomposer food chain than does the undisturbed plant.

Consumer Efficiencies

Within-Trophic-Level Efficiencies

The energy consumed by the herbivores forms the basis for their existence and also that of the secondary and tertiary consumers of the grazing food chains. We can follow the efficiency with which energy is transferred further in the ecosystem by many other ratios, both within and between trophic levels.

The basic within-trophic-level ratio is **ecological growth efficiency:**

$$\frac{\text{New biomass}}{\text{Gross energy intake}}.$$

But this is less informative than its two component ratios, **assimilation efficiency** and **tissue growth efficiency.**

There are two ways in which energy is lost between energy intake and the addition of new biomass (in the form of growth or reproduction). One is the egestion of undigested, unabsorbed, or unassimilated food. Assimilation efficiency measures the efficiency at this stage, and is

$$\frac{\text{Assimilation}}{\text{Gross energy intake}}.$$

Energy is also lost through use in the animal's maintenance, and we may write tissue growth efficiency as

$$\frac{\text{New biomass}}{\text{Assimilation}}.$$

The efficiency with which the animal assimilates food depends on the nature of the food and, consequently, varies over a wide range. Cockroaches fed on yeast and sugar assimilate 100% of their gross energy intake but some detritivores, eating food containing compounds highly resistant to digestion, may have assimilation efficiencies below 10% (LaMotte and Bourlière 1983). Browser and grazer efficiencies seem to run 30 to 60%. Carnivores average higher, 60 to 90%.

Tissue growth efficiency varies with the energy used in maintenance. This includes the energy for running from enemies, foraging, digesting food, and performing all the other activities of just staying alive. Steers in a feedlot may turn more than half of the energy they assimilate into beef, whereas elephants on their natural range may have an efficiency of just over 1% (Petrides and Swank 1965). The feedlot efficiency is achieved largely by substituting human and fossil fuel energy for maintenance activities that, in nature, the animals have to provide for themselves.

Many factors affect the percentage of assimilated energy that an individual animal spends in maintenance, including activity, size, and age. Insects that gather nectar or seeds will probably have to spend more energy on foraging than ones that suck sap or eat leaves, and predators that run down their prey probably have to use more energy than ones that hunt from ambush. Although many factors interact, it is, nevertheless, true that a single factor accounts for a large share of the variation in tissue growth efficiency: Homoiotherms are much less efficient than poikilotherms.

One reason for the difference is the substantial amount of energy homoiotherms use just in maintaining their body temperature. For poikilotherm populations, tissue growth efficiencies are highly variable but usually between 20 and 50%. Most poikilothermic vertebrates, such as reptiles and amphibians, seem to be within this range, though at the low end (for example, Lamotte 1972). For homoiotherms, that is, birds and mammals, production efficiencies are only 1 to 3%. Social insects, whose activities in maintaining colony homeostasis are energy demanding, seem to have low efficiencies (Golley and Gentry 1964, Peakin and Josens 1978).

Combining the efficiencies with which assimilation and production occur gives us the ecological growth efficiency, with which we started, that is, the ratio between new biomass production and gross energy intake. For the reasons we have discussed, these tend to run about as follows:

Herbivores	Homoiotherms 0.3–1.5%
	Poikilotherms 9–25%
Carnivores	Homoiotherms 0.6–1.8%
	Poikilotherms 12–35%

Because the bulk of the energy flow in most ecosystems is through poikilotherms, mostly invertebrates, ecological growth efficiencies for whole trophic levels tend to run around 10 to 20%.

Between-Trophic-Level Efficiencies

Several between-trophic-level ratios have been described (Kozlovsky 1968). We can compare gross energy intake at trophic level n relative to trophic level $n - 1$. Alternatively, we can compare new biomass (production) at the two trophic levels. The first of these has been called **Lindeman's efficiency,** after R. L. Lindeman. The second is **trophic-level production efficiency.**

In the laboratory, these comparisons can be made. For well-defined food chains in nature, the necessary information can be obtained, although with difficulty. But calculating such ratios for whole ecosystems is very difficult. The reason is the practical problem of delimiting trophic levels, discussed in the next section. Suppose we study a wren that eats insects that eat grass. It is, in this respect, a secondary consumer. It also eats similar-sized spiders. The spiders are predators in a food chain that starts with dead grass and proceeds through bacteria and fungi, collembola and mites, and small carnivorous insects, which are spider food. Along this pathway, the bird is a quintary (fifth-level) consumer.

At present, the most we can say is that if each consumer trophic level totally utilizes the net production of the preceding consumer level, both Lindeman's efficiency and trophic-level production efficiency are likely to be between 10 and 20%. Higher nonpredatory losses (adverse weather, disease, accidents) will produce lower efficiencies. It may be that one or both ratios tend toward a constant within particular ecosystems or throughout nature. Or it may be that efficiencies tend to increase at higher trophic levels in some or all ecosystems. If any of these patterns exist, they have yet to be substantiated.

Food Chains and Food Webs

A large and complex literature has grown up on these subjects in the past several years. Dealt with are such questions as how many trophic levels exist and why, how food chains are interconnected into webs, and the relationships between food web structure and community stability.

The field has certain problems; one involves the databases used. These consist of assemblages of published diagrams or descriptions of communities, not necessarily gathered with the expectation that they would be used in this way. Thorough descriptions that include the identity of the organisms and their feeding relationships (with quantitative detail) are difficult and time-consuming. As a consequence, all published food webs are incomplete, some absurdly so (Paine 1988, Hall and Raffaelli 1991). They also suffer from great variability in resolution—how much lumping has been done (Polis 1991). For example, one top predator may be identified to species and another may be "several species of fish-eating birds," or vertebrates may be given separately and the various lepidopteran larvae listed simply as "caterpillars." Upon this somewhat shaky foundation has been erected theoretical structures of considerable complexity but, in some cases, mind-numbing ambiguity.

Here, we deal with just one question, What determines the length of food chains? Most food chains consist of two to five species (Pimm, 1982, Hall and Raffaelli 1991). For most food webs, it would be possible to imagine longer energy pathways, and simulated food chains constructed by seemingly realistic rules tend to be longer than their real-life counterparts. Accordingly, it is appropriate to ask, as G. E. Hutchinson (1959) did, Why are food chains so short?

Stuart Pimm (1982) listed several possible explanations:

1. Energy flow. Because of the great loss of energy for each organism's maintenance, the energy base for a predator at the quaternary or quintary consumer level is very small. There

may be so little that (1) patches of habitat are rarely large enough to support viable populations of these animals or (2) the animal requires too much time or energy to cover sufficient area to gather enough calories.

If insufficient energy alone sets the limit to food chain length, several predictions follow. The simplest is that highly productive ecosystems ought to have more trophic levels (more links in the food chains) than unproductive ones. It does seem true that extremely unproductive systems, such as Antarctic lakes having a NPP of about 10 g/m^2/year, lack a third (secondary consumer) trophic level. Otherwise, though, evidence relating food chain length to productivity seems ambiguous.

2. Design constraints. It is difficult to imagine a predator large, fierce, and fast enough to make a specialty of preying on adult peregrine falcons, but evolution has produced a lot of unlikely creatures. Design constraints is a hard hypothesis to test, but it seems possible that, in conjunction with other factors, biological impracticability may prevent the addition of another trophic level in some systems.
3. Optimal foraging. Three trophic levels are easy to understand: plants use sunlight; herbivores eat plant material; and carnivores eat herbivores. Why should a fourth trophic level exist? Carnivore flesh is not much different from herbivore flesh, and the energy per hectare available for a top predator, specializing on other carnivores, is only about one tenth that available if it ate herbivores.

   An evolutionary viewpoint provides perspective. Suppose that a three-trophic level system exists and a new carnivore arrives on the scene. Two strategies are open to this animal: It can compete with the already existing secondary consumers for abundant herbivore biomass or it can specialize on scarcer carnivore biomass for which there is no competition. Which route will be the fitter will depend on various circumstances. Among them are what the new predator is like in terms of size and speed, for example, and how much secondary consumer biomass is currently going to decomposer food chains through natural deaths. It is clear that the evolutionary answer will not always be for the new predator to fit in as another secondary consumer. There is no need to change the question we started with to, Why are food chains so long?
4. Dynamical constraints. The crux of this hypothesis is that certain models predict that long chains take longer to return to equilibrium when perturbed. The longer time away from equilibrium tends to increase the possibility of extinction of one of the species (thus shortening the food chain). For example, a perturbation reduces one population's density and while it is slowly rebuilding, random changes in the environment kick it below the threshold for recovery.

   A certain amount of real-world evidence can be interpreted as supporting this viewpoint. For example, treeholes with water have three to four trophic levels in subtropical rain forest in Australia whereas similar ones in England have only two trophic levels (Pimm 1988). The treeholes are similar in many ways; one obvious way they differ is the much greater seasonality in leaf fall (the food base for the system) and weather (in temperate regions the water in treeholes freezes).

More than one of these hypotheses may be true. Also, these do not necessarily include all possible explanations. For example, there is some suggestion that, among the assemblage of food webs, three-dimensional ecosystems such as forests and lakes have longer food chains than two-dimensional ones such as prairies and streams (Briand and Cohen 1987). It is not clear that this result, if it is true, would be predicted by any of the four proposed explanations.

So, a seemingly simple question, What determines the length of food chains? has proved to be complex or else has not yet been looked at in the right way. It seems likely that the path to an eventual answer will involve more evolutionary thought than has informed the discussion to this point.

Difficulties With Trophic Levels

Trophic levels are a useful way to think about energy in ecosystems, but in carrying out actual studies, we soon run into various practical difficulties. It is impossible to place most species into a single trophic level. Green plants form a fairly uniform producer group, but few animals are strictly herbivores. Sparrows eat seeds, true, but they feed their young almost exclusively on insects. What of ruminants that gain some energy from the grass they eat and some from the microorganisms in their gut? The higher levels are even worse. Where do we find a secondary consumer, that is, a carnivore that eats only herbivores, or harder yet, a tertiary consumer, a carnivore that eats only secondary consumers? Some species move from one trophic level to another as they get older and bigger.

Also, what should we do with dead organic matter? Is it still a part of the trophic level where it was produced? In practice, such dead material as wood, hair, and feathers, still attached to the organism, is generally considered a part of that trophic level. Dead leaves, fallen tree trunks, and dead birds generally are not.

The organisms that get their energy from such materials, in detritus food chains of soil or water, are also hard to fit into trophic levels. Where are we to put a clam or caddis fly that filters water to remove algae, leaf fragments, and other particles that potentially include contributions from the dead bodies and feces of every other organism in the stream? Detrital energy flow that involves energy storage in reduced sulfate, nitrate, or iron, described in the next section, poses even knottier problems.

As a result of such problems, few recent energetic studies have attempted a pure trophic-level approach. Rather, the tendency has been to use a "compartment" approach in which the compartments are whatever can be realistically studied. Both the amount of energy present in a compartment at a given time and flow rates into and out of the compartment need to be measurable. The compartments can be populations of important species, amounts of litter, etc.

The trophic-level concept remains useful as a simplifying way of looking at ecosystems. It retains usefulness for more serious analytical purposes if we view trophic levels abstractly, as energy that has traveled a certain number of steps from its arrival in the biosphere as sunlight. The biomass of some species will have to be partitioned into several trophic levels. In this view, it is clear that energy that is stored in dead organic tissue belongs to the trophic level where it was produced until it is assimilated by other organisms. The energy in a leaf fragment in the feces of an earthworm, for example, is still part of the producer trophic level.

Chemolithoautotrophs

We tend to use "autotroph" and "green plant" synonymously. However, there are two categories of autotrophs, only one of which comprises photosynthetic organisms (including both plants and bacteria). Chemosynthetic autotrophs, or chemolithoautotrophs, are the other category. In most ecosystems they are of small importance in energy flow, but recent studies point to greater importance in certain aquatic systems discussed in the following sections.

Lakes and Coastal Marine Ecosystems

Oxygen is absent only a few millimeters deep in the sediments of lakes and coastal marine waters. Consequently, the decomposition of detritus through most of the sediment depth has to be anaerobic. The process is performed by various heterotrophic bacteria. In conjunction with their oxidation of (that is, removing electrons from) the organic molecules being respired, something has to be reduced. The electron acceptors are usually carbon dioxide, sulfate, nitrate, or ferric iron.

If carbon dioxide is the electron acceptor, methane is produced. The methane, which bubbles up out of the sediments, still has energy to yield and does so, ordinarily to methane-oxidizing bacteria in the lake water.

If sulfate, nitrate, or iron is the electron acceptor, a more reduced form of sulfur, nitrogen, or iron is the product. Sulfur-reducing bacteria, for example, break down carbohydrate in the presence of sulfate to yield carbon dioxide, water, and hydrogen sulfide (H_2S). In the process, the bacteria gain about 25% of the energy stored in the compounds they respire. The other 75% of the energy is stored in the bonds of the H_2S.

Several things can happen to this energy, and the relative importance of the different pathways is not yet clear. The H_2S can diffuse upward and be reoxidized chemically, with the result that the energy is lost from the system as heat. Another possibility is the use of the reduced sulfur compounds by chemosynthetic, sulfur-oxidizing bacteria. These grow as mats on the sediment surface. They convert carbon dioxide and water to carbohydrate using the energy they obtain by oxidizing H_2S to elemental sulfur and elemental sulfur to sulfate. The summary reaction is

$$CO_2 + H_2S + O_2 + H_2O \rightarrow [CH_2]_n + H_2SO_4$$

Of the energy stored in the reduced sulfur compounds, 20% or more may go to produce new bacterial biomass (Howarth 1984).

The mats of sulfur-oxidizing bacteria are fed upon by bottom-feeders, such as crabs and fish. The unique feature of this sequence of events is the temporary uncoupling of energy flow from carbon metabolism. The energy is stored for a time in H_2S (or some other reduced sulfur compound such as iron pyrite, FeS_2) and then re-enters the community in the new biomass of chemosynthetic bacteria. Although the chemosynthesizers are autotrophs, the new biomass is not primary production. The energy originally entered the ecosystem as sunlight (Figure 11-16). The new biomass, accordingly, is secondary production, just as the new biomass of a heterotroph is.

The process may be ecologically important in several ways. Much of the energy in sediments is nearly unavailable to most bottom-feeders. It may be tied up for years in cellulose and other relatively indigestible compounds. Also, it is in small pieces, mixed with mineral materials, so that efficient use of it is difficult for most animals. The activities of sulfate-reducing bacteria followed by sulfur-oxidizing bacteria convert this energy into concentrated, high-quality food for clams and fishes.

These activities of the sulfate, nitrate, and iron-reducing bacteria also slow the rate at which lakes fill in by removing resistant organic sediments from the lake bottom (Rich and Wetzel 1978).

How much energy takes these pathways is still uncertain. Howarth (1984) suggested that sulfate reduction may account for 25 to 70% of total sediment respiration in near-shore marine sediments and 70 to 90% in salt-marsh sediments. The remaining sediment respiration is by other anaerobic bacteria, producing methane and, near the sediment surface, by aerobic heterotrophic bacteria (although Canfield and Des Marais, 1991, indicate

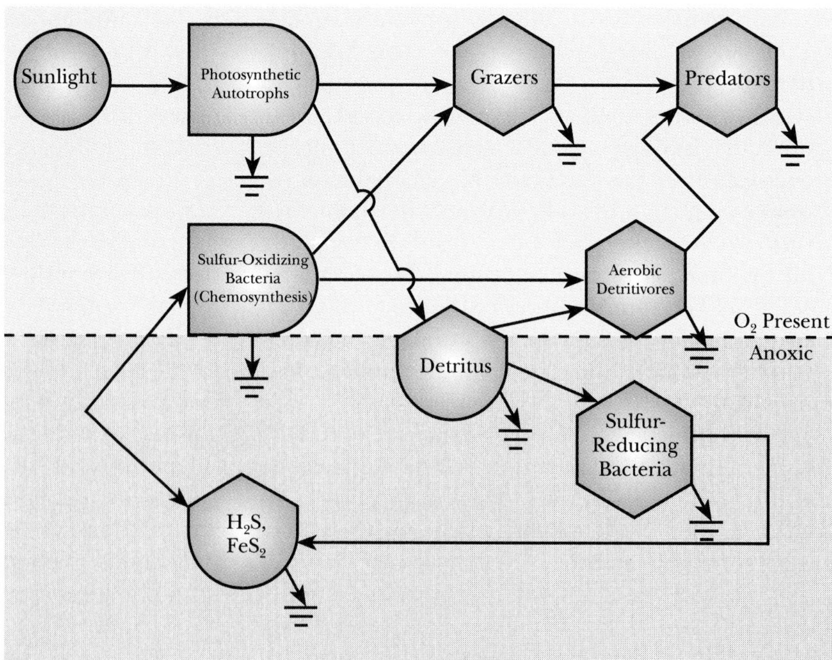

Figure 11-16

Simplified diagram of energy flow in coastal marine ecosystems showing energy storage in reduced sulfur compounds. (Based on Howarth 1984.)

that in some situations sulfate reduction occurs even in the upper, oxygenated zones). The percentage of energy reintroduced to the community by the sulfide-oxidizing bacteria would, necessarily, be much less, perhaps 3 to 18% of total sediment respiration (Howarth 1984). In lakes, methane production is relatively more important and production by chemolithoautotrophs is small (Wetzel 1983).

Hydrothermal Vents

An unusual community was first discovered about 30 years ago around hot springs (vents) associated with seafloor spreading zones. Ordinarily, the bottom fauna 2000 to 3000 m deep in the ocean is sparse because energy input from the surface (as detritus) is low. These vent communities, however, are dense beds of giant white-shelled clams and blood-red tube worms (Pogonophora) up to 3 m long. Many other animals including polychaete worms, crabs, shrimp, limpets, siphonophores, isopods, and fish are also present (Grassle 1985). The ecosystem was first observed in 1977 along the Galápagos Rift east of the Galápagos Islands. Since then, deep-sea research submersibles have found similar groupings of organisms in several other localities.

The energy base, or a large part of it, for this ecosystem is chemolithoautotrophy by sulfur-oxidizing bacteria. These behave much as do similar bacteria described in the preceding section: They obtain energy by oxidizing H_2S and use this energy to combine carbon dioxide and water to yield carbohydrates. Several kinds of bacteria are present (Jannasch and Wirsen 1979), and it is likely that some of them have other modes of nutrition (Jannasch and Mottl 1985).

The sulfide comes from the reduction of sulfate in seawater as it percolates through rock formations and magma (lava before it comes to the surface) at temperatures up to several hundred degrees centigrade. Another source is leaching of sulfide from the hot basalt. Some distance away, the hot water, charged with sulfides and other reduced materials including methane and various metals, is expelled through vents around which the animals cluster in patches occupying many square meters. Close to the outlet the vent water is above 20°C. Away from the vents, the water temperature at the ocean bottom is 1.8°C.

Most of the organisms of the vent ecosystems probably live off the chemosynthetic bacteria in one way or another. The mussels, among others, probably feed by filtering bacteria out of the water. Some organisms, such as crabs, may graze on bacterial mats. Others, including at least one of the fish, are predators.

The tube worm apparently has formed a symbiotic union with sulfur-oxidizing bacteria. Pogonophora have no digestive tract. How other species obtain food is not entirely clear, but this one evidently shares in the carbohydrate produced by the bacteria. The bacteria live in an organ called the **trophosome,** which means, reasonably enough, "food body." The worms apparently extract sulfide and oxygen from the water coming from the vent and convey this through their blood to the trophosome (Arp and Childress 1983). There the symbiotic bacteria carry out reactions like those of the free-living forms. The giant clam has symbiotic sulfide-oxidizing bacteria in its gills.

This ecosystem is remarkable in several ways. It is one of the few types of ecosystems where the primary production is not based on sunlight. As described in the preceding section, the sulfide used by chemosynthetic bacteria of marshes, estuaries, and waterlogged soils comes from the activities of sulfur-reducing bacteria. The energy stored in the sulfide and, later, in the biomass of the sulfur-oxidizing bacteria of these ecosystems came, several steps back, from sunlight. This is not so for the hydrothermal vent bacteria. The energy source is geothermal; the sulfide reduction is accomplished by heat generated within the earth.

Although the energy base is geothermal, the hydrothermal vent ecosystems, nevertheless, depend on photosynthesis. The sulfide bacteria as well as the associated animals are aerobes, requiring the oxygen produced by plants. That this strange ecosystem of the sea bottom depends on oxygen liberated far away in the ocean or on land is a measure of the interconnectedness of the biosphere—and also of the overriding importance of sunlight to it.

Another interesting feature of the vent ecosystems is the great heat tolerated by some of the bacteria. The water welling up out of the springs is milky blue. Analysis showed that the turbidity is due, not so much to colloidal sulfur as was initially expected, but mainly to bacteria in vast numbers, millions or even billions of cells per milliliter (Karl et al. 1980, Corliss et al. 1979). But this water is hot. Some of the bacteria—mostly members of the archaebacteria, or archaeotida—are living and even growing at temperatures above the boiling point of water at sea level (Pool 1990). This indicates unusual high-temperature adaptations of proteins and nucleic acids but just what these consist of is not yet known. At vents where the water is even hotter, 300 to 400°C, no microbial cells were found. These hot-water vents occur at the top of chimneys several meters tall, which may spurt water black with sulfides (Rise Project Group 1980). The zone next to the chimneys lacks clams and tube worms, but animals do live close to, if not in, the extremely hot water. Most noticeable is a polychaete, the Pompeii worm, that forms white snowball-like masses.

Finally, the communities seem temporary. We are accustomed to thinking of deep-sea communities as highly stable, but these vents have a life measured in years or tens of years (Grassle 1985). Old ones disappear and new ones appear, perhaps many kilometers away. Vents studied on one expedition already showed declines when revisited two or three years later. The organisms of the vent ecosystems are fugitive species like the weeds of forest clearings or the beetles and flies of carrion. Most marine organisms, with their motile larval stage, have good dispersal abilities, but the specific adaptations to life in these temporary habitats will be interesting to learn as research continues. The plumes of hot water coming from the vents are, themselves, a feature that larvae can take advantage of, riding several hundred meters above the ocean floor and, from there, being moved by the deep-sea currents (Mullineaux et al. 1991).

Regulatory Processes and Energy Quality

The amount of energy flow through the birds (Wiens 1977) or even the bison of prairie is minute compared with the energy flow through the nematodes or the mites. This is true if we compare energy intake and, because of homoiothermy, even more true if we look at production.

Does this mean that these species are unimportant in the ecosystem? It may be that some species in an ecosystem are virtual supernumeraries, like an extra toe, and that if they disappeared, the only change in the ecosystem would be a trifling increase in energy flow along some other food chain. But it is clear that many species that process relatively few kilocalories, nevertheless, play important roles in the sense that if they disappeared, the structure or function of the ecosystem would change appreciably.

These roles are generally regulatory; for example, predators may regulate prey numbers, reducing the likelihood that the herbivores will overexploit the producers. They also weed out the slow, the unwary, and the diseased. The herbivores do such things as stimulating biomass production within a season and fertilizing, helping to maintain productivity. They may also do some planting, as in the case of squirrels or blue jays storing nuts, and some tillage, as the case of ground squirrels and earthworms. We can think of many other examples; in fact the chapters on coactions and reactions are full of them, looked at from a different perspective.

S. R. Carpenter, J. F. Kitchell, J. R. Hodgson, and coworkers (Carpenter et al. 1985, 1987) have suggested that many aspects of the structure of lake ecosystems depend on the degree of fish predation. Specifically, they cite evidence that a high level of predation by piscivorous (fish-eating) fish, such as bass or pike, results in a lowered biomass of plankton-eating fish (such as minnows). In what has been termed **cascading trophic interactions,** the lowered biomass of planktivorous fish results in lower predation on and, thereby, a higher biomass of zooplankton. This in turn results in a lower biomass of phytoplankton. We are speaking here of the standing crop of the phytoplankton. The turnover rate and, hence, the productivity of the phytoplankton may well be higher.

Other investigators have connected this view with ideas of S.D. Fretwell (1977) and related both food chain length and the relative importance of competition or predation to productivity (Persson et al. 1988). Figure 11-17 illustrates these ideas. In areas of low productivity, only two trophic levels may be supported, in which case, it is suggested, the phytoplankton may be limited by predation and the zooplankton, with no predators, limited by competition. More productive systems support three

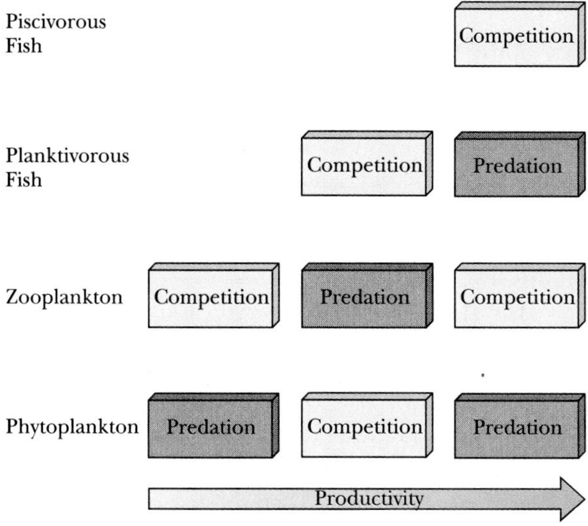

Figure 11-17

Increasing productivity may lengthen food chains. If so, the top trophic level in any particular system may be structured by competition while predation from it structures the trophic level below it. (Based on Persson et al. 1988.)

trophic levels, with the zooplankton now predator limited (by planktivorous fish) and the phytoplankton (with lessened grazing) competition limited. In still more productive systems, the planktivorous fish are limited by predation from piscivorous fish, and the structuring coactions flip-flop on down the line.

Regulatory activities are work applied to the ecosystem, corresponding to the work that humans do when they plant, till, weed, and harvest in agricultural ecosystems. In an energy systems diagram, we would represent them as feedback loops, as in Figure 11-18. These loops do not imply that the energy re-enters the metabolism of the organisms shown; the energy does, however, do work on the trophic levels indicated. An analogous situation is the energy we apply to the steering wheel of a car; it does not become involved in the process of combustion going on in the engine, but it keeps the car on the road.

H. T. Odum (1977) has suggested that in complex systems every kilocalorie is not like every other kilocalorie; rather the energy is of higher quality depending on the number of steps (with the accompanying energy losses) required to produce a given stage, process, or organism. For example, it takes a lot of energy being processed by producers and first- and second-level consumers for an ecosystem to produce a peregrine falcon or a timber wolf. Even with the best of intentions, we cannot avoid these energy losses and produce one on grass.

Each calorie in such an organism, Odum claims, is of higher quality than a calorie in lower trophic levels. The higher quality is reflected in a greater ability to do work on the system—for example, by regulating secondary consumer numbers that regulate herbivore numbers, etc. An analogous situation would be the difference in energy required to produce a skilled and experienced surgeon versus a hospital parking lot attendant and the corresponding difference in their ability to affect health.

It is not certain that thinking of differences in energy quality is the best way of looking at this situation. Neither is it certain that regulatory processes "cascade" down the trophic ladder just as shown. In many types of ecosystems, trophic levels are probably not distinct enough for things to work out this neatly.

It is, nevertheless, clear that organisms at higher trophic levels exert various types of regulatory effects on lower levels, and this suggests some

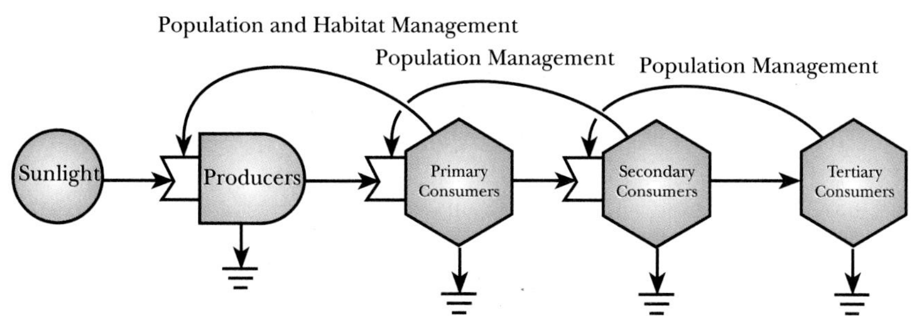

Figure 11-18

The feedback of energy from higher to lower trophic levels in the form of work.

powerful strategies for ecosystem management. By relatively small changes at the highest trophic level, produced by relatively small expenditures of time and money, we may be able to exert large effects at lower levels. Humans have, in fact, probably already done this on a large scale in our elimination of many top predators. By going the other way and restoring some of these, we may be able to improve matters. For example, stocking piscivorous fish (thereby reducing planktivorous fish and phytoplankton) could become an important tool in rehabilitating artificially eutrophied lakes (Shapiro and Wright 1984).

Energy in Agriculture

Basic Patterns

The basics of energy flow in agriculture are clear. Sunlight strikes a field on which humans have planted a crop and from which we have eliminated competing vegetation. Some of the sun's energy is fixed in photosynthesis and stored in new protoplasm. The energy in the edible portions of the plant is then passed on to us. When we eat meat, another step is added to the food chain, with cattle, pigs, or chickens eating the plants and we obtaining the energy stored in the new animal protoplasm.

This picture of energy flow is shown in Figure 11-19. The figures are growing season approximations for 1 acre of Illinois cornfield producing 100 bushels of corn an acre. Of the sunlight striking the acre (2043 million kcal), 1.6% is used in photosynthesis. Corn has the efficient C_4 pathway so that relatively little energy is used in plant respiration; 80% of the 33 million kcal of gross production winds up as net production. Of this total production, only the grain is used directly as food. If humans function as a primary consumer, then approximately 8.2 million kcal is potentially available, disregarding indigestible parts of the grain and processing waste.

If the corn is instead fed to livestock, placing us at the carnivore level, there are losses in such forms as material wasted or not digested by the livestock, respiration by the animals, and processing waste in the slaughterhouse. The result is that the human food in the form of beef represents about 5% of the kilocalories of grain fed (Heichel 1976). If the grain is fed to chickens or to pigs the loss is less, with human food in the form of poultry or pork containing about 10% of the energy in the grain. By American and western European dietary standards, a human needs about 1 million kcal per year, meaning that 1 acre of corn can feed about eight persons. When corn is converted to beef, it provides only enough food for one person for about five months.

These figures emphasize the reason why high populations of humans must be mainly vegetarian (herbivorous), but they are not necessarily an argument against raising livestock. Meat may be a requirement for optimal human growth and development, although not in the quantities that Americans consume. Much of the world's cattle production is on rangeland that for one reason or another, such as low or erratic rainfall, should not be cultivated. Ruminants such as cattle can live on the grass produced, whereas humans cannot. Pigs and chickens are often raised on scraps and garbage although this is no longer true of commercial pork and poultry production. Also, as we shall see, low-intensity meat production is as efficient (comparing food calorie output with petroleum calorie input) as is most high-intensity vegetable farming (tomatoes and lettuce, for example) and fruit production (grapefruits and pears).

Energy Flow in Food Systems

Figure 11-20 diagrams energy flow in the human food system. For hunter-gatherer societies, the food chains included in the two lower pathways predominate. Pastoralists, like the shepherds of the Bible, the Ngisonyoka (Coughenour et al. 1985), and the yak-herding natives of the Tibetan Plateau, substitute livestock for wild herbivores in native ecosystems. Where crop tillage is practiced, the upper two pathways are added. These pathways dominate the food system of modern technological societies.

Every animal expends energy to get food. For hunter-gatherers, pastoralists, and persons practicing shifting cultivation (such as slash-and-burn), this is mostly the energy of their metabolism. Hunter-gatherers have to find and catch prey; pastoralists herd and milk their animals and drive off predators; subsistence farmers hoe, plant, weed, and harvest.

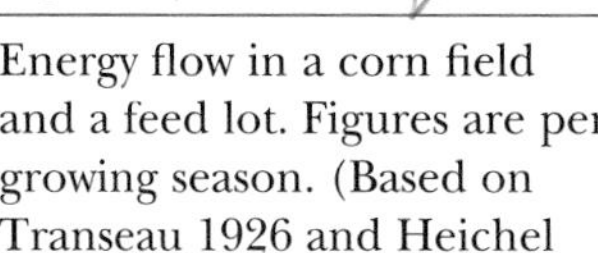

Figure 11-19

Energy flow in a corn field and a feed lot. Figures are per growing season. (Based on Transeau 1926 and Heichel 1976.)

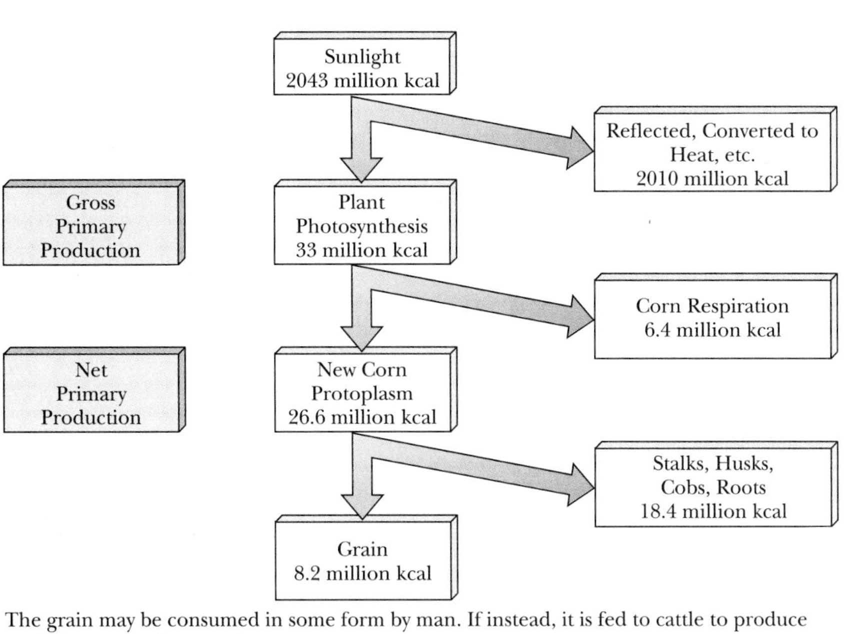

The grain may be consumed in some form by man. If instead, it is fed to cattle to produce beef, the energy flow continues:

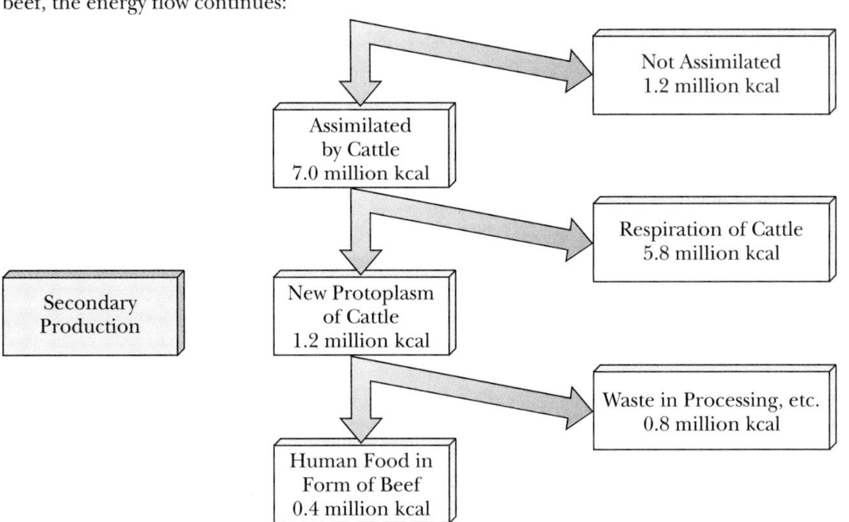

With the introduction of draft animals such as oxen and horses, humans began to use energy from outside their own metabolism to do work on the ecosystems supporting them. Draft animals use mostly foods that are unsuitable for humans and, thus, likely to go to detritus food chains. The animals, consequently, convert energy of little direct benefit to humans to useful agricultural work such as plowing, transportation, and fertilizing.

Modern mechanized agriculture uses large amounts of energy from outside sources, mostly fossil fuels. Many of these energy inputs are obvious to anyone—the fuel to run the tractors and combines; the energy needed to make pesticides and fertilizer; gasoline and electricity for irrigating, harvesting, and drying. Some other inputs may not come so readily to mind—the energy to build and maintain tractors, silos, and other equipment; and the energy to produce the hybrid seed corn at the seed companies. Some energy inputs most of us would fail to think of at all but, nevertheless, continue to eat up fossil fuel. Included here are such things as the en-

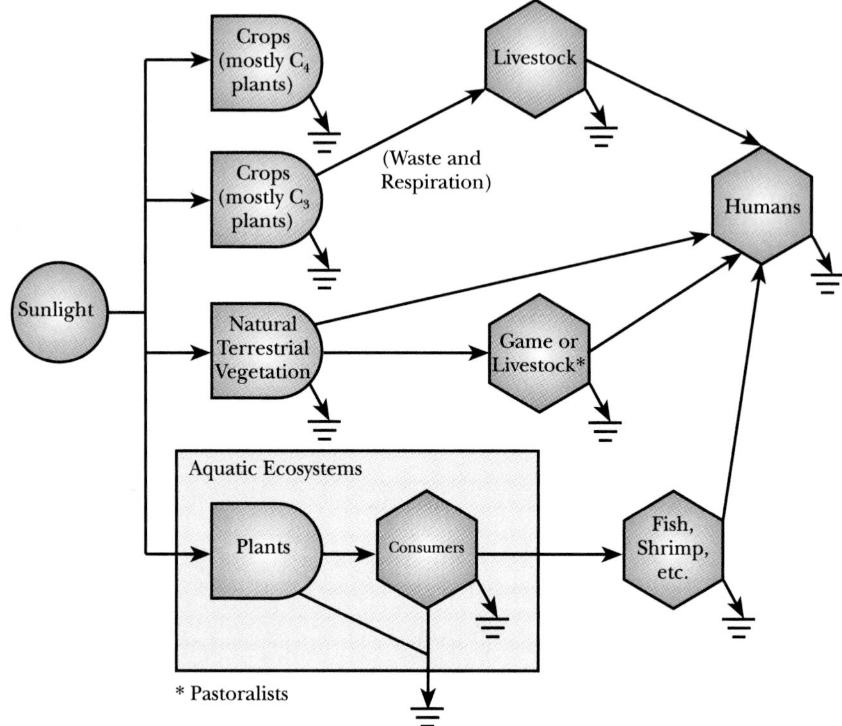

Figure 11-20

Energy flow to humans.

ergy required to support veterinarians' offices (and veterinarian schools) and the energy expended by seed companies, agricultural colleges, and county extension agents in developing, testing, and promoting new plant varieties.

The fossil fuel subsidy to agriculture has increased greatly over the 20th century. In 1945, according to an analysis by David Pimentel and colleagues (1973) of the New York State College of Agriculture, it took a fossil fuel input of somewhat less than 1 million kcal to raise 1 acre of corn yielding about 3.4 million kcal of food energy (primarily for livestock) (Table 11-7).

Currently, farmers spend over 4 million kcal of fossil fuel energy per acre. They were also producing more, but not proportionally more, and so the ratio of kilocalorie return to kilocalorie input dropped from 3.7 to 2.7.

Table 11-7 gives a breakdown of the cultural energy inputs to corn production. "Cultural" in this sense is as opposed to biological—the 33 million kcal of photosynthetic input—and includes the fossil fuel subsidy plus human labor (Heichel 1973). From these figures we can see what the energy goes for in everyday terms, but what, ecologically, does all this energy expenditure accomplish? Aside from the energy used simply to harvest, which even the hunter-gatherer needed to expend, it performs important ecological functions:

1. It continually sets back succession by plowing, cultivation, and herbicide usage. A cornfield is an analog of a pioneer community. Without human intervention any such field of annuals would be rapidly invaded by perennial herbs and shrubs or trees and proceed within a few years to a community of perennial grassland or forest, depending on climate.
2. It performs various maintenance and regulatory functions. For example, in a natural ecosystem, plants tend to be protected from heavy herbivory by their own protective devices, such as secondary compounds, and by predators eating the herbivores. Crops usually lack protective devices; they have been bred out to channel more production to human yield. Predators are often absent or ineffective because of pesticides, lack of alternative prey,

Table 11-7
Cultural Energy Inputs in Corn Production

Figures are kilocalories per acre. 1990 represents current practice; 1945 is included to represent somewhat lower (although still high) technology and energy subsidies. Based mainly on Pimentel et al. 1973 and Pimentel 1991.

Input	*1945*	*1990*
Labor	12,500	2300
Machinery	180,000	410,000
Gasoline and diesel	543,400	403,000
Nitrogen	58,800	1,287,000
Phosphorus	10,600	191,000
Potassium	5,200	97,000
Seeds for planting	34,000	210,000
Irrigation	19,000	716,000
Insecticides	0	40,000
Herbicides	0	161,000
Drying	10,000	496,000
Electricity	32,000	36,000
Transportation	20,000	36,500
Other	—	54,000
Total inputs	925,500	4,127,500
Corn yield (output)	3,427,200	11,088,000
kcal return/input kcal	3.70	2.69

or lack of cover. Accordingly, such regulatory functions have to be supplied from the outside.

Crop plants are, in effect, plants adapted to living in a certain peculiar ecosystem. Adaptation, itself, is a regulatory process requiring large amounts of energy; that is, natural selection works through the production and destruction of a great deal of biomass. Crop breeding is human-directed adaptation that is similarly energy demanding.

It seems likely that maintenance costs are not, in fact, being fully met. Many agricultural lands are suffering erosion and degradation of soil structure that would be prevented by the plants and litter of a natural ecosystem (Larson et al. 1983). On some lands, fertilizer applications do not have the function of increasing soil fertility (as below), but simply of bringing it up to fertility levels of ordinary soil before agricultural degradation (e.g., Cox 1984b).

3. It improves certain soil traits affecting productivity. About the only natural ecosystems showing periodic enrichment comparable to agricultural fertilization are lands subject to flooding. Floods add nutrients every year or few years, allowing heavier growth than on similar, unfertilized sites. Flooding, too, represents an energy subsidy, but one derived from the hydrological cycle rather than from fossil fuel.

Agricultural fertilization improves crop yield, up to a point. Although most crops never approach the net primary production of natural ecosystems of the region even when heavily fertilized, some, such as corn, may do so (Figure 11-21).

Irrigation has effects on productivity similar to fertilization in regions where soil moisture is limiting.

In raising corn we get back 3 to 5 kcal in grain for every 1 kcal spent as petroleum. Another C_4 plant, sugarcane, does equally well (Table 11-8). Most grains, fruits, and vegetables yield more in usable calories than they consume in petroleum. Intensively cultivated fruits and vegetables, however, return less than a calorie of food for each calorie of energy subsidy.

Animal products all tend to take more energy than they return. Certain kinds of meat require incredible amounts of fossil fuel. Feedlot beef, for example, requires 15 cal of fossil fuel for every calorie in meat. Range-fed beef (in Colorado) returns a calorie of beef for every 1.7 cal of fossil fuel, the latter for operating a pickup truck to check the herd (Ward et al. 1977). Nutritionally, the importance of meat is not so much calories as protein. Even as a protein source, meat production is wasteful of fossil fuel energy—although again scarcely more so than some intensively cultivated fruits and vegetables (Table 11-9). Most seafoods require heavy energy inputs for travel and refrigeration.

Obviously, if we tried to live on any of the foods below the breakeven point (ones that return less than a kilocalorie of food for a kilocalorie of cultural energy input), and had only the energy of our own metabolism, we would starve to death in short order. The U.S. agriculture system as a whole is below the breakeven point. Overall we use somewhat

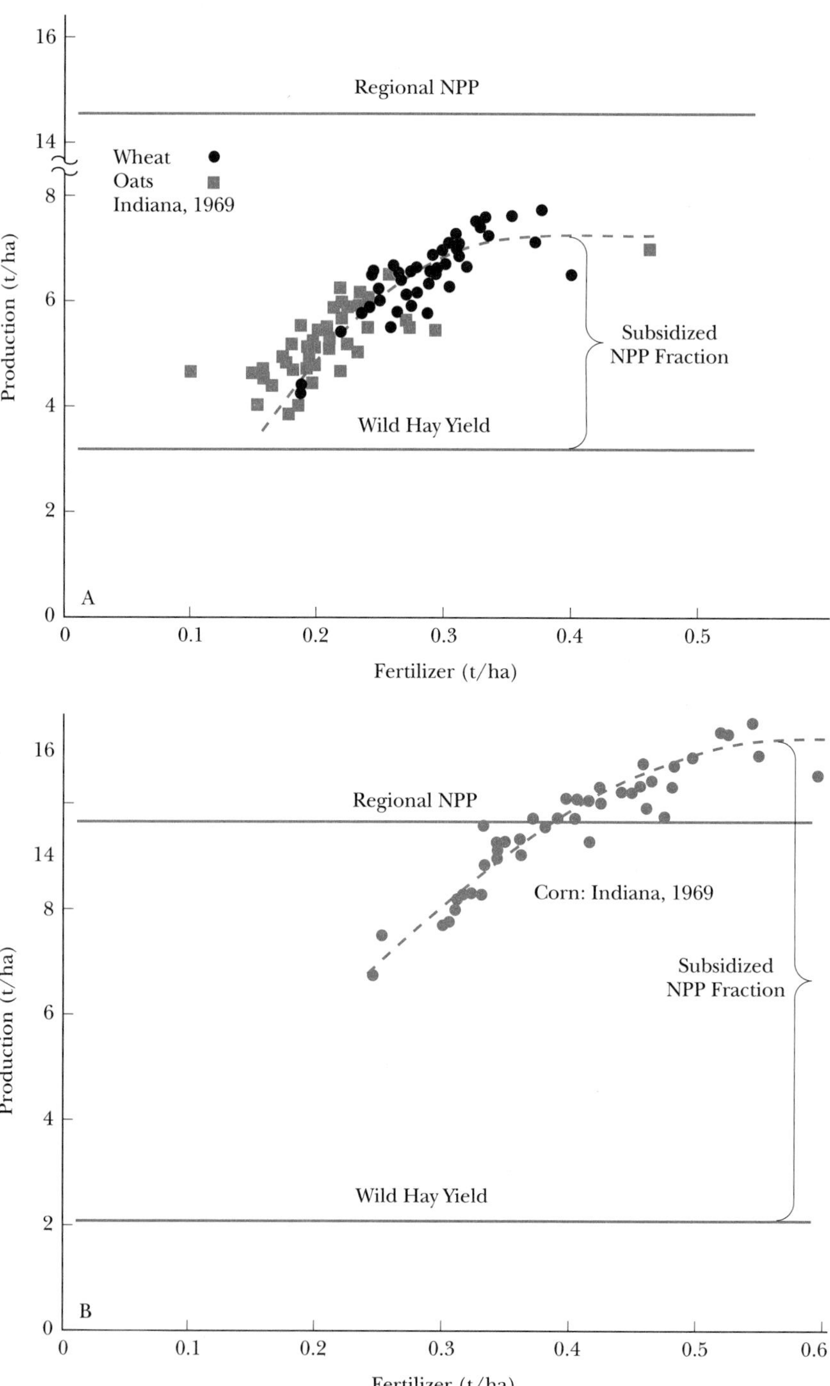

Figure 11-21

(A) Top. This curve relates wheat and oats production to fertilizer use. "Production" includes not just the grain but also the stems and leaves. The lower horizontal line is the yield from wild (unfertilized) hay fields for the same year. Consequently, the difference between the wheat-oats curve and the wild-hay line is a rough measure of the increase in productivity achieved by fertilization. The top horizontal line is regional net primary production of natural vegetation. This includes belowground as well as aboveground production but, since wheat and oats have skimpy root systems, adding in the weight of their roots would not bring the wheat-oats curve up to the regional NPP line. (Note that the left-hand scale is broken; wheat and oats level off below 8 t/ha whereas regional NPP is close to 15.) *(B) Bottom.* This figure is like *A* except that it's for corn. Here fields receiving small amounts of fertilizer are not as productive as native ecosystems but heavily fertilized cornfields are more productive. As before, this is without adding in root production. Adding in the roots might run heavily fertilized corn productivities up to 20 t/ha. (Adapted from Cox 1984.)

more than 2 kcal of oil to produce 1 kcal of food (Cox and Atkins 1979).

Even this does not tell the whole story of our dependence on fossil fuel. The energy used in the food system between harvest and our bellies has to be added in to provide a full estimate. Included is the energy used for processing, packaging, transporting, storing, advertising, and cooking. Figure 11-22 is an energy flow diagram that includes some of the major energy components of the whole food

Table 11-8
Fossil Fuel Subsidies for the Agricultural Production of Various Kinds of Food

This is the energy required to grow the crop on the farm, to raise the livestock until ready for slaughter, or to catch the fish or shrimp and bring them back to port. Included is the fossil fuel energy for such things as laor, fuel, fertilizer, and depreciation of machinery. The figures are ratios of fossil fuel to food energy. For example, every kilocalorie of milk ready to go to the dairy requires the use of 2.8 kilocalories of fossil fuel on the farm.

Crop	*Fossil Fuel Energy/Energy in Food*
C_4 plants	
Sugarcane	0.2
Corn	0.2
C_3 plants	
Wheat	0.3
Soybeans	0.4
Rice	0.7
Barley	0.4
Peanuts	0.7
Potatoes	0.7
Apples	0.8
Sugar beets	0.8
Grapes	1.0
Pears, peaches	1.0
Most vegetables	2.0
Oranges, grapefruit	3.0
Animal products	
Perch	1.3
Milk	2.8
Eggs	6.0
Tuna	8.1
Chicken	10.0
Pork	10.0
Beef	15.0
Shrimp	60.0

From Brewer, R., and McCann, M. T., *A Laboratory and Field Manual of Ecology,* Philadelphia, Saunders College Publishing, 1982.

Table 11-9
Protein Yield Related to Cultural Energy Input for Various Crops

Grams of Protein per 1 kcal Input			
<0.01	*0.01–0.04*	*0.05–0.19*	*0.20–0.39*
Feedlot beef	Chicken	Peanut	Soybean
Catfish	Eggs	Barley	Alfalfa
Milk	Rangeland lamb	Oats	
Pork	Rangeland beef	Wheat	
Ocean fish	Brussel sprouts	Corn	
Green beans	Intensive rice	Sorghum	
	Tomato		
	Potato		

Compiled from Heichel, G. H., "Agricultural production and energy resources," *Am. Scientist,* 64(1):64–72, 1976; Pimentel, D., Dritschilio, W., Krummel, J., and Kutzman, J., "Energy and land constraints in food protein and production," *Science,* 190: 754–761, 1975; Leach, G., *Energy and Food Production,* Guildford, England, IPC Sci. and Tech. Press, 1976.

system from sunlight to you. Attempts to put figures to all these flows have produced estimates of the total fossil fuel energy subsidy that vary over a wide range. The estimates of the ratio of kilocalorie input to the entire food system to food calories on the table range from 6.4 (Hirst 1964) to about 10 (Steinhart and Steinhart 1974, Cox and Atkins 1979) to 112 (Hannon and Lowman 1978).

Although agricultural production can and should be made more energy efficient, the basic aim of the application of energy to agriculture seems well directed. For example, corn production as practiced by North American Indians, based on photosynthesis and a certain amount of work done by humans, had a small net yield. Application of fossil fuel subsidies has greatly increased the yield. The natural output has been amplified (Odum and Odum 1981), as a public address system amplifies sounds allowing a message to be carried to thousands instead of tens of persons.

Some of the other energy inputs in the larger food system also seem justified from an energetic standpoint. For example, canning, drying, and, to a degree, freezing and refrigeration preserve energy for humans that would otherwise be shunted off to decomposer food chains. Some other energy uses seem dubious. Fossil fuel energy used in heavily processed convenience foods may, for example, amount to 40 times the food energy. The large amount of energy required for transportation and refrigeration to provide stores with fresh lettuce, strawberries, or apples every day of the year can

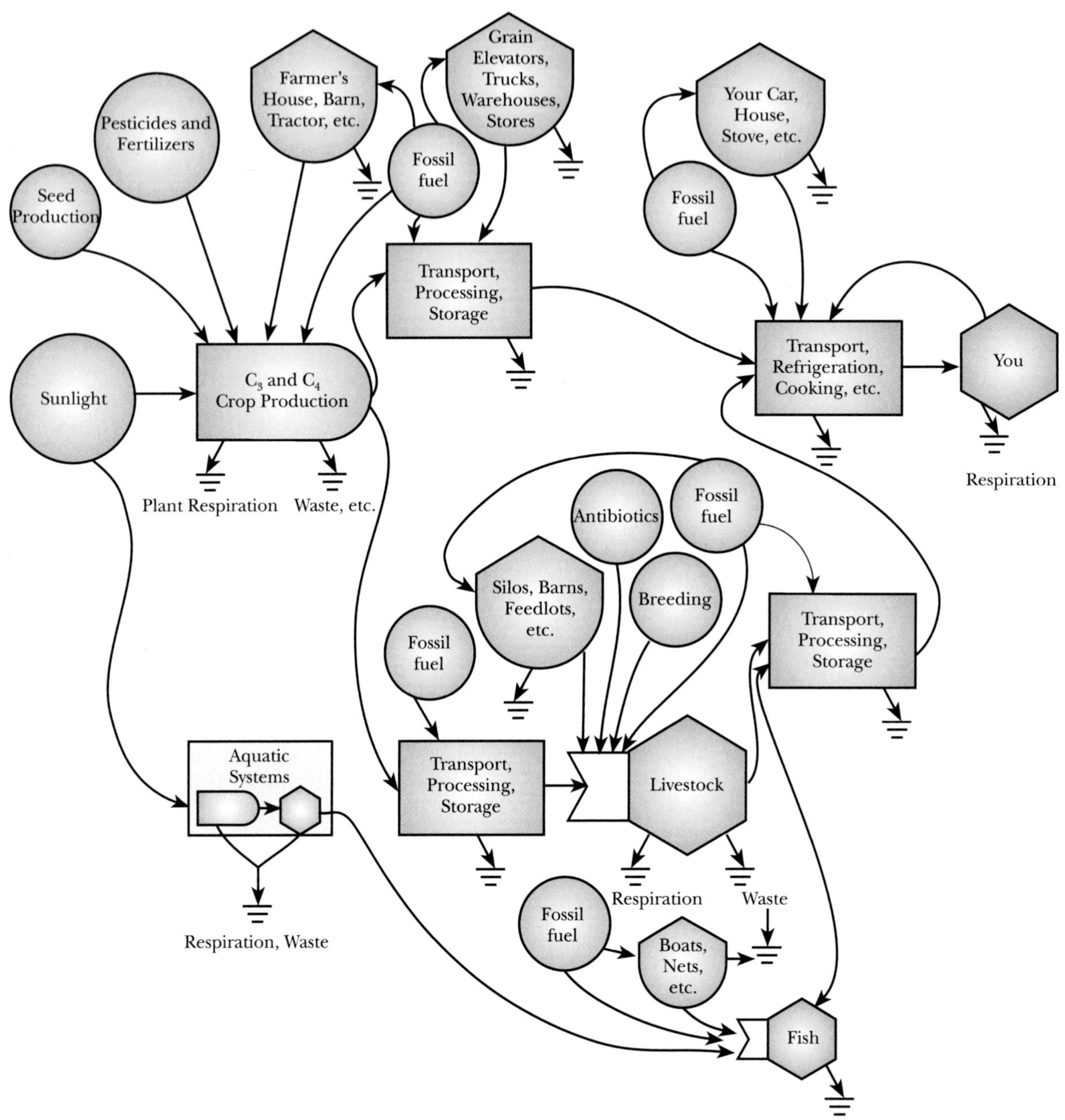

Figure 11-22

Simplified representation of the whole food system from which you obtain your food. (Adapted from Brewer & McCann 1982.)

only be supported in a era of abundant, cheap energy.

Human Expropriation of the Earth's Primary Production

One measure of the dominance of an organism is the extent to which it monopolizes and directs the energy flow of the system to which it belongs. If we think of the biosphere as one large ecosystem, we can make such calculations for humans and get some idea of human impact on the earth and the implications for the other species that are members of earth's biotic community.

Table 11-10 gives an estimate of the total NPP of the earth (from Vitousek et al. 1986, based on Atjay et al., 1979 for terrestrial systems and DeVooys 1979 for aquatic systems). Total annual NPP for the earth amounts to about 225 million petagrams. A petagram (pg) is 10^{15} g or 10^{9} metric tons (a metric ton is about 2200 pounds). This is the energy base on which all organisms above the producer trophic level depend—birds, mammals, insects, bacteria, and all the other heterotrophs, several million species in all.

Table 11-10
Surface Area and NPP Contributed by the Earth's Major Ecosystems

Type	*Surface Area ($\times 10^6$ km^2)*	*NPP (Pg)*
Forest	31	48.7
Woodland, grassland, and savanna	37	52.1
Deserts	30	3.1
Arctic-alpine	25	2.1
Cultivated land	16	15.0
Human area	2	0.4
Other terrestrial (chaparral, bogs, swamps, and marshes)	6	10.7
Subtotal terrestrial	**147**	**132.1**
Lakes and streams	2	0.8
Marine	361	91.6
Subtotal aquatic	**363**	**92.4**
Total	510	224.5

From Vitousek et al. 1986, based on Atjay et al. 1979 and De Vooys 1979.

Of this production, the human population **uses directly** a fairly small amount for food, fiber, firewood, and construction. The estimate, as of 1986, was about 3.2% (7.2 pg). To understand human impact fully, however, we have to add two other categories—co-opted and missing production (Vitousek et al. 1986).

Co-opted production is the production (minus that used directly for food, wood, etc.) on lands that are strictly dedicated to human use; included is the production on cultivated land in such forms as straw and corn stalks and on forested land in slash and stumps. All the production on such basically inedible sites as golf courses and lawns is included here. Also under co-opted production is the energy in biomass burned on grazing lands and in slash-and-burn agriculture and the energy lost in vegetation and litter on areas cleared for agriculture or development.

Missing production is the energy that would have been there except for human activities. The calculations of Vitousek and his coworkers (1986) are probably conservative in making these calculations. They concentate on the declines (compared with the regional natural vegetation) resulting from four processes: (1) agriculture (for example, having cornfields instead of tallgrass prairie in central Illinois), (2) conversion of forest to pasture, (3) conversion of natural vegetation to malls, highways, factories, university research parks, etc., and (4) the desertification that has accompanied human occupation of dry savanna and grassland. Omitted, as too hard to calculate, are declines in productivity from acid rain and other atmospheric pollution and soil erosion and degradation. Also, no attempt was made to calculate declines in productivity in aquatic systems.

Co-opted production does support some other species besides humans. Vesper sparrows and moles live on golf courses, weeds and weevils live in crop fields, and desert species occupy desertified areas; such groups are supported by this energy flow.

The upshot of this string of calculations is that humans currently expropriate by direct use, co-optation, or reduction of productivity an estimated 38.8% of the potential NPP of the land and 2.2% of the aquatic production. The latter is small; if the figure is basically accurate, it suggests, that human exploitation of marine resources is not having massive effects on marine ecosystems much beyond the directly targeted whales, tuna, cod, swordfish, anchovies, shrimp, crabs, lobsters, oysters, and a few hundred other species and close coactors with them.

For terrestrial situations, the conclusion is less optimistic. Nearly 40% of the earth's primary productivity is now required to support humans, rats, and house sparrows (and other human commensal species). All the other species of the world have to get by on the remainder. Suppose that the earth's human population doubles, which, if present trends continue, will happen in about 40 years. A simple proportionality suggests that more than 75% of the earth's production would then have to go to support human needs. Many creatures have been lost to extinction in past centuries as humans have extended their hegemony across the globe. The losses yet to come, if current trends continue, are almost unimaginable.

Community and Ecosystem Ecology: *Reactions and Biogeochemistry*

OUTLINE

- Reactions on Land
- Reactions on Air
- Reactions in Fresh Water
- Reactions in the Ocean
- Nutrient Cycling
- The Hydrological Cycle
- Watershed Studies
- Biological Control of the Composition of the Atmosphere and Oceans

12

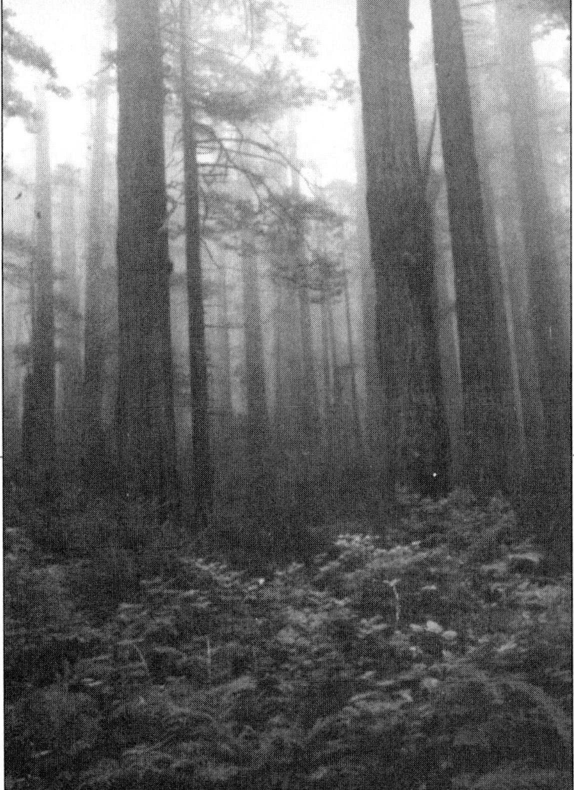

Old-growth temperate rain forests of the western United States sequester large amounts of carbon in their biomass. (*Stan Osolinski/Dembinsky Photo Associates.*)

Visualize a large area of bare dirt, perhaps a construction site, and then consider how it would be different if a forest, including not just trees but smaller plants, animals, and microorganisms, were suddenly there. The community would produce shade so that the light intensity would be lowered and temperatures would be moderated. The trees would act as barriers to wind and sound. As leaves fell and earthworms tunnelled, soil would be built. Plant leaves would intercept rainfall and allow it to re-evaporate, and humus would soak up water so that runoff and erosion would be lessened. Such effects are all alterations of the physical environment produced by a forest. The effects of organism on habitat we have referred to as reactions. The organisms of grasslands, lakes, and other communities produce their own reactions.

We described the immediate effects of habitat on organism in Chapters 2 and 3. Individual organisms produce reactions, and so we could have included a part of this chapter's discussion in that section of the book. For example, beavers, by their metabolic activities, keep the temperature warm inside their lodges through the winter (Buech et al.

1989). Generally, though, the important reactions are those produced by the whole community or, at least, large segments of it. The community accumulates or integrates effects that, produced by an isolated individual, would be insignificant, local, or transitory. The different effects of shade from an isolated tree and from a forest canopy illustrate the distinction.

Most nonecologists tend to think of the earth, including its ocean and its atmosphere, as the independent variable on which the organisms of the planet depend. This view is inaccurate; the world, to a considerable degree, is a biotic product. The composition of the earth's atmosphere, crust, and oceans is the result of biological activity. The involvement of organisms in the cycling of such chemicals as carbon dioxide, oxygen, and nitrogen is of global importance and is dealt with in the latter part of this chapter. To begin with, we will concern ourselves with effects more nearly confined to the ecosystem where they occur. Reactions are so numerous and varied that we can only sample them under headings corresponding to the medium (land, air, or water) that they affect.

Reactions on Land

Soil Formation

In Chapter 3, we discussed the broad features of soil formation as background for discussing soil as a "factor" in the ecology of individual organisms. Here we deal specifically with the roles of organisms in soil formation. Organisms, first of all, contribute to weathering, by which large particles are turned into small ones. They are involved in mechanical weathering by such processes as the growth of roots in rock cracks. They are involved in chemical weathering by the production of acids, both organic acids and carbonic acid from carbon dioxide production. Snails in the Negev desert of Israel rasp away the surface of rocks to feed on the endolithic lichens (growing inside the rocks) (Shachak et al. 1987). In so doing, the snails produce fine mineral material amounting to about one metric ton per hectare per year (about 5000 pounds per acre).

One of the major processes in soil formation is the addition of organic matter to the soil and the subsequent conversion of it to humus (Figure 12-1). The amount of organic material added to the surface of a piece of land depends largely on the productivity of the vegetation. The general range of litter fall in temperate ecosystems is 250 to 450 g/m^2 per year, or 1 to 2 tons per acre (Jordan 1971). In some desert ecosystems, however, litter fall may be almost nil, and in some tropical rain forests amounts above 2000 g/m^2 have been reported.

Most organic matter is added to the top of the soil but in a few ecosystems, notably grassland, large amounts are added by the death of roots. The thick, black topsoil of the tallgrass prairies is largely the result of this process; however, there is little quantitative information for this mode of addition of organic material to the soil.

The rate at which litter decays varies with temperature, moisture, soil pH, and the chemical composition of the litter (Table 12-1). It is faster at higher temperatures, higher moisture levels (but not under anaerobic conditions), and a neutral or slightly alkaline pH. Lignified and resinous materials decay slower than others (Figure 12-2). Especially in moist soils, humus formation may convert

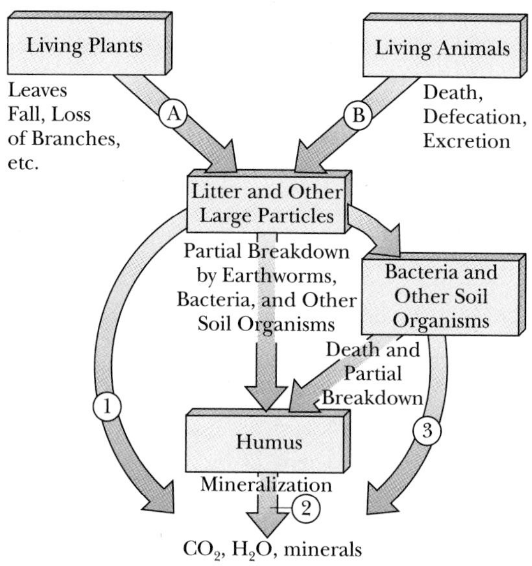

Figure 12-1

Litter and humus formation and breakdown. If the organic matter of the soil is at a steady state, the yearly amount added along paths *A* and *B* will be equal to the amount of lost along paths 1, 2, and 3.

Table 12-1
Half-Lives of Organic Matter in Different Kinds of Forest

The half-life is the time required for half the litter to disappear.

Type of Forest	*Organic Matter Half-Life (Years)*
Tropical forest	0.36
Temperate forest hardwoods	1.01
Temperate forest conifers	4.86
Boreal forest conifers	11.23

Calculated from Table 5 in Perala, D. A., and Alban, D. H., "Rates of forest floor decomposition and nutrient turnover in aspen, pine, and spruce stands on two soils," USDA Forest Serv. Res. Paper NC-227, pp. 1–5, 1982.

a part of the soil organic matter into certain humic acids or other compounds that are highly resistant to decay. These may persist in the soil for hundreds of years before they are eventually broken down to carbon dioxide, water, and minerals (Anderson and Coleman 1985).

In mature ecosystems, the breakdown of organic matter evidently proceeds at approximately the same rate as its addition. Earlier in succession, decomposition does not keep pace, and thus most ecosystems have some amount of organic matter stored as litter or humus, or both.

Oak and pine forests tend to grow on acid soils (which they, in part, create) in cooler temperatures; their lignified foliage, low in calcium, is relatively decay resistant. Consequently, they tend to build up deep litter and **mor** humus (Table 12-2). Temperate forests of maple and basswood and other trees with thin leaves of high base content tend to have litter on the forest floor only during the fall and winter. By midsummer most of the litter has decayed into **mull** humus. Decay is even more rapid under the hot, moist conditions of tropical rain forest.

Most of the chemical breakdown of organic materials in the soil occurs through the action of bacteria, actinomycetes, and fungi; however, soil animals are also important. Several experiments (such as Santos et al. 1981) have shown that litter decomposition goes much more slowly if earthworms, isopods, and other such animals are kept out. Both the mechanical processes of chopping big pieces of litter into small ones, giving more surface for bacteria to occupy, and also digestive processes occurring in the body of the soil animals seem important. The animals also help in dispersing fungal spores and bacteria to fresh litter (Stevenson and Dindal 1987).

A close look at the soil of a maple or basswood forest late in the summer, after the leaves have all

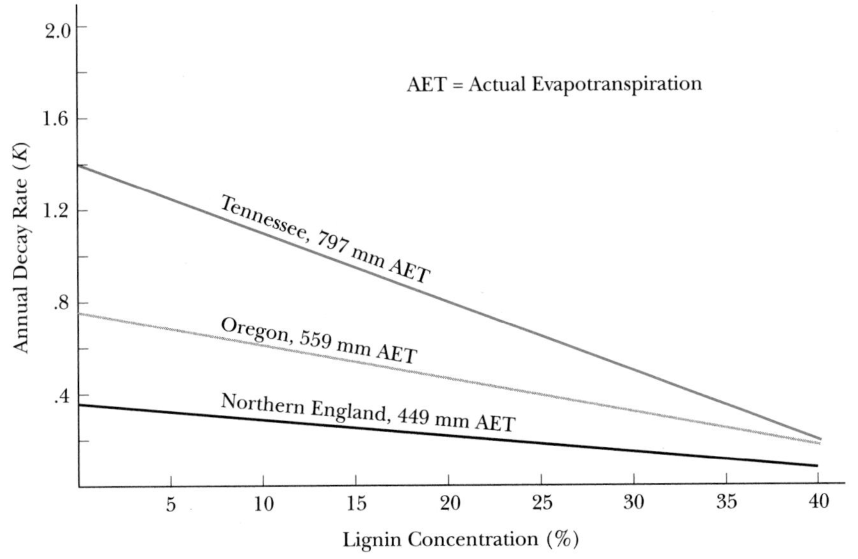

Figure 12-2

Decomposition rate of forest litter as affected by lignin content and climate. Decomposition rate is fastest for materials with little lignin under a warm, moist climate. An annual decay rate of about 0.7 (approximately 70%) gives a half-life of about a year. (Adapted from "Macroclimate and lignin control of litter decomposition rates," by V. Meentenmeyer, *Ecology*, 1978, *59:* 465–472. Copyright © by Ecological Society of America. Reprinted by permission.)

Table 12-2
General Characteristics of Mor and Mull Humus Types

Trait	*Mor*	*Mull*
Vegetation type	Usually needle-leaved	Usually broad-leaved
pH	Acid (usually <4.5)	Less acid (usually >5.0)
Boundary with mineral soil	Distinct	Gradual
Nitrification	Usually absent	Occurs
Decomposition rate	Slow	Faster
Litter	Thick	Thin
Agents of decomposition	Fungi important	Bacteria important
Soil animals	Relatively few	Many, including earthworms

disappeared, can be a little unsettling. The surface consists entirely of earthworm castings, that is, soil that has been passed through the gut of the worm. Such soil, consisting of fine mineral particles cemented into crumblike aggregates by the action of the worms, has excellent structure, highly conducive to plant growth. The worms combine in close association organic and inorganic fractions of soil; they add mineral materials from lower parts of the soil profile to the surface; and they pull leaves down into their burrows, adding organic material to subsurface portions of the soil. They also provide channels in the soil through which air and water can move.

The importance of earthworms in soil processes has, of course, been known for a long time. Darwin studied the ecological role of earthworms; his last book, in 1881, was called *The Formation of Vegetable Mould through the Action of Worms.* In it, he suggested that worms bring a layer about 0.2 inch deep to the top of the ground every year in favorable (moist, circumneutral) soils, amounting to 10 or more tons per acre (23,000 kg/ha). "When we behold a wide, turf-covered expanse," Darwin wrote, "we should remember that its smoothness is mainly due to all the inequalities having been slowly levelled by worms. The plough is one of the most ancient and most valuable of man's inventions; but long before he existed the land was in fact regularly ploughed, and still continues to be thus plowed by earthworms. It may be doubted whether there are many other animals which have played so important a part in the history of the world."

Earthworms are not the only live plows. All burrowing animals move earth. The burrowing and mound building of pocket gophers (Geomyidae) may displace 15 tons of earth per year in an acre of mountain rangeland (34,000 kg/ha) (Teipner et al. 1983). They also harvest and take below ground large amounts of vegetation. These activities tend to displease humans as much as the activities of earthworms please them, and a great deal of money is spent every year to kill gophers on agricultural and forest lands.

Topography

Darwin mentioned the leveling effect of earthworms on the land, but plants and animals may also have the opposite effect, of increasing topographical relief. In sandy areas plants, by offering resistance to the wind, are responsible for soil deposition and dune building. Animals may cause large local effects by making and using trails, rooting, and wallowing. Explorers and early settlers found buffalo trails and wallows to be prominent features of the Great Plains—understandably so, since buffalo weighed half a ton apiece and there were 50 million of them. The trails were narrow, except in areas where several converged, and sometimes were worn 25 cm deep. The wallows (Figure 12-3) were dish-shaped depressions about 4 m across that formed temporary ponds (Keim 1885).

Wild pigs are native to the forests of Eurasia but not to North America. They have, however, been brought to this country by hunting groups and allowed to escape. Now they occur in 11 states (Singer et al. 1984) and are spreading. They entered Great Smoky Mountain National Park in the 1940s and

Figure 12-3

A buffalo wallow in Yellowstone National Park, August 14, 1988. (Photo by the author.)

currently occupy much of the park. Comparisons of areas of heavy pig use with undisturbed sites show that rooting and trampling reduce herb cover, increase bare ground, expose tree roots, grind litter up and mix it with the upper soil horizons, accelerate leaching of minerals, and lower cation exchange capacity. Some of the immediate biological results are known, such as declines in populations of soil invertebrates and certain small mammals; however, the long-term consequences to the ecosystem, such as effects on tree growth and reproduction, are still unclear.

The uneven topography of many forested lands is the result of **windthrow** (Brewer and Merritt 1978, Beatty and Stone 1985). When a large tree is tipped over by wind storms, the root mass retains the soil around it, leaving a hole in the ground (Figure 12-4). Eventually, when the tree has decayed, all that remains are the mound, resulting from the dirt around the roots, and the pit, often occupied by a vernal pool. Such pit-and-mound topography dominates many old-growth forests, representing windthrows occurring over several centuries. Shallow rooting, resulting from impermeable soil layers or a high water table, predisposes trees to windthrow.

Somewhat similar terrain in unforested areas of western North America and various other regions, including eastern Africa (Cox and Gakahu 1983), is the result of animal activity. Mima (pronounced "my-ma") mounds are round earth mounds up to about 2 m high and 20 m or more in diameter. San Diego State University ecologist G. W. Cox (1984a) has presented evidence that they result from the earth-moving activities of burrowing rodents—in North America, pocket gophers. As with windthrow mounds, they are most noticeable in areas with a seasonally high water table or an impenetrable soil layer not far below the surface. Once a pocket gopher burrow is established, the animal moves the dirt from new tunnels back toward the center of its burrow system and deposits it on top of the ground. This tends to produce a mound, surrounded by a lower area. In well-drained areas of deep soil, any topographical effects produced by one animal might be erased a few years later by another that placed its burrow in a different spot. In shallow soil, however, mounds are the most suitable sites for burrows, and the activities of the pocket gophers make them more so; the mounds tend to be continuously occupied and, thus, to be maintained and added to. On wetter ground, the resulting low spots be-

Figure 12-4

(Left) A windthrown beech in Warren Woods, Michigan. The soil from the pit is clinging to the roots but erosion of the soil mass will produce a mound alongside the pit. *(Right)* A close-up of the base of a wind-throw, showing the water-filled pit.

tween the mounds may be occupied by spring pools.

Soil Moisture

In general, the growth of vegetation has a moderating influence on soil moisture, tending to make areas that would otherwise be dry, moister, and making wet areas less so. For example, trees planted in arid areas may precipitate moisture from fog. Plants remove enormous quantities of water from the soil in transpiration, and there are many accounts going back to ancient times of forested areas that, when cleared, became marshes through the resulting rise of the water table (Daubenmire 1959).

There is, however, no guarantee that the effect of vegetation will be moderating. Peat moss, once started on a site, has such a large water-holding capacity that it can cause the soil to become waterlogged. In Alaska, Sitka spruce forest may be converted to muskeg in this way (Lawrence 1958). Contrariwise, alfalfa, which is deep rooting, when grown in arid regions can deplete the soil moisture to such depths that growing more shallowly rooted crops may be impossible for one or more years afterward.

Reactions on Air

Solar Radiation

Shading is not so much a reaction on the air as on the space above the ground, but it is convenient to discuss it here. Vegetation reduces the light intensity (or irradiance) beneath it and also changes the composition of the light by filtering out more in some parts of the spectrum than others.

There are two aspects to how shady an area is: (1) how dim the light in shady patches is and (2) how much ground area is shaded. The reduction of light intensity in the shady spots is related in a general way to the number of leaf layers through which incoming solar radiation passes. The amount of unshaded ground is related to the spacing of the plants or leaves. Vegetation varies in its overall shadiness from extreme desert where sparse plants leave large areas unshaded and cast only thin shade beneath them to tropical forest where several layers of trees absorb much of the light and provide few holes for direct transmission of sunlight to the ground. Savanna areas may have dense shade under individual trees but bright sunlight away from them at a height of 1 or 2 m. At ground level, however, light intensity in savannas and grasslands may be low even away from trees because of shading by the herbaceous vegetation.

Completely satisfactory comparisons of the shadiness of different vegetation types are difficult (Anderson 1966); nevertheless, it is clear that there are very shady communities and not-so-shady communities (Table 12-3). Among the very shady ones are temperate mesic forest and tropical rain forest, in both of which general light intensities may be below 1% of full sunlight.

The sunlit parts of the ground caused by holes in the forest canopy are called sunflecks (Figure 12-5). These have been measured at about 5% of the ground area in Illinois maple forest (Park 1931) and about 2% of Nigerian rain forest (Evans 1966).

Any one spot on the forest floor will, of course, be in a given sunfleck for only a brief period of the day, when the spot, the gap in the canopy, and the sun are aligned. Sections of the floor that are in a sunfleck around noon receive intense sunlight, but areas illuminated by early morning or late afternoon sunflecks receive only low light intensities.

The duration of a sunfleck at a particular point depends on the size of the canopy gap and the height of the canopy. Most last 2 minutes or less; nevertheless, a substantial fraction of the photosynthesis carried out by understory plants may occur during these brief periods of high light intensity (Chazdon and Pearcy 1991). For sugar maple seedlings in northern hardwood forests, about 35% of the summer carbon gain was found to occur during sunflecks (Weber et al. 1985). Understory herbs are

Table 12-3
Reduction of Overall Light Intensity by Different Types of Vegetation

Values are (light intensity under vegetation/light intensity in the open) × 100

Vegetation Type	*Locality*	*Light Intensity (% Full Sunlight)*	*Source*
Leafless oak forest	Michigan	54–78	Brewer et al. 1973
Leafless hardwood forest	Various	55	Shirley 1945
Pine-oak woodland	Arizona	43* 34†	Whittaker and Marks 1975
Field of sunflowers 130–150 cm tall	England	35†	Anderson 1966
Prairie, summer	Nebraska	25‡ 5†	Weaver and Flory 1934
Staghorn sumac stand, summer	New York	14	Shirley 1930
Young oak-pine forest, summer	New York	13* 6†	Whittaker and Marks 1975
Loblolly pinewoods	Florida	9	Hailman 1979
Red pine forest (25–30 years old)	Minnesota	7	Shirley 1945
Tropical rain forest	Nigeria	5–6	Evans 1939
Ponderosa pine forest	Colorado	4	Hailman 1979
Mature spruce-fir forest, summer	Tennessee	4* 1†	Whittaker and Marks 1975
Tamarack forest, summer	Michigan	2§	Brewer, original data
Oak forest, summer	Michigan	2–6	Brewer et al. 1973
Oak forest, summer	Tennessee	1–7	Whittaker 1966
Beech-maple forest, summer	Michigan	1§ <1†	Brewer, original data
Cove forest	Tennessee	<1	Whittaker and Marks 1975
Two-stratum tropical forest	Panama	<1	Hailman 1979
Secondary tropical forest	Nigeria	<1	Evans 1939

* Below trees
† Near ground
‡ Median vegetation height
§ 3m

Figure 12-5

A sunfleck on the floor of a beech-maple forest. (Photo by the author.)

able to increase photosynthetic rate rapidly in response to brief periods of bright light; however, it is not yet clear how different they are from plants of other habitats in this ability (Chazdon and Pearcy 1991).

The light distribution resulting from sunflecks has been suggested as the physical basis for the height—in fact, the existence—of the subcanopy tree stratum in temperate deciduous forests. Just below the canopy, a given point will be in constant shade if it is not in a gap. If it is in a gap, it will be in sunlight when the sun is directly in line with it and in shade the rest of the day.

Going down, a level is eventually reached in certain kinds of forest (oak, for example) where a given point is illuminated by sunflecks through most of the day, first by a sunfleck through a gap on the east, then by one from more nearly overhead, and finally by the next gap to the west. The height of the subcanopy trees seems to correspond fairly well with this level in the forest—7 to 12 m in several forests in the southeastern United States (Terborgh 1985). Trees that go above this level are growing into a more variable light regime (unless thay can reach the canopy); shorter trees are apt to be shaded by neighbors that grow up to this level.

Evergreen forests, whether broadleaf or needleleaf, are shady all year round. Deciduous forest has a leafless period, during either a cold season or a dry one, when light intensity is much higher, generally 50 to 70% full sunlight (Figure 12-6).The length of the light phase of such forests varies; in temperate America, oak forests may leaf out a month later than maple forests.

Although some radiation across the range from ultraviolet to infrared is transmitted through (and reflected from) leaves, much of the light energy is absorbed. Because red and blue light are absorbed by photosynthetic pigments, we might suppose that the light in forests would be green. We would have "green shade" because of the transmission and reflection of the green portion of the spectrum. This is true but only for forests composed of broadleaf trees (Coombe 1957). Needleleaf canopies tend to act like neutral filters. Needleleaf trees are said to cast "blue shade."

Broadleaf vegetation absorbs most infrared radiation beyond 2000 nanometers but absorbs very little of the near infrared from about 700 to 1100 nm (Figure 12-7). This is transmitted or reflected (the large amount reflected is why broadleaf vegetation looks white in infrared photographs). If this

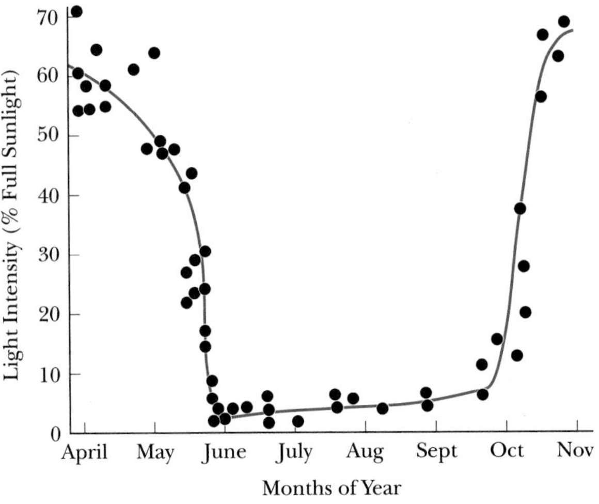

Figure 12-6

Light intensity from spring through fall in Ontario deciduous forests. Each dot is an average for a single forest stand. From November to April, with the trees bare, light intensity is 55–70% full sunlight. (Adapted from Sparling 1964.)

radiation were absorbed (and nothing else changed), leaf temperatures of plants receiving direct sunlight would be higher, and overheating would presumably be a problem. Sunlight contains a considerable amount of energy at these wavelengths; it is likely that this radiation is important in the energy budgets of plants and animals in the understory of broadleaf forests.

Some ecological effects of shading are obvious. Organisms sort out in successional sequences from plants intolerant of shade in pioneer stages to highly tolerant plants in climax stages (see Table 3-2). The canopy trees of a forest produce, to a considerable degree, the light conditions under which all the other organisms must operate; consequently, the understory plants of temperate mesic forest are either **sciophytes,** plants capable of living under very low light intensities, or else plants that evade shading, such as the spring ephemeral herbs that do virtually all their growing and food making prior to canopy closure.

Many other effects of shading are yet to be worked out with certainty. For example, do the dif-

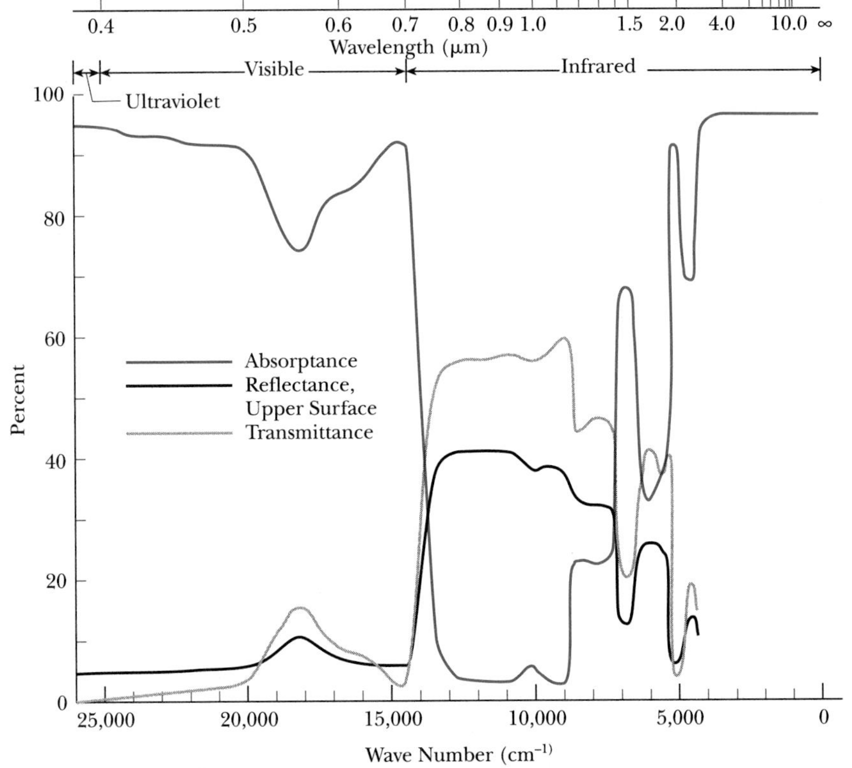

Figure 12-7

Cottonwood leaves have high absorptance into the near infrared (the dip around .55 micrometers is green light which, of course, is reflected and transmitted to a greater degree than red, orange, blue, indigo, and violet light). Tree leaves absorb very little infrared light of wavelengths between 0.75 and 1.2 micrometers. (Reprinted from Gates 1980 by permission.)

ferences in spectral composition of the light in broadleaf versus needleleaf forest—green shade versus blue shade—have physiological or behavioral effects of ecological importance? Is the sharp decrease in absorption of solar radiation at 700 nm by broadleaf plants (Figure 12-7) the result of evolution or simply physics? It is a fertile field for study.

Temperature, Humidity, Rainfall, and Wind

Vegetational cover moderates temperature extremes below it by absorbing solar radiation and acting as insulation. Average temperatures may not be changed much. Humidity is higher within vegetation than in bare areas and higher in forest than in grassland. Wind velocity is reduced; the reduction 245 m within a beech-maple forest in Ohio was 90% in the summer and 75% when the trees were leafless (Williams 1936).

Rainfall within the forest is reduced because of the amount intercepted and re-evaporated by the foliage (Figure 12-8). How big the reduction is varies, depending on the vegetation and whether the rainfall is light and brief or heavy and sustained. In regions with frequent, heavy fog, on the other hand, precipitation can be increased as a result of condensation. Precipitation under an old-growth Douglas fir forest in Oregon was nearly 30% more than in an adjacent clear-cut, evidently as the result of fog drip from the trees (Harr 1982). In such a region, stream flow might be reduced by clear-cutting.

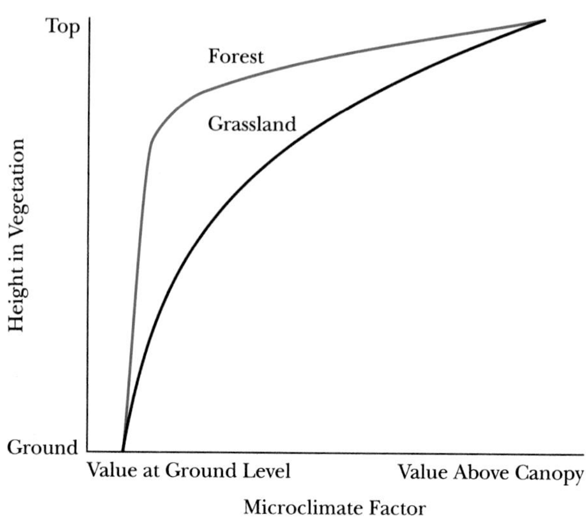

Figure 12-9

Vertical gradients in vegetation. In forests, reactions on solar radiation, temperature, etc. are produced mostly by the canopy, so rapid change occurs there and little occurs below. In grassland, the change is more gradual.

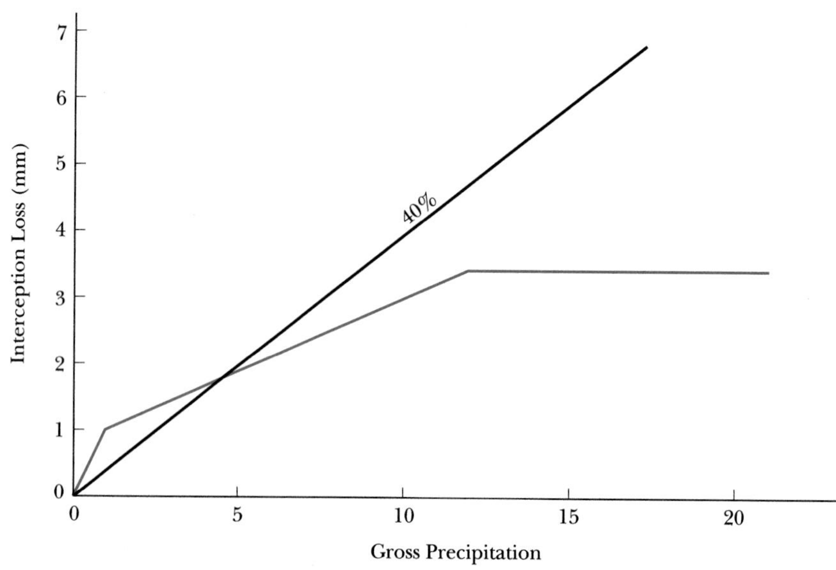

Figure 12-8

Interception loss in a Danish oak stand (trees about 17 m tall) in relation to the total amount in single showers. The dashed line represents 40% interception loss. Small showers of up to 4 or 5 mm, total, have interception losses above this level. Only about 3.5 mm are intercepted in a shower of 12 mm—about 30%. In heavier rains, no additional interception loss occurs, so the percentage steadily drops. (Adapted from Rasmussen & Rasmussen 1984.)

Temperature, humidity, and wind, as well as light, may show vertical gradients, which tend to be more pronounced in shrubby or herbaceous vegetation than in forest. In forest, most of the change (in temperature or light intensity, for example) occurs between the top and the bottom of the canopy (Figure 12-9). In grassland, the changes occur gradually from the top of the vegetation to the ground.

Vegetation affects not just its own microclimate but also the regional climate. In the 15th century Christopher Columbus noted that after the forests were cut on the Canary, Madeira, and Azores Islands, there was less mist and rain (Shukla and Mintz 1982). One way in which biological processes may affect the regional rainfall, by producing condensation nuclei, is discussed later in connection with hydrological cycle.

Reactions in Fresh Water

Ponds and lakes tend to fill in; some of this filling is from sediments washed into the basin, but another part is through the addition of organic matter from the aquatic plants. In many ponds and lakes, this may be in the form of leaves, fruits, and rhizomes of water lilies and cattails. In more northerly regions, especially, basins may fill with peat from the remains of sphagnum moss and sedges. Bogs and the reactions of sphagnum are discussed in more detail in Chapter 18. Hydroseres, the succession from open water to more or less dry land (Chapter 18), are based in part on this filling in of the basin. Other reactions of the vegetation including transpiration and shading may also be important.

Plants growing in water lower the light intensity below them just as terrestrial plants do. This produces conditions that are favorable for some organisms and unfavorable for others. In hypereutrophic lakes, the productivity (which is high) is restricted to the upper level of the water because of the absorption of light by the dense populations of algae; in a highly productive Ethiopian soda lake, the zone where photosynthesis occurred was less than 0.6 m thick. In less fertile lakes, photosynthesis may occur many meters into the lake.

Plants, animals, and bacteria exchange gases and minerals with the water in which they live and, in so doing, cause both local and global effects. Locally, diatoms which require silica for their shells (tests) may deplete the silica concentration to a point where they decline. The green algae that then increase may deplete nitrogen concentrations to a point where only nitrogen-fixing blue-green algae can thrive.

The deposition of marl, a form of calcium carbonate, on rocks and submerged vegetation in lakes and streams with hard water, is largely the result of plant activity. Inorganic carbon in water exists in an equilibrium as given below:

$$Ca\ (HCO_3)_2 \rightleftharpoons CaCO_3\downarrow + H_2O + CO_2$$

(this fits into the carbon cycle as shown later). During rapid photosynthesis by algae and other plants, carbon dioxide may be rapidly withdrawn, causing the reaction to proceed to the right. At the site where this occurs, calcium carbonate accumulates as deposits.

The flooding and subsequent effects resulting from dam building by beavers are among the most striking reactions by animals. Beavers dam streams (generally where the water is the swiftest) with trees they fell, along with mud, stones, and sticks (Figure 12-10). Terrestrial vegetation in the pond dies, and aquatic succession begins. Eventually the beavers leave (usually, under natural conditions, because they exhaust their winter food supply), the dam deteriorates and breaks, and the valley is returned to dry land but with an altered soil consisting of the old lake bed.

A large part of the primeval landscape in many regions must have been influenced by beaver activity, current or past. Clements and Shelford commented in 1939 that "today the reaction is practically confined to remote regions." The situation has not changed much since. The scarcity of beavers is not, however, the result of their unwillingness to live around humans. Rather, it is because beavers are systematically removed when they begin operations in regions where they might interfere with agriculture or other human activities.

Reactions in the Ocean

Aside from the enormous role that ocean organisms play in the processes that control the inorganic

Figure 12-10

The beaver dam in the foreground, across a stream in Algonquin Provincial Park, has produced this large pond. (Photo by the author.)

composition of the ocean and the atmosphere, perhaps the outstanding reaction of marine organisms is the building of coral reefs. Reefs grow through the buildup of the calcareous skeletons of stony corals with similar contributions from other organisms, especially red algae. The total weight of the limestone contributed by algae evidently is greater than that of the corals. The reefs grow in warm, shallow waters along continents or islands or atop submarine heights. The reef provides habitats for a diverse grouping of organisms that include unicellular algae that live within the coral polyps, filamentous algae that occur as bands in the coral skeleton, and a variety of invertebrates—sponges, sea cucumbers, worms, crabs, mollusks—and brightly colored fishes that live on the surface of the reef or in its crevices.

The swamps of mangroves that occupy portions of subtropical and tropical coasts are important in decreasing erosion from tropical storms. By trapping mud and adding organic matter they may also build up the shoreline; how much ocean is really turned into land is unclear. Both reefs and mangrove swamps are discussed further in Chapter 18.

The bottom of the ocean deeper than about 2000 m consists mostly of the shells (tests) of marine organisms, mostly protists (King 1968). The deposits tend to be a few hundred meters deep and are called "ooze." Calcareous ooze, mostly from Foraminifera, covers about one half of the area of the ocean depths. Siliceous oozes make up about 15% of the bottom and are from diatoms (in cooler waters) or radiolarians (in the tropical Pacific). Slightly less than 40% of the area of the ocean bottom consists of inorganic sediments called "red clay."

Nutrient Cycling

The energy talked about in Chapter 11 is, by and large, in the form of chemical bonds of organic compounds made up of such elements as carbon, hydrogen, oxygen, nitrogen, sulfur, and about 20 (or probably more) others—the essential elements of protoplasm. These elements move from the abiotic to the biotic portion of the ecosystem as plants take in carbon dioxide from the air and water and minerals from the soil and use these to produce carbohydrates, fats, and proteins. As the consumers eat the plants and obtain the organic compounds of which the plant body is made, these elements are passed on to the consumers. They are returned to the abiotic portion of the ecosystem through excretory processes of producers, consumers, and decomposers. This, then, is another aspect of ecosystem function—the flow of materials. Energy flow, as we have already seen, is a one-way process. Energy enters the ecosystem as sunlight and leaves as heat that is dissipated in the universe. However, materials (matter) move in more or less complete circular systems known as **biogeochemical cycles.** This may be an unnecessarily long name, but it emphasizes the global interconnectedness of organisms (bio) and the rest of the earth (geo). Different organisms, different habitats, and different parts of the globe affect one another through the movement of these essential chemical elements.

The word "cycle" emphasizes the circularity of the system, the dependence of each segment on the one preceding it. Although it is clear to all of us that animals depend on plants, it is easy to fall into the error of thinking that plants, able to synthesize complex organic molecules from simple inorganic materials, are relatively self-sufficient. However, most of the "simple inorganic materials" that plants use—carbon dioxide, ammonia, sulfates—are the products of other organisms. There is no starting point for biogeochemical cycles. They are not for our benefit or even for the benefit of green plants any more (or less) than they are for the benefit of iron or sulfur bacteria.

The Carbon Cycle

Everyone is familiar with biogeochemical cycling in at least a vague way, through the processes of photosynthesis and respiration. Figure 12-11 illustrates the **carbon cycle.** Carbon dioxide in the air (or, for aquatic plants, dissolved in the water) is taken up by plants and used in photosynthesis. As organic compounds are used by the plants themselves, some carbon dioxide is returned to the environment, but much of the carbon is retained in the plant bodies. Primary consumers obtain their carbon when they eat plants and higher-level consumers when they use lower-level consumers as food. Decomposers act on the dead bodies of plants and animals when they die and obtain their carbon from these. All consumers and decomposers carry on respiration that returns carbon dioxide to air or water.

This is the basic part of the cycle, but there are some further ramifications. There is a rather complicated chemical system involving carbon dioxide in water. When carbon dioxide dissolves, some of it combines with water to form carbonic acid. Bicarbonates and carbonates may, in turn, be formed. Carbonates are not very soluble and may precipitate and be deposited as sediments on the bottom of lakes and oceans. All of these are reversible reactions; the overall effect is to buffer the carbon dioxide content of the air. If there is a local depletion of carbon dioxide in the air, a reversal of the reactions just described occurs, resulting in the liberation of carbon dioxide from water to air. If there is an increase in carbon dioxide in the air, more dissolves in water, more is converted to carbonic acid, and so forth. Material that becomes incorporated in sediments probably returns to the system slowly, but these sediments may be brought back into circulation by geological processes such as volcanic activity or by the uplift of land and subsequent weathering of limestone (calcium carbonate).

Another complication is the storage of fossil carbon. Some dead organic matter such as the peat in bogs escapes decomposition. The amount of new organic material formed without subsequently decomposing seems to be small, but a fairly large amount has accumulated throughout the period of life on earth, much of it over a period of about 65 million years ending 280 million years ago. This period, the Carboniferous, was the time during which most of our coal, oil, and gas were stored. Before the advent of humans these materials probably decomposed slowly. There are, for example, petroleum bacteria that are able to use oil as their carbon and energy source (that is, as their food).

With our discovery of the use of this material as an energy source, the release of the stored carbon as carbon dioxide has greatly increased. In 1900, the addition of carbon dioxide through burning of these fossil fuels to the atmosphere by hu-

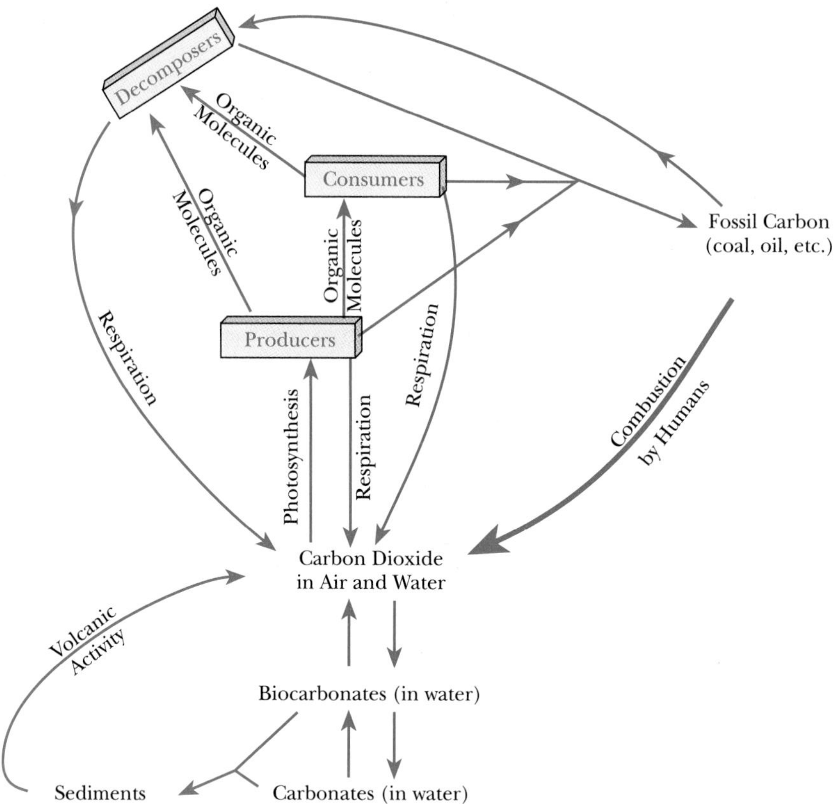

Figure 12-11

Main features of the carbon cycle in the biosphere.

mans was about 1 billion tons, in 1955 it was about 2 billion tons, and in 1980 it was above 5 billion tons, enough to raise the carbon dioxide content of the air by more than 2.3 parts per million each year if none were removed. The carbon dioxide content of the air has increased but not by this amount. Instead it increased by a little less than 1 part per million per year during the 1960s and by a little more than 1 part per million per year in the 1970s.

Figure 12-12 shows the increasing carbon dioxide concentration of the atmosphere. The graph also shows a regular yearly trend, rising from a September low to a May peak. This pattern probably is biological, resulting from the predominance of photosynthetic use of carbon dioxide during the summer and the predominance of respiratory production of carbon dioxide by animals, decomposers, and plants during the rest of the year.

What has happened to the additional carbon released from fossil fuels but not now in the atmosphere? Some of it has gone into the oceans, but by no means all, as recent studies have made plain (Tans et al. 1990, Quay et al. 1992). Currently about 7 billion tons of carbon per year are entering the atmosphere from fossil fuel combustion and tropical deforestation. Slightly more than 3 billion tons are being added per year to the atmosphere. Ocean uptake is now estimated at 1 or 2 billion tons per year, leaving 1 or 2 billion tons unaccounted for (Kerr 1992).

The most likely additional sink is increased plant biomass. For plants that have plenty of water and sunlight and a favorable temperature, carbon dioxide concentration can be a limiting factor. Consequently, high carbon dioxide levels create the potential for increased plant growth. The additional carbon could, then, be locked up in more or bigger trees or other forms of biomass, especially in the tropics (Lugo 1983).

Forest regrowth in temperate regions could be another place where carbon is being stored. Forest

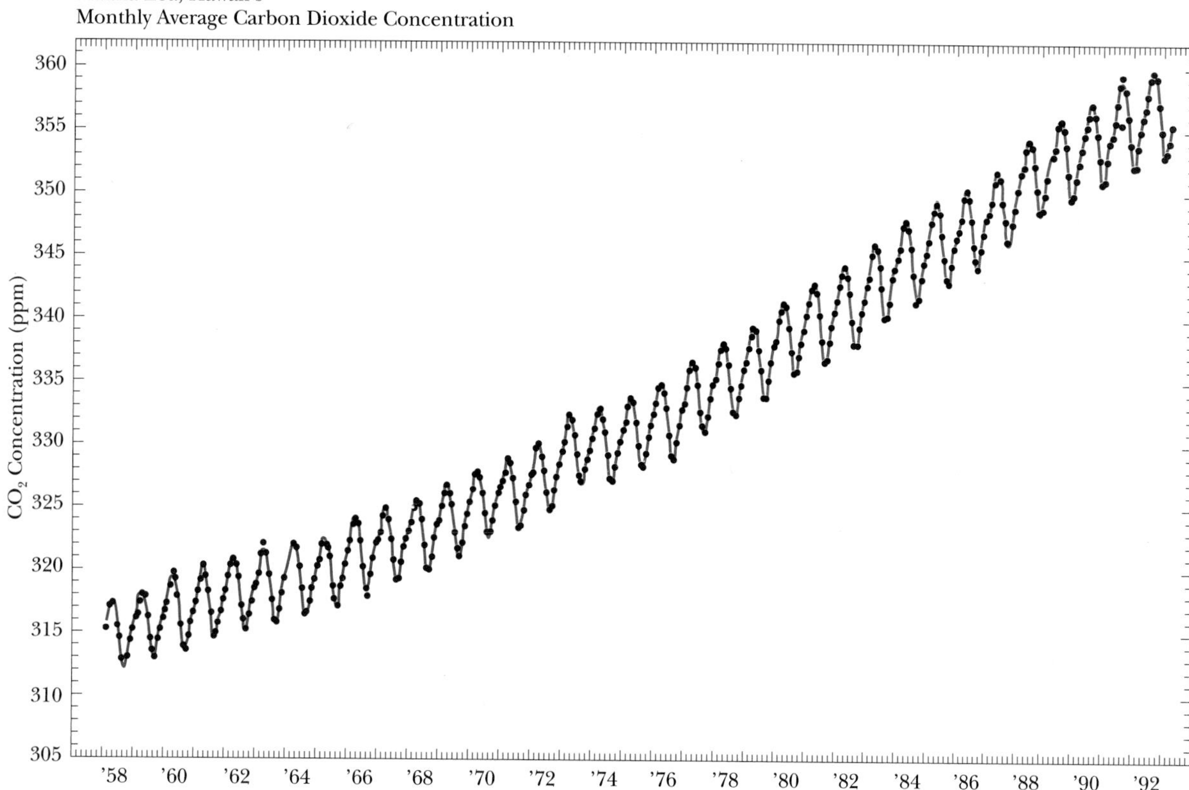

Figure 12-12

Atmospheric carbon dioxide levels (parts per million of dry air) 1958 through 1992, as measured at the Mauna Loa observatory, Hawaii. A smooth curve is fitted to monthly averages as given by the dots. Mauna Loa Observatory is operated by the U. S. National Oceanic and Atmospheric Administration. The measurements were obtained in a cooperative program of NOAA and the Scripps Institution of Oceanography. (From C. D. Keeling, D. J. Moss, and T. P. Whorf, ''Measurements of the Concentration of Atmospheric Carbon Dioxide at Mauna Loa Observatory, Hawaii, 1958–1986,'' Final Report for the Carbon Dioxide Information and Analysis Center, Martin–Marietta Energy Systems Inc., Oak Ridge, Tennesee, subcontract 19X-89696C, issued in April 1987. Updated by C. D. Keeling, R. B. Bacastow, A. F. Carter, S. C. Piper, T. P. Whorf, M. Heimann, W. G. Mook, and H. Roeloffzen, ''A Three Dimensional Model of Atmospheric CO_2 Transport Based on Observed Winds: Observation Data and Preliminary Analysis,'' Appendix A, in Aspects of *Climatic Variability in the Pacific and the Western Americas,* Geophysical Monograph, American Geophysical Union, vol. 55, 1989; and personal communication C. D. Keeling.)

acreage was probably at its minimum in eastern North America in the early years of the 20th century. Wood volume has evidently increased in Europe in the past 20 years (Kauppi et al. 1992). It is widely believed that forest area has been increasing in the United States. This was not true in Kalamazoo County, Michigan, between 1967 and 1983. Young forest increased but mature forest was lost. Also increasing were old field, thicket, and other successional stages as cropland was taken out of cultivation (Brewer et al. 1992). It seems unlikely that the amount of carbon stored in wood, perennial roots, and litter in these ecosystems and in young forest is greater than that released when mature forest was destroyed.

Some scientists believe that world biomass has declined rather than increased in the past two or three decades (Woodwell et al. 1978, Houghton et

al. 1983). There has been a great deal of deforestation in the tropics as well as continued loss of high quality forest in North America. The oxidation, by burning or decay, of the trunks, roots, litter, and humus from these areas has released large amounts of carbon dioxide.

The fact remains that the carbon dioxide content of the atmosphere has increased considerably early 19th century levels. This has the potential to increase global temperatures, as discussed in the section on the greenhouse effect (Chapter 20).

The Nitrogen Cycle

Plants use nitrogen in the production of proteins and many other compounds. Although the large reservoir of nitrogen in the air is not directly available to plants, its presence facilitates the global cycling of the element. Most plants must take in nitrogen as ammonia or nitrate (Figure 12-13). Animals produce their proteins from proteins in food, and decomposers make their proteins from those in dead animal and plant tissue.

When proteins are broken down in respiration, a waste product containing nitrogen is produced. The usual cellular waste is ammonia, which some organisms excrete directly; however, many convert it to another, less toxic compound. Mammals, including humans, convert ammonia to urea. Urea and similar compounds contain energy that decomposers can obtain in breaking them back down to ammonia.

Ammonia is retained in soils (as the ammonium ion) by the cation exchange system. Plants can use it directly, so the basic cycle is complete with the excretion of ammonia by animal or decomposer. There are, however, important subsystems involving various groups of microorganisms.

Nitrogen usable by plants can be lost from ecosystems in several ways. Under aerobic conditions in soil and water, nitrite bacteria convert am-

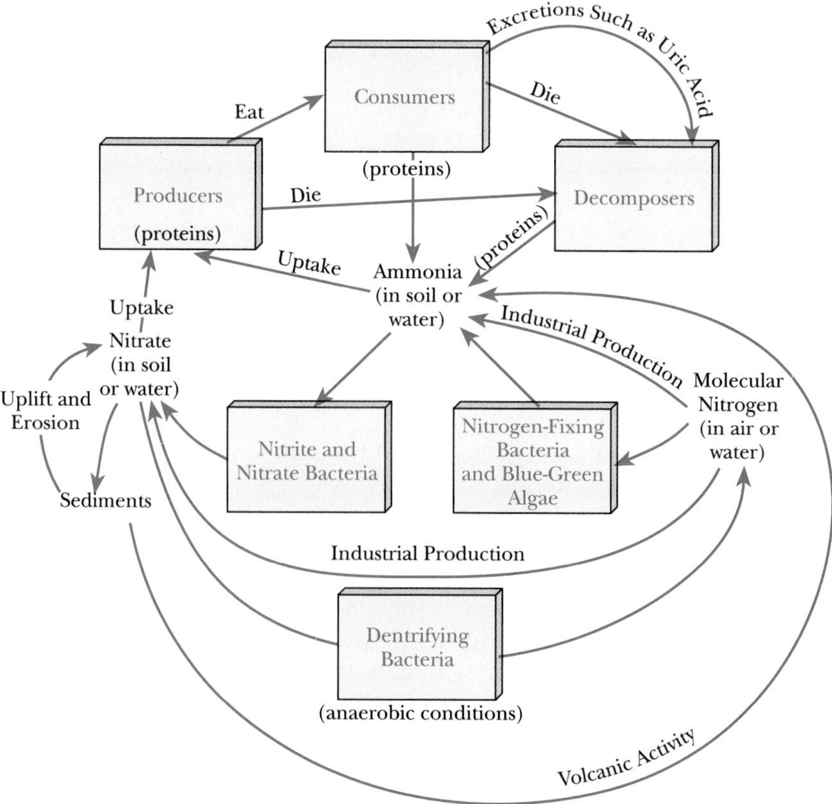

Figure 12-13

Main features of the nitrogen cycle in the biosphere.

monia to nitrite, and nitrate bacteria convert nitrite to nitrate. The process, called **nitrification,** yields energy; the bacteria involved are chemosynthetic autotrophs. Nitrate is highly soluble and, hence, is readily leached from soils; nevertheless, nitrogen tends to be tightly conserved within the ecosystem. This is because nitrate tends to be speedily taken up by plants for new growth. Also, nitrogen within the plant tends to be conserved by retranslocation; that is, nitrogen tends to be shifted out of plant parts that are dying and stored within roots or other living parts (Cole and Rapp 1981).

Even so, ecosystems can lose appreciable amounts of nitrate at times. Organisms that cut down leaves before they're ready to fall, such as squirrels (Skinner and Klemmedson 1978), caterpillars, and maple petiole borers, short-circuit the translocation process, and may result in detectable rises in the loss of nitrate from the ecosystem (Swank et al. 1981). Nitrogen loss can be very high in some circumstances. The first year after a forest is clear-cut, the nitrogen output may rise 3 to 50 times over the precut output (Vitousek et al. 1979) as foliage, twigs, bark, and litter decompose.

Nitrogen-fixing organisms convert the molecular nitrogen of the atmosphere into ammonia. Nitrogen fixation is an expensive process energetically, but a large number of organisms do it, including both photosynthetic and heterotrophic bacteria in both aerobic and anaerobic environments. The process may occur symbiotically or nonsymbiotically. The classic case of symbiotic nitrogen fixation is by *Rhizobium* bacteria that live in nodules on the roots of legumes, such as soybeans and black locust trees. The legume gets a plentiful supply of ammonia from the bacterium for protein production. From the plant, the bacterium receives photosynthate and a suitable microhabitat.

Various nonlegume trees and shrubs, such as alders, also have nodules. The nitrogen-fixing organisms in these associations are actinomycetes. An interesting aquatic symbiosis is known between the aquatic fern *Azolla* and the nitrogen-fixing cyanophyte *Anabaena* (Peters et al. 1986). The alga lives in pores in the fern fronds and is passed on to the next generation in the spores.

Nonsymbiotic nitrogen fixation is carried out by free-living cyanophytes and bacteria in soil and water and also by ones living epiphytically on trees and other plants. Mutualistic associations between these organisms and non-nitrogen fixers probably occur but have not been thoroughly studied.

What nitrogen fixers do, denitrifying bacteria undo. They convert nitrate to molecular nitrogen, which may then escape from the soil or water as a gas. The bacteria are using nitrate in place of oxygen as electron donors in their respiration of organic compounds. The process occurs under anaerobic conditions at the bottom of lakes and estuaries and in wet soils.

Denitrifiers seem perverse from a human viewpoint: How unfortunate that all that perfectly good nitrate is being turned into an almost inert gas. It would be a typical human reaction to decide that denitrifying bacteria, like mosquitos and ragweed, ought to be controlled. The special lesson of ecology is that we need to look beyond the immediate effect and see how things are interconnected. Denitrification is the main process returning molecular nitrogen to the atmosphere. Without denitrifiers, our atmosphere would slowly be depleted of nitrogen, which would tend to accumulate as nitrate in the ocean.

The Phosphorus Cycle

Phosphorus is a component of DNA, RNA, and ATP. As such, it is essential to the functioning of every living cell. Phosphorus also plays other roles varying by organism. It is, for example, a constituent of the vertebrate skeleton. Even though phosphorus is a small part of the bulk of most organisms, it is often a limiting nutrient because of its relative unavailability. Plants contain only about 3% phosphorus, but the soil solution that they draw on contains only about 0.000003% (Foth 1978).

Plants usually take up phosphorus as orthophosphate ($H_2PO_4^-$). In terrestrial situations, mycorrhizae are generally important in the process. Animals obtain most of the phosphate they need in their diet. Excretions of animals return phosphorus as phosphate to soil or water. Dead bodies of plants and animals are acted upon by decay organisms, regenerating phosphate.

The basic biotic portion of the phosphorus cycle is relatively uncomplicated (Figure 12-14). Complications are provided by the abiotic portion. These come from the relative insolubility of most forms of phosphate. This, rather than its scarcity in

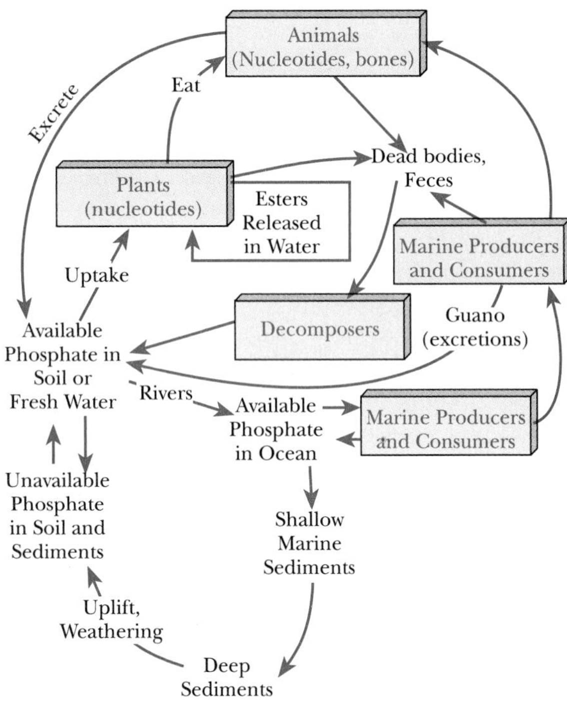

Figure 12-14

Main features of the phosphorus cycle in the biosphere.

most soils or sediments, is the key to its limiting role for plant growth.

Orthophosphate resulting from weathering of rock particles, if not taken up by plants (or bacteria), tends to be locked up abiotically. This is especially true below pH 5.5, through the formation of relatively insoluble iron and aluminum compounds, and above 7.0, through the formation of other unavailable forms, including complex calcium phosphates.

Phosphorus in usable form liberated by organisms is, as we would expect, rapidly recycled through plant uptake. We can get a hint of this by the depletion of phosphorus that sometimes occurs in lake waters in the summer (Vallentyne 1974). But how long does an average phosphate ion stay in the water before it is taken up and reincorporated into an organism? Limnologists have studied this problem by adding orthophosphate labeled with radioactive phosphorus. By measuring how much of the radioactivity was left after various intervals, they were able to specify turnover time. The shortness of the period is astonishing. The average turnover time of phosphate in the epilimnion of lakes in the summer is a few minutes (Wetzel 1983). Winter turnover time is longer, usually several hours. This is about the same turnover time measured for the open ocean.

Such recycling is not 100% effective, and phosphorus tends to accumulate in relatively insoluble compounds in lake sediments. These may contain phosphorus at concentrations hundreds of times higher than in the water. Certain processes such as the formation of ferrous sulfide under anaerobic conditions at the lake bottom (when iron phosphates react with hydrogen sulfide) return some phosphorus to the water. Nevertheless, the main road runs to the sediments. Phosphates in fresh water tend to be carried to the oceans and accumulate eventually in marine sediments.

Terrestrial ecosystems seem to recycle phosphorus more efficiently than do aquatic ones. In the northern hardwood forest drained by Hubbard Brook, New Hampshire, about 10 kg of phosphorus per hectare per year are released in decomposition. Weathering of soil minerals releases another 1.5 to 1.8 kg. Despite this, only 0.007 kg of dissolved inorganic phosphorus per hectare per year is carried away in the stream waters. By collecting soil water at various depths, it was found that although some phosphate was trapped "geochemically" by iron and aluminum compounds in the B horizon, most of the phosphate never got that deep. Instead, it was recycled biotically in the upper portions of the forest floor (Wood et al. 1984). Similar tight conservation of organic phosphorus has been suggested for tropical forests (Went and Stark 1968, Odum 1983). Saprophytic and mycorrhizal fungi are probably involved, but the details are still unclear.

Return of phosphorus from ocean to land is evidently extremely slow. Because phosphate is so insoluble, windblown sea salt contains little. There seems to be no important gaseous compound of phosphorus to mediate such a return, although some speculation has centered on phosphine in this role. At present, it seems that the main way that phosphorus returns from the ocean is through the landward migration of animals that feed there, such as salmon, seals, and seabirds.

These animals return phosphorus in their bodies, especially their skeletons. Also, their excretions contain phosphate. The most spectacular examples of this aspect of the return of phosphate to the land are the guano deposits on the dry coasts and islands of Africa, Australia, and South America. Hundreds or thousands of years of occupancy of some sites produced deposits of seabird droppings up to 50 m deep. Various kinds of birds are involved. The deposits on the Chincha Islands of Peru, the most famous guano deposits, are from the Guanay cormorant.

Guano was an exceptional fertilizer ("a name to conjure with," according to the article on manures in the 11th edition of the *Encyclopaedia Britannica*). It contained most nutrients needed by plants in usable forms and in good proportions. Although guano is still mined and used, the best deposits were depleted in a period of only about half a century beginning in 1840.

Humans are now involved in the phosphate cycle, as in others, on a large scale. The mining of phosphate-containing rocks and the subsequent use of phosphate as fertilizers and in detergents, among other uses, has speeded up enormously the slow movement from rock to soil to fresh water to ocean.

One result has been lake eutrophication, discussed in Chapter 18. One might expect, knowing the characteristics of phosphate cycling, that if phosphate fertilization is halted, the system would return to "normal" quickly. This is usually true. The addition of phosphate to a lake causes a jump in algal production, but if the phosphate is cut off, productivity drops within weeks or months as the phosphate is lost to the sediments.

The Sulfur Cycle

Sulfur is an essential component of protoplasm, occurring in various amino acids, enzymes, and also certain other compounds including those that produce the odors of garlic and skunks. Sulfur is rarely a limiting nutrient for plant growth under natural conditions, being generally abundant relative to phosphorus or nitrogen.

Uptake by plants is mostly as sulfate from soil (or water, for aquatic plants; Figure 12-15). Some plants can use sulfur dioxide (SO_2) directly from the air. Animals ordinarily satisfy their sulfur requirements in their food. The decay of plant and animal tissue by various bacteria and fungi eventually yields sulfates or hydrogen sulfide. Sulfate is regenerated from H_2S, mostly in water, by chemosynthetic and photosynthetic sulfur bacteria. Oxidation of H_2S to sulfate also occurs abiotically in water and in the atmosphere. Sulfate is returned to land from the sea as windblown sea salt.

Some additional details are shown in Figure 12-16. The activities of sulfur-reducing and sulfur-oxidizing bacteria are discussed in the section on chemolithoautotrophs in Chapter 11.

Human effects have become heavy in this cycle. Globally, more than one fourth of the sulfur going into the atmosphere is sulfur dioxide from the burning of fossil fuels (Kellogg et al. 1972); in North America the fraction is above 90% (Galloway and Whelpdale 1980). Chemical reactions in the atmosphere yield sulfuric acid, producing acid rain, discussed in more detail in Chapter 20.

Nutrient Limitation of NPP

Individual species populations can be limited by any of a variety of nutrients, depending on physiological requirements. For example, potatoes have a high potassium requirement and do poorly if grown in organic soils, which are potassium deficient. The growth of diatoms, with their siliceous outer walls, can be limited by the silicon content of the water. When silicon concentrations are low, however, other species of phytoplankton tend to replace diatoms, and net primary production (NPP) of the phytoplankton as whole is left pretty much unaffected.

Nutrient limitation at this level, that is, of NPP in natural settings, is usually by nitrogen or phosphorus. There are situations, of course, in which NPP is not limited by any mineral nutrient. In the desert, moisture limits NPP. Only if adequate moisture is provided does a nutrient—usually nitrogen—become limiting.

Peter Vitousek and Robert Howarth (1991) summarized a variety of evidence indicating that phosphorus is commonly the element limiting NPP of lakes, at least in the temperate zone. Limitation of marine primary production evidently varies by habitat. Nitrogen is widely important but phosphorus seems implicated in some lagoons and estuaries, and iron may limit productivity at times in the open ocean. Nitrogen seems to be the limiting nutrient

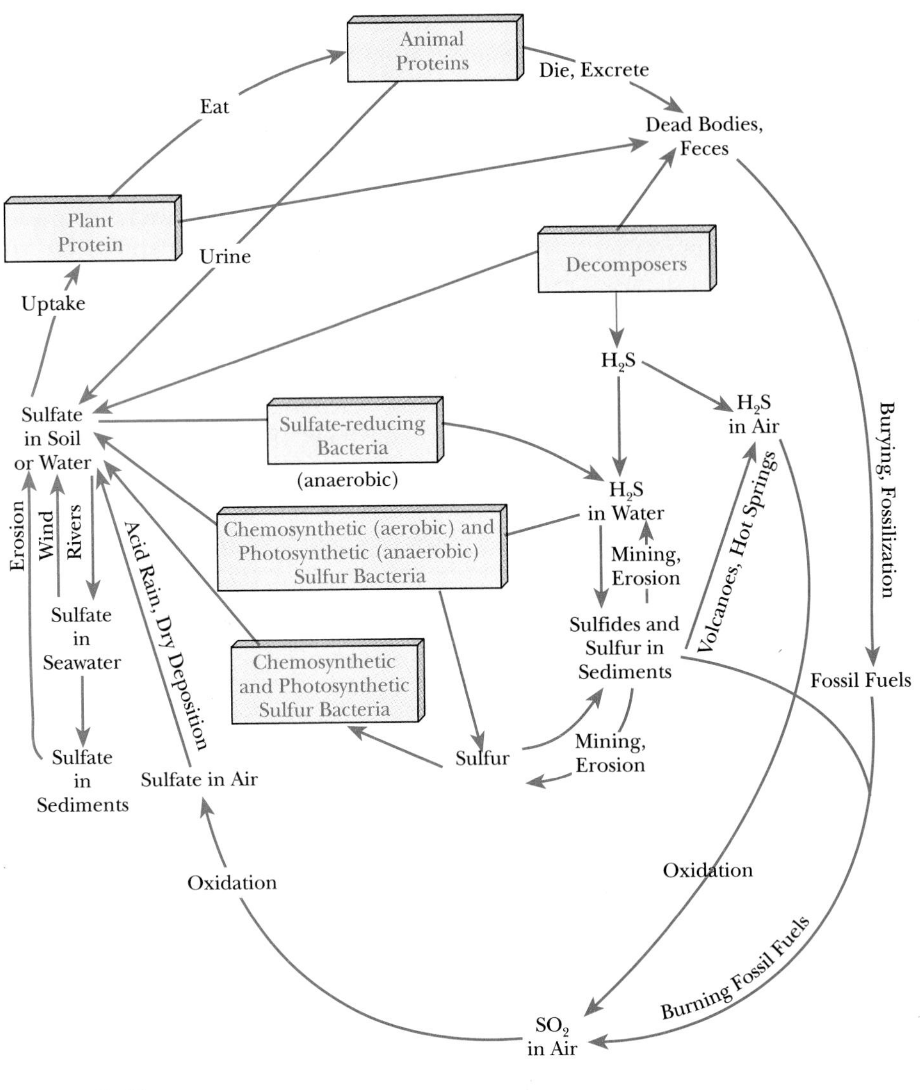

Figure 12-15

Main features of the sulfur cycle in the biosphere.

on terrestrial sites, with just a few exceptions. On very old soils, such as in the tropics and ancient, unglaciated landforms in temperate regions, phosphorus may be limiting.

The conclusion that nitrogen limitation of NPP is widespread in marine and terrestrial habitats is puzzling (Vitousek and Howarth 1991). Why, if nitrogen is scarce, don't nitrogen-fixing organisms increase in these habitats, just as the algae with low silicon requirements increase in lakes? They could thrive forever, or at least until they built up the nitrogen supply to a point where it was no longer limiting to other plants. It is a good question, with probably more than one answer. Possibly the most general explanation for terrestrial sites is something like this: On low nitrogen sites, the litter that is produced has a high carbon : nitrogen ratio and, hence, is slow to decompose. That is, the decomposers quickly become nitrogen limited. They then take up such nitrogen as is available from the soil,

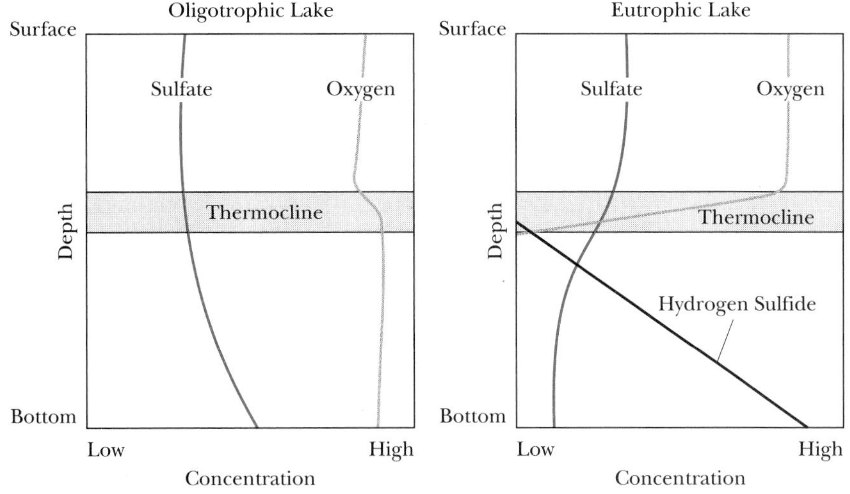

Figure 12-16

Diagrammatic representation of sulfate and hydrogen sulfide in unproductive (oligotrophic) and productive (eutrophic) lakes. Hydrogen sulfide is readily oxidized to sulfates in the presence of oxygen. Only under conditions of low oxygen as at the bottom of thermally stratified lakes does the hydrogen sulfide generated from decomposition build up. It is in such sites that green and purple sulfur bacteria use the hydrogen sulfide in photosynthesis. (Adapted from Wetzel 1983.)

further depleting the supply available for plant growth, and may tie it up, or "immobilize" it, for long periods.

This is one way in which the nitrogen available for plant growth can become depleted but it would seem to set the stage for invasion of nitrogen fixers. This may not happen, in temperate forests at least, because of the interplay of the high energetic cost of nitrogen fixation and the deep shade, hence low energy availability, in the understory. That is, even though the producers as a whole are nitrogen limited, no nitrogen fixers can invade and prosper in the closed forest because of the high energy requirements for nitrogen fixation. Consequently, the understory (including tree seedlings that may form a future canopy) consists of plants that both tolerate shade and have low nitrogen requirements.

Suppose we retrace the origin of such a community through successional time. Often early successional stages include nitrogen fixers, such as alders and legumes. These communities can be invaded by non-nitrogen fixers that make use of the nitrogen but do not have to pay the price of producing it. An example would be a grove of black locusts invaded by more shade-tolerant trees. These invading species eventually produce the low light community in which nitrogen fixers find it difficult to survive. This process by which the nitrogen fixers set the stage for their own replacement is a good example of the central paradigm of how succession proceeds as conceived by Frederic E. Clements (1916) and was, in fact, used as an example by him.

The Hydrological Cycle

The seaward movement of materials such as phosphorus and calcium occurs in connection with one of the most basic of material cycles on the earth, the **hydrological cycle** (Figure 12-17). This is the movement of water between ocean, earth, and atmosphere. The basic pattern of circulation is that precipitation falls on the earth, mainly as rain, although snow and even fog are important in some areas. Some of this water is returned to the atmo-

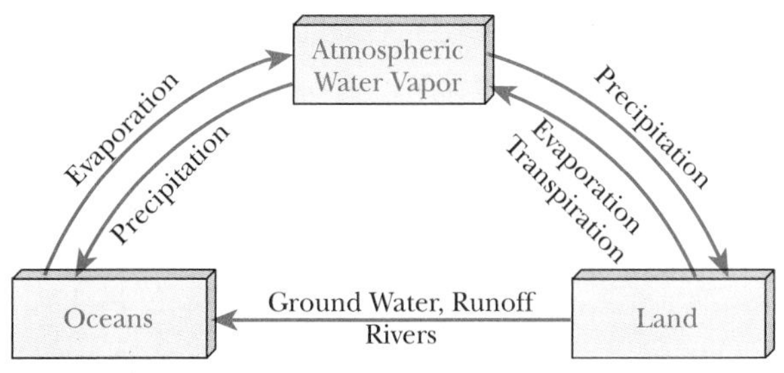

Figure 12-17

Main features of the earth's hydrological cycle.

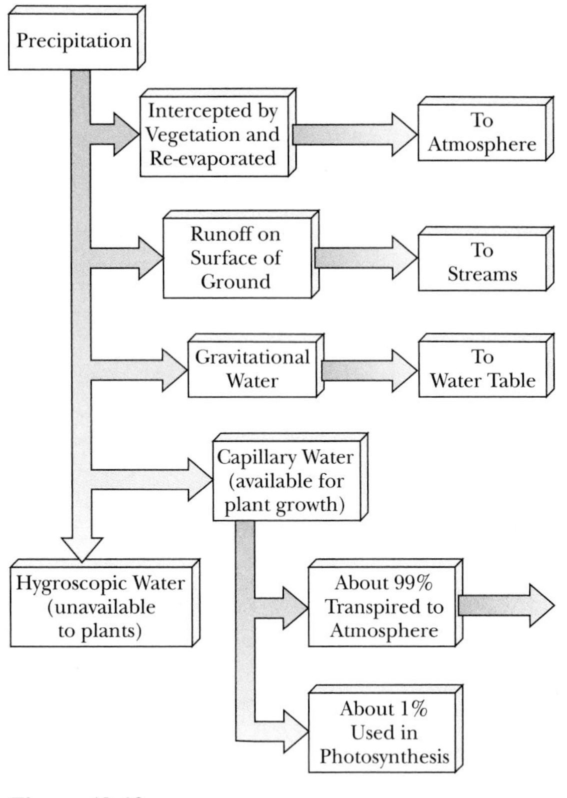

Figure 12-18

The pathways taken by rain falling on vegetated land.

sphere through evaporation and transpiration (Figure 12-18). For the land areas as a whole, less water is returned in this way than falls as rain; the excess soaks into the ground or runs off. Whatever is not used somewhere along the way eventually makes its way through streams or groundwater to the ocean.

About 97% of the earth's available water is contained in the oceans. Evaporation from the ocean's surface, as well as evaporation and transpiration from the land, adds water vapor to the air. The water vapor in the air at any one time amounts to only about 0.001% of the earth's water, equal to about 2.5 cm of rain over the whole earth's surface if it were all precipitated at once. Because the annual rainfall averaged over the earth as a whole is about 80 cm (Furon 1967), this implies a rapid turnover of atmospheric moisture.

About five sixths of the water evaporated in the hydrological cycle comes from the oceans, but only about three fourths of global precipitation falls on them. The difference is exported as rainfall to the land, later returning to the ocean in rivers. Without this asymmetry, the land would be a drier place.

The hydrological cycle is powered by solar energy in the evaporation of water. In the downhill part of the cycle, as water falls from the air and continues its descent toward sea level, the stored energy is converted to work. The continents are carved, and materials are moved from higher elevations to lower. Humans make use of some of the energy, now mostly to generate hydroelectric power, but in earlier times directly for such purposes as grinding grain and sawing lumber.

Much of the rain falling on a vegetated area is intercepted by the vegetation and re-evaporated without reaching the ground. Some of the rain that reaches the ground runs off on the surface. Generally, surface runoff on an undisturbed soil does not begin until the litter and soil have absorbed a fair amount of water. Some of the water absorbed by the soil will move on downward through the ground to the water table **(gravitational water).** The water remaining after the rain is over and the gravitational water has drained away is generally available for plant growth.

Gravitational water moves downward until it reaches a level below which all the pore spaces are filled. The water residing here is **groundwater,** and the upper surface of it is the **water table.** The water table is generally described as a "subdued replica" of the topography above it (Figure 12-19). Low spots in the land may be below the water table, in which case they contain streams or lakes. Water in the ground flows—slowly, a few thousand feet a year or less—and most of it eventually enters the regional stream flow on its way back to the ocean.

Groundwater is not important as a water source for most plants and animals, although certain plants (phreatophytes) may tap it. It is, however, exceedingly important to humans for recharging lakes and streams and for tapping directly by wells. A rock formation or other geological material such as glacial outwash that yields water in usable quantities is termed an **aquifer.** The material must be sufficiently porous that the sideways flow of groundwater replenishes that which is pumped.

In discussing carbon and nitrogen we traced parts of the pathways of these elements within the organisms. For water, however, we can virtually ignore such things. This is not because water is un-

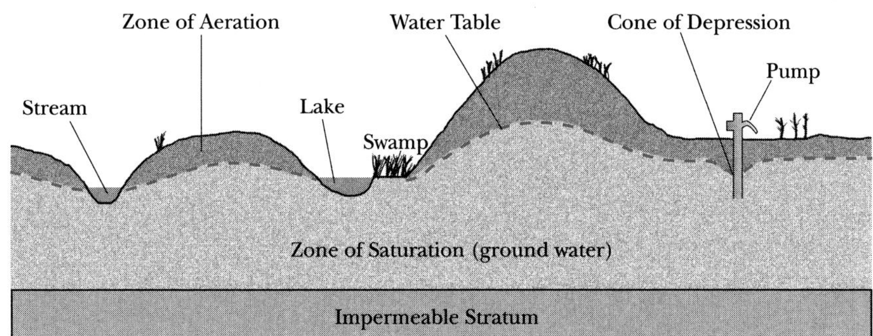

Figure 12-19

Diagrammatic view of a section through the earth's surface showing the relationships of the water table. The water table is lowered as the result of pumping, forming a "cone of depression" around a well.

important to the organisms but because more than 99% of the water taken in by the roots of a plant is transpired again as water. Less than 1% becomes involved in the complicated biochemical pathways of photosynthesis. The situation is comparable in animals; although water serves a variety of functions, it leaves the body mostly as water and in about the quantities in which it was taken in.

Organisms modify rates of water movement by slowing runoff and increasing the water storage capacity of soils. They also affect the distribution of rainfall and of wet and dry soil in ways described earlier in this chapter. In terms of the global hydrological cycle, nevertheless, organisms have seemed to play less fundamental roles than in most biogeochemical cycles. This appearance may be misleading. Recent studies suggest that organisms have additional roles that are not yet fully understood.

Rain forms as the result of condensation that begins around **nucleation particles** in the atmosphere. In the past, these have been casually referred to as "dust." Only recently has it become clear that the particles are primarily of biological origin. The complete range of biological materials that serve as such nuclei is not yet known. Pollen can do so, and so can bacterial cells (Vali et al. 1976) that occur as plant leaf epiphytes, pathogens, and saprobes. These are much more abundant in the air over vegetated areas than over bare soil (Lindemann et al. 1982). Only a few species of bacteria are "ice-nucleation active," and, even within these, not all strains have the necessary properties.

Cloud condensation nuclei are also produced by aerosol sulfate particles. Some of these come from air pollution resulting from human release of sulfur compounds. Another important source is the release to the atmosphere of dimethyl sulfide produced by certain marine phytoplankton, such as the coccolithophores (Falkowski et al. 1992).

One team of scientists (Vali et al. 1976) working on the source of ice nuclei commented that this connection between the biota and atmospheric processes "is a fascinating new manifestation of the interdependences of our natural world." And so it is. What the patterns of cloudiness and rainfall would be like on an earth devoid of life—or even of a few kinds of bacteria and marine phytoplankton—is, at this time, impossible to say.

Watershed Studies

Carbon and nitrogen are both elements that occur in a gaseous form that constitutes a fair part of the atmosphere. In these cases it is easy to see how movement of the elements from place to place over the globe can occur, but what of other elements such as phosphorus and calcium that are not present in significant amounts in the atmosphere? E. P. Odum (1983) referred to the first group as having **gaseous cycles** and the second as having **sedimentary cycles.** Elements of the latter type are released from rocks by weathering and become available for plant use. They tend to be leached from soil and carried into streams and eventually to the ocean, where they may be deposited as sediments. Will these materials eventually all be lost from the soil to the oceans? In the geological long term it is expected that there will be uplift of some areas now covered by ocean (and subsidence of some areas that are now dry land) as in the past. Soil will then be made from the material accumulated as sediments, and the minerals will again become available for use by organisms.

Studies of nutrient relations within small watersheds have led to a clearer picture of the cycling of these elements, as well as those with a gaseous form. Figure 12-20 diagrams the basic watershed relationships that need to be studied. A great deal can be learned just by measuring atmospheric inputs and stream outputs. If the watershed is underlain by impermeable material, as at Hubbard Brook, New Hampshire, nutrient losses by groundwater flow will be slight; otherwise, this output would have to be measured and added to that picked up in the stream. For elements with a gaseous form, of course, inputs and outputs as gases will have to be measured.

Watershed studies by the U.S. Forest Service and its co-operators have shown that nutrient loss from natural ecosystems may be small. From a hectare (about 2½ acres) of fairly mature northern hardwood forest around Hubbard Brook about 14 kg of calcium (Figure 12-21), 3 kg of magnesium, and 2 kg of potassium were lost over the period of a year in stream water (Likens et al. 1977).

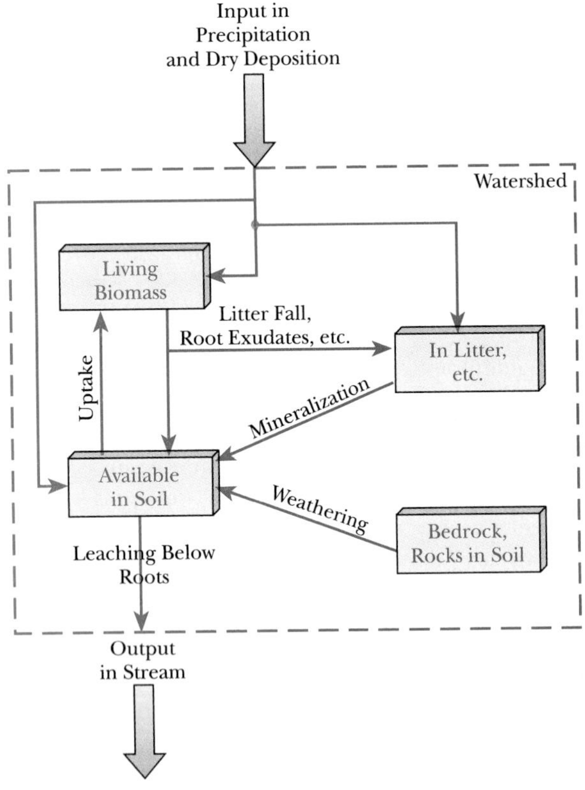

Figure 12-20

Nutrient relationships of watersheds for elements having sedimentary cycles.

Another finding is that considerable quantities of nutrients are returned to ecosystems from the atmosphere. The nutrients get into the atmosphere mainly from the oceans, through wave action and spray, and by windblown dust from soils. For several elements, such as sulfur, nitrogen, and various trace elements, human activities, especially burning fossil fuels, have become important.

Atmospheric input to a watershed is either in rainfall, fog, or some other form of wet or dry deposition. The latter includes processes ranging from dust settling on foliage or soil to the absorption by plants of gases, such as sulfur dioxide. Probably most published estimates of atmospheric input are too low because of the difficulty of measuring some forms of dry deposition. Table 12-4 compares wetfall and dryfall at two southeastern U.S. sites. The large amount of dryfall at Walker Branch is at least partly derived from fly ash from coal-fired power plants in that region (Swank 1984).

At Hubbard Brook, atmospheric inputs were measured as 2.2 kg/ha per year for calcium, 0.6 for magnesium, and 0.9 for potassium. Net losses, then, subtracting output from atmospheric input, were 11.7 kg for calcium, 2.7 for magnesium, and 1.5 for potassium.

Presumably such losses are balanced by the weathering of rock minerals. For calcium, as an example, about 65,000 kg is available in the rocks of a hectare of the Hubbard Brook watershed, just in the top 45 cm. This is about 5500 times the annual loss. Another 11,000 kg/ha is in smaller soil particles, living biomass, and litter.

Ecosystems such as crop fields, where there is little vegetation, litter, or humus, may lose large amounts of nutrients. They are, in fact, veritable sieves, passing through large quantities of nitrogen fertilizers. Losses can also occur from forest areas immediately after they are clear-cut or burnt, although here most of the output is from nutrients that had been stored in vegetation and litter. At Hubbard Brook, one small watershed was clear-cut and then doused with a herbicide. Nitrate loss increased nearly 50 times (Likens et al. 1970). Elevated nitrate loss after commercial logging

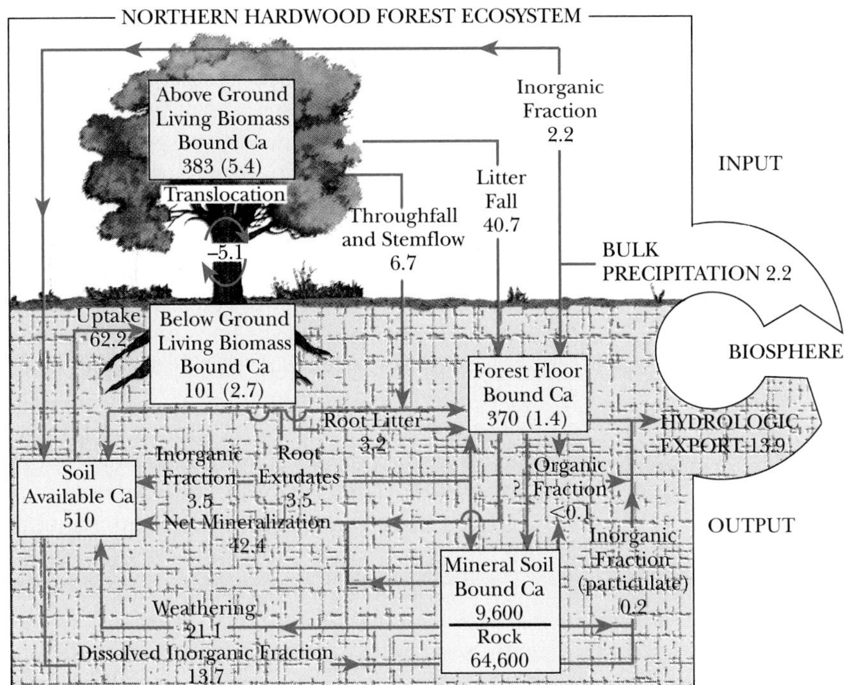

Figure 12-21

Major features of the calcium cycle for a forest ecosystem at Hubbard Brook, New Hampshire. Compartment amounts are in kg/ha. Numbers in parentheses are annual increases. Flow rates are in kg/ha/year. (From Likens et al. 1977.)

operations is usually restricted to the first year or so and is much less than the 97 kg/ha per year found in the Hubbard Brook experiment (Vitousek et al. 1979). Usually, other elements show only small increases in output after disturbance.

Unlike crop fields, mature ecosystems are tight; they do not waste nutrients. The plants themselves conserve certain materials by removing them from leaves prior to leaf fall, and there is conservative recycling of other materials from litter back to plant growth via decomposers and mycorrhizae. It should not be concluded from this, however, that the streams flowing out of watersheds covered by mature forest or other climax ecosystems will contain no nutrients or even that the nutrients will be lower than in the stages just preceding. Logically, this is not possible, as W. B. Leak (1974) of the U.S. Forest Service has observed. Fully mature forests are at a steady state, not adding new biomass or accumulating litter. If there is any input of nutrients by rainfall or by weathering, there is no place for them to be retained; consequently, an amount equal to the input must be carried away.

Successional communities, on the other hand, which are accumulating biomass and litter will retain large amounts of the minerals they use in making protoplasm so that the output will be less than the input (atmospheric plus weathering.) At Coweeta, for example, where the oak-hickory forests are still "aggrading," that is, adding biomass, several times as much calcium, potassium, and nitrogen are stored in new biomass as lost in streamflow (Table 12-5).

Table 12-4
Comparison of Wet Deposition and Dry Deposition at Two Southeastern U.S. Sites

Figures are kilograms per hectare per year.

	Coweeta, NC		*Walker Branch, TN*	
Element	*Wetfall*	*Dryfall*	*Wetfall*	*Dryfall*
Calcium	3.92	0.96	9.91	5.82
Potassium	1.11	0.51	1.09	1.90
Phosphorus*	0.17	0.02	0.06	0.49

* As PO_4^{---}

From Swank, W. T., "Atmospheric contributions to forest nutrient cycling," *Water Res. Bull.*, 20:313–321, 1984.

Table 12-5
Comparison of Input (Bulk Precipitation), Output (Streamflow), and Storage in New Biomass at Coweeta, North Carolina

Figures are kilograms per hectare per year.

Element	*Input*	*Output*	*Stored in Biomass*
Calcium	4.8	7.7	23.0
Potassium	2.1	5.6	13.0

From Swank, W. T., and Waide, J. B., "Interpretation of nutrient cycling research . . . in a management context," in *Forests: Fresh Perspectives from Ecosystem Analysis,* R. H. Waring (ed), Oregon State University Press, Corvallis, 1980, pp. 137–158; and Swank, W. T., "Atmospheric contributions to forest nutrient cycling," *Water Res. Bull.,* 20:313–321, 1984.

P. M. Vitousek and W. A. Reiners (1975) put these ideas into a model of how watershed outputs should be related to the maturity of a community. In early development there should be essentially no output of limiting nutrients such as nitrogen and phosphorus. Elements that are not essential for plant growth may be simply passed through. Some, however, could be stored on clay or humus. In the mature ecosystem, output should equal input for all elements. If an ecosystem is heavily disturbed, as by clear-cutting or fire, output of the limiting nutrients may rise sharply as nutrients formerly stored in biomass are carried away.

Increased outflow in streams or ground is not, of course, the only way that logging removes nutrients from a site. The trunks contain nutrients, and all logging operations remove these. Recently, machines have been introduced that can harvest all the woody vegetation on a site and turn it into chips (for use as fuel or making paper). Such **whole-tree harvest** is an additional drain on the nutrients of a site. Forests differ, but whole-tree harvest tends to increase biomass harvest by roughly 30% while increasing nutrient removal 50 to 150% (Table 12-6). Foliage, in particular, adds little biomass to a harvest but removes large quantities of nutrients.

Perhaps these losses can be made up by atmospheric input and weathering on fertile sites, given a reasonably long rotation period. It is possible, though, that disruption of nitrogen cycling, through decreased fixation or increased denitrification, could reduce productivity on whole-tree sites or even on ones where conventional logging is practiced (Swank and Waide 1980). On the other hand, many shrubs with nitrogen-fixing ability exist. In some localities, these may counteract trends toward nitrogen depletion (DeBell et al. 1985). It is still too early to predict how complete a disaster whole-tree harvest is likely to be.

Table 12-6
Biomass and Nutrient Content of Merchantable Trunks Compared with the Foliage, Branches, and Small Stems

Foliage, branches, and small stems are removed along with trunks in whole tree harvest. A 41-year loblolly pine plantation in South Carolina and a Tennessee mixed hardwood forest are compared. Figures are percentages. Biomass of the hardwood site was about 1.5 times that of the pine plantation, and the amounts of nutrients were 1.5–2.7 times higher.

	Biomass		*Phosphorus*		*Nitrogen*		*Potassium*	
Component	*Pine*	*Hardwood*	*Pine*	*Hardwood*	*Pine*	*Hardwood*	*Pine*	*Hardwood*
Trunks	78%	78%	42%	59%	50%	62%	63%	50%
Foliage	3	2	25	16	23	17	11	29
Branches, small stems	19	20	33	25	27	21	26	21
Total	100	100	100	100	100	100	100	100

Data from Phillips, D. R., and Van Lear, D. H., "Biomass removal and nutrient drain as affected by total-tree harvest in southern pine and hardwood stands," *J. Forestry,* 82:547–550, 1984.

Biological Control of the Composition of the Atmosphere and Oceans

The Early Earth

It has taken the effects of pollution to make the general public realize the power humans have to alter the earth, its oceans, and its atmosphere. But it is only this realization and not the effects that are new. As ecologists have long realized, the earth and the life on it have evolved together, each affecting the other. The atmosphere in which humans evolved is a biological product.

Current theory suggests that the earth was formed somewhat less than 5 billion years ago. The first atmosphere, consisting of elements that are gases at moderate temperatures, was evidently lost, probably by escaping from the earth's gravitational field. This conclusion was reached because the earth, in comparison with the rest of the solar system, is deficient in most of the light elements such as hydrogen, helium, neon, and argon.

The second atmosphere of the earth is believed to have been produced by the release of gases from the earth's interior, mainly by volcanic activity. This process is still going on. The secondary atmosphere presumably consisted of compounds that are found in volcanoes today, along with compounds formed from them. Water, carbon monoxide, hydrogen, nitrogen, carbon dioxide, sulfur dioxide, and hydrochloric acid were probably present. So also may have been methane and ammonia. This is debated by students of the earth's early history, but there is general agreement that the atmosphere was a reducing atmosphere—that is, there was no free oxygen but instead an excess of hydrogen.

With the accumulation of water, oceans would begin to form, containing dissolved gases and salts. It is generally believed that life originated in the primitive oceans. It may have occurred in a fashion similar to that originally proposed by A. I. Oparin and J. B. S. Haldane (Wald 1964) although the details of their reconstruction (as well as more recent modifications) may well be incorrect.

The details of the origin of life are not of major concern here. The model basically supposes chance collisions of inorganic molecules occurring in the presence of a suitable energy source.* Some of these collisions would produce more complicated molecules. The probability of producing a compound as complicated as an amino acid in this way in one swoop is low, but the probability of producing less complicated molecules is higher; once these are formed, the probability of their coming together to form still more complicated molecules is reasonably high. Given a glass flask containing water and a methane, ammonia, and hydrogen atmosphere, and with energy supplied in the form of heat and an electric spark, you might suppose that it would take years, or even forever, to produce measurable quantities of amino acids, fatty acids, and sugars. In fact, in an experiment performed by S. L. Miller in 1953 (Miller 1955), it took a week.

Since then, many other similar experiments have been performed under various conditions including lower temperatures, ultraviolet radiation as an energy source, and many combinations of gases. In the presence of a reducing atmosphere the production of many kinds of organic compounds occurs rapidly.

The step from organic compounds to something that could be called an organism is a long one, but there was a long time available for the step (or rather, the very many small steps) to have occurred. The first organisms are believed to have been some sort of single-celled heterotrophs. Their food, which they broke down through anaerobic respiration, consisted of the various organic compounds produced in the way already described, which had made the ocean a dilute soup by this time.

The World's Greatest Reaction

With the increase in numbers of these primitive heterotrophs, competition would develop. The competition would be for food—for the rather complex

* This might have been ultraviolet radiation, which would have been more intense at this period. Today most of the ultraviolet radiation is screened out before reaching the earth's surface by an ozone layer in the upper atmosphere. But the ozone layer is produced by the action of sunlight on oxygen. Another possibility is the energy from natural radioactivity, which would also have been more intense; 4 billion years ago there was 32 times as much uranium-235 as today.

organic molecules that had been produced in the millions of years of chemical evolution. N. H. Horowitz (1945) reasoned that a change allowing the use of some simpler compound by one of the primitive heterotrophs would give it an advantage, because it could then exploit a food source not available to others. As these simpler materials became depleted, further changes allowing the use of still simpler materials would be favored. The eventual result would be the development of organisms able to manufacture all their required organic materials from inorganic materials. In this sequence of events lies the origin of autotrophs.

Let us summarize the probable trophic categories at this time, the Golden Age of the anaerobic world. Probably there were

- Anaerobic heterotrophs, including, among others, fermenters, sulfate reducers, and methane producers
- Chemosynthetic autotrophs, such as iron and manganese sulfide oxidizers
- Photosynthetic autotrophs, resembling today's green and purple sulfur bacteria

The organisms in the last category use light energy to reduce carbon dioxide and produce carbohydrates but do not produce oxygen. This is because their hydrogen source is hydrogen sulfide rather than water:

$$CO_2 + H_2S \rightarrow (CH_2O)_n + H_2O + S$$

Presumably the photosynthesizers that liberated oxygen arose along this line, with that trait a side effect of the ability to use as an electron donor the abundant H_2O rather than the relatively scarce H_2S.

With the evolution of this new group of photosynthetic autotrophs (probably ancestors of what we call blue-green algae, or cyanobacteria, today) the world began to change. Oxygen accumulated slowly in the atmosphere. Some of it oxidized metals such as iron; the earliest "red beds" indicating deposition of iron-containing sediments with appreciable oxygen in the water are dated at about 1.9 billion years ago (Knoll 1992). As oxygen accumulated in the air, an ozone layer would be produced at high altitudes, decreasing the amount of ultraviolet radiation reaching the earth's surface and making the upper levels of the oceans (and eventually the land) habitable.

The accumulation of oxygen in the atmosphere made a conversion from anaerobic to aerobic respiration possible. Aerobic respiration yields 19 times as much energy per molecule of sugar broken down as does anaerobic respiration, so that those organisms that developed mutations allowing aerobic respiration began to win out over their anaerobic relatives. The evolution of plants and animals as they are known today could then begin.

The point is that our present atmosphere is of biological origin (Berkner and Marshall 1964). The development of the oxygen-liberating brand of photosynthesis marked the beginning of the greatest reaction in the earth's history, a revolution that made possible the development of advanced life. It was, at the same time, a catastrophe of unparalleled proportions for the anaerobes that had dominated the world up to this time. They lived on, but in today's world they are out of sight, in the muck and ooze of swamps and estuaries, the bottoms of lakes, the guts of animals, and a few other oxygen-free habitats.

The evolution of the eucaryote cell, another important development in the switch from the anaerobic, moneran-dominated world, was discussed in the section "The World's Greatest Symbiosis."

Biological Controls

We can use the conditions on Mars and Venus to judge what the earth would be like if no life were present here (Lovelock 1979). The earth's atmosphere would be mostly carbon dioxide, and because of the greenhouse effect, surface temperatures would be much higher than our current approximately 13°C. Nitrogen would make up about 2%, rather than 79%, of the atmosphere, and there would be trace amounts of oxygen, instead of 21%. These differences between what is and what would be are the result of organisms acting in the ways we have described in previous sections. For example, photosynthesizers convert carbon dioxide to organic compounds and also cause the precipitation of carbon as insoluble carbonates. In photosynthesis, oxygen is liberated. Nitrogen is returned to the air by denitrying bacteria.

Earth's atmosphere is odd in many other ways (Margulis 1981). For example, methane is present in low but regular quantities. In an atmosphere with such abundant oxygen, the methane would be quickly lost by conversion to carbon dioxide and water if it were not steadily replenished, which, of course, it is. Organisms including anaerobic bacteria in swamps and the guts of herbivores release a million tons of it every day (Erhalt 1985).

The oxygen concentration of the atmosphere has been fairly stable for approximately the past 400 million years, and the concentration of many materials in the oceans has shown similar long-term stability (Redfield 1958). This suggests some sort of steady-state control. Why has the atmosphere stopped at 21% oxygen? Why have the oceans not gotten steadily saltier?

A. C. Redfield, an ecologist at Woods Hole Oceanographic Institute, was struck by the fact that phosphorus and nitrogen are available in seawater in the same proportions (1:15) in which they are used by phytoplankton in photosynthesis. Furthermore, the decomposition of phytoplankton containing 1 atom of phosphorus (and 15 of nitrogen) requires about 240 atoms of oxygen, which corresponds well with the oxygen content of saturated seawater. Redfield argued that these ratios are produced by biological control processes and sketched some possibilities.

Full understanding of the control processes is yet to be achieved. To consider just the question of oxygen concentration in ocean and atmosphere, the following seem reasonable possibilities: At any one time, the oxygen content of seawater is largely determined by the oxygen content of the atmosphere. The oxygen content of the atmosphere is determined by a balance between the release of oxygen in photosynthesis and the uptake of oxygen in respiration (plants, animals, and decomposers). These two processes, however, balance. If no other processes were involved, oxygen would never have accumulated in the atmosphere. Any oxygen present would, in fact, be slowly removed through combination with reduced materials in the earth's crust, as they become exposed by erosion and other geological processes.

The oxygen in the atmosphere is largely the result of slightly more organic matter being produced in photosynthesis than is respired. This material in the form of living biomass, litter, humus, peat, coal, and petroleum represents a kind of oxygen debt. If it were all burnt or respired, the oxygen would have to be repaid by depleting the atmosphere.

But this does not get at the question of why the oxygen content is at a steady state. At 21% atmospheric oxygen, anoxic environments in the ocean are few (Redfield 1958). Aerobic processes predominate; there is little burial and consequently little long-term storage of undecomposed organic matter. Suppose, now, that the oxygen content of the atmosphere were to drop. The oxygen content of seawater would also drop, and anoxic environments would increase (Figure 12-22). There would be more burial of undecomposed organic matter and, therefore, a larger imbalance of photosynthesis over respiration. The result would be a shift back toward a higher oxygen content (Walker 1984).

Suppose, on the other hand, that the oxygen content of the atmosphere increased. The result would be still fewer anoxic environments than today. More complete decomposition of recently produced organic matter would occur, as well as aerobic decomposition of organic matter stored in formerly anoxic environments. The use of oxygen in this process would tend to lower the atmospheric oxygen content.

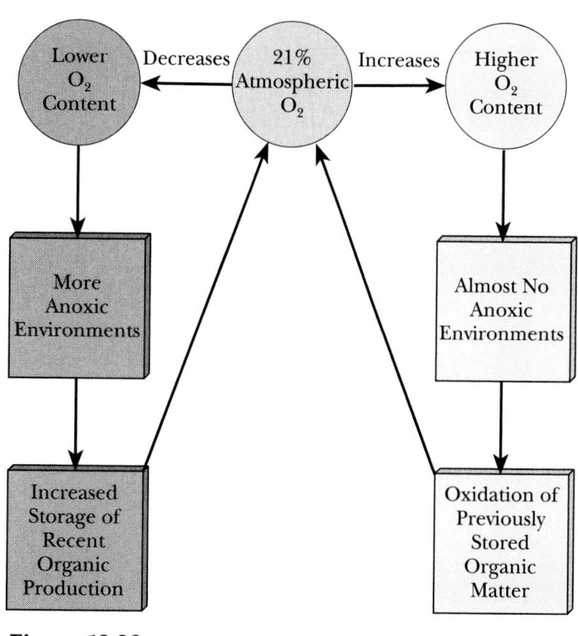

Figure 12-22

One possible mechanism in the steady-state control of the atmospheric oxygen content.

Another factor that may have the same homeostatic tendency is the activity of sulfate-reducing bacteria (Redfield 1958). These organisms prosper under anoxic conditions. They release carbon dioxide, which becomes available for photosynthesis, which liberates more oxygen, which can diffuse into the atmosphere. (Sulfide is produced in the process, and it has to be withdrawn into bottom sediments or otherwise would consume oxygen equivalent to that liberated.) The combined result of these two processes and possibly others is the 21% oxygen content that the earth's atmosphere has enjoyed for many millions of years.

The Gaia Hypothesis

Recently an English inventor, J. E. Lovelock, expressed views similar to Redfield's but more sweeping. In what he termed the Gaia hypothesis, Lovelock (1979) suggested that the biosphere is a self-regulating entity that has evolved to keep our planet healthy by controlling the physical and chemical environment. The name refers to the Greek idea of the earth itself as an enormous goddess, Gaia (or Gaea or Ge), who gave rise to all the other gods and goddesses and, of course, people.

It is easy to make light of the Gaia hypothesis. There is, first of all, the name. Although it is undoubtedly appropriate, how do you pronounce it—"gay-uh," "guy-uh," or even "gee-uh"? Then there is the fact that the book was written in admitted ignorance of Redfield's work and apparent ignorance of the long history of holistic thought in ecology from Clements and Cowles to Aldo Leopold and the Odum brothers. T. C. Chamberlin, Cowles' mentor, wrote in 1916, "the most important public interest lies in convincing as many of our citizens as possible that this earth of ours is not a dead planet, passed on to us from the past, but is a living, active organism" (Engel 1983). (*Gaia's* Chapter 8 begins, "Some of you may have wondered how it has been possible to travel so far in a book about relationships amongst living things with only a brief mention of ecology," but the explanation seems obvious.)

The most serious problem, though, is the evolutionary one. The idea that the homeostatic mechanisms seen in the biosphere evolved with the specific function of maintaining a steady state is, to most evolutionary biologists, group selection pushed to an absurd extreme. What Lovelock sees as Gaia's tendency to optimize conditions for life, most biologists would see as a network of individually evolved interactions. "Gaia" is a by-product—in effect, an accident.

But it would be unwise to write the Gaia hypothesis off completely. It is at least a charming metaphor. It has been successful in bringing home to many persons—including many scientists outside the field of ecology—the ecological message that life and environment are an interacting system rather than separate realms. It is, also, too early to state that no credible mechanism will ever be conceived whereby control for, as well as by, the biosphere might evolve. That would be closed-mindedness.

Community and Ecosystem Ecology: *Community Change*

Studying old-field succession in a class from Western Michigan University, 1965. (*Photo by the author.*)

OUTLINE

- Types of Community Change
- Replacement Changes
- Fluctuation
- Succession
- The Climax Community
- Stability

Types of Community Change

Communities and ecosystems are dynamic; they change constantly. One tree dies and a sapling will grow up to take its place. A forest is a different place from night to day and from spring to fall. Winter comes as a prelude to another season of growth. Such changes as these, however, do not alter the community permanently; they are called **nondirectional** changes. Included are the daily and seasonal cycles that have already been discussed in Chapter 10. Included also are the replacement changes associated with maintaining a steady state in stable communities. Fluctuations in communities associated with fluctuations in physical factors, generally related to climate, are nondirectional changes.

Some other types of community change do result in permanent alteration of the community. They are **directional** changes. Included are the long-term changes resulting from long-term climatic change, such as the shifts of plants and animals that occurred with glacial advance and recession. Changes that occur on an even longer time scale which involve the evolution and extinction of species, such as the changes in land communities as birds and mammals evolved, are directional. So also is succession, the process of community development which may produce a climax community. It is a directional change that occurs even though climate and the species pool available do not change.

The community, of course, has no regard for ecologists' categories. The decrease of a certain xerophyte may be part of a climatic fluctuation in which rainfall is higher, part of a long-term climatic trend toward greater rainfall, or part of succession toward a more mesic community. All three of these types of community change may be happening at once.

Replacement changes, fluctuation, and succession are discussed in the following sections of this chapter. Long-term community change is discussed in Chapter 14, Paleoecology.

Replacement Changes

In stable communities, organisms tend to be long-lived, but none is immortal. In a climax forest, an individual tree may spend a century or two in the canopy, but eventually it dies and is replaced. A species at the outside extreme is bristlecone pine; individuals of this species seem regularly to reach ages over 4000 years (Harlow and Harrar 1968). Clones of vegetatively reproducing herbs and shrubs, such as may-apple and pawpaw of the eastern deciduous forest, may also be extremely long-lived, though there is little good information on this point. Creosote bush clones in the Mohave Desert often reach a few hundred years of age, and one about 16 m across may have been as much as 11,700 years old (Vasek 1980).

Some other constituents of the community are replaced on a much shorter time scale. The average individual of most bird species lives no more than a few years, and even the larger mammals live only a little longer. Their replacements, more so than the plants, are likely to come from another location, brought in by dispersal coupled with habitat selection.

The net result of these changes may be that if we had visited a mature forest 20 years ago and returned today, it would seem unchanged. It is still dominated by large beeches and maples; ovenbirds and Acadian flycatchers still sing; red-backed salamanders wiggle away when we turn over a log. But if we had a detailed accounting, we would see that the seeming constancy is an illusion; the community has been maintained by change. The log we turned over 20 years ago has rotted away, and the one we turn over today was a canopy tree sometime in the past.

Cyclic Replacement

In some communities, replacement processes are cyclic (Watt 1947). The scale involved varies from small patches to landscapes. The cycling may be dependent mainly on the dynamics of the community itself, but often some outside agency, such as fire, is important. Some cyclic systems originally interpreted as being endogenous, controlled by the life cycles of the plants involved, are now seen as being more strongly influenced by outside factors. The cyclic regeneration of *Calluna* heathland in England is one example of such a reinterpretation (Marrs 1987).

A classic example of a small-scale cycle evidently dependent on the characteristics of the plants themselves is the "hummock-and-hollow cycle" of English bogs (Kershaw 1964). Small pools are invaded by pointy-leaved sphagnum and, in turn, by less hydrophytic species (Figure 13-1). Nipple sphagnum forms hummocks on which heather grows. The hummock continues to grow, the stems of heather and other plants helping to provide support. After a time, the top of the hummock becomes too dry for the sphagnums and is invaded by lichens. Eventually, the heather dies, and the hummock surface erodes. As this occurs, and as surrounding hummocks grow up, a small pool is reconstituted. The bog surface consists of a mosaic formed of the different phases of this cyclic system (Godwin and Conway 1939).

A similar cyclic system can be seen in North American peat bogs with other heath shrubs, generally leatherleaf, taking the place of heather. North America, however, has a number of larger shrubs and trees, notably tamarack and black spruce, that can invade, leading to a successional conversion of the open bog to thicket or forest. The

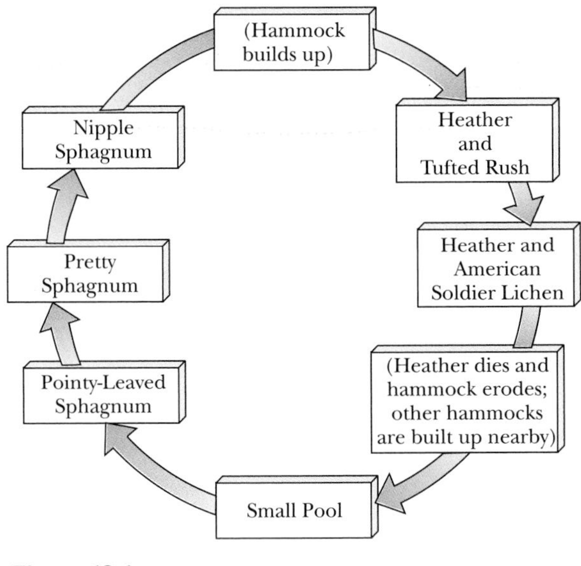

Figure 13-1

Cyclic replacement in an English bog.

hummock-and-hollow cycle is based on the ecological features of the plants, but its importance in England is related to the impoverished nature of the British flora.

In the Chihuahuan Desert, two important species, creosote bush and the Christmas tree cholla, seem connected in a cyclic replacement system (Yeaton 1978). Here the cycle is partly driven by animals in the community. Creosote bushes serve as perches and refuges for birds and mammals, among which are individuals that have eaten fruits of the cholla cactus. The seeds, which pass unharmed through the digestive tract, are deposited beneath the creosote bush. Here they encounter relatively good soil as the result of wind accumulation of finer particles by the creosote bush. The cholla germinates beneath the creosote bush and eventually replaces it, at least in part because its root system occupies the top 10 cm of soil, above the root system of the creosote bush (Figure 13-2).

This is a formula for the competitive replacement of creosote bush by Christmas tree cactus, not the coexistence of the two. What completes the cycle? The cholla bushes are preferred sites of various desert rodents, such as kangaroo rats. Their burrowing and root gnawing are harmful. Also the shallow root system may, in the absence of the creosote bush, be disadvantageous, being easily exposed by wind and water erosion. The cholla dies, leaving an open space. This is potentially open to invasion by either plant, but the heavier seed production by creosote bush and the coarse soil, relatively unfavorable for the cactus, make creosote bush the usual colonizer.

Several other cyclic replacement systems that function at a larger scale are described in other sections of the book. Examples are chaparral (Chapter 16), ponderosa pine savanna (Chapter 15), longleaf pine savanna (Chapter 3), and sand-pine scrub (Chapter 15). In all these, fire initiates a series of changes that eventually culminates in another fire. The community after the fire may be slightly different from that before it, as in ponderosa and longleaf savannas, or substantially different, as in chaparral.

Canopy Replacement in Forest

Trees are lost from the canopy in two general ways. They either die in place and fall to pieces, rapidly in the case of beech or elm, more slowly in the case of oak or chestnut. Or they are removed from the canopy by wind (sometimes coupled with other factors, such as ice storms). They may be windthrown, that is, the tree is blown over with the roots intact, creating a mound of earth and an adjoining pit at the base of the tree, or they may be snapped off, leaving a stump that is generally a meter or 2 tall or taller.

In any case, a gap is created—a space is left in the canopy where full sunlight is available. Occasionally, through the action of such agents as hurricanes and tornadoes, large gaps, amounting to many tree crowns, are produced. Timber cutting and forest fires may kill most or all of the canopy trees over considerable areas, but for most forests, these are rare events compared with the loss of one or a few trees. If the average time a tree lives in the canopy is 150 years, and there are 1500 canopy individuals in a 40-acre woodlot, then 10 trees (1500/150) will die or be killed in an average year. The estimated yearly loss of canopy trees in Warren Woods, a climax beech-maple forest in Michigan, was 0.2 trees per hectare from windthrow alone (Brewer and Merritt 1977). This amounts to about three trees in 40 acres.

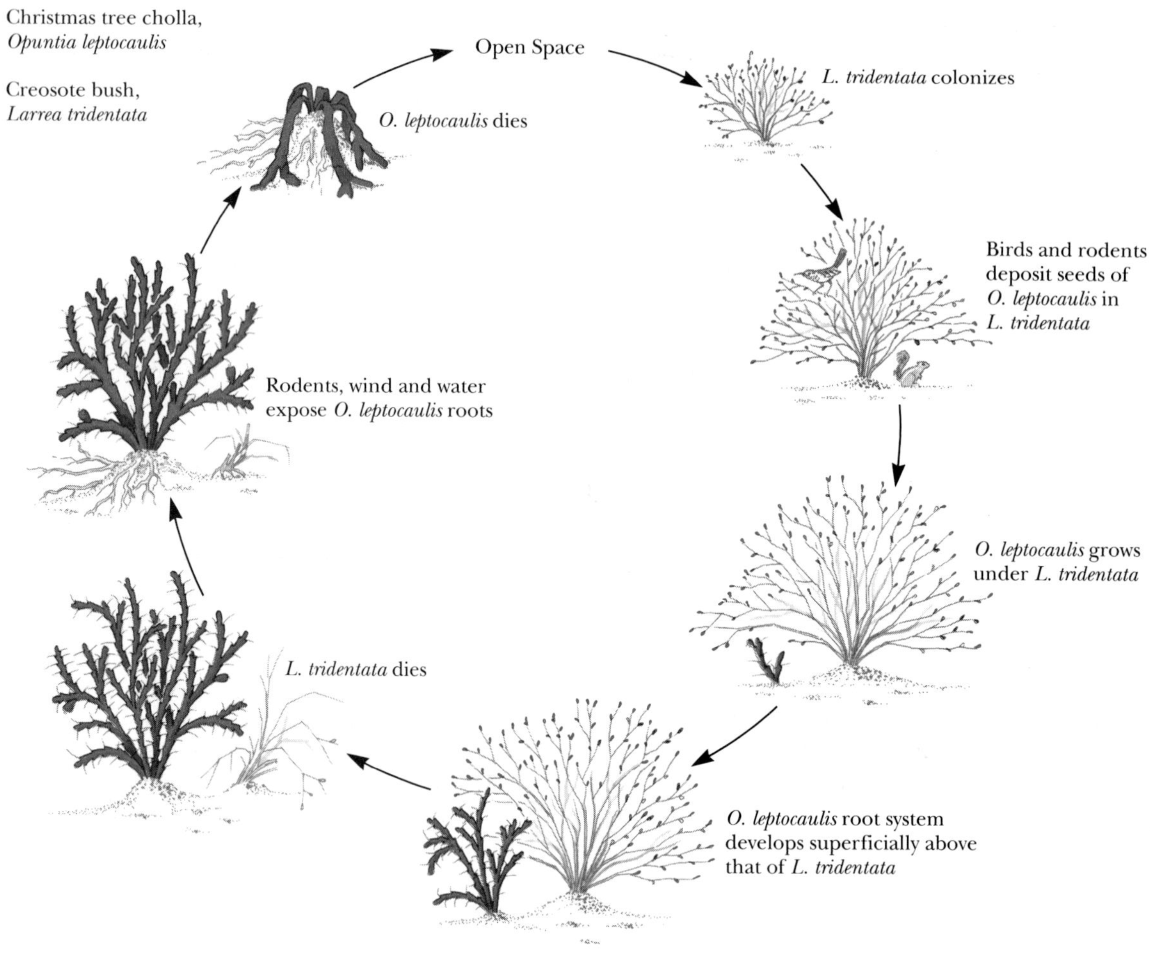

Figure 13-2

Suggested pattern of cyclic replacement of creosote bush and Christmas tree cholla in the Chihuahuan desert. (From Yeaton 1978.)

Most canopy replacement, then, involves filling in a space in the canopy the size of one to a few tree crowns (gap-phase replacement). Just how this occurs has been the subject of many recent studies, but there is still much to learn. If tree density stays the same, each tree is, on the average, replaced by one tree. This does not mean that this is the outcome of every gap; certainly some gaps are filled in by surrounding canopy individuals expanding their crowns.

In other cases, a tree not yet in the canopy fills in the space. Which individual this is depends on position in the gap, height, and rate of upward growth, among other things. Chance factors may be involved; the best situated individual may be damaged or killed when the canopy tree falls. Few studies have evaluated the relative importance of these factors in determining canopy replacements.

There appears to be a fundamental difference between many temperate forests and many tropical forests in gap-phase replacement. For temperate forest, as just discussed, gaps tend to be filled by suppressed, subcanopy individuals, generally already many years old. The literature on tropical forests suggests that the smallest canopy gaps there may be filled by already established young trees. Most gaps, however, seem to be filled by relatively light-demanding species whose germination occurs subsequent to gap formation (Whitmore 1978).

If we can specify the probability with which a given species of tree will be replaced by other species, then we can study the replacement process mathematically by the use of transition matrices (Waggoner and Stephens 1970, Horn 1975). Suppose, for example, that for a forest currently dominated by black oak and white oak, we are able to determine replacement probabilities as shown in Table 13-1. It takes no mathematics and only modest insight to tell that this hypothetical forest will eventually be dominated by red maple and that black oak will probably disappear.

Finer details of composition are a bit harder to discern; however, the table forms a transition matrix that, if certain conditions are met, can be used to discover these details. It can be used to project canopy composition at various times in the future and to determine an eventual steady-state canopy composition (if the transition matrix is a constant). How to use transition matrices in such a way is explained in *A Laboratory and Field Manual of Ecology* (Brewer and McCann 1982: pp. 181 to 186).

One of the interesting questions about forests is why the one most shade-tolerant tree species does not eventually exclude all the rest. The two general answers are that (1) recurrent sizable disturbances, such as fires or hurricanes, prevent forests from reaching a competitive equilibrium and (2) the group of species may in one of several ways satisfy the Lotka–Volterra conditions for the coexistence of competing species (Chapter 8).

That recurrent disturbance is an important factor in maintaining diversity in many forests is now widely recognized (Heinselman 1973, Brewer 1980). To what level diversity will eventually drop in the absence of large-scale disturbance is still unclear. Does gap-phase replacement occur in such a way that a fair number of species retain canopy membership?

Several studies (examples are Barden 1980 and Runkle 1981) suggest that diversity will be maintained at some level above a single canopy species. Cyclic replacement in which a heavily reproducing, faster-growing, shorter-lived tree (such as sugar maple) and a less prolific, slower-growing, longer-lived tree (such as beech) tend to replace one another may be involved (Fox 1977). What level of diversity is maintained—two tree species, a few, or many—still requires a great deal of study. It is an interesting question and an important one, bearing on how best to conserve our old-growth forests.

Table 13-1
Hypothetical Replacement Probabilities for a Hypothetical Black Oak-White Oak Forest

The probability of a canopy white oak being replaced by a red maple, for example, is 0.7.

	Canopy Species			
Replacers	*Black Oak*	*White Oak*	*White Pine*	*Red Maple*
Black oak	0.0	0.0	0.0	0.0
White oak	0.2	0.1	0.2	0.1
White pine	0.2	0.2	0.2	0.1
Red maple	0.6	0.7	0.6	0.8

Fluctuation

Fluctuation refers to alterations in community structure or composition that occur as the result of shifts in habitat factors. With a drop in the level of the water table, for example, a shallow basin may change from a pond surrounded by a typical zonation of water lilies and cattails to a wet meadow. With a return to a higher water table, a pond is reestablished. Habitats like this, in which fluctuations are expectable, often have "shifts" of species that flourish only during one part of the oscillation. Some plant species may not be seen for years until favorable conditions recur. Presumably they pass the intervening periods as seeds or underground organs. An example of such a shift is certain coastal plain species that also occur on the sandy shores of midwestern lakes. In the latter habitat they are virtually absent except in years of unusually low water, when they may be abundant (Pierce 1974).

The Great Drought

One of the best studied examples of the "fluctuation" type of community change was the change in grassland vegetation that occurred during and following the Great Drought of the 1930s in the Great Plains. The effects of the drought were studied by the grassland ecologist John Weaver and his coworkers in Nebraska, Iowa, and Kansas. Rainfall for six years preceding 1933 had been 5 inches (13 cm) above average; during the early years of the drought there was a deficit of 7 inches (18 cm) below the long-term average of 23 inches (58 cm) a year. The percentage of ground covered by vegetation in ungrazed shortgrass prairie was over 85% before the drought (Figure 13-3). This percentage steadily declined as the grasses died back or failed to come up in the spring—it was 65% by 1935 and down to 20% in 1940. Litter disappeared, decomposing faster than it was added.

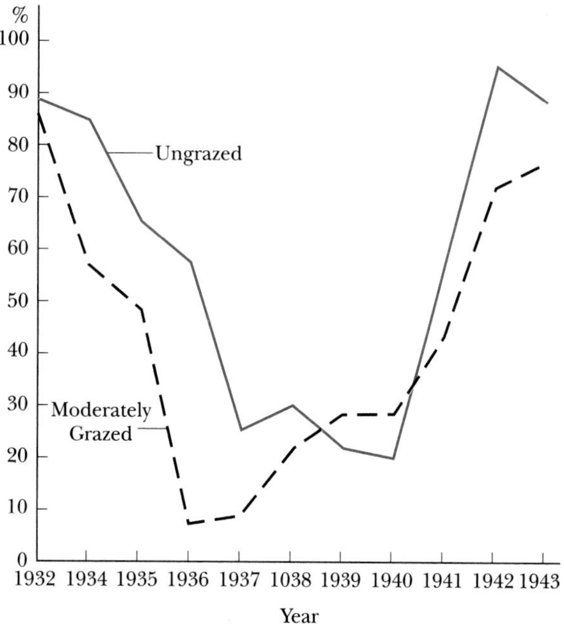

Figure 13-3

Changes in basal cover during the Great Drought. Although there were occasional moist periods such as April–May 1938, the main reason for the increases between 1936 and 1939 was the increasing coverage by side-oats and blue grama. The period 1940–43 shows the results of the return to higher rainfall. (Adapted from Weaver & Albertson 1956.)

Not all species of plants declined. Although the cover of little and big bluestem dropped from 61 to 3% between 1932 and 1939, the grama grasses increased from a negligible percentage to 27% (Weaver and Albertson 1956). If these arid-land grasses from the Southwest had not increased, the ground would have been even more bare. Another grass that increased was western wheatgrass, a plant that makes most of its growth in the spring when moisture is most available and which has a high percentage of its roots deep in the soil. Some of the shallowly rooted forbs such as coneflower and pussytoes disappeared in the first two years of the drought, but several forbs, such as heath aster and Missouri goldenrod, increased for a while as competition from the grasses declined.

With the ground bared, soil began to blow (Figure 13-4). Blowing of topsoil was worst in the southwestern part of the Great Plains where Oklahoma, Kansas, Texas, Colorado, and New Mexico come together. Row-crop farming, mainly wheat, had expanded in this region onto lands that should have been left in their native shortgrasses (Figure 13-4). With the coming of the drought this became the Dust Bowl (Figure 13-5), an area of 50 million acres by 1936. Farm families ruined by crop failures began to flee, mostly to the "Promised Land" of California. John Weaver studied one face of the drought; the story of another was told by John Steinbeck in *The Grapes of Wrath* and by Woody Guthrie in such songs as "Dust Bowl Refugees" and "Dust Pneumonia Blues." Their writings are less scientific than Weaver's but of no less ecological interest.

It was not until the winter of 1941–1942 that the drought really ended. Rainfall increased, and it was found that many grasses still lived as roots and underground stems, even though they had not shown themselves aboveground for several years. Now they sprouted, cover and shade increased, and by 1943 a mulch of leaves and stems from the preceding year began to cover the ground again. Forbs also began to sprout from underground stems or from seeds that had lain dormant in the soil.

The end of the fluctuation was not the events of the first part in reverse, like a videotape on rewind. By 1945, several species of forbs and shrubs were abundant on sites that the grasses had held before the drought. Silverleaf psoralea, typically found on moist, level land, had spread over the dri-

Figure 13-4

Blowing soil in Beadle County, South Dakota, October 16, 1935. The vegetation in the foreground is the annual Russian thistle, a tumbleweed that has invaded on the bare soil. (Photo courtesy U.S. Soil Conservation Service.)

est ridgetops, and redroot seedlings covered whole hillsides. Not until 1959–1960 did forb distribution on Weaver's study areas return to its predrought condition (Weaver 1961). Western wheatgrass, one of the xerophytic grasses, was the slowest to decline because its early spring growth allowed it to resist encroachment by other grasses. Shading by big bluestem finally caused its decline.

One trap that both ecologists and farmers must avoid is thinking of the wet years just preceding the 1930s as the normal condition. Drought, no less than years of good rain, is normal for these semiarid regions. Droughts had come before—an especially bad one in the 1890s gave rise to the slogan "In God we trusted, in Kansas we busted"—and they came afterward (the Dust Bowl area experienced another drought in the mid-1950s). They will come again. Land use in such an area must be designed for living with dry as well as moist periods. An ecologist studying the area must realize that western wheatgrass is much a member of the ecosystem as is big bluestem.

Succession

If a new area essentially free from life is produced, for example by a river depositing a new sandbar or by humans strip-mining an area for coal, the community that initially develops there is a **pioneer community,** which usually does not persist very long. In the course of time various species are lost and others invade. These, in turn, may disappear and still others may enter. Some tens, hundreds, or thousands of years later, a community may develop that is stable, or at a steady state known as the **climax community.** This process, in which the same area is successively occupied by different communities, is referred to as **succession.**

Pioneer communities tend to be made up of relatively few species, those that are able, first, to immigrate quickly and, second, to live under the particular environmental conditions, often extreme, of the new area. The earliest plants to grow may be either those with windblown seeds or those

Figure 13-5

An approaching dust storm in Texas, August 1, 1936. (Photo courtesy U.S. Soil Conservation Service.)

that are carried long distances in or on birds or mammals. Alternatively, they may be plants whose seeds live for many years in the soil in the seed bank. They are plants that can withstand direct sunlight and various other harsh conditions. Legumes, able to grow in nitrate-poor soil because of the nitrogen-fixing bacteria associated with their roots, may be prominent.

For the first one or two years annual herbs may be most important, but they are soon replaced by perennial herbs. One obvious factor involved in this replacement is that of occupancy. Once a perennial has established itself, it is able to hold onto an area and, usually, to spread by vegetative means. Because annuals have to grow from seed each year, they tend to be crowded out quickly.

This perennial herb stage might be replaced by a community dominated by shrubs, followed by a community dominated by a forest of light-tolerant trees. Some of these changes represent differences in the time of invasion of various species, while others represent merely the different time needed to reach dominance. The trees may invade about the same time as the shrubs, but for several years they are no taller; later they begin to overtop the shrubs and change the look of the community.

The light-tolerant trees are also generally shade intolerant and are unable to reproduce in the shade they cast. Beneath them shade-tolerant trees invade, and of course there are also shifts from shade-intolerant to shade-tolerant shrubs and herbs. At this point we may have a community in which the canopy trees are mainly shade-intolerant species whereas all the lower layers are shade-tolerant species. The seedlings of shade-tolerant trees grow up, and as the shade-intolerant trees die, a community of shade-tolerant trees develops. The community is at or near climax.

The whole sequence from bare ground to climax community is called a successional series or a **sere,** with the various communities existing at different points in time called **seral stages.** For the sere just discussed the following seral stages might be recognized: annual herb, perennial herb, shrub, early forest, followed by the climax forest.

Succession was one of the major topics of study by early ecologists in the United States. It does not now occupy such a central position in ecological thought, but as a framework for vegetation management, succession has assumed increasing importance (Lukens 1990). For example, the only way we will ever have sizable blocks of forest in many parts of the United States is by restoring forest on abandoned fields and other open areas (McPeek and Brewer 1991). How to achieve this rapidly and efficiently is a question in applied succession.

Many early ecologists overestimated the orderliness of succession, at least in statements for pedagogical purposes. Most current students of vegetation would cite the roles of different levels of herbivory, varying availability of seed sources, climatic fluctuations, and certain other stochastic events as affecting the timing and sequence of events in succession. They would probably agree with H. A. Gleason (1927) who wrote:

> *Does vegetation succession proceed in a systematic and orderly fashion? Obviously not, for all detailed studies of succession show that the various associations in a region stand in the most complicated successional relations to each other, and that an extraordinary number of different routes may be followed in their progression.*

Pinewoods on the North Carolina piedmont derived from old fields abandoned before 1900 have shortleaf pine at a much higher frequency than loblolly pine, but this dominance is reversed for fields abandoned after 1930 (Christensen 1989). Old field succession seemingly took a somewhat different path in these two time periods, but why? Several possibilities exist:

1. Fields abandoned in the late 1800s were surrounded by other fields; consequently, pine dispersal had to be over a long distance. Shortleaf pine has small, widely dispersed seeds. Fields abandoned in the decades following 1930 were apt to be surrounded by earlier abandoned fields that had reached the pine stage, or even by pine plantations. Loblolly pine is more widely used in silviculture, hence, its seeds became increasingly available. Also, on all but the poorest sites, it grows more rapidly and may outcompete shortleaf pine.
2. Nitrogen and phosphate fertilizers were rarely used on 19th century fields. Shortleaf pines may have had the advantage on the wornout fields of that era, loblolly on the later fields with some residual fertilizer.
3. Probably the order in which fields were abandoned was related to quality of the soil, poorer sites being abandoned earlier. This would likewise lead to shortleaf pine having the advantage in the pre-1900 period.
4. Climate was different during these two periods, the latter part of the 19th century being cold and the period 1930 to 1955 being relatively warm (Brewer 1991). This, or some other aspect of climate, could have favored different species in the two time periods.

One or more of these factors may have influenced the difference in successional pathways, in this case, of old-field succession. There are other possibilities in general, whether or not they apply to this case in particular. For example, herbivores, seed predators, and seed dispersers could vary enough in abundance from one time period to another to produce differences in details of the sere. Immigrations and extinctions could be important. On neutral to slightly alkaline bare areas in the Midwest, such as old gravel quarries, spotted knapweed is an important successional species. This member of the Asteraceae was, however, introduced to this country from Europe fairly recently. It arrived in southwestern Michigan about 1928 (Kenoyer 1933).

Types of Succession

Seres can be classified according to the water relations of the site where they begin. A successional sequence beginning in water is termed a **hydrosere.** One beginning where there is a deficiency of water, such as bare rock, a talus slope, or dry sand, is a **xerosere.**

Succession may begin in many different ways. In the examples discussed so far, it started on a totally bare area, an area that had never before supported a community. When succession occurs in such a place, on a new sandbar or mudbar in the river or on a new lava flow, it is called **primary** succession, and the sere is a **prisere. Secondary** succession **(subsere)** occurs on areas that have already supported a community, for instance, on abandoned cropland allowed to revegetate or on a piece of burned-over forest land. Usually the rate of secondary succession is faster than that of primary succession because a developed soil and some organisms are already present.

The dividing lines separating secondary succession, certain kinds of fluctuations, and certain kinds of replacement changes are not sharp. That a fire in a forest sets off a series of changes is a matter of fact; how we classify the changes is a matter of definition.

Microseres

Within any community, seral or climax, particular microhabitats may support a sequence of populations making up a **microsere.** It is difficult to give a brief definition of "microsere," but the idea is easily grasped. A forest tree blows over. The sequence of changes, including the various insects and other organisms involved, as the tree disintegrates, decays, and eventually becomes indistinguishable from the forest floor is one microsere. Microseres, thus, are minor successional sequences that occur within the framework of a larger community. They have no separate climax stage of their own but, at the end, become indistinguishable from the larger community.

The term was coined by V. E. Shelford and first used by C. O. Mohr (1943), an entomologist at the Illinois Natural History Survey, in his classic study of the dung microsere. Besides logs and dung, other microseres that have been studied occur in rotting fruit, seasonal ponds, and pond infusion cultures. Probably most studied has been the carrion microsere (Table 13-2). One minor branch of applied ecology is the use of the carrion microsere in forensics, specifically in medicocriminal investigations to establish time of death (Catts and Goff 1992). Entomologists specializing in blowflies get most of these jobs.

Causes of Succession

Under succession we are talking about the directional community change that occurs given a basically unchanging climate and a certain species pool. Of course, climates fluctuate and species come and go. The sere occurring on any real piece of land is affected by such changes; the route or the destination may be different than it would have been, had no changes occurred.

Even with no climatic shifts, forces exterior to the community may cause some successional changes. Such factors in succession are termed **allogenic.** In succession on bare rock (a very slow process, by the way) some of the breakdown of the rock is by organisms, but some is by physical and chemical weathering. These latter processes are allogenic. In pond succession, the filling-in by sand, silt, and clay carried from surrounding lands is an allogenic factor.

Autogenic factors in succession are ones resulting from the community or its constituent organisms. The filling-in of a pond by rhizomes and organic detritus is an autogenic factor. Autogenic factors include (1) immigration, (2) growth, (3) reactions, and (4) coactions. Reactions are especially significant; Frederic E. Clements spoke of reactions as being the driving force of succession.

The Role of Reactions

Generally in the course of time the reactions of the dominant organisms of each stage make the area relatively less favorable for themselves; they may also make it more favorable for organisms of the next stage. The process is seen in the sequence of more and more shade-tolerant organisms paralleling the deeper and deeper shade produced during the course of forest succession, and in the addition of nitrate to the soil by plants with associated nitrogen fixers allowing the invasion of organisms with higher nitrate requirements. The process continues to the climax, which is composed of organisms able to tolerate their own reactions.

One semicontroversy has involved the role of immigration in succession. F. E. Egler (1954) suggested two extreme possibilities: initial floristic composition and floristic relays (Figure 13-6). Looking at the floristic relay model in terms of old-

Table 13-2
Some Features of the Carrion Microsere

In an acre of forest (or grassland or marsh) several fair-sized animals die every day. Relatively few of these are seen because they are quickly eaten by opossums or vultures or buried by those remarkable insects called burying beetles. If this does not occur, the carcass goes through a series of changes like those described below. The particular study on which this table is based was done by M. D. Johnson (1975), who put out medium-sized carcasses (mostly squirrels) in hardware-cloth cages in a red oak-sugar maple-basswood forest in northern Illinois.

Stage	***Duration (Spring and Summer)***	***Characteristics***
Fresh	3 hours–2 days	Blowflies may visit the carcass to lay eggs within hours or even minutes of death.
Bloat	4–19 days	This stage begins when the carcass bloats from the gases produced by anaerobic decomposition within. Blowfly maggots hatching from eggs feed on the flesh. Predaceous beetles such as rove and hister beetles enter the carcass to prey on the blowfly larvae.
Decay	13–23 days	The burrowing of maggots and other insects allows the entrance of air into the carcass, halting putrefaction and allowing aerobic decomposition to occur. The maggots contribute to decomposition by secreting digestive enzymes; bacteria and fungi are also important in decay. This is the period of greatest insect abundance—the carcass may swarm with fly larvae. Blowflies peak about the beginning of this stage and then leave the carcass to pupate; however, other kinds of scavenging flies remain. Beetles, both scavengers and predators, are common during this stage.
Dry	30–60 days	By the end of the decay stage the carcass is a hollow shell of hard, dry skin. The lower side is disintegrated and hair and bones are scattered around the carcass. Insect populations are much diminished; some fly larvae of a few kinds remain, but beetles, mainly scavengers, are more numerous. An example is the skin beetle, so called because of its occurrence at this stage. By the end of this stage characteristic carrion species are gone and the invertebrates found are typical soil and litter forms such as millipedes and daddy longlegs.

field succession, there is a pioneer group of annual herbs into which invades a group of perennial herbs, into which invades a group of shrubs, etc. The initial floristic composition model suggests that the species that are going to occupy the site are there from the beginning. The physiognomic stages are the result of the annuals being pinched out by the perennials, the early perennials being overtopped and shaded out by the shrubs, and the shrubs being overtopped in turn by the trees. This view is a kind of ecological analogue of the preformation idea of embryology in which each egg was believed to hold a complete, miniature animal that simply grew.

Although the floristic relay model has sometimes been pictured as the traditional view, Clements commented in 1916 that

> *Secondary areas such as burns, fallow fields, drained areas, etc., contain a large number of germules often representing several successive stages. In addition, some stages owe their presence to the fact that certain species develop more rapidly and become characteristic or dominant, while others which entered at the same time are growing slowly.*

Perhaps neither model is perfectly right for any real sere. Tree and shrub seedlings—elm, aspen, box-elder, cherry, sumac, dogwood—do begin to appear within a few years in old fields, on road cuts, and on other similar sites. The shrub stage in this instance may simply be the period when the sumacs overtop the goldenrods, and the young-tree stage when the elms or cherries overtop the shrubs.

To this degree, the initial floristic composition

***Frederic E. Clements** (American, 1874 to 1945)*

Clements was born on the western frontier, in Lincoln, Nebraska, when the surrounding prairie was still intact and while the Plains Indians were still being subjugated. He entered the University of Nebraska when he was 16 and got his Ph.D. there in 1898. His major influences at Nebraska were Charles E. Bessey, who probably had the best understanding of ecology of any of the older generation of U.S. botanists, and Roscoe Pound, who had graduated from the department a few years earlier and, at this time, was head of the Nebraska Botanical Survey.

Clements stayed on at Nebraska until 1907. He then moved to the University of Minnesota as chairman of its botany department. In 1917, largely on the strength of his monograph on plant succession, he was appointed Research Associate in Ecology with the Carnegie Institution in Washington, D.C. He stayed in this position the rest of his life, at first spending winters in Tucson, Arizona, and summers at his Alpine Laboratory on Pike's Peak, Colorado, and later moving his winter headquarters to Santa Barbara, California.

The image of Clements from his wife Edith's semibiography *Adventures in Ecology* is of someone totally occupied with understanding the continent's vegetation. Besides being preoccupied, Clements was sickly, a diabetic, for the latter part of his life, and Edith, called "Cherié," took care of him and served as "chauffeur, photographer, secretary, mechanic, and field assistant." According to Pound, the coauthor of Clements' important early publication *The Phytogeography of Nebraska,* Clements was cooperative, conscientious, devoutly religious, and full of enthusiasm.

Clements is probably still the one person most responsible for the shape of ecological thought in the United States and, perhaps, throughout the world. His straightforwardly holistic approach is not the only philosophical theme in American ecology, but it is still an important one. His views on the importance of succession in vegetation and its preeminence as a basis for classification, on the climax community as an integrated system, and on the role of competition in structuring communities have been modified but not lost. He invented methods and coined terms that are part of the permanent fabric of ecology. He gave an early direction to several fields, such as physiological plant ecology and paleoecology, that we may not today think of as lying within Clements' domain.

Clements was a theoriest, but his theorizing was grounded in experimentation and unmatched field observation. He was, furthermore, interested in the practical applications of ecology to such topics as range management, regeneration of forests, and reclamation of the Dust Bowl.

Clements' influence acted largely through his writings and, to a degree, through the hiring of young ecologists for summer work on Carnegie Institution projects. Although Clements spent 18 years in academic life, it is not as an academic that he is remembered. He was, in his words, an "escaped professor."

One of the voices through which Clements spoke was the textbook *Plant Ecology,* published in 1929, revised in 1938, and still the most widely used textbook in U.S. ecology courses through the early 1950s. Although J. E. Weaver—about the only student of Clements most of us could name—was senior author, the book is pure Clementsian ecology. It is still worth reading for that reason and for the sound ecological observations it reports.

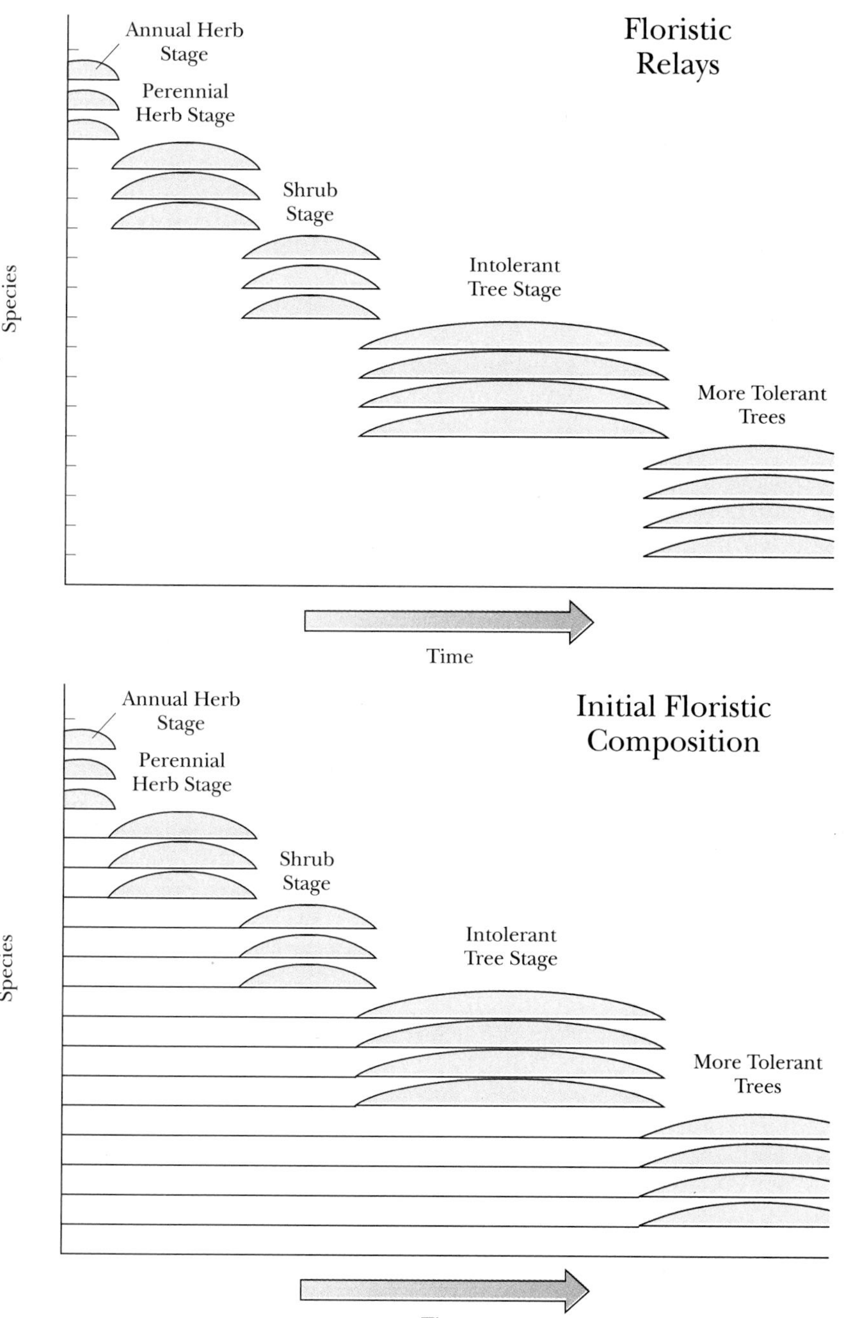

Figure 13-6

Two models of succession: initial floristic composition and floristic relays.

model has some validity for old field succession, the sere for which it was proposed. No one who has carefully examined early successional stages on expanses of bare rock, river bars, or sand dunes believes there are beech or sugar maple seedlings growing on those sites. Still less are there the herbs and animals that are part of the later stages; for these groups, the relay model is close to being correct.

John VanKat (1991) of Miami University examined the initial floristic composition model directly by compiling lists of the flora (and also the species present in seedbanks) at five stages of old-field succession in Ohio. He used old fields 2, 10, and 50 years postcultivation, a 90-year-old maple-elm forest and an old-growth beech-maple forest undisturbed since pioneer settlement (more than 200 years). Of the seven common trees and shrubs of the old-growth forest, none was present in the two or ten-year-old old fields. Of the nine common herbs of the old-growth forest, one was present but scarce in the ten-year-old field and none was present in the two-year-old field. Similarly, only 1 of 15 common species in the 90-year-old forest was present in the 10-year-old field and none in the two-year-old field.

In many subseres following surface fires or timber cutting, there may be little invasion of new species. Cutting the large trees of a beech-maple forest leaves a more open forest of beech and maple, along with individuals of a few additional intermediate or shade-intolerant species that were poorly represented in the canopy of the original forest. Any invasion consists of rare individuals of successional species such as staghorn sumac or cottonwood. In such seres there is no pioneer community, merely an initial one not much different in floristic composition from the predisturbance community (Clements 1916, VanKat and Carson 1991).

When different species arrive in succession and how they increase are worthy subjects for study. Such matters are of practical importance from the standpoint of vegetation management (Kenfield 1966). Relatively few trees are able to invade well-established thickets. In old-field succession, as mentioned, shrubs are shaded out by trees that invaded about the same time as the shrubs. If these trees are eliminated, an extremely persistent shrubland may be produced that could be a relatively maintenance-free cover for certain sites, such as highway rights-of-way.

Another question is the degree to which early species in succession make conditions more suitable for the later successional species. Most ecologists agree that the reactions of the early species tend to modify the environment in ways that make it less suitable although not necessarily uninhabitable, for themselves. As to later species, Connell and Slatyer (1977) distinguished three possibilities:

1. Reactions of the earlier species make the environment more suitable for later successional species. This they termed the **facilitation model.**
2. Reactions of the earlier species have little or no effect on the growth of the later species **(tolerance model).**
3. Reactions of the earlier species make the environment less suitable for later species **(inhibition model).**

All three situations occur; it is quite possible that all three occur in the same sere. It is worth considering what we would expect in evolutionary terms.

An Evolutionary View of Succession

Visualize a situation in which no succession occurs; one or more of a small group of species can invade bare sites and retain possession indefinitely. Two opposing evolutionary trends are likely. On the one hand, genetic variations in any species that allow it to displace the others will be advantageous. These include differences in life form, whereby perennials, by monopolizing space and perhaps other means, displace annuals; shrubs by shading, litter production, etc., displace perennial herbs; and trees, by similar mechanisms, displace shrubs. The ability to make better use of the conditions created by the early species and, hence, to outcompete them, should be favored.

Just the ability to tolerate the conditions set up by the early species may be enough to allow replacement. Individuals of the later species with such abilities will be present, ready to take over when older individuals of the first species (which has modified conditions to its own disadvantage) die.

At the same time, any features that give the earlier species the ability to hang on and continue to reproduce will be advantageous to them. From this tendency we would expect cases in which earlier species somehow inhibit the entrance or growth of later species. Many such cases are known, and many more can doubtless be found. Poverty grass dominates the old-field sere in Oklahoma for several years, replacing the annual weeds and eventually being replaced by little bluestem. The question is, How does this short annual hold out so long against the taller perennial grass? The answer appears to be that by allelopathy it inhibits nitrogen-fixing bacteria (Rice 1974). The result is that the soil stays nitrogen poor, which is favorable for poverty grass and unfavorable for little bluestem.

When Douglas fir forests of the Pacific Northwest are cut over, succession may take several routes. If the nitrogen-fixing red alder enters in good numbers, it grows faster than most of the young conifers and may delay conifer dominance for a long period (Franklin 1979).

White spruce dominance can be delayed on burnt-over lands where paper birch or trembling aspen precedes it. Here the inhibition occurs by simple mechanical means. Wind-whipped branches of the birch and aspen tend to knock off the new growth of the spruce, removing the apical bud and retarding its height growth (Rowe 1953).

Even for those many cases in which one species does modify the environment in a way that makes it more favorable for other species, we must not fall into the mistake of thinking that the earlier species has as its "function" paving the way for the later species. There is no evidence that ragweed is altruistic. Presumably the earlier as well as the later organisms are doing as well as they can for themselves. Succession is a matter of the later species being able to invade and come to dominance under the conditions created by the unavoidable activities of the earlier ones.

The Role of Coactions

Coactions play important roles in succession. Competition is often important in the replacement of earlier by later successional species as we have already seen. Many of the changes in plant composition during the course of a sere are based on competition for light or space. Probably many changes in animal species are also competitively based. For example, there is a great deal of evidence that the successional loss of grassland rodents to forest rodents results from competition (Grant 1972a, Fox and Pople 1984).

Noncompetitive coactions are also important. The presence of animals depends, in part, on a favorable microclimate (a reaction of the plants), but also on the provision by plants of elevated song perches, trees suitable for nest cavities, and appropriate nesting material—among many other coactions.

The role of noncompetitive coactions by animals and fungi in determining what the dominant organisms are and the rate of their replacement is probably substantial. The invasion of certain trees into an area is known to depend on mycorrhizal fungi. Most of the spread of oaks, beeches, hickories, and many pines into an area is dependent on dispersal by animals such as squirrels and jays.

Darwin (1859) mentioned that the heathlands near Farnham in Surrey when enclosed quickly grew up to Scotch fir (actually a pine, *Pinus sylvestris*). When he examined the open areas, he found a multitude of little trees, 32 in a square yard, "perpetually browsed down by the cattle." Since then, a variety of enclosure (or exclosure) studies have verified the importance of grazing or browsing—and other studies have shown that in other communities or locations they are of little importance.

The kind of plants that animals eat, or do not eat, can affect the following stages. In pastures the cattle or horses do not generally eat thorny plants. Consequently, a savanna of hawthorns or other thorny species tends to develop. Other woody plants develop within the protection of these "nurse" plants. Rabbits have their own preferences for browse, and it may be their partiality for oaks, nearly as much as immigration and growth rates, that allows black cherries and elms to dominate the early tree stage of some seres. Insects play similar roles and perhaps more pervasively, although there seem to be few studies directed at this question. One reason shadbush is rare in beech-maple forest is that caterpillars find them unusually attractive compared with maple saplings and spice bush, as I have learned by trying to grow shadbush in such a forest.

That coactions affect succession in such ways is well known; how important the effects are in general and for particular ecosystems is not.

Examples of Succession

Sand dune succession and old-field succession are discussed in the following pages. Hydroseres are discussed in Chapter 18.

Sand Dune Succession

Along its southern and eastern shores Lake Michigan is surrounded by sand deposited by the lake waters and thrown up into dunes by the action of the wind. These sand areas extend inland a considerable distance from the presentday shoreline because at times during the period when the glaciers were retreating the lake level was higher than it is now. To the casual visitor the distribution of communities on the sand areas may seem chaotic, but there is a pattern, worked out by Henry Chandler Cowles (1899) and Victor E. Shelford (1908), in some of the earliest ecological studies done in the United States. The intellectual and social background from which the early ecological studies of the dunes emerged is described in the book *Sacred Sands: the Struggle for Community in the Indiana Dunes* (Engel 1983).

Lower and Middle Beaches

The area adjoining the lake is divided into the lower, middle, and upper beaches (Figure 13-7). The lower beach includes the area washed by summer storms. The middle beach is the section from this point to the limit of wave action of winter storms. A line of driftwood and debris usually marks the upper edge of the middle beach.

Because of its instability the lower beach supports no vegetation. However, the water-filled spaces between the sand grains a few centimeters below the surface may support bacteria and a large invertebrate fauna of protozoa, rotifers, copepods, and tardigrades. The middle beach provides a relatively stable area during the summer, with the result that annual plants are able to grow and reproduce there. Perennial plants may start growth in one summer but tend to be washed away or buried under new sand in winter. The annuals that exploit this specialized habitat are succulents, a trait usually associated with deserts. In fact, the middle beach is desert-like despite the abundance of soil moisture. The high temperatures near the surface and the constant winds make loss of water by evaporation high enough that water balance is a problem. Some of the succulent annuals occurring on the middle beach are sea rocket (Figure 13-8), bugseed, and a spurge.

Relatively few animals live permanently on the middle beach. Most of these are either scavengers or predators on the scavengers. The scavengers make their living off the bodies of animals washed up on shore. Some of these are fish, as might be expected, but there are also large numbers of insects carried over the water, drowned, and washed onto shore, and also, during spring and fall, migrating birds that are killed by storms or become exhausted while over the lake. The scavengers include such invertebrates as blowflies, dermestid and rove beetles, and also vertebrates such as opossums, raccoons, and grackles that come to the beach from other inland communities. Tiger beetles and spiders are predators on the invertebrate scavengers.

The lower and middle beaches are ecologically interesting, but their biotas are totally dependent on the lake. Only if the lake level drops for several years does the current middle or lower beach begin to undergo successional changes such as described here as beginning with the vegetation of the upper beach. Such drops in lake levels do occur. Large and long-continued drops occurred at times during glacial recession and lesser ones in more recent times including the past few years following high lake levels of the 1970s and early 1980s.

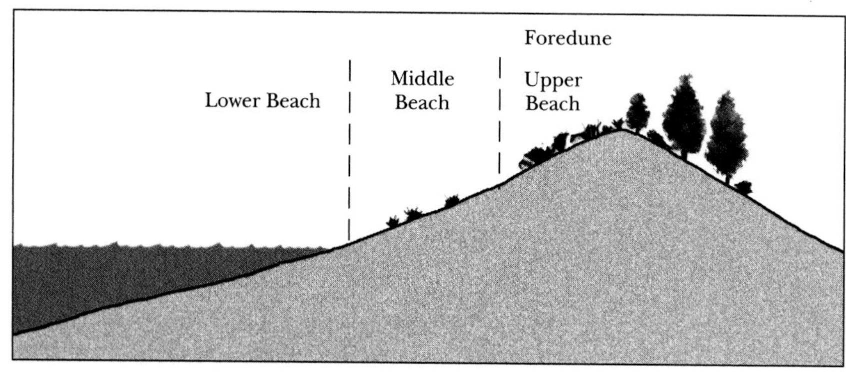

Figure 13-7

Zonation on a Lake Michigan beach. The lower beach, washed by the waves of summer storms, supports no plants. Sparse vegetation, mostly annuals, grows on the middle beach. The upper beach, outside the reach of the highest waves of winter, supports permanent vegetation, such as marram grass and cottonwood.

***Henry Chandler Cowles** (American, 1869 to 1939)*

Cowles grew up in Connecticut, but it is as a Midwesterner that we think of him. After earning a B.A. at Oberlin College in Ohio, he went to the newly founded University of Chicago for graduate work. He never left. After getting a Ph.D. in 1899, he remained in the botany department, becoming head in 1925 and retiring in 1934.

Originally aiming to study geology, Cowles fell under the influence of J. M. Coulter and turned to botany; however, a sound basis in geology became a major feature of Cowles' approach to community ecology.

His dissertation on sand dune succession, published in the university's *Botanical Gazette* in 1899, became one of the cornerstones of ecology in the United States. His works on the plant societies and physiographical ecology of the Chicago region, his 1911 paper on the causes of vegetational cycles, and his textbook, also published in 1911, were influential. But as F. E. Egler (1951) pointed out, "Cowles wrote relatively little, devoting his energies mainly to the training and developing of his graduate students."

Cowles was small, Irish, witty, and good-humored. The year after he retired, the Ecological Society of America published a special issue of *Ecology* in his honor (the same was done for F. E. Clements but only after he was dead). In the introduction, W. S. Cooper wrote, "A man may be a great scientist and a great teacher and yet inspire little affection or none at all. With Cowles it is far otherwise."

Cowles was the author of the second volume of the Chicago *Text Book of Botany,* dealing with autecology. In it, Cowles displayed a nonevolutionary, physiological approach to plant structure and life history. Plant features were not to be called adaptations without good reasons for thinking they had evolved in the service of the function they now performed.

Like Clements, Cowles saw succession as an important organizing process in ecology. He looked at succession from a less abstract height than Clements did, however, and his characterization of it as a variable approaching a variable is still a useful metaphor.

So Cowles and Clements, America's two great ecological pioneers, were very different men—one holist, one reductionist, one a theorist on a grand scale, the other interested in the details and the exceptions. Perhaps in this accident of history, the science has been lucky.

Dune Formation

The upper beach is out of reach of the waves. Typically we find a dune, the fore-dune, occupying the upper beach; the dune is built as a result of plant activity. Occasionally, through some catastrophic cause, a flat area will be produced in the upper beach zone, and here embryonic dunes will begin to form (Figure 13-9).

Dunes are built by sand being deposited around something that presents an obstacle to the wind; the wind is slowed and consequently drops the sand it is carrying. Small, temporary dunes can be formed by snow fences or even by a pile of trash, but only plants are able to build large, permanent dunes. Three traits make a plant a good dune former: (1) it must be perennial so that it holds the sand from year to year; (2) it must be able to grow upward, surviving continual burial as sand is deposited around it; and (3) it must be able to spread laterally so that the dune can grow in width as it grows in height.

Two species of grass, beach, or marram, grass and dune grass, are the best dune formers and generally occupy much of the fore-dune (Figure 13-10).

Figure 13-8

Sea rocket, a succulent annual on the middle beach, Leelanau County, Michigan. (Photo by author.)

Figure 13-9

An embyonic dune formed by a small clump of marram grass, Ottawa County, Michigan. (Photo by the author.)

Figure 13-10

Sand dune succession. A marram grass-covered foredune is shown in *A*. With high lake levels, dunes that were deposited in years of lower levels may be eroded by wave action. In *B*, the waves have cut away a foredune, exposing the spreading rhizomes that make marram grass such an effective dune former. In *C*, a blowout has produced a wandering dune which is moving away from the lake *(left)* across an area of bearberry, juniper, and jack pine. In the sheltered valleys produced by the slopes of old dunes (*D*), basswood is one of the important tree species. (Photos by the author.)

Both are perennial, can put out new roots from buried stems while sending up new shoots above the sand, and can send out long underground stems, rhizomes, from which new aboveground stems sprout. The two grasses have differences as well as similarities. The differences tend to favor long-distance dispersal and success in areas of active sand by marram grass and good local dispersal and success where sand deposition is reduced in the case of dune grass. Compared with dune grass, marram grass produces few seeds, but they float for an average of a week versus a day for dune grass (Maun 1985). During periods of shore erosion, marram grass is also spread by rhizomes being washed out of the sand and carried along the shore. Dune grass rhizomes do not float.

Growing rhizomes of marram grass can shoot out several meters a year (Figure 13-11), rapidly colonizing new bare areas, whereas dune grass tends to form a spreading clump that grows only a few tenths of a meter per year. Although burial by a small amount of sand increases the growth of both

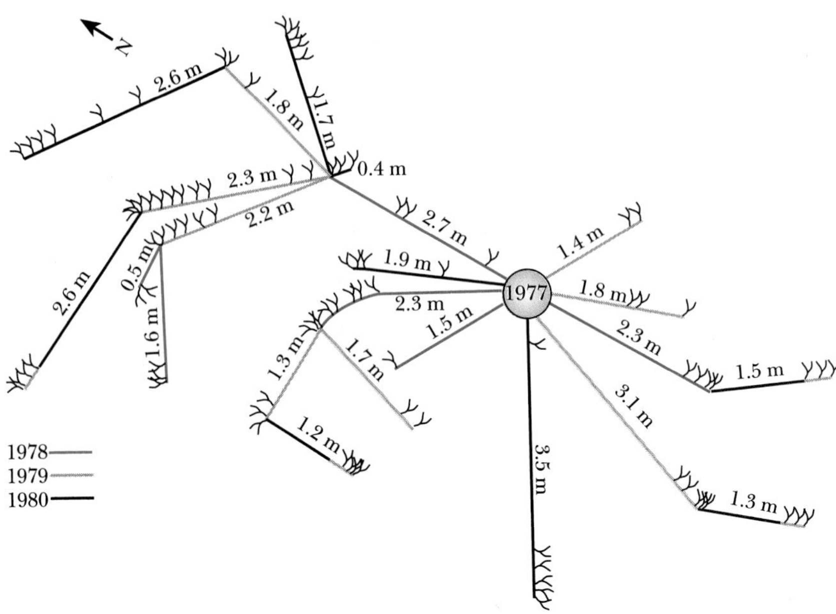

Figure 13-11

The spread of rhizomes from one small marram grass transplant. The planting was in 1977. The spread during the next three years is shown. Locations of above-ground shoots are indicated along the rhizomes. (Adapted from Maun 1985.)

grasses, marram grass can tolerate deeper burial. With little or no sand deposition, vigor apparently declines rapidly in marram grass, resulting in a tendency for dune grass to increase on sites where sand deposition has decreased.

Some other grasses and a few woody plants are also important members of the flora of the upper beach. Possibly the most important dune-forming woody plant is a small shrub, the sand cherry, that forms large patches by sprouting from its spreading roots. Cottonwood is also important; it can survive burial by continuing to grow upward and putting out new roots, but it does not spread laterally as do the grasses and sand cherry.

If we look at succession beginning at an embryonic dune being formed around a clump of grass, we can consider the pioneer stage in the sere to be a herbaceous stage dominated by the plants mentioned above and some others that generally invade, although they contribute little to dune formation. One is yellow beach (or Pitcher's) thistle, a species that occurs only on the sand dunes around Lakes Michigan, Huron, and Superior.

A fore-dune is built, with the sand stabilized by these plants. At this point succession may proceed through other stages (described below), and eventually a forest may occupy this fore-dune area. Sometimes, however, the dune covered by dune and beach grass may become a **wandering dune;** the vegetation loses its grip on the sand, and the wind begins to move the dune back across the landscape.

Blowouts and Wandering Dunes

Blowouts result from some disturbance that breaches the plant cover and allows the wind to get at the sand. The wind blows out the exposed sand, undermining the vegetation at the edges of the opening. The blowout enlarges rapidly as the vegetation on all sides of the hole topples in through the action of gravity.

Such disturbance probably occurred only rarely in primeval times. The importance of blowouts and wandering dunes in the current landscape is "an artifact of postsettlement trauma" (Wilhelm 1990). Agents for breaking the plant cover include "dune schooners," from which tourists can view the dunes without having to walk up and down hill, motorcycles and other recreational vehicles, construction of houses, and heavy human foot traffic.

Ian McHarg (1969) characterized the primary dunes along the ocean coasts, which correspond to the Great Lakes fore-dunes, as absolutely intolerant of trampling. "As a consequence, no development should be permitted on the primary dune, no walking should be allowed and it should not be breached at any point." Access to the beach should be by bridges over the primary dune. Much the same points can be made about the Great Lakes fore-dunes, but human acceptance of this fragility comes only slowly.

The wandering dune (Figure 13-10) moves back across the landscape through what might be called the dune complex. It may cover pine, oak, or even beech-maple forests, and fill in ponds. Eventually a point is reached where the sand is no longer being moved very much by the wind. Reduction of the wind may occur partly through distance from the lake, but the formation lakeward of other dunes, including a new fore-dune, is also important.

Plant Succession

At this point the dune, which during its wanderings has been simply bare sand, can now be captured by vegetation again. On the windward side of the dune the invading vegetation tends to be similar to that of the upper beach. Marram and dune grass invade, along with several other plants. This stage may be replaced by a shrub stage in which patches of bearberry, junipers, and various cherries, dogwoods, and willows are intermixed with vegetation of the preceding stage.

A cottonwood stage may occur at this point, but when it does, cottonwood does not represent a newly invading species. Cottonwood germinates only on wet sand, so that the usual site for a cottonwood to begin is on a blowout that removes sand all the way down to the water table. When cottonwood trees appear on a dune in succession, then, they have invaded long before and become dominant by being able to grow higher and shade the other plants.

Especially on flat, inter-dunal sites, the shrub stage may be invaded by jack pines, and occasionally other species of pines. One characteristic landscape consists of scattered jack pines (or, in northern Michigan, red or white pines) with large saucer-shaped clumps of ground juniper separated by sparse low vegetation of bearberry, little bluestem, and, to the north, creeping juniper. Thick patches of pine may occur, but usually both the cottonwood and jack pine communities are open enough for oaks to germinate and grow. Around the south end of Lake Michigan, a low, open black oak forest or savanna seems to be the usual end point of dune succession. The sterility of the sand and its poor moisture-retaining capacity along with the lack of soil improvement by the oaks have been suggested as reasons why these areas are not eventually occupied by the more mesophytic trees of the region (Olson 1958); however, it seems likely that recurrent fire has been at least as important (Henderson and Pavlovic 1986). The presence of highly flammable prairie and the generally east-west orientation of the dunes contribute to spread of fires.

In certain protected pockets in this region and generally further north along the eastern shore of Lake Michigan, oak forest (usually red oak) is invaded by basswood (Figure 13-10), sugar maple, beech, and hemlock. Fire has been less of a factor along the eastern shore; there is little prairie, and, given prevailing westerly winds, fires would have to start near the lake and spread across the dunes, rather than along them, as at the south end of the lake.

These seem to be some of the sequences occurring on the windward slope of stabilized dunes, sequences involving some hundreds of years. Succession is faster on leeward slopes. Physical conditions are more favorable, and usually the leeward slope is immediately adjacent to an already developed forest.

If we return now to a fore-dune occupied by marram grass and sand cherry, and suppose that it does not suffer any disturbance that might convert it to a wandering dune, it may undergo a succession similar to that just described. It is not unusual to find a section of forest-covered fore-dune next to one occupied by marram grass. But even forested dunes are not immune to blowouts; human disturbance or possibly a tornado may open up a section of forest enough so that even a dune occupied by beech-maple-hemlock forest may be converted into a wandering dune.

Within this basic framework, most of the communities of the **dunes complex** can be understood. The dunes complex is a mosaic of embryonic dunes, blowouts, wandering dunes, and stabilized dunes in all stages of succession from marram grass to beech-maple forest. But this is not the whole story. Some additional seres are involved, such as those around the interdunal ponds.

The sand dune sere described is based on reactions of the dominant plants, mainly shading and soil development (Table 13-3). Other reactions that they produce include moderated temperature extremes, a decrease in wind, an increase in relative

Table 13-3
Physical Features of Sand Dune Successional Stages

Disregarding blowouts, successional changes in the dominant plants on the sand dunes are caused mainly by shading and soil development. These effects are reactions, exerted by the plants themselves. The plants also produce other effects on the physical environment.

	Successional Stage				
Physical Feature	*Marram Grass*	*Cottonwood*	*Jack Pine*	*Oak*	*Beech Maple*
Midsummer light intensity 10 cm from the ground (expressed as % of full sunlight)	96	—	37	2	1
Soil moisture (weight of water expressed as % of dry weight of soil)	1	—	2	5	24
Summer evaporation rate (ml water per day)	—	21	11	10	8
Soil pH	7–8	—	5.5–7.0	5.5–7.0	5.5–6.0

Evaporation data from Fuller, G. D., "Evaporation and plant succession," *Bot. Gaz.*, 52:193–208, 1911; other data original.

humidity, and decreased evaporative power of the air. On the basis of these and other changes, the subdominant plants and the animals also change throughout the sere.

Water Levels

Sand dune formation as described here is a feature of depositing shores. Higher lake levels, such as occurred in the 1970s and early 1980s, lead to a preponderance of erosion over deposition. The lake eats into the fore-dune producing a steep bluff above a narrow beach, or none at all. The cliff edges closer and closer to houses built to look out over the lake and eventually undercuts them.

Lakefront property owners tend to regard these higher lake levels as a catastrophe from which the government ought to protect them. It is now clear, however, that fluctuating lake levels have been a feature of the Great Lakes throughout their post-glacial history. Early, extreme changes from the 230-foot (140 m) Chippewa stage of 9000 years ago to the 605-foot (184 m) Nipissing stage of 4000 years ago were the result of changing outlets and drainageways to the sea. More recently the changes are based on rainfall and other aspects of climate.

A partial chronology of lake-level changes has been developed. It suggests high water levels between 1600 and 1200 B.P. (before present), 950 and 750 B.P., and 450 and 185 B.P. Doubtless, smaller-scale fluctuations occurred within the major phases. The 580-foot (176 m) level of the early 1900s that we tend to think of as normal is "a low phase on a naturally fluctuating trend" (Larsen 1987).

Animal Succession

This generalized account of changes in animal communities with time is based on investigations by Shelford (1913, 1963), Kendeigh (1974), Lowrie (1948, spiders), Park (1930, beetles), Strohecker (1937, Orthoptera), Talbot (1934, ants), Van Orman (1976, birds), and Olson (1978, mammals). Factors important in driving the succession vary across the groups, but widely influential are

- Changing life forms of the dominant vegetation
- Disappearance of bare sand
- Increase in leaf and woody litter and logs
- Increase in standing dead trees
- Declining temperature maxima and light intensity
- Increasing relative humidity and soil moisture

Marram Grass-Cottonwood

Invertebrates. Burrowers are especially important in the bare, open sand of this community. "This is preeminently the stage of digger wasps,'' according to V. E. Shelford (1963). Their holes, where they store flies for their larvae, pock the open sand in large numbers. The unstable sand is less suitable for long-lived ant colonies, and the only conspicuous ants are crater-forming species. Web-weaving spiders are few in this area of high winds; however, there are some interesting non-web-builders, notably Wright's earthwolf that hunts mostly at night and spends the heat of the day in a silk-lined burrow. Shelford's study of tiger beetle succession on the dunes was one of the earliest contributions to the study of animal ecology. He found the white tiger beetle to be the marram-grass representative of this group of bright-colored, active predators. Grasshoppers are the only Orthoptera.

Birds and Mammals. The characteristic small mammals are the prairie deer mouse and thirteen-lined ground squirrel. Several species of birds nest, including the vesper sparrow, a grassland species; however, the avifauna is mostly forest-edge. Characteristic species include field sparrow, brown thrasher, and eastern kingbird. In the past few decades red-winged blackbirds have come to occupy this habitat in substantial numbers.

Jack Pine

Invertebrates. Many of the invertebrates present in the marram grass-cottonwood community continue in the more open and grassier areas of jack pine. Several species of ants are added including ones needing litter or logs. Grasshoppers remain numerous and some new species are added. With reduced wind velocities and more stable anchorages, the number of web-building spiders increases; however, numbers of such non-web-builders as the jumping spiders (Salticidae) and crab spiders (Thomisidae) also increase. The bronze tiger beetle occurs here and continues in the oak stage.

Birds and Mammals. The most prominent small mammals are the eastern chipmunk, a forest ground dweller of broad ecological amplitude, and the red squirrel, an inhabitant of conifer forests in the northern United States. Many of the same forest-edge birds that live in the cottonwoods occur in the pines. The prairie warbler is particularly characteristic of the open areas where large clumps of ground juniper occur among scattered pines. Where larger trees occur, such species as northern oriole, great crested flycatcher, and black-capped chickadee are added. Recently, the house finch has expanded into the jack-pine areas of the dunes.

Black Oak

Invertebrates. For most groups of organisms, the biggest turnover in species composition occurs between the jack pines and the black oaks. Species that require open sand decline as do those dependent on open savanna landscape or conifers. Several of the grasshoppers drop out, and species living in litter and under logs, such as crickets and wood roaches, become prominent. Several log- and snag-inhabiting ants are added. There is a rich spider fauna of diverse microhabitats and foraging styles. The orbweavers (Argiopidae) first become common in this stage, but overall, ''vagabond'' species, such as wolf and jumping spiders (Lycosidae and Salticidae), predominate. Despite an abundance of sand, ant-lions are rare in the sand dune sere except in stable bare areas of the oak stage.

Birds and Mammals. If a substantial area of oak forest is present, mammal and bird species are similar to oak woodlots inland. White-footed mice, eastern chipmunks, and fox squirrels are common. With an increase in the soil fauna, masked and short-tailed shrews appear. In addition to orioles, crested flycatchers, and chickadees, such species as scarlet tanager, ovenbird, red-eyed vireo, and eastern wood-pewee are prominent.

Beech-Maple or Northern Hardwood Forest

Invertebrates. Large areas of forest tend to have the same species as mesic forests of inland sites. Diversity among both ants and Orthoptera is low in the unbroken forest. Only two ant species, both log dwellers, are common. The Pennsylvania wood roach, a holdover from the oak forest, is common. Added are camel crickets under the logs and a tree-dwelling katydid. Spider diversity is also lower in unbroken beech-maple forest than in any preceding stage; nevertheless, more than 50 species are present. Numbers of some web-builder groups increase, especially the small sheet-web weavers (Linyphiidae), which build delicate and complex webs among herbaceous stems. There is a general decrease in the ''vagabond'' spiders. The tiger beetle seen flying ahead across the bare soil of the summer forest is the green tiger beetle.

Birds and Mammals. Both the bird and mammal faunas are richer than earlier stages. In general, the species occupying oak forest are retained, and other species are added. Gray squirrels may come in, and red squirrels reappear, making use of tulip tree cones as they do the cones of pines. New among the birds may be the Acadian flycatcher, American redstart, and hooded warbler. Where hemlock or white pine are added to the forest in Michigan, such northern mixed forest species as the black-throated green and blackburnian warblers may occur.

Old-Field Succession

Succession on abandoned cropland is termed "old-field succession" (Keever 1950). Nothing quite like old-field succession occurred in North America prior to the beginning of colonial agriculture in the 1600s, nor anyplace else in the world earlier than perhaps 10,000 or 12,000 years ago. The seral stages are probably recent combinations of species drawn from various habitats such as sandbars, lake margins, and forest openings. Eurasian weeds are frequent in the early stages. For these reasons, it is unclear what old-field succession can tell us about other kinds of succession. Given the enormous fraction of our landscape occupied by croplands, it is, nevertheless, of interest in its own right. We look at it here as an example of secondary succession.

Old fields generally have soil that is low in organic matter from years of cultivation. Buried seeds of the weedy species that are able to persist with the crops are present. Vegetative propagules are scarce, although some, such as common milkweed, seem often to remain. The soil is generally infertile; however, some fields may have a fertilizer residue that declines in the year or two just after abandonment.

The earliest stages of old-field succession show considerable similarity over a fairly wide geographical area. Annuals such as crabgrass, horseweed, and ragweed are common and seed profusely (Figure 13-12). Such sites are excellent foraging sites for seed-eating animals such as doves, sparrows, and mice.

The second year generally finds a denser vegetation. The annuals are declining, at least in size, the biennials flowering, and perennials such as goldenrods and asters becoming noticeable. By the third year, perennials are generally dominant; broom sedge, a perennial grass, is usually important.

Figure 13-12

The end of the first growing season in old field succession. The dominant plants in this Kalamazoo County, Michigan, old field were horseweed (middle of photo) and annual ragweed (foreground).

In some parts of the prairie region, such as Oklahoma, the annual poverty grass replaces the earlier annual grasses and forbs and persists for 9 to 13 years. Little bluestem, a relative of broom sedge, eventually replaces it. The bluestem stage may last for 20 years or more but, it is thought, other grasses and forbs characteristic of climax prairie will eventually enter. In other parts of the prairie, perennial prairie species may be frequent within five years of abandonment (Buss 1956) (Figure 13-13).

In the forested parts of the eastern and midwestern United States, tree and shrub seedlings tend to appear, at least in small numbers, during the first few years. In the Southeast, the trees are generally pines, but elsewhere deciduous trees such as aspen and elm tend to be the earliest species. Woody plants are usually noticeable within 10 years; by 20 to 40 years, a young forest may be present.

These are broad generalizations; the exact sequences may be affected by such factors as time of the last cultivation, herbicide treatments, soil type, and availability of seed sources. As just one example of such differences, a field in southern Illinois that is undisturbed after the last crop is removed tends to be dominated in the first year by winter annuals. These are plants that germinate in the autumn of the last crop year, overwinter as rosettes, and grow up to flower in the first year after abandonment. Examples are fleabane, horseweed, and shepherd's purse. Under them occur large numbers of spindly

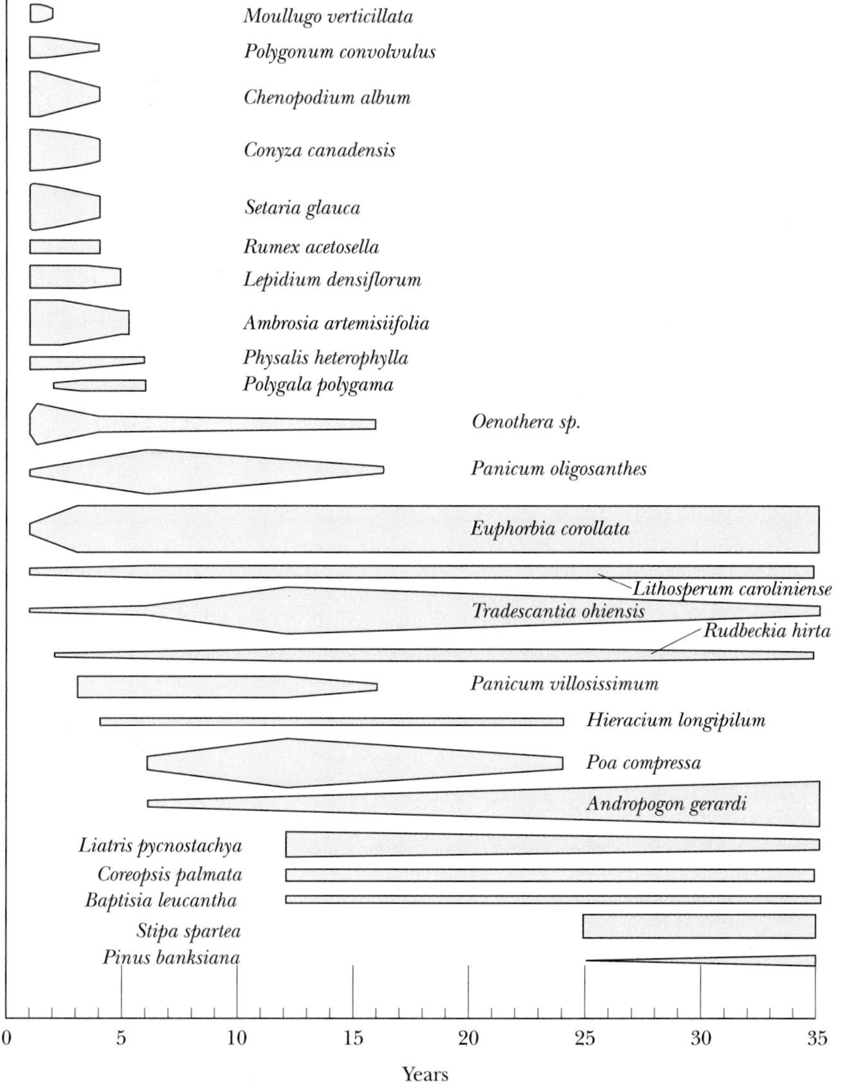

Figure 13-13

Old-field succession on sandy areas in the prairie-forest border region of western Wisconsin. A few of the weedier perennial prairie species are present in the first year. By 5 years, the vegetation is essentially a species-poor prairie and by 12 years, a richer one. By 35 years, the vegetation is similar to native prairie relicts in the region, though lacking some of the rarer species; however, jack-pine has begun to invade in sites adjacent to older jack-pine stands. (Adapted from Buss 1956.)

summer annuals such as ragweed and lamb's-quarters. If, however, the field is plowed after the last crop, that fall or the following spring, the first-year field is dominated by vigorously growing summer annuals. The same effect can be achieved experimentally by removing the rosettes of the winter annuals (Raynal and Bazzaz 1975).

In general, plant species richness is high in the first two years when many alien species are present (Bazzaz 1968, Mellinger and McNaughton 1975). Richness is low during the perennial grass stage, but woody plants are being added and, later, plants associated with the shade they provide. The very high species richness shown for the 50-year field in Ohio (Table 13-4)may be idiosyncratic, based on the specific 50-year field that was studied. Just where the peak in species richness occurs along the sere is uncertain; however, the slight drop in richness in the long-undisturbed (200+-year) site is probably typical. The low species diversity (that is, the combination of richness and equitability) in the ten-year field is the result of concentration of dominance in a few species (in this case, Canada goldenrod and meadow fescue) (VanKat and Snyder 1991).

Productivity may be high in the first year of old-field succession, if residual fertilizer is present (Odum 1960). After a few low years, there may be a general rise in net production with plateaus corresponding to the physiognomic stages.

Early successional species contribute to a seedbank (Table 13-4), but species typical of mature or climax forest generally do not. There are in the soil, of course, rhizomes, bulbs, corms, and other vegetative parts of many of the mature forest species. Many early successional herbaceous species remain common in the seedbank for years after they have disappeared above ground; most will sprout readily in disturbed areas. Their seeds do not last forever, though, and the seedbanks of sites where the canopy has been closed for several decades have few early successional species still represented (VanKat 1991).

The Climax Community

The climax community is in equilibrium with its environment, of which climate is an especially important factor. The climax is at a steady state of species competition, structure, and energy flow. "Steady state" indicates the dynamic nature of the climax; changes occur continually but they are changes that tend to perpetuate the community rather than alter it.

Table 13-4
Some Features of the Old Field Sere in Southern Ohio

The fourth line of the table is the percent of species in the flora of each stage that is represented in its own seedbank (for example, the percent of 90-year-forest species found in the seedbank of the 90-year forest). The fifth line of the table is the percent of the pioneer (2-year species) flora found in the seedbank of all the different stages. Based on VanKat and Snyder 1991 and VanKat 1991.

	Community				
Feature	*2-Year Field*	*10-Year Field*	*50-Year Field*	*90-Year Forest*	*200+-Year Forest*
Number of species	69	67	109	95	84
Diversity (inverted Simpson's index)	4.2	2.3	7.3	5.4	4.7
Percent annuals in flora	42	24	19	10	7
Percent of species represented in seedbank	54	31	29	6	8
Percent of species in 2-year-field represented in seedbank	54	42	41	19	9

Climax and other relatively mature ecosystems differ, at least in degree, from pioneer and other immature ecosystems. The following seem to be valid statements about the climax community, although individual exceptions may occur:

1. The climax community is able to tolerate its own reactions.
2. The climax tends to be mesic for the climate in which it occurs; many pioneer communities are relatively xeric or hydric.
3. The climax community tends to be more highly organized; for example, stratification may be more complex and there may be more, and more complex, coactions (more complex food webs, more mutualism).
4. Mature communities have more species (higher diversity) than immature ones, which tend to have high populations of a few species. This latter situation is to be expected where a relatively harsh environment permits only a few species to occur. The larger number of species in the climax community is related to the more complex organization, thus allowing more niches. However, it is now clear that there may be some loss of diversity in climax forests, as the species list drops to those best able to tolerate the low light and other conditions (Brewer 1980). Normally this loss of diversity is reversed at intervals by processes that "disturb" patches of the forest—such things as severe wind storms and infrequent surface fires (Figure 13-14). These processes rarely eliminate the climax species but allow some other species to persist or to invade again.
5. The organisms composing the climax community tend to be long-lived, relatively large, and with a low biotic potential (*K*-selected), whereas species of earlier successional stages tend to be smaller, shorter-lived, with a higher biotic potential (*r*-selected). Most croplands are man-made pioneer ecosystems, and most crops are annuals or short-lived perennials. Most game animals, such as bobwhite and rabbits, are species of relatively early successional stages; they produce a large number of young each year, many of which, when not harvested by humans, fall to predators and otherwise die. The mourning dove, for example, is a species of earlier successional stages. A given female lays only two eggs at a time but renests several times a year. Ecologically it is an appropriate game species, at least in the South. The related passenger pigeon, however, was a forest bird living on beechnuts and acorns and produced only a single egg a year. It should not have been subjected to hunting pressure, and when it was (mainly market hunting), it quickly became extinct.
6. Gross primary production tends to be greater than community respiration in immature stages, whereas energy is at a steady state (net community production is zero) in climax communities. This implies that energy is being stored in immature systems, and, consequently, the amount of plant or animal tissue or dead organic matter (such as litter or humus) is increasing. In immature systems this excess can be harvested by humans. A cornfield starts with seeds weighing pounds and ends up the season with ears and stalks weighing tons.
7. The stability of climax ecosystems is high. This is obviously true in the sense that immature ecosystems are temporary, but it also seems often to be true in the senses that immature ecosystems show broader changes than mature ones and are less resistant to invasion by other species.

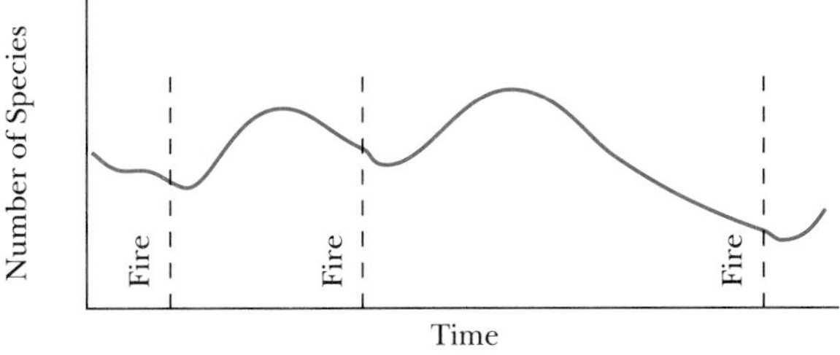

Figure 13-14

In this graphical representation of O. Loucks' model of species diversity relations in northern forests, "random perturbations" such as forest fires allow the growth and establishment of species which lose out in the stretches between perturbations. (Adapted from Loucks 1970.)

All these trends seem statistically true. The fundamental distinction between climax and earlier stages, however, seems to be the first one: The climax community consists of organisms able to tolerate the conditions that this community sets up. Another way of saying almost the same thing is that succession proceeds until there are no more species left in the regional species pool that are able to make more efficient use of the conditions prevailing in the mature communities.

The actual composition of the climax community in a region is influenced also by the kinds and frequencies of disturbance. It makes no sense to think of the regional climax as consisting of only the most shade-tolerant trees if fires, windthrow, or other disturbances recur at intervals short enough that species of intermediate tolerance are regular members of the canopy.

Monoclimax and Polyclimax

Frederic E. Clements was a clear observer who liked to fit complicated matters into systems. His view of succession and the climax community, now called the **monoclimax hypothesis,** was that within a given region all land surfaces eventually tend to be occupied by a single kind of community which is climax. The climax is determined by the regional climate. Given a stable climate the climax community is stable indefinitely.

There are several types of observations that seem to conflict with this hypothesis. For example, there is evidence that even under primeval conditions it was difficult to find large areas of uniform vegetation. In southwestern Michigan at the time of settlement there was about as much oak-hickory forest as beech-maple forest, and there was also prairie, oak savanna, marsh, swamp forest, and sand barrens (Figure 13-15). This area had about 15,000 years (after the last glaciers) for the climax community to establish itself. Should we not recognize several different communities as climax on various sites (the **polyclimax concept**)?

Clements' answer would probably have been that replaceability, not persistence, is the key to the status of a community. The drier, less shade-tolerant vegetation on the tops of hills may last for a very long time, but as a mature soil develops, it will be replaced by the climax vegetation; if not then, at least when the hills erode down further (as, of course, they surely must). It is clear to anyone, Clements would have said, that vegetation is a veritable mosaic, consisting of sites supporting the climax, sites that are undergoing rapid succession, and sites where succession is proceeding slowly or is stalled. The solution is not to tamper with the concept but to find appropriate names for the various nonclimax stages that are seen. Clements (1916) gave them names such as subclimax, serclimax, and disclimax.

Few present-day ecologists would accept this solution. They would argue that a climax community ought to be in harmony with the whole environment, not just climate. Climate is not very stable; the climatic climax in southern Michigan, in the 15,000 years since the glaciers withdrew, has been tundra, boreal forest, moist deciduous forest, and dry deciduous forest. Why, then, choose climate as the one factor determining *the* climax? There are also some other, more abstract, objections.

The solution of the polyclimax school is also basically terminological. After sorting out the obviously seral communities, they would recognize different kinds of climaxes, such as edaphic (where soil factors are important), topographic (elevation, slope), and fire.

A third hypothesis, called by R. H. Whittaker (1953), the **climax pattern concept** rejects the classification approach. In this view the undisturbed vegetation of southwestern Michigan or the Smoky Mountains of Tennessee and North Carolina would be treated not by recognizing a bunch of communities but by studying and describing the spatial changes in the vegetation of the region. In the climax pattern concept there is, in a sense, only one big community that changes according to soil, slope, and other habitat factors. The climax pattern approach considers it more useful and closer to reality to describe this pattern of variation.

These three views are not as different as they might at first seem. On the level of describing the vegetation and its environmental relations, proponents of all three might agree on most of the fundamentals. For example, Leak (1982) used a polyclimax approach in describing variants of mature northern hardwood forests of the White Mountains of New Hampshire that correlate primarily with soil. One could also describe these variations using the climax pattern concept. But Clements would not have found the variations an argument against the monoclimax concept. His view of the climax was broad. The subtle variations of northern hardwood

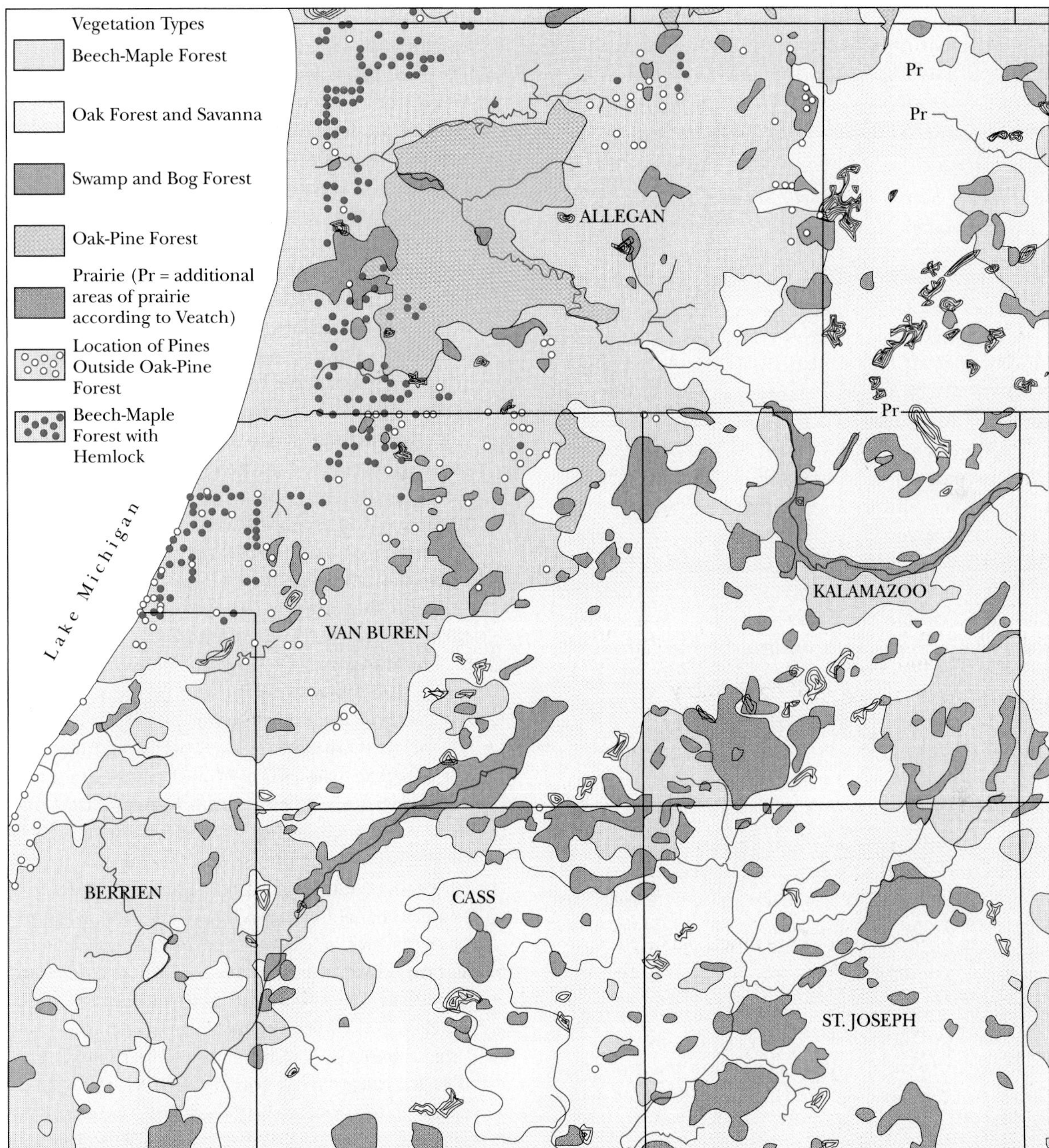

Figure 13-15

This map of southwestern Michigan shows that the primeval vegetation was far from uniform. The map was mostly constructed from information contained in the notes of the original land survey. As land was opened up for settlement, government surveyors laid out the lines for township, range, and section. Part of the procedure involved marking and recording two or four trees at each section corner and halfway down some section lines. Consequently, there is information available on the forest or other vegetation at least every mile and sometimes at shorter intervals. (From Brewer 1980.)

forest exist, and it is interesting to know something of their environmental basis, but to Clements they would all have been representatives of one climax community.

Is There Such a Thing as a Climax Community?

As a part of a recent trend toward questioning the existence of equilibrium states in ecology, the concept of climax is sometimes avoided or, at least, circumlocuted. Long-undisturbed stands of forest showing no directional changes in species composition—climax forests—are now often referred to as "old-growth forests," even in the Midwest and the Northeast. The term arose in the Pacific Northwest, for an understandable reason. Some of the important species of the temperate rain forests there, such as the long-lived Douglas fir and coastal redwood, are less shade tolerant than, hence potentially replaceable by, western hemlock and western red cedar. Even stands undisturbed for hundreds of years may not be "climax," in that the latter two species may be increasing.

There is no situation comparable to this in the East. Probably the closest approach was the occasional stand found by settlers where giant tulip trees or red oaks formed a discontinuous upper layer above the beeches and sugar maples. No, or virtually no, old-growth forests of this sort still exist, although some mature, or even climax, forests do. Warren Woods in southwestern Michigan is a notable example.

The term "old-growth forest" is unlikely to be dislodged soon, because it has spread widely in the conservation biology literature and it avoids confronting the issue of climax. The fundamental question is not terminology but whether processes occur that tend to lead to equilibrium communities. The following statements seem defensible.

1. In the absence of large-scale disturbance, succession tends to produce a relatively stable end product on most sites. Species composition varies according to site conditions (Leak 1991). Forests that result from long freedom from disturbance tend to consist of the most shade-tolerant trees and other biota associated with heavy shade. They have lower diversity than the communities that early ecologists in North America considered to be climax (Brewer 1980, Martin 1960).
2. Under natural conditions, recurrent disturbances, such as fires and large wind storms, interacting with site factors, such as water table, tend to interrupt the autogenic processes mentioned in (1). One result is to produce a landscape having greater diversity than the individual sites referred to in (1). Slightly moister or drier sites have slightly different species than medium sites. Recently burned sites have species that are rare in or absent from sites that have long escaped fire. Sites disturbed a fair number of years previously have some of these species plus the most shade-tolerant species that declined on or were eliminated from areas of most severe disturbance. Furthermore, transitory factors affecting dispersal and establishment early in succession can produce idiosyncrasies of composition that may be detectable for a long time afterwards (McCune and Allen 1985).

Much of this site-to site variation within a landscape was well within the broad scope of the climax concept as developed by F. E. Clements. Concerning the role of chance factors, for example, Weaver and Clements (1938) wrote, "It sometimes happens that one dominant occupies an area so completely as to exclude the others simply because it invaded first, although the habitat was equally suitable to all." Where the Clementsian view of climax fell short most seriously was in assigning complete primacy to climate and, consequently, undervaluing disturbance as a natural process. Even natural and predictable "disturbances," such as fire, were regarded as flies in, rather than constituents of, the ointment. They were considered mere impediments on the road to the climatic climax.

The idea of an **equilibrium landscape** was probably first stated by W. S. Cooper (1913), one of Henry Chandler Cowles' early students, based on observations on the boreal forest (largely balsam fir, paper birch, and white spruce) of Isle Royale. Windfalls are frequent because of the brittleness of the balsam firs. "The result in the forest in general is a mosaic or patchwork which is in a state of continual change. The forest as a whole remains the same, the changes in various parts balancing each other.'' Similar conclusions for various other ecosystems were reached over the years by many observers. For example, F. H. Bormann and G. E. Likens (1979) from their studies of the Hubbard Brook watershed suggested that the presettlement northern hardwood landscape was ''a collection of ecosystems in various stages of development.''

3. It is probably trivially true that few sites are actually at equilibrium. Almost every landscape unit shows some deviation: Perhaps there is an underrepresentation of hemlock because of a plague of white-tailed deer some winter a century ago. There may yet be forests in the Appalachians where the series of changes in canopy trees resulting from the elimination of American chestnut by chestnut blight 50 years ago (Woods and Shanks 1959) is still going asymptotically to completion. Chestnut was absent from the presettlement forests of most of Michigan, not because it wouldn't grow but because it hadn't gotten that far yet in its postglacial range expansion. In that way the deciduous forests of presettlement Michigan were nonequilibrium. They were probably also nonequilibrium as a result of several climatic changes in the preceding few hundred years.

   Douglas Sprugel (1991) suggested that a high percentage of the old-growth forests of the Pacific Northwest represent stages of regeneration dating from a warmer period 600 to 700 years ago when fires were more frequent than before or since. (As already mentioned, these old-growth forests would not be considered equilibrium communities in the strict Clementsian sense either and, in fact, Weaver and Clements [1938] referred to the climax association of this area as the cedar-hemlock forest.)

It is important to be aware of factors that may keep a community from being fully in harmony with the current environment. We need to realize that major climate changes can occur within the lifetime of the canopy trees, that unique events can cast long shadows, that catastrophes can affect areas larger than a small watershed (Sprugel 1991). If, however, processes occurring on a site or in the landscape are in the direction of an equilibrium, we are likely to advance our understanding by recognizing and studying these processes.

Stability

Knowing what makes a population or a community stable is obviously of great practical importance. People would prefer their crops and game to be stable at high levels and pests and diseases to be stable at low levels. Unfortunately the question, Is this population or community stable? has turned out to be complex or, really, to be several questions that may not be very closely related.

In one sense climax communities are stable and successional communities are unstable; given a reasonably stable environment, the climax community persists whereas successional communities disappear or change. Of course, there are examples of very persistent subclimax communities, and climax communities themselves may slowly change, but to a practical individual the distinction between stable climax communities and changeable successional communities is valuable.

What happens to a climax community if there is a disturbance from the outside—for example, a temporarily changed climate? If stability implies an unchangeable nature, the community should retain the same species in the same numbers, despite drought or flood. Usually, this does not happen, although if the climatic change is short compared with the lives of the adult dominants (such as trees in a forest), there may be little evident change. The response to change usually seems to be more like that described for prairie (earlier in this chapter). In a period of drought some prairie species decreased and others increased. With an end to the drought, the community returned to a predrought structure and composition. Timber cutting in a climax forest produces a similar response. Some shade-intolerant species increase temporarily or invade, but later trends are toward restoring prelogging conditions.

On the community level, then, there seems to be little tendency to resist change. Rather, the community undergoes internal shifts in response to external change, which may maintain some stability (although not constancy) of such traits as cover and productivity. Climax communities also show stability in the sense that they tend to shift back toward their earlier state once the external disturbance has ceased; this is generally expressed by saying that they recover from the perturbations. The tendency to remain unchanged in the face of perturbations is sometimes referred to as **resistance stability.** The tendency to return to the former state is then called **resilience stability.**

What would be the evolutionary advantages of resistance stability of individual species? Very wide swings in numbers will be selected against in the sense that a species that does fluctuate widely may go to extinction, so that only species having those traits that tend to moderate such swings (at least on the down side) continue to be with us. Beyond this, the advantages of population stability to individuals, as affecting their life span and reproduction, need to be considered.

When this is done, it seems clear that stability, in the sense of constancy, could rarely be selected for. Regulation of the size of local populations (see Chapter 4) could be selected because mechanisms for recognizing and preventing overcrowding would enhance survival and reproduction. But crowding is relative to resources, at least over a very broad range of densities. Consequently, if conditions become unusually favorable for a species, any genotypes of that species unable to respond to the increased favorability by increasing in numbers will be selected against. We would anticipate that ordinarily species will respond to a better environment by increasing in numbers and, when actual cases are considered, this is usually what we see.

Although the stability of the community is based on the individual organisms composing it, it is worth repeating that some community traits may show more stability than those shown by individual species. If half the species increase and half decline, the net effect may be that the community shows a stable productivity.

Constancy, as useful as it might be to humans in their use of ecosystems, seems to be a kind of accident resulting when a species has an effectively constant environment. Species that fluctuate in numbers show no less stability, in any important ecological sense, than ones that do not, as long as the fluctuations tend to be up when the population is low and down when it is high. A little thought will make it clear that this restriction is no restriction at all, because it must apply to every species that does not become extinct.

Much of the interest in stability stemmed from a small book by the English animal ecologist Charles Elton, *The Ecology of Invasions by Animals and Plants* (1958). Elton summarized several lines of evidence that suggested to him that stability was greater in more diverse habitats or ecosystems. He, and later followers, pointed out that the predator-prey models of Lotka and Volterra predicted oscillations in numbers, and laboratory experiments seem to confirm that oscillations and even extinction are usual results in very simple systems. Furthermore, simple natural systems, such as the tundra, do show wide fluctuations in animal numbers, such as with the snowshoe hare and the lynx. Outbreaks of pest species are most common in simple ecosystems, especially in the very simple monoculture systems of crops favored by modern agriculture. Such outbreaks seem very uncommon in the complex natural ecosystems of the tropics. There is also the fact that a high proportion of the species that have become extinct in historical times have been species of the simplified ecosystems of islands.

The particular kind of stability in which Elton was most interested is a special kind of resistance stability—resistance to invasion. An alien species arrives: What happens? In this limited meaning, stability does seem strongly correlated with diversity. Diverse communities, or, at least, intact natural communities that are diverse, seem much harder to invade than low-diversity areas produced by either human disturbance or natural agents, such as lava flows.

Another leg to the diversity-stability stool was a short paper by Robert MacArthur (1955) suggesting that community stability was related to the number of links in the food web. In a more diverse ecosystem, a perturbation, perhaps an unusually heavy production of food by one plant species, would not be passed along as simply a great increase in one herbivore followed by a great increase in one carnivore, as might occur in a very simple system. Instead, the food would be spread among several herbivore species, each of which was fed upon by several carnivores, so that the large bulge at the plant level would be dispersed into many scarcely detectable bulges at the carnivore level.

A direct connection between diversity and stability was attractive for several reasons, one being that it provided a practical argument against the extreme simplification of ecosystems favored by recent technology. Unfortunately, the connection has proved difficult to confirm. For one thing, some of the evidence was probably misinterpreted. One example already mentioned is that it now seems clear that oscillations and fluctuations are not necessarily indications of instability.

Observation has sometimes confirmed that diverse communities are more stable than less diverse ones, but there are very simple communities that are apparently very stable, such as those of hot springs. At the same time there is some evidence that species of insects with broad diets may fluctuate more in numbers than those with narrow ones (Watt 1968).

Mathematical analyses have suggested that, for systems in general, complexity tends to generate instability (May 1974), confirming everyday experience that simple devices break down less often than complicated ones. Of course, mathematical modeling does not tell us that there cannot be a connection between diversity and stability in real ecosystems that, because of evolutionary processes, may not correspond to expectations for "systems in general."

We have already mentioned (in Chapter 10) the fact that instability, in the particular sense of recurrent disturbance, seems to generate or maintain diversity in some situations. The higher diversity of wetlands vegetation where water levels shows year-to-year fluctuations is one such instance.

It seems clear, nevertheless, that some connections between diversity and stability exist. Extremely simple systems, simplified beyond any natural ecosystem, are unstable. Laboratory cultures of predator and prey usually end with the extinction of one or the other until spatial heterogeneity is added (Huffaker 1958). There is a great deal of observational evidence and several experiments indicating that crop monocultures are more prone to outbreaks of pests than natural communities. The reasons for the outbreaks seem to be related to (1) the high density of the host plant, allowing rapid spread and exponential growth of the specialized pest species; and (2) the absence from the system of such complications as habitat patchiness and alternative hosts, which would maintain satisfactory populations of competitors and predators.

Outbreaks and extinction are the two types of instability that seem to have ecological significance as well as practical importance to humans. Some evidence seems to point toward the possibility that these may be discouraged—and stability thereby favored—by spatial diversity. In promoting stability, complexity within the community may be no more important than complexity in the landscape.

Community and Ecosystem Ecology: *Paleoecology*

OUTLINE

- Uniformitarianism
- The Paleocommunity
- Autecology
- Population Ecology
- Community Ecology
- Applying Paleoecological Knowledge

Dawn redwood, *Metasequoia,* perhaps the most characteristic tree of the Arcto-Tertiary forests. (*Photo by the author.*)

Unlike the molecular biologist, the chemist, or the physicist, the ecologist often needs to know more than the current state of a system. The past is important. This may be the recent past—a game biologist may need to know what happened to a population last winter to understand what is likely to happen to it this summer. Or it may go back a little farther—plant ecologists studying the prairie may reach some wrong conclusions unless they know whether the past few years have been wet or dry. Understanding why communities are organized or distributed as they are may require a knowledge of events that occurred thousands or millions of years ago. Paleoecology is the study of the environmental relationships of organisms in the geological past. It is obviously a hybrid subject, partly biology and partly geology.

Uniformitarianism

The key assumption of paleoecology, as of geology in general, is **uniformitarianism.** This principle, which was stated by the British geologists E. Hutton and C. Lyell, is given in the catchphrase, "The present is the key to the past." It holds that processes have been uniform through time, so that we can understand the past geology of the earth—its mountains and rock strata, for example—by studying current processes such as erosion and sedimentation.

In paleoecology, application of the principle to three specific topics is important:

1. The ecological rules or principles observable today, it is assumed, held true in the past. Ancient organisms, just like modern ones, had tolerance ranges, they competed, and they became involved in food chains.
2. The environmental relations of fossil organisms resembled the environmental relations of their closest relatives today. For example, ancient clams lived in water, and a genus of clam that today lives only in the ocean probably also lived in salt water in the past.
3. Morphology—which is basically all that is available to study for fossil organisms—was adapted to the environment in which the organism lived. Plants with morphological features for restricting water loss, for example, probably lived in dry habitats.

Assumption 1, we will hope, is always true. It is, in fact, the assumption on which all of science rests—that the universe is orderly and we can, at least in principle, understand its rules. Assumptions 2 and 3 may fail or mislead us in special cases. One species of a marine genus could invade fresh water; seemingly xerophytic features might be those of a bog plant.

The Paleocommunity

Using the fossil remains of organisms that we find associated, the paleocommunity, we attempt to determine the number and kinds of organisms that formed the actual biotic community that existed at some time in the past. If we include a consideration of the physical environment, we could substitute the terms "ecosystem" and "paleoecosystem" for "community" and "paleocommunity." Knowing the numbers of different kinds of organisms, we can then infer other characteristics of the community—food chains, biomass pyramids, etc.

Unfortunately, between the death of organisms and their eventual recovery as fossils, many events occur that may hamper a precise and accurate reconstruction of the ancient community from which they came (Figure 14-1).

The main problems in reconstructing and studying paleocommunities are these:

1. Many organisms have soft bodies that rarely if ever fossilize, so organisms without bones, shells, wood, etc., are underrepresented (or unrepresented) in the paleocommunity.
2. Each paleocommunity is a sample from the real community from which it was derived. Many things may make the sample unrepresentative. If the sample was preserved by burial from a volcanic eruption, for example, some kinds of organisms—birds, perhaps—might have escaped. Insect-pollinated plants are poorly represented in fossil pollen samples.
3. Many habitats have no situations in which fossilizing conditions occur. The remains of organisms from these communities, if they are

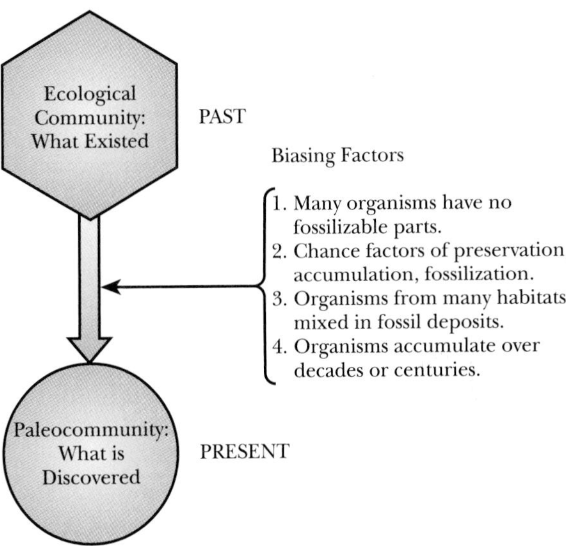

Figure 14-1

Relationships of the paleocommunity to the ecological community from which it was derived.

found at all, are mixed in fossil deposits with organisms from other habitats. Most fossil beds are derived from deposits in the shallow water of lakes or seas. Into these sites are washed plants and animals from the upland habitats of the region. An early step has to be sorting the organisms into their presumed habitats.

4. Precision in separation of events by time decreases as we deal with older material, to the point where a paleocommunity of some pre-Pleistocene period may represent thousands of years of accumulation (Knoll 1986). The actual community present at any given time might have been different in many details from our reconstruction. We could, for example, overestimate diversity if species came and went with range contractions and expansions.
5. Evidences of life history, physiology, and coactions must nearly always be incomplete. Only occasionally are paleoecologists lucky enough to find an insect preserved with an intact load of pollen, for example. Generally they must rely on inference, reasoning that animals with plant-eating teeth were probably eaten by the contemporaneous animals of the right size with flesh-tearing teeth.

Some frequently studied types of fossils used for paleoecological reconstructions include

1. Fossil pollen. The pollen of many kinds of plants has a decay-resistant layer, the exine, and so accumulates in bog and lake sediments. By identifying and counting pollen grains, palynologists can reconstruct vegetation of earlier periods and make inferences concerning the climate of those times. Most of the reconstructions of vegetation during glacial and postglacial times in the "Community Ecology" section beyond are based on the analysis of fossil pollen. The cold, acid water of bogs also helps to preserve other fossils, such as stems and nuts of plants and skeletons of animals.
2. Plant macrofossils. One problem with pollen is that it is rarely identifiable to species. A pollen grain can be identified to genus, as pine, spruce, or birch or, for some plants, only to family, as sedge or composite. Often macrofossils, such as cones, seeds, or needles, can be identified to species, allowing us to tell whether the birch pollen is dwarf birch that invades tundra just ahead of spruce, or paper birch, a tree that's a member of the northern forest. Consequently, paleobotanists increasingly use techniques to recover macrofossils as well as pollen from the deposits they study.
3. Bones. Bones can be dug up out of bogs where the animals became mired down and died. Animals perish in the same way in tar pits at Rancho La Brea in Los Angeles County, California. The tar pits are places where asphalt deposits come to the surface. Animals get stuck, like Br'er Rabbit, when they come to drink at water holes or to prey on some other animal that is already stuck. They die, and their bones are eventually found, beautifully preserved in the asphalt. Caves are another rich source of bones accumulated by vultures, owls, hyenas, or primitive humans.
4. Petrified stems, shells, bones, etc. The three preceding categories are particularly useful in studying Pleistocene paleoecology. In older strata, the hard parts of organisms tend to be encased in rock, and the tissues may be partially or wholly converted to mineral material.

Autecology

Paleoecology can be practiced at the level of the individual, the population, or the community or ecosystem. Most climatic reconstructions are based on autecology, with past climates inferred from the current climatic requirements of various fossil species. To take a simple example, the pond slider, a turtle, was found in the kitchen midden (garbage dump) of an Indian village that flourished on the shore of Saginaw Bay at the base of the thumb in Michigan 1500 to 2000 years ago (Cleland 1966). Today this turtle occurs no farther north than northwestern Indiana. The inference is of a warmer climate and communities containing greater numbers of southern species at that time.

The sauropod dinosaur *Diplodocus* from the Jurassic of Colorado and Wyoming was an enormous, long-necked, long-tailed dinosaur. It stood 4 m tall and was nearly 30 m long. Using details of its morphology, we can draw conclusions about the habitat in which it occurred and its role in its community. Because its teeth were pencil-like and not deeply rooted, it was probably not a carnivore, but just what kinds of plants or detritus it ate is unclear.

Diplodocus was probably not terrestrial but, instead, aquatic or amphibious. This is indicated by several features of its skeleton. For example, its nostrils were merged into a single opening on an upward-projecting process on top of its head. Its eyes were also set high on the head. The usual reconstruction is of an animal submerged up to its nose, like a hippopotamus, and this may well be true. Another possibility is that the position of the nostrils on top of the head allowed it to feed on shallow aquatic vegetation without interruptions for breathing, even if the whole body was not submerged. A 10-ton (9000 kg) animal living on water weeds probably needed to devote as much time to foraging as it possibly could.

Population Ecology

The Swedish ecologist Bjorn Kurtén (1958, 1976) investigated the population ecology of the cave bear, *Ursus spelaeus,* which lived during the later part of the Pleistocene and became extinct early in the postglacial period. Bones and teeth of this species are among the most common fossils in European caves. The size and shape of its teeth compared with the teeth of other bears suggest that it was vegetarian. It was a large animal, about the size of a grizzly, its closest North American relative. Males probably weighed 500 to 1000 pounds (225–450 kg) and females, about one half that. The bears hibernated in caves and perhaps also slept in them when not hibernating. Cubs were born during hibernation.

The fossil teeth and jaws can be sorted into groups corresponding to ages (a high percentage of the mortality evidently occurred during hibernation, so that jaws with an amount of wear intermediate between, say, three and four years, are uncommon). The sexes can be separated by size, among other traits. Making the assumption that the teeth give the ages of animals dying natural deaths, Kurtén constructed a Pleistocene cave bear life table (Table 14-1). The life table is an age-specific one, but composite.

In the life table, we see a high rate of juvenile mortality. Many of the cubs born in the cave died without ever leaving it. Less than one half the cohort remained to re-enter the cave the next winter. Mortality dropped in the second year, though it was still high, and continued to drop through age seven years. The period from about four to ten years of age had a fairly steady, low mortality rate of 13 to 16%. Then mortality climbed again; only 1% of the original cohort was left by 16 years of age.

If we assume that the cave bear population was approximately stable in size, we can calculate an age structure (L_x) column that will tell us its age distribution and compare this with the age distribution calculated by John Craighead and his coworkers (1974) for grizzly bears in Yellowstone National Park, Montana. The comparison is given in Table 14-2. The biggest difference seems to be the much larger percentage of adult grizzlies. The grizzlies seem to have a greater potential longevity; one of the individuals that Craighead studied lived more than 25 years.

Were there enormous Pleistocene predators to pick off the cave bears, preventing them from reaching a ripe old age? Probably not; there were lions closely related to the African lion of today, but they could have found easier prey than a 500-pound cave bear. The most likely explanation is less exciting. The cave bears seem to have died of bad teeth. Vegetable material, especially foliage, is hard on teeth, and though the cave bears' teeth were better adapted for a vegetarian diet than the teeth of other bears, they were not the grass grinders of cows or horses. Older bears show a high percentage of badly worn teeth, cavities, and abscesses. These conditions, in themselves, might not cause death very often, but they would have interfered with feeding and caused the bears to enter hibernation in condition too poor to last the winter.

There are very large numbers of cave bear fossils; remains representing about 50,000 individuals were recovered from Mixnitz Cave in Austria. Does this mean bears were around in droves? The population size in the vicinity can be calculated if we know the mortality rate and how long the cave was occupied. Overall mortality rate in Table 14-1 is approximately 20% per year. Mixnitz Cave was occupied during what we in the United States call Wis-

Table 14-1
Life Table for the Pleistocene Cave Bear

The basic data were mortality rates, calculated as $q_x = a/(a + b)$, where a = teeth of age x and b = total of older teeth. $a + b$ then, represents the total number of individuals—teeth, actually—entering age x and a represents those that died during the interval.

Age (Years), x	*Survivorship,* l_x	*Age Structure,* L_x *(%)*	*Deaths,* d_x	*Mortality Rate (Year),* q_x	*Life Expectation,* e_x
0–.5	1000	23.94	191	0.500	3.48
0.5–1	809		309		3.75
1	500	14.14	113	0.226	4.75
2	387	11.14	76	0.196	5.00
3	311	8.97	59	0.190	5.09
4	252	7.40	39	0.155	5.17
5	213	6.32	30	0.141	5.03
6	183	5.74	25	0.137	4.77
7	158	4.69	21	0.133	4.44
8	137	4.08	18	0.131	4.05
9	119	3.51	18	0.141	3.58
10	101	2.97	16	0.158	3.13
11	85	2.42	18	0.212	2.63
12	67	1.85	18	0.269	2.20
13	49	1.28	17	0.347	1.83
14	32	0.80	13	0.406	1.53
15	19	0.45	9	0.474	1.24
16	10	0.22	6	0.600	0.90
17	4	0.06	4	1.000	0.50

Recalculated (following Brewer and McCann 1982) from data in Kurtén, B., *The Cave Bear Story,* New York, Columbia University Press, 1976.

Table 14-2
Age Structure in the Pleistocene Cave Bear Compared with the Grizzly Bear

Age	*Cave Bear (%)*	*Grizzly Bear (%)*
Cubs	23.9	18.6
Yearlings	14.1	13.0
Two-year-olds	11.1	10.2
Three- and four-year-olds	16.4	14.7
Adults	34.5	43.7

Cave bear figures from the L_x (%) column of Table 14-1. Grizzly data from 1959–1967, Table 1, in Craighead et al. (1974). Based on a similar table in Kurtén (1976).

consin glaciation and also during the preceding interglacial, or for about 100,000 years in all. So, 50,000 deaths in 100,000 years implies 0.5 deaths per year, or one every other year.

From a population of what size will an annual mortality rate of 20% give 0.5 dead animal per year? We can set that up as

$$0.2 \times \text{population size} = 0.5$$
$$\text{Population size} = 0.5/0.2 = 2.5 \text{ bears}$$

So the cave bear was a relatively rare animal. The vicinity of each cave probably supported only one family at a time.

It is natural to wonder why the extinct cave bear became extinct. The climate was changing as the glaciers receded but since the bear had already weathered one interglacial period, climatic change seems unpromising as a complete explanation. Could humans have been involved? Cave bears and Neanderthals coexisted for quite a while, but this was the time and place of the arrival of Cro-Magnons from the south. Perhaps Cro-Magnons hunted the cave bears to extinction.

That is a possibility, but there is little, if any, evidence showing that the cave bear was hunted. A better possibility is competition with humans for caves. Caves are a limited resource. The cave bear evidently monopolized the caves against the brown bear where these two species occurred together on the continent of Europe. In the British Isles, where the cave bear did not occur, the caves are full of brown bear remains. But humans are tenacious; they might well have outcompeted the cave bear for caves. If human numbers were increasing, the bears may have been forced to spend their winters in fewer and fewer and less and less satisfactory sites. They may have been pushed to extinction in a process not much different from the impersonal extermination humans are visiting upon the animals of the tropical rain forests today.

But that is just a possibility. The best we can say, for now at least, is that the reason for the extinction of the cave bear is uncertain. But we can and should admire the human ingenuity involved in giving us figures on mortality and population size for an organism extinct for 10,000 years.

Community Ecology

Because of the difficulties listed earlier in reconstructing the original ecological community, community-level studies dealing with pre-Pleistocene communities are scarce. Those that have been attempted are valuable, particularly when they confirm—or fail to confirm—ecological concepts developed on the basis of existing communities. Here we will focus our discussion on the history of the North American biota since about 65 million years ago.

There are three main episodes in this history. To begin with, climates were milder than today and tropical forests grew in much of what is now the United States. Climates then began to get cooler and dryer and the vegetation changed, so that much of the United States was occupied by temperate forest and grassland. Finally the Ice Age arrived, with glaciers advancing and receding, and the vegetation changed greatly with each advance and recession.

The Early Tertiary

Early in the Tertiary (Table 14-3) the climate in North America was probably warmer, and there was probably less difference between summer and winter. Ralph W. Chaney (1947) and Daniel I. Axelrod (1958) have described three great geofloras that occurred in North America at this time, the Neotropical-Tertiary, the Arcto-Tertiary, and the Madro-Tertiary Geofloras. The southern part of the continent, extending north to what are now the states of Washington and Colorado, was occupied by the Neotropical-Tertiary Geoflora (Figure 14-2). The fossil leaves and fruits found in rocks laid down at this time show that the plants present were similar to those now found in tropical and subtropical forest and savanna.

The northern part of the continent was occupied by communities of the Arcto-Tertiary Geoflora. At this time there was still a land connection between North America and western Europe across the North Atlantic, and this geoflora ranged all across North America and Eurasia at higher latitudes. In its southern portions this was temperate deciduous forest, evidently similar to the mixed mesophytic forests of the present day. Many of the same kinds of trees were present, probably not the same species but close relatives. There were maples, beech, chestnut, ash, walnut, oaks, basswood, elms, and many others. To the north, perhaps in what is now northern Canada, conifers such as pine and spruce increased.

One of the most characteristic members of the Arcto-Tertiary forests was the dawn redwood, or *Metasequoia.* This plant, related to the present-day sequoia and redwood of the West Coast, was known only as a fossil for many years and then was found in China in 1946 (Chaney 1948). It was growing in valleys with chestnut, oak, sweet gum, cherry, beech, and spice bush, or, in other words, in an association much the same as found in fossil beds 65 million years old.

Table 14-3
Geological Timetable

Era	*Period*	*Epoch*	*Years Ago*
	Quaternary	Holocene	10,000
		Pleistocene	2,000,000
		Pliocene	7,000,000
Cenozoic		Miocene	26,000,000
	Tertiary	Oligocene	38,000,000
		Eocene	54,000,000
		Paleocene	65,000,000
	Cretaceous		136,000,000
Mesozoic	Jurassic		190,000,000
	Triassic		225,000,000
	Permian		280,000,000
	Carboniferous (Pennsylvanian and Mississippian)		345,000,000
Paleozoic	Devonian		395,000,000
	Silurian		430,000,000
	Ordovician		500,000,000
	Cambrian		570,000,000
Precambrian			5,000,000,000

The third geoflora was a newly developing one, the Madro-Tertiary Geoflora, in the drier, cooler highlands of the southwest United States and Mexico. The plants included such types as live oaks, pines, junipers, yucca, and mesquite.

The Later Tertiary

That was the situation in the Paleocene and Eocene. By the end of the Oligocene, 26 million years ago, a climatic deterioration had begun, with climates becoming cooler and drier. This produced two general trends. First, there was a southward shifting of communities, so that by the end of the Miocene the Neotropical-Tertiary Geoflora had retreated south of the United States and the Arcto-Tertiary Geoflora had spread southward and occupied much of North America. We need to guard against an anthropomorphic view of these geofloras packing their bags and heading south. The actual movements occurred because conditions at the north edge of the range of the tropical species became unfavorable so that these species could no longer survive or reproduce. As these died out, more northerly species found conditions favorable and were able to grow, thereby extending their range southward.

The second effect resulting from the cooling and drying climatic trend was the spreading out of the Madro-Tertiary Geoflora from its place of origin, and the development of various vegetation types such as chaparral, woodland, and desert, adapted to dry to very dry conditions.

Sometime in the Miocene temperate deciduous forest derived from the Arcto-Tertiary Geoflora was extremely widespread, occupying large areas of North America and Eurasia (Mai 1991). This is a major source of the similarities between the forests of western Europe, eastern Asia, and eastern North America; they were once part of a continuous deciduous forest vegetation type. Subsequent events in the last 20 million years have led to the fragmentation of the deciduous forest, complete extinction of some lines, and extinction of other lines in some regions but not in others. For example, horsechest-

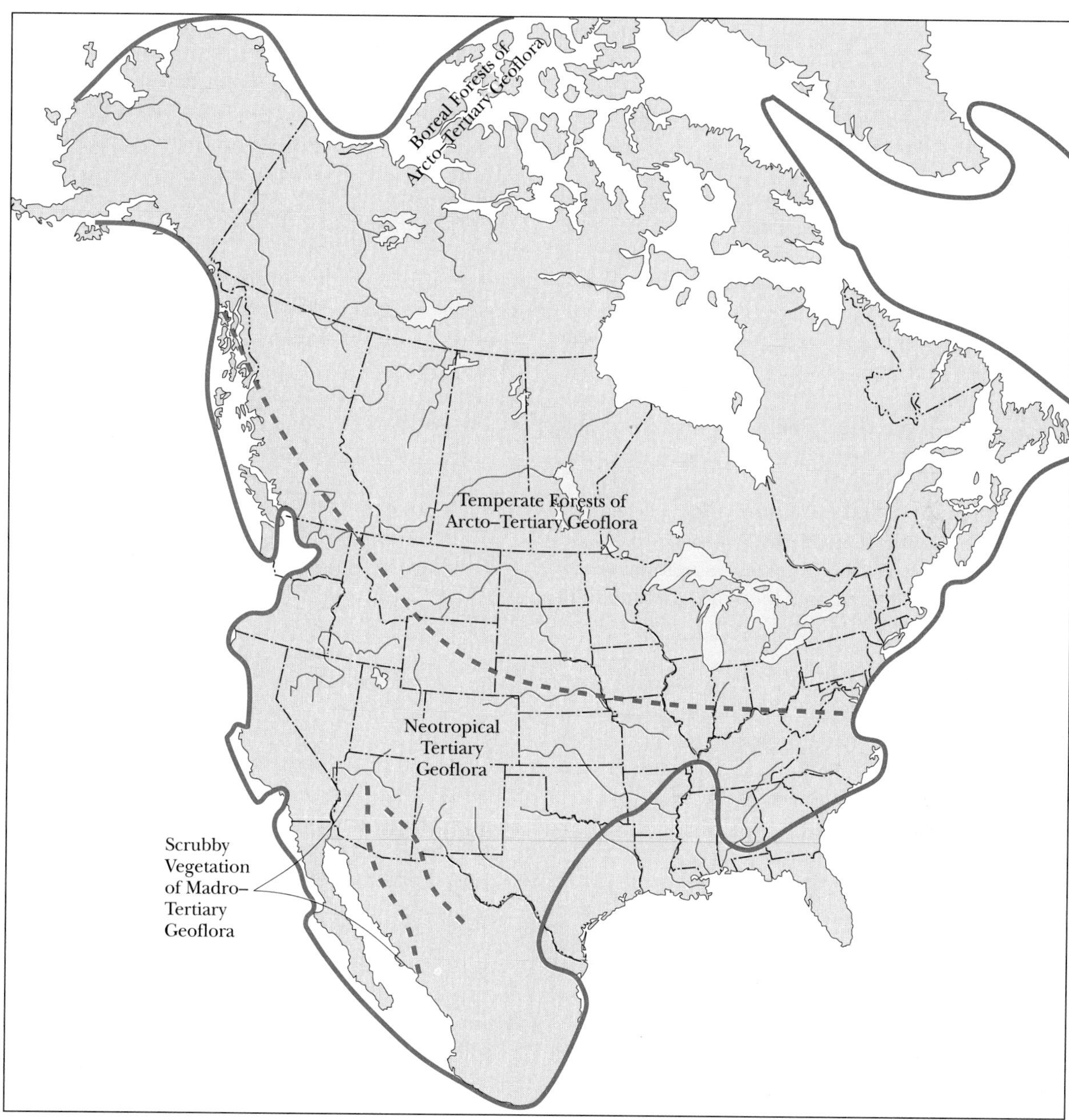

Figure 14-2

Diagrammatic representation of the vegetation of North America in the Paleocene, approximately 65 million years ago. At this time there was still a North Atlantic connection with Eurasia through Greenland but the Bering Straits connection had not yet been established. The Central American isthmus connecting North and South America was not yet in existence but the southeast coast of North America and the northeast coast of South America were fairly close together. The coastal plain of the southeast United States, north to about Cairo, Illinois, was still under water and, not long before, the area of the Great Plains had been submerged under a broad, shallow sea.

nut, sweet gum, sour gum, magnolia, hickory, walnut, and hackberry grew in western Europe in the Miocene. All these genera still occur in North America, but in Europe the first three went extinct in the Pliocene and the rest in the Pleistocene (Critchfield 1980).

Continued cooling and drying led to the development of forests dominated by evergreens in the northern part of the continent and the further spread of arid-land communities, including grassland, into the interior of North America. The trend of greater dryness in western North America was produced by the uplift of mountains in the West; these mountains produced a rain shadow in the interior. Areas of the Great Plains changed from forest to savanna to grassland over millions of years, and the animals present also changed, by emigration, immigration, and evolution.

The evolution of horses is a favorite example for the evolution section of beginning biology textbooks. Here is the place and the time in which that evolution occurred. In the Eocene, horses were four-toed little animals about the size of fox terriers that browsed on tree leaves. There were several lines of evolution, one of which, by the early Pleistocene, led to the big creatures we ride or bet on, having a single toe with a hoof on it and teeth suitable for cropping and chewing up grass.

The Pleistocene

By the beginning of the Pleistocene, the distribution of vegetation may have been similar to that found now. There were certainly differences—probably no tundra existed (Frenzel 1968)—but at first glance a biome map of North America in a late Pliocene ecology text might have looked much like the one in Chapter 14. Many of the same, or very similar, species were involved, but the Pleistocene communities were probably richer. This was definitely true among the larger animals.

The vertebrates of the La Brea, California, tar pits studied by Hildegard Howard (1955), Chester Stock (1961), and their coworkers provide a good example. Some of the large animals, now extinct, that lived in the Los Angeles valley in the late Pleistocene included herbivores, such as 4-foot-tall ground sloths, horses, camels, mammoths, mastodons, antelopes, and peccaries, and predators and scavengers such as the saber-toothed cat, dire wolf, short-faced bear (larger than the grizzly bear), and vultures with a wing span of 14 feet. Large animals of these species or others also occurred elsewhere in North America throughout most of the Pleistocene but then disappeared, rather abruptly, 10,000 to 11,000 years ago.

The climatic deterioration that began in the Oligocene reached a climax with the continental glaciation of the Pleistocene. Four or more times glaciers came down from the north as ice sheets hundreds to thousands of feet thick. Communities occupying the land over which they advanced were obliterated. Plants and animals were shifted south as the result of the cooler climates. We know relatively little about the early glaciations. About the last period of glaciation, the Wisconsin, we know much more.

Wisconsin glaciation began about 70,000 years ago and was at its peak about 18,000 years ago (Figure 14-3). After a period of relative stability in area covered, retreat began about 16,000 B.P. (before the present) (Schoonmaker and Foster 1991). The front of the ice was sitting in southern Michigan possibly 15,000 years ago. By 9000 or 10,000 years ago the glacier was north of the Great Lakes basin, and by 4000 or 5000 years ago it was gone from the North American continent. Some geologists believe that we are now in an interglacial period, with the next glaciation due somewhere between a few hundred to a few thousand years hence.

The great student of North American deciduous forest, Lucy Braun (1950), believed that plants south of the ice sheet remained in the associations in which we now see them, with only some slight southward shifting. Many more fossil pollen sites have been studied since Braun wrote, and they suggest otherwise. Specifically, it seems likely that

1. Shifts in the ranges of many plant and animals were extensive. Details are given a little later.
2. The changes in vegetation associated with glaciation were not a simple southward displacement of biomes; instead, species reassorted into new communities. The great similarity between the mixed mesophytic forest of today and the temperate forests in the Tertiary has long been thought to indicate great coherence in time. Although the organisms form-

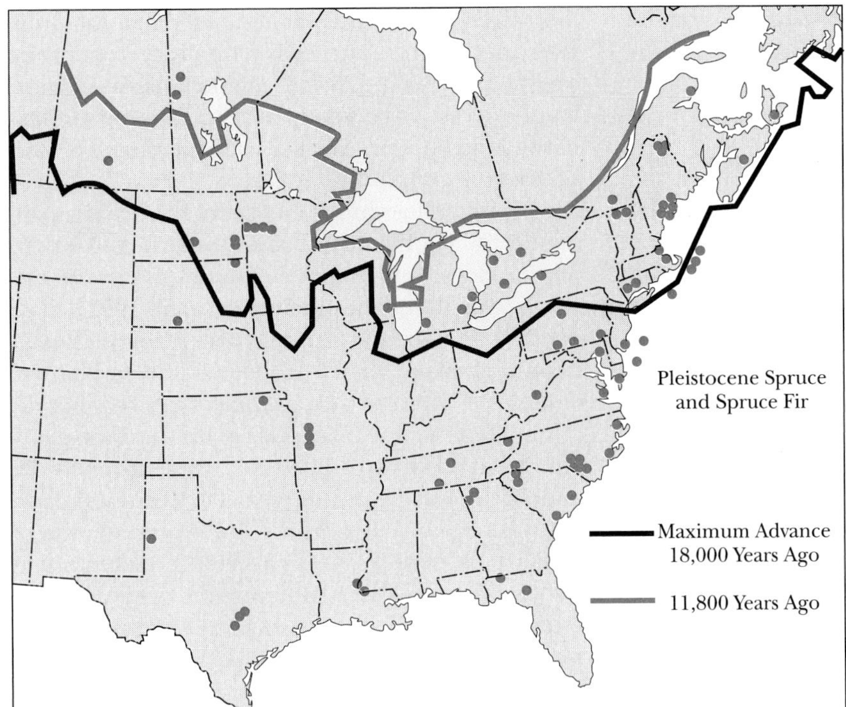

Figure 14-3

Shown here is the maximum advance of Wisconsin glacial ice about 18,000 years B.P. (before the present) and the approximate extent of the ice sheet 11,800 years B.P., or about the time of the Valders readvance. Dots are sites of spruce occurrence. (Adapted from Gordon 1985.)

ing the rich mesophytic forests may have continued in association in some localities through the Wisconsin, the fossil pollen record suggests that large areas of the landscape were at times occupied by combinations that would be considered unusual today. In some cases, a similar assemblage still occurs but is now strongly restricted geographically or by habitat. Other cases involve the co-occurrence of species with widely separated ranges today.

Although combinations of species that are peculiar by contemporary standards could result from a variety of factors including soil, migration patterns, and coactions, climatic differences are usually considered a major cause. Many of the cases of "disharmonious" faunas are explained by changes in **climatic equability,** that is, in how much the seasons differ (Hibbard 1960). An example is the Ladds quarry fauna of northwestern Georgia radiocarbon dated as 10,000 to 11,000 years old (Holman 1985). The known fauna includes about 90 vertebrate species, of which 20 do not occur in the region today. Of these, ten are northern and ten are southern. The northern species include the spruce grouse, today occurring no closer than Ontario; southern species include the round-tailed muskrat that does not now occur north of central Florida as well as extinct species of tropical affinities.

The "equability" argument goes like this: If certain northern species have their southern range limits set by hot summers and if certain southern species have their northern range limits set by cold winters, then a more equable climate, with milder winters and cooler summers, will allow these species to expand and, possibly, overlap. This situation, it is thought, prevailed at the time of the Ladds quarry fauna. With a subsequent decrease in equability—colder winters and hotter summers—the ranges have separated.

From the limited perspective of human history, glaciation was an unusual event, a catastrophe of fabulous proportions. But, in fact, glaciation has been the usual condition of the Pleistocene; interglacial periods have made up less than 20% of the Pleistocene's million or so years (Davis 1976). The biomes of today are not the norm from which the glacial communities were derived as temporary aggregations; the reverse is at least partly true.

Full-Glacial Plant and Animal Distribution

At full glacial, say 20,000 years B.P., with the ice sheet sitting on Long Island, New York, near Kent, Ohio, and just south of Charleston, Illinois, what was the distribution of plants and animals? A zone of tundra or very open spruce woodland evidently occurred south of the glacier, in some localities over a broad zone. Spruce woodland and forest grew south of the tundra. Spruce occupied the Great Plains at least as far west as Kansas (Watts 1983).

Both tundra and spruce forest extended south in the Appalachians; spruce grew south at least to central Georgia and northern Florida (Figure 14-3). Beyond the spruce forest grew pine, in the Southeast evidently jack or red pine or both. For example, the prevailing vegetation around Gainesville in central Florida from 24,000 to 18,500 years B.P. consisted of open pine forest (Watts and Stuiver 1980). High values of herb pollen suggest sandhill vegetation and prairie of some type in the vicinity.

All this implies a great southward shifting of ranges of northern species. The current southern boundaries of jack and red pine are in the middle of Michigan, or, in other words, at least 1200 km north of where they were thriving around Gainesville 20,000 years ago.

Animal distribution in full-glacial times is less well known, but there is evidence of considerable southern displacement. Bones of muskox have been found in Nebraska, caribou in Tennessee, and the collared lemming in southern Pennsylvania (Lundelius et al. 1983). The spruce grouse in northwestern Georgia was already mentioned. Clearly, tundra and boreal forest animals were displaced far to the south of where they now occur. On the other hand, one of the most noteworthy features of the full-glacial fauna of the Southeast was the large proportion of tropical species, some now occurring in Central America but others extinct. Included were giant land tortoises, armadillo, capybaras, and large cats (Lundelius et al. 1983).

Given the great southern shift of plants and animals, where was temperate deciduous forest? The most likely explanation seems to be that most of the tree species that currently compose it were far south, occupying favored microclimates—for example, moist valleys—in the generally pine-dominated landscape in Florida, elsewhere in the Gulf states, and perhaps in Mexico.

A pollen diagram from Camel Lake in the panhandle of Florida had a lower section (40,000 to 30,000 B.P.) with many deciduous forest species along with abundant pine (Watts et al. 1992). Oak, beech, elm, ash, and chestnut were present. Pollen deposited from 30,000 to 14,000 B.P., which includes the Wisconsin glacial maximum, was heavily dominated by pine, with only scant hardwood pollen. From 14,000 to 12,000 B.P., pine dropped off, and hardwood pollen, especially beech and hickory, rose again. Whether the low representation of hardwoods from 30,000 to 14,000 B.P. means that the major populations of temperate deciduous forest trees had moved somewhere else or just that the populations became very small and restricted in distribution is not certain.

A surprising number of plants of the forests of eastern North America also occur in the humid forests of Mexican mountains. Examples are beech (and its root parasite, beechdrops), sweet gum, and white pine (Graham 1973). Forest animals showing the same disjunct (that is, separated) distributional pattern are fewer but do occur. The red-bellied snake and the eastern flying squirrel are two (Martin and Harrell 1957). Fossil eastern chipmunks from central Texas (Edwards Plateau) cave deposits (Graham 1984) suggest temperate deciduous forest in that region between 9000 and 10,000 B.P. This, of course, is well past full glacial, but it implies that chipmunks were in the vicinity—or, more likely, farther south—at full glacial.

Another possible location for temperate deciduous forest species 20,000 years ago was on exposed portions of the continental shelf. A large fraction of the earth's water was tied up in the glaciers, with the result that large areas that are now submerged were exposed as a broad coastal plain. The drop in sea level was more than 100 m, so in some places the added strip of land was many miles wide. The land added to the west coast of Florida averaged more than 25 miles (40 km) wide and the strip below Louisiana and east Texas nearly as wide. Deciduous forest species may have occurred on these exposed shelves from Florida across the Gulf and down into Mexico, but little is yet known about their biota. Further north, along the Atlantic coast from Virginia to Boston, dredging operations have brought up mastodon and mammoth remains from depths up to 90 m.

So the picture of temperate deciduous forest persisting in sheltered valleys in the shadow of the glacier seems wrong. Also probably wrong, however, is the idea that nothing temperate persisted near the glacial border. Several plant species are endemic to (that is, occur only in) the area just south of the glacial border in the Southeast or in the Driftless Area of southwestern Wisconsin (Braun 1951, Fassett 1931). Most of these are rock-ledge species such as two species of Sullivantia, one of which occurs only on wet cliffs in southern Ohio, southeastern Indiana, and northern Kentucky and the other only in the Driftless Area and south to northeastern Missouri. It is easier to believe that these species have persisted where they are since preglacial times than to believe that they marched south and countermarched north with the fluctuations of the ice sheet, without ever leaving behind any relicts in the ravines of the South.

*Late-Glacial and Postglacial Changes**

The retreat of the glaciers and the accompanying warming set off major changes in plant and animal distribution (Figure 14-4). Melting of the glaciers left large areas of bare earth (Figure 14-5). This land was presumably occupied first by pioneer vegetation, mostly species from the nearby tundra and spruce woodland, blown in or brought in by animals. If we can judge from the succession around Glacier Bay, Alaska (Lawrence 1958), nitrogen fixers such as alders and Dryas might have been early pioneer plants.

Succession probably proceeded quickly to a community similar to the tundra-like vegetation that had occurred just south of the glacial border. Thereafter, most of the marked vegetational changes that occurred were presumably the result of a combination of climatic changes and migration patterns of the various species (Figures 14-6 to 14-9). It is not always easy to separate the two effects.

Spruce woodland tended to follow tundra and then to develop into spruce forest. Around 10,500 years ago, over a broad area, spruce gave way to pine forest (Ogden 1967). Pines gave way to hardwoods 8000 or 9000 years ago at the level of southern Michigan, earlier to the southeast.

Not every part of eastern North America followed this pattern, however. In New England, for example, tundra persisted for a long time. In southern New England, it gave way to spruce woodland about 12,500 B.P., and in northern New England, it gave way to mixed forest about 10,000 B.P. (Davis 1983). In much of Indiana and Illinois the pine stage is absent. Neither jack pine nor white pine dispersed into those areas ahead of the hardwoods.

The Hypsithermal Interval: A Postglacial Warm Period Although hardwood forest has been the prevailing vegetation type over a wide belt across the eastern United States and southern Canada for about 9000 years, the vegetation has not been static. Prairie expanded during a warm, dry period, early on in Kansas, Nebraska, and western Minnesota and later further east (about 8500 B.P. in Illinois). Grassland evidently reached its greatest extent about 7000 B.P. when the Prairie Peninsula (Transeau 1935) reached fingers at least as far east as Ontario and eastern Ohio.

It is likely that some of the expansion of prairie in the eastern part of the Prairie Peninsula was not so much the result of an invasion of species from the West but a matter of the increase on favorable sites of species already in the flora. Probably the tundra-like community of full-glacial or late-glacial times itself had prairie elements, especially those characteristic of the northern prairies. A cave deposit from southern Pennsylvania dated about 11,300 B.P. contained bones of sharp-tailed grouse and thirteen-lined ground squirrels (Guilday et al. 1964), species today characteristic of grassland, not tundra.

In the drier part of the Hypsithermal, also, xeric forest probably expanded at the expense of mesic. At Volo Bog in Lake County, Illinois, near the Wisconsin border, the switch from a mesic forest cover to mostly oak occurred about 7900 B.P. (King 1981). In areas like southern Michigan where settlers in the 19th century found a mosaic of beech-maple forest, oak forest, oak savanna, and prairie, it is likely that the proportions were tipped

* Although the timing of events differed by region, late glacial is approximately the period 14,000 to 10,000 years B.P. and postglacial, the period from 10,000 years to the present.

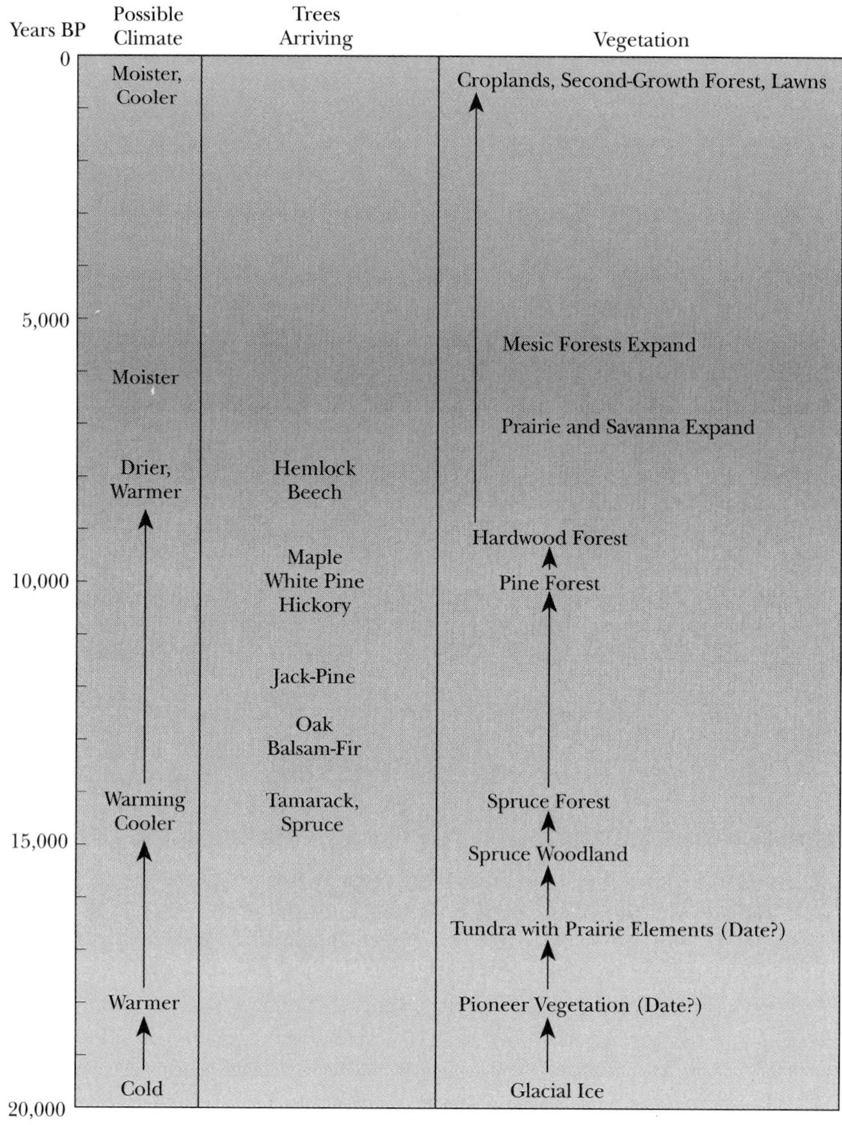

Figure 14-4

This is a generalized sequence of the vegetation from glacial times to the present for the northern Midwest. Arrival dates of trees are compiled from Davis (1983). Other features are from a variety of sources.

toward prairie and oak savanna. Past 7000 B.P., the prairie-forest border in Minnesota and Wisconsin slowly moved back westward, so this may mark the end of the driest part of the Hypsithermal.

During the Hypsithermal many southern, as well as western, species expanded their ranges beyond the limits we see today. White pine, for example, grew further north in Canada 5000 years ago than it does today (Terrasmae and Anderson 1970). The pond slider from Saginaw Bay, Michigan, is an example of an animal whose range extended further north 2000 years ago (Cleland 1966). A prairie vole from 730 B.P. at Sleeping Bear dunes in northwestern Michigan (Pruitt 1954), 180 miles (288 km) north of its range in historic times, was probably a holdover from these warmer, drier conditions.

The Last Thousand Years About 1000 years ago, a trend toward generally moister and cooler conditions began. The ranges of some southern and western species retreated, in some cases leaving relict populations in favorable microhabitats (Smith

Figure 14-5

The bare glacial drift left by a receding glacier is shown in this view of Portage glacier, Kenai peninsula, Alaska. The ice sheet is about 30 m thick at this edge. (Photo by Michael Gaule.)

1957). Some northern species reoccupied ground given up thousands of years earlier. Spruce, for example, reappears in the upper parts of the pollen profiles of some northern bogs.

The most striking feature of the top part of pollen diagrams is the arrival of European settlers. This event is clearly marked by an enormous jump in ragweed pollen, along with a corresponding drop in tree pollen, as the forests were cleared.

Pollen analysis is only one of many lines of evidence for reconstructing the climate of the past thousand years. Others include archeological remains, tree-ring measurements, the use of written accounts of exploration and farming, and, of course, for the past couple of hundred years, actual weather records. Combining these, we see a pattern for eastern North America and western Europe as follows (Bryson and Murray 1977):

A.D. 900–1130. Still relatively warm and moist. Vineyards flourished in England, the Vikings colonized Iceland and Greenland, and Indians in northwestern Iowa lived in a landscape with abundant elk and deer (which require trees) as well as bison.

1130–1380. Cooler and dry.

1380–1550. Evidently variable.

1550–1850. Cold. The "Little Ice Age." Advances of valley glaciers in the Alps.

1850–1960. Increasing temperatures. The U.S. Weather Bureau was not founded until 1872, but European weather records go back further. Mid-20th century compared to mid-19th century records indicate a two- to three-week longer growing season in England and an average temperature 1.4°C (2.5°F) higher in Denmark. The white spruce altitudinal tree line shifted upward a few tens of meters along the east coast of Hudson Bay (Payette and Filion 1985).

1960–present. At least in the eastern United States, the period from about 1955 or 1960 to about 1985 saw a return to cooler temperatures (Diaz and Quayle 1980). There were notably cold winters in the late 1970s. More recently still, in the late 1980s and early 1990s, global temperatures have risen.

What the future will bring climatically is uncertain. Possible future climates and their effects are discussed further in the last section of this chapter and in Chapter 20 under the greenhouse effect.

Pleistocene Changes Elsewhere

The foregoing discussion has focused mostly on eastern North America, but changes associated with the glacial period occurred through much or all of the world. A little farther westward in the United States, in central Texas, sphagnum bogs preserve pollen of spruce, birch, basswood, maple, and alder

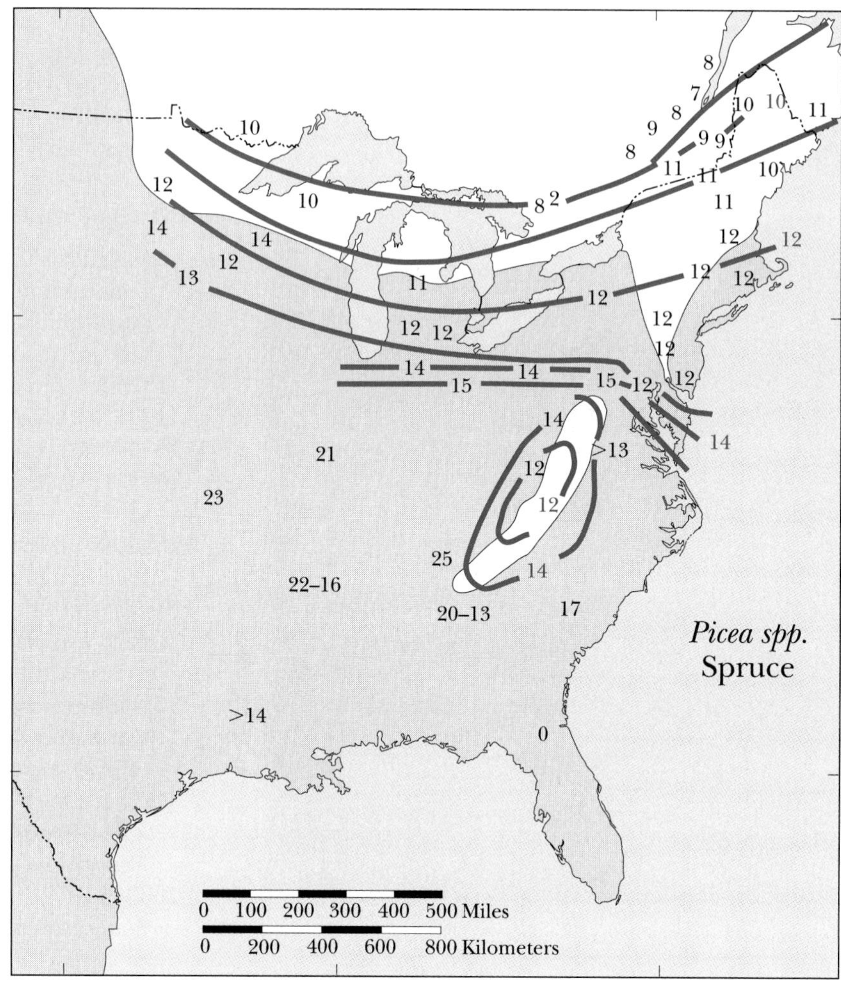

Figure 14-6

Migration patterns following deglaciation were highly individualistic. The white area in this and the next three maps represents the current geographical range. Small numbers indicate the arrival times (in thousands of years B.P.) at individual sites. The heavy lines connect points of equal age. Spruce, shown in this map, evidently occurred widely in the South and Midwest at full glacial and moved northward on a broad front following deglaciation. (Adapted from M. B. Davis 1983.)

from 15,000 B.P. (Gehlbach 1991), suggesting a landscape rather like northern Michigan. In lateglacial times (14,000 to 10,000 B.P.), spruce disappeared, and probably central Texas was largely occupied by temperate deciduous forest plants and animals.

With continued warming and drying in the period from 10,000 to 6,000 B.P., continuous deciduous forest came to be restricted to the eastern part of this region, tallgrass prairie began to occupy the adjoining lands, and westward, on the highlands of the Edwards Plateau, oak-juniper woodland began to develop. This general distribution of vegetation and animals—zones from west to east of evergreen woodland, tallgrass prairie, and temperate deciduous forest—persists today in central Texas, although with a few complexities added by a dry period beginning about 2000 B.P.

Farther west (Baker 1983, Beiswenger 1991), reconstructions are complicated by the scarcity of pollen analyses and by the mountains, where organisms with widely differing climatic requirements live within short distances of one another but at different elevations. Available evidence indicates major shifts downward, on the order of 500 to 1000 m (Figure 14-10). There were also southward shifts and a generally moister (pluvial) climate. The desert areas of southern Nevada at full glacial supported woodlands of pinyon pine and juniper, which now occur mostly above 2000 m. That the changes were not a simple shifting of zones is suggested by the fact that the desert shrub shadscale

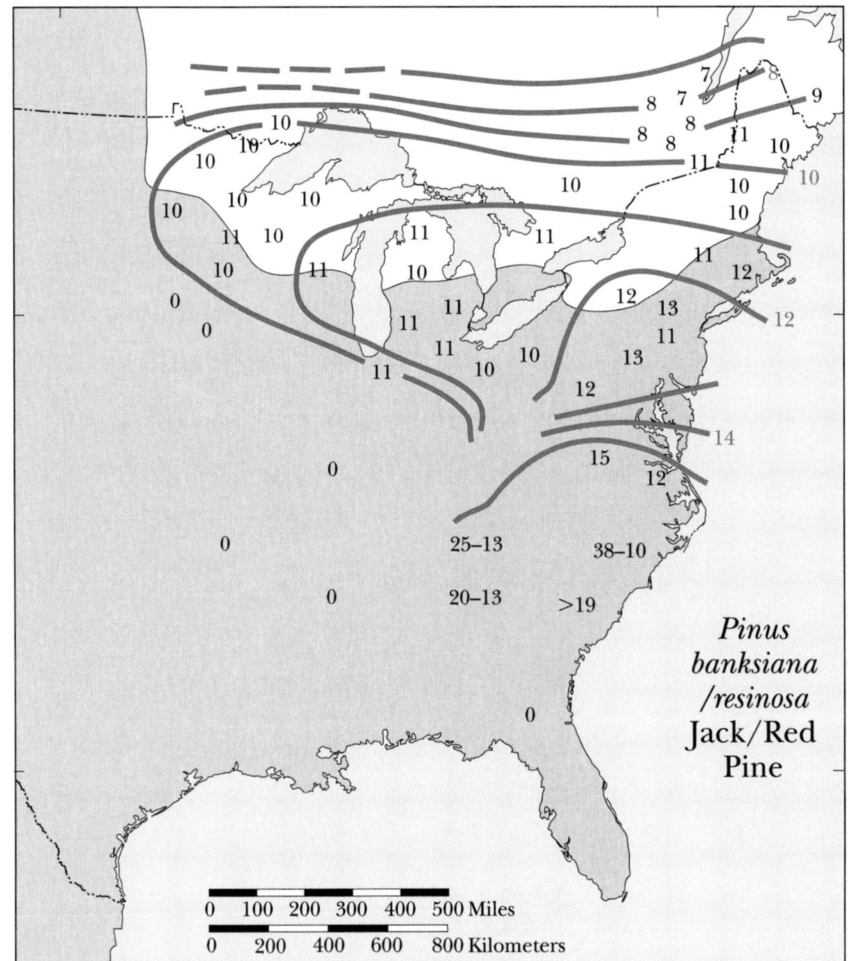

Figure 14-7

Jack and red pine apparently were restricted to the Southeast at full glacial but moved rapidly northward and, later, westward. (Adapted from Davis 1983.)

was able to extend its range *upward.* Also, ponderosa pine, which now forms a lower montane zone above pinyon-juniper, was seemingly absent from the region.

Accumulating information on South America suggests climates that were generally humid and cool during the early part of Wisconsin glaciation and cold and dry from about 21,000 to 14,000 B.P. (van der Hammen 1991). Vegetational changes have not yet been worked out in detail, but tropical rain forest must have contracted, and recent evidence suggests that the Amazonian forest was subdivided (with savanna intervening) into large northeastern and southwestern parts plus various smaller refugia.

In Europe, virtually the whole land mass north of the Alps supported only tundra at full glacial. Even around the Mediterranean, much of the landscape was not forest but cool steppe, similar to that found today in central Asia (Davis 1976).

Extinction of the Pleistocene Megafauna

In the Pleistocene, large animals (the megafauna) roamed North America. Almost all of them—mastodons, mammoths, giant beaver, woodland muskoxen, even giant tortoises and the Shasta ground sloth—disappeared rather suddenly 10,000 to 11,000 years ago. Their departure left an essentially modern fauna of deer, raccoon, teal, coot, and mallard—the animals that we see in Indian kitchen middens and that the settlers found in the 1700s. A

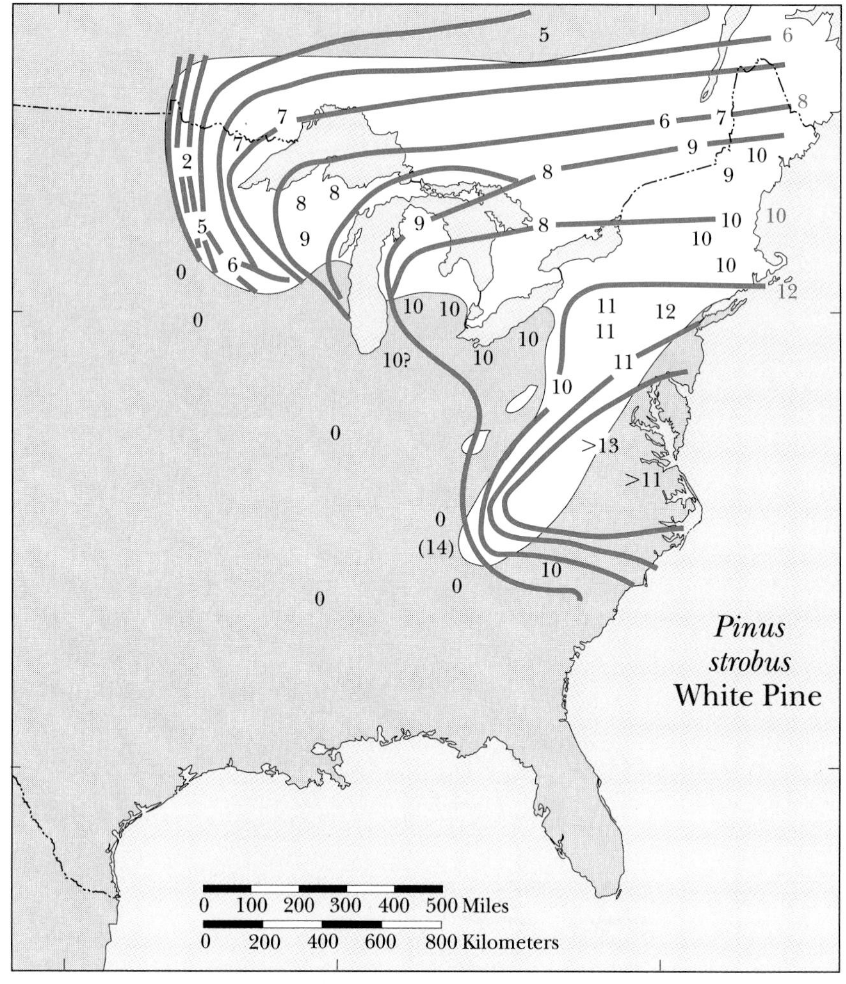

Figure 14-8

White pine showed a pattern somewhat similar to jack pine but did not reach as far south. At full glacial, it apparently occurred mainly on the now-submerged continental shelf. It tended to arrive at a given site in the Northeast or northern Midwest after jack pine but slightly ahead of most hardwoods. (Adapted from Davis 1983.)

few of the big animals survived—bison, the California condor, and the grizzly bear are examples. But they are a remnant.

What happened to the large Pleistocene animals? The University of Arizona paleoecologist P. S. Martin (1973) has pointed to two seemingly peculiar features of their extinction. (1) Unlike the usual course of extinction, it did not involve replacement by other kinds of organisms having similar niches. (2) Mainly large animals were involved.

Martin has assembled data suggesting that large-scale extinctions occurred on all the continents about the time primitive humans appeared there. His conclusion is that primitive humans, in our case Paleoindians, caused the extinctions, primarily through hunting the animals for food. Large predators might not have been hunted, but they, along with scavengers and commensals, might have become extinct when the large herbivores were exterminated.

Some have had difficulty accepting the picture of a small number of primitive humans exterminating millions of mastodons, mammoths, and other large animals. Extinction, however, is the outcome predicted by computer models based on reasonable assumptions about population sizes of hunter and hunted, migration and kill rates, and population growth rates (Mosimann and Martin 1975; also see Whittington and Dyke 1984). Humans presumably arrived in North America across

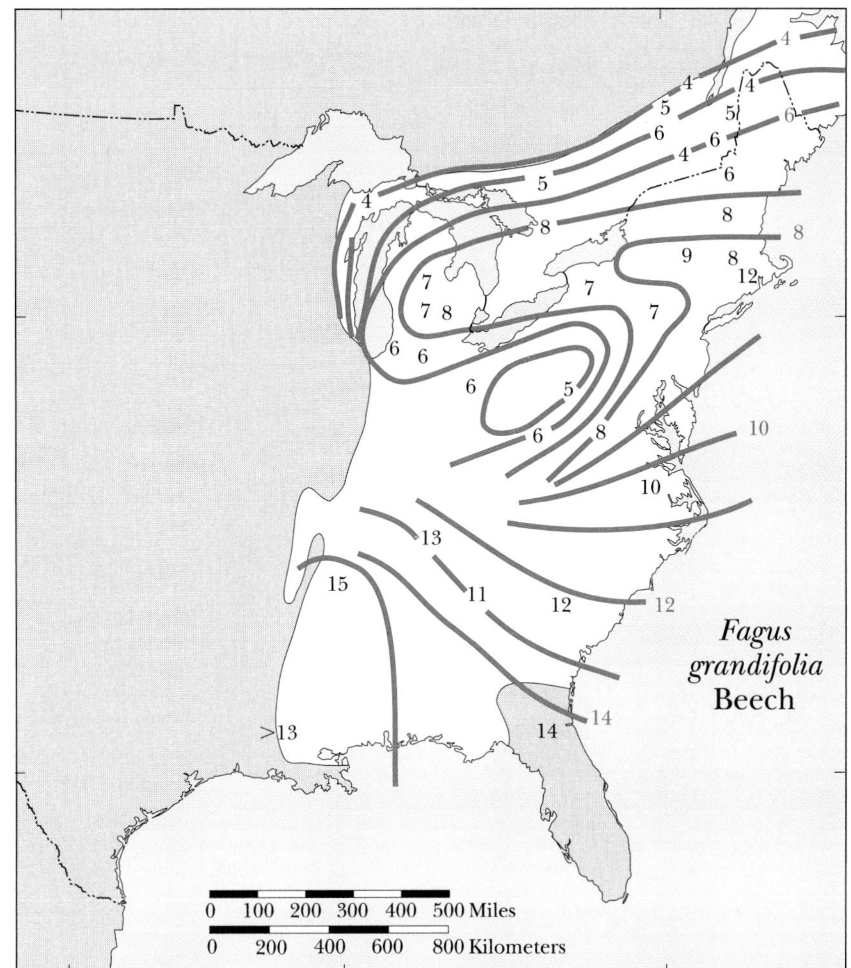

Figure 14-9

Beech apparently occurred widely, although at low densities, in the far South at full glacial. Relative to most of the hardwoods, it shows a peculiarity in its northward movement; it enters the region we now think of as the Prairie Peninsula very late (Benninghof 1964); white pine and hemlock show a similar inability to invade south of the Great Lakes. (Adapted from Davis 1983.)

the Bering land bridge. The models begin with a band of 100 men and women who followed the ice-free corridor east of the Canadian Rockies from Alaska to southern Alberta, arriving there about 11,500 years ago. According to the models these hunters and their descendants formed a front of extinction that spread southward across North America, reaching the Gulf of Mexico only 300 to 500 years later. At that time the human population along the front, running from New England to Louisiana and into Mexico, might have numbered as high as a few hundred thousand. In its spread southward it would have killed on the order of 100 million mammoths and other big game.

We have already pointed out that countless animals and plants became extinct before *Homo sapiens* appeared on earth. The overkill hypothesis has been criticized on several grounds—it does not seem to account for the large-scale extinctions of birds at about the same time, and there is some evidence suggesting that humans may have been in North America as early as 20,000 or 30,000 years ago. Certainly the period 10,000 to 11,000 years ago was one of climatic and vegetational changes that might have contributed to the extinctions. Humans may have produced extinctions by effects other than hunting, such as competition and habitat modification. But it seems clear that primitive humans, like modern humans, were a potent ecological force.

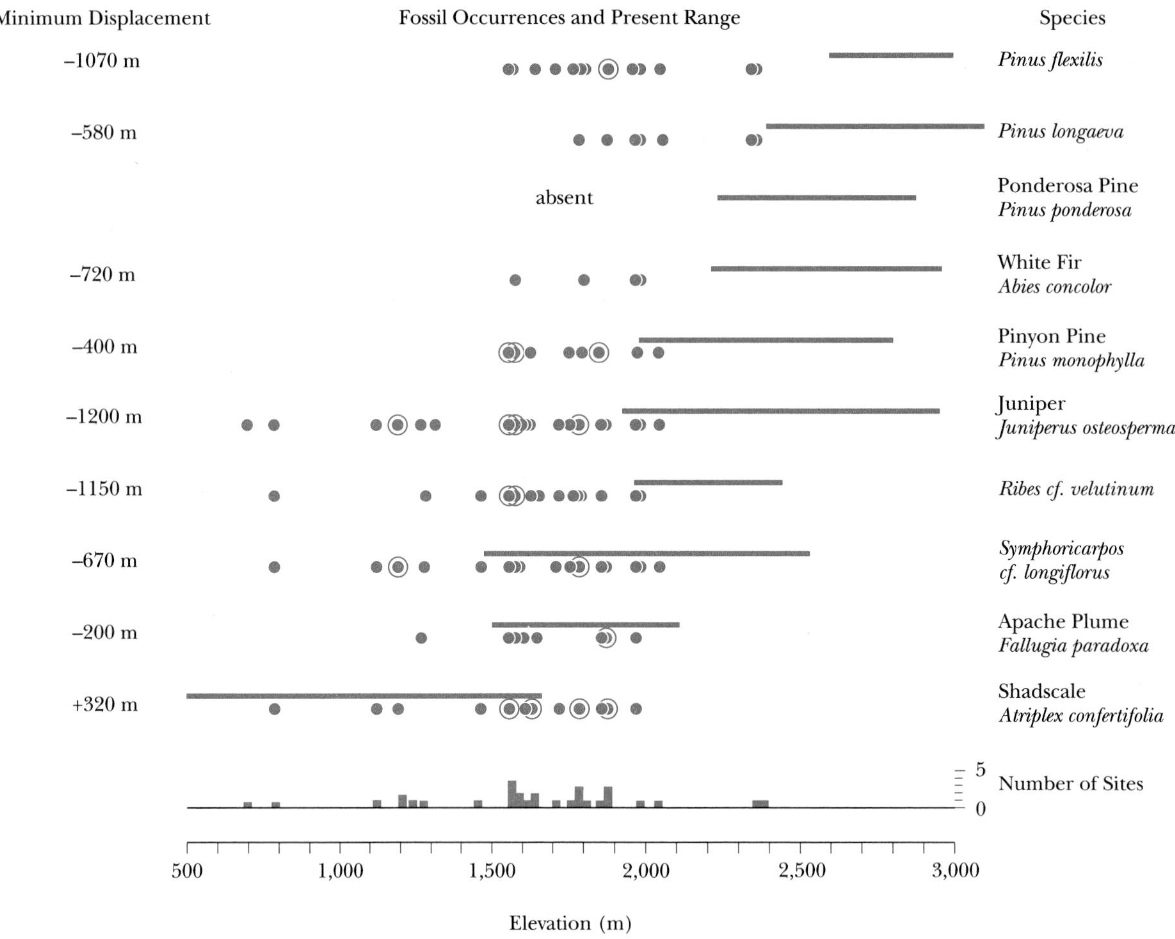

Figure 14-10

The current elevational range of various trees and shrubs compared with the elevations at which they have been found as Wisconsin period fossils in southern Nevada. Current ranges are given as the solid lines; fossil occurrences (11,000 years B.P. or older) are given by dots. Circled dots mean occurrence at more than one site at that elevation. (Adapted from Spaulding et al. 1983.)

Applying Paleoecological Knowledge

There are two possibilities for climate in the future: It can change or it can stay the same. Almost nobody except a few politicians is predicting stability, but there are two divergent opinions about the direction of change. Some geologists predict a new glacial period. Based on the history of the Pleistocene, we are overdue for a return of continental glaciation. Calculations from the most widely accepted model of glaciation—orbitally forced oscillations—yield the same prediction, although conditions as severe as the worst of the Wisconsin are not due for a few thousand years (Schneider and Londer 1984).

On the other hand, many atmospheric scientists foresee global warming as the result of the greenhouse effect (Chapter 20). The interaction of factors that cause climatic fluctuations is too poorly understood to make a firm prediction, but there is a developing consensus that the next several decades are likely to bring substantially warmer temperatures and accompanying changes in other climatic factors.

Paleoecological studies of biotic effects of past climatic changes are the best available evidence for predicting effects in the future (Davis 1991). The message from paleoecology is, however, mixed. The rate of postglacial range expansion for most trees ran from a little less than 100 to nearly 300 m per year (Schoonmaker and Foster 1991). Seemingly realistic estimates of the effects of carbon dioxide concentration on climate suggest that the climatic limits of tree ranges will be shifted northward at the rate of 7000 to 9000 m per year in the next century (Gear and Huntley 1991). It seems obvious that few trees will be able to expand their ranges at that rate, so plant ranges and community composition will fall into disequilibrium with climate.

Besides, Pleistocene range shifts occurred across natural landscapes. In coming Warm Ages, they will have to occur across the unfavorable terrain of cropfields, highways, and shopping centers. It seems certain that the spread of many species will be slower under these conditions, but paleoecology has little to tell us on this score.

On the other hand, the shifting assemblages seen through the Pleistocene are somewhat encouraging. We will expect to see new assortments—new communities—as organisms move and die at different rates. Catastrophic deterioration of ecosystems is unlikely. Whether the new combinations of climate and soil will support organisms able to satisfy current human needs and desires for food, fiber, and aesthetics is, however, hard to predict.

Community and Ecosystem Ecology: The Biome System, Temperate Deciduous Forest, and Southeastern Evergreen Forest

Perhaps the most characteristic bird of the eastern temperate deciduous forest is the red-eyed vireo. *(Alan G. Nelson/Dembinsky Photo Associates)*

OUTLINE

- Physiognomy
- The Biome System
- Landscape Ecology
- Temperate Deciduous Forest
- Deciduous Forest Edge
- Southeastern Evergreen Forest
- Animals

This chapter and the next two describe the plant-animal communities of North America and, less comprehensively, other geographical regions. In this chapter, the biome system is introduced, and temperate deciduous forest and southeastern evergreen forest are described.

Physiognomy

Ecosystems tend to have a certain look based on broad features of the structure of the vegetation, as well as on the shape and features of the land—the dirt and rocks. Physiognomy is a term that refers to the overall appearance of the vegetation. Plant characteristics that interact to determine physiognomy (Dansereau 1957) include

1. Growth form—whether the dominant species are trees, shrubs, or herbs
2. Function—deciduous or evergreen
3. Size
4. Coverage—which determines how much of the landscape is rock, dirt, water, or ice

5. Leaf size and shape—for example, broad-leaved or needle-leaved
6. Leaf texture—for example, succulent, thin, or hard

Using these six categories, each with a few subdivisions, yields several thousand physiognomic vegetation types. Many, however, are only trivially different from others, and many do not exist. There are several physiognomic types that almost anyone, including nonbiologists, would recognize:

Forest—a landscape dominated by trees close enough together for their crowns to touch.
Savanna—scattered trees in a grassy or shrubby area.
Thicket—tall shrubs or small trees growing so thickly that they are difficult to walk through.
Grassland—grass and other herbs dominating a landscape from which trees are scarce or absent.
Desert—a landscape in which plants are sparse and often scrubby; here the sand and rock may be more important in the landscape than the vegetation.

Another dozen physiognomic types added to these suffice for most everyday ecological purposes, although specialists in biogeography may require a more detailed classification. The following sections on the biomes and a later section on wetlands describe and discuss most types.

Climate is a major determinant of vegetation types. The relationship between climate and physiognomy was one of the earliest stated of all ecological principles, beginning with the observations of the early 19th century plant geographers, mostly Germans such as Alexander von Humboldt and August Grisebach. They were impressed by the tendency for vegetation of a single physiognomic type to occupy a given climatic region and, further, for areas that were remote geographically but similar climatically to have vegetation of similar physiognomy. The dry areas of the different continents, they found, supported deserts that looked much the same, although the flora—the plant species—differed. Grisebach (1838) applied the term **formation** to a physiognomic vegetation type, and he and later plant geographers such as Oscar Drude and A. F. W. Schimper constructed classifications of world vegetation with the formation as the basic category. Note that the relationship between climate and physiognomy is a community relationship. There are grasses, shrubs, and trees, as well as xerophytes, mesophytes, and hydrophytes in virtually every climate. How they are combined in communities is what determines physiognomy.

Most systems relating vegetation and climate would put southern Michigan in some moist forest category, but we have seen that the first white settlers found, in addition to moist forest, dry and wet forest, and also native grassland, bogs similar to tundra, and sand areas similar to desert. Today at least patches of these occur along with still other human-produced landscape types, emphasizing the fact that other factors in addition to temperature and precipitation can be important, at least locally, in determining vegetation type. These other factors include fire and special features of soil and drainage.

Among the more motile animals such as birds and mammals physiognomy seems to be one of the main bases for habitat selection (Odum 1945). In the eastern United States a savanna, or forest-edge, landscape can be formed in many different ways and a variety of different plants can be involved. There can be a wholly human-produced community of exotic trees set in mowed lawn with imported shrubbery, or an old-field area with goldenrods, blackberries, and cherry trees, or a bog with tamarack trees, blueberries, and sphagnum moss (Brewer 1967). Many of the birds in these three situations—such as song sparrows, catbirds, and robins—are the same, indicating that their occurrence is related to physiognomy rather than to plant species.

The Biome System

Frederic E. Clements agreed with many of the European plant geographers that regional climate is primary in determining regional vegetation. He realized, however, that what climate does is set a regional potential. At any one time, various local areas may not support this type of vegetation because they have been recently denuded or because the

Victor E. Shelford *(American, 1877–1968)*

Shelford was born in Chemung County in southern New York state. He attended the University of West Virginia but received B.S. and Ph.D. degrees from the University of Chicago, the latter in 1907. Shelford was a zoologist, working on his doctoral research with C. B. Davenport and C. M. Child, but probably his strongest ecological influence of this period was the plant ecologist Henry Chandler Cowles. Shelford followed Cowles' lead and studied the sand dune sere, initially concentrating on tiger beetles. His first book, published in 1913 while he was on the faculty at Chicago, was *Animal Communities in Temperate America,* a slightly inflated title since the book actually dealt with an area 30 by 50 miles at the southern end of Lake Michigan.

Shelford's last book (in 1963), *The Ecology of North America,* was a lineal descendant of the first. In it he attempted to describe the continent as it was in 1500, before European settlement. The attempt to understand the distribution and functioning of the continent's primeval ecosystems was one of the important themes of Shelford's work. Connected with it was his interest in preserving natural areas. Beginning in 1917 he headed a committee of the Ecological Society of America charged with the task of listing all "preserved and preservable" natural areas. The result was the valuable *Naturalist's Guide to the Americas* (1926). A later similar committee gave rise to the Nature Conservancy.

In 1914, Shelford moved to the University of Illinois, where he remained, retiring in 1946. While at Illinois, he must have been influenced by S. A. Forbes, but the influence did not include Forbes' evolutionary approach to ecology. Shelford, like Cowles, was uninterested in how the habitat relations of organisms evolved. That was speculation. What interested him was explaining community and habitat relations, as they existed, in terms of physiology and behavior.

Physiological ecology was Shelford's second important theme. What gave strength to his approach was, first, his realization that for ecology the important questions come from observations of the community; autecology, as such, is uninteresting. A second strength was his insistence, dating at least from *Animal Communities in Temperate America,* on the need to study community and habitat relations by experimentation. "Nature supplies the problems," Shelford wrote, "to be solved in the field and laboratory experiments."

Shelford was, by all odds, the most influential animal ecologist in the world for the first half of the 20th century. At midcentury, the chances would have been high that an animal ecologist picked at random in the United States would be one of Shelford's students or else a student of one of Shelford's students.

In 1939, Shelford and Frederic E. Clements collaborated on the book *Bioecology.* An effort by the leading animal ecologist and the premier plant ecologist to see ecology as a whole instead of as separate taxonomic disciplines seemed promising, but it was only partially successful. The book turned out to be less a fusion and synthesis of the two men's ideas than a grafting of some of Shelford's ideas onto Clements' basic framework. "We didn't agree on anything except the grassland," Shelford later said, "and not very well on that."

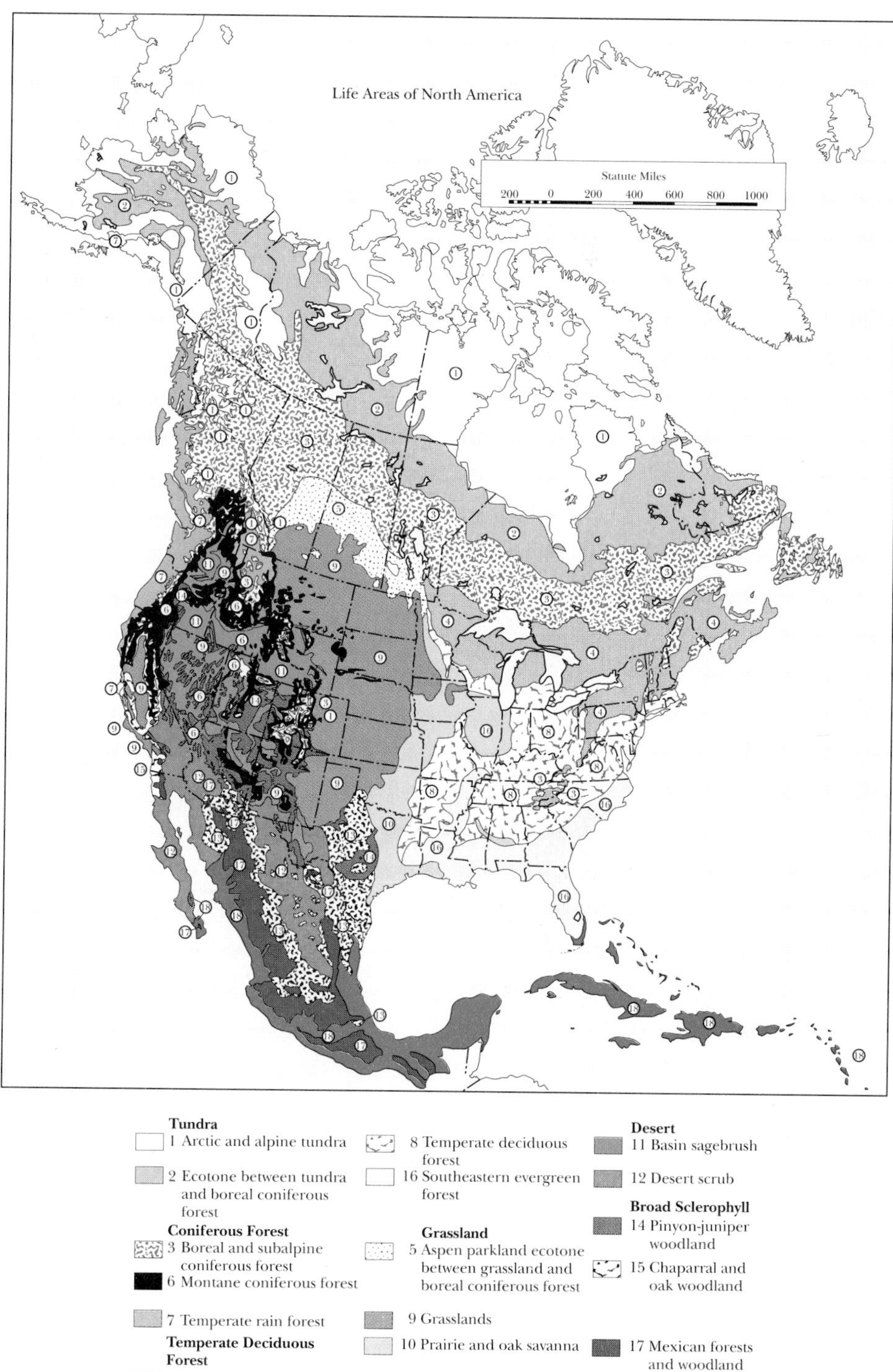

Figure 15-1

Biomes of North America. (Based on map prepared by Dr. John W. Aldrich, U.S. Bureau of Sport Fisheries and Wildlife Poster 102, Fish and Wildlife Service, U.S. Department of Interior, Washington, D.C., 1966. Terminology slightly modified.)

soil is unusually rich or unusually sterile, or wet or dry.

Clements adopted the European term "formation" but gave it a subtly different meaning. A formation in his classification was a plant community of geographical extent—one of the great landscape divisions of the continent—characterized by distinctiveness of life form in the climax dominants. The distinction is the word "climax": Nonclimax communities within the formation, no matter how much area they occupied, were a part of the formation. The extensive aspen forests of the north, although deciduous, are a part of the evergreen boreal forest formation.

This brings us to the biome system. Clements and V. E. Shelford (1939) (p. 428) recognized the desirability of a community classification based on a biotic approach rather than one based just on vegetation, which was usual, or just on animals, which had occasionally been tried (Kendeigh 1954). Their solution was the plant-animal formation, or **biome** (pronounced with slightly greater emphasis on the second syllable). As it turned out, the biomes are Clements' formations with the animals added in. Because physiognomy is such a powerful factor in animal habitat distribution, the correspondence between plant community, physiognomically defined, and animal community is tolerably close.

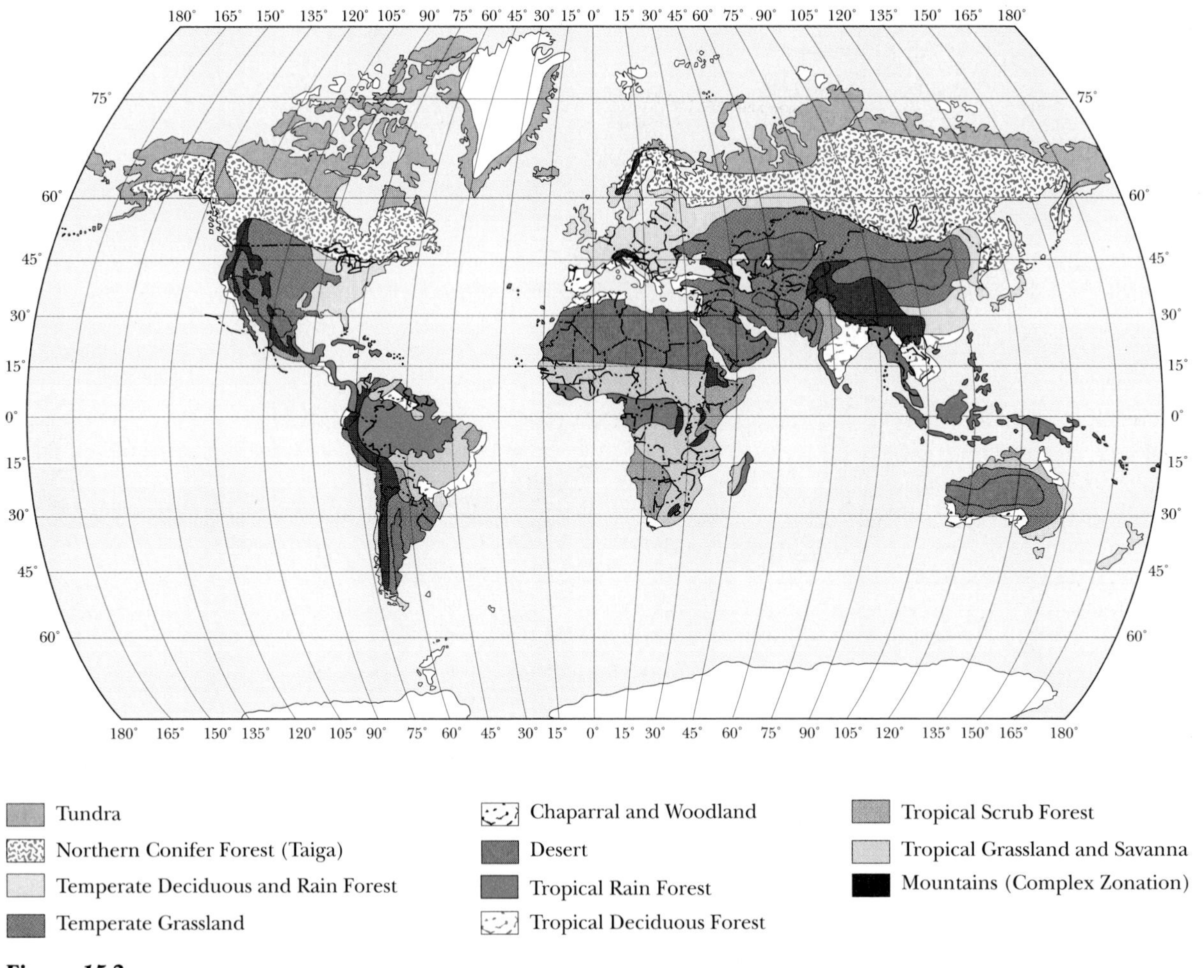

Figure 15-2

World distribution of biomes, highly diagrammatic. (From Odum 1971.)

A biome, then, is a biotic community of geographical extent characterized by a distinctive physiognomy based on the climax dominants. Although plant life forms are important as distinguishing traits of biomes, the corresponding features for animals, as S. C. Kendeigh (1954) pointed out, are the special physiological and behavioral features adapting them to the landscape.

Similar plant-animal communities in different parts of the world are considered to belong to the same biome type but not to the same biome. The biome, then, has some degree of geographical and paleoecological specificity.

The major biomes in North America are temperate deciduous forest, boreal coniferous forest, subalpine and montane coniferous forests, broad sclerophyll communities of woodland and chaparral, grassland, tundra, and desert (Figure 15-1, p. 429). In more tropical regions some additional biomes are present, including tropical savanna, seasonal (deciduous) forest, and rain forest (Figure 15-2). The main features of the major biomes are discussed in the following sections.

Landscape Ecology

Recently a new subdivision of ecology, **landscape ecology,** has grown up that emphasizes the interconnections among ecosystems of a region (Moss 1987, Turner 1987, Turner and Gardner 1991). The value of landscape ecology is, first, its emphasis on larger land areas of interacting ecosystems. That is, it focuses on the next higher level of organization above the local ecosystem. Its second value is as a reaction to the tendency to compartmentalize. We study a stand of forest or a lake, but landscape ecology considers the connections between them, such as the fact that herons that forage in the lake nest in the forest. As one aspect of their daily routine, the herons move nutrients from water to land.

Although landscape ecologists sometimes see their approach as a break from some earlier approaches to community ecology, many of the early ecologists used the term *community* broadly. Certainly the biome system is landscape based. In describing the mixed prairie association, Clements and Shelford (1939) included not just the areas dominated by mid- and short grasses, but also bottomlands of small rivers occupied by tall grasses, the riparian woodlands of larger rivers, and various other plant-animal communities, such as those occupying sandhill areas.

Charles Elton's (1966) ecological survey of Wytham Woods at Oxford, England, provided a relatively early and well-thought-out rationale for what has come to be called landscape ecology. Elton was setting out to study in a scientific way, he said, what the traveller, the naturalist, and the poet talk about as "the face of nature."

We briefly visited the field of landscape ecology in Chapter 13, where we talked about the question of equilibrium landscapes—the situation where at the scale of the local patch, disturbance (such as windthrow or fire) might cause major changes in structure, composition, or function, but at the landscape level, the percentage of patches in a given phase or at a given stage might approach a steady state. We will also see how the landscape approach is useful in conservation ecology in Chapter 19.

Temperate Deciduous Forest

Location and Climate

Temperate deciduous forest biomes occupy eastern North America, western Europe, Japan, eastern China, and Chile. The Northern Hemisphere temperate deciduous forest biomes contain many of the same or very similar organisms, but South American deciduous forest is not closely related.

The climate for the biome is one of a well-defined warm summer and cool to cold winter and of fairly high precipitation well distributed through the year (Table 15-1). Most of the region has a positive water balance in which the surplus is carried away by generally permanent streams; however, at the western edge of the biome, precipitation and potential evapotranspiration may be closely balanced. North-south trends within the biome are for the south to have higher precipitation, warmer winters, less snowfall, and a longer growing season. The importance of fire in the primeval forest varied greatly, being greatest in the Midwest and South.

Table 15-1
Generalized Climatic Data for Three Sections of North American Temperate Deciduous Forest and Southeastern Evergreen Forest

	Forest Type (and Location)			
Climatic Factor	*Northern Hardwoods (Northern Michigan)*	*Mixed Mesophytic (Central West Virginia)*	*Oak-Pine (North-Central Georgia)*	*Southeastern Evergreen (Southern Georgia)*
July temperature (°C)	19	23	26	27
January temperature (°C)	−7	1	6	12
Growing (frost-free) season (days)	130	175	210	260
Annual precipitation (cm)	76	114	132	127
Potential evapotranspiration (cm)	60	70	80	80
Percent of precipitation for April–September	53	53	48	62
Annual snowfall (cm)	178	89	8	Less than 2

Based on USDA, *Climate and Man; Yearbook of Agriculture* (House Document No. 27), 77th Congress of the United States, 1941.

Soils

Soil development under this vegetation and climate tends to produce a gray-brown podzol. Such soils are acid owing to the loss of calcium through leaching. The leaching is, however, less intense than under coniferous cover, and the A_2 horizon is gray-brown rather than the pale ash color of the podzols.

In the southern part of the biome, gray-brown podzols give way to red podzols. In this warmer region, bacterial action is rapid, and little humus is present to darken the soil. Leaching of silica occurs, and the red of oxidized iron is the predominant color of the soil. Under the U.S.D.A. Comprehensive System, the gray-brown podzols are alfisols, specifically udalfs, and the red podzols are alfisols (ustalfs) or ultisols (udults).

Plants and Vegetation

The dominants of temperate deciduous forest are tall trees with broad leaves that are shed each fall. The types of trees forming the canopy of the climax forest vary geographically, but oaks and maples are widespread. There may be a subcanopy tree layer in which such species as flowering dogwood, blue beech, and hop hornbeam occur.

The shrub layer with spicebush, witch hazel, and gooseberry is usually sparse. Often there is a rich herb layer. Some species of herbs come into flower all through the growing season, but the greatest number bloom in the spring and the herb type *par excellence* is the spring ephemeral (Figure 15-3). In the deep shade of the undisturbed climax forest, the herb flora may be nearly reduced to these plants which grow and flower early in the spring and die back by the time the trees are fully leafed (Table 15-2, Figure 15-4). Most of the spring ephemerals are perennials, living through ten months of the year as corms, bulbs, or roots. There are only a few annual plants, and most of these behave as spring ephemerals; blue-eyed mary and false mermaid weed are examples. The touch-me-nots, however, grow through the spring and summer and bloom from July to September.

Net primary production (NPP) is usually in the range of 1 to 2 kg/m^2 per year (Cannell 1982). Two thirds to three fourths of this is stored in wood and is thus not immediately available for herbivore consumption. Biomass in mature forests accumulates to around 300 to 400 metric tons per hectare but is

Figure 15-3

Warren Woods, a beech-maple climax forest in spring (*A*) and summer (*C*). Spring ephemerals, such as trout lily (*B*), grow in great numbers before the trees have leafed out but very few herbs grow up and flower in the dense shade of the late spring and summer. Several herbs and shrubs mature fruit in the summer and many of these are fleshy, animal-dispersed fruits. Most are red or blue but these doll's-eyes (*D*) are white. (Photos by the author.)

surprisingly variable, from less than 100 in short, dry oak forests to more than 700 in rich maple. Roots make up 20% or slightly more of the total biomass in oak forest and slightly less than 20% in maple forest.

Generally 3 to 4 metric tons of foliage per hectare are present in the summer. Some 5% of this is consumed by herbivores, mostly caterpillars, while the leaves are on the trees. The rest is not consumed until they fall to the ground. In some deciduous forests, decomposition of one year's leaf production is complete in a year or less, so that the litter layer has disappeared by the time of the next autumn leaf fall. This is usually true of forests dom-

Table 15-2
Number of Herb Stems in Spring, Early Summer, and Late Summer in a Climax Beech-Maple Forest

As shown by these original data from Warren Woods in southwestern Michigan, the forest floor in spring is densely covered by herbs. Many of these are spring ephemerals which sprout, make their vegetative growth, and flower before shading is at its heaviest; they then die back for the rest of the year. Most spring ephemerals are perennial, but a few (such as false mermaid weed in the table) are actually annuals that compress their life cycle into the spring and exist as seeds for the rest of the year. Wild leek is different from the other perennials in that it makes only its vegetative growth in the spring; weeks after its leaves have died back it sends up bare flower stalks from its bulbs (Figure 15-4). By mid-June there are only about 9 plants per square meter, and by mid-September the floor is practically bare, with only about two stems per square meter. Most species that are numerous in June and September were already present in April. Beechdrops, a root parasite of beech trees, is one exception; it does not sprout and flower until late in the summer. Not all the spring-flowering plants die back completely in the summer; hepatica, phlox, and ginger are some that do not.

	Number of Stems (per square meter)		
Species	*Late April*	*Mid-June*	*Mid-September*
Spring beauty	123	—	—
Yellow trout lily	107	—	—
Squirrel corn	25	—	—
False mermaid weed	19	—	—
Wild leek	9	—	<0.5
Canada mayflower	1	4	<0.5
Smooth yellow violet	3	3	Present
Sweet cicely	<0.5	1	Present
Pennsylvania sedge	—	1	<0.5
New York fern	—	<0.5	—
Beechdrops	—	—	1
Wild ginger	<0.5	—	<0.5
Hepatica	—	—	<0.5
Plantain-leaved sedge	<0.5	<0.5	<0.5
Blue phlox	—	—	<0.5
All species	About 307	About 9	About 2

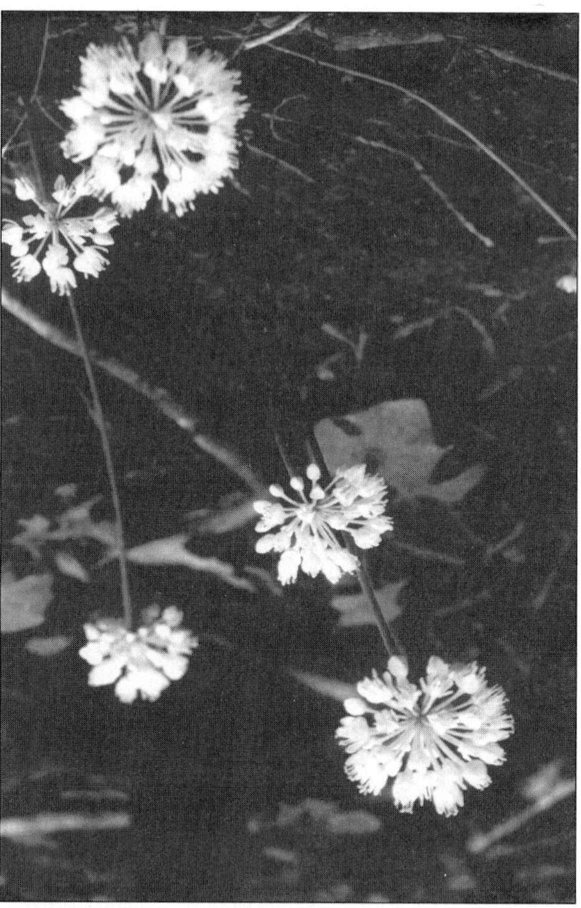

Figure 15-4

Wild leek is a card-carrying spring ephemeral with respect to its foliage. Its flowering, however, occurs in summer, in June or July, long after its leaves have died back and disappeared. (Photo in Michigan by the author.)

inated by sugar maple or basswood. In oak forests, however, decomposition and consumption of one year's production take more than a year, so that a thick litter layer tends to develop.

Subdivisions of Temperate Deciduous Forest

Maples and associated species tend to be more important as climax tree species in the northern part of the biome and oaks in the southern part. In oak-dominated regions, maple forests may, nevertheless, occupy certain habitats, generally ones that are cooler, moister, or have better soils. Similarly, oak forests may grow on dry or poor soils in regions where maple-dominated forests are the prevailing type.

Seral forests vary geographically and according to whether the starting point of the sere is relatively wet or dry. Elms are widely important on wetter sites and aspen and pine on drier ones. Oaks may be important in successional forests in maple-dominated regions.

E. Lucy Braun (1950) recognized nine subdivisions of the biome (Figure 15-5) (including southeastern evergreen forest).

Mixed Mesophytic Forest

In this centrally located region on the Cumberland Plateau and adjacent areas, the terrain is diverse. So, too, are the climax forests. Dominance may be shared by beech, tulip tree, white basswood, sweet buckeye, sugar maple, red and white oaks, and several other deciduous species. Hemlock, an evergreen, is also important on some sites. Relative abundances shift from site to site; at higher elevations, for example, tulip tree drops out and yellow birch becomes more frequent (Braun 1950). Above about 1250 m, red spruce comes in, but there is no development of a subalpine forest in this region comparable to that of the Great Smoky Mountains to the east.

Western Mesophytic Forest

The western mesophytic forest lies west of the mixed mesophytic region. Both regions are mostly unglaciated and thus have ancient land surfaces that have supported vegetation continuously for many millions of years. The range of elevations is

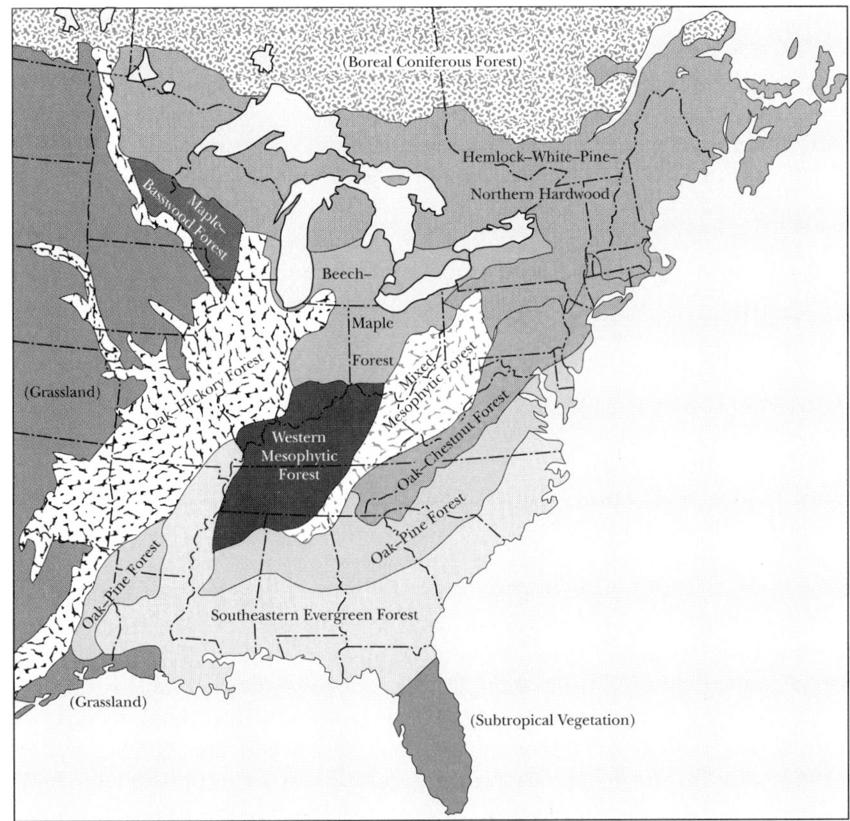

Figure 15-5

E. L. Braun recognized nine subdivisions of eastern deciduous forest, as shown on this map. The tree species making up the climax forests were the main criteria for deciding regional boundaries. (Based on E. Lucy Braun, *Deciduous Forests of Eastern North America,* Philadelphia, Blakiston, 1950.)

E. Lucy Braun (American, 1889–1971)

Lucy Braun (pronounced Brown) was born in Cincinnati, took all three of her degrees (Ph.D., 1914) at the University of Cincinnati, where she taught until 1948, and died in Cincinnati, aged 82. One of her strongest ecological influences, however, came during a year away at the University of Chicago in 1912, where she studied with Henry Chandler Cowles.

Lucy Braun was the great student of the temperate deciduous forest, studying its distribution, successional relationships, and its origin and evolution in geologic time. Her views are an interesting amalgam of Cowles and Clements. Like Cowles, she was interested in the differences in communities based on minor habitat differences—small differences in species composition that Clements would have found insignificant. Like Clements, she was willing to try to put things together in a grand picture, reconstructing the temperate forests of the Tertiary and attempting to understand how the forests of eastern North America were derived from them. Her forest studies culminated in the 1950 book *Deciduous Forests of Eastern North America.*

The preservation of natural areas was a practical matter to which Braun devoted a great deal of time and energy. She was a member of the Ecological Society of America's Committee on Natural Areas beginning in 1917, and later of the Society's Committee for the Preservation of Natural Conditions. She was an important force in land preservation in Ohio through out her life.

Braun made her way during a time when women needed a special strength to survive as professional scientists. This may help to explain the mixed sentiments with which she is remembered. One former colleague remembered her as "a mild-mannered, gentle person, confident and sure, a woman of steel, a master of her craft, confident, absolute." Another expressed agreement with everything except the "gentle."

Some years after her retirement, it is told, Lucy Braun and her sister Annette attended a lecture in the Department of Geology at the University of Cincinnati. (The two sisters lived together. Lucy, as breadwinner, was boss, even though Annette was five years older and also held a Ph.D. Annette did the cooking and housecleaning, as well as continuing her studies of lepidoptera.) The lecturer, a palynologist, presented evidence from pollen analysis suggesting that one of Lucy's conclusions about the deciduous forest— that it had survived *in situ* in the southern Appalachians throughout Pleistocene glaciation—was probably mistaken. After the talk, Lucy rose and delivered a ferocious attack. The palynologist gulped and thanked her for her opinion. To the students, most of whom had never heard of Lucy Braun and to whom the whole issue was uninteresting, the scene (one biographer noted) must have seemed straight out of *Alice in Wonderland.*

lower here than in the mixed mesophytic region, but in other ways this region is more diverse. The Ohio and Mississippi Rivers have provided additional land forms, and the shift from gray-brown to red podzols occurs in this region.

As a result of these factors plus those relating to plant migration, the forests of the region form a complex mosaic. Eastward, many sites are occupied by forests little different from those of the mixed mesophytic region. Westward, these become restricted to moist but well-drained slopes and bottomlands. Certain species, notably white basswood and sweet buckeye, drop out. Drier habitats tend to be occupied by oak and oak-hickory forest.

In southern Illinois, for example, ravine bottoms and lower slopes have forests of sugar maple, beech, tulip tree, red oak, and white oak (Voigt and Mohlenbrock 1964). Red buckeye, which bears showy red flowers in the spring, is a characteristic shrub. A rich herbaceous flora is present, with the familiar spring ephemerals.

Midslope forests in southern Illinois are mostly white, red, and black oak. Few of the herbs are spring ephemerals; grasses are prominent.

The ridgetops are also oak dominated. Post and blackjack oak are characteristic, although white and black oak also occur. Red cedar also occupies upland sites, especially where the soil is thin around rock outcrops. Ground cover is scant and dominated by grasses such as poverty oats grass and broom sedge. On poor, dry, or rocky soil, the growth may be savanna rather than forest. **Cedar glades** in which prairie species, mosses, and lichens are intermingled occur on such sites (Kurz 1981, Aldrich et al. 1983). Also, small patches of prairie **(hill prairies)** dominated by little bluestem, June grass, and side-oats grama may occur on south and west faces of bluffs.

Lowlands in the western mesophytic region support swamps and bottomland forests similar to those of the southeastern evergreen forest (discussed later). Various hickories occur regularly in upland or lowland sites but are rarely among the leading two or three species.

Oak-Hickory Forest

This region, as recognized by Braun, forms a band running southwest from Wisconsin and Indiana to Texas. In the southern part of the region, forest composition is similar to that described in the preceding section. North of the glacial boundary, many of the species of oaks and hickories drop out, leaving mostly those of wide ecological amplitude. White and black oaks and pignut and shagbark hickories are widespread. Red oak and bitternut hickory tend to be restricted to the most mesic sites.

Flowering dogwood is an important understory species. On many sites, oak and oak-hickory forest were maintained by recurrent fire, and more shade-tolerant trees such as black gum and sugar and red maples are well represented in the understory (Downs and Abrams 1991). Beggar's-ticks of several species and also woody vines such as Virginia creeper and poison ivy may be important in the ground layer.

Much of the area that Braun included within the oak-hickory forest region in the northern Midwest, especially Illinois and Indiana, was a mosaic of vegetation in which prairie occupied more land than forest. Forest was restricted mostly to floodplains and areas to the lee of firebreaks such as rivers and scarps (steep hills). Areas that burned frequently supported tallgrass prairie or savannas with the fire-resistant bur oak. Savannas of white or yellow chestnut oak seemingly occurred on lands of intermediate fire frequency. Fire suppression after settlement quickly led to the conversion of these open, parklike landscapes to forest (Brewer and Kitler 1989).

Oak-Chestnut Forest

Lying east of the mixed mesophytic forest and also extending northward to New York and Massachusetts, this region occupies diverse topography from the Great Smoky Mountains to the coastal plain. American chestnut, red oak, chestnut oak, and tulip tree were the original climax dominants; however, the chestnut has been virtually eliminated by the chestnut blight, a fungus parasite accidentally imported from Japan around 1900. The place of chestnuts in the canopy seems to be taken mostly by oaks. An understory of heath shrubs, such as the flame azalea in the south, but blueberries and huckleberries in the north, occurred on many sites.

The vegetation of the Great Smoky Mountains, a part of this region, is highly diverse (Whittaker 1956). Oak-chestnut forests tended to occupy most sites below about 1300 m; however, mixed mesophytic forest occupies the coves (small valleys) and lower north slopes. A deciduous montane forest occurs above this elevation to about 1800 m. This forest resembles the northern hardwoods in having sugar maple, beech, and yellow birch well represented but also contains southern species such as sweet buckeye. The red spruce of the subalpine forest comes in at 1600 or 1700 m and Fraser's fir about 1800 m.

In the forests, but especially on mountaintops with little tree cover *(heath balds),* rhododendron forms dense thickets that provide a spectacular flowering display in late May or early June. *Grassy balds*—eastern subalpine meadows—also occur, dominated by mountain oat grass.

Oak-Pine Forest

This region occupies most of the Piedmont Plateau and also occurs in the northern part of the Gulf states. Topography is rolling to flat. The climax forests are dominated by oaks, particularly white oak, and hickories; however, most of the upland areas were cleared for farming. Old-field succession here proceeds to a broom-sedge stage as in the oak-hickory and mixed mesophytic regions to the north, but in the oak-pine region this grassland stage is invaded not by broad-leaved trees but by pines, especially loblolly and short-leaf. After 40 years the pine stand is middle-aged, and by 70 to 80 years a secondary canopy of oaks and hickories is present (Oosting 1942). In the absence of fire, succession proceeds to an oak-hickory forest in 150 to 200 years.

The Pine Barrens of New Jersey, an extensive area dominated by woodlands of pitch pine or oaks, are included in this region by Braun (1950).

The oak-pine region is transitional to the southeastern evergreen forest, treated separately in a later section.

Beech-Maple Forest

This region occupies the southern Great Lakes region. The climax forests resemble those of the mixed mesophytic region but with the dominant species reduced to two, beech and sugar maple (Figure 15-3). These two highly shade-tolerant species seem to be able to coexist indefinitely. Sugar maple's great rate of reproduction and ability to grow rapidly in canopy openings are balanced by beech's longevity, its ability to continue growth toward the canopy despite deep shade and, in the primeval forest, its ability to root-sprout after fire. Third place in the canopy is usually occupied by basswood, although tulip tree, red oak, white ash, or elm can outnumber it.

Maple-Basswood Forest

In a small area centered on the Driftless (unglaciated) Area of southwestern Wisconsin and northwestern Illinois occur forests similar to beech-maple but lacking beech. Basswood largely takes its place though red oak may also be prominent. Much of the land, however, is occupied by prairie and oak openings.

Hemlock-White Pine-Northern Hardwoods Forest

This extensive region, running across the northern United States and southern Canada from Minnesota to the Atlantic, is an ecotone between temperate deciduous and boreal coniferous forest. Fair to good soils of medium moisture content support northern hardwoods—that is, sugar maple, beech, and yellow birch—with more or less hemlock. White pines occur in such forests mostly as large, old trees, relicts from earlier stages of succession.

On drier, usually sandier soils, infrequent fires in this region may yield a landscape dominated by white pine with scrubby oaks. These pineries once formed a magnificent timber resource, stretching discontinuously from Maine to Minnesota, but every sizable area of large white pines was cut over in a brief period centered around 1890 (Figure 15-6). Peaks of timber harvest were about 1850 in New York, 1860 to 1870 in Pennsylvania, 1880 in Michigan, 1890 in Wisconsin, and 1900 in Minnesota (Steers 1948, Holbrook 1957). And then the pine-

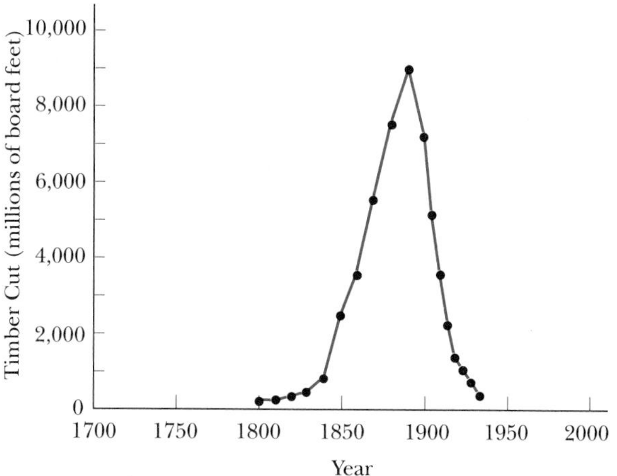

Figure 15-6

The white pine harvest in the northern hardwood region, 1799–1935. Virtually all of this magnificent resource was cut in a 90-year span from 1840 to 1930. One board foot is an amount of rough lumber 1 foot square and 1 inch thick. (Data from Steer 1948. Figures for 1799–1859 were calculated by assuming the softwood cut in the eastern United States was 63% white pine.)

lands were gone, turned into stumplands. At the peak, more than 3,500,000,000 board feet of white pine per year was cut in Michigan, all in the form of large trees. White pine is still a part of the timber industry. Currently the yearly production in Michigan is about 30,000 board feet (Smith et al. 1990).

The logging debris supported hot slash fires later (Haines and Sando 1969). Most of the replacement forests in the Great Lakes states are aspen or oak (Figure 15-7). On many sites, enough white pine reproduction (that is, small trees) is available that restoration of the white pine ecosystem would occur through natural processes within a few decades (McCann 1991). However, most of these forests are on state or federal land and are cut for timber or "wildlife management" on a rotation too short for this to occur.

In some cases, very hot slash fires on sterile soil produced persistent open landscapes called "stump prairies." Into these areas spread open country birds such as the prairie chicken, formerly restricted to the prairies further south (Brewer 1991).

Figure 15-7

The stump is a relict of the white pine forest cut for lumber in Allegan County, Michigan, in the late 1800s. Charcoal on the stump as well as the multiple-trunked black oak to the left are indicative of the slash fire that followed lumbering. (Photo by the author.)

Some poor sandy areas in the Great Lakes region, particularly higher, colder sites, supported jack pine forests and savannas. Clear-cutting here, also, sometimes led to persistent open areas (Abrams et al. 1985). Jack pine is a fire-adapted species with serotinous cones and tends to grow in even-aged stands that burn every 20 to 40 years. The endangered Kirtland's warbler occurs almost exclusively in open stands of orchard tree-sized jack pines. In poorly drained areas throughout the northern hardwoods region occur tamarack, black spruce, arborvitae, and white pine.

Northward, coniferous forests become more permanent; white spruce and balsam fir occupy mesic sites that, farther south, would support northern hardwoods. East-west differences also exist, although these are surprisingly few considering that the region extends across 20° of longitude. The climax hardwood forests seem to grow depauperate in both the east and the west. Basswood drops out to the east, beech and hemlock to the west. Red spruce becomes important in the Adirondacks and eastward. It may appear with the hardwoods in mature forests and also is important in some successional forests.

Most of the northern hardwoods region of New England was cleared for farming. In many areas farming was unprofitable and forest was allowed to return, but much of it has been cut over again for timber. Pin cherry, paper birch, red maple, and red spruce tend to be important successional species, varying by habitat (Leak 1982). Climax forest also varies by soil type; in the White Mountains area of New Hampshire, rocky areas and outwash plains tend to be occupied eventually by spruce-hemlock forests in which hardwoods are a minor element. Areas of till, however, tend to support northern hardwoods with few, if any, conifers.

On the higher mountains in New England and in the Adirondacks of New York, a subalpine red spruce-balsam fir forest occurs. Above it on the tallest peaks occurs alpine tundra.

Animals

The special conditions of life for animals in the deciduous forest are mostly related to the presence of trees, shade, and the strong seasonality in climate, cover, and food.

Trees and Shade

Trees are widely used for nests, den sites, food sources, and perches. Ground-nesting birds are scarce. Besides the gallinaceous birds, about the only species that nests on the ground is the ovenbird. A relatively large number of birds nest not on the trees, but in them, either in cavities they excavate themselves—woodpeckers and chickadees, for example—or in natural cavities and old woodpecker holes. Species in the latter category include great crested flycatchers (Figure 15-8), wood ducks, and horned owls. Probably predation is the big factor favoring these patterns of nesting; losses to predators drop in the sequence ground nest to open tree nest to cavity nest.

Burrowing is common among the mammals of deciduous forest, but of those that do not burrow, most are arboreal. This is well known for squirrels, but other mammals that we may tend to think of as ground dwellers climb to get food or to escape being food for others. Examples are black bear, gray fox, and white-footed mouse. The last obtains a good share of its food from trees and vines during some parts of the year (Gosling 1977).

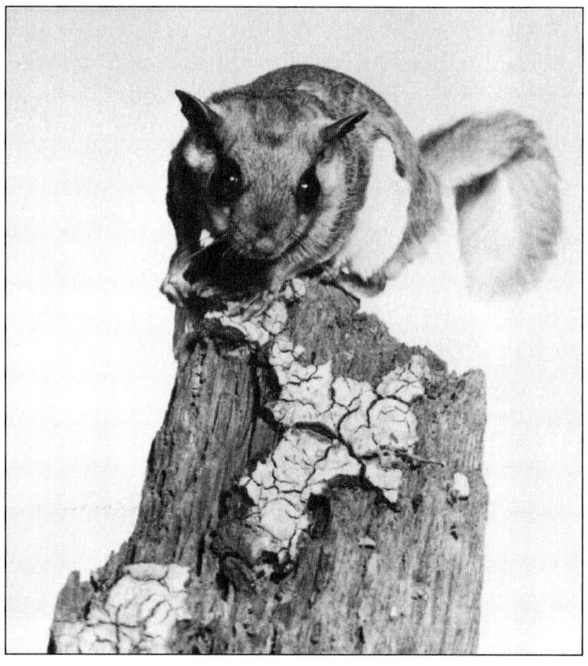

Figure 15-9

The nocturnal, cavity-nesting southern flying squirrel occurs throughout the eastern temperate deciduous forest. (Peter J. Van Huizen, U.S. Fish and Wildlife Service.)

Figure 15-8

A great crested flycatcher in front of her nest cavity, Kalamazoo County, Michigan. (Photo by the author.)

Certain reptiles and amphibians also show arboreal abilities. The black rat snake, rough green snake, and tree frogs are excellent climbers.

Morphological adaptations for climbing include prehensile tails (as in the white-footed mouse and opossum), adhesive toe pads (tree frogs), and skin flaps used in parachuting (flying squirrels, Figure 15-9). Wings fit birds, insects, and bats for life in the forest but also have many other uses.

Most birds give their territorial song from trees. About the only one with a flight song is the ovenbird, although it, too, usually sings from shrubs and trees. Voice and hearing are well developed in many forest animals—birds, insects, squirrels, chipmunks—presumably because the branches and leaves restrict long-distance communication by vision.

Seasonality

The forest alternates between easy and hard times, but the times do not correspond for every inhabitant. For the foliage insects, such as caterpillars

(Figure 15-10) and leafhoppers, the easy time is the spring and early summer when there is a new crop of tender leaves. Soon the leaves harden, and a few months later, they fall. This begins a time of plenty for the litter invertebrates, bacteria, and fungi that is interrupted by the cold of winter but resumed in the spring.

At higher trophic levels, the foliage insects provide food for a large number of insectivorous birds. The ground invertebrates fall prey mostly to larger ground invertebrates, to amphibians (Figure 15-11) and reptiles, to mammals such as shrews, and to only a few birds such as the ovenbird. A third important source of food is standing dead trees and dead limbs on live trees. They provide food for fungi and insects and, in turn, woodpeckers.

Figure 15-10

Caterpillar larvae, like this Cecropia, are important herbivores of forest leaves. (Photo by George and Betty Walker.)

Most cold-blooded animals and a few mammals, such as woodchucks and jumping mice, hibernate. Winter bird populations drop by three fourths as most of foliage gleaners—vireos, wood warblers, and tanagers—and flycatchers leave for the south. The resident species—chickadees, ruffed grouse, woodpeckers, and the white-breasted nuthatch—are augmented by a few species from the boreal coniferous forest such as golden-crowned kinglet and brown creeper. Most of the birds that reside in the forest in winter are flexible—able, during this hard time, to make do with hidden-away eggs and pupae and with seeds, fruits, and nuts. The small birds often move in flocks, probably gaining both foraging efficiency and predator protection in this habitat that has undergone a drastic switch in both food availability and accessibility by predators.

Although seasonal changes are great, year-to-year changes are slight. Catastrophes like fires and hurricanes occur but, by and large, the forest this year is much like the forest last year. The same trees and logs are present. Although drought and insect peaks can cause fluctuations in the density of the canopy, a crop of leaves is there year after year. The evolutionary result is that forest birds are *K*-selected, mesic forest birds most of all. Many of the most typical climax forest birds have a clutch size of only three eggs and raise only one brood a year (Brewer and Swander 1977).

A large number of animals occur virtually through the deciduous forest biome, giving it considerable uniformity. Table 15-3 lists most such species and also lists several that are more characteristic of the northern or southern parts of the biome. In addition to these species, most of which are nearly confined to eastern deciduous forest, several other species are (or were) found throughout the biome but also occur widely outside it. Examples include American toad (Figure 15-11), spring peeper, red-tailed hawk, great horned and screech owls, wild turkey, several woodpeckers, black-capped chickadee, American crow, black bear, raccoon, and bobcat.

Breeding-bird populations tend to be 150 to 250 pairs per 40 ha with lower populations in the drier forests. Red-eyed vireo is usually the leading species. Ovenbird is often second in abundance in the North, wood thrush or Acadian flycatcher in the South.

Figure 15-11

The American toad is a heavy nocturnal consumer of insects in forests. (Photo by the author.)

There is a rich invertebrate fauna. Highest densities are in the litter where very large numbers of springtails, mites, millipedes, centipedes, snails, sowbugs, and beetles live. Most of these also occur in large numbers in the soil, along with earthworms and nematodes. The foliage invertebrates of the herbs, shrubs, and trees include mosquitos, flies, moths, butterflies, leafhoppers, thrips, and spiders.

Human Occupation

In the Time-Life Book *The World We Live In* (New York 1955) there is a good chapter on forests in which temperate deciduous forests are referred to as "the woods of home." The phrase has a nice ring to it and a certain amount of truth, but it is incorrect in several ways. Temperate deciduous forest biomes are home to a great many people in western Europe, eastern Asia, and eastern North America, but other biomes are home to such people as Eskimos, Arabs, southern Europeans, and Polynesians. Even to those of us in North America, temperate deciduous forest is rarely home in the sense that we live in one. Forest edge, or temperate savanna, seems to be the landscape we prefer. Where our ancestors settled in forest they cut openings in it; when eventually they moved out onto the prairie they planted trees. The Indians were similarly uninterested in climax deciduous forest as a place to live. Through the summer it was too buggy, and the nuts and berries gathered and the game hunted were found in forest edge and open oak forest (Table 15-4).

Deciduous Forest Edge

Today much of the forest is gone, and so are many of the forest animals, gone altogether like the passenger pigeon or gone from large sections of their former range. Some of the more obvious examples of the latter situation are the wapiti (elk), which

Table 15-3
Species Occurring Virtually Throughout the Temperate Deciduous Forest Biome

Species marked "S" are absent from the northern part of the biome, and species marked "N" are absent from the southern part.

Reptiles and Amphibians	*Birds*	*Mammals*
Box turtle	Wood duck	Opossum (S)
Wood turtle (N)	American woodcock	Short-tailed shrew
Five-lined skink	Ruffed grouse (N)	Least shrew
Broad-headed skink (S)	Red-shouldered hawk	Eastern pipistrelle
Common water snake	Barred owl	Eastern chipmunk
Timber rattlesnake (S)	Whippoorwill	Gray squirrel
Black rat snake	Red-bellied woodpecker (S)	Southern flying squirrel
Spotted salamander	Eastern pewee	Pine vole
Marbled salamander (S)	Acadian flycatcher (S)	White-footed mouse
Red-backed salamander (N)	White-breasted nuthatch	Gray fox
Gray tree frog	Carolina chickadee (S)	
	Wood thrush	
	Red-eyed vireo	
	Yellow-throated vireo	
	American redstart	
	Hooded warbler (S)	
	Louisiana waterthrush	
	Kentucky warbler (S)	
	Cerulean warbler (N)	
	Ovenbird (N)	
	Scarlet tanager (N)	
	Summer tanager (S)	
	Rose-breasted grosbeak (N)	

once occurred virtually through the deciduous forest biome; timber wolf and a southern relative, red wolf; black bear; bobcat; mountain lion; and gray fox.

Where the forest remains, it is mostly as isolated woodlots of 40 acres or a little more or less. This is less than one family's home range for many of the larger species. A family of gray foxes, for example, range over about a square mile, 640 acres, during spring and summer (Lord 1961). For some of the animals, the discrepancy between what is needed and what is left is laughable—or weepable. Mountain lions, for example, probably hunted over areas of 300 square miles and up—an area 5000 times as big as a 40-acre woodlot.

In the early biological surveys of the midwestern states, a switch in abundance between gray squirrels and fox squirrels was usually mentioned (Figure 15-12). The fox squirrels originally inhabited the savannas and forest edges, but as the forests were fragmented, they spread throughout the remaining woodlots (Allen 1942). It is not easy to decide whether the switch occurred simply because of the changed habitat—one can stand in the middle of a 40-acre woods and see to the edges—or whether competition with the larger fox squirrel displaced gray squirrels from woods that they would otherwise still occupy (Brown and Batzli 1985).

The effects of forest fragmentation are also intermixed with those of direct human persecution. The postsettlement history of North America has been a history of overhunting and predator eradi-

Table 15-4
Some Important Food Plants of Indians in the Upper Great Lakes Region by Habitat

R. A. Yarnell has listed about 120 plants used by Indians at the time they first came into contact with Europeans. Although virtually every habitat contained some food plants, the partial lists in this table indicate correctly that a greater number of species and more important species were in marsh and forest edge. The climax beech-maple forests had relatively few important species; the same was true of coniferous forests.

Habitat	*Food Plant*
Beech-maple forest	Sugar maple, wild leek, spring beauty, basswood, toothwort, prickly gooseberry, beech, paw-paw
Oak and oak-hickory forest	Bracken fern, dryland blueberry, shagbark hickory, white oak, black oak
Swamp and bog forest	Silver maple, yellow birch, jack-in-the-pulpit, black huckleberry, wild currant, black walnut, skunk cabbage, hackberry
Coniferous forest	White pine, creeping snowberry, bunchberry
Marsh, bog, and lake	Great bulrush, American lotus, yellow pond lily, marsh marigold, fragrant pond lily, arrowhead, wild rice, cranberry, highbush blueberry, cinnamon fern
Prairie and bur oak savanna	Butterfly weed, wild strawberry, hazelnut, bur oak, wild pea, dewberry
Forest edge	Common milkweed, common elder, raspberry, blackberry, chokecherry, black cherry, plum, nannyberry, wild grape, hawthorn, bittersweet

Species list from Yarnell, R. A., "Aboriginal relationships between culture and plant life in the Upper Great Lakes region," Ann Arbor, University of Michigan Museum of Anthropology, Anthropological Paper No. 23, 1964.

cation. We will probably never know whether mountain lions or timber wolves could make a living in the eastern deciduous forest landscape of today; they would be shot before they had a chance to starve.

The passenger pigeon nested in the beech-maple forests, in immense colonies that shifted from place to place depending on where the beechnut and acorn crop of the preceding summer had been big enough. As Aldo Leopold (1949) put it, "Whatever Wisconsin did not offer him gratis today, he sought and found tomorrow in Michigan, or Labrador, or Tennessee." You can almost take your pick of whether overhunting or forest loss caused the demise of the species. The slaughter was incredible, adults and squabs alike, but the forests that they needed to live were disappearing. And if they had learned to eat soybeans like their cousins the mourning doves, descending on the field in vast flocks, our agricultural ancestors would have exterminated them for that.

Studying the diminished forests of today, expurgated of pigeons and predators, probably misleads us. What, for example, were the consequences of the overfertilizing by the pigeons of their nesting grounds on biogeochemical cycling and on tree and herb composition? Perhaps we will never know for sure, but the pigeon, like the bison of the prairie, must have had profound effects on the functioning of its ecosystem.

Today the common animals of the biome are not the mature forest species, not even the small ones that escaped our forefathers' attentions. Rather, they are species of the forest edge. These are animals that lived along the actual edges of the forest—next to the prairie, along streams, marshes, and cliffs—and also in such sites as holes in the forest from windthrow and Indian clearing and successional stages after forest fires and other disturbances. We have created a landscape in which once-rare forest-edge species have become common and common forest species rare. Through most of the primeval biome, for example, robins must have been scarce and wood thrushes common.

Table 15-5 lists some important species of deciduous forest edge. Many of these species seem to require patchiness in the vegetation—for example, they may nest in trees but forage in grassy areas.

Figure 15-12

The eastern gray squirrel (top) is typically a species of unbroken deciduous forest. As forests were cut over and fragmented, gray squirrels decreased and the larger, savanna-inhabiting eastern fox squirrel (bottom) increased. (Top, E. P. Haddon, U.S. Fish and Wildlife Service; bottom, Dave Menke, U.S. Fish and Wildlife Service.)

Table 15-5
Some Important Forest-Edge Species in the Deciduous Forest Biome

Reptiles and Amphibians	*Birds*	*Mammals*
Six-lined racerunner	American kestrel	Eastern mole
Common garter snake	Northern bobwhite	Short-tailed shrew
Hognose snake	Mourning dove	Eastern cottontail
Blue and black racers	Common nighthawk	Woodchuck
	Red-headed woodpecker	Fox squirrel
	Common flicker	Red fox
	Eastern kingbird	Striped skunk
	American crow	White-tailed deer
	Blue jay	
	House wren	
	Gray catbird	
	Brown thrasher	
	Mockingbird	
	Eastern bluebird	
	American robin	
	Cedar waxwing	
	Warbling vireo	
	Blue-winged warbler	
	Yellow warbler	
	Yellowthroat	
	Yellow-breasted chat	
	Brown-headed cowbird	
	Northern oriole	
	Common grackle	
	Indigo bunting	
	Northern cardinal	
	Rufous-sided towhee	
	American goldfinch	
	Chipping sparrow	
	Field sparrow	
	Song sparrow	

Figure 15-13

The northern cardinal, a widespread forest-edge bird. (Photo by George and Betty Walker.)

The group is far from uniform, however; each species has different habitat requirements reflecting various aspects of the primeval vegetation. Cardinals (Figure 15-13) and indigo buntings seem to have been originally birds of forest openings; they sing from good-sized trees but nest in a tangle of vegetation. Such birds as common flicker, red-headed woodpecker, blue jay (Figure 15-14), warbling vireo, and northern oriole seem to be savanna birds, nesting where there is an open growth of fair-sized bigger trees. Shrub-carr birds, nesting in the dense thickets near wet areas, include goldfinch, yellowthroat, song sparrow, and rufous-sided towhee.

Forest-edge communities occur in the river-fringing woods far out into the Great Plains, and the riparian communities of the West share many of the same species.

Southeastern Evergreen Forest

This region occupies much of the coastal plains along the Atlantic and Gulf states. It has hot summers and a long growing season (Table 15-1). Winters are mild enough that there is no period in which the ground is frozen or snow covered. Damaging tropical storms, including hurricanes, recur

Figure 15-14

The blue jay is a bird of wide occurrence but large numbers of it now occupy forest edge. (Photo by George and Betty Walker.)

on a time scale of years or tens of years (Nelson and Zillgitt 1969). Thunderstorms are frequent, occurring on 50 or more days per year, and are not necessarily accompanied by rain. Such "dry lightning" sets frequent fires. Fire occurrence rates run from high to very high (282 or more fires per million acres per ten years; Nelson and Zillgitt 1969). This rate is similar to that for the adjacent oak-pine forest but much higher than the rates for most other parts of the deciduous forest biome. Soils are mostly ultisols, spodosols and, on the extensive floodplains, inceptisols.

Like the oak-pine region of the deciduous forest biome, the uplands of this region are dominated by pine stands maintained by fire (Figure 15-15). If lands are protected from fire, pines are replaced by hardwoods. Perhaps the most convincing demonstration of the inexorable succession to hardwoods on fire-protected land comes from a study on a loblolly-shortleaf pine stand in southern Arkansas (Cain and Yaussey 1984).

In 1951, all hardwoods more than 9 cm diameter at breast height (dbh) on four test plots (total size 1 acre, 0.4 ha) and for 400 m all around were girdled. Smaller hardwoods on the plots down to 2.5 cm were cut and the stubs treated with a herbicide (AMS). Hardwoods less than 2.5 cm were sprayed with herbicide. The next year, 1952, all hardwood sprouts were sprayed with herbicide. For the next ten years, every hardwood seedling or sprout was grubbed out by hand. In the spring of 1962, the total number of hardwood stems in the entire acre was 19, which were grubbed out. The hardwood control measures were then discontinued.

Eighteen years later, the plots were re-examined. Table 15-6 summarizes the results. With continued fire protection, hardwood forest is just a tree generation away even after this treatment.

Vegetation

Upland Hardwoods

The composition of the mature fire-protected community varies by region and site. Northward, it is oak-hickory. Southward and also on favorable sites in the North, it is a mixed hardwood forest in which beech is important. Magnolia is usually a codominant, and other typical trees are spruce pine, water oak, sweet gum, and pignut hickory (Platt and Schwartz 1990). Magnolia, American holly, and several other species are evergreen components in a largely deciduous forest type in northern Florida (Figure 15-16). Among the shrubs and herbs of these forests are several that also occur in more northern mesophytic forests. Examples include hophornbeam, flowering dogwood, Christmas fern, and Indian pipe.

By mid-Florida, many of the deciduous species become less important or drop out and fire protection yields a forest that is mostly broad-leaved evergreens. A. M. Laessle (1942) recommended restricting the term "hammock" to broad-leaved evergreen forests, but in current usage all hardwood forests in Florida are apt to be called hammocks, including the mostly deciduous forests of the northern part of the state. Mid-Florida hammocks on mesic sites are dominated by magnolia and American holly but several other trees, mostly

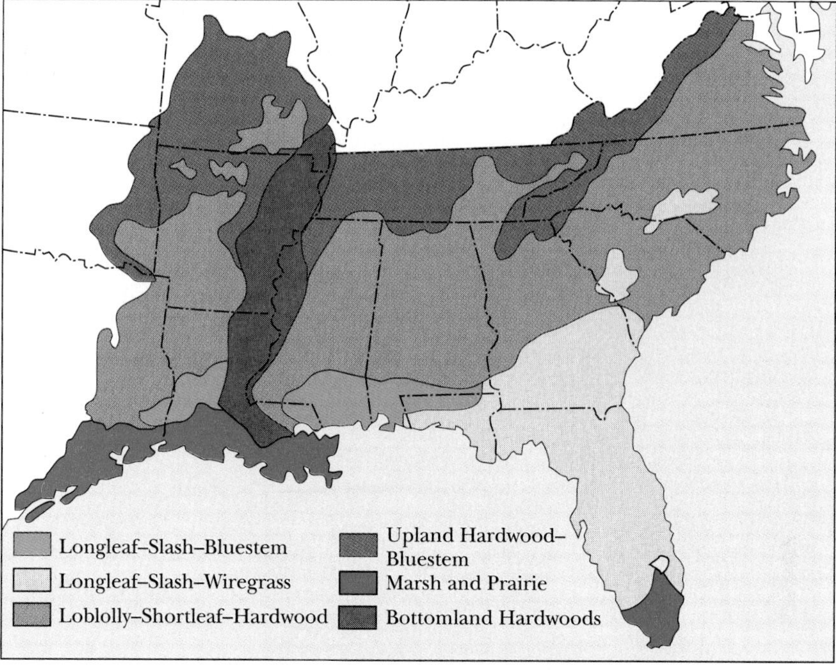

Figure 15-15

This map of range types in the southeastern U.S. includes not just the Southeastern Evergreen Forest but also oak-pine forest and the other adjacent southern subdivisions of Temperate Deciduous Forest (see Figure 14-4). (Adapted from Grelen & Hughes 1984.)

evergreen but some deciduous, may be present. Examples are laurel oak and sweet gum. Saw palmetto and staggerbush are characteristic shrubs. Herbs are sparse.

Evergreen forest of live oak (xeric hammock) occupies drier sites and occurs also along a narrow belt close to the southern Atlantic and Gulf coasts. Other trees include laurel oak and cabbage palm.

Table 15-6
Understory in a Loblolly-Shortleaf Pine Forest in Southern Arkansas 18 Years After Complete Eradication of Hardwoods on the Site

Species	*Stems per Acre*
Red oak	480
Pines	273
Red maple	262
White oak	150
American holly	65
Elm	58
Other hardwood	491
Total	1649

Based on Cain, M. D., and Yaussey, D. A., "Can hardwoods be eradicated from pine sites?" *South. J. Appl. Forestry,* 8:7–13, 1984.

Shrubs are numerous, and grasses are important in the understory. The epiphyte Spanish moss is characteristic of these woodlands, although it occurs in deciduous communities also.

One of the striking changes with latitude along the Florida peninsula is the jump in numbers of epiphytic species southward. Besides Spanish moss, there is a single epiphytic bromeliad in the Florida panhandle but a dozen species just north of the Everglades, 300 miles (500 km) south (Clewell 1985, Wunderlin 1982). The panhandle has one epiphytic orchid whereas 300 miles (500 km) south there are about ten. Most species of epiphytes grow in wet forests—hydric hammock and cypress swamps.

The southern third of Florida is a mixed bag including many tropical species of plants and animals, with new ones arriving daily—or, at least, frequently. Tropical hammocks on south Florida rocklands (that is, limestone ridges) include many plants of Caribbean affinities that do not occur elsewhere on the North American continent. Gumbo limbo, the legume wild tamarind, pigeon plum (whose dark red fruits are much prized by white-crowned pigeons), and strangler fig are characteristic trees, but tropical hammocks have a diverse tree flora of more than 150 species (Snyder et al. 1990).

Figure 15-16

Mesophytic hammock in northern Florida. The large tree center left is American beech and the one center right is southern magnolia. American holly, sweetgum, and pignut hickory are also important members of the canopy. (Photo by the author at Woodyard Hammock, Leon County, Florida.)

Pinelands

There are two major types of pinelands. Well-drained soils tend to support sandhill vegetation, savannas or open forests of longleaf pine and turkey oak. If fires are frequent, the very fire-resistant pine tends to predominate. Logging, as formerly practiced, tended to favor the oaks. The herbaceous layer is dominated by wire grasses.

Poorly drained uplands support flatwoods vegetation (Figure 15-17). Seasonally dry flatwoods that tend to be burned annually or nearly so support longleaf pine with wire grass and various xeromorphic shrubs. Flatwoods that are more likely to remain wet year-round and have less frequent fires tend to be occupied by slash pines.

Figure 15-17

A recently burned area of slash pine flatwoods with cabbage palms and small oaks in Citrus County, Florida. (Photo by the author.)

Prior to 1700, what is currently called the ''longleaf pine ecosystem'' occupied a high percentage of the coastal plain from North Carolina to east Texas. This broad ecological unit includes all sites with longleaf as the dominant tree, upland and lowland, forest and open. Included are some ''sa-

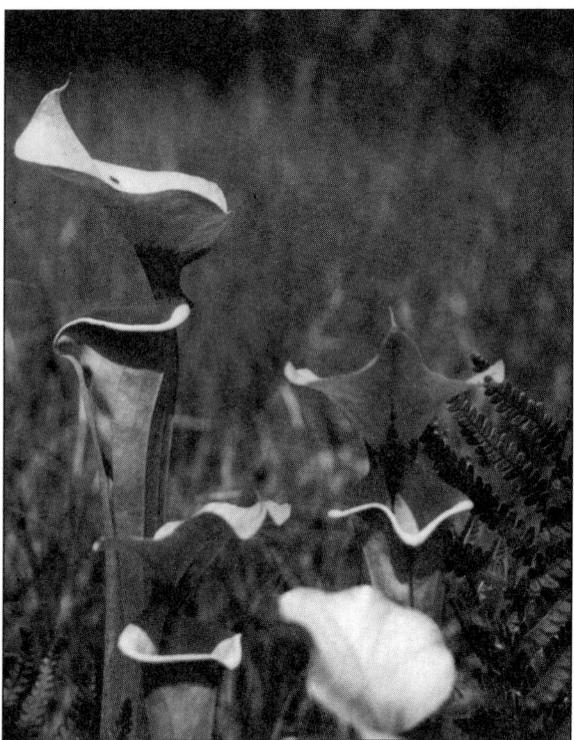

Figure 15-18

Trumpets are a southern species of pitcherplant that grow in grassy bogs of the southeastern coastal plain. (Photo by the author.)

vanna'' sites so open that they would be called prairies in the North and some so wet, with sphagnum, pitcherplants (Figure 15-18), and pipewort (''hatpins''), that they would be called bogs.

Like the white pine ecosystem of the northern hardwood forest, the longleaf pine ecosystem was decimated following European settlement, but the details are different. A considerable amount of clearing for agriculture went on in the Colonial period, yielding cotton and tobacco fields. The pines were tapped for pitch and turpentine production, a process that was deleterious to vast acreages all through the longleaf region (Frost 1991). Most virgin longleaf pine stands were logged out between 1870 and 1920.

Beginning in the 1940s, natural pinelands began to be converted into slash and loblolly plantations, cut on short rotation, and subjected to intensive site preparation that removes much of the existing vegetation. Currently, less than 3% remains of natural upland vegetation types in the original longleaf pine range. Recently, however, interest has been shown in restoring the longleaf pine ecosystem and managing it on ecological principles with broad conservation aims, rather than to maximize short-term timber production (Myers 1991, Neel 1991). Natural regeneration, single-tree selection for cutting (rather than clear-cutting), and the use of prescribed fires every other year (Figure 15-19) are involved. More discussion of fire in the longleaf pine ecosystem is given in Chapter 3.

This region is one of the important timber-growing areas of the United States; however, in recent years, a general decrease in pine growth rate has occurred (Sheffield et al. 1985). This has occurred throughout the Southeast, on the Piedmont and in the mountains as well as on the coastal plain. Figure 15-20 suggests a 30 to 50% reduction in growth rate of slash pine on the Georgia coastal plain, and reductions have also been measured for loblolly, shortleaf, and longleaf pine. There are several possible causes. One is ''atmospheric deposition,'' that is, acid rain and related types of air pollution that could be acting directly or making the forests more susceptible to other adverse effects, such as disease or drought stress.

Hydric Communities

In this low-lying region where the Mississippi and other rivers run to the sea, extensive wetlands exist. Some of these are marshes and bogs, including extensive grassy marshes called ''prairies'' in the region.

The Everglades is a vast, grassy marsh that once occupied a large section of the Florida peninsula south of Lake Okeechobee (Kushlan 1990). Today it is much more restricted, less than one half the original size, as the result of water diversion and conversion of land to other purposes, sugar cane near Lake Okeechobee and winter vegetables such as tomatoes and strawberries further south. The dominant natural vegetation is a tall sedge, called sawgrass, but islands of trees occur on slightly higher ground (limestone ridges) and also slightly lower ground (George 1988) (Figure 15-21). There is no ground that is more than slightly higher or lower.

The basis for the functioning of the Everglades lay in a combination of factors. In the summer, the Everglades was an 80-km-wide shallow river carrying

Figure 15-19

A prescribed burn in longleaf pine forest, specifically the Wade Tract in southern Georgia. The seeming narrowing of tree trunks at the fire line is an artifact of the photographic process. (Photo by Todd Engstrom.)

the overflow from Lake Okeechobee. In the dry season, the flow dropped, and the vast sawgrass areas became meadows that were susceptible to fire which tended to keep out invading woody species. This natural schedule has been completely disrupted by the U.S. Corps of Engineers and the South Florida Water Management District in concert with agricultural and commercial development interests of the region (Boucher 1991). An added complication is pollution of the water reaching the Everglades by high levels of phosphorus from the agricultural lands. Cattail tends to replace sawgrass in these overfertilized areas.

W. B. Robertson, Jr., one of the great students of Everglades ecology, commented (1959) that what we see today is "only a remnant and a memory of what was one of the great wildlife spectacles of the North America." William Bartram in 1774 and John James Audubon in 1832 described egrets, storks, and other birds in such numbers that they could not credit their eyes; the noise from the immense flocks kept Bartram awake. "Much water has passed down the Everglades since the days of Bartram and Audubon," Robertson wrote, "and much that should have come down has flowed elsewhere." The outlook for the Everglades seems more optimistic now than for many years. Strong conservation measures have been proposed that, if fully implemented, should halt the degradation (Boucher 1991, Kanamine 1991).

Forested wetlands, or swamps, vary in size from small gum ponds and cypress domes set in a matrix of dryland vegetation up to enormous wetlands like the Okefenokee of southern Georgia-northern Florida. Swamps can be classified as river swamps and stillwater swamps. Structure and species composition of stillwater swamps are related to hydroperiod (how long the soil is water saturated) and fire frequency (Ewel 1991). Fires tend to occur during the dry season of dry years, and a high fire frequency would be one fire per decade. Pond cypress is the most common tree in stillwater swamps, but a variety of others including pines, palms, and hardwoods occur.

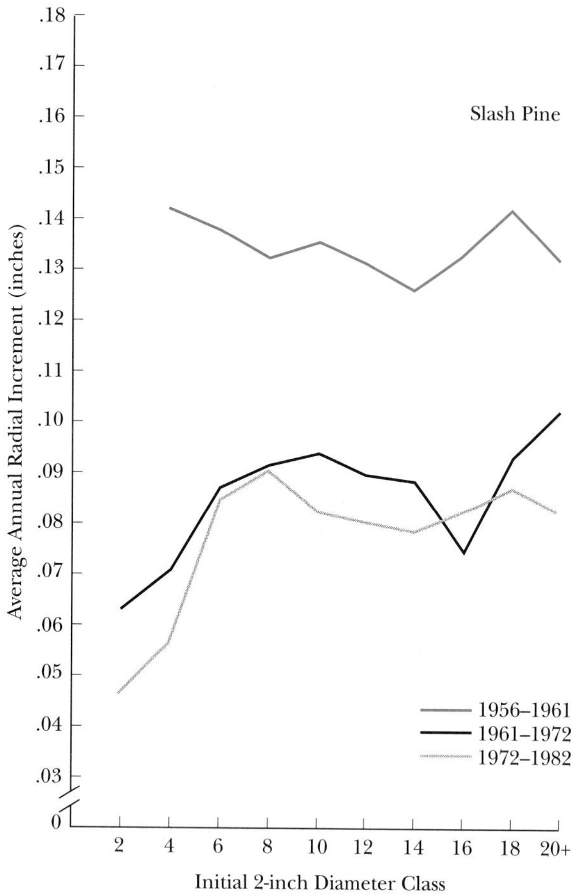

Figure 15-20

Growth rate of slash pine on the Georgia coastal plain during three time periods. Recent growth has been much slower for trees of all sizes compared with the period 1956–1961. (Adapted from Sheffield et al. 1985.)

Along the larger rivers, four types of lowland forest are recognizable:

1. Stream margin communities. These communities support black willow, sycamore, hackberry, and pecan.
2. Swamp forests. These occur inland from the natural levee that usually runs alongside the stream. The land is usually underwater throughout the year. Bald cypress and tupelo gum characteristically dominate these forests (Figure 15-22).
3. Hardwood bottoms. Higher areas that are flooded for varying lengths of time in the spring are very extensive and support sweet gum, red maple, shingle oak, overcup oak, and several other tree species. Live oak becomes important on such sites near the Gulf coast.
4. Ridge bottoms. Still higher areas on the floodplain, flooded only during high-water years, support sweet gum, oaks, shagbark hickory, and pecan. These sites supported formerly dense growths of cane, but most have been cut for fishing poles or destroyed in logging operations.

Scrub

Sand-pine scrub (Figure 15-23) is a distinctive vegetation type occurring on well-drained, sterile sand in the interior of Florida (Laessle 1958). It is a two-layered community with sand pine above and often dense evergreen shrubs, including several scrubby oaks, staggerbush, fetterbush, and semaphore cactus (Figure 15-24) below. The ground cover is largely mosses and lichens. The community is fire dependent, but fires are infrequent because of the sparse understory. When fires come, they are crown fires that kill off most of the vegetation (Laessle 1942). Most of the shrubs root-sprout readily. Sand pine reseeds quickly from serotinous cones (Myers 1985), but for a portion of the fire cycle it is a negligible part of the cover. The scrub jay is a characteristic bird occupying a small patch of habitat more than a thousand miles disjunct from the bird's main range in the southwestern United States.

Animals

The upland broad-leaved forests of this region tend to have a temperate deciduous forest fauna, although with a southern flavor. Such species as the northern parula, hooded warbler, blue-gray gnatcatcher, Acadian flycatcher, Carolina wren, and cardinal are as apt to be among the leading species as the red-eyed vireo. Many deciduous forest species do not reach this far south (Table 15-3), but several do, including gray squirrel, southern flying squirrel, gray fox, and bobcat. A few distinctive species are added. Examples are the golden mouse, a small tree-nesting rodent; the extremely rare red hills sal-

Figure 15-21

The Everglades, a vast "river of grass," with scattered islands of trees. The saw grass dominating the landscape is actually a sedge. (Photo by the author.)

Figure 15-22

Swamp forest in an ox-box lake, Alexander County, Illinois. Both bald cypress seen at the right and tupelo gum in the background have a similar growth form with expanded trunk bases. The water is covered with duckweed. (Photo by the author.)

Figure 15-23

Sand pine scrub, Ocala County, Florida. Palmettos and scrubby oaks are seen in the foreground, sand pine in the background. (Photo by the author.)

Figure 15-24

Semaphore cactus in sand-pine scrub, Ocala National Forest, Florida. (Photo by the author.)

amander that makes burrows in mesic ravines in a small section of southern Alabama (Mount 1975); and the Seminole bat that roosts in Spanish moss (Golley 1962).

Although the lowland forests also have many deciduous forest species, they have a larger proportion of distinctive southeastern species. This is true for the mammals, which include the nine-banded armadillo (formerly occurring no further east than Texas but now widespread in the Southeast), swamp rat, and cotton mouse. Among the birds, also, are some distinctively southern species, although many of these have suffered greatly from human activities. The rare and declining Bachman's and Swainson's warblers are canebrake species. The Carolina parakeet, last observed in the wild around 1900, nested in the floodplain forests in cavities in large trees. It traveled in large, noisy, colorful flocks, feeding on seeds, nuts, and fruits.

The ivory-billed woodpecker of the southeastern United States lasted until the 1940s but is all but extinct now. A small population of a different subspecies still exists in Cuba. Like the parakeet, it centered its activities in large swamp-forest trees, from which it roamed over an enormous home range of perhaps 8 km^2, feeding primarily on the insects of standing dead trees.

More than the other groups, however, it is the reptiles and amphibians that give the southeastern evergreen region its distinctive fauna. Some deciduous forest biome species are important, but there are many species not found further north. Alabama and Florida, for example, have 130 to 140 species of reptiles and amphibians, whereas Illinois has slightly fewer than 100 (Mount 1975) and Michigan has only about 50.

Some of the species prominent in the lowland forests are listed in Table 15-7. More aquatic habitats—marshes, ponds, and lakes—have an even larger number of distinctive forms including the American alligator (Figure 15-25) and a great variety of turtles and frogs. Among the characteristic birds of open wetlands are the snail kite with its specialized apple-snail diet, gallinules, the limpkin, and several herons and ibises.

The pinelands have a mixture of grassland, deciduous forest edge, and distinctive pineland species, the relative contributions varying with the openness of the stand. A few species that we tend to think of as more characteristic of continuous deciduous forest than of edge also occur. The only common summer grassland bird species is the eastern meadowlark. Edge species include northern bobwhite, mourning dove, Carolina wren, white-eyed vireo, northern cardinal, and rufous-sided towhee. Species of deciduous forest include red-bellied woodpecker, downy woodpecker, tufted titmouse, Carolina chickadee, and summer tanager. Pineland species are pine warbler, brown-headed nuthatch, Bachman's sparrow, and the endangered red-cockaded woodpecker (Figure 15-26).

This last species once occurred widely in the longleaf pine area. It has now been reduced to a few locations most of which are in North and South Carolina, Georgia, and Florida. The species requires large living pines, 70 years old or older, in which it drills holes for nesting and roosting. It seems not to tolerate hardwoods, deserting cavities in which a hardwood understory reaches the height of the entrance. The former wide distribution of

Table 15-7
Some Characteristic Mammals, Amphibians, and Reptiles of Lowland Forests and Pinelands in the Southeastern Evergreen Forest

Lowland Forests	*Pinelands*
	Mammals
Opossum	Fox squirrel (usually where oaks are also present)
Nine-banded armadillo	Southeastern pocket gopher (sandhills)
Swamp rabbit	
Gray squirrel	
Cotton mouse	
Wood rat	
Raccoon	
	Amphibians
Bird-voiced tree frog	Squirrel tree frog (flatwoods)
Mole salamander	Little grass frog (flatwoods)
Dusky salamander	Flatwoods salamander (flatwoods)
Slimy salamander	Dwarf salamander (flatwoods)
	Reptiles
Mud snake	Indigo snake
Red-bellied watersnake	Pine woods snake
Banded watersnake	Coral snake
Cottonmouth	Diamondback rattlesnake
Timber rattlesnake	Gopher tortoise (sandhills)

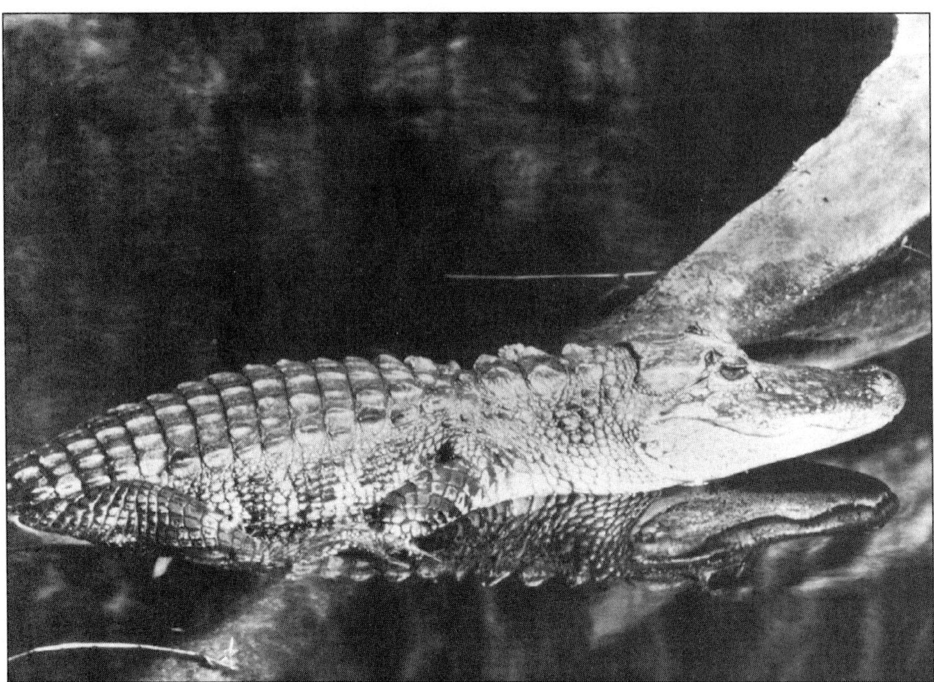

Figure 15-25

American alligator, Sanibel Island, Florida. (Photo by George and Betty Walker.)

the species is a testimony to the effectiveness of fire in excluding hardwoods in the region.

It is not hardwood encroachment, however, that has led to the endangerment of the red-cockaded woodpecker. Southern pines are cut for pulp at about 25 years of age or for sawtimber at about 40, and so few pines in managed stands—and there are not many nonmanaged ones—ever reach the size that the woodpeckers require. For many years, including the decade after the red-cockaded woodpecker was placed on the endangered species list, the U.S. Forest Service resisted changes in timber practices that would have protected the birds. The result was continued decline (Conner and Rudolph 1989, AOU Committee 1991, James 1991). Recently, the U.S. Forest Service has been persuaded that conservation values need to be included along with economic benefit to the timber industry in their management considerations. A developing cooperative approach involving many agencies and interested parties may yet preserve the species from extinction. The approach may include preservation of remaining old-growth longleaf pine, lengthening rotations and other measures to allow development of more old growth, and developing techniques that allow the retention of red-cockaded woodpeckers in a predominantly second- and third-growth landscape (Lennartz 1988, Hooper et al. 1991).

The mode of growth of pines rarely leads to the formation of natural cavities, and the red-cockaded woodpecker is the only pineland species in the Southeast capable of excavating in live wood (Harris and Skoog 1980). In this warm, moist climate standing dead trees, which might provide opportunities for cavity excavation by other species, disintegrate rapidly. As a consequence, the red-cockaded woodpecker is a keystone species for the other pineland animals that depend on cavities, such as other woodpeckers, flying squirrels, and bats. In fact, usurpation by red-bellied woodpeckers and southern flying squirrels of the laboriously drilled cavities is one of the red-cockaded woodpecker's problems. It is a problem, rather than simply an interesting coaction, because of the limited amount of remaining habitat acceptable to the red-cockaded woodpecker.

Figure 15-26

The red-cockaded woodpecker, an endangered species characteristic of older stands of longleaf pine. (U.S. Department of Agriculture.)

Figure 15-27

The gopher tortoise is characteristic of areas of longleaf pine and sandpine scrub. With its broad, flattened forelimbs, it digs burrows that are also inhabited by a variety of other vertebrates. (Gary Meszaros/Dembinsky Photo Associates.)

A number of distinctive mammals, reptiles, and amphibians inhabit the pinelands (Table 15-7). Two sandhill species dig burrows that have important effects on the habitat and other species. These are the gopher tortoise (Figure 15-27), called simply "gopher" by the locals, and the southeastern pocket gopher—called "salamander" by the locals. The good-sized burrows serve as homes for many other animals, among them burrowing owls, raccoons, opossums, the Florida mouse, gopher frogs, pine snakes, and diamondback rattlers (Cox et al. 1987). The indigo snake occurs mostly in wet areas but is able to invade the dry sandhills in localities where tortoise burrows offer protection from desiccation (Jackson 1985). The sandy soil deposited around the burrows returns leached nutrients to the surface and provides a different habitat for plant invasion. Both tortoise and pocket gopher are herbivores, the mammal specializing on underground plant parts and the reptile on leaves and fruits. The foliage and plumlike fruits of the gopher apple are favorite foods of the tortoise, which disperses the seeds in its feces (Laessle 1942). Tortoise populations have been decimated within the past couple of decades and, with them, populations of several of the species that depend on their burrows.

Breeding-bird populations vary by forest type. Pinelands have low populations, 100 to 200 pairs per 40 ha, the higher populations occurring in more open stands. Upland oak-hickory forests have similar densities. Mesic hardwoods run 200 to 300 pairs per 40 ha. Wet forests may have high populations, 300 to 400 pairs per 40 ha.

Few breeding species leave the region for the winter, but many northern birds migrate in. As a result, many communities have a greater number of species and often higher populations and biomass in the winter compared with summer. Some of the more abundant winter additions include Savannah sparrow, short-billed marsh wren, and palm warbler in open habitats and cedar waxwing, ruby-crowned kinglet, yellow-rumped warbler, and hermit thrush in forested ones. Red and hoary bats also winter in the Gulf states.

Community and Ecosystem Ecology:
Biomes of the High Latitudes and Elevations

The polar bear is emblematic of the Far North. *(Gerry Atwell, U.S. Fish and Wildlife Service.)*

OUTLINE

- Arctic Tundra
- The Antarctic
- Boreal Coniferous Forest
- Mountain Zonation
- Alpine Tundra
- Montane and Subalpine Forests
- Shrublands and Woodlands

Here we continue descriptions of the world's biomes. This chapter concentrates on the biomes of the high latitudes—arctic tundra and boreal coniferous forest. The high elevation homologues of these communities, alpine tundra and subalpine forest, are also described, as are conifer-dominated communities of lower elevations including montane forest and temperate rain forest. A variety of shrublands and woodlands that typically lie between montane forest and the grassland or desert of the flatlands are also briefly considered.

Arctic Tundra

Distribution and Climate

The Russian word "tundra" has been adopted for the vegetation that occurs at high latitudes beyond the tree line in Eurasia, North America, Greenland, and Iceland (Figure 16-1). Tundra in all these localities is similar, sharing many species of plants and

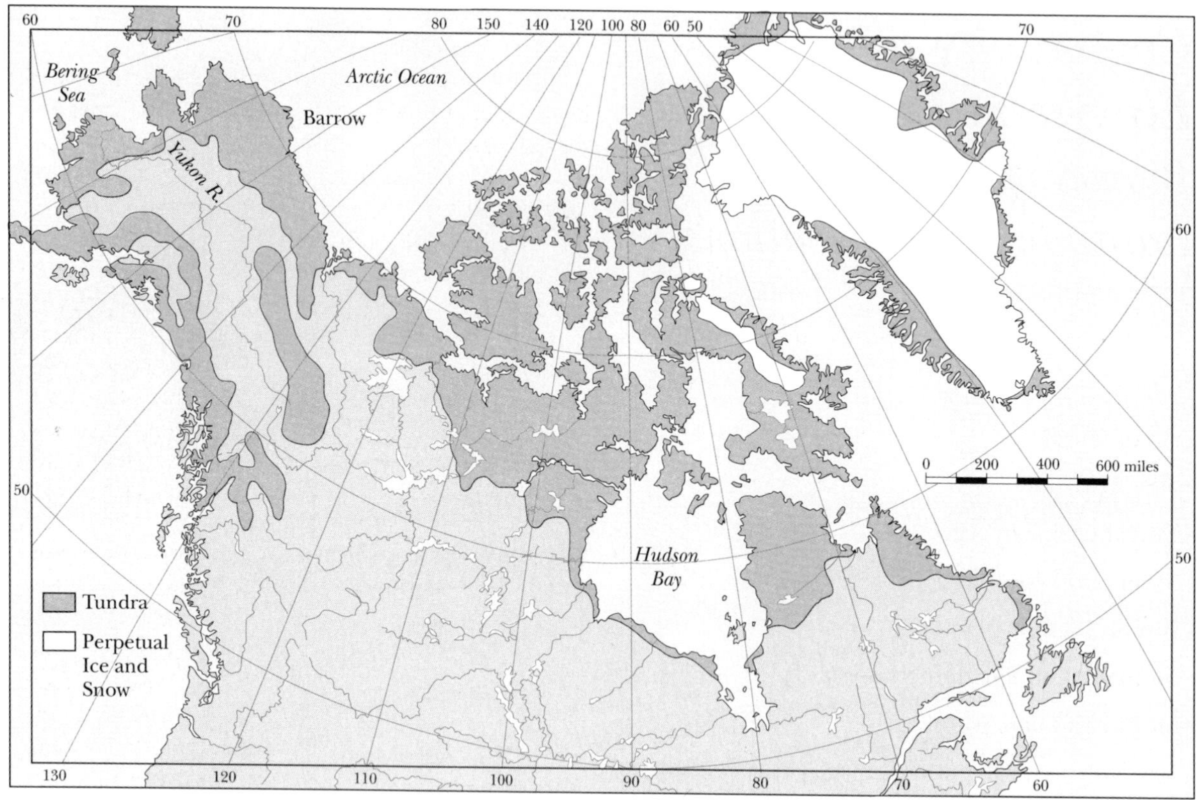

Figure 16-1

Distribution of arctic tundra in North America. There is no clear distinction between arctic and alpine tundra in parts of Alaska and western Canada. (Based mainly on the Department of Forestry map in W. E. Godfrey 1979; Shelford 1959; and Zasada 1976. Base map copyright McKnight and McKnight, Bloomington, Illinois.)

animals. From about 70° north latitude to the North Pole is the high arctic; the more southerly portions are low arctic (Freuchen and Salomonsen 1958).

Tundra growing seasons are short, 100 days or less. Winters are long and cold. Most of us who live in the temperate regions have experienced temperatures as cold as those of the tundra regions—but not for eight months of the year. December in Barrow, Alaska, averages below −20°C, and so do January, February, and March (Barry et al. 1981).

Precipitation is surprisingly low, less than 25 cm per year and less than 10 cm in the high arctic. Most of it falls as rain in summer and autumn. Average winter snow depths are only 10 to 20 cm, although deeper drifts as well as bare patches are created by occasional strong winds.

The short summers are cool; the southern edge of tundra corresponds well with an average temperature of 10°C for the warmest month (usually July). Average daily temperatures move above freezing sometime in May or June, but spring comes slowly because of the frozen soil and continuing cold nights. Freezing temperatures can occur on any day of the year (Figure 16-2).

Photoperiods are long in summer and short in winter. The weather is not a particularly interesting topic of daily conversation in the arctic; although one summer can be warm and another cold, day-to-day changes in weather are slight (Myers and Pitelka 1979). Light intensities are low even in the summer because of cloudiness and the oblique angle at which the sun's rays pass through the atmosphere.

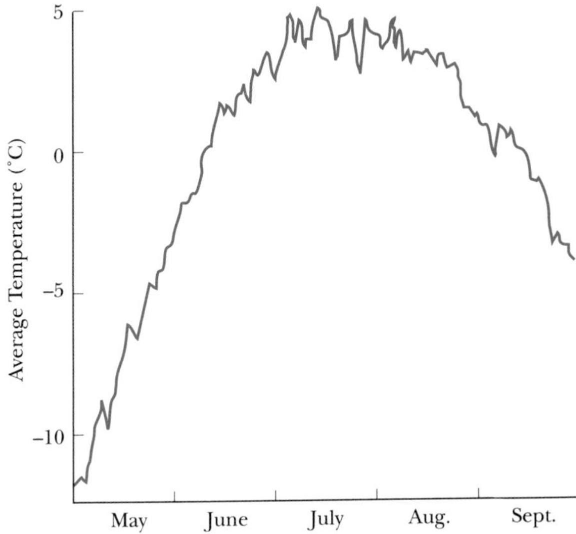

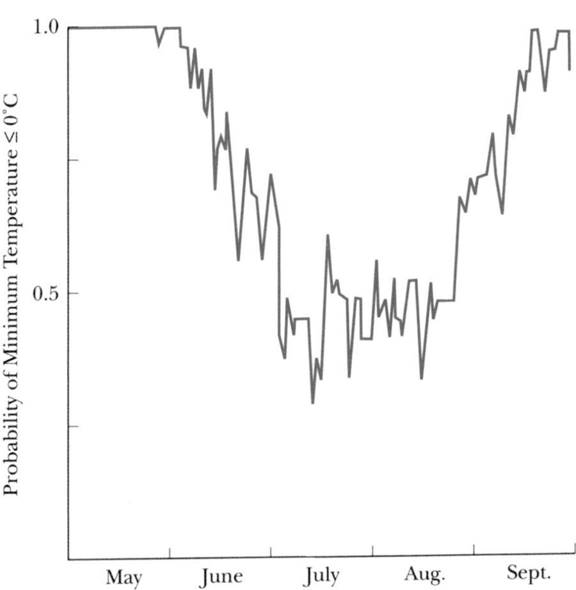

Figure 16-2

Temperatures in tundra. The top graph gives the daily mean temperature from May 1 to September 30 for Barrow, Alaska, located on the Arctic Ocean about 71° north latitude. Average daily temperatures rise above freezing early in June and stay there until early in September. As shown by the bottom graph, however, the possibility of the temperature dropping below freezing is fairly high throughout the summer. The lower graph plots the probability that the daily minimum temperature will be 0 degrees or colder. (From Myers & Pitelka 1979.)

Soil

Although upper layers of tundra soils thaw, lower layers remain frozen throughout the year, forming **permafrost.** As a consequence, drainage is poor and many soils are waterlogged in summer. Ponds, lakes, and bogs (muskegs) are numerous. Wetter soils are histosols; other tundra soils are mostly entisols or inceptisols. A mat of peat generally covers the ground surface. Soils tend to be nitrogen deficient (Everett et al. 1981), presumably because of low microbial activity.

Many tundra soils have a sticky subsurface layer called **glei,** which rhymes with, and means, clay. Glei is blue-gray from reduced iron compounds (because of poor aeration) and amorphous because it gets squeezed between permafrost and the refreezing surface layers every fall. High-arctic soils consist mainly of little-weathered mineral particles.

One of the most striking features of tundra seen from the air is **patterned ground,** rough polygons a few to many meters wide produced by frost action (Drew and Tedrow 1962). The edges of the polygons tend to be raised and composed of coarser materials. The center of the polygon may be low or raised but is composed of finer soil particles. Adjacent polygons are separated by troughs. The total relief within a group of polygons tends to be a few centimeters to a meter but this, combined with the texture differences, is enough to produce obvious differences in vegetation.

Vegetation

Tundra vegetation is short, mostly below 20 cm, with grasses, sedges, mosses, and lichens predominating (Figure 16-3). Herbs, when they occur, often have large, bright flowers. Because of the short growing season, completion of a life cycle in one season is difficult, so there are almost no annuals. Woody plants are mostly chamaephytes such as dwarf willows (Figure 16-4) and birches.

Of course, the tundra begins where the trees end. Although treeline in mountains may be abrupt, the zone between boreal forest and tundra over which trees thin out and become more dwarfed may be many kilometers wide. Several factors could conspire to keep trees out of the tundra: They might not be able to withstand the sandblast-

Figure 16-3

Tundra in Denali National Park, Alaska. A female grizzly bear and her two-year-old are seen in the foreground feeding on blueberries. A grove of black spruce grows on the poorly drained soil on the lowland. (Photo by Michael Gaule.)

ing by windblown snow. Tall plants may lose heat too rapidly, even in the summer, to maintain a favorable year-round energy balance. Or slow water uptake from cold or frozen soils may result in dehydration and death of exposed plant parts. As yet, it is unclear how these possible limiting factors interact.

Types of tundra vegetation tend to occur in a mosaic based on drainage and snow cover. Major types are summarized in Table 16-1.

The high arctic has fewer plant species than the low arctic, about 360 species of vascular plants compared with 600 (Bliss 1981). Even the southern part of the tundra has low diversity by temperate zone standards. There are, for example, more than 1700 species of vascular plants just in Kalamazoo County, Michigan (Hanes and Hanes 1947). Ground coverage is low in the high arctic. Bryophytes and lichens are more important northward and may make up the largest part of the cover in the arctic desert of the far north.

Important plant adaptations to the tundra climate (Savile 1972) include low growth forms—mats, cushions, and tussocks—that lower wind velocities and trap warm air, making favorable energy and water balances easier to achieve. There are many trailing and creeping plants, both woody and herbaceous, perhaps because they can "forage" across the patchy ground for suitable microhabitats (Sonesson and Callagan 1991). Self-pollination is frequent, presumably because insect pollination is undependable. Vegetative reproduction is widespread. Seed dispersal is largely by wind, though some animal dispersal of fleshy-fruited plants occurs.

Net annual primary production is low, generally below 250 g/m^2. Rocky areas of the high arctic may have much lower productivities. Much of the year's production is stored in bulbs and roots. Even at the height of the growing season, 80 to 90% of the plant biomass in sedge-grass tundra is below ground (Wielgolaski et al. 1981). Daily rates of pro-

Figure 16-4

A dwarf willow, pistillate plant in flower, Glacier Bay National Park, Alaska. (Photo by R. Bruce Moffett.)

duction during the short growing season are comparable to those of temperate regions.

Animals

In tundra, the lemmings and voles eat the sedges, and everything else eats the lemmings and voles. That is an oversimplification, of course, but these small rodents are a main support of an astonishing variety of predators from weasels to wolves (Figure 16-5).

Reindeer moss and other lichens are a primary winter food of the caribou, whose main enemy is the wolf, but which is also preyed upon—mainly in the calf stage—by many other carnivores. Where conditions permit willows and other shrubs to be present, other food chains that include ptarmigans and common redpolls are added. Ptarmigans are a prime food of snowy owls, though the owls also eat lemmings.

Detritus food chains, based on the thick litter that accumulates in many tundra habitats, include dense populations of enchytraeids—small, white relatives of earthworms—and insects, mostly flies. Grazing insects are rare. Flies, as larvae during most of the summer and as adults during their brief period of great abundance, are the main food of the Lapland longspur, the snow bunting, and several

Table 16-1
Types of Tundra Vegetation

Tundra Type	*Characteristics*
Sedge-grass tundra	Matlike vegetation of poorly drained areas where various members of the sedge genus *Carex* usually dominate.
Herbaceous tundra	Forb-dominated vegetation, mostly in mountainous areas. Included are saxifrages (*Saxifraga* sp.), cinquefoils (*Potentilla* sp.), and fireweeds (*Epilobium* sp.).
Tussock tundra	Characterized by clumps (tussocks) of cottongrass *(Eriophorum vaginatum)* in areas of moderate drainage. Other plants, such as dwarf birch *(Betula nana)* and Labrador tea *(Ledum palustre)*, are frequently present.
Mat and cushion tundra	Well-drained areas usually dominated by one or another species of a nitrogen-fixing member of the rose family, mountain avens (*Dryas* spp.).
Shrub tundra	Shrub-dominated tundra is taller than the others, having an overall height up to 1.5 m and taller along streams. The most important shrubs are dwarf species of willows *(Salix)*, birch *(Betula)*, bilberry *(Vaccinium)*, and other members of the heath family, along with crowberry *(Empetrum nigrum)*.

Classification based on Viereck and Dyrness (1980).

species of shorebirds. The first two also eat seeds. The insects and other soil invertebrates tend to have long life cycles relative to temperate zone relatives; crane flies, for example, take four years to develop at Barrow, Alaska (MacClean 1980).

The food chains of other groups of animals start in aquatic habitats. Lakes, ponds, and marshes support plants and insects (including hordes of blackflies and mosquitos) that, in turn, support vast numbers of shorebirds and waterfowl. Eiders, snow geese, and the tundra (whistling) swan are characteristic tundra waterfowl.

The sea provides the food base for a coastal arctic fauna that includes many species of colonial seabirds in the auk family—ecological equivalents of the Southern Hemisphere penguins. Polar bears feed on seals at the end of a long food chain beginning with marine algae.

The period in which food and favorable temperatures are available for birds in the tundra is short. Virtually all of them migrate in, raise their young during the brief flush of plant and insect life, and then retreat. Many species arrive already paired (Hoffmann 1973). Migratory birds in general molt in the interval between rearing young and southward migration. That interval is short or nonexistent for tundra birds, and many of them delay their postnuptial molt until they reach their wintering grounds (Morehouse and Brewer 1963, Holmes 1966). In several species of shorebirds, the adults of whichever sex is least involved in taking care of the young begin their southward migration before

Figure 16-5

Simplified energy flow diagram for arctic tundra.

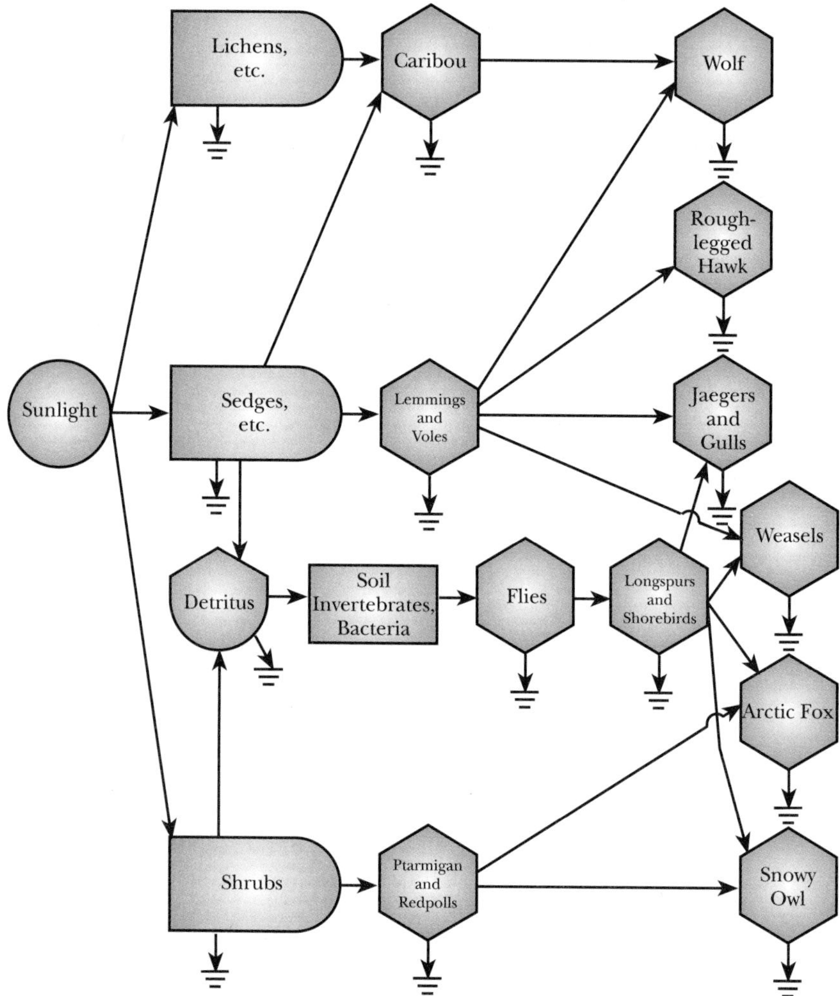

the young are grown—before they are even hatched in some species (Pitelka 1959). Male pectoral sandpipers spend less than a month on the nesting grounds at Barrow, Alaska, and female red phalaropes less than three weeks (MacClean 1980).

Winter bird life in the tundra is sparse. One ornithologist took a Christmas bird count at Bethel in western Alaska and found only three species—267 common ravens, 18 hoary redpolls, and 4 snow buntings. The next year he tried again and found nothing but ravens (Baxter 1984).

Ptarmigan and snowy owls are about the only other birds that stay through the winter. Even these few hardy species depend on special situations, such as food derived from human settlements for the raven and a good supply of exposed willow buds for the ptarmigan. Vast stretches of tundra in the winter have no birds at all. In bad years even the snowy owl and ptarmigan move south. The snowy owl is known as one of the irruptive species that appear far south of the tundra during years of low lemming populations.

Few mammals migrate. The barren-ground caribou (Figure 16-6) winters at the northern edge of the boreal forest, and wolves tend to move with them (Hoffmann 1973), but otherwise there is little mammal migration. Hibernation, likewise, is rare. There is nowhere for them to go. The single mammalian hibernator is the arctic ground squirrel that occurs in western North America and also eastern Siberia. It makes use of one of the rare situations in which the ground does not freeze down to the permafrost—well-drained slopes along streams where the permafrost is fairly deep and where snowdrifts keep the surface soils from freezing deeply.

Neither reptiles nor amphibians live in the tundra. The closest approach is the wood frog, which does cross the Arctic Circle but only in the wooded MacKenzie River Valley in Canada's Northwest Territories (Cook 1984). The short period favorable for growth and the absence of hibernacula are the most likely limiting factors.

Most tundra mammals remain active through the year, depending on various kinds of adaptations to cold to see them through the winter. Excellent insulation is one such adaptation, exemplified by the incredible coat of the muskox that is topped with a dense layer of guard hairs a foot long.

The arctic fox increases its fur depth by about 200% from summer to winter, resulting in a lower critical air temperature—the temperature at which metabolic rate has to be increased to compensate for increased heat loss—somewhere below −40°C. In effect, this means that arctic foxes never encounter conditions under which they must increase heat production to maintain body temperature (Prestrud 1991). The ptarmigan and snowy owl are also unusually well covered including feathers over their feet, which are bare in most birds.

Small animals can use the snow as insulation, living beneath it as do the lemmings and voles (and the weasels that prey on them) or diving into it for roosting, as does the ptarmigan. Willow ptarmigan may remain in burrows in the snow up to 21 hours per day, decreasing their 24-hour energy requirements by about one third (Andreev 1991). Do not think, though, that it is warm under the snow. The temperature may be 10 or 20 or even 50°C warmer than on top of the snow, but it will still be well below freezing (Table 16-2). When lemmings are not generating heat by running about searching for food, they evidently stay warm by huddling together in nests of dry grass and fur (Hoffmann 1973). Reproduction in lemmings occurs under the snow through much of the winter, as well as during the summer, but stops during snow melt and freeze-up.

Female polar bears sleep in caverns or tunnels in the snow near the seacoasts and, like other bears, give birth to their cubs at this season. Males may remain active, feeding on ringed seals that come to breathing holes in the sea ice (Freuchen and Salomonsen 1958).

White coats presumably serve to camouflage animals against the snow; whether they adapt their wearers to arctic conditions in other ways is still unclear. The ermine and other weasels, the arctic fox, the collared lemming, and the various ptarmigan are white in the winter and some shade of brown in summer. Polar bears, snowy owls, snow geese, and the gyrfalcon are among the species that stay white year-round. Most of these are high-arctic dwellers. High-arctic populations of the arctic hare, which in the south turns brown in the summer, remain white all year.

Lemming populations are cyclic, and this has ramifying effects on the ecosystem, many of which are still not well understood. At the peaks, which come every three to six years, populations may be 25 lemmings per hectare (at Barrow, Alaska); at the

Figure 16-6

Caribou at a tundra pond, Denali National Park, Alaska. (Photo by Michael Gaule.)

Table 16-2

Typical Temperature Gradients in Winter and Summer in Tundra

Temperatures above freezing are emphasized in boldface.

Height	*Winter (February)*	*Summer (July)*	
8 m (air)	−25°C	**8°C**	
1 m (air)	−31°	**9°C**	
Snow surface (about 0.3 m)	−35°	—	
Soil surface	−22°	**11°C**	
−0.25 m (soil)	−19°	**1°C**	
−1 m (soil)	−16°	− 3°C	Permafrost
−4 m (soil)	−10°	− 8°C	Permafrost

Based on Figure 3.2 in Barry, R. G., Courtin, G. M., and Labine, C., "Tundra climates," in *Tundra Ecosystems,* Bliss, L. C., Heal, D. W., and Moore, J. J., (eds.), Cambridge, Cambridge University Press, 1981, pp. 81–114.

lows there may be less than 1. At the highs, the lemmings may devastate the vegetation, consuming half the aboveground production in the summer and continuing their destruction in the winter. Such activities speed up nutrient cycling and, by removing the insulating litter, allow the ground to thaw faster and deeper.

Numbers, social structure, and reproduction of many of the predators are keyed to lemming populations. During highs, dense and diverse communities of predators are present (Figure 16-7). During lows, many of the predators do not breed, or even desert the region.

Biogeographical Relations

It is reasonable to consider the whole circumpolar tundra belt as a single biome. As H. A. Gleason and Arthur Cronquist (1964) note, "The same things occur monotonously over acre after acre, from Ungava to Point Barrow and on around the pole."

They were speaking of the flora, but the similarities among the animals are also very high. The polar bear, arctic fox, arctic hare, muskox, caribou (called reindeer in Europe), snowy owl, rough-legged hawk, dunlin, and snow bunting are just a sampling of animal species that are widely distributed throughout North America and Eurasia. The antarctic tundra, discussed in the next section, is not closely related.

Human Occupation

Eskimos (Inuits) have inhabited the New World arctic for perhaps 10,000 years, and Eurasian tribes have made use of the arctic in the Old World for a longer period. The Eurasian peoples, such as the Lapps, however, have remained tied to the boreal forest, living in the tundra only in the summer. The Eskimo developed a culture uniquely attuned to tundra, including such features as a diet almost exclusively of meat. Most hunter-gatherers get most of their calories from plants.

Although hunting and trapping have been an important factor in the lives of many arctic mammal species for decades, western technology has been slow in reaching the tundra. The cold, the six-month-long winter nights, and the problems of constructing roads and buildings on the unstable permafrost-based soils have acted as barriers. As late as 1940 the population of Alaska was less than 75,000. In 1990, it was more than half a million and growing at the rate of 4% a year.

The discovery of oil at Prudhoe Bay on the North Slope in 1968 led to one of the major environmental battles of the early 1970s over the construction of a trans-Alaska pipeline across 800 miles of tundra and boreal forest. Construction was finally authorized in 1973, with the addition of several environmental safeguards (Figure 16-8). How effective the safeguards have been is still a subject of debate. There has been some disruption of drainage and melting of permafrost (Walker et al. 1987). One study (Smith and Cameron 1985) found that caribou on their way to the seacoast for relief from mosquitos were reluctant to cross under an elevated section of oil pipeline. Instead they tended to move parallel, often traveling long distances in apparent attempts to get around it.

It is clear that the trans-Alaska pipeline is only the first of a series of drastic changes that will come to the tundra as western technology invades this fragile ecosystem. Humans have a chance to be more careful stewards of the arctic than they have been of much of the rest of the world. Certain signs are hopeful, such as the Alaska Lands Act of 1980 whereby the United States set aside 56 million acres as wilderness. There were some stirrings of an environmental consciousness in Alaska following the oil spill from the *Exxon Valdez* in Prince William Sound at the southern terminus of the oil pipeline (Lemonick 1989).

Less encouraging were the attempts of the Reagan and Bush administrations to allow oil production in the Arctic National Wildlife Refuge, a 7.7 million-ha preserve east of Prudhoe Bay. The environmental impact statement done for oil development in this national park projected construction of 160 km of pipeline, salt-water treatment

Figure 16-7

Long-tailed jaeger on tundra near Bethel, Alaska. Although related to the gulls, jaegers are predators and, like most predators in tundra, they eat lemmings. When lemmings are scarce, jaegers may not breed at all; when lemmings are abundant, they breed in large numbers and may even have larger clutches. (Photo by Lawrence H. Walkinshaw.)

Figure 16-8

The Alaska oil pipeline, here running through spruce forest northeast of Valdez. (Photo by R. Bruce Moffett.)

plants, gravel roads covering about 2300 ha of coastal terrain, and four airfields, with possible adverse effects on muskox, caribou, polar bear, and snow geese (Walker et al. 1989), not to mention the Indians and Eskimos (Inupiat) native to the region (Yukon Executive Council Office 1988).

The Antarctic

On the high mountain peaks of the arctic and in the Arctic Ocean, we eventually reach zones of permanent ice and snow; at the other pole, however, most of the Antarctic is ice. Half of the 14 million km^2 labeled on a globe as "Antarctica" is the frozen ocean extending out from the land. About 70% of the earth's total fresh water is tied up in the antarctic ice cap (Grotta and Grotta 1992). Only one fortieth of the continent's land mass is visible, as scattered mountains reaching up through the 2-km-thick ice and as narrow bands along parts of the coast, most notably the Antarctic Peninsula, which projects up toward South America.

Rainfall is mostly below 30 cm per year and most of Antarctica has no month with an average temperature above freezing. The *highest* temperature ever recorded at the Russian research station at Vostok is −21°C. "Inhospitable" is a mild word for most of Antarctica—"Great God! This is an awful place," the British explorer Robert Falcon Scott wrote (Hodgson 1990). No aboriginal humans lived in this harsh and isolated land, but there are now permanent residents—at the research stations maintained by about 26 countries—and many sorts of visitors including about 3500 tourists in 1991 (Grotta and Grotta 1992).

Conditions are more favorable on the Antarctic Peninsula and especially some of the antarctic and subantarctic islands. The islands have mild to cool damp summers and winters that, because of the effect of the ocean, have temperatures averaging only a few degrees below freezing. At these sites, vegetation physiognomically similar to arctic tundra occurs (French and Smith 1985). South Georgia Island, at 54° south latitude, has 6 species of ferns and 16 native flowering plants (Stonehouse 1972).

On the Antarctic Peninsula occur only two species of flowering plants, a pink *(Colobanthus quitensis)* and a grass in a genus *(Deschampsia)* that is also widespread in arctic and alpine tundra. Both species have short, dense stems forming a cushion or tussock (Komarkova et al. 1985). Algae, lichens, and mosses form the base of food chains involving mites and collembola (Llano 1962).

The vertebrates of the region are virtually all parts of a highly productive marine food web (Stonehouse 1972). Fish and krill (small crustaceans) form the food of seals, whales, and birds such as penguins and petrels. Most of these birds nest on land in large colonies, where they may be preyed upon by skuas. The wattled sheathbill, one of two species of a peculiar Southern Hemisphere shorebird family, breeds on the Antarctic Peninsula and as far north as South Georgia Island. It walks about, rarely flying or swimming, looking rather like a white pigeon in the park, and eating virtually anything. A pintail duck and a pipit occur on South Georgia Island but not further south.

What happens to the krill is crucial to the vertebrates of the Antarctic, but krill population trends aren't clear. There is some suggestion that, because of the heavy overharvesting of whales, krill have increased, allowing an increase of certain other krill-eating animals. Chinstrap penguins, for example, tripled in numbers between the 1940s and 1980s. Alternative explanations for the increase have, however, been proposed (Carey 1991).

Krill are also harvested by humans, mostly by fleets from Japan and countries of the former Soviet Union. The prevailing opinion seems to be that no more than local overfishing has occurred as yet, but new ways to market krill for human consumption (such as "krillburgers") are being actively pursued. A new concern is the effect of the Antarctic ozone hole on the krill, on the algae on which they feed, and on other organisms. The hole, first detected in 1986, develops in the Southern Hemisphere spring (September to November). The specific concern is the effects of increased ultraviolet radiation, no longer intercepted by the ozone shield.

What is happening to the krill is uncertain, but human effects on several other organisms are well known to have been disastrous. Fur and elephant seals were virtually exterminated in the 19th century, but have increased somewhat in the past few decades. Blue and fin whales were decimated by the mid-1960s (Stonehouse 1972). Currently, heavy harvesting of Antarctic cod, other fish, and squid, particularly by Russian fishing fleets, is of concern (Hodgson 1989).

The 1959 Antarctic Treaty (ratified in 1961 and up to this time signed by 39 countries) provided for protection of the antarctic flora and fauna, but did not cover the seas and pack ice regions. Mineral exploration and development were also not covered. More recent agreements have plugged the second gap, at least partially. The Madrid Protocol of 1991, signed by 31 nations, banned oil and gas exploration for 50 years (Grotta and Grotta 1992).

The unique status of the Antarctic is widely understood. It is a place for international cooperative research on relatively undisturbed ecosystems, on earth history, glaciation and climatology, and as a place for preservation as the last real wilderness. It is debatable whether this understanding will, in the face of ever-increasing human populations, prevent mineral development and the depletion of the seas for human food.

Boreal Coniferous Forest

Distribution and Climate

Coniferous forest occurs as a broad band across northern North America (Figure 16-9) and northern Eurasia (called *taiga* in Russia) and also extends southward at higher elevations in the mountains as subalpine forest.

The climate within this biome is characterized by cold winters and short, mild summers. Growing seasons are about 120 days in the south and shorter northward. The coldest parts of the winter can be very cold indeed, so the annual range of temperatures is large. The warmest month (usually July) has an average temperature of 10 or even 15°C; the coldest month (usually January) has an average temperature of −15 to −40°C. Although precipitation is low, 40 to 100 cm per year, correspondingly low evaporation rates produce a humid climate. Snowfall is heavy.

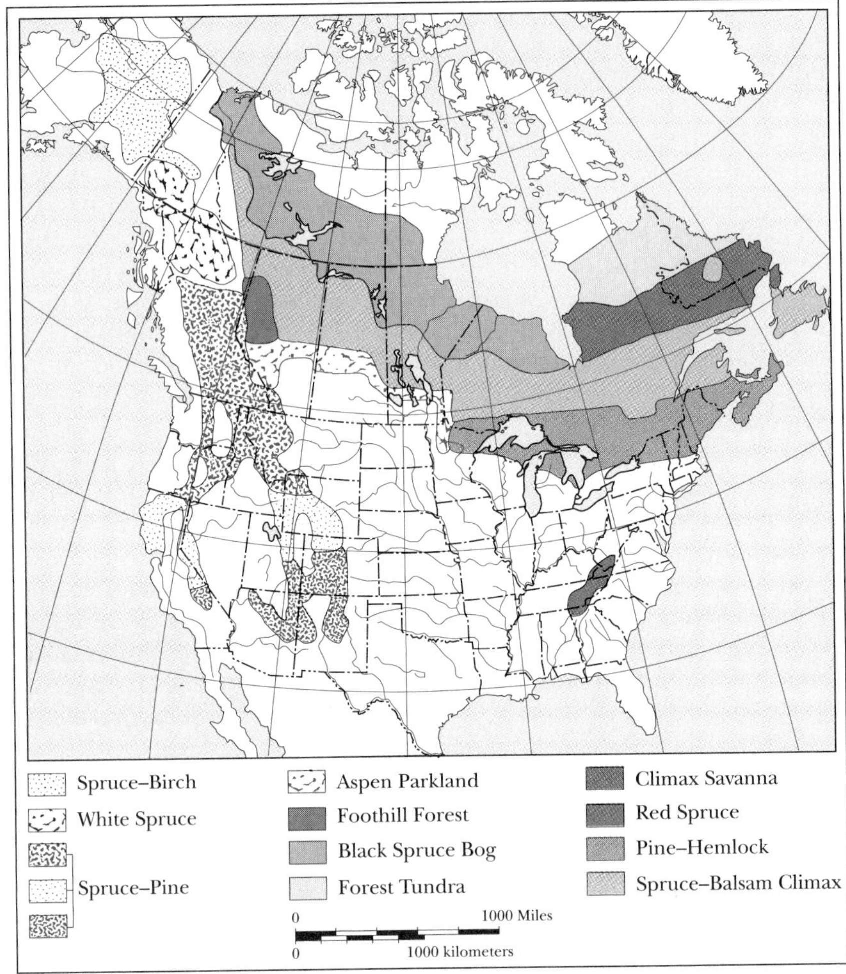

Figure 16-9

The distribution of northern conifer forests is shown in this map. Besides boreal forest, southern extensions of conifers as subalpine forest are shown in the western mountains and the southern Appalachians. (From *The Ecology of North America,* by V. E. Shelford, University of Illinois Press, 1963. Copyright © Board of Trustees, University of Illinois.)

Soil

This is the region for which the term "podzol" was coined. A combination of climatic and vegetational factors produces soils with a thick layer of mor humus that gives way abruptly to a dark gray layer of mineral soil below that looked, to Russian soil scientists, like ashes (*zola,* in Russian; *pod* simply means "ground"). Podzols are acid and infertile, leached of calcium, magnesium, and potassium. Under the U.S.D.A. Comprehensive System, such soils are called spodosols.

Although we tend to think of permafrost as a tundra characteristic, the northern reaches of the boreal forest are also underlain with permanently frozen soil. Southward, permafrost becomes restricted to the coldest microhabitats and disappears altogether around 50° north latitude.

Lakes, ponds, and wet boggy areas with extensive peatland development (muskegs) are common (Figure 16-10). In mountainous areas, soils may be shallow and rocky.

Plants

Boreal forests are usually short, 5 to 10 m, and of low diversity. One or two species of trees may dominate vast expanses. White spruce and balsam fir are often codominants in the eastern half of the biome. Black spruce and white cedar are most often found

Figure 16-10

The boreal forest landscape near Fawcett, Alberta. (Photo by Lawrence H. Walkinshaw.)

in wet areas, where tamarack is a common seral species.

The features of conifers that allow them to dominate the boreal region are probably related to their evergreenness and their needle-leavedness. Being evergreen conveys two related advantages: They do not need to invest energy in a new set of leaves every spring, and they are able to begin full-scale photosynthesis as soon as temperatures warm up in the spring—or any other time. Needles shed snow, especially when the tree itself is spire shaped. Unseasonable snowstorms can do heavy damage to broad-leaved trees.

Disturbance is frequent in the boreal forest. One major source is the spruce budworm. When high populations of this moth build up, they defoliate and kill a large percentage of the trees on a site. Balsam fir is affected more severely than spruce. Secondary succession restores spruce-fir forest on these areas through growth of the surviving trees. It is possible that budworm damage helps maintain spruce as a component of the forest in the face of the aggressively reproducing and very shade-tolerant fir (Gordon 1985).

Retention of dead lower branches and heavy litter accumulation both predispose boreal forest to fires. The vegetation following fire may include grassy areas and savannas or forests dominated by jack pine or broad-leaved trees such as balsam poplar, quaking aspen, or paper birch. The forests that eventually develop may be even-aged, that is, all the canopy trees may be of about the same age.

Features that are advantageous in an environment where fire recurs on a time scale of a few decades include the ability to root-sprout, well developed in aspen, and cones that shed their seeds only after they have been heated. This is seen in black spruce (LeBarron 1939) and, even more markedly, in jack pine. Such cones are termed "serotinous," which has nothing to do with fire or heat but simply means "late"; that is, the seeds are not shed right away.

In the climax forests, shrubs are scarce. Those that do occur, here or in seral stages, generally have fleshy fruits. Examples are cherries, blueberries, gooseberries, and a dogwood only a few inches tall, called dwarf cornel.

Along with dwarf cornel in the herb layer of the climax forest are such species as wild sarsaparilla, beadlily, twinflower, ferns, and mosses; however, the herb layer, like the shrub layer, may be sparse in the heavily shaded, even-aged forests.

Despite the short period between spring and fall frosts, productivity may be fairly high because the evergreen habit allows photosynthesis on any day when conditions (including air and soil temperatures) are even temporarily favorable. Net annual production may be as high as 2000 g/m^2 but more often is 500 to 750 (Gordon 1985). A large part of the annual production accumulates in wood and litter.

Figure 16-11

Moose, one of the characteristic large mammals of boreal forest. (Photo by Michael Gaule.)

Animals

V. E. Shelford called the boreal forest the "spruce-moose" biome, and the moose is one of the characteristic large herbivores (Figure 16-11). To picture it as a permanent inhabitant of unbroken spruce forest is mistaken, however. Twigs of deciduous trees and shrubs from successional and wet areas seem to be the preferred winter food; in summer, leaves of these plants and of hydrophytes from bogs and ponds are added (Shelford and Olson 1935). Woodland caribou is more nearly restricted to the climax forests where, like the barren-ground caribou of the tundra, it specializes on lichens but also eats bark and grass.

White-tailed deer are more characteristic of seral stages. Populations have increased greatly with lumbering to the extent that deer are now the primary large-game animal in many areas where the moose was once prevalent. In the areas where deer have recently entered, moose populations tend to be reduced or eliminated because of the brain worm, a nematode parasite that does little harm to deer but is fatal to moose.

Grizzly bears, cougars, and timber wolves were the main predators of these large herbivores. Smaller herbivores include several mice and voles, red squirrels, the porcupine (a bark specialist), and the snowshoe hare. The last is the prime food of the lynx. Both hare and lynx show "snowshoe" adaptations for travel over the surface of deep snow (Murray and Boutin 1991).

The fisher preys on snowshoe hares, red squirrels, and porcupines. It manages to kill this relatively invulnerable animal by worrying the porcupine's face with quick, small bites (Roze 1989). Repeated wounds bewilder the porcupine enough that the fisher can flip it over and attack the unprotected belly. The predator-prey connection between porcupine and fisher is strong enough that reintroduction of the fisher to areas from which it was extirpated apparently has led to substantial reductions in porcupine abundance.

Another porcupine predator of the North Woods is the wolverine, which looks like a miniature bear although it is actually a giant (15 kg) weasel. Details of the wolverine's food habits seem poorly known. Morris (1965) stated that it eats hare and small rodents taken from ambush, but most of

Murie's (1961) observations of it were as a scavenger.

Wolverine fur has been claimed not to collect and hold moisture, which could be a useful feature for an animal living in a cold and wet climate. Because of this reputation, wolverines have been heavily trapped for trimming parka hoods. Murie (1961) commented, "It so happened that the wolverine fur trimming my own parka hood collected considerable frost, but possibly a wolf or dog trim would have collected even more."

The boreal forest has a diverse avifauna. Breeding populations are high, regularly 300 to 500 birds per 40 ha (Kendeigh 1974, Crawford 1985), but most species migrate out in winter. Year-round residents include the northern goshawk; ruffed and spruce grouse; the great gray, boreal, and hawk owls; three-toed and black-backed woodpeckers (Figure 3-24); the common raven; the gray jay; the boreal chickadee (Figure 16-12); and red and white-winged crossbills. Some of these show substantial southern movement in years of relative food shortage.

Summer populations (Figure 16-13) are swelled by finches, thrushes, and, especially, wood warblers (Figure 16-14). Nearly two dozen warbler species breed, although not all are characteristic of the climax forest. Some species, notably bay-breasted, Cape May, and Tennessee warblers, increase greatly on plots of high spruce budworm infestations (Kendeigh 1947). At subepidemic levels, several species, including blackburnian warbler, golden-crowned kinglet, Nashville warbler, and white-throated sparrow, may consume a large proportion of the budworm population. It seems possible that forest management practices that provided habitat for the largest number of these species—basically some form of selective cutting—could prevent budworm outbreaks (Crawford 1985).

Expectably, arboreal adaptations are well developed among birds and mammals. Included is the

Figure 16-12

Boreal chickadee at a nest cavity, Chippewa County, Michigan. (Photo by Lawrence H. Walkinshaw.)

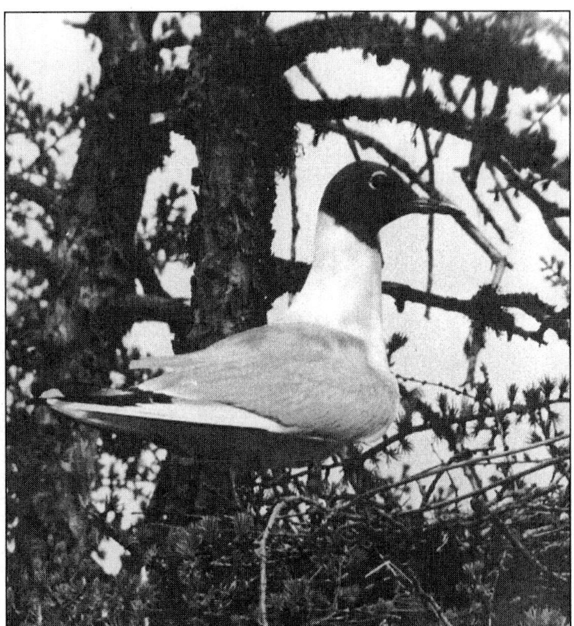

Figure 16-13

One of the more surprising nesting species of the boreal forest is a gull, the Bonaparte's gull, which occurs in conifer forest from Alaska to western Quebec (formerly to Michigan). (Photo by L. H. Walkinshaw, Fawcett, Alberta.)

Figure 16-14

The most abundant birds of the eastern boreal forests are the wood warblers. Here a myrtle, or yellow-rumped warbler, feeds three newly fledged young. (Photo by L. H. Walkinshaw.)

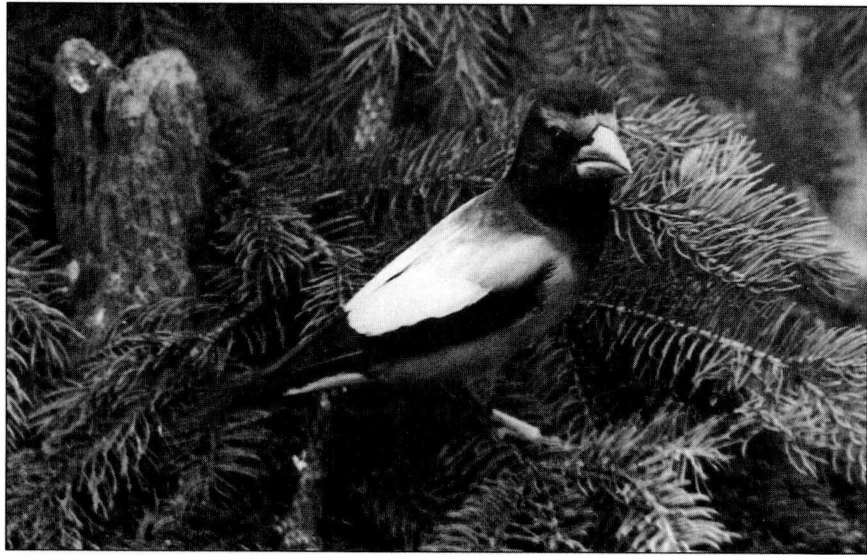

Figure 16-15

Evening grosbeaks have massive beaks with which they can crack cherry pits and other hard fruits and seeds to remove the kernels. The bird is seen here in a white spruce in Alger County, Michigan. (Photo by Lawrence H. Walkinshaw.)

use of trees for nesting (Figure 16-12 and 16-13) and singing. Bird songs tend to be high pitched. Other adaptations include crossed bills for feeding on conifer seeds and massive beaks for cracking cherry pits and other drupes (Figure 16-15). A few species, notably the snowshoe hare and some of the weasels, wear white fur in the winter and brown in the summer. Hibernation is scarcely better developed in boreal forest than in the tundra; the jumping mice hibernate, and various other species such as bears and chipmunks spend much of the winter sleeping. Many mammals are active under the snow, which forms a thicker and more complete insulating layer than in the tundra.

One animal that specializes in hunting through the snow is the great gray owl (Norberg 1987). It has a large head and large facial ruffs that serve to focus sound to the ears, allowing it to pinpoint the undersnow rustlings of the small rodents and shrews on which it feeds. It is the largest owl (about 1 kg) specializing on such small mammals, its size giving it the mass to plunge deeply into the snow and the reach to make the capture.

There are few species of salamanders, although the American toad, boreal chorus frog, northern leopard frog, and wood frog are widespread, and the mink frog is nearly restricted to this biome. Reptiles drop out at still lower latitudes, few reaching even 50° north. One exception is the common garter snake, which occurs through most of the biome and lives beyond 60° north latitude in the Northwest Territories of Canada.

Insects may be abundant. Included are the defoliators such as budworms and the larch sawfly and the various bark beetles that make use of the standing dead timber that these leave. Mosquitos, blackflies, biting midges (no-see-ums), and other flies reach densities at which they become important factors in the movements and behavior of the large ungulates—and of humans.

In a book describing a trip through boreal forest and tundra in western Canada, Ernest Thompson Seton (1911) devoted an early chapter wholly to mosquitos. His aim was to get the subject out of the way so as to avoid mentioning them on every page through the rest of the book. "At Salt River," he wrote, "one could kill 100 with a stroke of the palm, and at times they obscured the color of the horses. A little later they were much worse."

Biogeographical Relations

The boreal forests of North America and Eurasia resemble one another slightly less closely than tundra of the two regions but more closely than temperate forest. Similarities are strong on the generic level. Spruce, fir, larch, and pine dominate the Eur-

Table 16-3
Relatedness Among the Birds and Mammals of North American and Eurasian Boreal Forest

Below are some examples in which North American and Eurasian animals are considered to belong to the same species (conspecific) and others in which the populations are considered to be separate but closely related species.

Mammals	Birds
Conspecific	
Moose (called elk in Europe)	Great gray owl
Wolverine	Hawk owl
Timber wolf (gray wolf in Europe)	Goshawk
Lynx	Bohemian waxwing
	Red crossbill
	White-winged crossbill
	Pine grosbeak
Closely Related Species	
Elk or wapiti (red deer in Europe)	Gray jay (Siberian jay)
American marten (stone marten in Europe; sometimes considered conspecific)	Golden-crowned kinglet (goldcrest)
	Ruby-crowned kinglet (firecrest)

asian forests just as they do those of North America. Birch is important in seral communities, and there are many shared understory plants.

Among animals, many species occur in both regions, or closely related species replace one another (Table 16-3, p. 475). Each region, however, has its own specialties. There is nothing in North America quite like the capercaille, an enormous black grouse with red wattles over its eyes, and the songbird fauna of Eurasia is comparatively drab, lacking the wood warblers.

Ecotones

At the northern border of the boreal forest is an extensive ecotone with tundra (Figure 16-16). Several birds seem more characteristic of this ''forest-tundra'' ecotone than of either of the separate vegetation types (Kendeigh 1974). Included are hawk owl, great gray owl, northern shrike, Bohemian waxwing, gray-cheeked thrush, common redpoll, tree sparrow (Figure 16-17), white-crowned sparrow, and Harris' sparrow.

Along its southern edge, boreal forest in the east forms an ecotone with deciduous forest, the already-mentioned hemlock-northern hardwood forest (Figure 16-9). Westward, from Minnesota to Montana but best developed in Manitoba, Alberta, and Saskatchewan, occurs aspen parkland, considered a transition zone between prairie and boreal forest. Quaking aspen is the most common tree. The fauna is largely deciduous forest edge, although grassland and coniferous forest species are also present. Snowshoe rabbits may be abundant and, by girdling trees and shrubs, affect the composition of the vegetation (Bird 1930). American goldfinch and yellow warbler are abundant birds. Predators include great horned owl, goshawk, and coyote.

Further west and also along the Appalachian chain, boreal forest merges into subalpine forests, discussed in a later section.

Figure 16-16

''Drunken forests'' of black spruce seen in permafrost areas in taiga and boreal forest. The trees are rooted in the shallow layer of soil that thaws in the summer; they are tipped this way and that by creep or flow of this layer. (Photo by Michael Gaule.)

Figure 16-17

The tree sparrow, seen here at its nest near Bethel, Alaska, is a characteristic breeding bird of the wide ecotone between boreal forest and tundra. It winters to the south and is a familiar winter bird in much of the U.S. (Photo by Lawrence H. Walkinshaw.)

Mountain Zonation

The general pattern for vegetational zonation on high mountains in temperate regions is a zone of alpine tundra at the top descending through zones of subalpine and montane forest. In relatively moist regions, as in the Appalachians, montane forest merges with the regional flatland vegetation. In dry regions, a woodland or savanna zone occurs below the montane forest, and nonforest vegetation, such as desert or grassland, occurs at the lowest elevations of the mountain and as the regional flatland vegetation (Figure 16-18).

In arid regions, consequently, there are two timberlines. The upper one is a ''cold timberline,'' related to inadequate summer warmth. The relationship is complex, but generally a two-month period with no more than light frosts is necessary for tree growth (Arno 1984). As a rule of thumb, sites having a warmest-month (July in the Northern Hemisphere) mean temperature below 10°C (50°F) are above the line where trees can grow.

The lower timberline is a ''dry timberline,'' below which moisture is insufficient to support tree growth. This implies, and it is a fact that, without the mountains, trees would not grow in these arid climates. The mountains provide conditions moist enough for forest or savanna growth because of increased humidity and rainfall on the slopes. These both result from lower temperatures in the mountains.

The drop in temperature with elevation is the general basis for mountain vegetation zones and the distributional limits of individual plant and animal species in mountains. Temperature does not act alone, nor is it necessarily the immediate cause. For example, the increased rainfall in mountains occurs primarily because of the expansion and resulting cooling of an air mass as it rises (recall the gas law from chemistry class). As the air cools below

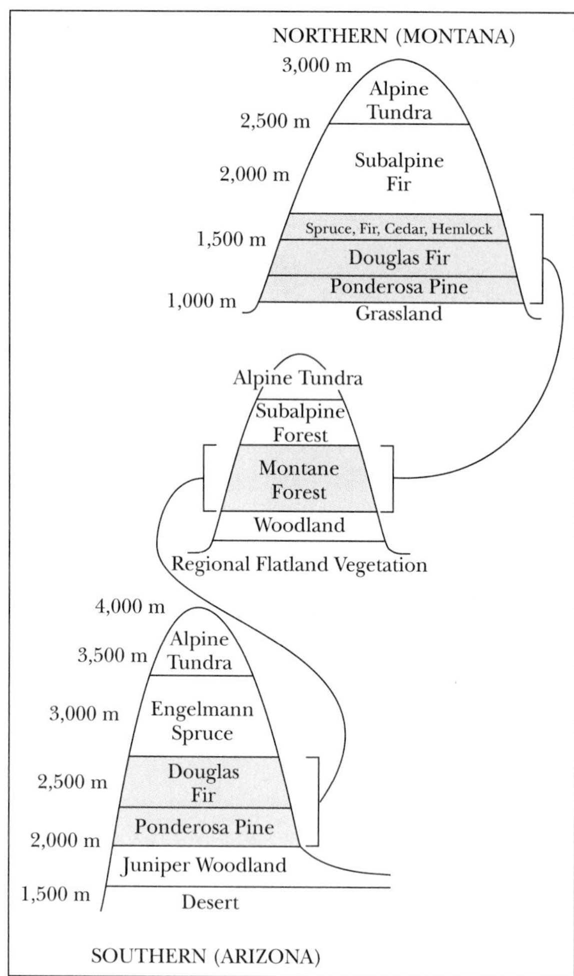

Figure 16-18

Zonation in the Rocky Mountains. A generalized zonation is shown in the middle and more detailed diagrams are given for northern and southern areas. The bracketed areas are the montane zone in each case.

the dewpoint, the moisture in the air condenses, clouds form, and rain falls. This effect is most important on the windward side of mountains. On the lee side, the descending air, from which much of the moisture has been removed, eventually warms above the dewpoint. The belt of dry climate leeward of mountains is called a **rain shadow.**

The relationship between altitude and temperature, whether in mountains or away from them, is complex. The general decline in temperature with altitude (the lapse rate) occurs at the rate of about 6.4°C per kilometer (3.6°F per 1000 ft). This is true up to the tropopause, which in temperate regions, is about 8 miles (12.5 km) up, or higher than we need to worry about in talking about mountain zonation. The underlying basis for the decrease in temperature with elevation is the fact that the atmosphere is warmed very little by incoming solar radiation, heavy in the short wave lengths, but is heated mainly from the earth, particularly by long-wave terrestrial radiation (Barry and Chorley 1987). The water vapor and dust that absorb this radiation are, of course, more concentrated in the denser lower air.

The lapse rate varies between day and night, between seasons, and in different climates (Barry 1981). In mountain ranges, temperature at a given altitude is also affected on a daily basis by the direct heating of soil and vegetation by solar radiation and by the passage of weather fronts, among other things. These other things include slope exposure. Northern and eastern slopes tend to be cooler and moister, with consequent differences in the elevation of vegetation zones compared with southern and western slopes. For example, both upper and lower timberlines tend to be lower on north and east slopes. In arid regions, the only tree growth may be on the cooler, moister slopes (Figure 16-19).

Vegetation zonation is also affected by recent history, including fire, and by specific features of topography, as between ridges and valleys (Arno 1984). Cold air drainage into valleys results in their being much cooler than adjacent ridges; consequently, vegetational zones tend to run higher up ridges. Wind and snow cover may counteract this effect in some cases; for example, ridges accumulate less snow and provide no protection from the sandblasting effect of the wind, and so some ridge sites are unable to support tree growth.

Alpine Tundra

The herb- and shrub-dominated zone above the treeline on mountains is alpine tundra (Figure 16-20). The largest area is on and around the Tibetan Plateau. Conventional map projections exaggerate the size of land areas near the poles so much that you may need to look at a globe to realize that this

Figure 16-19

Ponderosa pine occurs only on cooler, moister sites on these mountains in short-grass prairie at the National Bison Range, Montana. (Photo by the author.)

is, in fact, the largest single block of tundra anywhere, including the Arctic.

In North America, alpine tundra occurs at high elevations in the western mountains into Mexico but only into New England in the Appalachians. There is also a sizable area of alpine tundra in the Alps. Tropical and Southern Hemisphere upper treelines and alpine zones differ from Northern Hemisphere ones in many ways. Timberline trees are generally broad-leaved, and tussock grasses are often prominent in the alpine zone (Arno 1984).

Treeline varies with latitude, being about 900 m in the Mackenzie Mountains of the Yukon (65° N latitude), about 2100 m in Glacier National Park (49° N latitude), and about 3300 m in Rocky Mountain National Park (40° N latitude) (Arno 1984). Treeline on Popocatapetl in Mexico (19° N latitude) is about 4200 m. Other factors besides latitude also influence the elevation at which treeline occurs. In some Southern Hemisphere areas, cold-hardy trees are absent and so treeline is low; treeline near Rio de Janeiro, Brazil, at 22° S latitude, is below 2000 m.

Climate and Soil

Alpine tundra in temperate regions is similar to arctic tundra in having a short growing season and a long, cold winter. Otherwise, the climates are very different. Alpine tundra has more precipitation (mainly as snow), less extreme photoperiods, stronger winds, and great day-to-night temperature fluctuations (Figure 16-21). In contrast to arctic tundra, the weather is very changeable. The Appalachian Mountain Club (1979) guide for Mount Washington, New Hampshire, warns: "The appalling and needless loss of life on this mountain has been due largely to the failure of robust hikers to realize that wintry storms of incredible violence oc-

Figure 16-20

Alpine tundra at Logan Pass, Glacier National Park, Montana. (Photo by the author.)

Figure 16-21

Frost formed on the cinquefoil and other low plants of this subalpine meadow at Yellowstone during a mid-August night. (Photo by the author.)

cur at times even during the summer months. Rocks become ice-coated, winds of hurricane force exhaust the strongest hikers and, when they stop to rest, a temperature below freezing completes the tragedy."

Alpine tundra also differs from arctic tundra in having lower oxygen concentrations and higher ultraviolet radiation.

Soils of alpine tundra generally lack permafrost but are thin and immature, classified mostly as cryic (cold) entisols or inceptisols. They are better drained than arctic soils, and, in fact, late-summer drought often limits plant growth (Billings and Mooney 1968).

Boulder fields and other rocky sites with little plant cover are common (Figure 16-22). Solifluction, the downhill flow of water-saturated soil over bedrock or a deeper frozen layer, occurs in both arctic and alpine tundra but is especially prominent in the mountains because of sloping ground. It tends to produce an undulating landscape.

Vegetation

Alpine tundra resembles arctic tundra physiognomically and taxonomically. Similarities in the flora are great in the East (Bliss 1963), but the tundra of the Sierra Nevada shares only 15 to 20% of its species with the arctic (Van Kat 1979).

The upper limit of the zone where subalpine trees grow to normal size is sometimes called the **timberline.** Trees grow above this to the **tree limit,** but are increasingly dwarfed and misshapen (flagged, with the branches all on the side away from the wind). Even above the tree limit, some species may grow as **krummholz,** or elfin timber (Figure 16-23). They grow low and matlike, looking "as if they had been cultivated by overly ambitious bonsai gardeners" (Zwinger and Willard 1972).

The gardener is the wind, pruning off any sprouts not covered by snow. Although the complete explanation for treeline may be complex, abrasion of leaves by windblown snow, leading to

Figure 16-22

Boulder fields in Rocky Mountain National Park, Colorado. (Photo by Gerald Martin.)

Figure 16-23

Tree growth forms at timberline in Glacier National Park. The tree is subalpine fir. (Photo by the author.)

desiccation and death, is a large part of it (Hadley and Smith 1986).

Bilberry and other heath shrubs are important in the alpine tundra of the eastern United States but not in the West. Overall, alpine tundra tends to have more grass, sedge, and forb and less shrub and lichen than arctic tundra. Extensive grass-dominated areas are called **alpine meadows.**

Many of the herbs are dwarf forms, saxifrages and buttercups that can be covered with a penny—belly plants that you have to lie down to get a good look at. Cushion plants, short with dense-packed stems coming from a taproot, are common, especially in the thin soil between rocks in fellfields (Figure 16-24). Vegetative reproduction among the forbs is less common than in arctic tundra, and cross-pollination, mainly by flies, is more common.

As in the arctic, vegetation is a mosaic related to drainage and snow cover. Because there is more snow and also more topographic opportunities for deep drifts, some snowbed sites may not be uncovered until late summer (Kudo 1991), or not at all in a cool summer following a winter of heavy snow. Vegetational zonation related to the length of snow cover is characteristic of snowbed sites.

Sites of soil disturbance, such as pocket gopher "gardens," have a distinctive but temporary community of bright-flowered forbs such as skypilots and alpine avens.

Animals

Arctic and alpine tundra share few animals. Most of the species that inhabit alpine tundra are more characteristic of other vegetation types at lower elevations. There are no distinctive alpine tundra birds in the eastern United States, where the avifauna consists mostly of common raven, dark-eyed junco, and white-throated sparrow. The western mountains have two distinctive tundra birds, the white-tailed ptarmigan and the rosy finch. Both may stay year-round or migrate to lower elevations for the winter.

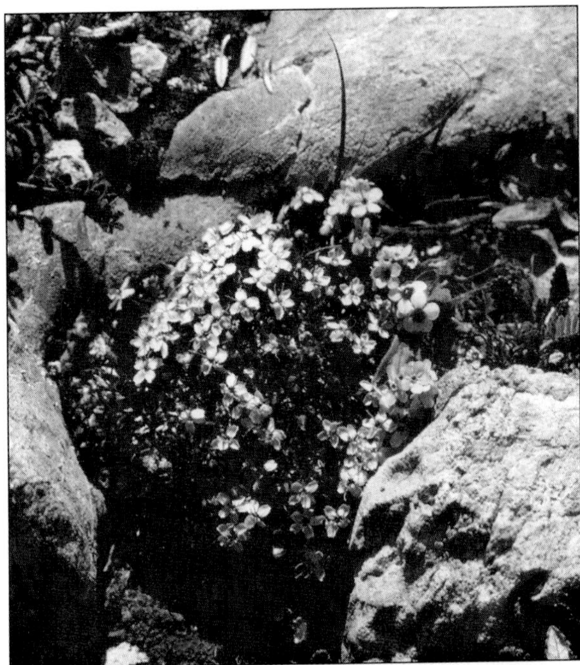

Figure 16-24

These "belly plants" are typical of the flora of alpine tundra. The clump is mostly pink mountain heather. At the right edge of the clump are also cream-colored flowers of rock jasmine, and blue flowers of alpine forget-me-not. These are the largest flowers in the scene, about the size of the head of a thumbtack. (Photo by Marcella Martin, between Banff and Jasper National Parks, Alberta, Canada.)

Other birds that summer in alpine tundra in the West include horned larks, water pipits, rock wrens, and white-crowned sparrows. Mountain bluebirds may forage in tundra but nest in tree cavities in the subalpine zone. Migrating rufous hummingbirds visit alpine flowers in the late summer. Raptorial birds, including golden eagles, red-tailed hawks, prairie falcons, and American kestrels, frequent the alpine zone in the summer but usually nest at lower elevations.

There is a larger number of distinctive alpine mammals. Pikas are among the most characteristic. The common pika occurs widely in the western United States, and the collared pika northward, and they have several relatives in the mountains of Asia. They nest in rock crevices and are perhaps best known for drying and storing hay, on which they live during the winter. Another characteristic tundra animal, the marmot, is a hibernator (Figure 16-25). Yellow-bellied and hoary marmots are colonial versions of the woodchuck (pikas are colonial, also).

Other herbivorous mammals inhabiting alpine areas of the West typically include a chipmunk, a ground squirrel, a pocket gopher, and one or more species of vole (such as the heather vole and the montane vole), plus other small rodents (Verner and Boss 1980, Reichel 1986). Large herbivores include the mule deer, mountain (bighorn) sheep (Figure 16-26), mountain goat (Figure 16-27), and elk. Mammalian predators include the shrew, weasel, badger, red fox, bobcat, coyote, and grizzly bear.

Figure 16-25

A hoary marmot in alpine tundra in Glacier National Park, Montana. (Photo by Marcella Martin.)

Figure 16-26

Mountain (bighorn) sheep in the northern Rockies between Banff and Jasper National Parks. (Photo by Marcella Martin.)

Few of the larger mammals live permanently in the alpine tundra of North America. Some, such as the mule deer and red fox, summer there, and others are regular visitors.

Amphibians and reptiles are scarce. So are bees, but flies, beetles, ants, and spiders are numerous. Many of the tundra insects are wingless, but many winged insects, such as grasshoppers, are blown up from lower elevations (Kendeigh 1974).

Organisms of extremely high elevations, beyond the limits of plant growth, are discussed in "The Aeolian Biome" in Chapter 18.

Human Relations

The natives of the Tibetan tundra are permanent occupants of alpine tundra. Their life is tied to the yak, an oxlike animal as furry as a muskox but larger and with long horns. Tibetans use it for transportation, milk, and meat, and use the dung for fuel, like buffalo chips. In the past 30 years overgrazing by domesticated yaks and overhunting have begun to degrade this once remote and sparsely populated region. Most of the alpine tundra elsewhere in the world has been subject only to occasional or seasonal visits, but the impact of these visits has increased greatly.

Just under 3 million ha of alpine tundra exist in the western United States (Johnston and Brown 1979). A 1976 survey showed 12% of this was disturbed, with additional disturbance occurring at the rate of about 15,000 ha per year. Sheep grazing had affected the biggest area. Recreation and mining were also important causes of disturbance. Recreation increases year by year. It includes many things, from off-road recreational vehicles and snowmobiles, which can heavily disrupt the thin tundra soil and vegetation, to hiking by sincere nature lovers whose lugged boots may, nevertheless, cause damage that takes years to repair.

The alpine tundra is, of course, worth protecting for its own sake, but there is a strikingly practical reason for taking good care of it. It is a prime source

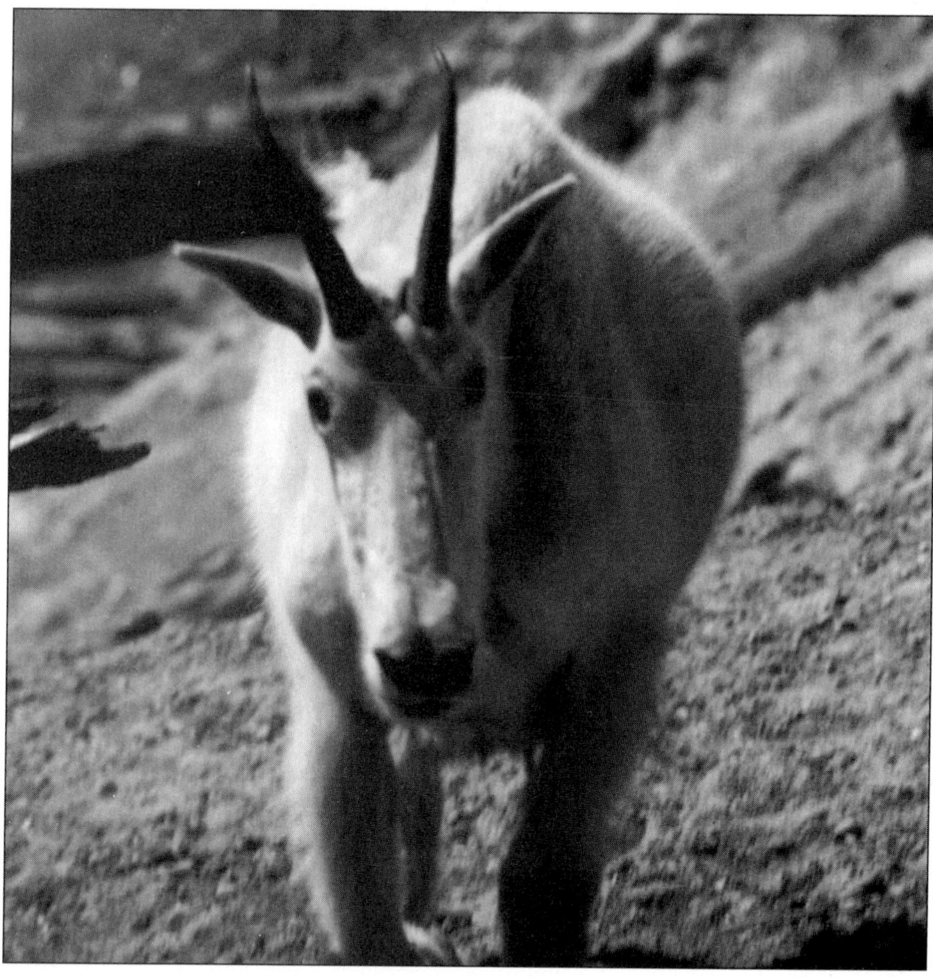

Figure 16-27

Mountain goat at a mineral lick between Banff and Jasper National Parks. (Photo by Gerald Martin.)

of the streamflow on which the life and economy of most of the West depends. Two examples are cited by Johnson and Brown (1979): Alpine areas of Colorado making up 3.5% of the land generate 20% of the streamflow. In Utah, 10% of the land in the highest parts of the Uinta and Wasatch Mountains produces 60% of the streamflow.

Because the alpine areas melt late and over a prolonged period, they prevent or temper floods and maintain streamflow during the peak water-demand period of summer (Figure 16-28). The quality of water coming from alpine tundra is high, although disturbance can change that. Sediments from erosion and mining can be harmful. Baring the soil also exposes acid-yielding minerals such as iron pyrites. In the acid runoff, aluminum and other metals can reach toxic levels, destroying aquatic communities for several miles downstream from surface mines.

The man in Aesop's fable was dissatisfied with only one golden egg a day, and, similarly, some have wondered whether still more streamflow might not be pulled out of the alpine tundra. Proposals include cloud seeding (to increase snowfall) and glacier building. However, the operation most likely to be put into practice is the use of snow

Figure 16-28

This stream running through the alpine tundra in Glacier National Park, Montana, is fed primarily by the snow fields at higher elevations still present in late summer. (Photo early August by the author.)

fences (or "terrain modification" with bulldozers) to trap snow more efficiently. More and bigger snowbeds would affect the vegetation, probably producing larger areas of the bare soil characteristic of late-melting snowfields.

Montane and Subalpine Forests

The highest forests, extending down from the tree limit on mountains, are termed subalpine. Virtually always dominated by some species of spruce or fir, they are homologues of boreal coniferous forests. Forests at lower elevations are montane. In the North, conifer forest in the mountains is continuous with the surrounding forest. Southward in the Appalachians and the mountains of western North America, conifer forest becomes discontinuous and increasingly restricted to higher elevations.

The situation is somewhat different in Eurasia, different because of the general lack of north-south mountain chains; nevertheless, the mountains of temperate latitudes, such as the Alps, support conifer forests similar to those of the North.

Compared with boreal forest, montane and subalpine conifer forests have shorter days in the growing season and longer ones in winter. The length of the growing season is similar, generally 120 days or less. Amount and seasonal distribution of precipitation vary regionally. The amount of precipitation is high both in the Appalachians, above 90 cm per year (Lull 1968), and in the coniferous forests near the Pacific Ocean, but some eastern slope areas in the Rockies may get no more than 40 or 50 cm per year (Arno 1979).

Snowfall is high, though regionally variable. In the spruce-fir forests of the East, 250 cm is a typical figure (Lull 1968), but even the drier, eastern slope areas of the Rockies regularly get more than 100 cm.

Temperature varies from the warmer, lower-elevation conifer forests to the tree limit. The mean temperature of the warmest month (July) tends to be between 15 and 20°C but may be as low as 10°C near timberline. Summer frosts are frequent in subalpine forest. The coldest month (January) may be as warm as −2°C in the ponderosa pine zone of central Arizona, or as cold as −12°C in the subalpine forests of higher elevations in New England or Montana.

Soils tend to be shallow, rocky, acid, and well drained. Occasional areas of deeper soils are on glacial drift or volcanic ash. Podzols and brown podzols are important soil types. Under the U.S.D.A. Comprehensive System, such soils are generally classified as spodisols or alfisols; however, cryic (that is, cold) entisols and inceptisols are also common in these regions.

Appalachians

No coniferous montane forests occur in the eastern United States, but spruce-fir subalpine forest occurs above 760 m in New England (Harvey 1903, Leak 1982) and above 990 m in the Adirondacks. Forests below this level are northern hardwoods of beech, yellow birch, and maple (Figure 16-29). Two conifer species, white pine and hemlock, occur in some of these forests but generally at lower elevations, not adjacent to the spruce-fir forests. The lower limit of spruce-fir forest in the Smoky Mountains of Tennessee and North Carolina is 1400 to 1800 m.

The predominant spruce in the Appalachians is red, although black occurs at high elevations and is the krummholz tree on Mount Washington (Leak and Graber 1974). Balsam fir extends south to the Adirondacks but is replaced by Fraser's fir in the southern Appalachians. Paper birch and mountain ash are deciduous components of the subalpine forest in New England.

Fire is an infrequent event in these forests, but hurricanes and avalanches may cause large-scale disturbance in New England. Spruce budworm epidemics south of Maine were formerly rare or nonexistent but more recently have been recorded in Vermont forests (Walker 1985).

Subalpine forests may show a peculiar regeneration pattern that begins with wave mortality. Canopy fir trees die in an advancing line; new trees, usually of the same species, grow up in the zone behind. Once a wave has begun, it may be perpetuated by increased exposure of live trees to wind that can damage both foliage and roots (the latter from the swaying of the trees in shallow soil over rocks) (Marchand et al. 1986). What additional fac-

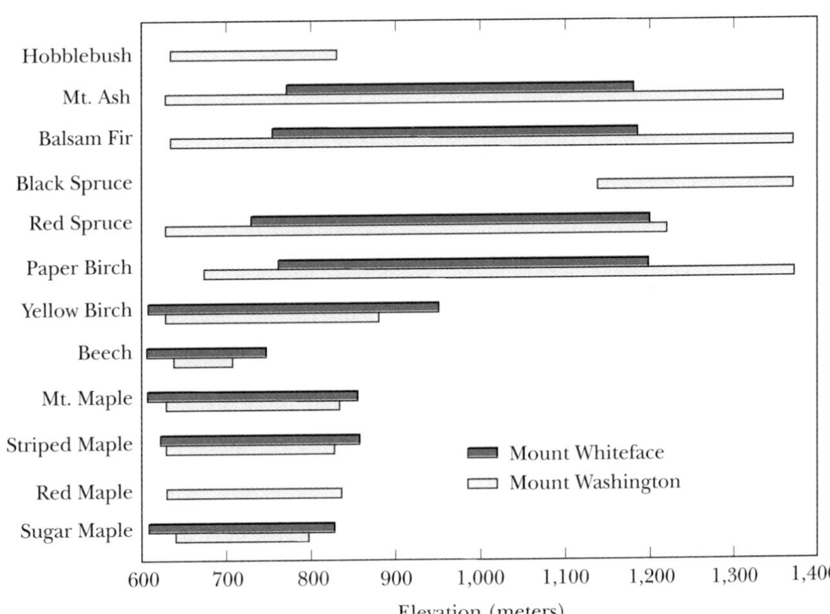

Figure 16-29

Elevational ranges of woody species at two localities in the White Mountains, New Hampshire. The transect on Mt. Whiteface ran from 610 to 1201 m, that on Mt. Washington from 632 to 1373 m. The greater range of conifers and paper birch on Mt. Washington is probably related to past logging there; these species play a successional role in the northern forests of the region. (Adapted from Leak & Graber 1974.)

tors are involved in the phenomenon is unclear. Unclear also is the importance of wave mortality as a normal regeneration process.

The northeastern United States is experiencing a "red spruce decline" that involves slowed growth beginning about 1960 in some areas and increased mortality, especially at high elevations. Possible explanations for the decline include (Van Deusen et al. 1991, Joslin et al. 1992):

1. Acid deposition or other types of air pollution.
2. The "stand history" hypothesis. This supposes that some historical event, such as the hurricane of 1938, synchronized stands in such a way that many are reaching a stage of slower growth or higher mortality about the same time.
3. Climatic hypotheses that, in one form, suggest that warm temperatures of the early to mid part of the 20th century encouraged rapid growth and that cold winters beginning in 1959 caused a slowdown.

The fauna is an attenuated version of the boreal fauna with some additions from neighboring vegetation types. Moose, wolverine, and snowshoe hare once occurred well down the Appalachian chain but are now greatly reduced in many areas. Smaller mammals include the red-backed mouse, which is also widely distributed in conifer forest across North America and southward in the Rockies, and the rock vole (Shelford 1963). Both occur south to the Smokies.

Many characteristic boreal forest species breed in the subalpine forest as far south as New Hampshire. Included are spruce grouse, black-backed three-toed woodpecker, gray jay, boreal chickadee, ruby-crowned kinglet, and pine and evening grosbeaks. None of these extends as far as the Smoky Mountains, but other boreal species, such as brown creepers, winter wrens, golden-crowned kinglets, and dark-eyed juncos, do. These four species may reach very high densities in the Smokies, making up three fourths of the breeding avifauna (Kendeigh 1974).

The Appalachians have a rich herpetofauna, and several species reach up into the spruce-fir zone. The red-cheeked, or Appalachian woodland, salamander for example, occurs at the highest elevations in the Smokies (Huheey and Stupka 1967).

Rocky Mountains and the High Plateaus

Conifer forest occurs in the mountains of this region from Alaska through eastern Washington, Idaho, and western Montana south to the southern Rockies in New Mexico and the mountains of the Colorado Plateau in Arizona, with some continuation on into Mexico.

Below alpine tundra grow subalpine forests (Figure 16-30), typically of Engelmann spruce or subalpine fir. Joining these species near the actual tree limit may be alpine larch and whitebark pine (Figure 16-31) in the North, where the tree limit is near 2500 m (northern Montana) (Pfister et al. 1977). In the South, where the tree limit is around 3300 m (Arizona; Merriam 1890), foxtail pine and bristlecone pine may grow with Engelmann spruce.

Forests of Douglas fir often form the upper part of the montane zone (Figure 16-32) and ponderosa pine forest or savanna, the lower part. Not every location shows just this zonation. Depending on geography, soil, and climate, a variety of other species may occur. The upper montane forest is especially variable. There, such species as western hemlock, western red cedar, grand fir, and blue spruce may join or replace Douglas fir.

The understory vegetation in these forests is not very diverse, although special habitats such as boulder fields and stream banks add species. Many species also live, or have close relatives, in the boreal and northern hardwood forests of the East. Dry, open sites in montane forest may have an understory of bunch grasses or other plants related to the savannas or grasslands below. On moister sites with a complete canopy, some herbs and shrubs are concentrated in either montane or subalpine forest, but many occur over a wide range of elevations (Alexander 1988, Cooper et al. 1991). Among these are western yew, the thorny and immense-leaved Devil's club, queencup beadlily, twin flower, and beargrass (Figure 16-33). The long, hard, grasslike leaves of this member of the lily family are said to be used by grizzly bears for nests; the creamy white flower spikes are eaten by bears, deer, and elk.

Fires are frequent occurrences in conifer forests of this region. They are particularly frequent in ponderosa pine, recurring under natural conditions every five to ten years. Fires are fewer at higher

Figure 16-30

Narrow spires of spruce or fir characterize the upper subalpine zone. Here, in the Mission Mountains of Montana, the predominant tree is subalpine fir. (Photo by the author.)

Figure 16-31

The upper limits of subalpine forest in the northern Rockies of the U.S. often includes alpine larch, the light-colored trees in this autumn photograph, and white-bark pine. (Photo at about 2700 m in the Anaconda-Pintlar range, Montana, courtesy U.S. Forest Service.)

Figure 16-32

Douglas fir. (Photo in Montana by the author.)

Figure 16-33

Beargrass, a striking herb of montane and subalpine zones from British Columbia to California and Montana. (Photo by the author.)

elevations, with fire-free periods longer than 100 years in some subalpine forests (Arno 1980).

The effect of frequent fire in ponderosa pine is to maintain an open grassy stand. Few of the characteristic understory species are susceptible to fire, and only the small pines are killed. Biswell et al. (1974) suggest a patchy regeneration cycle from meadow to grove to meadow with the eventual death of a grove occurring as the combined result of lightning, insects, disease, and wind. Protection from fire leads to denser stands, the buildup of litter, and the invasion of other, fire-susceptible species (Figure 16-34). Crown fires in such stands kill trees over many square kilometers.

Pure, even-aged stands of lodgepole pine may grow in either montane or subalpine forest elevations. These seem to be the result of fire. Like its relative jack pine of the boreal forest, lodgepole has serotinous cones. With fire restriction, lodgepole pine on most sites tends to be replaced by less fire-adapted conifers. Pure stands of quaking aspen also seem to have a fire history.

Cascade Range and Sierra Nevada

The Cascades of Oregon and Washington and the Sierra Nevada of California occupy similar geographical positions, running north-south parallel to the Pacific coast but separated from it by a 200- to 400-km belt occupied by the Coast Range and an intervening valley.

Air rising over the Coast Range, and especially the taller Cascades and Sierra Nevada, is cooled, so that it can hold less moisture. The surplus is deposited as fog, rain, or snow. The result on the leeward side of the range is a rain shadow, where conditions are much drier. The Coast Range produces a modest rain shadow, and the Cascades and Sierra

Nevada produce one that is much more pronounced (Figure 16-35).

The wet western slopes of these mountains show affinities with the forests of the Coast Range and the lower elevations near the Pacific coast, discussed in the next section, whereas the dry eastern slopes show affinities with the forests of the Rockies (Figure 16-36). Zonation in and around the Cascades is shown in Figure 16-36.

Conifer zones in the Sierra Nevada typically include a lower montane forest of ponderosa pine, into which oaks of the woodlands of lower elevations may penetrate. Ponderosa pine tends to dominate between 600 and 1500 m. Higher in the montane zone, between 920 and 1850 m, is a mixed-conifer forest in which ponderosa pine is joined by several species including Douglas fir, sugar pine, white fir, and incense cedar. Still higher, from 1850 to 2740 m, is red fir forest, sometimes considered to be lower subalpine but denser and taller than typical subalpine forest. Clearly subalpine are many stands of lodgepole pine and also mixed stands including whitebark pine and mountain hemlock. These are typically open stands that become increasingly interrupted by subalpine meadows, with the tree limit around 3300 m.

It is in the mixed-conifer forest of the Sierra Nevada that the giant sequoia grows (Figure 16-37). It now occurs in 75 stands or groves at elevations mostly between 1360 and 2270 m (Figure 16-38). These trees, the most massive organisms in the world, reach heights greater than 80 m, diameters at the base of 10 m, and weights of more than 400 metric tons dry weight (Hartesveldt et al. 1975). Considerable lumbering occurred between 1856 and the mid-1950s, although sequoia is not a particularly useful timber species. Most of the remaining area is now in public ownership and, presumably, protected.

The giant sequoias are fire resistant. In their

Figure 16-34

Here Douglas fir is invading a ponderosa pine stand that has not experienced a recent fire. The Douglas fir are the small Christmas-tree-like trees. (Photo near Missoula, Montana, by the author.)

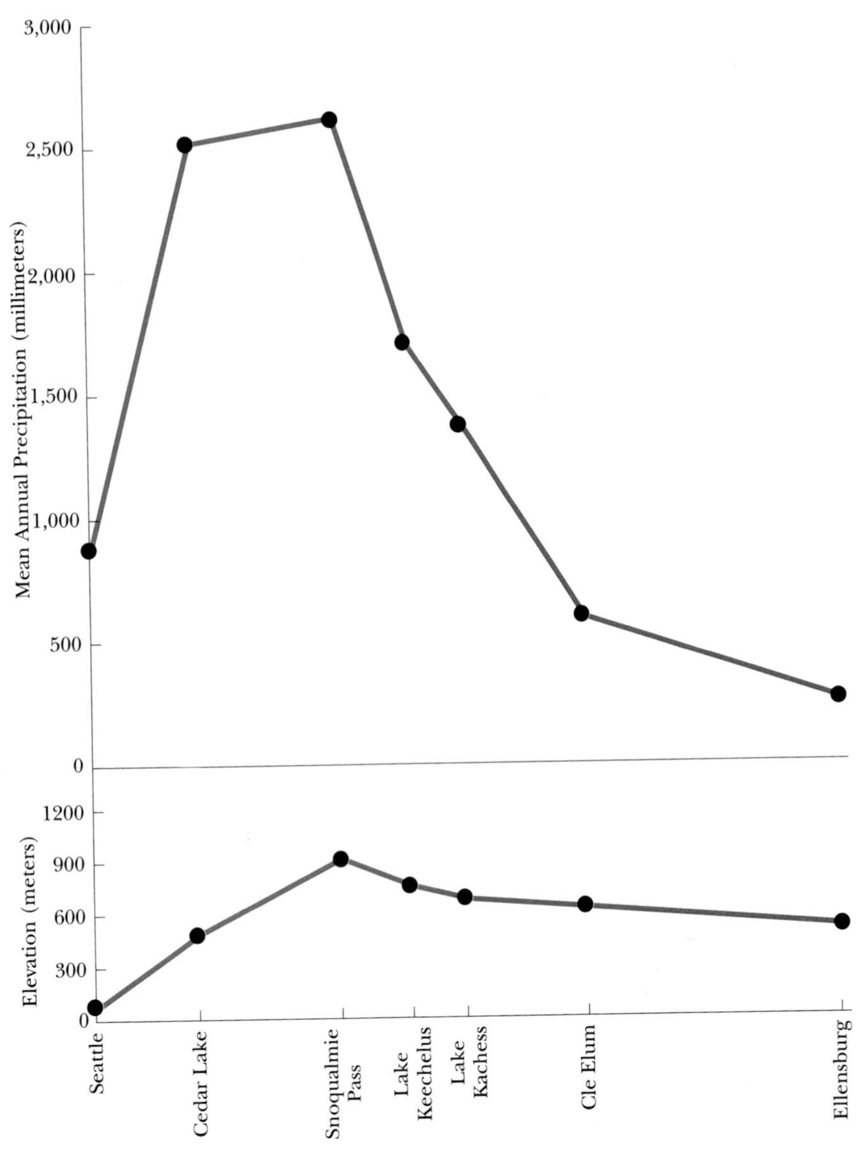

Figure 16-35

The rain shadow cast by the Cascade Range is illustrated here. Shown is a 150 km transect running southeast from Seattle to Ellensberg, Washington. Rainfall in Seattle is a moderate 85 cm. Going up in the Cascades, it rises to above 250 cm, about 100 inches. The east slope, however, is dry and Ellensberg gets only about 20 cm, most of it in the winter as snow. (Adapted from Franklin & Dyrness 1969.)

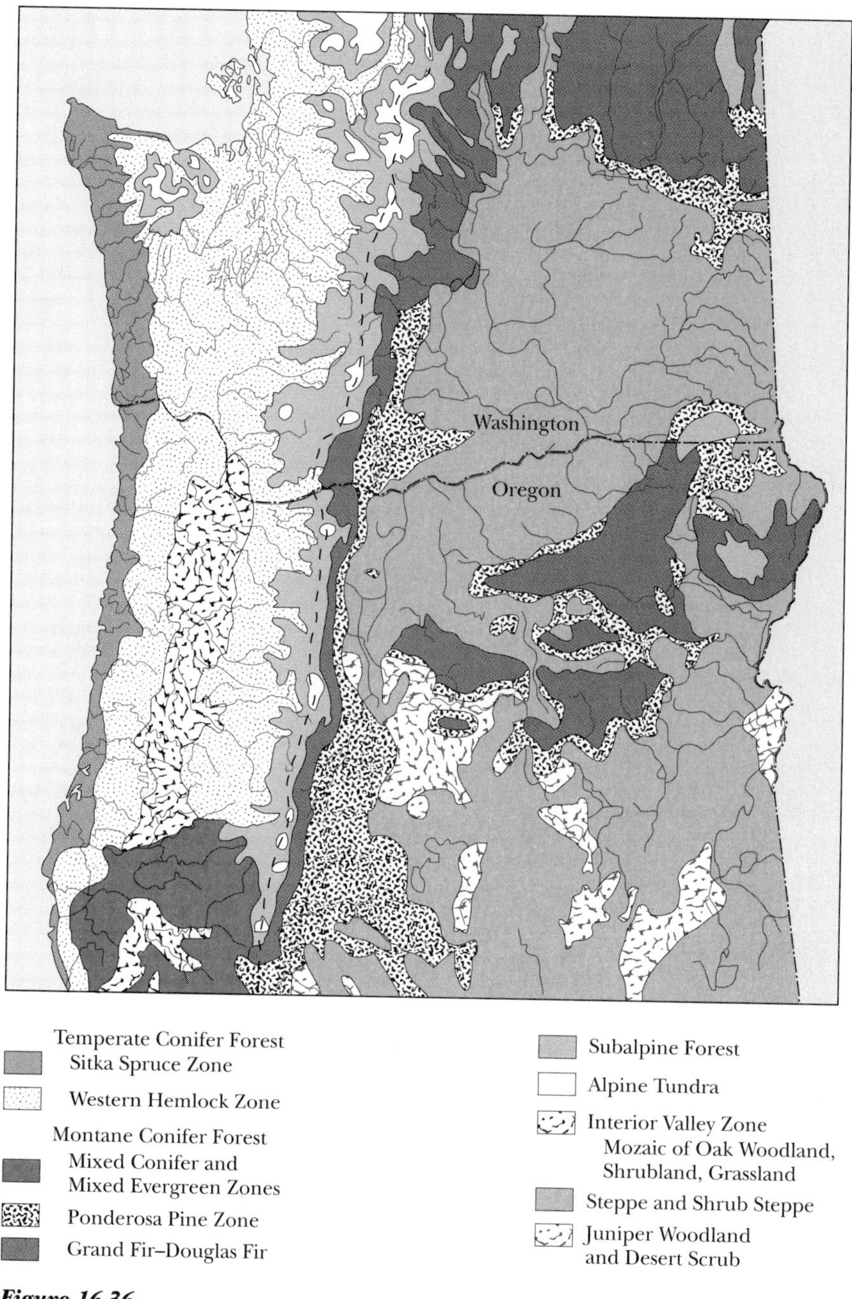

Figure 16-36

The vegetation of Oregon and Washington is a mosaic related to temperature and rainfall. The north-south broken line is the crest of the Cascades. The vegetation to the west drops down to the temperate conifer forests of western hemlock, western redcedar and, along the coast, Sitka spruce. The vegetation going down the east slopes is similar to the zonation in the northern Rockies. (Adapted from Franklin & Dyrness 1969.)

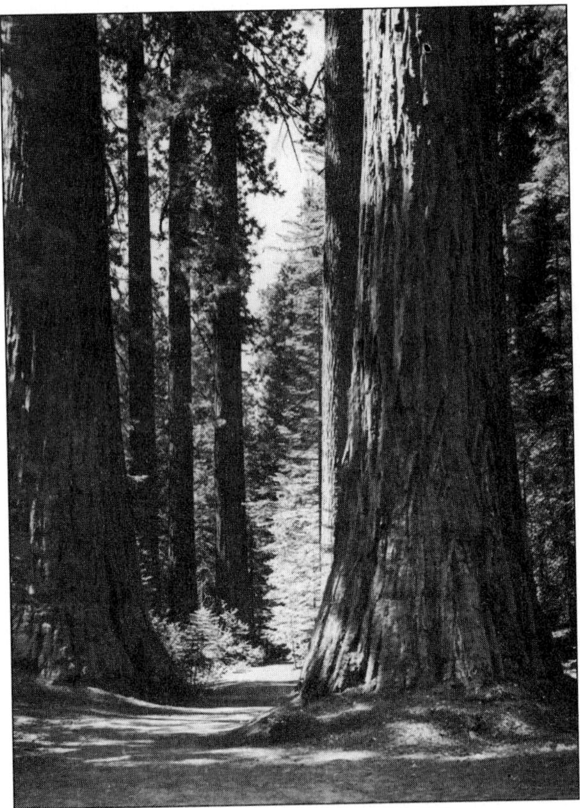

Figure 16-37

The Calaveras grove of giant sequoia in 1935. (Photo by Fred E. Cunham, courtesy U.S. Forest Service.)

life span, which may be more than 3000 years, the trunks become heavily fire scarred (Figure 16-39). Seed production is heavy, and germination evidently proceeds best on bare, mineral soil that fire provides. Apparently the bottleneck that accounts for the rarity and restricted distribution of the species is the seedling period, when moisture requirements are narrow. Depredations by animals seem of little importance. The seeds, in fact, are ignored by most of the mammals of the community. The chickaree or Douglas squirrel eats the fleshy scales of the cones, rather than the seeds, and may aid dispersal.

The giant sequoia is considered intermediate in shade tolerance (Table 3-2). In stands that are long protected from fire, more shade-tolerant species, especially white fir, increase. For trees that can live 3000 years, though, even very infrequent fires can serve to maintain them in the community.

Temperate Rain Forests of the Pacific Northwest

The most magnificent forests in the world grow in a belt up to 200 km wide along the Pacific Coast from California to Alaska (Figure 16-40). Temperate rain forests with few or no related species also occur along the coasts of Chile, New Zealand, and Australia (Billings 1964). Canopy heights of 60 to 70 m are usual, and the coastal redwoods of California regularly reach 100 m. The record height for a redwood is about 112 m, the height of a 28-story building. This is tall. None of the species approach the giant sequoia in girth, but diameters of 100 to 200 cm dbh (diameter at breast height) are not unusual (Franklin et al. 1981).

Characteristic members of the forest are Douglas fir, western hemlock, and western red cedar (Table 16-4). Sitka spruce joins these in a narrow zone (about 20 km wide in Washington) next to the coast. Southward, in California, the coast redwood is important in a comparable zone (Figure 16-38). Many redwood stands are on stream terraces, on sites subject to occasional flooding and silt deposition.

The climate in which these forests grow is maritime and influenced by the cool Alaskan current. Winters are wet and mild; summers are warm and dry. Seattle, Washington, in the valley between the Coast Range and the Cascades, has an average January temperature of 5°C and a July temperature of 19°C. Seattle gets less than 90 cm of rain per year, but precipitation varies considerably within the vegetation type. On the Olympic Peninsula and the western slope of the Cascades, it is above 300 cm. Snowfall and snow cover are also variable within the vegetation type.

Fog is frequent. Seattle has fog on about 50 days of the year, but Cape Disappointment on the Pacific Ocean at the mouth of the Columbia River averages over 2550 hours of heavy fog annually (Ludlum 1971). Summer fog has been suggested as an important factor in the occurrence of these forests. The coastal redwood stands, especially, occur in a

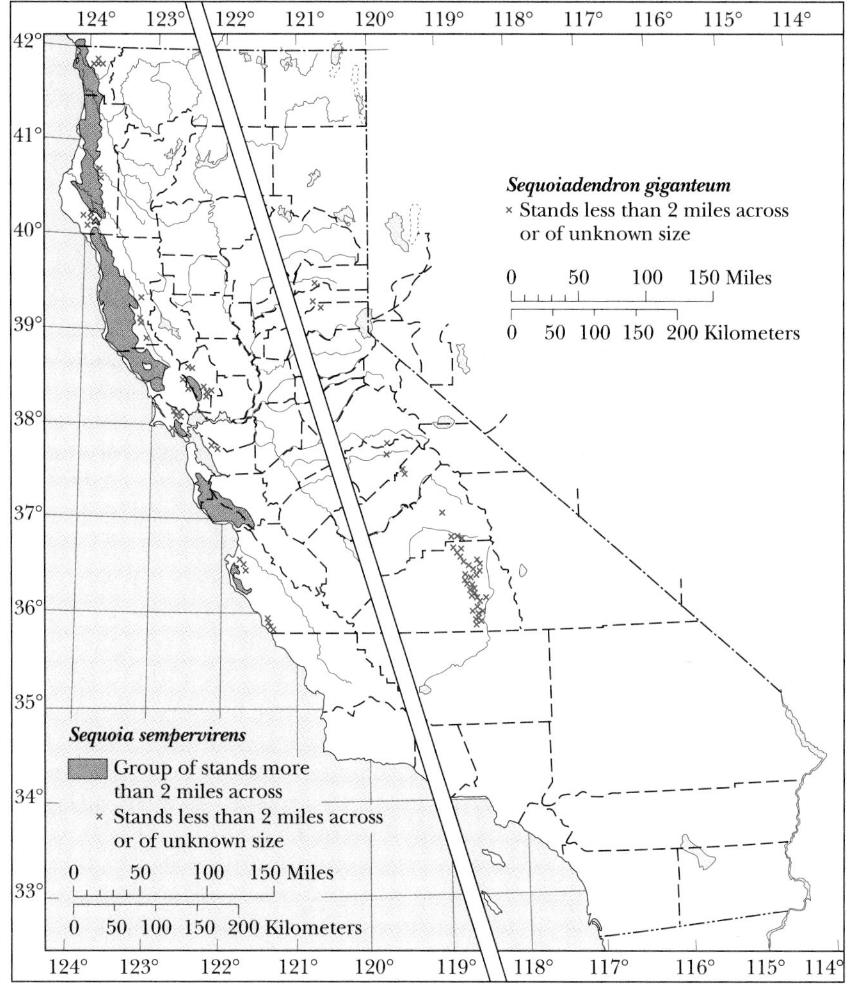

Figure 16-38

Distribution of giant sequoia and coast redwood. The giant sequoia occurs mostly in the mixed conifer zone of the montane forests of the Sierra Nevada. The coast redwood occurs in the southern portion of the temperate coastal conifer forest. (Adapted from Griffin & Critchfield 1972 [repr. 1976].)

Mediterranean climate that, without the fog, would be extremely dry in summer. San Francisco has basically no rain from July to September and only 10 cm for the seven months between April and October.

The scanty summer rainfall is the most obvious difference between this climate and temperate climates supporting broad-leaved trees. Whether this is the only factor and just how it operates to favor needle-leaved trees are uncertain. Whatever factors are influential, they seem to have been in operation for a long time; similar forests grew in the region in the early Pleistocene, before glaciation (Waring and Franklin 1979).

Succession after wildfire or after clear-cutting and burning of slash begins with a four- to five-year herbaceous stage (Franklin 1979). In the shrub stage that follows, dominance is shared by several species left over from the earlier forest and invaders such as willows and the nitrogen-fixing snowbush.

The shrubs are overtopped by young trees within 10 to 25 years. The young tree stand may be even-aged Douglas fir. On other sites, the more shade-tolerant species such as western hemlock and western red cedar may assume immediate importance. Elsewhere, dense stands of red alder take over. This nitrogen fixer grows rapidly enough to

Figure 16-39

The fire-scarred base of a giant sequoia in Sequoia National Park, California. (Photo by R. Bruce Moffett.)

overtop most young conifers and may delay conifer dominance on these sites.

A Douglas fir forest 125 years old is still young, with densely growing Douglas fir 30 to 60 cm dbh and western hemlock generally of smaller diameters. Stands 175 to 250 years of age take on "old-growth" characteristics. Trees of all sizes are present, forming a multilayer canopy. Density of the larger trees is low. Nitrogen-fixing foliose lichens grow high in the canopy on trunks and branches, and the nitrogen is carried by rainfall and fog drip to the soil below.

A lush green understory of mosses, ferns, and herbs is usually present in the old-growth forests. Certain saprophytic herbs find optimum habitat here, including phantom orchid and candystick. Standing dead trunks and logs are numerous, providing habitat for a rich variety of animals, mosses, lichens, plus tree seedlings.

Biomass in old-growth Douglas fir forest (350 to 750 years old) may be more than 100 kg/m^2, easily ten times that of an average temperate deciduous forest. Coastal redwood stands run even higher, 200 to 400 kg/m^2. An additional large amount of organic matter occurs as standing dead trees, logs, and litter (Figure 16-41). Biomass of standing dead trees and logs totaled almost one third as much biomass as living trunks in a 450-year-old Douglas fir forest in the Cascade Mountains (Waring and Franklin 1979).

Annual net primary productivity is high but not remarkably so; 2000 g/m^2 is probably an average figure for young stands on good sites (Waring and Franklin 1979). Rather, the high biomass results from the great longevity of the trees and their ability to continue height growth even when very old. Douglas fir is known to reach 1200 years of age, coastal redwood 2200 years, and many of the species live to more than 500 years. One climatic factor related to this longevity is the lack of hurricanes and typhoons along this coast.

The term "old-growth," rather than "climax," forest has been applied to these forests because neither Douglas fir nor redwood tolerates shade as well as the very tolerant western hemlock and western red cedar (Table 3-2). Both Douglas fir and redwood show fire adaptations (such as thick bark; redwood also has a low resin content and root-sprouts readily). Redwood also puts out adventitious roots when flooding deposits silt around it (Hewes 1981). In an environment without fires or flooding, these two species would tend to be pinched out by the more shade-tolerant species. Where fires and floods do occur, trees having life spans of 1000 or 2000 years seem as well adjusted to maintaining themselves as shorter-lived but more shade-tolerant species.

Coastal redwood is a desirable timber tree because of its large size (Figure 16-42), workable wood, and resistance to insects and decay. The demolition of the redwood stands is another sad chapter in the history of the North American continent. And it is not ancient history. Redwoods were not

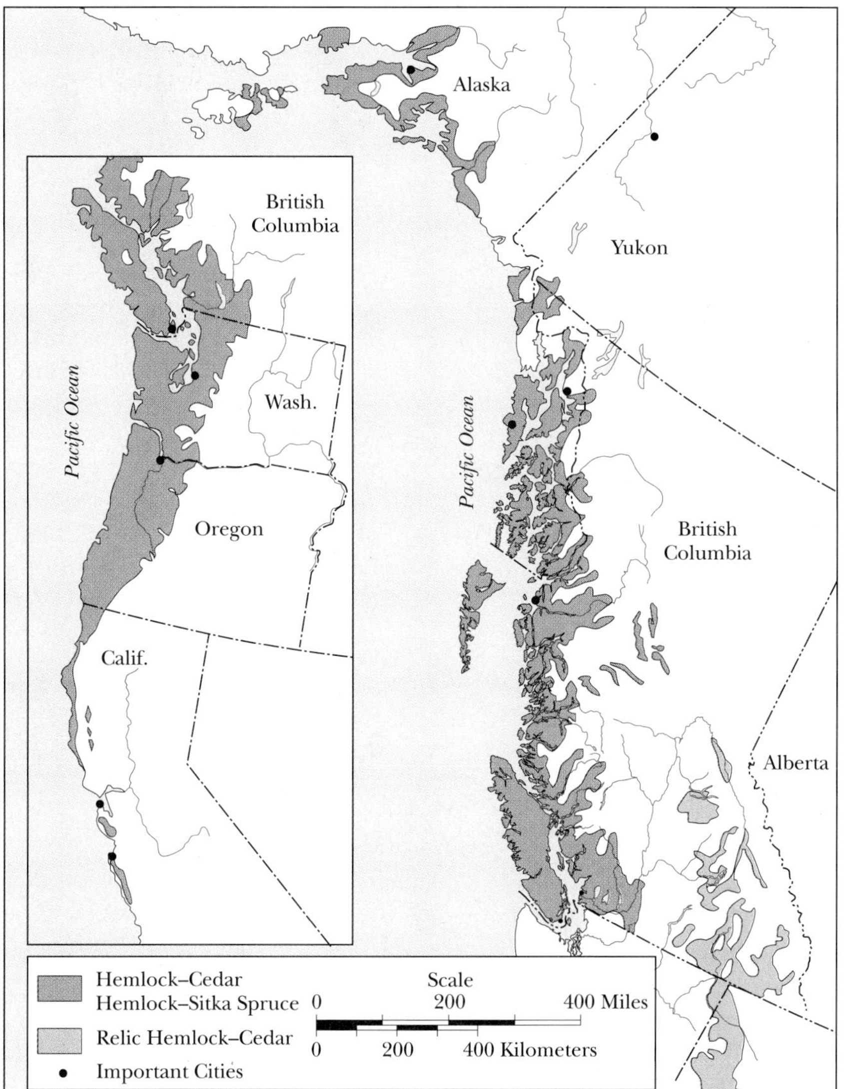

Figure 16-40

Distribution of the temperate evergreen forests of the Pacific Northwest (temperate rain forest). Western hemlock and western redcedar are climax dominants. Near the coast, Sitka spruce is also important from Alaska to northern California and coastal redwood occurs in the southern tail of the forest in central California. Douglas fir is also widely important. (From *The Ecology of North America,* by V. E. Shelford, University of Illinois Press, 1963. Copyright © Board of Trustees, University of Illinois.)

Table 16-4
Important Tree Species of Temperate Conifer Forests of the Coastal Range and Adjacent Lowlands

	Zone		
Species	*Western Hemlock*	*Sitka Spruce*	*Coastal Redwood*
Douglas fir	X	X	X
Western hemlock	X	X	X
Western red cedar	X	X	
Sitka spruce		X	
Coastal redwood			X
Grand fir			X
Tan oak			X*

* Understory

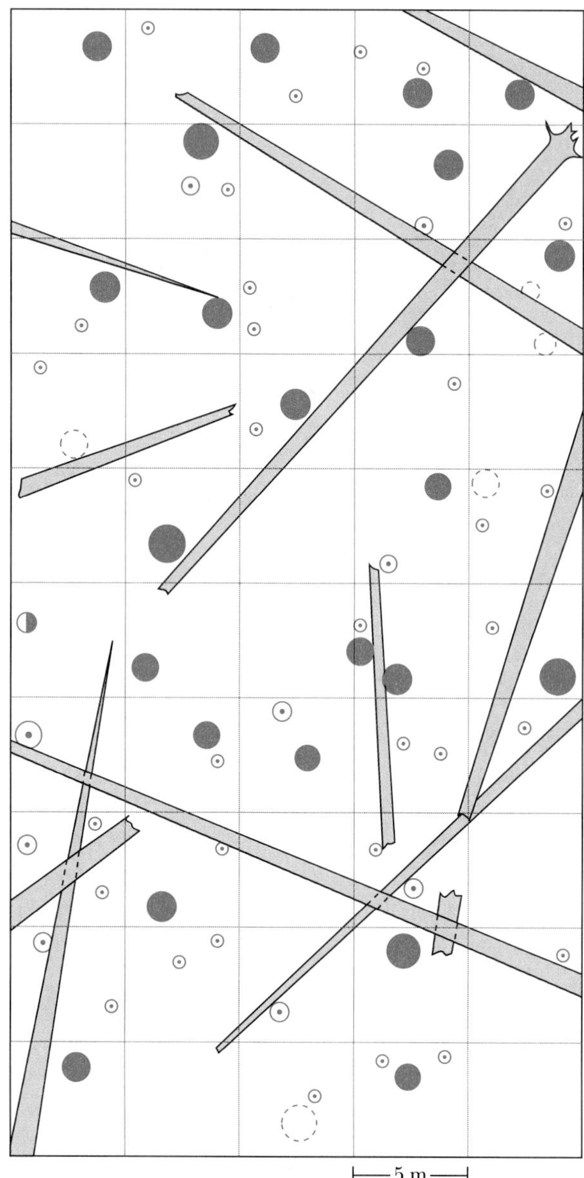

Douglas Fir
Western Hemlock
Standing Dead Tree

Figure 16-41

A map of a 25 × 50 meter area in a 250-year-old stand of Douglas fir. The fallen trunks are shown. Live trees are represented by circles whose diameter indicates relative tree sizes. (Adapted from Franklin et al. 1981.)

cut for timber until well into the 1800s, and as late as 1925 probably two thirds of the original redwood forests were still intact (Hewes 1981). The heaviest destruction came with the building boom in the years following World War II, when for many years about a billion board feet were cut annually.

Although a California state park protecting the redwood was established as early as 1901, timber interests repeatedly prevented protection of many areas that merited it. Only about 4% of the original redwood acreage has been protected, and some of this is subject to damage resulting from lumbering operations on adjacent land. Essentially no virgin redwood stands now exist outside the public lands. Cutover stands do, of course, regenerate redwood, but cutting continues to outpace regrowth. Old-growth Douglas fir forests have been pushed far along the same route to oblivion.

Animals of the Western Coniferous Forests

Many species that are widespread in the boreal forest also occur in the conifer forests of the West (Table 16-5, Figure 16-43). In addition, many western specialties exist that spread little if at all into the transcontinental boreal forest. The grizzly bear ranged over a variety of habitats but was centered in these forests. Large herbivores associated with the western conifers include the mule deer, big-

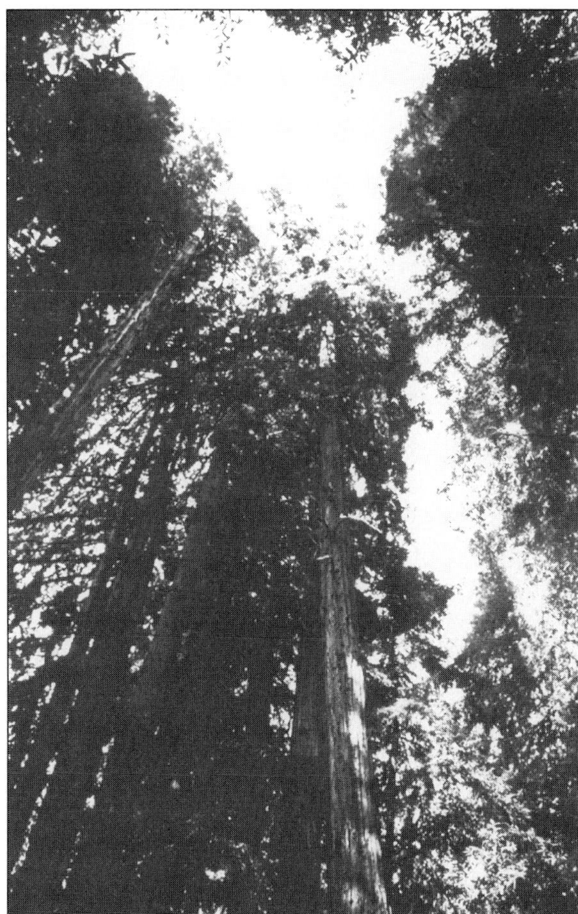

Figure 16-42

Coastal redwoods in Muir Woods, California. (Photo by R. Bruce Moffett.)

Table 16-5

Examples of Vertebrate Species Occurring Both in Boreal Forests and in Conifer Forests of the Western United States

Mammals	*Birds*
Snowshoe hare	Goshawk
Northern flying squirrel	Gray jay
Red-backed vole	Red-breasted nuthatch
Porcupine	Brown creeper
Marten	Winter wren
Wolverine	Swainson's thrush
Timber wolf	Golden-crowned kinglet
	Ruby-crowned kinglet
	Yellow-rumped warbler*
	Pine siskin
	Red crossbill
	Pine grosbeak
	Olive-sided flycatcher†
	Nashville warbler†
	Lincoln's sparrow†

* Western populations formerly considered a separate species called Audubon's warbler.

† Species of edge or successional vegetation.

horn sheep, and mountain goat. The elk, or wapiti, is now virtually restricted to the western mountains but formerly occurred widely in other biomes, including temperate deciduous forest in the East (Shelford 1963).

The fauna is rich. The vegetation itself is varied, ranging from savanna to dense forest. Added to this is the diversity produced by streams with their riparian vegetation, ponds and subalpine meadows, and cliffs and talus slopes. Examples of characteristic birds are given in Table 16-6.

One obvious difference between these forests and the boreal forest is the scarcity here of wood warblers. Although the West has several distinctive species of wood warblers, most are adapted to other vegetation types. Townsend's and hermit warblers, two western species that are related to the black-throated green warbler of the East, are conifer forest species.

The foliage-gleaning niches of the warblers are largely occupied in these forests by species of Old World affinities, such as chickadees and kinglets, which tend to reach higher densities here than in boreal forest. Overall bird densities are generally lower than in boreal forest.

Relative to boreal forest, a much higher proportion of the birds in these forests are resident throughout the year. Several species that do move out for the winter simply drop down to lower elevations (altitudinal migration). Examples are Townsend's solitaires, pine siskins, Steller's jays, (Figure 16-44), and dark-eyed juncos. Although most of the flycatchers, warblers, vireos, and tanagers that will spend the winter in the tropics depart in August, the same period sees an upslope migration into subalpine forest and even alpine tundra by individuals of several species (Small 1974). Presumably taking advantage of the brief, late-summer

Figure 16-43

The olive-sided flycatcher occurs in boreal coniferous as well as the subalpine and upper montane forests of the West. (This photograph by Lawrence H. Walkinshaw was taken in Schoolcraft County, Michigan.)

bonanza of food, blue grouse, dark-eyed juncos, fox sparrows, and American robins move in for a few days or weeks. Green-tailed towhees from chaparral at still lower elevations may also spend the late summer here.

An interesting coevolved mutualism exists between whitebark pine, mentioned earlier as a timberline tree in the western mountains, and the Clark's nutcracker, a noisy gray bird related to the crows and jays (Tomback et al. 1990). Whitebark pine seeds are large and wingless and, hence, not wind dispersed like most pines. The seeds are eaten and cached primarily by red squirrels and Clark's nutcrackers (Hutchins 1990). Seeds that fall or are experimentally placed on the ground are quickly eaten by rodents. For various reasons the large caches made by the squirrels rarely lead to whitebark pine germination; virtually all whitebark pine germination is from seeds stored but not retrieved by nutcrackers. The nutcrackers subsist almost entirely on the seed caches from November until well into the summer, which includes the nesting period. They disseminate the seeds widely, allowing whitebark pine to be a successional species in new burns in subalpine areas and also allowing it to regenerate itself in rocky areas and the edges of meadows at treeline.

Most larger mammals, such as the elk, mule deer, and bighorn sheep, show altitudinal migration, ranging higher in the summer than winter. H. S. Graves and E. W. Nelson (1919) summed up the situation for the elk this way:

Table 16-6
Characteristic Birds of Montane and Subalpine Conifer Forests in Western North America

Species Occurring Widely in Both Montane and Subalpine Forest	
Groshawk	Golden-crowned kinglet
Hammond's flycatcher	Yellow-rumped warbler
Gray jay	Townsend's warbler
Brown creeper	Pine siskin
Hermit thrush	Red crossbill
Townsend's solitaire	Dark-eyed junco
Species More Common in Subalpine Forest	
Williamson's sapsucker	Evening grosbeak
Black-backed woodpecker	Pine grosbeak
Clark's nutcracker	Cassin's finch
Mountain chickadee	
Species More Common in Montane Forest	
Blue grouse	American robin
Great gray owl	Varied thrush
White-headed woodpecker	Hermit warbler
Steller's jay	Western tanager
Red-breasted nuthatch	

The lower reaches of the montane zone also include many woodland and savanna species of wide distribution. The pygmy nuthatch is a distinctive ponderosa pine species of this zone.

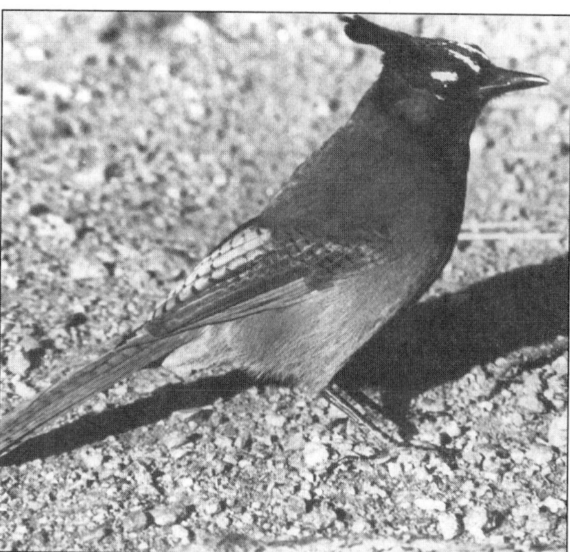

Figure 16-44

Steller's jay, a close relative of the blue jay, is a characteristic montane forest species. (Photo by George and Betty Walker.)

In fall they drifted out of the mountains ahead of the storms and snow, scattering over the bordering open valleys and plains where the snowfall was light and nourishing dry grasses were plentiful. They often worked their way more than 200 miles from their summer feeding grounds. In spring they followed the melting snow back to the high mountains, above the zone of the annoying flies, where the climate was cool and refreshing, the forest offering grateful cover, and where fresh and succulent feed abounded. Then came the settlers.

Fields replaced the winter range of the elk, fences prevented the fall and spring migrations, and livestock ate the elk's winter food. "Farther and farther settlement advanced into the mountain regions," Graves and Nelson wrote. "More and more restricted became the area on which the elk could spend the winter months." These remnant herds survive today, aided by artificial winter feeding in some localities.

As with the birds, many of the mammals are shared with the transcontinental boreal forests (Table 16-5). This is particularly true of squirrel-sized and larger mammals. The red-backed vole is often the most abundant small mammal (Raphael 1987). Some important species not shared with the boreal forest include several species of chipmunks and ground squirrels and the western jumping mouse.

Several species of reptiles and amphibians occur in the altitudinal range of montane and even subalpine forests, but most are associated with special habitats such as streams, ponds, and rock outcrops. Species with the least affinities with lower altitudinal vegetation include the long-toed salamander, Mount Lyell salamander, tailed frog, Cascades frog, and western terrestrial garter snake.

The cold, fast mountain streams are home to the American dipper. Dippers are year-round residents, nesting on rock ledges and rarely straying far from the stream. They feed on invertebrates of the benthos by "flying through the water," that is, propelling themselves along the bottom with their wings. Related species occur in the Andes and all across Eurasia but, curiously, not in the Appalachians. The beaver is a characteristic mammal of the mountain streams, producing through its activities some of the subalpine meadows.

A varied invertebrate fauna inhabits these western conifer forests (Blake 1945). Because of their importance to timber production, tree-inhabiting insects have been extensively studied (Furniss and Carolin 1977). About 30 species, mostly bark beetles, sawflies, and various moths including the western spruce budworm, are considered major insect pests. They are pests, of course, only in the context of our desire to use the forests exclusively for timber production. The insects are part of the natural functioning of the ecosystem, although some have reached higher-than-normal abundance as the result of human activities, especially fire suppression.

The coastal forests of Douglas fir and associated species have a diverse avifauna (Table 16-7). They share many birds with the coniferous forests of higher elevations but also have some distinctive features. One of the most unexpected inhabitants of these forests is the marbled murrelet (pronounced mer-lut and meaning "little murre"), a member of a family of mostly cliff-nesting seabirds. The marbled murrelet also gets its food from the ocean, but lays its single egg high on a thick tree limb in these forests, up to 10 km from the coast (Figure 16-45).

Figure 16-45

The enigmatic marbled murrelet is a member of a marine family and gets its food from the sea, but nests high on a limb in old-growth temperate rain forests. (T. Hamer/VIREO.)

Table 16-7
Important Birds of the Temperate Rain Forests

Marbled murrelet	Western flycatcher
Blue grouse	Steller's jay
Ruffed grouse	Chestnut-backed chickadee
Band-tailed pigeon	Brown creeper
Northern pygmy owl	Winter wren
Spotted owl	Varied thrush
Allen's hummingbird	Hermit thrush
Vaux's swift	Swainson's thrush
Pileated woodpecker	Golden-crowned kinglet
Hairy woodpecker	Wilson's warbler
Western wood-pewee	Hermit warbler
	Purple finch

Based mainly on MacNab 1958, Larrison and Sonnenberg 1968, Small 1974, and breeding-bird censuses published in *American Birds* by B. Milton, S. Murray, K. Darling, E. Akers, and R. Hansen (1971–1975) and R. Judah (1982–1983).

Adult, egg, and young all are camouflaged in this habitat; for example, the egg is green, matching the moss-covered limb (Binford et al. 1975). Some people (Abate 1992) have suggested that the marbled murrelet is a more precise indicator of old-growth temperate rain forest than the spotted owl (discussed below). It is, in any case, another species that is harmed by clear-cutting these forests (Quinlan and Hughes 1991) and has been proposed for listing as a threatened species in Washington, Oregon, and California.

The Vaux's swift, unlike its relative the chimney swift, has remained true to its original habitat, nesting in large, hollow trees of grand fir or coastal redwood. Pileated woodpecker drillings are important for access to the trees (Bull and Cooper 1991).

In most regions of the United States, relatively few animals are known to require old-growth or climax forest; many seem able to make do with younger forest, although they may need substantial acreages to do well. A case has been made that the old-growth coastal forests are an exception (Franklin et al. 1981). Certain birds, such as the marbled murrelet and the Vaux's swift, may require old-growth forest, and so also may various mammals that depend on cavities, logs, or epiphytes and fungi. Included are species of bats, the red tree vole (Figure 16-46), and the coast mole. The characteristic habitats and associated organisms of streams in such forest are also dependent on conditions created by logs and other debris from these forests.

Relatively few people become excited about the possible extinction of bats or the red tree vole, although that is not a flaw of the bats or the vole. The red tree vole is, in fact, a fascinating small mouse that eats only conifer needles and may spend its entire life in the tops of the Douglas firs—the most specialized vole in the world (Carey 1991).

Many more people have become concerned about the potential loss of the northern spotted owl, a large, secretive bird related to the more familiar barred owl. Recent studies have shown that this bird is very closely tied to the old-growth forests (Gutierrez and Carey 1985) and, further, that it requires large areas of such forest. In an Oregon study that used radiotracking, the smallest individual home range was over 1900 acres (777 ha), of which 75% (583 ha) was old growth (Carey et al. 1991). The average amount of old growth included in an individual home range was about 2200 acres (875 ha). In mixed landscapes, most foraging, roosting, and nesting occurred in the patches of old growth rather than in the 100-year-old and younger patches.

Exactly why the spotted owl requires old-growth forest has not been proved. One strong possibility is its need for cool, shady sites for daytime roosts; even within the old-growth forests, cooler micro-

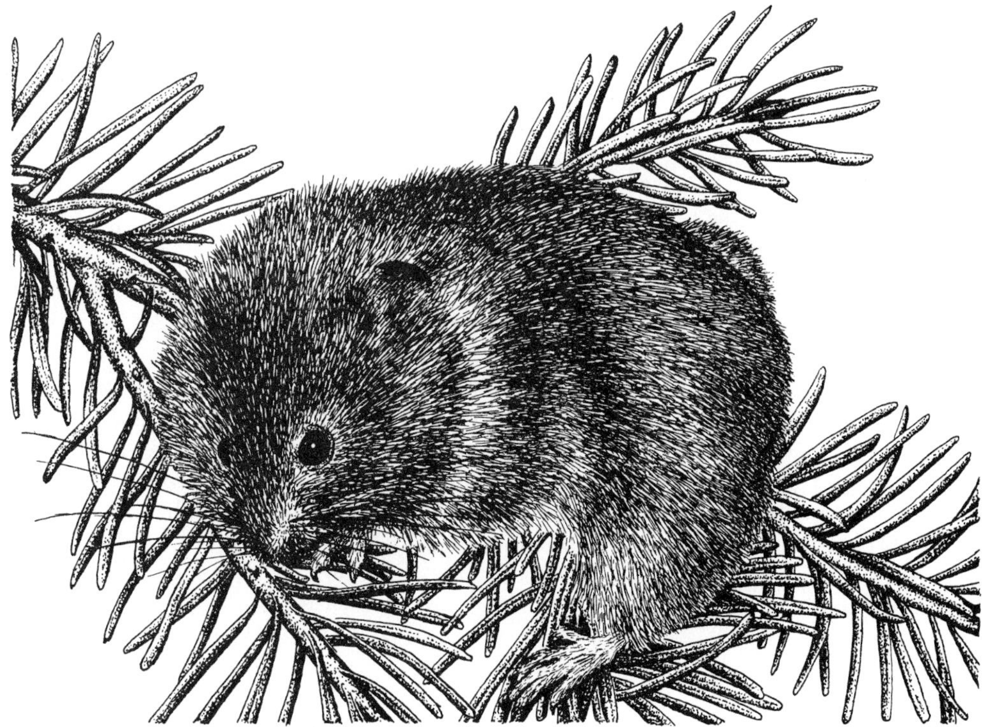

Figure 16-46

The red tree vole is a small, nocturnal, arboreal rodent characteristic of the old-growth forests of the Pacific Northwest. (Drawing by Brigitta Van Der Raay from Carey 1991.)

habitats are sought. Probably several features are important, and it may be that greater exposure to predation (by the larger great horned owl) and unavailability of the spotted owl's prey (mostly northern flying squirrels and red tree voles) make cutover areas unsuitable.

The best estimate of primeval old-growth acreage is probably that of Booth (1991), who calculated that there were originally 19.5 million acres (7.9 million ha) in Washington and Oregon. Timber cutting began early, even before the land was surveyed for sale to settlers. By 1986, about 3.4 million acres (1.4 million ha) was left or, in other words, more than 80% was gone. Logging reduced the amount of old growth on private lands rapidly in the 1960s and 1970s (from 470,000 to 88,000 ha between 1961 and 1984, Simberloff 1987), and timber companies increasingly turned to National Forest lands, where about 1.5 million acres of old growth supported about 1200 pairs of northern spotted owls.

Worries about the survival of the nothern spotted owl began to be voiced in the early 1970s, and it was designated as a threatened species in Oregon in 1973. In 1976, the National Forest Management Act required that the National Forests be managed so as to maintain adequate habitat for existing vertebrate species. It also required that "management indicator species"—ones having the potential to indicate serious loss or degradation of a major biological community—be given special attention. The spotted owl is such a species for the old-growth forest. It has since been listed as nationally threatened. Common sense might have suggested that a simple and direct approach would have been to stop cutting old-growth timber in the National (that is, the citizens') Forests, preserving not only the spotted owl but the other organisms making up this remarkable ecosystem.

The U.S. Forest Service bureaucracy did not see it that way. In a 1981 draft plan and a 1984 Final Regional Guide, the Forest Service proposed to manage (that is, allow lumbering in) the forests in such a way as to reduce the spotted owl population to 375 pairs. Even this figure was misleadingly high, because the Forest Service assumed that spotted owls needed only 400 ha of old growth per territory, about half the average actually used.

Environmental groups objected to the Guide and gained considerable support from scientists. The Forest Service produced a revised plan in 1987, including a Preferred Alternative that was possibly sufficient for retaining about 550 pairs. In other words, the Forest Service was sponsoring a 50% reduction in the population of a threatened species.

The battle has continued in succeeding years, in Congress, and in the courts. The federal agencies involved formed an Interagency Scientific Committee that produced (May 1990) *A Conservation Strategy for the Northern Spotted Owl.* The gist of the plan is a network of habitat conservation areas (HCAs) of around 60,000 acres (25,000 ha) apiece on which no timber cutting except for spotted owl management purposes is permitted (Wood 1991). This amounts to a sizable piece of real estate—about 7.7 million acres (about 3 million ha)—but when the acreage already protected in national parks and wilderness area or not loggable for other reasons is subtracted, the amount of land actually removed from timber cutting is considerably less than half that figure (Interagency Scientific Committee 1991). This plan may be adequate to save habitat for 1750 pairs of spotted owls. Whether it preserves as much of the old-growth forests as we should preserve is another question.

The main opponents to preserving old-growth forest are timber companies (who find it highly profitable to sell old-growth trees to Japan), residents of the region who depend on forest jobs, certain politicians, and, to a much lesser degree than earlier, the Forest Service hierarchy. It is impossible not to be sympathetic to persons facing unemployment; however, the impact of forest preservation on jobs has been exaggerated (Sample and LeMaster 1992). Projections of declining timber production, hence loss of jobs, if the cut of old growth was not decreased were made as early as 1963 and repeated in 1969 and 1976. It is, nevertheless, a fact that the average yearly cut of Douglas fir *increased* from 4400 million board feet in the early 1980s to 6320 million board feet in the late 1980s (data in Warren 1991). During this period employment declined, much of the loss coming from automation and computerization in the lumber industry (Stiak 1990, Sample and LeMaster 1992).

Many persons see the northern spotted owl as a test case for the nation's commitment to conservation. The issues are clear: The spotted owl requires old-growth forest to survive as a species. For economic reasons, business interests want to remove old growth at a rate that will almost certainly lead to its extinction and probably to the extinction of other species. Legislation is in place that should serve to protect the owl, but the willingness of the federal government to live up to its obligations has not yet been demonstrated. As Daniel Simberloff (1987) commented, "If this nation intends to pay more than lip service to the principle of conserving biotic diversity, now is the time." And, it might be added, the old-growth forests are the place.

Shrublands and Woodlands

In the western United States and elsewhere, a variety of shrublands and woodlands occur that resemble one another physiognomically but that seem to depend on several environmental factors whose roles vary from place to place. (**Woodland** refers to areas dominated by small trees spaced far enough apart that they do not form a closed canopy.) A "Mediterranean climate" of winter rain and summer drought often supports such communities, but similar vegetation may also grow where summers are wetter. Other factors that may be involved are soil, fire, and animal effects including past overgrazing. In mountainous regions these vegetation types tend to occur below the lower montane zone of ponderosa pine and above dry grassland or desert.

Here we discuss only California and Arizona chaparral and pinyon-juniper woodland; however, several other vegetation types that are physiognomically related also exist. Examples are the oak woodlands and savannas of California valleys and foothills and the chaparral (petran bush) occurring northward in the Rockies and adjacent highlands. More remote but still physiognomically similar are the shinnery, oak areas of sandhills and river dunes in Texas, Oklahoma, and New Mexico (Weaver and Clements 1938, Garrison et al. 1977), and the sand-pine scrub of central Florida.

California Chaparral

This dense shrubland occurs on dry slopes and ridges from southern Oregon to Baja California; however, the main distribution is in California, from Shasta County southward near the coast and on the west side of the Sierra Nevada (Munz and Keck 1968) (Figure 16-47). Elsewhere in the world, similar vegetation occurs around the Mediterranean Sea (maquis or garrigue), on the southern coast of Australia (mallee scrub), at the southern tip of Africa (cape scrub or fynbos), and in central Chile (matorral). Yearly rainfall is usually moderate, 30 to 75 cm, but summer rainfall is below 5 cm and often virtually nil.

Temperatures are generally mild in winter and warm to hot in summer; January and July averages for Los Angeles are 13 and 23°C. Fog and, in heavily populated areas, smog are frequent, especially in summer.

Soil types vary depending on parent material, slope, moisture regime, and erosion. Degraded red clay soils called *terra rossa* are characteristic of much of the Mediterranean area.

A dense thicket of shrubs 1 to 3 m tall with little other vegetation is the characteristic physiognomy. The shrubs tend to have root systems possessing both an extensive, shallow, lateral portion and a deep, penetrating portion (Sampson 1944). The latter taps water as deep as the rocky parent material during the summer drought period. Most growth, however, occurs during the rainy period.

On hot and dry sites, chamise often forms nearly pure stands (Paysen et al. 1980). Western mountain mahogany occupies more mesic sites (Figure 16-48). The manzanitas tend to occur at higher elevations. Other widespread species include California scrub oak, "California lilac," and buckthorns.

Under natural conditions, a stand of chaparral tends to be burnt every 30 to 40 years (Figure 16-49). The largest fires occur in late fall and early winter in connection with hot, dry Santa Ana winds (Countryman 1974). Fire protection by humans produces stands with excessive flammable material and, accordingly, more severe fires when they finally occur. The loss of plant and litter cover, together with fire-induced changes in soil, leads to rapid runoff and erosion if heavy rains precede regrowth of vegetation. Floods and mudslides (DeBano et al. 1979) may occur but are not an inevitable consequence of fire. Mudslides are encouraged by construction practices that cut off the tops and through the sides of hills.

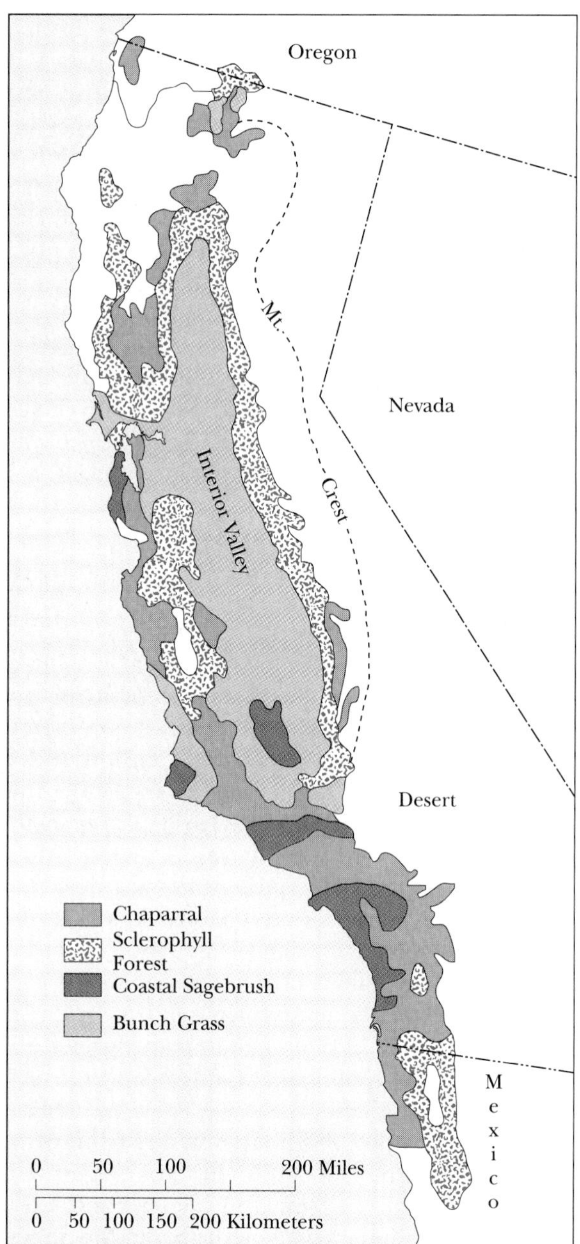

Figure 16-47

Distribution of California chaparral and related vegetation types. (From *The Ecology of North America,* by V. E. Shelford, University of Illinois Press, 1963. Copyright © Board of Trustees, University of Illinois.)

Figure 16-48

Mountain mahogany, a bushy shrub in the rose family, is a characteristic chaparral species. (Drawing by A. E. Hoyle from Judd 1962.)

Many, though not all, of the chaparral shrubs root-sprout within a few months after the fire. Not for several years, however, do these species resume dominance. A great variety of annual and perennial herbs spring up, many from a seedbank, and briefly dominate the scene. The base-rich ash may be important to these plants. Heat (or charred wood) stimulates germination in some herbs and non-root-sprouting shrubs, such as manzanitas and California lilacs. The postfire herbs tend to have showy fragrant flowers and include lobelias, phacelias (Figure 16-50), and whispering bells (Sampson 1944). Herbs are generally scarce in or absent from the mature chaparral, perhaps because of allelopathic effects by chamise and other shrubs.

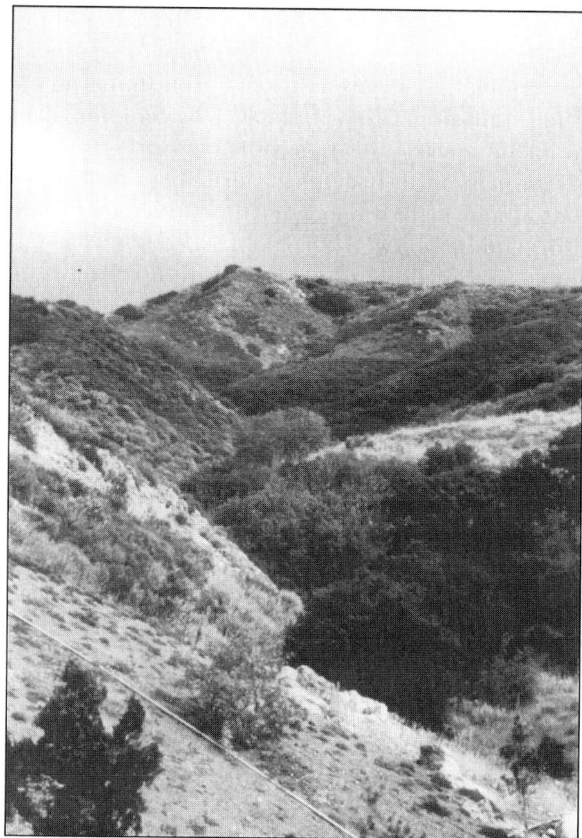

Figure 16-49

Pre- and post-fire scenes in California chaparral. One photo is from June, before the fire. The other is shortly after the fire, in October. (Photos in the Santa Monica Mountains, northern Los Angeles County, by Janet Vail.)

In time, the dense, low thickets are regenerated. By 20 years after the fire, 10 to 15% of the branches in the shrub crowns are dead, and by 40 to 50 years, the percentage is 35 to 40 (Green 1982). The stage is set for the next fire in the cyclic chaparral system.

Aboveground biomass in chaparral varies with time since the last fire. Older stands run mostly 3 to 6 kg/m^2 (Riggan and Dunn 1982), or something less than one fourth that of temperate deciduous forest. Below-ground biomass may be high. The root-sprouting shrubs have massive burls (lignotubers) that may weigh three or four times the shoot biomass (Kummerow 1982).

Growth is rapid after a fire, although some of this may be from energy stored in the burl. General productivity is rather low, perhaps 625 g/m^2, in part because of low moisture in the summer. Much of the stored energy in older chaparral stands is tied up in litter and dead stems and is later lost in fires. During the period of high oil prices in the 1970s, consideration was given to harvesting chaparral for fuel, and this idea will doubtless resurface in the future.

Chaparral is the climax vegetation over much of the range it occupies; however, lumbering and burning of low-elevation coniferous forest can give rise to chaparral, as can overgrazing of oak woodland. Although chaparral is fire dependent, too frequent fires can kill the chaparral shrubs, leading to grassland.

The fauna consists largely of species that also

Figure 16-50

Caterpillar phacelia, one of the annuals that grows profusely after fire in chaparral. (Photo in northern Los Angeles County by Janet Vail.)

occur in desert or montane forest. Among mammals, the brush rabbit is nearly restricted to chaparral and oak woodland (Verner and Boss 1980). The mule deer is common in winter, having summered at higher elevations. A variety of other mammals is present including many species of small rodents.

Small mammal populations drop drastically after a fire. How much of this is attributable to death from the fire, how much to their abandoning the area, and how much to later death through increased exposure to predators is uncertain. Wood rats, which are common in unburned sites, are especially vulnerable to fire because of their above-ground nests of dead twigs. Survivors and early reinvaders include kangaroo rats and deer mice. Wood rats begin to return in numbers after about two years.

Breeding-bird diversity is low, reflecting the low vegetational diversity. Perhaps the one most characteristic species is the wrentit (Figure 16-51). The small, pale-eyed bird "is so partial to dense chaparral that it will usually refuse to cross firebreaks cut through the brush" (Small 1974). The short wings, long tail, and dull brownish coloration of the wrentit also characterize several other of the typical chaparral species. Many of these forage on the ground between the shrub bases. Examples are rufous-sided, brown, and green-tailed towhees; California thrasher; sage sparrow among others; and California quail. Anna's hummingbird nests here.

Overall breeding-bird density is moderate, generally 200 to 300 pairs per 40 ha. Winter densities tend to be as high or higher, including many of the same species plus others, particularly from higher elevations. Winter additions include yellow-rumped warbler, ruby-crowned kinglet, and hermit thrush.

Although populations of chaparral birds such as California quail and brown towhee drop after a burn, grassland species and wide-ranging predatory

Figure 16-51

The wren-tit, a small, grayish-brown bird with a long tail, is characteristic of the Pacific coast chaparral. (P. La Tourrette/VIREO.)

species increase (Lawrence 1966). Examples of the former are mourning dove and western meadowlark. The overall result is a slight increase in population density.

Lizards and snakes are common in chaparral. Among the most characteristic species are the southern alligator lizard and the striped racer.

A diverse insect fauna exists. Bees are the most important pollinators (Force 1982). The postburn herbs draw large numbers of flower-visiting insects from outside the burn, even though many of the herbs are able to self-pollinate.

In the Mediterranean region this climate and general vegetation type was the important early center of civilization. (In the United States for the past 60 years it has attracted an increasing percentage of the population; however, its status as a center of civilization is not yet clear.) From the descriptions by classical writers of the "thick woods and gigantic trees" of southeastern Spain, of "wooded Samothrace," and similar allusions, it is evident that parts of the Mediterranean region were more heavily wooded 2000 to 3000 years ago. Timber cutting for shipbuilding and other purposes, clearing for cultivation (followed by abandonment as Rome declined), fires, goats, soil erosion, and climatic changes may have been involved in the conversion of such forests to the scrub that now occupies much of this region.

It would make sense to leave alone most of the chaparral and the adjacent lands subject to flooding. We could observe the functioning of a remark-

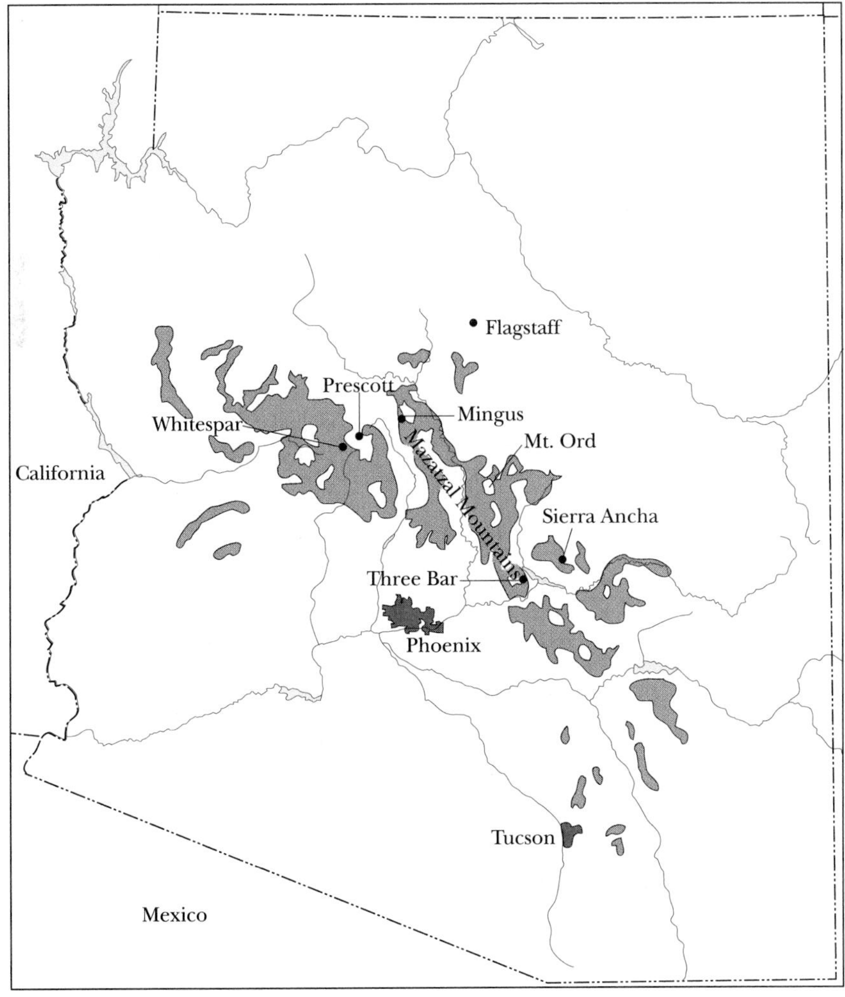

Figure 16-52

Distribution of the chaparral vegetation type in Arizona. (Adapted from Carmichael et al. 1978.)

able cyclic ecosystem, which could serve such useful functions as watershed protection, recreation, and game production, as Arthur Sampson (1944) long ago pointed out. But this has never been the case. By the 1880s, ranchers were running cattle on chaparral, burning it frequently to "control brush." This century has seen a steady migration of suburbanites into the chaparral, where they build flammable houses.

Arizona Chaparral

This vegetation type is closely related to the better-known California chaparral. It forms a discontinuous band running southeast across the state at elevations between 915 and 1830 m (Carmichael et al. 1979) (Figure 16-52, p. 509). Below it lies desert or desert grassland, and above, pinyon-juniper woodland or ponderosa pine.

Annual precipitation is similar to that for California chaparral, but there are two rainy periods, December to March and July to August, and two dry periods. Soils are entisols, youthful and coarse textured, derived from the weathering of granite and other rocks.

Shrub live oak is widely important (Figure 16-53). Mountain mahoganies, manzanitas, and ceanothus are also important, including species that also occur in the California chaparral. Chamise is absent.

Plants show fire adaptations similar to those of the California chaparral; however, the role of fire has not been as thoroughly studied here. There is some evidence that historically stands were more open, with grama grasses, dropseeds, and muhlies interspersed (Cable 1975). Overgrazing by cattle around the turn of the century evidently led to an increase in dominance by shrubs.

Currently efforts are being made to convert the Arizona chaparral to cattle (and goat) range through the use of fire and herbicides followed by seeding with exotic grasses. According to one set of Forest Service authors (Hibbert et al. 1974), conversion to a grass-dominated community "increases forage production, improves wildlife habitat, and provides more aesthetically pleasing landscapes that have a reduced fire hazard." The obvious conclusion is that the natural vegetation of the region was simply one of Mother Nature's mistakes.

Figure 16-53

Shrub live oak is widely important in Arizona chaparral. (Drawing from Judd 1962.)

Pinyon-Juniper Woodland

This is the most extensive woodland type, occupying a large, although discontinuous, area below the lower montane forests (usually ponderosa pine) from Mexico to Oregon and Montana (Figure 16-54). In California, pinyon-juniper occurs on the

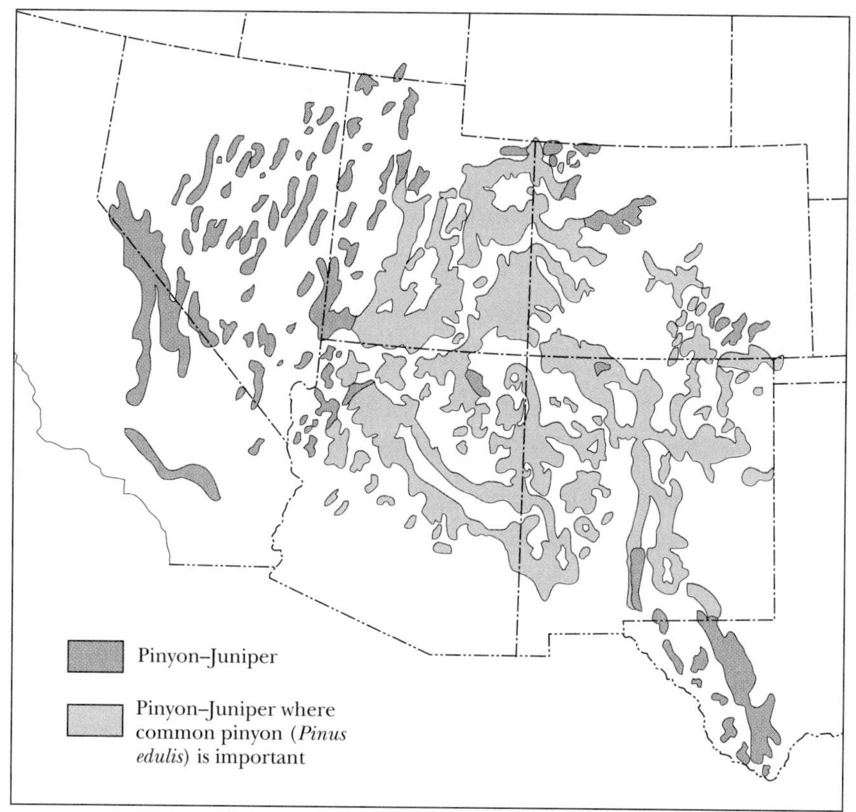

Figure 16-54

Distribution of pinyon-juniper in the southwestern U.S. Similar vegetation, though with pinyon decreasingly important, continues further north (see, for example, Figure 16-36). (Adapted from Barger & Ffolliott 1972.)

eastern side of the Sierra and various oak woodlands occur further west.

The physiognomy is a savanna of small (usually less that 10 m) evergreen trees with grasses and shrubs between. Pinyons are slow-growing, short-needled pines whose large seeds were a staple food of the Southwestern Indians (Little 1977). Both the pinyon pines and the junipers, which have fleshy fruits, are adapted for animal dispersal. Three species of pinyon and about five species of juniper are important constituents in different geographical areas. North of southern Idaho, pinyon drops out and the vegetation type is dominated by juniper. Also, juniper tolerates drier conditions, so that there is generally an altitudinal gradient with juniper more important at lower elevations and pinyon at higher. From Arizona and New Mexico southward, oak becomes increasingly important (Pieper 1977).

Shrub and herb associates are mostly species that also occur in lower, drier vegetation types. Sagebrush and bitterbrush are important shrubs in the North and various oaks and monocots in the South. The herb layer is dominated by gramas, muhlies, and other grasses.

Although physiognomy is relatively uniform over a wide area, species composition varies greatly, and so do climate and soil. Rainfall tends to be lower than in adjacent areas of montane forest but may range from 25 to 50 cm annually. A summer dry period is typical but not universal. Summers are hot and winters cold.

Soils are generally shallow, rocky, and infertile. Under the U.S.D.A. Comprehensive System, most soils are mollisols and aridosols. Under the older system, most were classified as brown and chestnut soils or regosols (Dealey et al. 1978).

Fire is clearly less important in this ecosystem than in chaparral. In the absence of fire, or with overgrazing, junipers spread into sagebrush (Dealey et al. 1978) and grass areas (Baxter 1977). Both juniper and pinyon are generally killed by fire, yielding grass- and shrub-dominated successional communities. Re-entry of the trees is by animal dispersal, and 100 to 150 years may be required for re-establishment of pinyon-juniper dominance (Koniak 1985). Nurse plants, such as shrubs and, later, small junipers and pinyon, are important in the process, supplying perching sites for juniper-dispersing birds, cover for juniper-dispersing mammals, and shade for developing seedlings.

Pinyon pines are dispersed mostly by pinyon jays; the two are involved in a mutualistic system similar to that already described for Clark's nutcrackers and whitebark pine (Ligon 1978). The nutcrackers also eat and disperse pinyon pine where they co-occur, in the southern part of the nutcracker's range (Vander Wall and Balda 1977).

Grazing, or overgrazing, has been the fate of most pinyon-juniper areas since the arrival of white settlers. Increasingly, management of the vegetation type has involved wholesale removal of the trees and conversion to grassland by seeding.

The mammal species most closely tied to pinyon-juniper are the pinyon mouse, an arboreal mouse that makes its nests in hollow junipers and uses juniper berries as its winter food (Armstrong 1972), and woodrats (especially *Neotoma lepida* and *Neotoma mexicana*). Woodrats build nests composed of several bushels of twigs, pinyon nuts, and juniper berries around the bases of pinyons or junipers.

A number of other small mammals occur. The deer mouse, a wide-ranging grassland relative of the pinyon mouse, is often common. Other species of grassland, desert, and coniferous forest are also present.

Pinyon-juniper, like chaparral, is an important winter-spring habitat for mule deer (Figure 16-55) which browse on the juniper and various shrubs. This is also one of the habitat types occupied by the introduced wild horse. Large predators include the mountain lion and coyote, both of which prey on mule deer and, like them, tend to move to higher elevations in summer.

Probably the most distinctive bird species is the pinyon jay, a noisy, social bird that caches pinyon nuts. Bushtits, gray vireos, plain titmice, gray fly-

Figure 16-55

Mule deer are characteristic large mammals of woodland and chaparral. In some regions they occupy these communities throughout the year; elsewhere they spend the summer at higher elevations in coniferous forest and come down for the winter. These deer were photographed in the Malheur Refuge, Oregon, by R. C. Erickson. (Photo courtesy U.S. Fish and Wildlife Service.)

catchers, and greater pewees (Coues' flycatchers) are characteristic, but a great many other species that occur in a variety of woodland types or in desert or montane forest may also nest.

In southwestern pinyon-juniper, breeding-bird populations are moderate, between 200 and 300 pairs per 40 ha (Balda 1969). Although such summer residents as the back-chinned and broad-tailed hummingbirds and black-throated gray warbler depart for the winter, other birds arrive from higher elevations, so that winter populations may be fairly high. Included are a number of species that make use of the juniper berries as food, including mountain bluebird, American robin, and cedar waxwing.

In the northeastern corner of the biome, in South Dakota, the avifauna is mostly forest edge, with mourning dove, rufous-sided towhee, chipping sparrow, and black-capped chickadee as important species (Sieg 1991). Long-eared owls and black-billed magpies also nest. The winter avifauna, as in the Southwest, includes species that eat juniper berries; examples are American robin, Townsend's solitaire, and Bohemian waxwing.

Amphibians and ground-dwelling invertebrates that require high soil moisture are scarce. Several insectivorous lizards with desert affinities are present at lower elevations; the short-horned toad, however, is more numerous at higher elevations and is also common in the ponderosa pine forest above (Frischknecht 1975). Ants are common. A number of insects utilize pinyon, but relatively few attack juniper (Smith 1977).

Community and Ecosystem Ecology: *Grassland, Desert, and Tropical Biomes*

Characteristic of drier grasslands and shrub-steppe is the pronghorn. *(E. P. Haddon, U.S. Fish and Wildlife Service.)*

OUTLINE

- Temperate Grassland
- Desert
- Tropical Biomes

This chapter concludes the discussion of the major biomes. Temperate grassland, desert, tropical forest, and tropical savanna are considered.

Temperate Grassland

Distribution and Climate

The climate of regions occupied by grassland is variable (Table 17-1). Summers are usually hot, but winters may be cold to mild. The growing season varies from 120 days in the northern plains to 300 days in the coastal prairie of Texas.

Precipitation is one of the least variable features, being usually between 30 and 85 cm per year. Usually, also, there is a dry period in late summer, fall, or winter. This predisposes grassland areas to spring or fall fires. Some of these may be lightning set, but humans have been responsible for much of the burning of the prairie.

Table 17-1
Generalized Climatic Data for Three Sections of North American Grassland

	Prairie Type (and Location)		
Climatic factor	*Tallgrass (Central Illinois)*	*Desert Plain (Southwest Texas)*	*Palouse (Southeast Washington)*
July temperature (°C)	24	30	19
January temperature (°C)	−4	12	−2
Growing (frost-free) season (days)	170	270	150
Annual precipitation (cm)	86	55	40
Number of days with 1 inch (2.5 cm) or more snow cover	45	<1	50

Based on USDA, *Climate and Man: Yearbook of Agriculture* (House Document No. 27), 77th Congress of the United States, 1941.

Most grassland regions occur in the interior of continents. In central North America the eastern tallgrass portion was called prairie and the western portion, plains. There has been a recent tendency to refer to all native American grassland as prairie. Elsewhere temperate grassland biomes occur in eastern Europe (in Hungary, termed *puszta*), central and western Asia (in Russia, *steppe,* a term also used generally to refer to any shortgrass grassland), Argentina *(pampas),* and New Zealand.

Soils

The soils that develop under grassland are, like grassland climates, variable (Figure 17-1). Typical soil formation is seen as occurring under rainfall too low to leach clay colloids and positive cations but high enough to support an annual productivity of 500 g/m^2 of new biomass, much of which is added to the soil each year. The result is a soil that is neutral to basic, fertile, and high in organic matter. However, soils at the eastern margin of the prairie may have about the same pH and be leached to the same degree as adjacent forest soils, although they have a dark, humus-rich A horizon. At the opposite extreme, the driest grassland soils may, like adjacent desert soils, have a shallow calcium carbonate hardpan and little organic matter accumulation.

In the older soil terminology, grassland soils were termed "prairie soils" (brunizems), "chernozems," or "chestnut" and "brown soils" along a geographical and moisture gradient running from the moist areas of the east to the arid grasslands adjoining desert. Under the U.S.D.A. Comprehensive System, most grassland soils are mollisols.

Although some local grassy areas, such as the hill prairies of Illinois (Evers 1955), are on slopes, most of the grassland biome is flat to rolling. "The prairie," Wallace Craig (1908) wrote after 2 years in North Dakota, "offers a minimum view of the earth and a maximum view of the sky. The prairie view is, in this particular, precisely similar to a view in mid-ocean."

Plants

The main body of the biome was a sea of grass—a vast expanse of grassland occupying more than a million square miles in the middle of North America (Figure 17-2). Through much of this region from Illinois to Colorado and Alberta to Texas, the grass ran uninterrupted except for an occasional strip of forest along the rivers. "And these rivers," Wallace Craig (1908) wrote, "lie score of miles apart. One could fence off an area of North Dakota plains of the size and shape of Delaware without enclosing or even touching a river."

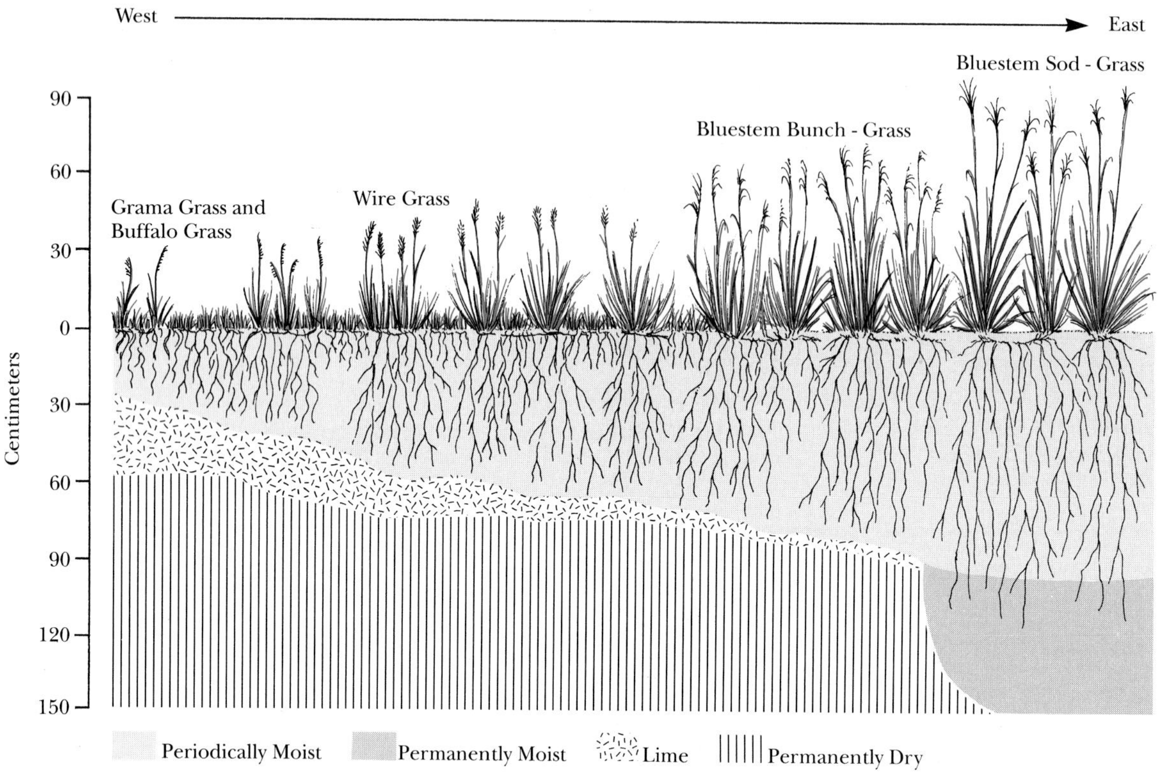

Figure 17-1

Soil-plant relationships in a generalized transect from west to east in North American grassland. (Adapted from Kendeigh 1961; after H. L. Shantz.)

Eastward the primeval grassland was interrupted more and more by trees, joining them in oak savanna and giving way to them in prairie groves. By Indiana and Michigan, the forest became the matrix and the grassland the inclusions. Westward, also, the grassland becomes discontinuous, forming parts of the vast mosaic of vegetation types between the Great Plains and the Pacific.

Yearly productivity varies greatly within the grassland biome, aboveground production running from below 100 to more than 1500 g/m^2. Grasses generally contribute 60 to 90% of this. Productivity is strongly dependent on moisture (Table 11-4), the highest levels occurring on prairie relicts within the deciduous forest. At any one site, increases of 100% between a dry and wet year are not unusual (Mueggler and Stewart 1980). In dry years, many species are dwarfed or, if the spring is dry, do not even appear above ground.

Although perennial grasses form the bulk of prairie biomass, much of the species diversity comes from forbs, the nongrasslike herbs. Composites (Figure 17-3), legumes, and mints (Figure 17-4) are especially well represented. In Wisconsin prairie, grasses make up only about 10% of the total number of species (Curtis 1959). In the more arid grasslands of the West where grass numbers are greater and forbs fewer, nearly one half the flora may be grasses or sedges.

Grasses are classified as short, mid, or tall, depending on their height at flowering (Table 17-2). In spring even the vegetation of tallgrass prairies is only a few centimeters high because, more than virtually any other community, the aboveground vegetation of prairie starts over every year. There is a seasonal progression of flowering heights in the prairie (Figure 17-5), the early spring flowers, such as bird's-foot violet, being short and the late summer ones, such as compass plants and rosinweed, being tall.

The progression results from each species tending to stay in the rising canopy of the vegetation

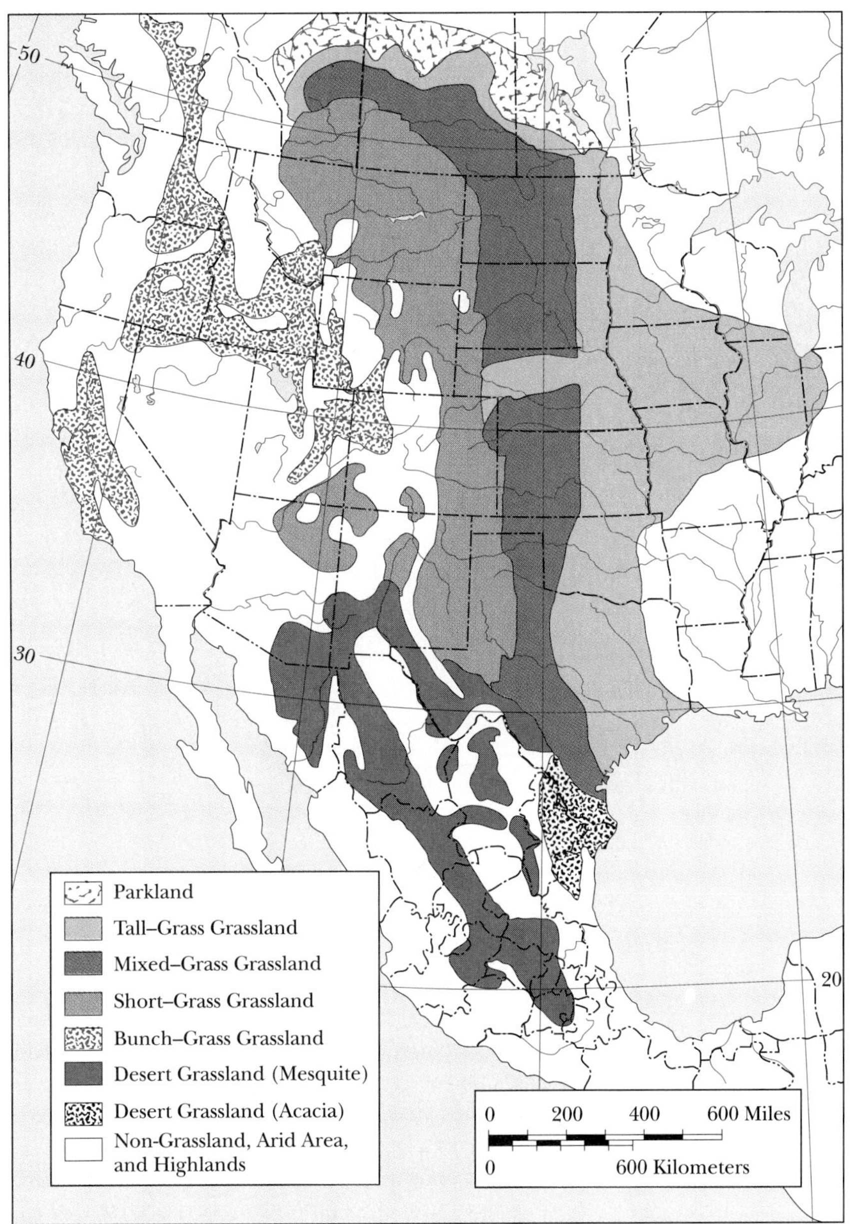

Figure 17-2

Distribution of grassland and related vegetation types in North America. (From *The Ecology of North America,* by V. E. Shelford, University of Illinois Press, 1963. Copyright © Board of Trustees, University of Illinois.)

(mostly set by the taller grasses) until it flowers and then ceasing growth (Brewer 1985). A few wind-pollinated species tend to flower above the general vegetation height. Also, in midsummer, when the greatest number of species are flowering, several species tend to shoot above the rising grass cover, perhaps gaining an advantage in attracting insect pollinators.

Fire has been an important environmental influence in all parts of the grassland. It is clear that,

Figure 17-3

Prairie coneflower, a prairie forb in the family Asteraceae. (Photo by the author.)

Table 17-2
Examples of Tallgrasses, Midgrasses, and Shortgrasses in North American Prairie

In general, tallgrasses reach heights of 1 to 3 m, shortgrasses 20 cm or less, and midgrasses in between.

Tallgrasses	*Midgrasses*	*Shortgrasses*
Big bluestem	Needlegrasses	Buffalo grass
Indian grass	Dropseeds	Gramas
Slough grass	Wheatgrass	Pennsylvania sedge*
Sand bluestem	June grass	
Switchgrass	Little bluestem	
	Wild ryes	

* Not a grass.

at least at the margins of the grassland with both forest and desert, fire suppression leads to invasion by woody plants. Recurrent fires kill most woody species above ground or altogether while doing little harm to the prairie herbs. Many grasses make better growth in the year following a burn than after a few years without fire (Petersen 1983). Most prairie species flower more vigorously following a burn; the effect is especially striking with certain species, such as purple coneflower and prairie dropseed (Roosa and Smith 1991), which may scarcely flower at all after a few seasons without fire.

Grazing by large herbivores is another important factor in the life of prairie plants. Grasses are well adapted to recover from grazing, having their growing points at the base of the plant rather than the tip. Although grazing removes biomass, it is beneficial to the grasses in some ways. Perhaps this explains why few grasses are poisonous, although

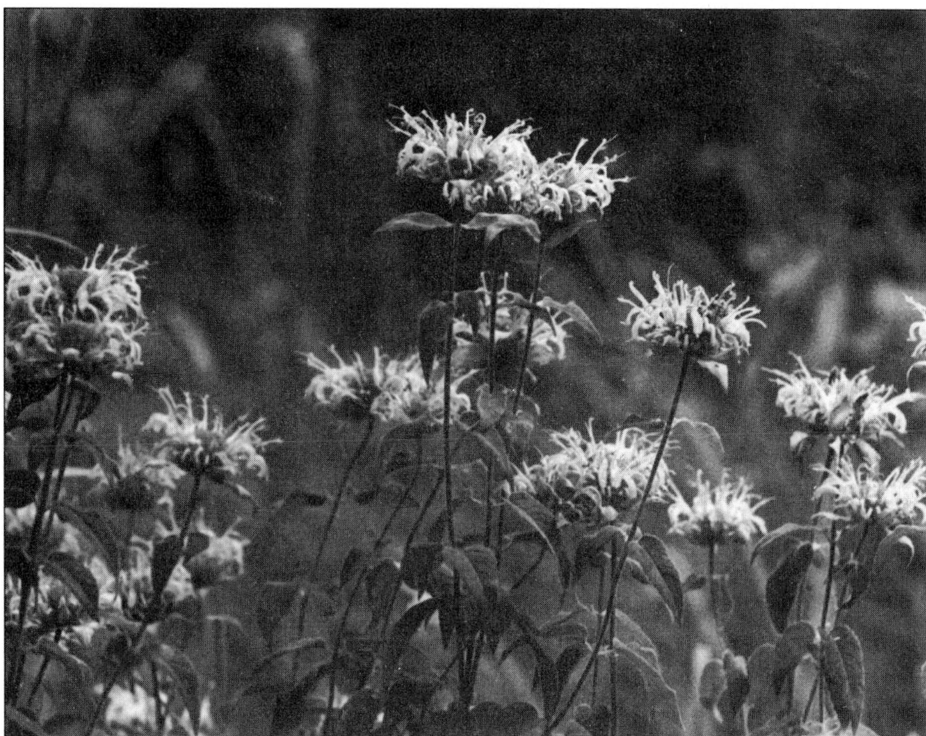

Figure 17-4

Wild bergamot, a prairie forb in the family Lamiaceae. (Photo by the author.)

some are less palatable at one season than another. Many prairie grasses have spines or barbs on the fruits that discourage grazers from eating their reproductive structures.

Forbs tend to have their growing points at the top of the stem and, thus, tend to be harmed more by grazing. The list of poisonous grassland forbs is long. Among the best known are the locoweeds of the western United States. Horses affected by locoweed tend to stagger along in a straight line until they run into something. They may also suffer hallucinatory visual disturbances, causing them to leap or shy at nothing.

Within limits, grazing helps to control woody plants, especially if fire keeps invaders small and tender. Overgrazing leads to the entry of woody plants and weeds by reducing the competitive ability of the grasses, by opening up bare and compacted ground for invasion, and by reducing the ability of the vegetation to sustain a fire.

Types of Grassland

Six subdivisions of the grassland biome (Figure 17-2) are recognized here:

Tallgrass Prairie

This is the easternmost section dominated by grasses, notably big bluestem, that by late summer stand 2 to 3 m tall (Figure 17-6). The forb flora is rich (Table 17-3).

At the time of settlement, much of this area was poorly drained. Drain tiles and ditches put in as the prairie was converted to cornfields have lowered the water table to a degree that a traveler today cannot appreciate how wet large sections of the tallgrass prairie were in early spring.

Although a few shrubs such as New Jersey tea occurred in the prairie, it was basically treeless. Trees were restricted mostly to floodplains and to

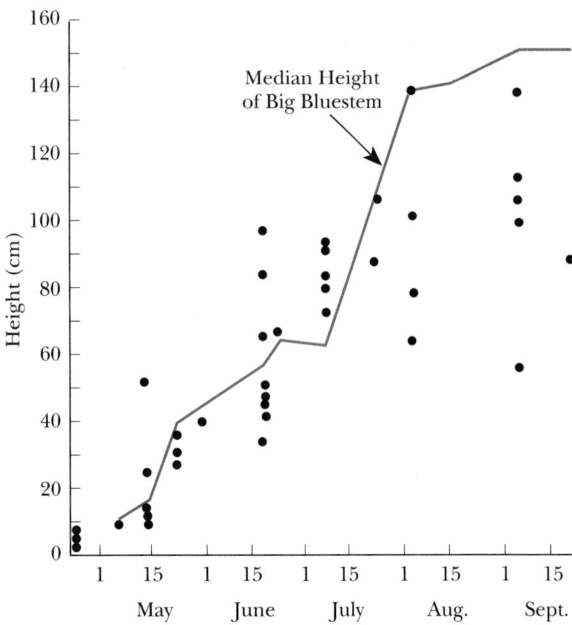

Figure 17-5

Flowering height in prairie tends to increase through the growing season, roughly keeping pace with the increasing general vegetational height. In this graph for a Michigan prairie, median height of big bluestem is used as an indicator of general vegetation height. Dots indicate height on the date of first flowering of other prairie species. (Adapted from Brewer 1985.)

topographical sites that formed firebreaks, such as steep slopes (scarps).

Tallgrass prairie called coastal prairie occurs along the warm, moist Gulf Coast of Texas. It is floristically rich, containing species of southeastern affinities as well as ones typical of other prairies to the north and west (Clements and Shelford 1939). Silver beard grass and Texas needlegrass are among the dominant grasses.

Mixed Prairie

This is a transition zone in which mid and tall grasses, usually of eastern affinities, tend to grow on moister sites and short grasses, often of western affinities, on drier ones. Tallgrass prairie has been largely converted to corn and soybeans; mixed prairie became the Wheat Belt.

Shortgrass Prairie

A little past the 100th meridian, the West begins with the shortgrass plains (Figure 17-7). Row crops are restricted to irrigated areas. These have increased over the past few decades but, with the depletion of groundwater in the Ogallala aquifer, may be expected to decrease again.

Rangeland is the primary landscape. Here the native vegetation is retained in a more or less degraded state depending on the level of overgrazing. Characteristic species of the less disturbed areas include buffalo grass, blue grama, needle-and-thread, and galleta (Figure 17-8). Widespread forbs include a blazing star *(Liatris punctata)* and prairie coneflower.

Figure 17-6

Flowering stalks of big bluestem, the most important tall grass of the prairie, seen in the late summer. (Photo by the author.)

Table 17-3
Some Characteristic Plants of Tallgrass Prairie Arranged by the Season in Which Flowering Begins

Approximate heights at flowering time (in centimeters) are given in parentheses.

	Spring
Prairie buttercup (5)	Shooting star (25)
Bird's-foot violet (10)	Downy phlox (30)
Penn sedge (10)	Lupine (40)
Prairie smoke (15)	Spiderwort (40
	Early Summer
Butterfly weed (60)	Wild bergamot (80)
Needlegrass (60)	Wild indigo (90)
New Jersey tea (75)	Prairie coneflower (105)
Leadplant (75)	Compass plant (150)
	Late Summer
Whorled milkweed (60)	Rigid goldenrod (100)
Purple prairie clover (60)	Culver's root (110)
Little bluestem (75)	Big bluestem (150)
Rattlesnake master (95)	Prairie dock (160)
	Fall
Silky aster (50)	Riddell's goldenrod (90)
Bottle gentian (60)	Smooth aster (105)
Rough blazing star (70)	Indian grass (130)

Based mainly on Brewer, R., "Seasonal change and production in a mesic prairie relict in Kalamazoo County, Michigan," *Mich. Bot.*, 24:3–13, 1985; Nichols, S., and Entine, L., *Prairie Primer,* Madison, University of Wisconsin Extension, 1976, pp. 1–43; and Schulenberg, R. F., Notes on the Propagation of Prairie Plants (mimeographed).

Bunchgrass Prairie

This grassland is also called Palouse prairie, after the Palouse River region of southeastern Washington, but the vegetation type extends to adjacent states and British Columbia. Occurring mostly west of the Rocky Mountains and east of the Cascades, it is relatively isolated from the great midcontinent grasslands. The grasses do not form a sod but grow in bunches or clumps. Bluebunch wheatgrass and two species of fescue are especially characteristic. Forbs are numerous. Sagebrush is frequently a member of the community (Figure 17-9); the drier sites in the region are occupied by shrub-steppe or desert. Overgrazing and the absence of fire have converted large portions of this region from grass dominance to sagebrush dominance (Clements and Shelford 1939). Cheatgrass, spotted knapweed, and thistle also increase with overgrazing (Mueggler and Stewart 1980).

California Prairie

The Central Valley of California, between the Coast Ranges and the Sierra Nevada, was originally grassland. Native perennial bunchgrasses, notably purple needlegrass, dominated, but the striking feature of this grassland was the March-to-May profusion of annual forbs, knee-high, with showy blossoms of yellow and purple. It was still a 400-mile-long (640 km) bee pasture, "one smooth countinuous bed of honey-bloom" when John Muir (1961) arrived in the spring of 1868. Since then, overgrazing, mostly by sheep, has obliterated the California prairie as completely as cultivation did the tallgrass prairie of Illinois. The region is now dominated by weedy annuals introduced from Europe, including wild oats and cheatgrass.

Desert Grassland

This warm and arid grassland fringes the hot desert in Mexico, Arizona, New Mexico, and Texas. Although much of it is on broad intermountain basins, it also occurs in a zone above desert and below woodland on the mountains.

Much of the area is savanna rather than real grassland. The woody plants may be oaks or desert shrubs such as creosote bush, yucca, and prickly pear and other cacti. By far the most important woody plant, however, is mesquite, a small leguminous tree which, like the shrubs, is also widespread in the Chihuahuan Desert.

Just how common mesquite was before the arrival of whites is unclear; clearly, though, it has spread widely as a result of overgrazing that began in the late 1800s. Three hundred head of cattle would be run on range that might support 20 to 40 on a sustainable basis (Cottle 1931). Weeds, cacti, and, especially, mesquite increased in these degraded habitats.

Mesquite is marvelously suited to take advantage of such a situation. Its branches are spiny and its foliage unpalatable (Humphrey 1958), but its pods are honey-sweet and nutritious. The seeds

Figure 17-7

Shortgrass prairie in western South Dakota with grama grasses and the low shrub soapweed (a yucca). (Photo by the author.)

within are hard coated and pass unharmed through the digestive tract and are, thereby, dispersed widely. Once established, mesquite is a formidable competitor for water. It resembles chaparral shrubs in having a two-phase root system consisting of a dense mat of shallow roots and a tap root that may reach groundwater 15 m deep.

In pre-cattle days, fires presumably kept the woody vegetation at bay in desert grasslands, though mesquite root-sprouts readily (S. C. Martin 1983). Fire also kept cactus density low, in a way that may not be immediately obvious: Fire burns off the spines, after which the cacti are speedily eaten by cattle, jackrabbits, and deer. It may also be that prairie dogs, prior to their near elimination by ranchers and the federal government, assisted by nipping off young mesquite shoots.

The grasses of desert grassland include many genera of short bunchgrasses. The gramas, three-awns, and tobosa grass are widely important. Forbs mostly of southwestern origin are also numerous (Clements and Shelford 1939).

Animals

Grassland animals show adaptations to several features of vegetation, terrain, and climate.

Vegetation

The brown, streaked backs of such birds as meadowlarks and sparrows (Figure 17-10) serve as camouflage. Burrowing is another way of escaping predators, though weasels and ferrets may follow the animals into their burrows.

The large herbivores cannot burrow, but in these flat, open landscapes, neither can they hide. Herding is one of their antipredator devices. Another is good eyesight coupled with speed afoot. In

Figure 17-8

Galleta is a perennial grass characteristic of shortgrass prairie and also present in desert grassland and other open arid vegetation. "Galleta" means biscuit, and I have no idea what the connection is between biscuits and *Hilaria jamesii.* (Drawing based on Judd 1962.)

Figure 17-9

Shrub-steppe in Yellowstone National Park. (Photo August 1988 by the author.)

large ungulates, the speed has been provided by a series of morphological changes that include running on their toes, long legs that move back and forth but not sideways, and reduction in number of toes. The ruminant habit, in which grass is eaten and sent to one section of the stomach, then regurgitated and chewed at leisure before transport to another section for chemical digestion, is also an adaptation to foraging in a situation open to enemies.

The teeth of ungulates show adaptations to grass eating; the molars are large and flat and show various adaptations to accommodate the wear that the abrasive grass causes. The front teeth and lips are specialized for cropping off grass. A section of the digestive tract containing symbionts able to digest cellulose—a part of the stomach in bison, pronghorn, and their relatives, and the caecum and large intestine in horses and zebras—is an adaptation to the use of foliage as food.

Grass-eating adaptations also occur in small mammals, mainly voles in the northern grasslands and cotton rats in the southern ones. These rodents, like the ungulates, have flattened molars and a digestive system with features for cellulose processing, such as greatly enlarged caeca (Baker 1971). Usually, a particular grassland habitat has only one species of these grass-eating, runway-making mice but several species of seed-eaters such as white-footed mice, harvest mice, and pocket mice. Seed-eating species are also common among the birds, especially the sparrows and their relatives.

The fluctuations in cover and productivity from season to season in grasslands have repercussions for the animals. A site may be undercrowded, relative to resources, in one year and overcrowded the next. Grassland birds show much more flexible patterns of occupancy than do birds of most habitats. They evidently keep track of other areas of appropriate habitat in the vicinity by flying off their ter-

Figure 17-10

Streaked-backed birds, like this grasshopper sparrow, are characteristic of grassland. This bird, at a nest in Calhoun County, Michigan, was photographed by Lawrence H. Walkinshaw.

ritories frequently and looking things over, and they may shift to a new site even within a breeding season (Brewer and Swander 1977). Polygyny is also a response to variations in habitat quality (Verner and Willson 1969).

In this treeless habitat, birds nest on the ground, usually in grass clumps (Figure 17-10). Many species, such as the bobolink and lark bunting, have flight songs, though they also sing from grass clumps and forb stalks. Perches added to grassland, such as utility wires or wooden posts, are heavily used as song posts (Harrison 1974), although the addition of perches may not raise the densities of the species using them (Harrison and Brewer 1979).

Climate

The windiness of the prairie makes it advantageous for animals to be strong fliers, as are most birds and such insects as grasshoppers; however, another way of coping with the wind is to avoid flying, and the loss of wings has been one evolutionary trend among grassland insects (Hayes 1927). No birds are flightless in the North American grassland, though some of the gallinaceous species probably do most of their traveling by foot.

In tropical grasslands, flightless species have evolved, though it is not certain exactly how climate, vegetation, and other factors have combined to make this way of life advantageous. The ostrich of the African savanna, the rhea of the grasslands and savannas of Brazil and Argentina, and the emu of the Australian grasslands are all large, flightless species. They show certain convergences with the large grassland mammals. For example, they live in groups and show a reduction in toe number, although in none has the reduction proceeded all the way to a single toe as in the horse.

In the northern grasslands, winter is harsh—cold, windy, and snowy. Virtually all the birds migrate out except for a few wide-ranging predatory species that, like the resident red fox and coyote, prey mostly on small mammals. Buffalos, though showing some seasonal migration, stayed the winter throughout most of the northern grassland, accompanied by their "constant and numerous attendants," the timber wolves (Coues 1893). Prairie dogs and ground squirrels hibernate. The carnivore burrowers, badgers and skunks, do not, but they stay below ground for long periods in winter.

In view of the close association of fire with grassland, it is curious that so few evolutionary adjustments to it are evident among animals. Burrowing animals, of course, are preadapted for surviving fires, but burrowing serves many other functions including predator protection, low-temperature survival in the winter, and water conservation and avoidance of high temperature in summer. Voles evidently survive fires in their burrows, but then over the next two to three days they move into unburned areas (Schramm 1970). Some of these same individuals return to their original home ranges as regrowth occurs.

The larger mammals and birds can flee the oncoming prairie fire; however, they do not always escape. Alexander Henry wrote of the prairies along the Red River of the North in November 1804 (Burroughs 1961):

> *Plains burned in every direction and blind buffalo seen every moment wandering about. At sunset we arrived at the Indian camp, having made an ex-*

traordinary day's ride, and seen an incredible number of dead and dying, blind, lame, singed, and roasted buffalo.

Effects of Animals

Animals may have strong effects on the habitat in grasslands. Burrowing improves soil by aiding the incorporation of organic matter and returning mineral nutrients to the top (Gross 1969). Buffalo trails, by which the buffalo visited streams and water holes, were areas of soil compaction and erosion. Some were worn 25 cm deep (Keim 1885). Buffalo wallows were dish-shaped depressions about 4 m across in low areas. They might hold water well into the summer.

Also prevalent are direct effects of animals on the vegetation. Trampling, grazing, and other effects of bison and other grazers helped prevent woody invasion and allowed for the continued presence of forbs. The impact of a herd of bison on its preferred food plants must have been severe.

Burrow entrances and dirt piles of pocket gophers, prairie dogs, and ground squirrels also helped prevent total grass dominance by providing habitats for annuals and other forbs (Figure 17-11). Burrows also provide habitat for animals that do not, themselves, dig them, including burrowing owls, cottontails, and rattlesnakes. Apparently prairie dog grazing can, in some circumstances, maintain shortgrass prairie. In moister areas, grazing by bison or cattle is necessary to prevent tallgrass dominance, which will lead to the demise of the prairie dog town (Koford 1958).

Figure 17-11

Here a black-tailed prairie dog, one of the keystone species of mixed and shortgrass prairie, sits at a burrow entrance in western South Dakota. Bare ground around the entrance supports plant species rare in or absent from the grassland away from the prairie dog towns. (Photo by the author.)

Animal Distribution

Only a few species occurred throughout the North American grassland, and these tended to be ones of wide habitat distribution like the mourning dove, bull snake, and coyote. A few species closely tied to grassland are widely distributed, including the thirteen-lined ground squirrel, badger, bison (in the past), western meadowlark, bobolink, grasshopper sparrow (Figure 17-10), and vesper sparrow.

A list of characteristic vertebrates of mixed prairie is given in Table 17-4. Food chains of the undisturbed prairie are diagrammed in Figure 17-12. Hunting and the conversion of tallgrass and mixed prairie to row crops have eliminated many of the most characteristic animals. Bison, once one of the most widespread of North American mammals (Figure 17-13), now are reduced to small herds in a few parks and reserves. The range of pronghorn, which extended east to Missouri, has now contracted virtually to the Rocky Mountains. In numbers, bison declined from 60 million to a current few thousand and pronghorn from an estimated 40 million to a current 365,000 (Evans and Probasco 1976).

The timber wolf is all but extinct in the United States. The black-footed ferret became extinct in the wild in 1985 but has recently been reintroduced with animals from captive breeding populations. Several other species, including Swainson's and ferruginous hawks, the prairie falcon, the swift fox, and greater and lesser prairie chickens have declined greatly in the past few decades.

Two of the most common grassland species today are new arrivals. The ring-necked pheasant is an Asian bird introduced to this country in the 19th century. The red-winged blackbird, although long present in prairie marshes, has spread into upland grasslands only in the past half century (Brewer 1991).

Prairie marshes and ponds—potholes—have long been among the major producers of waterfowl in North America. Half or more of the young ducks fledged each year are from the northern prairie region. Mallards, pintail, blue-winged teal, shovelers, and gadwall are the most numerous species (Evans and Probasco 1976).

Some species characteristic of mixed prairie extend into the tallgrass region (Table 17-4); to these are added a few species of eastern affinities, such as the American toad, eastern hognose snake, sedge wren, eastern meadowlark, Henslow's sparrow, and woodchuck. Most of the species of tallgrass prairie also occupy seral grasslands in the deciduous forest biome. Greater prairie chickens inhabited tallgrass prairie originally but spread into mixed prairie with early agriculture. One subspecies, Attwater's prairie chicken, was characteristic of coastal prairie.

In the desert grasslands of central Mexico, breeding birds include several species widespread in northern grasslands including horned lark, eastern meadowlark, grasshopper sparrow, red-tailed hawk, American kestrel, and burrowing owl (Webster 1956). More distinctive are the white-tailed hawk, a rodent eater that is attracted to grass fires, and Botteri's and Cassin's sparrows, two pale seed-eaters whose ecology and life history are still poorly known. Where mesquite is present in the grassland, some scrub species, such as brown towhee and black-chinned sparrow, are added.

Breeding-bird densities vary with moisture, geographically, and between years. In an average year, tallgrass prairies run 120 to 140 territorial males per 40 ha, mixed prairie 80 to 100, and shortgrass prairie 50 or fewer. Because red-winged blackbirds, bobolinks, dickcissels, and most gallinaceous birds are polygynous, merely doubling these numbers underestimates total population size.

Table 17-4
Some Characteristic Vertebrates of Mixed Prairie in North America

Reptiles and Amphibians	*Birds*	*Mammals*
Ornate box turtle*	Upland sandpiper*	White-tailed jackrabbit
Slender glass lizard*	Greater prairie chicken*	Black-tailed jackrabbit
Common garter snake*	Ring-necked pheasant*	Thirteen-lined ground squirrel*
Plains garter snake*	Sharp-tailed grouse	Franklin's ground squirrel*
Ribbon snake*	Long-billed curlew	Black-tailed prairie dog
Lined snake	Red-tailed hawk*	Plains pocket gopher
Western hognose snake	American kestrel*	Plains pocket mouse
Bull snake*	Swainson's hawk	Deer mouse*
Prairie rattlesnake	Ferruginous hawk	Northern grasshopper mouse
Great Plains toad	Prairie falcon	Meadow vole*
	Burrowing owl	Prairie vole*
	Horned lark*	Coyote
	Western meadowlark*	Timber wolf*
	Red-winged blackbird*	Swift fox
	Bobolink*	Badger*
	Dickcissel*	Black-footed ferret
	Brown-headed cowbird*	Pronghorn
	Lark bunting	Bison
	McCown's longspur	
	Chestnut-collared longspur	
	Grasshopper sparrow	
	Baird's sparrow	

* Species also important in tallgrass prairie.

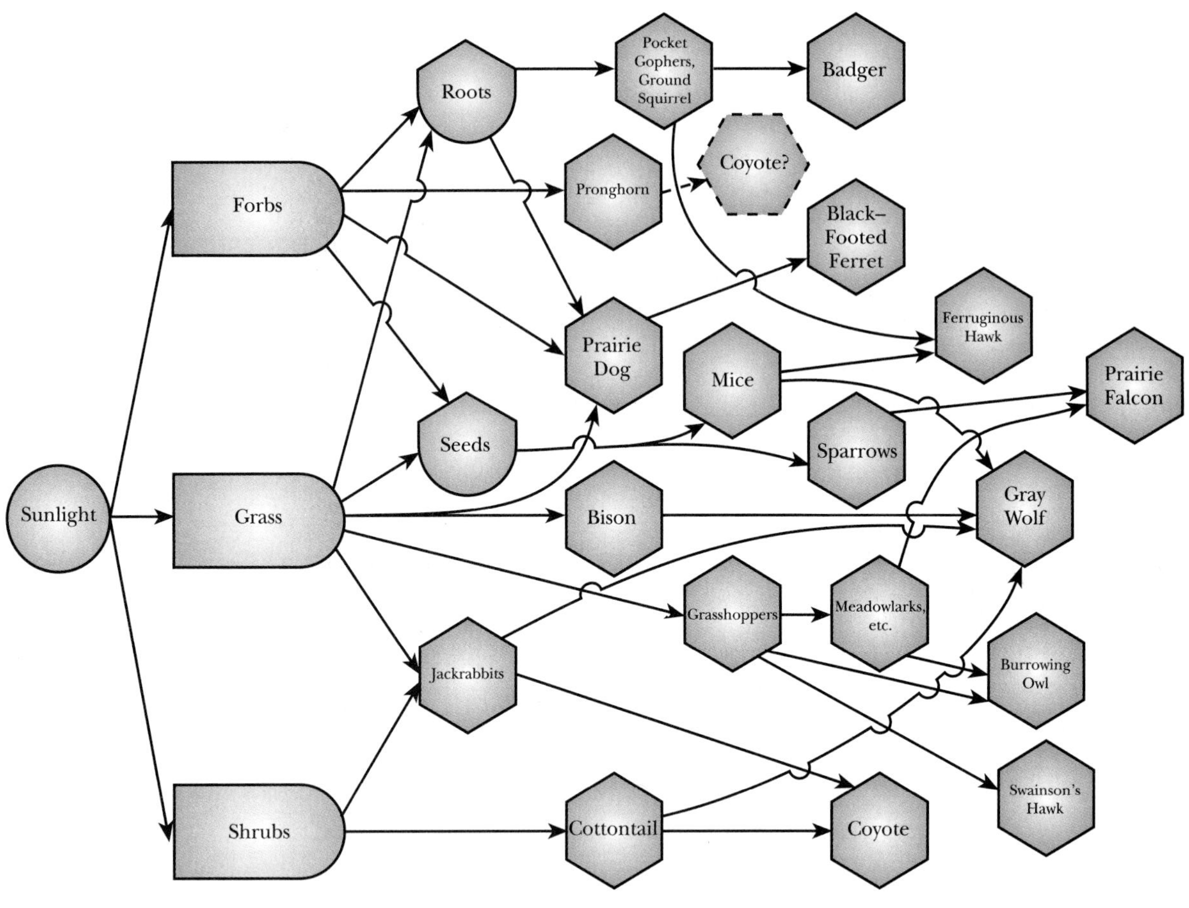

Figure 17-12

A generalized diagram of energy flow in primeval prairie.

Oak Openings

In the ecotone between tallgrass prairie and temperate deciduous forest occurred oak savannas, called **oak openings** (Curtis 1959). The settlers distinguished bur oak openings (Figure 17-14), which usually occurred fringing mesic prairie, from just plain oak openings, in which the trees were yellow, black, or white oak (Hodler et al. 1981). Both savanna types, like the prairies themselves, were maintained by fire.

The bur oak openings shared most of the species of mesic prairie, but the second type was more variable. On sandy sites it thinned out into dry prairie and sand barrens. On sites where fires were infrequent, such as the east side of rivers, it graded into oak forest. Hazelnut was a characteristic shrub in both types of oak openings. To the more strictly grassland animals were added in the oak openings a variety of other species that depended on trees or brush. Examples include eastern kingbird, common yellowthroat in the wetter areas, song sparrows, red-headed woodpecker, eastern bluebird, fox squirrel, and raccoon.

Prairie Restoration and Reconstruction

More so than almost any other ecosystem, the mixed and tallgrass prairies are gone, obliterated

Figure 17-13

More than any other animal, the bison is a symbol of the native North American grassland—and its disappearance. The animals here were photographed on the National Bison Range, Montana. (Photo by H. W. Henshaw, courtesy U.S. Fish and Wildlife Service.)

by humans. Of an original 240 million acres (97 million ha) of tallgrass prairie, about 3% remains today (Smith 1992). The plants persist, mostly in pitiful remnants a few meters wide along railroads and in pioneer cemeteries (Figure 17-15). The destruction of the tallgrass prairie took less than 100 years, from the small eastern prairies of Ohio, settled about 1825, to the draining of the last wet prairies of Iowa, about 1920; in fact, most of the prairie disappeared under the plow in just 50 years between 1830 and 1880.

The past 30 years have seen a remarkable widespread movement to restore at least small stands of prairie to the midwestern landscape. Reconstructing a beech-maple forest in a few years is not feasible. It would cost large amounts of money to plant even modest-sized trees. The leaf litter, logs, and herbaceous plants could be obtained only by damaging some other forest site. Because the dominant prairie species are herbaceous and can be readily grown from seed, it is, however, possible to produce a community resembling native prairie within a few years. So the reasoning goes, and the results have been encouraging, if not perfectly satisfying.

Probably the first attempt at prairie restoration was at the University of Wisconsin Arboretum, another of Aldo Leopold's good ideas. The first plantings were made in 1936 using employees from the Civilian Conservation Corps, one of the federal make-work projects of the Depression (Sperry 1983). Another early effort was at the University of Illinois Brownfield Woods Preserve where 8 ha were

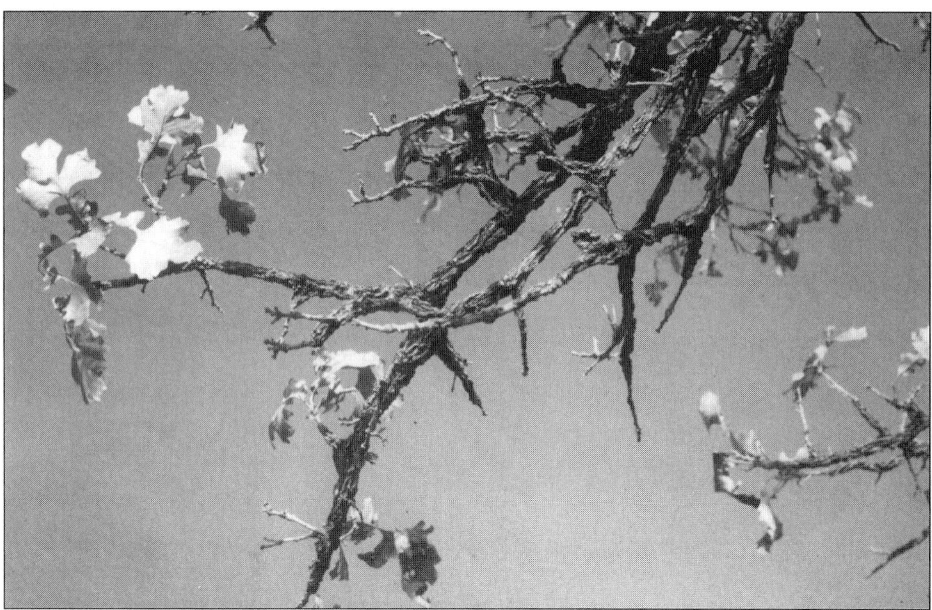

Figure 17-14

Bur oak, fire protected by its thick, corky bark and ability to root-sprout, formed savannas at the fringe of the mesic prairies. (Photo by the author.)

planted to prairie grasses in 1949 by V.E. Shelford and S.C. Kendeigh. One of the most successful restorations and one whose methods have been widely copied was begun at the Morton Arboretum south of Chicago by R. F. Schulenberg (1967) in 1963.

The usual procedures are to remove as much of the nonprairie vegetation of the site as possible, perhaps by planting row crops for a year or more. Into this site are then introduced prairie species. It is usual to gather seeds from local prairie relicts, the idea being to preserve the genetic constitution of the local populations. After a cold treatment which suffices to break dormancy in most prairie species, either the seeds are planted on the reconstruction site or, if an especially quick response is desired, some may be planted in the greenhouse in early spring and later transplanted. The latter method is usually reserved for forbs. On larger sites, the grasses may be planted using a mechanized seed drill in the same fashion as wheat.

For approximately the first two years, it is desirable, although time consuming, to weed the plantings. Weed seedlings are pulled or cut out with a sharp linoleum knife. As soon as enough grass litter has accumulated to support a fire, annual burning (in early spring, if possible) should be begun (Figure 17-16). After a few years, a switch to burning half the restoration one year and the other half the next may help maintain characteristic prairie invertebrates, if any (Schramm 1992).

Such reconstructed prairies ranging from a few square meters to a few tens of hectares have been developed at universities and nature centers at many locations in the prairie biome and outside it. They are important sites for education and research, and they preserve populations of plants and animals that are otherwise being pushed closer and closer to extinction. An analysis of the Wisconsin Arboretum reconstruction after 40 years concluded that it was a creditable replication of a tallgrass prairie but also found truth in J. E. Weaver's claim that a prairie, once destroyed, takes a thousand years to be restored to a completely natural condition (Sperry 1983). Even after 40 years, much of the planting pattern was still evident, rather than having been erased by internal processes of the community. It is also usual for reconstructed prairies to

Figure 17-15

In late summer, big bluestem flowering stalks shoot up around the gravestones in old prairie cemeteries, as in this one on Prairie Ronde, Kalamazoo County, Michigan. (Photo by the author.)

have certain species more abundant than is ever seen in relict prairie.

Another problem with reconstructions is the absence of prairie animals. This results partly from a failure to naturalize animals on the sites—a harder job than propagating plants. A second cause is that most of the reconstructions are too small to maintain viable populations of animals, especially the more wide-ranging ones. It may be that no reconstructions will reproduce the dynamics of the original prairie satisfactorily without the characteristic animals. The next step in prairie reconstruction, consequently, may be to increase prairie size and to begin to bring in insects, rodents, and, eventually, bison.

Desert

Distribution and Climate

Desert occurs in arid areas in the rain shadow of mountain ranges, along coasts next to cold ocean currents, and deep in the interior of continents. Extensive deserts occur in intermountain and southwestern North America, western South America, Arabia and northern Africa, southern Africa, central Asia, and central Australia.

Rainfall is typically less than 25 cm (10 inches) per year, and the evaporation rate is high. In fact, the potential evaporation is usually much more than rainfall actually supplies. In Phoenix, Arizona, for example, potential evapotranspiration is 130 cm (52 inches) per year, whereas precipitation is only 18 cm (7.2 inches).

The warm deserts of the Southwest are sunny and hot. Frosts are rare, and the frost-free season is 200 or even 300 days. The cold desert of the Great Basin has more winter cloudiness, and frost may occur from September to May. Average January and July temperatures are 10 and 32°C for Phoenix in the warm desert. For Boise, Idaho, toward the northern edge of the Great Basin, corresponding temperatures are −2.5 and 24°C. Large day-to-night differences in temperature are usual.

Nighttime temperatures often drop below the dewpoint, causing moisture to condense on soil,

Figure 17-16

The eastern prairies were maintained by fire, and fire is a necessary management tool in prairie preservation and restoration. Here is seen a fall burn in a restored prairie at the Kalamazoo Nature Center, Kalamazoo County, Michigan. (Photograph by Robert J. Pleznac.)

rocks, and plants. Dew is probably an important source of moisture for many small animals and certain plants, some of which are able to absorb it through their stems or leaves (Larson 1970). In coastal deserts, such as the Viscaíno in North America (Figure 17-17), fog may be an important moisture source. Ball-moss, an epiphyte in the same genus as Spanish moss, grows on the cactuses here, probably supported by moisture supplied by the fog.

Topography and Soil

In the desert, rainfall comes in brief, heavy thunderstorms, the runoff from which moves a large amount of soil, but not very far. The water evaporates fast, so that streams are short, tending to end at or before the low point of each subbasin. The result is a topography in which the basins are filled with alluvium sloping gently from the mountains toward flat areas, **playas,** occupied by lakes, often temporary, from which water evaporates rather than running to the sea (Figure 17-18). The streambeds running from the mountains, called **arroyos,** carry water only after snowmelt in the mountains and after thunderstorms.

The long, sloping alluvial fans reaching out of the mountains are called **bajadas.** The soil here is coarse, neutral or only slightly alkaline, and low in organic matter. The same calcification processes that produce grassland soils operate in the desert but in more extreme form. Rainfall does not generate enough groundwater to leach out the basic cations or soil colloids. Calcium and magnesium remain in the upper portions of the soil and, often, also form a carbonate hardpan, called *caliche,* in the B horizon.

The old names were gray desert soils and sierozems for the cold desert and red desert soils for

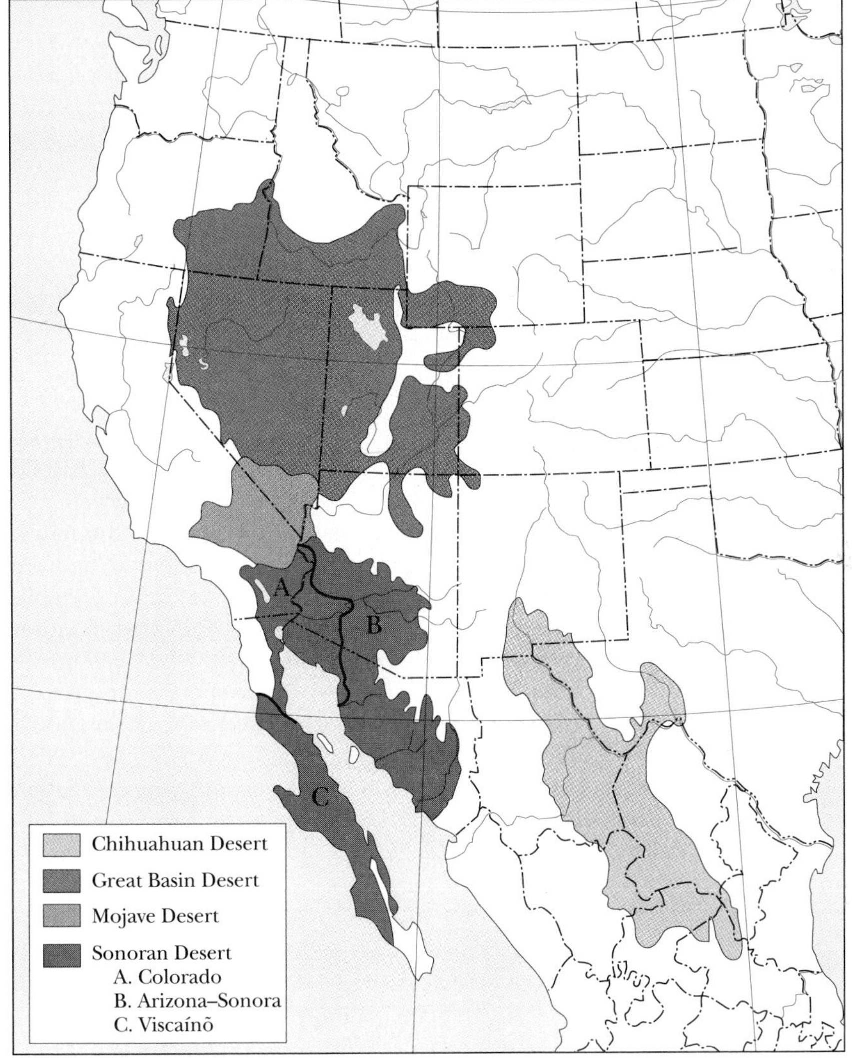

Figure 17-17

Distribution of desert vegetation in North America. (Adapted from *Botanical Review* 8(4), 195–246, Forrest Shreve, 1942. Used by permission of the New York Botanical Garden.)

the warm desert. In the U.S.D.A. Comprehensive System, they are now mostly aridosols and, in the northern regions of the Great Basin with more grass, mollisols.

It is windy in the desert, and wind and water erosion may remove the fine particles from an area, leaving only the stones, forming desert pavement, on which little grows. Sand dunes are frequent and may be bare or stabilized by vegetation. Evaporation from the soil surface tends to produce a crust that limits blowing of the soil.

The soils toward the bases of the bajadas and in the playas are finer textured. The playas are full of salts carried to the bottom of the basin in the runoff from the slopes.

Plants

Desert vegetation is characterized by sparse xerophytes of several types. The prevailing landscape is of wide-spaced, small-leaved shrubs a meter or so tall. Many of the plants are spiny as a defense against browsers though, if the spines are dense

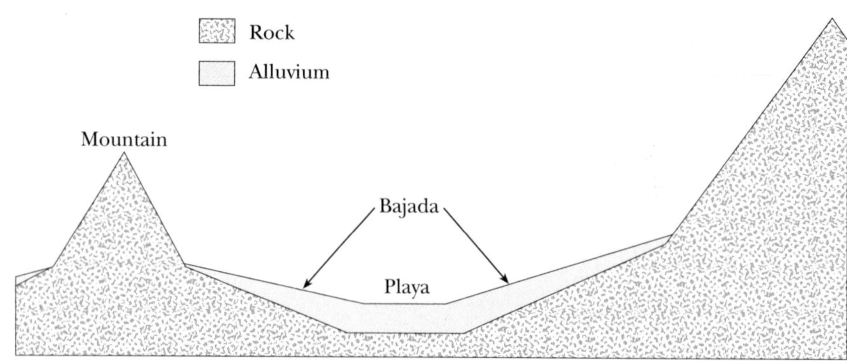

Figure 17-18

Generalized topography in desert.

enough, they may also cast shade useful for the rest of the plant. Halophytes occur in saline areas.

Although most of the desert is dry, there are moist sites dominated by mesophytes and hydrophytes. One such community is floodplain forest **(riparian woodland)** of willows and cottonwoods, among other species. Marsh (tule) vegetation of bulrushes and cattails is another.

Annuals that are structurally mesophytic grow and flower after rains heavy enough to ensure the completion of their life cycle. Although such annuals form a large percentage of the species list of most deserts, they may not appear at all in dry years.

Succession occurs slowly, if at all, in the drier parts of the desert, because the community's reactions on the physical environment are slight. Areas studied by botanists early in this century were much the same, often with the same individual plants, when restudied 30 years later. Even the establishment of an individual of one of the characteristic desert species may depend on a particularly favorable combination of factors. What you see is what you get is true for the future as well as the present in desert.

Cold Desert (Basin Sagebrush)

Cold desert occurs in Nevada and Utah with extensions into adjacent states (Figure 17-17). It lies in the rain shadow of the Sierra Nevada and the Cascades (Figure 16-35) at elevations mostly above 1200 meters. The most characteristic plant is sagebrush, a bushy shrub with gray-green leaves and small, wind-pollinated flowers (Figure 17-19). In the northern part of the Great Basin and in Oregon and Idaho, sagebrush tended to be associated with other shrubs, such as bitterbrush and perennial bunchgrasses, forming **shrub-steppe.**

Overgrazing has drastically reduced the native grasses (and, along with fire, the sagebrush and bitterbrush) in favor of cheatgrass and other alien invaders (Franklin and Dyrness 1969). This situation goes back many years; Aldo Leopold (1949) wrote an essay "Cheat Takes Over" some time around 1940. By the 1980s, the replacement of sagebrush and other shrubs by cheatgrass was a widespread problem from the shrub-steppe of Oregon through the sagebrush areas of the Great Basin to the Mohave desert of Nevada and California (Young and Tipton 1990).

The general pattern seems to be that overgrazing removes native grasses and allows the invasion of cheatgrass. Cheatgrass provides a continuous litter layer that encourages frequent and extensively spreading fires. Natural vegetation supported fires at a frequency of once every 60 to 110 years (Whisenant 1990), whereas sites with cheatgrass burn about every five years. Most of the native shrubs are killed by fire and do not resprout, so desert and shrub-steppe areas are converted to low-diversity grasslands dominated by cheatgrass and other exotic annuals (Table 17-5).

Leopold (1949) wrote, "I listened carefully for clues whether the West has accepted cheat as a necessary evil, to be lived with until kingdom come, or whether it regards cheat as a challenge to rectify its past errors in landuse. I found the hopeless attitude almost universal. There is, as yet, no sense of shame in the proprietorship of a sick landscape."

Sagebrush is replaced by shadscale on soils of high salt content. In such salt-desert shrub vegetation, grasses are relatively unimportant. Although some additional shrubs such as winterfat and Mormon tea are often present, much of the ground is bare. Similar vegetation occurs on sites with fine-textured but nonsaline soil where rainfall is very low (below 15 cm) and also on saline sites outside the Great Basin where grassland would otherwise be the expected vegetation.

A few species of prickly pear *(Opuntia)* occur in cold desert, but cacti as a group are unimportant. Annuals may be present in the spring following the winter rains but are inconspicuous here compared with the warm desert (Gleason and Cronquist 1964).

Aboveground biomass is low, 200 to 400 g/m^2 (Pearson 1965, Holgren and Brewster 1972). Below-ground biomass is twice to several times higher. Total biomass is probably generally 1–2 kilograms per square meter, or 1/15 to 1/30 that of temperate deciduous forest. Net annual aboveground productivity probably varies from 30 to 100 grams per square meter. Below-ground productivity is substantial and brings total productivity up to 250 to 300 gm/m^2, however, very dry sites may be lower. Decomposition is slow, and, although dead material tends to accumulate both above and below ground, most soils have little humus.

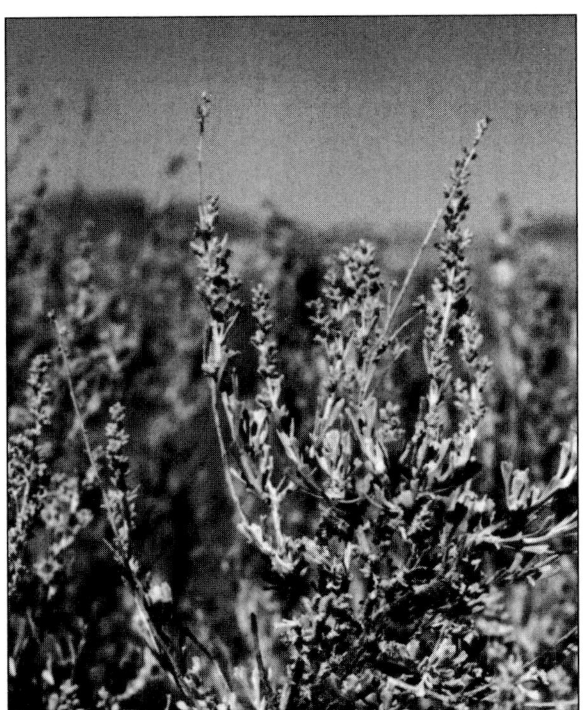

Figure 17-19

Sagebrush, an aromatic, silvery gray shrub in the composite family, occurs widely in arid and semiarid habitats in the American West. (Photo by the author.)

Warm Desert (Desert Scrub)

Three subdivisions of warm desert are commonly recognized. The Mohave Desert occupies southern Nevada, southeastern California, and northwestern Arizona (Figure 17-17). It is transitional between cold desert and the Sonoran Desert which occupies southern Arizona, adjacent Mexico, a small corner of southwestern California, and most of Baja California. The Chihuahuan Desert lies to the east, mostly in Mexico but also extending into southern New Mexico and Texas.

What sagebrush is to the cold desert, creosote bush is to the warm desert. A branching shrub a meter or 2 tall with small, shiny, evergreen leaves

Table 17-5

Species Richness (Number of Species in 100 0.1-m^2 Plots) and Typical Dominant Plants on Sites With Different Fire Frequencies in Idaho Sagebrush-Steppe

Fire Frequency (Fires/Year)	*Species Richness*	*Typical Dominants*
0*	18	Sagebrush, needlegrass, wheatgrass, and other native bunchgrasses
0.03	16	Sagebrush, needlegrass, wheatgrass, and other native bunchgrasses
0.44	9	Cheatgrass, gray rabbit-brush, some native bunchgrasses
0.61	7	Cheatgrass, stork's-bill, Russian thistle, Jim Hill mustard

Based on Whisenant 1990.

* More than 100 years since last fire.

and nearly black stems, it bears large yellow flowers in the spring. Creosote bush occurs sparsely over vast areas, each bush 5 or 10 m from the next, in some places almost as regularly spaced as fruit trees in an orchard. Forming a lower shrub stratum may be bur sage in the western part of the hot desert and tarbush in the eastern part (Shelford 1963).

Most of the Mohave Desert is mile after mile of creosote bush and bur sage. The Joshua tree is considered the trademark of the Mohave, but it grows mostly at higher elevations, above 1000 m, in a transition zone between desert and pinyon-juniper woodland (Figure 17-20).

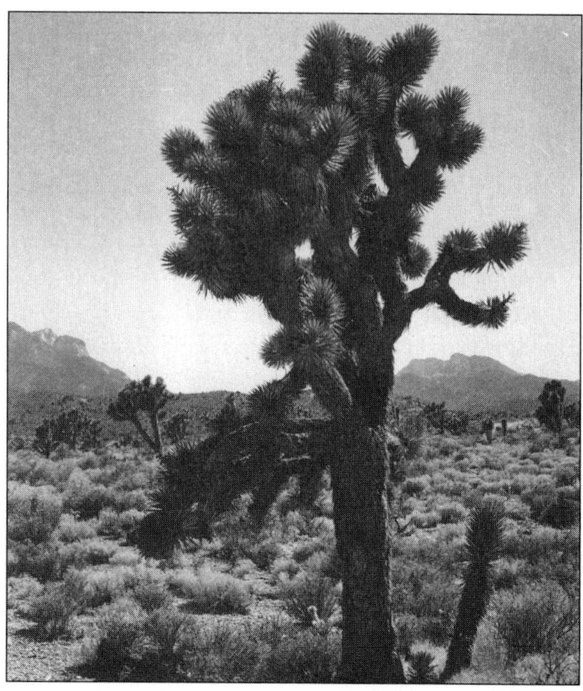

Figure 17-20

The Joshua tree is the most characteristic plant of the Mohave Desert. It is a yucca and, like others, its flowers are pollinated by yucca moths. At 8 to 10 m it is the tallest plant of the community, and it provides a nesting site for birds such as the Scott's oriole. (Photo by P. S. Bieler in Nevada National Forest, courtesy U.S. Forest Service.)

Here, as elsewhere in the warm desert, two distinct groups of annuals exist: Winter annuals germinate if fall or winter rains are sufficient, grow slowly through the winter, and bloom following a period of rapid growth in the spring (Beatley 1974). Their life cycle is not much different from that of various temperate winter annuals.

Rains when the temperature is high do not trigger the germination of the desert's winter annuals but are the proximate factor in the germination of another, smaller set of species. These summer annuals are the real desert ephemerals. They germinate if 25 mm or more rain comes in August or September and flower and fruit within a month. Flowering of perennial herbs is generally in spring, but is based on autumn rains sufficient to trigger new vegetative growth.

Timing and success of reproduction of heteromyid rodents (such as kangaroo rats and pocket mice) and, consequently, their population size later in the year seem to be dependent on the winter annual crop (Beatley 1969). The main effect may be the provision of green forage, but perhaps other specific nutrients are involved.

The Sonoran Desert is the lowest and hottest of North American deserts. Rainfall may be extremely low but tends to be balanced between winter and summer. Drainage is into the Gulf of California, and, in contrast to the other desert subdivisions, soil salinity is rarely an important factor in plant distribution.

In southern Arizona and northern Mexico, creosote bush and bur sage tend to dominate the lower bajada but the upper bajada supports a complex and diverse vegetation (Figure 17-21) that includes saguaro, ocotillo, and paloverde. Saguaro is the tall (up to 15 m) cactus with "arms." Ocotillo is an odd, skinny plant that grows a new crop of small leaves in each moist period and loses them in each dry one. Paloverde is a small tree with green bark, which allows it to photosynthesize without having to grow new leaves if conditions are otherwise suitable. This is not such a remarkable situation when we realize that it is what cacti also do, except that most of them never grow leaves.

The concept of keystone species becomes muddied, but the concept of interdependence shines through when we consider the nexus to which the saguaro (and its larger Mexican sister, the cardon) belongs. Without these large cacti, the Gila woodpecker and the northern flicker (gilded form) would have no place to excavate cavities. Without their abandoned cavities, no homes would be available for elf owls, treehole mosquitos, or various hole-dwelling mammals. The fleshy fruits of the cacti are an important food source for several of the desert mammals and birds.

Among the mammals that disperse cactus seeds are fruit-eating bats, which also pollinate these large cacti. The face of the lesser long-nosed bat fits into the cone-shaped flowers like a key in a lock. Saguaro flowers open at night and stay open till the next afternoon; cardon and organ-pipe flowers close in the morning and so are more strictly bat pollinated. Birds, such as white-winged dove, and the introduced honey bee, also pollinate saguaro. Bats are declining—the lesser long-nosed bat is on the U.S. endangered species list—and fruit set of some of the cacti is low in some areas (Tuttle 1991).

The decline of large saguaros within Saguaro National Monument (Anon. 1991) has a different cause. As older plants die, there are few or no middle-aged plants to replace them. The absence of this age class seems to be from a combination of factors stemming from misguided policies of federal agencies—policies that continued long after the problems had been diagnosed by desert ecologists (Larson 1970). Important among the causes were overgrazing on the vulnerable young saguaros and also on nurse plants (Figure 17-22) and predator control that led to high populations of rodents.

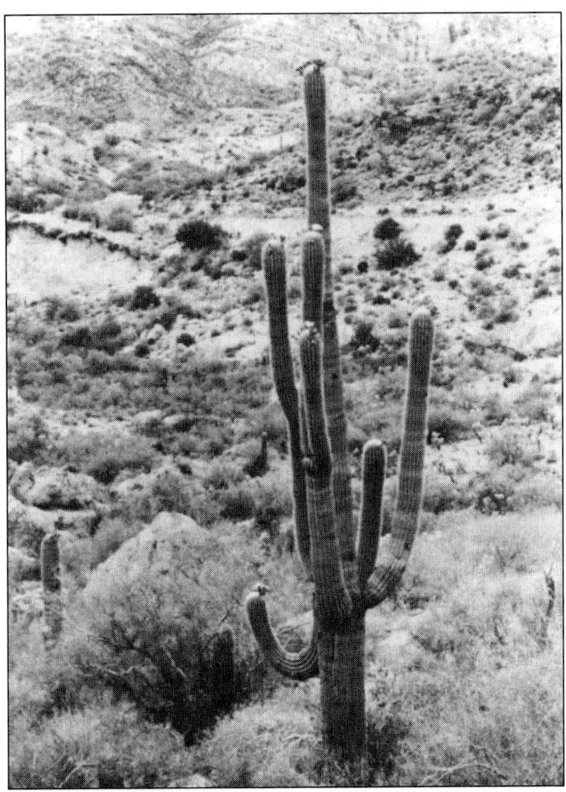

Figure 17-21

A diverse vegetation, including saguaro, grows on the upper bajada in the Sonoran Desert. Below, the vegetation is less diverse. The saguaro seen here, an ancient specimen probably more than a hundred years old, is in bloom. (Photo by George and Betty Walker.)

The Sonoran Desert is diverse, and some botanists recognize several subdivisions (Figure 17-17). Extreme southeastern California and adjacent areas are sometimes separated out as the Colorado Desert. Slightly more distinctive is the Viscaíno Desert of Baja California that, among other features, has several endemic species of reptiles including an odd, pink, earthworm-like burrowing lizard (Stebbins 1985).

The Chihuahuan Desert lies on the east side of the Sierra Madre Occidental. It is higher and generally moister and cooler than the Sonoran Desert, and more of the precipitation comes in the summer.

Creosote bush, ocotillo, and mesquite, all common in the western deserts, also occur here (Shelford 1963). Although saguaro does not occur in the Chihuahuan Desert, other cacti, including prickly pears of tree stature, are abundant. One among the many small cacti that have their center of distribution here is mescal or peyote, whose fruits contain the psychedelic alkaloid mescalin. Spineless and bluish, the peyote cactus tends to grow under a cover of mesquite along the Rio Grande in Texas and southward into Mexico.

Also abundant are yuccas—shrubby members of the lily family with tough, pointed leaves bristling out from a short woody base—and agaves. A couple of dozen species of agave grow in the southwestern United States and more in Mexico, although not all are in desert. Some become good-sized plants a couple of meters wide that may be grown in plantations for fiber and alcohol. Of the beverages produced from agave, the most familiar to U.S. citizens is tequila.

Some of the large agaves are textbook examples of nonannual but semelparous organisms. That is, they reproduce only once and then die but do so only after several years of growth. When the flow-

Figure 17-22

Several saguaros of different ages have managed to get started under this palo verde "nurse tree." Saguaros do not begin to grow arms until they are 5 to 8 m tall, by which time they are several decades old. (Photo by R. Bruce Moffett.)

ering stalk appears, it shoots up rapidly and has been known to poke a hole in the roof of the greenhouse.

Animals

Aridity and high summer temperatures are the environmental conditions that desert animals must survive. Reptiles are preadapted for success in the desert, being poikilothermic, well protected against desiccation, and conservative of water through the use of uric acid as their nitrogenous waste product. They are diverse and common in the deserts. Amphibians, as well as moist-skinned invertebrates, are scarce away from permanent water.

Birds as a group are reasonably well suited to desert life by uric acid production, a high body temperature (so that they do not have to shift to a ventilation mode as soon as mammals), and good motility (allowing them to get to shade and water).

Mammals do not start out so well adapted to life in the desert, but many species have, nevertheless, been outstandingly successful there. Large mammals are not common in American deserts, although the mule deer, collared peccary (a piglike animal a couple of feet tall), and mountain sheep occur. The pronghorn is widespread at low densities (generally below $0.5/km^2$) in basin sagebrush; however it needs drinking water throughout the year and is more characteristic of shortgrass prairie (Kindschy et al. 1982).

Elsewhere in the world's deserts, however, certain large mammals have done very well. Camels are the most familiar example. This line arose in North America, spreading to the Old World in the Pliocene, but the last North American camel became extinct in the Pleistocene. The one-humped camel of the hot desert of Africa and Arabia is now extinct in the wild, but a few million domesticated animals still exist (Gauthier-Pilter and Vagg 1978).

The camel shows an imposing array of features that fit it to life in the desert, including adaptations for feeding on sparse, thorny, and salty food; walking in sand; tolerating long periods of little food or water; and surviving high temperatures. Some of these features are related directly to its large size. For example, an animal that weighs 450 kg can absorb a lot of heat without its body temperature rising greatly. Added to this is the fact that the camel's body accepts a few extra degrees as normal and does not begin to lose water by sweating until body temperature has risen to nearly 41°C, or about 105°F.

The approach of small mammals to a desert existence has been quite different. They spend the day in burrows and come out at night when it is cool and the relative humidity is high. Physically, burrowing is feasible for small mammals but, also, avoidance of high temperatures is essential because their body water would be quickly depleted if they had to use it in thermoregulation (Clousdley-

Thompson 1965). Medium-size mammals fall between these two adaptive modes and are scarce in desert.

Even nonburrowing animals, such as birds, curtail activity and seek the shade during the hottest part of the day. Other adaptations shown by desert animals include coloration matching the substrate, often pale grays or tans. Oddly, black coloration is not uncommon even in areas of pale soil. It may be an adaptation for heat gain in the cool morning hours in animals for which predation is not a significant factor (poisonous species, for example).

Adaptations for locomotion in sand are also seen, including the camel's large flat feet and the fringes on the toes of desert lizards. Some reptiles "swim" through the sand, either in a straightforward way or by the peculiar sidling movement of the sidewinder, a rattlesnake of the Mohave and Sonoran Deserts.

Many animals are inactive in burrows during the driest part or parts of the year. They are said to be aestivating. Examples are common in the invertebrates but also include vertebrates, such as ground squirrels. Couch's spadefoot toads may aestivate 10 to 11 months of the year, appearing above ground only after summer storms fill pools with water (Cornejo 1985). The proximate factor in their emergence seems to be the ground vibrations caused by pounding rain. Mating and egg laying occur immediately, and only nine days are required for the development of the miniature toadlets.

Animal Communities in North American Desert

Certain animals occur widely through both cold and warm desert in North America. Included are certain species restricted to arid regions and others that also occur widely in other habitats (Table 17-6). Many species are more or less restricted to either the warm desert (desert scrub) or the cold desert (basin sagebrush).

The list for desert scrub is longer. Included are a variety of small granivorous rodents such as pocket mice and kangaroo mice and rats. Kangaroo mice and rats hop, as do many other unrelated small desert rodents elsewhere in the world. Possibly hopping is a very good way to get around if there is not much vegetation in the way.

There is a long list of distinctive bird species (Table 17-7). Although some are seed-eaters, a large percentage are insectivores or carnivores, both foods providing a better percentage of water than do seeds.

Predaceous birds include Harris' hawk, a slim, dark bird with striking chestnut patches in the wings and white in the tail—an odd, social hawk, nesting in saguaro if available. In saguaro, also, nests the elf owl. The roadrunner (Figure 17-23), the speedy, ground-dwelling cuckoo of cartoon fame, is a predator on lizards, snakes, and large insects. Black and turkey vultures and the crested caracara are largely carrion-feeders.

Breeding-bird densities are low, mostly below 100 pairs per 40 ha and often below 50 pairs. Many birds are year-round residents, and to them may be added some birds from higher elevations or latitudes, so that winter densities are similar to or even higher than those of summer.

There are also many distinctive desert scrub reptiles (Figure 17-24), mostly lizards and snakes (Table 17-8), although one of the most distinctive is the desert tortoise, a large, burrowing turtle.

Although there are no native large mammals characteristic of American desert, one introduced species is present. This is the burro, or donkey, the domesticated version of the wild ass native to the deserts of eastern Africa. Released by discouraged

Table 17-6
Some Animals Widely Distributed in Both Cold and Warm Desert in North America

Primarily Desert Species	
Black-tailed jackrabbit	Black-throated sparrow
Desert woodrat	Long-nosed leopard lizard
Spotted skunk	Side-blotched lizard
	Desert horned toad
Species of Wide Ecological Amplitude	
Coyote	Red-tailed hawk
Deer mouse	Great horned owl
	Mourning dove
	Gopher snake

prospectors or otherwise escaped into the wild, these have built up populations in the Grand Canyon, Death Valley, and various other sites to the point where they are seriously degrading the vegetation.

In the cold desert, grassland animals join the widespread desert species mentioned earlier. There are also some distinctive species but fewer than in the warm desert. Many of the characteristic species have "sagebrush" as part of their name. Examples are the sagebrush vole and sagebrush lizard.

To the characteristic cold-desert birds listed in Table 17-9 are added such grassland species as western meadowlark, lark bunting, and vesper sparrow. Overall breeding-bird density is generally below 50 pairs per 40 ha. There is a fair amount of migration from the northern reaches of the Great Basin in

Table 17-7
Some Characteristic Warm Desert Bird Species in North America

	Primarily Seed-Eaters
Gambel's quail	Inca dove
Masked bobwhite	Pyrrhuloxia
White-winged dove	Abert's towhee
	Primarily Insectivores
Gila woodpecker	Verdin
Ladder-backed woodpecker	Cactus wren
Brown-crested flycatcher	Curve-billed thrasher
Vermilion flycatcher	LeConte's thrasher
	Phainopepla

Figure 17-23

The greater roadrunner lives in the desert and adjacent arid habitats and is a predator on reptiles and larger insects. (Luther C. Goldman, U.S. Fish and Wildlife Service.)

winter but less in the south, where populations may be swelled by such birds as rosy finches from the mountaintops and lapland longspurs from the tundra.

Ants, grasshoppers, and tenebrionid beetles are conspicuous invertebrates of desert. Arachnids, including tarantulas and scorpions, both venomous predators, are also conspicuous.

Many animals in the desert are restricted to specific physiographic sites. In treeless areas, for example, cliffs and caves may supply virtually the only nesting or roosting sites for various birds, including several of the large raptors, such as golden eagles, and for several species of bats. Rocky areas or talus slopes are the favored habitats for certain species such as the collared lizard, chuckwalla (a large lizard that puffs up, making it hard for predators to pull it out of rock crevices), and rock wren. Sand dunes have their characteristic fauna including several lizards and snakes and the kit fox (Maser et al. 1979).

Table 17-8
Some Distinctive Lizards and Snakes of Warm Desert in North America

Banded gecko	Western leaf-nosed snake
Chuckwalla	Rosy boa
Desert iguana	Western shovel-nosed snake
Gila monster	Mohave rattlesnake

Riparian and oasis habitats tend to have an abundant fauna including species in four main categories: (1) desert species from the surrounding area, (2) woodland and chaparral species, (3) species restricted to the region but mostly absent from the drier habitats, and (4) a large contingent of species typical of deciduous forest, savanna, or

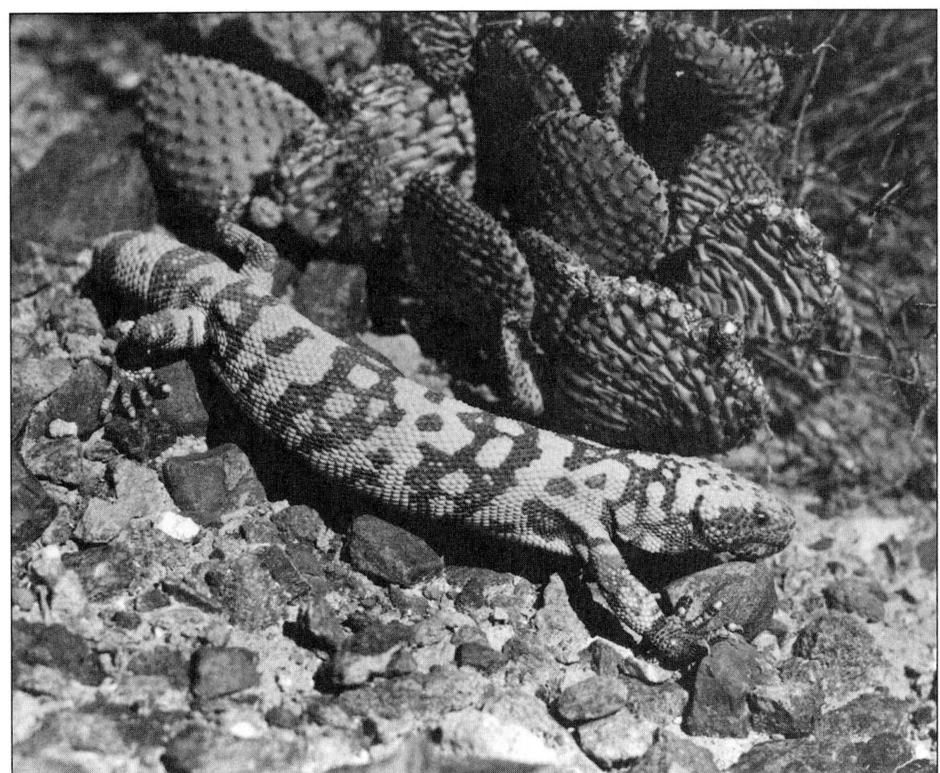

Figure 17-24

This handsome lizard, the Gila monster, is characteristic of rocky areas in the hot desert but is much reduced in numbers. (E. P. Haddon, U.S. Fish and Wildlife Service.)

even grassland in the eastern and southern United States.

Examples of category 3—distinctive riparian species—include the vermilion flycatcher, ladder-backed woodpecker, and Lucy's warbler. Some of the many avian examples of category 4 are northern flicker, yellow warbler, summer tanager, cardinal, and song sparrow (Szaro and Jakle 1985). Small mammals in category 4 include such species as deer mouse, white-footed mouse, and cotton mouse (Blair 1940).

Ponds and lakes in the desert are isolated, resulting in restrictions on gene flow for the fish populations occupying them. As a result, evolution has produced many species of restricted range, perhaps a single spring-fed pool, and many of these species are now threatened, endangered, or already extinct.

Not all the endangered species, however, have such restricted ranges. The Sonoran topminnow has been reduced to endangered status in Arizona, although it was once widespread and common over most of the southern part of the state. Some of the factors driving it and several other species toward extinction include the following (Minckley 1991):

- Habitat destruction by development
- Damming and diversion of streams, creating reservoirs and concrete-lined channels
- Overgrazing, logging, and other abuse of watersheds
- Pollution from mines, agriculture, domestic sewage, etc.
- Drying up of ponds and streams resulting from wells lowering the water table, typically by 50 to 200 m
- Competition and predation by alien species

The mosquito fish was introduced in the early 1920s—based on ecological ignorance, like many introductions. It was brought in for "mosquito control" but occupies the same food niche as the Sonoran topminnow and is also a "notorious predator on small fish" (Minckley 1991). The Monkey Spring pupfish in southern Arizona was eliminated within a few months by introduced largemouth bass. The original fish fauna of the Colorado River consisted of 32 species, but there are more than 80 present now, including a wide variety of predators.

The now-isolated populations of Sonoran topminnows have lost genetic diversity, presumably as the result of inbreeding and genetic drift, and so may be less able to adapt to ecological changes of the future (Vrijenhoek et al. 1985). In northern Mexico, where humans have not yet had such severe impacts on the desert, the species is still abundant.

Table 17-9
Some Characteristic Birds of Cold Desert in North America

Sage grouse	Green-tailed towhee
Sage thrasher	Sage sparrow
Poorwill	Brewer's sparrow

Human Relations

The human population of the southern deserts has exploded since 1950 and is still exploding. The population of Arizona has increased nearly 200% since 1960, Nevada more than 50% just since the 1980 census. This expansion has been possible only through the provision of water from outside the region, mostly through massive water projects such as diversions of the Colorado River. It has also been possible only by provision of outside money, mostly taxpayer dollars from the rest of the country, financing the water projects.

"Making the desert bloom" has been a favorite slogan over the years. Of course the desert has always bloomed, but with cacti, ghost flowers, and panamint daisies. Making it bloom with crops entails several problems. Irrigation is expensive, so much so that if farmers of these regions had to pay the full price for it, in the same way farmers have to pay for fertilizer, their crops would cost more than the same things grown in Michigan or Missouri.

Another problem is the increase in salts in downstream irrigation water and in irrigated croplands. The salinization has already led to the abandonment of some irrigated lands in the American West and declines in yields on one fourth of them (Anon 1980). Drinking water, if it is derived from groundwater, may also become saline.

Fields and citrus groves and subdivisions have obliterated large sections of desert, and grazing has degraded others. There are still substantial areas of relatively undamaged desert left, but the threats, in addition to the spreading development, are many. Cactus thieves steal and sell plants that may not be replaced for decades in this harsh environment (Nabhan 1991). Then there are the off-road recreational vehicles whose passage through the desert causes longer-lasting destruction than a bulldozer does in temperate regions (Stebbins and Cohen 1976).

The situation in the Sahara region of North Africa is even bleaker. The Sahara alternated between desert and somewhat moister, semiarid conditions through the Pleistocene (Le Houérou 1992). A rich fauna persisted until the first half of the 20th century, when the extinction rate increased drastically. Extirpated or nearly so are several large mammals (which, however, still persist in equatorial Africa); examples are elephant, hippopotamus, lion, cheetah, and wildebeest. It is likely that biological diversity will be further reduced in the next few decades by desertification at the edges of the Sahara and by increasing human pressure in the mountainous areas that form refuges for rarer animals and plants.

The desertification is the result of occasional droughts combined with effects of exponentially growing human populations (Le Houérou 1992), such as overgrazing (Skarpe 1991). Droughts are not new, but in this century, human populations have increased from 20 to 112 million in the northern Sahara and from 10 to 56 million in the southern Sahara. At the fringes of the Sahara, grassland and savanna are being converted to desert at a rate such that the area of desert is increasing 1 to 2% annually. Few protected reserves exist and in those that do, "wildlife is often persecuted by those who are supposed to protect it" (Le Houérou 1992).

Tropical Biomes

The tropics are the part of the globe "where winter never comes," to use Marston Bates' (1952) phrase. A January (or, for the Southern Hemisphere, July) isotherm somewhere between 15 and 20°C (or 60 to 70°F) seems to fit that description well enough. Within the tropics are a variety of climates distinguished mainly by the length and severity of the dry season. As in the arctic, the weather is a dull subject for conversation because it tends to be the same, day after day, for long periods.

Tropical Rain Forest

Location and Climate

Tropical rain forest occurs primarily within the equatorial zone, within 10° of the equator. The largest areas are in the Amazon Basin of South America, western Africa, and Indonesia; however, many smaller areas occur on the rainy leeward slope of mountains elsewhere in the tropics (Figure 17-25).

Annual precipitation is very high, regularly over 200 cm and in some localities over 1000 cm. There may be a dry season of a month or less, corresponding with the "cool" season. Seasonal variation in temperature is very slight, however, the monthly means remaining close to 27°C throughout the year.

Variations in temperature and humidity during the course of a day may be greater than the differences among the monthly averages. At dawn in a clearing in rain forest, the foliage is dew covered but the air is fresh and cool—perhaps 22°C (72°F). As the sun rises higher, the temperature goes up and the birds fall silent. By noon, it is hot—although rarely above 32°C (90°F)—and muggy. Clouds develop, and there is a brief thunderstorm with heavy, driving rain. After the sun comes out, "the freshness only lasts for a short period. The humidity again becomes unbearable, as in a hothouse. Relief comes only with the evening when one can regain strength after sundown. The same daily sequence is repeated virtually unchanged throughout the whole year. This makes a prolonged stay in the humid tropics nearly unbearable for people from the temperate zone" (Walter 1971).

Away from clearings, daily climatic conditions are more nearly constant. Light intensity near the ground is very low, there is little daily temperature variation, and the humidity is constantly near saturation.

Soils

Most soils in the tropics are old, products of a very long period of weathering. High rainfall and temperature promote the leaching of all the soluble constituents of the original rocks, leaving behind a latosol, a red or yellow soil composed mainly of aluminum and iron oxides. The soils are acid, show little in the way of horizon development and, in this

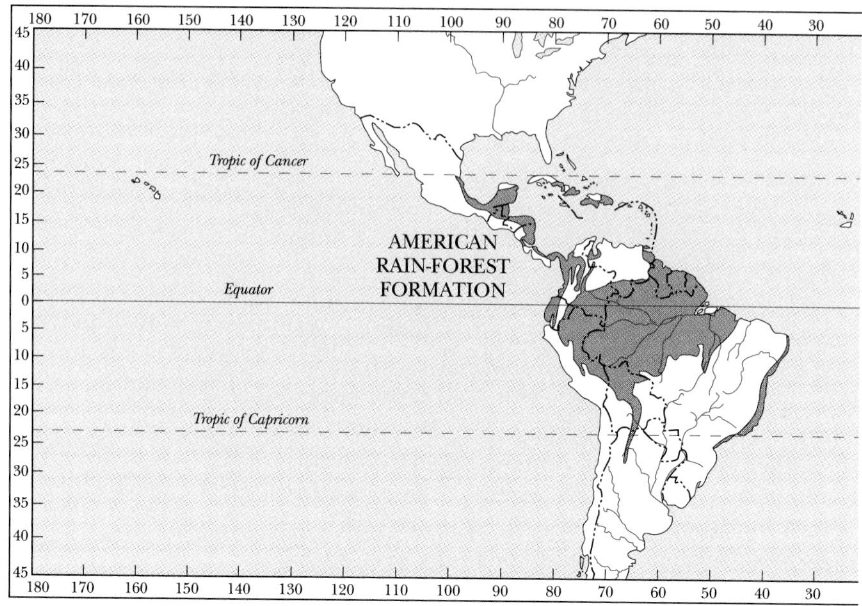

Figure 17-25

Distribution of tropical rain forest throughout the world. (From *The Tropical Rain Forest: An Ecological Study,* by P. W. Richards. Cambridge University Press, 1952. Reprinted with permission.)

climate of rapid decay, have no humus-rich layers. Such soils in the U.S.D.A. Comprehensive System are termed oxisols.

Any addition of nutrients to the system occurs via rainfall, and cycling occurs mainly within the biotic portion of the ecosystem. The soil contains few nutrients and lacks the cation exchange capacity to retain them if they are added. A mat of roots and mycorrhizae lying on or near the soil surface (Jordan 1982) quickly returns to living plants minerals freed in leaching and decomposition. Invertebrates, especially ants and termites, also quickly remove plant litter to their nests where they use it, or the fungi that grow on it, as food.

Because the soils themselves contain few plant nutrients, logging tends to reduce a site's productivity. Nutrients are removed in the wood, the nutrient cycling system is disrupted, and leaching carries nutrients away. On some sites, the resulting infertile soil is converted to laterite, a hard, red material which, when dry, can be used as bricks but which supports only scrubby vegetation.

Plants

Popular books often use the word "jungle" to refer to tropical forest. This is not necessarily wrong but, to many, the word carries a connotation of dense tangles requiring a machete (or as we call it in the Midwest, a corn knife) to penetrate. Dense vegetation does occur in early seral stages following forest clearing (Table 17-10) and along the banks of rivers where sunlight can create a wall of lianas and small trees. But typical rain forest is, by virtue of the low light intensity at the forest floor, open and easy to walk through. "Cathedral-like" is a recurring term in descriptions of the climax forest.

Rain forest is a tall forest composed of broad-leaved evergreens, often forming three layers (Figure 17-26). The upper layer consists of trees 45 to 55 m tall with round or umbrella-shaped crowns (Holdridge et al. 1971). These trees, "emergents," do not form a canopy. They are straight and unbranched to near the crown. Although the trees are tall, they are not often big in diameter (Figure 17-27). The bark of many of the species is thin, smooth, and light colored. Buttress roots, apparently an adaptation promoting stability in these shallowly rooted plants, are frequent (Figure 17-28). Leaves are typically elliptical, smooth margined, leathery, and dark green.

The second layer is 30 to 40 m tall, filling in the spaces between the trees of the upper layer. Trees of the second layer, whose crowns are below the strongest action of the wind, are generally unbuttressed.

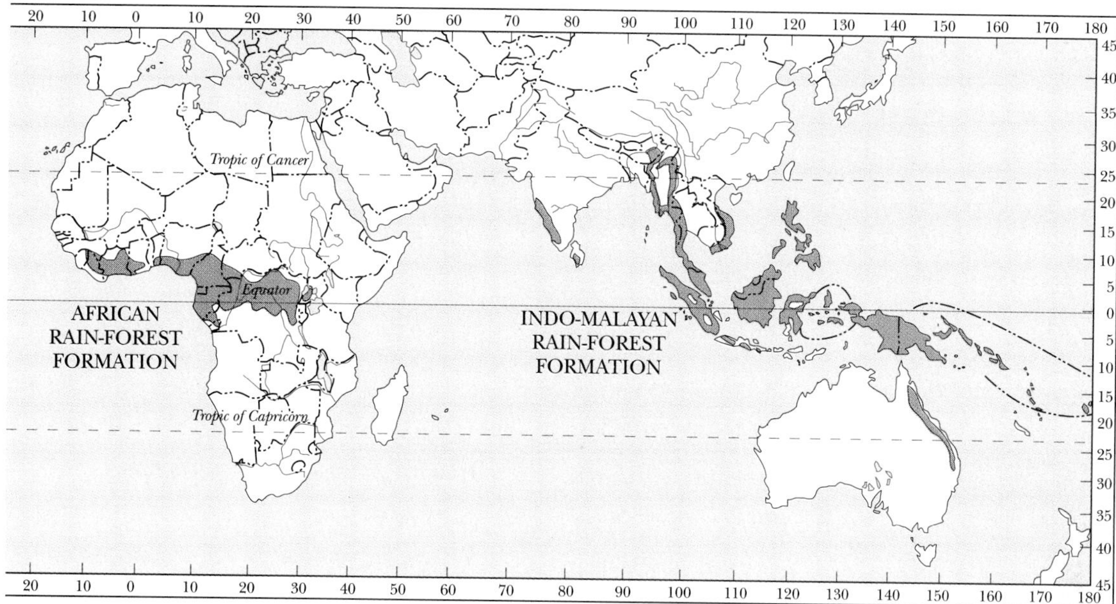

The third layer of trees is only 10 to 25 m tall. Crowns are narrow and conical, and the trunks are often leaning, twisted, or crooked. Cauliflory, the situation in which flowers and, later, fruits grow directly from the trunk rather than from small, leafy branches, is a trait of many of these trees (Figure 17-29). Leaves and leaflets are larger than those of trees of the upper strata, and "drip tips," long, pointed tips, are characteristic (Figure 17-30).

The shrub layer includes dwarf palms and giant herbs with banana-like leaves. Within the forest are few other herbs, and the ground may be virtually bare. In this perpetually shady community, their place in the flora is taken by two other types of plants that are able to find a place in the sunshine. These are **lianas,** vines that are rooted in the soil but climb into the canopy, and **epiphytes,** plants that grow on other plants (Figure 17-31).

Leslie A. Kenoyer, a Western Michigan University plant ecologist who was an early student of the American rain forest, compared the flora of Barro Colorado Island, Panama, with that of a temperate area, Kalamazoo County, Michigan. Temperate lianas exist—Virginia creeper and grape are examples—but lianas make up only about 1% of the Kalamazoo County flora compared with 10 to 15% of the rain forest flora (Kenoyer 1929). Tropical lianas include vanilla and philodendrons (whose name means "tree loving").

The temperate flora contained no epiphytes, whereas over 10% of the Barro Colorado Island vascular plants were in this category. Included are large, striking plants such as orchids, bromeliads, and cacti (the familiar Christmas cactus, for example, is a Brazilian epiphyte). But there are also many smaller epiphytes, such as algae and lichens, that grow on the evergreen leaves themselves.

Strangler figs are "hemiepiphytes." They germinate on a branch in the canopy, but their roots slowly grow down the supporting trunk to the ground. Once they are there, with a better supply of water and nutrients, the fig plant grows more rapidly. Eventually the host tree may be killed, probably by shading rather than strangulation, and the fig is left as a hollow trunk (Figure 17-32).

Perhaps the most striking community-level feature of the tropical rain forest is its diversity and consequent lack of dominance by one or a few species (Kenoyer 1929). About 2500 species of trees are known just from the Malay Peninsula, as compared with about 800 for all of North America north of Mexico. Forty species per hectare is usual and 100 per hectare occasional (Walter 1971). The richest mesic forest in Michigan would be dominated by 2

Table 17-10
Characteristics of Successional Communities and the Climax in Tropical American Humid Forests

	Pioneer	*Early Secondary*	*Late Secondary*	*Climax*
Age of communities	1–3 years	5–15 years	20–50 years	More than 100 years
Height, meters	5–8	12–20	20–30, some reaching 50	30–45, some up to 60
Number of woody species	Few, 1–5	Few, 1–10	30–60	Up to 100 or a little more
Number of strata	1, dense	2, well differentiated	3, increasingly difficult to discern	4–5, difficult to discern
Lower stratum	Dense, tangled	Dense, large herbaceous species frequent	Relatively sparse, includes tolerant species	Sparse with tolerant species
Life span, dominants	Very short, less than ten years	Short, 10–25 years	Usually 40–100 years	Very long
Tolerance to shade, dominants	Very intolerant	Very intolerant	Tolerant in juvenile stage, later intolerant	Tolerant, except in adult stage
Dispersal of seeds of dominants	Birds, bats, wind	Wind, birds, bats	Principally wind	Gravity, mammals, rodents, birds
Seed viability	Long, in seed bank	Long, in seed bank	Short to medium	Short
Leaves of dominants	Evergreen	Evergreen	Many deciduous	Evergreen
Epiphytes	Absent	Few	Prominent, but few species	Many species and life forms
Vines	Abundant, herbaceous species few	Abundant, herbaceous species few	Abundant, but few large	Abundant, including very large woody species

Based on Budowski, G., "The distinction between old secondary and climax species in tropical central American lowland forests," *Trop. Ecol.,* 11:44–48.

or 3 species and would be unlikely to have as many as 20 species in a hectare.

Net primary productivity in tropical rain forest is high, probably regularly over 2000 g/m^2. Much of this is stored in wood. However, the percentage so stored is lower than in temperate forests (Jordan 1983), meaning that a relatively large amount is available in leaves and litter fall for herbivores. Yearly litter fall in tropical forests at the equator averages about 10 metric tons per hectare, about a kilogram per square meter. In the middle of the eastern deciduous forest, say 40° N latitude, it is about one half that (Lugo and Brown 1991).

Standing crop is the highest of any ecosystem except for the temperate rain forests. Plant biomass is about 420 metric tons per hectare in the African rain forest (Tucker et al. 1985). This amounts to more than 90 pounds (40.5 kg) of plant material per square meter. Biomass in Amazonian rain forest is apparently lower (Brown and Lugo 1992). Tropical forests seem to accumulate biomass rapidly, whereas temperate forests may not reach steady-state biomass levels until they are 100 or more years old (Lugo and Brown 1991). One implication of this is that temperate zone forests remain functioning as carbon sinks long past the age at which tropical forests have quit sequestering carbon.

One recent development in the understanding

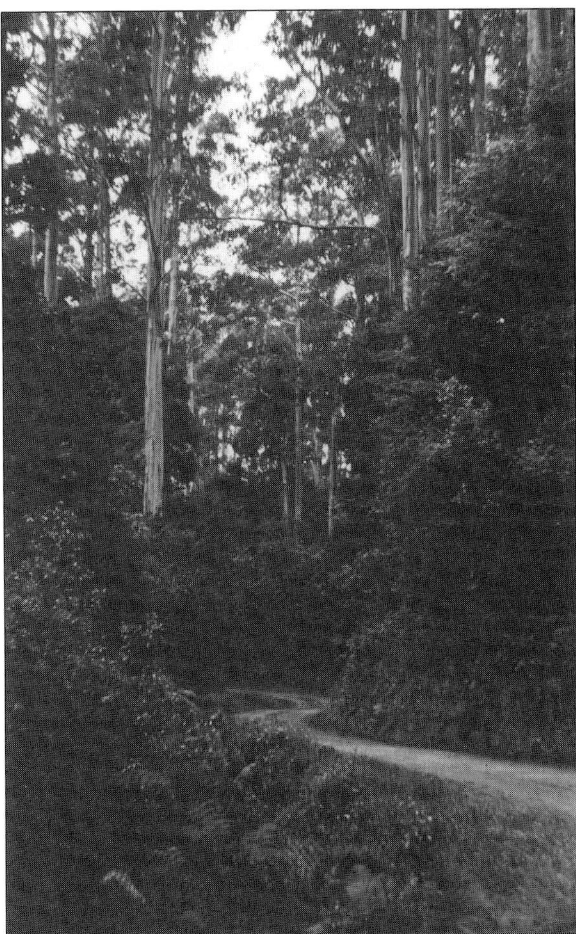

Figure 17-26

The tall, straight, pale trunks of the emergent trees in tropical rain forest are seen here, as is something of the stratification. (Photo in a eucalyptus forest in Victoria, Australia, by Joseph G. Engemann.)

of how tropical forests function is a realization of the importance of disturbance. It is now realized that fires, although infrequent, do occur (Sandford et al. 1985). Changes in river channels are important disturbances in lowland forest. The late 1980s brought into prominence the role that hurricanes can have on forest dynamics. Early data have suggested that openings caused by hurricanes may be filled in by primary forest species, often by resprouting of downed trees, rather than by invasion of typical pioneer trees (Boucher 1990). Tall emergent trees seem to survive hurricane winds better than small and medium-size trees. The upshot of these recent observations on fire and hurricanes is that it is a rare hectare in a given rain forest landscape that has gone more than 200 years without substantial disturbance (Gentry 1991).

Animals

Tropical forests are rich in animal species (Bourlière 1983), although there are regional differences. The Amazon rain forests are richer in bird species than those of the Congo, but even the latter have twice the species of temperate forest (Amadon 1973). Densities and biomass of each species tend to be low.

A high proportion of the animal life is concentrated in the upper layers of the forest, all but invisible to ground-dwelling humans. Many species of snakes, monkeys, squirrels, and of course insects, birds, and bats live permanently in the canopy. The same is true of various frogs that are able to make use of water-filled treeholes and bromeliad leaf bases and so do not have to descend to lay their eggs in ponds. Lizards live in the tree canopy too, but most lay eggs in the ground; however, the flying gecko glues its eggs to tree trunks.

Arboreal adaptations are even more highly developed than in the temperate forests. A number of species have skin flaps or flattened bodies that are used in gliding (Eisenberg 1983). These include species of flying squirrels, including ones much larger than the temperate zone species; three other groups of mammals, including marsupials (phalangers) that live in the eucalyptus-dominated rain forest of eastern Australia; and also "flying" snakes, frogs, and lizards.

Some of these gliders inhabit the emergent trees above the main canopy, as does that strange creature with the Jimmy Durante face, the proboscis monkey (Ayensu 1980). Most monkeys and related primates, however, live lower, in the continuous canopy or the understory trees, where they run along limbs and swing from branch to branch with their long arms (called brachiation). The New World monkeys, although not those of Africa or Asia, have prehensile tails by which they can dangle and which are also useful for holding onto food (Figure 17-33). Many other animals in tropical rain forest have prehensile tails, including lizards and

Figure 17-27

Relative numbers of trees of different diameter classes in virgin tropical forest at La Selva, Costa Rica, and climax beech-maple forest at Warren Woods, Berrien County, Michigan. A majority of the Warren Woods trees are larger than 33 cm; a majority of La Selva trees are below 17 cm. (Redrawn from "Treefalls and Forest Dynamics" by G. S. Hartshorn, pp. 617–634 in *Tropical Trees as Living Systems,* P. B. Tomlinson and M. H. Zimmerman, eds. Copyright © Cambridge University Press. Reprinted with permission.)

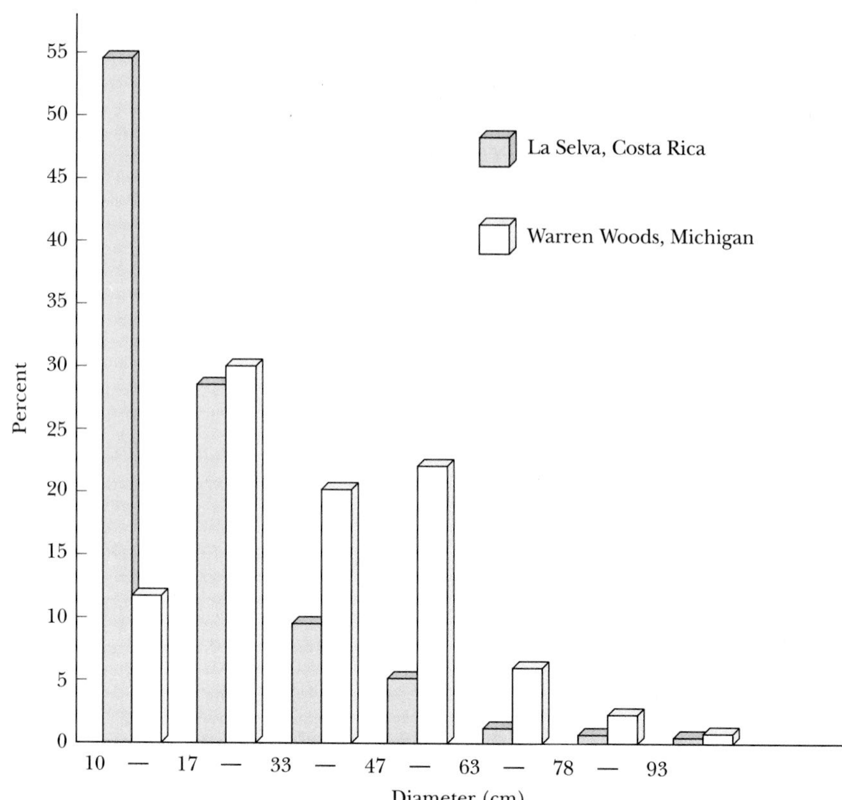

Figure 17-28

Buttress roots are a feature of many tropical forest trees. Here the base of a large fig is seen in a cleared area in tropical dry forest at Tikal, Guatemala. There are several hypotheses as to the function of the buttress roots; the most satisfactory seems to be that, for these tall trees lacking tap roots, the buttresses help the tree resist being blown over by acting like guy wires. The leaves along the trunk are herbaceous vines. (Photo by Richard W. Pippen.)

Figure 17-29

Bearing flowers and, later, fruits directly on the trunk (cauliflory) is widespread in the tropics but almost absent elsewhere. (Photo by Richard W. Pippen.)

snakes. Snakes and lizards are often camouflaged, and they hunt from ambush, shooting out their body or, in the case of some of the lizards, their tongue, to grab a passing prey.

One of the best-known canopy dwellers is the sloth. There are seven species, all New World, and similar in that they spend their lives hanging upside down from tree branches. Long claws that function like the top of a coat hanger and fur that grows from the belly toward the back fit the sloths to their habitat. They do not move fast—''slow'' and ''sloth'' come from the same root—but they do not need to. They are vegetarians, feeding on tree leaves, and if they hustled around to do it, they might attract the attention of predators, such as the harpy eagles that nest in the emergents and hunt all through the upper tree levels. In an illustration of the symbioses that seem particularly common in the rain forests, algae grow on the sloth's fur and are fed on by a moth that lays its eggs in sloth dung (Beazley International Limited 1981).

Birds are, of course, present in profusion in the upper levels of the rain forest. Toucans (Figure 17-34) in the New World and hornbills in Africa and Asia nest in cavities in the emergent trees. Scattered through the canopy are other fruit-eaters, such as parrots, and nectar-feeders, such as hummingbirds in the New World and sunbirds in the Old. Only in the tropics are nectar and fruit year-round resources. The opportunity for specializing on these foods is part of the reason for the high diversity of bird species in the tropics. One or another species of mannakin—small, polygynous frugivores in the genus *Pipra*—ranked as either the first or second most common bird (based on mist-net captures) in four New World tropical forests that have been intensively studied (Karr 1991).

Also present, of course, are insectivores and carnivores in great variety. Despite the great variety of flycatchers in some habitats in the tropics, this group is not well represented in the moist forests.

Little food making goes on below the lower layer of tree foliage, and there are few ground-dwelling herbivores. Most areas of rain forest, however, have a fairly common, rather long-legged mammal slightly bigger than a large rabbit that escapes predators by bounding. This mammal, however, may be either a rodent, a deer, or a tiny antelope, depending on the region. The whole situation is one of the better examples of ecological equivalents. In the African rain forest, elephants and the odd, semistriped okapi stand on the ground and eat foliage from the lower tree branches—or, in the case of the elephants, the branches or trees themselves.

There is plenty of food on and below the forest floor in the form of insects and other invertebrates and roots, tubers, and other underground plant parts. There are ecological equivalents among the animals living here too, such as the Old World pigs and the New World peccaries, both of which are omnivorous rooters. Ants and termites (Figure 17-35) are present in great numbers and, like fruits and nectar, are a resource to which several species have become adapted. Examples are anteaters, armadillos, and pangolins. Some of these are arboreal, making use of the tree-dwelling ants and termites.

A number of birds feed mainly at the level of the forest floor. New World examples include the tinamous and trumpeters. Most of the large flightless birds are grassland or savanna species, but the

Figure 17-30

"Drip tips," elongated leaf tips off which the water tends to run, are characteristic of many plants of the lower strata of tropical rain forest. They are also typical of many well-known house plants, which are such because they can grow in deep shade. (This photograph is of a plant in the forest in Tikal, Guatemala, by Richard W. Pippen.)

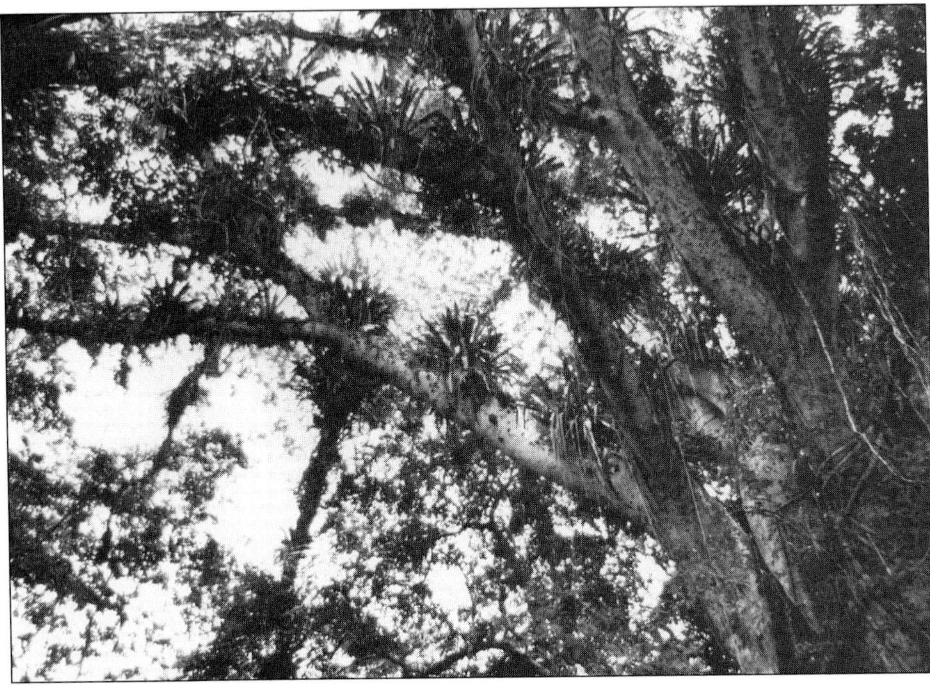

Figure 17-31

Epiphytes are prominent in tropical rain forests and a few other habitats with steadily high temperatures and humidities. (Photo by Richard W. Pippen.)

Figure 17-32

Strangler figs are hemiepiphytes. The tree that once supported this one in the Masai Mara region of Africa has died and rotted away. (Photo by R. Bruce Moffett.)

cassowaries of the New Guinea region are forest dwellers. Despite their size—a meter and a half tall—and the bright colors of their head and neck, cassowaries are not easily observed. They can run at high speeds "through all sorts of obstructions, leaping prodigiously, and even swimming with a facility that must be seen to be believed. When cornered, cassowaries attack by leaping feet first, striking with the powerful, heavily muscled legs" (Gilliard 1958). One of the toes has a "long straight murderous nail, which can sever an arm or eviscerate an abdomen with ease. There are many records of natives being killed by this bird."

The three birds just mentioned tend to be herbivores, feeding largely on fruits. Trumpeters are said to be commensals, following groups of howler monkeys, parrots, and coatis around and feeding on the fruits and nuts dislodged or dropped (Gilliard 1958). Other birds, such as the red jungle fowl—the ancestor of the domestic chicken—and many small birds have insects and other invertebrates as a large part of their diet. Included are various species in the antbird family that follow army ant columns, not so much to eat the ants as to feed on the other invertebrates that scurry away from the line of march.

Many predator species hunt the forest floor—poisonous and constricting snakes, hawks and owls in several sizes, and cats from the size of the margay to the jaguar (and in Asia, the tiger, which is twice as big as the jaguar). All of the cats are now rare because of human overhunting, but through long stretches of evolutionary time they and the other predators have been a potent selective force. Forest-floor herbivores and omnivores tend to be armored, possessed of vicious claws or tusks, extremely wary, or social—or some combination.

As in temperate forest in summer, communication here at any distance is accomplished better by sound than by vision. The jungle can be a noisy place. William Beebe (1921) wrote: "In a tropical jungle the birds and the frogs, the beasts and the insects are sending out their messages so swiftly, one upon the other, that the senses fail of their mission and only chaos and a great confusion are carried to the brain." Nevertheless, some voices can be picked out. Monkeys tend to be vocal creatures, none more so than the howlers (Alouattinae) of the South American forests that defend group territories vocally with morning and evening choruses that can be heard for miles. Basically diurnal voices include most birds, including such ear-splitting contributors as the macaws.

There are relatively few nocturnal birds, although the voice of the potoo is said to resemble the cry of a midnight murder victim (Gilliard 1958). But many frogs and insects, especially cicadas, can be heard calling at night throughout the year.

Destruction of the Rain Forest

Rain forests and other natural ecosystems of the tropics are being destroyed at an almost unbe-

Figure 17-33

The prehensile-tailed howler monkey is a characteristic mammal in New World tropical forests. It is seen here in Venezuela. (Animals Animals © 1994 Raymond A. Mendez.)

lievable pace. Current estimates are that something more than 40 million acres (more than 160,000 square kilometers) of tropical forest are being cleared each year (Lovejoy 1991). This is an area, every year, about the size of the state of Washington. Simply projecting this rate would mean that the last hectare of rain forest on the globe would be cut or burned about the middle of the 21st century; however, some areas are going fast and others, by virtue of low human populations or inaccessibility, more slowly.

Environmentalists have suggested possibly serious consequences to climate of eliminating the earth's largest source of photosynthesis and evapotranspiration and of adding the carbon contained in its organic matter to the atmosphere. They have suggested that the shrub savanna that replaces much of the forest will not be productive enough to support dense populations of humans, and that soil deterioration will lower productivity further, leading, on some sites, to virtual deserts. The extinction of thousands or perhaps a million species of plants and animals has been mentioned as a consequence of the continued destruction of tropical habitats. It is impossible to give a firm number, because vast numbers of species in the tropics have not even been described and given scientific names. It seems likely that a few thousand species are already being exterminated every year. This represents a substantial loss of our heritage. Of course, it is not just tropical species that are likely to be snuffed out. Many birds that breed in temperate America winter in the American tropics—the "Neotropical migrants," and there is a similar group of Paleotropical migrants.

Then there is the plight of the forest-dwelling native peoples, who "get their food, medicines, building materials, and spiritual meaning from their natural environment" (Hardman 1991).

The governments of developing countries

Figure 17-34

Toucans are frugivores of the New World tropical forests. The keel-billed toucan occurs north into Mexico and is seen here in Yucatan. (Animals Animals © 1994 C. C. Lockwood.)

Figure 17-35

An arboreal termite nest in Cosco Verde National Park, Costa Rica. The earth tunnels by which the termites go to the ground and higher in the tree are visible on the trunk. (Photo by R. Bruce Moffett.)

have, in the past, been generally unsympathetic to such concerns. Several tropical countries have established nature preserves; however, they tend to regard suggestions that their nations should, in effect, become one large sanctuary for wildlife and indigenous peoples as being another tactic in an attempt by the developed nations to retain dominance. They must industrialize, they believe, and to do that they need cash.

Within the past ten years, many conservationists have begun to suggest additional tactics in the battle to preserve the tropical forests. It seems certain that a cornerstone must be the preservation of large areas of relatively undisturbed forest. One way of encouraging such preservation can be "debt-for-nature swaps."

Many of the Third World countries borrowed large amounts of money to pay for weapons and development. They are now having a very difficult time making their loan payments. Conservation groups in more affluent countries raise money, buy the paper from the First World banks, and sell or trade it to countries involved. The bankers are happy to sell to the conservation groups at a substantial discount because the prospect for full repayment is slim for many of the loans. Concessions the conservation groups get from the Third World governments in return for the money vary but are usually promises to set aside land in preserves or promises to put money into special conservation funds. The World Wildlife Fund has probably been the most active player in debt-for-nature swaps up to now, and Costa Rica, Ecuador, and the Philippines have benefited the most (Williams 1992).

Other approaches that could encourage the preservation of natural communities include the following:

- Developing markets for rain forest products, such as fruits and seeds, that can be harvested without destroying the forest (Bennett 1992)
- Selling licenses to drug companies to test plants and animals for compounds that may have medical uses
- Developing tourism based on visits to large nature preserves (Norris 1992)
- Developing a sustainable forestry plan based on logging rotations using secondary forests, rather than simply cutting the primary forests until they are gone (Brown and Lugo 1990).

We must hope that an overall conservation strategy employing these and other devices will preserve a high percentage of the natural ecosystems that still exist in the tropics. For that to happen, the Third World countries will have to show a stronger commitment to conservation than the developed countries, including the United States, have shown toward their own land. In the meantime, the rain forests continue to go the way of the North American prairie.

Other Tropical and Subtropical Forests

A variety of forest types exists in the tropics, depending on such factors as rainfall, drainage, soil type, and elevation. The familiar comparisons between temperate forests and tropical moist forests do not necessarily extend to other tropical forest types (Lugo and Brown 1991).

Zonation on mountains differs in different geographical regions (Walter 1972); however, certain features are general. The lower mountain slopes have submontane forest that is similar to the flatland rain forest around it. Changes accumulate with altitude. The lower part of the montane forest may be similar in structure to, although floristically different from, the surrounding rain forest. Tree ferns, for example, tend to be common, and palms, uncommon. There is usually a montane belt of near-constant cloud or mist (cloud forest) where epiphytes are at their most abundant. At higher altitudes, flowering epiphytes drop out and the tree branches are covered with mosses, ferns, and lichens (mossy forest).

There are also altitudinal trends in tree layers, tree height, leaf size, and number of species. All decline. The decline in species, however, is not generally accompanied with strong dominance of one or two species (Richards 1957). On the higher mountains, alpine tundra occurs and, above that, perpetual ice and snow.

Rain forest is replaced by seasonal forest in regions with a dry season (Figure 17-36). Where it is short—less than two months—and not very dry, the forest may be little different from rain forest. With increasing aridity, the top layer of emergent trees comes to be made up of deciduous species. Eventually this layer disappears altogether, and the lower strata also become deciduous. The trees lose their leaves at the beginning of the dry season, at which time many of them come into flower. They leaf out again prior to the beginning of the rainy season. Epiphytism, buttress roots, and cauliflory all decline in frequency.

Tropical Savanna and Grassland

Location and Climate

With increasing aridity, tree cover declines, and we pass through various woodland and savanna types and eventually to steppe and desert. Savanna grasses tend to be bunchgrasses, 2 m or more tall in Africa but rarely more than 1.5 m in South America (Huber and Prance 1986). The trees are often legumes and tend to be less than 10 m tall, flat topped, and thorny.

The largest area of tropical savanna is in Africa, in two broad belts on each side of the equator all across the continent. Substantial areas also occur in southern Asia, Australia, Venezuela *(llanos),* and Brazil *(cerrados).*

Tree density varies in tropical savanna, but much of it is so open that it would be considered grassland or prairie by temperate zone standards (Figure 17-37). As in temperate regions, the relative roles of climate and fire in restricting tree growth are not precisely known; however, there is ample evidence that human-set fires are responsible for the maintenance of open savanna in many of the less arid regions (Figure 17-38). There is some suggestion that the widespread presence of the umbrella-shaped acacias in the African savanna is the result of a brief period of heavy recruitment following the introduction of rinderpest—cattle plague—in the late 19th century (Sinclair and Norton-Griffiths 1979). This disease, brought in with domestic livestock, decimated several native ruminants and perhaps resulted in a temporary decline in grazing/browsing pressure. The overall role of grazing in controlling woody growth in savannas is not clear; to the degree it has importance, it probably works in conjunction with fire (Scarpe 1991). Shallow or seasonally waterlogged soils may locally favor savanna over forest (Beard 1953).

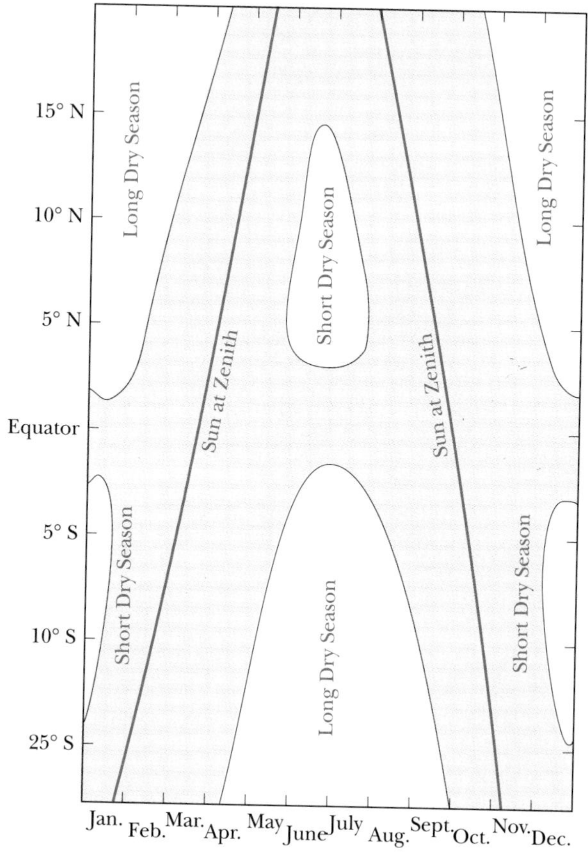

Figure 17-36

A simplified view of wet and dry seasons in the tropics by latitude. The shaded section is the wet season. (From *The Tropical Rain Forest: An Ecological Study,* by P. W. Richards, Cambridge University Press, 1952; after E. de Martonne.)

There tend to be three seasons in savanna—warm and rainy, cool and dry, and hot and dry. The main portions of the biome type occur under a yearly rainfall of 90 to 150 cm, but the driest three months of the year generally have less than 5 cm. The grasses dry up, and in many savanna areas the trees lose their leaves; however, phreatophytes that stay green year-round are common in some regions. As in seasonal forests, most trees flower and fruit in the dry season. The switch from dry season to rainy season occurs rapidly; in the course of a few days, the trees leaf, grasses sprout, and the insectivorous birds begin to nest. The large ungulates calve near the beginning of the rainy season (Talbot and Talbot 1963).

Soils

Soils are generally acid and strongly leached, although with a slightly better nutrient supply than forest soils. Laterization is the predominant soil-forming process. Organic matter is low, and evidently no soils occur comparable to the chernozems of temperate regions (Montgomery and Askew 1983). Most soils are oxisols in the U.S.D.A. Comprehensive System.

Plants and Animals

The large mounds built by certain termite species are a prominent feature of most savannas (Figure 17-39). Frequently standing 2 to 4 m tall and being four or five times as wide at the base, they are formed of soil brought from below the surface, cemented together with saliva. The number is highly variable, but most areas probably have between 10 and 150 per hectare (Josens 1983).

The termite colony lasts a few years or a decade or two, after which the mound begins to break down. In the process, fine soil particles are added to the soil surface. A study in Zaire calculated that the material in the mounds, if spread evenly over the surface, would be 20 cm deep (Montgomery

Figure 17-37

Much tropical savanna is treeless or only sparsely wooded as in the view in Masai Mara, Africa. (Six common zebras and two hartebeests are seen in this photo by R. Bruce Moffett.)

Figure 17-38

A fresh burn in tropical savanna in Belize. The trees in the background are oaks. (Photo by Richard W. Pippen.)

and Askew 1983). It is clear that termite activities are an important factor in the development of the soil profile, particularly in rocky or gravelly areas.

The termite mounds also tend to be richer in nutrients than the adjacent soil and so probably counteract the effects of leaching as well as forming a microhabitat occupied, following death of the colony, by a different association of plants. The mounds are used for sunning and as lookout posts by a wide variety of animals (Brown 1972). Some Australian kingfishers excavate nests in active termite nests.

The evolutionary advantages of mound building are still under debate, but clearly mounds aid in oxygen-carbon dioxide exchange for the colony and avoid seasonal flooding. In some areas, the mounds are oriented with their long axis north-south, presumably hastening the warming of the mound in the morning while avoiding noontime overheating (Jacklyn 1991).

Termites are the main processors of litter in the savanna, one third or more of the dead plant material being processed by them (Josens 1983). They are also a major food source for all sorts of animals. Some species feed on them throughout the year; the aardvark and aardwolf (*aard* is Afrikaans for "earth") are examples. But when the reproductive swarms appear at the beginning of the rainy season, many other animals also switch to them. One observer recorded over 150 species of birds feeding at termite swarms, including frugivores and raptors.

Figure 17-39

A termite mound in tropical savanna, Samburu Game Preserve, Africa. (Photo by R. Bruce Moffett.)

Net annual production in tropical savanna varies with climate, from below 150 to above 2000 g/m^2. Much of it goes into foliage that supports herds of large grazers and the large carnivores that prey upon them.

Flora and fauna vary from one region to another. South American savannas have no large grazers, and none of the other savanna biomes approaches Africa in this regard. African savanna is the home of many of the large animals most familiar to us from zoos and fables—grazers like zebras (Figure 16-37), wildebeests, and gazelles; browsers like elephants, giraffes, and gerenuks (Figure 17-40); scavengers such as hyenas (Figure 17-41); and carnivores such as lions (Figure 17-42), leopards, and cheetahs (Figure 17-43).

The bird fauna is rich. Areas without trees have more species than does temperate grassland, and the fauna is richer still where the tree layer is more conspicuous. Familiar African birds are the secretary bird, a 150-cm hawk that runs through the grass and kills snakes by stamping on them, and the ostrich, also cursorial, that goes in herds like, and sometimes with, the ungulates. Ecologically equivalent species to the ostrich are the rheas of South American savannas and the emu of Australian ones.

Among amphibians, only frogs and toads are important, and they do not occur far from water. Reptiles also seem to play a lesser role here than they do in forest on the one hand or desert on the other.

Migration is an important factor in tropical savanna for both birds and larger mammals. Wildebeests and zebras move to areas of dry-season moisture. Some birds are able to avoid the dry season altogether; the dry-land stork, for example, breeds in savannas north of the equator in their wet season and winters in the savannas south of the equator in

Figure 17-40

A browser typical of tropical savanna, the gerenuk. (Photo in the Samburu Game Preserve by R. Bruce Moffett.)

Figure 17-41

Striped hyenas at Masai Mara, Kenya, Africa. (Photo by R. Bruce Moffett.)

their wet season. Many North American and European species winter in Latin America or Africa, mostly in various types of woodland or savanna rather than in tropical rain forest. Most of these species do not go below the equator and so (Figure 17-36) are present in these areas during the dry season. The ecological and evolutionary ramifications of this situation are largely unknown (Fry 1983).

Loss of the African Megafauna

Several wildlife sanctuaries have been set up in African savanna. The 5600-square mile (14,560-km^2) Serengeti National Park in Tanzania is probably best known, but there are some two dozen others. Latin American countries have lagged, but recently Bolivia established a 585-square mile (1521-km^2) sanctuary, the Beni Biological Reserve.

Outside the African parks, only a remnant of the once-magnificent fauna exists. Rinderpest brought in with domestic livestock at the end of the 19th century affected several native ruminants. Big-game and ivory hunters killed many animals. Still, most of the decline has occurred just since the end of World War II, with the rising tide of human population.

The human population of Tanzania (then called Tanganyika), for example, was less than 6 million at the time of Ernest Hemingway's visit (1933 to 1934) described in *The Green Hills of Africa.* By 1965 it was more than 10 million, and today it is more than 25 million. Hunting for trophies and for the fur trade has continued. Native populations, needing protein, poach. But agriculture is perhaps most damaging. Farmers kill herbivores that threaten their crops and carnivores that threaten their livestock. Habitat and traditional migration

Figure 17-42

Male African lion on the Samburu Game Preserve, Africa. (Photo by R. Bruce Moffett.)

Figure 17-43

Female cheetah at Masai Mara, Kenya, Africa. (Photo by R. Bruce Moffett.)

routes have been lost because of overgrazing, conversion of rangeland to crops, and the development of large, privately owned farms. These processes, like many of the projects most destructive of tropical rain forest, have often been financed by loans from large international banks, such as the World Bank and the Inter-American Development Bank.

Community and Ecosystem Ecology: *Aquatic Communities and Special Habitats*

OUTLINE

- Lakes and Ponds
- Streams
- The Oceans
- Freshwater Wetlands
- Caves, Phytotelmata, and Other Special Habitats

Water striders are members of the neuston, living at the interface between water and air. *(Dennis Drenner.)*

Generalizations concerning the ecology of aquatic organisms and situations have been dealt with in appropriate places in other chapters. Here aquatic ecology is treated specifically. Major features of fresh- and salt-water habitats and their communities are described, but the details of aquatic ecology form a large and specialized field beyond the scope of this book. Caves and other special habitats are also briefly discussed.

Lakes and Ponds

The fraction of the earth's surface covered by fresh water is trifling, about 2%. The percentage of the total water of the biosphere that is in lakes and rivers is even less impressive, less than 0.2%. Because of people's interest in inland waters, **limnology,** the science of fresh water (physical, chemical, and biological aspects), has received what might seem a disproportionate amount of study. People are interested in lakes and rivers for their scenic beauty; for boating, fishing, and related recreation; for dumping sewage in and getting water out for drinking; for industry; and for irrigation. People are also interested in lakes and rivers scientifically as links in the hydrological cycle and as homes of a set of organisms with modes of life very different from those of terrestrial organisms.

This section deals with lakes and ponds—or, in other words, areas of standing fresh water. (The classical term used to refer to standing, or still, water is **lentic**.) There is no clear distinction between lakes, ponds, and marshes. Basically, ponds are small, shallow lakes. Marsh refers to a vegetation type, the floating-leaved and emergent vegetation that grows in shallow water that often occurs as a fringe around lakes.

Origins of Lakes

Lakes are holes or depressions in the ground that are filled with water. They originate in many ways (Cole 1979).

Glacial action is responsible for the formation of many types of lakes. The kettle lake, perhaps the most common kind of glacial lake, forms when a retreating glacier leaves a large block of ice buried in the drift. The typical lake in Michigan or Minnesota is a kettle lake.

In mountainous areas, glaciers can dam valleys with rock debris producing, when the glacier is gone, a kind of reservoir sometimes called a tarn. During an ice age, sea level is lower. Valleys that form near sea level during a period of glacial advance become flooded with seawater when sea level rises. Such estuaries are termed fjords.

Oxbow lakes are formed when a meandering river straightens its course, leaving a large loop of the old river channel cut off.

Caldera and maar lakes form in a dormant volcano's crater. If Mount Saint Helens becomes dormant, a lake may form in the caldera left by the explosion of 1980.

In areas where the bedrock is soluble and close to the surface, karst lakes may form when caverns, cut by water in the bedrock, collapse and dam the underground stream. These flooded sinkholes are common in Florida, southern Indiana, and other areas with limestone bedrock.

Impoundments are human-produced lakes that result from the damming of a water course. They range from farm ponds to massive reservoirs.

There are still other modes of lake formation; basically lakes form wherever there is a source of water and a depression to collect it. Although how a lake starts may influence its development chemically or in other ways, lakes have many similar characteristics everywhere.

Physical Factors in Lakes

Water as a Habitat

Water as a habitat has a number of features that need to be mentioned. It has a high specific heat; that is, a large amount of heat must be added to raise the temperature of water. This means that daily temperature changes are slight and seasonal changes slow compared with terrestrial habitats.

For most lakes most of the heat input is through solar radiation. Some fraction of the sunlight that hits a lake surface is reflected away. As it passes through the water, the remaining light energy is absorbed and converted to heat. About one half of the total light energy is absorbed in the first meter, but some light penetrates great depths. The rate at which light is absorbed by the water varies by wavelength; 95% of the red light is lost in the first 6.5 m, but 95% of the blue light is not lost until a depth of about 550 m. Light penetration can, of course, be very much reduced in turbid waters, those containing suspended materials. The upper zone of the lake in which oxygen production in photosynthesis exceeds the use of oxygen is called the **trophogenic** (or **euphotic**) **zone.** The depth at which oxygen release and oxygen absorption balance is the **compensation depth;** below it lies the **tropholytic zone.** At the compensation depth, light intensity is reduced to about 1% of full sunlight. This depth in feet or meters varies greatly depending on turbidity and water color.

Water is about 775 times as dense as air. This allows it to provide the buoyancy that makes swimming possible, but it also means that water provides much greater resistance to movement than does air. Unlike most materials, fresh water is densest at a temperature above its freezing point—ice floats—

and this is a major feature of water as a habitat. It would be a different world if lakes froze from the bottom up.

Thermal Stratification and Oxygen Depletion

The temperature-density relationship of water is important in producing patterns of temperature change with depth, referred to as **thermal stratification** (Figure 18-1). The pattern for most of the deeper lakes of the temperate regions of the earth is as follows: In the spring, sunshine and warm air temperatures melt the ice cover and bring the upper layers of the lake to the same temperature as that of the lower, about 4°C. When this occurs, winds can cause the whole lake to circulate. During the period of this spring overturn, the lake temperature may stay around 4°C, or it may rise several degrees higher.

Usually during a warm, calm period the temperature of the surface waters is raised a few degrees higher than that of the lower portions of the lake. This warmer, lighter water tends to circulate by itself on top of the cooler heavier water. From this develops the typical summer stratification, with an upper **epilimnion** of warmer, circulating water, and the lower **hypolimnion** of cold water that has little circulation. The area between, in which temperature change with depth is rapid, is called either the **metalimnion** or the **thermocline.** The temperature

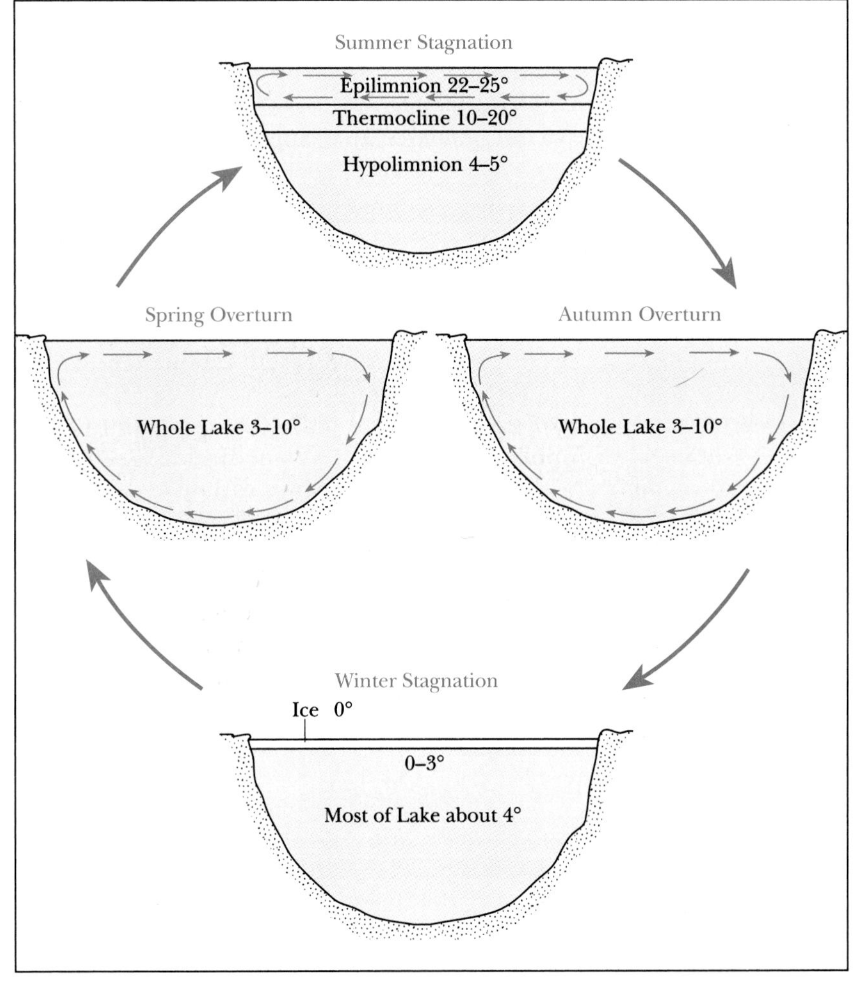

Figure 18-1

Temperatures (°C) with depth for a fairly deep, temperate-zone lake for four seasons. The small arrows show water circulation caused by wind blowing across the lake.

of the hypolimnion may be 4°C, but it is often higher and may increase through early summer.

In the autumn the lake loses more heat than it gains. The surface waters cool to the temperature of the upper part of the metalimnion so that, in effect, more and more of the lake is drawn into the epilimnion. Eventually the autumn overturn occurs; the upper waters cool to about the same temperature as the lower waters and the whole lake circulates.

As air temperatures drop through the early winter, the lake temperatures also drop, but circulation continues. As surface waters cool below 4°C, they no longer sink, but the density differences in the range between 0 and 4°C are so slight that even small amounts of wind can keep the lake circulating. Once the lake has frozen, however, winter stratification can occur. Here the lighter, colder water of about 0 to 2°C lies on top of the warmer water of around 4°C.

Lakes with two overturns per year, as just described, are called **dimictic** lakes (meaning two mixings). In other-than-temperate climates, other patterns exist involving a single overturn, or several, or none.

The concentration of oxygen is lower in water than in air and is also very variable from time to time and from place to place. In general, it is the times and places of very low oxygen concentrations that are of the most ecological interest. One situation in which oxygen may become very low is in ponds or shallow lakes with rooted hydrophytes covering the whole bottom. At the end of the summer, when these die back, their rapid decomposition may lower oxygen levels through the whole lake enough to kill much of the fauna. Die-offs of fish under such circumstances are called "summer kills."

Oxygen depletion in the hypolimnion during the summer is likely to occur if the hypolimnion is below the zone of effective light penetration, thus having little or no photosynthesis occurring in it. In this situation, where the hypolimnion is sealed off from the oxygen of the air and no oxygen is being added by plants, the respiratory activities of the animals and bacteria can exhaust the oxygen content of the hypolimnion by late summer. Oxygen depletion during this summer stagnation period is more apt to occur in highly productive lakes where there is more organic material for decay. The third situation in which oxygen may be severely depleted is in winter when there is a heavy snow cover over the ice. Light reduction slows or stops photosynthesis in the algae; if the snow cover lasts several weeks, decomposition and animal respiration can reduce the oxygen supply below the tolerance limits of some organisms, with "winter kills" of fish as a result.

Meromictic Lakes

Meromictic lakes (meaning partially mixed) are more or less permanently stratified, usually because the lower stratum is so chemically dense that even if it becomes warmer than the upper stratum, it still won't rise and cause the water in the lake to circulate. The differences in dissolved materials, usually salts, that produce meromixis stem from a variety of processes, but meromictic lakes are all similar in having virtually permanent anaerobic conditions at the bottom.

Occasionally some event causes a meromictic lake to circulate. That apparently was what happened at Lake Nyos, a caldera in the Cameroons, the evening of August 21, 1986 (Kerr 1986, Kling 1987). Perhaps an earthquake, a landslide, or a small eruption caused an overturn. Carbon dioxide bubbled out of the depressurized water and spread across the landscape. Around 2000 people in three villages downslope died of suffocation in the course of a few hours.

The circulation removed about one third of the hypolimnion's store of carbon dioxide (Kerr 1989). In the five years following, carbon dioxide built back up to nearly half the predisaster level (McCann 1991), presumably coming from warm, mineral- and carbon dioxide-rich springs in the bottom of the lake.

Communities and Organisms

Pond and Lake Regions

For descriptive purposes, three regions of lakes are usually recognized (Figure 18-2). These are the **littoral zone,** the shallow water around the shore occupied by rooted vegetation; the **limnetic zone,** the open water beyond the littoral zone; and the **profundal zone,** the bottom below the limnetic zone.

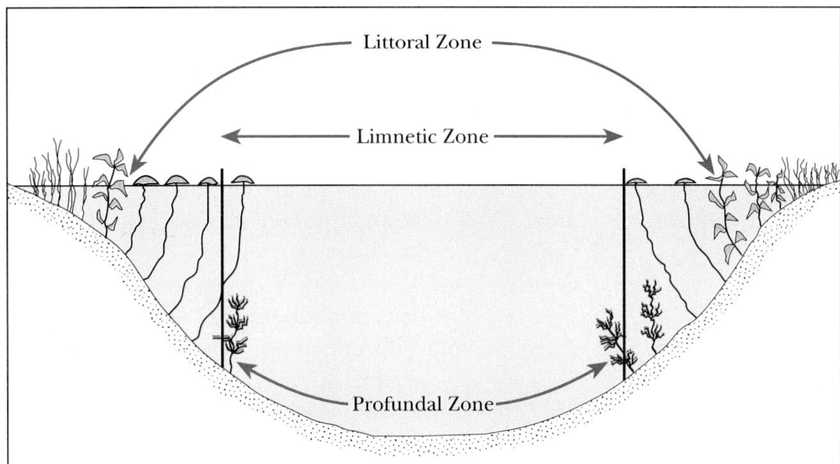

Figure 18-2

Littoral, limnetic, and profundal zones in a lake.

Sometimes the deeper parts of the open water are included with the profundal zone. In shallow lakes the whole basin may be occupied by rooted vegetation, in which case neither limnetic nor profundal zones exist.

Habitat Groups

There are basically five life habit types among the organisms of lakes: benthos, periphyton, plankton, nekton, and neuston (Welch 1952).

Benthos. The **benthos** is composed of the bottom-dwelling organisms. The rooted plants of the littoral zone are in a sense benthos but are usually considered separately as aquatic macrophytes. Although there may be large numbers of microscopic invertebrates in the benthos, the more noticeable animals are such forms as sponges, flatworms, annelid worms, crustaceans such as crawfish (or crayfish), mollusks, and aquatic insect larvae. Bacteria are abundant. The food of the bacteria and most of the benthonic animals consists of the dead organic material, or detritus, that accumulates on the bottom.

Species diversity is high in the benthos of the littoral zone and much lower in the profundal zone, probably because of the low habitat diversity there. In eutrophic lakes, another factor is the extremely low oxygen levels that bottom dwellers of the profundal region have to endure during the summer stagnation period. Different animals show different mechanisms for dealing with this situation. Some larvae avoid it by migrating from the profundal zone. The worms *Tubifex* and *Limnodrilus* and also the midge fly larva *Tendipes* have blood that contains hemoglobin, the same oxygen-carrying pigment found in mammals that is especially efficient at transporting oxygen at lower concentrations. The tendipedid larvae have the common name "bloodworms." Some bacteria of the profundal zone are true anaerobes, requiring no oxygen in their respiration. Many animals of the profundal zone accumulate an oxygen debt during periods of little or no oxygen. Without oxygen to break glucose down to carbon dioxide, they break it down instead to lactic acid. The process is only about 5% as efficient as respiration in the presence of oxygen, but it does provide some energy for the animal's needs. The lactic acid may be excreted, with a considerable loss of energy, or, later when oxygen is available again, it may be treated in such a way as to break it down to carbon dioxide (and water).

Periphyton. The **periphyton** is made up of the microscopic or near-microscopic plants and animals that live attached or clinging to aquatic macrophytes, larger animals, rocks, or other surfaces projecting into the water. Included are diatoms and other types of algae, and such animals as hydras and the protozoans *Stentor* and *Vorticella.*

Plankton. The **plankton** consists of the mostly microscopic plants (phytoplankton) and animals (zooplankton, rhymes with mo', po', plankton) that float in the water. Although most of the animals and some of the algae are capable of locomotion, their movements are determined largely by water currents and the wind-driven cycling of the epilimnion. Most planktonic organisms are slightly denser than water and, accordingly, tend to sink following Stokes' law. According to Stokes' law:

$$S = [K\,r^2\,(b - m)]/v$$

where S = sinking rate, K = constant determined by the shape, r = radius, b = density of the organism, m = density of the medium, and v = viscosity of the medium.

If m and v are assumed constant, an organism's sinking rate could be altered by changing its size (r), shape (K), or density (b). Many planktonic organisms seem to employ these and other tactics to keep their sinking rate low but positive. Most of the plankton are very small: A reduced size (r) equates to a lower sinking rate. Some plankters exude a gelatinous envelope that reduces the overall density (b) of the organism. During the course of the season some organisms exhibit changes in their shape (K).

The name given to seasonal changes in shape is **cyclomorphosis.** Typically, organisms will grow or expand translucent or transparent projections of their body shell during the warmer parts of the year.

The evolutionary bases of cyclomorphosis are not fully agreed upon. In phytoplankton, the changes have been shown to influence buoyancy, and it may be that the ultimate factor is a calibration of sinking rate to viscosity changes of the water. In zooplankton, the growth of spines may somehow be related to predation (Wetzel 1983). One hypothesis is that by growing the transparent helmets and other projections, the animal can store energy without increasing its apparent size and, hence, conspicuousness to predators. Another likely possibility is that the projections foil attacks of the smaller (invertebrate) predators.

The phytoplankton compose one of the two important producer groups in lakes (the other being the macrophytes); diatoms, cyanobacteria, and green algae are the important groups. The zooplankton organisms are mostly grazers on the phytoplankton, the most abundant being rotifers, copepods, ostracods, and water fleas.

Many of the limnetic zooplankton, especially the water fleas, show an interesting pattern of daily vertical migration. They migrate several meters each day, toward the surface at night and toward the bottom in daylight. The rate of ascent is usually a few centimeters per hour but may be a meter or more. The proximate factor in the vertical migration is light; the ultimate factor or factors are uncertain. It has been suggested that by coming up to feed at night the zooplankton (1) feed at the time when protein content of the algae is highest; (2) feed when predation is least; and (3) feed in waters of high temperature, allowing rapid feeding, while spending the rest of the day in waters of lower temperature, allowing more efficient growth.

The phytoplankton of lakes of temperate regions often show a characteristic pattern of abundance. After a winter minimum there is a spring peak of numbers corresponding with increasing day length and light intensity of spring. An increased nutrient supply in the epilimnion as a result of the circulation associated with the spring overturn and rising temperatures may also be involved. Diatoms are often important in the spring peak. A decline in late spring may result from a decrease in the nutrient supply through nutrients being tied up in dead bodies of plants and animals and deposited on the bottom. In the case of diatoms the limiting factor may be silica, of which their shells or tests are made.

In nutrient-rich lakes a later summer "bloom" of blue-green algae (cyanobacteria) often occurs, but in nutrient-poor lakes algal populations may be low until fall. Another peak occurs about the time of the autumn overturn, when nutrients again are replenished in the upper layers of the lake. During the winter, ice forms on the surface of the lake, and the epilimnion does not circulate. At this time motility is advantageous, and flagellated algae do, in fact, dominate during the winter. Individual species of algae may have yearly patterns of abundance different from these overall patterns.

Nekton. The **nekton** consists of the strongly swimming organisms, mostly fish but also including a few amphibians (such as the mud puppy) and birds (such as mergansers and loons). Depending on the species, fish may feed on detritus, plankton, small invertebrates, or other fish. One of the interesting features of fish communities, causing them to be more diverse ecologically than they would seem merely on the basis of species number, is the tendency for different sizes of the same species to use different foods. Small perch under 10 cm may eat mainly small zooplankton, larger perch may eat large zooplankton and some invertebrates from the bottom, and perch over 20 cm may eat mainly small fish (Allen 1935). Changes in the specific habitat occupied may also occur, for example from littoral to limnetic.

Neuston. The **neuston** is a group of remarkable organisms that are as they are because of another unusual property of water—it has the highest surface tension of any liquid except mercury. On land this allows the capillary rise of water in soils. In lakes, streams, and the ocean it produces a special habitat at the air-water interface, inhabited by the neuston. One set of neuston organisms lives on top of the surface film. Included are the long-legged water striders that actually walk on the water and the whirligig beetles that have two-parted eyes, the lower part suitable for the refractive index of water and the upper for the refractive index of air. Some other organisms use the underside of the water film, such as hydra, water fleas, and insect larvae such as mosquito wigglers.

Energy Flow

The heterotrophs of lakes depend on three main sources of energy. These are production of organic matter in the lake by algae of the plankton and periphyton, production by macrophytes, and the import of organic matter from the surrounding basin. The imported material may vary from dissolved organic matter (DOM) such as amino acids or carbohydrates to leaves, twigs, and tree trunks. One obvious type of food chain in the lake ecosystem begins with phytoplankton and ends with some top carnivore such as a trout. In between may be zooplankton and small fish. But here, just as in many terrestrial ecosystems, such grazing food chains are relatively unimportant in the total energy metabolism of most lakes. Most organic material, whether originating as phytoplankton or macrophytes in the lake or as terrestrial plants carried into the lake, ends up on the bottom. Here it and the invertebrates and bacteria that work on it form the food chains through which most of the lake's energy input flows.

Succession and Eutrophication

The general fate of lakes is to be filled in by sediments from the surrounding uplands and also by the deposition of organic remains, and so to become dry land. The process may, however, be very slow. In much of the temperate part of the world, lakes tend to be unproductive, or oligotrophic, when they are first formed. Characteristically, they are deep and have low phytoplankton populations. With only a small amount of organic debris settling into a large volume of hypolimnion, decomposition is not sufficient to exhaust the oxygen supply; thus, the benthos is diverse, and there are cold-water fish, such as char and whitefish, living in the deep waters.

Some lakes tend to become more productive, or **eutrophic,** as they age. As sediment fills the lake, the epilimnion and the thermocline scarcely change in depth, but the hypolimnion becomes so greatly reduced in size that the reserve of oxygen stored in the hypolimnion runs out before the fall overturn. Eutrophic lakes tend to have high algal populations, especially blue-greens, to show summer oxygen depletion in the hypolimnion and, consequently, to have few or no fish in the hypolimnion and a bottom fauna made up of only the forms tolerant of very low oxygen concentrations. It should be realized that the descriptions of oligotrophic and eutrophic lakes as given are like parts of a spectrum, and intermediate lakes do exist.

Natural eutrophication is related to an increased nutrient supply for the algae, especially of phosphorus and possibly also of nitrogen. The idea that lakes in general show increasing eutrophy with age is no longer widely held, mostly because of paleolimnological evidence to the contrary (Moss 1988). G. E. Hutchinson (1973) suggested that after a short oligotrophic phase a steady state, or trophic equilibrium, tends to be established. At Linsley Pond, Connecticut, studied by Hutchinson and colleagues, a steady state seemed to persist from 7000 or 8000 years ago until the time of European settlement.

The steady state of productivity is probably related to the nutrient input from the drainage basin. Although an initial condition of oligotrophy may characterize some lakes, others may be well supplied with nutrients from the start. Changing conditions in the watershed, such as deforestation or changes in the water table, may lead to shifts back and forth along the oligotrophy-eutrophy spectrum.

The one most general trend in lake productivity with time is that a high percentage of lakes in human-influenced watersheds show evidence of artificial eutrophication.

Ponds and lakes receiving sewage and runoff from fields, fertilized lawns, or other high phosphate/nitrogen sources commonly show heavy "blooms" of algae, especially blue-greens, in the summer. These may look and smell unpleasant and may interfere with swimming and boating (Figure 18-3). The other features of eutrophic lakes, such as oxygen depletion in the hypolimnion, may be accentuated. The large amount of organic debris produced tends to accumulate on the bottom, and presumably the succession to dry land is speeded up. This extreme eutrophication connected with human activities may have few natural parallels. In any case, it is clear that most lakes affected by increased nutrient input from human activities can still be restored to approximately their previous productivity simply by stopping the input of sewage, fertilizer, or phosphate detergent.

A well-studied case in point is Lake Washington, next to the city of Seattle (Edmondson and Lehman 1981). By the early 1960s, the lake was heavily eutrophied from sewage pollution. When sewage input to the lake was stopped in the mid-1960s, the phosphate level quickly dropped, and many of the symptoms of eutrophication, such as dense algal growths, had disappeared by 1969.

The concentric zones of vegetation around the edges of lakes are usually regarded as representing successive seral stages in the last stages of succession

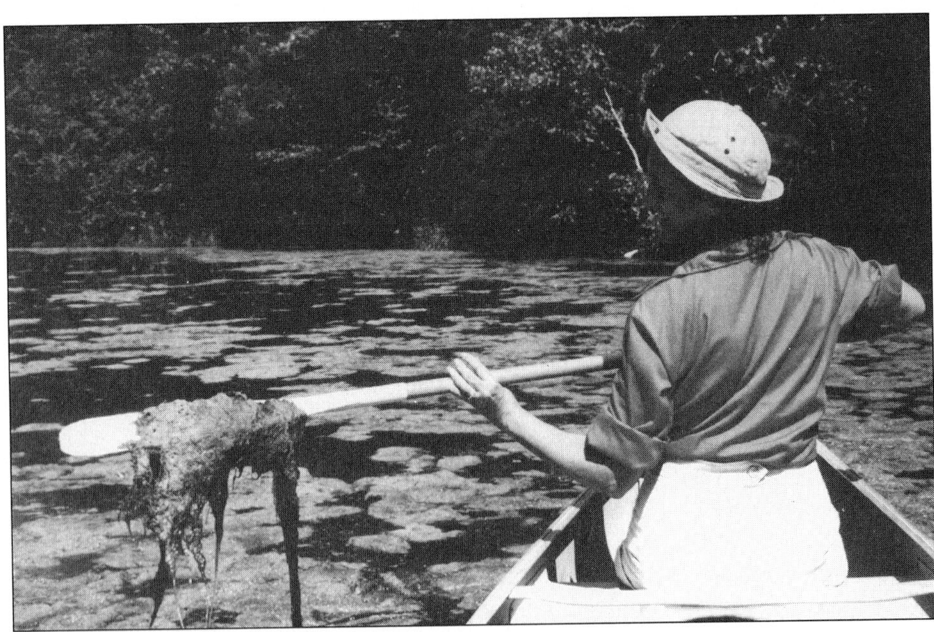

Figure 18-3

Eutrophication in the Pinnebog River, Huron County, Michigan. The algae is *Cladophora.* (Photo by Joseph G. Engemann.)

from shallow water to land. The zones, from lakeward to landward, or from earlier to later, are often the following (Figure 18-4):

Submerged vegetation, with waterweeds, pondweeds, eelgrass, *Chara*, and bladderwort.
Floating vegetation, with water and pond lilies, pickerel weed, certain pondweeds, and duckweed.
Emergent vegetation, with cattail, reeds, rushes, bur reeds, arrowheads, and wild rice. This zone generally occupies the edge of standing water, from about 1.5 m in depth onto the areas where the soil is merely damp.
Swamp shrub vegetation, with buttonbush, willows, and dogwoods.

The zonation seems to be controlled primarily by water depth. Consequently, as the lake gets more shallow, each zone tends to move outward and occupy the area previously held by the preceding one. For example, floating-leaved plants move out into areas earlier occupied by submerged vegetation, while the area formerly dominated by floating-leaved plants is invaded by emergent vegetation. Marshes and other wetlands are discussed further in a later section.

Streams

Streams flow. That is not all that there is to know about them, but it is by far the most important thing. The current sets the conditions under which stream organisms compared with other aquatic organisms must live. The water that is here today is mostly far away tomorrow; whatever the organisms living in or along a stream do to the water is continually exported downstream.

Streams come in all sizes, from temporary trick-

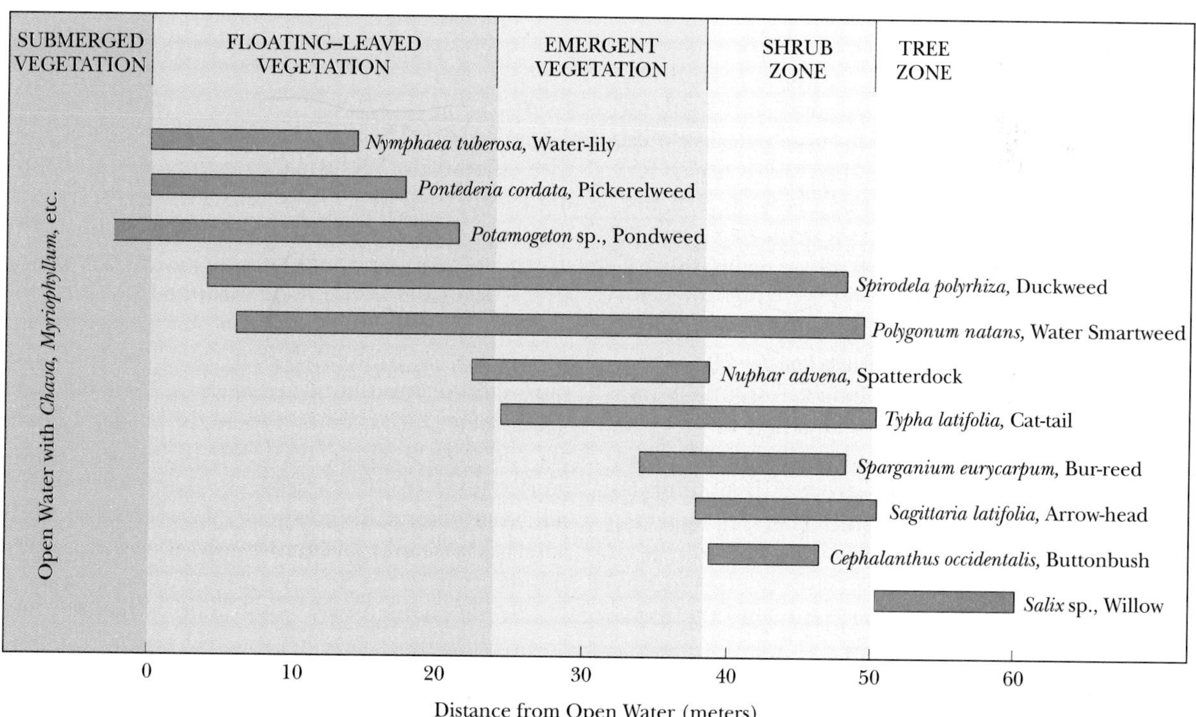

Figure 18-4

Zonation around a pond in Allegan County, Michigan.

les that flow only after rains to the miles-wide stretches of the great rivers such as the Mississippi and the Amazon. The classical term for flowing water is **lotic.** A basic subdivision is between creeks or brooks, which are less that 3 m wide, and rivers, which are 3 m or more. Streams of most sizes show two basic habitats, rapids or riffles and pools. Although there may be some distance separating pools from rapids, commonly the two habitats are close together. In a meandering river the water at the outside of the bend moves rapidly and cuts into the bank, whereas the water at the inside moves slowly and deposits silt and sand, thus slowly increasing the meander. The two habitats here are at the same place in the river, but on different sides.

Riffles

It is the riffle that is the true stream habitat. Here the water is swift flowing over rocks or pebbles. Because of the turbulent flow, air is continually mixed with the water, so the oxygen level is always high. The plants of the riffles are mostly algae that grow attached to the rocks. They may be filamentous forms that grow as cushion-like or crustlike mats, or they may be sessile diatoms that live as periphyton on other algae or attached to rocks and pebbles. The vegetation in streams that most people would call "moss" is usually algae, but there are a few aquatic mosses that are important in some streams.

Animals of the riffles keep from being swept away by the current through three main mechanisms: They attach themselves to solid surfaces, swim strongly, or avoid the current. Many animals show some combination of these methods.

Sponges and bryozoans behave essentially like plants—they are sessile, living attached to rocks or to dead trees in the water. However, many other invertebrate animals move around but cling to rocks or vegetation using hooks, claws, or suckers. The limpet is a characteristic swift-water animal, a peculiar snail that clings to rock surfaces by means of its muscular foot, and also diminishes the force of the current by its low conical shape that resembles a Chinese coolie's hat. Water pennies, the larvae of a certain kind of beetle, are almost flat; they cling to the surface of rocks where the current is slow. Blackfly larvae spin silk threads to which they attach themselves by a circle of hooks on their posterior end. Mayfly naiads and many other immature insects occurring in the riffle are streamlined to reduce drag, and have legs adapted for clinging to the substrate. Many caddis flies live in heavy cases constructed of sand, pebbles, or twigs cemented together.

In addition to having shapes that reduce drag, animals may avoid the current by actually going where there is little or no current. They may go under rocks or even down into the rubble on the bottom. Also, within dense mats of vegetation, the current may be practically zero. Some organisms, for example turtles and aquatic birds, like the ouzel, simply leave the water to avoid the current.

Several kinds of invertebrates such as crawfish and amphipods can swim, but fish are the stream organisms that can swim strongly enough to maintain their position readily. Combined with the ability to swim, fish show an orientation response to current, or rheotaxis. Fish and many stream invertebrates will face into and move against a current. Some fish seem to use landmarks along the shores as cues for maintaining their position in the stream, but darters, perhaps the most characteristic riffle fishes, seem to use tactile stimuli from the bottom.

Pools

The current in certain parts of the stream, including shallow margins, backwaters, and pools, is much reduced. Pools ordinarily have water entering at the upstream high side and spilling out at the downstream low side, with no real through-flow. With the lowered velocity, sand, silt, and clay carried by the stream are deposited, so that these areas tend to have sand or mud bottoms. Many of the organisms living there, especially in the mud-bottomed pools, are the same as those occurring in ponds. This includes some of the same marsh plants such as cattail, arrowleaf, and smartweed. Watercress is a characteristic stream emergent rarely, if ever, found in ponds.

Some characteristic pool animals include burrowing mayfly naiads, tendipedid larvae, tubificid or limnodrilid worms, and various clams, in the benthos. The nekton consists mostly of fish, some the same species as occur in ponds but with several distinctive stream fish. A neuston of water striders and whirligig beetles is often well developed.

Plankton in general is poor in streams, except in the larger rivers. Here sizable populations of plankton may develop in a mass of water as it moves toward a large lake or the ocean. Most characteristic are certain diatoms, rotifers, water fleas, and copepods. Plankton is rare in smaller streams; a plankton net put into the water will sieve out only a few true planktonic organisms. It will, however, pick up drifting bottom dwellers such as mayfly naiads and the larvae of caddis flies, blackflies, and midges.

Downstream Drift

It is now clear that such **downstream drift** is so prevalent that it must be a phenomenon of fundamental importance in streams. Although much is still uncertain, it is known that drift is more pronounced at night than in the day, and in spring and summer rather than at other seasons (Waters 1972). For some organisms, such as mayflies, drift seems to be part of cyclical, migratory phenomena in which the immature stages drift downstream and the adults, upon emergence, fly upstream and lay their eggs in the headwaters, allowing the cycle to continue. The evolutionary advantage of drift apparently is that it may bring the immature insects to less crowded situations. One effect of this in the functioning of the ecosystem is that areas of a stream depopulated by floods are quickly recolonized. The same is true of sections of streams made barren by heavy pollution.

Energy Flow

In small, shaded forest streams as much as 99% of the organic material may be imported (allochthonous) and 1% or less produced by instream photosynthesis (autochthonous) (Fisher and Likens 1977). The high populations of animals that the streams sustain are parts of food chains based mainly on imported organic matter rather than on the small amount of plant matter produced in the stream. The imported organic matter enters the stream usually as leaves but also as twigs, fruits, and other plant parts. The large pieces of organic matter are quickly colonized or attacked by bacteria, fungi, and animals that K. W. Cummins (1974) has called "shredders." Included are various caddis flies, crane flies, and stone flies. Other organisms use the smaller detritus fragments that result from the activities of these organisms and organic compounds leached out of the leaves. Many of the detritus-feeders may obtain their nutrition as much from the bacteria they ingest as from the leaves or leaf fragments. There are, of course, predators that feed on the detritus- and plant-feeders; among these are dobsonflies, some stone flies, and certain fish such as sculpins.

The disappearance of leaves in streams has been studied by methods similar to those used in terrestrial situations—packets are anchored in the stream to be gathered and weighed at intervals. Weight loss is much more rapid than in terrestrial systems. Judging from several studies (such as Horton and Brown 1991 and McArthur and Barnes 1988), the half-life for tree leaf mass in temperate streams tends to be in the range from about 0.05 to 0.20 year (about two weeks to two and one half months), to be compared with a half-life of about one year for leaves on the soil of temperate hardwood sites (Table 10-1). Shredders are important in the processing of the leaves. When the leaf packet is enclosed in fine mesh to exclude organisms such as stoneflies and isopods, the rate of loss is slowed (Barnes et al. 1986) (Figure 18-5). It is

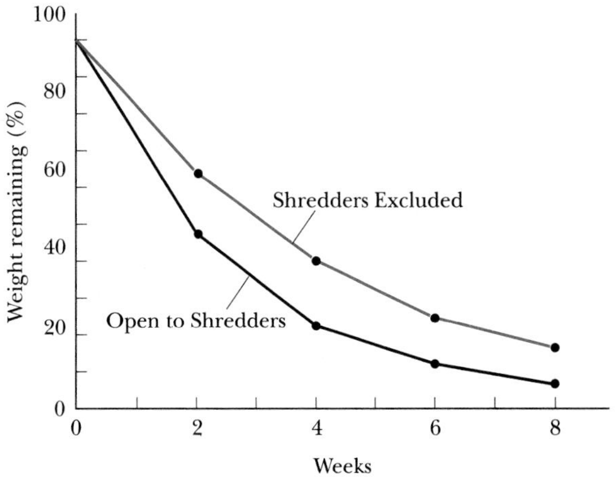

Figure 18-5

Loss of weight of box elder leaves in a small spring-fed stream in cold desert (shrub-steppe) in Washington. Insects and other shredders were excluded from one set of leaf packets by fine screen (1 mm) whereas the other set was left open. The beginning date was August 2. (Data from Barnes et al. 1986.)

interesting that fresh green leaves tend to be processed more rapidly than leaves that have fallen naturally through leaf abscission. The most likely reason is that, prior to leaf fall, the tree has removed nutrients useful for the shredder invertebrates, bacteria, and other organisms.

As a stream widens into a river, autochthonous production by filamentous algae and rooted vascular plants becomes the dominant source of energy in the system, and the terrestrial influence is diminished (Figure 18-6). This means, among other things, that large streams in similar climatic regions are relatively similar, but small streams may be very different depending on the surrounding habitat.

In still larger rivers there is a gradual return to

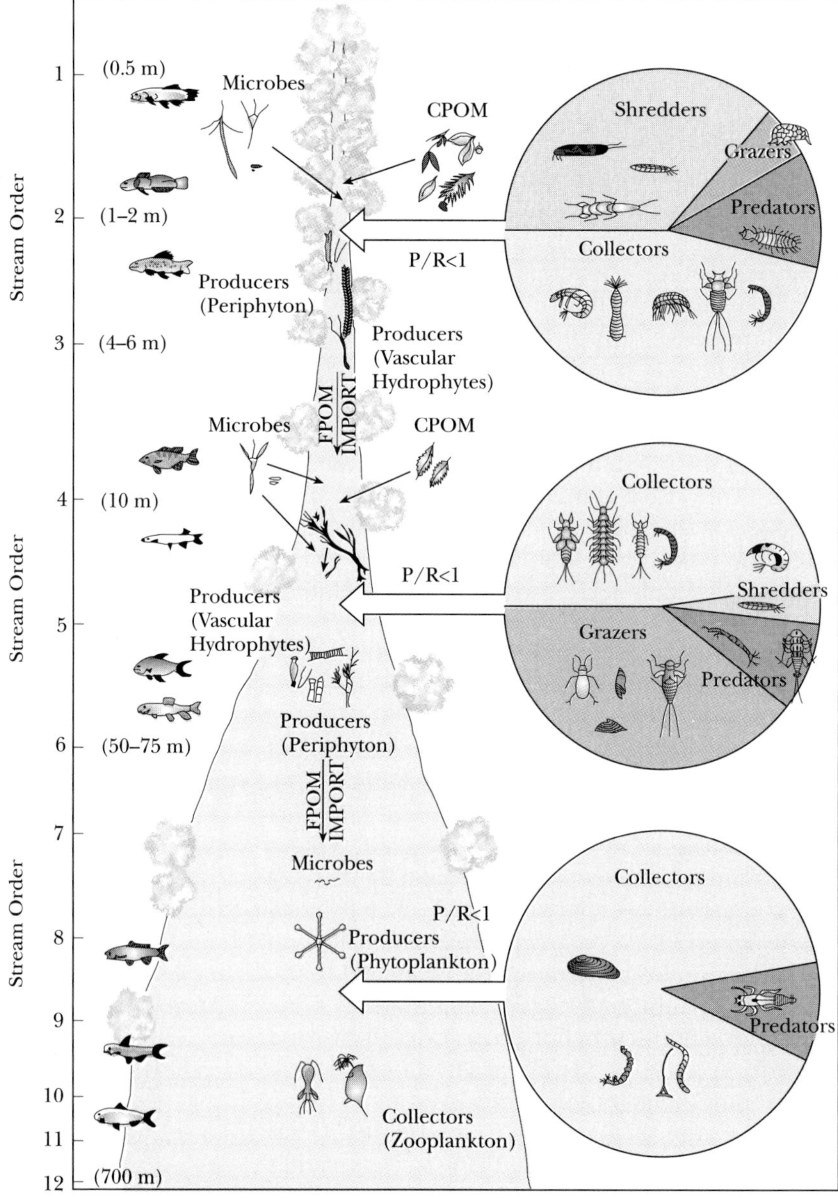

Figure 18-6

The longitudinal zonation of a stream is shown here. Small headwaters are stream order 1 and the Mississippi River at its mouth is stream order 12. There are trends with stream order in production/respiration (P/R) ratios, feeding types among animals, and also species composition. (Adapted from Cummins 1977.)

heterotrophy. Here it seems neither instream production nor import from the terrestrial vegetation along the banks is very important. Turbidity holds down production within the stream, and the amount of overhanging vegetation is small compared with the volume of water. Much of the energy base in these larger rivers is imported from upstream and from sloughs and backwaters.

Physiographic Succession

An immature landform has few streams and is poorly drained. As the land ages in geological time, the headwaters of the streams move upslope and the streams cut deeper into the land. At any one time, the stream system consists of intermittent streams in the furthest reaches of the headwaters and progressively more mature sections downstream to the oldest section of the stream system near base level where it flows into a major river, a large lake, or the ocean.

Youthful streams tend to have narrow, V-shaped valleys and few tributaries. Water flow is swift. In older sections, the valleys are wider, the gradient more gentle, the flow accordingly slower, and tributaries more numerous. Near base level, the river flows slowly and meanders through a broad floodplain.

In 1911, Victor E. Shelford noted that each stream consists of a longitudinal series of communities. Within any one region, such as the North Shore region of Lake Michigan above Chicago, the order in which the communities appeared was essentially identical, although the actual distance from mouth or source varied.

For fish, as an example, the pools closest to the source of the stream were characterized by a single species, the creek chub. The next species to appear was the red-bellied dace. Along the course of the stream, other fish were added and some dropped out. Near the mouth of the streams that had a long sluggish section near the base level of Lake Michigan were added several new species such as pike, chub sucker, and two species of shiners.

Thus, a longitudinal zonation of species, as well as of trophic characters (described in the preceding section), exists. As the stream ages, a particular habitat and its attendant community move upstream, a process referred to as physiographic succession. Physiographic succession, as the name implies, is an almost totally allogenic process.

The earliest stage, the intermittent stream, has few or no strictly aquatic animals. If the gully is deep enough to contain water for a few days or weeks, a few invertebrates such as blackfly larvae may be found during that period (Shelford 1913).

The deeper parts of the intermittent stream may have temporary pools that hold water for several weeks. Here crawfish and other crustaceans and aquatic snails may be permanent inhabitants, surviving the dry periods by burrowing.

When downcutting brings the pools to the level of the water table, the pools become permanent pools. This is the habitat where the creek chub enters. To some of the invertebrates mentioned for the temporary pool are added neuston types such as water striders as well as other invertebrates that live within the pool, such as dytiscid beetles.

The next major stage is the swift stream community, and the last is the sluggish stream community. Hynes (1970) summarizes some other classifications of longitudinal zonation in streams.

Although longitudinal zonation and its basis in physiographic succession are useful concepts, the study of any real stream system reveals additional complexities. A stream may switch from sluggish to swift and back depending on such factors as the presence of lakes and the susceptibility of the substrate to erosion. The upstream parts of Jordan Creek, running through the cornfields of central Illinois are slow flowing and silted, whereas the downstream portion runs through a forested floodplain and is fast flowing over gravel and bedrock. Here the usual distribution of fish is reversed, with such sluggish water species as largemouth bass and sand shiners occurring upstream and swift water darters being more common downstream (Larrimore et al. 1952).

Also, spring-fed streams and ones that start from snowfields in mountains have upper reaches with different features than streams that arise from runoff on relatively flat ground. Streams may be rejuvenated and set off on another cycle of downcutting if the base level toward which they are flowing is lowered. Furthermore, stream habitats are always subject to disturbance resulting from the scouring of floods or sedimentation.

May T. Watts (1975), following Henry Chandler Cowles (1911), gives a readable description of changes in the terrestrial vegetation on the rock walls and other habitats next to a stream in relation to stream age.

The Oceans

The oceans are vast, and so is oceanography, the science of the oceans. Oceans occupy just over 70% of the earth's surface and have an average depth of about 4000 m, or something more than 2 miles. The deepest areas are trenches in the western Pacific more than 11,000 m deep.

One of the important ways that the oceans differ from fresh-water habitats is in salinity, or saltiness. Ocean water contains about 3.5% salt by weight, generally expressed as 35 parts per thousand (rather than 3.5 parts per hundred) and written 35‰. The salinity of fresh water, by contrast, is mostly below 0.3‰. About 78% of sea salt is sodium chloride.

To an inlander the saltiness of the ocean might seem to make it an extreme environment, to be populated by a biota of low diversity, but in fact the opposite is the case. Life originated in the oceans, and evolution went on there more than 2 billion years before plants and animals began to evolve the traits allowing them to use such inhospitable habitats as dry land and fresh water. Only a few groups are better represented outside the ocean than in it. There are whole phyla such as the comb jellies, brachiopods, chaetognaths, and echinoderms that do not occur in fresh water at all. There are not, however, a large number of species in the sea. The total number of species is only about 160,000 (Thorson 1971), to be compared with well over a million known terrestrial and fresh-water species.

One reason for the low number of species is the virtual absence from the sea of two very successful groups, the seed plants and the insects. Other reasons have to do with the lower habitat diversity compared with land and the fact that the sea is continuous rather than being isolated like the various land masses. The opportunity for geographical isolation and, accordingly, the opportunity for speciation, is thereby reduced.

Ocean water temperatures vary from about −2°C, its approximate freezing point, to about 30°C, or higher in the hottest regions such as the Persian Gulf. Salt water becomes denser the colder it gets, in contrast to fresh water, which is densest at 4°C. Sea ice floats, of course, but that is because ice forms through the freezing of pure water, leaving behind a saltier brine.

Water density is also affected by salinity, water that is less saline floating on top of water that is more saline. The density of typical fresh water is about 97% that of typical seawater at the same temperature.

Ocean Life Zones

Very broadly, the ocean consists of three zones (Figure 18-7): a **littoral zone** that is approximately the area between the lines of high and low tide, a **neritic zone** that is the portion of the ocean over the continental shelf with depths down to about 200 m and, by far the largest, the open ocean, the **oceanic zone.**

These terms apply to the water and its organisms. The bottom with its benthos is subdivided into the **littoral zone,** the **sublittoral zone** (occupying the continental shelf), the **bathyal zone** of the continental slope, and the **abyssal zone** of the abyssal plain. This zone between 2000 and 6000 m occupies large portions of the ocean and, in fact, about one half of the globe. There are also hills and ridges on the ocean bottom and trenches that may be as deep as 11,500 m. The portion of the bottom in these trenches has been called the **hadal zone,** from "Hades."

One other way of subdividing the ocean ought to be mentioned. There is an upper **photic** zone and a lower **aphotic** zone. Photosynthesis occurs in the upper part of the photic zone, to a depth of perhaps 150 m, depending on turbidity, season, etc. Below this level, however, down to about 500 to 550 m in very clear water, enough light penetrates for vision.

Littoral Zone

The littoral zone of the ocean is a habitat of physical extremes, changing from aquatic to terrestrial as the tide goes out, twice a day (Figure 18-8). Two

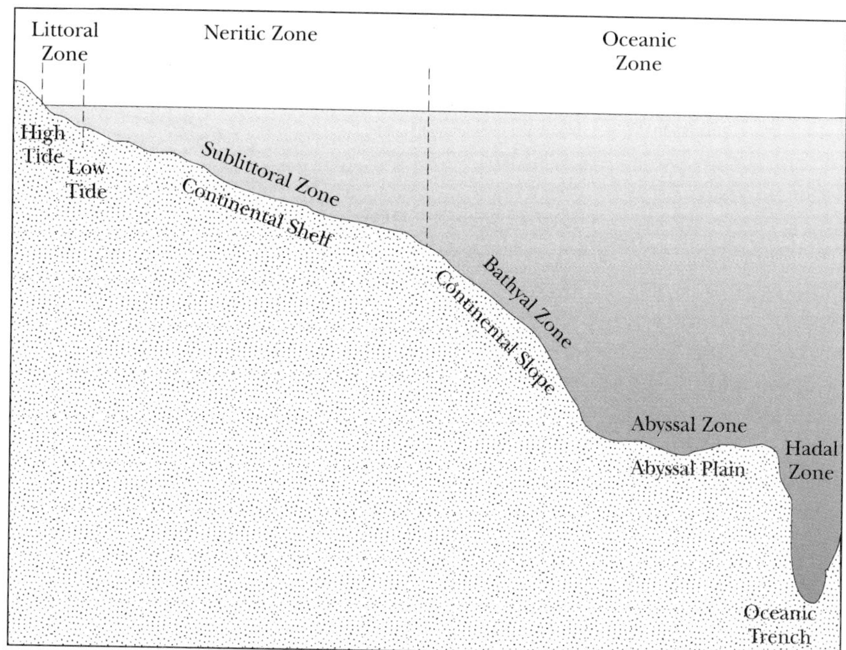

Figure 18-7

Zonation in the ocean.

Figure 18-8

A rocky shore and a sandy shore are seen here in close proximity. (Photo at Clifton Beach, Tasmania, by Joseph G. Engemann.)

high and low tides per day is what ought to happen based on the position of the moon, the single strongest influence on the earth's tides. Because of the particular characteristics of each basin, not every shore shows this regular pattern. From Louisiana to western Florida, for example, the Gulf Coast tends to have only one high and low tide per day (Britton and Morton 1989). There are also large variations in tidal range (how high and low the water gets) among localities and at different times in the 28-day lunar cycle.

Accompanying the shifts from high to low tide and back and again are changes in temperature and many other factors, both abiotic and biotic. At low tide, for example, organisms in the zone left by the receding water may be exposed to direct sunlight and to terrestrial predators. The organisms are subjected to the pounding of incoming waves and the scouring by the water as the waves recoil.

Within the littoral, it is usual to recognize three zones on both sandy and rocky shores. These are the **intertidal** zone itself, a **supratidal** zone above it that is covered only by the highest tides or influenced by spray, and the **subtidal** zone below the line of average low tide. The organisms inhabiting the three zones are almost entirely different on rocky and sandy shores.

Rocky Shores. Many of the organisms of rocky shores are sessile, permanently attached to the rock (Figure 18-9). The supratidal zone is a spray region where black-colored blue-green algae grow on the rocks and are fed on by small, heavy-shelled snails called periwinkles. Sea slaters, which are large isopods, are also characteristic, feeding on seaweed washed up by the waves.

In a widely used classification (Stephenson and Stephenson 1972), the lower part of this lichen-periwinkle zone is referred to as the **supralittoral fringe.**

Below, the **intertidal zone** has a large and varied fauna, despite the harsh physical conditions. Barnacles, snails, limpets, oysters, mussels, sea anemones, chitons, and sea urchins are some of the ani-

Figure 18-9

Zonation on a rocky shore at Cole's Bay, Tasmania. Barnacles form the light zone at the top of the intertidal zone. Below is a zone where algae are prominent. Black attached mussels in the genus *Mytilus* form the lower part of the intertidal zone. (Photo by Joseph G. Engemann.)

mals that occur attached to rocks, in crevices, or in the masses of brown and red algae. These animals are either sessile, like the barnacles, or else very good at holding on. Even the motile forms may actually move very little. Duke University ecologist A. S. Pearse (1939) mapped the locations of several sea anemones in a beach pool and found, when he came back two weeks later, that they were all in the same spots. The littoral fauna are mostly gill-breathers but are necessarily adept at avoiding desiccation. When the tide is out, they close down, generally inside some sort of shell-like structure that helps against predation as well as water loss (Figure 18-10). This barnacle-limpet zone is basically the **eulittoral zone** of Stephenson and Stephenson (1972).

Below the average low-tide line, in the **subtidal zone,** grow large brown algae called kelps. Here too are a great variety of animals, including many that move up into the intertidal zone when it is water covered. Typical animals of the subtidal zone include sea anemones, starfish, sea urchins, polychaete worms, sea cucumbers, nudibranchs, and crabs.

The kelp itself is covered with organisms. The fronds support other smaller algae, such as diatoms, and colonies of small animals, such as bryozoans and one of our chordate relatives, the golden-star tunicate (Carson 1955). Even more animals are associated with the holdfast, the rootlike structure by which the kelp is attached to the rocks. Within the branches of the holdfast live representatives of most of the benthic groups of the subtidal zone—sponges, brittle stars, sea squirts, and many more.

The upper part of the region dominated by the attached algae is referred to as the **sublittoral fringe** in the Stephenson and Stephenson (1972) classification.

The actual elevations or altitudes occupied by the various zones depend on how exposed the shore is. Headlands facing the open sea take the full force of the waves; the effect is to move the zones of organisms upward compared with coves (Figure 18-11). Although the basic zones remain, some organisms are more characteristic of exposed and some of sheltered shores. For example, mussels of the genus *Mytilus* (Figure 18-9) and barnacles in the genus *Chthamalus* are characteristic

Figure 18-10

A pool in the intertidal zone of a rocky shore near Snug, Tasmania. Chitons, snails, and the tubes of a polychaete worm can be seen. (Photo by Joseph G. Engemann.)

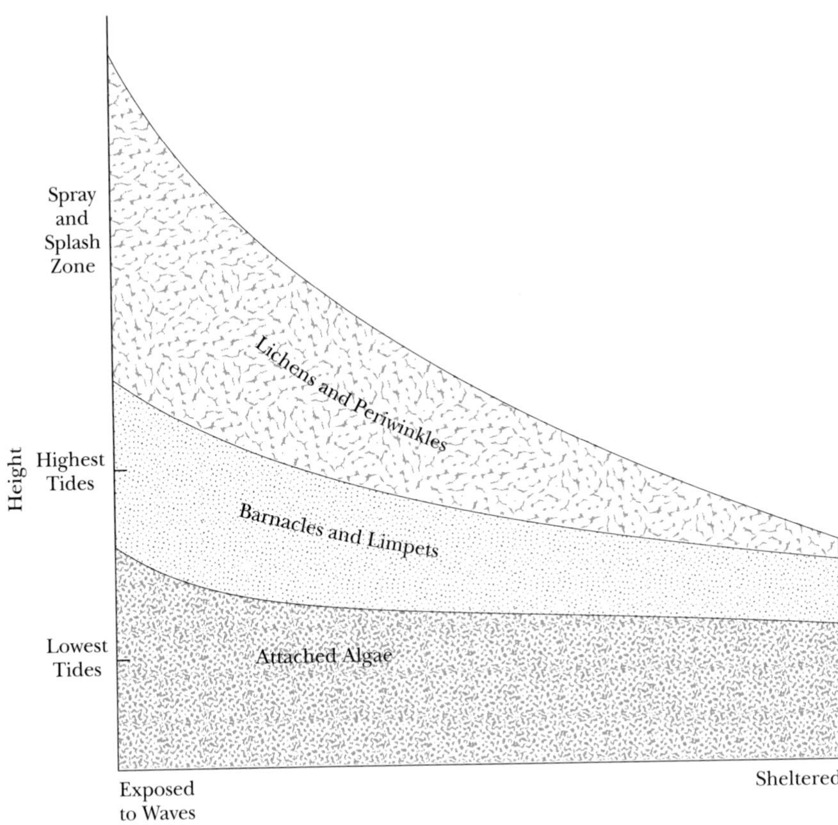

Figure 18-11

Relationship between rocky shore zonation and exposure. On exposed shores, where splash and spray from waves are important, the zones reach higher and are more spread out.

of exposed shores in north temperate regions (Fincham 1984).

The generalization is often made that vertical distribution of sessile organisms on rocky shores is determined by physical factors at the upper limit and biotic factors at the lower limit (e.g., Connell 1961). In other words, the upper limit for a given species is set by its tolerance to such factors as desiccation or temperature. The lower limit is set by competition or predation. However, habitat selection by larvae may be more important as a proximate factor in zonation than generally thought (Underwood and Denley 1984).

Sand and Mud Shores. On sandy (and muddy) shores the substrate is unstable; here burrowing rather than attachment is the predominant mode of life. Animals of the supratidal zone include ghost crabs and amphipods which shift their activities into the intertidal zone when the tide is out. In the intertidal zone are large populations of clams, polychaete worms, isopods, amphipods, and other crustaceans. Most of these animals are hidden in burrows when the tide is out but are active when submerged. The burrows made by the larger burrowing animals are also homes during low water for many smaller forms. In the subtidal zone occur fish such as flounders, and killifish, some of which move up into the intertidal zone at high tide, and sand dollars, clams, snails, and blue crabs.

Crabs are present in great variety on sand and mud shores, from the predominantly subtidal blue crab to the intertidal fiddler crabs to the supratidal but mostly nocturnal ghost crab. G. Thorson (1971) had the experience of seeing a herd of about a hundred ghost crabs abroad during the day. He at first took the group for a flock of sandpipers. "They all moved like lightning in the same direction with military precision and with responses so rapid that they could hardly be followed by the human eye." When he approached, they rapidly dug themselves in, leaving only their large eyes visible. As he approached closer, the eyes also disappeared and then the antennae. "I then tried to dig out the crabs one by one, but failed; it was also impossible to chase them as they ran faster than I could." Ghost crabs are primarily predators, digging out mole crabs and coquina clams in the surf, but they also make use of the often abundant carrion in the form of offshore biota stranded on the beach by high waves and storms.

On sandy beaches, unlike rocky beaches, there is little production of organic matter by plants. Consequently grazing and plankton feeding are not very important; most of the organisms are detritus-feeders.

Estuaries. **Estuaries** are coastal bodies of water connected with the ocean, such as tidal marshes, tidal rivers, and the bodies of water behind barrier islands. Typically, estuaries are highly productive at both the producer and consumer levels, for a variety of reasons. One is the action of the tides, bringing nutrients and carrying away waste products. Another reason for the high productivity is the wide variety of producers that can live in the estuarine habitats, from phytoplankton to seaweeds to the very important emergent marsh grasses, such as cordgrass.

Some of the organic matter produced in estuaries is exported, enriching the adjacent ocean waters. Many of the fish and shellfish of the coastal waters spend a part of their life history in estuaries. The evidence is fairly strong that fishery production is higher along coasts with extensive estuaries (Figure 18-12). Probably the nutritive value of the cordgrass and other detritus, in the marshes as well as exported, and also the function of the salt marshes as refuges from predation for the juveniles are both important factors (Boesch and Turner 1984).

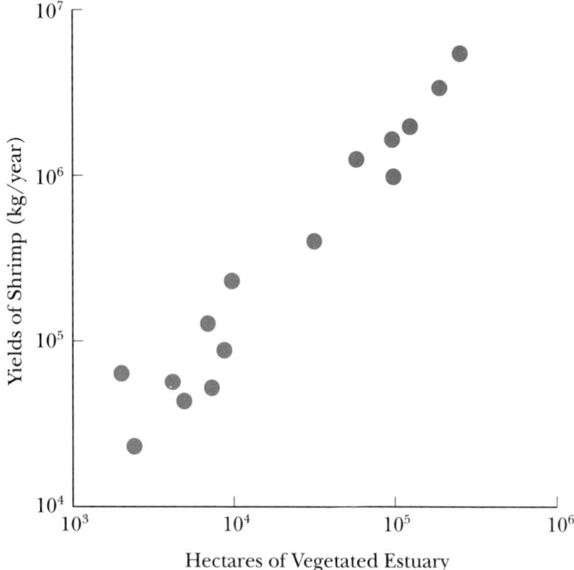

Figure 18-12

The relationship between yearly inshore shrimp yields and the area of vegetated habitat within the estuaries. Data from the northern Gulf of Mexico. (Adapted from Boesch and Turner 1984.)

The great value of estuaries and other coastal wetlands has been recognized by nonbiologists only recently. As with inland wetlands, business, government, and the average citizen regarded them as mosquito-breeding wastelands. The inclination, often supported by tax money, was to fill them, dredge them, dike them, or drain them to produce something useful like a pasture, an industrial site, a football stadium, or a marina. The degradation of San Francisco Bay was described by F. H. Nichols and his coworkers (1986) with the U.S. Geological Survey. They point out that one of the reasons that the problems of San Francisco Bay are perceived as less severe than those of many other estuaries is "that many of the major changes (such as the disappearance of wetlands, introduction of exotic species, and loss of many commercial and sports fisheries) occurred decades ago and have been forgotten." It is a sad thought, recalling Aldo Leopold's comment that one of the disadvantages of an ecological education is that one lives alone in a world of wounds.

Coastal wetlands have suffered, not just in the United States, but over much of the world. For example, the Fenlands, marshes formed in a bay of the North Sea on the east coast of England, were drained in one of the first "reclamation" projects during the 17th and 18th centuries. The natives of the Fenlands were said to have led an amphibious life, wading, rowing, and even using stilts to get around in the marshes. They lived on the abundant and diverse fish and game, raised livestock on marsh grass pasturage and hay, burned peat for cooking and warmth, and raised geese as a cash crop. Besides selling the geese, which they drove by road to London for Michaelmas, they also sold down, feathers, and quills for pens.

As the fens were drained under the orders of the king, the area was divided up among the companies and individuals connected with the projects and turned into farmland. The fenmen, early environmental activists, fought the projects over a period of centuries. One incident in 1642 involved a group of dispossessed fenmen near the River Witham. They "took arms, and in a riotous manner fell upon the Adventurers, broke the sluices, demolished their houses, and forcibly retained possession of the land" (quoted in Darby 1956).

But the Fenlands were drained, and the marshes with their rich fauna disappeared. The resulting farmland has been productive, although what the future holds, as the peat is used up and blows away and the surface drops lower, is uncertain.

The destruction of coastal wetlands in the United States has been slowed considerably in the past 20 years but scarcely stopped. In Louisiana over 700 ha of coastal wetlands were removed in 1981, mostly for oil and gas exploration, extraction, and transportation (Boesch and Turner 1984). Even where filling and draining have stopped, many less direct effects continue, including canal construction and chemical pollution. Attempts have begun, in small ways, to restore some estuarine sites that were disturbed but not wholly obliterated; an example is the attempt to restore a diked pasture on the Oregon coast to salt marsh (Frenkel and Marlan 1991).

Coral Reefs. Eugene Odum (1971) referred to coral reefs as oases in a desert ocean, but that is an understatement. Besides being beautiful (see, for example, *The Coral Seas* by H. W. Fricke, 1973), coral reefs are marvels of productivity, diversity, and complex interdependencies.

Although individual coral species occur more widely, coral reefs develop only in warm water. Good development occurs only in waters above 20°C; 23 to 25°C is optimal. This means that reefs occur mostly between 30° North and 30° South latitude and that they are absent along the western coasts of continents where offshore winds cause the upwelling of cold water (Figure 18-13). In all, coral

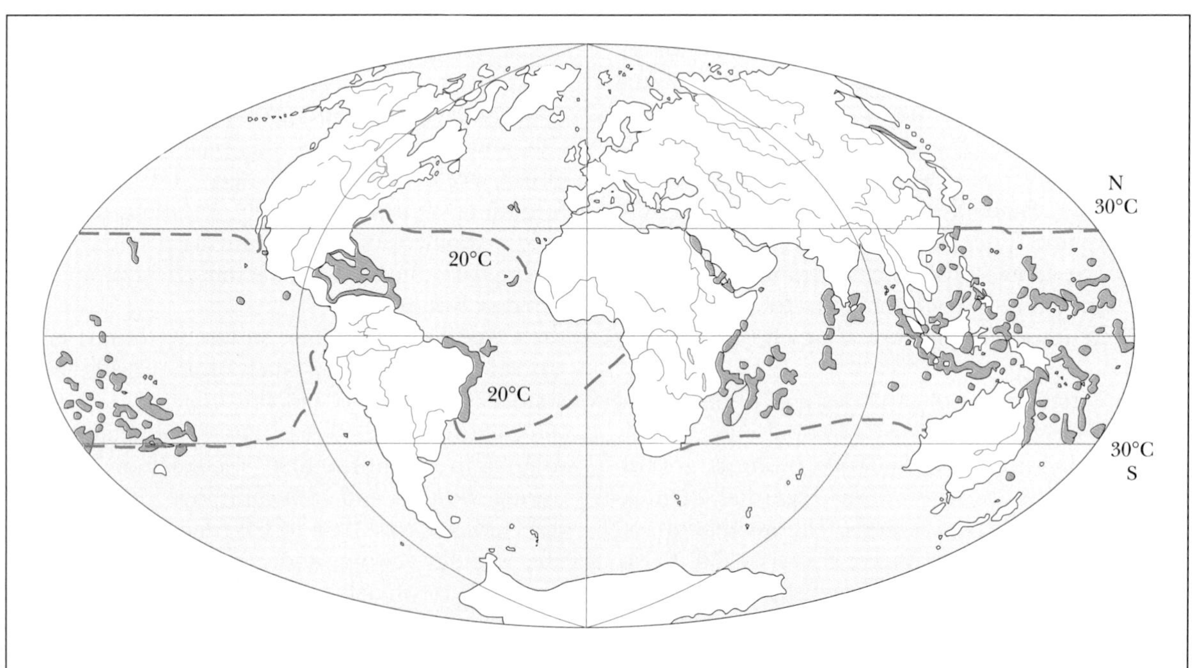

Figure 18-13

The world distribution coral reefs (blue shading) and the 20°C isotherm for the coldest month of the year (broken line). (Adapted from Clarke 1954.)

reefs cover about 600,000 km^2 or 0.17% of the ocean surface (Crossland et al. 1991).

The living reef consists of corals—small colonial animals, coelenterates of the order Anthozoa, which secrete a skeleton made up of calcium carbonate. Many other organisms also contribute to the reef, and many others live on it. The living portion of the reef, however, forms only a thin skin over the reef surface. The bulk of the reef consists of dead skeletons formed in past years and centuries.

Living coral occurs generally in waters from slightly below low tide down to about 50 m. If sea level rises, upward growth can occur. If sea level recedes, leaving the reef exposed as an island, the upper layers of the coral die, and the skeletons begin to disintegrate. Small islands are also formed by storms throwing up coral debris on top of reefs. Mangroves or land organisms, such as coconut palms, can invade these islands.

The relative elevations of land and sea are responsible for the development of the three reef types—fringing, or shore, reefs; barrier reefs, in

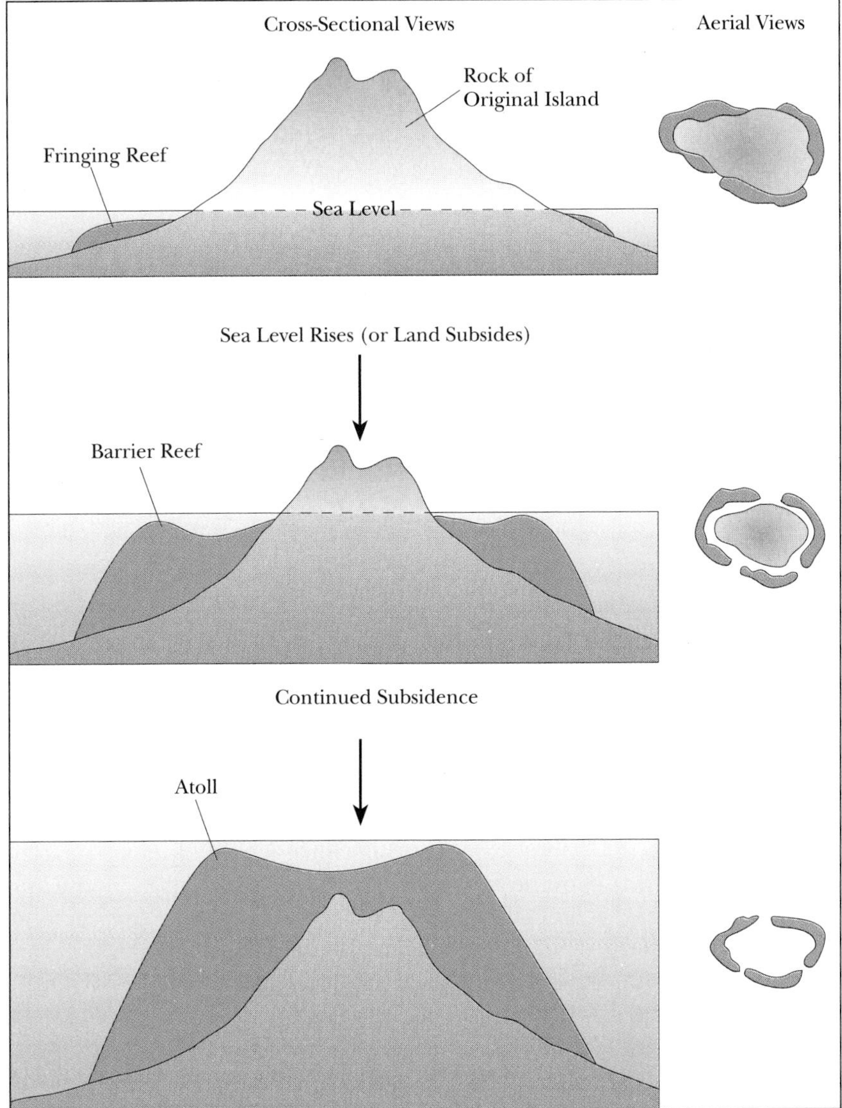

Figure 18-14

The development of barrier reefs and atolls from fringing reefs.

which a channel lies between the reef and the shore; and atolls, in which the reef forms a more or less complete ring with no central island visible.

Charles Darwin correctly outlined the basic pattern of development of reefs in 1842. Coral grows around the base of an island or along a continent at the appropriate depth. The land subsides or sea level rises, and the coral grows upward. The reef-forming organisms do better in the surf, and also silt and fresh water from the land are inhibitory to the coral next to the shore. As a result, a barrier reef develops (Figure 18-14, p. 581). With further subsidence, the top of the original land mass disappears below the sea while the coral continues upward growth. We are left with a ring, the atoll.

We speak of coral reefs, but several other kinds of organisms participate in reef formation. Other invertebrates and green algae may deposit calcium carbonate. Bryozoans and, especially, red algae help to cement the skeletons together, forming the actual reef. There are also processes eating away at the older parts of the reef. The surf pounds it and, in addition, animals such as clams, polychaetes, and sponges bore into it and sea cucumbers process the broken-up coral fragments. Parrotfish *(Scarus)* (Figure 18-15) gnaw away living portions of the coral, ingesting the living portions and letting the calcium carbonate fall as a fine sediment (Fricke 1973).

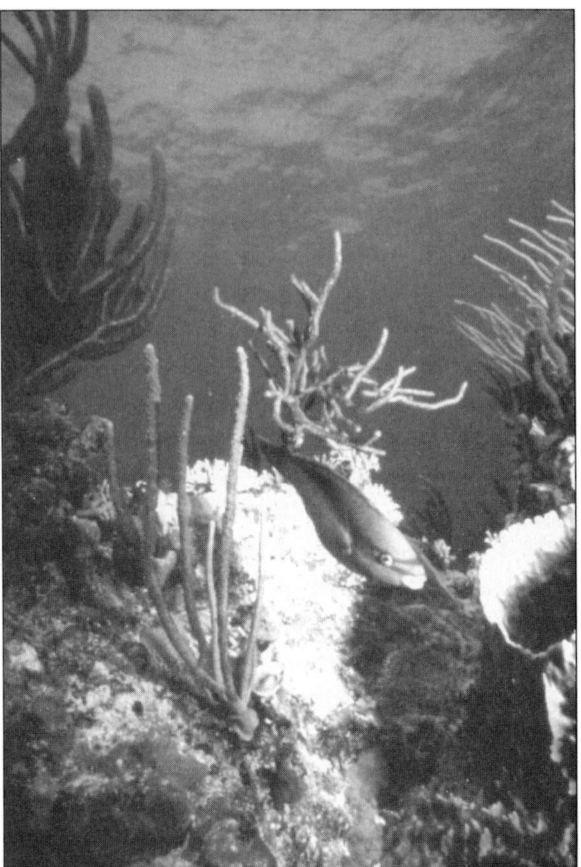

Figure 18-15

Parrotfish (family Scaridae) have beaklike jaws with hundreds of teeth. They bite off chunks of coral, digest the polyps and the algae. The ground-up coral is discharged from the cloaca as a cloud of white sand. (Susan Blanchet/Dembinsky Photo Associates.)

A variety of organisms, beyond those directly involved in reef formation, are part of the reef community. The giant clam, a meter across, is one of the best-known animals. Other invertebrates include sea urchins, sea fans, sponges, brittle stars, and shrimp that prey on starfish.

Fishes are incredibly numerous and brightly colored. Many are territorial. A tall, thin body form is common, apparently allowing maneuverability through the cracks and branches of the coral. Long, pursed lips evidently act like forceps to pick food out of crevices (Figure 18-16). Large predators include moray eels which live in recesses in the reef, and sharks and barracudas that move back and forth between the reef and the more open water.

There are many striking and complex coactions in the reef community. This is the setting of the cleaning symbiosis in which certain small fish swim into the mouth and even between the gills of large fish, removing ectoparasites, mostly copepods (Eibl-Eibesfeldt 1955, Limbaugh 1961). The cleanees, or customers, which include such predators as groupers and barracudas, adopt characteristic postures that lead the cleaners, such as the cleaner wrasse (the "w" and "e" are silent), to approach. Cleaner species tend to have blue longitudinal stripes and perform a kind of dance in front of the customer. The customer opens its mouth and gill covers and the cleaner swims in. "Involuntarily the observer fears for the life of the cleaner, as it swims unperturbed into the gaping jaws—as it were directly into the lion's mouth," writes H. W. Fricke (1973).

When the customer snaps its mouth or gill covers partially shut and reopens them and shakes its

Figure 18-16

Many coral reef fish, like this banded butterfly-fish of the Caribbean, have tall, thin bodies that allow maneuvering between the coral branches and long lips that can pluck food items out of crevices. (Susan E. Blanchet/Dembinsky Photo Associates.)

body laterally, it is an indication that the cleaning session is over, and the cleaner exits. The cleanees are said to give these signals even if they are startled and depart suddenly (Eibl-Eibesfeldt 1955).

The cleaners do not travel with the cleanees. Rather they live hidden in the reef, and the larger fish stop by when cleaning is in order. They may even queue up. Eibl-Eibesfeldt (in Fricke 1973) described seeing three sea bass, one grouper, and several impatient butterfly fish waiting their turn at a wrasse's "barber shop."

Cleaner mimics also exist. As Eibl-Eibesfeldt watched, another small blue-striped fish swam up with the characteristic cleaner's approach movements. One of the sea bass "welcomed him with open mouth." But the newcomer did not clean. It was a saw tooth blenny, a cleaner mimic, and it took a bite of the bass and made its escape.

How much of the cleaning symbiosis is innate and how much the result of learning is still uncertain, but certain elements seem not to require previous experience. H. Hediger (in Trivers 1971) raised a grouper from infancy for six years, by which time it was over a meter long. He dropped a wrasse into the tank, and the grouper immediately opened its mouth and spread its gill covers—the latter behavior Hediger had never seen before. The wrasse went through the cleaning operation, although the grouper had no parasites. Hediger added two more wrasses, which also cleaned. After a few days, the grouper grew restless and began shaking itself and hiding in the corner but did not attempt to eat the wrasses.

The cleaning symbiosis is mutualistic. The customers have potentially harmful parasites removed, and the cleaners get a dependable supply of food and security against predation. Another mutualism is, however, much more basic to the functioning of the reef community. This is the symbiosis between the corals and a group of dinoflagellates, single-celled algae, that live symbiotically within the bodies of the corals. They are often so abundant that the coral appears yellow-brown. The roundish non-flagellated dinoflagellate cells are called zooxanthellae (rhymes with Jojo Mantelli).

Although there are probably additional details to be added, the basics of the mutualism seem to be that the algae get protection, carbon dioxide, and a better supply of nutrients than they could garner by themselves. The corals get oxygen and some photosynthate from the zooxanthellae. The corals also produce new skeletons at a faster rate when zooxanthellae are present. This is apparently because the removal of carbon dioxide in zooxanthellae photosynthesis favors the production of the relatively insoluble calcium carbonate from calcium bicarbonate.

These corals can live without zooxanthellae by catching and eating zooplankton. Or they can live without zooplankton on the products of the zooxanthellae. In nature they continue to catch and eat zooplankton, and the suggestion has been made that getting scarcer nutrients such as phosphorus and nitrogen is as important as caloric gain. The nutrients can be recycled back and forth between coral and zooxanthellae within the body of the animal. In this way, a high concentration of nutrients is available for primary production, despite the generally low nutrient level of the sea.

This is one factor responsible for the high productivity of the reef. Others are favorable temperature, good water supply, and year-round high-intensity sunlight. Because of the variety of producers and the intimate connections between producer and consumer (for example, the giant clams also have zooxanthellae), precise measurements of reef productivity are difficult. Obviously reefs are very productive, perhaps having a yearly net primary production (NPP) of 2500 g/m^2, or more. The producer biomass is probably less than a kilogram per square meter (Odum and Odum 1955), to be compared with perhaps 45 kg for tropical forest (Table 11-3). Although primary production is high, respiration within the coral reef community

Figure 18-17

The well-named crown of thorns starfish specializes in eating coral polyps. (Shark Song/M. Kazmers/Dembinsky Photo Associates.)

is also high; consequently, net community production is low (Crossland et al. 1991). In fact, it is not clear that there is any consistent net increase or export of biomass; the reefs seem very close to a trophic steady state.

One of the interesting ecological events of the late 1960s was the population explosion of the crown of thorns starfish (Figure 18-17) in the Pacific Ocean. The starfish, which specializes in eating coral polyps, reached densities of thousands per hectare on the Great Barrier Reef of Australia (Done 1986). When the coral polyps are killed—and 90% devastation was not uncommon—the skeletons become overgrown with algae and some of the characteristic reef fishes and other organisms decline or disappear.

In the 1970s the crown of thorns declined in most areas and rapid re-establishment of coral followed; then in the early 1980s another outbreak occurred. It is not clear whether the outbreaks and consequent reef destruction were triggered by some sort of human disturbance, such as pollution. Neither is it clear which changes in the reefs are reversible and which ones are permanent.

Mangrove Swamps. Salt marshes dominated by grasses and sedges tend to dominate mud shores in temperate regions. In the tropics, mangrove forest and scrub covers 60 to 70% of the shoreline. Mangrove grows best on protected shores, behind coral reefs or barrier islands or within bays. Although mangroves grow on coral reefs and sand, the best development is on muddy shores in the intertidal zone. Only the crowns of the trees project above water at high tide and, in fact, lower leaves may be immersed. At low tide, the roots, which project above the soil, are exposed.

These are generalities about the distribution of mangrove vegetation. What, though, are mangroves? They are broad-leaved evergreen trees and shrubs that are obligate halophytes (that is, they grow poorly or not at all in nonsalty soil). Most have "breathing roots," and are viviparous. Some species have stilt roots that, as the name suggests, support the stems well above the soil (Figure 18-18). Others have pneumatophores that project up above the soil from undergound roots. A gas-exchange function has been demonstrated for the breathing roots, their lenticels allowing oxygen and carbon

Figure 18-18

A red mangrove swamp in south Florida. Note the stilt roots. (Photo by Joseph G. Engemann.)

dioxide exchange with the air. Vivipary consists of the seed germinating within the fruit, often while still attached to the plant. In mangroves, the fruit or seedling then falls to the mud below and may grow there or be carried considerable distances by the sea currents.

There are about 90 species of mangroves, and what is remarkable is that they belong to about ten not-very-closely-related families (Chapman 1970). Mangroves are, as P. W. Richards (1957), the great student of tropical forest, remarked, perhaps the most remarkable of all examples of ecological equivalents—of the evolutionary convergence of unrelated species living under similar environmental conditions.

The advantage of having aboveground breathing roots for a plant living in waterlogged soil is evident. The advantage of vivipary is possibly less so. As A. C. Joshi (1933–1934) pointed out, it is probably related to the difficulty of seed germination in water of high osmotic concentration. Once established, mangrove plants are able to limit salt uptake through their roots (Scholander 1968). Also, several species of mangroves (for example, the black mangrove) excrete salt via salt glands on the leaves.

Typically a zonation of mangrove species occurs, based on some combination of factors including frequency of flooding, soil type, and salinity (that is, how much fresh water is mixed with the seawater) and also the competitive interactions of the various species. The approximate zonation on south Florida shores (Davis 1940) runs from red mangrove on the outer fringes where the soil is almost continually submerged, through black mangrove, a more open community than the dense and tangled red mangrove zone (Figure 18-19). Salt marsh plants fill in the openings. The white mangrove, where it occurs as a separate zone, is beyond the reach of the tide much of the time, and button mangrove is still more landward. In some parts of the world, particularly the western Pacific, many more species occur, and the zonation is more complicated.

On many shores, mangroves aid in soil building, the encroachment of the land on the sea, by trapping and holding sediments either washed from the land or brought in by waves and currents. In this situation, a succession occurs, corresponding more or less with the zonation. For example, in south Florida, red mangrove invades continuously submerged areas by seedling or roots, raises the soil level, and allows the eventual entrance of black mangrove.

In other places, the zonation seems stable with little evidence of succession. Shores suffer natural disturbances, of course, such as hurricanes, and the natural functioning of the ecosystem includes recovery from such events.

As well as trapping soil and all sorts of other debris, the stilt roots of red mangrove form surfaces and crevices for oysters and many other marine animals. The mud surface is home for such detritus-feeders as fiddler crabs and is visited at low tide by more terrestrial forms, such as raccoons, that come down as predators and scavengers. Many birds nest in mangrove forests including herons, ibises, cormorants, and anhingas. The mangrove cuckoo is a characteristic species of the mangroves around Florida. So also is the black-whiskered vireo, which is very much like the red-eyed vireo of the temperate deciduous forests but has a larger bill.

The energy base for most of the biota consists of the mangroves, including mangrove seedlings; however, periphyton also makes a contribution (Lugo 1990). Mangrove swamps are productive ecosystems; yearly NPP is probably 30 or more metric tons per hectare (over 3000 g/m^2). Although herbivorous insects consume leaf biomass, most consumption is delayed until the leaves are shed.

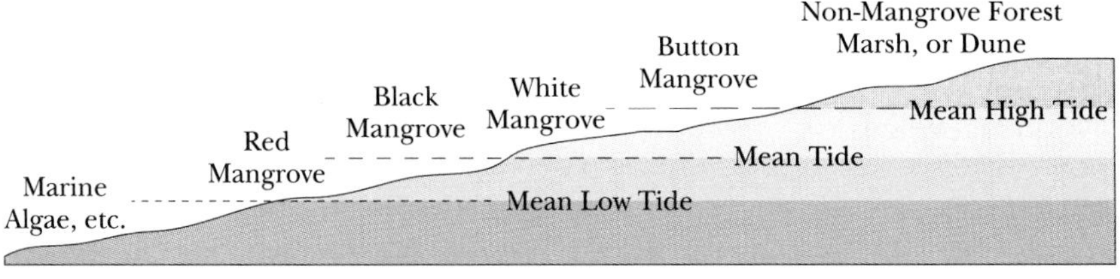

Figure 18-19

Simplified diagram of zonation in mangrove swamps in south Florida.

Yearly leaf fall is high, in the range from 5 to 10 metric tons per hectare (Lugo 1990, Robertson 1991). As a comparison something under 4 metric tons per hectare would be a reasonable figure for leaf fall in temperate oak forest.

Anywhere from one fourth to three fourths of the mangrove leaf biomass is exported to other areas of the estuary or to the ocean. The substantial amount of leaf litter retained in the mangrove swamps is utilized by various detritivores. In well-studied Australian mangrove ecosystems, the largest percentage was processed by crabs in the genus *Sesarma,* which are also serious seed predators of the mangroves.

New wood makes up 20–50% of the yearly NPP in mangrove swamps (Robertson 1991). In the Australian mangrove systems, the most important consumers of dead wood were the burrowing mollusks called shipworms. The half-life of a unit of dead wood is about two years, most of the loss being the work of the shipworms.

Mangrove swamps have been reduced over much of the earth from a variety of human disturbances. Local use of the wood for construction and cooking has been important, but also large areas have been lost to urban development, drainage for agriculture especially sugarcane, and repeated defoliation in warfare (in Vietnam).

Bathyal, Abyssal, and Hadal Zones

The bathyal zone is the slope between the continental shelf and the seabed. The slope is furrowed with submarine canyons and has a fairly rich and abundant fauna. The glass sponges, with skeletons of silicon, come in at these depths. Crinoids, or sea lilies—sessile, stalked relatives of the starfish—are characteristic, as are shrimp, clams, brittle stars, and the odd, big-eyed rabbitfish (Chimaera).

Life thins out at greater depths, in the abyssal zone. Both numbers and kinds are reduced, and many of the species have large or even cosmopolitan geographical ranges. The physical environment is, of course, very stable and uniform over most of the globe, with temperatures of 2 to 3°C and constant darkness. The bottom is either ooze or red clay mud (Chapter 12), and the animals are divided between those that eat their way through the substrate for its organic content and their predators.

Sea cucumbers, watery transparent echinoderms, are the most common detritus-eaters, but there are also other echinoderms, mollusks, crustaceans, and polychaetes. Bacteria also process the detritus. The predators include brittle stars and fishes. Here is where such bizarre fish as the gulpers, swallowers, angler fish, and tripod fish occur.

Except for the hydrothermal vent ecosystems discussed in Chapter 11, the organisms of the sea bottoms are dependent on imported energy, detritus, that falls in a slow rain from the photosynthetic zone. Not much is left by the time it gets there, and the biomass supported on the bottom is very low, below 5 g live weight per square meter (Thorson 1971). Considering that the organisms at this depth are mostly water, this translates to 0.1 g or less dry weight.

The hadal zone seems to be like the abyssal zone but more so.

It is worth noting that currents occur throughout the ocean and, at least partly as a result, oxygen depletion such as at the bottom of lakes is rare in the sea. The fauna is not one adapted to low oxygen levels. Some of the more prominent adaptations are to the soft nature of the bottom, such as long arms, fins, or spines for support, and to the constant darkness. Many of the fish and squids of the depths have luminescent organs where light is produced in the same way as in a lightning bug's abdomen. The light is used in various ways. Some organisms seem to have searchlights, others have luminous baits to lure prey, and in some cases the luminous structures may serve for sex or species recognition. There are few obvious special adaptations to pressure, despite the fact that the pressure at an average depth of the ocean bottom is more than 350 atmospheres. One feature shared by the organisms of the depths is the absence of gas-filled cavities such as swim bladders in fish. Beyond this, it may be sufficient that pressures within and outside the organisms are the same.

Neritic and Oceanic Zones

The important producers are phytoplankton, except in the shallowest parts of the neritic zone. Diatoms, especially in cooler waters, and dinoflagellates are numerous. It now seems clear, though, that a considerable fraction of the production is by extremely small organisms, called nannoplankton or picoplankton, too small to be taken in plankton nets and, in fact, small enough to pass through 1-μm filters (Li et al. 1983). Cyanobacteria (blue-green algae) are probably one of the groups involved.

Phytoplankton is fed upon by zooplankton, and smaller zooplankton may be eaten by larger. Bacteria ("bacterioplankton") are also important, utilizing dead cells, organic excretions of the other plankton, and other dissolved organic matter (DOM). It is uncertain whether these minute organisms are eaten by microflagellates that are then eaten by protozoa, and so on up, joining the food chains of the net plankton. There is some evidence (Ducklow et al. 1986) that the bacterioplankton and/or the microflagellates that graze on them (and on the autotrophic picoplankton) may not link up in this way but instead contribute to energy transported downward. Whether or not this is true, these organisms contribute to the liberation of inorganic nutrients, which become available for more photosynthesis.

Marine zooplankton is diverse, especially in neritic regions where it is enriched by the motile larvae of sedentary animals of the bottom and sessile animals of the littoral zone. Examples of the animals that live only temporarily in the plankton, as larvae, are barnacles, sea urchins, polychaete worms, and crabs. The permanent plankton is diverse in itself and includes more larger forms than are ordinarily important in fresh water. Some of the important groups are copepods, the good-sized crustaceans called "krill," jellyfish and combjellies, and protozoans such as foraminiferans and radiolarians.

One of the problems facing plankton is sinking. This is particularly a problem for the phytoplankton, which need to stay in the light, but is probably also a problem for some of the heterotrophic organisms that must avoid sinking below the zone where sufficient DOM is available. One way in which organisms of this zone avoid sinking is by being very small. A spherical organism 0.1 mm in diameter sinks much more than 100 times faster than one that is 0.001 mm in diameter. A sand grain 1 mm across sinks over 8000 m in a day (Clarke 1954) or, in other words, it would fall to the bottom of all but the deepest parts of the ocean in one day. An organism 0.001 mm (that is, 1 μm) sinks at the rate of about 0.1 m per day. Almost any sort of locomotor ability or even very modest currents could suffice to keep such an organism from sinking. There may be other selective advantages to being small (Colinvaux 1986). And, as we pointed out earlier, there are other ways of avoiding sinking, including increasing the surface area with long extensions. This is probably at least part of the explanation for the spines of some foraminiferans and the bizarre antennae and other appendages of certain marine copepods and other invertebrates.

Nekton, mostly feeding on the zooplankton, consists of such fish as herring, menhaden, and sardines as well as squid and baleen (or whalebone) whales. Higher-order carnivores among the nekton include toothed whales, sharks, and birds. Pelagic birds are mostly restricted to albatrosses, shearwaters, petrels, and penguins; in neritic regions are added gulls, terns, auks, pelicans, gannets, and many others.

The depth to which photosynthesis exceeds plant respiration varies, but it is probably rarely more than about 30 m in the more cloudy waters of the neritic zone nor more than about 200 m in the clearest waters of the pelagic zone. Organisms at greater depths are dependent on import of organic matter originally produced above. In shallow waters much of this import may occur by the sinking of dead bodies of plankton and other organisms. In deeper waters the organic remains of organisms from near the surface probably decay before they reach the bottom. The way in which the organic matter that supports the benthos and the deep-water nekton reaches the bottom is not wholly understood. The aggregation of DOM into particles, activities at all levels of plankton able to use dead organic material, and transport along the bottom from shallower regions may all be important.

Productivity

The popular literature that used to talk about the "infinite resources of the sea" has now disappeared. It has become clear that fish and even algae as food are no more infinite than the blue whale that has been hunted virtually to extinction. Even so, it still surprises most people when they discover just how unproductive most of the ocean is. A reasonable figure for NPP in the open ocean is probably 125 to 150 gm/m^2 per year, similar to semidesert communities on land.

The factor limiting primary production in the open ocean appears to be nutrient supply, generally phosphorus. The problem is the continual loss of nutrients from the photosynthetic zone. This occurs mainly through organisms dying and sinking, eventually to the bottom. Furthermore, the bottom is far from the top, and there is nothing that corresponds exactly to the overturns of fresh-water lakes which replenish the nutrient supply at the top. If all nutrient regeneration were postponed until dead organisms reached the bottom of the ocean, the nutrient supply would probably be much poorer than it actually is. In fact, considerable decay goes on as these dead bodies slowly sink. Furthermore, there is direct use by phytoplankton of zooplankton excretions, such as ammonia, and the phytoplankton itself is leaky, allowing regeneration of nutrients by bacterioplankton.

Secondary production at the levels that interest humans as food is low because of long food chains and also probably because of complications such as bacterial heterotrophy. A food chain in the open ocean might have five steps (small phytoplankton → small zooplankton → larger zooplankton → still larger zooplankton → planktivorous fish) before reaching the piscivorous fish, the top carnivores such as tuna and salmon that humans tend to prefer as food.

In a few situations in the ocean nutrient levels are kept high, and these areas are highly productive. Of particular importance to humans are areas of upwelling, coastal areas where nutrient-rich waters from lower levels circulate upward by the action of currents carrying surface waters away from shore. Probably the best known of such areas is that of the Peru current. Here, in an area of the Pacific Ocean about the same size as Lake Michigan, humans harvest some 10.5 million metric tons of anchovies per year. The net annual production by the phytoplankton, especially diatoms, at the base of the food chains here is clearly very high, possibly on the order of 1000 or 1250 gm/m^2 per year. Estuaries and coral reefs may have even higher levels of productivity.

Wetlands

Not long ago wetlands were considered wastelands. The best thing to do was drain them, so the land could be farmed, or fill them in and build houses. Of course, the foundations cracked and the septic tanks ran fecal coliforms into the groundwater. Today, as a result of the increased awareness of the environment that came about in the 1960s and 1970s, the value of wetlands is more widely recognized. More so than most natural ecosystems, they are protected by law for the benefits they provide society.

Because it is now a legal matter, we should be specific about what we mean by a wetland. The U.S. Fish and Wildlife Service says, "Wetlands are lands where saturation with water is the dominant factor determining the nature of soil development and the types of plant and animal communities living in the soil and on its surface. Wetlands are lands transitional between terrestrial and aquatic systems where the water table is usually at or near the surface or the land is covered by shallow water" (Cowardin et al. 1979).

Five major types of fresh-water wetlands—marsh, bog, fen, shrub-carr, and swamp—are discussed in the following sections; marine wetlands were discussed earlier. The categories are modified slightly from a wetlands classification for Ontario (Jeglum et al. 1974).

Marsh

Marsh is a wetland dominated by graminoids, that is, by grasslike plants that may, in fact, be grasses but may also be cattails, sedges, rushes, or other plants with more or less grasslike leaves. Marsh may occur as a zone around a lake or pond or alongside a river, but may also occur away from any water bodies in areas where the water table is high.

Standing water is present year-round or during the wetter seasons. Depth may be up to a couple of meters, although it is usually less. At greater depths one tends to find only floating-leaved or submerged vegetation. The soil is usually mineral, but organic content may be high. pH is usually near neutral to slightly alkaline.

Marshes where the soil surface is dry by midsummer tend to be occupied by a type of marsh called wet meadow. Sedges in the genus *Carex* are the usual dominants, so "sedge meadow" is a virtual synonym.

Marsh vegetation tends to be patchy, the patchiness based in part on water depth, in part on successional processes, and probably, in part, on chance events of colonization. A cattail marsh, for example, may have deeper holes where water lilies such as spatterdock grow, sections where such species as sweet flag or Phragmites dominate, shrub-dominated areas (shrub-carr), and sedge meadow on higher areas.

Successional processes tend to cause marshes to be replaced by shrub-carr and eventually by swamp forest. High water levels, caused by natural fluctuations in rainfall or by beaver flooding, and low water levels, leading to fires, tend to kill woody plants and to maintain (or re-establish) marsh.

Marsh animals include many of the pond invertebrates when standing water is present. Small ponds and marshes are important breeding sites for frogs and toads. Bird populations are high, often 300 to 400 pairs per 40 ha. Characteristic species include rails (Figure 18-20), bitterns, ducks, marsh wren, red-winged blackbird, and swamp sparrow (Aldrich 1943). Sandhill cranes usually nest in marshes, although they may feed in drier areas. Among mammals, the herbivorous muskrat and the mink, which preys on it, are characteristic (Errington 1957).

Figure 18-20

A Virginia rail at its nest in a cattail-bulrush marsh. (Photographed at Munuscong Bay, Chippewa County, Michigan by Lawrence H. Walkinshaw.)

Bog

A bog is a landform characterized by the accumulation of peat. Peat-filled depressions are common in northern latitudes, especially glaciated regions; but shallow bogs, peat-covered surfaces, occur throughout the world in areas of high humidity. The discussion here centers on northern bogs.

Bog Succession

Typical bog succession begins around the margin of a lake or pond (Figure 18-21). Although submerged, floating-leaved, and emergent vegetation may be present, there is, in addition, a plant capable of growing out over the water as a floating mat. Usually this is hairy-fruited sedge or water willow (Figure 18-22). Onto the mat invade sphagnum moss and other plants (Table 18-1). The accumulated biomass of these plants, as peat, begins to fill in the basin, allowing the floating mat to be extended further toward the center (Figure 18-23).

Eventually, after some hundreds or thousands of years, the encroaching mat may meet itself in the middle, enclosing the pond beneath. In the meantime, bog shrubs and trees have spread onto the mat. Continued biomass accumulation and settling may eventually fill the basin with peat, and trees, such as tamarack or black spruce, may extend from shore to shore.

Habitat Conditions in Bogs

Once sphagnum has invaded the bog, it sets, by its reactions, most of the conditions under which other organisms must live (Curtis 1959). The waters of bog lakes are often near neutral; however, the plants growing on the bog mat live in an acidic medium as the result of activities of sphagnum (Table 18-2). At least two activities are important: Sphagnum adsorbs cations, such as calcium, and releases hydrogen (Table 18-2). It also releases hydrogen in the dissociation of organic acids, especially polygalacturonic acid (PGA) (Kilham 1982). The pH within the bog is generally below 5.0 and often below 4.0 (Boelter and Verry 1976).

1. Cross section of ice block stranded and buried in gravel 11,000 years ago.

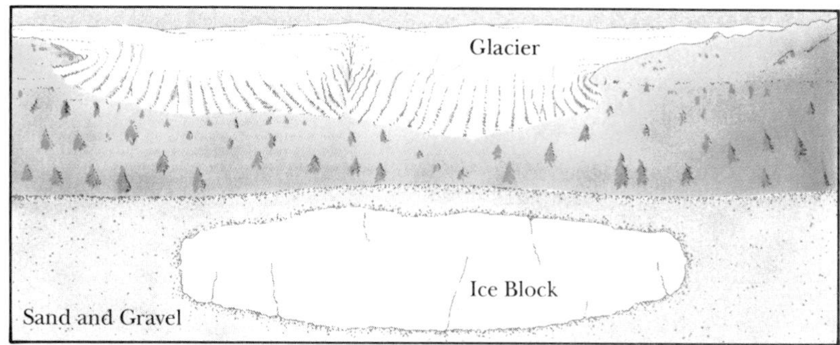

2. Cross section of water-filled "kettle" left when the ice block melted.

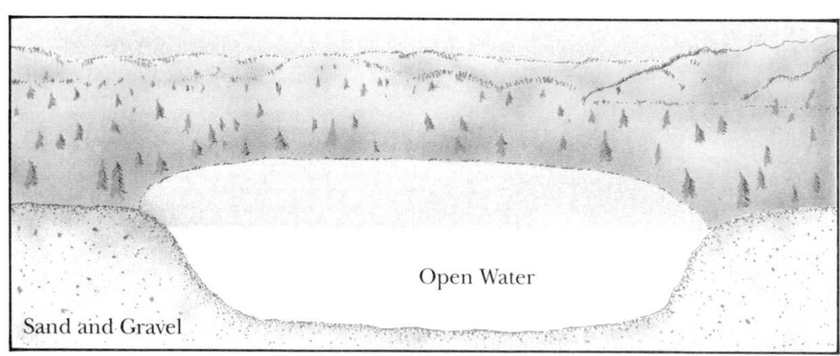

3. Cross section of kettle as bog mat advances from the edges over the open water.

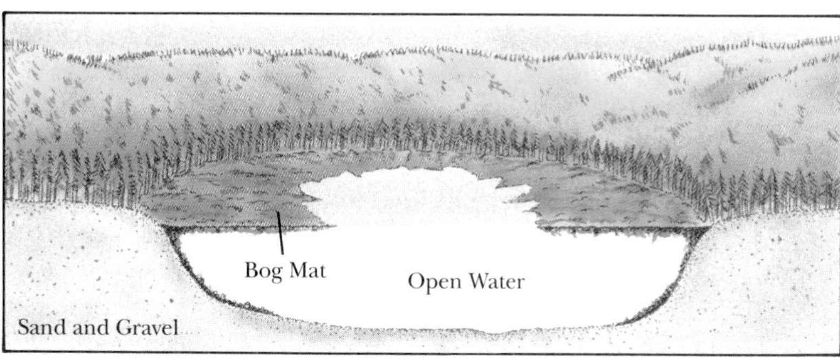

4. Cross section of kettle bog at a later stage. The mat completely covers the water and trees have started to close in.

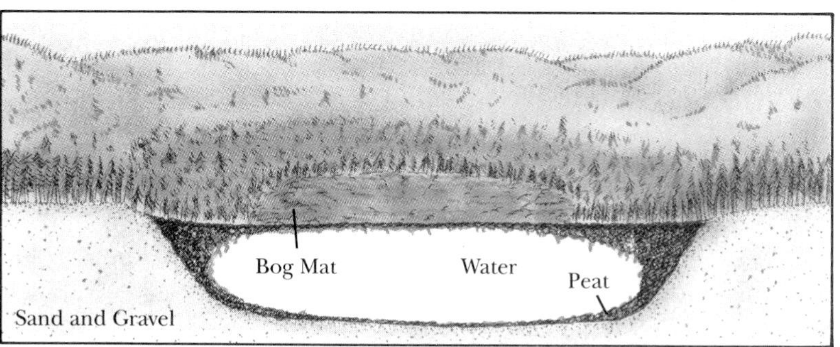

Figure 18-21

The development of a bog in a kettle lake. (Adapted from "Spruce Bog Boardwalk," text by Dan Strickland, drawings by Howard Coneybeare. Algonquin Provincial Park trail guide published by The Friends of Algonquin Park.)

Figure 18-22

The edge of a bog mat in Algonquin Provincial Park, Ontario, Canada. The vegetation at the left is dominated by hairy-fruited sedge and the leaves sticking up from the water among the waterlilies are from rhizomes of this species that have grown outward from the mat. (Photo by the author.)

Table 18-1
Important Bog Plants in the Northern Midwest

Floating mat
Hairy-fruited sedge
Water willow
Bladderwort
Spikerush
Three-way sedge
Open wet bog
Sphagnum spp.
Cranberry
Round-leaved sundew
White beak rush
Cotton grass
Open dry bog
Leatherleaf
Bog rosemary
Labrador tea
Sphagnum spp.
British soldier lichen
Pitcher plant
Blueberry
Haircap moss

Living sphagnum as well as peat can absorb 15 to 20 times its own weight in water (Figure 18-24). The water can be readily squeezed out of living sphagnum or fresh fibrous peat, but partially decomposed peat holds the water much more tenaciously. The "soil" of bogs is continuously waterlogged, although the upper few centimeters will dry out if the water table drops below the level to which capillary action can supply the underground parts of the live plants (about 33 cm).

As a result of the high water content and the high specific heat of water, together with the insulative action of the peat, temperature changes in bogs are slow. This is complicated by the fact that bogs generally lie in basins, so that cold air drains into them from the surrounding higher lands. Consequently, bogs warm very slowly in the spring but do not show a corresponding slowness to cool in the fall. In fact, frosts are possible any night of the year at the latitude of Wisconsin, and night temperatures 10 or 15°C lower than the surrounding uplands are not unusual (Curtis 1959).

The dark color of the sphagnums along with the openness of bog vegetation encourages high

Figure 18-23

An aerial view of a kettle lake bog in Kalamazoo County, Michigan. The bog mat is dominated by sphagnum and leatherleaf; the conifer zone is tamarack. Beyond the tamarack is deciduous swamp forest of red maple. The trees with darker crowns are on the upland. (Photo by Clayton D. Alway.)

Table 18-2
Water Characteristics of Streamflow From Bog and Fen in Itasca County, Minnesota

Water Characteristic	*Bog*	*Fen*
pH	3.6	6.5
Color units	336	100
Organic nitrogen (mg/l)	0.69	0.33
Sulfate (mg/l)	4.6	6.0
Calcium (mg/l)	2.4	16.6
Magnesium (mg/l)	0.97	2.88
Specific conductance (μmho at 25°C)	51	125

Based on Verry, E. S., "Streamflow chemistry and nutrient yields from upland-peatland watersheds in Minnesota," *Ecology*, 56:1149–1157, 1975.

temperatures at the surface on sunny days when snow cover is absent. Low temperatures at root level, perhaps coupled with the acidity of the water, together with high temperatures at leaf level have favored plants with morphological traits adapted to water conservation such as thick leaves with hairy lower surfaces and in-rolled margins—xeromorphic traits, in other words. Labrador tea and bog rosemary are examples.

Low temperatures, acidity, and the waterlogged, and, hence, oxygen-poor, soil all discourage decomposition. The basin fills in with peat, and the bog forms.

These conditions also inhibit the microbial activity that would result in nitrogen fixation. The bog material is low in most nutrients, anyway. The bog mat is largely isolated from groundwater, because its surface is above the water table. The surface at the edge of the bog may be lower, forming a moat, or lagg, around the bog where growth of the peat-

Figure 18-24

Sphagnum holds several times its own weight in water. The plant projecting out from the handful of sphagnum is the miniature shrub cranberry. (Photo by the author.)

forming plants is inhibited by nutrient-charged groundwater. Phragmites, Virginia chain fern, and bluejoint are characteristic lagg species.

Nutrient input, then, is mainly atmospheric. **Ombrotrophic,** meaning "rain fed," is the technical term. Even these scant nutrients may be unavailable for most plants because of sphagnum's tendency to absorb cations.

Some organic nitrogen doubtless enters atmospherically as the result of fixation by lightning, and blue-green algae and perhaps lichens also produce some, but the habitat is nitrate poor. One group of plants has found another avenue for supplying their nitrogen requirements. These are the insectivorous plants.

The insectivorous plants of North American bogs are sundews, pitcher plants, and bladderworts. The sundews catch insects in a sticky material secreted by hairs on their leaves, which enfold the insect for digestion (Figure 18-25). Pitcher plants digest insects in a specialized, hollow, water-filled leaf—the "pitcher" (Figure 18-26). It is possible that the pitcher in some species also serves the function of catching rain-borne nutrients ahead of the sphagnum.

The pitcher plant sarcophagus, as it is called, is a burial chamber only for some insects. A mosquito, *Wyeomeia smithii,* undergoes its larval development within the water of the pitcher plant, as does a midge. Rotifers, other invertebrates, and a variety of ciliates and flagellates live in the small pond that the pitcher plant leaf forms.

Bladderworts grow in water at the edges of bog lakes and catch aquatic invertebrates in small bladders that expand when they are touched, sucking the animal in, and close with a trapdoor.

Bog Forest

Open bog can be invaded by bog shrubs, such as mountain holly, chokeberry, and blueberry or by tamarack (Figure 18-27). Northward, black spruce may also enter directly, or it may follow tamarack. In these regions black spruce forest of short stature may occupy such sites for long periods. South of the range of black spruce, tamarack tends to be re-

Figure 18-25

Round-leaved sundew is one of the insectivorous plants of bogs. (Photo by the author.)

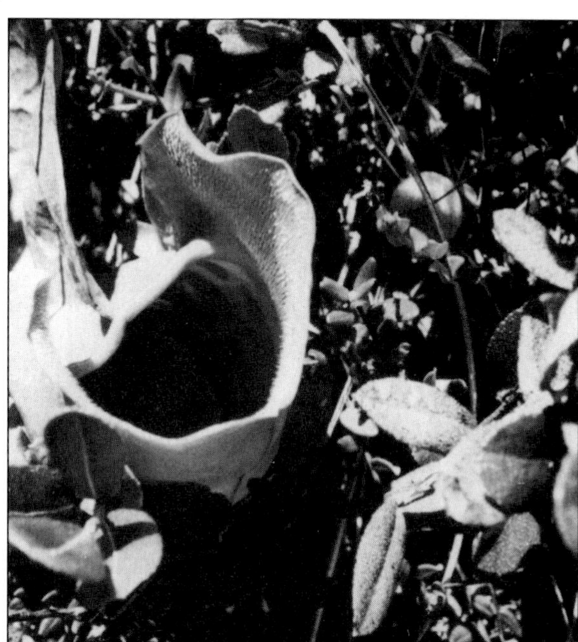

Figure 18-26

Looking down into a pitcherplant sarcophagus. The round fruit to the right is a cranberry. (Photo by the author.)

placed by a mixed bog forest in which red maple, white pine, and black ash share dominance (Brewer 1966).

These are expectable trends in the absence of disturbance; however, disturbance is a constant feature of bogs. Climatically induced or beaver-induced changes in water level, fire, and insect depredations (as by the larch sawfly) frequently counteract successional trends (Brewer 1966, Schwintzer 1979).

Bogs as Boreal Islands

Vernon Bailey, in a paper read before the Biological Society of Washington in 1896, referred to bogs as "boreal islands." There is a certain amount of truth to the picture of bogs as relicts, enclaves of northern species, left behind as the glaciers receded and climate warmed. Several typical bog plants have northern affinities, among them tamarack, black spruce, cotton grass, and some of the heath shrubs. Some northern animals evidently occupied bogs well south of their continuous ranges. The snowshoe hare is an example.

Figure 18-27

Open bog dominated by heath shrubs (here mostly leatherleaf) growing in sphagnum peat is typically invaded by a conifer tolerant of water-logged soil, in this case tamarack. (Photo by the author.)

For birds, the idea is valid only for a narrow range of latitudes, corresponding roughly to the upper half of Michigan (Figure 18-28). Tamarack-spruce bogs here contain a good collection of boreal forest and boreal forest-edge breeding species (Ewert 1982). Northward the bogs are not much more boreal than the rest of the countryside and, southward the boreal species drop out. The bogs of southern Michigan and northern Indiana and Ohio tend to be occupied by a few deciduous forest and deciduous forest-edge species of wide ecological amplitude (Brewer 1967).

Although there are northern plants in bogs, there are also species of other affinities. Pitcher plant, for example, is not boreal. Its closest relative occurs in the shallow bogs and acid sands of the Atlantic coastal plain, where also grow several species of sundews. Despite the elegant integration of the bog ecosystem, it is probably a newcomer, assembled in the Pleistocene of components from a variety of sources including arctic and alpine tundra, boreal forest, and the wet acid-soil communities of the coastal plain.

Fen

A fen is an ''alkaline bog,'' meaning that it is a mineral-rich peatland. Usually, fens occur at the base of slopes in the path of mineral-charged groundwater that results in a near-neutral to slightly alkaline pH (Table 18-2). Fens are **minerotrophic** (''mineral-fed''), in contrast to the ombrotrophic bogs.

Sphagnum is usually uncommon in fens, although some other bog plants, notably pitcher plant, may be prominent. Sedges are usually dominant, but the flora is rich. Fens share many species

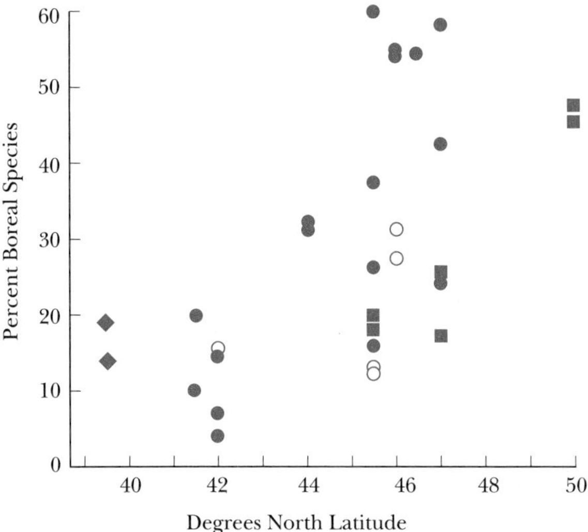

Figure 18-28

Bogs are "boreal islands" but how boreal they are depends on their geographical location and on the group of organisms considered. The percent of boreal species among the birds, shown here, increases with latitude but only around 43° to 45° north latitude is the bog avifauna much more boreal than that of the upland plant communities. For a given latitude western bogs (squares) are less boreal and Appalachian bogs (diamonds) are more boreal than ones in Michigan, Wisconsin, or Ontario. (Adapted from Brewer 1967.)

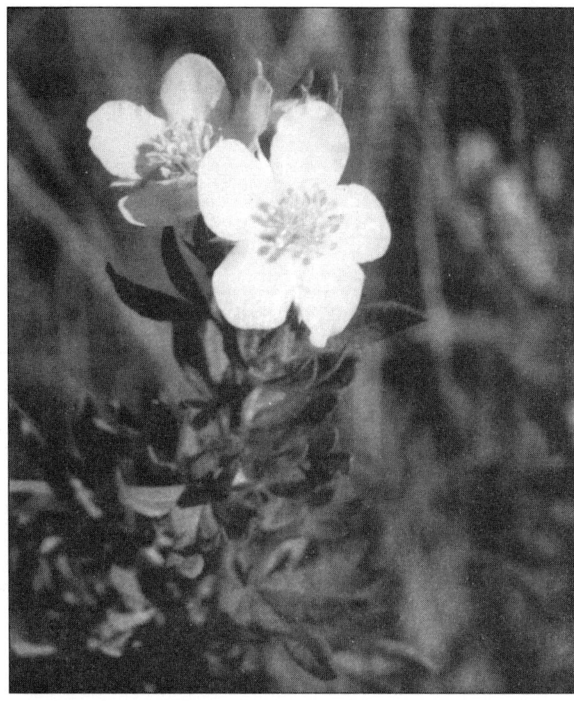

Figure 18-29

Shrubby cinquefoil, one of the most characteristic plants of the minerotrophic wetlands called fens. (Photo by the author.)

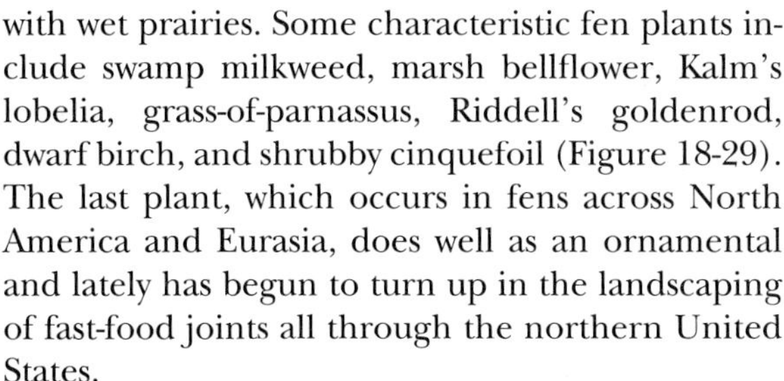

with wet prairies. Some characteristic fen plants include swamp milkweed, marsh bellflower, Kalm's lobelia, grass-of-parnassus, Riddell's goldenrod, dwarf birch, and shrubby cinquefoil (Figure 18-29). The last plant, which occurs in fens across North America and Eurasia, does well as an ornamental and lately has begun to turn up in the landscaping of fast-food joints all through the northern United States.

Tamarack may invade fens. It may be replaced by white cedar, or white cedar may enter without a tamarack stage. The resulting white cedar swamp may be long persistent. Cedar swamps have few distinctive bird species. Boreal species may be present if spruce also occurs; otherwise, the avifauna seems to be that of the northern hardwood forest and forest edge (Brewer 1967, Ewert 1982). White cedar swamps are a favored winter habitat for white-tailed deer (Dahlberg and Guettinger 1956).

Shrub-Carr

Shrub-carr is the wetlands variety of thicket. Shrubby vegetation may invade marshes, fens, or bogs. Dogwoods, willows, buttonbush, and birches are the most common kinds of shrubs in marsh. Alders are common along stream margins, especially northward.

Many of the characteristic plants of the more open stages of wetlands disappear, at least above ground, following the development of the dense shrub-carr clones. Several vines are characteristic of the shrub-carr stage and persist into the following swamp forest. Poison ivy, grape, and Virginia creeper may grow with great vigor, but several species of much more limited occurrence include virgin's bower, wild cucumber, and moonseed.

If this habitat adjoins marsh, many of the bird species listed for that habitat occur, and the total density may be 500 or more pairs per 40 ha. The added species are forest-edge birds such as song sparrow, American goldfinch, and gray catbird. Especially characteristic are alder or willow flycatchers, yellow warbler, and common yellowthroat.

Swamps

A swamp is a wooded wetland. Several types of swamps have already been mentioned in connection with the type of open wetland they replace. The major swamp types, each of which can be subdivided, include the following:

Northern Conifer Swamps. These are wetlands, usually on peat, where the forest is composed of tamarack, black spruce, or white cedar.

Hardwood Swamps. Wetland forests of elms, ashes, red maple, and birches occur across much of eastern North America. Oaks and hickories become increasingly important southward. These forests occupy a variety of wetland sites, including peatlands south of the conifer swamp region. Floodplain forests, although generally similar to the regional wetland forests away from rivers, show some differences. In most of the eastern United States, for example, species with southern affinities tend to be better represented in floodplain forests.

Cypress Swamps. Southward from southern Illinois and southern Indiana in the Mississippi embayment and on the coastal plain occur swamps dominated by cypress and gum, and usually including a variety of other southern species. Such swamps near rivers usually include bald cypress and tupelo gum *(Nyssa aquatica);* away from rivers, pond cypress *(Taxodium ascendens)* and water gum *(Nyssa biflora)* may dominate.

Vertebrate communities tend to be similar to those of forests of similar physiognomy on upland sites; however, diversity and density are usually higher. Certain species seem more numerous in, or restricted to, swamp forests; this is particularly true in the South. Two such species that extend northward along the floodplains are prothonotary warbler and yellow-throated warbler (one race of which was formerly called by the ecologically apt name of "sycamore warbler"). Breeding colonies of many of the large waders, such as great blue herons, tend to be located in swamp forest.

Wetlands Productivity

Productivity in marshes may be as high as in any ecosystem on earth. Westlake (1963) suggests an annual NPP of 38 metric tons per hectare for "reedswamp" on fertile sites in temperate climates; this amounts to 3800 g/m^2. Tropical marshes are probably still more productive.

Bogs are not very productive. Aboveground production probably averages less than 50 g/m^2 per year, and considerably less where vascular plants are sparse (Bartsch and Moore 1985). Fens may be more productive.

Biomass accumulates in the rhizomes of marsh plants, but this is slight compared with the biomass accumulation in the peat of bogs and fens. Decomposition of below-ground plant parts must be extremely slow, although there is little information on the subject. Above-ground parts in a poor fen decomposed at rates below 12% per year for sphagnum and 13 to 22% for sedges (Bartsch and Moore 1985). Decomposition rates vary with nutrient content, so it is likely that the least fertile, most heavily sphagnum-dominated sites accumulate peat faster, relative to production, than do graminoid fens.

Caves, Phytotelmata, and Other Special Habitats

In any region, certain habitats support communities different from those characteristic of the general landscape. Even though temperate forest may be the prevailing vegetation (for example), there may be wetlands, rock outcrops, caves, and other special habitats occupied by a different set of organisms.

Some of these habitats may be geographically restricted, but others recur widely throughout the world. Some may be the result of recent technological activity, such as the areas around a smelter or a radiation source. These usually have an impoverished flora and fauna consisting of whatever local species are able to tolerate the strange, new conditions. Other special habitats such as rock outcrops are, as a type, very ancient. The opportunity for adaptation by organisms has existed for a long time, allowing for communities highly integrated with these special circumstances.

In this section we discuss only three special habitats, caves, phytotelmata and the "aeolian biome," but others have been described earlier, such as hydrothermal vent ecosystems and communities of dung and carrion.

Caves

A cave is a hollow extending below the surface of the ground. The most abundant and best-studied type of cave is that produced in limestone by the underground movement of water. An underground stream may still be present, year-round or seasonally, or the cave may be basically dry.

The single most important feature of the cave habitat is, of course, the absence of light; however, caves also differ from most habitats in other ways. Except for variations in streamflow, the physical environment is nearly constant. Humidity is always high. Temperature shows almost no day-to-night change and little seasonal variation, being always near the average annual aboveground* temperature. As we would expect, water in caves tends to be hard and alkaline.

One of the consequences of the lack of light is that no green plants live in caves, and, therefore, food chains are based on energy brought in from the outside. The cave ecosystem, thus, is heterotrophic; food chains start with bacteria and scavengers. Water carries in both dissolved and particulate organic matter. If the holes to the surface are big enough, particles of substantial size—leaves, twigs, earthworms, and insects—can be carried in.

The other main source of organic matter is from animals that spend some time on the outside. Bats, of course, fly in and out of caves. Many other animals make use of caves as dens for sleeping, raising young, or hibernating. Vultures, wolves, and bears are obvious examples; humans are another. All such animals bring in various kinds of plant and animal debris which, along with their own dead bodies and guano, make up sizable imports of energy.

The bats, bears, and other cave dwellers we have mentioned retain direct connections with the epigean world. Others, however, are permanent inhabitants, never venturing into the aboveground world. The word, coined long ago, for such permanent residents is **troglobite** (Racovitza 1907).

Troglobites usually are related taxonomically to the epigean biota of the region where they live (Packard 1894). Not all the aboveground groups are represented, however. There are no green plants and, consequently, no grazers like caterpillars or antelope. There are no leaping or flying troglobites. The flying animals that do make use of caves use sonar. Bats are the obvious example, but two kinds of birds that nest and roost in caves, the oilbird and the swift, also use echolocation.

The important groups of troglobites are similar throughout the world. Generally planaria, isopods, amphipods, crawfish, cavefish (amblyopsids), and salamanders inhabit pools and streams in caves. Terrestrial residents usually include cave crickets, millipedes, spiders, collembola, and carabid beetles. Spiders and cavefish tend to be the top carnivores of the cave ecosystem.

Troglobites generally mature late and lay few but large eggs. These are traits we associate with *K*-selection and may well be a response to the great stability of the cave environment. One way in which many caves are stable is that there is always very little food, so it is possible that some of these traits may be more directly adapted to extreme food limitation. Troglobites are, thus, "tolerators," in Grime's (1979) terminology (Chapter 5). Adaptation to a sparse food supply may account for the lower metabolic rate of troglobites compared with their above-ground relatives.

One other feature that many troglobites share is more obvious than a low metabolic rate; many have no eyes or pigment (Eigenmann 1909). Neither is of much good to an organism living in perpetual dark. The debate still goes on (Culver 1982) about whether the loss is positively adaptive (by saving energy that would otherwise be wasted on useless structures) or whether it is, in effect, the result

* Although it may seen unnecessary to us surface swellers, students of cave life use a special term, *epigean,* to refer to the aboveground environment. The corresponding term for the cave environment is *hypogean.*

of a series of genetic accidents that go uncorrected by natural selection.

A greater development of structures associated with other senses would be handy in a habitat where sight is of no help, and many cave dwellers show such adaptations—long antennae on crawfish, for example.

Phytotelmata

The aquatic microcosms that form in or on certain plants are called **phytotelmata** (Varga 1928), based on the Greek word *telma* meaning, approximately, "pond." The water may consist of plant secretions or be derived from precipitation or both. Important types of phytotelmata include the hollow leaves of the pitcher plants in which the water contains digestive enzymes (discussed earlier under Bogs) and the "tanks" of many bromeliads, consisting of expanded leaf bases. Many other kinds of plants in 20-odd families show similar, although less spectacular, modifications. Bromeliad and similar pond-bearing plants are most abundant in the tropics. Tree holes that hold water are another distinct phytotelma type.

For some plants, certain functions of the little ponds are clear. Pitcher plants get, at least, nitrogen from the insects that die in the pitchers. The epiphytic bromeliads trap nutrient-containing rainfall that would otherwise be lost by flowing on down the stem on which they are situated. In many plants rooted in mineral soil, such as the teasels, the role of the phytotelma in the life of the plant is less clear, and it seems likely that trees derive little, if any, benefit from water-filled rot cavities.

From the standpoint of other organisms, the phytotelmata constitute a reasonably permanent aquatic habitat in an otherwise terrestrial setting. The bulk of the studies have concentrated on insects, especially mosquitos. Nearly every type of phytotelma studied includes at least one characteristic mosquito (Fish 1983). Some disease vectors such as mosquitos that transmit eastern equine encephalitis are treehole breeders (Engemann 1982). So, also, is the Asian tiger mosquito whose recent introduction to the United States has been highly publicized by tax-supported insect control agencies.

Other characteristic insect groups of phytotelmata include helodid beetles and dragonflies. Invertebrates other than insects include various protozoans which may be extremely common, flatworms, oligochaetes, ostracods, copepods, and rotifers. Crabs, such as the bromeliad crab of Jamaica, may inhabit the larger bromeliads and one vertebrate, a frog, *Hyla brunnea,* breeds in the bromeliads (Laessle 1961). Other frogs may visit the phytotelmata, perhaps to feed upon the insects there. Dogs quickly learn the locations of accessible treeholes with a dependable supply of water, and it seems likely that other forest vertebrates such as opossums and raccoons do also.

Phytotelmata in sunny situations may have an abundant algal flora. Such microcosms tend to show strong diurnal fluctuations in carbon dioxide and oxygen (because of photosynthesis by the algae) and in temperature. Ostracods and chironomid larvae tend to be important in such microcosms (Laessle 1961). In such situations, algal grazing may form the base of important food chains. Most phytotelmata, however, are shaded. Here food chains are largely detritus based. The detritus is from dead tissue of the plant itself, from animals that have died in the phytotelma, and from leaves that have fallen into it.

Some organisms may feed on living tissue of the host plant; the larvae of hispine beetles living in the bracts of Heliconia inflorescences scrape tissue from the plant bracts (Seifert 1982). Some of this tissue falls into the water, becoming food for bacteria and other organisms living on detritus.

Phytotelmata undergo changes during the course of their existence that can be thought of as constituting microseres. One difference from microseres such as dung and carrion is that, although the individual phytotelma has a short existence, generally of the order of months or years, new ones may be produced in virtually the same location for a long period. The dispersal requirements of organisms occupying this sort of temporary habitat are probably different from those occupying microhabitats that recur in spots that are widely separated spatially.

Few phytotelmata successions have yet been thoroughly studied. In some a general progression is seen from microorganisms such as protozoa to larger filter feeders such as mosquito larvae; around the same time, larger detritus-feeders and then predators also come in. A characteristic sequence of species within some of these groups may also occur.

Phytotelmata are attractive units for community-level research. They are natural, potentially coevolved systems that exist on a scale that makes field and laboratory experimentation feasible. They may help provide answers to important questions of succession and community organization.

The Aeolian Biome

Climbers on the high mountains of the earth eventually reach elevations at which not even isolated plants grow. Flowering plants and mosses drop out somewhere around 5500 to 6000 m, and not even lichens reach past 6700 m (Halloy 1991). Among suggested limiting factors are the great temperature fluctuations of the surface, low soil temperatures, and low available soil moisture.

In a few locations, unique plant-animal communities exist on volcanoes at elevations well above the general limits of plant growth. These "warmspots" are heated by geothermal energy. On such sites high on the Socompa volcano on the border between Chile and Argentina occur thick carpets of mosses and liverworts (Halloy 1991). Insects (springtails, earwigs, and flies), a leaf-eared mouse, and a small bird, the greenish yellow-finch, were observed around the warmspots. These communities are presumably autotrophic, based on green plant production at the site; however, part of the energy base for the animals is probably detritus blown up from ecosystems at lower elevations.

Such organic debris carried upward by air currents is the main source of food for the animals that live away from these rare high-altitude warmspots. L. W. Swan (1992) termed the high-altitude area above the limit of plant growth, the aeolian zone and suggested that the aeolian zones throughout the earth could be grouped into the "aeolian biome." "Aeolian" is pronounced "eolian" and means wind produced.

The rock fields and crevices above the plant line accumulate organic debris—insects, leaves, seeds, spores, etc.—on which live springtails and flies which are, in turn, fed on by spiders. There may also be a variety of scavengers including birds, mice, and lizards that live mostly on the dead or injured insects swept up to these heights. Such detritus-based ecosystems are, of course, heterotrophic.

There are also animals characteristic of the ice and snow of these heights. There is a strange small snow annelid (in the genus *Mesenchytraeus*) that can survive in solid ice "and that seemingly moves along crystalline boundaries within the ice" (Swan 1992). With small thaws, it feeds on detritus particles in the rivulets. This worm occurs in North America on glaciers from Washington to Alaska.

In Nepal occurs a flightless midgefly (genus *Diamesa*) that spends its entire life in the snow and ice of the high Himalayas (Kohshima 1984). The larvae live in melt-water drainage channels where they feed on blue-green algae and bacteria. Adults were observed active at temperatures down to −7°C and in experiments remained active at temperatures down to −16°C. They displayed a corresponding sensitivity to even moderately warm temperatures, becoming immobilized from heat when placed on a human palm.

The Practical Ecologist: *Conservation Biology*

America's premier endangered mammal, the black-footed ferret. *(Dean Biggins, U.S. Fish and Wildlife Service.)*

OUTLINE

- The Idea of Conservation
- Conservation Biology
- Wildlife Management
- Extinction
- Preserving Ecosystems and Landscapes
- The Preservation of Natural Areas
- Why Preserve Biodiversity?

> *"The world is not going to survive for very much longer as humanity's captive. Does that need explication?"*
>
> *Daniel Quinn,* Ishmael

In earlier chapters we discussed applications of ecological knowledge where appropriate, and in the last three we focus specifically on putting ecology to practical use. This chapter deals with a field of environmental science that is very close to the heart of ecology—conservation biology. In this chapter, we deal with the principles of conservation biology. Many topics that pertain to conservation are discussed elsewhere. Among them are management of the longleaf pine ecosystem and constituent organisms, such as the red-cockaded woodpecker (Chapter 15), the use of fire in managing ecosystems (Chapter 3), preservation of the spotted owl and old-growth temperate rain forest (Chapter 16), prairie restoration and reconstruction (Chapter 17), loss of tropical habitats and their biota (especially in Chapter 17), endangered fish of the desert biome (Chapter 17), conservation problems in the antarctic (Chapter 16), and paleoecological lessons in the study of global change (Chapter 14).

Much of the material in Chapters 20 and 21 can be considered "conservation biology," in the broadest interpretation of the term. Chapter 20 considers one specific aspect of human relationship with the environment—pollution. It includes, for example, pesticide pollution of the Great Lakes and oil pollution of marine shorelines. Chapter 21 deals with a broad range of environmental topics to which ecology makes important contributions, including human overpopulation.

The Idea of Conservation

Although we may believe that the American Indians or other native peoples were conservationists through their approach to the land and the biota, conservation as an idea belongs to the 20th century. It owes much to Theodore Roosevelt. He realized that the wildlife, forests, and ranges are renewable organic resources that can last forever if used wisely. He understood also that conservation is a public responsibility and that science is the tool for discharging that responsibility (Leopold 1933).

A 1947 textbook by a quartet of Cornell University scientists (Gustafson et al. 1947) gave the following definition for natural resource conservation. The definition is wordy but typical of the idea of conservation as the wise use of the environment for the sake of humans. Conservation, they wrote, is "the wise use of the existing supplies, the husbanding of those that remain for the benefit of future generations, the restoration and careful management of renewable resources, and the establishment of a workable program that will make them all serve the people as a whole, perpetually and to the fullest advantage."

Conservation Biology

By the 1940s conservation was a well-established field, with such subdivisions as soil conservation, wildlife and fisheries management, and forestry. Its concerns included soil erosion, degradation of grazing lands, declining stocks of fish and game, and the destruction of our forests by fire. Conservation writings of the era have an optimistic, rather Soviet sound to them. "A new era is at hand. The day of wasteful use is gone and cannot return," wrote Gustafson and his colleagues (1947).

By the 1960s, the more perceptive writers on the environment had realized that the separate problems of soil, water, wildlife, and forests were all manifestations of a single big problem—human degradation of the biosphere. Conservation texts began to be replaced by or to converge with environmental science texts; they had chapters on pesticides and population and concluded, not with a section on conservation of the nonmetallic minerals, but with speculations on the future of Planet Earth.

We might date conservation biology as a separate field to the year 1980 and the publication of the book *Conservation Biology—an Evolutionary-Ecological Perspective,* edited by Michael Soulé and Bruce Wilcox. In 1987, the Society for Conservation Biology was formed and began to publish a journal in May of that year.

Why, after old-style conservation had merged with a broader appreciation of environmental concerns, did it prove useful to resurrect "conservation biology"? There are probably several reasons. It was evident to biologists by the late 1970s that the world was on the verge of a tidal wave of extinctions as a direct result of human insensitivity to their actions and responsibilities. Conservation biology emphasized the need for conserving species and habitats, subjects of which many of the more humanistically oriented environmentalists of this era were oblivious.

Secondly, subfields of old-style conservation were still in operation in such forms as wildlife management and forestry; however, their practice fell short of what advocates of conservation biology saw as desirable. For one thing, wildlife management had been largely co-opted by the hunting industry and forestry by the timber industry. The science practiced was often concerned with minutiae about timber species and game species. Nongame organisms and habitats as ecosystems got scant attention.

Conservation biology focused on the big ecological picture, not on biological resources as commodities. Also, it brought into play recent advances in population ecology and genetics and computer modeling that the holdover "conservation" had failed to incorporate.

The goal of conservation biology can be simply stated. It is to provide the principles and tools for preserving biological diversity (Soulé 1985). Preserving biodiversity involves preventing extinctions of species, but not just that. A measure of biodiversity has already been lost if a species has been reduced to a few scattered populations, so that it is no longer an important constituent of the ecosystems to which it formerly belonged. Genetic diversity is important in its own right and for its contribution to short- and long-term survival of species. The preservation of characteristic ecosystems and landscapes is a part of biodiversity.

We can target a select group of organisms, such as spotted owls or Kirtland's warblers, and work hard to learn enough about each to protect it from extinction, but we cannot do this for every bird, insect, moss, or fungus. The only way to retain all these species is to make sure that their characteristic ecosystems or habitats are preserved on the scale necessary for proper functioning.

In the following section, we review the field of wildlife conservation, or wildlife management, to set the stage for a consideration of the major topics of conservation biology.

Wildlife Management

Wildlife management is the craft that applies ecological principles to produce sustained yields of wild animals. Under the broadest interpretation, wildlife management is virtually synonymous with natural resource conservation. As currently practiced, wildlife management does, in fact, include certain activities with broad environmental aims. Programs for the preservation of endangered species, for example, are usually housed in governmental units devoted to the management of wildlife. Usually, however, the term "wildlife management" is far too generous. Most of what occurs under the name is still "game management" with hardly a nod to nongame animals, plants, or the ecosystem as a whole.

Game management differs from other sorts of wildlife management in one fundamental way. Its aim is to produce a shootable surplus. Animals equivalent to this surplus are then removed in recreational hunting. The ecological basis for hunting is that it acts as a replacement for natural agents of mortality. In particular, hunting is viewed as surrogate predation, taking the place of the wolves, mountain lions, owls, and falcons that we have lost.

An appropriate game species, from a biological viewpoint, is one that has a biotic potential high enough to produce a yearly shootable surplus (Figure 19-1). Animals with a low biotic potential can make satisfactory game species if their habitat is abundant, but the take must be small relative to the population and numbers should be closely monitored because they cannot recover quickly from decimating factors such as overhunting and weather disasters.

Hunting in the United States

Hunting is an ancient human practice, and probably most primitive societies practiced some form of game management. Some practices may have been explicitly for management, but often they were in the form of taboos or religious rules with no conscious management connection. Widely cited as the earliest written game law is the biblical rule on nest robbing: It's OK to take the eggs or young, but let the hen go (Deuteronomy 22:6).

A society without game laws would be an oddity. One such society was colonial America. Most of the colonists came from Europe where game belonged to the king. Poaching brought severe penalties, such as blinding or death. In medieval England, the penalty for causing a stag to pant was a year's loss of liberty (Graham 1947). Later laws reduced the severity of the sentences, but the average colonist arriving in the 1600s from Europe had no hunting experience. This must be one of the main reasons for the privations suffered by many early settlements, despite an abundance of game, fish, and other wild foods. The settlers had to learn from the Indians how to hunt and fish, just as they had to learn to grow corn.

The colonists became hunters, but they had no traditions that had evolved to assure the perpetuation of game populations, and they did not adopt the Indian's limitations. So they were launched into a land of seemingly inexhaustible game with no checks on their conduct. The result was an era of unbridled killing that began on the eastern sea-

tinction of any other species tied to it in an obligate way (its obligate parasites, for example).

Many described cases of recent and incipient extinction seem to involve a two-stage process (Figure 19-8):

1. Deterministic events reduce the species to small and isolated populations.
2. Stochastic events drop some and eventually all the populations below the threshold for recovery.

We will explore the ramifications of this model in the next sections.

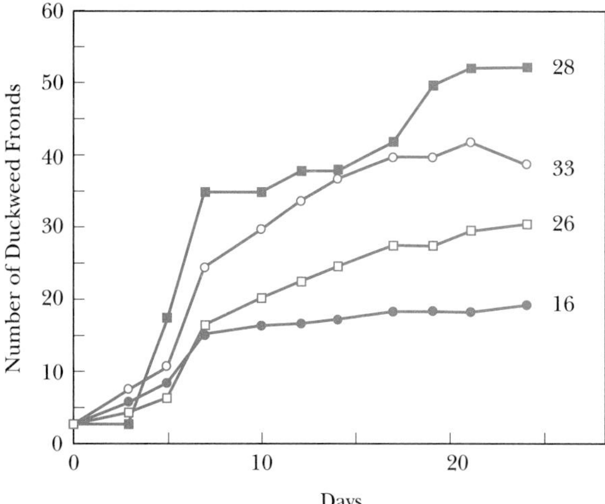

Figure 19-9

Demographic stochasticity. Duckweed population growth in four cultures started at the same time with three fronds and maintained under the same conditions. Numerals are code numbers of individual cultures.

Deterministic and Stochastic Models

In earlier parts of this book, nearly all the models used were deterministic. In the logistic model of population growth, for example, if we specify one value for r, one value for N_0, and one value for K, we always get the same answer about how many individuals will be present on any given day in the future. The model is deterministic.

In fact, as any student of biology knows, organisms are more variable than that. The likelihood of giving birth or of dying varies among individuals based on their genotypes and their past experience and on minor changes or differences in environment.

As an example, Figure 19-9 shows population growth in four separate duckweed cultures. Each was set up in a large test tube beginning with a group of three fronds and was maintained under the same conditions of constant light and temperature (see Brewer and McCann 1982). Each of the populations followed a unique path, and so did 30-odd others begun at the same time. For the whole collection of cultures, we can calculate average values of r and K, but these do not tell us what any one of the cultures did or what a new culture will do.

Although each culture was different, there was not infinite variability in the numbers of duckweed on day 5 or on day 24, nor in the value of r calculated from the cultures. All these formed approximately normal distributions, allowing us to state probabilities. For example, the mean r for these samples was 0.28 per day, and the standard deviation was 0.082. This means that approximately 67% of the samples had r between 0.20 and 0.36.

A stochastic model allows for the variation that is inevitably encountered. To predict what is likely to happen in one duckweed culture—or what would be likely to happen if we started a new Kirtland's warbler population in a previously unoccupied stand of jack pine—we would have to put together a set of calculations for the important steps and then have to run through these calculations many times, changing the important features, such as r (or birth rates and death rates). Our answer for the Kirtland's warbler would not be in the form that the population would survive or not but in the form of probabilities of survival for varying lengths of time.

We earlier used "random" as a one-word definition of stochastic, but "probabilistic" may be a better one. In choosing a value for r or for birth rates or death rates, we would not pick out numbers wholly randomly. Rather, we would pick them randomly from the appropriate probability distribution. If we were picking rs for duckweed, we would program our computer so that, on the average, 67% of the values picked would be between 0.20 and 0.36. But this is just on average: If we actually did this ten times, or if we started ten cultures, it could occur by chance that all ten were below 0.20 or above 0.36. This illustrates what is probably the best way of thinking about stochasticity—it consists of sampling errors, errors that have increasingly strong effects as populations become smaller.

Stochasticity in Extinction

Virtually every population that becomes extinct first becomes small. If 5% of the females of a given species are sterile, it is unlikely that this will have a detectable effect on the growth of a population of 8000 individuals of that species. If, instead, we are dealing with a population of eight it becomes a possibility that all four females will, by chance, be sterile some year in the near future.

It is useful to recognize three types of stochasticity: demographic, genetic, and environmental.

Demographic stochasticity refers to chance variation in demographic features of the population, such as births and deaths. The examples of stochasticity used so far have all been demographic stochasticity.

Genetic stochasticity refers to chance changes or variations in the genotypes represented in the population. Of most importance in terms of biodiversity is the loss of alleles from a population.

Environmental stochasticity refers to the uncertainty of the weather—a drought, a blizzard—and other aspects of environment. The abiotic and biotic components of the organism's environment are ordinarily lumped together here, although some additional insights could be gained by separating them.

Demographic Stochasticity

Among demographic features that can vary are age at which breeding begins, clutch or litter size, survivorship, dispersal rate, and sex ratio. Most models of the relation between demographic stochasticity and the probability of population extinction have concluded that demographic stochasticity alone is of importance only in very small populations (MacArthur 1972, Leigh 1981).

Some of these simplified models have suggested that fluctuations in population size from demographic uncertainty are roughly proportional to $1/\sqrt{K}$, where K is carrying capacity in the sense of maximum population size. If so, fluctuations in a population of 10 will be three times as great, proportionally, as in a population of 100 ($1/\sqrt{10} = 0.32$ versus $1/\sqrt{100} = 0.1$).

Population numbers in these models are effective population size, rather than actual population size. If, in a population of 20, 5 individuals are too young to reproduce, 3 too old, and 2 too unattractive, effective population size may be only 10. Or if, in a population of 60 individuals, there are only 5 females, the effective population size will be about 10.

A primary reason that demographic stochasticity, by itself, rarely leads to extinction is density dependence. If one female fails to produce offspring one year, the offspring of the females that do reproduce have enhanced survivorship. If several adults die, all the immature animals may survive where, in an average year, many would perish.

Genetic Stochasticity

At generation x, different alleles have certain frequencies in the gene pool. The gene pool of the next generation is the result of a complicated sampling process (involving pairing of individuals, egg and sperm production, zygote formation, and the production and survival of young). In a small population, it may occur by chance that none of the individuals with allele A′ survive to mate, or it may be by chance that none of the eggs or sperm with allele A′ happen to fuse to form a zygote.

The loss of alleles by this process of sampling error is called **genetic drift.** In large populations, sampling errors will rarely result in any allele (except an initially very rare one) being lost entirely. In small populations, however, drift leads inexorably to the loss of genetic variation, as one allele after another is lost. Eventually, many loci come to be represented by only a single allele; that is, all the individuals are homozygous for that allele. Genetic drift results in loss of alleles and, consequently, loss of heterozygosity (Figure 19-10).

Making some simplifying assumptions (specifically, that we are dealing with random fusion of gametes in and among self-compatible hermaphrodites), it is possible to predict the rate at which heterozygosity is lost from a population. From the

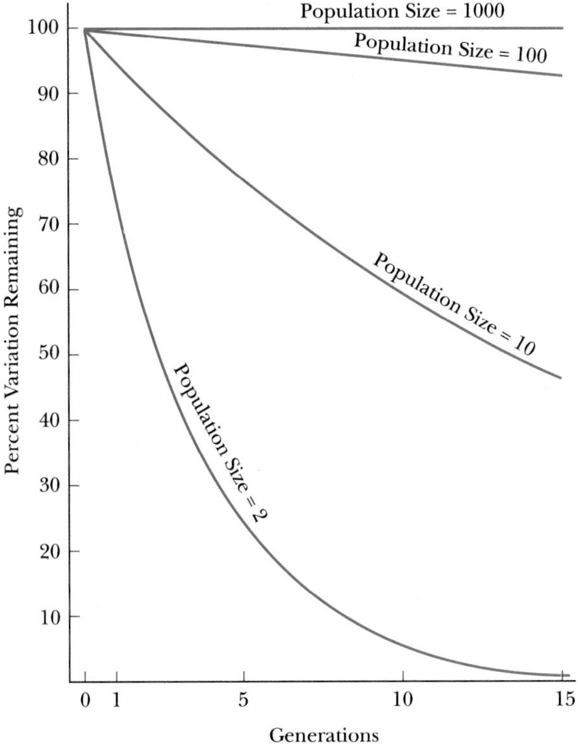

Figure 19-10

Genetic variability is lost from small populations at a rapid rate. Losses are estimated for different effective population sizes from the formula $V = (1 - 1/[2N])^t$, where V is remaining variability (heterozygosity), N is effective population size, and t is number of generations (Roughgarden 1979).

Hardy–Weinberg law, the probability of obtaining homozygotes in a given generation is about ½ N (N is effective population size for the breeding generation). Consequently, heterozygosity tends to be lost at the rate of 1−(½N) (Roughgarden 1979).

Here also, the same comments about effective population size being less than actual population size apply. There are, in fact, some additional important limitations, such as nonrandom mating—if the creatures tend to mate within their neighborhood, this increases the discrepancy between actual and effective population size. (For genetic drift, effective population size, N_e, is defined as the size of an ideally behaving population that shows the same increase in homozygosity as the observed population [Falconer 1981]. Formulas are available.)

Genetic drift is the result of a kind of inbreeding. Another likely consequence of small populations is actual mating of close relatives, resulting in **inbreeding depression.** This is defined as the reduction of fitness in inbred lines. Examples in humans and domestic animals include smaller size, smaller litters, and lowered milk production (Falconer 1981). In the short term, there seem to be two main disadvantages to inbreeding: (1) Increasing homozygosity increases exposure of deleterious recessive alleles and (2) any enhanced fitness as the result of **heterosis** (hybrid vigor, overdominance, and heterozygote superiority are other terms expressing much the same concept) is lessened (Allendorf and Leary 1986).

A **bottleneck** is a temporary steep reduction in population size. If conditions change such that the population again expands, it will have suffered some reduction in genetic diversity. However, bottlenecks of only a generation or two tend to cause only modest reductions, as can be seen from Figure 19-10. If during a prolonged bottleneck, most individuals homozygous for deleterious recessives die or fail to reproduce, the expanded population seen later could have low heterozygosity but would not show a loss in fitness associated with these deleterious alleles. In other words, the inbreeding process would have purged from the gene pool the most disadvantageous recessives.

Although this is logical, highly inbred lines nevertheless seem to have low fitness, perhaps because of their general lack of heterosis. The cheetah is well known as a highly homozygous species that has probably suffered such a bottleneck or series of bottlenecks. Among other problems, cheetahs have high levels of abnormal sperm and seem highly sensitive to disease (Ballou 1989).

One optimistic note about genetic drift has to do specifically with the fact that it is random. If the species occurs as a metapopulation—that is, a series of subpopulations—allele A′ may be lost in one population, allele A in another, and allele A″ in a third. It turns out that only a small amount of interchange—individuals dispersing among populations and joining the breeding population at their new home—can greatly reduce the loss of diversity. A widely cited figure based on analytical models is that two new subpopulations will be expected to retain the diversity they start with (ignoring mutation and selection) as long as they exchange one individual per generation (Falconer 1989, Lande and Barrowclough 1987).

Evolutionary Potential Although the loss of genetic diversity may lower the current fitness of a population, the most serious consequence is for the future by reducing adaptive response. When environments change—for example, climatic change or the arrival of a new predator or parasite—one way in which populations cope is by changing genetically, that is, by evolving. Most of this change, as we have pointed out earlier, is based on pre-existing genetic variability, not on new mutations. Loss of genetic variability means a restriction of the field over which adaptation can occur.

Environmental Stochasticity

Environmental stochasticity, often interacting with demographic and genetic stochasticity, seems likely to be the major agent in the demise of small populations. Models suggest that the long persistence times for small populations when only demographic stochasticity is involved are drastically shortened when the main source of stochasticity is environmental (Goodman 1987). From the standpoint of the models, the main reason is that demographic stochasticity acts through individuals; if birth rate or survivorship is by chance low for one individual, it may be high for another. However, an abnormally hot summer or cold winter will affect many of the individuals in the population in much the same way, so that the overall rate of increase of the population may be severely affected.

Figure 19-11 plots persistence times for populations with various rates of increase and various population ceilings (which for our purposes can be considered a population size a little bigger than *K*). A population with between-individual (demographic) variance = 1 (no environmental variance), with a low population ceiling, and a low *r* is likely to go extinct soon, but a population with a medium ceiling and a moderate *r* is likely to last a very long time. Persistence times shown on the graphs are relative values; consider them years if

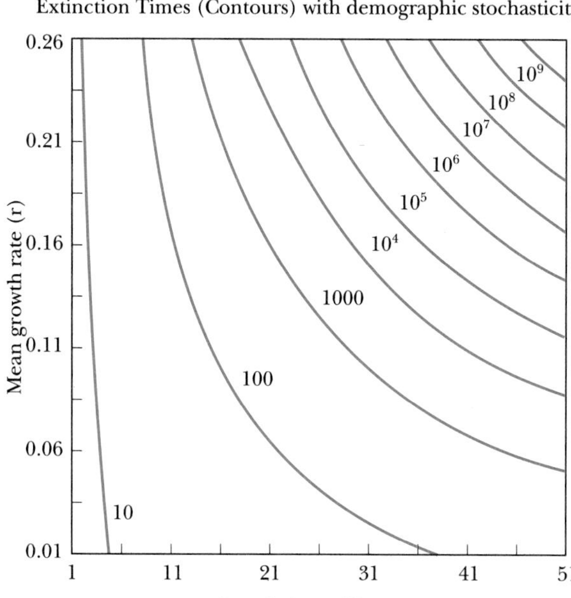

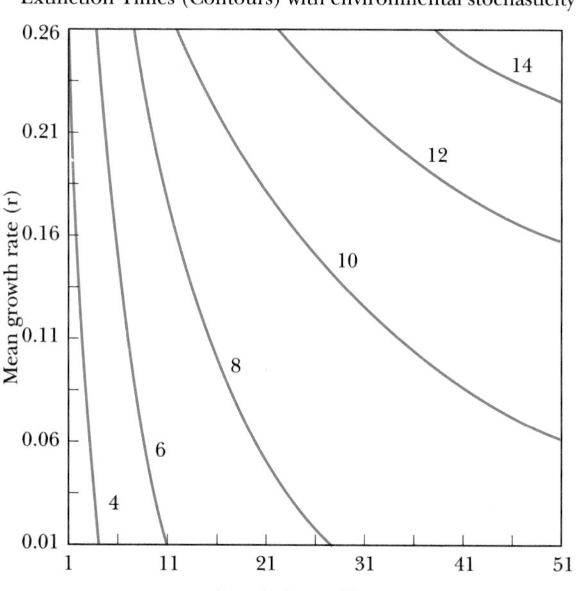

Figure 19-11

These graphs plot mean persistence times for populations, with demographic but no environmental stochasticity in the top graph and environmental but no demographic stochasticity in the bottom graph. In both cases, time to extinction increases with increasing population ceiling (approximately equal to *K*, carrying capacity) and mean growth rate (equal to *r*, intrinsic rate of natural increase). However, much lower times to extinction are predicted when environmental stochasticity is present. Details of the model are given in Goodman 1987, from which these figures are adapted.

you like. A similar graph with environmental variance = 1 (no between-individual variance) predicts short persistence times even with large *r* and sizable populations.

Sometimes catastrophes, like the winter storm of March 1993 in the eastern United States, are considered separately from ordinary environmental stochasticity, such as the usual run of wet, dry, hot, or cold years. The reason for this is that the probability distribution of such very rare events is unlikely to be known; however, if it is known or can be reasonably approximated, catastrophes can be considered as just the rare end of stochastic environmental events.

The Threshold of Extinction

If the last individuals of a population perish of starvation after an ice storm or from a disease, the immediate cause of extinction is clearcut. Often, however, the processes outlined in the preceding sections do not end with every animal dead. What, then, constitutes the threshold below which recovery is unlikely or impossible?

If various factors reduce a species to a single male or a single female, that is the effective end to the population, even though (as in the case of the health hen discussed in Chapter 4) that individual may linger on for years. Even if stochastic factors leave us with ten individuals but they are all males or all females, the population is doomed.

Some populations may be—with a high probability—doomed even if the foregoing processes leave a few individuals of both sexes. In highly motile and not-very-social species, it is possible that none of the few remaining males will find any of the few remaining females. In social species, numbers may be too low to trigger successful reproductive activities. Or, numbers may be too low to satiate predators. The last has been suggested as a reason for the final descent of the passenger pigeon (Halliday 1980).

A Case Study of Extinction: The Black-Footed Ferret

The black-footed ferret is the size and shape of a small mink, but pale colored and with a black mask. It once occupied grasslands from southern Canada to the middle of Texas. Its geographical range pretty much corresponded to the range of the two prairie dog species (black-tailed and white-tailed). The black-footed ferret lived in abandoned prairie dog burrows, and prairie dogs in nearby active burrows formed its main food. The ferret is nocturnal and secretive, and no one knows how common it was at the time of European settlement of the Great Plains.

The species was unknown to science until 1851, when one of John James Audubon's collectors sent him a specimen from "the lower waters of the Platte River." Audubon named it and painted it, but was unaware of the prairie dog connection; he had the ferret hovering over a bird nest (Audubon and Bachman 1854). Although unknown historically until 1851, the black-footed ferret has been found in fossil deposits back to 100,000 years old.

By the 1950s, it was evident that the black-footed ferret was nearly extinct. A small population found in South Dakota in 1964 was studied intensively; it disappeared in 1974. The species was not seen again until 1981, when a rancher's dog near Meeteetse, Wyoming, killed one. This population was again studied intensively, but with caution (Clark 1987). Major features of black-footed ferret demography and life history are summarized in Table 19-1.

The Meeteetse ferrets were found to be occupying 37 white-tailed prairie dog colonies that totaled about 3000 ha (7500 acres) scattered over 200 km^2 (77 square miles). Numbers of ferrets fluctuated but were around 100 in 1983 and 1984 (Forrest et al. 1988). Most of this population perished from canine distemper in 1985; however, 18 ferrets were captured at Meeteetse in 1986 and 1987 (the last in February 1987) and have been used for captive breeding by the Wyoming Fish and Game Department and the U.S. Fish and Wildlife Service. No natural wild populations are known, but reintroductions of captive-reared animals have been made recently.

The major deterministic factors pointing the ferret on the road to extinction stemmed from the near-eradication of the prairie dog by ranchers and the federal government. The stated reason for the extermination campaign, which has spanned decades and covered much of the West, was that prairie dogs compete with cattle for grass. It seems likely, however, that prairie dogs have little effect on cattle production on range that is not already overgrazed

Table 19-1
Some Features of Life History and Demography of the Black-Footed Ferret, Based Primarily on Information From the Meeteetse, Wyoming, Population (Forrest et al. 1988, Harris et al. 1989)

Feature	*Comments*
Breeding age	Possibly 85% of yearlings and a higher proportion of older females breed every year; however, the data are scanty
Sex ratio	Past juvenile stage seems to be female biased; this could be partly from higher predation on males and partly because males tend to disperse longer distances (hence are lost and not replaced by immigration to an isolated population)
Mating season	Probably late February–April
Litter size	3.38, standard deviation 0.86
Development of young	Juveniles appear above ground in company of mother in early July; become self-sufficent in about a month
Dispersal	Dispersal by juveniles occurs beginning in late August; more long-distance movements by males
Survival rate	Probably density dependent and varies by age and sex

(Robinson and Bolen 1989). The eradication efforts were relatively successful; the once-abundant and widespread prairie dogs were reduced to scattered, mostly small colonies. Eradication efforts continue: Some Kansas counties require landowners to control prairie dogs. If they don't do it, the county does and bills the owners (Cohn 1991). At Pine Ridge Indian Reservation in South Dakota the Bureau of Indian Affairs recently spent 5 million dollars on prairie dog eradication.

As a result of the prairie dog extermination, black-footed ferrets were deprived of their food base and their homes. The ferret population must have been reduced and fragmented. It is likely that the ferrets always occurred as separate, rather sparse populations. Some 36 ha of prairie dog town are needed per breeding ferret female. Under natural conditions, ferrets probably died out or moved on when they overate the local prairie dog food base. Later, when conditions improved, this area could be repopulated by dispersal from other ferret populations.

Increasingly, this normal population event—the restocking of a depleted ferret population by immigration—must have become impossible. Other populations were too few and too far away. The normal dispersal of young animals became simply another way of dying.

The stochastic event that pushed the Meeteetse population over the edge was environmental. Specifically, it was evidently an epidemic of canine distemper, perhaps caught from coyotes or some other local carnivore. Although we are saying that environmental stochasticity was the agent, it was mediated through demography. That is, the population dropped because disease resulted in high death rates.

Distemper was diagnosed in two animals live trapped and removed from the population in October 1985; however, the population had already declined by 50% or more between 1984 and 1985, so the disease had probably been present for several months at this time. About six animals remained in November 1985 (Forrest et al. 1988). Fourteen individuals (ten young in two litters) were present in the summer of 1986 (Clark 1987). Most of these were captured for the captive breeding program, and afterwards no ferrets were observed at Meeteetse.

The captive breeding effort has been successful (Cohn 1991) and beginning in 1991, ferrets have been reintroduced into the wild. The first two years, the ferrets were released in white-tailed prairie dog colonies in the Shirley Basin, Wyoming. Few persons believe that a fully self-sustaining ferret population in the wild is likely; monitoring and various forms of intervention, such as transport to maintain genetic variability will have to occur. It is not yet clear where the eventual balance will lie between a nearly self-sustaining wild population propped up by occasional input from reproduction ofcaptive animals and a showpiece wild population largely maintained by captive breeding.

There is further discussion of the management of small populations based on principles from the preceding pages in the sections How Big Is Big Enough? and Advantages of Multiplicity.

Preserving Ecosystems and Landscapes

The emphasis by traditional management on single species, such as white-tailed deer or slash pine, is obviously misplaced. These species are members of ecosystems and an approach that manipulates the landscape to maximize the production of these commodity species is apt to distort it to the detriment of many other species.

Similarly, the emphasis on the management of a few endangered and threatened species is misplaced, and few conservation biologists would claim otherwise. We can apply heroic efforts and spend millions of dollars to restore the black-footed ferret to the wild, and that is good. But there are too many endangered species for each to receive similar treatment. In the long run, the black-footed ferret will live or die based on the presence of healthy Great Plains grasslands supporting a healthy biota including well-distributed prairie dog colonies. Those grasslands will save other at-risk species—including many that are so poorly known that we have not yet recognized their risk. In the same way, maintaining and restoring large blocks of other ecosystems is the eventual route of choice to saving their endangered and threatened species.

The Preservation of Natural Areas

In southwestern Michigan the first settlers claimed lands at the edges of the small prairies (Peters 1970). They valued the open land for crops and hay and not for massasaugas or prairie flowers; they valued the oaks for lumber and firewood and acorns for their pigs and not for the orioles that nested there. Next the settlers claimed either grassland or open oak forest. The beech-maple climax forests were called ''heavy timber'' and were settled last. These forests were thought of mainly as obstacles to raising the crops that the settlers needed to survive, to make a living, and to become successful. The forests were laboriously cleared, at which time the area was said to have ''come out of the woods.''

Forested land is now scarce over much of our country. Prairie is almost totally gone, having been cultivated continuously since settlement. The oak areas, transformed to farm woodlots, are very different communities from those of primeval times. Bogs and marshes have been drained or filled. Sand dunes have been mined or covered with vacation houses. If we had the chance we might develop our continent more thoughtfully than did our forefathers. We might, but there are no second chances in such matters. The best we can do now is spend our time, energy, and money for preserving and restoring the natural landscapes that remain.

A **natural area,** according to the Indiana ecologist A. A. Lindsey and his coworkers (1969), is ''any outdoor site that contains an unusual biological, geological or scenic feature or else illustrates common principles of ecology uncommonly well.'' No areas on the earth are undisturbed by humans, given our use of pesticides, spread of radioisotopes and greenhouse gases through the atmosphere, alteration of water tables, and many other far-reaching actions. However, most natural areas are sites on which our influence has been slight relative to our influence on the rest of the landscape. A beech-maple forest in which no timber was ever cut would be a natural area. So too could be a beech-maple forest in which many trees had been cut 40 years previously. At the present time it might illustrate principles of succession, and in the future it might show the workings of a mature ecosystem.

A preserved natural area is one that has been legally dedicated for such a purpose. Natural areas not specifically set aside are lost every year as landowners interested in preservation die and their property passes to heirs not sharing this interest, or as university administrators decide that a ''research park'' (read, ''industrial compound'') represents a higher and better use for a piece of land than as an outdoor laboratory for biology classes.

The past 30 years have seen an accelerating trend toward preserving natural areas. After good progress on federal lands in the 1970s, federal preservation efforts stalled almost completely between 1980 and 1992. Several states have an oustanding record of recognizing their responsibility for accomplishing this public good; Indiana and Florida are two that come to mind. Even some counties, townships, and cities have lived up to their responsibilities; however, many others, either because of financial or philosophical shortfalls, have allowed land that should have been preserved to fall to development.

Overall, the land preservation record for government needs to be improved. The bright spot in the past few years has been the private conservation groups usually referred to as conservancies or land trusts. The best known such organization is the Nature Conservancy, an outgrowth of a natural areas committee of the Ecological Society of America. As the Nature Conservancy has grown, so has the scope of its projects. It is now interested primarily in "landscape"-level projects—essentially very large, pristine blocks of habitat.

Although such projects are essential for the health of the continent, smaller sites are also deserving of preservation—a 16.19-ha woodlot full of spring wildflowers, the pond at the end of the road where the frogs congregate to breed. Sometimes these are preserved by concerned owners giving them to nature centers, Audubon societies, or similar nature organizations. Increasingly, however, such conservation tasks tend to fall to the local land trusts.

These numbered about 50 in 1950. By 1975, there were more than 300 and by 1989, nearly 750 (Chown 1991). Currently, about a thousand exist. Dedicated to protecting natural areas or other open space (sometimes including historical or agricultural sites), the land trusts acquire lands by purchase or gift and also protect natural areas by acquiring conservation easements. A **conservation easement** is a legal agreement in which a land owner conveys development rights to the land trust. For example, the owner of 16 ha of beech-maple forest might wish to keep the property in the family but make sure that it was never developed. He or she could give a conservation easement to a local land trust and achieve this aim while also achieving income and property tax savings.

The Design of Nature Preserves

If it is the case that not all land that merits protection is likely to be protected, one then has to ask questions about the types of land and the sizes of perserves. Most thoughtful persons would agree that each region should preserve examples of every important natural ecosystem, and most agencies and organizations involved with natural areas have adopted this position. Species ought also to be preserved, and landscape diversity ought to be maintained.

Beyond these general principles, many questions remain as to the number, size, shape (Blouin and Connor 1985), and arrangement of preserves. One basic question is, given limited money for acquiring land, will we preserve more diversity with one big area or several little ones. This could be known as the JOBOVALOLO controversy—Just One Big One Versus A Lot Of Little Ones (however, it has more often been called the SLOSS controversy—Single Large Or Several Small).

For a variety of practical reasons, the question rarely comes into play in the real-life preservation of nature (Soulé and Simberloff 1986); nevertheless, exploring answers to the question provides insights into several important topics.

To begin with, two opposing trends in the way nature is constructed make a simple answer difficult. The first is the species-area relationship (discussed in Chapter 10). As area is increased, species number goes up, but the rate of increase drops, so that, at a fairly small size, adding more area increases diversity only slightly. From this standpoint, several small areas—of prairie, for example—each with its own species pool, will often perserve the largest number of prairie plants (Simberloff and Gotelli 1983, Quinn and Harrison 1988).

Presumably this is for two main reasons: (1) Each area includes slightly different habitats, hence, additional species and (2) not all the rare species are likely to be found as a part of the sample that one site constitutes, even if the site is fairly large.

The opposing trend is the habitat island effect, or area sensitivity. Although only a few new species are added as we increase the size of a stand, these are often the rare species and the ones most highly adapted to the specific features of the community. To sum up, one large area will probably not preserve as many species as several smaller ones totaling the same area (Figure 19-12). The one large preserve may, however, include species that are impossible to retain on the smaller preserves. We will delve into the effects of small preserve size and habitat fragmentation in general in the following sections.

Habitat Fragmentation

Even under natural conditions, most landscapes were a mosaic, not one long sweep of homogeneous habitat. The fragmented landscapes that human

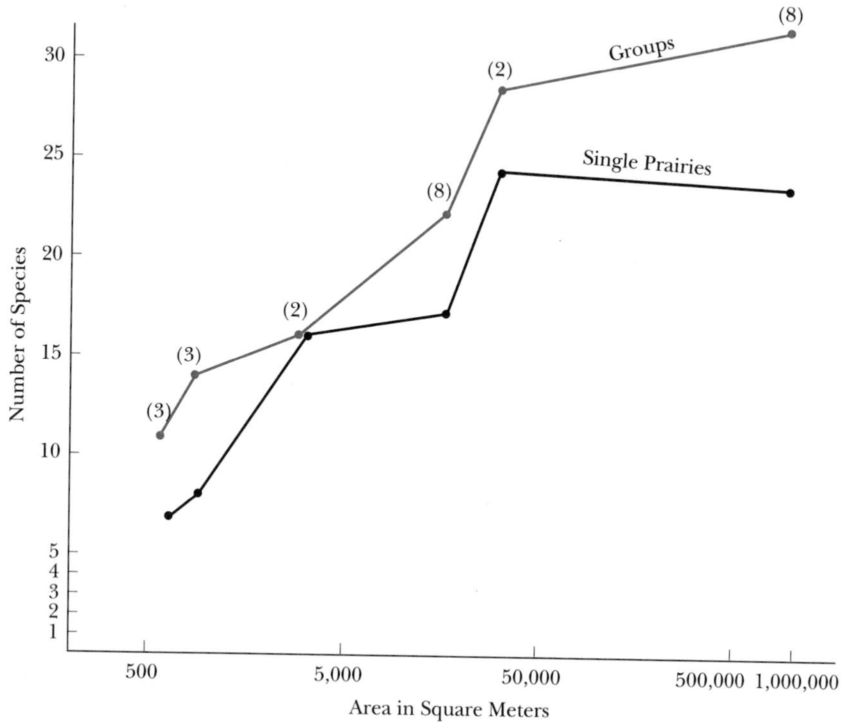

Figure 19-12

Numbers of species of three plant families found in prairie relicts in Iowa and Minnesota. The dashed line gives the number in single relicts of various sizes. The solid line gives the number in groups of smaller relicts added together. The numbers combined to produce a certain area are shown in parentheses. Thus, three smaller prairies adding up to about 500 m^2 contained a total of 11 species. (Data of William Glass, from Simberloff & Gotelli [1983], Table 1.)

disturbance produces, however, often lack the transitions and exchanges—for example between forest and savanna—of the natural landscape. Humans leave relics of the natural landscape isolated in a sea of row crops, parking lots, and suburbs (Figure 19-13). Two interrelated ways in which this fragmentation has undesirable effects on biodiversity are area effects and edge effects.

Area Effects There is now abundant evidence that some of the characteristic species of forest birds are absent from small tracts of forest (Bond 1957, Galli et al. 1976, Robbins 1979, Terborgh 1989, Askins et al. 1990). As you go from large forest stands to small—from 64 ha to 8, perhaps—you lose ovenbirds and Acadian flycatchers (Figure 19-14) and gain house wrens and song sparrows (Table 19-2).

For some species, loss in population is about proportional to loss of habitat. For **area-sensitive species** (those that are scarce or absent in small habitat patches), however, removing 50% of the forest will drop their populations by 50% plus some additional fraction based on the pattern of fragmentation and how small a patch of habitat they tolerate. We can approximately characterize the decline in non-area-sensitive and area-sensitive species thus:

Non-area sensitive species

% decline in bird population = % loss of habitat

Area-sensitive species

% decline in bird population	=	% loss of habitat	+	% of remaining habitat not inhabited because of area effects

Many of the area-sensitive species are small land birds occupying the forest interior, have small clutches, and have open nests placed on or near the ground. Evidence is increasing that some species of grassland, marsh, and shrubland habitats also show area sensitivity.

There are several suggestions as to why forest interior birds tend to be scarce in or absent from small forests. One broad explanation may be that these *K*-selected forest species (Brewer and Swander 1977) have a combination of life history traits that predispose them to local extinction and slow re-establishment.

Figure 19-13

Forest fragmentation. In a formerly forested region of northwestern Ohio only small woodlots are left surrounded by cultivated fields. (Photo by the author.)

Figure 19-14

The Acadian flycatcher is lost from smaller forests. It is seen here on a nest in a beech tree in Calhoun County, Michigan. (Photo by L. H. Walkinshaw.)

Table 19-2
The Effect of Forest Stand Size on Types of Birds Living There

Because of the way our land was surveyed, in townships, sections, and quarter-sections, isolated woodlots of 40 acres (16.19 ha) are frequent. Several of the species in the third column are rare in or absent from such stands. The species in the middle column are about equally likely to be in a forest of, for example, 8 ha as one of 24, but as the size of the woodlot decreases to 4 or 2 or 1 ha, even these begin to drop out. About the only species that may be more common in smaller forests are the forest-edge species of the first column.

Species More Likely to be Found in Forests of 14 ha or Less	*Species Not Much Affected by Forest Size*	*Species More Likely to be Found in Forests of 24 ha or More*
Downy woodpecker	Red-bellied woodpecker	Pileated woodpecker
Common flicker	Eastern wood pewee	Acadian flycatcher
House wren	Black-capped chickadee	Wood thrush
American robin	White-breasted nuthatch	Blue-gray gnatcatcher
Blue jay	Red-eyed vireo	Black-and-white warbler
Northern cardinal	Northern oriole	Worm-eating warbler
Indigo bunting	Brown-headed cowbird	Hooded warbler
European starling	Rose-breasted grosbeak	Cerulean warbler
		Ovenbird
		American redstart
		Scarlet tanager

Adapted from material in Bond, R. R., "Ecological distribution of breeding birds in the upland forests of southern Wisconsin," *Ecol. Monogr.,* 27:351–384, 1957; Galli, A. E., Leck, C. F., and Forman, R. T. T. "Avian distribution patterns in forest islands of different sizes in central New Jersey," *Auk,* 93:356–364, 1976; and Whitcomb, R. F., et al., "Effects of forest fragmentation on avifauna of the eastern deciduous forest," in *Forest Island Dynamics in Man-dominated Ecosystems,* Burgess, R. L., and Sharpe, D. M. (eds.), New York, Springer-Verlag, 1981, pp. 125–205.

Another possible factor in the absence of bird species from small woodlots (and small patches of other types of habitat) is that the habitat factors they key on are absent or behaviorally insufficient. A small woodlot might, for example, be too brushy in the subcanopy layer. For some birds, size of the habitat patch or factors correlated with size (such as the sound of other members of the same species singing in the distance) may be a component of the habitat selection process. The birds visit the small woodlot briefly and leave, or do not even stop to visit. For large birds, such as the pileated woodpecker or barred owl, it is especially likely that 40, 80, or even 320 acres (16, 33, or 130 ha) will not supply their long-term requirements for nesting sites, escape cover, or food. The average home range of one pair of pileated woodpeckers in northeastern Oregon forests is about 890 acres (360 ha) (Bull and Holthausen 1993).

The general regional population level of the species interacts with area. For a common species, a small patch of habitat may be occupied by the overflow individuals that failed to become established in large blocks (Brewer 1963). As general population size falls, most of the remaining birds accumulate in the most favorable areas—for forest interior birds, perhaps those that are darkest, least windy, least brushy, and where good habitat quality is suggested by the presence of other members of their species in the vicinity.

Edge Effects As forests are fragmented, the ratio of border to interior rises. Very small plots may be "all edge," that is, most of the stand is subjected to higher light intensities, more wind, and various biotic factors that are associated with the perimeter, the boundary between forest and grassland or shrubland.

Game managers have a love affair with "edge," dating back to Aldo Leopold's (1933) book *Game Management.* Leopold made a fairly narrow point: The potential density of certain game species, such as ruffed grouse, is increased by increasing the amount of edge. Over the years, the premier technique for managing almost anything has come to be cutting holes in the forest (Figure 19-15).

The idea of the "edge effect" later ballooned to include the idea that species diversity is also increased by increasing edge (for example, Odum 1959). This is true up to a point, the point at which forest interior species begin to be lost. The species added to a forest tract by adding edge are the species that are today common anywhere we stop our car, along every roadside and powerline, such as white-tailed deer and song sparrow.

Today when the term "edge effect" is used in the conservation literature, it is rarely with favorable connotations. Rather, it usually refers to the fact that many forest species are harmed by edge in one or both of two ways. Many avoid the space for a few to many meters next to an edge (Kroodsma 1984) so that 4 ha cut from a tract of forest actually removes from use a considerably larger amount. Secondly, reproductive success is often adversely affected because nest-parasitizing cowbirds and many predators, such as blue jays, raccoons, foxes, and domestic cats tend to enter the forest via edges (Wilcove 1985, Sandström 1991, Robinson 1992). For particularly vulnerable species, such as ground nesters, forests with large amounts of edge may be "ecological traps"—attractive habitat, but deadly to them or their eggs and young (Gates and Gysel 1978).

Figure 19-15

Clear-cutting whether for silvicultural or wildlife management purposes produces edge that is avoided by many forest-interior species. (Photo in the White Mountains of New Hampshire by the author.)

Brown-headed cowbirds, blue jays, and raccoons are examples of species that are uncommon in unbroken forest, but have increased tremendously in the chopped-up landscapes that humans have created. Brown-headed cowbirds forage in open country, so it is likely that under primeval conditions, large populations were present only around the large grasslands from Illinois westward (Mayfield 1965).

In the eastern Midwest, these social parasites were probably relatively scarce birds, occurring mostly around the small prairies. In Michigan, for example, cowbirds were probably restricted to the vicinity of the prairies in the southwest part of the state. With deforestation, they spread and multiplied. As they spread north, they put pressure on the Kirtland's warbler, a species hitherto unexposed to this species (Mayfield 1977).

Another population of birds that has undergone a precipitous decline seemingly connected with cowbird parasitism is the least Bell's vireo (Goldwasser et al. 1980). In California, the low riparian thickets favored by this species have been reduced, so some of the drop in the vireo is probably related to habitat loss. Much of the decline, though, seems connected with the invasion of cowbirds, which were probably absent or nearly so from the valleys of California before irrigated agriculture.

For a few species, competition from forest-edge species overflowing from the surrounding countryside may adversely affect forest species in small woodlots (Ambuel and Temple 1983). There seem to be few credible examples of this, but European starlings taking nest holes of red-headed or red-bellied woodpeckers are a possibility.

Most research on the adverse effects of edge has been done on birds, but it is likely that other animals and plants are affected. For example, the Canada yew plant has been nearly eliminated from fragmented northern hardwood forests. The reason

Figure 19-16

Few remaining stands of forest in the eastern United States are large enough that any part is inaccessible to white-tailed deer. (Photo at the Kalamazoo Nature Center by the author.)

is browsing by white-tailed deer (Alverson et al. 1988), which also tend to overbrowse various kinds of trees and herbs, preventing regeneration (Anderson and Katz 1993). White-tailed deer are rare in extensive areas of unbroken forest. As forests become increasingly fragmented, no area is far enough from an edge to escape browsing (Figure 19-16).

Deforestation and Other Changes on the Wintering Grounds

Many of the birds that have declined in the past two decades (Finch 1991) are **Neotropical migrants,** that is, they breed in North America but winter in central or South America (Figure 19-17). Although they are, in one sense, members of the Nearctic avifauna, many spend considerably more time away from their breeding grounds than on them. The yellow-bellied flycatcher arrives on the bogs and muskegs of the boreal forest region in June and departs for Central America in August. It spends less than one third of the year on its breeding grounds (Walkinshaw 1991). Many other Nearctic flycatchers, warblers, vireos, and orioles spend much more time in the tropics than on their breeding grounds.

It is logically possible that events on the wintering grounds are instrumental in the decrease of some of these species, and the destruction of natural vegetation in the tropics in recent decades lends credence to the possibility (Hall 1984, Terborgh 1989). One of the first declines for which such an explanation was proposed was the grassland-nesting dickcissel (Fretwell 1977). The postulated explanation for this species is slightly more complicated than a simple loss of winter habitat. It was hypothesized that replacement of small-seeded herbaceous plants by planted sorghum in the wintering grounds on the grasslands of Venezuela reduced the food supply of the female dickcissel.

The male dickcissels are larger and should be better able to use the big sorghum seeds. This situation could result in a shift in the sex ratio toward more males and fewer females, a highly disadvantageous outcome for this polygynous species (Zimmerman 1991). The decline of the dickcissel has continued since 1966, based on breeding-bird surveys (BBS).† Although there have been occasional brief reversals, overall populations have dropped at an annual rate of 1.7% (Peterjohn and Sauer 1993).

For several other species, a fairly compelling circumstantial case can be made that deteriorating conditions on the wintering grounds are the cause of their decline. Terborgh (1989) sets forth the reasons for thinking that Bachman's warbler, an endangered, or possibly already extinct, bird is in this category. The bird evidently once nested in swamp forest over much of the Southeast. Although large amounts of swamp forest have been cut over or cleared throughout its breeding range, fair quantities remain.

Apparently Bachman's warbler wintered mainly in Cuba in dense evergreen forest (Hamel 1986). Much of this was cleared during the first half of the 20th century, mainly for sugar cane plantations. Reduced winter habitat would increase winter mortality. The declining numbers of survivors would head back each spring to the scattered swamps and narrowed floodplain forests distributed over some 300,000 square miles (800,000 km^2) of breeding range. This dilution of the breeding population, making it harder and harder for birds to find one another, may have been the *coup de grace* (Terborgh 1989). This is, of course, an example of the Allee effect in action.

Figure 19-17

The ovenbird is a ground-nesting Neotropical migrant. (Carl R. Sams II/Dembinsky Photo Associates.)

So, the declines of some Neotropical migrants are probably the result of habitat destruction (perhaps combined with heavy pesticide use or other influences) in the Neotropics. Nevertheless, the evidence for wintering ground problems as a general cause for the current declines in North American breeding species is less compelling than evidence for loss, fragmentation, and deterioration of habitat on the breeding range. A recent study analyzed BBS data for migratory songbirds in the eastern and central United States (Böhning-Gaese et al. 1993). In this analysis, population declines tended to be significantly related to breeding-range factors such as cowbird and predator vulnerability but not to whether the birds were long-distance (Neotropical) migrants versus short-distance migrants (southern United States) or residents.

† Much of the information on trends in bird populations is derived from the BBS database. Once each breeding season, cooperators with the U.S. Fish and Wildlife Service and Canadian Wildlife Service take roadside counts of birds at 50 stops along predetermined 24.5-mile (39 km) routes. Some 2000 BBS routes are run each year in the United States and Canada.

Other studies using the same BBS database have, however, implicated Neotropical migrant status as an important risk factor (Robbins et al. 1989). It is unlikely that the massive deforestation going on in the tropics could be having no effects on the North American breeding species that winter there. The questions are, Which species? How big are the effects? What are the mechanisms?

It is, of course, certain that many tropical resident species are suffering from deforestation and outright killing or capture of birds (parrots for the pet trade, for example). Tropical preserves are necessary for these species as well as our migrants. For the Nearctic-Neotropical migrants, the migratory pathways are another link in the chain; suitable habitat for migration stopovers must be maintained, where necessary by preserves.

How Big Is Big Enough?

In preserving a tract of forest or prairie, the desire is to preserve the integrity of the whole ecosystem, including its form, function, and species diversity. Even though this is true, the area requirements of particular species may provide useful guidelines for the size of preserve required (Soulé and Simberloff 1986). In general, if we hope to preserve the large animals, we will need large preserves. We can probably never have cougars unless we are willing to commit several hundred square kilometers to a refuge (Beier 1993).

An analysis using the same model (Goodman 1987) as in Figure 19-11 related the population size needed to avoid extinction to area or size of preserve. The connection was made by use of known regularities in energy requirements and density for different-sized mammals (Belovsky 1987). Although the actual values resulting from this deterministic model may underestimate extinction probabilities in a given situation, the trends are probably correct. Big mammals need more space, but small mammals need higher populations (Figure 19-18). Carnivores, at the top of the energy flow pyramid, need more space than herbivores. And tropical mammals, with their generally lower densities, need more space than temperate mammals.

Large creatures, especially the large carnivores such as cougars, grizzly bears, wolves, and the large raptorial birds, are sometimes referred to as "umbrella species." The implication is that a preserve for them will also shelter many other plants and animals that use the same habitat.

Small preserves—say 20 acres (8 ha) or less—may play no role in preserving large mammals and often possess only forest-edge birds rather than ones characteristic of the preserved habitat. This does not mean that such sites are worthless, however. Plants, especially long-lived vegetatively reproducing ones, may be preserved indefinitely in a small strip of land between a highway and a railroad. Some of the prairie forbs, such as compass plants, are examples. Small preserves may also serve as stepping stones in dispersal of larger animals, or as migratory stopovers.

The keystone species of a system need to be preserved, if there are any. In some systems the extinction of one species may change the functioning or structure of the whole system or, at the least, cause substantial changes in abundance of several other species. If we are aiming to preserve Sonoran desert, we need to have preserves of a size to retain saguaro cactus and its bat pollinators. The saguaro provides homes and food for many other organisms, and the bats are useful in perpetuating the saguaro. In longleaf pine, we ought to have preserves big enough to keep red-cockaded woodpeckers and gopher tortoises. In both cases, the holes they dig become homes to many other species.

In other ecosystems, no particular keystone species are evident. Each of several species is more or less interchangeable as far as major function goes, or any keystone species are among the most common species of the system, requiring no special management.

The process of figuring out how big a population of a particular species has to be to survive indefinitely has been called **population viability assessment** or **population vulnerability analysis** (in either case, **PVA**). With this figure and knowing characteristic density, we can calculate the area required for survival. Our discussion of extinction earlier in this chapter makes it clear that "survival" has to be couched in terms of probabilities over some time span (Hooper 1971).

In a pioneer study, M. L. Shaffer (1981) defined the **minimum viable population (MVP)** for a given species in a given habitat as "the smallest isolated population having a 99% chance of remaining extant for 1000 years despite the foreseeable effects of demographic, environmental, and genetic sto-

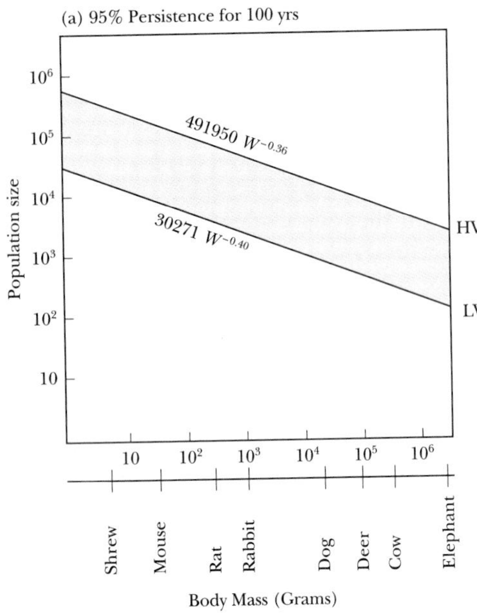

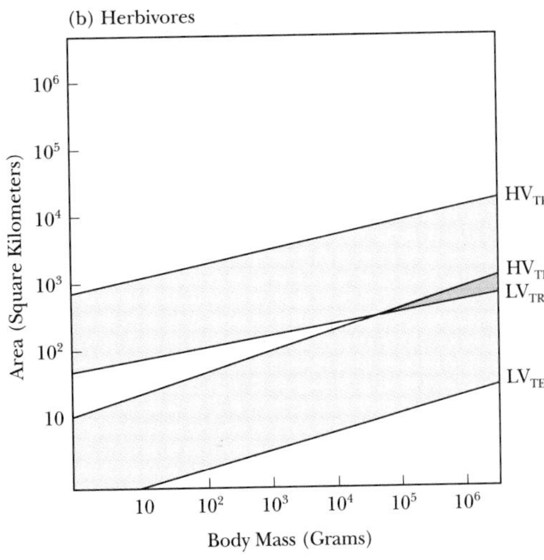

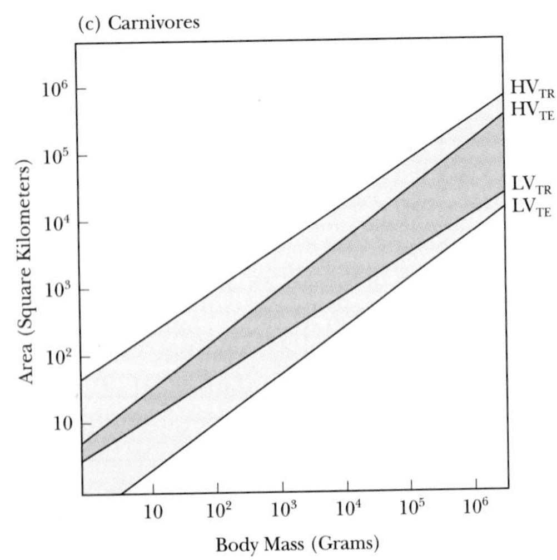

chasticity, and natural catastrophes.'' The considerably less stringent requirements of a 95% chance of survival for 100 years are now more generally used.

Using the 95%/100 years criteria, Shaffer's (1983) analysis for the grizzly bear (Figure 19-19) based on data from Yellowstone National Park (Craighead et al. 1974) suggested that the MVP was 50 to 90 bears and the minimum area required to support such a population lay between 1000 and 13,500 km^2, depending on habitat quality. Yellowstone contains a little more than 9000 km^2 (3472 square miles). During the years of the Craigheads' study, the average population of grizzlies in Yellowstone was 177. Shaffer's study did not consider genetic factors, which would probably indicate that a considerably higher population is needed to provide the genetic diversity for long-term adaptability (see point 4, below).

Although the idea of an MVP served a useful development function, MVP estimates should probably be discontinued. This is true for several reasons.

1. The PVA, giving probabilities of survival for different lengths of times under different sets of assumptions, provides more information than a single MVP number.
2. Once an MVP has been put forth, it becomes grist for the political mill. It will be hard to revise upward, even if later study shows that the earlier analysis was flawed. Furthermore, a stated MVP above the carrying capacity of a particular proposed reserve can be used as an argument against establishing the reserve, despite other values the reserve may have.
3. For the species likely to be involved in MVP discussions, that is, rare species, it is almost al-

Figure 19-18

The top graph (A) gives population sizes required to give a 95% chance of persisting for 100 years for different-sized mammals. The bottom two graphs translate these figures into area, or size of preserve, required for herbivores (B) and carnivores (C) and for tropical (TR) and temperate (TE) habitats. HV refers to high variance—habitats where environmental stochasticity is great. LV is low variance—habitats with little environmental stochasticity. The mammal names are indicative of sizes but were not actually used in calculations. (Based on Belovsky 1987.)

Figure 19-19

A grizzly bear in Yellowstone National Park. (William S. Keller, U.S. Fish and Wildlife Service.)

ways going to be the case that more will be better. If a tract of federal land contains 1200 animals of some threatened species, an MVP estimate of 800 may well lead to pressure to cut over, develop, or otherwise remove 33% of the habitat. This reduces the margin for error in the original estimate and removes flexibility in such forms as having a surplus for transporting animals to other sites.

4. MVP is not a simple concept. A distinction has to be made between crude population size—what we can go and count on the ground—and effective population size. It appears that most genetic variability can be maintained indefinitely with an effective population size of around 500 individuals (Lande and Barrowclough 1987). However, when we take into account age structure, sex ratio, variations in family size, neighborhood effects in mating, and fluctuations in population size (in bad climatic years, for example), an area having a carrying capacity of a few thousand is

likely to be necessary to provide this effective population size for most animals (Soulé 1987).

One more concept is worth discussing under the general heading of How big? Buffer zones can increase the value of small reserves. Although there is little research directly on the topic, it seems reasonable that buffers can ameliorate effects of small size in two general ways: (1) They can increase the apparent and, in some cases, the effective size of the preserve for area-sensitive species. For example, a forest-interior species might be willing to settle in a small reserve of mature forest surrounded by successional forest. It might nest only within the mature core but might be affected in its habitat selection by the larger visual size of the block and might also make use of the younger forest for some of its activities, such as feeding. (2) The buffer might decrease certain types of edge effects. For example, a zone of grassland or shrubland, rather than lawn, around a marsh might reduce erosion into the marsh, might cut down on phosphate input, and might provide singing posts for some of the marsh birds.

If the buffer zones are not owned as part of the reserve, it may nevertheless be possible to manage them through cooperative arrangements or by conservation easements.

Advantages of Multiplicity

Capturing area-sensitive species and reducing edge effects are good reasons for large preserves. So also is the fact that large populations are less vulnerable to stochastic processes that may take them near or to the extinction threshold. There are, however, several lines of evidence suggesting benefits of multiple, rather than just one, preserve. The upshot, as stated by Michael Soulé and Daniel Simberloff (1986), is that "Nature reserves should be as large as possible, and there should be many of them." This conclusion is absolutely correct, although the implementation of it poses practical problems today and for the foreseeable future.

In this section we review some of the thinking and evidence supporting the desirability of multiple reserves.

The first reason for multiple reserves is reflected in the aphorism about not putting all your eggs in one basket. The aphorism represents good sense even if you have a large basket. Local extinction of a rare species is more likely on a small preserve, but if the whole population of a species is on a single preserve, the population may be totally eliminated by some extreme environmental variation, such as a catastrophic storm or a new disease.

The detrimental effect of environmental stochasticity will be moderated by multiple preserves as long as they are separated enough or different enough that the same environmental variations do not affect all of them in the same way. In other words, if a drought lowers reproduction in some patches, the total population will not be affected as much if (a) some of the preserves are outside the region of the drought or (b) some of the preserves have microclimates where the effects of the drought are not felt as severely. If all the preserves endure the same environmental variation, a population gains little relief from the effects of environmental stochasticity. This is like putting your eggs in several baskets and then dropping all the baskets.

An eggs-in-baskets model incorporating environmental stochasticity showed that subdividing a species of 1000 individuals into almost any number of subpopulations decreases the probability of species extinction (all the subpopulations extinct) over the short term (Quinn and Hastings 1987). At the time horizon is extended, however, some intermediate number of subpopulations was more effective in preventing species extinction than either a large number of subpopulations or no subdivision at all (Figure 19-20).

The model just described was based on the assumption that no interchange occurred among the subpopulations. For species with very low mobility, such as plants with ineffective long-distance dispersal mechanisms, this assumption may be realistic. For many animals, however, interchange occurs among isolated populations.

In these metapopulations (Levins 1970), the same mitigation of environmental stochasticity is obtained, and there are often other benefits. We have already mentioned the fact that a subdivided population tends to retain more population-wide genetic variability than a nonsubdivided one. This is because (1) different alleles may go to fixation (from inbreeding) in different subpopulations and (2) different selective pressures in the different lo-

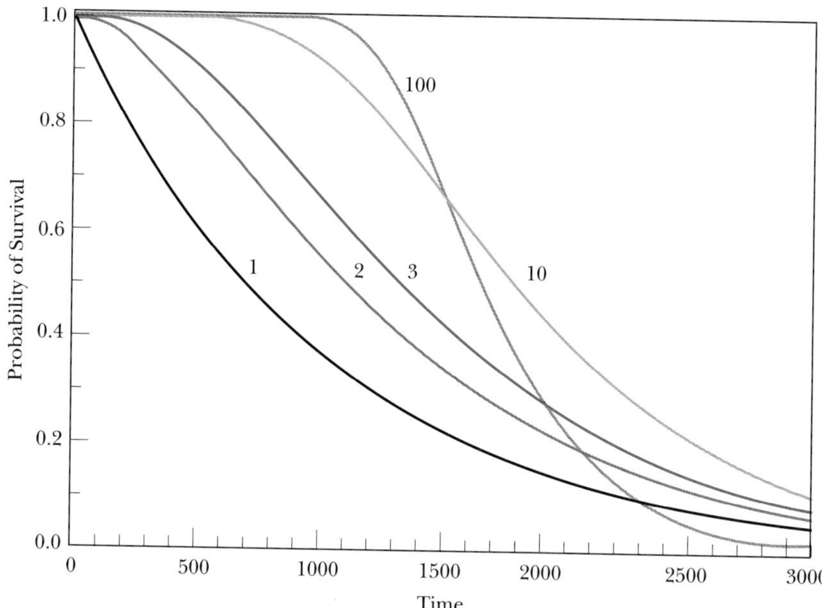

Figure 19-20

The graph plots the probability of survival of one or more populations against time when the population (N = 1000) is not subdivided (1) or subdivided into 2, 3, 10, or 100 subpopulations. (Details of the model are given in Quinn and Hastings 1987, on which this is based.)

cales may lead to different gene frequencies (Hedrick et al. 1976).

When the local population occupying a patch of habitat (such as an isolated preserve) goes extinct, it may be replenished by immigrants from another patch. Many cases of local extinction followed (or sometimes, not followed) by recolonization have been reported (examples are given in Taylor 1991 and Verboom et al. 1991). Suppose that a certain butterfly goes extinct in a single 8-ha maple woodlot every ten years on the average. If four such forest fragments are close enough to one another that dispersal can occur among them, the likelihood that the butterfly would go extinct in the four-woodlot system is greatly reduced. The likelihood that extinction would occur simultaneously in all four is, in fact $(1/10)^4$. In other words, only once in 10,000 years, on the average, would the butterfly go totally extinct in the four-woodlot system. This example is oversimplified in various ways, especially the possibility that the same cause of extinction could affect all four woodlots at the same time if they were close together.

For highly motile organisms, like birds, local extinction and reinvasion may happen on a time scale so short that we do not detect the extinctions unless we are dealing with a closely studied marked population. This rapid replenishment has been called the "rescue effect" (Brown and Kodric-Brown 1977).

Michael Gilpin (1987) suggested three rules derived from island biogeography theory as a start on understanding how the extinction/colonization process works in a metapopulation.

1. Large patches produce more emigrants
2. The probability of a patch being colonized falls exponentially with distance from the nearest source population
3. The probability of local extinction is inversely related to the size, or carrying capacity, of the patch

A study of European nuthatch occurrence in forest patches in the agricultural landscape of western Europe suggests that these generalizations are broadly correct. In three regions possessing 135 forest fragments studied from two to three years, 23 extinctions and 14 colonizations were recorded (Verboom et al. 1991). Extinction frequency was significantly related to carrying capacity of the patch (Figure 19-21) and also to patch quality (more mature forest = better); it was unrelated to an index of probable immigration volume. This index combined distances to and occupancy status (including number of territories) of adjacent patches. Thus, it combined information from

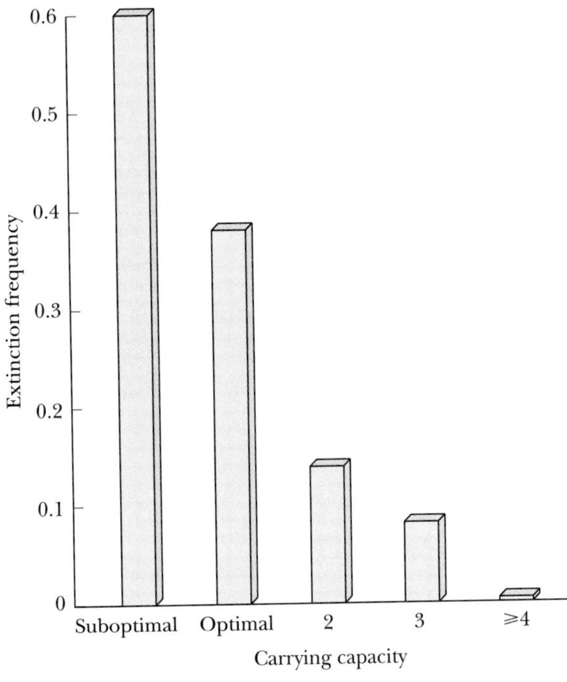

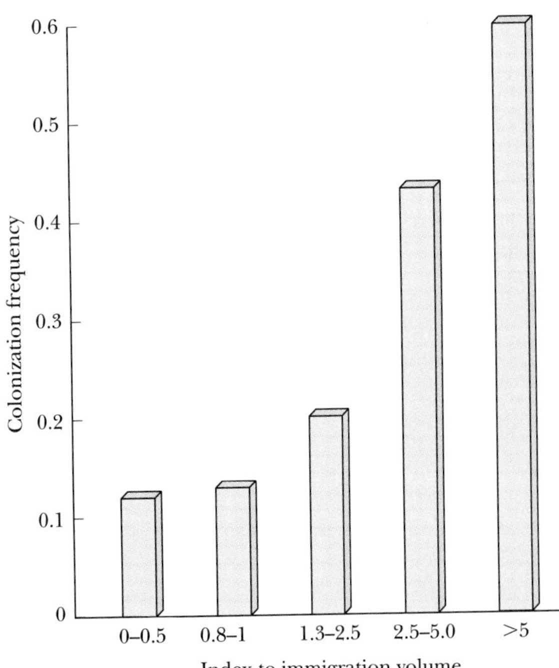

points 1 and 2 in Gilpin's list. Colonization frequency was significantly related to this index but not to carrying capacity or habitat quality (Figure 19-21).

A set of terms related to metapopulation structure is "source" and "sink" to refer to habitats or populations (Van Horne 1983, Pulliam 1988). Some patches may house local populations that are nearly always "sources," that is, birth rate is greater than death rate, leading to surplus individuals that disperse from the patch. Other patches, smaller or of marginal habitat quality, may be nearly always "sinks." Birth rate is less than death rate; populations in these patches tend to go extinct unless replenished by immigration (Figure 19-22).

What is the significance of sink populations? Can they contribute to the extinction of a species by draining individuals away from patches of good habitat (Buechner 1987)? Would we be doing the forest-interior birds a favor by cutting down all the woodlots less than 16 ha?

A complete answer as to the effects of sink habitats would be complex, but the short answer seems to be that they are more likely to contribute to the survival of a species than to hinder it (Howe et al. 1991). The term "sink" may be misleading, perhaps carrying an implication that nothing gets out once it enters. In fact, reproduction goes on in most sink habitats, independent young are produced, and some of these emigrate to other patches, including source habitats. Even adult animals in sink habitats may move on to other patches, especially if they have poor breeding success (Roth and Johnson 1993).

This "storehouse effect" by habitats of marginal quality may be important to the persistence of rare species. The jack pine areas resulting from controlled burns in northern Michigan, although

Figure 19-21

Metapopulation dynamics in European nuthatches. (The top graph relates extinction frequency (per year) to size of the habitat patch. Also for small patches optimal (mature deciduous forest) and suboptimal (young or mixed forest) are compared. The bottom graph relates colonization frequency to an index of probable immigration volume that includes distances to and occupancy status (including number of territories) of adjacent patches. (Based on Verboom et al. 1991.)

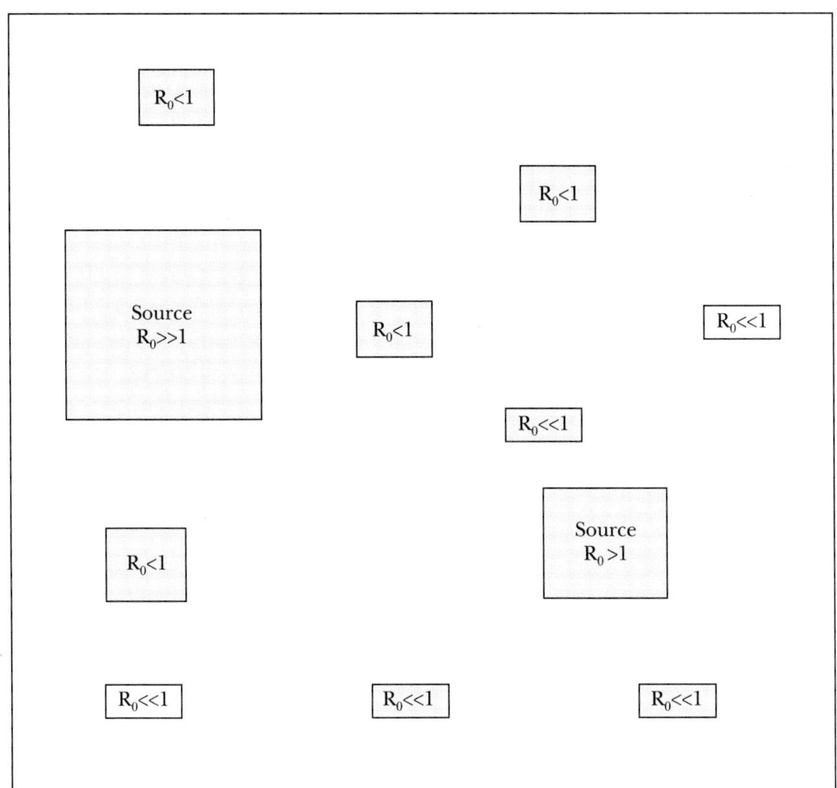

Figure 19-22

A hypothetical metapopulation structure. Shown are two source habitat patches (net reproductive rate above 1) and nine sink patches (net reproductive rate below 1). Two of the sink habitats (the unshaded patches) are currently unoccupied.

not optimal habitat, may have served as storehouses for the Kirtland's warbler during a period in which good habitat was scarce or even nonexistent. A sink to which dispersing black-footed ferrets from the Meeteetse population could have been funneled might have helped preserve a population of that species in the wild.

Among the probable benefits to a source-sink system—compared with a single source patch—are (Howe et al. 1991)

1. A larger overall population size
2. A larger source population
3. Extended survival time of declining populations

One of the groups of organisms for which a source-sink population structure is probably important is the forest-interior birds. It is likely that the continued occurrence of several of these species—such as ovenbirds and scarlet tanagers—in small woodlots in much of the eastern United States depends on immigration from a few source habitats. Often, reproductive rates in these woodlots seem clearly to be below the probable mortality rates (for example, Robinson 1992). The identification, preservation, and proper management of these source habitats are of great importance for the survival of these birds (Brewer 1992).

Two final points having to do with biological variability of species should be made. The same patch or preserve may be a source for some species and a sink for others. Also worth remembering is that the same landscape that is a series of habitat patches—sources and sinks—for small animals of limited movement may simply be a heterogeneous habitat with scattered foraging, nesting, and escape cover for large, wide-ranging species.

Looking beyond the species level, multiple reserves even if designed to preserve the same ecosystem type, such as mesic forest or wetland, are likely to preserve additional diversity above that of a single preserve. We have already indicated that they will probably have more species in total. Usually, they will also capture additional variability in such community-level traits as successional status and variations in structure and biota related to moisture and other gradients.

Corridors

A corridor by the dictionary is a narrow passageway. Corridors in conservation biology are linear connections between two or more patches of wildlife habitat that are meant to serve as conduits for animals (Figure 19-23). Although major symposia and books have dealt with corridors (Saunders and Hobbs 1991), very little is known about how well they function in getting organisms from one patch to another (Simberloff et al. 1992).

It seems logical that connecting isolated patches with a strip of somewhat similar vegetation would facilitate the interchanges that we have talked of as being valuable in moderating unfavorable stochastic effects. There are, however, also logical reasons for doubting that interchange would be enhanced and also for thinking that corridors could have disadvantages. Among the reasons for doubting the effectiveness of corridors is the fact that narrow corridors would be mostly edge. Individuals of forest species might be no more willing to traverse them than to fly from tree to tree across a field. Also, travel through the corridor might be hazardous because of the forest-edge predators frequenting the corridor. If the corridors were long and winding, the exposure to predators might be enough that successful movement through the corridor would be unlikely.

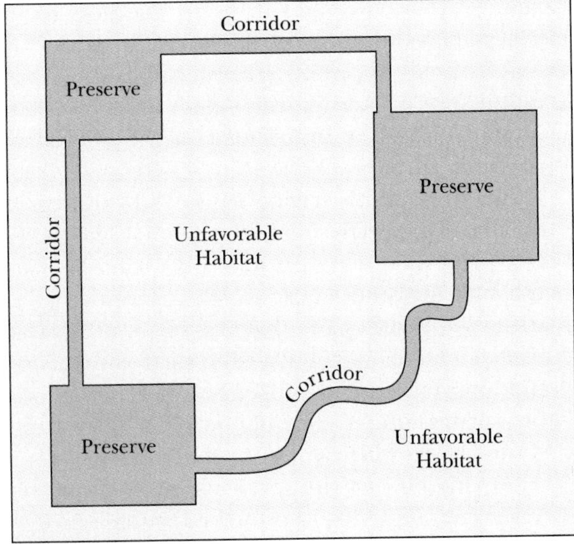

Figure 19-23

Three preserved areas linked by corridors. As shown, the three corridors together make up somewhat more than 10% of the area of the smallest preserve.

Among active disadvantages of corridors might be that they could themselves serve as highly unfavorable sink habitats. For example, forest animals might be attracted to them for breeding but suffer the edge effects of cowbird parasitism, predation, and possibly unfavorable microclimatic effects. Conceivably, they could serve as conduits for disease and introduced pests. Corridors also take money that might—or might not—be spent more profitably for preserving other pieces of land.

If corridors that are meant to allow dispersal of organisms from one large patch to another are ever useful, one group that would seem predisposed to profit most from them would be the large predators. They routinely move considerable distances and would be at very little risk from predation themselves. Observations of cougar movements suggest that these animals could and would make use of corridors (Beier 1993), although often the corridors most valuable to them will be across prime development land that would be very expensive to preserve (Simberloff et al. 1992, Beier 1993).

Intrarange corridors, or travel paths, meant to facilitate movement within the normal home range of the animals living within a patch are clearly of benefit to individual animals. An example is the highway underpasses that allow amphibians to move to and from breeding ponds and cougars simply out on a night's hunt to avoid getting run over by cars. Often these travel paths must raise survivorship and/or reproduction for the affected species, and overall it seems reasonable that they will improve persistence time of the species in the patch where these amenities are provided.

Landscapes

When resources will allow very large reserves, these can incorporate some of the advantages of multiplicity and add others. We will refer to this as preservation at the landscape level. Here a large area, bigger than a small watershed, is being protected—

a few thousand hectares, for example, that contain a mosaic of vegetation types or ecosystems. In Montana, as an example, an area of this size might contain short-grass grassland, ponderosa pine savanna, ponderosa pine forest, ponderosa with invading more shade-tolerant trees, and a stretch of stream with riparian forest. It might also contain cutover areas of ponderosa pine and overgrazed and degraded grassland.

If we can protect areas on the landscape scale, we have the potential to do several things:

1. Provide homes for some of the large, wide-ranging animals
2. For the small animals and plants, preserve several patches among which interchanges can occur
3. Provide a setting in which many large-scale processes, such as windthrow and fire, can occur naturally without compromising the value of the preserve. That is, a tornado passing through a small forest preserve can blow down many trees and reduce its value for preserving some birds, for the public as an example of a mature forest, and for certain research purposes. A much larger preserve—one with several forest stands or one very large one—would not suffer these ill effects. The tornado-devastated area would be a patch of interest in its own right.
4. Illustrate landscape-level ecological processes. These include the role of disturbance, such as tornadoes and fire, in producing landscape patches and the presence (or absence) of a tendency toward an equilibrium landscape. Other landscape-level processes include spatial transitions between ecosystems and flows between ecosystems or patches—the movement of organisms and the transport of organic matter and minerals.

Several plans for large nature reserve systems have been put forth in recent years. The **greater ecosystem concept** is one such plan (Grumbine 1990). The greater ecosystem concept stems from the realization by F. C. Craighead, Jr. (1979) that Yellowstone National Park by itself cannot provide the permanent needs of a viable grizzly bear population. A Greater Yellowstone Ecosystem has been proposed that includes Yellowstone National Park plus much of the Gallatin, Shoshone, Bridger-Teton, Targhee, and Gallatin national forests plus Red Rock Lakes National Wildlife Refuge (Reese 1984). In all, perhaps 8000 km^2 would be involved.

Preservation and Management

There is an old dichotomy, still trotted out occasionally by the wildlife managers and foresters: Preservation versus conservation. Preservationists are the bad guys. They want to preserve some piece of land—lock it up and leave it alone. Conservationists, on the other hand, adhere to Theodore Roosevelt's principles; conservation is "wise use." This view of conservation may still be defensible if "wise use" is broadened to include no use, or at least no commodity use. But the dichotomy is faulty and should be packed away in the same trunk with such outmoded concepts as "overmature" for the old trees that woodpeckers and flying squirrels like.

Another dichotomy is of more importance, because it still leads to dissension among those who share the viewpoint that nature is worth preserving for its noncommodity values. Everyone in this category might agree that preserving diversity among the herbs of a prairie is a desirable aim. Where they may part company is whether this aim is best achieved by allowing natural processes to take their course or by active management of preserved areas.

The primary argument for management is that in our preserved fragments, many important natural processes have been altered or suspended. In the case of prairie preservation, repeated observations and studies have made it clear that most prairies will be converted to brush in a few years if management is not practiced. The most desirable form of management is frequent controlled burning, mimicking the wildfires of the past.

Because many preserves are isolated from other examples of the same ecosystem, little or no dispersal may occur among them. For some of the characteristic but rare species, this lack will lead to loss of genetic diversity and even extirpation. For such species, active transport of individuals is probably a desirable intervention. In Illinois, greater prairie chickens numbering in the millions once danced and boomed on the tall-grass prairies. Now there are 40 birds in three isolated populations. Low egg fertility and high embryo mortality today compared with the 1930s suggest a loss in genetic diversity. Attempts are being made to counteract this by interchanging eggs between nests among the populations (Westemeier 1992). For birds this procedure of using resident adults as foster parents has a greater likelihood of success than transporting independent young or adults.

One of the most pervasive problems today is the scarcity of top predators in the landscape. In some cases, we probably cannot even imagine the ways that our preserved "natural" areas would be different if we had the predators back. In other cases, the consequences of their absence are all too clear. One consequence is the overabundance of white-tailed deer in much of the eastern United States and the resulting depauperization of the forest understory (Diamond 1992, Anderson and Katz 1993).

Most preserved natural areas ban hunting, just as they ban digging up flowers, and that is part of the problem on these preserves (Girard et al. 1993). One nature center in Michigan (the Chippewa Nature Center near Midland) recently faced up to the issue and invited bow hunters in to reduce the deer herd. Many persons who are supporters of nature centers oppose hunting, either in general or under the sanctuary conditions of a nature center. Bow hunters were presumably favored because the bow is more natural and less disruptive than the thunder-stick.

Even those conservation biologists most strongly committed to the need for managing preserves would probably agree that (1) management practices should be as unintrusive as possible; (2) they should mimic the natural processes that have been suspended or altered; and (3) reintroduction of the natural processes, if possible, would be the best management option.

Even in our largest parks, such as Yellowstone and Glacier, a totally hands-off approach seems unlikely to be the best option. Even these large reserves suffer from predator lack, as shown by the overabundance of elk in Yellowstone, for example (Diamond 1992). It is likely that, big as it is, Yellowstone cannot sustain a population of grizzly bears over the long term (Shaffer 1991, Dennis et al. 1991). Timber wolves were exterminated from both parks long ago. Reintroduction is desirable, but convincing the local hunters, ranchers, and shopkeepers of this will be at least as difficult as convincing nature center members of the desirability of permitting deer hunting.

One more issue needs discussion. In managing an ecosystem, what is our target? We can probably agree on the aim of preserving natural form, function, and diversity. But community change is a constant, as droughts and glaciers come and go, as species evolve and go extinct, as ranges expand and contract, so we must decide on what earlier natural time we are going to target (Sprugel 1991).

Suppose that a new preserve is acquired, such as the land for the Indiana Dunes National Lakeshore. Among the parcels were some with relatively undisturbed vegetation and others with buildings and plantings, roads now to be abandoned, and other heavily disturbed areas (Hiebert and Pavlovic 1987). Even some of the more natural areas included nonnative plants and animals that were maintaining themselves or spreading.

What earlier natural time should management of the Indiana Dunes National Lakeshore and other such areas attempt to emulate? The most frequent decision has been that the target, or model, should be the landscape and communities of the time immediately before widespread European-style settlement. We will try to figure out what things were like then—say, about 1800 for northern Indiana—and manage to encourage those conditions.

There is much to recommend this decision, but it is not the only one possible. We could choose a slightly earlier period, before the fur trade began. We might even wish to choose a time prior to any human influence, perhaps 20,000 years ago, but would probably reject this option because of the extinction of the Pleistocene megafauna that followed the arrival of humans in North America. We might feel that the absence of mastodons, mammoths, saber-tooth tigers, and giant predatory birds would compromise our restoration efforts even more seriously than the absence of the passenger pigeon and Carolina parakeet, which disappeared following European settlement.

Having made the choice of time period, we still have many practical questions to contend with. One set involves the alien, or exotic, species. Most conservation biologists believe that aliens should be eradicated from nature reserves (although some, such as Lugo 1992, hesitate). Even if we are quite sure that we want purple loosestrife out of the marshes and common buckthorn out of the oak openings, how to accomplish this may pose a tough management problem (Thompson 1991).

Restoration

In many regions of the world, few traces of the natural landscape remain. In much of eastern North America—and the situation is much worse in most of Europe—were we to preserve every remaining patch of natural vegetation, we and our children and grandchildren would still live in a depauperized landscape. If central Illinois is ever to have a square mile of native prairie to marvel at, it will come about by setting aside a section of cornfield and restoring the tall grasses and prairie forbs that once grew there.

Restoration ecology is a new field that grew out of just such efforts as bringing back Midwestern prairie (Jordan 1981). **Restoration** can be defined as the return of an ecosystem to its predisturbance form and function. The aim, according to a National Research Council Committee on Restoration of Aquatic Ecosystems (1992), "is to emulate a natural, functioning self-regulating system that is integrated with the ecological landscape in which it occurs." It is probably best to use the term restoration in a broad sense, while retaining additional terms that describe kinds of restoration or related types of construction and reconstruction of ecosystems.

Restoration efforts up to now have mostly been concerned with prairie and aquatic systems. Prairie reconstructions (Pleznac 1983) have often been successful in recreating the plant composition and look of native grassland (Schramm 1992). Wetland restoration and reconstruction have had variable results (National Research Council on Restoration of Aquatic Ecosystems 1992). Wetland **mitigation** efforts seem often to have failed to achieve the aim of "a natural, functioning self-regulating system." "Mitigation" simply means moderating; in this context, it refers to moderating the effects of destroying wetlands—for a new golf course, highway, or housing development. The mitigation often consists of digging out an area of comparable or larger size somewhere else and letting it fill with water.

If some restoration efforts have been less than successful, this probably means simply that more care is going to have to be taken—more time for research on hydrology and ecosystem function, more money for study, finding and buying the right site, introducing the appropriate elements, and management. Restoration, even more than preserving remaining pristine sites, is the wave of the future for many regions. If southern Michigan is to have even one moderately large beech-maple forest—say 160 acres (64 ha), it will be through restoration in some sense, because no such forest plot exists to preserve.

As McPeek and Brewer (1991) pointed out, large areas in the eastern half of the United States have been taken out of agriculture in past decades and have undergone succession to shrub or tree stages. It should be a major priority for conservation organizations or government to buy large blocks of such land on dry, wet, and mesic sites and allow the natural processes of succession to restore the corresponding forest types. The process can be accelerated by seeding, transplanting, introducing various animals, especially the less motile ones, and judicious removal of single individuals of successional tree species.

One landscape to which serious restoration efforts have begun to be directed is the longleaf pine of the Southeast. Restoration measures appropriate to the level of degradation have begun on several sites by the Nature Conservancy. One particularly challenging case is a large tract in northern Florida where a heavily degraded area of slash pine plantation is being converted back to wiregrass and longleaf (Seamon 1992).

Another landscape that needs to be restored is the white pine ecosystem of the northern Midwest (Brewer 1992). These pineries are part of the historic heritage of the Great Lakes area but were obliterated in the timber era of the late 19th century. Natural successional processes could restore the native white pine landscape, but this has been prevented by forestry practices on state and federal lands. Timber has been cut on a short rotation that has kept the land in oak and aspen.

Note that in both these cases, the landscape level is the appropriate scale for restoration. Although the usual phrase has been restoration of the "longleaf ecosystem," the "longleaf ecosystem" actually consists of several ecosystems from dry sandhills to open wet savannas. Similarly, the "white pine ecosystem" consisted in primeval times of forest, savanna, areas with more and less oak, and areas in various stages of recovery from fire.

Why Preserve Biodiversity?

Aesthetic and Practical Reasons

It is time to consider why we should bother to preserve nature. "In what way does the sooty tern serve mankind?" someone asked a friend of mine and waited with great seriousness for the answer.

Why should a stretch of shoreline be set aside permanently as a natural area, when it might well be the site of a new condominium complex? Why should we worry about the disappearance of a species? Do we really need whooping cranes or California condors? There are many arguments to be made, but they tend to belong to three broad categories—practical (including prudence), aesthetic, and moral or ethical.

It is a more pleasant world with forests, marshes, and prairies, with birds singing, lions and black rhinos (Figure 19-24) roaming the veldt, and spotted owls hooting in the old-growth forests. So there are aesthetic reasons for preserving nature, and these are a powerful argument for many persons. The few areas of remaining natural longleaf pine in southern Georgia were preserved by the plantation owners partly for the practical reason that they provide good bobwhite hunting but also for the beauty of this landscape. Warren Woods in southern Michigan—probably the best remaining stand of beech-maple forest in the United States—was bought and preserved by Edward K. Warren "for its great natural beauty" (Brewer 1980), and there are similar examples in every region.

One practical argument is that many natural areas are irreplaceable, at least in the short run, and species are irreplaceable on any time scale. If a hardware store lies in the path of a new highway, the owner can, although he may not want to, build a new store somewhere else. But beech-maple forest destroyed in highway construction is, for practical purposes, permanently lost. To produce a new beech-maple forest starting with bare ground would take hundreds of years of succession (Table 19-3).

Figure 19-24

The black rhinoceros is an endangered species of southern Africa. Poaching to obtain the horns, which are sold as aphrodisiacs in China, Taiwan, Korea, and Yemen, is a major cause of mortality. (Fritz Polking/Dembinsky Photo Associates.)

If we get rid of all the old-growth forest this year, we cannot have some back next year even if that turns out to be highly desirable.

Natural areas are places for scientific research on the whole ecosystem, individual plant and animal species, geology, and soil science. They are places for education, not only in ecology, zoology, and conservation, but in geography, literature, and history. It is impossible to convey the sense of wonder felt by the settlers at their first sight of the prairie if the nearest thing to a prairie still available is an oat field.

By providing a place in which species incapable of living on lawns and roadsides can continue to exist, these species with their unique gene pools will be preserved. The potential practical benefits of this are enormous. The uses of native organisms—plant, animal, or microbe—for medicines and crops, in biological pest control, and for using in processing sewage and other wastes have only begun to be investigated.

May-apple, with the chemical podophyllum used in treating skin cancer and venereal warts, and

Table 19-3
Estimated Replacement Times for Communities in Illinois

One aspect of deciding how land should best be used is the "cost" of replacing it, measured in the number of years required to develop an ecosystem once it is destroyed. The right-hand column, "years of replacement time," is the number of years to get back some typical or medium-aged example of the community after disturbance. For example, cutting the large trees in a beech-maple forest produces disturbance, but this may be largely erased in 35 to a few hundred years. If, however, the forest is cut and cleared and the land planted to crops, an additional amount of "successional lead-in time" would have to be added. The cost of replacing each community starting at bare ground, then, is given by adding the two columns together.

Gross Habitats	*Years of Successional Lead-in Time*	*Years of Replacement Time*
Bottomland forest		
Oak-gum-cypress	100–150	20–600
Elm-ash-cottonwood by age 5–29 years	35	5–29
(willow-cottonwood) 30–59 years	35	30–59
(willow-cottonwood-maple) 60–99 years	35	60–99
(hackberry-gum) 100+	135–600	100–500
(hackberry-gum, elm-oak-hickory, and succession to climax)		
Upland forest by age 10–29 years	25	10–29
(black cherry-elm-hawthorn, elm-persimmon-sassafras) 30–59 years	50	30–59
(elm-oak-hickory) 60–99 years	100	60–99
(oak-hickory) 100+	100+	100–500
(oak-hickory with possible succession to maple-beech)		
Maple-beech	150–200+	35–500+
Aspen	5	5–39
Pine forest by age		
10–39 years	25	10–39
40+	25	40–100+
Shrub areas	3	3–30
Residential habitat		1–100+
Marsh, natural	1,000+	600+
Marsh, man-made	3	3–100+
Prairie	10–15	10–30+
Ungrazed and fallow fields		1–10
Pastures		1–10
Hayfields		1–3
Small-grain fields		1
Row-crop fields		1

From Graber, J. W., and Graber, R. R., "Environmental evaluations using birds and their habitats," Ill. *Nat. Hist. Surv. Biol.* Note No. 97, pp. 1–39, 1976.

Pacific yew, source of taxol useful for ovarian cancer, are two cases of medicinal use (Hamburger and Hostettmann 1991). Other plants have compounds that are promising in themselves or as a starting point for systematic structural modification in the pharmaceutical laboratories for treating other types of cancers, human immunodeficiency virus, stroke, malaria, as contraceptives, and in many other applications. When we lose species, we, of course, lose the opportunity even to test for potentially useful compounds. But the loss of every natural area also diminishes this resource. We lose sources for harvesting such chemicals, and also we lose genetic variability. This diminishes the chance of finding strains of plants that lend themselves to cultivation or that produce unusually high yields of the chemicals.

Two scientists attempted to place a dollar value on the consequences of plant extinction in the United States, based just on potential medical use (Farnsworth and Soejarto 1985). Comparing the number of species that are used in medicine today (about 25% of all prescriptions use compounds derived from higher plants) with the number that have been examined for therapeutic effects, they obtained a ratio of 1/125. With this as a starting point and with data on extinction rates and the dollar value of prescriptions, they concluded that the loss from higher plant extinctions in the United States by the year 2000 would be about $3.25 billion. A more complete economic analysis adding in such things as lost work time and premature deaths stemming from the lack of the lost drugs would increase that figure.

There are many other types of practical uses of organisms and ecosystems. Organisms are sources of genes that can be transferred to crop plants or bacteria. Natural ecosystems are the best way of storing this genetic variability. Native African grazers may do a better job of producing protein on the savannas than imported cattle. Some wetlands are good sewage treatment facilities. The study of natural ecosystems may provide insight into how human-influenced ecosystems can be made more stable.

There is a large category of what, from an anthropocentric point of view, are "ecosystem services" (Odum 1989, Ehrlich and Ehrlich 1992)—things that ecosystems and their constituent organisms do for us. We have already met many of these services in their basic ecological contexts earlier. Included are the roles organisms and communities play in maintenance of the gaseous composition of the atmosphere, biogeochemical cycling, soil formation, watershed management, pest control, and pollination.

In analyzing the effects of natural areas in the landscape on humans, it is difficult to know where to draw the line between practicality and aesthetics. Nearly everyone will agree that a forest or an undeveloped lake is valuable as scenery and as a peaceful place to escape from the pressures of civilization. Some would go further and argue that humans are the product of some several hundred thousands of years of evolution, which occurred not in apartment buildings and parking lots but in nature. As adaptable as humans are, it may nevertheless be true, as H. H. Iltis (1969) claims, that ". . . for our physical and mental well-being nature in our daily life is an indispensable biological need."

Moral and Ethical Reasons

There are aesthetic reasons for preserving nature. If aesthetics is the only criterion, though, many people could readily dispense with various unspectacular landscapes and ugly and insignificant species. We could compile a list of organisms that the average citizen could get along without: Who needs bats, chiggers, spiders, slugs, or snakes? Surely no aesthetic argument could be made for preserving the proboscis monkey. Dispensable, too, might be many swamps, ponds covered with scum (that is, algae), and native prairie, which looks a lot like a weed field.

There are many practical reasons for preserving nature, of which we have listed a few. These do not, however, make persuasive arguments for the *permanent* preservation of species or communities. If marshes are preserved only because they process sewage cheaper than man-made tertiary treatment plants, then the marshes must go when sewage technology improves and land prices go up. If some species of plant is thoroughly tested by the phytochemists and found to have no promising bioactive compounds, it can be discarded.

Many of the practical arguments are those of prudence. Some species or ecosystems may have an importance that we cannot foresee (Farnsworth

***Aldo Leopold** (American, 1887–1948)*

Leopold was born into affluence in Burlington, Iowa, a rail and manufacturing center on the Mississippi River in the southeastern corner of the state. After prep school in New Jersey, he went to Yale University and stayed for a master's degree (1909) at Yale's new Forestry School. The early part of his career was with the U.S. Forest Service, first in the Arizona and New Mexico Territories, where he married into a prominent Spanish family.

After four evidently frustrating years (1924–1928) as Associate Director of the Forest Products Laboratory at Madison, Wisconsin, he became an agent for the Sporting Arms and Ammunition Manufacturers' Institute, promoting game management. Following publication of his book *Game Management* (1933), a position in that field was created for him at the University of Wisconsin, where he remained until his death, aged 61.

Leopold's connection with game was not only academic. He was a hunter from his earliest days to his last. In the first two weeks of a trip to the Delta Colorado, in 1922 (described in *Round River*), he and his brother killed 35 Gambel's quail, 8 Canada geese, 8 doves, 3 teal, 2 mallards, 2 wigeon, 2 coots, 3 avocets, 1 willet, 1 coyote, 1 bobcat, and 1 peregrine falcon, at which I stopped counting. There is probably more, rather than little, connection between his hunting experience and the ideas he eventually formulated.

Leopold published many scientific papers, was president of the Ecological Society of America and the Wildlife Society, and will be remembered for his pioneering efforts to put wildlife management on a scientific footing. He made many personal contributions to conservation. He was, for example, instrumental in getting the first wilderness area set aside in a national forest (in Gila National Forest, New Mexico, in 1924). He was one of the founders, in 1935, of the Wilderness Society. In the last years of his life, from 1943 on, he was a member of the Wisconsin Conservation Commission. There, he and his fellow commissioners engaged in interminable battles among themselves and with the legislature and certain segments of the public on how to manage the state's deer herd. In retrospect, it is an odd and peculiarly American spectacle: Aldo Leopold arguing game management policy with barbers and lawyers—and generally losing.

More than anything else, Leopold will be remembered as the author of *Sand County Almanac*. After several years of failing to attract a publisher, the book was accepted by Oxford University Press in April 1948. A week later Leopold died of a heart attack while fighting a grass fire on land adjoining the Sauk County farm that figures in many of the book's essays.

Sand County Almanac was published in 1949, but it was not until the 1960s that it came into its own. The spokesmen and prophets of the environmental movement of those times were well intentioned, but many displayed a distressing ignorance of elementary facts about the universe. Leopold's vision, although revolutionary, was sound. Recognizing other organisms as kin and the biosphere as a partnership to which humans belong requires a reorientation of thought that may take a few minutes or many years. For those who achieve the reorientation, the land ethic, as Leopold called these ideas, has provided a satisfying way of looking at our place in nature.

and Soejarto 1985). Some prairie plant may provide a cure for cancer. Prudence would suggest that we not waste genetic material that is the end product of millions of years of evolutionary work.

This is an important argument, but it is not a very convincing one to the money-minded (Ehrenfeld 1976). Economists, who are practical people, think very little of such values. In their terminology, they discount them. Who knows what the future may bring? It might be nice to have a cure for cancer 30 years from now, but by that time cancer may be preventable and a cure would be obsolete. In 30 years, we may all be dead from a nuclear holocaust. The slim chance of a plant curing cancer tomorrow has to be weighed against the very real profit to be made in weed eradication today. In less practical persons, such a view might be called shortsighted.

In many ways, the decision not to destroy a species or a stand of forest is a decision similar to the decision not to murder someone. If someone stands in the way of our profit, pleasure, or convenience, few of us solve the problem by murder. It is possible that practical and aesthetic considerations come into play to some degree in the decision. Practically, it might require a lot of planning and we might get caught. Aesthetically, we may shrink from blood and violence.

But most of us avoid murder for basically ethical reasons. It is not the right thing to do. Killing another human is a proscribed act, either by ethical or religious teaching or because of peer pressure. If biodiversity is to be preserved, it will be through the development of similar rules or taboos against being the agent of the demise of forests and marshes and of populations of animals and plants.

The religious concept of **stewardship** is one such reason for preserving nature. The biblical ark is a part of this tradition. People were put upon the earth as, among other things, stewards. They are guardians for the flora and fauna. Unfortunately, stewardship has never had a very high priority in the Judeo-Christian tradition. The relevant passage

> *Be fruitful, and multiply, and replenish the earth, and subdue it: and have dominion over the fish of the sea, and over the fowl of the air, and over every living thing*

has more often been interpreted as saying that earth and its creatures were given to humans for their use (Ogburn 1966). The stewardship approach to conservation and the environment in general is well set forth in the book *Earthkeeping: Christian Stewardship of Natural Resources* (Wilkinson 1980).

Another ethical approach to preserving nature is the view that humans are a part of nature, rather than apart from it. The semiheretic St. Francis of Assisi came close to this idea when he preached to Brother Swallow and Brother Wolf just as to a human audience (White 1967). The belief held by American Indians and many pagan societies that birds, trees, and rocks have souls or guardian spirits is another version. Various eastern ways of thought, such as Taoism and Zen Buddhism, also hold a view of nature in which man is "something more than a frustrated outsider" (Watts 1958).

There is no scientific reason for preserving nature. If we show that by exterminating a species, we will ourselves go extinct or the biosphere will die, this is simply a subcategory of the practical argument. The scientific-technological mainstream would tend to see this situation, not as a fundamental constraint, but as a challenge. Once we find a technological way to counteract the effects of the loss of this super-keystone species, we are free to exterminate it. Although there is no scientific reason, as such, for preserving nature, there is a scientific basis—apart from religious, magical, or superstitious bases—for an ethical concern for nature. This is the land ethic developed by Aldo Leopold (1949).

In *Sand County Almanac,* Leopold wrote, "A thing is right, when it tends to preserve the integrity, stability, and beauty of the biotic community. It is wrong when it tends otherwise." The basis for this ethical statement is a sense of community that extends to the land, including the other organisms. We humans are members of the biosphere, not the bosses or the owners of it.

Another essay in *Sand County Almanac* was written for the dedication of a monument to the extinct passenger pigeon. If we had gone extinct, Leopold noted, the passenger pigeons would not have mourned us. Their approach would have been entirely practical, in the economist's and businessperson's sense of the word. The ability to mourn the loss of another species, to think in other than purely practical terms—"in this fact, lies objective evidence of our superiority over the beasts."

But why should the loss of a species concern us? Because, Leopold said, we are all kin:

> *It is a century now since Darwin gave us the first glimpse of the origin of species. We know now what was unknown to all the preceding caravan of generations: that men are only fellow-voyagers with other creatures in the odyssey of evolution. This new knowledge should have given us, by this time, a sense of kinship with fellow-creatures; a wish to live and let live; a sense of wonder over the duration and magnitude of the biotic enterprise.*

We have the responsibility for protecting the earth because every species is kin but we are the only one that knows it.

The land ethic is not humanistic or anthropocentric. It is biospheric. If we adopt it, we adopt a view of conservation that is far removed from the human-centered "wise use" version with which conservation began the 20th century.

The Practical Ecologist: *Pollution*

Coal-fired electric generating plants pollute the air.
(Adam Jones/Dembinsky Photo Associates.)

OUTLINE

- The Touchstone Formulation for Environmental Damage, Destruction, Degradation, and Deterioration
- Pollution as a Global Problem
- Types of Pollutants
- Biological Magnification
- Pollution in the Environment
- Recycling

20

The Touchstone Formulation for Environmental Damage, Destruction, Degradation, and Deterioration

The topic that springs to mind first when the general public hears the word "environment" is pollution. It is reasonable that this should be true; we can see filthy water and belching smokestacks, and the health problems associated with pesticides, smog, and radioactive wastes bring home the seriousness of pollution in a very personal way.

In discussing pollution or any other environmental problem, it is easy to bog down in specifics about pesticides, solid waste, loss of open space, and a thousand other issues. It is well to have a simple formulation that can serve as a touchstone, something to bring things back in focus when they begin to get fuzzy or lost in detail. The touchstone formulation is

$$D = \overline{E}P$$

where D = environmental damage, destruction, degradation, and deterioration; $\overline{E}$ = the average effect of one individual on environmental quality; and P = population.

This formulation is certainly oversimplified, but it is basically realistic. Among the quick lessons we can learn are

1. Every person has some adverse effects on the environment. The average is composed of small effects of those who tread lightly and large effects of the consumers and exploiters. In general, $\overline{E}$ in less-developed countries is lower than in more-developed countries because of the amplification in the latter of harmful activities by technology feeding on heavy subsidies from fossil or nuclear energy.
2. Human numbers are a basic term in any calculation of environmental damage.
3. We can achieve an improvement in the environment (lowering D) by reducing $\overline{E}$ or by reducing P; however, we cannot achieve permanent improvement by reducing $\overline{E}$ while allowing P to increase. For example, if the average damage done by an individual to environmental quality is reduced by 25%, but population size doubles,

$$D = 0.75\overline{E} \times 2P,$$

the overall effect on the environment is a decline in quality of 50%.

In reading this chapter and the next, stop occasionally and recast your thinking about specific issues in terms of the touchstone formula.

Pollution as a Global Problem

In Chapter 2, we defined pollution as the unfavorable modification of the environment as a result of human activity. This is the same thing as saying that pollution consists of the reactions of humans. In this sense, pollution is a natural occurrence. Being natural is not necessarily the same thing as being good. For example, yeasts carrying on fermentation—their way of obtaining energy—produce ethyl alcohol. Ecologically, this addition of ethyl alcohol to their environment is a reaction. Yeasts cannot tolerate alcohol concentrations higher than about 16%, however, which is why there are no naturally produced wines with an alcohol content higher than this. When the yeasts have produced an alcohol concentration around 16%, they begin to die as a result of their own modification of the environment.

Humans have always modified their environments locally (by setting fires, for example), but especially in the past half century human reactions have become globally serious. This is for four main reasons:

1. Humankind has become a cosmopolitan species, occurring over the whole surface of the globe; no other species is so widely distributed.
2. Humans are large in both physical size and numbers. Their physical size has always allowed them to make substantial local alterations. Their numbers have increased from about a billion in 1850 at the beginning of the Industrial Revolution to more than 5 billion now.
3. By utilizing energy subsidies, mainly from fossil fuel, humans exert effects on the environment many times greater than a comparable animal that used only the energy from its own metabolism.
4. Humans have produced new chemicals with which they and the biosphere have had no evolutionary experience. Chemists have been known to show naive audiences a long list of the organic chemicals occurring in, say, coconut milk or apple cider. Their message is that we're exposed to organic chemicals all the time so we shouldn't worry about the ones that the chemical companies make to kill insects or plants. But the whole point of a successful pesticide or herbicide is its evolutionary novelty. Humans have had immemorial generations of experience with the chemicals that occur naturally in common foods. Humans with genotypes predisposing them to get cancer from such chemicals got cancer long ago and left few or no children, with the result that human genotypes today are ones that, by and large, get along okay with most naturally occurring organic compounds. Most new chemicals are useful precisely because they are evolutionarily unique, and this may result not only in their killing organisms that

Landmarks in Environmental Studies and Protection

Included are a selection of some of the most important events. Only constructive events have been included, although important in their own way were such occurrences as the death of the last Carolina parakeet in 1914, the first dust storm of the Great Drought in 1933, the "killer fog" in London in 1952, and the Chernobyl nuclear disaster in 1986.

1605. A scheme to drain 120,000 ha of marshland on the east coast of England was foiled by the Fenmen, early environmental activists.
1854. H. D. Thoreau published *Walden.*
1864. *Man and Nature, or Physical Geography as Modified by Human Action,* by George Perkins Marsh appeared.
1870. The first official wildlife sanctuary in the United States was established by the state of California at Lake Merritt.
1872. Yellowstone, the first national park, was created.
1892. The Sierra Club was formed.
1898. The Royal (British) Commission on Sewage disposal was set up.
1900. The Lacey Act, a landmark in federal wildlife protection, was signed.
1903. The first federal wildlife refuge, Pelican Island, Florida, was created by Teddy Roosevelt.
1908. The White House Conference on Natural Resources, which led to the formation of most state conservation departments, was called by Teddy Roosevelt.
1916. The U.S. National Park Service was created.
1918. The Migratory Bird Treaty protecting birds that migrate between the United States and Canada was signed.
1920. James Ritchie published *The Influence of Man on Animal Life in Scotland.*
1926. Victor E. Shelford edited *Naturalist's Guide to the Americas,* the earliest attempt to compile a list of natural areas. Calvin Coolidge assembled an international conference on oil pollution in the ocean.
1933. Civilian Conservation Corps and Soil Erosion Service (later called the Soil Conservation Service) were created.
1935. *Deserts on the March* by Paul B. Sears was published. The Wilderness Society was founded.
1943. Publication of *An Agricultural Testament* by Albert Howard, a treatise on organic farming, hence one of the sources of the environmental movement.
1948. *Malabar Farm* by Louis Bromfield was published.
1949. *Sand County Almanac* by Aldo Leopold appeared (however, an earlier version of its most important chapter, "The Land Ethic," had been published in 1933).

they are not designed to kill, such as us, but also in their building up in the biosphere because no organisms have evolved to use them as an energy source.

Neither pollution nor the scientific study of it is new. What is new is a change in attitude on the part of a great many people. This change has involved the realization, first, that pollution is harmful to humans and the natural systems of the earth and, second, that pollution is for the most part preventable. Finally, it has involved the realization that the prevention of pollution should be a cost of doing business, like paying salaries and buying raw materials.

When businesses do not pay the cost of preventing pollution, the general public has to pay in medical bills for respiratory ailments, in taxes for ever-bigger sewage treatment plants, and in a great many ways that are hard to place a dollar value on, such as having streams with sewage worms instead of trout. Economists call such effects **negative ex-**

1949. The (British) Nature Conservancy was established.
1956. The proceedings of a symposium (1955), *Man's Role in Changing the Face of the Earth,* edited by W. L. Thomas, appeared. Federal Water Pollution Control Act passed by Congress.
1962. Rachel Carson's *Silent Spring* was published.
1964. The Wilderness Act was signed into law.
1965. The conference Future Environments of North America was held by the Conservation Foundation (proceedings published in 1966).
1966. Endangered Species Act passed.
1968. *The Population Bomb* by Paul R. Ehrlich appeared.
1968. Garrett Hardin published *The Tragedy of the Commons.*
1969. Ian L. McHarg's ecological approach to environmental planning, *Design With Nature,* was published.
1969. A catalytic book of readings selected by Paul Shepard and Daniel McKinley and called *The Subversive Science* (after an essay by Paul Sears) was published.
1970. The National Environmental Policy and Clean Air Acts became law.
1970. *Ecotopia* by Ernest Callenbach was published.
1970. First Earth Day was held.
1970. Creation of UNESCO's "Man and the Biosphere Program," which has taken the lead in creating biosphere reserves.
1972. United Nations Conference on the Human Environment in Stockholm.
1972. *The Limits to Growth* by Donella H. Meadows, Dennis L. Meadows, Jørgen Randers, and William W. Berhens III appeared.
1975. *The Monkey-Wrench Gang* by Edward Abbey was published.
1980. Publication of the *World Conservation Strategy* by the International Union for the Conversation of Nature and Natural Resources (IUCN). Publication of the *Global 2000 Report to the President* by the Council on Environmental Quality.
1983. Conference on the long-term worldwide biological consequences of nuclear war was held in Washington, D.C.
1992. U.N. Conference on Environment and Development held in Rio de Janeiro. Publication of *Ishmael* by Daniel Quinn.

ternalities. When businesses assume the duty of preventing them, as they have increasingly been forced to do, the costs do not thereby disappear; they are paid for instead by higher prices for the company's product and, possibly sometimes, by smaller profits to the company's shareholders. However, it often turns out that, after the immediate changeover costs (such as buying the necessary equipment), pollution control is not very expensive.

Types of Pollutants

Pesticides

Pesticides are materials used to kill pests; pests are organisms that interfere with our profit, convenience, or welfare. The definition of pest is, of course, absolutely anthropocentric. Mosquitos are not pests of the swamps; rather, they are pests to humans living near the swamps. There is no sharp line between what is and what is not a pest, but no really exact distinction is necessary, because most pesticides cannot distinguish between a beetle that is or isn't a pest nor, for that matter, between a beetle and a cat.

Early pesticides were the familiar inorganic or organic poisons such as arsenic and nicotine. New organic poisons began to be synthesized beginning in the 1930s and especially in the 1940s. At present these are of four main types (Watterson 1988), in the order in which they were widely introduced into the environment.

Chlorinated hydrocarbons—examples of this group, which tends to be persistent in the environment, include DDT, DDD, aldrin, dieldrin, heptaclor, lindane, and chlordane.

Organic phosphorus compounds—examples include malathion, parathion, and diazinon.

Carbamates—generally short-lived compounds such as aldicarb, carbaryl, Zectran, and carbofuran (Furadan).

Synthetic pyrethroids—mimics of the active compounds of the pyrethrum plant; they tend to be moderately persistent and highly toxic; Bifenthrin and Fenvalerate are examples.

There are, in addition, the fungicides and herbicides.

So much has been written and spoken about the dangers of pesticides that you would think that everyone would avoid contact with them if at all possible. If you believe this, a visit to your local garden store some Saturday morning will convince you otherwise. Evidently a great many people have not heard of the extreme toxicity of some of these compounds, do not believe what they have heard, or just cannot comprehend it. The last is a distinct possibility. Most people know that arsenic is fairly poisonous and have no desire to have it around, but the fact that parathion is 140 times as poisonous as lead arsenate may be an idea, like the idea of a billion dollars, that is incomprehensible to many people.

Given the potential dangers, surprising amounts of pesticides are used for what can only be regarded as friviolous purposes, such as growing fancy roses and killing harmless but annoying insects. However, most pesticide usage is in agriculture and the control of potentially disease-carrying insects. Although many spokesmen for agriculture-business speak lightly of the dangers of pesticides, most farmers have a realistic idea of the personal hazards of pesticide use. Usually farmers use pesticides because they believe they must, not to raise crops (their fathers grew many of the same crops they grow today without these pesticides), but to compete with other farmers.

The competition is of two forms, the first being yield. If a farmer using pesticides can produce 50 bushels a hectare and one not using them can produce only 45, then the pesticide user can sell his crop at a lower price, assuming the pesticide is fairly cheap. If the nonuser has to sell at the same price as the user, he may not be able to make a living.

The second kind of competition is for appearance. For instance, pesticides have made possible the routine production of apples of a perfection that, 40 years ago, was rarely found outside the basket of Snow White's stepmother. The unsprayed apple is often smallish with an occasional worm. These were the sort of apples that everybody once ate and made jelly and applesauce from, but most people, if given a choice, usually pick the large shiny, pesticide-produced fruit. Commercial buyers and governmental agencies now have rules as to "defects" that may make agricultural products raised without pesticides essentially unmarketable.

The chance of one of us starting up from the breakfast table and falling dead of poisoning from the pesticide residue on our food is slight. Why, then, do environmentalists oppose many of the current practices in pesticide use?

Human Toxicity

To begin with, people do die of pesticide poisoning; a 1985 U.N. report estimated a million poisonings and 20,000 fatalities per year (Wright 1990). Most of these are agricultural workers or other persons having direct contact with concentrated forms of pesticide. Research in the Philippines led M. E. Loevinsohn (1987) to conclude that previous figures were underestimates and that tens of thousands of persons per year die from pesticide poisoning in the rice-growing areas of Asia alone.

***Rachel Carson** (American, 1907–1964)*

Silent Spring, published in 1962, crystallized the environmental movement. It was able to do this for several reasons. It appeared at the right time. Many people were already concerned with the reckless addition of synthetic chemicals to the environment and with the general deterioration of the quality of life that was occurring in the name of progress. Other books and articles had made many of the same points as *Silent Spring*, but they had sunk under ridicule from the chemical industry. Much of the educated public had been convinced that worrying about such things was unscientific, effete, and even slightly shameful.

Rachel Carson made concern with the environment respectable. Her biology background and skill as a writer allowed her to draw together the developing technical literature into an utterly convincing indictment of the hazards of pesticides. As a prize-winning author on natural history, her work had a credibility and an audience unavailable to most others who had written about toxic chemicals.

But the feature that made *Silent Spring* the trigger of the environmental movement was its vision. Carson did not call her book *Our Poisoned Food;* that was not what she was writing about. *Silent Spring* is thoroughly ecological. It recognized indiscriminate pesticide use as an attack on the biosphere itself, not just another human health problem.

Rachel Carson was born in Springdale, Pennsylvania, and majored in biology at Pennsylvania College for Women (now Chatham College) in Pittsburgh. She did graduate work at Woods Hole Marine Biological Station and at Johns Hopkins, receiving an M.A. in 1932. Her thesis was on a subject typical for the period, the embryology of the kidney of the catfish. Most of her professional life was spent as a writer and editor for the U.S. Fish and Wildlife Service. She resigned to write full time in 1952 after *The Sea Around Us* had become a best-seller.

Silent Spring was begun in 1958 and completed after Carson was aware that she had developed cancer. The book was an immediate success. Most reviews in important magazines were favorable, although some tended to be merely jocular and tolerant (and sexist). A few were definitely hostile, the hostility increasing the closer the link to the chemical industry. A review in *Chemical and Engineering News* was titled "Silence, Miss Carson" and suggested that the book would appeal to "organic gardeners, the antifluoride leaguers, the worshippers of 'natural foods,' those who cling to the philosophy of a vital principle, and pseudoscientists and faddists."

Despite the chemical industry's attacks, *Silent Spring* led to important changes in ecological research, in laws, and in people's ways of looking at things. It would be nice to say that the earth's problems of chemical contamination were thereby solved; but they were not. The fight goes on.

A. Watterson (1988) concluded that pesticide usage is not a "high-risk hazard well controlled but rather a high-risk hazard out of control."

Effects on Nontarget Organisms

Ideally a pesticide should kill only the pest for which it is applied, but most are not very selective. The die-offs of robins (Wallace et al. 1961) from spraying elm trees with DDT (to kill the elm bark beetle that carries Dutch elm disease) became so notorious that they helped to begin the revolt against indiscriminate pesticide use.

A group of wildlife biologists from the Illinois Natural History Survey (Scott et al. 1959) studied the results of an application of dieldrin (1.3 kg/ ha, spray and granules) for Japanese beetle control. Doubtless some Japanese beetles were killed, but the researchers also found dead in the treated areas a variety of animals, ranging from pheasants, brown thrashers, meadowlarks, ground squirrels, and cottontails, to 14 farm cats.

Of course, that was 1959. The pesticides of today are much more specific in their action, right? Well, not exactly. Although the use of dieldrin was restricted by the Environmental Protection Agency (EPA) in 1974, certain applications such as mothproofing and ground insertion for termite control are still permitted (Regenstein 1982). Shell Chemical Company has continued to manufacture dieldrin outside the United States and to promote and sell it widely, especially in the Third World countries. Many other broad-spectrum pesticides that we sometimes think of as banned have merely been restricted in their usage, label restrictions that may or may not have an effect on how the chemicals are actually used.

What about the various biocides that are claimed to affect only certain target organisms? A test of 12 biocides claiming to be narrow spectrum found that none had effects that were strictly limited to their target groups (Ingham 1985). For example, carbofuran (Furadan), which is supposed to be effective against nematodes and microarthropods, was; but it also reduced populations of field crickets, earthworms, and the nitrogen-fixing fungus Rhizobium as well as decreasing nitrogen fixation by blue-green algae.

The pink-purple granules of Furadan were, nevertheless, spread by the ton through the 1970s and 1980s over corn and rice fields and garden crops. Birds killed by carbofuran poisoning became commonplace, and by 1989, several groups had recommended that the EPA cancel the pesticide's registration (West 1992). After several bald eagles in Virginia died from eating carbofuran-killed smaller birds, Virginia's Pesticide Control Board banned granular carbofuran from use in the state, and the EPA finally negotiated an agreement with the manufacturer of Furadan (FMC Corporation of Philadelphia) to phase out use of the chemical in the United States over a period of years.

Persistence and Accumulation

There is, of course, a very good reason why cats died in the dieldrin-treated areas in Illinois. They are carnivores, at the top of the food web, where the concentration of pesticides that do not break down rapidly is greatest. Although some recent pesticides are less persistent than the extremely long-lived chlorinated hydrocarbons, many "hard" pesticides are still in wide use. One such is endrin with a half-life of around a decade. In 1981, more than 100,000 ha of croplands in Montana, Wyoming, South Dakota, and Colorado were sprayed with endrin. Migrating waterfowl picked up body loads three to four times the level permitted in commercial poultry (Regenstein 1982). Problems of persistence and biological magnification are discussed in more detail later in this chapter.

Problems of Spreading

Pesticides do not stay where they are applied. If one farmer sprays his land, his neighbors also get sprayed. When Key Largo sprays for mosquitos, which it does, twice a week in the rainy season, John Pennecamp Coral Reef State Park also gets a dose of the organophosphate Naled (Ward 1990). Furthermore, persistent pesticides travel long distances in water and air. DDT and various other harmful organic chemicals are found in the fat of organisms throughout the world, even though the chemicals have never been used there; for example, penguins in Antarctica had DDT levels up to 18 parts per million (ppm) as early as 1964.

Many chemicals now banned or restricted in the United States are still widely used elsewhere in the world. Apart from the ethical indefensibility of encouraging pesticide use in the Third World, such persistent and dangerous pesticides as dieldrin, DDT, chlordane, and benzene hexachloride return to us regularly in residues on bananas, coffee, chocolate, and many other imports.

Migratory animals also transport pesticides. Double-crested cormorants reached a low level in the United States in the years before and just after the DDT ban in 1972. In recent years, populations have increased markedly (Brewer 1981, Ludwig 1984). One reason, evidently, is that most double-crested cormorants winter within the United States. Some other species whose numbers had dropped but which winter in the tropics where DDT is still heavily used have not shown similar population rebounds.

Sublethal Effects

Pesticides do not have to kill an organism to harm it. Trout are more susceptible to environmental stresses such as temperature extremes if they are carrying a burden of DDT in their tissues. There is now a large amount of evidence linking DDT to reproductive failure in several kinds of birds, including falcons and pelicans. More than one effect may be involved, but an important one is that DDT and its breakdown products cause birds to lay eggs with shells so thin that they cannot hold up under normal incubation (Figure 20-1). With the banning of DDT in the United States and the subsequent slow reduction in body loads, reproductive success has improved in several species such as the bald eagle (Grier 1982) and the osprey.

If you are average, you are carrying around in your fat DDT, aldrin, dieldrin, and small amounts of many other pesticides. On your food, every day, you probably eat a few tenths or hundredths of micrograms of lindane, dieldrin, malathion, and other pesticides. A microgram is a very small amount, one millionth (0.000001) of a gram, and, in general, the amounts you eat are below the World Health Organization's acceptable daily intake levels. Are you being harmed by these pesticides taken in day after day in small amounts?

Are they depressing your immune system, making you more susceptible to infectious diseases? Are they affecting your behavior? Were your SAT scores lower than they should have been? Are they causing mutations? Are they causing cancer?

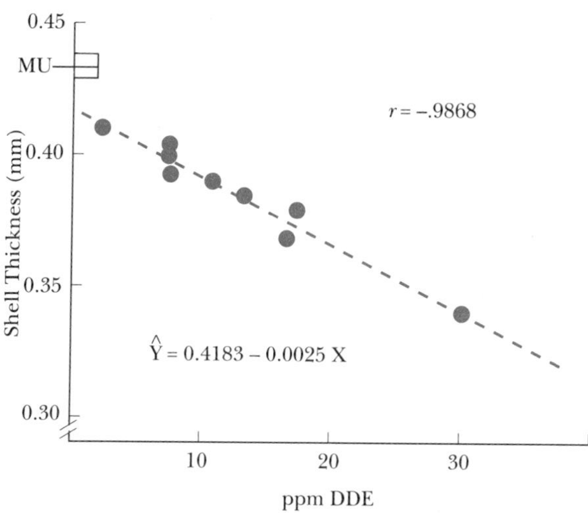

Figure 20-1

The relationship between eggshell thickness and DDE (a DDT breakdown product) concentration in double-crested cormorant eggs. The samples were taken in 1965 from the northcentral United States and adjacent southern Canada. "MU" indicates shell thickness in museum specimens from the same geographical area but from years prior to the use of DDT. (Adapted from D. W. Anderson et al. 1969.)

There are epidemiological studies suggesting that a pesticide connection exists and other studies that fail to show a connection. For example, there are studies that show no increase in cancer among persons exposed to various pesticides and other studies that suggest increased rates of leukemia, myeloma, lymphoma, sarcoma, and cancers of the prostate, brain, skin, lung, nervous system, esophagus, and ovaries (Watterson 1988). A writer in the *British Medical Journal* (Coggon 1987) commented that "epidemiological data on pesticides are few, and those that there are give a clouded picture." Unfortunately, exactly the same sentence could have been written 20 years ago. One problem is that new pesticides are introduced frequently, so that data on the effects of human exposure may become available about the time widespread use is being phased out.

A related problem is the length of time required for a definite answer, especially for questions such as cancer-producing effects. Doesn't the fact that people are living longer today constitute a general proof that pesticides, pollution, and all the rest are not really hurting us? No it doesn't, for several reasons. The old folks of today, dying at a ripe age of 75 or 80 years, were born early in this century; they never met a DDT molecule until they were 35 or 40 years old. Will the life span of someone born in 1970 or 1975 also be 75 years of age? It is an interesting question, interesting in a more personal way to someone of 20 years than someone of 65 years.

Synergistic Effects

As we pointed out in discussing limiting factors in Chapter 2, some environmental factors act more strongly in combination than would be suspected from their effects singly. Malathion is considered a relatively safe pesticide because it is rapidly detoxified by the liver; however, any chemical that interfered with the detoxification function of the liver would increase the danger of malathion poisoning. Just such a chemical is another pesticide, EPN. Malathion was made 50 times more poisonous to dogs when EPN was given at the same time.

There are a great many human-made chemicals

in use (half a million, according to the editors of *The Ecologist*) in pesticides, food additives, hair sprays, medicines, paints, gasoline additives, and many other things. The chance of unsuspected synergistic effects increases as the number of such chemicals increases.

Ecosystem Effects

The direct effects of a pesticide are rarely confined to one species, and furthermore, the overall effects are rarely confined just to the species that are harmed directly. The decrease in abundance of one species has ramifying effects on other populations at the same and at other trophic levels. Frequently, pesticides are more effective at removing predators of pests than at removing the pests themselves. This occurs because the predators eat a lot of pests, each containing pesticides, and because predators are generally rare relative to their prey, so that a given application of pesticide is likely to kill practically all the predators while leaving a fair number of the pests. If so, the remaining pests can increase, virtually unchecked by the predators. There are now many cases in which applications of pesticides have been followed by an initial reduction in pest density followed by an explosive increase. Chemical treatments often generate new pests; the European red mite became an important pest in orchards only after the widespread use of chemical pesticides, probably because of their effect on its predators (and possibly competitors).

Resistance to Pesticides

Most pests have a short generation length and high populations—otherwise they would not be pests. But these two traits favor rapid evolution and consequently rapid development of resistance to pesticides. Recent counts show that 84 species of weeds are resistant to herbicides that once killed them, more than 100 species of fungi are resistant to one or more formerly effective fungicides, and the numbers of insects and mites known to be resistant must now be well above the 500 documented in 1986 (Gould 1991).

The answers to this problem have been, first, to increase the dose. Initially parathion was used against the walnut aphid at a dosage of 0.11 kg/380 L of water, but seven years later the dosage was increased to 3.8 to 5.7 kg/380 L. The second step is the development and introduction of a new pesticide. The third and fourth steps are the first and second repeated. So are the fifth and sixth.

Mackinac Island, Michigan, is a summer resort where cars are prohibited and transportation for tourists is provided by about 500 horses. Stable flies and houseflies that develop in the manure have been a nuisance that, with the development of DDT, was met by wholesale spraying. By the early 1950s, DDT-resistant fly populations had evolved, and so the city and state employees, anxious to protect the tourists, switched to the organophosphate malathion.

By the early 1960s, the flies were also resistant to malathion. The next step was dimethoate, under the brand name Cygon (pronounced "Saigon"). By the 1970s it was also ineffective (Coulson and Witter 1984). In this situation, as in many others, better sanitation, the use of screens and flypaper, and a higher threshold of tolerance by humans for other organisms may be the long-term solutions.

Not many insects that are predators or parasitoids of pest species are known to have developed resistance to pesticides, for several reasons. As we have pointed out, populations of the predator are usually smaller than those of the prey. Because a new pesticide is generally effective at first in killing a very large proportion of the pest, any resistant predator individuals that survive the pesticide may die of starvation before they can reproduce.

There is no question that pesticides can result in a short-term decrease in the numbers of the target organism, if it has not evolved resistance. If your house is full of cockroaches, the exterminator can get rid of them. Whether it is worth the risk to you, your cat, and the environment is dubious. Being careful to put food away and using nonpoisonous control measures, such as boxes lined with glue that traps the roaches, may do as well, and perhaps better in the long run.

Cases in which pesticides have exacerbated the control problem are now numerous, as are cases in which, after spending many millions of taxpayer dollars, government agencies have decided to give up on attempts at chemical control. The gypsy moth, once the object of large-scale spraying efforts, is an example. Although some midwestern agriculture departments still spray, the governmental agencies in the areas most familiar with the problem have concluded that chemical control is ineffective (Batts 1981). Given the heavy promotion of pesticides by the chemical industry and the natural reluctance of most people to admit that they have been wrong, this change of direction is especially convincing.

Herbicides

Biocides that are aimed at plants rather than animals are called **herbicides.** They are used for such purposes as keeping down weeds in cropfields, but the most famous of all herbicides was put to a different use. Agent Orange, a combination of 2,4-D and 2,4,5-T, was used by the United States in Vietnam to defoliate trees, making the enemy more visible, and to destroy crops that might be used by the Vietcong. Over 2 million hectares in that small country were sprayed with over a hundred metric tons in the 1960s. Some of the ecological effects were described by Tschirley (1969).

It is unnecessary here to repeat the horror stories of Vietnam (Whiteside 1970)—the careless handling of the material claimed by Dow Chemical to be as safe as aspirin, the Air Force C-123s laying down 15.2 L a second in a swath 72 m wide with or without people on the ground. The spraying was stopped in December 1970 after the U.S. National Academy of Science concluded that the spraying was causing birth defects among the Vietnamese. It is unnecessary, likewise, to dwell on the aftermath—the cancers, diseased livers and kidneys, and other illnesses of the Vietnam veterans, or the out-of-court settlement in 1984 between the chemical companies (which admitted no culpability) and 15,000 veterans for $180 million.

A dioxin, TCDD, widely considered the most toxic chemical known (Roberts 1991), has been a frequent contaminant of 4,5,5-T (among other things including pentachlorophenol, a wood preservative, and hexachlorophene, a widely used antiseptic). How much of the effects of Agent Orange and 2,4,5-T in general are attributable to the dioxin is uncertain.

2,4,5-T and silvex, a similar product, were sprayed widely over western forestlands in the 1970s, until the EPA issued a temporary ban on them as imminent threats to human health. Oddly enough, the EPA has allowed their continued use on rangeland and rice fields. In the Northwest, they were used by timber companies in conjunction with the U.S. Forest Service to kill broad-leaved vegetation and, thereby, favor the economically more important evergreens, such as Douglas fir. There seems to be little published research on whether the increased growth actually occurred. The destruction of nitrogen-fixing plants, increased runoff, and various other detrimental effects may offset any reduction of competition from hardwoods.

Herbicides, including 2,4-D and 2,4,5-T, are extremely widely used chemicals today; in total tonnage, they far outstrip the insecticides and other animal poisons. They are used in virtually every part of the country to control vegetation along roads, along powerline rights-of-way, in cropfields, and even in parks and people's lawns to get rid of innocuous dandelions or crabgrass. Low or no-till agriculture, in which plowing is omitted, has increased, and this uses large amounts of herbicide as the alternative to cultivation for weed control. An environmental benefit here, of course, is a lower loss of topsoil through erosion.

It is unclear how serious the threats from herbicides are to human health and to the functioning of individual ecosystems and the biosphere as a whole. A book on naturalistic landscaping published in 1966 advocated herbicides as the main tool (Kenfield 1966). "The view-hiding thicket, the poisonous-ivy, the rank weeds, will vanish. In their places will emerge the landscape that you yourself choose to sculpture." Those who think that herbicides are "poisons" are uninformed, the book claims. "The chlorophenoxy herbicides [that is, 2,4-D and 2,4,5-T] are nonpoisonous to animal life and man as here suggested. Applied discriminately, they only rootkill the plants they touch. Then the chemical is utilized as food by soil bacteria, and vanishes."

It seems possible that similar claims of safety for other herbicides in current wide use may also prove false. The National Cancer Institute recently published data showing a sixfold increase in the rate of a particular kind of lymph cancer (non-Hodgkin's lymphoma) among farmers who regularly use 2,4-D compared with nonusers.

The Economics of Pesticide Use

Chemical-based agriculture has costs and benefits for the individual farmer and for society. For each, the accounting is complex. The farmer gains productivity and convenience. He no longer needs, for example, to rotate crops to reduce pest populations and replenish soil nitrogen. He can grow whatever crop he guesses will make the most money in a given year. The need for human labor is reduced, so he may not need a hired hand, and the children can go off to the city.

Studies have shown that the expenditure of $1 on pesticides raises gross agricultural output by $3 to $6.50 (Zilberman et al. 1991), seemingly a handsome return on investment. Most of these studies neglect several considerations, however. Productivity may be increased for a crop that we do not need any more of. This is not necessarily a concern of the individual farmer; if he can grow a crop more cheaply, he may be able to fulfill the need for a crop at a profit, while his less-efficient neighbor merely produces a nonsalable surplus.

Still, the studies neglect other things that are possible costs to the individual farmer. The studies do not include, for example, the future cost of cleaning up sites of pesticide storage or of pesticide-contaminated groundwater. They do not include the costs of the pesticide-related illnesses we have talked about in earlier sections. For large agricultural operations, the farm owners may not have to bear this burden; rather, health costs fall on the field workers (in the forms of sickness and shortened lives) and on the taxpayers (to pay for Medicaid and related welfare items). Cesar Chavez (in many issues of *Food and Justice,* published by the United Farm Workers of America) pointed to pesticide use by the California table grape industry as a serious threat to the health of grape workers.

For the United States as a whole, approximately 500,000 tons of 600 different pesticides are applied to crops each year at a cost of just over $4 billion dollars (Pimentel et al. 1992). Approximately $16 billion in crop production is saved, it is estimated. David Pimentel and associates at Cornell University have estimated that the environmental and social costs of this pesticide usage amount to about $8 billion. These additional costs, largely not borne by the makers or users of pesticides, include

- Human health effects
- Animal poisonings and contaminated products—crops that have to be destroyed because they have pesticide residues that are too high (and that happen to be detected)
- Destruction of beneficial natural control agents
- Various types of nonindustry funded research, such as studies to detect pesticide-induced mortality in songbirds
- Bee poisonings and reduced fruit- and seed-set as a result of lowered bee populations
- Crop losses from such causes as pesticide drift into adjacent fields or reduced yield because of residual pesticides in the soil
- Remediation of pesticide-contaminated groundwater
- Deaths or contamination of birds, mammals, and fish
- Governmental monitoring of pesticide levels and education of pesticide applicators.

So, the societal balance sheet may be a $16 billion increase in crop production compared with $4 billion in direct costs (production, sales, application) plus $8 billion in indirect costs (Pimentel et al. 1992), or a $12 billion total cost. Even these figures may suggest a more favorable balance than actually exists. The future costs of chemical cleanup at manufacturing facilities and in sediments and groundwater are going to be high and are probably going to be largely paid by the taxpayer. Human health effects may be more serious than any persons except the alarmists are currently saying. Some of the current declines of animals may involve pesticides. Endangered or threatened species of restricted range might be pushed over the edge by some pesticide incident.

One justification for using pesticides is the need to feed the steadily increasing world populations. A case could be made that in spending money on pesticides, we are losing the opportunity to spend it on population control.

Radioactive Materials

All creatures on this earth are subjected to naturally occurring low levels of high-energy radiation (called **background radiation**) in the form of radioactive material in the earth's crust and cosmic radiation from space. This background radiation may be several times higher on one part of the earth than another. There is no evidence that these natural differences have any ecological effect; there may, however, be health effects. Statistical correlations suggest a higher incidence of congenital malformations (such as harelip and clubfoot) and certain types of cancer (Wilkison 1986) in areas of high background radiation. One aspect of background radiation that has received recent attention is the filtration of radon gas (from soil and bedrock) into basements. Modern "tight" houses allow radon concentrations to build up to potentially troublesome levels in areas of granitic rock.

We are concerned here primarily with the human release of radioactive materials to the environment. This could occur in two ways: By the testing or actual use of nuclear weapons, or from peaceful applications of nuclear energy, particularly nuclear power plants (Table 20–1).

Radiation of the sort we are talking about is thought to exert its harmful effect by ionizing materials in the cells—that is, by adding or removing electrons. Radioactive materials emit three types of ionizing radiation: (1) alpha radiation, which has relatively little penetrating power, so that generally a source has to be within the body of the organism to do harm; (2) beta radiation, which can penetrate a couple of centimeters of tissue; and (3) gamma radiation (similar to X-rays), which has great penetrating power. There are several units of measurement of radiation and radioactive materials; we will mention only the rad and the rem. Both are measures of the amount of energy absorbed (a rad is an absorbed dose of 100 ergs per gram of tissue and is equivalent to the older term "roentgen"; the rem is similar but is adjusted for the relative biological effects of the different kinds of radiation).

The biological effects of ionizing radiation are of three general sorts:

1. Acute, "radiation sickness" effects resulting from brief exposures to high doses. For humans, the LD_{50} is about 500 rads. (LD stands for "lethal dose" and 50 for 50%. An LD_{50}, then, is the dosage at which half those exposed to the radiation live and half die within hours or days.) The 50% surviving generally suffer some ill effects such as cataracts, sterility, or leukemia. Sensitivity to acute exposure varies from one type of organism to another, with mammals among the most sensitive and bacteria among the least.
2. Chronic exposure to low-level radiation may have several effects, of which the best studied is the increased tendency toward cancer.
3. Related to the chronic effect is the production of mutations in sperm and eggs. Geneticists agree that there is no threshold for radiation in its role of increasing mutation. That is, any slight increase in radiation increases mutation rate at least slightly. Most mutations that are detected are harmful, so that this effect of radiation increases the percentage of children bearing harmful mutations.

The problems involving radioactivity associated with nuclear power plants are of three general sorts: (1) the routine release of small amounts of radioactive materials into the air and water, (2) problems associated with transport, storage, and disposal of wastes, and (3) the possibility of some catastrophe affecting the plant.

Low-level wastes routinely released vary by reactor type, but they include tritium and radioisotopes of iodine, xenon, krypton, strontium, cesium, cobalt, manganese, iron, chromium, molybdenum, and zinc. The release of these materials raises the general level of background radiation. Also troublesome is the possibility that the low-level wastes discharged will become high-level wastes through biological magnification.

High-level wastes, such as spent fuel rods, require storage, transportation, possibly reprocessing, and disposal. Although some countries have dumped sealed containers of these wastes into the oceans, the United States has its wastes in temporary storage. Through 1985, spent fuel rods equal to 32,300,000 curies (the equivalent of about 36 tons of radium) were in temporary storage in swimming pool-like water basins at power plants (Oak Ridge National Laboratory 1986). Most high-level nuclear wastes, however, have come from the weapons industry. Wastes in the amount of 1,430,000,000 curies are stored at Department of Defense facilities.

Just where the wastes can be put for long-term storage (sometimes referred to as disposal) is puzzling, because they must be kept for a very long time. Because of the long half-life* of such elements as plutonium, unreprocessed spent fuel rods must be stored for about 7000 years before they reach the "innocuous" level of uranium ore (Klingsberg and Duguid 1982).

The site chosen should then be geologically sta-

* The half-life of a radioactive substance is the amount of time it takes for half of the radioactive material originally present to decay (transmute).

Table 20-1
Approximate Doses and Effects of Different Kinds of Release of Radiation to Our Natural Environment

Source	*Kinds of Radiation*	*Duration of Exposure*	*Dose to Environment*	*Secondary Activity Induced?*	*Heat and Blast Effects?*	*Total Area Involved*	*Direct Effects?*	*Incorporated in Natural Cycles?*
Natural (background) radiation	Alpha, beta, gamma	Several billion years	0.1 to 0.5 roentgens per year	No	No	The earth	No	Yes
Medical and occupational	Not normally delivered to human's environment							
Gamma fields up to 4000 curies cobalt-60	Gamma	Chronic, several years	Up to several thousand roentegens per hour	No	No	Thousands of acres	Yes	No
Shielded reactor	Mixed gamma-neutron	Intermittent	Up to several times background	Negligible	No	Acres	No	Negligible
Unshielded reactors	Mixed gamma-neutron	Intermittent	Up to 100,000 rads per hour	Yes	No	Hundreds of acres	Yes	Negligible
Reactor effluents	Alpha, beta, gamma	Continuous	Above background	No	No	Hundreds of square miles	No	Yes
Waste disposal	Alpha, beta, gamma	Continuous	Slightly above background	No	No	Hundreds of acres	No	Potentially
Accidental explosions	Alpha, beta, gamma	Acute	Up to several thousand rads per hour	No	Yes	Acres	Yes	No
Nuclear detonations	Alpha, beta, gamma	Acute	Up to several million rads per hour	Yes	Yes	Hundreds of square miles	Yes	No
Fallout from above two sources	Alpha, beta, gamma	Chronic, thousands of years	Up to several times background	No	No	The earth	No	Yes
Nuclear war (projected)	Alpha, beta, gamma	Acute	Up to hundreds of millions rads per hour	Yes	Yes	Thousands of square miles	Yes	No
Fallout from nuclear war (projected)	Alpha, beta, gamma	Chronic, thousands of years	Up to several roentgens per hour	No	No	The earth	Yes	Yes

From Platt, R., "Ecological effects of ionizing radiation on organisms, communities and ecosystems," *in* Schultz, Vincent, and Klement, A. W., eds., *Radioecology*, New York: Reinhold, 1963, p. 248.

ble for at least this length of time, because it would not do to have the material dispersed by an earthquake or a volcano. A geologically stable site may be easier to find than a politically stable one. Dating from the Declaration of Independence, the United States has been in existence just over 200 years, or about one one-hundredth of the half-life of plutonium. In fact, the 24,000-year half-life of plutonium would take us back to a time when much of North America was covered with glaciers, when there were still mammoths and saber-toothed cats; and when human culture was still Old Stone Age (in Europe; primitive humans had probably not reached North America yet).

Alvin Weinberg (1972), director of Oak Ridge National Laboratory, has written: "We nuclear people have made a Faustian bargain with society. On the one hand, we offer—in the catalytic burner—an inexhaustible source of energy .... But the price we demand of society for this magical energy source is both a vigilance and a longevity of our social institutions that we are quite unaccustomed to."

Most opponents of nuclear power believe that such a bargain would be a bad one even if only this generation were affected, and that it is still less desirable to force such a bargain upon future generations. The increased tempo of social change in the past decades and the possibility of instability as populations continue to rise and resources to decline suggest that our institutions may lack not only the vigilance or longevity but eventually even the technological means of keeping nuclear wastes contained indefinitely.

The safety record of nuclear power plants was fairly good until 1979. There had been a few near-misses, such as the Brown's Ferry plant in Alabama that came close to a core meltdown in 1975 when its emergency cooling system malfunctioned. Then, on March 28, 1979, came the incident at Three Mile Island, Pennsylvania. A partial meltdown was accompanied by the venting of radioactive materials, some more than a year after the accident. The facility is still inoperative, and the cleanup cost has most recently been estimated at 2 billion dollars.

Other more or less near-misses occurred, and then the most serious accident at a power plant to date occurred at Chernobyl, near Kiev, Ukraine, on April 25, 1986. Partial meltdown occurred followed by an explosion and a fire leading to the spewing of radioactive krypton, xenon, iodine, cesium, and cobalt—among other things—into the atmosphere. The rest of the world learned of the disaster two days later when Swedish monitors picked up a large rise in radioactivity as the cloud of radioactive material drifted over them from the southeast.

Several people at the Chernobyl plant died within the first few hours, and others have died since. Four months after the accident, the toll stood at 36, but others will die of radiation-induced cancers in the years ahead. Much of the cropland near the plant will be too heavily radioactive to be used again for years or decades, and most of the residents of the area have been permanently relocated. The disaster is likely to be a continuing story with more details coming out over a long period.

This has been the result in a relatively uninhabited farming region. A similar accident at the Fermi-II plant near Detroit or the Indian Point plant near New York City would probably kill and dislocate many more people.

Nuclear trouble does not have to be accidental. Terrorism could occur at several levels. The possibility exists for the theft of the 5 kg of plutonium needed to make an atomic bomb, which could then be used to blackmail a city, state, or possibly the federal government. However, there really would be no need to go to the trouble of constructing an atomic device. The theft of virtually any fair-sized quantity of concentrated nuclear waste and the threat of blowing it up with conventional explosives would be frightening enough.

Nuclear War

At this point, "peaceful" uses of atomic energy and warfare begin to converge. The period following the signing of the nuclear test ban treaty of 1963 was a relatively optimistic time. Only France and Communist China continued to conduct atmospheric tests. Most ecological concern focused on radiation releases from peaceful applications.

In the renewed Cold War atmosphere of the early years of the Reagan administration, consideration began again to be given to the military use of nuclear force. The literature of the 1950s and early 1960s on the cataclysmic effects of nuclear war once more became relevant. Examples of these assessments include Comar's (1962) "Biological Aspects of Nuclear Weapons" and the 1963 symposium of the Ecological Society of America on the ecological effects of nuclear war (Woodwell 1965).

New scenarios were developed that envisaged a nuclear exchange between the United States and the Soviet Union involving 5000 to 6500 megatons (a megaton is the equivalent of a million tons of TNT). Blast, heat, and the initial ionizing radiation resulting from these rocket attacks would kill outright about 1.1 billion people (Harwell and Grover 1984). Added to this total would be as many more that would require skilled, lengthy, and labor-intensive treatment in by-then mostly nonexistent hospitals.

Although ecologists had long speculated about ecosystem effects such as loss of vegetation, increased erosion, and disruption of food chains and biogeochemical cycling (Odum 1964), some new considerations were added in the 1980s. The so-called TTAPS study (Turco et al. 1983) and later versions (such as Turco et al. 1990) put forth models and data suggesting that dust and soot from fires following a nuclear exchange would block out the sun. In the pale light of this "nuclear winter," photosynthesis would be reduced, temperatures would drop, and much of the vegetation would die. So would animals, forced to forage in the dark for a dwindling food supply. Current agricultural crops, with their heavy dependence on technology for seed production, fertilizer, and pesticides, would not fare very well in the postwar world, even after the smoke cleared. It was projected that more people would die in India from starvation than died in the United States and the Soviet Union from the direct effects of the explosions (Pittock 1986).

The 1990s seem to be the beginning of a new era of optimism. As a result of peace initiatives by Mikhail Gorbachev, followed by the economic and political collapse of the Soviet Union, nuclear war on a massive scale currently seems unlikely. Use of nuclear weapons on a smaller scale is perhaps more likely, owing to the spread of nuclear technology to many nations, such as Pakistan and Israel, and instability as the result of ethnic tensions around the world.

Other Toxic Materials

For some time laws have been in effect that require testing before commercial use of pesticides, drugs, and food additives, and that require safety precautions in the handling of radioactive wastes. These safeguards may not always prove adequate but the legislation does exist. A great many other potentially hazardous chemicals do not fall into any of the categories listed but may enter the environment and do much the same harm as pesticides. Examples are asbestos, which enters the environment from automobile brake bands, talcum powders, and mining wastes; mercury, used as a fungicide, in silent electrical switches, in dental work, and in the paper industry; and polychlorinated biphenyls (PCBs), extremely stable compounds related to DDT and once widely used in industry in large electrical insulators, as plasticizers, and in "carbonless" carbon paper. PCBs seem to be at least as harmful as DDT in much the same ways and, furthermore, seem to be as widespread in the bodies of organisms, despite the fact that they were never purposely added to the environment. In 1976 Congress finally passed legislation (the Toxic Substances Control Act) requiring testing and safeguards in handling these sorts of chemicals, but full implementation of the law has been slow in coming.

The potential for trouble that exists is illustrated by another compound, polybrominated biphenyl (PBB). Like PCB, PBB is a very stable hydrocarbon with a variety of industrial uses. One of these was in a mixture manufactured by the Michigan Chemical Corporation as a fire retardant and sold under the name "Firemaster." The company also manufactured a grain supplement called "Nutrimaster" to be added to the food of dairy cattle. In the summer of 1973, Farm Bureau Services received a shipment of Firemaster, rather than Nutrimaster, from the Michigan Chemical Corporation and began to mix it into livestock feed that went to 1100 or more farms (Stadtfeld 1976). Cattle sickened and died, but the source of the problem was not identified for several months. By May 1974, PBB had been identified, and when the Michigan Department of Agriculture, after much foot dragging, finally began to investigate, it found PBB in milk and meat throughout much of the state.

The full extent of the problem is still not clear several years later. One difficulty was the virtual lack of research on PBB's biological effects. The farm families who received the contaminated feed, through drinking the milk from their own herds and eating meat from butchered animals, have high levels of PBBs in their body. Preliminary medical research has shown that the farm families show a high frequency of certain symptoms, including those of brain damage and a reduced ability to resist infection. In 1981, a sampling showed that 97% of the state's residents had at least trace amounts of PBBs in their bodies. To what degree they have suffered the same or other effects may not be known for a long time.

Residents of other states can take no encouragement from Michigan's misfortune; they have their own toxic substance crises. Besides, only a few hundred pounds of PBB were involved in the Michigan incident, but the Michigan Chemical Corporation manufactured 2400 tons of PBB in 1974 and much more in previous years. It was shipped widely and is undoubtedly a contaminant of the whole country, if not the whole world.

With new reports of carcinogenicity and other chemical hazards arriving frequently in the news, it is easy to fall into the trap of thinking that every chemical is hazardous, and so there is no point in worrying about it. It is true that even water is a danger if it is above our nose, but most of us continue to try not to drown.

As we pointed out earlier, most of the chemicals that a species has evolved with tend not to be harmful in the amounts it is likely to encounter naturally. As of 1980 about 7000 chemicals had been tested for carcinogenicity in the laboratory. Only about 500, less than 8%, caused cancers. A few natural materials are included in the list, but these are things that, under natural conditions, are usually dilute or inaccessible, such as asbestos and cadmium.

The fact is that many chemicals are relatively safe, but some are hazardous to us or to other parts of the biosphere. Most of the toxic, carcinogenic, and mutagenic chemicals are the novel compounds, new in evolutionary history, that the chemical companies synthesize, promote, and spread in the environment.

Biological Magnification

There is an old saying among sanitary engineers that "dilution is the solution to pollution." Historically, this is one reason why cities and industries tended to locate on rivers, lakes, or the ocean: They could run their wastes into the public waters. Unfortunately, several factors have combined to make "dilution" unsatisfactory as a method of waste disposal. First there is too much waste being produced; a city or industry of any size just does not have enough water available to it to dump wastes without harming water quality. Second, there are so many people that the solution to one person's pollution may come out of the next person's faucet. If there is only one town on a river, sewage dumped into the water may be decomposed within 20 or 30 miles and the water may be clean again but if there are towns every 10 miles, this recovery may not occur.

Another reason why dilution no longer works is slightly more complex. Because of biological magnification (also called bioconcentration and bioaccumulation), some materials may be put into the environment dilute and come back concentrated. Two classes of materials showing such magnification are some pesticides and other human-made chemicals and radioactive isotopes of certain elements. Two factors are involved. One is that the material persists in the bodies of organisms; the organism takes it in but does not readily break it down or excrete it. The second is the structure of food chains and trophic levels.

Some human-made chemicals are extremely persistent—they are not broken down quickly. One reason for this is their novelty; they are new to life and few organisms have evolved the metabolic pathways for decomposing them.

DDT (and many of the other chemicals we are concerned about) is soluble in fat but not in water. Consequently, when DDT is in or on the food eaten by a herbivore, it tends to be stored in the fat of the animal, rather than excreted. If the concentration of the pesticide on the plant is some dilute level that we can represent as 1, a particular herbivore may eat 100 plants to stay alive and the concentration in the herbivore may build up to 100 times (100×) the low level in plants. In the same way, a carnivore may eat 100 herbivores and the pesticide is magnified to something like 10,000× the concentration at the plant level.

Top carnivores are of course particularly susceptible to being harmed by pesticides as the result of biological magnification. Prime examples are some of the hawks and large seabirds, in which reproductive abnormalities have been shown to be caused by high pesticide levels.

Most of us are familiar with the population declines in the bald eagle (Figure 20-2) and some other raptors that occurred through pesticide poisoning. With the phasing out of some of the more persistent chemicals, some of these birds have begun to rebound, but we should not be fooled into thinking that biological magnification is a problem of the past. Researchers from the Illinois Natural History Survey studying hawk movements by means of radiotracking found a recent compelling example. An adult male peregrine falcon showed abnormal signs of tameness and weakness in its southward migration through Wisconsin and Illinois (Cochran et al. 1992). His hunting effort and success dropped through mid-October. On October 20, 1989, the falcon was tracked to a farmyard and about dark was observed creeping through the grass to hide under a brushpile.

During the night, the falcon was killed and partially eaten, probably by a raccoon or skunk. Chemical analysis of tissues of the bird showed high levels of pesticides including dieldrin, DDE (a DDT breakdown product), PCBs, and mercury.

Radioisotopes are concentrated in a fashion similar to persistent pesticides. Here the materials that are especially dangerous are mainly the elements essential for some bodily process and that are relatively scarce in the environment. The body tends to retain as much of these materials as possible. Phosphorus, for example, is rare in nature but is part of certain essential cellular compounds (such as ATP, the P standing for phosphate) and is also concentrated in such places as birds' eggs. One study has shown that radioactive phosphorus in

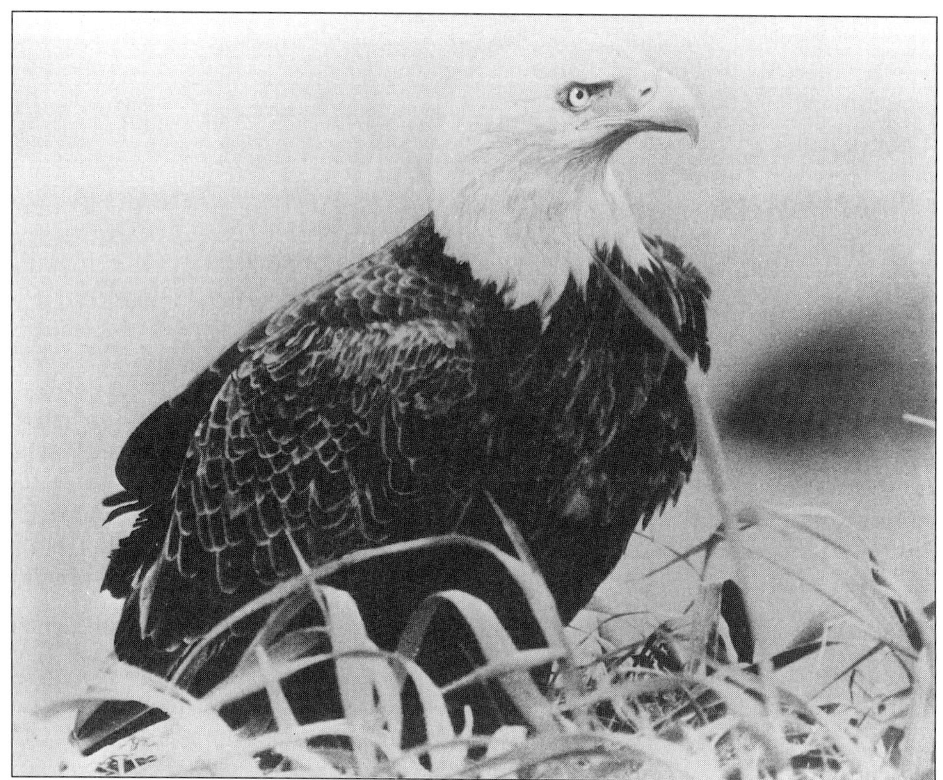

Figure 20-2

The bald eagle, our national emblem, is at the top of its food chains and, accordingly, accumulates heavy concentrations of persistent chemicals, such as DDT/DDE and PCBs. (Karl Kenyon, U.S. Fish and Wildlife Service.)

duck and geese flesh was 7.5× the concentration in the water where they fed; the concentration in their eggs was 200,000× the concentration in the water.

Radioactive iodine is another candidate for concentration. Iodine is rare in nature but necessary for producing thyroxin, the hormone of the thyroid gland. Individual organisms concentrate iodine in their thyroids, and further concentration may occur up the food chain, through biological magnification.

Some other concentration factors recorded for various radioisotopes have been zinc, 1000× in adult whitefish; iron, 100,000× in algae; strontium, 3900× in muskrat bone; and cesium, 250× in waterfowl muscles.

Neither cesium nor strontium is essential in metabolism, but both can be concentrated, cesium being taken up like potassium and strontium like calcium. Both were released in considerable quantities in the early 1960s in the testing of nuclear weapons, and both were found concentrated in human tissue as well as in that of other animals.

Pollution in the Environment

In this section, we focus on three aspects of pollution—air, water, and the environmental release of genetically engineered organisms. Other adverse effects of humans on the environment could be considered. Neglected but important issues, for example, are noise pollution and the loss of darkness, as light pollution progressively eliminates the night.

Water Pollution

Certain populations of humans have the habit of putting their wastes into water. As discussed in earlier sections, this has the effect of diluting the wastes and, especially in streams, getting them carried out of the vicinity of the producer. Types of pollutants include:

- Sewage and other organic wastes (such as wood fibers from paper mills)
- Pathogens (mostly bacteria and viruses) from human wastes
- Toxic materials (discussed in preceding sections of this chapter) such as salt, mercury, pesticides, and petroleum
- Chemicals (usually phosphorus or nitrogen) important in eutrophication (mentioned in the section on limiting factors in Chapter 2 and in the section on lakes and ponds in Chapter 18)
- Waste heat (thermal pollution)

Any one source of pollution may include materials in several of these categories.

For a variety of reasons, large amounts of time and money have gone into studying and attempting to correct the problem of water pollution. These include the fact that water pollution is often readily identifiable by sight or odor and is thus more easily noticed by the public (Figure 20-3). Second, clean water has obvious economic values (in the price of waterfront property, for example) and public health importance. Third, there seem to be technological solutions to water pollution problems such as building sewer lines and designing large and sophisticated sewage treatment plants. Many other important ecological problems that are more subtle or whose solutions depend on something other than engineering have so far received disproportionately small shares of public attention (and money).

The primary effect of stream pollution by sewage and other organic wastes is in providing large amounts of food for decomposer organisms, mainly bacteria and fungi. The sewage fungus *Sphaerotilus* characteristically appears in organically polluted streams. The respiratory activity of the decomposers removes oxygen from the water, and many characteristic stream animals cannot tolerate the resulting low oxygen levels (Figure 20-4). Some animals, however, such as sewage worms, may become very abundant.

The term **biochemical oxygen demand** (BOD) refers to the amount of oxygen needed to break down (oxidize) organic materials to carbon dioxide, water, and minerals. It is considered that each person generates 0.07 kg of BOD in a day. Other types of organic pollution can be expressed directly in terms of BOD or converted to **population equivalents** (PE) by dividing BOD by 0.07. By far the greatest amounts of organic pollution in the United States come from the food industry, including rais-

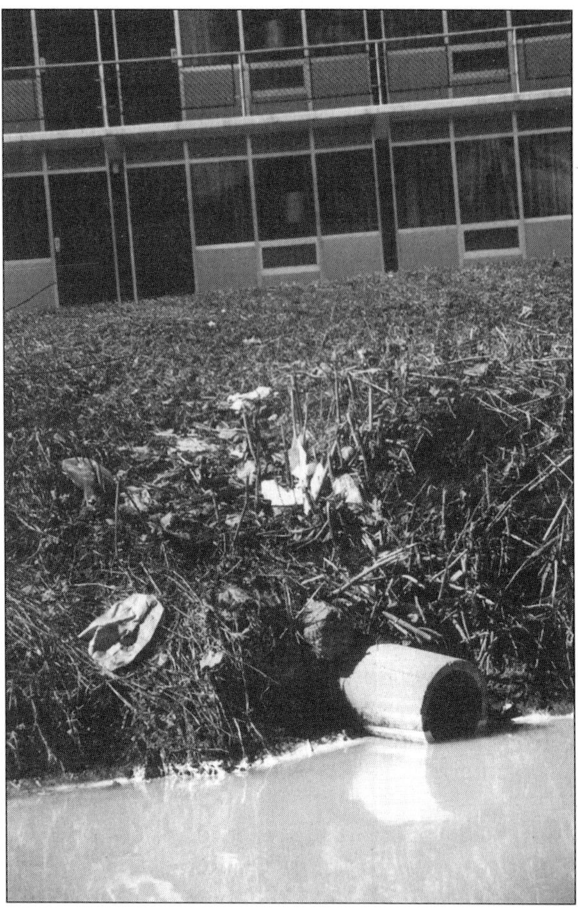

Figure 20-3

The sight and stink of polluted water draws attention. The milky color of Portage Creek in Kalamazoo, Michigan, in the 1960s was caused by paper mill wastes including wood fibers and clay. It would be impossible to duplicate this photograph today. Because of pollution control measures, this stream and many others over the country are much cleaner. (Photo by Gerald Martin.)

ing livestock and processing meat and other foods. The water pollution resulting from a slaughterhouse per ton of slaughtered cattle is 100 to 200 PEs—that is, the same BOD as produced by 100 to 200 people in a day.

The addition of waste heat to water is mainly from electrical generating plants, both conventional and nuclear. Water is taken in to cool the generators and then returned to a lake or stream. Although local modifications of the biota occur in the vicinity of the outfall, it is still unclear how seriously the addition of waste heat is to be regarded (Sharitz and Gibbons 1981). A good-sized power plant located on a small, deep lake might lower the thermocline and extend the summer stagnation period to such a degree that certain types of fish might be harmed. Hot water added to streams tends to eliminate some species of animals and to decrease diversity. In other situations addition of warm water may have somewhat desirable effects—for example, providing open water in winter where waterfowl may feed. The current approach to thermal pollution is to use cooling towers that dissipate the heat to the atmosphere rather than into lakes or streams.

The conclusion of the National Academy of Science (Spilhaus 1966) that "pollution is often a resource out of place" describes organic and thermal pollution of water very well. The eventual solutions to both problems should be aimed at *using* the organic materials (to enrich soil, for example) and the waste heat, rather than simply trying to dissipate them.

The use of deicing salt, often mixed with sand, on roads in cold climates in winter is a kind of pollution that needs careful study. About 4.5 million tons per winter is used in the United States (EPA 1980). Potentially serious effects are salination of the groundwater and effects on nearby streams. Deicing salt also kills susceptible vegetation along the roadside while allowing the entry, in some cases, of halophytes (Reznicek 1980).

At sites of severe salt pollution, invertebrate species diversity declines. At lower concentrations, changes result from increased sedimentation. This comes both from roadside erosion following death of vegetation and from the sand applied to the road. Sedimentation seems to cause a seasonal drop in invertebrate biomass and a shift to a higher proportion of oligochaetes (Molles 1980). Both probably result in less food for trout. The sedimentation itself may directly harm eggs and young of spring-spawning trout.

One lesson of the past several years is that a large fraction of the pollutants in water (more than half) come from "nonpoint" sources, rather than from a pipe. Examples of nonpoint sources include

Urban storm-water runoff, with its oil residues, lead from gasoline, asbestos from brake linings, and chemicals from people's lawns.

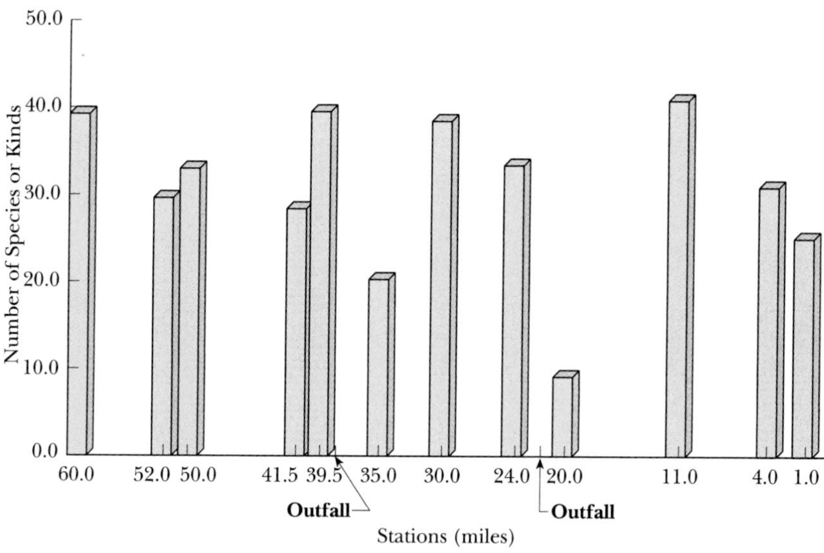

Figure 20-4

In this graph, the Mad River, one of Ohio's few trout streams, flows from left to right (the distances are miles above Dayton). At the time of the survey in the 1950s there were two main sources of pollution, Urbana, just below mile 39.5, and Springfield, just below mile 24.0. Most of the pollution was sewage and paper mill waste. The characteristic decline in diversity just downstream from both outfalls is obvious. Species of cleanwater forms such as stoneflies and caddis flies dropped out whereas the pollution-tolerant species such as sewage worms and bloodworms increased. (Adapted from Gaufin 1958.)

Construction, which generates erosion that runs sediments into streams and lakes.

Forestry, which has similar effects plus heavy inputs of herbicides and pesticides.

Mining, which produces sediments and may release harmful chemicals from ores (such as acid from iron pyrites).

Septic systems, which, when located near lakes, may run sewage, bacteria, and household chemicals almost directly into the lake water.

Atmospheric deposition, which, although hard to evaluate, is clearly an important source of various highly toxic organic compounds, such as toxaphene and PCBs, and, of course, of acids.

Agriculture, which generates sediments from soil erosion, organic matter from feedlots, pesticides, herbicides, and phosphate and nitrate from fertilizers. About half of the total nonpoint pollution has been estimated to come from agriculture.

One important pollutant coming from atmospheric deposition is mercury. Mercury has been detected in fish from lakes in many localities with no nearby human source of mercury contamination. Studies of lake sediments in Minnesota and Wisconsin showed that annual deposition of atmospheric mercury has increased by a factor of more than 300% (from 3.7 to 12.5 $\mu g/m^2$) from 1850 (that is, the preindustrial era) to the present (Swain et al. 1992). The fact that the increase was similar among all seven lakes studied suggests regional or even global sources for the atmospheric mercury being deposited.

Pollution in the Great Lakes

Many of the themes of earlier sections of this chapter come together in a consideration of chemical pollution of the Great Lakes (Figure 20-5). This marvellous resource, bounded by eight U.S. states and one large Canadian province, Ontario, was greatly abused in the past. In recent years, the dis-

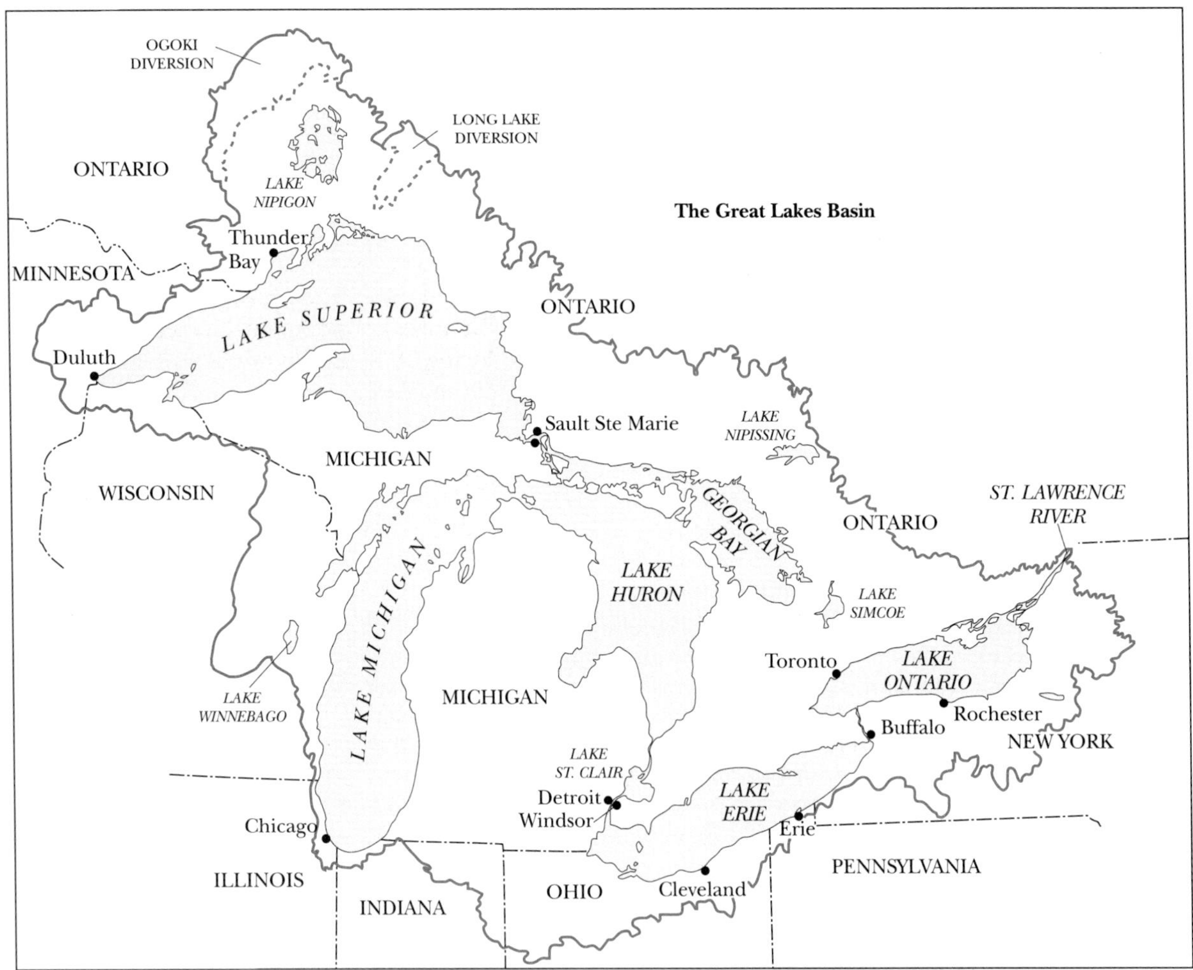

Figure 20-5

The Great Lakes. The dark blue line shows the watershed or drainage basin. The main flow is from Lake Superior to Huron to Erie to Ontario and out the St. Lawrence River to the Atlantic Ocean, a distance (from the west end of Superior) of about 1700 miles (2700 km). (Based on Great Lakes Water Quality Agreement 1978. International Joint Commission, Windsor, Ontario.)

charge of many kinds of pollutants has decreased. Sewage and waste water containing phosphate detergents are examples. Use of the more persistent pesticides has also dropped in the watershed.

Eutrophication has been slowed or reversed, especially in Lake Erie, which was seriously eutrophied in the 1960s. Erie is the shallowest of the Great Lakes and has a large throughflow from Huron to Ontario. Only three years is required to replace its water volume completely, so flushing out of pollutants is relatively rapid. Lake Michigan, a cul de sac with essentially no southern outlet, has a water replacement time of 70 years. In all the lakes, persistent chemicals remain in the sediments—and in the organisms, especially the bottom fauna and other organisms having bottom dwellers in their food chains. Chemicals continue to be added to the lakes, a large fraction now coming from nonpoint sources; examples are atmospheric PCBs and mercury.

The problem in the Great Lakes is one of biological magnification. The water itself is relatively clean, as we see in Table 20-2. The same table shows high concentrations of pesticides in a top carnivore fish, rainbow trout. Along a trout food chain in Lake Ontario we might find concentrations of PCB (Bishop and Weseloh 1990) in a series like this:

Plankton	1
Crustaceans	32
Smelt	166
Lake trout	564

Although top carnivore fish have high levels, top carnivore birds tend to have even higher ones. For example, the lake trout concentration given above might translate to 5.64 μg PCB/gram wet weight, but herring gulls may have ten times as much (Bishop and Weseloh 1990). The main reason is that the birds are homoiotherms and thus require more food intake to add a given amount of biomass. Organisms that use top carnivore fish as food, and consequently, suffer the next step up in biomagnification, include bald eagles and humans.

Several types of reproductive abnormalities have been documented in colonial waterbirds nesting around the Great Lakes (Gilbertson et al. 1991). Species involved include herring and ring-billed gulls, double-crested cormorants, Caspian, common, and Forster's terns, and black-crowned night herons. Observed abnormalities include eggshell thinning, high embryo mortality, slow development, altered liver and thyroid function, and high levels of chick deformities including crossed bills and malformed feet.

This syndrome has been termed **GLEMEDS (Great Lakes embryo mortality, edema, and deformities)** and has been equated with a disease of poultry attributed to pesticide poisoning. Exactly which of the Great Lakes pollutants are most important in producing GLEMEDS is not definitely known but the list of possibilities includes DDT and its breakdown products, dieldrin, PCBs, and dioxin.

Several studies have evaluated the effects of eating Great Lakes fish on humans, especially human reproduction (Swain 1991). The general approach has been to compare the offspring of women who eat more, less, or no fish. A high level of consumption would be someone who ate lake fish at least twice a month and had done so for three or more years. Attention was focused on PCB levels. Among maternal health problems associated with high levels of fish consumption (that is, PCB exposure) were edema, anemia, and susceptibility to infection.

The developing infant is exposed to PCBs and similar contaminants both (1) in the uterus (by transfer across the placenta) and (2) by drinking the mother's milk. Evidently intrauterine exposure is generally more important. Conditions associated with increasing PCB levels include low birth weight, shortened gestation period, and several sorts of behavioral abnormalities or problems in infancy and early childhood. For example, when children from one of the studies were tested at four years of age, those with greater uterine PCB exposure scored poorer on standardized tests for verbal ability and memory.

For these reasons as well as the carcinogenicity of these chemicals, public health advisories have been issued for the Great Lakes states (Figure 20-6). The 1992 Michigan Department of Natural Resources Fishing Guide has two columns under Lake Michigan: "Restrict Consumption" and "No Consumption." The no-consumption category includes larger individuals of whitefish, lake trout, chinook salmon, and brown trout, and all carp and catfish. The restrict-consumption category indicates that no more than one meal per week of the listed fish should be eaten, but it also has this footnote for

Table 20-2
Concentrations of Some Organic Compounds in Lake Ontario Rainbow Trout Compared With Lake Ontario Water

	Concentration (μg/g)		
Compound	*Water*	*Fish*	*Bioconcentration Factor*
Lindane	0.0003	0.3	1000×
Trichlorobenzene	0.0005	0.6	1200×
PCB 18	0.00002	13	650,000×
PCB 153	0.00003	250	8,000,000×
Mirex	0.000008	110	14,000,000×

***Bioconcentration factor = concentration in fish/concentration in water.**

Calculated from Oliver, B. G., and Niimi, A. J., "Bioconcentration factors of some halogenated organics for rainbow trout: Limitations in their use for prediction of environmental residues," *Envir. Sci. Tech.*, 19:842–849, 1985.

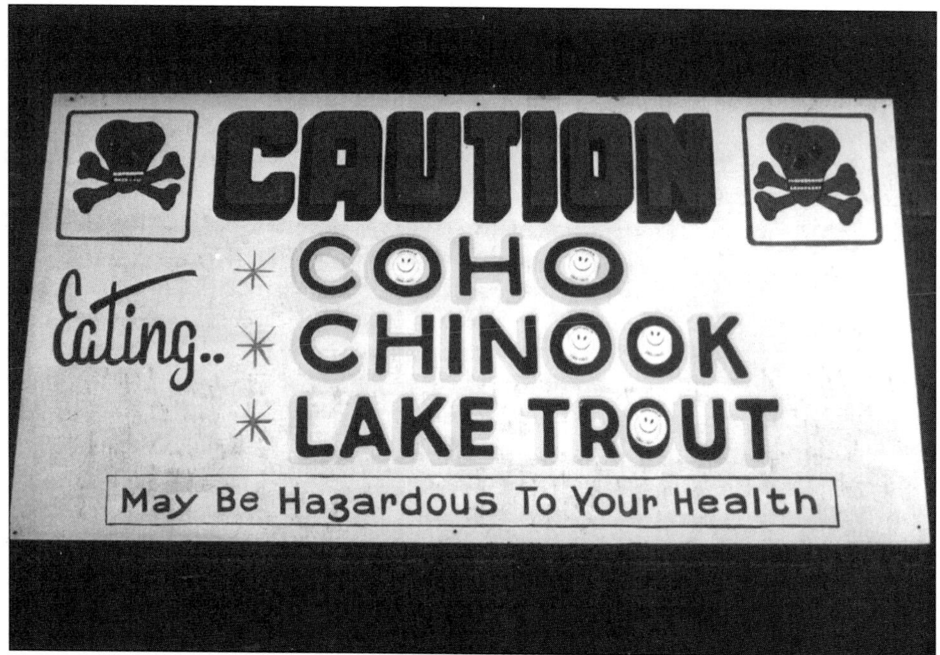

Figure 20-6

Potentially dangerous levels of PCBs, heavy metals, and other chemicals may be present in several species of Great Lakes fish, as this sign along the Lake Michigan shore indicates. (Photo by Marcella Martin.)

most of the fish: "Nursing mothers, pregnant women, women who intend to have children, and children under age 15 should not eat these fish."

These preceding paragraphs show that some of our mistakes of the past are still with us, a continuing drain on resources, including the human resource. A lesson we might draw is that we need to revise the belief that because we can do something, we might as well do it. We have not yet reached the necessary level of caution in our release of materials into the environment.

As far as Great Lakes pollution goes, it is possible to conclude on a semioptimistic note. Things are generally better than they were. Lake Michigan DDT, dieldrin, and PCB levels have been monitored for many years in the bloater, a small deepwater fish that feeds mainly on zooplankton (Hesselberg et al. 1990). Both DDT and PCB have shown a fairly steady decline beginning in the 1970s (Figure 20-7). Dieldrin (not shown in the graph), although banned in 1974, peaked in the later 1970s and has since declined.

Although chemical contamination is still harming the birds that depend on the lake for food, they

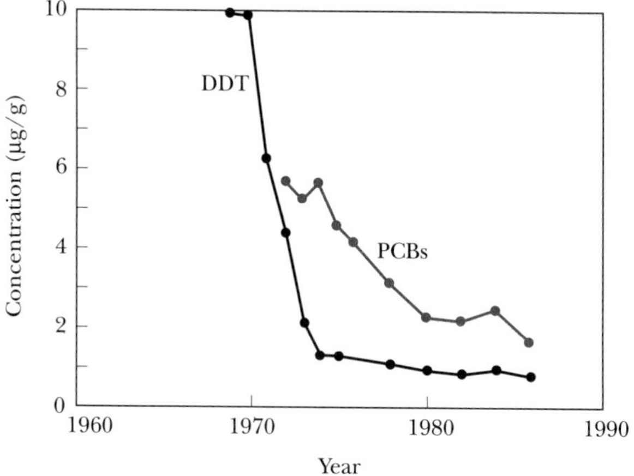

Figure 20-7

Concentrations of DDT and PCBs in Lake Michigan bloaters. Figures are mean concentrations in whole fish, wet weight. DDT was banned in the Great Lakes states in 1969 to 1971 and federally in 1972. PCBs were the subject of voluntary controls beginning in 1972 and were banned in 1976. PCBs were not measured prior to 1972. (Data from Hesselberg et al. 1990.)

have shown signs of recovery. Double-crested cormorants disappeared from the waters around Michigan in the 1950s; the last known nesting was in Lake Superior in 1961 (Ludwig 1991). Then in 1977 with dropping DDT levels, there were eight successful nests in Lake Michigan's Green Bay. Colonies totaling more than 2000 nests now exist at more than 25 localities in the state of Michigan, and similar increases have been noted in Wisconsin and Ontario.

The Exxon Valdez *Oil Spill*

In the middle of the night of March 23–24, 1989, the oil tanker *Exxon Valdez,* Joseph Hazelwood, captain, went aground on Bligh Reef in Port William Sound, on the southern shore of Alaska (Church 1989). Through holes in the single-thickness hull something more than 10 million gallons of petroleum poured out into the water. The oil was from the bitterly contested Alaska pipeline, whose southern terminus is at the port of Valdez.

It was the largest oil spill in North America and the largest in the arctic—in each case, so far—but it was not an unusual event. In the ten months between March 1989 and January 1991, there were four other oil spills of more than a million gallons and more than 60 spills of more than 100,000 gallons (Holloway 1991). These occurred from the Arabian Sea to Texas to the Philippines and from South Africa to Norway.

The response on the part of Alyeska, the pipeline consortium, and the Exxon Corporation was slow (Hodgson 1990). The oil eventually spread over a linear distance of about 750 km (to the southwest) (Piatt et al. 1990) and fouled 5300 km of shoreline (Estes 1991) (Figure 20-8). Faster action might have contained more of the oil with booms and sucked it off the ocean for disposal.

There was immense mortality of many kinds of organisms, much of it undocumentable in any quantitative way. For some of the more conspicuous animals, figures are available. More than 30,000 dead birds of 90 species were actually recovered from polluted areas by the beginning of August (Piatt et al. 1990). Murres and their relatives suffered the greatest losses. Among the relatives were several hundred marbled murrelets, a species al-

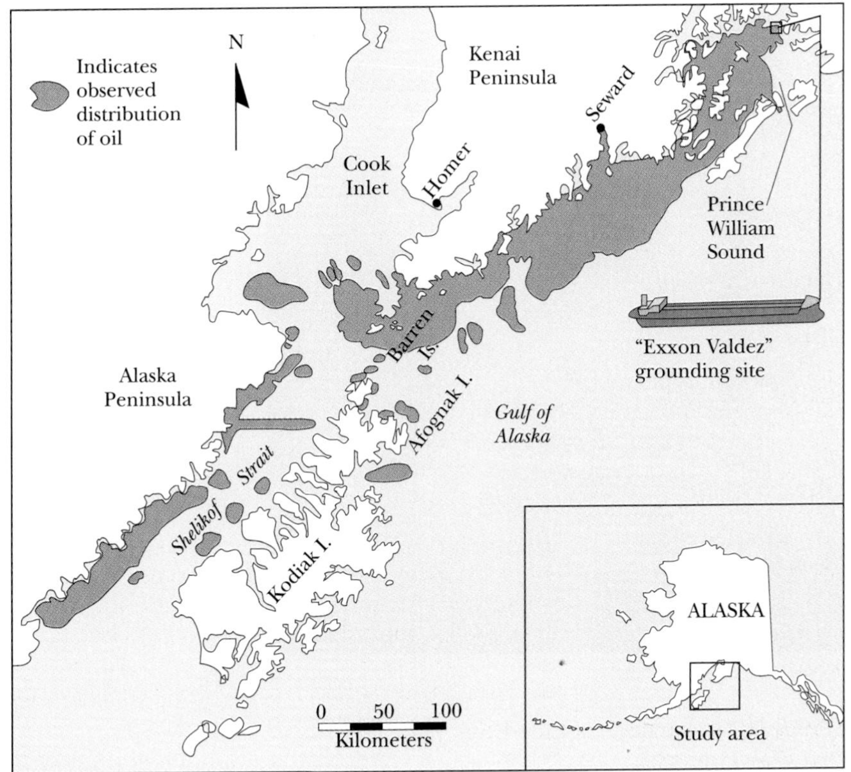

Figure 20-8

Area of the *Exxon Valdez* oil spill between March 24 and May 10, 1989. (Based on Piatt et al. 1990.)

ready at risk because of its ties to old-growth temperate rain forest. The 30,000 dead birds retrieved were estimated to be between 10 and 30% of the actual fatalities—most of the carcasses sank, were carried away by currents, or suffered other fates. Probably the actual number of seabirds killed was 100,000 to 300,000.

Eight hundred seventy-eight sea otter carcasses were found and 357 live otters were captured and taken to rehabilitation facilities (Estes 1991). As with the marine birds, the number of sea otters killed but not retrieved was probably high. Of the 357 living otters brought in, 160 died or were too badly injured to return to the wild. Of the 197 animals returned to the wild, survival rate was apparently low.

In Prince William Sound as in many other oil spill sites, there was an expensive cleanup effort (Figure 20-9). The Exxon Company focused on beach cleanup, although it did not get started until six days after the spill (Church 1989). A variety of methods were used, ranging from removal by rakes and shovels to blasting the oil off rocks with pressurized heated water (Hodgson 1990). Also employed was "bioremediation" in which a fertilizer mixture containing nitrogen and phosphorus was sprayed on shores. The aim was to supply limiting nutrients to naturally occurring petroleum bacteria (which, of course, had an ample carbon supply in the form of the oil).

Considering this spill and others studied before, certain lessons are clear and others are yet to be learned in final form. One lesson is that the more drastic shoreline cleanup procedures, such as hot water blasting, tend to produce as much damage to the biota as the original spill and probably should be omitted (Holloway 1991). Another is that the rehabilitation efforts for most of the wildlife species are expensive and ineffective (Estes 1991). Restricting rehabilitation efforts to endangered and threatened species and doing a superb job with those would make more sense than attempting to save every oiled bird or seal.

These conclusions were re-enforced by the *Exxon Valdez* incident, but they are not new. How soon they will be implemented is doubtful. Corporations want to seem to be doing everything possible so as to improve their image and to position themselves favorably for the lawsuits that will follow. Individuals do not take well to the idea of standing

Figure 20-9

Trying to clean up the shoreline of Prince William Sound after the *Exxon Valdez* oil spill, 1989. (Rob Stapleton/Dembinsky Photo Associates.)

around; they want to be doing something, even if it is cleaning the plumage of a bird that is going to die from the oil it ingested.

A much more effective way to spend time and money is in preventing spills. Energy conservation, which will reduce the amount of petroleum used, hence shipped, is one way in which everyone can help. Requiring the use of double-hulled tankers is desirable, although the *Exxon Valdez* rammed Bligh Reef with such force that the inner wall would probably also have been breached. "Hydrostatic loading" should certainly be instituted; this involves loading tankers to a shallower depth (up to about the water line). When this is done, the pressure exerted by the oil at the bottom of the hull is less than the water pressure; if a hole is punched in the hull, water enters rather than oil running out (Holloway 1991).

Perhaps not even these measures would be as effective in preventing spills as criminal prosecutions with jail time for CEOs of corporations responsible for spills.

Groundwater

Contamination of groundwater is an environmental problem that came to public attention in the 1980s. Who would have thought a decade or two ago that all that stuff we were putting in landfills would move through the ground and start coming out in our drinking water? Actually, most people, including the government units in charge of the dumps, didn't think about it at all, but even those who did underestimated the rate of movement through the ground and overestimated rates of breakdown and adsorption on soil particles.

The estimate is that about 1% of U.S. aquifers are badly polluted (Abselson 1984); however, because that 1% is mostly in heavily populated regions, the percentage of the population exposed to potentially dangerous water is much higher.

The landfills with their burdens of trichloroethylene, carbon tetrachloride, benzene, PCBs, and many other poisons from industries and homes are one obvious source of groundwater contamination. Not many years ago, a prime use of wetlands was as landfills, so the hazardous chemicals did not even have to leach down to the water table; they were put directly into it. Landfills occur everywhere and are a source of some of the most potentially dangerous contaminants. They are not, however, the largest source of groundwater contamination; that is the petroleum industry.

In Michigan, about 13% of the known groundwater contamination sites are from landfills; at least 30% and possibly over 40% are from the petroleum industry. We start with the brine and various organics such as benzene from well-drilling and move on to losses from pumping, storage, and pipeline transportation to the leakage from underground storage tanks along with the constant, day-after-day spillage that goes on at filling stations. These are just some of the obvious, larger losses that will eventually reach the water table.

Other sources of groundwater contamination include on-site storage of raw materials and wastes from industry, deicing salt storage and application, laundromats, and car washes. Another source of which we will hear more in the next several years is injection wells. Industry is increasingly disposing of hazardous waste by injecting it under high pressure into underground formations a thousand or more feet deep. Although the technology is favored by the EPA's Office of Drinking Water, regulation has been weak and several states have acted to ban deep-well injection altogether.

The sorts of problems that have occurred in connection with injection wells include spillage from aboveground storage ponds and tanks and, in an Ohio case, the production of toxic clouds that drifted across nearby farmland (Culver and Audette 1985). Leaks through the well casings have occurred, allowing direct contamination of groundwater. And a major uncertainty is whether the hazardous wastes, injected below any underground source of drinking water, will, in fact, stay there or whether they will instead migrate sideways or upward.

About half the U.S. population depends on groundwater for drinking water; however, of the 60 gallons of water an average American uses per day, the water he or she actually drinks amounts to only about 0.4%, about a fourth of a gallon. Further, home use accounts for only 4% of the water usage in the United States. Agriculture (mostly irrigation) and industry are the big users, accounting for about 43% and 48% of total water use, respectively (Miller 1982).

Groundwater pollution seems certain to increase in the years ahead as toxic chemicals continue their spread through aquifers. Given this fact, the serious surface-water pollution that exists in many localities, and the figures on usage in the last paragraph, it seems possible that more and more communities will give up the battle to provide drinking-quality water at the faucet. Individuals in many localities have already decided that the battle has been lost and switched to bottled water.

Atmospheric Pollution

Historically, most air pollution has resulted from burning fossil fuels in homes, factories, power plants, and automobiles. From one or all of these sources have come such things as soot, ash, carbon monoxide and carbon dioxide, oxides of sulfur and nitrogen, sulfuric acid, unburned hydrocarbons, formaldehyde, arolein, peroxycetylnitrate, lead, and bromine. Some of these are secondary pollutants resulting from irradiation of the primary pollutants by sunlight, producing photochemical smog.

Direct effects of air pollution on human health seem to range from simple irritation of eyes and nostrils through bronchitis and emphysema to lung cancer. During episodes of severe air pollution, death may occur in short order. During the five-day "killer fog" of early December 1952, 4000 more persons died in London than normally would have died in such a time span. Large-scale conversion from high-sulfur coal to natural gas as a fuel supply

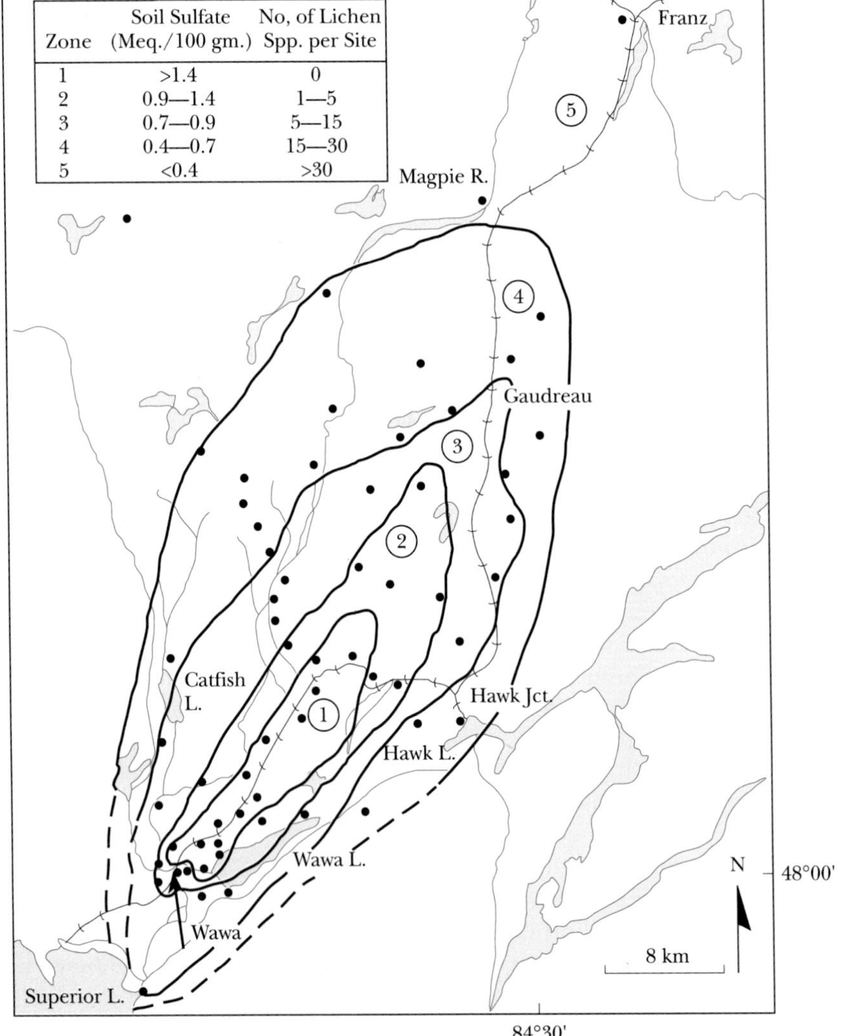

Zone	Soil Sulfate (Meq./100 gm.)	No, of Lichen Spp. per Site
1	>1.4	0
2	0.9—1.4	1—5
3	0.7—0.9	5—15
4	0.4—0.7	15—30
5	<0.4	>30

Figure 20-10

Air pollution zones around an iron-sintering plant at Wawa, Ontario. As a result of sulfur dioxide pollution (measured here by soil sulfate levels), no lichens grew on tree trunks and branches for more than 16 km downwind and diversity was reduced for 50 km. (Adapted from Rao & LeBlanc 1967.)

in London seems to have made these catastrophes a thing of the past—as long as the natural gas supply holds out.

The same pollutants affecting humans also harm plants and other animals, although there has been little work done on their effects on wild animal populations. Certain types of crops and trees cannot be grown successfully in the vicinity of large cities or heavily traveled highways. Automobile exhaust fumes are known to be harmful to the invertebrate animals living in forest litter. Lichens are sensitive to air pollution, especially to levels of sulfur dioxide and thus are valuable as indicator species (Figure 20-10).

Air pollution has ecosystem effects in addition to direct effects on organisms. Humans are discharging large enough quantities of some elements that pollution has become an important biogeochemical process for the whole biosphere. Combustion of fossil fuels does not add as much carbon to the atmosphere as the respiration of organisms, but it adds more than such sources as volcanoes. Losses from burning fossil fuels and other human activities now add more mercury to the environment than does the weathering of the earth's crust.

Worldwide atmospheric pollution can come in less conspicuous packages than the smokestacks of Pittsburgh. Studies suggest that the chlorofluorocarbons **(CFCs)** used as refrigerants are moving to the upper atmosphere, where they are causing ozone to break down to ordinary oxygen molecules. This is significant because the ozone layer, which lies in the zone from about 15 to 35 kilometers out from the earth, screens out most of the ultraviolet radiation in sunlight. Ozone depletion is discussed in more detail a little later.

A study committee of the National Academy of Sciences recommended that the use of CFCs as propellants in spray cans be stopped, and this has been done in most of the world. This is enouraging because it is one of the few cases concerning atmospheric pollution in which the attempted solution has been something other than what environmentalists have referred to as a "technological fix," a situation in which a problem is attacked at the engineering level rather than through a reorientation of attitude or behavior (Figure 20-11). For example, the solutions up to now for automotive pollution have been to add on "pollution control" devices such as catalytic converters. Technology has been and will continue to be a powerful tool, but purely technological solutions often turn out to be new problems in themselves. Catalytic converters and other pollution control devices, for example, are expensive, may cause engines to run less efficiently—resulting in higher levels of hydrocarbon pollutants and the wasting of gasoline—and may generate more nitrogen oxides, contributing to the acid rain problem (discussed below). Solutions using other than technological approaches (for example, banning private passenger cars from urban areas) have been unpopular but may be the eventual solution.

Figure 20-11

The technological fix as applied to a stream. Real streams meander, taking up more space than they need, occasionally flood, and have other traits that may not be convenient for humans. One approach is to fix the river, as shown in this channelized section of Battle Creek, Calhoun County, Michigan. (Photo by Gerald R. Martin.)

Acid Rain

Acid rain is likely to be one of the serious environmental problems of the next decade and beyond. The harmful effects of acid deposition on ecological systems were first identified in Scandinavia in

1959, when acid rain was connected with the acidification of lake waters and a decline in fish production.

Rain, snow, fog, or dew is made acid by the presence of sulfuric and nitric acids that result when sulfur and nitrogen oxides combine with oxygen and water from the atmosphere. Dry deposition of particles yielding acids also occurs. The bulk of the sulfur and nitrogen oxides in the air now comes from the burning of fossil fuels. For eastern North America, more than ten times as much sulfur enters the atmosphere from pollution than from natural sources (Galloway and Whelpdale 1980).

In the United States, most sulfur dioxide comes from electrical power generation, but elsewhere in the world other industry plays a large role. Nitrogen oxides come mainly from the internal combustion engine, which means, generally, the automobile; however, industry and electrical generation also contribute large amounts (Postel 1984). The sulfur in sulfur dioxide is from the fossil fuel itself, and largely from coal; nitrogen oxides, however, come mainly from the reaction at high temperature of oxygen and nitrogen from the air.

The pH of normal rainwater, without industrial contamination, can be as low as 5.6 to 5.7 because of carbonic acid resulting from the natural process of water taking up atmospheric carbon dioxide. As recently as 1955 to 1956, rain in North America with a lower pH was restricted to a corridor running northeast from the industrialized Ohio River Valley through Ontario to New England. By the early 1970s virtually all of eastern North America was regularly receiving rains below 5.0 pH. Rain in southern Michigan had an average pH of 3.9 to 4.7 in 1983 (Schneider 1986).

Much lower pHs have been recorded at times; the record low I have seen is 1.4 in Wheeling, West Virginia. Because the pH scale is logarithmic, a pH of 5.0 is four times as acid as one of 5.6, and a pH of 1.4 is 15,800 times as acid.

Aquatic Systems Naturally acid waters are not common, and relatively few organisms are adapted to live under such conditions. Virtually all fish and frogs die in waters below 4.5 pH, and reproduction generally stops at even higher levels. Naturally occurring acid lakes, such as bog lakes, are not devoid of life. Algae, crustaceans, and insects live in them, but not mollusks with their calcareous exoskeletons. A few fish survive in lakes below pH 5, such as the sphagnum sunfish (Patrick et al. 1981).

An early contribution to the biology of naturally acid waters was by a pioneer female ecologist, Minna E. Jewell (Jewell and Brown 1924). A University of Illinois Ph.D., Jewell spent most of her career at Thornton Junior College near Chicago. She studied a small bog lake in northern Michigan with a pH of 4.4 and found four species of fish living there. Some of the organisms that live in such naturally acid waters may not be able to survive in human-produced acid lakes because the heavy metals (which tend to be chemically held, or chelated, by humic compounds in bog lakes) may be toxic.

Evidence of damage to lake ecosystems accumulated steadily during the 1970s. By 1979, 200 lakes in the Adirondack Mountains of New York were said to be fishless as the result of acidification. A recent paleolimnological study (Sullivan et al. 1990) confirmed that acid deposition was important in the acidification of lakes in this region. Another recent study (Baker et al. 1991) covering the whole United States concluded that atmospheric deposition was the dominant source of acid anions in 75% of some 1180 U.S. lakes and 47% of about 4670 streams. A high percentage of the rest of the acid streams were ones carrying acid mine wastes.

Some, mostly persons connected by politics or employment with the mining or electrical power industries, have continued to argue that the fishlessness of acid lakes is caused by something other than acidification. That argument cannot be made against a long-term study begun by the Freshwater Institute of Canada in 1974.

In a region of virgin boreal forest set aside for the experimental study of lake ecology, one lake, Lake 223, was chosen and studied for baseline information for two years. Then beginning in 1976, the lake was acidified by adding sulfuric acid. What happened in this lake was followed in comparison with nearby control lakes. Some of the events resulting from progressive acidification are summarized in Table 20-3.

It is clear that no one factor has done all the damage in Lake 223 or other acid waters. Some of the possibly important factors include the disturbance of normal ionic balance, the lower pH, and changes in species composition. (1) The first factor, disturbance of normal ionic balance, is probably pervasively important and may explain such phe-

Table 20-3
Changes Occurring With Acidification in Lake 223, Northwestern Ontario, Canada

Year	*pH*	*Changes*
1976	6.8	pH unchanged but alkalinity dropped. Unimportant biological changes.
1977	6.1	Brown algae decreased, green increased. Other changes slight.
1978	5.9	Opossum shrimp declined greatly. Fathead minnow failed to reproduce. Heavy emergence of midges.
1979	5.6	Thick mats of the alga *Mougeotia,* not seen previously, began to grow. Crawfish exoskeletons stayed soft. Fathead minnow declined in numbers. Phytoplankton production and midge emergence remained high.
1980	5.6	Brown algae still dominant but blue-greens increased. An acidophilic diatom, *Asterionella ralfsii,* became abundant. A third species of copepod followed two earlier ones to apparent extinction in this lake. The pearl dace increased. Midge emergence reached an all-time high. Crawfish were heavily infested with parasites and their egg masses were often infested with fungus. No successful crawfish reproduction noted. No successful lake trout reproduction.
1981	5.0	Brown algae declined greatly, blue-green increased. Several species of rotifers increased; most copepods continued to decrease. White sucker reproduction was no longer successful although growth was unimpaired. Crawfish population declined. The pearl dace began to decline.
1982	5.1	Their spawning sites covered with *Mougeotia,* lake trout attempted to spawn in locations not previously used. Lake trout condition poor. No young-of-the-year of any fish species was observed.
1983	5.1	Algal populations had largely shifted from edible to inedible forms (e.g., with gelatinous sheaths). Lake trout still numerous but in bad shape. Crawfish, leeches, and Hexagenia mayflies absent by fall.

Based on Schindler, et al., "Long-term ecosystem stress: The effects of years of experimental acidification on a small lake," *Science,* 228:1396–1401, 1985.

nomena as the failure of the crawfish exoskeletons to harden and perhaps the cessation of reproduction in some species. (2) The lower pH provides a favorable habitat for some organisms and an unfavorable one for others. The enhanced conditions for fungal growth (and poorer conditions for bacteria), together with physiological stress, may account for some diseased conditions, such as the fungal infestation of crawfish eggs. (3) Changes in species composition may, in turn, produce a variety of further effects, such as the covering by *Mougeotia* of trout spawning beds and the disruption of food-chain relationships through the replacement of edible by inedible species.

In "natural" acidification, damage from heavy metals and aluminum, which are more soluble in acid waters, is a possible source of reproductive failure or death. In Lake 223, because the acid was added directly to the lake rather than coming in through the soil, this effect probably was less important.

Terrestrial Systems Effects of acid deposition on terrestrial systems were slower to be recognized. Recent tree die-offs in Europe *(Waldsterben)* and the northeastern United States on a scale unlike anything known historically have been attributed to acid deposition. The Black Forest area of Germany has been especially hard hit; in 1983 more than 75% of the area covered by fir forest showed damage such as crown dieback, deformed shoots, and premature death (Postel 1984). Declines in forest biomass resulting from the death of larger trees have been reported for the red spruce forests on Camels Hump Mountain, Vermont, and declines in density for the hardwood forests at lower elevations (Vogelmann et al. 1985).

The precise mechanisms of damage to terrestrial systems are uncertain. Soil acidification, leading to the loss of essential nutrients such as calcium, and the breakdown of normal soil structure may be involved. The increased availability of toxic metals, especially aluminum, is a possibility. Loss of other organisms such as important soil bacteria, most of which do not tolerate high acidity, or mycorrhizal fungi may be important. Direct effects on the foliage of the plants may be important. Acid rain need not be acting alone: Interactions with other atmospheric pollutants, such as ozone, may be involved (Hinrichsen 1986).

Vulnerable Regions The damage that acid rain does to lakes depends on an interaction between the acid of the rain and the buffering capacity of the soils

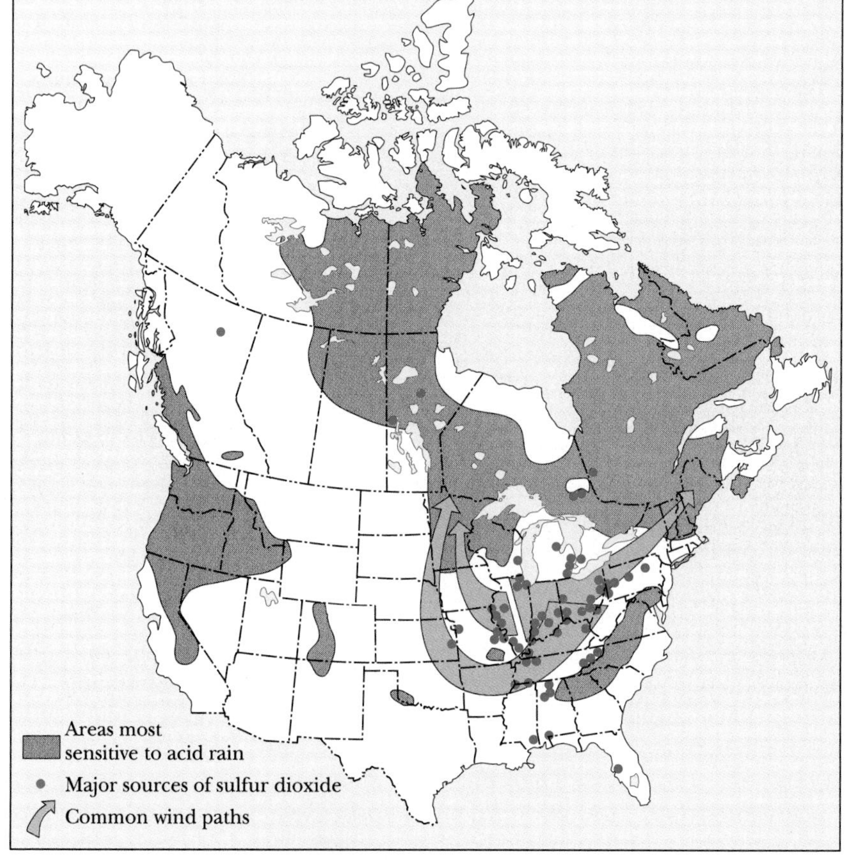

Figure 20-12

The shaded areas are those parts of North America most vulnerable to acid rain. Major sources of sulfur dioxide (more than 100 kilotons per year) are shown as dots. (Adapted from Borchard 1984; after Galloway and Cowling.)

and lakes of a given region. Damage has been greatest where the soil is thin and contains little calcium or magnesium. Lakes in such regions quickly use up their buffering capacity, and their pH drops. Some of the areas with a geology that makes them sensitive to acid rain include Scandinavia, virtually all of eastern Canada, much of the northeastern United States, and the Appalachians (Figure 20-12).

Much of the pollution that affects these regions is generated outside, upwind from them. For example, much of the pollution that has dropped the pH of Adirondack lakes from 5.5 to below 5.0 in the past 40 years comes from the Midwest. Much of Canada's acid rain problem is also imported from the United States, although its own industry is not blameless. The Sudbury, Ontario, smelter of International Nickel has for years been the largest single source of atmospheric sulfur in the world. The United States is off the hook, though, for Scandinavia's acidification problem; most of its pollution comes from Great Britain and Germany.

One may wonder why acid rain has become a serious problem only in the past 30 years, even though fossil-fuel combustion products have added to the atmosphere over a much longer period. Probably part of the reason was simply a failure to recognize the problem. Of course, the local effects of sulfur dioxide emissions from coal-burning power plants, smelters, steel mills, and similar industrial plants have been recognized for a long time. The surrounding areas were sorry places with bronchitis sufferers living in grimy neighborhoods where little could grow except ailanthus trees and rats. At the worst sites, it was not just the most sensitive plants, like lichens, that died. The areas around certain smelters, such as those at Copperhill, Tennessee (Odum 1971:25), or Queenstown, Tasmania (Figure 20-13), were all but devoid of vegetation.

Effects further away from the sources of pollution and especially aquatic effects may have simply been missed. But the problem has undoubtedly increased recently, and some probable causes seem evident.

Figure 20-13

Emissions from the copper smelter located at Queenstown, Tasmania, have killed vegetation on the surrounding hills. Of course, the United States has similar human-produced wastelands such as the one at Copperhill, Tennessee. (Photo by Joseph G. Engemann.)

1. Although sulfur dioxide emissions have been with us since the beginning of the Industrial Revolution, they did, in fact, increase from the mid-1950s to the mid-1960s (Figure 20-14). Also, nitrogen oxide levels, which stayed fairly level for a long time, started a sharp rise about 1950, presumably resulting from the spread of the private automobile.
2. A second factor may be simply an exhaustion of buffering capacity. Many systems may have had sufficient buffering capacity (alkalinity) to neutralize acid rain input for a time without showing major pH changes, but eventually their resilience was used up.
3. The third is related to the technological solutions to local air pollution that began to be introduced in the industrial nations in the 1950s and 1960s. Laws, such as the Clean Air Acts of 1965 and 1970, were passed to help the citizens of the heavily industrialized areas, such as Pittsburgh and the Gary-Michigan City area.

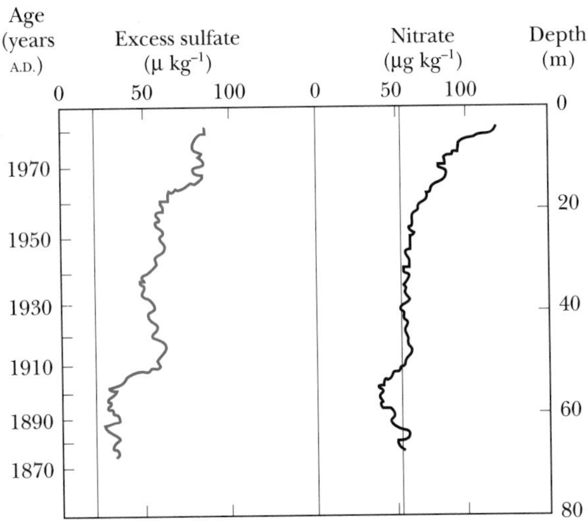

Figure 20-14

Graphed here are sulfate and nitrate concentrations in an ice core from southern Greenland. The figures give a measure of sulfur and nitrogen compounds added to the air from human sources and thus a measure of the acidifying ability of the atmosphere. Air masses from both North America and Eurasia cross Greenland. The big increase in excess sulfate starts a little past 1900 but for nitrate it does not start until about 1955. The vertical lines represent preindustrial ("old ice") levels. (From "Sulfate and nitrate concentrations from a south Greenland ice core," by P. A. Mayewski et al., *Science* 232:975–977, 1986. Copyright 1986 by the AAAS.)

The response to this legislation tended to be various sorts of antipollution technology. To reduce local air pollution, for example, industry built taller and taller smokestacks. The average height of a smokestack in 1956 was about 60 m; by 1976 the average was close to 240 m, and some stacks more than 360 m tall had been built (Patrick et al. 1981). These put the discharges higher in the air, allowing wind to carry them out of the immediate vicinity. Unfortunately, tall stacks also increase the time the sulfur and nitrogen compounds stay in the air and, consequently, the size of the area subjected to acidification.

Precipitators to remove particulates were another technological device adopted by coal-fired power plants in the United States. A good many were in use by 1960, and particulate control was required by law beginning in 1975 (Patrick et al. 1981). These cut down on the rain of soot near the plants, but they also removed alkaline-earth metals that, before, had helped neutralize the acid components of the emissions.

The overall result of such measures, it seems, has been that the air in Pittsburgh and other highly industrialized areas is better, but lakes and forests hundreds of miles downwind are suffering. Generating problems more severe than the one solved is a frequent result when problems stemming from technology are attacked by grafting on more technology instead of meeting them head-on.

The Future Even for persons who do not fish and have never taken a walk in the woods, the impact of acid rain on fisheries production and on the lumber industry is likely to hurt economically. Also, outdoor works of art, bridges, buildings, and highways are being damaged. Sulfur dioxide damage to paint in the Boston area has been estimated at \$11 to \$12 per person per year.

Other problems potentially affecting humans may not yet be appreciated. Will crop yields be affected? Will mercury, lead, asbestos, or cadmium reach dangerous levels in drinking water (Mc-

Donald 1985)? Such questions are just coming under study.

There are some piecemeal approaches to the acid rain problem. The acidification of lake waters can be counteracted by adding lime. This is sometimes effective in allowing the survival or restocking of sport fish. Liming would have to continue as long as acid rain continues. It is an expensive process costing \$55 to \$420 per hectare (Schneider 1986).

There, however, is a simple and direct solution: Greatly reduce sulfur and nitrogen oxide emissions. Several studies have indicated that effects of acidification are reversible in many lakes and soils if sulfate input is drastically reduced (Wright and Hauhs 1991). Unfortunately, the solution is tied to the world's energy demands, and the coming decade may well see an overall increase in sulfur and nitrogen emissions (stemming especially from Russia and eastern Europe) rather than a decrease.

The Greenhouse Effect and Global Climatic Change

As shown earlier in Figure 12-12, the carbon dioxide content of the earth's atmosphere has increased substantially from the burning of fossil fuels and deforestation. The carbon dioxide content is now more than 30% higher than in the immediate preindustrial era. Higher levels of carbon dioxide can have direct effects on organisms. For plants with plenty of sunlight and water, increased carbon dioxide concentration increases photosynthesis, more so in C_3 than C_4 plants (Sveinbjornsson 1984). At least, this is the result in short-term experiments. Long-term exposure also affects leaf surface area, branching patterns, the effects of photoperiod, and nitrogen-fixation rates. Most such effects seem positive considered individually, but the net result on productivity is difficult to predict.

The aspect of rising carbon dioxide levels that has received the most attention, however, is the "greenhouse effect." Carbon dioxide is transparent to light but absorbs much of the energy in the infrared, or heat, wavelengths. Consequently, when sunlight hits the earth and warms it (that is, when the light energy is converted to heat energy), the heat radiation from the ground is absorbed by the carbon dioxide of the air. The result is that the air is warmed and so is the earth, by reradiation from the air.

The greenhouse effect is not some recent invention of radical environmentalists. The term was first used in the 19th century by the famous chemist Arrhenius. Carbon dioxide is a normal constituent of the earth's atmosphere; its presence within certain limits is what keeps the earth habitably warm. The concern—first expressed by Arrhenius in 1896—is that increasing levels of carbon dioxide generated by human activities, especially burning fossil fuels, will amplify the greenhouse effect causing global temperatures to climb and leading to other changes that will be at least troublesome and possibly dangerous.

It is now realized that human actions are increasing other "greenhouse gases" (that is, gases that operate like carbon dioxide to trap heat radiation from the earth). Included are natural components of the atmosphere, such as methane, and wholly new compounds such as the CFCs. Most of these gases are, per molecule, much more potent than carbon dioxide (Table 20-4), but in terms of overall effect carbon dioxide is still by far the most important greenhouse gas.

Some persons deny that any substantial global climatic change is likely to occur. They suggest that the excess carbon dioxide will be incorporated into more phytoplankton in the ocean and bigger trees on land. They argue that at the same time we've been putting carbon dioxide into the atmosphere, we've also been adding solid particles and droplets (of sulfuric acid, for example). These increase the *albedo,* or shininess of the earth. This will act counter to the greenhouse effect, reducing sunlight reaching the earth and tending to lower temperatures.

These phenomena may help counteract global warming trends, but other changes may tend to speed it along. For example, warming temperatures will decrease winter snow and ice cover, thereby decreasing albedo, increasing absorption of sunlight, and further raising temperatures. Melting of permafrost, it is believed, will result in more methane production from the water-logged soils. The main agents for removing methane from the atmosphere are methane-oxidizing bacteria in aerobic soils. There is some suggestion that uptake of methane by such soils is inhibited by certain pollutants, especially high nitrogen levels.

Although we have been lucky so far in our thoughtless alterations of the earth, it is unlikely

Table 20-4
Important Greenhouse Gases

Compound	*Anthropogenic Sources*	*Contribution to Increased Greenhouse Effect*	
		Per Molecule	*Overall (%)*
Carbon dioxide	Fossil fuel combustion, deforestation	1	60
Methane	Anaerobic decomposition of manure, paddy rice production, flatulence of ruminants, natural gas leaks	25	15
Nitrogen oxides	Fossil fuel combustion, catalytic converters, agricultural fertilizers	200	5
Ozone (in troposphere)	Photocatalysis of automobile pollutants	2000	8
CFCs	Industrial chemicals used as refrigerants, blowing agents in manufacture of plastic foam; no longer used in spray cans in the United States	12,000–15,000	12

Contribution to Greenhouse Effect figures from Rodhe 1990.

that all these various forces will balance one another so as to maintain the status quo. The most satisfactory models of global climate (**general circulation models** [GCMs]) suggest that a doubling in atmospheric carbon dioxide will lead to an increase in average global surface temperature of 1.3 to 5.2°C (Pearman 1988, Gribbin 1990). Just how global climates are determined is, nevertheless, still poorly understood, and it may be that important factors are not yet being included in the models. Probably most ecologists and atmospheric scientists today lean towards the view that temperatures are going to go up, and a good many believe that global warming is already to be seen in the record warm temperatures of the past few years (Henderson-Sellers and McGuffie 1986, Kerr 1990).

There are optimists who see a grand opportunity in every disaster. One physicist, seeing the possibilities for increased productivity, wrote (Idso 1985) that the increased carbon dioxide concentration "appears not to be the climatic catastrophe which has so often been portrayed in the popular press but rather a beneficent blessing in which all Mankind may share."

It is, perhaps, too early for rejoicing. Climatic models predict lowered rainfall and increased evaporation in the North American wheat belt. Sea level will rise, partly just because warm water takes up more room than cold but mainly because of melting of the polar ice caps. Complete melting would result in a rise of about 70 m (Henderson-Sellers and McGuffie 1986). Considerably smaller rises would still produce mind-boggling dislocations. With a 15-m rise, nothing would be left of Florida south of Sarasota—and, of course, Sarasota would be gone too, because only the interior part of Florida is above 15 m, even in the north.

A sea-level rise of only 10 m would flood 10 million km^2 worldwide, a land area about the size of China. Many of the world's population centers, including cities such as Beijing, New York, and London as well as densely populated agricultural areas, are near sea level. Rises of only a few meters would cause millions of people to attempt to relocate to higher ground—ground already more or less fully occupied by other people.

Among the changes that would come from global climate alterations stemming from the greenhouse effect could be

Northward shifts of many plant and animal ranges.

Other shifts in range resulting from changed patterns of precipitation and other climatic factors, including fire.

Shifts in the distribution of biomes (Figure 20-15). For example, the Goddard Institute for Space Studies (GISS) GCM was applied to northern North America postulating a doubling of atmospheric carbon dioxide (Rizzo 1988). The model predicted that tundra would shrink northward, the taiga-boreal forest zone would greatly compress especially in the West, and temperate grassland and forest would come to occupy all of southern Canada, reaching to the southwest shore of Hudson Bay.

Shifts in community composition. It is unlikely that all species will be able to disperse at the same rate and also unlikely that all dispersers will find suitable conditions. For example, some temperate forest plants and invertebrates dispersing northward may not be able to tolerate the soils developed under boreal forest.

Possible extinctions. It has been suggested that the jack pine would be lost from the high sandy plains of northern Michigan, perhaps leading to the extinction of the Kirtland's warbler. This rare bird nests on the ground and it may be that, even if individuals moved northward, they could not nest successfully in the heavier soils of southern Ontario (Cohn 1989). Another example is that the less motile species among the animals and plants of timberline and alpine tundra may be pinched off the top of the western mountains.

Anachronistic nature preserves. In most regions, nature preserves occupy only a small fraction of the landscape and are usually established to capture a particular habitat or ecosystem. With climatic change, the vegetation of these preserves will change in complicated ways. They will not become valueless, because all preserved land serves many useful functions. Nevertheless, a piece of land set aside to protect some northern orchid living at the southern edge of its range may become obsolete for that purpose. Rare species mostly confined to preserves are additional candidates for extinction because current preserves may become useless before the species can become re-established in others. These are, of course, not arguments against preserving land; they are arguments for having many more preserves, so that stepping stones are available in all directions for such species.

Altered patterns of agricultural production. Getting rid of paddy rice production and cutting back drastically on cows and other ruminants would be useful measures to reduce methane emissions, hence reducing the future rates of temperature increase. Whether such voluntary changes are made or not, changes will be dictated by new patterns of temperature and rainfall. Up to this time, the agricultural establishment seems to have made little headway in thinking about the coming Warm Ages (Rawlins 1991). It seems likely that the optimal range for some crops will shift across political boundaries with resulting changes in standards of living and national power. Political instability is a likely result of such events, as it is from the displacement of populations by rising sea levels.

This list is indicative of some possible changes but is far from complete. Many of the changes will result in impoverishment of the biosphere, but others are merely inconveniences for humans. From a humanistic or anthropocentric perspective, some changes will be desirable. For example, year-round navigation on the Great Lakes might become possible (Assel 1991).

What are our options in dealing with global climatic change and its adjuncts? They are three and, if global changes approach the magnitude suggested in preceding paragraphs, we will try all of them at various times and places.

1. Fight the changes. In forestry and agriculture, we could, for example, irrigate and fertilize and try to keep growing the same things in the same places. We can try to hold back the sea; we can dyke and pump water out; we can pump sand in to save beaches (Titus et al. 1991). There is little doubt that these things

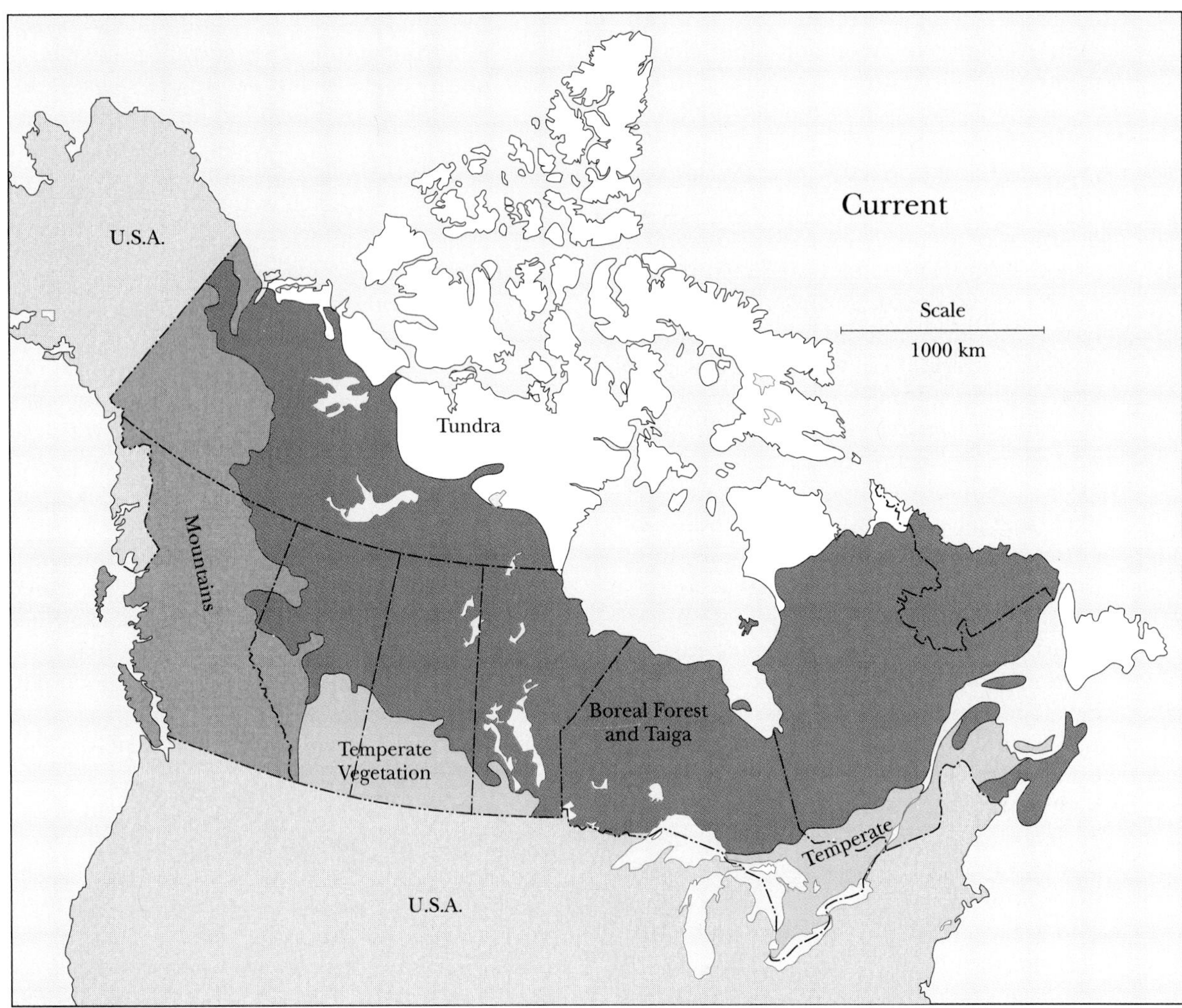

Figure 20-15

The taiga-boreal forest zone of North America currently and after doubling of atmospheric carbon dioxide. The zone moves northward and shrinks, based on one model of the resulting climates. Vegetation north of the taiga-boreal forest zone is tundra; vegetation to the south is temperate forest or grassland. (Modified from Rizzo 1988.)

will be tried, probably at taxpayer expense, for coastal cities and resort areas.

2. Reduce the changes (mitigation). We can introduce energy conservation measures to lower the amount of carbon dioxide going into the atmosphere and we can phase out the use of CFCs (Rubin et al. 1992). We can halt deforestation and begin reforestation. Phasing out paddy rice and cows fits in this mitigation category.
3. Adjust and accept. Canada will, perhaps, become the major corn producer of the world, and the United States will grow more okra than ever before. We can try to develop crops that grow well in the heat. We can accept more crowded conditions resulting from re-

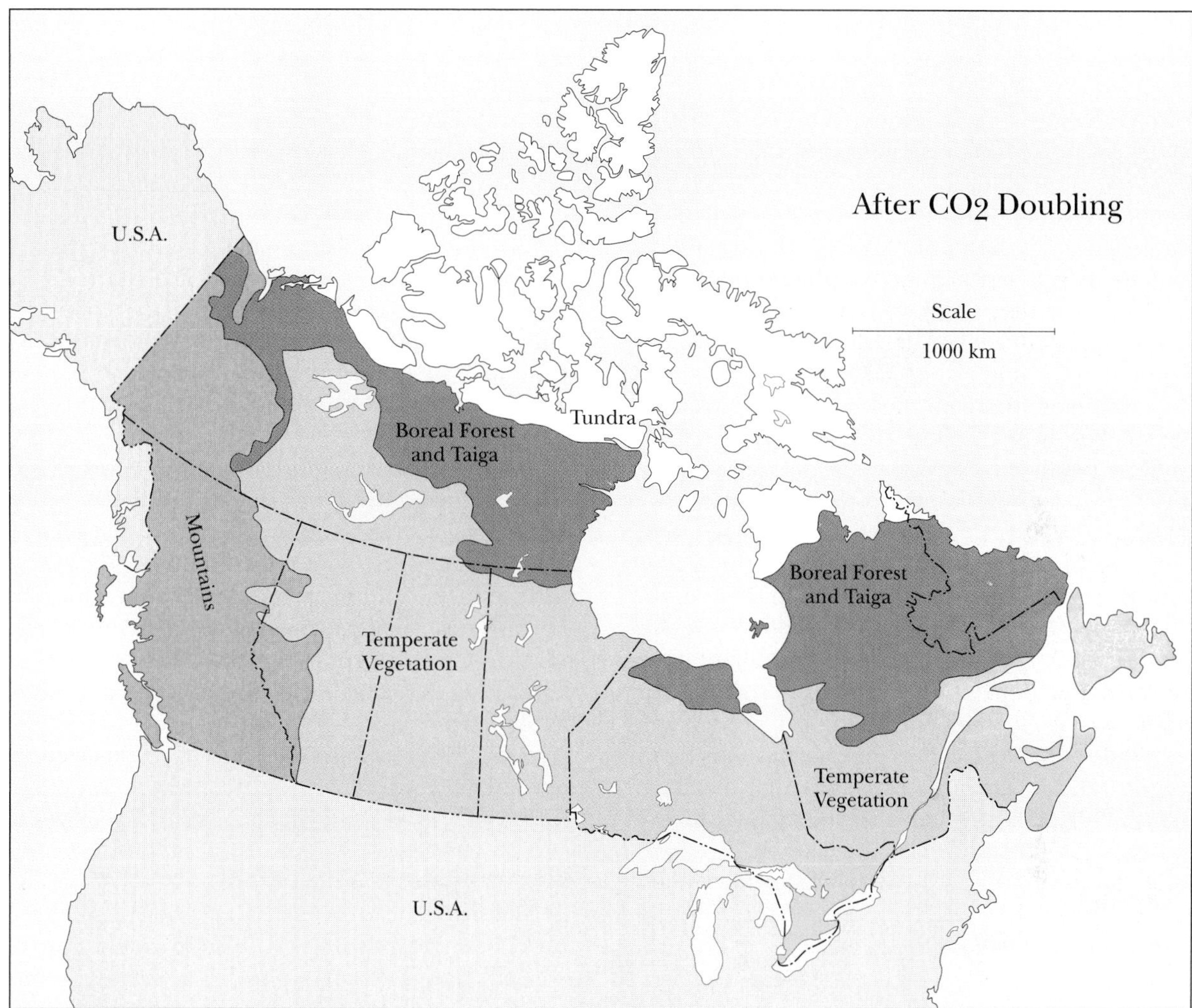

Figure 20-15 (*cont.*)

duced land area. We can take an active role in assisting plants and animals to get to newly suitable habitats.

The Hole in the Ozone

CFCs do another thing besides contributing to the greenhouse effect. They destroy ozone (O_3). This might seem like a good thing because we all know ozone is a pollutant, with many damaging effects. But that is ozone in the lower atmosphere (troposphere). Ozone in the stratosphere, 15 to 35 km up, performs the useful biospheric process of absorbing ultraviolet radiation. The absorption of ultraviolet radiation warms this layer, which is important to the stability of the troposphere. Also, the amount of ultraviolet radiation reaching the surface of the earth is greatly reduced.

The ozone belt of the stratosphere was in approximate equilibrium with about as much being produced as being destroyed. Then humans began liberating or losing certain chlorine-containing compounds to the air. Some of the more reactive of these are broken down near the ground but others are carried into the stratosphere. The very nonreactive CFCs and relatives are the main culprits.

In the stratosphere, through a complicated set of reactions, chlorine (or bromine) is removed from these molecules and catalyzes the breakdown of ozone to molecular oxygen (O_2). Note that the action of the chlorine is catalytic; at the end of the reaction it reappears, able to break down more ozone. Eventually, the chlorine is removed from the stratosphere in the form of hydrochloric acid and, in time, falls to earth. However, the residence time in the atmosphere of the CFCs is long, tens of years or even a few hundred depending on the specific compound.

The most dramatic decrease in stratospheric ozone has been around the South Pole. Atmospheric monitoring stations were set up in Antarctica as a part of the International Geophysical Year (1957 to 1958). Their measurements of stratospheric ozone drifted lower through the mid-1970s and then began to drop rapidly (Rowland 1989). Currently, an area of the stratosphere above Antarctica about the size of the United States shows severe ozone depletion (70% lost) during the Antarctic spring (September to October) (Anderson et al. 1991). Similar but less severe thinning of the ozone layer is already occurring over the Arctic (Austin et al. 1992), and ozone depletion over other parts of the world has also been detected. The mid-latitude loss over North America was estimated at 1 to 3% per decade a few years ago; more recently the loss has been occurring at 4 to 5% per decade (Kerr 1991).

Why should we worry about the increased ultraviolet radiation we and other organisms are now encountering and will increasingly be subjected to in the future? Personally we should worry because ultraviolet radiation is implicated in skin cancer including the very dangerous malignant melanoma, in cataracts, and in suppression of the immune response. We should also worry because there are observations indicating that motile marine phytoplankton avoid ultraviolet exposure by going deeper in the water. This screens out the ultraviolet light but it also screens out sunlight, so that productivity drops. Consequently, higher ultraviolet levels may inhibit one avenue for removing excess carbon dioxide from the air.

And, of course, we should also say that higher ultraviolet levels will have other effects, some of which probably no one has yet imagined.

In one sense, the handling of the CFC-ozone depletion link has been one of the environmental movement's success stories. The United States banned CFCs as aerosol propellants effective 1978 and so did Canada and a few European countries. Little then happened until the discovery of the Antarctic ozone hole. The Montreal Protocol signed by many of the developed countries in 1987 specified slight reductions in CFC emissions by 1994 and an overall 50% reduction (compared with 1986) by 1999. As knowledge of the seriousness of the problem grew, so did concern, and the protocol was renegotiated in 1990. The aim is to stop production and use by the industrialized countries by the year 2000.

All this is encouraging. What is discouraging is that because of the great stability of CFCs, stratospheric levels will continue to increase beyond 2000, even if production is halted then. This means that their contribution to the greenhouse effect will increase (as they move up through the troposphere) and that ozone depletion will continue to increase as they reach the stratosphere and slowly break down.

As CFCs are phased out, they are replaced by new chemicals (Rhodes 1991). Many of these are simply less potent ozone depleters or greenhouse gases. Examples are the hydrochlorofluorocarbons (HCFC) (Kroeze and Reijnders 1992). One of the biggest problems of the next few decades will be the rise in refrigeration and air conditioning in tropical countries. The Montreal Protocol commits the developed countries to help finance CFC reduction in Third World countries. Whether switching to less potent greenhouse gases and ozone depleters in refrigeration units while greatly multiplying the number of the units will actually result in a decrease in the overall atmospheric effects remains to be seen.

Genetically Engineered Organisms in the Environment

In the early 1970s, humans developed the technology to alter the genetic makeup of individual organisms. The best-developed approach currently is through recombinant DNA techniques. A piece of donor DNA that codes for the desired trait, such as insulin production, is coupled to a vector, such as plasmids or a bacteriophage. The resulting recombinant DNA molecule is inserted into a host cell, such as the bacterium *Escherichia coli,* where replication occurs. Details are available in any recent microbiology text.

Early researchers using the technique showed considerable restraint (Fisher 1985). In subsequent years, as no serious problems became evident, guidelines were relaxed. At the present time, none of the disasters originally envisioned have occurred, so far as is known. There has been no cancer outbreak from the spread of recombinant *Escherichia coli* organisms carrying animal tumor viruses nor have microbes engineered to produce insulin infected people and caused an epidemic of endocrine imbalance.

In the 1980s individual researchers and biotechnology companies began to propose field release of genetically engineered organisms. The first such test proposed was to spray genetically altered forms of *Pseudomonas syringae* on crop plants with the commercial objective of making them more frost resistant.

Specifically, the attempt was to lower the temperature at which frost forms on plants. The *Pseudomonas* forms were ones lacking the ability to produce the proteins that are important in ice nucleation. *Pseudomonas* lives naturally on the leaves of many plants but the naturally occurring forms ordinarily possess this protein, leading to ice formation on the leaves near 0°C. In laboratory tests, frost did not form on the leaves inhabited by non-ice-nucleating (non-IN) strains until the temperature was lowered to about −5°C (Crawford 1987).

Such a test was opposed in court by environmentalists who claimed that introducing such an organism into the environment has a potential for damage that needs to be carefully examined. Might the non-IN genotype replace the normal type? Since the IN properties of *Pseudomonas syringae* are what give it an important role in precipitation, might such an event alter world weather patterns? Clearly, no one can be sure, although the proponents of the field experiment noted that a small proportion of the naturally occurring *Pseudomonas* lack the gene for the protein. The field trial would involve a concentration of these organisms unlike that found normally but would not be a qualitatively new event.

A California judge dismissed the environmentalists' suit, stating that they had failed to show that irreparable harm would result. The field trial, using a strawberry field somewhat less than 100 m^2, went forth in April 1987 with, so far as can now be seen, no spectacular results for either the climate or strawberries.

The business community, led by the drug and agricultural industries, sees biotechnology as having the potential for enormous profits. They expect, for example, to sell seeds that will yield plants that are tolerant to herbicides and pesticides, so that fields can be zapped to kill everything but them. Similarly, they expect to make money off plants resistant to viruses, that have enhanced cold tolerance, and that have more (or less) sugar, starch, or particular amino acids (Kessler et al. 1992, Wrubel et al. 1992). In chemical factories, microbes will produce hormones, cocoa, coffee, and tea.

Some environmentalists see the introduction of genetically engineered organisms into the environment as being of such incalculable risk that we should simply avoid the technology. We cannot assess the risks to the biosphere without field tests and after the wrong one it may be too late to put the cork back in the culture bottle. Why not simply quit trying to grow crops in localities too cold for them or else accept an occasional frost as a cost of doing business?

Many of the other problems that introductions of genetically engineered organisms are supposed to solve also have similar, non- or low-technology solutions. For example, time and money are being spent engineering canteloupe and squash to be resistant to mosaic viruses. Such resistance could increase production; also, virus-infected plants have fruits that do not have the coloration preferred by marketers of produce. There is no overall shortage of canteloupe and squash, and increasing canteloupe and squash production is not going to feed the starving masses of the world. But, of course, these are not the relevant questions in a free-enterprise society. The questions are, can a corporation make money selling such seeds, and can a farmer who uses them make more money than a farmer who does not.

Although some uses of biotechnology are likely to be answers to nonexistent problems, other uses may be valuable to humanity. Even for frivolous uses, simple opposition will be ineffective. Human cultures have almost never achieved prudential restraint from new technology (Phillips 1971). Consequently, most ecologists accept the proposition that tests and, later, widespread use of biotechnology will occur and that their role as scientists is to try to advise on the likelihood and seriousness of environmental effects (Pimentel et al. 1989 is an example).

For this task, ecologists have two important assets: Knowledge of the ecological effects of other introductions and a systems perspective. The accumulated knowledge of what happened when such organisms as the chestnut blight and house sparrow (and, of course, the camel and coturnix, both of which failed) were introduced in regions of natural and human-modified ecosystems is probably the best source we have for assessing what may happen as the direct results of releasing genetically engineered organisms (Tangley 1985).

Ecology also teaches its practitioners to suspect that preventing frost on strawberries (if it does) may not be the only effect of spraying altered *Pseudomonas* into the environment. It may change the weather or do any of several other things. Likewise, plants with a gene for an insect toxin may (or may not) be protected from pests but they may also kill the earthworms around them or perhaps the birds that eat any pest insects that happen to tolerate the toxin. Effects on nitrogen fixation and many other aspects of nutrient cycling are among the other possibilities for effects far removed from the original intent (Flanagan 1987). Attempting to identify and warn about such potential effects will be one of ecology's important roles in the coming years.

Recycling

One of the oddly charismatic topics of the past few years has been solid waste. Many people have devoted time to the question, What do we do with our garbage and trash? It is a reasonable question because each of us in the United States generates close to 1 metric ton of solid waste, on the average.

The solution from time immemorial has been individual or community garbage dumps. Until recently many of these were placed in wetlands. To the average municipal official, this was killing two birds with one stone: Getting rid of the trash and filling in a piece of worthless wasteland.

Landfill sites are harder to come by now. Even when placed on upland sites, they have often polluted the water table, so that wells in the vicinity become unusable. Landfills can no longer be put in wetlands; rather the site must have a geology such that there is clay or some other impermeable layer that will keep the leachate from the landfill away from the water table. No one wants a landfill in their neighborhood; people see no reason to put up with its pollution, smells, noisy trucks, and circling gulls and vultures.

There have been several types of responses to the solid waste problem. One business approach has been for firms connected with the moribund nuclear power plant industry to promote solid waste incinerators, which they design and build. Several aspects of these incinerators trouble environmentalists. One is that the incinerators, constructed at great cost, demand a constant waste stream, especially if they have been sold as having the dual functions of waste disposal and power generation. This requirements inhibits efforts to cut down on the waste stream by recycling and other ways.

Another argument against incinerators is that they tend to spew out toxic materials contained in the waste, such as mercury and cadmium. The residue after incineration tends to have high enough concentrations of some materials to qualify as toxic waste if federal guidelines for waste in general were followed. Disposing of incinerator wastes properly would be very expensive.

Environmentalists favor recycling, of course. They see it as the analog in the human ecosystem of biogeochemical cycling, without which stability and sustainability are impossible. Thoughtful environmentalists put one other option before recycling; this is waste prevention or reduction. Recycling is environmentally sounder than landfilling or incineration for ordinary solid wastes, but recycling is not without problems. Time, energy, and organization on the part of individuals and governments are required to save, sort, and collect the materials. There are not always ways to make a profit turning the old items into new ones, so the cycle may not be completed efficiently.

Most such problems are soluble, and many have been solved in some states or countries; nevertheless, avoiding them altogether by waste prevention or reduction is more efficient. Environmental handbooks from the 1960s to today provide lists of ways in which individuals can reduce their contribution to the waste stream. Most of them fall under the headings of a slogan popular a few years ago—Make it do, wear it out, do without. We can buy compact fluorescent light bulbs that last years. We can buy well-made, durable clothes in classic styles and wear them for a long time. We can use cloth grocery bags. We can simply avoid a lot of the junk that the mall and the TV set tell us we need.

Most such options can be put under the heading of *frugality*. In general, they have the virtue, not just of reducing garbage which, despite its current popular appeal, is a minor environmental problem. They also help with energy conservation, greenhouse warming, and ozone depletion which are much more serious issues.

The Practical Ecologist: *Energy, Food, Health, Population, and Land*

Sustainable agriculture. *(Carl R. Sams II/Dembinsky Photo Associates.)*

OUTLINE

- Ecology and Global Change
- Energy
- Food Production and Organic Farming
- The Role of Environment in Health and Disease
- Population
- Land Use
- Radical Environmentalism
- Spaceship Earth

In this third chapter of "The Practical Ecologist" are discussed a variety of topics having to do with the application of ecological knowledge to matters important to human life. The primary topics involve energy, food production, health and disease, and land use. Along the way, understanding the role of humans in the biosphere is given some additional consideration.

Ecology and Global Change

Ecologists in the past few years have increasingly become willing to think globally. "Ecological aspects of global change" is one of the three research topics of the Sustainable Biosphere Initiative (Ecological Society of America Committee 1991). The global change that first leaps to mind is usually global warming and other climatic changes expected to result from increased levels of greenhouse gases. But this is only one type of global change. Global warming may be coming, but many other types of global change are well underway.

Most have ecological causes and will have ecological consequences.

All types of human-produced global change begin with local actions. Some, such as the release of carbon dioxide from fossil-fuel combustion and deforestation, are translated into global effects by the circulation of the atmosphere or the oceans. Others, such as forest fragmentation, achieve global stature by their ubiquity and by producing such problems as the worldwide decline in biological diversity. A partial list of changes of clearly global stature includes

- Increased carbon dioxide levels as an ecological factor apart from the greenhouse effect (Chapter 20).
- Ozone depletion (Chapter 20).
- Changes in atmospheric chemistry, such as acid rain (Chapter 20).
- Deforestation and habitat fragmentation (Chapters 11 and 15 through 19).
- Other processes creating heavily disturbed natural systems (Chapters 11 and 15 through 19).
- Rising human population, the driving force behind many other types of change (Chapters 4 and 21).
- The triumph of capitalism, bringing with it, among other things, worldwide increases in the demand for consumer goods.

Energy

Humans are uniquely dependent on energy subsidies. A paragraph from H. T. Odum's (1971) *Environment, Power, and Society* states the relationship between humans and the availability of energy beyond that of our own metabolism in this way:

> *Most people think that man has progressed in the modern technological era because his knowledge and ingenuity have no limits—a dangerous partial truth. All progress is due to special power subsidies, and progress evaporates whenever and wherever they are removed. Knowledge and ingenuity are the means for applying power subsidies when they are available, and the development and retention of knowledge are also dependent on power delivery.*

In historic and prehistoric times, human life was affected by the use of fire and currents, by waterwheels, sailing ships, and windmills, and by animals domesticated as beasts of burden or draft animals. But daily life as we know it is based on abundant, cheap fossil fuel, a development of the 19th century.

A period of a little less than ten years starting with the Oil Producing Exporting Countries (OPEC) petroleum price increase of 1973 gave us a slight taste of what life will be like when the era of cheap fossil fuel energy is over. Since then, the OPEC oil cartel has failed, oil prices have fallen, and energy is no more expensive (allowing for inflation) than it was in the 1950s.

We can learn many lessons from that period of relative energy scarcity. The first is that energy is a unique resource. The average economist does not worry much about resource scarcity. When one resource gets scarce and, thereby, expensive, some replacement becomes economic and is substituted for it. When we have used up the big trees, we build things out of plywood, and when plywood becomes too expensive, we turn to particleboard. When energy becomes too expensive, we can turn to . . . energy. Of course, but, specifically, when petroleum becomes expensive, we will turn to oil shales, nuclear power, solar cells, etc.

What this sort of conventional economic analysis misses is that it takes energy to get energy, especially energy in the right form and place. For example, the energy of sunlight is plentiful but dilute. It is great for drying clothes, but to run a boiler you have to concentrate it, and that takes energy. A controlled nuclear reaction gives you plenty of heat, but you can't bake a cake with it; it's too hot. For that you first have to generate electricity and that process uses, and loses, energy. You also have to provide safety precautions; that takes energy. And you have to store the wastes; that is not just a current cost but is also a mortgage that will require energy payments for thousands of years into the future.

Energy systems analysts use the term "net energy" to describe the useful energy that a process yields above and beyond the energy that goes into the process (Odum and Odum 1981). A hundred years ago, drilling into an oil pool a couple of feet below the surface of the ground yielded a lot of net energy. Today's oil wells, involving such energetically expensive operations as drilling deep into the seabed, yield much less. Many of the processes the economists have assumed to be substitutes for petroleum will yield little net energy. Some would not even break even; for every joule we got, we'd have to spend more than one.

One of the best examples of such a losing process is the production of ethanol for use as a fuel. A gallon of ethanol yields 19,152 kcal when we burn it in our car or elsewhere. To produce a gallon of ethanol, however, requires slightly more than 33,000 kcal of fossil fuel energy input (Pimentel 1991). The largest amounts go to the agricultural production of the corn (for fertilizers, tractor fuel, drying, etc.) and operation of the factory for fermentation and distillation. The upshot is that we could drive our car 1.7 times as far on the petroleum used in producing ethanol as on the ethanol itself.

All this means that, in the future, energy will again become expensive and that it behooves us to make plans. Actually, the solutions to the world's coming energy problems are easy to see. We must stop population growth, improve the efficiency of our energy use, and find feasible alternatives to fossil fuels. Only the implementation is difficult.

Another of the lessons of the 1970s energy crunch is how readily conservation practices take hold when energy prices go up. That is a heartening discovery. In the United States, energy use peaked in 1979 and declined through 1983, by which time we were using less energy than we had been ten years earlier (World Resources Institute 1986). Since then, of course, energy consumption has begun to increase again.

Conservation took many forms. Smaller, more energy-efficient cars—and less driving—were a big part of it. Better insulated houses and commercial and industrial buildings were also important. People were less extravagant with lighting and air conditioning. Greater use of solar power, wood burning, and other alternatives helped lower fossil fuel use.

Many of these conservation measures are permanent, such as better insulation, more efficient machinery, and perhaps even a taste for cooler indoor winter temperatures. Other things have shifted back during the recent cheap energy interlude. People are driving more and faster, and the gas-hog type of large car is again popular among a certain segment of the population. But there are many conservation measures still available, which the energy crunch of the 1970s didn't last quite long enough to see implemented.

Better control of heating and lighting, especially in commercial buildings, is one place where large savings could still be made. Replacing inefficient lights, air conditioners, motors, appliances, and smelters with energy-efficient ones could, according to energy experts Amory and Hunter Lovins (1980), reduce energy demands in those applications by three fourths.

Conservation, then, retains a great deal of potential for solving the energy problem. Eventually, though, all fossil fuels must become scarce. As that happens, they will become too valuable for the manufacture of plastics and other chemicals to be burnt as fuel. The alternatives to oil, gas, and coal are

Nuclear power
Hydroelectric power, including both falling water and tidal installations
Solar energy, including the burning of wood and other recent organic matter, the use of sun and wind for such things as windmills and drying clothes, the use of solar cells to generate electricity, and various "solar heating" applications for houses, hot-water heaters, and so forth
Geothermal power, making use of steam or hot water from the earth

Nuclear Power

The federal government has heavily subsidized the development of a nuclear power industry that, after more than 40 years of protection and nurturing, supplies about 22% of our electricity (Ahearne 1993). This translates to about 7% of our total energy use. The discrepancy between four decades of claims and 1990s reality is one of the reasons why nuclear power has faded to insignificance in America's power future.

About 110 nuclear power plants are currently in operation, 9 are under construction, and 60 proposed plants have been cancelled. Many of the plants now operating could be eliminated. Because of conservation measures and with relatively cheap improvements in efficiency still available, there is plenty of reserve electrical generating capacity in the United States for the foreseeable future.

As a result of energy conservation by the citizenry, the electrical utilities projections of generation capacity have regularly been too high. In 1974, the industry council was forecasting a peak electrical demand of 700,000 megawatts by 1982 (Ahearne 1993). Each year since then the projection has been lowered. The actual peak demand in 1982 was about 450,000 megawatts, and the peak demand currently is still well below 500,000.

Other problems with nuclear power have been cost and performance. John F. Ahearne, chairman of the Nuclear Regulatory Commission from 1979 to 1981, recently (1993) pointed out that utility planners need to know the answers to three questions: "How much will the plant cost? When will the plant be ready? How well will it run? The usual answers for nuclear power are: It will cost a lot, it will take a long time to build, and it won't run well."

To be added to high costs of building and operating nuclear plants are the costs of decommissioning them. When each is taken out of service 30 or 40 years after being licensed, it will have to be cleaned up and then made secure against entry by the public for a long period.

The 110 nuclear plants operating in the United States present various dangers. A Chernobyl-style disaster could occur. They are vulnerable to terrorist attack. Great quantities of radioactive materials are stored on the sites in containers of doubtful durability. As these plants are taken off line—something that will gain momentum in the next few years—no generation of nuclear plants will replace them. The energy contribution of, and the threat from, nuclear power plants will be declining in the United States.

We have been talking here about power plants in which fission of the radioactive isotope uranium-235 (^{235}U) produces heat that is used to generate electricity. Two other types of nuclear reactors need to be mentioned. The breeder reactor uses ^{235}U, a relatively scarce material, to convert ^{238}U into plutonium-239, that, like ^{235}U, is fissionable.

The past justification for breeder reactors has been that we were going to run out of ^{235}U, but these predictions were based on the earlier overestimates of electric generating capacity needs. President Jimmy Carter tried to kill the proposed Clinch River breeder reactor in the late 1970s and Congress finally did in the project about ten years later.

Breeder reactors will work; a few countries are going in for them in a big way. France is one (Zaleski 1980). There are questions, however, of practicality, cost-effectiveness, and safety. The reactors are enormously expensive, and it takes tens of years after start up for breeding potential to be realized. There are considerable doubts that reprocessing centers capable of maintaining what Americans see as necessary safety and security precautions can be economically feasible. And, of course, the plutonium from breeder reactors is usable for bombs by both the governments operating the facilities and terrorists.

The fusion reactor, the third type, would use deuterium, plentiful in seawater, or tritium to produce helium. In this process, which is essentially that used in the sun and the hydrogen bomb, vast amounts of energy would be liberated. Fusion reactors would presumably be less of a pollution problem than the other two types, but this might depend on the specific technical details in commercial application of the process. Any commercial application, however, is far in the future; in fact, there is no assurance that it will ever happen.

Our Energy Future

At the present time, solar energy seems to be the obvious long-term solution to our energy needs. It is abundant and nonpolluting, and it does not add to the heat budget of the earth. It will take a variety of forms. Solar architecture, such as solar greenhouses for heating and solar chimneys for cooling, will be a major part; it is hard to understand why anyone would build a house today without incorporating such features. Electrical generation from solar cells **(photovoltaic)** is economic today for certain applications (such as calculators and watches), and the price per kilowatt-hour is bound to decrease. Windmills, wood burning, and hydroelectric energy are also solar applications.

The future will see a diversity of energy sources used, rather than the current dependence on electricity from large power companies. The exact mix will depend on the locality and the individual.

Sunlight's calories are spread out in time and

space, and a great deal of energy will have to be expended just to make it usable for some applications. Accordingly, the future may not be a time of abundant energy. That need not make it a gloomy time. H. T. Odum and E. C. Odum (1981) spoke of "The Hopeful Future, A Steady State," and suggest that a moderate energy future can be a pleasant time in which, among other features, there is little advertising, government budgets are balanced, pride is taken in efficiency and workmanship, and everything is smaller scale and more attuned to the local environment. We will all have a garden, ride a bicycle, and learn to play the acoustic guitar.

This sort of future may be ours if we plan for an energy subsidence, or we might have a catastrophic decline to a less pleasant future if we do not plan. On the other hand, suppose that fusion or some unforeseen technology gives us abundant, cheap energy. Most environmentalists would agree that this sort of future would probably be the bleakest of all. A continually growing population and technology, feeding on abundant energy, would disastrously alter the land, oceans, and atmosphere.

Food Production and Organic Farming

There is a wide range of farming practices, from routine heavy usage of herbicides, pesticides, and fertilizers—at present the most common forms of commercial agriculture in the United States—through various intermediates to totally "organic" farming, in which no chemicals are employed for weed or insect control and the only fertilizers used are organic matter, such as manure and crushed limestone or other rocks. Although the practice of organic gardening and farming has grown in popularity in the past few years, some nutritionists and agricultural experts scoff at organic farming as being hard, unprofitable, and pointless, in that organically grown foods are no more healthful than any other. Advocates of organic farming claim that the current U.S. agricultural system is poisoning us with chemicals, depriving us of, at least, useful trace minerals and possibly other, unknown healthful factors, and leading us down a topsoil-depleted, petroleum-dependent road to ruin.

The facts are far from being well defined. The question of pesticides has already been discussed; prudent consumers will wish to reduce their intake of chemical residues to as low a level as possible. This is no less true for meat and milk than for fruits and vegetables. Those who consume meat or milk from organically raised livestock, which do not receive food containing synthetic additives, will be safe from the effects of these additives and will also be protected from the effects of mistakes, such as the addition of PBB to livestock feed, discussed in the "Firemaster" episode (Chapter 20).

It is worth noting that the mere addition of organic matter to soil does not assure its fertility. If neither the original soil nor the organic material added contains much calcium, for example, then calcium will have to be added in some other way. Does it make any difference whether the calcium, phosphate, or nitrogen is supplied "organically" or as manufactured chemicals? It seems unlikely; the plant absorbs the calcium as an ion from the clay or humus. One calcium ion is much like another.

Are organically grown foods more healthful because of some unknown factors that plants take from the soil or manufacture? Nutritionists correctly state that there is no evidence that this is so. The fact is that there seems to be little evidence at all; the experiments have not been done. It would be poor science to accept the claims of the organic food faddists on this question, but it is equally poor science to say that they are absurd. Science is not based on closed-mindedness.

Of course, environmentalists have suggested that the real food faddists are the food industries. Forty-five years ago most persons ate meat, milk, and vegetables that would be considered "organic" today. Except for details, most foods were grown, cooked, and stored just as they had been through human generation after generation. It is primarily since World War II that our diets have come to consist of fast-food meals, refined and processed foods, emulsifiers, anticaking agents and shelf-life extenders, spray-dried vegetable oil, sodium sulfite, monosodium glutamate, artificial color, artificial flavor, potassium sorbate, and calcium disodium EDTA (to preserve freshness). This, environmentalists say, is the fad. It is one in which many of us have been caught up. An encouraging countercurrent of the past few years has been renewed attention to safe and healthful foods.

The questions of costs and benefits in organic farming are complex. Several studies over the past 20 years (Lockeretz et al. 1975, National Research Council 1989) have shown that well-managed organic farms are as profitable as the currently "conventional" factory farm. Although some farms on which pesticide and synthetic fertilizer use is eliminated or greatly decreased experience modest drops in yields of some crops, this tends to be more than offset by lower costs (Scully et al. 1992). This is the conclusion reached even in this period of cheap energy. As energy becomes more expensive, such practices as crop rotation, integrated pest management, greater use of human and animal labor, and less dependence on mechanization may be even more profitable.

Why, then, hasn't agriculture become green again? One reason is that the pesticide-fertilizer model has been heavily promoted by chemical companies and agricultural college extension services for the past 45 years. Many farmers have had no other sources of information. Only recently, with the availability of some federal money earmarked for low-input agriculture, have some of the agriculture schools begun to consider alternative farming methods.

Other reasons for the persistence of current farming methods include rules, practices, and attitudes that may make it difficult to market some organically produced crops if they have cosmetic defects. Also, there is no question that organic farming is labor intensive. The labor may not be available, and the farmers themselves may wish to spend their time another way—by farming more and more acres or by holding a second job.

Another problem is that bugaboo of high-school students—peer pressure. Neighboring farmers may claim that the organic farmer is growing pests that are eating their crops or forcing them to spray more often.

Finally, it is probably true that many of the costs of current agricultural practice are not being paid by the farmer. If the long-term deterioration of agricultural soils, eutrophification and siltation of our waters, disruption of social patterns, and widespread chemical contamination are included, the price of current agriculture may already be more than we would have voted to pay.

Let us not lose sight of the fact that U.S. agriculture has had a remarkable history of productivity, if we look no farther than bushels, pounds, or calories per hectare (Figure 21-1). Much of the increase, however, has come from increasing the energy subsidy to food production, and it appears that the U.S. food system is well along on the curve of diminishing returns.

The "green revolution" beginning in the late 1960s was an attempt to increase food production in undeveloped countries by exporting to them a crop production system similar to that of the United States. New varieties of wheat, rice, and corn were developed that responded better than traditional varieties to fertilization, irrigation, and chemical pest and weed control. Yields were increased substantially, particularly in Asia and parts of Latin America. Along with the increase in food production came an increased dependence on energetically expensive technology—for producing such items as the hybrid seeds, fertilizers, pesticides, and tractors—and changes in social and political structures (Dahlberg 1979).

By the 1980s, yield increases had stalled, and the social and environmental dislocations were evident even to the World Bank. Talk turned to **sustainable agriculture,** defined by the Consultative Group on International Agricultural Research (CGIAR) as agriculture that satisfies changing hu-

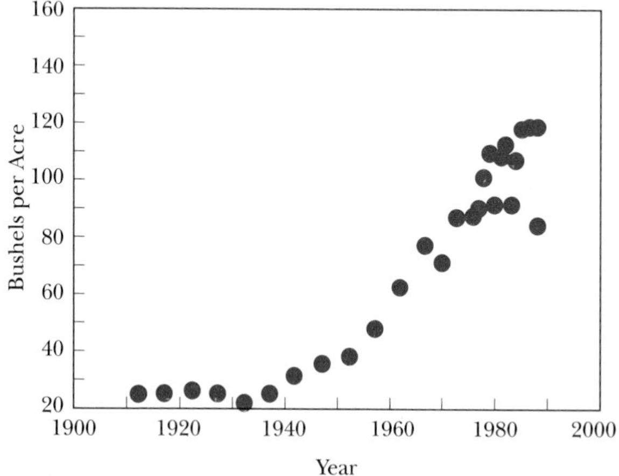

Figure 21-1

Corn production in bushels per acre in the United States from 1909 to 1989. This illustrates the increase in productivity beginning in the early 1940s and the leveling-off evident by the 1980s. (Data from U.S. agricultural censuses.)

man needs while maintaining or enhancing the quality of the environment and the natural resource base.

Africa is a special case. For the world as a whole, increases in food production have slightly outstripped population growth during the past quarter-century. This has not been true for much of Africa. For the continent as a whole, per capita grain production in the early 1980s was about 70 kg below the 180-kg-per-person subsistence level attained during three years of the 1960s (World Resources Institute 1986). The reasons are many, although the one big reason is unrelieved population growth at the rate of more than 3% per year. Other important factors are African governments' use of resources for industrialization rather than farming and their concentration on cash crops such as coffee, tea, and cotton, instead of food. Political instability has also contributed to the abysmal state.

The seminomadic agricultural system of Africa's past was well attuned to the continent's climates and vegetation. With drought, herdsmen moved to wetter areas either seasonally or between years. With rising populations and changes in political structures, African governments have discouraged nomadism in favor of the permanent farms typical of the moist and fertile developed countries. Low food production, malnutrition, and starvation have been the result. One geographer was able to list 157 ways in which traditional societies in Kenya coped with drought. For the same societies today, only two mechanisms are left: Moving to the city and prayer (Wijkman and Timberlake 1985).

There has been some recent movement toward expanding what is left of the "green revolution" to include agroforestry, greater use of native crops, and other practices that have the potential to be less destructive socially and environmentally (Walsh 1991). However, the vision to produce a "sustainable" agriculture, ecologically based and in tune with local environment and culture, seems to be a scarce commodity anywhere in the world (Dahlberg 1991). This is particularly so in that most discussions begin with talk about increasing production to meet the needs of increasing human populations. The first premise of a truly sustainable agriculture has to be bringing population size in line with long-term carrying capacity.

The Role of Environment in Health and Disease

In Chapter 7, we pointed out that health and disease are relationships between an organism and its environment. In that chapter we explored infectious diseases, diseases produced by other organisms, and pointed out that most of us harbor microbes capable of producing disease but show no symptoms for many of them. The disease produced by one of these organisms may be precipitated by various environmental events that lower the host's resistance—poor diet, for example, or fatigue or emotional stress. Lowered resistance can often be translated as "suppression of the immune system." There is now clear evidence that stressful environmental events can impair various aspects of the immune response.

There are several other kinds of conditions generally considered as disease, all with important ecological ramifications. There are **deficiency diseases** such as pellagra and scurvy, produced by improper diet that is, in itself, an environmental matter. This usually depends, in turn, on some sort of cultural dislocation relative to the environment; primitive peoples living on the diets that they have evolved rarely show deficiency diseases. It was not until corn was introduced into southern Europe and Africa that the burning, weakness, and eventual near-mummification caused by pellagra appeared. It was easier to grow enough calories using corn rather than other grains, but corn is deficient in niacin.

There are inherited **metabolic disorders** such as diabetes and phenylketonuria. It is not known whether something in the environment causes the underlying metabolic problem, but the health of the individual *is* entirely an environmental matter. In environments where it is possible to obtain the hormone insulin in the case of diabetes and to reduce the amino acid phenylalanine in the case of phenylketonuria, the individual having these conditions may develop and live normally. In a similar fashion a hay fever sufferer may show no symptoms of allergy in an environment where ragweed does not grow.

Cancer may be considered as another disease category. Since 1971, billions of dollars have been spent in the war on cancer. Most of the money has gone to study cellular and biochemical aspects of cancer. This research has produced a great deal of basic scientific information on cells but little in the way of methods for curing cancer. Cancer rates continue to rise; approximately one fourth of all deaths in the United States are from cancer. Although the cure rate (five-year survival after diagnosis) has risen for certain types of cancer, notably cancer of the cervix, for most types it has been static over the past few decades (Epstein 1990).

It now seems clear that the majority of cancers, possibly as much as 90% (Doll and Peto 1981), are environmentally produced (Figure 21-2). Cancer-producing (carcinogenic) agents are various: X-rays, cigarette smoke, food additives, viruses, and workplace chemicals, among others. Curing cancer is difficult, but it is "essentially a preventable disease" (Epstein 1990). Although not easy to achieve, an environment allowing us to avoid carcinogens almost completely would nearly eliminate the problem of cancer.

Stress-related diseases such as hypertension and gastric ulcers have a strong environmental component. The environmental stress usually is some kind of social pressure. Many biologists have been impressed with the parallels in both physiology and behavior between crowded rats and crowded people. Stressed humans and stressed rats both respond physiologically with such things as arteriosclerosis and ulcers. It is possible, although still not certain, that increases in crime, alcoholism, divorce, and mental illness in crowded inner cities are the same responses on the human level as are increased aggression, abnormal sexual behavior, and breakdown of normal social functioning in crowded rats.

Poisoning, stemming from toxic materials in the environment, has been around for a long time. In the old days, humans might be poisoned by metals, acids, or plant alkaloids. Workers that silvered the backs of mirrors and hatmakers, for example, suffered from mercury poisoning. They became mad as hatters, trembling so much that they couldn't lift a teacup to their lips and breaking into a dancing trot when they tried to walk.

Such poisons are still with us. The most pathetic recent example of mercury poisoning was Minamata disease, named for Minamata Bay in Japan. There and in several other coastal areas, people depended heavily on clams and fish for food, but in the 1950s and 1960s, these came to contain high concentrations of methylmercury as a result of bacterial conversion of industrial discharges of inor-

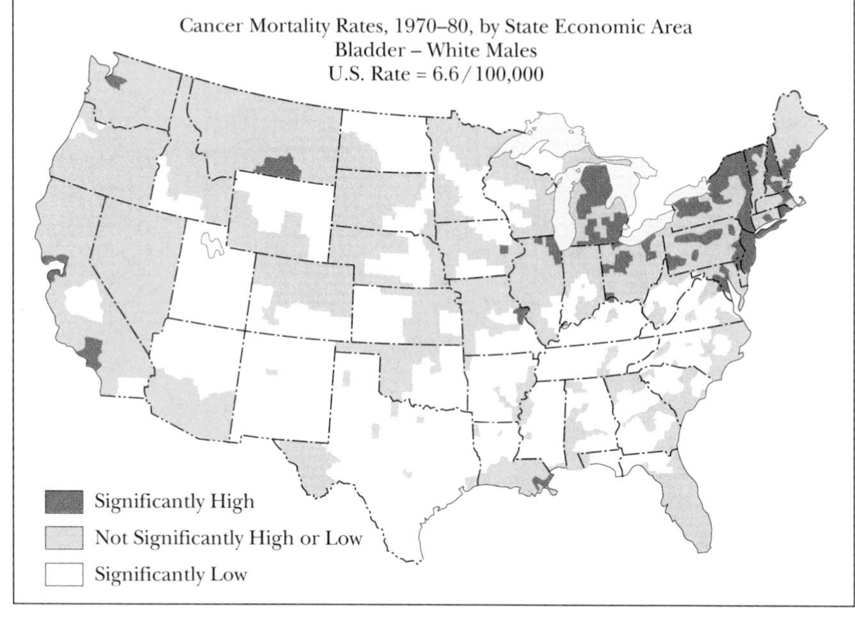

Figure 21-2

Relative death rates (by state economic area) from cancer of the bladder among white males, 1970–1980. This shows one line of evidence suggesting an environmental cause for many kinds of cancer. The high rates are found mostly in regions with industries manufacturing or using organic chemicals. Occupational exposure is thought to be important. (From Pickle et al. 1987.)

ganic mercury salts (Pier and Bang 1980). Sickness and death and a 50% rate of cerebral palsy among the newborn children were the results.

To add to these poisons, we have produced a vast array of new chemicals that we can meet in our food, our water, at work, or just walking down the street. Like the older poisons, they can affect our behavior, making us dull, anxious, or irritable, make us physically sick, or kill us.

Parkinson's disease, characterized by a fixed, masklike expression, pill-rolling tremors of the hands, and a stooped, shuffling gait, results from degeneration in the substantia nigra of the midbrain. The condition was unknown prior to the industrial revolution. A Canadian research team found a strong correlation between the incidence of Parkinson's disease and levels of pesticide usage in rural areas of Quebec (Lewin 1985). Of course, pesticides are only one class of human-produced chemicals that might be responsible for damage to this part of the nervous system; the same researchers found a higher incidence of the condition in persons living in the industrialized sections of Montreal than in the nonindustrialized sections.

Holistic Medicine

Holistic medicine is an approach to maintaining good health that considers whole organisms in relation to their environment. Western, and especially American, medical thought has been dominated by the "biomedical" model: People get sick. They go to a physician who diagnoses their illness based on symptoms and laboratory tests. A treatment is prescribed, generally pharmacological or surgical. As a result, the patients recover—or so we hope. Birth, death, and a variety of other normal life history events have been forced into the same model. Health insurance contracts sometimes read, "Pregnancy is considered like any other illness."

The biomedical approach has had its successes and may even have been a necessary stage in the development of health care. Its successes, however, have been less than we sometimes think. The decline in mortality from infectious disease beginning about 1750 in Europe has stemmed mostly from improved hygiene and better diet. One author (McKeown 1979) attributed only 10% of the decline in death rates to improved medical practice, including modern drugs such as antibiotics.

Something of the limits of standard medical practice is shown by a study of childhood mortality rates in Boston (Wise et al. 1985). Access to intensive tertiary health care is extremely widespread for children and pregnant women in Boston. For example, 93% of all births occurred in facilities with directly associated neonatal intensive-care units (a neonate is a child less than 28 days old). Despite the high level of availability—and use—of such facilities, childhood death rates were much higher for low-income compared with high-income families (Figure 21-3). Why should this be so?

The reasons are cultural and environmental. Children from low-income families tended to be undersized at birth, presumably because of factors such as nutrition and tobacco and alcohol use by the mothers. However, even normal-sized infants from low-income groups had higher mortality rates from bronchopneumonia and sudden infant death syndrome. Accidents, especially fires, were a much higher source of mortality among low-income families. Remarkably, adolescent death rates were sim-

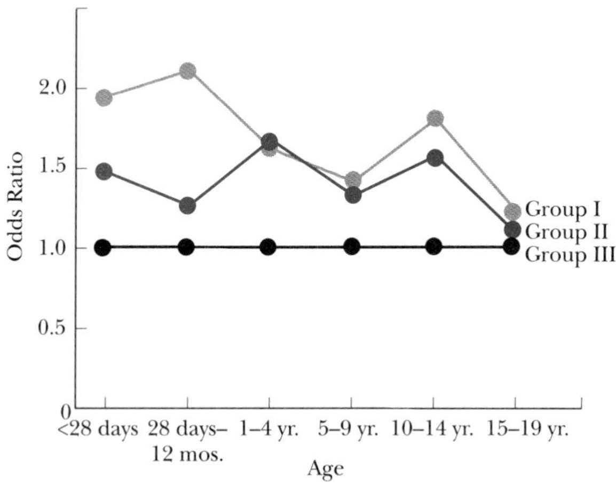

Figure 21-3

Odds ratio for childhood mortality by income group in Boston 1972–1979. Group III were children of higher income families, Group II, medium, and Group I, low. Group III is used as the standard (odds ratio = 1.0). (From Wise et al. 1985. Reprinted by permission of the *New England Journal of Medicine* 313: 361, 1985.)

ilar among the various income groups (Figure 21-3), automobile-related deaths among the affluent balancing homicides among the poor.

It is clear that the disparity in survivorship between the affluent and the poor will not be removed by improvements in medical practice. The hope for improvement lies in thinking of the children and their families in connection with their whole environment. Improved nutrition, less drug usage, decreased population density—these are the kinds of changes that will bring improvement. And they are, of course, a difficult prescription.

One of the most troubling sets of statistics from current medical practice has to do with iatrogenic—literally, physician-caused—illness. Such problems now constitute a high proportion of physician visits. A study of 815 consecutive patients at the Boston University Medical Center Hospital found that over one third suffered at least one iatrogenic illness (Steel et al. 1981), and 15% had two to seven such medically induced problems. The problems were severe enough to be potentially disabling or life-threatening for 9% of the patients. Thirty of the 290 patients with iatrogenic problems did, in fact, die.

Iatrogenic illness is sometimes the result of sloppiness on the part of physician, nurse, or laboratory. All of us have heard the horror stories (if not, see Couch et al. 1981): A surgeon removes the healthy rather than the diseased kidney. But generally the problems are expectable, statistically predictable effects of the biomedical approach. A baby is delivered by Caesarean section rather than vaginally because malpractice suits against obstetricians are more frequent for normal births. Staphylococcus bacteria present in the hospital cause an infection in the woman. She is given a broad-spectrum antibiotic that successfully controls the staph infection but also kills most of the rest of her normal body flora, resulting in enterocolitis and a vaginal yeast infection. There are, of course, medical treatments for these problems too.

This approach is the medical equivalent of the "technological fix" that our society is prone to apply to more general environmental problems. Each solution begets another problem that, in turn, is given another technological solution.

Holistic medicine emphasizes staying healthy. It starts from a basis of appropriate diet, exercise, reduction of stress, a healthful environment, and health-promoting behavior. Surgery and pharmacology are useful adjuncts along with a variety of new and old techniques such as relaxation responses and acupuncture. At least, this is what holistic medicine ought to be. One problem with the term is that it has sometimes been appropriated by practitioners of some bizarre techniques, like diagnosis based on your irises or treatments consisting of high-colonic irrigation, that even to the open-minded seem unlikely to be medically useful.

J. M. May (1960) described a situation in northern Vietnam illustrating the importance of ecological factors in health. In the delta region at the time he described, people built their houses on the ground, with the stables on one side and an outdoor cooking area on the other. In the hills, people built their houses on stilts, with animals kept in the space under the house and meals cooked inside. Anopheles mosquitos, which can carry malaria, occur in the hills but not in the delta region. These mosquitos rarely fly higher than 8 or 9 feet above the ground, however, so they usually only encounter the livestock. If they do fly as high as the living quarters, smoke from the cooking fires tends to repel them.

The rich delta lands are crowded. When people from the delta relocate in the hills, they retain their delta culture. Living at ground level, they are bitten by the mosquitos and contract malaria. The delta people ascribe their troubles to demons in the hills. We know they are wrong, but if we attribute the disease to plasmodium, the malaria parasite, we are only a little way closer to the truth.

May concluded his article, "The Ecology of Human Disease," like this:

> *The ancient formula of one ill, one pill, one bill, which seems to have been the credo of physicians for many generations should be abandoned. Disease is a biological expression of maladjustment. This is what should be taught to our students in medical school, and this phenomenon against which they are going to fight all their lives cannot be understood without an ecological study in depth [that includes] the environment, the host, and the culture.*

May's article appeared in 1960. Thirty-odd years later, this approach seems somewhat nearer.

Population

Given a limited world, populations cannot continue to grow forever. Even if food, water, energy, pollution levels, or some other factor does not become limiting first, present growth rates will fill the land area with people standing next to one another in just under 500 years. One can envision housing everyone in skyscrapers or strange underground cities and adding floating or submarine communities in the lakes and oceans, but they would merely postpone matters, even if they were possible; a few hundred years more would be sufficient for the human population to weigh more than the earth. In addition to the logical difficulties of infinite growth in a finite world, our experience with other species shows that the absence of population limitation would be ecologically unique.

Optimal Population Size

If growth cannot continue forever, the appropriate question is, When should it stop? Or a better way of putting the same question, What human population size is optimal?

There is no simple answer, but we can begin by realizing that it is a different question from, What is the maximum population that the earth can support? The earth could probably support a higher population living in poverty and distress than in comfortable circumstances, but few would claim that the former are optimal conditions. A second step is in realizing that there is no one answer, no one optimal population, because human preferences are involved in parts of the equation. For example, we might be able to specify an optimal population for a world population living on an almost exclusively vegetarian diet (*almost,* because it may be true that optimal development of children requires fatty acids or other materials found in meat). We might consider safeguarding the land against overuse, preventing pollution of air and water, and take into account the caloric and other nutritional requirements of individuals—and then perhaps we might say that the earth could comfortably support so many billions of people. To allow some fairly substantial part of the world population to include meat in their diet would necessitate a lower population than could be supported—optimally, in purely nutritional terms—on plant material. The freedom to choose a way of life, whether as to diet, religion, or hairstyle, would be held by many persons to be an essential part of an optimal population size.

Local and National Populations

Although we have been talking on a worldwide scale, three different population sizes are important to humans. The population size of the whole earth is important, but so also is the population size of individual nations and of local communities. Determining and achieving optimal populations at each level are separate, although interrelated, problems.

For local communities an optimal size may be one that is large enough to provide such things as a consumer base for specialized shops and businesses and a social life and clientele for cultural events, and one small enough to avoid the higher crime rates, excessive highway building, destruction of farmland and open space, congestion, and other disadvantages of urban life (Sale 1978). Exactly what this optimum is has not yet been clearly defined. Athens in its golden age had a population of about 40,000, and London in Shakespeare's time was about 180,000. Today in the United States it appears that an urban area having 100,000 persons can provide a sufficient audience for most cultural and sporting events, and that per capita taxes for police, education, and welfare go up sharply at a population somewhere between 100,000 and 500,000 (Figure 21-4).

At some point between 1 million and 5 million population, this trend seems to produce a situation in which a city can no longer support itself. The city's problem then becomes a problem of the rest of the state or of the nation. The recurring financial problems of Detroit and New York City are cases in point. Although most large cities are not "sustainable," there are two advantages in having them: (1) Certain professions and activities may be so specialized that only cities of a few million provide a population base big enough to support them; and (2) by concentrating millions of people in a few square kilometers, many of their adverse environmental effects are localized. For these reasons,

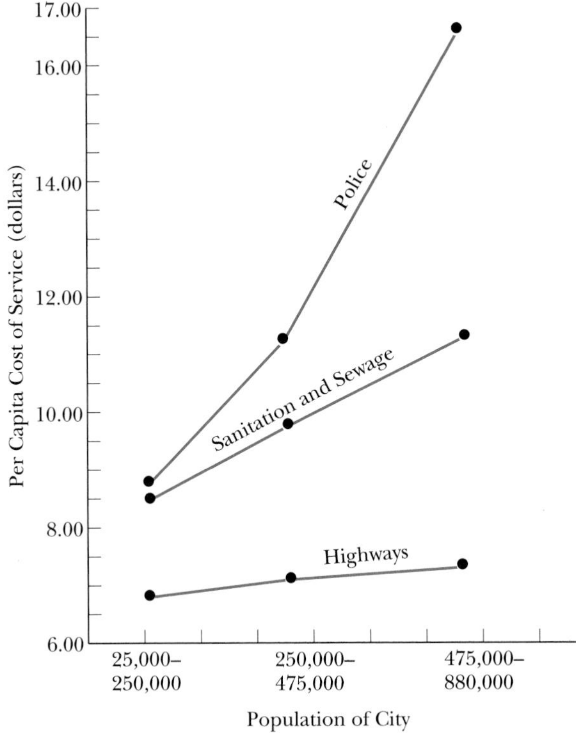

Figure 21-4

The diseconomies of scale. The per capita costs of police, sanitation and sewage, and highways in Ohio cities are graphed against size of the city. Such services do not just cost more as the size of the city rises; they cost more for each individual, despite the fact that there are more people paying taxes. The biggest city in Ohio is Cleveland with a population of 876,000, but the upward trends continue at higher populations. For Los Angeles, with a population of 2.8 million, per capita cost of police protection was $22.39 (Data from Advisory Commission on Intergovernmental Relations, 1968.)

those of us who live elsewhere may, nevertheless, wish to help subsidize a few large cities.

Under any circumstances, city and countryside are interdependent in many ways. The city imports food and raw materials for manufacturing, and its residents use the open spaces of the farms, forests, streams, and lakes in various ways. The countryside is dependent on the city for such things as manufactured goods and services and cash for crops. (There are other sorts of exports, also: The city exports air pollution, after importing the fuels that produce it, and the countryside exports chemical residues in food after importing the pesticides and other compounds.) Here, as previously, the freedom of opportunity to live in rural, small town, or city surroundings is important. So too is the retention of diversity in the landscape so that dwellers in any one of these environments can make use of the unique features of the others.

Controlling local population size has proved particularly difficult in the United States for several reasons, despite its importance. Historically, mobility has been an asset for industrial development, but such mobility may cause rapid growth of communities through immigration. When local communities have attempted to limit growth, the legal methods used (such as zoning) have sometimes been considered discriminatory by the courts. Probably most important, however, is that most local communities want growth—or to put it more precisely, local leaders want growth. (This matter is dealt with further in the section on land use.)

At the national level, recent fertility levels of about two children per family would allow a stabilized population of something less than 300 million (versus today's 250 million) early in the 21st century (ignoring immigration; see the following section). At that point, a family size of 2.1 children would suffice to keep the population constant. Nationally we seem, then, to have achieved the lowering of the reproductive rate that 25 years ago (say, in 1968, when P. R. Ehrlich's book *The Population Bomb* was first published) seemed unlikely or impossible.

Is 300 million, or even 260 million, Americans optimal? Probably most environmentalists—at least those having direct knowledge of the natural world—would guess that it is too high. Land is being eaten up for new housing developments, new roads, new jails, and research and business "parks." Declining farmland acreages are being farmed too intensively to sustain yields without heavy inputs of energy, fertilizers, and pesticides. Natural areas and open spaces are disappearing, taking with them the animal species that need more than 8 or 16 ha of forest or marsh for living space.

It would be easy for a well-intentioned person to drive through some of the less heavily populated parts of the United States—Wyoming, for example—and conclude that we still have plenty of space to spare. What they do not realize is that the Wyoming grassland with its sparse herds of cattle and pronghorns is just as fully occupied as Cleveland or

Baltimore. Wyoming rainfall is low, productivity is low, and it is thoroughly possible that some of these lands that we glimpse driving along I-80 or I-90 may already be overgrazed even at these low densities. It is possible to irrigate these lands and occasionally we see the green, relatively lush patches of alfalfa or other crops. But irrigation is expensive, energetically and in dollars. It cannot be done without subsidies from taxpayers all over the United States. And when it is done, it produces the environmentally undesirable consequences of all such water projects.

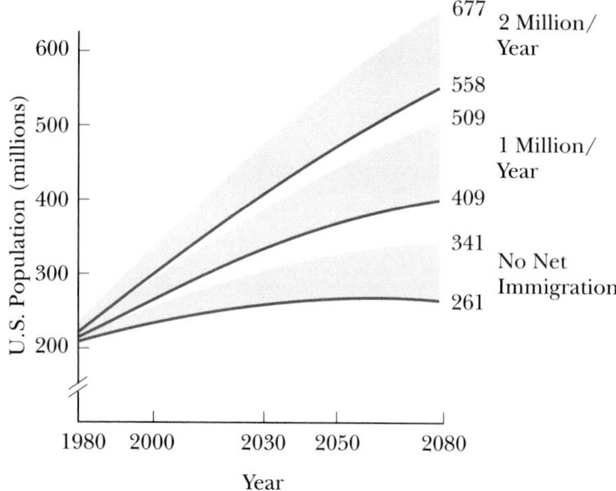

Figure 21-5

These projections of U.S. population size made by The Environmental Fund (now Population-Environment Balance) are a powerful illustration of the impact of immigration on population. Made more than ten years ago, they are close to being on target and have been considerably more realistic than official U.S. Census Bureau predictions of the past decade. A total fertility rate of 2.0 children per woman is assumed; the shaded areas above the curves show the added growth if fertility rates are up to 10% higher (2.2 children). (Adapted from TEF Data [The Environmental Fund], June 1982.)

Immigration

Although the current fertility rates of citizens would have allowed stabilization of the U.S. population no later than halfway through the 21st century, that is no longer a realistic possibility because of the high level of immigration. In 1992, the U.S. Census Bureau revised its population projections for the year 2050 (Day 1992). Just three years earlier, in 1989, it had been predicting a 2050 population of 302 million (Spencer 1989). Now, using somewhat more realistic immigration figures, it is predicting a 2050 population at 383 million and rising fast. Two factors that may make even this figure an underestimate are, first, that a high proportion of current immigrants, legal and illegal, are in their peak reproductive years and, second, that many come from cultures in which large families are favored.

Of course, this prediction of 383 million, like the Ghost of Christmas Future, is what may be rather than what must be. A policy in which only as much immigration to the United States is permitted as is balanced by emigration might yet lead to a stabilized population at a tolerable level (Figure 21-5). But such a future will require an understanding of the consequences of immigration that currently is much better developed among the citizenry than among politicians, government bureaucrats, and the media.

Most of us in the United States are descendants of immigrants, and some thoughtful persons have felt that closing the door now that we're here is selfish. They portray opposition to unrestrained immigration as somehow not respectable, rather the way the chemical industry portrayed opposition to unrestrained pesticide use, before Rachel Carson. If a liberal immigration policy was good then, they say, it's good now. But things change. One of ecology's many lessons is that what is good varies with the stage of a system. Early in succession, for example, *r*-selected species do well, but it is a strategy for failure in mature systems. A high level of immigration, however useful in the 1880s, is a recipe for environmental disaster in the 1990s and beyond.

But isn't this pretty tough on the countries with populations rising beyond the resources to support them? It has been in the self-interest of the governments of these countries to send as many of its citizens as possible to the United States, Canada, or western Europe. By doing so, hard decisions on the economy, land reform, birth control, and many other topics can be postponed. Although it is in the self-interest of the governments, it may not be in the best long-term interests of the countries. The decisions need to be made but, more than that, those fleeing the countries are often those with the initiative, the hope for a better future and, very often, the training and skills that might make reform in their own country a possibility.

Let us not lose sight of the global picture. If the areas of the world with the highest birth rates can steadily export part of their population to somewhere less crowded, it is much more likely that their population growth will continue unabated, and the earth's population will continue inexorably to rise.

World Populations

Almost no one believes that the earth can support its present 5.3 billion at American levels of affluence. We eat about 3000 calories worth of food a day, including some 65 or 70 g of animal protein. For the two thirds of the world population in what the food scientist Georg Borgstrom (1973) called the Hungry World, the figures average 2200 calories and 11 g of animal protein per day (Scully et al. 1992). There is about 1 ha of tilled land per person in the United States. At this ratio of cropland to population, the world could support about 1.4 billion persons at U.S. levels. Using similar reasoning but different aspects of life and more complicated formulas, several persons have come up with estimates for an optimal world population of around 1 billion.

This leads to one of basically four approaches to the world population problem: One approach would somehow involve reducing world population by about 75% in order to achieve a high standard of living for everyone. A second approach would be to stabilize world population—or allow it to continue to grow—but to allocate food and other resources evenly so that everyone had a low standard of living. The third approach would be to count on the development of abundant cheap energy (as, for example, by fusion, should it prove technically feasible and inexpensive). If this were to happen, and if the problems of pollution from power plants, transportation, food processing, and so forth could be dealt with, and if unhealthily crowded areas could be reduced without destroying too much of the remainder of the earth, a world population possibly as high as it is now could be supported at a high standard of living.

The fourth approach is that of the present: Each sovereign state looks to its own self-interests and uses as much of the earth's resources as it can obtain. The United Nations Conference on Environment and Development held in Rio de Janeiro in 1992 perhaps contains the possibility of a small step forward. Although the Rio Declaration's principle 2 stated the right of a nation to exploit its own resources in accordance with its own policies, added was the clause "without harming the environment elsewhere" (Parson et al. 1992). Overpopulation, however, was scarcely mentioned at the conference. Principle 8 of the Rio Declaration espoused sustainable development that promotes "appropriate demographic policies," which could be anything.

Land Use

Many environmental problems—population, pollution, natural area preservation—converge in the issue of land use. There are two possible approaches to land use: Public planning and what most communities have now.

The land-use pattern where you live may be a pleasant patchwork of villages, fields, and forest, or it may be a ghastly mess of smoke-belching factories (Figure 21-6), ancient and dirty tenements, and misbegotten high-rise apartment projects. Whether it is one of these or something else, it probably got that way largely by accident. Even so, it would be a mistake to say that land-use planning does not occur. It does, but the planning is mainly in such places as the offices of land developers, paid or unpaid local booster organizations, and road commissions.

Land-use decisions—like everything else in the environment—rarely have only a single effect. A decision to build a new factory causes the highway commission to widen the road past the site to four lanes. Motorists begin to use the good, new road as a major artery. No one wants to have his children and dogs next to the speeding cars, the noise, and the fumes, so the zoning board is asked to rezone the neighborhoods along the road for apartment buildings and hamburger stands. These generate more traffic, causing the highway department to put in a six-lane highway that is so convenient as an access into the central city that housing develop-

Figure 21-6

The industrial landscape.

ments spring up around the end of it in the surrounding countryside.

To deal with this sort of ramifying effect and, when necessary, to prevent it, regions and communities must have a well thought out plan for land use and evaluate each proposed development on the basis of its eventual as well as its immediate effects.

Phases in Environmental Planning

In his textbook, *Ecological Systems and the Environment,* T. C. Foin Jr., (1976), identified five phases in environmental planning.

1. *An environmental resource inventory:* This should include all environmentally relevant features such as soil types, natural communities, rare and endangered species of plants and animals, human population distribution, air quality, archeological and historical sites, and topography. Aerial photographs and remote sensing data from satellites such as LANDSAT are valuable in the inventory but are not a substitute for actual field work by biologists and other scientists.
2. *Value inputs and interpretations:* Ecology and environmental science can carry us only so far in this matter. The recharge areas for groundwater can be identified as areas that should not be paved over, the remaining bogs, clean streams, and climax forests can be identified as areas that should be preserved, but in the end a great deal of latitude for judgment is left. It may be a fact that the citizenry is unhealthy without open space, but preferring good health over ill is still a matter of taste. Ideally this phase of planning should include considerations of the total amount of industrialization and commercialization allowable and the total population that would be tolerated. It is important to learn the opinions of the citizens at this stage.
3. *Production of maps:* Maps of the area are the most straightforward way to assemble the information gathered under Phase 1. A **Geographic Information System (GIS)** approach, in which data are assembled in computer files on a common base map, is an appropriate

technology. Otherwise, transparent overlays can be prepared by hand. The aim is the integration of information to produce new maps that indicate, for example, the degree of suitability of various regional areas for different uses.

4. *Identification of development constraints and the regional plan:* The plan is now produced. It should indicate such things as the areas to be preserved, the areas to be retained as agricultural land, the areas on which one residence per hectare can be tolerated, and the areas for high-density housing.
5. *Implementation:* Implementation of the plan involves such things as the various legal devices for setting it up, cooperation with state and federal agencies and private groups or persons, such as land conservancies and local philanthropists, and any necessary education of the citizenry.

One of the pioneers and leading practitioners of ecologically based land-use planning, as described here, has been Ian McHarg (1969). His book *Design with Nature* has several case histories that, besides showing the method, are full of ecological insights.

A few areas have a rationally developed plan for the future that has been developed by some process similar to that just described, but most do not. Most areas have some type of land-use plan or at least a zoning system, but up to now such plans have been mostly arbitrary, without the factual basis of a resource inventory and with a value basis having, as its main principle, "growth is good."

Although the past few years have seen several communities decide that they were large enough and would grow no further, most local governments still seem to believe that *any* growth of industry or population is desirable because it "increases the tax base." This ignores the evidence (mentioned under "Population") that beyond some size per capita tax rates increase, despite the fact that there are more people to pay them. Few local governments thoroughly analyze each proposed new business to determine whether the benefits in the form of taxes and payrolls outweigh the costs of such factors as pollution (direct plus indirect, in a form such as auto emissions from employees' cars), loss of open space, provision of utilities, and the tendency of new industry to attract immigration.

Of course, growth is not the only thing valued by local governments. Another important rule seems to be "better us than them." If a new shopping mall is to be located in a region, many local governments are willing to violate their own land-use plans so it can be constructed under their jurisdiction rather than in the adjoining city, township, or county. This sort of competition needs to be lessened, but whether this can be achieved through regional and state land-use plans or by some method of reducing the benefits from winning the competitive game is not clear.

If the U.S. population were 50 million, perhaps even if it were 100 million, land-use decisions of this sort could be tolerated. If we did not like what was happening to the neighborhood we could simply move to where it was less crowded and the air was better, just as our ancestors did all through the 1800s. But our population is 250 million, heading toward 260. For most of us there is no longer anywhere to move.

Radical Environmentalism

Members of the Reagan and Bush administrations used the term "environmental extremist" for persons who had earlier thought of themselves simply as "environmentalists" or, even, "conservationists." It is instructive to try to identify extremist positions—pro-environment and also anti-environment—on various issues. What, for example, are the two extreme positions on natural areas?

One nomination for the pro-environmental extreme would be that all lands now retaining near-natural communities ought to be preserved as such, protected from any threat of development. But surely that is not really a very extreme position. It is easy to think of stronger ones. For example, is it not time to begin restoring some of the areas that have been environmentally degraded but that could recover? Examples would be lakes whose shores have been turned into housing developments, the ocean beaches, and many of the grazing lands of the West.

Such proposals have been seriously made (Foreman 1981):

> *Pick an area for each of our major ecosystems and recreate the American wilderness—not in little pieces*

of a thousand acres but in chunks of a million or ten million. Move out the people and the cars. Reclaim the roads and plowed lands. It is not enough any longer to say no more dams on our wild rivers. We must begin tearing down some dams already built.

And what would be the anti-environmental extremist position on natural areas? Most students of whom I have asked the question come up with a statement to the effect that land-use decisions would be based solely on economic considerations. It is not a very difficult matter to decide how close American policy and practice are to striking a moderate compromise between these two extremist positions.

Some persons who wish to protect the environment and correct past abuses have felt that most of the older, mainstream environmental organizations, such as the Audubon Society, Sierra Club, Nature Conservancy, and Wilderness Society, have tended to adopt positions that are too moderate. Such organizations do so, perhaps, in hopes of ingratiating themselves with the public and potential financial contributors. They believe in the old maxim that you can catch more flies with honey than vinegar.

The belief that we should be striving for what is right for the biosphere, rather than what will not displease the city fathers or the large corporate donors, has resulted in the formation of new environmental organizations that are characterized by being willing to take a more confrontational approach. Depending on the organization, they may be primarily interested in stopping pollution, stopping whale killing, or preserving wilderness, among other goals. These are what might be called the radical environmentalists.

The best-known such group is Greenpeace (Figure 21-7). Formed in 1970, it adopted as its main tactic the Quaker practice of "bearing witness." This is a nonviolent form of protest that involves gathering at the scene of the environmental crime. A catchy banner and good press coverage also help.

Among environmentally damaging activities that Greenpeace has tried to stop have been atmospheric and undersea testing of nuclear weapons by France (which refused to sign the Nuclear Test Ban Treaty of 1963), whaling, the killing of Newfoundland harp seal pups for fur, the dumping of toxic and nuclear waste in Europe's North and Mediterranean seas, and the development of Antarctica (Brown and May 1991). In many of these causes and others, they have had a fair measure of success, often in conjunction with the activities of other groups.

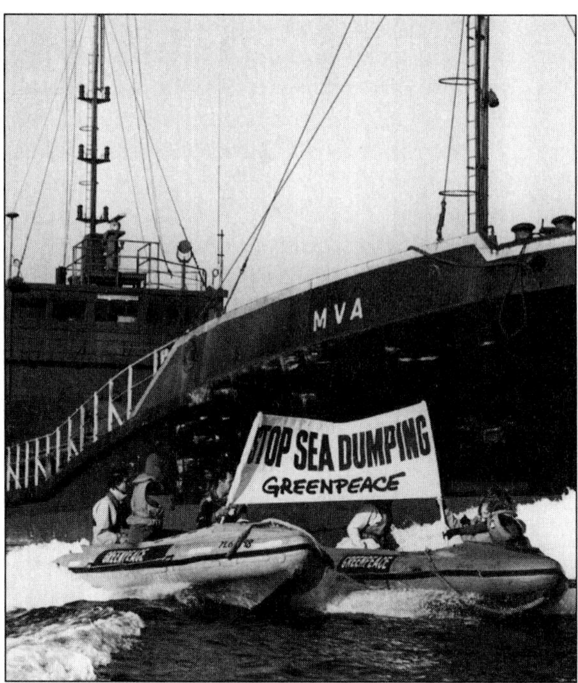

Figure 21-7

Greenpeace activists intercept a ship headed from England to the North Sea to dump fly-ash waste from a coal-burning power plant. (© Greenpeace/Walker 1987.)

Although Greenpeace's approach is nonviolent, its members have had violence directed toward them. The most serious was the bombing of the Greenpeace vessel the *Rainbow Warrior* on July 10, 1985. The vessel was anchored at Auckland, New Zealand, on its way to protest French nuclear testing scheduled at Moruroa, a Pacific atoll. Two bombs planted by a member of the French spy agency *Direction Général des Services Extérieurs* sank the vessel, drowning a Greenpeace photographer.

At the end of September, after nearly three months of denial by France, the French prime minister admitted the government's involvement. Two

French spies arrested earlier by New Zealand authorities pleaded guilty to charges of manslaughter and wilful damage. They were sentenced to serve ten years.

One of the most colorful of the radical environmental groups has been Earth First!, founded in 1980 not as a formal organization but an informal movement by Dave Foreman and several other conservationists (Foreman 1981). Their general position was summed up by one of their slogans, "No compromise in defense of Mother Nature." One controversial aspect of the movement was its seeming encouragement of "monkey-wrenching," or ecosabotage (Foreman and Haywood 1987). The name is derived from the novel *The Monkey Wrench Gang* by Edward Abbey.

Earth First! was particularly interested in stopping logging in the Western wilderness. Persons inspired by Earth First! principles would chain themselves to trees, sit down in front of bulldozers, and commit similar acts of annoyance to persons in the timber industry, including U.S. Forest Service bureaucrats. In some cases, the Earth First!ers would "spike," or claim to spike, trees in areas scheduled for clear-cutting. The "spikes" are large nails such as are used in framing houses.

A chain or mill saw encountering such a nail might be damaged and might injure the person using or operating the saw. Persons doing such spiking would contact the media proclaiming what they had done and would often paint spots on trees they had spiked (and probably some others too). The practice of spiking has been widely condemned. One person was injured in an Oregon sawmill when the saw ran into a nail. It was claimed by timber interests that the nail was from an Earth First! spiking operation. Sawmills tend to be dangerous workplaces, and when one Earth First!er, Judi Bari, discovered how badly safety was neglected, she began efforts to organize the millworkers (Rosen 1990).

On May 24, 1990, Bari and a fellow Earth First! advocate were injured when a pipe bomb went off in their auto. They had been on their way to a "Redwood Summer" rally in Santa Cruz, California, where they would have sung and spoken on the evils of destroying the old-growth forests. The immediate response of the Oakland, California, police was to arrest the two Earth First!ers, the charge being transporting an explosive device. The charges were later dropped. No other arrests have been made.

The confrontational tactics of the radical environmentalists obviously stir strong feelings. Some persons believe that such tactics are ineffective or even counterproductive. The views espoused are seen as extremist, and the tactics as threats to an orderly society. Radical environmentalism, it is cautioned, may do more harm than good.

Studies of similar popular movements, such as the Civil Rights movement, that confronted entrenched practices do not support this view (Haines 1988, McAdam 1992). A radical fringe is helpful to such movements for several reasons. For one, it provides an extreme view—in this case, an extreme pro-environmental view—as a counterbalance to the extremists at the other end of the spectrum. In so doing, it enlarges the area of debate. This may allow the position of mainstream environmental organizations to be perceived correctly as moderate—the radical environmentalists would call them already compromised—and to have a better chance of being adopted.

Spaceship Earth

"We travel together, passengers on a little spaceship . . ." wrote Adlai Stevenson, and since then, the image of "Spaceship Earth" has become a popular one. On short spaceflights such as those to the moon food can be taken along and wastes stored until the end of the voyage. On longer trips this arrangement would be impossible because of space and weight limitations. The spaceship would have to be complete in itself, producing food and oxygen by the use of photosynthetic plants and recycling wastes as nutrients for plant growth. In the small closed system of the spacecraft the interdependence of the various parts, living and nonliving, is obvious. The spacecraft is, of course, a miniature model of the earth; however, it made a striking image to turn the comparison upside down and think of earth as a spaceship.

The image emphasizes the limitation of space and resources; it is very different from the traditional American idea of moving on to a new frontier further west if we make too big a mess of things here. The image emphasizes interdependencies; we are all here together, and earth, air, and water are part of a system to which we ourselves belong. Our continued well-being is dependent on the continued functioning of the whole system.

So the spaceship analogy has its value, but we need to remember that we are not going anyplace; we *live* on "Spaceship Earth." If we find the regimented, claustrophobic conditions of spaceship living a poor sort of home, it is up to us to keep the model right-side up. That is, it is up to us to keep human impact small compared with the earth's capacity to absorb it, so that fields remain fields and oceans remain oceans, rather than becoming "life-support systems." We are travelers together, but there is no end of the trip to look forward to when we will get out and stretch, drink clean water, and eat home cooking again.

Bibliography

Aaron, W. I., and S. H. Smith. 1971. Ship canals and aquatic ecosystems. *Science* 174:13–20.

Abelson, P. H. 1984. Groundwater contamination. *Science* 224:673.

Abrahamson, W. G., and D. C. Hartnett. 1990. Pine flatwoods and dry prairies. Pp. 103–149 in *Ecosystems of Florida,* R. L. Myers and J. J. Ewel, eds. University of Central Florida Press, Orlando.

Abrams, M. C., D. G. Sprugel, and D. I. Dickmann. 1985. Multiple successional pathways on recently disturbed jack pine sites in Michigan. *For. Ecol. Manage.* 10:31–48.

Abrams, P. 1983. The theory of limiting similarity. *Annu. Rev. Ecol. Syst.* 14:359–376.

Acree, F., R. B. Turner, H. K. Gouck, M. Beroza, and N. Smith, 1968. L-lactic acid: A mosquito attractant isolated from humans. *Science* 168:1346–1347.

Adams, C. C., H. A. Gleason, O. McCreary, and M. M. Peet. 1909. *An ecological survey of Isle Royale.* Rep. Board Geol. Surv. Michigan 1908.

Addicott, J. F., J. Bronstein, and J. Kjellberg. 1990. Evolution of mutualistic life cycles: Yucca moths and fig wasps. Pp. 143–161 in *Genetics, evolution, and coordination of insect life cycles,* F. Gilbert, ed. Springer Verlag, London.

Advisory Commission on Intergovernmental Relations. 1968. *Urban and rural America: Policies for future growth.* U.S. Government Printing Office, Washington, D.C.

Ahearne, J.F. 1993. The future of nuclear power. *Am. Sci.* 81:24–35.

Aho, J. M. 1990. Helminth communities of amphibians and reptiles: Comparative approaches to understanding patterns and processes. Pp. 157–195 in *Parasite communities: Patterns and processes.* G. W. Esch, A. O. Bush, and J. M. Aho, eds. Chapman and Hall, New York.

Alatalo, R. V., L. Gustafsson, A. Lundberg, and S. Ulfstrand. 1985. Habitat shift of the willow tit *Parus montanus* in the absence of the marsh tit *Parus palustris. Ornis Scand.* 16:121–128.

Albrecht, L. 1991. Die Bedeutung des toten Holzes im Wald. *Forstw. Cbl.* 110:106–113.

Aldous, C. M. 1937. Notes on the life history of the snowshoe hare. *J. Mammal.* 18:46–57.

Aldrich, J. R., J. A. Bacone, and M. D. Hutchison. 1983. Limestone glades of Harrison County, Indiana. *Proc. Indiana Acad. Sci.* 91:480–485.

Aldrich, J. W. 1943. Biological survey of the bogs and swamps of northeastern Ohio. *Am. Midl. Nat.* 30: 346–402.

Alexander, R.R. 1988. Forest vegetation on national forests in the Rocky Mountain and intermountain regions: Habitat types and community types. *USDA For. Serv. Gen. Tech. Rep. RM*-162.

Allee, W. C. 1958. *The social life of animals.* Rev. ed. (originally published 1938). Beacon Press, Boston.

Allee, W. C., A. E. Emerson, O. Park, T. Park, and K. P. Schmidt. 1949. *Principles of animal ecology.* W. B. Saunders Co., Philadelphia.

Allen, D. L. 1942. Populations and habits of the fox squirrel in Allegan County, Michigan. *Am. Midl. Nat.* 27:338–379.

Allen, D. L. 1962. *Our wildlife legacy.* Rev. ed. Funk & Wagnalls, New York.

Allen, K. R. 1935. The food and migration of the perch *(Perca fluviatilis)* in Lake Windermere. *J. Anim. Ecol.* 4:199–273.

Amadon, D. 1973. Birds of the Congo and Amazon forests: A comparison. Pp. 267–277 in *Tropical forest ecosystems in Africa and South America: A comparative review,* B. J. Meggers, E. S. Ayensu, and W. D. Duckworth, eds. Smithsonian Institution Press, Washington, D.C.

Ambuel, B., and S. A. Temple. 1982. Songbird populations in southern Wisconsin forests: 1954 and 1979. *J. Field Ornithol.* 53:149–158.

Andreev, A. V. 1991. Winter adaptations in the willow ptarmigan. *Arctic* 44:106–114.

Anderson, D. W., and D. C. Coleman. 1985. The dynamics of organic matter in grassland soils. *J. Soil Water Conserv.* 40:211–216.

Anderson, D. W., et al. 1969. Significance of chlorinated hydrocarbon residues to breeding pelicans and cormorants. *Can. Field. Nat.* 83:91–112.

Anderson, F. K., and M. Treshow. 1984. Responses of lichens to atmospheric pollution. Pp. 259–290 in *Air pollution and plant life,* M. Treshow, ed. John Wiley & Sons, Chichester.

Anderson, J. G., D. W. Toohey, and W. H. Brune. 1991. Free radicals within the Antarctic vortex: The role of CFCs in Antarctic ozone loss. *Science* 251:39–52.

Anderson, M. C. 1966. Some problems of simple characterization of the light climate in plant communities. Pp. 77–90 in *Light as an ecological factor,* R. Bainbridge, G. C. Evans, and O. Rackham, eds. Blackwell Scientific, Oxford.

Anderson, R. C. 1983. The eastern prairie-forest transition—An overview. Pp. 86–92 in *Proceedings of the Eighth North American Prairie Conference,* R. Brewer, ed. Western Michigan Department of Biology, Kalamazoo.

Anderson, R. C., and A. J. Katz. 1993. Recovery of browse-sensitive tree species following release from white-tailed deer *Odocoileus virginianus* Zimmerman browsing pressure. *Biol. Conserv.* 63:203–208.

Anderson, R. M. 1982. Directly transmitted viral and bacterial infections of man. Pp. 1–37 in *The population dynamics of infectious diseases: Theory and applications,* R. M. Anderson, ed. Chapman and Hall, London.

Andrewartha, H. G., and L. C. Birch. 1954. *The distribution and abundance of animals.* University of Chicago Press, Chicago.

Andrews, R. V. 1968. Daily and seasonal variation in adrenal metabolism of the brown lemming. *Physiol. Zool.* 41:86–94.

Anonymous. 1980. Low salt diet prescribed for irrigation. *EPA Water Qual. Manage. Bull.* March 1980:11–13.

Anonymous. 1984. Mack Lake: Burning gets scorched. *J. For.* 82:85.

Anonymous. 1991. Saguaro expanded to aid vanishing cacti. *National Parks* 65 (September-October):11–12.

AOU Committee for the Conservation of the Red-cockaded Woodpecker. 1991. The red-cockaded woodpecker: On the road to oblivion? *Auk* 108:200–201.

Appalachian Mountain Club. 1979. *The A.M.C. White Mountain guide.* Appalachian Mountain Club, Boston.

Armstrong, D. M. 1972. Distribution of mammals in Colorado. *Univ. Kans. Mus. Nat. Hist. Monogr.* 3:1–415.

Arno, S. F. 1979. Forest regions of Montana. *USDA For. Serv. Res. Pap. INT* 218:1–39.

Arno, S. F. 1980. Forest fire history in the northern Rockies. *J. For.* 78:460–465.

Arno, S. F. 1984. *Timberline.* The Mountaineers, Seattle, Wash.

Arp, A. J., and J. J. Childress. 1983. Sulfide bonding by the blood of the hydrothermal vent tube worm *Riftia pachyptila. Science* 219:295–297.

Arrhenius, S. 1896. On the influence of carbonic acid in the air upon the temperature on the ground. *Philos. Mag.* 41:237.

Askins, R. A., J. F. Lynch, and R. Greenberg. 1990. Population declines in migratory birds in eastern North America. *Curr. Ornithol.* 7:1–57.

Assel, R. A. 1991. Implications of CO_2 global warming on Great Lakes ice cover. *Climatic Change* 18:377–395.

Atjay, G. L., P. Ketner, and P. Duvigneaud. 1979. Terrestrial primary production and phytomass. Pp. 129–182 in *The global carbon cycle,* B. Bolin, E. T. Degens, S. Kempe, and P. Ketner, eds. John Wiley and Sons, New York.

Audubon, J. J., and J. Bachman. 1989. *The quadrupeds of North America* (originally published 1854). Wellfleet Press, Secaucus, N.J.

Austin, J., N. Butchart, and K. P. Shine. 1992. Possibility

of an Arctic ozone hole in a doubled-CO_2 climate. *Nature* 360:221–225.

Axelrod, D. I. 1958. Evolution of the Madro-Teritiary geoflora. *Bot. Rev.* 24:433–509.

Ayensu, E. S., ed. 1980. *Jungles.* Crown, New York.

Baird, D. J., L. R. Linton, and R. W. Davies. 1986. Life-history evolution and post-reproductive mortality risk. *J. Anim. Ecol.* 55:295–302.

Baker, F. S. 1949. A revised tolerance table. *J. For.* 47: 179–181.

Baker, J. R. 1938. The evolution of breeding seasons. In *Evolution: Essays on aspects of evolutionary biology,* G. R. de Beer, ed. Oxford University Press, New York.

Baker, L. A. et al. 1991. Acidic lakes and streams in the United States: The role of acidic deposition. *Science* 252:1151–1154.

Baker, R. G. 1983. Holocene vegetational history of the western United States. In *Late-Quaternary environments of the United States,* Vol. 2, H. E. Wright, Jr., ed. University of Minnesota Press, Minneapolis.

Baker, R. H. 1971. Nutritional strategies of myomorph rodents in North American grasslands. *J. Mammal.* 52:800–805.

Balda, R. P. 1969. Foliage use by birds of the oak-juniper woodland and ponderosa pine forest in southeastern Arizona. *Condor* 71:399–412.

Ballegaard, T. K., and E. Warncke. 1985. The age distribution of a *Cirsium palustre* population in a spring area in Jutland, Denmark. *Holarctic Ecol.* 8:59–62.

Barbour, M. G., R. B. Craig, F. R. Drysdale, and M. T. Ghiselin. 1973. *Coastal ecology: Bodega Head.* University of California Press, Berkeley.

Barden, L. S. 1980. Tree replacement in a cove hardwood forest of the southern Appalachians. *Oikos* 35: 16–19.

Barger, R. L., and P. F. Folliott. 1972. Physical characteristics and utilization of major woodland tree species in Arizona. *USDA For. Serv. Res. Pap. RM* 83.

Barnes, D. M. 1987. AIDS: Statistics but few answers. *Science* 236:1423–1425.

Barnes, J. R., J. V. McArthur, and C. E. Cushing. 1986. Effects of excluding shredders on leaf litter decomposition in two streams. *Great Basin Nat.* 46:204–207.

Barnett, C. 1977. Aspects of chemical communication with special reference to fish. *Biosci. Commun.* 3: 331–392.

Barrett, S. W. 1980. Indians and fire. *West. Wildlands* Spring 1980:17–21.

Barry, R. G. 1981. *Mountain weather and climate.* Methuen, London.

Barry, R. G., and R. J. Chorley. 1987. *Atmosphere, weather and climate.* 5th ed. Routledge, London.

Barry, R. G., G. M. Courtin, and C. Labine. 1981. Tundra climates. Pp. 81–114 in *Tundra ecosystems,* L. C. Bliss, D. W. Heal, and J. J. Moore, eds. Cambridge University Press, Cambridge.

Barta, R. M., L. B. Keith, and S. M. Fitzgerald. 1989. Demography of sympatric arctic and snowshoe hare populations: An experimental assessment of interspecific competition. *Can. J. Zool.* 67:2762–2775.

Bartholomew, B. 1970. Bare zone between California scrub and grassland communities: The role of animals. *Science* 170:1210–1212.

Bartsch, I., and T. R. Moore. 1985. A preliminary investigation of primary production and decomposition in four peatlands near Schefflerville, Quebec. *Can. J. Bot.* 63:1241–1248.

Bates, H. W. 1862. Contributions to an insect fauna of the Amazon valley. Lepidoptera: Heliconidae. *Trans. Linn. Soc. Lon.* 23:495–515.

Bates, M. 1952. *Where winter never comes.* Scribner's, New York.

Batlerton, J. C., G. M. Bough, and F. Matsumura. 1972. DDT: Inhibition of sodium chloride tolerance by the blue-green alga *Anacystis nidulans. Science* 176: 1141–1143.

Batts, H. L., Jr. 1981. *The gypsy moth in Cooper Township.* Kalamazoo Nature Center, Kalamazoo, Mich.

Baudvin, H. 1979. Taille des pontes et reussite des nichées chez la Chouette Effraie, *Tyto alba* en Bourgogne. *Alauda* 47:13–16.

Baver, L. D. 1940. *Soil physics.* Wiley, New York.

Baxter, C. 1977. A comparison between grazed and ungrazed juniper woodland. Pp. 25–27 in *Ecology, uses, and management of pinyon-juniper woodlands. USDA For. Serv. Gen. Tech. Rep. RM* 39.

Baxter, R. 1984. Bethel, AK. *Am. Birds* 38:472.

Bazzaz, F. A. 1968. Succession on abandoned fields in the Shawnee Hills, southern Illinois. *Ecology* 49:924–936.

Beard, J. S. 1953. The savanna vegetation of northern tropical America. *Ecol. Monogr.* 23:149–214.

Beatley, J. C. 1969. Dependence of desert rodents on winter annuals and precipitation. *Ecology* 50:721–724.

Beatley, J. C. 1974. Phenological events and their environmental triggers in the northern Mojave Desert. *Ecology* 48:745–750.

Beatty, S. W., and E. L. Stone. 1985. The variety of soil microsites created by tree falls. *Can. J. For. Res.* 16: 539–548.

Beazley International Limited. 1981. *The international book of the forest.* Simon and Schuster, New York.

Beck, D. E. 1977. Twelve-year acorn yield in southern Appalachian oaks. *USDA For. Serv. Res. Note SE* 244: 1–8.

Beck, E., M. Senser, R. Scheibe, H. Steiger, and P. Pongratz. 1982. Frost avoidance and freezing tolerance in Afroalpine "giant rosette" plants. *Plant Cell Environ.* 5:215–222.

Beebe, W. 1921. *Edge of the jungle.* Duell, Sloan and Pearce, New York.

Beebe, W. 1945. *The book of naturalists.* A. A. Knopf, New York.

Begley, S., G. C. Lubenow, and M. Miller. 1986. Silent spring revisited? *Newsweek* July 14:72–73.

Beier, P. 1993. Determining minimum habitat areas and habitat corridors for cougars. *Conserv. Biol.* 7:94–107.

Beiswenger, J. M. 1991. Later-Quaternary vegetational history of Grays Lake, Idaho. *Ecol. Monogr.* 6:165–182.

Bejcek, V., and K. Stastny. 1984. The succession of bird communities on spoil banks after surface brown coal mining. *Ekol. Pol.* 32:245–259.

Belsky, A. J. 1986. Does herbivory benefit plants? A review of the evidence. *Am. Nat.* 127:870–892.

Bennett, B. C. 1992. Plants and people of the Amazonian rainforests. *BioScience* 42:599–607.

Benninghoff, W. S. 1963. The prairie as a filter barrier to postglacial plant migration. *Proc. Indiana Acad. Sci.* 72:116–124.

Bensch, S., and D. Hasselquist. 1991. Nest predation lowers the polygyny threshold: A new compensation model. *Am. Nat.* 138:1297–1306.

Bergerud, A. T. 1967. The distribution and abundance of Arctic hares in Newfoundland. *Can. Field Nat.* 81: 242–248.

Berkner, L. V., and L. C. Marshall. 1964. The history and growth of oxygen in the earth's atmosphere. Pp. 102–126 in *The origin and evolution of atmospheres and oceans,* D. J. Brancazio, ed. Wiley, New York.

Berndt, R., and W. Winkel. 1975. Gibt es beim Trauerschnäpper *Ficedula hypoleuca* eine Prägung auf den Biotop des Geburtsortes? *J. Ornithol.* 116:195–201.

Berryman, A. A. 1992, The origins and evolution of predator-prey theory, *Ecology* 73:1530–1535.

Bertram, B. C. R. 1980. Vigilance and group size in ostriches. *Anim. Behav.* 28:278–286.

Bierbaum, R. M., and S. Ferson. 1986. Do symbiotic pea crabs decrease growth rate in mussels? *Biol. Bull.* 170:51–61.

Billings, W. D. 1978. *Plants and the ecosystem,* 3rd ed. Wadsworth Pub. Co., Inc., Belmont, Calif.

Billings, W. D., and H. A. Mooney. 1968. The ecology of arctic and alpine plants. *Biol. Rev.* 43:481–529.

Binford, L. C., B. G. Elliott, and S. W. Singer. 1975. Discovery of a nest and the downy young of the marbled murrelet. *Wilson Bull.* 87:303–319.

Bingham, B. L., and C. M. Young. 1991. Influence of sponges on invertebrate recruitment: A field test of alleopathy. *Marine Biol.* 109:19–26.

Birch, L. C. 1953. Experimental background to the study of the distribution and abundance of insects. *Ecology* 34:698–711.

Bird, R. D. 1930. Biotic communities of the aspen parkland of central Canada. *Ecology* 11:356–442.

Bishop, C., and D. V. Weseloh. 1990. Contaminants in herring gull eggs from the Great Lakes. *State of Environment Canada Fact Sheet no.* 90-2:1–12.

Biswell, H. H., H. R. Kallander, R. Komarek, R. J. Vogl, and H. Weaver. 1973. *Ponderosa fire management: A task force evaluation of controlled burning in ponderosa pine forests.* Tall Timbers Research Station Miscellaneous Publication No. 2, Tallahassee, Fla.

Björkman, E. 1960. *Monotropa hypopitys* L.—An epiparasite on tree roots. *Physiol. Plant.* 13:308–327.

Blackman, F. F. 1905. Optima and limiting factors *Ann. Bot.* 19:281–295.

Blair, W. F. 1940. *A contribution to the ecology and faunal relationships of the mammals of the Davis Mountain region, southwestern Texas.* Miscellaneous Publication of the Museum of Zoology University of Michigan Publication No. 46:1–39.

Blake, I. H. 1945. An ecological reconnaissance in the Medicine Bow mountains. *Ecol. Monogr.* 15:208–242.

Bliss, L. C. 1963. Alpine plant communities of the Presidential Range, New Hampshire. *Ecology* 44:678–697.

Bliss, L. C. 1981. North American and Scandinavian tundras and polar deserts. Pp. 8–24 in *Tundra ecosystems,* L. C. Bliss, O. W. Heal, and J. J. Moore, eds. Cambridge University Press, Cambridge.

Block, W. 1990. Cold tolerance of insects and other arthropods. *Philos. Trans. R. Soc. London [Biol.]* 326: 613–633.

Blouin, M. S., and E. F. Connor. 1985. Is there a best shape for nature reserves? *Biol. Conserv.* 32:277–278.

Boelter, D. H., and E. S. Verry. 1977. Peatland and water in the northern lake states. *USDA For. Serv. Gen. Tech. Rep. NC* 31:1–22.

Boesch, D. F., and R. E. Turner. 1984. Dependence of fishery species on salt marshes: The role of food and refuge. *Estuaries* 7:460–468.

Bolin, B. 1977. Changes of the land biota and their importance for the carbon cycle. *Science* 196:613–615.

Bond, R. R. 1957. Ecological distribution of breeding birds in the upland forests of southern Wisconsin. *Ecol. Monogr.* 27:351–384.

Bonner, J. 1962. The upper limit of crop yield. *Science* 137:11–15.

Booth, D. E. 1991. Estimating prelogging old-growth in the Pacific Northwest. *J. For.* 89:25–29.

Borchard, M. 1984. *Acid rain in Minnesota.* Minnesota

Department of Education and Minnesota Pollution Control Agency.

Borgstrom, G. 1973. *The food and people dilemma.* Duxbury Press, North Scituate, Mass.

Bormann, F. H., and G. E. Likens. 1967. Nutrient cycling. *Science* 155:424–429.

Bormann, F. H., and G. E. Likens. 1979. *Pattern and process in a forested ecosystem.* Springer-Verlag, New York.

Bouchard, R. J., Jr., D. T. Lykken, M. McGue, N. L. Segal, and A. Tellegen. 1990a. Sources of human psychological differences: The Minnesota study of twins reared apart. *Science* 250:223–228.

Bouchard, T. J., Jr., D. T. Lykken, M. McGue, N. L. Segal, and A. Tellegen. 1990b. When kin correlations are not squared. *Science* 250:1498.

Boucher, D. H. 1990. Growing back after hurricanes. *BioScience* 40:163–166.

Boucher, N. 1991. Smart as gods. *Wilderness* 55:11–21.

Bourlière, F. 1983. Animal species diversity in tropical forests. Pp. 77–91 in *Tropical rain forest ecosystems,* F. B. Golley, ed. Elsevier, Amsterdam.

Bourlière, F., ed. 1983. *Ecosystems of the world: Tropical savannas.* Elsevier Scientific Publishing Co., New York.

Boyer, W. D. 1978. Heat accumulation: An easy way to anticipate the flowering of southern pines. J. For. 76:20–23.

Braun, E. L. 1950. *Deciduous forests of eastern North America* (reprinted 1964). Hafner Pub. Co., New York.

Braun, E. L. 1951. Plant distribution in relation to the glacial boundary. *Ohio J. Sci.* 51:139–146.

Braun-Blanquet, J. 1932. *Plant sociology: The study of communities* [translation of Pflanzensoziologie, 1928]. McGraw-Hill, New York.

Bray, O. E., J. J. Kennelly, and J. L. Guarino. 1975. Fertility of eggs produced on territories of vasectomized red-winged blackbirds. *Wilson Bull.* 87:187–195.

Brett, J. R. 1956. Some principles in the thermal requirements of fishes. *Q. Rev. Biol.* 31:75.

Brewer, R. 1960. A brief history of ecology. Part I—Pre-nineteenth century to 1919. *Occas. Pap. C. C. Adams Cent. Ecol. Stud. West. Mich. Univ.* 1:1–18.

Brewer, R. 1963. Ecological and reproductive relationships of black-capped and Carolina chickadees. *Auk* 80:11.

Brewer, R. 1963. Stability in bird populations. *Occas. Pap. C.C. Adams Cent. Ecol. Stud. West. Mich. Univ.* No. 7:1–12.

Brewer, R. 1966. Vegetation of two bogs in southwest Michigan. *Mich. Bot.* 5:36–46.

Brewer, R. 1967. Bird populations of bogs. *Wilson Bull.* 79:371–396.

Brewer, R. [1980.] *Vegetation of southwestern Michigan* (map). Western Michigan University, Department of Biology, Kalamazoo.

Brewer, R. 1980. A half-century of changes in the herb layer of a climax deciduous forest in Michigan. *J. Ecol.* 68:823–832.

Brewer, R. 1981. The changing seasons. *Am. Birds* 35: 915–919.

Brewer, R. 1985. Seasonal change and production in a mesic prairie relict in Kalamazoo County, Michigan. *Mich. Bot.* 24:3–14.

Brewer, R. 1991. Original avifauna and postsettlement changes. Pp. 33–58 in *The atlas of breeding birds of Michigan,* R. Brewer, G. A. McPeek, and R. J. Adams, Jr., eds. Michigan State University Press, East Lansing.

Brewer, R. 1992. *Management recommendations for the Allegan State Game Area.* Western Michigan University Department of Biological Sciences, Kalamazoo.

Brewer, R. 1992. A deficiency of credulousness. *BioScience* 42:123–124.

Brewer, R., et al. 1992. *Trends in Kalamazoo County natural areas, 1967–1987.* Unpublished manuscript.

Brewer, R., A. Raim, and J. D. Robins. 1969. Vegetation of a Michigan grassland and thicket. *Occas. Pap. C.C. Adams Cent. Ecol. Stud. West Mich. Univ.* No. 18: 1–29.

Brewer, R., D. A. Boyce, J. R. Hodgson, J. O. Wenger, M. H. Mills, and M. M. Cooper. 1973. Composition of some oak forests in southwest Michigan. *Mich. Bot.* 12:217–234.

Brewer, R., and G. A. McPeek. 1991. Priorities and perspectives in conservation. Pp. 87–94 in *The atlas of breeding birds of Michigan,* R. Brewer, et al., eds. Michigan State University Press, East Lansing.

Brewer, R., and L. Swander. 1977. Life history factors affecting the intrinsic rate of natural increase of birds of the deciduous forest biome. *Wilson Bull.* 89:211–232.

Brewer, R., and M. T. McCann. 1982. *Laboratory and field manual of ecology.* Saunders College Publishing, Philadelphia.

Brewer, R., and P. G. Merritt. 1978. Wind throw and tree replacement in a climax beech-maple forest. *Oikos* 30:149–152.

Brewer, R., and S. Kitler. 1989. Tree distribution in southwestern Michigan bur oak openings. *Mich. Bot.* 28:73–79.

Briand, F., and J. E. Cohen. 1987. Environmental correlates of food chain length. *Science* 238:956–960.

Briggs, L. J., and H. L. Shantz. 1912. The relative wilting coefficient for different plants. *Bot. Gaz.* 53:229–235.

Britton, J. C., and B. Morton. 1989. *Shore ecology and the Gulf of Mexico.* University of Texas Press, Austin.

Brodie, E. D. 1980. Phenologic relationships of model and mimic salamanders. *Evolution* 35:988–994.

Bromfield, L. 1948. *Malabar Farm.* Harper & Brothers, New York.

Bronstein, J. L. 1991. The nonpollinating wasp fauna of *Ficus pertusa:* Exploitation of a mutualism? *Oikos* 61: 175–186.

Brookmeyer, R. 1991. Reconstruction and future trends of the AIDS epidemic in the United States. *Science* 253:37–42.

Brower, J. V. Z. 1958. Experimental studies of mimicry in some North American butterflies. Part I. The Monarch, *Danaus plexippus,* and the Viceroy, *Limenitis archippus archippus. Evolution* 12:32–47.

Brower, J. V. Z. 1960. Experimental studies of mimicry: Part IV. *Am. Nat.* 96:297–308.

Brower, L. P. 1969. Ecological chemistry. *Sci. Am.* 220: 22–29.

Brower, L. P., and J. V. Z. Brower. 1964. Birds, butterflies and plant poisons, a study in ecological chemistry. *Zoologica* 49:137–159.

Brown, B. W., and G. O. Batzli. 1984. Field manipulations of fox and gray squirrel populations: How important is interspecific competition? *Can. J. Zool.* 63:2134–2140.

Brown, C. R. 1986a. Cliff swallow colonies as information centers. *Science* 234:83–85.

Brown, F. B. 1986b. The evolution of Darwin's theism. *J. Hist. Biol.* 19:1–45.

Brown, J., P. C. Miller, L. L. Tieszen, and F. L. Bunnell. 1980. *An arctic ecosystem: The coastal tundra at Barrow, Alaska.* Dowden, Hutchinson, & Ross, Inc., Stroudsburg, Penn.

Brown, J. H., and A. Kodric-Brown. 1977. Turnover rate in insular biogeography: Effect of immigration on extinction. *Ecology* 58:445–449.

Brown, J. H., and D. W. Davidson. 1986. Reply to Galindo. *Ecology* 67:1423–1425.

Brown, J. H., D. W. Davidson, and O. J. Reichman. 1979. An experimental study of competition between seed-eating desert rodents and ants. *Am. Zool.* 19:1129–1143.

Brown, J. L. 1964. Territorial behavior and population regulation of birds. *Wilson Bull.* 81:293–329.

Brown, L. 1972. *The life of the African plains.* McGraw-Hill, New York.

Brown, M., and J. May. 1991. *The Greenpeace story.* Dorling-Kindersley, New York.

Brown, R. B., E. L. Stone, and V. W. Carlisle. 1990. Soils. Pp. 35–69 in *Ecosystems of Florida,* R. L. Myers and J. J. Ewel, eds. University of Central Florida Press, Orlando.

Brown, S., and A. E. Lugo. 1990. Tropical secondary forests. *J. Trop. Ecol.* 6:1–32.

Brown, S., and A. E. Lugo. 1992. Aboveground biomass estimates for tropical moist forests of the Brazilian Amazon. *Interciencia* 17:8–18.

Brown, W. L., Jr., and E. O. Wilson. 1956. Character displacement. *Syst. Zool.* 5:49–64.

Bryan, W. C., and P. P. Kormanik. 1977. Mycorrhizae benefit survival and growth of sweetgum seedlings in the nursery. *South. J. Appl. For.* 1:21–23.

Brylinsky, M. 1980. Estimating the productivity of lakes and reservoirs. Pp. 411–453 in *The functioning of freshwater ecosystems,* E. D. Le Cren and R. H. Cowe-McConnell. Cambridge University Press, Cambridge.

Bryson, R. A., and T. J. Murray. 1977. *Climates of hunger.* University of Wisconsin Press, Madison.

Buech, R. R., D. J. Rugg, and N. L. Miller. 1989. Temperature in beaver lodges and bank dens in a near-boreal environment. *Can. J. Zool.* 67:1061–1066.

Buechner, H. K., and F. B. Golley. 1967. Preliminary estimation of energy flow in Uganda kob. Terrestrial. Pp. 243–254 in *Secondary productivity of ecosystems,* K. Petrusewicz, ed. Polish Academy of Science.

Buechner, M. 1987. Conservation in insular parks: Simulation models of factors affecting the movement of animals across park boundaries. *Biol. Conserv.* 41: 57–76.

Bugbee, B., and O. Monje. 1992. The limits of crop productivity. *BioScience* 42:494–502.

Bull, E. L., and H. D. Cooper. 1991. Vaux's swift nests in hollow trees. *Western Birds* 22:85–91.

Bull, E. L., and R. S. Holthausen. 1993. Habitat use and management of pileated woodpeckers in northeastern Oregon. *J. Wildl. Manage.* 57:335–345.

Bünning, E. 1967. *The physiological clock.* 2nd ed., rev. Springer-Verlag, Berlin.

Burroughs, R. D. 1961. *The natural history of the Lewis and Clark expedition.* Michigan State University Press, East Lansing.

Burton, M. 1969. *Animal partnerships.* Fredrick Warne & Co. Inc., New York.

Busnel, R. G., and J. F. Fish, eds. 1979. *Animal sonar systems.* NATO Advanced Studies Institute Series, Series A.

Buss, I. O. 1956. Plant succession on a sand plain, northwest Wisconsin. *Trans. Wisconsin Acad. Sci. Arts Lett.* 45:11–19.

Butler, L. 1953. The nature of cycles in populations of Canadian mammals. *Can. J. Zool.* 31:242–262.

Cable, D. R. 1975. Range management in the chaparral type and its ecological basis. *USDA For. Serv. Res. Pap. RM* 155:1–30.

Cain, M. D., and D. A. Yaussey. 1983. Reinvasion of hardwoods following eradication in an uneven-aged pine stand. *USDA For. Serv. Res. Pap. SO* 188.

Cairns, J., Jr., D. W. Albaugh, F. Busey, and M. D. Chanay. 1968. The sequential comparison index—A simplified method for non-biologists to estimate relative differences in biological diversity in stream pollution studies. *J. Water Pollut. Control Fed.* 40: 1607–1613.

Cairns, J., Jr., J. R. Pratt, and B. R. Niederlehner. 1985. A provisional multispecies toxicity test using indigenous organisms. *J. Test. Eval.* 13:316–319.

Calhoun, J. B. 1963. The social use of space. Pp. 1–187 in *Physiological mammalogy,* Vol. 1; W. Mayer and R. van Gelder, eds. Academic Press, New York.

Canfield, D. E., and D. J. Des Marais. 1991. Aerobic sulfate reduction in microbial mats. *Science* 251:1471–1473.

Cannell, M. G. R. 1982. *World forest biomass and primary production data.* Academic Press, New York.

Carey, A., and S. Carey. 1989. *Yellowstone's red summer.* Northland Publishing, Flagstaff, Ariz.

Carey, A. B. 1991. The biology of arboreal rodents in Douglas-fir forests. *USDA For. Serv. Gen. Tech. Rep. PNW-GTR* 276.

Carey, A. B., J. A. Reid, and S. P. Horton. 1991. Spotted owl home range and habitat use in southern Oregon coast ranges. *J. Wildl. Manage.* 54:11–17.

Carey, F. G. 1990. Further observations on the biology of the swordfish. Pp. 103–122 in *Planning the future of billfishes,* R. H. Stroud, ed. National Coalition for Marine Conservation, Savannah, Ga.

Carey, J. 1991. Hot science in cold lands. *Natl. Wildl.* 29: 5–12.

Carl, E. A. 1971. Population control in arctic ground squirrels. *Ecology* 52:395–413.

Carlquist, S. 1974. *Island biology.* Columbia University Press, New York.

Carmichael, R. S., O. D. Knipe, C. P. Pase, and W. W. Brady. 1978. Arizona chaparral: Plant associations and ecology. *USDA For. Serv. Res. Pap. RM* 202.

Carnes, B. A., and N. A. Slade. 1982. Some comments on niche analysis in canonical space. *Ecology* 63: 888–893.

Carpenter, S. R., J. F. Kitchell, and J. R. Hodgson, 1985. Cascading trophic interactions and lake productivity. *BioScience* 35:634–639.

Carpenter, S. R., J. R. Kitchell, J. R. Hodgson, P. A. Cochran, J. J. Elser, M. M. Elser, D. M. Lodge, D. Kretchmer, X. He., and C. N. von Ende. 1987. Regulation of lake primary productivity by food web structure. *Ecology* 68:1863–1876.

Carson, R. 1955. *The edge of the sea.* Houghton Mifflin, Boston.

Carson, R. 1962. *Silent spring.* Houghton Mifflin, Boston.

Case, T. J., and E. A. Bender. 1981. Testing for higher-order interactions. *Am. Nat.* 113:705–714.

Castellano, M. A., and J. M. Trapp. 1985. Mycorrhizal associations of five species of Monotropoideae in Oregon. *Mycologia* 77:499–502.

Catts, E. P., and M. L. Goff. 1992. Forensic entomology in criminal investigations. *Annu. Rev. Entomol.* 37: 253–272.

Chaney, R. W. 1947. Tertiary centers and migration routes. *Ecol. Monogr.* 17:139–149.

Chaney, R. W. 1948. The bearing of the living Metasequoia on problems of Tertiary paleobotany. *Proc. Natl. Acad. Sci. USA* 34:503–515.

Chapman, R. N. 1931. *Animal ecology.* McGraw-Hill, New York.

Chapman, V. J. 1970. Mangrove phytosociology. *Trop. Ecol.* 2:1–24.

Charnov, E. L. 1976. Optimal foraging, the marginal value theorem. *Theor. Popul. Biol.* 9:129–136.

Charnov, E. L., and W. M. Schaffer. 1973. Life-history consequences of natural selection: Cole's result revisited. *Am. Nat.* 107:791–793.

Chasan, R. 1991. Molecular biology and ecology: A marriage of more than convenience. *Plant Cell* 3:1143–1145.

Chazdon, R. L., and R. W. Pearcy. 1991. The importance of sunflecks for forest understory plants. *BioScience* 41:760–766.

Chelazzi, G., and R. Calzolai. 1986. Thermal benefits from familiarity with the environment in a reptile. *Oecologia (Berl)* 68:557–558.

Cherfas, J. 1989. Hope for AIDS vaccines. *Science* 246: 23–24.

Chikishev, A. G., ed. 1965. *Plant indicators of soils, rocks, and subsurface waters.* Consultants Bureau Enterprises, Inc., New York.

Chitty, D. 1960. Population processes in the vole and their relevance to general theory. *Can. J. Zool.* 38: 99–113.

Chitty, D. 1967. The natural selection of self-regulatory behavior in animal populations. *Proc. Ecol. Soc. Aust.* 2:51–78.

Chown, G. A. 1991. Protecting natural land and community character with conservation easements. *Planning Zoning News* 9:5–10.

Christensen, M. 1989. A view of fungal ecology. *Mycologia* 81:1–19.

Christensen, N. L. 1989. Landscape history and ecological change. *J. For. Hist.* 33:116–124.

Christian, J. J. 1963a. The pathology of overpopulation. *Military Med.* 128:571–603.

Christian, J. J. 1963b. Endocrine adaptive mechanisms and the physiologic regulation of population growth. In *Physiological mammalogy*. Vol. 1. W. Mayer and R. van Gelder, eds. Academic Press, New York.

Christian, J. J., and D. E. Davis. 1964. Endocrines, behavior, and population. *Science* 146:1550–1560.

Church, G. J. 1989. The big spill. *Time* April 10:38–41.

Clark, L., and C. A. Smeraski. 1990. Seasonal shifts in odor acuity by starlings. *J. Exp. Zool.* 255:22–29.

Clark, L., and J. R. Mason. 1988. Effect of biologically active plants used as nest material and the derived benefit to starling nestlings. *Oecologia (Berl)* 77:174–180.

Clark, T. W. 1987. Black-footed ferret recovery: A progress report. *Conserv. Biol.* 1:8–10.

Clarke, G. C. 1954. *Elements of ecology*. Wiley, New York.

Clayton, J. L. undated. Nutrient supply to soil by rock weathering Pp. 75–96 in *Proceedings of the Impact of Intensive Harvesting on Forest Nutrient Cycling.* USDA Forest Service, Washington, D.C.

Cleland, C. E. 1966. The prehistoric animal ecology and ethnozoology of the Upper Great Lake Region. *Univ. Mich. Mus. Anthropol. Pap.* 29:1–294.

Clements, F. E. 1916. Plant succession. *Carnegie Inst. Wash. Publ.* 242:512.

Clements, F. E. 1920. Plant indicators. *Carnegie Inst. Wash. Publ.* 290.

Clements, F. E. 1936. Nature and structure of the climax. *J. Ecol.* 24:252–284.

Clements, F. E., and V. E. Shelford. 1939. *Bio-ecology*. McGraw-Hill, New York.

Clements, F. E., J. E. Weaver, and H. C. Hanson. 1929. Plant competition. *Carnegie Inst. Wash. Publ.* 398.

Clewell, A. F. 1985. Guide to the vascular plants of the Florida panhandle. University Presses of Florida. Tallahassee.

Cloudsley-Thompson, J. L. 1965. *Desert life*. Pergamon Press, Oxford.

Coale, A. J. 1974. The history of the human population. *Sci. Am.* 231:41–51.

Cochran, W., S. Wood, A. Raim, and A. in't Veld. 1992. A peregrine tale: Part two. *Illinois Nat. Hist. Surv. Rep. No.* 313:4–5.

Coggon, D. 1987. Are pesticides carcinogenic? *Br. Med. J.* 294:725.

Cohen, J. E. 1976. Schistosomiasis: A human host-parasite system. Pp. 237–256 in *Theoretical ecology, principles and applications,* R. M. May, ed. W. B. Saunders, Philadelphia.

Cohn, J. P. 1989. Gauging the biological impacts of the greenhouse effect. *BioScience* 39:142–146.

Cohn, J. P. 1991. Ferrets return from near-extinction. *BioScience* 41:132–135.

Cole, D. W., and M. Rapp. 1981. Elemental cycling in forest ecosystems. Pp. 341–410 in *Dynamic properties of forest ecosystems,* D. E. Reichle, ed. Cambridge University Press, Cambridge.

Cole, G. A. 1979. *Textbook of limnology*. 2nd ed. The C. V. Mosby Co., St. Louis.

Cole, L. C. 1951. Population cycles and random oscillations. *J. Wildl. Manage.* 15:233–252.

Cole, L. C. 1954. The population consequences of life history phenomena. *Q. Rev. Biol.* 29:103–137.

Coley, P. D., J. P. Bryant, and F. S. Chapin, III. 1985. Resource availability and plant antiherbivore defense. *Science* 230:895–899.

Colinvaux, P. 1986. *Ecology*. Wiley, New York.

Colwell, R. K. 1982. Community biology and sexual selection: Lessons from hummingbird flower mites. Chap. 24 in *Community ecology*, J. Diamond and T. J. Case, eds. Harper and Row, New York.

Colwell, R. K., and D. W. Winkler. 1984. A null model for null models in biogeography. Pp. 344–359 in *Ecological communities,* D. R. Strong, Jr., et al., eds. Princeton University Press, Princeton, N.J.

Comar, C. L. 1962. Biological aspects of nuclear weapons. *Am. Sci.,* 50:339–353.

Commoner, B. 1971. *The closing circle.* A. A. Knopf, New York.

Connell, J. H. 1961. Effects of competition, predation by *Thais lapillus,* and other factors on natural populations of the barnacle *Balanus balanoides. Ecol. Monogr.* 40:49–78.

Connell, J. H. 1978. Diversity in tropical rain forests and coral reefs. *Science* 199:1302–1310.

Connell, J. H. 1980. Diversity and the coevolution of competitors, or the ghost of competition past. *Oikos* 35:131–138.

Connell, J. H. 1983. On the prevalence and relative importance of interspecific competition: Evidence from field experiments. *Am. Nat.* 122:661–696.

Connell, J. H., and R. O. Slatyer. 1977. Mechanisms of succession in natural communities and their role in community stability and organization. *Am. Nat.* 111: 1119–1144.

Conner, R. N., and D. C. Rudolph. 1989. Red-cockaded woodpecker colony status and trends on the Angelina, Davy Crockett and Savine National Forests. *USDA For. Serv. Res. Pap. SO* 250:1–15.

Connor, E. F., and D. Simberloff. 1979. The assembly of species communities: Chance or competition. *Ecology* 60:1132–1140.

Connor, E. F., and D. Simberloff. 1984. Neutral models of species co-occurrence patterns. Pp. 316–331 in *Ecological communities: Conceptual issues and the evidence,* D. R. Strong, et al., eds. Princeton University Press, Princeton, N.J.

Conover, D. O., and D. A. Van Vorhees. 1991. Evolution of a balanced sex ratio by frequency-dependent selection in a fish. *Science* 250:1556–1558.

Cook, F. R. 1984. Introduction to Canadian amphibians and reptiles. *Natl. Mus. Nat. Sci. (Ottawa) Publ. Zool.*

Cook, R. E. 1977. Raymond Lindeman and the trophic-dynamic concept in ecology. *Science* 198:22–26.

Coombe, D. E. 1957. The spectral composition of shade light in woodlands. *J. Ecol.* 45:823–830.

Cooper, C. F. 1961. The ecology of fire. *Sci. Am.* 204: 150–160.

Cooper, S. M., and N. Owen-Smith. 1985. Condensed tannins deter feeding by browsing ruminants in a South Africa savanna. *Oecologia (Berl)* 67:142–146.

Cooper, S. V., K. E. Neiman, and D. W. Roberts. 1991. Forest habitat types of northern Idaho: A second approximation. *USDA For. Serv. Gen. Tech. Rep.* INT-236.

Cooper, W. E. 1990. Chemical detection of predators by a lizard, the broad-headed skink (*Eumeces laticeps*). *J. Exp. Zool.* 256:162–167.

Cooper, W. S. 1913. The climax forest of Isle Royale, Lake Superior, and its development. I, II, and III. *Bot. Gaz.* 55:1–44, 115–140, 189–235.

Corliss, J. B., et al. 1979. Submarine thermal springs on the Galápagos rift. *Science* 203:1073–1083.

Cornejo, D. 1987. For the spadefoot toad, rain starts a race to metamorphosis. *Smithsonian* 17:98–105.

Cossins, A. R., and K. Bowler. 1987. *Temperature biology of animals.* Chapman and Hall, London.

Cottle, H. J. 1931. Studies in the vegetation of southwestern Texas. *Ecology* 12:105–155.

Couch, N. P., N. L. Tilney, A. A. Rayner, and F. D. Moore. 1981. The high cost of low-frequency events. *N. Engl. J. Med.* 304:634–637.

Coues, E., ed. 1965 (originally published 1893). *History of the expedition under the command of Lewis and Clark.* Dover, New York.

Coughenour, M. B., et al. 1985. Energy extraction and use in a nomadic pastoral ecosystem. *Science* 230: 619–625.

Coulson, R. N., and J. A. Witter. 1984. *Forest entomology.* Wiley, New York.

Countryman, C. M. 1974. Can southern California wildland conflagrations be stopped? *US For. Serv. Gen. Tech. Rep. PSW* 7.

Courtney, S. P., and T. T. Kibota. 1989. Mother doesn't always know best: Selection of hosts by ovipositing insects. Pp. 161–188 in *Insect-plant interactions.* Vol. 2. E. A. Bernays, ed. CRC Press, Boca Raton, Fla.

Cowardin, L. M., V. Carter, F. C. Golet, and E. T. LaRoe. 1979. *Classification of wetlands and deepwater habitats of the United States.* US Dep. Inter. Fish and Wildlf. Serv., Washington, D.C.

Cowles, H. C. 1899. The ecological relations of the vegetation of the sand dunes of Lake Michigan. *Bot. Gaz.* 27:95–117, 167–202, 281–308, 361–391.

Cowles, H. C. 1911. The causes of vegetative cycles. *Bot. Gaz.* 51:161–183.

Cowles, R. B., and C. M. Bogert. 1944. A preliminary study of the thermal requirements of desert reptiles. *Bull. Am. Mus. Nat. Hist.* 83:261–296.

Cox, G. W. 1984a. The distribution and origin of mima mound grasslands in San Diego County, California. *Ecology* 65:1397–1405.

Cox, G. W. 1984b. The contribution of external inputs and intrinsic state conditions to agroecosystem productivity. *Options méditerranéennes* 147–160.

Cox, G. W., and C. G. Gakahu. 1983. Mima mounds in the Kenya highlands: Significance for the Dalquest-Scheffer hypothesis. *Oecologia (Berl)* 57:170–174.

Cox, G. W., and M. D. Atkins. 1979. *Agricultural ecology.* Freeman, San Francisco.

Cox, G. W., and R. E. Ricklefs. 1977. Species diversity and ecological release in Caribbean land bird faunas. *Oikos* 28:113–122.

Craig, W. 1908. North Dakota life: Plant, animal, and human. *Bull. Am. Geogr. Soc.* 40:321–415.

Craighead, F. C., Jr. 1979. *Track of the grizzly.* Sierra Club Books, San Francisco.

Craighead, J., J. R. Varney, and F. C. Craighead, Jr. 1974. A population analysis of the Yellowstone grizzly bears. *Bull. Mont. St. Univ. For. Conserv.* 40:1–20.

Crawford, H. S. 1985. Effects of silvicultural practice on bird predation. Pp. 173–175 in *Spruce-fir management and spruce budworm. USDA For. Serv. Gen. Tech. Rep. NE* 99.

Crawford, M. 1987. California field test goes forward. *Science* 236:511.

Crawley, M. J. 1987. Benevolent herbivores? *Trends Ecol. Evol.* 2:167–168.

Critchfield, W. B. 1980. Origins of the eastern deciduous forest. Pp. 1–14 in *Proc. dendrology in the eastern deciduous forest biome.* V. Polytechn. Inst. State Univ. Res. Publ. FWS 2-80.

Croll, N. A. 1973. *Parasitism and other associations.* Pitman Medical, London.

Crossland, C. J., B. G. Hatcher, and S. V. Smith. 1991. Role of coral reefs in global ocean production. *Coral Reefs* 10:55–64.

Culver, A., and R. M. Audette. 1985. Danger's in the well. *Environ. Action,* March/April, pp. 15–19.

Culver, D. C. 1982. *Cave life: Evolution and ecology.* Harvard University Press, Cambridge, Mass.

Cummins, K. W. 1974. Structure and function of stream ecosystems. *BioScience* 24:631–641.

Cummins, K. W. 1977. From headwaters to rivers. *Am. Biol. Teach.* 39:305–312.

Cummins, K. W., and J. C. Wuycheck. 1971. Caloric equivalents for investigations in ecological energetics. *Int. Assoc. Theor. Appl. Limnol.* 18:1–158.

Cundall, A. W., and H. T. Lystrup. 1987. *Hamilton's guide to Yellowstone National Park.* Hamilton Stores, Inc., West Yellowstone, Mont.

Cunningham, W. J. 1954. A nonlinear differential-difference equation of growth. *Proc. Natl. Acad. Sci. USA* 40:708–713.

Currie, D. J. 1991. Energy and large-scale patterns of animal- and plant-species richness. *Am. Nat.* 137:27–49.

Currie, D. J., and V. Paquin. 1987. Large-scale biogeographical patterns of species richness in trees. *Nature* 329:326–327.

Curtis, J. T. 1959. *The vegetation of Wisconsin.* University of Wisconsin Press, Madison.

Dahlberg, B. L., and R. C. Guettinger. 1956. The white-tailed deer in Wisconsin. *Wis. Conserv. Dep. Tech. Bull.* 14:282.

Dahlberg, K. A. 1979. *Beyond the green revolution: The ecology and politics of global agricultural development.* Plenum Press, New York.

Dahlberg, K. A. 1991. Sustainable agriculture—Fad or harbinger? *Bioscience* 41:337–340.

Dansereau, P. 1957. *Biogeography: An ecological perspective.* Ronald, New York.

Darby, H. C. 1956. *The draining of the fens.* 2nd ed. Cambridge University Press, Cambridge.

Darnell, R. M. 1964. Organic detritus in relation to secondary production in aquatic communities. *Verh. Int. Ver. Limnol.* 15:462–470.

Darwin, C. 1859. *On the origin of species by natural selection.* Murray, London.

Darwin, C. 1881. *The formation of vegetable mould through the action of worms* (1896 reprint). D. Appleton and Co., New York.

Daubenmire, R. F. 1947. *Plants and environment* (2nd ed., 1959). Wiley, New York.

Daubenmire, R. F. 1968. Ecology of fire in grasslands. *Adv. Ecol. Res.* 5:209–266.

Davidson, D. W., J. H. Brown, and R. S. Inouye. 1980. Competition and the structure of granivore communities. *BioScience* 30:233–238.

Davidson, D. W., R. S. Inouye, and J. H. Brown. 1984. Granivory in a desert ecosystem: Experimental evidence for indirect facilitation of ants by rodents. *Ecology* 65:1780–1786.

Davidson, J., and H. G. Andrewartha. 1948a. Annual trends in a natural population of *Thrips imaginis* (Thysanoptera). *J. Anim. Ecol.* 17:193–199.

Davidson, J., and H. G. Andrewartha. 1948b. The influence of rainfall evaporation and atmospheric temperature on fluctuations in the size of a natural population of *Thrips imaginis* (Thysanoptera). *J. Anim. Ecol.* 17:200–222.

Davies, L. 1987. Long adult life, low reproduction and competition in two sub-Antarctic carabid beetles. *Ecol. Entomol.* 12:149–162.

Davis, D. W., C. A. Engelkes, and J. V. Groth. 1990. Erosion of resistance to common leaf rust in exotic-derived maize during selection for other traits. *Phytopathology* 80:339–342.

Davis, J. H., Jr. 1940. The ecology and geologic role of mangroves in Florida. *Carnegie Inst. Wash. Publ.* 517:303–314.

Davis, J. W. 1983a. Snags are for wildlife. Pp. 4–9 in Snag habitat management: Proceedings of the symposium. *USDA For. Serv. Gen. Tech. Rep. RM* 99.

Davis, M. B. 1976. Pleistocene biogeography of temperate deciduous forests. *Geosc. Man* 13:13–26.

Davis, M. B. 1983b. Holocene vegetational history of the eastern United States. In *Late-Quaternary environments of the United States,* Vol. 2, H. E. Wright, Jr., ed. University of Minnesota Press, Minneapolis.

Dawkins, R. 1976. *The selfish gene.* Oxford University Press, New York.

Day, J. C. 1992. *Population projections of the United States, by age, sex, race, and hispanic origin: 1992–2050.* U.S. Bureau of Census Current Population Report P25-1092.

Dayan, T., D. Simberloff, E. Tchernov, and Y. Yom-Tov. 1989. Inter- and intraspecific character displacement in mustelids. *Ecology* 70:1526–1539.

Dean, J., D. J. Aneshansley, H. E. Edgerton, and T. Eisner. 1990. Defensive spray of the bombardier beetle: A biological pulse jet. *Science* 248:1219–1221.

DeBach, P. 1964. *Biological control of insect pests and weeds.* Reinhold, New York.

DeBach, P., and R. A. Sundby. 1963. Competitive displacement between ecological homologues. *Hilgardia* 34:105–166.

DeBano, L. F., R. M. Rice, and C. E. Conrad. 1979. Soil heating in chaparral fires: Effects on soil properties, plant nutrients, erosion, and runoff. *USDA For. Serv. Res. Pap. PSW* 145.

DeBell, D. S., C. D. Whitesell, and T. H. Schubert. 1985. Mixed plantations of Eucalyptus and leguminous trees enhance biomass production. *USDA For. Serv. Res. Pap. PSW* 175.

Deevey, E. S., Jr. 1947. Life tables for natural populations of animals. *Q. Rev. Biol.* 22:283–314.

De Jong, G. 1976. A model of competition for food. I. *Am. Nat.* 110:1013–1027.

Dennis, B., P. L. Munholland, and J. M. Scott. 1991. Estimation of growth and extinction parameters for endangered species. *Ecol. Monogr.* 61:115–143.

Department of Energy. 1986. *Annual energy review.* Energy Information Administration, Washington, D.C.

DeVooys, C. G. N. 1979. Primary production in aquatic systems. Pp. 259–292 in *The global carbon cycle,* B. Bolin, E. T. Degens, S. Kempe, and P. Ketner, eds. John Wiley and Sons, New York.

DeVos, A. 1964. Recent changes in the ranges of mammals in the Great Lakes region. *Am. Midl. Nat.* 71: 210–231.

Diamond, J., and T. J. Case. 1986. Overview: Introductions, extinctions, exterminations, and invasions. Pp. 65–79 in *Community ecology*, J. Diamond and T. J. Case, eds. Harper and Row, New York.

Diamond, J. 1992. Must we shoot deer to save nature? *Nat. Hist.* August 1992:2–8.

Diamond, J. M. 1969. Avifaunal equilibria and species turnover rates on the Channel Islands of California. *Proc. Natl. Acad. Sci. USA* 64:57–63.

Diamond, J. M. 1972. Biogeographic kinetics: Estimation of relaxation times for avifauna of southwest Pacific islands. *Proc. Natl. Acad. Sci.* USA 69:3199–3200.

Diamond, J. M. 1975. Assembly of species communities. Pp. 342–344 in *Ecology and evolution of communities*, M. L. Cody and J. M. Diamond, eds. Harvard University Press, Cambridge, Mass.

Diaz, H. F., and R. G. Quayle. 1980. The climate of the United States since 1895: Spatial and temporal changes. *Monthly Weather Rev.* 108:249–266.

Dindal, D. T. 1971. *Ecology of compost.* SUNY College of Environmental Science and Forestry, Syracuse.

Dixon, K. R., and R. C. Juelson, 1987. The political economy of the spotted owl. *Ecology* 68:772–776.

Dodson, S. 1989. Predator-induced reaction norms. *BioScience* 39:447–452.

Doll, R., and R. Peto. 1981. *The causes of cancer.* Oxford University Press, New York.

Done, T. J. 1986. The significance of the crown of thorns starfish. *Oceanus* 29:58–59.

Done, T. J. 1987. Simulation of the effects of *Acanthaster planci* on the population structure of massive corals in the genus *Porites:* Evidence of population resilience? *Coral Reefs* 1–16.

Downs, J. A., and M. D. Abrams. 1991. Composition and structure of an old-growth versus a second-growth white oak forest in southwestern Pennsylvania. Pp. 207–223 in Proceedings of the 8th central hardwood conference, L. H. McCormick and K. W. Gottschalk, eds. *USDA For. Serv. Gen. Tech. Rep.* NE-148.

Dreistadt, S. H., D. L. Dahisten, and G. W. Frankie. 1990. Urban forests and insect ecology. *BioScience* 40:192–198.

Drew, J. V., and J. C. F. Tedrow. 1962. Arctic soil classification and patterned ground. *Arctic* 15:109–116.

Dublin, L. I., and A. J. Lotka. 1925. On the true rate of increase as exemplified by the population of the United States, 1920. *J. Am. Statist. A.* 20:305–339.

Dubos, R. 1955. Second thoughts on the germ theory. *Sci. Am.* 192:31–35.

Ducklow, H. W., D. A. Purdie, P. J. LeB. Williams, and J. M. Davies. 1986. Bacterioplankton: A sink for carbon in a coastal marine plankton community. *Science* 232:865–869.

Dueser, R. D., and H. H. Shugart. 1979. Niche pattern in a forest-floor small-mammal fauna. *Ecology* 60: 106–118.

Duffy, J. E., and M. E. Hay. 1990. Seaweed adaptations to herbivory. *BioScience* 40:368–375.

Dumond, D. E. 1975. The limitation of human population: A natural history. *Science* 187:713–721.

Dunning, D. C. 1968. Warning sounds of moths. *Z. Tierpsychol.* 25:129–138.

Dunning, D. C., and K. D. Roeder. 1965. Moth sounds and the insect-catching behavior of bats. *Science* 147:173–174.

Du Rietz, G. E. 1930. Classification and nomenclature of vegetation. *Sv. Bot. Tidskr.* 24:333–428.

Ecological Society of America Committee. 1991. The sustainable biosphere initiative: An ecological research agenda, *Ecology* 72:371–412.

Economou, A. N. 1991. Is dispersal of fish eggs, embryos and larvae an insurance against density dependence? *Environ. Biol. Fishes* 31:313–321.

Edmondson, W. T., and J. T. Lehman. 1981. The effect of changes in the nutrient income on the condition of Lake Washington. *Limnol. Oceanogr.* 26:1–29.

Edwards, J. 1984. Patterns of wolf predation and effects on moose feeding habits. In *Research in forest productivity, use, and pest control.* Proceedings of a Symposium in Recognition of the Contribution by Women Scientists.

Egler, F. E. 1954. Vegetation science concepts. I. Initial floristic composition, a factor in old-field development. *Vegetatio* 14:412–417.

Ehrenfeld, D. W. 1976. The conservation of non-resources. *Am. Sci.* 64:648–656.

Ehrlich, P. R. 1968. *The population bomb.* Ballantine, New York.

Ehrlich, P. R., and A. H. Ehrlich. 1992. The value of biodiversity. *Ambio* 21:219–226.

Ehrlich, P. R., and P. H. Raven. 1965. Butterflies and plants: A study in coevolution. *Evolution* 18:586–608.

Eibl-Eibesfeldt, I. 1955. Über Symbiosen, Parasitismus und andere besondere zwischenartliche Beziehungen bei tropischen Meeresfischen. *Z. Tierpsychol.* 12: 203–219.

Eigenmann, C. H. 1909. Cave vertebrates of America: A study in degenerative evolution. *Carnegie Inst. Wash. Publ.* 104:1–241.

Einarsen, A. S. 1945. Some factors affecting ring-neck pheasant density. *Murrelet* 26:7.

Eisenberg, J. F. 1983. Behavioral adaptations of higher vertebrates to tropical forests. Pp. 267–278 in *Tropi-*

cal rainforest ecosystems, F. B. Golley, ed. Elsevier Scientific Pub. Co., New York.

Elder, W. H., and W. W. Dierker. 1981. Do birds follow Hopkins's bioclimatic law? *Inland Bird Banding* 53: 33–38.

Elgar, M. A. 1989. Predator vigilance and group size in mammals and birds: A critical review of the empirical evidence. *Biol. Rev.* 64:13–33.

Elkin, A. P. 1964. *The Australian aborigines.* Natural History Library, Doubleday, Garden City, New York.

Elton, C. S. 1927. *Animal ecology.* Sidgwick and Jackson, London.

Elton, C. S. 1942. *Voles, mice, and lemmings.* Clarendon Press, Oxford.

Elton, C. S. 1946. Competition and the structure of ecological communities. *J. Anim. Ecol.* 15:54–68.

Elton, C. S. 1958. *The ecology of invasions by animals and plants.* Methuen & Co., London.

Elton, C. S. 1966. *The pattern of animal communities.* Wiley, New York.

Emslie, S. D. 1987. Age and diet of fossil California condors in Grand Canyon, Arizona. *Science* 237:768–770.

Engel, J. R. 1983. *Sacred sands: The struggle for community in the Indiana dunes.* Wesleyan University Press, Middletown, Conn.

Engelmann, M. D. 1961. The role of soil arthropods in the energetics of an old field community. *Ecol. Monogr.* 31:235.

Engemann, J. G., ed. 1982. *Eastern equine encephalitis and public health in southwestern Michigan.* Science for Citizens Center, Western Michigan University Kalamazoo.

Enright, J. T. 1976. Climate and population regulation. The biogeographer's dilemma. *Oecologia (Berl)* 24: 295–310.

Environmental Protection Agency. 1980. *Groundwater protection.* Water Planning Division, Washington, D.C.

Epstein, S. S. 1990. Losing the War Against Cancer: Who's to blame and what to do about it. *Int. J. Health Sci.* 20:53–71.

Epstein, S. S., L. O. Brown, and C. Pope. 1982. *Hazardous waste in America.* Sierra Club Books, San Francisco.

Erhalt, D. H. 1985. Methane in the global atmosphere. *Environment* 27:6–12, 30–33.

Erickson, M. M. 1938. Territory, annual cycle, and numbers in a population of wren-tits *(Chamaea fasciata). Univ. Calif. Pub. Zool.* 42.

Errington, P. L. 1945. Some contributions of a 15-year local study of the northern bobwhite to a knowledge of population phenomena. *Ecol. Monogr.* 15:1–34.

Errington, P. L. 1957. *Of men and marshes.* Macmillan, New York.

Errington, P. L. 1962. Of man and the lower animals. *Yale Rev.* 51:370–383.

Errington, P. L. 1967. *Of predation and life.* Iowa State University Press, Ames, Iowa.

Estes, J. A. 1991. Catastrophes and conservation: Lessons from sea otters and the *Exxon Valdez. Science* 254:1596.

Evans, G. C. 1939. Ecological studies on the rain forest of southern Nigeria II. *J. Ecol.* 27:436–482.

Evans, G. C. 1966. Model and measurements in the study of woodland light climates. Pp. 77–90 in *Light as an ecological factor,* R. Bainbridge, G. C. Evans, and O. Rackham, eds. Blackwell Scientific Pub., Oxford.

Evans, K. E., and G. E. Probasco. 1977. Wildlife of the prairies and plains. *USDA For. Serv. Gen. Tech. Rep. NC* 29.

Evans, W. G. 1966. Perception of infrared radiation from forest fires by *Melanophila acuminata* De Geer (Coleoptera: Buprestidae). *Ecology* 47:1061–1065.

Everett, K. R., V. D. Vassiljevskaya, J. Brown, and B. D. Walker. 1981. Tundra and analogous soils. Pp. 139–179 in *Tundra ecosystems,* L. C. Bliss, O. W. Heal, and J. J. Moore, eds. Cambridge University Press, Cambridge.

Evers, D. C. 1991. Black-backed woodpecker. Pp. 270–271 in *The Atlas of breeding birds of Michigan,* R. Brewer, G. A. McPeek, and R. J. Adams, eds. Michigan State University Press, East Lansing.

Evers, R. A. 1955. Hill prairies of Illinois. *Bull. Ill. Nat. Hist. Surv.* 26:368–446.

Ewel, K. C. 1990. Swamps, Pp. 281–323 in *Ecosystems of Florida,* P. L. Myers and J. J. Ewel, eds. University of Central Florida Press, Orlando.

Ewert, D. 1982. Birds in isolated bogs in central Michigan. *Am. Midl. Nat.* 108:41–50.

Faegri, K., and L. van der Pijl. 1979. *The principles of pollination biology.* 3rd ed. Pergamon Press, Oxford.

Falconer, D. S. 1989. *Introduction to quantitative genetics.* 3rd ed. Wiley, New York.

Falkowski, P. G., Y. Kim, Z. Kolber, C. Wilson, C. Wirick, and R. Cress. 1992. Natural versus anthropogenic factors affecting low-level cloud albedo over the north Atlantic. *Science* 256:1311–1313.

Farner, D. S. 1955. Bird banding in the study of population dynamics. Pp. 397–449 in *Recent studies in avian biology,* A. Wolfson, ed. University of Illinois Press, Urbana.

Farnsworth, N. R., and D. D. Soejarto. 1985. Potential consequences of plant extinction in the United States on the current and future availability of prescription drugs. *Econ. Bot.* 39:231–240.

Feeny, P. 1970. Seasonal changes in oak leaf tannins and nutrients as a cause of spring feeding by winter moth caterpillars. *Ecology* 51:565–581.

Finch, D. M. 1991. Population ecology, habitat requirements, and conservation of Neotropical migratory birds. *USDA For. Serv. Gen. Tech. Rep. RM* 205:1–26.

Fincham, A. A. 1984. *Basic marine biology.* Cambridge University Press, Cambridge.

Fine, P. E. M. 1982. The control of infectious disease group report. Pp. 121–127 in *Population biology of infectious diseases,* R. M. Anderson and R. M. May, eds. Springer-Verlag, New York.

Fish, D. 1983. Phytotelmata: Flora and fauna. Pp. 1–27 in *Phytotelmata: Terrestrial plants as hosts for aquatic insect communities,* J. H. Frank and L. P. Lounibos, eds. Plexus Publ., Medford, N.J.

Fisher, E. 1985. The management and assessment of risks from recombinant organisms. *J. Hazard. Mater.* 10:241–261.

Fisher, J., N. Simon, and J. Vincent. 1969. *The red book: Wildlife in danger.* Collins, London.

Fisher, R. A. 1930. *The genetical theory of natural selection.* Clarendon Press, Oxford.

Fisher, S. G., and G. E. Likens. 1977. Energy flow in Bear Brook, New Hampshire: An integrative approach to stream ecosystem metabolism. *Ecol. Monogr.* 43:421–439.

Fitzgerald, S. M., and L. B. Keith. 1990. Intra- and interspecific dominance relationships among arctic and snowshoe hares. *Can. J. Zool.* 68:457–464.

Fitzpatrick, J. W., and G. E. Woolfenden. 1986. Demographic routes to cooperative breeding in some New World jays. Pp. 137–160 in *Evolution of behavior,* M. Nitecki and J. Kitchell, eds. University of Chicago Press, Chicago.

Flanagan, P. W. 1986. Genetically engineered organisms and ecology. *Bull. Ecol. Soc. Am.* 67:26–30.

Fleming, T. H. 1979. Do tropical frugivores compete for food? *Am. Zool.* 19:1157–1172.

Fleming, T. H. 1991. Fruiting plant-frugivore mutualism: The evolutionary theater and the ecological play. Pp. 119–144 in *Plant-animal interactions,* P. W. Price, T. M. Lewinsohn, G. W. Fernandes, and W. W. Benson, eds. John Wiley & Sons, New York.

Flesness, N. R. 1989. Mammalian extinction rates: Background to the black-footed ferret dilemma. Pp. 3–9 in *Conservation biology and the black-footed ferret,* E. S. Seal, et al., eds. Yale University Press, New Haven, Conn.

Flower, S. S. 1938. Further notes on the duration of life in animals. IV. Birds. *Proc. Zool. Soc. Lond. Series A,* 108:195–235.

Foin, T. C. 1976. *Ecological systems and the environment.* Houghton Mifflin, Boston.

Forbes, L. S., and R. C. Ydenberg. 1992. Sibling rivalry in a variable environment. *Theor. Popul. Biol.* 41: 135–160.

Forbes, S. A. 1880. On some interactions of organisms. *Bull. Ill. Lab. Nat. Hist.* 1:3–17.

Forbes, S. A. 1887. The lake as a microcosm. *Bull. Sci. A. Peoria* 1887:77–87.

Forbes, S. A., and R. E. Richardson. 1919. Some recent changes in Illinois river biology. *Ill. Nat. Hist. Surv. Bull.* 13:140–156.

Forbes, S. A., and R. E. Richardson. 1920. *The fishes of Illinois.* 2nd ed. Illinois State Natural History Survey, Urbana.

Force, D. C. 1982. Postburn insect fauna in southern California chaparral. Pp. 234–240 in *Dynamics and management of Mediterranean-type ecosystems. USDA For. Serv. Gen. Tech. Rep. PSW* 58.

Foreman, D. 1981. Earth first. *Progressive* 45:39–42.

Foreman, D., and B. Haywood. 1987. *Ecodefense: A field guide to monkeywrenching.* A Ned Ludd Book: Tucson, Ariz.

Forkman, B. 1991. Some problems with current patch-choice theory: A study on the Mongolian gerbil. *Behaviour* 117:243–254.

Forrest, S. C., et al. 1988. Black-footed ferret (*Mustela nigripes*) population attributes, Meeteetse, Wyoming, 1981–1985. J. Mammal. 69:261–273.

Forsman, E. D., and E. C. Meslow. 1985. Old-growth forest retention for spotted owls—how much do they need? Pp. 58–59 in *Ecology and management of the spotted owl in the Pacific Northwest,* R. J. Gutierrez and A. B. Carey, eds. *USDA For. Serv. Gen. Tech. Rep. PNW* 185.

Foth, H. D. 1978. *Fundamentals of soil science.* 6th ed. Wiley, New York.

Fox, B. J., and A. R. Pople. 1984. Experimental confirmation of interspecific competition between native and introduced mice. *Aust. J. Ecol.* 9:323–334.

Fox, J. F. 1977. Alternation and coexistence of tree species. *Am. Nat.* 108:268–289.

Frank, S. A., and M. Slatkin. 1992. Fisher's fundamental theorem of natural selection. *Trends Ecol. Evol.* 7: 92–95.

Frankel, O. H., and M. E. Soulé. 1981. *Conservation and evolution.* Cambridge University Press, Cambridge.

Franklin, J. F. 1979. Vegetation of the Douglas-fir region. Pp. 93–112 in *Forest soils of the Douglas-fir region,* P. E. Heilman, et al., eds. Washington State University, Pullman, Wash.

Franklin, J. F., and C. T. Dyrness. 1969. Vegetation of Oregon and Washington. *USDA For. Serv. Res. Pap. PNW* 80:1–216.

Franklin, J. F., et al. 1981. Ecological characteristics of

old-growth Douglas-fir forests. *USDA For. Serv. Gen. Tech. Rep. PNW* 118.

French, D. D., and V. R. Smith. 1985. A comparison between northern and southern hemisphere tundras and related ecosystems. *Polar Biol.* 5:5–21.

French, J. L. 1937. *The pioneer west.* Garden City Pub. Co., Inc., New York.

Frenkel, R. E., and J. C. Morlan. 1991. Can we restore our salt marshes? Lessons from the Salmon River, Oregon. Northwest Environ. J. 7:119–135.

Frenzel, B. 1968. The Pleistocene vegetation of northern Eurasia. *Science* 161:637–649.

Fretwell, S. D. 1972. Populations in a seasonal environment. *Princeton Mon. Pop. Biol.* 5:1–217.

Fretwell, S. D. 1977a. The regulation of plant communities by the food chains exploiting them. *Perspect. Biol. Med.* 20:169–185.

Fretwell, S. D. 1977b. Is the Dickcissel a threatened species? *Am. Birds* 31:923–932.

Fretwell, S. D., and H. L. Lucas, Jr. 1970. On territorial behavior and other factors influencing habitat distribution in birds. I. Theoretical development. *Acta Biotheor.* 19:16–36.

Freuchen, P., and F. Salomonsen. 1958. *The arctic year.* Putnam, New York.

Fricke, H. W. 1973. *The coral seas.* Putnam, New York.

Frischknecht, N. C. 1975. Native faunal relationships within the pinyon-juniper ecosystem. Pp. 55–65 in *Proceedings of the pinyon-juniper symposium.* Utah State University, Logan.

Fritschen, L. J., and L. W. Gay. 1979. *Environmental instrumentation.* Springer-Verlag, New York.

Frost, C. C. 1991. Four centuries of changing landscape patterns in the longleaf pine ecosystem. *Abstract, The longleaf pine ecosystem: Ecology, restoration and management. 18th Tall Timbers Fire Ecology Conference,* Tallahassee, Fla.

Frost, S. W. 1959. *Insect life and insect natural history.* 2nd rev. ed. Dover Publications, New York.

Fry, C. H. 1983. Birds in savanna ecosystems. Pp. 337–357 in *Ecosystems of the world: Tropical savannas,* F. Bourlière, ed. Elsevier Scientific Pub. Co., New York.

Fryxell, J. M., D. J. T. Hussell, A. B. Lambert, and P. C. Smith. 1991. Time lags and population fluctuations in white-tailed deer. *J. Wildl. Manage.* 55:377–385.

Fullard, J. H., M. B. Fenton, and J. A. Simmons. 1979. Jamming bat echolocation: The clicks of arctiid moths. *Can. J. Zool.* 57:647–649.

Fuller, G. D. 1911. Evaporation and plant succession. *Bot. Gaz.* 52:193–208.

Furniss, R. L., and V. M. Carolin. 1977. Western forest insects. *USDA For. Serv. Misc. Publ.* 1339:1–654.

Furon, R. 1967. *The problem of water: A world study.* Elsevier, New York.

Gabriel, W., and M. Lynch. 1992. The selective advantage of reaction norms for environmental tolerance. *J. Evol. Biol.* 5:31–59.

Gadagker, R. 1985. Kin recognition in social insects and other animals—a review of recent findings and a consideration of their relevance for the theory of kin selection. *Proc. Indiana Acad. Sci. Anim. Sci.* 94: 587–621.

Gaddy, L. L., P. W. Suckling, and V. Meentemeyer. 1984. The relationship between winter minimum temperatures and spring phenology in a southern Appalachian grove. *Arch. Metereol. Geophys. Bioclimatol. Ser. B.* 34:155–162.

Gadgil, M. 1971. Dispersal: Population consequences and evolution. Ecology 52:252–261.

Gajdusek, D. C. 1977. Unconventional viruses and the origin and disappearance of kuru. *Science* 197:943–960.

Galef, B. G., Jr. 1991. Information centers of Norway rats: Sites for information exchange and information parasitism. *Anim. Behav.* 41:295–302.

Galindo, C. 1986. Do desert rodent populations increase when ants are removed? *Ecology* 67:1422–1423.

Galli, A. E., C. F. Leck, and R. T. T. Forman. 1976. Avian distribution patterns in forest islands of different sizes in central New Jersey. *Auk* 93:356–364.

Galloway, J. N., and D. M. Whelpdale. 1980. An atmospheric sulfur budget for North America. *Atmos. Environ.* 14:409–417.

Gansner, D. A., and O. W. Herrick. 1985. Host preference of gypsy moth on a new frontier of infestation. *USDA For. Serv. Res. Note NE* 330:1–3.

Garcia, R., L. E. Caltagirone, and A. P. Gutierrez. 1990. Comments on a redefinition of biological control. *BioScience* 38:691–694.

Gargett, V. 1978. Sibling aggression in the black eagle in the Matopos, Rhodesia. *Ostrich* 49:57–63.

Garrison, G. A., A. J. Bjugstad, D. A. Duncan, M. E. Lewis, and D. R. Smith. 1977. Vegetation and environmental features of forest and range ecosystems. *USDA For. Serv. Agricultural Handbook* No. 475.

Gates, D. M. 1968. Transpiration and leaf temperature. *Annu. Rev. Plant Physiol.* 19:211–239.

Gates, D. M. 1980. *Biophysical ecology.* Springer-Verlag, New York.

Gaufin, A. 1958. The effects of pollution on a midwestern stream. *Ohio J. Sci.* 58:197–208.

Gause, G. F. 1934. *The struggle for existence.* Williams & Wilkins, Baltimore.

Gauthier-Pilter, H., and A. I. Dagg. 1961. *The camel.* University of Chicago Press, Chicago.

Gear, A. J., and B. Huntley. 1991. Rapid changes in the range limits of Scots pine 4000 years ago. *Science* 251:544–547.

Gehlbach, F. R. 1991. The east-west transition zone of terrestrial vertebrates in central Texas: A biogeographical analysis. *Texas J. Sci.* 43:415–427.

Gentry, A. H., ed. 1991. *Four neotropical rainforests.* Yale University Press, New Haven, Conn.

George, J. C. 1988. *Everglades wildguide.* USDI National Park Service Official Handbook 142.

Gerard, G. T., B. D. Anderson, and T. A. DeLaney. 1993. Managing conflicts with animal activists: White-tailed deer and Illinois nature preserves. *Nat. Areas J.* 13:10–17.

Gibbons, J. W. 1990. Sex ratios and their significance among turtle populations. Pp. 171–182 in *Life history and ecology of the slider turtle,* J. W. Gibbons, ed. Smithsonian Institution Press, Washington, D.C.

Gibbs, H. L., P. J. Weatherhead, P. T. Boag, B. N. White, L. M. Tabak, and D. J. Hoysak. 1990. Realized reproductive success of polygynous red-winged blackbirds revealed by DNA markers. *Science* 250: 1394–1397.

Gibson, D. O. 1974. Batesian mimicry without distastefulness? *Nature* 250:77–79.

Gilbert, F. F. 1974. *Parelaphostrongylus tenuis* in Maine: II—prevalence in moose. *J. Wildl. Manage.* 38:42–46.

Gilbertson, M., T. Kubiak, J. Ludwig, and G. Fox. 1991. Great Lakes embryo mortality, edema, and deformities syndrome (GLEMEDS) in colonial fish-eating birds: Similarity to chick-edema disease. *J. Toxicol. Environ. Health* 33:455–520.

Gilliard, E. T. 1958. *Living birds of the world.* Doubleday, Garden City, New York.

Gillon, Y. 1971. The effect of bush fire on the principal acridid species of an Ivory Coast savanna. Pp. 419–471 in The proceedings of the Tall Timbers Fire Ecology Conference, Tallahassee, Fla.

Gilpin, M. 1987. Spatial structure and population vulnerability. Pp. 125–139 in *Viable populations for conservation,* M. E. Soulé, ed. Cambridge University Press, Cambridge.

Gilpin, M. E., and F. J. Ayala, 1976. Schoener's model and Drosophila competition. *Theor. Popul. Biol.* 9: 12–14.

Ginzburg, L. R., and H. R. Akçakaya, 1992. Consequences of ratio-dependent predation for steady-state properties of ecosystems. *Ecology* 73:1536–1543.

Glass, G. E., ed. 1973. *Bioassay techniques and environmental chemistry.* Ann Arbor Science Pub., Inc., Ann Arbor, Mich.

Gleason, H. A. 1922. On the relation between species and area. *Ecology* 3:158–162.

Gleason, H. A. 1926. The individualistic concept of the plant association. *Bull. Torrey Bot. Club* 53:1–20.

Gleason, H. A. 1927. Further views on the succession-concept. *Ecology* 8:299–326.

Gleason, H. A. 1939. The individualistic concept of the plant association. *Am. Midl. Nat.* 21:92–108.

Gleason, H. A., and A. Cronquist. 1963. *Manual of vascular plants of the northeastern United States and adjacent Canada.* Van Nostrand-Reinhold, New York.

Gleason, H. A., and A. Cronquist. 1964. *The natural geography of plants.* Columbia University Press, New York.

Gleick, J. 1987. *Chaos. Making a new science.* Viking Penguin, Inc., New York.

Glendinning, J. I., and L. P. Brower. 1990. Feeding and breeding responses of five mice species to overwintering aggregations of the monarch butterfly. *J. Anim. Ecol.* 59:1091–1112.

Godfrey, W. E. 1979. *The birds of Canada.* National Museums of Canada, Ottawa.

Godwasser, S., D. Gaines, and S. R. Wilbur. 1980. The least Bell's vireo in California: A de facto endangered race. *Am. Birds* 34:742–745.

Godwin, H., and V. M. Conway. 1939. The ecology of a raised bog near Tregaron, Cardiganshire. *J. Ecol.* 27:313–359.

Gogan, P. J. P., and R. H. Barrett. 1988. Lesson in management from translocations of tule elk. Pp. 275–287 in *Translocations of wild animals,* L. Nielsen and R. D. Brown, eds. Wisconsin Humane Society and Caesar Kleberg Wildlife Research Institute, Milwaukee, Wis.

Golley, F. B. 1962. *Mammals of Georgia.* University of Georgia Press, Athens.

Golley, F. B. 1983. *Tropical rain forest ecosystems.* Elsevier, Amsterdam.

Golley, F. B. 1991. The ecosystem concept: A search for order. *Ecol. Res.* 6:129–138.

Golley, F. B., and J. B. Gentry. 1964. Bioenergetics of the southern harvester ants, *Pogonomyrmex badins. Ecology* 45:217–225.

Good, N. E., and D. H. Bell. 1980. Photosynthesis, plant productivity, and crop yield. Pp. 3–51 in *The biology of crop productivity,* P. S. Carlson, ed. Academic Press, New York.

Goodall, J. 1968. The behaviour of free-living chimpanzees in the Gombe Stream Reserve. *Anim. Behav. Monogr.* 1:165–301.

Goodman, D. 1987. The demography of chance extinction. Pp. 11–34 in *Viable populations for conservation,* M. E. Soulé, ed. Cambridge University Press, Cambridge.

Gordon, A. G. 1985. "Budworm! But what about the forest?" Pp. 3–29 in *Spruce-fir management and spruce budworm. USDA For. Serv. Gen. Tech. Rep. GTR NE* 99.

Gosling, N. 1977. Observations on the three-dimensional range of *Peromyscus leucopus noveboracensis. Jack Pine Warbler* 55:43–44.

Gould, S. J., and R. C. Lewontin. 1979. The spandrels of San Marco and the Panglossian paradigm: A critique of the adaptationist programme. *Proc. R. Soc. Lond. [Biol.]* 205:581–598.

Graber, R. E., and D. F. Thompson. 1978. Seeds in the organic layers and soil of four beech-birch-maple stands. *USDA For. Serv. Res. Pap. NE* 401:1–8.

Graham, A. 1964. Origin and evolution of the biota of southeastern North America: Evidence from the fossil plant record. *Evolution* 18:571–585.

Graham, A. 1973. History of the arborescent temperate element in the northern Latin American biota. Pp. 301–314 in *Vegetation and vegetational history of northern Latin America,* A. Graham, ed. Elsevier Sci. Pub. Co., Amsterdam.

Graham, E. H. 1947. *The land and wildlife.* Oxford University Press, New York.

Graham, R. W. 1984. Paleoenvironmental implications of the quaternary distribution of the eastern chipmunk (*Tamias striatus*) in central Texas. *Quat. Res.* 21:111–114.

Grant, P. R. 1972a. Interspecific competition among rodents. *Annu. Rev. Ecol. Syst.* 3:79–106.

Grant, P. R. 1927b. Convergent and divergent character displacement. *Biol. J. Linn. Soc.* 4:39–68.

Grant, P. R., and D. Schluter. 1984. Interspecific competition inferred from patterns of guild structure. Pp. 201–231 in *Ecological communities,* D. R. Strong, et al., eds. Princeton University Press, Princeton, N.J.

Grassle, J. F. 1985. Hydrothermal vent animals: Distribution and biology. *Science* 229:713–717.

Graves, H. S., and E. W. Nelson. 1919. *Our national elk herds.* USDA, Dept. Circular 51.

Green, L. R. 1982. Prescribed burning in the California Mediterranean ecosystem. Pp. 464–471 in *Dynamics and management of Mediterranean-type ecosystems. USDA For. Serv. Gen. Tech. Rep. PSW* 58.

Greene, H. W., and R. W. McDiarmid. 1981. Coral snake mimicry: Does it occur? *Science* 213:1207–1212.

Greenwood, P. J., and P. H. Harvey. 1982. The natal and breeding dispersal of birds. *Annu. Rev. Ecol. Syst.* 13:1–21.

Grelen, H. E., and R. H. Hughes. 1984. Common herbaceous plants of southern forest range. *USDA For. Serv. Res. Pap. SO* 210.

Gribbin, J. 1990. *Hothouse earth.* Grove Weidenfeld, New York.

Grier, C. C. 1975. Wildfire effects on nutrient distribution and leaching in a coniferous ecosystem. *Can. J. For. Res.* 5:599–607.

Grier, J. W. 1982. Ban of DDT and subsequent recovery of reproduction in bald eagles. *Science* 218:1232–1234.

Griffin, D. R. 1953. Bat sounds under natural conditions, with evidence of echolocation of insect prey. *J. Exp. Zool.* 123:435–465.

Griffin, J. R., and W. B. Critchfield. 1972 (reprinted 1976). The distribution of forest trees in California. *USDA For. Serv. Res. Pap. PSW* 82:1–118.

Grimble, D. G., and F. B. Lewis. 1985. Symposium Proceedings: Microbial control of spruce budworms and gypsy moths. *USDA Gen. Tech. Rep. NE* 100:1–175.

Grime, J. P. 1979. *Plant strategies and vegetation processes.* Wiley, Chichester.

Grinnell, J. 1917. The niche-relationships of the California thrasher. *Auk* 34:427–433.

Grisebach, A. 1838. Über den Einfluss des Klimas auf die Begrenzung der näturlichen Floren. *Linnaea* 12 (Repr. in *Gesammelte Abhandlungen und Kleinere Schriften zur Pflanzengeographie.* Leipzig 1880).

Grmek, M. D. 1990. *History of AIDS,* R. C. Maulitz and J. Duffin, transl. Princeton University Press, Princeton, N.J.

Gross, A. O. 1928. The heath hen. *Mem. Boston Soc. Nat. Hist.* 6:491–588.

Gross, J. E. 1969. The role of small herbivorous mammals in the functioning of the grassland ecosystem. Pp. 268–278 in *The grassland ecosystem: A preliminary synthesis,* R. L. Dix and R. G. Beidleman, eds. Colorado State Range Department Science Series No. 2.

Grotta, D., and S. Grotta. 1992. Antarctica: Whose continent is it anyway? *Popular Sci.* 240(1):62–67.

Grover, H. D., and G. F. White. 1985. Toward understanding the effects of nuclear war. *BioScience* 35: 552–556.

Grumbine, E. 1990. Protecting biological diversity through the greater ecosystem concept. *Nat. Areas J.* 10:114–119.

Guilday, J. E., P. S. Martin, and A. D. McCrady. 1964. New Paris No. 4: A Pleistocene cave deposit in Belford County, Pennsylvania. *Bull. Nat. Speleol. Soc.* 26: 121–194.

Gutierrez, R. J., and A. B. Carey, eds. 1985. Ecology and management of the spotted owl in the Pacific Northwest. *USDA For. Serv. Gen. Tech. Rep PNW* 185.

Hackett, T. 1989. Fire. *The New Yorker,* October 2:50–73.

Hadley, E. B., and B. J. Kieckhefer. 1963. Productivity of

two prairie grasses in relation to fire frequency. *Ecology* 44:389–395.

Hadley, J. L., and W. K. Smith. 1986. Wind effects on needles of timberline conifers: Seasonal influence on mortality. *Ecology* 67:12–19.

Haeckel, E. 1870. Ueber Entwickelungsgang und Aufgabe der Zoologie. *Jenaische Z.* 5:353–370.

Hagan, J. M., T. L. Lloyd-Evans, and J. L. Atwood. 1991. The relationship between latitude and the timing of spring migration of North American landbirds. *Ornis Scand.* 22:129–136.

Hagen, D. W., and L. G. Gilbertson. 1973. Selective predation and the intensity of selection acting upon the lateral plates of threespine sticklebacks. *Heredity* 30:273–287.

Hailman, J. P. 1979. Environmental light and conspicuous colors. Pp. 289–345 in *The behavioral significance of color,* E. H. Burtt, Jr., ed. Garland STPM Press, New York.

Haines, D. A., and R. W. Sando. 1969. Climatic conditions preceding historically great fires in the north central region. *USDA For. Serv. Res. Pap. NC* 34.

Haines, H. H. 1988. *Black radicals and the civil rights mainstream.* University of Tennessee Press, Knoxville.

Hairston, N. G. 1965. On the mathematical analysis of schistosome populations. *Bull. WHO* 33:45–62.

Hairston, N. G., F. E. Smith, and L. B. Slobodkin. 1960. Community structure, population control, and competition. *Am. Nat.* 94:421–425.

Halfpenny, J. C., and R. D. Ozanne. 1989. *Winter. An Ecological handbook.* Johnson Books, Boulder, Colo.

Hall, E. T. 1969. *The hidden dimension.* Anchor Books, Garden City, New York.

Hall, S. J., and D. Raffaelli. 1991. Food-web patterns: Lessons from a species-rich web. *J. Anim. Ecol.* 60: 823–842.

Halley, E. 1694. An estimate of the degrees of the mortality of mankind, drawn from curious tables of the births and funerals at the City of Breslaw with an attempt to ascertain the price of annuities upon lives. *Philos. Trans. R. Soc. Lond.* 17:596–610.

Halliday, T. R. 1980. The extinction of the passenger pigeon *Ectopistes migratorius* and its relevance to contemporary conservation. *Biol. Conserv.* 17:157–162.

Halloy, S. 1991. Islands of life at 6000 m altitude: The environment of the highest autotrophic communities on earth (Socompa Volcano, Andes). *Arct. Alp. Res.* 23:247–262.

Halverson, H. G., and J. L. Smith. 1979. Solar radiation as a forest management tool: A primer of principles and application. *USDA For. Serv. Gen. Tech. Rep. PSW* 33:1–13.

Hamburger, M., and K. Hostettmann. 1991. Bioactivity in plants: The link between phytochemistry and medicine. *Phytochemistry* 30:3864–3874.

Hamel, P. B. 1986. *Bachman's warbler: A species in peril.* Smithsonian Institution Press, Washington, D.C.

Hamilton, A. C. 1982. *Environmental history of East Africa. A study of the Quaternary.* Academic Press, London.

Hamilton, W. D. 1963. The evolution of altruistic behavior. *Am. Nat.* 97:354–356.

Hamilton, W. D. 1971. Geometry for the selfish herd. *J. Theor. Biol.* 31:295–311.

Hamilton, W. D., and R. M. May. 1977. Dispersal in stable habitats. *Nature* 269:578–581.

Hanes, C. R., and F. N. Hanes. 1947. *Flora of Kalamazoo County, Michigan.* The Anthoensen Press, Portland, Me.

Hannon, B. M., and T. G. Lowman. 1978. The energy cost of overweight in the United States. *Am. J. Public Health* 68:765–767.

Hanson, A. D., and C. E. Nelsen. 1980. Water: Adaptation of crops to drought-prone environments. Pp. 77–152 in *The biology of crop productivity,* P. S. Carlson, ed. Academic Press, New York.

Hansson, L., and H. Henttonen. 1985. Regional differences in cyclicity and reproduction in Clethrionomys species: Are they related? *Ann. Zool. Fennici* 22: 277–288.

Harborne, J. B. 1982. *Introduction to ecological biochemistry.* Academic Press, London.

Hardin, G. 1960. The competitive exclusion principle. *Science* 131:1292–1297.

Hardin, G. 1977. *The limits of altruism.* Indiana University Press, Bloomington.

Hardman, M. G. 1991. Indigenous peoples. Pp. 108–111 in *1992 Earth Journal.* Buzzworm Books, Boulder, Colo.

Hardy, G. H. 1908. Mendelian proportions in a mixed population. *Science* 28:49–50.

Harlow, F. C., and E. S. Harrar. 1980. *Textbook of dendrology.* 5th ed. McGraw-Hill, New York.

Harper, J. L. 1977. *Population biology of plants.* Academic Press, London.

Harr, R. D. 1982. Fog drip in the Bull Run municipal watershed, Oregon. *Water Resour. Bull. Am. Water Resour. Assoc.* 18:785–789.

Harris, M. 1977. *Cannibals and kings: The origins of cultures.* Random House, New York.

Harris, R. B., T. W. Clark, and M. L. Shaffer. 1989. Extinction probabilities for isolated black-footed ferret populations. Pp. 69–82 in *Conservation biology and the black-footed ferret,* E. S. Seal, et al., eds. Yale University Press, New Haven, Conn.

Harrison, K. G. 1977. Perch height selection by grassland birds. *Wilson Bull.* 89:486–487.

Harrison, K. G., and R. Brewer. 1979. The role of elevated perch sites and mowing in the distribution of grassland and savanna birds. *Jack-Pine Warbler* 57: 179–183.

Hartesveldt, R. J., H. T. Harvey, H. S. Shellhammer, and R. E. Stecker. 1975. *The giant sequoia of the Sierra Nevada.* US Dep. Int., Natl. Pk. Serv., Washington, D.C.

Hartl, D. L. 1981. *A primer of population genetics.* Sinauer Associates, Inc., Sunderland, Mass.

Hartshorn, G. S. 1978. Treefalls and forest dynamics. Pp. 617–634 in *Tropical trees as living systems,* P. B. Tomlinson and M. H. Zimmerman, eds. Cambridge University Press, Cambridge.

Harvey, L. H. 1903. A study of physiographic ecology of Mount Ktaadn, Maine. *Univ. Maine Stud.* 5:1–50.

Harwell, M. A., and H. D. Grover. 1985. Biological effects of nuclear war I: Impact on humans. *BioScience* 35:570–575.

Haskell, E. F. 1949. A clarification of social science. *Main Curr. Modern Thought* 7:45–51.

Haukioja, E. 1980. On the role of plant defenses in the fluctuation of herbivore populations. *Oikos* 35:202–213.

Haukioja, E., J. Kapiainen, P. Niemlea, and J. Tuomi. 1983. Plant availability hypothesis and other explanations of herbivore cycles: Complementary of exclusive alternatives. *Oikos* 40:419–432.

Haukioja, E., and T. Hakata. 1979. On the relationship between avian clutch size and life span. *Ornis Fennica* 56:45–55.

Hayes, W. P. 1927. Prairie insects. *Ecology* 8:238–250.

Hedrick, P. W., M. E. Ginevan, and E. P. Ewing. 1976. Genetic polymorphism in heterogeneous environments. *Annu. Rev. Ecol. Syst.* 7:1–32.

Heichel, G. H. 1973. Comparative efficiency of energy use in crop production. *Conn. Agric. Exp. St. Bull.* 739:1–26.

Heichel, G. H. 1976. Agricultural production and energy resources. *Am. Sci.* 64:64–72.

Heinrich, B. 1979. "Majoring" and "minoring" by foraging bumblebees, *Bombus vagans:* An experimental analysis. *Ecology* 60:245–255.

Heinselman, M. L. 1973. Fire in the virgin forests of the Boundary Waters Canoe Area, Minnesota. *Quat. Res.* 3:329–382.

Heller, H. C. 1971. Altitudinal zonation of chipmunks (*Eutamias*): Interspecific aggression. *Ecology* 52:312–319.

Henderson, N. R., and N. B. Pavlovic. 1986. *Primary succession on the southern Lake Michigan sand dunes* (abstract). First Indiana Dunes Research Conference, Indiana Dunes National Lakeshore.

Henderson, R. 1982. Vegetation-fire ecology of tallgrass prairie. *Nat. Areas J.* 2:17–26.

Henderson-Sellers, A., and K. McGuffie. 1986. The threat from melting ice caps. *New Sci.* 12:24–25.

Hensley, M. M., and J. B. Cope. 1951. Further data on removal and repopulation of the breeding birds in a spruce-fir forest community. *Auk* 68:483–493.

Henttonen, H., T. Oksanene, A. Jortikka, and V. Haukisalmi. 1987. How much do weasels shape microtine cycles in the northern Fennoscandian tundra. *Oikos* 50:353–365.

Hepper, C. M. 1984. Isolation and culture of VA mycorrhizal (VAM) fungi. Pp. 95–112 in *VA Mycorrhiza,* C. L. Conway and D. J. Bagyaraj, eds. Academic Press, New York.

Hepper, P. G., ed. 1991. *Kin recognition.* Cambridge University Press, New York.

Herrera, C. M. 1985. Determinants of plant-animal coevolution: The case of mutualistic dispersal of seeds by vertebrates. *Oikos* 44:132–141.

Hesselberg, R. J. 1990. Contaminant residues in the bloater (*Coreognus hoyi*) of Lake Michigan, 1969–1986. *J. Great Lakes Res.* 16:121–129.

Hewes, J. J. 1981. *Redwoods: The world's largest living trees.* Rand McNally & Co., Chicago.

Hewitt, G. B. 1980. Plant phenology as a guide in timing grasshopper control efforts on Montana rangeland. *J. Range Manage.* 33:297–299.

Hibbard, C. W. 1960. An interpretation of Pliocene climates in North America. *Annu. Rep. Mich. Acad. Sci., Arts., Lett.* 62:5–30.

Hibbert, A. R., E. A. Davis, and D. G. Scholl. 1974. Chaparral conversion potential in Arizona. Part 1. Water yield response and effects on other resources. *USDA For. Serv. Res. Pap. RM* 126.

Hickman, J. C. 1975. Environmental unpredictability and plastic energy allocation strategies in the annual *Polygonum cascadense* (Polygonaceae). *J. Ecol.* 63:689–701.

Hiebert, R. D., and N. B. Pavlovic. 1987. Role of past land use on succession at the Indiana dunes; Implications for management. Pp. 47–70 in *Proceedings of the First Indiana Dunes Research Conference: Symposium on plant succession,* K. L. Cole, R. D. Hiebert, and J. D. Wood, Jr., eds. National Park Service Science Publications Office, Atlanta.

Hildén, O. 1965. Habitat selection in birds. *Ann. Zool. Fenn.* 2:53–75.

Hill, J. E., and J. C. Smith. 1984. *Bats: A natural history.* University of Texas Press, Austin.

Hinkle, G. 1991. Status of the theory of the symbiotic origin of undulipodia (cilia). Pp. 135–142 in *Symbiosis as a source of evolutionary innovation,* L. Margulis and René Fester, eds. MIT Press, Cambridge, Mass.

Hinrichsen, D. 1986. Multiple pollutants and forest decline. *Ambio* 15:258–265.

Hirst, E. 1974. Food-related energy requirements. *Science* 184:134–138.

Hodgson, B. 1990a. Can the wilderness heal? *Natl. Geogr.* 177:5–42.

Hodgson, B. 1990b. A land of isolation no more: Antarctica. *Natl. Geogr.* 177:2–51.

Hodler, T. W., R. Brewer, L. G. Brewer, and H. A. Raup. 1981. *Presettlement vegetation of Kalamazoo County* (map). Western Michigan University, Department of Geography, Kalamazoo.

Hoffman, K. H. 1974. Wirkung von konstanten und tagesperiodisch alterienden Temperaturen auf Lebensdauer; Nahrungsverwertung und Fertilitat adulter *Gryllus . . . Oecologia (Berl)* 17:39–54.

Hoffman, R. S. 1973. Terrestrial vertebrates. Pp. 475–568 in *Arctic and alpine environments,* J. D. Ives and R. G. Barry, eds. Methuen, London.

Holbrook, S. H. 1957. *Holy old Mackinaw.* Macmillan, New York.

Holdridge, L. R., et al. 1971. *Forest environments in tropical life zones, a pilot study.* Pergamon Press, Oxford.

Holling, C. S. 1959. The components of predation as revealed by a study of small-mammal predation of the European sawfly. *Can. Entomol.* 91:293–320.

Holloway, M. 1991. Soiled shores. Sci. Am. October: 102–116.

Holman, J. A. 1985. New evidence on the status of Ladds Quarry. *Natl. Geogr. Res.* 1:569–570.

Holmes, J. C. 1961. Effects of concurrent infections on *Hymenolepis diminuta* (Cestoda) and *Moniliformis dubius* (Acanthocephala). General effects and comparison with crowding. *J. Parasitol.* 47:209–216.

Holmes, R. T. 1966. Molt cycle of the red-backed sandpiper *(Calidris alpina)* in western North America. *Auk* 83:517–533.

Holmgren, R. C., and S. F. Brewster, Jr. 1972. Distribution of organic matter reserve in a desert shrub community. *USDA For. Serv. Res. Pap. INT* 130.

Holt, R. 1977. Predation, apparent competition, and the structure of prey communities. *Theor. Popul. Biol.* 12:197–229.

Holt, R. D., and J. Pickering. 1985. Infectious disease and species coexistence: A model of Lotka-Volterra form. *Am. Nat.* 126:196–211.

Hooper, M. D. 1971. The size and surroundings of nature reserves. Pp. 555–561 in *The scientific management of animal and plant communities for conservation,* E. D. Duffey and A. S. Watt, eds. Blackwell, Oxford.

Hooper, R. G., D. L. Krusac, and D. L. Carlson. 1991. An increase in a population of red-cockaded woodpeckers. *Wildl. Soc. Bull.* 19:277–286.

Hopkins, A. D. 1920. The bioclimatic law. *J. Wash. Acad. Sci.* 10:34–40.

Hopkins, M. N., Jr. 1975. The birdlife of Ben Hill County, Georgia and adjacent areas. *Georgia Ornithol. Soc. Occasional Publ. No.* 5.

Horn, H. S. 1975. Markovian properties of forest succession. Pp. 196–211 in *Ecology and evolution of communities,* M. L. Cody and J. M. Diamond, eds. Belknap Press, Cambridge.

Horn, H. S. 1978. Optimal tactics of reproduction and life history. Ch. 14 in *Behavioral ecology,* J. R. Krebs and N. B. Davies, eds. Sinauer Associates, Sunderland, Mass.

Hornaday, W. T. 1913. Our vanishing wildlife. Charles Scribner's Sons, New York.

Horowitz, N. H. 1945. On the evolution of biochemical synthesis. *Proc. Natl. Acad. Sci. USA* 31:153–157.

Horton, R. T., and A. V. Brown. 1991. Processing of green American elm leaves in first, third, and fifth order reaches of an Ozark stream. *J. Freshwater Ecol.* 6:115–119.

Houghton, R. A., et al., 1983. Changes in the carbon cycle of terrestrial biota and soils between 1860 and 1980: A new release of CO_2 to the atmosphere. *Ecol. Monogr.* 53:235–262.

Houston, D. C. 1974. The role of griffon vultures *Gyps africanus* and *Gyps ruppellii* as scavengers. *J. Zool. Lond.* 172:35–46.

Howard, H. 1955. Fossil birds with especial reference to the birds of Rancho LaBrea. *Los Angeles County Museum Contributions to Science Series No.* 17.

Howard, W. E. 1960. Innate and environmental dispersal of individual vertebrates. *Am. Midl. Nat.* 63: 152–161.

Howarth, R. W. 1984. The ecological significance of sulfur in the energy dynamics of salt marsh and coastal marine sediments. *Biogeochemistry* 1:5–27.

Hrbacek, J., M. Dvorakova, V. Korinek, and L. Prochazkova, 1961. Demonstration of the effect of the fish stock on the species composition of zooplankton and the intensity of metabolism of the whole plankton assemblage. *Verh. Int. Ver. Theoret. Angew. Limnol.* 14:192–195.

Huber, O., and G. T. Prance. 1986. Tropical savannas. *Nature Conserv. News* 36:19–23.

Hudson, P. J. 1986. The effect of a parasitic nematode on the breeding production of red grouse. *J. Anim. Ecol.* 55:85–92.

Huey, R. B., and E. R. Pianka. 1981. Ecological consequences of foraging mode. *Ecology* 62:991–999.

Huffaker, C. B. 1957. Fundamentals of biological control of weeds. *Hilgardia* 27:101–157.

Huffaker, C. B. 1958. Experimental studies on predation: Dispersion factors and predator-prey oscillations. *Hilgardia* 27:343–383.

Hughes, R. D. 1975. Introduced dung beetles and Australian pasture ecosystems. Papers presented at a symposium during the meeting of the Australian and New Zealand Association for the Advancement of Science at Canberra in January 1975.

Huheey, J. E., and A. Stupka. 1967. *Amphibians and reptiles of Great Smoky Mountain National Park.* University of Tennessee Press, Knoxville.

Humphrey, R. R. 1958. The desert grassland. *Univ. Ariz. Agr. Exp. St. Bull.* 299:1–62.

Hunt, G. L., Jr., Z. A. Eppley, and D. C. Schneider. 1986. Reproductive performance of seabirds: The importance of population and colony size. *Auk* 103: 306–317.

Hutchins, H. E. 1990. Whitebark pine seed dispersal and establishment: Who's responsible? Pp. 245–255 in *Proceedings of the Symposium on whitebark pine ecosystems: Ecology and management of a high-mountain resource. USDA For. Serv. Gen. Tech. Rep. INT* 270.

Hutchinson, G. E. 1948. Circular causal systems in ecology. *Ann. NY Acad. Sci.* 50:221–246.

Hutchinson, G. E. 1957. Concluding remarks. *Cold Spring Harb. Symp. Quant. Biol.* 22:415–427.

Hutchinson, G. E. 1965. *The ecological theater and the evolutionary play.* Yale University Press, New Haven, Conn.

Hutchinson, G. E. 1973. Eutrophication. *Am. Sci.* 61: 269–279.

Hutto, R. L., S. M. Pletschet, and P. Hendricks. 1986. A fixed point count method for nonbreeding and breeding season use. *Auk* 103:593–602.

Huxley, J. S. 1942. *Evolution: The modern synthesis* (Reprint 1964.) Wiley, New York.

Hyman, J. W. 1990. *The light book.* Tarcher, Inc., Los Angeles.

Hynes, H. B. N. 1970. *Ecology of running waters.* University of Toronto Press, Toronto, Canada.

Iason, G. R., and R. T. Palo. 1991. Effects of birch phenolics on a grazing and a browsing mammal: A comparison of hares. *J. Chem. Ecol.* 17:1733–1742.

Idso, S. B. 1985. The search for global CO_2 etc. "greenhouse effects." *Environ. Conserv.* 12:29–35.

Iltis, H. H. 1969. A requiem for the prairie. *Prairie Nat.* 1:51–57.

Ingham, E. R. 1985. Review of the effects of 12 selected biocides on target and nontarget soil organisms. *Crop Protection* 4:3–32.

Inglis, I. R., and N. J. K. Ferguson. 1986. Starlings search for food rather than eat freely available, identical food. *Anim. Behav.* 34:30–38.

Interagency Scientific Committee. 1991. An owl conservation strategy that works. *J. For.* 89:23–25.

Ito, Yo. 1980. *Comparative ecology.* Cambridge University Press, Cambridge.

Izawa, K. 1975. Foods and feeding behavior of monkeys in the upper Amazon Basin. *Primates* 16:295–316.

Jacklyn, P. 1991. Evidence for adaptive variation in the orientation of *Amitermes* (Isoptera: Termitinae) mounds from northern Australia. *Aust. J. Zool.* 39: 569–577.

Jackson, D. R. 1985. Florida's "desert" tortoise. *Nature Conserv. News* 35:24–26.

Jackson, R. R. 1992. Eight-legged tricksters. *Bioscience* 42: 590–598.

Jacobs, J. 1984. Cooperation, optimal density, and low density thresholds: Yet another modification of the logistic model. *Oecologia (Berl)* 64:389–395.

James, F. C. 1991. Signs of trouble in the largest remaining population of red-cockaded woodpeckers. *Auk* 108:419–423.

James, F. C., and W. J. Boeklen. 1984. Interspecific morphological relationships and the densities of birds. Pp. 458–477 in *Ecological communities . . . ,* D.R. Strong et al., eds. Princeton University Press, Princeton, N.J.

Jameson, D. A. 1963. Responses of individual plants to harvesting. *Bot. Rev.* 29:532–594.

Jannasch, H. W., and C. O. Wirsen. 1979. Chemosynthetic primary production at East Pacific sea floor spreading centers. *BioScience* 29:592–598.

Jannasch, H. W., and M. J. Mottl. 1985. Geomicrobiology of deep-sea hydrothermal vents. *Science* 229: 717–725.

Janovy, J., Jr., M. A. McDowell, and M. T. Ferdig. 1991. The niche of *Salsuginus thalkeni,* a gill parasite of *Fundulus zebrinus. J. Parasitol.* 77:697–702.

Janzen, D. H. 1966. Coevolution of mutualism between ants and acacias in Central America. *Evolution* 20: 249–275.

Janzen, D. H. 1969a. Birds and the ant × acacia interaction in Central America, with notes on birds and other myrmecophytes. *Condor* 71:240–256.

Janzen, D. H. 1969b. Seed-eaters versus seed size, number, toxicity, and dispersal. *Evolution* 23:1–27.

Janzen, D. H. 1971. Escape of *Cassia grandis* L. beans from predators in time and space. *Ecology* 52:964–979.

Janzen, D. H. 1975a. Behavior of *Hymenea courbaril* when its predispersal seed predator is absent. *Science* 189: 145–147.

Janzen, D. H. 1975b. Seed predation by animals. *Annu. Rev. Ecol. Syst.* 2:465–492.

Janzen, D. H. 1977. Why fruits rot, seeds mold, and meat spoils. *Am. Nat.* 111:691–713.

Järvinen, A. 1986. Clutch size of passerines in harsh environments. *Oikos* 46:365–371.

Jarvis, J. U. M. 1981. Eusociality in a mammal: Cooperative breeding in naked mole-rat colonies. *Science* 212:571–573.

Jeglum, J. K., A. N. Boisseonneau, and V. F. Havisto.

1974 (reprinted 1979). Toward a wetland classification for Ontario. *Can. For. Serv. Dep. Environ. Info. Rep. O-X* 215:1–54.

Jenkins, D., A. Watson, and G. R. Miller 1963. Population studies on red grouse, *Lagopus lagopus scoticus* in north-east Scotland. *J. Anim. Ecol.* 32:317–376.

Jenkins, D., A. Watson, and G. R. Miller 1964. Predation and red grouse populations. *J. Appl. Ecol.* 1:183–195.

Jewell, M. E., and H. Brown. 1924. The fishes of an acid lake. *Trans. Am. Microsc.* Soc. 43:77–84.

Johnson, L. K., and S. P. Hubbell. 1975. Contrasting foraging strategies and coexistence of two bee species on a single resource. *Ecology* 56:1398–1406.

Johnson, M. D. 1975. Seasonal and microseral variations in the insect populations on carrion. *Am. Midl. Nat.* 93:79–90.

Johnson, M. L., and M. S. Gaines. 1990. Evolution of dispersal: Theoretical models and empirical tests using birds and mammals. *Annu. Rev. Ecol. Syst.* 21: 449–480.

Johnston, R. F. 1961. Population movements of birds. *Condor* 63:386–389.

Johnston, R. F., and J. W. Hardy. 1962. Behavior of the purple martin. *Wilson Bull.* 74:243–262.

Johnston, R. S., and R. W. Brown. 1979. Hydrologic aspects related to the management of alpine areas. Pp. 65–75 in *Special management needs of alpine ecosystems.*

Jones, D. A., and D. A. Wilkins. 1971. *Variation and adaptation in plant species.* Heinemann Educational Books Ltd., London.

Jordan, C. F. 1971. A world pattern in plant energetics. *Am. Sci.* 59:425–433.

Jordan, C. F. 1982. Amazon rain forests. *Am. Sci.* 70: 394–401.

Jordan, C. F. 1983. Productivity of tropical rain forest ecosystems and the implications for their use as future wood and energy sources. Pp. 117–136 in *Tropical rainforest ecosystems,* F. B. Golley, ed. Elsevier Scientific Pub. Co., New York.

Jordan, D. S. 1928. The distribution of freshwater fishes. *Ann. Rep. Smithson. Inst.* 1927:355–385.

Jordan, W. R., III. 1981. Restoration and management notes: A beginning. *Restoration Management Notes* 1: 2.

Josens, G. 1983. The soil fauna of tropical savannas. III. The termites. Pp. 505–524 in *Ecosystems of the world: Tropical savannas,* F. Bourlière, ed. Elsevier Scientific Pub. Co., New York.

Joshi, A. C. 1933–1934. A suggested explanation of the prevalence of vivipary on the sea shore. *J. Ecol.* 21: 209–212; 22:306–307.

Joslin, J. D., J. M. Kelly, and H. Van Miegroet. 1992. Soil chemistry and nutrition of North American spruce-fir stands: Evidence for recent change. *J. Environ. Quality* 21:12–30.

Kale, H. W. 1965. Ecology and bioenergetics of the long-billed marsh wren *Telmatodytes palustris griseus* (Brewster) in Georgia salt marshes. *Nuttall Ornithological Club Publ.* No. 5, p. 142.

Kanamine, L. 1991. Everglades in a struggle for survival. *USA Today,* July 12:6A.

Karban, R. 1986. Interspecific competition between folivorous insects on *Erigeron glaucus. Ecology* 67:1063–1072.

Karl, D. M., C. O. Wirsen, and H. W. Jannasch. 1980. Deep-sea primary production at the Galápagos hydrothermal vents. *Science* 207:1345–1347.

Karns, P. D. 1967. *Pneumostrongylus tenuis* in deer in Minnesota and implications for moose *J. Wildl. Manage.* 31:299–303.

Karr, J. R., S. C. Robinson, J. G. Blake, and R. O. Bierregaard, Jr. 1991. Birds of four Neotropical forests. Pp. 237–251 in *Four neotropical rainforests,* A. H. Gentry, ed. Yale University Press, New Haven, Conn.

Kaufman, D. W., E. J. Finck, and G. A. Kaufman. 1990. Small mammals and grassland fires. Pp. 46–80 in *Fire in North American tallgrass prairie,* S. L. Collins and L. L. Wallace, eds. University of Oklahoma Press, Norman.

Kauppi, P. E., K. Mielikäinen, and K. Kuusela. 1992. Biomass and carbon budget of European forests, 1971 to 1990. *Science* 256:70–74.

Keddy, P. A., and A. A. Reznicek. 1986. Great Lakes vegetation dynamics: The role of fluctuating water levels and buried seeds. *J. Great Lakes Res.* 12:25–36.

Keeley, J. E. 1987. Role of fire in seed germination of woody taxa in California chaparral. *Ecology* 68:434–443.

Keeley, J. E., and S. C. Keeley. 1989. Allelopathy and the fire-induced herb cycle. Pp. 65–72 in *The California chaparral: Paradigms reexamined,* S. C. Keeley, ed. Natural History Museum of Los Angeles County Science Series No. 34.

Keeling, C. D. 1984. Atmospheric and oceanographic measurement needed for establishment of data base. Pp. 11–22 in *The potential effects of carbon dioxide-induced climatic changes in Alaska,* J. H. McBeath, ed. *Univ. Alaska Sch. Agri. Land Res. Man., Misc. Publ.* 83–1.

Keever, C. 1950. Causes of succession in old fields of the Piedmont, North Carolina. *Ecol. Monogr.* 20: 229–250.

Keim, De B. R. 1937 (originally published 1885). General Sheridan hunts the buffalo. Pp. 284–295 in *The pioneer West,* J. L. French, ed. Garden City Publishing, Garden City, New York.

Keith, L. 1963. *Wildlife's ten year cycle.* University of Wisconsin Press, Madison.

Kellogg, W. W., et al., 1972. The sulfur cycle. *Science* 175: 587–596.

Kelly, C. A. 1986. Extrafloral nectaries: Ants, herbivores and fecundity in *Cassia fasciculata. Oecologia (Berl)* 69:600–605.

Kelsall, J. P., and W. Prescott. 1971. Moose and deer behavior in snow in Fundy Park, New Brunswick. *Can. Wildl. Serv. Rep. Ser.* 15:1–25.

Kendeigh, S. C. 1941. Territorial and mating behavior of the House Wren. *Ill. Biol. Monogr.* 18:1–120.

Kendeigh, S. C. 1945. Community selection by birds on the Helderburg Plateau of New York. *Auk* 62:418–436.

Kendeigh, S. C. 1947. Bird population studies in the coniferous forest biome during a spruce budworm outbreak. *Ont. Dep. Lands For. Biol. Bull.* 1:1–100.

Kendeigh, S. C. 1949. Effect of temperature and season on energy resources of the English sparrow. *Auk* 66:113–127.

Kendeigh, S. C. 1954. History and evaluation of various concepts of plant and animal communities in North America. *Ecology* 35:152–171.

Kendeigh, S. C. 1974. *Ecology with special reference to animals and man.* Prentice-Hall, Englewood Cliffs, N.J.

Kendrick, B. 1991. Fungal symbioses and evolutionary innovations. Pp. 249–261 in *Symbiosis as a source of evolutionary innovation,* L. Margulis and René Fester, eds. MIT Press, Cambridge, Mass.

Kenfield, W. G. 1966. *The wild gardener in the wild landscape.* Hafner Pub. Co., New York.

Kennedy, C. R. 1972. The effect of the cestode *Caryophyllaeus laticeps* upon production and respiration of its intermediate host. *Parasitology* 64:485–499.

Kennedy, C. R. 1975. *Ecological animal parasitology.* John Wiley and Sons, New York.

Kennedy, C. R., A. O. Bush, and J. M. Aho. 1986. Patterns in helminth communities: Why are birds and fish different? *Parasitology* 93:205–215.

Kennedy, J. S. 1950. Aphid migration and the spread of plant viruses. *Nature* 165:1024–1025.

Kenoyer, L. A. 1929. General and successional ecology of the lower tropical rain-forest at Barro Colorado Island, Panama. *Ecology* 10:201–222.

Kenoyer, L. A. 1933. Opportunities in plant ecology. Pp. 13–20 in *Opportunity for investigation in natural history by high-school teachers.* University of Michigan Press, Ann Arbor.

Kenoyer, L. A. 1934. Forest distribution in southwestern Michigan as interpreted from the original land survey (1826–1832). *Pap. Mich. Acad. Sci. Arts Lett.* 19: 211–217.

Kenward, R. E. 1978. Hawks and doves: Factors affecting success and selection in goshawk attacks on woodpigeons. *J. Anim. Ecol.* 47:449–460.

Kerr, R. A. 1985. Wild string of winters confirmed. *Science* 227:506.

Kerr, R. A. 1986. There may be more than one way to make a volcanic lake a killer. *Science* 233:1257–1258.

Kerr, R. A. 1989. Nyos, the killer lake, may be coming back. *Science* 244:1541–1542.

Kerr, R. A. 1991a. Global temperature hits record again. *Science* 251:274.

Kerr, R. A. 1991b. Ozone destruction worsens. *Science* 252:204.

Kerr, R. A. 1992. Fugitive carbon dioxide: It's not hiding in the ocean. *Science* 256:35.

Kershaw, K. A. 1964. *Quantitative and dynamic ecology.* Elsevier, New York.

Kerster, H. W. 1968. Population age structure in the prairie herb, *Liatris aspera. BioScience* 18:430–432.

Kessler, D. A. et. al. 1992. The safety of foods developed by biotechnology. *Science* 256:1747–1832.

Kettle, D. S. 1951. The spacial distribution of *Culicoides impunctatus* Goet. under woodland and moorland conditions and its flight range through woodland. *Bull. Entomol. Res.* 42:239–291.

Kettlewell, H. B. D. 1973. *The evolution of melanism: The study of a recurring necessity.* Clarendon, Oxford.

Khan, M. B., and J. B. Harborne. 1991. A comparison of the effect of mechanical and insect damage on alkaloid levels in *Atropa acuminata. Biochem. Syst. Ecol.* 19:529–534.

Kiester, E., Jr. 1984. A little fever is good for you. *Science* 84:168–173.

Kilham, P. 1982. The biogeochemistry of bog ecosystems and the chemical ecology of Sphagnum. *Mich. Bot.* 21:159–167.

Kindschy, R. R., C. Sundstrom, and J. D. Yoakum. 1982. Wildlife habitats in managed rangelands—the Great Basin of southeastern Oregon. *USDA For. Serv. Gen. Tech. Rep. PNW* 145.

King, C. A. M. 1968. *An introduction to oceanography.* McGraw-Hill Book Co., Inc., New York.

King, J. E. 1981. Late Quaternary vegetational history of Illinois. *Ecology* 51:43–62.

King, J. R., and L. R. Mewaldt. 1987. The summer biology of an unstable insular population of white-crowned sparrows in Oregon. *Condor* 89:549–565.

Kira, T., and T. Shidei. 1967. Primary production and turnover of organic matter in different forest ecosystems of the Western Pacific. *Jpn. J. Ecol.* 17:70–87.

Kirkpatrick, J. F., and J. W. Turner, Jr. 1991. Compensatory reproduction in feral horses. *J. Wildl. Manage.* 55:649–652.

Kleiber, M. 1961. *The fire of life. An introduction to animal energetics.* Wiley, New York.

Kling, G. W. 1987. Seasonal mixing and catastrophic degassing in tropical lakes, Cameroon, West Africa. *Science* 237:1022–1024.

Klingsberg, C., and J. Duguid. 1982. Isolating radioactive wastes. *Am. Sci.* 70:182–190.

Kluger, M. J., D. H. Ringler, and M. R. Anver. 1975. Fever and survival. *Science* 188:166–168.

Klump, G. M., E. Kretzschmar, and E. Curio. 1986. The hearing of an avian predator and its avian prey. *Behav. Ecol. Sociobiol.* 18:317–323.

Kluyver, H. N., and L. Tinbergen. 1953. Territory and the regulation of density in titmice. *Arch. Neerl. Zool.* 10:265–289.

Knipe, O. D., C. P. Pase, and R. S. Carmichael. 1979. Plants of the Arizona chaparral. *USDA For. Serv. Gen. Tech. Rep. RM* 64.

Knoll, A. H. 1986. Patterns of change in plant communities through geological times. Pp. 126–141 in *Community ecology,* J. Diamond and T. J. Chase, eds. Harper & Row, New York.

Koford, C. B. 1958. Prairie dogs, whitefaces, and blue grama. *Wildl. Monogr.* 3:1–79.

Kogan, M. 1986. Natural chemicals in plant resistance to insects. *Iowa State J. Res.* 60:501–527.

Kohshima, S. 1984. A novel, cold-tolerant insect found in a Himalayan glacier. *Nature* 310:225–227.

Kolata, G. 1985. Avoiding the schistosome's tricks. *Science* 227:285–287.

Komarek, E. V. 1969. Fire and animal behavior. *Proc. Annu. Tall Timbers Fire Ecol. Con.* No. 9, pp. 161–207.

Komarkova, V., S. Poncet, and J. Poncet. 1985. Two native Antarctic vascular plants, *Deschampsia antarctica* and *Colobanthus quitensis:* A new southernmost locality and other localities in the Antarctic peninsula area. *Arct. Alp. Res.* 17:401–416.

Koniak, S. 1985. Succession in pinyon-juniper woodlands following wildfire in the Great Basin. *Great Basin Nat.* 45:556–566.

Kormanik, P. P. 1983. Third-year seed production in outplanted sweetgum related to nursery root colonization by endomycorrhizal fungi. *Proc. 17th S. For. Tree Improve. Conf.*:49–54.

Kormanik, P. P., W. C. Bryan, and R. C. Schultz. 1977. The role of mycorrhizae in plant growth and development. *Proc. Symp. South. Sect. Am. Soc. Plant Physiol.*

Kozlovsky, T. T., and C. E. Ahlgren, eds. 1974. *Fire and ecosystems.* Academic Press, New York.

Kozlowski, D. G. 1968. A critical evaluation of the trophic level concept. 1. Ecological efficiencies. *Ecology* 49:48–116.

Kramer, P. J. 1969. *Plant and soil water relationships: A modern synthesis.* McGraw-Hill, New York.

Krantz, G. W. 1978. *A manual of acarology.* Oregon State University Bookstore, Corvallis.

Kroeze, C., and L. Reijnders. 1992. Halocarbons and global warming. *Sci. Total Environ.* 111:1–24.

Kroodsma, R. L. 1984. Effect of edge on breeding forest bird species. *Wilson Bull.* 96:426–436.

Kucera, C. L. 1960. Forest encroachment in native prairie. *Iowa State J. Sci.* 34:635–640.

Kuhn, T. S. 1970. The structure of scientific revolutions. 2nd ed, enlarged. In *Foundations of the unity of science.* Vol. 2. University of Chicago Press, Chicago.

Kullenberg, B. 1961. Studies in *Ophrys* pollination. *Zool. Bidr. Upps.* 34:1–340.

Kummer, H. 1968. *Social organization of the hamadryas baboons: A field study.* University of Chicago Press, Chicago.

Kummerow, J. 1982. The relation between root and shoot systems in chaparral shrubs. Pp. 142–147 in *Dynamics and management of Mediterranean-type ecosystems. USDA For. Serv. Gen. Tech. Rep. PSW* 58.

Kurtén, B. 1958. Life and death of the Pleistocene cave bear. *Acta Zool. Fenn.* 95:1–59.

Kurtén, B. 1976. *The cave bear story.* Columbia University Press, New York.

Kurtz, D. R. 1981. Flora of limestone glades in Illinois. *Proc. 6th North Am. Prairie Conf.:* 183–186.

Kushlan, J. A. 1990. Fresh-water marshes. Pp. 324–363 in *Ecosystems of Florida,* R. L. Myers and J. J. Ewel, eds. University of Central Florida Press, Orlando.

Lack, D. 1944. Ecological aspects of species formation in passerine birds. *Ibis* 86:260–286.

Lack, D. 1947. *Darwin's finches.* Cambridge University Press, Cambridge.

Lack, D. 1966. *Population studies of birds,* Clarendon Press, Oxford.

Lack, D. 1968. *Ecological adaptations for breeding in birds.* Methuen & Co., Ltd., London.

Laessle, A. M. 1942. The plant communities of the Welaka area. *Univ. Florida Pub. Biol. Sci. Ser. 4:*1–143.

Laessle, A. M. 1958. The origin and successional relationship of sandhill vegetation and sand-pine scrub. *Ecol. Monogr.* 28:361–387.

Laessle, A. M. 1961. A microlimnological study of Jamaican bromeliads. *Ecology* 42:499–517.

Laessle, A. M. 1965. Spacing and competition in natural stands of sand pine. *Ecology* 46:65–72.

LaGory, K. E. 1987. The influence of habitat and group characteristics on the alarm and flight response of white-tailed deer. *Anim. Behav.* 35:20–25.

Lahti, S., J. Tast, and H. Uotila. 1976. Pikkujyrsijöiden kannanvaihteluista Kilpisjävellä vuosina 1950–1975. *Luonnon Tutkija* 80:97–107.

Lamont, B. B., D. C. Le Matire, R. M. Cowling, and N. J. Enright. 1991. Canopy seed storage in woody plants. *Bot. Rev.* 57:277–317.

Lamotte, M. 1972. Bilan energḗtique de la croissance du male *Nectrophrynoides occidentalis* Angel, Amphibien Anoure. *C. R. Acad. Sci.* 274:2074–2076.

Lamotte, M. 1977. Observations preliminaires sur les flux d'energie dans un ecosysteme herbace tropical, la savane de Lamto (Côte d'Ivoire). *Geo-Eco-Trop* 1:45–63.

Lamotte, M., and F. Bourlière. 1983. Energy flow and nutrient cycling in tropical savannas. Pp. 583–603 in *Ecosystems of the world 13: Tropical savannas,* F. Bourlière, ed. Elsevier, Amsterdam.

Lande, R., and G. F. Barrowclough. 1987. Effective population size, genetic variation, and their use in population management. Pp. 87–124 in *Viable populations for conservation.* M. E. Soulé, ed. Cambridge University Press, Cambridge.

Landers, J. L., N. A. Byrd, and R. Komarek. 1989. A holistic approach to managing longleaf pine communities. Pp. 135–167 in *Proceedings of the Symposium on the management of longleaf pine,* R. M. Farrar, Jr., ed. *USDA For. Serv. Gen. Tech. Rep. SO* 75.

Landsberg, H. E. 1961. Solar radiation at the earth's surface. *Sol. Energy* 5:95–98.

Larrimore, R. W., O. H. Pickering, and L. Durham. 1952. An inventory of the fishes of Jordan Creek, Vermilion County, Illinois. *Ill. Nat. Hist. Surv. Biol. Notes* 29:1–26.

Larrison, E. J., and K. G. Sonnenberg. 1968. *Washington birds: Their location and identification.* Seattle Audubon Society, Seattle.

Larsen, C. E. 1986. Long-term trends in Lake Michigan levels, a view from the geologic record. Pp. 5–22 in *Proceedings of the 1st Indiana Dunes Research Conference* U.S. Department of the Interior, National Park Service, Atlanta, Ga.

Larson, P. 1970. *Deserts of America.* Prentice-Hall, Englewood Cliffs, N.J.

Larson, W. E., F. J. Pierce, and R. H. Dowdy. 1983. The threat of soil erosion to long term crop production. *Science* 219:458–465.

Lawrence, D. B. 1958. Glaciers and vegetation in southeastern Alaska. *Am. Sci.* 46:89–122.

Lawrence, G. E. 1966. Ecology of vertebrate animals in relation to chaparral fire in the Sierra Nevada foothills. *Ecology* 47:278–291.

Lawson, D. L., R. W. Merritt, M. J. Klug, and J. S. Martin. 1982. The utilization of late season foliage by the orange striped oakworm, *Anisota senatoria. Entomol. Exp. Appl.* 32:242–248.

Leach, G. 1976. *Energy and food production.* IPC Science and Technology Press, Guildford, England.

Leahy, C. 1984. *The birdwatcher's companion.* Bonanza Books, New York.

Leak, W. B. 1974. Some effects of forest preservation. *USDA For. Serv. Res. Note NE* 186:1–4.

Leak, W. B. 1982. Habitat mapping and interpretation in New England, *USDA For. Serv. Res. Pap. NE* 496: 1–28.

Leak, W. B. 1991. Secondary forest succession in New Hampshire, USA. *For. Ecol. Manage.* 43:69–86.

Leak, W. B., and J. R. Riddle. 1983. Why trees grow where they do in New Hampshire forests. *USDA For. Serv.,* NE For. Exp. Stn. NE-INF-37–79.

Leak, W. B., and R. E. Graber. 1974. Forest vegetation related to elevation in the White Mountains of New Hampshire. *USDA For. Serv. Res. Pap. NE* 299:1–7.

LeBarron, R. K. 1939. The role of forest fires in the reproduction of black spruce. *Proc. Minn. Acad. Sci.* 7: 10–14.

Leck, M. A., V. T. Parker, and R. L. Simpson. 1989. *Ecology of soil seed banks.* Academic Press, San Diego.

LeCren, E. D., and R. H. Lowe-McConnell. 1980. *The functioning of freshwater ecosystems.* Cambridge University Press, London.

Lee, R. E., Jr., and D. L. Denlinger. 1991. *Insects at low temperature.* Chapman and Hall, New York.

Le Houérou, H. N. 1992. Outline of the biological history of the Sahara. *J. Arid Environ.* 22:3–30.

Leigh, E. G., Jr. 1981. The average lifetime of a population in a varying environment. *J. Theor. Biol.* 90: 213–239.

Lemel, J. 1989. Habitat distribution in the great tit *Parus major* in relation to reproductive success, dominance, and biometry. *Ornis Scand.* 20:226–233.

Lemonick, M. D. 1989. The two Alaskas. *Time* April 17: 56–66.

Lennartz, M. R. 1988. The red-cockaded woodpecker: Old-growth species in a second-growth landscape. *Nat. Areas J.* 3:160–165.

Lennartz, M. R., and J. P. McClure. 1979. Estimating the extent of red-cockaded woodpecker habitat in the southeast. Selected Reprints from the 1979 Workshop on Forest Resources Inventories.

Leopold, A. 1933. *Game management.* Charles Scribner's Sons, New York.

Leopold, A. 1949. *A Sand County almanac and sketches here and there.* Oxford University Press, London.

Leopold, A., and S. E. Jones. 1947. A phenological record for Sauk and Dane counties, Wisconsin, 1933–1945. *Ecol. Monogr.* 17:81–122.

Leopold, A. S., M. Erwin, J. Oh, and B. Browning. 1976. Phytoestrogens: Adverse effects on reproduction in California quail. *Science* 191:98–100.

Leslie, P. H. 1945. On the use of matrices in certain population mathematics. *Biometrika* 33:183–212.

Leslie, P. H. 1966. The intrinsic rate of increase and the overlap of successive generations in a population of guillemots *(Uria aalge). J. Anim. Ecol.* 35:291–301.

Leslie, P. H., and R. M. Ranson. 1940. The mortality, fertility, and rate of natural increase of the vole *(Microtus agrestis)* as observed in the laboratory. *J. Anim. Ecol.* 9:27–52.

Leverich, W. J., and D. A. Levin. 1979. Age-specific survivorship and reproduction in *Phlox drummondi. Am. Nat.* 113:881–903.

Levins, R. 1970. Extinction. Pp. 75–108 in *Some mathematical questions in biology,* Vol. II. American Mathematical Society, Providence, R.I.

Levins, R., and R. Lewontin. 1985. *The dialectical biologist.* Harvard University Press, Cambridge, Mass.

Lewin, R. 1985. Parkinson's disease: An environmental cause? *Science* 229:257–258.

Li, W. K. W. 1983. Autotrophic picoplankton in the tropical ocean. *Science* 219:292–295.

Lieth, H., and J. S. Radford. 1971. Phenology, resource management, and synographic computer mapping. *BioScience* 21:62–70.

Lifjeld, J. T., T. Slagsvold, and H. M. Lampe. 1991. Low frequency of extrapair paternity in pied flycatchers revealed by DNA fingerprinting. *Behav. Ecol. Sociobiol.* 29:95–101.

Ligon, J. D. 1978. Reproductive interdependence of piñon jays and piñon pines. *Ecol. Monogr.* 48:111–126.

Likens, G. E., and F. H. Bormann. 1972. Nutrient cycling in ecosystems. In *Ecosystem structure and function,* J. A. Wiens, ed. Oregon State University Press, Corvallis.

Likens, G. E., et al. 1970. The effects of forest cutting and herbicide treatment on nutrient budgets in the Hubbard Brook watershed ecosystem. *Ecol. Monogr.* 40:23–47.

Likens, G. E., F. H. Bormann, R. S. Pierce, J. S. Eaton, and N. M. Johnson. 1977. *Biogeochemistry of a forested ecosystem.* Springer-Verlag, New York.

Lima, S. L., T. J. Valone, and T. Caraco. 1985. Foraging-efficiency—predation-risk trade-off in the grey squirrel. *Anim. Behav.* 33:155–165.

Limbaugh, C. 1961. Cleaning symbioses. *Sci. Am.* 205: 42–49.

Lindeman, R. L. 1942. The trophic-dynamic aspect of ecology. *Ecology* 23:399–418.

Lindemann, J., H. A. Constantinidou, W. R. Barchet, and C. D. Upper. 1982. Plants as sources of airborne bacteria, including ice nucleation-active bacteria. *Appl. Environ. Microbiol.* 44:1059–1063.

Lindén, M., L. Gustafsson, and T. Pärt. 1992. Selection on fledgling mass in the collared flycatcher and the great tit. *Ecology* 73:336–343.

Lindroth, R. L., and G. O. Batzli. 1986. Inducible plant chemical defenses: A cause of vole population cycles? *J. Anim. Ecol.* 55:431–449.

Lindsey, A. A., D. V. Schmelz, and S. A. Nichols. 1969. *Natural areas in Indiana and their preservation.* Indiana Natural Areas Survey, Lafayette.

Lipfert, F. W. 1985. Mortality and air pollution: Is there a meaningful connection? *Environ. Sci. Technol.* 19: 764–770.

Little, E. L., Jr. 1977. Research in the pinyon-juniper woodland. Pp. 8–19 in *Ecology, uses and management of pinyon-juniper woodlands. USDA For. Serv. Gen. Tech. Rep. RM* 39.

Llano, G. A. 1962. The terrestrial life of the Antarctic. *Sci. Am.* 207:213–230.

Lloyd, J. E. 1965. Aggressive mimicry in *Photuris:* Firefly femme fatales. *Science* 149:653–654.

Lloyd, J. E. 1980. Male *Photuris* mimic sexual signals of their female's prey. *Science* 210:669–671.

Lockeretz, W., et al. 1975. A comparison of the production, economic returns, and energy intensiveness of cornbelt farms that do and do not use inorganic fertilizers and pesticides. *Ctr. Biol. Nat. Syst. Rep. No.* CBNS-AE-4.

Loehle, C. 1987. Tree life history strategies: The role of defenses. Can. J. For. Res. 18:209–222.

Loehle, C. 1988. Problems with the triangular model for representing plant strategies. *Ecology* 59:284–286.

Loevinsohn, M. E. 1987. Insecticide use and increased mortality in rural central Luzon, Philippines. *Lancet* 1987:1359–1362.

Logan, K. T. 1965. Growth of tree seedlings as affected by light intensity. I. White birch, yellow birch, sugar maple and silver maple. *Can. Dept. For. Publ.* 1121: 1–16.

Logan, K. T. 1973. Growth of tree seedlings as affected by light intensity. V. White ash, beech, eastern hemlock, and general conclusions. *Can. For. Serv. Dept. Environ. Publ.* 1323:1–12.

Longhurst, W. M., A. S. Leopold, and R. F. Dasmann. 1952. A survey of California deer herds, their ranges and management problems. *Calif. Dept. Fish Game Bull.* No. 6.

Lopreato, J. 1984. *Human nature & biocultural evolution.* Allen & Unwin, Boston.

Lord, R. D. 1961. A population study of the gray fox. *Am. Midl. Nat.* 66:87–109.

Lotka, A. J. 1925. *Elements of physical biology.* Williams and Wilkens, Baltimore.

Loucks, O. L. 1970. Evolution of diversity, efficiency, and community stability. *Am. Zool.* 10:17–26.

Lovat, L., ed. 1911. *The grouse in health and disease.* Smith and Elder, London.

Lovejoy, T. 1991. Deforestation. Pp. 89–91 in *1992 Earth Journal.* Buzzworm Books, Boulder, Colo.

Lovelock, J. E. 1979. *Gaia: A new look at life.* Oxford University Press, New York.

Lovins, A. B., and L. H. Lovins. 1980. *Energy/war: Breaking the nuclear link.* Harper and Row, New York.

Lowrie, D. C. 1948. The ecological succession of spiders in the Chicago area dunes. *Ecology* 29:334–51.

Ludlum, D. A. 1971. *Weather record book.* Weatherwise, Inc., Princeton, N.J.

Ludwig, J. P. 1979. Present status of the Caspian tern population of the Great Lakes, *Mich. Acad.* 12:69–77.

Ludwig, J. P. 1984. Decline, resurgence and population dynamics of Michigan and Great Lakes double-crested cormorants. *Jack Pine Warbler* 62:91–102.

Ludwig, J. P. 1991. Double-crested Cormorant. Pp. 102–103 in *The atlas of breeding birds of Michigan,* R. Brewer et al., eds. Michigan State University Press, East Lansing.

Lugo, A., M. M. Brinson, and S. Brown. 1989. *The forested wetlands.* Elsevier Scientific Publishers, Amsterdam.

Lugo, A. E. 1983. Influence of green plants on the world carbon budget. Pp. 391–398 in *Alternative energy sources. V. Part E: Nuclear/conservation/environment.* Elsevier, Amsterdam.

Lugo, A. E. 1990. Fringe wetlands. Pp. 143–169 in *The forested wetlands.* Elsevier, Amsterdam.

Lugo, A. E. 1992. More on exotic species. *Conserv. Biol.* 6:6.

Lugo, A. E., and S. Brown. 1991. Comparing tropical and temperate forests. Pp. 319–330 in *Comparative analyses of ecosystems,* J. Cole, G. Lovett, and S. Findlay, eds. Springer-Verlag, New York.

Lukens, J. O. 1990. *Directing ecological succession.* Chapman and Hall, London.

Lull, H. W. 1968. *A forest atlas of the Northeast.* USDA For. Serv.

Lundelius, E. L., Jr., et al. 1983. Terrestrial vertebrate fauna. Ch. 16 in *Late-Quaternary environments of the United States,* S. C. Porter, ed. University of Minnesota Press, Minneapolis.

Macan, T. T. 1963. Freshwater ecology. Longmans, London.

MacArthur, R. H. 1955. Fluctuations of animal populations and a measure of community stability. *Ecology* 36:533–536.

MacArthur, R. H. 1957. On the relative abundance of bird species. *Proc. Natl. Acad. Sci. USA* 45:293–295.

MacArthur, R. H. 1958. Population ecology of some warblers of northeastern coniferous forests. *Ecology* 39:599–619.

MacArthur, R. H. 1972. *Geographical ecology.* Harper and Row, New York.

MacArthur, R. H., and E. O. Wilson. 1967. The theory of island biogeography. *Princeton Univ. Press. Monogr. Popul. Biol.* 1:1–203.

MacArthur, R. H., and E. R. Pianka. 1966. On optimal use of a patchy environment. *Am. Nat.* 100:276–282.

MacArthur, R. H., and R. Levins. 1967. The limiting similarity, convergence, and divergence of coexisting species. *Am. Nat.* 101:377–385.

MacDougall, W. B. 1931. *Plant ecology.* 2nd ed. (3rd ed. 1941). Lea and Febiger, Philadelphia.

MacLean, S. F., Jr. 1980. The detritus-based trophic system. Pp. 411–457 in *An arctic ecosystem,* J. Brown et al., eds. Dowden, Hutchinson & Ross, Stroudsburg Pa.

MacNab, J. A. 1958. Biotic aspection in the coast range mountains of northwestern Oregon. *Ecol. Monogr.* 28:21–53.

Mal, D. H. 1991. Paleofloristic changes in Europe and the confirmation of the Arctotertiary-Paleotropical geofloral concept. *Rev. Palaeobot. Palynol.* 68:29–36.

Malcolm, S. B. 1986. Aposematism in a soft-bodied insect: A case for kin selection. *Behav. Evol. Sociobiol.* 18:387–393.

Malcolm, S. B. 1990. Mimicry: Status of a classical evolutionary paradigm. *Trends Ecol. Evol.* 5:57–62.

Malcolm, W. M. 1966. Biological interactions. *Bot. Rev.* 32:243–254.

Malthus, T. R. 1798. *An essay on the principle of population.* London. (American ed. 1959. *Population: The first essay.* University of Michigan Press, Ann Arbor.)

Marchand, P. J., F. L. Goulet, and T. C. Harrington. 1986. Death by attrition: A hypothesis for wave mortality of subalpine *Abies balsamea. Can. J. For. Res.* 16:591–596.

Margalef, R. 1958. Information theory in ecology. *Gen. Syst.* 3:36–71.

Margulis, L. 1981. *Symbiosis in cell evolution.* Freeman, San Francisco.

Marks, P. L. 1974. The role of pin cherry (*Prunus pensylvanica* L.) in the maintenance of stability in northern hardwood ecosystems. *Ecol. Monogr.* 44:73–88.

Marquis, D. A. 1967. Clearcutting in northern hardwoods: Results after 30 years. *USDA For. Serv. Res. Pap. NE* 85:1–13.

Marquis, D. A. 1974. The impact of deer browsing on Allegheny hardwood regeneration. *USDA For. Res. Pap. NE* 308.

Marquis, R. J. 1991. Evolution of resistance in plants to herbivores. *Evol. Trends Plants* 5:23–29.

Marrs, R. H. 1987. Studies on the conservation of lowland Calluna heaths. *J. Appl. Ecol.* 24:163–175.

Marshall, J. T. 1963. Rainy season nesting in Arizona. *Proc. 13th Int. Ornith. Congr.* 2:620–622.

Marshall, L. G., et al. 1982. Mammalian evolution and the great American interchange. *Science* 215:1351–1357.

Martin, P. S. 1973. The discovery of America. *Science* 179:969–974.

Martin, P. S., and B. E. Harrell. 1957. The Pleistocene history of temperate biotas in Mexico and eastern United States. *Ecology* 38:468–480.

Martin, P. S., and R. G. Klein, eds. 1984. *Quaternary extinctions.* University of Arizona Press, Tucson.

Martin, R. E. 1983. Fire history and its role in succession. Pp. 92–98 in *Forest succession and stand development in the Northwest: Proceedings of a symposium,* J. E. Means, ed. USDA For. Serv., Corvallis, Oregon.

Martin, S. C. 1983. Responses of semidesert grasses and shrubs to fall burning. *J. Range Manage.* 36:604–610.

Martin, T. E. 1987. Food as a limit on breeding birds: A life history perspective. *Annu. Rev. Ecol. Syst.* 18: 453–487.

Martindale, S. 1982. Nest defense and central place foraging: A model and experiment. *Behav. Ecol. Sociobiol.* 10:85–89.

Marx, J. L. 1985. The immune system "belongs to the body." *Science* 227:1190–1192.

Maser, C., M. J. Geist, D. M. Concannon, R. Anderson, and B. Lovell. 1979. Wildlife habitats in managed rangelands—the great basin of southeastern Oregon. *USDA For. Serv. Gen. Tech. Rep. PNW* 99.

Mathews, J. 1988. Yellowstone reborn. Detroit News September 22, sec. F: 1.

Maun, M. A. 1985. Population biology of *Ammophila breviligulata* and *Calamovilfa longifolia* on Lake Huron sand dunes. I. Habitat, growth form, reproduction and establishment. *Can. J. Bot.* 73:113–124.

Maximov, N. A. 1931. The physiological significance of the xeromorphic structure of plants. *J. Ecol.* 19:273–282.

Maxwell, F. G., and P. R. Jennings, eds. 1980 *Breeding plants resistant to insects.* Wiley, New York.

May, J. M. 1960. The ecology of human disease. *Ann. N.Y. Acad. Sci.* 84:789–794.

May, M. 1991. Aerial defense tactics of flying insects. *Am. Sci.* 79:316–328.

May, R. 1974. Biological populations with nonoverlapping generations: Stable points, stable cycles and chaos. *Science* 186:645–647.

May, R. 1976. Simple mathematical models with very complicated dynamics. *Nature* 261:459–467.

May, R. 1989. Detecting density dependence in imaginary worlds. *Nature* 338:16–17.

May, R. M. 1974. Stability and complexity in model ecosystems. 2nd ed. *Princeton Monogr. Popul. Biol.* 6:1–265.

May, R. M. 1983. Parasitic infections as regulators of animal populations. *Am. Sci.* 71:36–45.

May, R. M., and R. M. Anderson. 1987. Transmission dynamics of HIV infection. *Nature* 326:137–142.

Mayewski, P. A., et al. 1986. Sulfate and nitrate concentrations from a south Greenland ice core. *Science* 232:975–977.

Mayfield, H. 1965. The brown-headed cowbird, with old and new hosts. *Living Bird* 4:13–28.

Mayfield, H. 1977. Brown-headed cowbird: Agent of extinction? *Am. Birds* 31:107–113.

Mayfield, H. F. 1983. Kirtland's warbler, victim of its own rarity? *Auk* 100:974–976.

Maynard Smith, J. 1964. Kin selection and group selection. *Nature* 201:1145–1147.

McAdam, D. 1992, Radicals and others. *Science* 255: 1448–1450.

McArthur, J. V., and J. R. Barnes. 1988. Community dynamics of leaf litter breakdown in a Utah alpine stream. *J. North Am. Benthol. Soc.* 7:37–43.

McCann, H. 1991. Ecologist strives to prevent Africa lake explosion. *Detroit News* December 23:2E.

McCune, B., and T. F. H. Allen. 1985. Will similar forests develop on similar sites? *Can. J. Bot.* 63:367–376.

McDonald, M. E. 1985. Growth of a grazing phytoplanktivorous fish and growth enhancement of the grazed alga. *Oecologia* 67:132–136.

McFall-Ngai, M. J. 1991. Luminous bacterial symbiosis in fish evolution: Adaptive radiation among the Leiognatid fishes. Pp. 380–409 in *Symbiosis as a source of evolutionary innovation,* L. Margulis and René Fester, eds. MIT Press, Cambridge, Mass.

McGinley, M. A., and T. G. Whitham. 1985. Central place foraging by beavers *(Castor canadensis):* A test of foraging predictions and the impact of selective feeding on the growth form of cottonwoods (*Populus fremontii*). *Oecologia (Berl)* 66:558–562.

McHarg, I. 1969. *Design with nature.* Doubleday, New York.

McIntosh, R. P. 1985. *The background of ecology.* Cambridge University Press, Cambridge.

McKeown, T. 1979. *The role of medicine: dream, mirage, or nemesis.* Princeton University Press, Princeton, N.J.

McLachlan, A. J., and M. A. Cantrell. 1980. Survival strategies in tropical rain pools. *Oecologia (Berl)* 47: 344–351.

McNeil, J. N., and R. E. Stinner. 1983. Seasonal biology in Quebec and prediction of phenology of the European Skipper, *Thymelius lineola* (Lepidoptera: Hesperiidae). *Can. Entomol.* 115:905–911.

McNeill, W. H. 1976. *Plagues and peoples.* Anchor Press/ Doubleday, Garden City, New York.

McPeek, G. A., and R. Brewer. 1991. Priorities and perspectives in conservation. Pp. 87–94 in *The atlas of breeding birds of Michigan,* R. Brewer, G. A. McPeek, and R. J. Adams, Jr., eds. Michigan State University Press, East Lansing.

McPherson, J. K., and C. H. Muller. 1969. Allelopathic effects of *Adenostoma fasiculatum,* "chamise," in the California chapparal. *Ecol. Monogr.* 39:177–198.

Medin, D. E., and W. P. Clary. 1990. Bird and small mammal populations in a grazed and ungrazed riparian habitat in Idaho. *USDA For. Serv. Res. Pap. INT* 425:1–8.

Meentemeyer, V. 1978. Macroclimate and lignin control of litter decomposition rates. *Ecology* 59:465–472.

Meffe, G. K., and A. L. Sheldon. 1990. Post-defaunation recovery of fish assemblages in southeastern blackwater streams. *Ecology* 71:657–667.

Mellinger, M. V., and S. J. McNaughton. 1975. Structure and function of successional vascular plant communities in central New York. *Ecol. Monogr.* 45:161–182.

Meredith, C. W., A. M. Gilmore, and A. C. Isles. 1984. The ground parrot (*Pezoporus wallicus* Kerr) in south-eastern Australia: A fire-adapted species. *Aust. J. Ecol.* 9:367–380.

Meredith, D. H. 1977. Interspecific agonism in two paraptric species of chipmunks *(Eutamias). Ecology* 58: 423–430.

Merriam, C. H. 1890. Results of a biological survey of the San Francisco Mountain region and desert of the Little Colorado, Arizona. *USDA North Am. Fauna* 3:1–136.

Messier, F., J. A. Virgl, and L. Marinelli. (1990) Density-dependent habitat selection in muskrats: A test of the ideal free distribution model. *Oecologia (Berl)* 84: 380–385.

Michod, R. E., and W. W. Anderson. 1980. On calculating demographic parameters from age frequency data. *Ecology* 61:263–269.

Miller, S. L. 1955. Production of some organic compounds under possible primitive earth conditions. *J. Am. Chem. Soc.* 77:2351–2361.

Miller, T. E., and W. C. Kerfoot. 1987. Redefining *indirect effects.* Pp. 33–37 in *Predation, direct and indirect effects on aquatic communities,* W. C. Kerfoot and A. Sih, eds. University Press of New England, Hanover, New Hampshire.

Mills, A. M. 1986. The influence of moonlight on the behavior of goatsuckers (Caprimulgidae). *Auk* 103: 370–378.

Milne, L. J., and M. Milne. 1976. The social behavior of burying beetles. *Sci. Am.* 235:84–89.

Minchella, D. J., and M. E. Scott. 1991. Parasitism: A cryptic determinant of animal community structure. *Trends Ecol. Evol.* 6:250–254.

Minckley, W. L. 1963. The ecology of a spring stream. Doe Run, Meade County, Kentucky. *Wildl. Monogr.* 11:5–124.

Minckley, W. L. 1991. Native fishes of arid lands: A dwindling resource of the desert Southwest. *USDA For. Serv. Gen. Tech. Rep. RM* 206.

Möbius, K. 1877. Die Auster und die Austernwirtschaft. Berlin. (Transl. 1880. The oyster and oyster culture. *Rep. US Fish Comm.,* 1880:683–751.)

Mock, D. W. 1984. Infanticide, siblicide, and avian nestling mortality. Pp. 3–30 in *Infanticide: Comparative and evolutionary perspectives,* G. Hausfater and S. B. Hrdy, eds. Aldine, New York.

Moffat, A. S. 1991. Research on biological pest control moves ahead. *Science* 252:211–212.

Mohr, C. O. 1943. Cattle droppings as ecological units. *Ecol. Monogr.* 13:280–281.

Molisch, H. 1937. *Der Einfluss einer Pflanze auf die andere-Allelopathie.* Fischer, Jena.

Molles, M. C., Jr. 1980. Effects of road salting on aquatic invertebrate communities. *Eisenhower Consortium Bull.* 10:1–9.

Montgomery, R. F., and G. P. Askew. 1983. Soils of tropical savannas. Pp. 63–78 in *Tropical Savannas,* F. Bourlière, ed. Elsevier, Amsterdam.

Moodie, G. E. E., J. D. McPahil, and D. W. Hagen. 1973. Experimental demonstration of selective predation on *Gasterosteus aculeatus.* Behaviour 47:95–105.

Moore, J. 1984. Parasites that change the behavior of their hosts. *Sci. Am.* 250:108–115.

Moreau, R. E. 1966. *The bird faunas of Africa and its islands.* Academic Press, New York.

Morehouse, E. L., and R. Brewer. 1968. Feeding of nestling and fledgling eastern kingbirds. *Auk* 85:44–54.

Morris, D. 1965. *The mammals.* Harper and Row, New York.

Morse, A. N. C. 1991. How do planktonic larvae know where to settle? *Am. Sci.* 79:154–167.

Morton, E. S. 1973. On the evolutionary advantages and disadvantages of fruit eating in tropical birds. *Am. Nat.* 107:8–22.

Mosimann, J. E., and P. S. Martin. 1975. Simulating overkill by Paleoindians. *Am. Sci.* 63:303–313.

Moss, M. R., ed. 1987. *Landscape ecology and management.* Polyscience Publications, Inc., Montreal, Canada.

Mount, R. H. 1975. *The reptiles and amphibians of Alabama.* Auburn Printing Co., Auburn, Ala.

Mueggler, W. F., and W. L. Stewart. 1980. Grassland and shrubland habitat types of western Montana. *USDA For. Serv. Gen. Tech. Rep. INT* 66.

Mueller, L. D., P. Guo, and F. J. Ayala. 1991. Density-dependent natural selection and trade-offs in life history traits. *Science* 253:433–435.

Mueller-Dombois, D., and H. Ellenberg. 1974. *Aims and methods of vegetation ecology.* Wiley, New York.

Muir, J. 1961. *The mountains of California* (originally published 1893). Doubleday, New York.

Mullen, D. A. 1969. Reproduction in brown lemmings (*Lemmus trimucronatus*) and its relevance to their cycle of abundance. *Univ. Calif. Publ. Zool.* 85:1–24.

Muller, C. H. 1965. Inhibitory terpenes volatilized from Salvia shrubs. *Bull. Torrey Bot. Club* 93:332–351.

Mullineaux, L. S., P. H. Wiebe, and E. T. Baker. 1991. Hydrothermal vent plumes: Larval highways in the deep sea? *Oceanus* 34:64–68.

Mundinger, P. C., and S. Hope. 1982. Expansion of the winter range of the House Finch, 1949–1979. *Am. Birds* 36:347–353.

Munger, J. C., and W. H. Karasov. 1989. Sublethal parasites and host energy budgets: Tapeworm infection in white-footed mice. *Ecology* 70:904–921.

Munn, R. E. 1970. *Biometeorological methods.* Academic Press, New York.

Munz, P. A., and D. D. Keck. 1968. *A California flora.* University of California Press, Berkeley.

Murie, A. 1961. *A naturalist in Alaska.* Doubleday, New York.

Murphy, G. I. 1968. Pattern in life history and the environment. *Am. Nat.* 102:391–403.

Murray, B. G., Jr. 1967. Dispersal in vertebrates. *Ecology* 48:975–978.

Murray, D. L., and S. Boutin. 1991. The influence of snow on lynx and coyote movements: Does morphology affect behavior? *Oecologia (Berl)* 88:463–469.

Muscatine, L., J. E. Boyle, and D. C. Smith. 1974. Symbiosis of the acoel flatworm *Convoluta roscoffensis* with the alga *Platymonas convolutae. Proc. R. Soc. Lond. [Biol]* 187:221–234.

Musselman, L. J., and W. F. Mann, Jr. 1978. *Root parasites of southern forests.* USDA Southern Forest Experiment Station.

Mutch, R. W. 1970. Wildland fires and ecosystems—a hypothesis. *Ecology* 51:1046–1051.

Myers, J. P., and F. A. Pitelka. 1979. Variations in the summer temperature patterns near Barrow, Alaska: Analysis and ecological interpretation. *Arct. Alp. Res.* 11:131–144.

Myers, N. 1983. Conversion rates in tropical moist forests. Pp. 289–300 in *Tropical rainforest ecosystems,* F. B. Golley, ed. Elsevier Scientific Pub. Co., New York.

Myers, R. L. 1985. Fire and the dynamic relationship between Florida sandhill and pine scrub vegetation. *Bull. Torrey Bot. Club* 112:241–252.

Myers, R. L. 1991. Restoring longleaf pine community integrity. *Abstract, The longleaf pine ecosystem: Ecology, restoration and management. 18th Tall Timbers Fire Ecology Conference, Tallahassee, Fla.*

Nabhan, G. 1991. Cryptic cacti on the borderline. *Orion* 10:26–31.

National Research Council. 1989. *Alternative agriculture.* National Resource Council, Washington, D. C.

National Research Council on Restoration of Aquatic Ecosystems. 1992. *Restoration of aquatic ecosystems.* National Academy Press, Washington, D.C.

Neel, W. L. 1991. An ecological approach to longleaf pine forestry. *Abstract, The longleaf pine ecosystem: Ecology, restoration and management. 18th Tall Timbers Fire Ecology Conference, Tallahassee, Fla.*

Neill, W. E. 1974. The community matrix and the interdependence of the competition coefficients. *Am. Nat.* 108:399–408.

Nelson, T. C., and W. M. Zillgitt. 1969. *A forest atlas of the South.* USDA Forest Service.

Newman, E. I., and R. D. Rovira. 1975. Allelopathy among some British grassland species. *J. Ecol.* 63: 727–737.

Nichols, F. H., J. E. Cloern, S. N. Luoma, and D. H. Peterson. 1986. The modification of an estuary. *Science* 231:567–573.

Nichols, S., and L. Entine. 1976. *Prairie primer.* University of Wisconsin Extension, Madison.

Nicholson, A. J. 1955. An outline of the dynamics of animal populations. *Aust. J. Zool.* 2:9–65.

Nilsson, S. G., and U. Wästljung. 1987. Seed predation and cross-pollination in mast-seeding beech (*Fagus sylvatica*) patches. *Ecology* 68:260–265.

Noble, E. R. 1960. Fishes and their parasite-mix and objects for ecological study. *Ecology* 41:593–596.

Noon, B. R. 1981. The distribution of an avian guild along a temperate elevational gradient: The importance and expression of competition. *Ecol. Monogr.* 5:105–124.

Norberg, R. A. 1987. Evolution, structure, and ecology of northern forest owls. Pp. 9–43 in *Biology and conservation of northern forest owls,* R. W. Nero, R. J. Clark, R. J. Knaption, and R. H. Hamre, eds. *USDA For. Serv. Gen. Tech. Rep. RM* 142.

Norris, R. 1992. Can ecotourism save natural areas? National Parks 66 (1–2):30–34.

Oak Ridge National Laboratory. 1986. *Integrated data base for 1986: Spent fuel and radioactive waste inventories, projections and characteristics.* DOE/RW-0006, rev. 2. U.S. Department of Energy, Washington, D.C.

O'Brien, W. J., H. I. Browman, and B. I. Evans. 1990. Search strategies of foraging animals. *Am. Sci.* 78: 152–160.

O'Connor, R. J. 1978. Brood reduction in birds: Selection for infanticide, fratricide, and suicide? *Anim. Behav.* 26:79–96.

O'Dowd, D. J., and M. F. Willson. 1991. Associations between mites and leaf domatia. *Trends Ecol. Evol.* 6: 179–182.

Odum, E. P. 1945. The concept of the biome as applied to the distribution of North American birds. *Wilson Bull.* 57:191–201.

Odum, E. P. 1953. *Fundamentals of ecology.* 3rd ed., 1971 W. B. Saunders, Philadelphia.

Odum, E. P. 1960. Organic production and turnover in old field succession. *Ecology* 41:34–49.

Odum, E. P. 1965. Summary of the ecological effects of nuclear war. Pp. 69–72 in *Ecological effects of nuclear war,* G. M. Woodwell, ed. Brookhaven Nat. Lab. Publ. No. 917:1–72.

Odum, E. P. 1983. *Basic Ecology.* Saunders College Pub., Philadelphia.

Odum, E. P. 1989. *Ecology and our endangered life-support systems.* Sinauer Associates, Sunderland, Mass.

Odum, E. P., and L. J. Biever. 1984. Resource quality, mutualism, and energy partitioning in food chains. *Am. Nat.* 124:360–376.

Odum, H. T. 1957. Trophic structure and productivity of Silver Springs, Florida. *Ecol. Monogr.* 27:55–112.

Odum, H. T. 1971. *Environment, power, and society.* Wiley, New York.

Odum, H. T. 1977. The ecosystem, energy, and human values. *Zygon* 12:109–133.

Odum, H. T., and E. C. Odum. 1981. *Energy basis for man and nature.* 2nd ed. McGraw-Hill Book Co., New York.

Odum, H. T., and E. P. Odum. 1955. Trophic structure and productivity of a windward coral reef community Eniwetok atoll. *Ecol. Monogr.* 25:291–320.

Ogburn, C., Jr. 1966. *The winter beach* (1971 reprint). Pocket Books, New York.

Ogden, J. G., III. 1967. Radiocarbon and pollen evidence for a sudden change in climate in the Great Lakes region approximately 10,000 years ago. Pp. 117–127 in *Quaternary paleoecology,* E. J. Cushing and H. E. Wright, eds. Yale University Press, New Haven, Conn.

Oliver, B. G., and A. J. Niimi. 1985. Bioconcentration factors of some halogenated organics for rainbow trout: Limitations in their use for prediction of environmental residues. *Environ. Science Tech.* 19:842–849.

Oliver, B. M. 1990. Metrification oversold? *Science* 250: 611–612.

Oliver, J. A. 1963. *Snakes in fact and fiction.* Doubleday Anchor Natural History Library, Garden City, New York.

Oliver, J. A., and C. G. Shaw. 1953. The amphibians and reptiles of the Hawaiian Islands. *Zoologica* 38:65–95.

Olson, J. 1978. Mammalian succession on Lake Michigan sand dunes. M.A. Thesis, Western Michigan University.

Olson, J. S. 1958. Rates of succession and soil changes on southern Lake Michigan sand dunes. *Bot. Gaz.* 199:125–170.

Oosting, H. J. 1942. An ecological analysis of the plant communities of the Piedmont, North Carolina. *Am. Midl. Nat.* 28:1–126.

Orians, G. H. 1969. On the evolution of mating systems in birds and mammals. *Am. Nat.* 103:589–603.

Owen, D. 1980a. *Camouflage and mimicry.* University of Chicago Press, Chicago.

Owen, D. F. 1980b. How plants may benefit from the animals that eat them. *Oikos* 35:230–235.

Owen, D. F., and R. G. Wiegert. 1976. Do consumers maximize plant fitness? *Oikos* 27:488–492.

Packard, A. S. 1894. On the origin of the subterranean fauna of North America. *Am. Nat.* 28:727–751.

Packer, C. 1985. Dispersal and inbreeding avoidance. *Anim. Behav.* 33:676–678.

Paige, K. N., and T. G. Whitham. 1987. Overcompensation in response to mammalian herbivory. *Am. Nat.* 129:407–416.

Paine, R. T. 1966. Food web complexity and species diversity. *Am. Nat.* 100:65–75.

Paine, R. T. 1988. Food webs: Road maps of interactions or grist for theoretical development? *Ecology* 69: 1648–1654.

Palca, J. 1991. The sobering geography of AIDS. *Science* 252:372–373.

Palca, J. 1992. Infection with selection: HIV in human infants. *Science* 255:1069.

Palmer, E. L., and H. S. Fowler. 1975. *Fieldbook of natural history.* McGraw-Hill, New York.

Park, O. 1930. Studies in the ecology of forest Coleoptera. *Ann. Entomol. Soc. Am.* 23:57–80.

Park, O. 1931. The measurement of daylight in the Chicago area and its ecological significance. *Ecol. Monogr.* 1:189–230.

Park, O. 1940. Nocturnalism—the development of a problem. *Ecol. Monogr.* 10:485–536.

Park, T. 1948. Experimental studies of interspecific competition. I. Competition between populations of the flour beetles *Tribolium confusum* Duval and *Tribolium castaneum* Herbst. *Ecol. Monogr.* 18:265–308.

Park, T. 1954. Experimental studies of interspecies competition. II. Temperature, humidity, and competition in two species of *Tribolium. Physiol. Zool.* 27: 177–238.

Parker, G. A. 1974. Courtship persistence and female-guarding as male time-investment strategies. *Behaviour* 48:157–184.

Parker, G. A., and R. A. Stuart. 1976. Animal behavior

as a strategy optimizer: Evolution of resource assessment strategies and optimal emigration thresholds. *Am. Nat.* 110:1055–1076.

Parker, M. A., and R. B. Root. 1981. Insect herbivores limit habitat distribution of a native composite, *Machaeranthera canescens. Ecology* 62:1390–1392.

Parson, E. A., P. M. Haas, and M. A. Levy. 1992. A summary of the major documents signed at the earth summit and the global forum. *Environment* 34(8): 12–15.

Patrick, R., V. P. Binetti, and S. G. Halterman. 1981. Acid lakes from natural and anthropogenic causes. *Science* 211:446–452.

Paul, V. J., and S. C. Pennings. 1991. Diet-derived chemical defenses in the sea hare *Stylocheilus longicauda* (Quoy et Gaimard 1824). *J. Exp. Marine Biol. Ecol.* 151:227–243.

Payette, S., and L. Filion. 1985. White spruce expansion at the tree line and recent climatic change. *Can. J. For. Res.* 15:241–251.

Payne, R. B., and L. L. Payne. 1989. Heritability estimates and behaviour observations: Extra-pair matings in indigo buntings. *Anim. Behav.* 38:457–467.

Paysen, T. E., et al. 1980. A vegetation classification system applied to southern California. *USDA For. Serv. Gen. Tech. Rep. PSW* 45:1–33.

Peakin, G. J., and G. Josens. 1978. Respiration and energy flow. Pp. 111–164 in *Production ecology of ants and termites,* M. V. Brian, ed. Cambridge University Press, Cambridge.

Pearl, R., and L. J. Reed. 1920. On the rate of growth of the population of the United States since 1790 and its mathematical representation. *Proc. Natl. Acad. Sci. USA* 6:275–288.

Pearl, R., L. J. Reed, and J. F. Kish. 1940. The logistic curve and the census count of 1940. *Science* 92:486–488.

Pearman, G. I., ed. 1988. *Greenhouse: Planning for climate change.* E. J. Brill, Leiden.

Pearse, A. S. 1939. *Animal ecology.* McGraw-Hill, New York.

Pearson, L. C. 1965. Primary production in grazed and ungrazed desert communities of eastern Idaho. *Ecology* 46:278–285.

Pemberton, R. W., and C. F. Turner. 1989. Occurrence of predatory and fungivorous mites in leaf domatia. *Am. J. Bot.* 76:105–112.

Perala, D. A., and D. H. Alban. 1982. Rates of forest floor decomposition and nutrient turnover in aspen, pine, and spruce stands on two soils. *USDA For. Serv. Res. Pap. NC* 227.

Pereira, M. S. 1982. The impact of infectious disease on human demography today. Pp. 53–64 in *Population biology of infectious diseases,* R. M. Anderson and R. M. May, eds. Springer-Verlag, New York.

Perry, L. M. 1936. A marine tenement. *Science* 84:156–157.

Persson, L., G. Andersson, S. F. Hamrin, and L. Johansson. 1988. Predator regulation and primary production along the productivity gradient of temperate lake ecosystems. Chapter 4 in *Complex interactions in lake communities,* S. R. Carpenter, ed. Springer-Verlag, New York.

Peterjohn, B. G., and J. R. Sauer. 1993. North American breeding bird survey annual summary 1990–1991. *Bird Popul.* 1:1–15.

Peters, B. C. 1970. Pioneer evaluation of the Kalamazoo County landscape. *Mich. Acad.* 3:15–25.

Peters, G. A., R. E. Toia, Jr., H. E. Calvert, and B. H. Marsh. 1986. Lichens to *Gunnera*—with emphasis on *Azolla. Plant Soil* 90:17–34.

Petersen, N. J. 1983. The effects of fire, litter, and ash on flowering in *Andropogon gerardii.* Pp. 21–24 in *Proceedings of the eighth North American prairie conference,* R. Brewer, ed. Western Michigan University, Department of Biology, Kalamazoo.

Petit, L. J. 1991. Experimentally induced polygyny in a monogamous bird species: Prothonotary warblers and the polygyny threshold. *Behav. Ecol. Sociobiol.* 29:177–187.

Petrides, G. A., and W. G. Swank. 1965. Population densities and the range-carrying capacity for large mammals in Queen Elizabeth National Park, Uganda. *Zool. Afr.* 1:209–225.

Pfeiffer, W. J., and R. G. Wiegert. 1981. Grazers on Spartina and their predators. Pp. 87–112 in *The ecology of a salt marsh,* L. R. Pomeroy and R. G. Wiegert, eds. Springer-Verlag, New York.

Pfennig, D. W., M. L. G. Loeb, and J. P. Collins. 1991. Pathogens as a factor limiting the spread of cannibalism in tiger salamanders. *Oecologia (Berl) 88:161–166.*

Pfister, R. D., B. L. Kovalchik, S. F. Arno, and R. C. Presby. 1977. Forest habitat types of Montana. *USDA For. Serv. Gen. Tech. Rep. INT* 34.

Phillips, C. S., Jr. 1971. The revival of cultural evolution in social science theory. *J. Dev. Areas* 5:337–370.

Phillips, D. R., and D. H. Van Lear. 1984. Biomass removal and nutrient drain as affected by total-tree harvest in southern pine and hardwood stands. *J. For.* 82:547–550.

Piatt, J. F. et al. 1990. Immediate impact of the ''Exxon Valdez'' oil spill on marine birds. *Auk* 107:387–397.

Pielou, E. C. 1974. *Population and community ecology.* Gordon and Breach, New York.

Pieper, R. D. 1977. The southwestern pinyon-juniper ecosystem. Pp. 1–6 in *Ecology, uses, and management of pinyon-juniper woodlands. USDA For. Serv. Gen. Tech. Rep. RM* 39.

Pier, S. M., and K. M. Bang. 1980. The role of heavy

metals in human health. Pp. 367–409 in *Environment and health,* N. M. Trieff, ed. Ann Arbor Science, Ann Arbor, Mich.

Pierce, G. J. 1974. The coastal plain floristic element in Michigan. M. S. Thesis, Western Michigan University.

Pimentel, D. 1991. Ethanol fuels: Energy security, economics, and the environment. *J. Agri. Environ. Ethics* 4:1–13.

Pimentel, D. et al. 1992. Environmental and economic costs of pesticide use. *BioScience* 42:750–760.

Pimentel, D., et al. 1973. Food production and the energy crisis. *Science* 182:443–449.

Pimentel, D., W. Dritschilo, J. Krummel, and J. Kutzman. 1975. Energy and land constraints in food protein production. *Science* 190:754–761.

Pimlott, D. H. 1967. Wolf predation and ungulate populations. *Am. Zool.* 7:267–278.

Pimlott, D. H., J. A. Shannon, and G. B. Kolenosky. 1969. The ecology of the timber wolf in Algonquin Provincial Park. *Can. Dep. Lands For. Res. Branch Res. Rep. (Wildlife)* No. 87.

Pimm, S. L. 1982. *Food webs.* Chapman and Hall, London.

Pimm, S. L. 1988. Energy flow and trophic structure. Chapter 13 in *Concepts of ecosystem ecology,* L. R. Pomeroy and J. J. Alberts, eds. Springer-Verlag, New York.

Pitelka, F. A. 1959. Numbers, breeding schedule, and territoriality in pectoral sandpipers of northern Alaska. *Condor* 61:233–264.

Pitelka, F. A. 1964. The nutrient-recovery hypothesis for arctic microtine cycles. Pp. 55–56 in *Grazing in terrestrial and marine environments,* D. J. Crisp, ed. Blackwell, Oxford.

Pittendreigh, C. S. 1960. Circadian rhythms and the circadian organization of living systems. *Cold Spring Harb. Symp. Quant. Biol.* 25:159–182.

Pittock, A. B. 1986. Rapid developments on nuclear winter. *Search* 17.

Platt, R. 1963. Ecological effects of ionizing radiation on organisms, communities, and ecosystems. Pp. 243–255 in *Radioecology,* V. Schultz and A. W. Klement, eds. Reinhold, New York.

Platt, W. J., G. W. Evans, and M. M. Davis. 1988. Effects of fire season on flowering of forbs and shrubs in longleaf pine forests. *Oecologia (Berl)* 76:353–362.

Platt, W. J., and M. W. Schwartz. 1990. Temperate hardwood forests. Pp. 194–229 in *Ecosystems of Florida,* R. L. Myers and J. J. Ewel, eds. University of Central Florida Press, Orlando.

Pleznac, R. 1983. Management and native species enrichment as an alternative to prairie reconstruction. Pp. 132–133 in *Proceedings of the Eighth North American Prairie Conference,* R. Brewer, ed. Western Michigan University Department of Biology, Kalamazoo, MI.

Polis, G. A. 1991. Complex trophic interactions in deserts: An empirical critique of food-web theory. *Am. Nat.* 138:123–155.

Polis, H. S., C. A. Myers, and R. D. Holt. 1989. The ecology and evolution of intraguild predation: Potential competitors that eat each other. *Annu. Rev. Ecol. Syst.* 20:297–330.

Pomeroy, L. R., and R. G. Wiegert. 1981. *The ecology of a salt marsh.* Springer-Verlag, New York.

Pool, R. 1990. Pushing the envelope of life. *Science* 247: 158–160.

Popper, K. 1968. *The logic of scientific discovery.* 2nd ed. Hutchinson, London.

Porter, K. G. 1976. Enhancement of algal growth and productivity by grazing zooplankton. *Science* 192: 1332–1334.

Porter, S. C. 1983. *Late-Quaternary environments of the United States. Vol. 1. The late Pleistocene.* University of Minnesota Press, Minneapolis.

Porter, W. P., J. W. Mitchell, W. A. Beckman, and C. B. DeWitt. 1973. Behavioral implications of mechanistic ecology. *Oecologia (Berl)* 13:1–54.

Postel, S. 1984. Air pollution, acid rain, and the future of the forests. *Worldwatch Pap.* 58:1–54.

Pough, F. H. 1980. The advantages of ectothermy for tetrapods. *Am. Nat.* 115:92–142.

Pound, R., and F. E. Clements. 1898. *The phytogeography of Nebraska.* (2nd ed. 1900.) University of Nebraska Botany Survey.

Powell, C. L., and D. J. Bagyaraj. 1984. *VA Mycorrhiza.* CRC Press, Inc., Boca Raton, Fla.

Powell, G. V. N. 1974. Experimental analysis of the social value of flocking by starlings *(Sturnus vulgaris)* in relation to predation and foraging. *Anim. Behav.* 22:501–505.

Power, M. E. 1990. Effects of fish in river food webs. *Science* 250:811–814.

Prestrud, P. 1991. Adaptations by the arctic fox *(Alopex lagopus)* to the polar winter. *Arctic* 44:132–138.

Primack, R. B., and M. ShiLi. 1991. "Safe sites" for germination using *Plantago* seeds: A repetition of a thrice-published experiment. *Bull. Torrey Bot. Club* 118:154–160.

Probst, J. R. 1991. Kirtland's warbler. Pp. 414–417 in *The atlas of breeding birds of Michigan,* R. Brewer, G. A. McPeek, and R. J. Adams, Jr., eds. Michigan State University Press, East Lansing.

Proctor, J., and S. R. J. Woodell. 1975. The ecology of serpentine soils. *Adv. Ecol. Res.* 9:255–366.

Prosser, C. L., ed. 1973. *Comparative animal physiology.* W. B. Saunders Co., Philadelphia.

Pruitt, W. O., Jr. 1954. Additional animal remains from under Sleeping Bear Dune, Leelanau County, Michigan. *Pap. Mich. Acad. Sci. Arts Lett.* 39:253–256.

Pulliam, H. R. 1973. On the advantages of flocking. *J. Theor. Biol.* 38:419–422.

Pulliam, H. R. 1988. Sources, sinks, and population regulation. *Am. Nat.* 132:562–661.

Putnam, A. R. 1983. Allelopathic chemicals. *Chem. Eng. News* (April 4):34–45.

Putnam, R. J., and S. D. Wratten. 1984. *Principles of ecology.* University of California Press, Berkeley.

Quay, P. D., B. Tilbrook, and C. S. Wong. 1992. Ocean uptake of fossil fuel CO_2: Carbon-13 evidence. *Science* 256:74–79.

Quinn, J. F., and A. Hastings. 1987. Extinction in subdivided habitats. *Conserv. Biol.* 1:198–208.

Quinn, J. F., and S. P. Harrison. 1988. Effects of habitat fragmentation and isolation on species richness: Evidence from biogeographic patterns. *Oecologia (Berl)* 75:132–140.

Quinlan, S. E., and J. J. Hughes. 1990. Location and description of a marbled murrelet tree nest site in Alaska. *Condor* 92:1068–1073.

Rabenold, K. 1984. Cooperative enhancement of reproductive success in tropical wren societies. *Ecology* 65: 871–885.

Rabotnov, T. A. 1969. On coenopulations of perennial herbaceous plants in natural coenoses. *Vegetatio* 19: 87–95.

Racovitza, E. G. 1907. Essai sur les problèmes biospéologiques. *Arch. Zool. Exp. Gen.* 6:371–488.

Rao, D. N., and F. LeBlanc. 1967. Influence of an iron-sintering plant on corticolous epiphytes in Wawa, Ontario. *Bryologist* 70:141–157.

Raphael, M. G. 1987. Nongame wildlife research in subalpine forests of the central Rocky Mountains. Pp. 113–122 in *Management of subalpine forests: Building on 50 years of research. USDA For. Serv. Gen. Tech. Rep. RM* 149.

Rasmussen, K. R., and S. Rasmussen. 1984. The summer water balance in a Danish oak stand. *Nord. Hydrol.* 15:213–222.

Raunkiaer, C. 1934. *The life forms of plants and statistical plant geography.* Clarendon Press, Oxford.

Rawlins, S. L. 1991. Global environmental change and agriculture. *J. Prod. Agri.* 4:291–293.

Raynal, D. J., and F. A. Bazzaz. 1975. Interference of winter annuals with *Ambrosia artemisiifolia* in early successional fields. *Ecology* 56:35–49.

Reader, R. J., K. C. Taylor, and D. W. Larson. 1991. Does intermediate disturbance increase species richness within deciduous forest understory? Chapter 17 in *Modern ecology: Basic and applied aspects,* G. Esser and D. Overdieck, eds. Elsevier, Amsterdam.

Redfield, A. C. 1958. The biological control of chemical factors in the environment. *Am. Sci.* 46:205–221.

Reese, R. 1984. *Greater Yellowstone.* Montana Geographic Series, no. 6. Montana Magazine, Helena.

Regenstein, L. 1982. *America the poisoned.* Acropolis Books, Ltd., Washington, D.C.

Rego, F.C., S. C. Bunting, and J. M. da Silva. 1991. Changes in understory vegetation following prescribed fire in maritime pine forests. *For. Ecol. Manage.* 41:21–31.

Reichel, J. D. 1986. Habitat use by alpine mammals in the Pacific Northwest, U.S.A. *Arct. Alp. Res.* 18:111–119.

Reiter, R. J. 1991. Neuroendocrine effects of light. *Int. J. Biometeorol.* 35:169–175.

Remmert, H. 1980. *Ecology: A textbook.* Springer-Verlag, Berlin.

Rey, J. R. 1984. Experimental tests of island biogeographic theory. Pp. 101–112 in *Ecological communities,* D. R. Strong et al., eds. Princeton University Press, Princeton, N.J.

Reynolds, J. C. 1985. Details of the geographic replacement of the red squirrel (*Sciurus vulgaris*) by the grey squirrel (*Sciurus carolinensis*) in eastern England. *J. Anim. Ecol.* 54:149–162.

Reznicek, A. A. 1980. Halophytes along a Michigan roadside with comments on the occurrence of halophytes in Michigan. *Mich. Bot.* 19:23–30.

Rhoades, D. F. 1979. Evolution of plant chemical defense against herbivores. Pp. 3–54 in *Herbivores: Their interaction with secondary plant metabolites,* G. A. Rosenthal and D. H. Janzen, eds. Academic Press, New York.

Rhodes, W. J. 1991. Stratospheric ozone protection: An EPA engineering perspective. *J. Air Waste Manage. Assoc.* 41:1579–1584.

Rice, E. L. 1974. *Allelopathy.* Academic Press, New York.

Rich, P. H., and R. G. Wetzel. 1978. Detritus in the lake ecosystem. *Am. Nat.* 112:57–71.

Richards, P. W. 1957. *The tropical rain forest: An ecological study.* Cambridge University Press, Cambridge.

Richardson, R. E. 1921. The small bottom and shore fauna of the middle and lower Illinois River and its connecting lakes. *Bull. Ill. Nat. Hist. Surv.* 13:363–522.

Richmond, R. C., M. E. Gilpin, S. P. Salas, and F. J. Ayala. 1975. A search for emergent phenomena: The dynamics of multi-species Drosophila systems. *Ecology* 56:709–714.

Riddiford, L. M., and C. M. Williams. 1967. Volatile principle from oak leaves: Role in sex life of the Polyphemus moth. *Science* 155:589–590.

Ridley, H. N. 1930. *The dispersal of plants throughout the world.* Reeve, Ashford, Kent.

Riggan, P. J., and P. H. Dunn. 1982. Harvesting chaparral biomass for energy—an environmental assessment. Pp. 149–157 in *Dynamics and management of Mediterranean-type ecosystems. USDA For. Serv. Gen. Tech. Rep. PSW* 58.

Rise Project Group. 1980. East Pacific Rise: Hot springs and geographical experiments. *Science* 207:1421–1433.

Rizzo, B. 1988. The sensitivity of Canada's ecosystems to climatic change. *Can. Committee Ecol. Land Class. Newsl.* 17:10–12.

Robbins, C. 1979. Effect of forest fragmentation on bird populations. Pp. 198–212 in *Management of northcentral and northeastern forests for nongame birds. USDA For. Serv. Gen. Tech. Rep. NC* 51.

Robbins, C. S., J. R. Sauer, R. S. Greenberg, and S. Droege. 1989. Population declines in North American birds that migrate to the neotropics. *Proc. Natl. Acad. Sci. USA* 86:7658–7662.

Roberts, L. 1991. Dioxin risks revisited. *Science* 251:624–626.

Robertson, A. I. 1991. Plant-animal interactions and the structure and function of mangrove forest ecosystems. *Aus. J. Ecol.* 16:433–443.

Robertson, W. B., Jr. 1959. *Everglades—the park story.* University of Miami Press, Miami.

Robhe, H. 1990. A comparison of various gases to the greenhouse effect. *Science* 248:1217–1219.

Robinson, R. K. 1972. The production by roots of *Calluna vulgaris* of a factor inhibitory to growth of some mycorrhizal fungi. *J. Ecol.* 60:219–224.

Robinson, S. K. 1992. Population dynamics of breeding Neotropical migrants in a fragmented Illinois landscape. Pp. 404–418 in *Ecology and conservation of Neotropical landbird migrants.* J. M. Hagan III and D. W. Johnston, eds. Smithsonian Institution Press, Washington, D. C.

Robinson, W. L., and E. G. Bolen. 1989. *Wildlife ecology and management.* 2nd ed. Macmillan, New York.

Roe, F. G. 1970. *The North American buffalo.* 2nd ed. University of Toronto Press, Toronto.

Roeder, K. D., and A. E. Treat. 1961. The acoustic detection of bats by moths. *Proc. 11th Intern. Cong. Entomol.* 3:7–11.

Romer, A. S. 1947. *Man and the vertebrates.* University of Chicago Press, Chicago.

Roosa, D., and D. Smith. 1991. Prairie fire: Points of view. Bison and Bluestem (Konza Prairie, Kansas State University) no. 12:1–2

Rosen, R. 1990. "This bomb had my name on it." *Progressive* 54:29–33.

Root, M. 1990. Biological monitors of pollution. *BioScience* 40:83–86.

Root, R. B. 1967. The niche exploitation pattern of the blue-gray gnatcatcher. *Ecol. Monogr.* 37:317–350.

Romme, W. H., and D. G. Despain. 1989. Historical perspective on the Yellowstone fires of 1988. *BioScience* 39:695–699.

Rosenthal, G. A. 1983. A seed-eating beetle's adaptation to a poisonous seed. *Sci. Am.* 249:164–171.

Rosenzweig, M. L. 1981. A theory of habitat selection. *Ecology* 62:327–335.

Rosenzweig, M. L., and R. H. MacArthur. 1963. Graphical representation and stability conditions of predator-prey interactions. *Am. Nat.* 46:209–219.

Roth, R. R., and R. K. Johnson. 1993. Long-term dynamics of a wood thrush population breeding in a forest fragment. *Auk* 110:37–48.

Rothschild, M., and T. Clay. 1961. *Fleas, flukes, and cuckoos.* Arrow Books, London.

Rouse, C. 1986. Fire effects in northeastern forests: Aspen. *USDA For. Serv. Gen. Tech. Rep. NC* 102:1–8.

Rowe, J. S. 1953. Forest sites—a discussion. *For. Chron.* 29:278–289.

Rowland, F. S. 1989. Chlorfluorocarbons and the depletion of stratospheric ozone. *Am. Sci.* 77:36–45.

Roze, U. 1989. *The North American porcupine.* Smithsonian Institution Press, Washington, D.C.

Rubin, E. S. 1992. Realistic mitigation options for global warming. *Science* 257:148–266.

Runkle, J. R. 1981. Gap regeneration in some old-growth forests of the eastern United States. *Ecology* 62:1041–1051.

Russell, M. J., T. Mendelson, and V. H. S. Peeke. 1983. Mothers' identification of their infant's odors. *Ethol. Sociobiol.* 4:29–31.

Ryan, M. J., and E. A. Brenowitz. 1984. The role of body size, phylogeny, and ambient noise in the evolution of bird song. *Am. Nat.* 126:87–100.

Ryel, L. A. 1981. Population change in the Kirtland's warbler. *Jack-Pine Warbler* 59:76–91.

Ryel, L. A. 1983. Status of the Kirtland's warbler, 1983. *Jack-Pine Warbler* 61:95–98.

Sabo, D. G., G. V. Johnson, W. C. Martin, and E. F. Aldon. 1979. Germination requirements of 19 species of arid-land plants. *USDA For. Serv. Res. Pap. RM* 210:1–26.

Saldarriaga, J. G., and R. J. Luxmoore. 1991. Solar energy conversion efficiencies during succession of a tropical rain forest in Amazonia. *J. Trop. Ecol.* 7: 233–242.

Sale, K. 1978. The polis perplexity: An inquiry into the size of cities. *Working Pap. New Soc.* 6:64–77.

Salk, J. 1987. Prospects for the control of AIDS by immunizing seropositive individuals. *Nature* 327:473–476.

Sample, V. A., and D. C. Le Master. 1992. Economic effects of northern spotted owl protection. *J. For.* 90: 31–35.

Sampson, A. W. 1944. Plant succession on burned chaparral lands in northern California. *Univ. Calif. Agric. Exp. Sta. Bull.* 685:1–144.

Sandberg, D. V., et al. 1979. Effects of fire on air. *USDA For. Serv. Gen. Tech. Rep. WO-O:* 1–40.

Sandford, R. L., Jr., J. Saldarriaga, K. E. Clark, C. Uhl, and R. Herrera. 1985. Amazon rain-forest fires. *Science* 227:53–55.

Sandred, K. B., and M. Emsley, 1979. *Rain forests and cloud forests.* Abrams, New York.

Sandstrom, U. 1992. Enhanced predation rates on cavity bird nests at deciduous forest edges—an experimental study. *Ornis Fennica* 68:93–98.

Santos, P. F., J. Phillips, and W. G. Whitford. 1981. The role of mites and nematodes in the early stages of buried litter decomposition in a desert. *Ecology* 62: 664–669.

Saunders, D. A., and R. J. Hobbs. 1991. *The role of corridors.* Surrey Beatty & Sons Pty Limited, Chipping Norton, Australia.

Saunders, D. S. 1977. *An introduction to biological rhythms.* Blackie, Glasgow.

Savile, D. B. O. 1972. Arctic adaptations in plants. *Can. Dep. Agric. Monogr.* 6:1–81.

Scaro, R. C., and M. D. Jakle. 1985. Avian use of a desert riparian island and its adjacent scrub habitat. *Condor* 87:511–519.

Schaffer, W. M. 1984. Stretching and folding in lynx fur returns: Evidence for a strange attractor in nature? *Am. Nat.* 124:798–820.

Schaffer, W. M. 1985. Order and chaos in ecological systems. *Ecology* 66:93–106.

Schaffer, W. M. 1988. Perceiving order in the chaos of nature. Pp. 313–350 in *Evolution of life histories of mammals,* M. S. Boyce, ed. Yale University Press, New Haven, Conn.

Schaller, G. B. 1963. *The mountain gorilla: Ecology and behavior.* University of Chicago Press, Chicago.

Schaller, G. B. 1972. *The Serengeti lion. A study of predator-prey relations.* University of Chicago Press, Chicago.

Schartz, R. L., and J. L. Zimmerman. 1971. The time and energy budget of the male dickcissel *(Spiza americana). Condor* 73:65–76.

Schimper, A. F. W. 1898. *Pflanzengeographie auf physiologischer Grundlage.* Jena. (Eng. transl. 1903. *Plant-geography upon a physical basis.* Oxford.)

Schindler, D. W., et al. 1985. Long-term ecosystem stress: The effects of years of experimental acidification on a small lake. *Science* 228:1395–1401.

Schmidt-Nielsen, K. 1972. *How animals work.* Cambridge University Press, London.

Schmidt-Nielsen, K., B. Schmidt-Nielsen, S. A. Jarnum, and T. R. Houpt. 1957. Body temperature of the camel and its relation to water economy. *Am. J. Physiol.* 188:103–112.

Schneider, J. C. 1986. "Acid rain": Significance to Michigan lakes and their fisheries. *Mich. Acad.* 18: 7–15.

Schneider, S. H., and R. Londer. 1984. *The coevolution of climate and life.* Sierra Club Books, San Francisco.

Schnell, D. E. 1976. *Carnivorous plants of the United States and Canada.* John F. Blair, Winston-Salem, N.C.

Schnitzler, H. V. 1978. Die Detektion von Bewegungen durch Echoortung bei Fledermausen. *Verh. Dtsch. Zool. Ges.* 1978:16–33.

Schoener, T. W. 1983. Field experiments on interspecific competition. *Am. Nat.* 122:240–285.

Schoener, T. W. 1986. Patterns in terrestrial vertebrate versus arthropod communities: Do systematic differences in regularity exist? Pp. 556–586 in *Community ecology,* J. Diamond and T. J. Case, eds. Harper & Row, New York.

Scholander, P. F. 1968. How mangroves desalinate seawater. *Physiol. Plant.* 21:251–261.

Schoonmaker, P. K., and D. R. Foster. 1991. Some implications of paleoecology for contemporary ecology. *Bot. Rev.* 57:204–245.

Schorger, A. W. 1955. *The passenger pigeon.* University of Wisconsin Press, Madison.

Schramm, P. 1970. Effects of fire on small mammal populations in a restored tallgrass prairie. Pp. 39–41 in *Proceedings of a symposium on prairie and prairie restoration,* P. Schramm, ed. Knox College, Galesburg, Ill.

Schramm, P. 1978. The do's and don'ts of prairie restoration. Pp. 139–151 in *Proceedings of the fifth midwest prairie conference.* D. C. Glenn-Lewin and R. Q. Landers, Jr., eds. Iowa State University, Ames.

Schramm, P. 1992. Prairie restoration: A twenty-five year perspective on establishment and management. Pp. 169–177 in *Proceedings of the twelfth North American Prairie Conference,* D. D. Smith and C. A. Jocobs, eds. University of Northern Iowa, Cedar Falls.

Schulenberg, R. F. 1967. *Notes on the propagation of prairie plants.* Morton Arboretum, Lisle, Ill.

Schullery, P., and D. G. Despain. 1989. Prescribed burning in Yellowstone National Park: A doubtful proposition. *Western Wildlands* 15:30–34.

Schultz, A. M. 1969. A study of an ecosystem: The Arctic tundra. Pp. 77–93 in *The ecosystem concept in natural resource management,* G. M. Van Dyke, ed. Academic Press, New York.

Schupp, E. W., and D. H. Feener, Jr. 1991. Phylogeny,

lifeform, and habitat dependence of ant-defended plants in a Panamanian Forest. Pp. 175–259 in *Ant-plant interactions,* C. R. Huxley and D. F. Cutler, eds. Oxford University Press, Oxford.

Schwintzer, C. R. 1979. Vegetation changes following a water level rise and tree mortality in a Michigan bog. *Mich. Bot.* 18:91–98.

Scott, J. A., N. R. French, and J. W. Leetham. 1979. Patterns of consumption in grasslands. Pp. 89–105 in *Perspectives in grassland ecology,* N. R. French, ed. Springer-Verlag, New York.

Scott, T. G., and W. D. Klimstra. 1955. Red foxes and a declining prey population. *South. Ill. Univ. Monogr.* 1:1–123.

Scott, T. G., Y. L. Willis, and J. A. Ellis. 1959. Some effects of a field application of dieldrin on wildlife. *J. Wildl. Manage.* 23:409–427.

Scott, V. E., K. E. Evans, D. A. Patton, and C. P. Stone. 1977. Cavity-nesting birds of North American forests. *USDA Agric. Handbook* 511:1–112.

Scully, J., T. Mast, and S. Morgan. 1992. Food. Pp. 33–66 in *Environmental Almanac,* Alan Hammond, ed. Houghton Mifflin, Boston.

Seamon, G. 1992. *Restoration at Apalachicola Bluffs and Ravines Preserve.* Unpubl. Rep., The Nature Conservancy, Tallahassee, Fla.

Searcy, W. A., and K. Yasukawa. 1989. Alternative models of territorial polygyny in birds. *Am. Nat.* 134: 323–343.

Seastedt, T. R., and R. A. Ramundo. 1990. The influence of fire on belowground processes of tallgrass prairie. Pp. 99–117 in *Fire in North American tallgrass prairie,* S. L. Collins and L. L. Wallace, eds. University of Oklahoma Press, Norman.

Seeley, T. D., and B. Heinrich. 1981. Regulation of temperature in the nests of social insects. Pp. 159–234 in *Insect thermoregulation,* B. Heinrich, ed. Wiley, New York.

Seifert, R. P. 1982. Neotropical Heliconia insect communities. *Q. Rev. Biol.* 57:1–18.

Semlitsch, R. D., R. N. Harris, and H. M. Wilbur. 1990. Paedomorphosis in *Ambystoma talpoideum:* Maintenance of population variation and alternative life history pathways. *Evolution* 44:1604–1613.

Seton, E. T. 1909. *Life histories of northern animals.* C. Scribner's Sons, New York.

Seton, E. T. 1911. *The Arctic prairies.* Scribner's, New York. (Harper and Row paperback, 1981.)

Shachak, M., C. G. Jones, and Y. Granot. 1987. Herbivory in rocks and the weathering of a desert. *Science* 236:1098–1099.

Shaffer, M. L. 1981. Minimum population sizes for species conservation. *BioScience* 31:131–134.

Shaffer, M L. 1983. Determining minimum viable population sizes for the grizzly bear. *Int. Conf. Bear Res. Manage.* 5:133–139.

Shantz, H. L. 1911. Natural vegetation as an indicator of the capabilities of land for crop production in the great plains area. *USDA Bureau of Plant Industry Bull.* 201.

Sharitz, R. R., and J. W. Gibbons. 1981. Effects of thermal effluents on a lake: Enrichment and stress. Pp. 243–259 in *Stress effects on natural ecosystems,* G. W. Barrett and R. Rosenberg, eds. Wiley, New York.

Shavyrina, A. V. 1965. On the possibility of using the geobotanical method in the search for freshwater in the southern deserts. Pp. 19–23 in *Plant indicators of soils, rocks, and subsurface waters,* A. G. Chikishev, ed. Consultants Bureau, New York.

Sheffield, R. M., N. D. Cost, W. A. Bechtold, and J. P. McClure. 1985. Pine growth reductions in the southeast. *USDA For. Serv. Res. Bull. SE* 83.

Shelford, V. E. 1908. Life histories and larval habits of the tiger beetles (Cicindelidae). *Linn. Soc. J. Zool.* 30:157–184.

Shelford, V. E. 1911. Ecological succession. *Biol. Bull.* 21:9–34.

Shelford, V. E. 1913 (2nd. ed. 1937). Animal communities in temperate America. *Bull. Geog. Soc. Chicago,* 5:1–368.

Shelford, V. E. 1929. *Laboratory and field ecology.* Williams and Wilkins, Baltimore.

Shelford, V. E. 1951. Fluctuations of forest animal populations in east-central Illinois. *Ecol. Monogr.* 21:183–214.

Shelford, V. E. 1963. *The ecology of North America.* University of Illinois Press, Urbana.

Shelford, V. E., and S. Olson. 1935. Sere, climax and influent animals with special reference to the transcontinental coniferous forest of North America. *Ecology* 16:375–402.

Sheppard, D. H. 1965. Ecology of the chipmunks *Eutamias amoenus luteiventris* (Allen) and *E. minimus orocetes* Merriam, with particular reference to competition. *Diss. Abstr.* 27:1663.

Sheppard, D. H. 1971. Competition between two chipmunk species *(Eutamias). Ecology* 52:320–329.

Shirley, H. C. 1930. A thermodynamic radiometer for ecological use on land and in water. *Ecology* 11:61–71.

Shirley, H. C. 1945. Light as an ecological factor and its measurement II. *Bot. Rev.* 11:497–532.

Shreve, F. 1942. The desert vegetation of North America. *Bot. Rev.* 8:195–246.

Shukla, J., and Y. Mintz. 1982. Influence of land-surface

evapotranspiration on the Earth's climate. *Science* 215:1498–1500.

Shy, M. 1982. Interspecific feeding among birds: A review. *J. Field Ornithol.* 55:370–393.

Sieg, C. H. 1991. Rocky Mountain juniper woodlands: Year-round avian habitat. *USDA For. Serv. Res. Pap. RM* 296.

Siegel-Causey, D., and G. L. Hunt, Jr. 1981. Colonial defense behavior in double-crested and pelagic cormorants. *Auk* 98:522–531.

Siivonen, L. 1954. Some essential features of short-term population fluctuation. *J. Wildl. Manage.* 18:38–45.

Silvertown, J. W. 1980. The evolutionary ecology of mast seeding in trees. *Biol. J. Linn. Soc.* 14:235–250.

Simberloff, D. 1970. Taxonomic diversity of island biotas. *Evolution* 24:23–47.

Simberloff, D. 1987. The spotted owl fracas: Mixing academic, applied, and political ecology. *Ecology* 68: 766–772.

Simberloff, D., and E. O. Wilson. 1970. Experimental zoogeography of islands: A two-year record of colonization. *Ecology* 51:934–937.

Simberloff, D., J. A. Farr, J. Cox, and D. W. Mehlman. 1992. Movement corridors: Conservation bargains or poor investments? *Conserv. Biol.* 6:493–504.

Simberloff, D., and N. Gotelli. 1983. Refuge design and ecological theory: Lesson for prairie and forest conservation. Pp. 61–71 in *Proceedings of the eighth North American Prairie Conference,* R. Brewer, ed. Western Michigan University, Department of Biology, Kalamazoo.

Simberloff, D., and T. Dayan. 1991. The guild concept and the structure of ecological communities. *Annu. Rev. Ecol. Syst.* 22:115–143.

Simmons, R. 1989. The Cain and Abel riddle in eagles and other birds. *Afr. Wildl.* 43:35–43.

Simpson, E. H. 1949. Measurement of diversity. *Nature* 163:688.

Sims, P. L., and J. S. Singh. 1978. The structure and function of ten North American grasslands. III. *J. Ecol.* 66:573–597.

Sinclair, A. R. E., and M. Norton-Griffiths, eds. 1979. *Serengeti—Dynamics of an ecosystem.* University of Chicago Press, Chicago.

Singer, F. J., W. Schrefer, J. Oppenheim, and E. O. Garton. 1989. Drought, fires, and large mammals. *BioScience* 39:716–722.

Singer, F. J., W. T. Swank, and E. E. C. Clebsch. 1984. Effects of wild pig rooting in a deciduous forest. *J. Wildl. Manage.* 48:464–473.

Siple, P. A., and C. F. Passel. 1945. Measurement of dry atmospheric cooling in subfreezing temperatures. *Proc. Am. Philos. Soc.* 89:177–199.

Skarpe, C. 1991. Impact of grazing in savanna ecosystems. *Ambio* 20:351–356.

Skinner, T. H., and J. O. Klemmedson. 1978. Abert squirrels influence nutrient transfer through litterfall in a ponderosa pine forest. *USDA For. Serv. Res. Note RM* 353.

Skutch, A. 1935. Helpers at the nest. *Auk* 52:257–273.

Slagsvold, T. 1980. Habitat selection in birds: On the presence of other bird species with special regard to *Turdus pilaris. J. Anim. Ecol.* 49:523–536.

Slatkin, M. 1983. Models of coevolution: Their use and abuse. Pp. 339–370 in *Coevolution,* M. H. Nitecki, ed. University of Chicago Press, Chicago.

Slatyer, R. O. 1967. *Plant-water relationships.* Academic Press, New York.

Sloan, A. W. 1979. *Man in extreme environments.* Charles C. Thomas Pub., Springfield, Ill.

Small, A. 1974. *The birds of California.* Winchester Press, New York.

Smith, B. P. 1988. Host-parasite interaction and impact of larval water mites on insects. Annual Review of Entomology 33:487–507.

Smith, D. C. 1979. From extracellular to intracellular: The establishment of a symbiosis. *Proc. R. Soc. Lond. [Biol]* 304:115–130.

Smith, D. D. 1992. Tallgrass prairie settlement: Prelude to demise of the tallgrass ecosystem. Pp. 195–199 in *Proceedings of the 12th North American Prairie Conference.* D. D. Smith, ed. University of Northern Iowa, Cedar Falls.

Smith, F. E. 1963. Population dynamics in *Daphnia magna* and a new model for population growth. *Ecology* 44:651–663.

Smith, J. S. 1935. The role of biotic factors in the determination of population density. *J. Econ. Entomol.* 28: 873–898.

Smith, J. W. 1915. Phenological dates and meterological data recorded by Thomas Mikesell at Wauseon, Ohio. *US Dep. Agric. Monthly Weather Rev. Suppl.* 2: 23–93.

Smith, P. W. 1957. An analysis of post-Wisconsin biogeography of the prairie peninsula region based on distribution phenomena among terrestrial vertebrate populations. *Ecology* 38:205–219.

Smith, S. A., and R. A. Paselk. 1986. Olfactory sensitivity of the turkey vulture *(Cathartes aura)* to three carrion-associated odorants. *Auk* 103:583–592.

Smith, T. 1977. Insects and diseases of pinyon-juniper. Pp. 20–21 in *Ecology, uses, and management of pinyon-juniper woodlands. USDA For. Serv. Gen. Tech. Rep. RM* 39.

Smith, W. B., A. K. Weatherspoon, and J. Pilon. 1990. Michigan timber industry—an assessment of timber product output and use, 1988. *USDA For. Serv. Res. Bull. NC* 121.

Smith, W. P. 1991. Ontogeny and adaptiveness of tail-flagging behavior in white-tailed deer. *Am. Nat.* 138:190–200.

Smith, W. T., and R. D. Cameron. 1985. Reactions of large groups of caribou to a pipeline corridor on the arctic coastal plain of Alaska. *Arctic* 38:53–57.

Snow, D. W. 1954. The habitats of Eurasian tits (*Parus* spp.). Ibis 96:565–585.

Snow, D. W. 1976. *The web of adaptation.* Demeter Press, New York.

Snow, D. W. 1981. Tropical frugivorous birds and their food plants: A world survey. *Biotropica* 13:1–14.

Synder, J. R., A. Herndon, and W. B. Robertson, Jr. 1990. South Florida parkland. Pp 230–277 in *Ecosystems of Florida,* R. L. Myers and J. J. Ewel, eds. University of Central Florida Press, Orlando.

Soil Survey Staff. 1990. *Keys to Soil taxonomy. 4th ed.* SMSS Technical Monograph no. 19, Blacksburg, Virginia.

Solomon, M. E. 1949. The natural control of animal populations. *J. Anim. Ecol.* 18:1–35.

Sonesson, M., and T. V. Gallaghan. 1991. Strategies of survival in plants of the Fennoscandian tundra. *Arctic* 44:95–105.

Soulé, M. E. 1987. Where do we go from here? Pp. 175–183 in *Viable populations for conservation,* M. E. Soulé, ed. Cambridge University Press, Cambridge.

Soulé, M. E., and D. Simberloff. 1986. What do genetics and ecology tell us about the design of nature reserves? *Biol. Conserv.* 35:19–40.

Sparling, J. H. 1964. Ontario's woodland flora. *Ont. Nat.* 2:18–24.

Spaulding, W. G., E. B. Leopold, and R. R. Van Devender. 1983. Late Wisconsin paleoecology of the American Southwest. Ch. 14 in *Late-Quaternary environments in the United States.* H. E. Wright and S. C. Porter, eds. University of Minnesota Press, Minneapolis.

Speakman, J. R. 1991. Why do insectivorous bats in Britain not fly in daylight more frequently? *Functional Ecol.* 5:518–524.

Speers, H. B. 1948. *Lumber production in the United States. USDA Forest Service,* Washington, D.C.

Spencer, G. 1989. *Projections of the population of the United States, by age, sex, and race: 1988–2080.* U.S. Bureau of the Census, Current Population Reports, series P-25, no. 1018.

Sperry, T. M. 1983. Analysis of the University of Wisconsin-Madison prairie restoration project. Pp. 140–147 in *Proceedings of the eighth North American Prairie Conference,* R. Brewer, ed. Western Michigan University, Department of Biology, Kalamazoo.

Spilhaus, A., ed. 1966. Waste management and control. *Nat. Acad. Sci. Publ.* 1400.

Springsett, B. P. 1968. Aspects of the relationships between burying beetles, *Necrophorus* spp. and the mite, *Poecilochirus necrophori* Vitz. *J. Anim. Ecol.* 37: 417–424.

Sprugel, D. G. 1991. Disturbance, equilibrium, and environmental variability: What is "natural" vegetation in a changing environment? *Biol. Conserv.* 58:1–18.

Stacey, P. B., and J. D. Ligon. 1991. The benefits-of-philopatry hypothesis for the evolution of cooperative breeding: Variation in territory quality and group size effects. *Am. Nat.* 137:831–846.

Stacey, P. B., and W. D. Koenig, eds. 1990. *Cooperative breeding in birds: Long-term studies of ecology and behavior.* Cambridge University Press, Cambridge.

Stadtfeld, C. K. 1976. Cheap chemicals and dumb luck. *Audubon* 78:110–118.

Stager, K. 1964. The role of olfaction in food location in the turkey vulture *(Cathartes aura). Los Ang. Cty. Mus. Contrib. Sci.* 81:1–63.

Stamatiadis, S., and D. L. Dindal. 1990. Coprophilous mite communities as affected by concentration of plastic and glass particles. *Exp. Appl. Acarol.* 8:1–12.

Stanchinsky, V. V. 1931. On the importance of biomass in the dynamic equilibrium of the biocenose. (In Russian.) *J. Ecol. Biocenol.* 1:88–98.

Stearns, S. C. 1976. Life history tactics: A review of the ideas. *Q. Rev. Biol.* 51:3–47.

Stearns, S. C. 1989. The evolutionary significance of phenotypic plasticity. *BioScience* 39:436–445.

Stebbins, R. C. 1985. *A field guide to western reptiles and amphibians.* Houghton Mifflin, Boston.

Stebbins, R. C., and N. W. Cohen. 1976. Off-road menace. A survey of damage in California. *Sierra Club Bull.* July/August.

Steel, K., P. M. Gertman, C. Crescenzi, and J. Anderson. 1981. Iatrogenic illness on a general medical service at a university hospital. *N. Engl. J. Med.* 304: 638–642.

Steinhart, J. S., and C. E. Steinhart. 1974. *Energy: Sources, use, and role in human affairs.* Duxbury Press, North Scituate, Mass.

Stephens, D. W., and J. R. Krebs. 1986. *Foraging theory.* Princeton University Press, Princeton, N.J.

Stephenson, T. A., and A. Stephenson. 1972. *Life between tide marks on rocky shores.* Freeman, San Francisco.

Stepney, P., H. R., and D. H. Power. 1973. Analysis of the eastward breeding expansion of Brewer's blackbird plus general aspects of avian expansion. *Wilson Bull.* 85:452–464.

Stevenson, B. G., and D. L. Dindal. 1987. Functional ecology of coprophagous insects: a review. *Pedobiologia* 30:285–298.

Stewart, F. M., and B. R. Levin. 1973. Partitioning of resources and the outcome of interspecific competition: A model and some general consequences. *Am. Nat.* 107:171–198.

Stiak, J. 1990. Old growth! *Amicus J.* 12:35–41.

Stock, C. 1961. Rancho La Brea, a record of Pleistocene life in California. *Los Ang. Cty. Mus. Ser.* 20:1–81.

Stoddard, H. C. 1936. Relation of burning to timber and wildlife. *Proc. North Am. Wildl. Conserv.* 1:1–4.

Stonehouse, B. 1972. *Animals of the Antarctica: The ecology of the far south.* Holt, Rinehart & Winston, New York.

Storey, K. B., and J. M. Storey. 1990. Metabolic rate depression and biochemical adaptation in anaerobiasis, hibernation and estivation. *Q. Rev. Biol.* 65: 145–173.

Stoutamire, W. P. 1974. Australian terrestrial orchids, thynnid wasps, and pseudocopulation. *Am. Orch. Soc. Bull.* 1974:13–18.

Strauss, S. Y. 1991. Indirect effects in community ecology: Their definition, study, and importance. *Trends Ecol. Evol. 6:*206–210.

Street, D. 1979. *The reptiles of northern and central Europe.* B. T. Bateford Ltd., London.

Strohecker, H. F. 1937. An ecological study of some Orthoptera of the Chicago area. *Ecology* 18:231–250.

Strong, D. R., Jr. 1980. Null hypotheses in ecology. *Synthese* 43:271–285.

Strong, D. R., Jr., et al. 1984. *Ecological communities: Conceptual issues and the evidence.* Princeton University Press, Princeton, N.J.

Sullivan, K. A. 1984. Information exploitation by downy woodpeckers in mixed-species flocks. *Behaviour* 91: 294–311.

Sullivan, T. J. et al. 1990. Quantification of changes in lakewater chemistry in response to acid deposition. *Nature* 345:54–57.

Svärdson, G. 1949. Competition and habitat selection in birds. *Oikos* 1:157–174.

Svärdson, G. 1957. The "invasion" type of bird migration. *Br. Birds* 50:314–343.

Sveinbjornsson, B. 1982. *Alaskan plants and atmospheric carbon dioxide.* Proceedings of a conference on the potential effects of carbon dioxide-induced climatic changes in Alaska. University of Alaska-Fairbanks, Misc. Pub.

Svejcar, T. J. 1990. Response of *Andropogon gerardii* to fire in the tallgrass prairie. Pp. 19–27 in *Fire in North American tallgrass prairies,* S. L. Collins and L. L. Wallace, eds. University of Oklahoma Press, Norman.

Swain, E. B., D. R. Engstrom, M. E. Brigham, T. A. Henning, and P. L. Brezonik. 1992. Increasing rates of atmospheric mercury deposition in midcontinental North America. *Science* 257:784–787.

Swain, W. R. 1991. Effects of organochlorine chemicals on the reproductive outcome of humans who consumed contaminated Great Lakes fish: An epidemiologic consideration. *J. Toxicol. Environ. Health* 33: 587–639.

Swan, L. W. 1992. The Aeolian biome. *BioScience* 42: 262–270.

Swank, W. T. 1984. Atmospheric contributions to forest nutrient cycling. *Water Resour. Bull.* 20:313–321.

Swank, W. T., and J. B. Waide. 1979. Interpretation of nutrient cycling research in a management context. Pp. 137–158 in *Fresh perspectives from ecosystem analysis,* proceedings of the 40th Biol. Colloq., Oregon State University Press.

Swank, W. T., J. B. Waide, D. A. Crossley, and R. L. Todd. 1981. Insect defoliation enhances nitrate export from forest ecosystems. *Oecologia (Berl)* 51:297–299.

Szaro, R. C., and M. D. Jakle. 1985. Avian use of a desert riparian island and its adjacent scrub habitat. *Condor* 87:511–519.

Talbot, L. M., and M. H. Talbot. 1963. The wildebeest in western Masailand, East Africa. *Wildl. Monogr.* 12: 1–88.

Talbot, M. 1934. Distribution of ant species in the Chicago region with reference to ecological factors and physiological toleration. *Ecology* 15:416–439.

Tangley, L. 1985. Releasing engineered organisms in the environment. *BioScience* 35:470–473.

Tanner, J. T. 1952. Black-capped and Carolina chickadees in the southern Appalachian mountains. *Auk* 69:407–424.

Tans, P. P., I. Y. Fung, and T. Takahashi. 1990. Observational constraints on the global atmospheric CO_2 budget. *Science* 247:1431–1438.

Tansley, A. G. 1911. *Types of British vegetation.* Cambridge.

Tansley, A. G. 1935. The use and abuse of vegetational concepts and terms. *Ecology* 16:284–307.

Taylor, A. D. 1991. Studying metapopulation effects in predator-prey systems. *Biol. J. Linn. Soc.* 42:305–323.

Taylor, C. E. 1986. Genetics and evolution of resistance to insecticides. *Biol. J. Linn. Soc.* 27:103–112.

Taylor, F. J. R. 1979. Symbionticism revisited: A discussion of the evolutionary impact of intracellular symbioses. *Proc. R. Soc. Lond. [Biol.]* 304:270–273.

Taylor, W. P. 1934. Significance of extreme or intermittent conditions in distribution of species and management of natural resources, with a restatement of Liebig's law of the minimum. *Ecology* 15:374–379.

Teague, R. D., ed. 1971. *A manual of wildlife conservation.* The Wildlife Society, Washington, D.C.

Teal, J. M. 1957. Community metabolism in a temperate cold spring. *Ecol. Monogr.* 27:298.

Teipner, C. L., E. O. Garton, and L. Nelson. 1983. Pocket gophers in forest ecosystems. *U.S. For. Serv. Gen. Tech. Rep. INT* 154.

Telfer, E. S. 1967. Comparison of moose and deer winter range in Nova Scotia. *J. Wildl. Manage.* 31:418–425.

Terborgh, J. 1985. The vertical component of plant species diversity in temperate and tropical forests. *Am. Nat.* 126:760–776.

Terborgh, J. 1989. Where have all the birds gone? Princeton University Press, Princeton, N.J.

Terrasmae, J., and T. W. Anderson. 1970. Hypsithermal range extension of white pine (*Pinus strobus* L.) in Quebec, Canada. *Can. J. Earth Sci.* 7:406–413.

Terres, J. K. 1980. *The Audubon Society Encyclopedia of North American birds.* A. A. Knopf, New York.

Thomas, W. R., M. J. Pomerantz, and M. E. Gilpin. 1980. Chaos, asymmetric growth and group selection for dynamical stability. *Ecology* 61:1312–1320.

Thompson, D. Q. 1991. History of purple loosestrife biological control efforts. *Nat. Areas J.* 11:148–150.

Thompson, J. N., and M. F. Willson. 1979. Evolution of temperate fruit-bird interactions: Phenological strategies. *Evolution* 33:973–982.

Thorson, G. 1971. *Life in the sea.* World University Library, New York.

Tilman, D. 1982. *Resource competition and community structure.* Princeton University Press, Princeton, N.J.

Tinbergen, L. 1960. The natural control of insects in pinewoods. I. *Arch. Néerl. Zool.* 13:265–336.

Tinbergen, N. 1951. *The study of instinct.* Oxford University Press, New York.

Titus, J. G. et al. 1991. Greenhouse effect and sea level rise: The cost of holding back the sea. *Coastal Manage.* 19:171–204.

Tomback, D. F., L. A. Hoffmann, and S. K. Sunde. 1980. Coevolution of whitebark pine and nutcrackers: Implications for forest regeneration. Pp. 118–129 in *Proceeding—Symposium on whitebark pine ecosystems: Ecology and management of a high-mountain resource. USDA For. Serv. Gen. Tech. Rep. INT* 270.

Toquenaga, Y., and K. Fuji. 1990. Contest and scramble competition in two bruchid beetles, *Callosobruchus analis* and *C. phaseoli* (Coleoptera: Bruchidae). I. Larval competition curves and interference mechanisms. *Res. Popul. Ecol.* 32:349–363.

Transeau, E. N. 1926. The accumulation of energy in plants. *Ohio J. Sci.* 26:7.

Transeau, E. N. 1935. The prairie peninsula. *Ecology* 16: 423–437.

Trappe, J. M., and R. D. Fogel. 1977. Ecosystematic functions of mycorrhizae. Pp. 205–214 in *The below-ground ecosystem: A synthesis of plant associated processes.* USDA Forest Service.

Travis J. 1989. The role of optimizing selection in natural populations. *Annu. Rev. Ecol. Syst.* 20:279–296.

Travis, J., J. C. Trexler, and M. Mulvey. 1990. Multiple paternity and its correlates in female *Poecilia latipinna* (Poeciliidae). *Copeia* 1990:722–729.

Trivers, R. L. 1971. The evolution of reciprocal altruism. *Q. Rev. Biol.* 46:35–57.

Trivers, R. L. 1972. Parental investment and sexual selection. Pp. 136–179 in *Sexual selection and the descent of man,* B. Campbell, ed. Aldine Publishing Co., Chicago.

Tschirley, F. H. 1969. Defoliation in Vietnam. *Science* 163:779–786.

Tucker, C. J., J. R. G. Townshend, and T. E. Goff. 1985. African land-cover classification using satellite data. *Science* 227:369–375.

Turco, R. P. 1990. Climate and smoke: An appraisal of nuclear winter. *Science* 247:166–176.

Turco, R. P., et al. 1983. Nuclear winter: Global consequences of multiple nuclear explosions. *Science* 222: 1283–1292.

Turlings, T. C. J., J. H. Tumlinson, and W. J. Lewis. 1990. Exploitation of herbivore-induced plant odors by host-seeking parasitic wasps. *Science* 250: 1251–1253.

Turner, M. G., ed. 1987. *Landscape heterogeneity and disturbance.* Springer-Verlag, New York.

Turner, M. G., and R. H. Gardner, eds. 1991. *Quantitative methods in landscape ecology.* Springer-Verlag, New York.

Turreson, G. 1922. The genotypical response of the plant species to habitat. *Hereditas* 3:211–350.

Tuttle, M. D. 1991. Bats: The cactus connection. *Natl. Geogr:* 130–140.

Udvardy, M. D. F. 1969. *Dynamic zoogeography.* Van Nostrand Reinhold, New York.

Underwood, A. J., and E. J. Denley. 1984. Paradigms, explanations and generalizations in models for the structure of intertidal communities on rocky shores. Pp. 151–180 in *Ecological communities,* Donald R. Strong et al., eds. Princeton University Press, Princeton, N.J.

Ungar, I. A. 1966. Salt tolerances of plants growing in saline areas of Kansas and Oklahoma. *Ecology* 46: 154–155.

Ungar, I. A. 1978. Halophyte seed germination. *Bot. Rev.* 44:233–264.

US Department of Agriculture. 1941. *Climate and man: Yearbook of agriculture.* House Document No. 27, 77th Congress of the United States.

USDA Soil Survey Staff. 1975. *Soil taxonomy.* Agriculture Handbook No. 436. USDA Forest Service.

USDA Soil Conservation Service. 1960. *Soil classification: A comprehensive system.* USDA, Washington, D.C.

US Department of Commerce. 1976. *Statistical abstract of the United States.* 97th ed. Bureau of the Census, Washington, D.C.

US Department of Commerce. 1985. *Statistical abstract of the United States.* 106th ed. Bureau of the Census, Washington, D.C.

US Department of Health and Human Services. 1986. *Vital statistics of the United States, 1981.* Public Health Service, Hyattsville, Md.

U.S. National Park Service. 1991. Draft environmental assessment for Wildland Fire Management Plan Yellowstone National Park. U.S. Dept. of Interior National Park Service.

Utida, S. 1953. Interspecific competition between two species of bean weevil. *Ecology* 34:301–307.

Vacanti, P.L., and K. N. Geluso, 1985. Recolonization of a burned prairie by meadow voles (*Microtus pennsylvanicus*). *Prairie Naturalist* 17:15–22.

Vali, G., M. Christensen, R. W. Fresh, E. L. Galyan, L. R. Maki, and R. C. Schnell. 1976. Biogenic ice nuclei. Part II: Bacterial sources. *J. Atmos. Sci.* 33:1565–1570.

Vallentyne, J. R. 1974. *The algal bowl—lakes and man.* Dep. Environ. Misc. Special Pub. 22, Ottawa, Canada.

Van Alstyne, K. L. 1988. Herbivore grazing increases polyphenolic defenses in the intertidal brown alga *Fucus distichus. Ecology* 69:655–663.

Van Beneden, P. J. 1876. *Animal parasites and messmates.* Appleton, New York.

Van Cleve, K., et al. 1983. Ecosystems in interior Alaska. *BioScience* 33:39–44.

Van der Hammen, T. 1991. Palaeoecological background: Neotropics. *Climatic change* 19:37–47.

Van Deusen, P. C., G. A. Rams, and E. R. Cook. 1991. Possible red spruce decline. *J. For.* 89:20–24.

Van Hook, R. I. 1971. Energy and nutrient dynamics of spider and orthopteran populations in a grassland ecosystem. *Ecol. Monogr.* 41:1–26.

Van Horne, B. 1983. Density as a misleading indicator of habitat quality. J. Wildl. Manage. 47:893–901.

Van Kat, J. 1991. Floristics of a chronosequence corresponding to old field-deciduous forest succession in southwestern Ohio. IV. *Bull. Torrey Bot. Club* 118: 392–398.

Van Kat, J., and W. P. Carson. 1991. Floristics of a chronosequence corresponding to old field-deciduous forest succession in southwestern Ohio. III. *Bull. Torrey Bot. Club* 118:385–391.

Van Kat, J. L. 1979. *The natural vegetation of North America.* Wiley, New York.

Van Orman, J. 1976. Avian succession on Lake Michigan sand areas. M.S. Thesis, Western Michigan University.

Van Valen, L. 1975. Life, death, and energy of a tree. *Biotropica* 7:263.

Vance, B. D., and C. C. Kucera. 1960. Flowering variations in *Eupatorium rugosum. Ecology* 41:340–345.

VanderMeer, J. H. 1969. The competitive structure of communities: An experimental approach with protozoa. *Ecology* 50:362–371.

Vander Wall, S. B., and R. P. Balda. 1977. Coadaptation of the Clark's nutcracker and the piñon pine for efficient seed harvest and dispersal. *Ecol. Monogr.* 47: 89–111.

Varga, L. 1928. Ein interessanter Biotop der Bioconose von Wasserorganismen. *Biol. Zentralbl.* 48:143–162.

Varley, J. D., and P. Schullery. 1991. Reality and opportunity in the Yellowstone fires of 1988. Pp. 105–121 in *The greater Yellowstone ecosystem: redefining America's wilderness heritage,* R. Keiter and M. Boyce, eds. Yale University Press, New Haven, Conn.

Vasek, F. C. 1980. Creosotebush: Long-lived clones in the Mohave desert. *Am. J. Bot.* 67:246–255.

Verboom, J., A. Schotman, P. Opdam, and J. A. J. Metz. 1991. European nuthatch metapopulations in a fragmented agricultural landscape. *Oikos* 61:149–156.

Vernadsky, W. I. 1929. *La biosphere.* Alcan, Paris.

Verner, J., and A. S. Boss, eds. 1980. California wildlife and their habitats: Western Sierra Nevada. *USDA For. Serv. Gen. Tech. Rep. PSW* 37.

Verner, J., and M. F. Willson. 1969. Mating systems, sexual dimorphism, and the role of male North American passerine birds in the nesting cycle. *Am. Ornithol. Union Ornithol. Monogr.* 9:1–76.

Verry, E. S. 1975. Streamflow chemistry and nutrient yields from upland-peatland watersheds in Minnesota. *Ecology* 56:1149–1157.

Viereck L. A., and C. T. Dyrness. 1980. A preliminary classification system for vegetation of Alaska. *USDA For. Serv. Gen. Tech. Rep. PNW* 106.

Viktorov, S. V., E. A. Vostokova, and D. D. Vyshivkin. 1965. Some problems in the theory of geobotanical indicator research. Pp. 1–4 in *Plant indicators of soils, rocks, and subsurface waters.* Consultants Bureau, New York.

Vitousek, P. M., J. R. Gosz, C. C. Grier, J. M. Melillo, and W. A. Reiners. 1979. Nitrate losses from disturbed ecosystems. *Science* 204:469–474.

Vitousek, P. M., P. R. Ehrlich, A. H. Ehrlich, and P. A. Matson. 1986. Human appropriation of the products of photosynthesis. *Bioscience* 36:368–373.

Vitousek, P. M., and R. W. Howarth. 1991. Nitrogen limitation on land and in the sea: How can it occur? *Biogeochemistry* 13:87–115.

Vitousek, P. M., and W. A. Reiners. 1975. Ecosystem succession and nutrient retention: A hypothesis. *Bioscience* 25:337.

Vogelmann, H. W., G. J. Badger, M. Bliss, and R. M. Klein. 1985. Forest decline on Camels Hump, Vermont. *Bull. Torrey Bot. Club* 112:274–287.

Vogt, K. A., C. C. Grier, C. E. Meier, and R. L. Edmunds. 1982. Mycorrhizal role in net primary production and mineral cycling in *Abies amabilis* (Dougl.) Forbes ecosystems in western Washington. *Ecology* 63:370–380.

Voigt, J. W., and R. H. Mohlenbrock. 1964. *Plant communities of southern Illinois.* Southern Illinois University Press, Carbondale.

Volterra, V. 1931. Variations and fluctuations in the number of individuals in animal species living together. Pp. 409–448 in *Animal ecology,* R. N. Chapman, ed. McGraw-Hill, New York.

von Frisch, K. 1967. *The dance language and orientation of bees.* Harvard University Press, Cambridge, Mass.

Vrijenhoek, R. C., M. E. Douglas, and G. K. Meffe. 1985. Conservation genetics of endangered fish populations in Arizona. *Science* 229:400–402.

Wade, M. J. 1978. A critical review of the models of group selection. *Q. Rev. Biol.* 53:101–114.

Wade, M. J. 1980. Kin selection: Its components. *Science* 210:665–667.

Wade, M. J. 1982. Group selection: Migration and the differentiation of small populations. *Evolution* 34: 799–812.

Waggoner, P. E., and G. R. Stephens. 1970. Transition probabilities for a forest. *Nature* 225:1160–1161.

Wald, G. 1964. The origin of life. *Proc. Natl. Acad. Sci. USA* 52:595–611.

Waldbauer, G. P., and W. E. LaBerge. 1985. Phenological relationships of wasps, bumblebees, their mimics, and insectivorous birds in northern Michigan. *Ecol. Entomol.* 10:99–110.

Walker, C. G. 1984. How life affects the atmosphere. *BioScience* 34:486–491.

Walker, D. A., et al. 1987. Cumulative impacts of oil fields on northern Alaskan landscapes. *Science* 238: 757–761.

Walker, D. A., et al. 1989. Impacts of petroleum development in the Arctic [letter] *Science* 245:765–766.

Walker, T. 1985. Introductory remarks for panel discussion: Managing spruce budworm in Vermont. P. 209 in *Spruce-fir management and spruce budworm. USDA For. Serv. Gen. Tech. Rep. NE* 99.

Walkey, M., and R. H. Meakins. 1970. An attempt to balance the energy budget of a host-parasite system. *J. Fish Biol.* 2:361–372.

Walkinshaw, L. H. 1991. Yellow-bellied Flycatcher. Pp. 280–281 in *The Atlas of breeding birds of Michigan,* R. Brewer, G. A. McPeek, and R. J. Adams, Jr., eds. Michigan State University Press, East Lansing.

Wallace, G. C., W. P. Nickell, and R. F. Bernard. 1961. Bird mortality in the Dutch elm disease program in Michigan. *Cranbrook Inst. Sci. Bull.* 41.

Waller, N. G., B. A. Kojetin, T. J. Bouchard, Jr., D. T. Lykken, and A. Tellegen 1990. Genetic and environmental influence on religious interests, attitudes, and values: A study of twins reared apart and together. *Psychol. Sci.* 1:138–142.

Walsberg, G. E. 1986. Thermal consequences of roost-site selection: The relative importance of three modes of heat conservation. *Auk* 103:1–7.

Walsh, J. 1991. The greening of the green revolution. *Science* 252:26.

Walter, H. 1971. *Ecology of tropical and subtropical vegetation.* Oliver & Boyd, Edinburgh.

Ward, G. M., P. Knox, B. Hobson, and T. P. Yorks. 1977. Energy requirements of alternative beef production systems in Colorado. Pp. 395–412 in *Agriculture and energy,* W. Lockeretz, ed. Academic Press, New York.

Ward, P., and A. Zahavi. 1973. The importance of certain assemblages of birds as "information centres" for food-finding. *Ibis* 115:517–534.

Waring, R. H., and J. F. Franklin. 1979. Evergreen coniferous forests of the Pacific Northwest. *Science* 204: 1380–1386.

Warming, E. 1895. *Plantesamfund.* Copenhagen. (Eng. transl. 1909. *Oecology of plants.* Oxford University Press, London.)

Warner, R. E. 1968. The role of introduced diseases in the extinction of the endemic Hawaiian avifauna. *Condor* 70:101–120.

Warren, D. D. 1991. Production, prices, employment and trade in Northwest forest industries, fourth quarter 1990. *USDA For. Serv. Res. Bull.* PNW-RB-187.

Washburn, J. O., D. R. Mercer, and J. R. Anderson. 1991. Regulatory role of parasites: Impact on host population shifts with resource availability. *Science* 253:185–188.

Waters, T. F. 1972. The drift of stream insects. *Annu. Rev. Entomol.* 17:253–272.

Watson, A., R. Moss, and R. Parr. 1984b. Effects of food enrichments on numbers and spacing behaviour of red grouse. *J. Anim. Ecol.* 53:663–678.

Watson, A., R., Moss, P. Rothery, and R. Parr. 1984a. Demographic causes and predictive models of population fluctuations in red grouse. *J. Anim. Ecol.* 53:639–662.

Watt, A. S. 1947. Pattern and process in the plant community. *J. Ecol.* 35:1–22.

Watt, K. E. F. 1968. *Ecology and resource management.* McGraw-Hill, New York.

Watterson, A. 1988. *Pesticide users' health and safety handbook.* Van Nostrand Reinhold, New York.

Watts, A. W. 1958 (1970 repr.). *Nature, man, and woman.* Vintage Books, New York.

Watts, M. T. 1975. *Reading the landscape of America.* Macmillan, New York.

Watts, W. A. 1983. Vegetational history of the eastern United States 25,000–10,000 years ago. Ch. 15 in *Late-Quaternary environments of the United States,* S. C. Porter, ed. University of Minnesota Press, Minneapolis.

Watts, W. A., B. C. S. Hansen, and E. C. Grimm. 1991. Camel Lake: A 40000-yr record of vegetational and forest history from Northwest Florida. *Ecology* 73:1056–1066.

Watts, W. A., and M. Stuiver. 1980. Late Wisconsin climate of northern Florida and the origin of the species-rich deciduous forest. *Science* 210:325–327.

Weaver, J. E. 1961. Return of midwestern grassland to its former composition and stabilization. *Occas. Pap. C. C. Adams Cen. Ecol. Stud.* 3:1–15.

Weaver, J. E., and E. L. Flory. 1934. Stability of climax prairie and some environmental changes resulting from breaking. *Ecology* 15:333–347.

Weaver, J. E., and F. E. Clements. 1938 (1st ed. 1929). *Plant ecology.* McGraw-Hill, New York.

Weaver, J. E., and F. W. Albertson. 1956. *Grasslands of the Great Plains.* Johnsen Pub. Co., Lincoln, Neb.

Webb, T., III, E. J. Cushing, and H. E. Wright, Jr. 1983. Holocene changes in the vegetation of the Midwest. Ch. 10 in *Late-Quaternary environments of the United States,* H. E. Wright, ed. University of Minnesota Press, Minneapolis.

Weber, J. A., T. W. Jurik, J. D. Tenhunen, and D. M. Gates. 1985. Analysis of gas exchange in seedlings of *Acer saccharum:* Integration of field and laboratory studies. *Oecologia (Berl)* 65:338–347.

Webster, J. D. 1956. Birds and grasslands in western Mexico. *Hanover Forum:* 34–45.

Weinberg, A. 1972. Social institutions and nuclear energy. *Science* 177:27–34.

Weinberg, W. 1908. Über den Nachweis der Vererbung beim Menschen. *Jahresh. Ver. Naturk. Wuerttemb.* 64: 368–382.

Weiser, C. J. 1970. Cold resistance and injury in woody plants. *Science* 169:1269–1278.

Welch, P. S. 1952. *Limnology.* McGraw-Hill, New York.

Weller, R. H., and L. F. Bouvier. 1981. *Population demography and policy.* St. Martin's Press, New York.

Welty, J. C. 1982. *The life of birds.* Saunders College Pub., Philadelphia.

Went, F. W., and N. Stark. 1968. Mycorrhiza. *BioScience* 18:1035–1039.

Weseloh, D. V., and R. T. Brown. 1971. Plant distribution within a heron rookery. *Am. Midl. Nat.* 86:57–64.

West, D. 1992. Taking aim at a deadly chemical. *Natl. Wildl.* 1992:38–41.

Westemeier, R. L. 1992. Prairie chicken egg exchanges. *Illinois Nat. Hist. Surv. Rep.* 316:1–2.

Westh, P., J. Kristiansen, and A. Hvidt. 1991. Ice-nucleating activity in the freeze-tolerant tardigrade, *Adorybiotus coronifer. Comp. Biochem. Physiol.* 99:401–404.

Westlake, D. F. 1963. Comparisons of plant productivity. *Biol. Rev.* 38:385–425.

Wetzel, R. W. 1983. *Limnology.* 2nd ed. Saunders College Pub., Philadelphia.

Wherry, E. T. 1948. *Guide to eastern ferns.* 2nd ed. University of Pennsylvania Press, Philadelphia.

Whisenant, S. G. 1990. Changing fire frequencies on Idaho's Snake River Plains: Ecological and management implications. Pp. 4–10 in *Proceedings—Symposium on cheatgrass invasion, shrub die-off, and other aspects of shrub biology and management. USDA For. Serv. Gen. Tech. Rep. INT* 267.

Whitcomb, R. F., et al. 1981. Effects of forest fragmentation on avifauna of the eastern deciduous forest. Pp. 125–205 in *Forest island dynamics in man-dominated landscapes,* R. L. Burgess and D. M. Sharpe, eds. Springer-Verlag, New York.

White, L. 1967. The historical roots of our ecological crisis. *Science* 155:1203–1207.

Whiteside, T. 1970. *Defoliation: What are our herbicides doing to us?* Ballantine Books, New York.

Whitmore, T. C. 1978. Gaps in the forest canopy. Pp. 639–655 in *Tropical trees as living systems,* P. B. Tomlinson and M. H. Zimmerman, eds. Cambridge University Press, Cambridge.

Whittaker, R. H. 1953. A consideration of climax theory: The climax as a population and pattern. *Ecol. Monogr.* 23:41–78.

Whittaker, R. H. 1956. Vegetation of the Great Smoky Mountains. *Ecol. Monogr.* 26:1–30.

Whittaker, R. H. 1966. Forest dimensions and production in the Great Smoky Mts. *Ecology* 44:176–182.

Whittaker, R. H. 1972. Evolution and measurement of species diversity. *Taxon* 21:213–251.

Whittaker, R. H. 1975. *Communities and ecosystems.* 2nd ed. Macmillan, New York.

Whittaker, R. H., and G. Woodwell. 1969. Structure, production, and diversity of the oak-pine forest at Brookhaven, New York. *J. Ecol.* 57:155–174.

Whittaker, R. H., and P. L. Marks. 1975. Methods of assessing terrestrial productivity. Pp. 55–118 in *Pri-*

mary productivity of the biosphere, H. Lieth and R. H. Whittaker, eds. Springer-Verlag, New York.

Whittaker, R. H., and P. P. Feeny. 1971. Allelochemics: Chemical interactions between species. *Science* 171: 757–770.

Whittington, S. L., and B. Dyke. 1984. Simulating overkill: Experiments with Mosimann and Martin model. Pp. 451–465 in *Pleistocene extinctions,* P. S. Martin and R. G. Klein, eds. University of Arizona Press, Tucson.

Wickler, W. 1968. *Mimicry in plants and animals,* R. B. Martin, trans. McGraw-Hill, New York.

Wiegert, R. G., and F. C. Evans. 1967. Investigations of secondary productivity in grasslands. Pp. 499–518 in *Secondary productivity of terrestrial ecosystems,* K. Petrusewicz, ed. Polish Academy of Science.

Wielgolaski, F. E., L. C. Bliss, J. Svoboda, and G. Doyle. 1981. Primary production of tundra. Pp. 187–225 in *Tundra ecosystems,* L. C. Bliss, O. W. Heal, and J. J. Moore, eds. Cambridge University Press, Cambridge.

Wiens, J. A. 1977. Model estimation of energy flow in North American grassland bird communities. *Oecologia (Berl)* 31:135–151.

Wijkman A., and L. Timberlake. 1985. Is the African drought an act of God or man? *Ecologist* 15:9–18.

Wilcove, D. S. 1985. Nest predation in forest tracts and the decline of migratory songbirds. *Ecology* 66:1211–1214.

Wilhelm, G. S. 1990. Special vegetation of the Indiana Dunes National Lakeshore. Indiana Dunes National Lakeshore Research Program Report 90-02:1–373.

Wilhelm, S. 1966. Chemical treatments and inoculum potential of soils. *Annu. Rev. Phytopathol.* 4:53–78.

Wilhm, J. F. 1975. Biological indicators of pollution. In *River ecology,* B. A. Whitton, ed. University of California Press, Studies in Ecology Vol. 2.

Wilkinson, L., ed. 1980. *Earthkeeping: Christian stewardship of natural resources.* Eerdmanns Publishing Company, Grand Rapids, Mich.

Wilkison, G. S. 1986. Gastric cancer in New Mexico counties with significant deposits of uranium. *Arch. Environ. Health* 40:307–312.

Williams, A. B. 1936. The composition and dynamics of a beech-maple climax community. *Ecol. Monogr.* 6: 318–408.

Williams, A. B. 1950. Census No. 7. Climax beech-maple forest with some hemlock. *Audubon Field Notes* 4: 297–298.

Williams, C. B. 1964. *Patterns in the balance of nature.* Academic Press, New York.

Williams, G. C. 1964. Measurement of consociation among fishes and comments on the evolution of schooling. *Mich. St. Univ. Mus. Pub.* 2:349–384.

Williams, G. C. 1966. *Adaptation and natural selection.* Princeton University Press, Princeton, N.J.

Williams, H., III. 1992. Banking on the future. *Nature Conserv.* 42:23–27.

Williams, K. S., and L. E. Gilbert. 1981. Insects as selective agents on plant vegetative morphology: Egg mimicry reduces egg laying by butterflies. *Science* 212:467–469.

Willson, M. F. 1983. *Plant reproductive ecology.* John Wiley & Sons, New York.

Wilson, D. S. 1980. *The natural selection of populations and communities.* The Benjamin/Cummings Pub. Co., Inc., Menlo Park, Calif.

Wilson, D. S. 1983a. The effect of population structure on the evolution of mutualism. *Am. Nat.* 121:856–869.

Wilson, D. S. 1983b. The group selection controversy: History and current status. *Annu. Rev. Ecol. Syst.* 14: 159–187.

Wilson, E. O. 1969. The species equilibrium. *Brookhaven Symp. Biol.* 22:38–47.

Wilson, E. O. 1975. *Sociobiology.* Harvard University Press, Cambridge, Mass.

Wilson, L. F. 1977. A guide to insect injury of conifers in the lake states. *US For. Serv. Agric. Handbook* 501: 1–218.

Wilson, O. 1967. Objective evaluation of wind chill index by records of frostbite in the Antarctic. *Int. J. Meteor.* 11:29–32.

Wise, P. H., M. Kotelchuck, M. L. Wilson, and M. Mills. 1985. Racial and socioeconomic disparities in childhood mortality in Boston. *N. Engl. J. Med.* 313:360–366.

Wolinsky, S. M., et al. 1992. Selective transmission of human immunodeficiency virus type-1 variants from mothers to infants. *Science* 255:1134–1137.

Wood, G. W. 1991. Owl conservation strategy flawed. *J. For.* 89:39–41.

Wood, T., F. H. Bormann, and G. K. Voight. 1984. Phosphorus cycling in a northern hardwood forest: Biological and chemical control. *Science* 223:391–393.

Woodel, S. R. J. 1985. Salinity and seed germination patterns in coastal plants. *Vegetatio* 61:223–229.

Woods, F. W., and K. Brock. 1964. Interspecific transfer of Ca-45 and P-32 by root systems. *Ecology* 45:886–889.

Woods, F. W., and R. E. Shanks. 1959. Natural replacement of chestnut by other species in the Great Smoky Mountains National Park. *Ecology* 40:349–361.

Woodwell, G. M., ed. 1965. Ecological effects of nuclear war. *Brookhaven Nat. Lab. Pub.* 917:1–72.

Woodwell, G. M., R. H. Whittaker, W. A. Reiners, and

G. E. Likens. 1978. The biota and the world carbon budget. *Science* 199:141–146.

Woolfenden, G. E. 1975. Florida scrub jay helpers at the nest. *Auk* 92:1–15.

Woolfenden, G. E., and J. W. Fitzpatrick. 1984. *The Florida scrub jay: Demography of a cooperatively breeding bird.* Princeton University Press, Princeton, N. J.

Wootton, J. T. 1991. Direct and indirect effects of nutrients on intertidal community structure: Variable consequence of seabird guano. *J. Exp. Marine Biol. Ecol.* 151:139–153.

Wootton, J. T., B. E. Young, and D. W. Winkler. 1991. Ecological versus evolutionary hypotheses: Demographic stasis and the Murray-Nolan clutch size equation. *Evolution* 45:1947–1950.

Worf, D. L., ed. 1980. *Biological monitoring for environmental effects.* Lexington Books, Lexington, Mass.

World Resources Institute. 1986. *World resources 1986.* Basic Books, New York.

Wright, A. 1990. *The death of Ramón Gonzáles: The modern agricultural dilemma.* University of Texas Press, Austin.

Wright, H. E., Jr., ed. 1983. *Late-Quaternary environments of the United States. Vol. 2, The Holocene.* University of Minnesota Press, Minneapolis.

Wright, R. F., and M. Hauhs. 1991. Reversibility of acidification: Soils and surface waters. *Proc. R. Soc. [Biol]* 97B:169–191.

Wrubel, R. P., S. Krimsky, and R. E. Wetzler. 1992. Field testing transgenic plants. *BioScience* 42:280–289.

Wunderle, J. M., Jr. 1991. Age-specific foraging proficiency in birds. *Curr. Ornithol.* 8:273–324.

Wunderlin, R. P. 1982. *Guide to the vascular plants of central Florida.* University Presses of Florida, Tampa.

Wynne-Edwards, V. C. 1962. *Animal dispersion in relation to social behavior.* Oliver and Boyd, Edinburgh.

Yager D. D., and M. L. May. 1990. Ultrasound-triggered, flight-gated evasive maneuvers in the flying praying mantis, *Parasphendale agrionina,* II. *J. Exp. Biol.* 152: 41–58.

Yarnell, R. A. 1964. Aboriginal relationships between culture and plant life in the Upper Great Lakes region. *Univ. Mich. Mus. Anthropol. Anthropol. Pap.* 23.

Yeaton, R. I. 1978. A cyclical relationship between *Larrea tridentata* and *Opuntia leptocaulis* in the northern Chihuahuan desert. *J. Ecol.* 66:651–656.

Yellowstone National Park. 1992. *Yellowstone National Park Wildland Fire Management Plan.* U.S. Department of Interior National Park Service.

Yorke, J. A., N. Nathanson, G. Pianigiani, and J. Martin. 1979. Seasonality and the requirements for perpetuation and eradication of viruses in populations. *J. Epidemiol.* 109:103–123.

Yorke, J. A., and W. P. London. 1973. Recurrent outbreaks of measles, chickenpox, and mumps. *Am. J. Epidemiol.* 98:469–482.

Young, H. 1968. A consideration of insecticide effects on hypothetical avian populations. *Ecology* 49:991–994.

Young, J. A., and F. Tipton. 1990. Invasion of cheatgrass into arid environments of the Lahontan Basin. Pp. 37–40 in *Proceedings—Symposium on cheatgrass invasion, shrub die-off, and other aspects of shrub biology and management. USDA For. Serv. Gen. Tech. Rep.* INT 267.

Young, T. P. 1990. Evolution of semelparity in Mount Kenya lobelias. *Evol. Ecol.* 4:157–171.

Young, T. P., and C. K. Augspurger. 1991. Ecology and evolution of long-lived semelparous plants. *Trends Ecol. Evol.* 6:285–289.

Yukon Executive Council Office. 1988. *The caribou are our life.* Government of the Yukon. Whitehorse, Yukon, Canada.

Zaleski, C. P. 1980. Breeder reactors in France. *Science* 208:137–144.

Zasada, J. C. 1976. Alaska's interior forests. *J. For.* 74: 333–337.

Zilberman, D., et al. 1991. The economics of pesticide use and regulation. *Science* 253:518–522.

Zimmerman, J. 1991. Dickcissel. Pp. 462–463 in *The atlas of breeding birds of Michigan,* R. Brewer, G. A. McPeek, and R. J. Adams, Jr., eds. Michigan State University Press, East Lansing.

Zucker, W. V. 1983. Tannins: Does structure determine function? An ecological perspective. *Am. Nat.* 121: 335–365.

Zwinger, A. H., and B. E. Willard. 1972. *Land above the trees.* Harper & Row, New York.

Glossary

This glossary is designed for the readers of this book. Consequently, most entries are in two categories: (1) Ecological terms that are repeated away from the main treatment of the subject and (2) technical terms from other fields that the student may have missed or forgotten. Meanings for unfamiliar words not in the glossary should be sought in the body of the book (see index), a comprehensive general biology text, or a good dictionary.

Abiotic. Nonliving.

Acclimation. Phenotypic adjustment (especially change in tolerance limits) to changed conditions.

Acid rain. Rain with a low pH because of the presence of sulfuric or nitric acid; more generally, any acidifying atmospheric deposition.

Acquired immunodeficiency syndrome (AIDS). A combination of pathological conditions resulting at least in part from infection with human immunodeficiency virus (HIV).

Adaptive. Tending to fit an organism to its environment.

Aestivation. A period of inactivity during a hot or dry season.

Age distribution. The proportions of individuals of different ages in a population.

Allee effect. The situation of higher population growth rate at a population density above the minimum.

Alleles. Alternative forms of the same gene; genes occupying the same position (locus) on homologous chromosomes.

Allelopathy. Chemical inhibition of one organism by another. Sometimes restricted to interactions between higher plants.

Allogenic. Caused by outside agents; generally refers to successional changes resulting from environmental change.

Allopatric. Occupying separate geographical areas.

Altruism. Helping others. Behavior that lowers one's fitness while raising the fitness of the organism to which the behavior is directed.

Amensalism. The coaction in which one organism harms another as a by-product of its activities.

Anaerobe. An organism able to live in the absence of free oxygen.
Annual. An organism whose life span is contained within a single year.
Anoxic. Lacking oxygen.
Aposematic. Conspicuous, usually referring to colors or patterns associated with poisonous or otherwise dangerous organisms.
Arboreal. Pertaining to or inhabiting trees.
Aspection. Seasonal change.
Assimilation. Conversion of nutrients to living material of the organism.
Association. (1) A community recognized on the basis of a characteristic combination of species; (2) a subdivision of a formation.
Autecology. The ecology of the individual organism.
Autogenic. Self caused; generally refers to succession based on reactions and other biotic effects.
Autotroph. An organism that can manufacture its food from inorganic materials: green plants and chemosynthetic organisms.

Bajada. The long, sloping alluvial fans at the base of mountain ranges in the desert.
Benthos. The bottom fauna of water bodies.
Biennial. A plant with a two-year life cycle, dying after it flowers in its second year.
Bioaccumulation. Synonym of biological magnification.
Biocenose. Community.
Biochemical oxygen demand (BOD). Oxygen required to break down organic material and to oxidize reduced chemicals (in water or sewage).
Biocide. A chemical that kills living things.
Bioconcentration. Synonym of biological magnification.
Biodiversity. Biological diversity—variety in landscapes, ecosystems, species, and genetics.
Bio-ecology. Ecology.
Biogeochemical cycle. A circular system through which some element is transferred between the biotic and abiotic parts of the biosphere.
Biological control. The use of predators or parasites to lower pest numbers to an economically acceptable level.
Biological magnification. The increase in concentration of some material in organisms compared with its concentration in the environment.
Biomass. The weight of living material (of a population, trophic level, etc.)
Biome. A community of geographical extent characterized by a distinctive landscape based on the life forms of the climax dominants.
Biosphere. The layer of the globe containing living organisms; also, the earth considered as an ecological system.
Biota. The organisms of an area.
Biotic. Pertaining to life.
Biotic potential. The inherent capacity of a population to increase in number.
Birth rate. Number of births in a population per unit time.
Blue-green algae. The group of procaryotic unicellular or filamentous organisms now generally termed the Cyanophyta or Cyanobacteria.
Bog. A type of wetland ecosystem in which organic matter accumulates as peat.
Boreal. Northern.

Canopy. The leafy top layer of a forest formed by the tree crowns.
Capitalization. Evolution of the ability to benefit from a coaction by a partner originally harmed in it.
Carcinogen. A cancer-producing agent.
Carrying capacity. The maximum number of a species that can be supported indefinitely by a certain area.
Cation. An ion having a positive charge, such as calcium (Ca^{+}), magnesium (Mg^{++}), and ammonium (NH_4^+).
Cation exchange capacity. The ability of a soil to store cations, a measure of potential fertility.
Chaparral. Dense sclerophyll shrublands occurring at lower elevations in the mountains of California; also similar vegetation elsewhere in the Southwest.
Character displacement. The situation in which two species are less similar where they occur together than in allopatry.
Circadian rhythm. A cycle of activity having an approximately 24-hour period.
Clear-cutting. Timber cutting that removes all the trees from a site.
Climax community. A community that is the stable end product of succession.
Coaction. The effect of one organism on another.
Coevolution. The simultaneous evolution of two or more coacting species.
Cohort. All the individuals in a population that began life in the same period.
Commensalism. The coaction in which one member is helped and the other neither gains nor suffers.
Community. A group of organisms occupying a particular area; the connotation is of a coacting system.

Competition. A combined demand in excess of the immediate supply.

Competitive exclusion. Restriction or elimination of one species from a habitat or area by another species through monopolization of shared resources.

Conservation. Maintenance of the biosphere's natural diversity; preservation and wise use of natural resources.

Conspecific. Belonging to the same species.

Continuum. A gradient of change in species composition related to habitat.

Crown. The upper, leafy part of a tree.

Culture. Socially transmitted behavior patterns, beliefs, language, institutions, etc., especially characteristic of human populations.

Cyanophyta. The group of unicellular or filamentous procaryotes also referred to as the blue-green algae.

Cyclic. Recurring periodically.

Death rate. Number of deaths in a population per unit time.

Definitive host. The host in which a parasite reaches sexual maturity.

Demography. The study of characteristics of populations (especially humans).

Denitrification. Reduction of nitrate or nitrite to molecular nitrogen or nitrogen oxide.

Density. Numbers per unit area.

Density-dependent factor. A factor affecting population size whose intensity of action varies with density.

Deterministic. Referring to models where chance variation in parameters is not considered.

Detritus. Dead particulate organic matter.

Diel. Daily.

Diploid. Having a 2N number of chromosomes.

Disaffiliation. The evolution by one member of a coaction of traits that terminate the association.

Dispersal. The movement of individuals from the homesite.

Dissolved organic matter (DOM). Organic carbon molecules in water resulting from leaching, secretion, decomposition, excretion, etc.

Disturbance. Interruption of the settled state.

Diurnal. Active in the daytime.

Diversity. Variety; usually refers to number of species in a community and often includes some measure of their relative abundance.

Dominance. Control exerted on the character and composition of an ecosystem by an organism.

Dominance hierarchy. Social organization in which high-ranking individuals have precedence in the use of resources.

Ecological amplitude. Range of tolerance.

Ecological efficiency. The percentage of energy taken in (as food) by one trophic level that is passed on as food to the next higher trophic level. Several other ratios of energy transfer may also be calculated.

Ecological equivalents. Unrelated species playing the same ecological role (occupying the same niche) in different geographical areas.

Ecological indicator. A species or community (or their response) that is a measure of environmental conditions.

Ecological niche. The role of a species in an ecosystem.

Ecology. The study of the relationships of organisms to their environment and to one another.

Ecosystem. The community plus its habitat; the connotation is of an interacting system.

Ecotone. A transition zone between two ecosystems.

Ecotype. A genetically differentiated population within a species, the differences having ecological significance.

Ectotherm. Synonym of poikilotherm.

Edaphic. Pertaining to soil.

Efficiency. The ratio of output to input.

Endangered species. A species in imminent danger of extinction.

Endotherm. Synonym of homoiotherm.

Energy subsidy. Energy other than its own metabolic energy used by an organism.

Environment. The surroundings of an organism.

Environmental impact statement (EIS). An examination of the potential environmental effects of a proposed action including a consideration of alternative courses of action, required for federal projects by the National Environmental Policy Act.

Environmental resistance. The effects of crowding that lower population growth below that potentially possible.

Epidemic. The spread of a communicable pathogen through a population.

Epiphyte. A plant that grows on another plant but is not parasitic.

Estuary. An arm of the sea where fresh and salt water mix.

Ethology. The study of the function, causation, and evolution of behavior.

Eucaryote. Organism having chromosomes included within a nuclear membrane.

Eusocial. Truly social. Populations characterized by

cooperative rearing of young, overlap of generations, and subdivision of labor.

Eutrophication. The increase of nutrients in lakes either naturally or artificially by pollution.

Evapotranspiration. Loss of water from a land surface through both transpiration from plants and evaporation.

Evolution. A change in gene frequencies with time; the adaptive modification of organisms through successive generations.

Exploitation. Competition directly for a resource.

Exponential population growth. Population growth at an ever-increasing rate; the situation in which population increase depends on population size and a constant per head growth rate.

Extirpation. Local extinction.

Extrinsic. Originating from the outside.

Fauna. The animals of an area.

Fecundity. The potential level of reproduction in a population. In human populations, for example, slightly more than one birth per reproductive-age female per year.

Fen. An alkaline bog.

Fertility. The realized level of reproduction in a population; the actual number of live births.

Fitness. In current evolutionary terminology, representation (of an individual, genotype, phenotype, or allele) in the next breeding generation.

Flight song. A territorial or mating song given on the wing.

Flood plain. The belt of flat, annually inundated land along a stream.

Flora. The species of plants in an area.

Fluctuation. An alteration, usually temporary, in community composition or structure resulting from a change in habitat factors.

Food chain. A sequence of organisms, each of which feeds on the preceding.

Food web. A trophic system composed of interconnected food chains.

Foraging. Searching for and securing food.

Forb. A herb that is not a grass, sedge, or rush.

Forest edge. The vegetational structure characteristic of the edge of a forest, a type of savanna with interspersed trees, thicket, and herbaceous areas.

Formation. One of the great subdivisions of the earth's vegetation (deciduous forest, tundra, etc.) having a distinctive physiognomy based on the life forms of the climax dominants.

Frass. Insect fecal pellets.

Frugivore. Fruit eater.

Fugitive species. Species characteristic of temporary habitats.

Gamete. A haploid reproductive cell; a sperm or egg.

Gap phase. The part of cyclic replacement in forests initiated by the loss of a canopy tree.

Gene. The basic unit of heredity.

Gene pool. The genes of all individuals of a population, collectively.

Generation. The average interval from the birth of parents to the birth of their offspring.

Genotype. Genetic makeup of an individual. For example, a human with two genes for blue eyes would have genotype *bb*.

Geographic range. The area of the earth's surface occupied by all the populations of a species.

Granivore. Subsisting on grain or seeds.

Gross primary production (GPP). Total energy storage by the autotrophs of an ecosystem.

Group selection. Changes in the gene pool as the result of differential survival or productivity of local populations.

Guild. A group of species that share a resource (have related niches) in a community.

Habitat. The specific set of environmental conditions under which an individual, species, or community exists. Sometimes restricted to conditions of the physical environment; sometimes restricted to individuals or species (in which case the habitat of a community is termed a biotope).

Habitat island. An isolated patch of a particular habitat.

Halophytes. Plants that typically grow in salty soil.

Haploid. Having a single set of chromosomes (1N).

Heavy soil. A soil with a high percentage of clay.

Herb. A plant without woody tissue.

Herbicide. A poison applied to land or water to kill plants.

Herbivore. An animal that feeds on plants.

Heritability. The proportion of phenotypic variation stemming from genetic variation.

Herpetofauna. Reptiles and amphibians.

Heterotroph. An organism that requires food in the form of complex organic molecules: Animals, fungi, and many bacteria.

Heterozygous. Having different alleles at a locus on the two homologous chromosomes.

Hibernation. A condition of winter torpor.

Holistic. Concerned with the properties of whole systems, such as organisms or ecosystems, rather than with the properties of their constituent parts; opposed to reductionistic.

Homeostasis. The maintenance by an organism of constant internal conditions in the face of changes in the environment.

Home range. The area routinely traversed by an animal.

Homoiotherm. Organism whose body temperature is maintained constant by automatic physiological responses.

Homozygous. The situation in which alleles at a locus are the same on both chromosomes.

Host. The organism that supports a parasite or commensal.

Human immunodeficiency virus (HIV). A retrovirus of humans tending to suppress immune function and believed to be important in producing AIDS.

Humus. Partially decomposed organic matter in soil.

Hydric. Wet.

Hydroperiod. In wetland, the portion of the year that soils are saturated.

Hydrophyte. A plant characteristically growing in water.

Indirect effects. Coactions once removed. Influence exerted by organism A on organism B as a consequence of A's direct effects on organism C.

Interference. Behavior that tends to deny access of another individual to a resource.

Intermediate host. A host in which parasite reproduction, if any, is asexual.

Interspecific. Among or between species.

Intraspecific. Within a species.

Intrinsic. Inherent in the organism or population.

Intrinsic rate of natural increase. Per head population growth rate when the population is uncrowded and has a stable age distribution; biotic potential.

Irruption. (1) An outbreak. (2) An incursion.

Iteroparous. Breeding more than once per lifetime.

Keystone species. A species upon which other species of a community depend.

Krummholz. Vegetation characterized by stunted, misshapen trees above the timberline of mountains.

***K*-selected.** Adapted to stable environmental conditions; hence with a low reproductive rate.

Lake. A large body of standing water.

Landscape. A substantial piece of terrain (several to many square kilometers) comprising a mosaic of ecosystems. Also, the look of an area or region.

Life form. The characteristic adult growth form of an organism (especially plants); usually does not refer to the Raunkiaer life-form system unless so specified.

Life table. Vital statistics, by age, of a population.

Light soil. A soil with high percentage of sand.

Limited. Confined at or below a certain level.

Limiting factor. The factor that, by its relative scarcity or unfavorability, limits a process or the numbers or range of an organism.

Limnology. The study of lakes, streams, and other inland waters.

Litter. Organic debris that has fallen to earth.

Littoral. Pertaining to the shore of lakes and oceans.

Loam. Soil in which both large and small particles are well represented.

Logistic curve. An S-shaped population growth curve in which growth rate begins to decline about halfway up.

Longevity. Length of life.

Marsh. Wetland dominated by emergent (herbaceous) vegetation.

Mediterranean climate. A weather regime characterized by cool, wet conditions interrupted by a hot, dry summer.

Megafauna. Large animals; especially the large reptiles, birds, and mammals present during the Pleistocene.

Mesic. Having or characteristic of a medium moisture content.

Metapopulation. A population of populations; a spatially subdivided population.

Microsere. Successional changes in a temporary microhabitat such as fallen logs or carrion.

Migration. Seasonal movements involving leaving an area and returning to it.

Mimicry. The resemblance of one species to another, not closely related species. The similarity is usually viewed as adaptive.

Monoculture. A single crop grown over a large area.

Montane forest. The belt of vegetation in mountains below subalpine forest but distinguishable from the regional flatland vegetation.

Mor. A type of acid humus that develops under most temperate conifers and oaks.

Mull. A type of slightly acid to neutral humus that develops under most temperate hardwoods.

Mutation. A heritable change in the genetic material.

Mutualism. An interaction between two species from which both derive benefit.

Mycorrhiza. A symbiotic association of vascular plant roots and fungus, important in nutrient uptake by the plant.

Natural area. A site (usually with minimal recent human disturbance) that contains an unusual biological or other natural feature, or that illustrates ecological principles unusually well.

Natural selection. The preservation from generation to generation of favorable individual differences; differential perpetuation of genes in successive generations caused by different degrees of adaptedness to the environment; survival of the fittest.

Net community production (NCP). Gross primary production minus respiratory energy loss at all trophic levels; thus, the increment (or decrement) of biomass in an ecosystem.

Net primary production (NPP). Energy storage by the autotrophs of an ecosystem after the energy used for autotroph maintenance is subtracted.

Net reproductive rate (R_o). The average number of daughters born to each female in a population; the capacity of the population to increase in one generation.

Nitrification. Oxidation of ammonium to nitrite or nitrate.

Nitrogen fixation. The conversion of atmospheric nitrogen to forms usable in biological processes (ammonia and nitrate).

Nocturnal. Active at night.

Old field. Abandoned cropland.

Old-growth. Pertaining to forests that have not been cut for decades or centuries (depending on the specific forest type). Mature. By forestry definition, overmature (in terms of timber production).

Open space. Land that is sparsely occupied by human construction; natural areas, parks, and farmland.

Optimal. Most favorable or advantageous.

Optimum. The most favorable portion within an organism's range of tolerance.

Organic farming. Agriculture in which synthetic chemicals are avoided and fertilizers in the form of composted organic matter and crushed rocks are employed.

Outwash. Material deposited by meltwater from glaciers so that a given site has soil particles of relatively uniform size.

Paleoecology. The study of the interactions of organisms and environment in the geologic past.

Parasitism. The coaction in which one organism (parasite) lives in or on another (host) and gains nourishment from it, but does not immediately kill it.

Particulate organic matter (POM). Detritus, especially detrital particles larger than 0.5 μm in aquatic habitats. Mostly derived by physical and biotic breakdown of dead plants.

Pathogen. An organism that causes disease.

Pattern. Any nonrandom spacing. Loosely, any spatial distribution, including random.

Perennial. A plant with a life span of more than two years.

Permafrost. Permanently frozen soil.

Pesticides. Poisonous chemicals marketed with the claim that they kill unwanted organisms; biocides.

Phenology. The study of seasonal change.

Phenotype. The morphological, physiological, or behavioral manifestation of a (genetically based) trait; the observable traits of an individual.

Phenotypic plasticity. Environmentally induced phenotypic variation.

Pheromone. A chemical messenger between members of a species.

Photoperiod. Day length.

Phreatophyte. A plant that taps the ground water for its water supply.

Physiognomy. The general structure or overall appearance of a community.

Phytogeography. Plant geography.

Phytoplankton. The passively floating plants, mostly algae, in water.

Phytosociology. The study of plant populations and communities (as contrasted with the study of the relations of plants to their abiotic environment).

Pioneer community. The initial community in succession.

Piscivorous. Fish eating.

Planktivorous. Plankton eating.

Plankton. The usually microscopic plants and animals that live free-floating in water.

Pleistocene. The geological epoch extending from about 2 million years ago to 10 thousand years ago; the Ice Age.

Poikilotherm. Organism lacking physiological means of maintaining a constant body temperature.

Pollution. The unfavorable modification of the environment by human activity.

Polygamy. The situation in which one individual has more than one mate.

Polygyny. The most frequent type of polygamy; the mating system in which a male has two or more female mates.
Pond. A small body of standing water, shallow enough to support rooted vegetation across the whole basin.
Population. A set of organisms belonging to the same species and occupying a particular area at the same time.
Postnuptial. Occurring after the breeding season.
Prairie. Native North American grassland; sometimes restricted to regions of tall-grass dominance.
Predator. An animal that kills and eats other animals.
Prey. The actual or potential victim of a predator.
Primeval. Original; especially pertaining to conditions when human influence was negligible.
Procaryote. Organism characterized by cells having the genetic material dispersed rather than in chromosomes and lacking a nuclear membrane.
Production. Energy storage; increase in organic matter.
Productivity. Rate of energy storage or increase in organic matter.
Proximate factor. The environmental trigger for a biological event.
P/R ratio. The balance between production and respiration in an ecosystem.

Quadrat. A sample plot (properly, rectangular).
Quaternary. The geological period from about 2 million years ago to the present. Includes the Pleistocene and Holocene epochs.

Rad. A measure of absorbed ionizing radiation, specifically, an absorbed dose of 100 ergs per gram of tissue.
Rain shadow. The region on the lee side of mountains where rainfall is lower than on the windward side.
Random. Haphazard; having no set pattern or design.
Reaction. The effect of organisms on their physical environment.
Recessive. A gene that is expressed phenotypically only when present in the homozygous state.
Regulation. Maintenance within circumscribed limits.
Relict. An individual, population, or community left over from some earlier period of more favorable conditions.
Resilience. The tendency for a population or an ecosystem to return to its original state after a perturbation.
Rhizome. A prostrate or underground stem, usually rooting at the nodes and serving as a means of vegetative spread.
Richness. Species diversity (number of species).
Riparian. Pertaining to or occupying river banks.
Rosette. A circle of basal leaves especially characteristic of the first year's growth of a biennial plant.
***r*-selected.** Adapted to temporary or unpredictable environmental conditions; hence with a high reproductive rate.
Ruminant. The cud-chewing hoofed mammals forming a large suborder of the order Artiodactyla.

Saline. Having a high salt concentration.
Sample. A portion chosen from a population. A **random sample** is one in which every individual (organism, point, plot) has an equal and independent chance of being chosen.
Saprobe. An organism that subsists on dead tissue.
Savanna. Vegetation type with trees widely spaced among herbaceous (usually) or shrubby vegetation.
Schistosomiasis. Infection with blood flukes of the genus *Schistosoma;* bilharziasis in older literature.
Sclerophyll. Plant with hard leaves, such as evergreen oaks and various desert shrubs.
Seed bank. Viable seeds that have accumulated in the soil.
Semelparous. Breeding once per lifetime.
Sere. The whole sequence of communities leading to the climax. **Seral stage,** any community in a sere; a successional community.
Serotinous. Referring to cones that retain their seeds until heated.
Shade tolerance. The ability of a plant to survive and grow in shade.
Sink. A habitat or population where natality does not balance mortality.
Snag. A standing dead tree.
Social. Pertaining to a grouping of organisms organized cooperatively.
Sociobiology. The study of the biological basis of social behavior.
Soil. The loose surface layer of the earth in which plants are rooted.
Source. A habitat or population having a reproductive surplus.
Species. A group of populations reproductively isolated from other such groups.

Species-area curve. A graph that plots cumulative number of species against area.

Spring ephemeral. Herbs that grow and (usually) flower during the period before canopy closure, die back, and are dormant the rest of the year.

Stable age distribution. The age structure assumed by a population growing at a constant rate.

Standing crop. Biomass (rarely, numbers) present at a given time.

Steady state. A dynamic equilibrium; that is, one in which relative stability is maintained by compensatory changes.

Steppe. Dry, short-grass grassland.

Stochastic. Pertaining to events randomly selected from a probability distribution.

Stratum. A layer of an ecosystem.

Subalpine forest. The belt of vegetation in mountains below the alpine (tundra) zone.

Succession. The change in numbers and kinds of organisms leading to a stable (climax) community. Replacement of communities, one by another, on an area.

Survivorship curve. A graph showing the number or proportion of a population still alive by age.

Swamp. Wooded wetland.

Symbiosis. Any intimate coexistence such as parasitism or mutualism.

Sympatric. Occupying the same geographical area.

Synecology. Community ecology, especially the habitat relations of communities.

Synergism. A combined effect greater than the sum of the effects of two agents acting separately.

Synusia. Plants of the same life form, often forming a layer.

Taiga. (1) Boreal coniferous forest. (2) The ecotone between boreal coniferous forest and tundra.

Talus. The accumulated rock fragments at the base of a cliff.

Territory. A defended area.

Tertiary. The geological period after the Cretaceous and before the Quaternary from about 65 million to about 2 million years ago.

Threshold. The intensity (of a stimulus) below which no response is produced.

Till. Glacial material deposited in such a way that mineral particles of various sizes are intermixed; found in terminal and ground moraines.

Timberline. The upper or poleward limit of the zone on mountains or in high latitudes where trees grow to normal size.

Tolerance, limits of. The lower and upper survivable levels of some physical factor for an organism or population.

Tolerance, range of. The set of conditions for some physical factor within which an organism or a population can survive.

Trade off. The situation in which something is gained by sacrificing something else.

Trophic. Feeding; thus, pertaining to energy transfers.

Tundra. A vegetation type occurring beyond the tree line and dominated by herbs, lichens and mosses, or dwarf shrubs.

Ultimate factor. The environmental feature that gives higher fitness to the organisms possessing a particular trait.

Understory. The vegetation below the canopy in a forest.

Ungulate. A hoofed mammal, such as a horse or cow.

Vector. An organism that spreads a parasite.

Vegetation. The plant cover of an area.

Virulence. Ability to produce disease; the mortality associated with a specific pathogen.

Wetlands. Areas where the soil is saturated with water for a significant part of the year.

Wilderness. Land unaffected, or not evidently changed, by human presence.

Woodland. (1) As a specific physiognomic type, vegetation dominated by small trees too widely spaced to form a closed canopy. (2) Any area with trees; timberland.

Xeric. Dry.

Xeromorphic. Having structural traits of plants characteristic of dry regions.

Xerophytes. Plants of dry areas.

Yield. Organic production used by humans.

Zero population growth (ZPG). A situation in which birth and immigration rates are balanced with death (and emigration) rates, so that a population neither increases nor decreases with time.

Zooplankton. Passively floating or weakly swimming animals living in water.

Scientific Names of Species Mentioned in Text

The Latin names of higher categories are generally not provided in the following list. Readers unfamiliar with them can find most in biology texts or unabridged dictionaries.

Trees

acacia, swollen-thorn, *Acacia* spp.
ailanthus, *Ailanthus altissima*
alder, red, *Alnus rubra*
apple, *Pyrus malus*
arbor vitae, *Thuja occidentalis*
ash, black, *Fraxinus nigra*
ash, green, *Fraxinus pensylvanica*
ash, mountain, *Sorbus americana*
ash, Oregon, *Fraxinus oregona*
ash, white, *Fraxinus americana*
aspen, bigtooth (or large-toothed), *Populus grandidentata*
aspen, quaking (or trembling), *Populus tremuloides*
bald cypress, *Taxodium distichum*
basswood, *Tilia americana*
basswood, white, *Tilia heterophyla*
beech (American), *Fagus grandifolia*
beech, blue, *Carpinus caroliniana*
birch, *Betula pendula*
birch, dwarf, *Betula glandulosa*
birch, paper, *Betula papyrifera*
birch, yellow, *Betula alleghaniensis*
buckeye, red, *Aesculus pavia*
buckeye, sweet, *Aesculus flava*
buckthorn, common, *Rhamnus cathartica*
butternut, *Juglans cinerea*
cassia, big, *Cassia grandis*
catalpa, *Catalpa bignonioides*
cedar, Alaska yellow, *Chamaecyparis nootkatensis*
cedar, incense, *Libocedrus decurrens*
cedar, Port Orford white, *Chamaecyparis lawsoniana*
cedar, red, *Juniperus virginiana*
cedar, western red, *Thuja plicata*
cedar, white (or northern white), *Thuja occidentalis*
cherry, black, *Prunus serotina*
cherry, pin, *Prunus pensylvanica*
chestnut, American, *Castanea dentata*
chinquapin, golden, *Castanea pumila*
chokecherry, *Prunus virginiana*
coffee tree, Kentucky, *Gymnocladus dioicus*

cottonwood, *Populus deltoides*
cypress, bald, *Taxodium distichum*
cypress, pond, *Taxodium ascendens*
dogwood, flowering, *Cornus florida*
dogwood, redosier, *Cornus sericea*
dwarf cornel, *Cornus canadensis*
elder, box, *Acer negundo*
elm, American, *Ulmus americana*
elm, rock, *Ulmus thomasii*
elm, slippery, *Ulmus rubra*
fig, *Ficus carica*
fig, strangler, *Ficus* spp.
fir, alpine, *Abies lasiocarpa*
fir, balsam, *Abies balsamea*
fir, Douglas, *Pseudotsuga taxifolia*
fir, grand, *Abies grandis*
fir, noble, *Abies procera*
fir, Pacific silver, *Abies amabilis*
fir, red, *Abies magnifica*
fir, scotch, *Pinus sylvestris*
fir, subalpine, *Abies lasiocarpa*
fir, white, *Abies concolor*
gum, sweet, *Liquidambar styraciflua*
gum, tupelo, *Nyssa sylvatica*
gum, water, *Nyssa biflora*
gumbo limbo, *Bursera simaruba*
hackberry, *Celtis occidentalis*
hawthorn, *Crategus* spp.
hazelnut, *Corylus americana*
hemlock, eastern, *Tsuga canadensis*
hemlock, mountain, *Tsuga mertensiana*
hemlock, western, *Tsuga heterophylla*
hickory, bitternut, *Carya cordiformis*
hickory, pignut, *Carya glabra*
hickory, shagbark, *Carya ovata*
hobblebush, *Viburnum alnifolium*
holly, American, *Ilex opaca*
hophornbeam, eastern, *Ostrya virginiana*
hornbeam, American, *Carpinus caroliniana*
Joshua tree, *Yucca brevifolia*
juniper, *Juniperus communis*
larch, alpine, *Larix lyalli*
larch, western, *Larix occidentialis*
laurel, California, *Umbellularia californica*
locust, black, *Robinia pseudoacacia*
locust, honey, *Glenditsia triacanthos*
madrone, *Arbutus menziesii*
magnolia, *Magnolia grandiflora*
mahogany, mountain, *Cercocarpus betuloides*
mangrove, black, *Avicennia nitida*
mangrove, button, *Conocarpus erecta*
mangrove, red, *Rhizophora mangle*
mangrove, white, *Laguncularia racemosa*
manzanitas, *Arctostaphylos* spp.
maple, bigleaf, *Acer macrophyllum*
maple, mountain, *Acer spicatum*
maple, red, *Acer rubrum*
maple, silver, *Acer saccharinum*
maple, striped, *Acer pensylvanicum*
maple, sugar, *Acer saccharum*
maple, vine, *Acer circinatum*
mesquite, *Prosopis juliflora*
oak, black, *Quercus velutina*
oak, blackjack, *Quercus marilandica*
oak, bur, *Quercus macrocarpa*
oak, California scrub, *Quercus dumosa*
oak, California white, *Quercus lobata*
oak, canyon live, *Quercus chrysolepis*
oak, chestnut, *Quercus prinus*
oak, laurel (uplands), *Quercus hemisphaerica*
oak, laurel (lowlands), *Quercus laurifolia*
oak, live, *Quercus virginiana*
oak, Oregon white, *Quercus garryana*
oak, overcup, *Quercus lyrata*
oak, post, *Quercus stellata*
oak, red, *Quercus rubra*
oak, shingle, *Quercus imbricaria*
oak, shrub live, *Quercus turbinella*
oak, turkey, *Quercus laevis*
oak, water, *Quercus nigra*
oak, white, *Quercus alba*
orange, osage, *Maclura pomifera*
palm, cabbage (Florida), *Sabal palmetto*
pawpaw, *Asimina triloba*
pecan, *Carya illinoiensis*
persimmon, *Diospyros viginiana*
pine, bishop, *Pinus muricata*
pine, bristle cone, *Pinus aristata*
pine, Coulter, *Pinus coulteri*
pine, digger, *Pinus sabiniana*
pine, eastern white, *Pinus strobus*
pine, foxtail, *Pinus aristata*
pine, jack, *Pinus banksiana*
pine, Jeffrey, *Pinus jeffreyi*
pine, knobcone, *Pinus attenuata*
pine, limber, *Pinus flexilis*
pine, loblolly, *Pinus taeda*
pine, lodgepole, *Pinus contorta*
pine, longleaf, *Pinus palustris*
pine, Monterey, *Pinus radiata*
pine, pinyon, *Pinus edulis*
pine, pitch, *Pinus rigida*
pine, pond, *Pinus serotina*
pine, ponderosa, *Pinus ponderosa*
pine, red, *Pinus resinosa*
pine, sand, *Pinus clausa*
pine, shortleaf, *Pinus echinata*
pine, slash, *Pinus caribaea*

pine, spruce, *Pinus glabra*
pine, sugar, *Pinus lambertiana*
pine, Virginia, *Pinus virginiana*
pine, western white, *Pinus monticola*
pine, white, *Pinus strobus*
pine, whitebark, *Pinus albicaulis*
pinyon, *Pinus edulis*
plum, *Prunus* spp.
poplar, balsam, *Populus balsamifera*
poplar, yellow, *Liriodendron tulipifera*
redbud, *Cercis canadensis*
redcedar, eastern, *Juniperus virginiana*
redcedar, western, *Thuja plicata*
redwood (coastal redwood), *Sequoia sempervirens*
sassafras, *Sassafras albidum*
sequoia, giant, *Sequoia gigantea*
spruce, bigcone, *Picea breweriana*
spruce, black, *Picea mariana*
spruce, blue, *Picea pungens*
spruce, Engelmann, *Picea engelmannii*
spruce, red, *Picea rubens*
spruce, sitka, *Picea sitchensis*
spruce, white, *Picea glauca*
sweetgum, *Liquidambar styraciflua*
sycamore, *Platanus occidentalis*
tamarack, *Larix laricina*
tamarind, wild, *Lysiloma latisiliqua*
tamarisk, *Tamarix* spp.
tanoak, *Lithocarpus densiflorus*
torreya, California, *Torreya californica*
tuliptree, *Liriodendron tulipifera*
tupelo, *Nyssa* spp.
walnut (English), *Juglans regia*
walnut, black, *Juglans nigra*
willow, *Salix* spp.
willow, black, *Salix nigra*
yew, pacific, *Taxus brevifolia*

Other Plants

agave, *Agave* spp.
alfalfa, *Medicago sativa*
anabaena, *Anabaena* spp.
apple, may, *Podophyllum peltatum*
arrowhead, *Sagittaria* spp.
arrowleaf, *Sagittaria* spp.
aster, silky, *Aster sericeus*
aster, smooth, *Aster laevis*
avens, alpine, *Geum rossii*
avens, mountain, *Dryas* spp.
azalea, flame, *Rhododendron calendulaceum*
azolla, *Azolla* spp.
ball-moss, *Tillandsia recurvata*
bamboo, *Arundinaria* spp. (and related genera)
barley, *Hordeum vulgare*
beadlily, *Clintonia borealis*
beadlily, queencup, *Clintonia uniflora*
bean, green, *Phaseolus vulgaris*
bearberry, *Arctostaphylos uva-ursi*
beargrass, *Xerophyllum tenax*
beardgrass, silver, *Bothriocloa saccharoides*
beechdrops, *Epifagus virginiana*
beet, sugar, *Beta vulgaris*
bellflower, marsh, *Campanula aparinoides*
bergamot, wild, *Monarda fistulosa*
berry, partridge, *Mitchella repens*
bilberry, *Vaccinium cespitosum*
bitterbrush, *Purshia tridentata*
bittersweet, *Celastrus scandens*
blackberry, *Rubus* spp.
bladderwort, *Utricularia* spp.
blazing star, *Liatris punctata*
blazing star, rough, *Liatris aspera*
blight, chestnut, *Endothia parasitica*
blueberry, dryland, *Vaccinium vacillans*
blueberry, highbush, *Vaccinium corymbosum*
blue-eyed mary, *Collinsia verna*
bluejoint, *Calamagrostis canadensis*
bluestem, big, *Andropogon gerardi*
bluestem, little, *Schizachyrium scoparium*
bluestem, sand, *Andropogon hallii*
bower, virgin's, *Clematis virginiana*
broomsedge, *Andropogon virginicus*
buffalo bur, *Solanum rostratum*
bugseed, *Corispermum* spp.
bulrush, *Scirpus* spp.
bulrush, great, *Scirpus validus*
bunchberry, *Cornus canadensis*
bush, creosote, *Larrea tridentata*
buttercup, *Ranunculus* spp.
buttercup, prairie, *Ranunculus fascicularis*
butterfly weed, *Asclepias tuberosa*
buttonbush, *Cephalanthus occidentalis*
cabbage, skunk, *Symplocarpus foetidus*
cactus, Christmas, *Zygocactus truncatus*
cactus, organ-pipe, *Cereus thurberi*
cactus, prickly pear, *Opuntia* spp.
cactus, semaphore, *Opuntia austrina*
candystick, *Allotropa virgata*
cardon, *Cereus pringelei*
carrot, tame, *Daucus carota* var.*sativa*
carrot, wild, *Daucus carota*
cat-tail, *Typha latifolia*
ceanothus, *Ceanothus* spp.
century plant, *Agave* spp.

chamise, *Adenostoma fasciculatum*
chara, *Chara* spp.
cheatgrass, *Bromus tectorum*
cherry, *Prunus* spp.
cherry, sand, *Prunus pumila*
chokeberry, *Aronia* spp.
cholla, christmas tree, *Opuntia leptocaulis*
cinquefoils, *Potentilla* spp.
cinquefoil, shrubby, *Potentilla fruticosa*
cladophora, *Cladophora* spp.
cleavers, *Galium aparine*
clover, *Trifolium* spp.
clover, purple prairie, *Dalea purpurea*
clover, red, *Trifolium pratense*
clover, white, *Trifolium repens*
clover, white sweet, *Melilotus alba*
columbine, *Aquilegia* spp.
compass-plant, *Silphium laciniatum*
coneflower, prairie, *Ratibida pinnata*
coneflower, purple, *Echinacea purpurea*
cordgrass, *Spartina alterniflora*
corn, *Zea mays*
corn, squirrel, *Dicentra canadensis*
cottongrass, *Eriophorum vaginatum*
crabgrass, *Digitaria* spp.
cranberry, *Vaccinium macrocarpon* or *V. oxycoccos*
creeper, Virginia, *Parthenocissus quinquefolia*
crowberry, *Empetrum nigrum*
cucumber, wild, *Echinocystis lobata*
currant, wild, *Ribes* spp.
daisy, seaside, *Erigeron glaucus*
dandelion, *Taraxacum officinale*
desmid, copperhead, *Ankistrodesmus falcatus*
Devil's club, *Oplopanax horridum*
dewberry, *Rubus flagellaris*
dock, prairie, *Silphium terebinthenaceum*
dodder, *Cuscuta* spp.
doll's-eyes, *Actaea alba*
dropseed, *Sporobolus* spp.
dropseed, prairie, *Sporobolus heterolepis*
duckweed, *Lemna* spp.
Dutch elm disease, *Ceratocystis ulmi*
eelgrass, *Zostera marina*
elder, common, *Sambucus canadensis*
elder, red-berried, *Sambucus racemosa*
fern, bracken, *Pteridium aquilinum*
fern, Christmas, *Polystichum acrostichoides*
fern, cinnamon, *Osmunda cinnamomea*
fern, New York, *Thelypteris noveboracensis*
fern, Virginia chain, *Woodwardia virginica*
fescue, Idaho, *Festuca idahoensis*
fescue, meadow, *Festuca elatior*
fetterbush, *Lyonia lucida*
fireweed, *Epilobium angustifolium*
fleabane, *Erigeron* spp.
flower, carrion, *Smilax herbacea*
flytrap, Venus, *Dionaea muscipula*
forget-me-not, alpine, *Eritrichium aretioides*
foxglove, false, *Aureolaria* spp.
galleta, *Hilaria rigida*
gentian, bottle, *Gentiana andrewsii*
gentian, fringed, *Gentianopsis crinita*
ginger, wild, *Asarum canadense*
goldenrod, Canada, *Solidago canadensis*
goldenrod, Riddell's, *Solidago riddellii*
goldenrod, rigid, *Solidago rigida*
goldthread, *Coptis trifolia*
gooseberry, *Ribes* spp.
gooseberry, prickly, *Ribes cynosbati*
gopher-apple, *Hilania michauxii*
grama, *Bouteloua* spp.
grama, blue, *Bouteloua gracilis*
grama, sideoats, *Bouteloua curtipendula*
grape, *Vitis* spp.
grass, buffalo, *Buchloe dactyloides*
grass, dune, *Calamovilfa longifolia*
grass, gama, *Tripsacum dactyloides*
grass, Indian, *Sorghastrum nutans*
grass, marram or beach, *Ammophila breviligulata*
grass, poverty, *Aristida oligantha*
grass, prairie cord, *Spartina pectinata*
grass, slough, *Spartina pectinata*
grass, tobosa, *Hilaria mutica*
grass, wire, *Aristida stricta*
grass of Parnassus, *Parnassia* spp.
hatpins, *Eriocaulon* spp.
heather, *Calluna vulgaris*
heather, pink mountain, *Phyllodoce empetriformis*
hepatica, *Hepatica americana* or *acutiloba*
holly, American, *Ilex opaca*
holly, mountain, *Nemopanthus mucronatus*
horseweed, *Conyza canadensis*
huckleberry, black, *Gaylussacia baccata*
indigo, wild, *Baptisia leucantha*
ivy, poison, *Toxicodendron radicans*
jack-in-the-pulpit, *Arisaema triphyllum*
jasmine, rock, *Androsace chamaejasma*
jimson-weed, *Datura stramonium*
Junegrass, *Koeleria pyramidata*
juniper, creeping, *Juniperus horizontalis*
juniper, ground, *Juniperus communis* var. *depressa*
knapweed, spotted, *Centaurea maculosa*
knotweed, *Polygonum aviculare*
Labrador-tea, *Ledum groenlandicum*
laurel, *Laurus nobilis*
leadplant, *Amorpha canescens*
leatherleaf, *Chamaedaphne calyculata*
leek, wild, *Allium tricoccum*

lettuce, *Lactuca sativa*
lichen, American soldier, *Caldonia sylvatica*
lichen, British soldier, *Caldonia cristatella*
lilac, California, *Ceanothus* spp.
lily, fragrant pond, *Nymphaea odorata*
lily, trout, *Erythronium americanum*
lily, water, *Nymphaea odorata*
lily, yellow pond, *Nuphar advena*
lily, yellow trout, *Erythronium americanum*
lobelia, *Lobelia* spp.
lobelia, Kalm's, *Lobelia kalmii*
locoweed, *Astragalus* spp.
loosestrife, purple, *Lythrum salicaria*
lotus, American, *Nelumbo lutea*
lousewort, *Pedicularis* spp.
lupine, *Lupinus* spp.
marigold, marsh, *Caltha palustris*
may-apple, *Podophyllum peltatum*
mayflower, Canada, *Maianthemum canadense*
mermaid weed, false, *Floerkea proserpinacoides*
milkweed, common, *Asclepia syriaca*
milkweed, swamp, *Asclepia incarnata*
milkweed, whorled, *Asclepia verticillata*
mint family, *Lamiaceae*
mistletoe, *Phoradendron flavescens*
moonseed, *Menispermum canadense*
morning glory, *Ipomoea* spp.
moss, haircap, *Polytrichum* spp.
moss, reindeer, *Cladonia rangiferina*
moss, Spanish, *Tillandsia usneoides*
moss, sphagnum, *Sphagnum* spp.
muhly, *Muhlenbergia* spp.
mullein, *Verbascum* spp.
nannyberry, *Viburnum lentago*
needle-and-thread, *Stipa comata*
needlegrass, purple, *Stipa pulchra*
needlegrass, Texas, *Stipa leucotricha*
oat, *Avena sativa*
oat grass, mountain, *Danthonia compressa*
oat, wild, *Avena fatua*
ocotillo, *Fouqueria splendens*
orchid, phantom, *Cephalanthera austineae*
paloverde, *Cercidium microphyllum*
passionflower, *Passiflora* spp.
pea, wild, *Lathyrus* spp.
pea, partridge, *Cassia fasciculata*
peach, *Prunus persica*
peanut, *Arachis hypogaea*
pear, *Pyrus communis*
peyote, *Lophophora williamsii*
phacelia, *Phacelia* spp.
philodendron, *Philodendron* spp.
phlox, *Phlox* spp.
phlox, blue, *Phlox divaricata*
phlox, downy, *Phlox pilosa*
phlox, Drummond's, *Phlox drummondii*
phragmites, *Phragmites communis*
pickerelweed, *Pontederia cordata*
pipe, Indian, *Monotropa uniflora*
pitcherplant, *Sarracenia purpurea*
pondweed, *Potamogeton* spp.
potato, *Solanum tuberosum*
primrose, evening, *Oenothera biennis*
psoralea, silver-leaf, *Psoralea* spp.
pussy-toes, *Antennaria* spp.
quarters, lamb's, *Chenopodium album*
rafflesia, *Rafflesia arnoldii*
ragweed, annual, *Ambrosia artemisiifolia*
raspberry, *Rubus* spp.
rattlesnake master, *Eryngium yuccifolium*
redroot, *Ceanothus americanus*
reed, bur, *Sparganium* spp.
rhododendron, *Rhododendron catawbiense*
rice, *Oryza sativa*
rice, wild, *Zizania aquatica*
rocket, sea, *Cakile edentula*
root, culvers, *Veronicastrum virginicum*
rosemary, bog, *Andromeda glaucophylla*
rosinweed, *Silphium integrifolium*
roundie, Schroeter's, *Sphaerocystis schroeteri*
rush, tufted, *Scirpus caespitosus*
rush, white beak, *Rhynchospora alba*
rye, wild, *Elymus canadensis*
sage, bur, *Ambrosia dumosa*
sagebrush (big), *Artemisia tridentata*
sagebrush, purple, *Salvia leucophylla*
saguaro, *Cereus gigantea*
samphire, *Salicornia* spp.
sand dropseed, *Sporobolus cryptandrus*
sarsaparilla, wild, *Aralia nudicaulis*
sausage tree, *Kigelia africana*
sawgrass, *Cladium jamaicense*
saw-palmetto, *Serenoa repens*
saxifrage, *Saxifraga* spp.
sea blite, *Suaeda depressa*
sea rocket (Great Lakes), *Cakile edentula*
sea rocket (Pacific coast), *Cakile maritima*
sedge, hairy-fruited, *Carex lasiocarpa*
sedge, penn or Pennsylvania, *Carex pensylvanica*
sedge, plantain-leaved, *Carex plantaginifolia*
sedge, three-way (arrowgrass), *Triglochin maritima*
shadscale, *Atriplex confertifolia*
sheep's sorrel, *Festuca ovina*
shepherd's-purse, *Capsella bursa-pastoris*
shooting star, *Dodeocatheon meadia*
skypilots, *Polemonium* spp.
smartweed, *Polygonum* spp.
smoke, prairie, *Geum triflorum*

snakeroot, white, *Eupatorium rugosum*
snakeweed, broom, *Gutierrezia sarothrae*
snowberry, creeping, *Gaultheria hispidula*
snowbush, *Ceanothus velutinus*
sorghum, *Sorghum bicolor*
sorrel, sheep, *Rumex acetosella*
soybean, *Glycine max*
spatterdock, *Nuphar advena*
spearscale, *Atriplex patula* var.*hastata*
sphagnum, nipple, *Sphagnum papillosum*
sphagnum, pointy-leaved, *Sphagnum cuspidatum*
sphagnum, pretty, *Sphagnum pulchrum*
spicebush, *Lindera benzoin*
spiderwort, *Tradescantia ohiensis*
spikerush, *Eleocharis* spp.
spring beauty, *Claytonia virginica*
sprouts, brussel, *Brassica oleracea* var.*gemnifer*
spurge, *Euphorbia* spp.
St.-John's-wort, *Hypericum perforatum*
staggerbush, *Lyonia mariana*
sugarcane, *Saccharum officinarum*
sundew, *Drosera* spp.
sundew, roundleaved, *Drosera rotundifolia*
sweet cicely, *Osmorhiza* spp.
sweet-flag, *Acorus calamus*
switchgrass, *Panicum virgatum*
tarbush, *Flourensia cernua*
tea, Labrador, *Ledum groenlandicum*
tea, Mormon, *Ephedra* spp.
tea, New Jersey, *Ceanothus americanus*
thistle, *Cirsium* spp.
thistle, Canada, *Cirsium arvense*
thistle, Russian, *Salsola kali*
thistle, yellow beach, *Circium pitcheri*
three-awn, *Aristida* spp.
toadflax, bastard, *Comandra umbellata*
tomato, *Lycopersicon esculentum*
toothwort, *Dentaria* spp.
touch-me-not, *Impatiens* spp.
trillium, purple, *Trillium* spp.
trumpet-creeper, *Campsis radicans*
tulip, *Tulipa sylvestris*
turtlehead, *Chelone* spp.
twin flower, *Linnea borealis*
vanilla, *Vanilla planifolia*
vine, *Dioclea meagacarpa*
violet, bird's foot, *Viola pedata*
violet, smooth yellow, *Viola eriocarpa*
violet, yellow, *Viola* spp.
watercress, *Rorippa nasturium-aquaticum*
wheat, *Triticum aestivum*
wheatgrass, bluebunch, *Elytrigia spicata*
wheatgrass, western, *Elytrigia smithi*
willow, water, *Decodon verticillatus*
winterfat, *Ceratoides lanata*
wintergreen, *Gaultheria procumbens*
wiregrass, *Aristida stricta*
witchhazel, *Hammamelis virginiana*
yew, western, *Taxus brevifolia*
yucca, *Yucca* spp.

Insects

ant, Argentine, *Iridomyrmex humilis*
aphid, peach, *Myzis persicae*
bee, honey, *Apis mellifica*
beetle, bombardier, *Brachinus* spp.
beetle, bronze tiger, *Cincindela lecontei*
beetle, bruchid, *Caryedes brasiliensis*
beetle, burying, *Necrophorus humator*
beetle, chestnut grain, *Tribolium castaneum*
beetle, Colorado potato, *Leptinotarsa decemlineata*
beetle, confusing grain, *Tribolium confusum*
beetle, Douglas fir, *Dendroctonus pseudotsugae*
beetle, dung, subfamily Scarabaeinae
beetle, dytiscid, family Dytiscidae
beetle, elm bark, *Scolytus multistriatus*
beetle, fire, *Melanophila acuminata*
beetle, green tiger, *Cincindela sexguttata*
beetle, Japanese, *Popillia japonica*
beetle, kidney bean, *Callosobruchus phaseoli*
beetle, retentive bean, *Callosobruchus analis*
beetle, vedalia, *Rodolia cardinalis*
beetle, whirligig, family Gyrinidae
beetle, white tiger, *Cincindela lepida*
blackfly, *Simulium* spp.
borer, maple petiole, *Caulocampus acericaulis*
budworm, spruce, *Choristoneura fumiferana*
bumblebee, *Bombus* spp.
butterfly, cabbage, *Pieris brassicae*
butterfly, monarch, *Danaus plexippus*
butterfly, viceroy, *Basilarchia archippus*
caddisfly, order Trichoptera
cecropia (moth), *Hyalophora cecropia*
cicada, 17-year, *Magicicada septendecem*
cockroach, family Blattidae
cricket, field, subfamily Gryllinae
dobsonfly, family Corydalidae
earthwolf, Wright's, *Geolycosa wrightii*
earwig, order Dermaptera
fly (Canada goldenrod gall), *Eurosta solidaginis*
fly, buffalo (hornfly), *Haematobia irritans*
fly, crane, family Tipulidae
fly, hessian, *Phytophaga destructor*
fly, house, *Musca domestica*

fly, stable, *Stomoxys calcitrans*
fly, stone, order Plecoptera
grasshopper, *Hesperotettix viridis*
mayfly, order Ephemeroptera
midge, pool-colonizing, *Chironomus imicola*
midge, Vanderplank's, *Polypedilum vanderplanki*
mosquito (pitcher plant), *Wyeomeia smithii*
mosquito, night-flying, *Culex pipiens fatigans*
mosquito, Sierra treehole, *Aedes sierrensis*
mosquito, yellow fever, *Aedes aegypti*
moth, calendula plume, *Platyptilia williamsii*
moth, catalpa sphinx, *Ceratomia catalpae*
moth, clear-winged, family Aegeriidae
moth, gypsy, *Porthetria dispar*
moth, noctuid, family Noctuidae
moth, peppered, *Biston betularia*
moth, polyphemus, *Telea polyphemus*
moth, sphinx (worm, tomato), *Protoparce quinquemaculata*
moth, tiger, family Arctiidae
moth, yucca, *Tegeticula alba*
roach, Pennsylvania wood, *Parcoblatta pennsylvania*
saddled prominent, *Heterocampa guttivita*
sawfly, European pine, *Neodiprion sertifer*
sawfly, larch, *Prisiphora erichsonii*
scale, California red, *Aonidiella aurantii*
scale, cottony-cushion, *Icerya purchasi*
scalesucker, golden-naveled, *Aphytis chrysomphali*
scalesucker, lingnan, *Aphytis lingnanensis*
spittlebug, meadow, *Philaenus spumarius*
tendipedid, family Tendipedidae
tumblebug, subfamily Scarbaeinae
wasp, digger, *Microbembex* spp.
wasp, fig, *Blastophaga psenes*
wasp, otherwise-spotted, *Neocatolaccus mamezophagus*
wasp, polistes, family Vespidae
water pennies, *Psephenus* spp.
water striders, family Gerridae
weevil, Chinese bean, *Callosobruchus chinensis*
weevil, four-spotted bean, *Callosobruchus quadrimaculatus*
worm, orange-striped oak, *Anisota senatoria*

Other Invertebrates

clam, coquina, *Donax* spp.
clam, giant, *Tridacna* spp.
clam, giant (hydrothermal vents), *Calypytogena magnificens*
coccidian, barbershop, *Adelina* spp.
crab, blue, *Callincectes sapidus*
crab, ghost, *Ocypode albicans*
crab, king, *Paralithodes camtschatica*
crab, mole, *Emerita* spp.
crab, pea, *Pinnotheres maculatus*
crab, stone, *Menippe mercenaria*
crawfish, broad-legged, *Potamobius astacus*
crawfish, long-legged, *Potamobius leptodactylus*
hookworm, *Necator americanus*
krill, *Euphausia* spp.
lambkin, Clark's, *Lambornella clarki*
limpet, *Ferrissia* spp.
lobster, *Homarus americanus*
lobster, spiny, *Panulirus argus*
malaria, *Plasmodium falciparum*
malaria, avian, *Haemoproteus* and *Plasmodium* spp.
mussel, zebra, *Dreissena polymorpha*
periwinkle, *Littorina* spp.
roundworm, *Ascaris lumbricoides*
schistosomiasis, *Shistosomiasis mansoni*
sea slaters, *Ligia* spp.
shipworm, family Teredinidae
shrimp, *Crago septemspinosus*
shrimp, opossum, order Mysidacea
sponge, glass, class Hexactinellida
sponge, loggerhead, *Spongia officinalis*
squid, *Loligo* spp.
starfish, crown-of-thorns, *Acanthaster plancii*
threadworm, *Trichostrongylus tenuis*
twister, Roscoff, *Convoluta roscoffensis*
waterflea, *Daphnia magna*
worm, brain, *Pneumostrogylus tenuis*
worm, Pompeii, *Alvinella pompejana*
worm, tube, *Riftia pachyptila*

Fishes

angler-fish, ceratioid, *Cryptopsaras couesi*
barbel, *Barbus barbus*
barracuda, *Sphyraena barracuda*
bass, large-mouthed, *Micropterus salmoides*
blenny, saw-tooth, *Aspidontus rhynorhynchus*
bloater, *Coregonus hoyi*
bream, *Abramis brama*
butterflyfish, *Heniochus nigrirostris*
butterflyfish, banded, *Chaetodon striatus*
catfish, *Ictalurus nebulosus*
charr (sea), *Salvelinus alpinus*
chub, *Leuciscus cephalus*
chub, creek, *Semotilus atromaculatus*
chub-sucker, *Erimyzon oblongus*
dace, pearl, *Semotilus margarita*

dace, red-bellied, *Chrosomus erythrogaster*
eel, *Anguilla* spp.
flounder, family Pleuronectidae
grouper, family Serranidae
grouper, tiger, *Mycteroperca tigris*
killifish, family Cyprinodontidae
lamprey, sea, *Petromyzon marinus*
minnow, fathead, *Pimephales promelas*
mosquito-fish, *Gambusia affinis*
parrotfish, *Scarus* spp.
perch, *Perca flavescens*
pike, *Esox lucius*
rabbit-fish, *Chimaera* spp.
remora, *Echeneis naucrates*
roach, *Rutilus rutilus*
salmon, coho, *Oncorhynchus kisutch*
salmon, Pacific, *Oncorhynchus* spp.
sculpin, *Cottus* spp.
seabass, family Serranidae
shiner, *Notropis* spp.
shiner, sand, *Notropis deliciosus*
squawfish, Colorado River, *Ptychocheilus lucius*
stickleback, three-spined, *Gasterosteus aculeatus*
sucker, white, *Catostomus commersonnii*
sunfish, sphagnum, *Ennaecanthus obesus*
tilapia, blue, *Tilapia aurea*
topminnow, Sonoran, *Poeciliopsis occidentalis*
trout, brown, *Salmo trutta*
trout, lake, *Salvelinus namaycush*
trout, rainbow, *Salmo gairdnerii*
trout, sea, *Salmo trutta*
tuna, *Thunnus thynnus*
whitefish, *Coregonus clupeaformis*
wrasse, family Labridae

Amphibians and Reptiles

alligator, American, *Alligator mississippiensis*
boa, rosy, *Lichanura trivirgata*
bullsnake, *Pituophis melanoleucus*
chuckwalla, *Sauromalus obesus*
cottonmouth, *Agkistrodon piscivorus*
frog, bird-voiced tree, *Hyla avivoca*
frog, boreal chorus, *Pseudacris triseriata*
frog, Cascades, *Rana cascadae*
frog, gopher, *Rana areolata*
frog, gray tree, *Hyla versicolor*
frog, little grass, *Hyla ocularis*
frog, mink, *Rana septentrionalis*
frog, northern leopard, *Rana pipiens*
frog, squirrel tree, *Hyla squirella*
frog, tailed, *Ascaphus truei*
frog, wood, *Rana sylvatica*
gecko, banded, *Coleonyx variegatus*
Gila monster, *Heloderma suspectum*
iguana, desert, *Dipsosaurus dorsalis*
lizard, burrowing, *Bipes biporus*
lizard, collared, *Crotaphytus collaris*
lizard, long-nosed leopard, *Gambelia wislizenii*
lizard, sagebrush, *Sceloporus graciosus*
lizard, slender glass, *Ophisaurus attenuatus*
lizard, southern alligator, *Gerrhonotus multicarinatus*
lizard, wide-blotched, *Uta stansburiana*
massasauga, *Sistrurus catenatus*
peeper, spring, *Hyla versicolor*
racer, black, *Coluber constrictor*
racer, blue, *Coluber constrictor*
racer, striped, *Masticophis lateralis*
racerunner, six-lined, *Cnemidophorus sexlineatus*
rattler, diamondback (eastern), *Crotalus adamanteus*
rattler (rattlesnake), timber, *Crotalus horridus*
rattlesnake, Mohave, *Crotalus scutulatus*
rattlesnake, prairie, *Crotalus viridis*
salamander, Appalachian woodland, *Plethodon jordani*
salamander, dusky, *Desmognathus fuscus*
salamander, dwarf, *Manculus quadridigitatus*
salamander, flatwoods, *Ambystoma cingulatum*
salamander, long-toed, *Ambystoma macrodactylum*
salamander, marbled, *Ambystoma opacum*
salamander, mole, *Ambystoma talpoideum*
salamander, Mount Lyell, *Hydromantes platycephalus*
salamander, red-backed, *Plethodon cinereus*
salamander, red-cheeked, *Plethodon jordani*
salamander, slimy, *Plethodon glutinosus*
salamander, spotted, *Ambystoma maculatum*
salamander, Texas blind, *Typhlomolge rathbuni*
skink, broad-headed, *Eumeces laticeps*
skink, five-lined, *Eumeces fasciatus*
skink, metallic, *Lygosoma metallicum*
skink, moth, *Lygosoma noctua*
slider, pond, *Chrysemys scripta*
snake, black rat, *Elaphe obsoleta*
snake, common garter, *Thamnophis sirtalis*
snake, coral, *Micrurus fulvius*
snake, eastern hognose, *Heterodon platyrhinos*
snake, hognose, *Heterodon platyrhinos*
snake, indigo, *Drymarchon corais*
snake, lined, *Tropidoclonion lineatum*
snake, mud, *Farancia abacura*
snake, pine, *Pituophis melanoleucus*
snake, pine woods, *Pituophis melanoleucus*
snake, plains garter, *Thamnophis radix*
snake, red-bellied, *Storeria occipitomaculata*
snake, ribbon, *Thamnophis sauritus*
snake, rough green, *Opheodrys aestivus*

snake, western hog-nosed, *Heterodon nasicus*
snake, western leaf-nosed, *Phyllorhynchus decurtatus*
snake, western shovel-nosed, *Chionactis occipitalis*
snake, western terrestrial garter, *Thamnophis elegans*
toad, American, *Bufo americanus*
toad, Couch's spadefoot, *Scaphiopus couchii*
toad, desert horned, *Phrynosoma platyrhinos*
toad, great plains, *Bufo cognatus*
toad, short-horned, *Phrynosoma douglassii*
tortoise, giant, *Geochelone* spp.
tortoise, gopher, *Gopherus polyphemus*
tortoise, Hermann's, *Testudo hermanni*
turtle, box, *Terrapene carolina*
turtle, ornate box, *Terrapene ornata*
turtle, wood, *Clemmys insculpta*
watersnake, banded, *Natrix sipedon fasciata*
watersnake, common, *Natrix sipedon*
watersnake, red-bellied, *Natrix erythrogaster*

Birds

apapane, Laysan, *Himatione sanguinea freethii*
auk, great *Pinguinus impennis*
bird, secretary, *Sagittarius serpentarius*
bittern, American, *Botaurus lentiginosus*
blackbird, Brewer's, *Euphagus cyanocephalus*
blackbird, red-winged, *Agelaius phoenicus*
bluebird, eastern, *Sialia sialis*
bluebird, mountain, *Sialia currucoides*
bobolink, *Dolichonyx oryzivorus*
bobwhite, masked, *Colonus virginianus ridgwayi*
bobwhite, northern, *Colinus virginianus*
brambling, *Fringilla montifringilla*
bullfinch, *Pyrrhula pyrrhula*
bunting, indigo, *Passerina cyanea*
bunting, lark, *Calamospiza melanocorys*
bunting, snow, *Plectrophenax nivalis*
bushtit, *Psaltriparus minimus*
capercaille, *Tetrao urogallus*
caracara, crested, *Polyborus plancus*
cardinal, northern, *Cardinalis cardinalis*
cassowaries, *Casuarius* spp.
catbird, gray, *Dumetella carolinensis*
chat, yellow-breasted, *Icteria virens*
chickadee, black-capped, *parus atricapillus*
chickadee, boreal, *Parus hudsonicus*
chickadee, Carolina, *Parus carolinensis*
chickadee, chestnut-backed, *Parus rufescens*
chicken, *Gallus domesticus*
chicken, Attwater's prairie, *Tympanuchus cupido attwateri*
chicken, greater prairie, *Tympanuchus cupido*
chicken, lesser prairie, *Tympanuchus pallidicinctus*
chicken, prairie, *Tympanuchus cupido*
chukar, *Alectoris chukar*
condor, California, *Vultur californianus*
coot, tristan, *Gallinula nesiotis nesiotis*
cormorant, double-crested, *Phalacrocorax auritus*
cormorant, guanay, *Phalacrocorax bouganvilli*
coturnix (common quail), *Coturnix coturnix*
cowbird, brown-headed, *Molothrus ater*
crane, sandhill, *Grus canadensis*
crane, whooping, *Grus americana*
creeper, brown, *Certhia americana*
crossbill, red, *Loxia curvirostra*
crossbill, white-winged, *Loxia leucoptera*
crow, American (or common), *Corvus brachyrhynchos*
crow, carrion, *Corvus corone*
cuckoo, black-billed, *Coccyzus erythropthalmus*
cuckoo, yellow-billed, *Coccyzus americanus*
curlew, *Numenius* spp.
curlew, long-billed, *Numenius americanus*
dickcissel, *Spiza americana*
dipper, American, *Cinclus mexicanus*
dove, Inca, *Columbina inca*
dove, mourning, *Zenaida macroura*
dove, white-winged, *Zenaida asiatica*
duck, black, *Anas rubripes*
duck, wood, *Aix sponsa*
dunlin, *Calidris alpina*
eagle, bald, *Haliaeetus leucocephalus*
eagle, golden, *Aquila chrysaetos*
eagle, imperial, *Aquila helica*
eider, *Somateria mollissima*
emu, *Dromaius novaehollandiae*
falcon, peregrine, *Falco peregrinus*
falcon, prairie, *Falco mexicanus*
fieldfare, *Turdus pilaris*
finch, medium ground, *Geospiza fortis*
finch, rosy, *Leucosticte arctoa*
finch, small ground, *Geospiza fuliginosa*
flicker, common, *Colaptes auratus*
flicker, northern, *Colaptes auratus*
flycatcher, Acadian, *Empidonax virescens*
flycatcher, alder, *Empidonax alnorum*
flycatcher, brown-crested, *Myiarchus tyrannulus*
flycatcher, gray, *Empidonax wrightii*
flycatcher, great crested, *Myiarchus crinitus*
flycatcher, Hammond's, *Empidonax hammondii*
flycatcher, olive-sided, *Contopus borealis*
flycatcher, pied, *Ficedula hypoleuca*
flycatcher, vermilion, *Pyrocephalus rubinus*
flycatcher, western, *Empidonax difficilis*
flycatcher, willow, *Empidonax traillii*
fowl, red jungle, *Gallus gallus*

gadwall, *Anas strepera*
gnatcatcher, blue-gray, *Polioptila caerulea*
godwit, *Limosa* spp.
goldfinch, American, *Carduelis tristis*
goose, Canada, *Branta canadensis*
goose, snow, *Chen caerulescens*
goshawk, *Accipiter gentilis*
goshawk, northern, *Accipiter gentilis*
grackle, common, *Quiscalus versicolor*
grosbeak, evening, *Coccothraustes vespertinus*
grosbeak, pine, *Pinicola enucleator*
grosbeak, rose-breasted, *Pheucticus ludovicianus*
grouse, blue, *Dendragapus obscurus*
grouse, red, *Lagopus lagopus scoticus*
grouse, ruffed, *Bonasa umbellus*
grouse, sage, *Centrocercus urophasianus*
grouse, sharp-tailed, *Tympanuchus phasianellus*
grouse, spruce, *Dendrogapus canadensis*
gull, herring, *Larus argentatus*
gull, ring-billed, *Larus delawarensis*
gyrfalcon, *Falco rusticolus*
hawk, ferruginous, *Buteo regalis*
hawk, Harris's, *Parabuteo unicinctus*
hawk, red-shouldered, *Buteo lineatus*
hawk, red-tailed, *Buteo jamaicensis*
hawk, rough-legged, *Buteo lagopus*
hawk, Swainson's, *Buteo swainsoni*
hawk, white-tailed, *Buteo albicaudatus*
hen, heath, *Tympanuchus cupido cupido*
heron, great blue, *Ardea herodias*
hummingbird, Allen's, *Selasphorus sasin*
hummingbird, Anna's, *Calypte anna*
hummingbird, black-chinned, *Archilochus alexandri*
hummingbird, broad-tailed, *Selasphorus platycercus*
hummingbird, ruby-throated, *Archilochus colubris*
hummingbird, rufuous, *Selasphorus rufus*
jaeger, long-tailed, *Stercorarius longicaudus*
jay, blue, *Cyanocitta cristata*
jay, Florida scrub, *Aphelocoma c. coerulescens*
jay, gray, *Perisoreus canadensis*
jay, pinyon, *Gymnorhinus cyanocephalus*
jay, scrub, *Aphelocoma coerulescens*
jay, Steller's, *Cyanocitta stelleri*
junco, dark-eyed, *Junco hyemalis*
kestrel, American, *Falco sparverius*
kestrel, Seychelles, *Falco araea*
killdeer, *Charadrius vociferus*
kingbird, eastern, *Tyrannus tyrannus*
kinglet, golden-crowned, *Regulus satrapa*
kinglet, ruby-crowned, *Regulus calendula*
kite, Everglade, *Rostrhamus sociabilis*
kittiwake, black-legged, *Rissa tridactyla*
lark, horned, *Eremophila alpestris*
longclaw, yellow-throated, *Macronyx croceus*
longspur, chestnut-collared, *Calcarius ornatus*
longspur, Lapland, *Calcarius lapponicus*
longspur, McCown's, *Calcarius mccownii*
magpie, black-billed, *Pica pica*
mallard, *Anas platyrhynchos*
meadowlark, eastern, *Sturnella magna*
meadowlark, western, *Sturnella neglecta*
moas, family Dinornithidae or Anomalopterygidae
mockingbird, *Mimus polyglottos*
murre, McKinley, *Uria iocosa*
nighthawk, common, *Chordeiles minor*
night-heron, black-crowned, *Nycticorax nycticorax*
nightingale, *Luscinia megarhynchos*
nutcracker, Clark's, *Nucifraga columbiana*
nuthatch, brown-headed, *Sitta pusilla*
nuthatch, red-breasted, *Sitta canadensis*
nuthatch, white-breasted, *Sitta carolinensis*
oilbird, *Steatornis caripensis*
oriole, northern, *Icterus galbula*
ostrich, *Struthio camelus*
ovenbird, *Seiurus aurocapillus*
owl, barn, *Tyto alba*
owl, barred, *Strix varia*
owl, boreal, *Aegolius funereus*
owl, burrowing, *Athene cunicularia*
owl, elf, *Micrathene whitneyi*
owl, great gray, *Strix nebulosa*
owl, great horned, *Bubo virginianus*
owl, hawk, *Surnia ulula*
owl, horned (or great horned), *Bubo virginianus*
owl, long-eared, *Asio otus*
owl, northern spotted, *Strix occidentalis caurina*
owl, screech, *Otus asio*
owl, Seychelles, *Otus insularis*
owl, snowy, *Nyctea scandiaca*
owl, spotted, *Strix occidentalis*
parakeet, Carolina, *Conuropsis carolinensis*
partridge, gray, *Perdix perdix*
parula, northern, *Parula americana*
penguin, chinstrap, *Pygoscelis antarctica*
pewee, eastern wood, *Contopus virens*
pewee, greater (Coues' flycatcher), *Contopus pertinax*
phainopepla, *Phainopepla nitens*
phalarope, red, *Phalaropus fulicaria*
pheasant, ring-necked, *Phasianus colchicus*
pigeon, band-tailed, *Columba fasciata*
pigeon, passenger, *Ectopistes migratorius*
pigeon, white-crowned, *Columba leucocephala*
pigeon, wood, *Columba palumbus*
pintail, northern, *Anas acuta*
pipit, water, *Anthus spinoletta*
poor-will, *Phalaenoptilus nuttallii*
potoo, *Nyctibius* spp.
ptarmigan, white-tailed, *Lagopus leucurus*

ptarmigan, willow, *Lagopus lagopus*
pyrrhuloxia, *Cardinalis sinuatus*
quail, California, *Callipepla californica*
quail, Gambel's, *Callipepla gambelii*
quail, scaled, *Callipepla squamata*
rail, sandwich, *Pennula sandwichensis*
raven, common, *Corvus corax*
razorbill, *Alca torda*
redpoll, common, *Carduelis flammea*
redpoll, hoary, *Carduelis hornemanni*
redstart, American, *Setophaga ruticilla*
rhea, family Rheidae
roadrunner, *Geococcyx californianus*
robin, American, *Turdus migratorius*
robin, European, *Erithacus rubecula*
sandpiper, pectoral, *Calidris melanotos*
sandpiper, upland, *Bartramia longicauda*
sapsucker, Williamson's, *Sphyrapicus thyroideus*
sapsucker, yellow-bellied, *Sphyrapicus varius*
scaup, *Aythya* spp.
sheathbill, American (white or wattled), *Chionis alba*
shoveler, *Anas clypeata*
shrike, northern, *Lanius excubitor*
siskin, pine, *Carduelis pinus*
snipe, common, *Gallinago gallinago*
solitaire, Townsend's, *Myadestes townsendi*
sparrow, Bachman's, *Aimophila aestivalis*
sparrow, Baird's, *Ammodramus bairdii*
sparrow, black-chinned, *Spizella atrogularis*
sparrow, black-throated, *Amphispiza bilineata*
sparrow, Botteri's, *Aimophila botterii*
sparrow, Brewer's, *Spizella breweri*
sparrow, Cassin's, *Aimophila cassinii*
sparrow, chipping, *Spizella passerina*
sparrow, field, *Spizella pusilla*
sparrow, fox, *Passerella iliaca*
sparrow, grasshopper, *Ammodramus savannarum*
sparrow, Harris', *Zonotrichia querula*
sparrow, Henslow's, *Ammodramus henslowii*
sparrow, house, *Passer domesticus*
sparrow, Lincoln's, *Melospiza lincolnii*
sparrow, sage, *Amphispiza belli*
sparrow, Savannah, *Passerculus sandwichensis*
sparrow, song, *Melospiza melodia*
sparrow, swamp, *Zonotrichia georgiana*
sparrow, tree, *Passer montanus*
sparrow, vesper, *Pooecetes gramineus*
sparrow, white-crowned, *Zonotrichia leucophrys*
sparrow, white-throated, *Zonotrichia albicollis*
starling, European, *Sturnus vulgaris*
stork, dry-land, *Ciconia abdimii*
stork, white, *Ciconia ciconia*
swallow, barn, *Hirundo rustica*
swan, trumpeter, *Cygnus buccinator*
swan, tundra (whistling), *Cygnus columbianus*
swift, chimney, *Chaetura pelagica*
swift, Vaux's, *Chaetura vauxi*
tanager, scarlet, *Piranga olivacea*
tanager, summer, *Piranga rubra*
teal, blue-winged, *Anas discors*
tern, Caspian, *Sterna caspia*
tern, common, *Sterna hirundo*
tern, Forster's, *Sterna forsteri*
thrasher, brown, *Toxostoma rufum*
thrasher, California, *Toxostoma redivivum*
thrasher, curve-billed, *Toxostoma curvirostre*
thrasher, Le Conte's, *Toxostoma lecontei*
thrasher, sage, *Oreoscoptes montanus*
thrush, gray-cheeked, *Catharus minimus*
thrush, hermit, *Catharus guttatus*
thrush, oahu, *Phaeornis obsurus oahensis*
thrush, varied, *Ixoreus naevius*
thrush, wood, *Hylocichla mustelina*
tinamou, family Tinamidae
tit, coal, *Parus ater*
tit, crested, *Parus cristatus*
tit, great, *Parus major*
tit, marsh, *Parus palustris*
tit, willow, *Parus montanus*
titmouse, plain, *Parus inornatus*
titmouse, tufted, *Parus bicolor*
towhee, Abert's, *Pipilo aberti*
towhee, brown, *Pipilo fuscus*
towhee, green-tailed, *Pipilo chlorurus*
towhee, rufous-sided, *Pipilo erythropthalmus*
trumpeter, *Psophia* spp.
turkey, wild, *Meleagris gallapavo*
verdin, *Auriparus flaviceps*
vireo, black-whiskered, *Vireo altiloquus*
vireo, gray, *Vireo vicinior*
vireo, least Bell's, *Vireo belli pusillus*
vireo, red-eyed, *Vireo olivaceus*
vireo, warbling, *Vireo gilvus*
vireo, yellow-throated, *Vireo flavifrons*
vulture, bearded (lammergeyer), *Gypaetus barbatus*
vulture, black, *Coragyps atratus*
vulture, griffon, *Gyps fulvus*
vulture, king, *Sarcorhamphus papa*
vulture, turkey, *Cathartes aura*
vulture, white-backed (Kenya), *Pseudogyps africanus*
warbler, Bachman's, *Vermivora bachmanii*
warbler, bay-breasted, *Dendroica castanea*
warbler, black-and-white, *Mniotilta varia*
warbler, blackburnian, *Dendroica fusca*
warbler, black-throated gray, *Dendroica nigrescens*
warbler, black-throated green, *Dendroica virens*
warbler, blue-winged, *Vermivora pinus*
warbler, Cape May, *Dendroica tigrina*

warbler, cerulean, *Dendroica cerulea*
warbler, hermit, *Dendroica occidentalis*
warbler, hooded, *Wilsonia citrina*
warbler, Kentucky, *Geothlypis formosus*
warbler, Kirtland's, *Dendroica kirtlandii*
warbler, Lucy's, *Vermivora luciae*
warbler, myrtle, *Dendroica coronata*
warbler, Nashville, *Vermivora ruficapilla*
warbler, palm, *Dendroica palmarum*
warbler, pine, *Dendroica pinus*
warbler, prairie, *Dendroica discolor*
warbler, prothonotary, *Protonotaria citrea*
warbler, Swainson's, *Limnothlypis swainsonii*
warbler, Tennessee, *Vermivora peregrina*
warbler, Townsend's, *Dendroica townsendi*
warbler, Wilson's, *Wilsonia pusilla*
warbler, worm-eating, *Helmitheros vermivorus*
warbler, yellow, *Dendroica petechia*
warbler, yellow-rumped, *Dendroica coronata*
warbler, yellow-throated, *Dendroica dominica*
waterthrush, Louisiana, *Seiurus motacilla*
waxwing, Bohemian, *Bombycilla garrulus*
waxwing, cedar, *Bombycilla cedrorum*
whip-poor-will, *Caprimulgus vociferus*
willet, *Catoptrophorus semipalmatus*
woodcock, American, *Scolopax minor*
woodpecker, black-backed, *Picoides arcticus*
woodpecker, black-backed three-toed, *Picoides arcticus*
woodpecker, downy, *Picoides pubescens*
woodpecker, Gila, *Melanerpes uropygialis*
woodpecker, hairy, *Picoides villosus*
woodpecker, ivory-billed, *Campephilus principalis*
woodpecker, ladder-backed, *Picoides scalaris*
woodpecker, pileated, *Dryocopus pileatus*
woodpecker, red-bellied, *Melanerpes carolinus*
woodpecker, red-cockaded, *Picoides borealis*
woodpecker, red-headed, *Melanerpes erythrocephalus*
woodpecker, white-headed, *Picoides albolarvatus*
wood-pewee, eastern, *Contopus virens*
woodpigeon, *Columba palumbus*
wren, Carolina, *Thryothorus ludovicianus*
wren, house, *Troglodytes aedon*
wren, marsh (or long-billed marsh), *Cistothorus palustris*
wren, rock, *Salpinctes obsoletus*
wren, sedge (or short-billed marsh), *Cistothorus platensis*
wren, stripe-backed, *Campylorhynchus nuchalis*
wren, winter, *Troglodytes troglodytes*
wren-tit, *Chamaea fasciata*
yellow-finch, greenish, *Sicalis olivascens*
yellowthroat, common, *Geothlypis trichas*

Mammals

aardvark, *Orycteropus afer*
aardwolf, *Proteles cristatus*
ant-eater, family Myrmecophagidae
antelope, *Capromeryx* spp.
armadillo, nine-banded, *Dasypus novemcinctus*
ass, wild, *Equus asinus*
baboon, hamadryas, *Papio mormom*
badger, *Taxidea taxus*
bat, hoary, *Lasiurus cinereus*
bat, horseshoe-nosed, *Rhinolophus ferrumequinum*
bat, Indiana, *Myotis sodalis*
bat, red, *Lasiurus borealis*
bat, Seminole, *Lasiurus seminolus*
bear, black, *Ursus americanus*
bear, cave, *Ursus spelaeus*
bear, grizzly, *Ursus horribilis*
bear, polar, *Thalarctos maritimus*
bear, short-faced, *Arctodus simus*
beaver, *Castor canadensis*
beaver, giant, *Castoroides ohioensis*
bison (American), *Bison bison*
bobcat, *Lynx rufus*
buffalo, *Bison bison*
camel (fossil), *Titanotylopus* spp.
camel, North American, *Camelops* spp.
camel, one-humped, *Camelus dromedarius*
caribou, barren-ground, *Rangifer arcticus*
caribou, woodland, *Rangifer caribou*
cat, saber-tooth, *Smilodon fatalis*
cattle, *Bos taurus*
cheetah, *Acinonyx jubatus*
chickaree, *Tamiasciurus douglasi*
chimpanzee, *Pan troglodytes*
chipmunk, eastern, *Tamias striatus*
chipmunk, least, *Eutamias minimus*
chipmunk, yellow pine, *Eutamias amoenus*
coati, *Nasua narica*
cottontail, eastern, *Sylvilagus floridanus*
cougar, *Felis concolor*
coyote, *Canis latrans*
deer, mule, *Odocoileus hemionus*
deer, red, *Cervus elaphus*
deer, white-tailed, *Odocoileus virginianus*
elephant (African), *Loxodonta africana*
elk (wapiti), *Cervus canadensis* or *C. elaphus*
ermine, *Mustela erminea*
ferret, black-footed, *Mustela nigripes*
fisher, *Martes pennanti*
fox, arctic, *Alopex lagopus*
fox, gray, *Urocyon cinereoargenteus*
fox, kit, *Vulpes fulva*

fox, swift, *Vulpes velox*
gazella, *Gazella* spp.
gerenuk, *Litocranius walleri*
giraffe, *Giraffa camelopardalis*
goat, mountain, *Oreamnos americanus*
gopher, plains pocket, *Geomys bursarius*
gopher, pocket, family Geomyidae
gopher, southeastern pocket, *Geomys pinetis*
gorilla, eastern mountain, *Gorilla gorilla beringei*
hare, arctic, *Lepus arcticus*
hare, varying (snowshoe), *Lepus americanus*
hartebeest, *Alcelaphus* spp.
horse, *Equus caballus*
hyena, striped, *Hyaena hyaena*
jackrabbit, black-tailed, *Lepus californicus*
jackrabbit, white-tailed, *Lepus townsendi*
jaguar, *Felis onca*
lamb, *Ovis aries*
lemming, collared, *Dicrostonyx groenlandicus*
lemming, European, *Lemmus lemmus*
lion, African, *Felis leo*
lion, mountain, *Felis concolor*
lynx, Canada, *Lynx canadensis*
mammoths, *Mammuthus* spp.
margay, *Felis tigrina*
marmoset, pygmy, *Cebuella pygmaea*
marmot, hoary, *Marmota caligata*
marmot, yellow-bellied, *Marmota flaviventris*
marten, American, *Martes americana*
marten, pine, *Martes martes*
marten, stone, *Martes foina*
mastodon, *Mammut americanum*
mink, *Mustela vison*
mole, coast, *Scapanus townsendi*
mole, eastern, *Scalopus aquaticus*
mongoose, *Herpestes* spp.
monkey, howler, subfamily Alouattinae
monkey, proboscis, *Nasalis larvatus*
moose, *Alces alces*
mouse, cotton, *Peromyscus gossypinus*
mouse, deer, *Peromyscus maniculatus*
mouse, golden, *Peromyscus nuttalli*
mouse, harvest, *Reithrodontomys* spp.
mouse, jumping, *Zapus hudsonius*
mouse, leaf-eared, *Phyllotis darwini rupestris*
mouse, northern grasshopper, *Onychomys leucogaster*
mouse, piñon, *Peromyscus truei*
mouse, plains pocket, *Perognathus flavescens*
mouse, pocket, *Perognathus* spp.
mouse, red-backed, *Clethrionomys gapperi*
mouse, western jumping, *Zapus princeps*
mouse, white-footed, *Peromyscus leucopus*
muskox, *Ovibos moschatus*
muskox, woodland, *Symbos cavifrons*
muskrat, *Ondatra zibethicus*
muskrat, round-tailed, *Neofiber alleni*
okapi, *Okapia johnstoni*
opossum, *Didelphis marsupialis*
orangutan, *Pongo pygmaeus*
pangolin, family Manidae
peccaries, *Platygonus* spp.
peccary, collared, *Pecari angulatus*
pig, domestic, *Sus scrofa*
pig, wild, *Sus scrofa*
pikas, *Ochotona* spp.
pipistrelle, eastern, *Pipistrellus subflavus*
porcupine, *Erethizon dorsatum*
prairie dog, black-tailed, *Cynomys ludovicianus*
prairie dog, white-tailed, *Cynomys leucurus*
pronghorn, *Antilocapra americana*
rabbit, brush, *Sylvilagus bachmani*
rabbit, European, *Oryctolagus cuniculus*
rabbit, swamp, *Sylvilagus aquaticus*
raccoon, *Procyon lotor*
rat, black, *Rattus rattus*
rat, cotton, *Sigmodon hispidus*
rat, kangaroo, *Dipodomys agilis*
rat, naked mole, *Heterocephalus glaber*
rat, Norway, *Rattus norvegicus*
rat, rice, *Oryzomys palustris*
rat, swamp (in Phillipines), *Rattus rattus*
rat, swamp, see round-tailed muskrat
rat, wood (eastern), *Neotoma floridana*
rhinocerus, black, *Diceros bicornis*
seal, elephant, *Mirounga angustirostris*
seal, fur, *Callorhinus ursinus*
seal, harp, *Pagophilus groenlandicus*
sheep, mountain (or bighorn), *Ovis canadensis*
shrew, least, *Cryptotis parva*
shrew, masked, *Sorex cinereus*
shrew, short-tailed, *Blarina brevicauda*
skunk, spotted, *Spilogale putorius*
skunk, striped, *Mephitis mephitis*
sloth, ground, *Megalonyx* spp.
sloth, Shasta ground, *Nothrotheriops shastensis*
squirrel, arctic ground, *Spermophilus* (or *Citellus*) *parryi*
squirrel, Douglas (chickaree), *Tamiasciurus douglasi*
squirrel, eastern flying, *Glaucomys volans*
squirrel, fox, *Sciurus niger*
squirrel, Franklin's ground, *Citellus franklini*
squirrel, gray, *Sciurus carolinensis*
squirrel, ground, *Citellus pygmaeus*
squirrel, northern flying, *Glaucomys sabrinus*
squirrel, red (Europe), *Sciurus vulgaris*
squirrel, red (North America), *Tamiasciurus hudsonicus*
squirrel, southern flying, *Glaucomys volans*

squirrel, thirteen-lined ground, *Citellus tridecemlineatus*
tiger, *Felis tigris*
vole, European meadow, *Microtus s. arvalis*
vole, field, *Microtis agrestis*
vole, gray-sided, *Clethrionomys rufocanus*
vole, heather, *Phenacomys intermedius*
vole, meadow, *Microtus pennsylvanicus*
vole, montane, *Microtus montanus*
vole, pine, *Pitymys pinetorum*
vole, prairie, *Microtus ochrogaster*
vole, red-backed, *Clethrionomys gapperi*
vole, red tree, *Phenacomys longicaudus*
vole, rock, *Microtus chrotorrhinus*
vole, sagebrush, *Lagurus curtatus*
vole, tundra, *Microtus oeconomus*
vole, tundra redback, *Clethrionomys rutilus*
wapiti, *Cervus elaphus* or *C. canadensis*
whale, blue, *Balaenoptera musculus*
whale, fin, *Balaenoptera physalus*
whale, killer, *Orcinus orca*
wildebeest, *Connochaetes taurinus*
wolf, *Canis lupus*
wolf, dire, *Canis dirus*
wolf, red, *Canis niger*
wolf, timber, *Canis lupus*
wolverine, *Gulo luscus* (or *G. gulo*)
woodchuck, *Marmota monax*
woodrat, *Neotoma* spp.
woodrat, desert, *Neotoma lepida*
zebra, common (or Burchell's), *Equus burchelli*

Index

Note: f following a page number indicates a figure or illustration; t indicates a table.

Abalone, 32
Abbey, Edward, 705
Abiotic factors, 14, 40–80
Aborigines, 154
Abyssal zone, 574, 575f, 587
Acacia, swollen-thorn, 161–162, 161f
Acacia, umbrella-shaped, 555
Acanthocephala, 221
Acclimation, 17–18
Acid rain, 450, 488, 673–679, 676f
Acquired Immunodeficiency Syndrome (AIDS), 214–216
Actions, 12f
Adams, Charles C., 2, 2f
Adaptation, 30–31, 176–177, 539
Adaptation and Natural Selection, 157
Adventures in Ecology, 384
Aeolian biome, 601
Aestival aspect, 274, 275f
Aestivation, 17, 539
Agave, 537–538
Age distribution, 130–133
  animal, 131
  human, 131, 132f
  plant, 131
  stable, 131
  stationary, 131
  U.S., 132t
Agent Orange, 655
Aggressive mimicry, 200–202
Aggressive resemblance, 200–202
Agriculture, 692–694
  African nomadism and, 694
  allelochemical use, 248
  biological controls, 202–203
  chemical use, 5, 655–656
  companion plantings, 203, 248
  crop rotation, 203, 248
  desert, 542–543
  energy flow in, 334–341
  fertilization, 337, 338f
  habitat destruction due to, 559–560
  herbicide use, 248, 655
  irrigation, 337, 542, 671
  pesticide use, 203, 650, 654–656
  phenology in, 203
  research, 248
  sustainable, 693
Agroforestry, 694
Ahearne, John F., 691
Air pollution, 646f, 672–686
  forest decrease and, 450, 488
  fossil-fuel-caused, 79
Albedo, 670
Alfalfa, 348
Algae
  amensalism and, 247
  in animal bodies, 251, 252f
  bloom, 566, 568
  DDT and, 21
  in lichens, 250–251
  phosphates and, 20
  photosynthetic, 25
Algae, blue-green, 181, 260, 566, 568
Algae, brown, 180
Algae, green, 252–253, 267
Algae, red, 32
Allee, W. C., 93
Allee effect, 93–95, 100–101, 190, 628
Allelochemicals, 203, 247, 255, 268–269
Allelopathy, 247–248, 255
Allen, Durward, 606
Alligator, American, 455, 456f
Allogrooming, 125
Alpine meadows, 482
Alpine tundra, 48, 478–486, 480f
  animals, 483–484
  climate and soil, 479–481
  human relations, 484–486
  vegetation, 481–482

Altruism, 151–157
reciprocal, 156
Amensalism, 183, 246–247, 266–267
Amphipods, 221–222
Andrewartha, H. G., 112, 114
Animal behavior, 28–29
Animal Communities in Temperate America, 428
Annelid, snow, 601
Annuals, 51, 141–144
Antarctic, 468–469
Antarctic Treaty of 1959, 469
Antbird, 551
Antibiotics, 133
Ants, 126
acacia trees and, 161–162
aphids and, 252, 253
plant coaction and, 182
as pollinators, 256–258
rodents and, 231, 266
Aphids, 32, 252, 253, 253f, 654
Aphotic zone, 574
Appalachians, 487–488
Applied ecology, 382
Aprovechados, 259
Aquatic habitat, 14
commensalism in, 222
dead organic matter in, 228
fresh-water, 14
marine or ocean, 14
mutualism in, 252–253
seasonal change, 275
Aquifer, 364
Arctic National Wildlife Refuge, 467–468
Arctic tundra, 416, 417, 459–468
animals, 463–466
biogeographical relations, 466–467
climate and temperature, 459–461, 461f, 466t
distribution, 459–461, 460f
energy flow, 464f
human occupation, 467–468
soil, 62t, 461
vegetation, 461–463, 463t
Area-sensitive species, 622, 623
Arena displays, 127
Aristotle, 143
Arizona chaparral, 509f, 510
Arroyos, 532
Asbestos, 660
Assembly rules, 288
Assimilation, 317, 318f
Assimilation efficiency, 326
Atmosphere, 369–372
first, 369
second, 369
Atmospheric pollution. *See* Air pollution
Attractants, 269
Audubon, John James, 451, 619
Audubon Society, 622, 704
Auk, great, 166
Autecology, 13–39, 389, 408–409
Automobiles, 673
Autotoxins, 269
Autotrophs, 23–24
chemosynthetic, 24, 251, 329–332, 359, 370
photosynthetic, 370
respiration, 324–325
Autumnal aspect, 275
Auxotrophism, 25
Axelrod, Daniel I., 411
Azolla, 359

Bacteria
carrion and, 223
green, 24
nitrifying, 24–25
nonsulfur purple, 24
photosynthetic, 262
purple, 24
Rhizobium, 359
sulfur, 25, 361
tolerance ranges, 30–31
in vent ecosystem, 331
Bacterioplankton, 588
Bailey, Vernon, 595
Bajadas, 532, 534f
Baker, J. R., 30
Balds, 437
Barbour, M. G., 14
Bari, Judi, 705
Barnacles, 576f, 577
Bartram, William, 451
Basic (intrinsic) reproductive rate, 211
Basidiomycetes, 254
Bass, large-mouthed, 17f
Basswood, 295, 295f, 391f, 339
Bates, H. W., 198
Batesian mimicry, 199, 200
Bathyl zone, 574, 575f, 587
Bats, 227
cave, 599
of desert, 537
moths and, 162–165
nocturnalism, 273, 274
as pollinators, 256, 257t, 258f, 629
sonar, 162, 164, 165
Beaches, 388, 388f
Bear, brown, 411
Bear, cave, 409–411, 410t
Bear, grizzly, 409, 410t, 462f, 630, 631f
Bear, polar, 459f, 465
Beargrass, 488, 490f
Bears, 46
Beater effect, 123
Beavers, 175, 268, 343, 353, 354f
Bee, honey, 126
hive, 119f
pollination, 256, 257t
thermoregulation, 125
waggle dance, 124
Bee, stingless, 245–246
Beebe, William, 198–199, 551
Beech, 423f, 449f, 624f
Beech-maple forest, 395–396, 433f, 434t, 438
reconstruction, 529, 640
species composition, 270, 271t
succession, 394t
Beetle, ambrosia, 226
Beetle, bark, 71–72, 226
Beetle, blister, 223
Beetle, bombardier, 170, 269
Beetle, bruchid, 109, 180
Beetle, burying, 223, 224–225, 383t
Beetle, chestnut grain, 108, 108f, 204, 220, 220t, 238, 239f, 265
Beetle, Colorado potato, 202
Beetle, confusing grain, 238, 239f, 265
Beetle, fire, 71
Beetle, Japanese, 651–652
Beetle, scarab, 225–226
Beetle, tiger, 395–396
Beetle, whirligig, 567
Behavior
evolution of, 148–160
agonistic, 28
caregiving, 28
maintenance, 28
selfish, 151–152
sexual, 28
Beltian bodies, 161f
Benthos, 565
Bergamot, wild, 519f
Bergerud, Arthur, 237
Bessey, Charles E., 384
Bioassay, 22
Biochemical oxygen demand (BOD), 663
Biodiversity, 604, 610, 640–644. *See also* Diversity
Bio-ecology, 3, 428
Biogeochemical cycles, 355
Biogeography model, 303–306
Biological control, 202–203, 370–372
Biological magnification, 661–663, 667
Biological weapons, 5
Biomass, 315–316
accumulation, 367, 368t, 368f
productivity and, 321–324, 322t, 323t
pyramid of, 310, 310f, 313, 315
world, 357
Biome, 427–431
defined, 431

North American, 429f
world, 430f
See also specific biomes
Biomonitoring, 22
Bioremediation, 670
Biosphere, 11, 320–321
homeostatic mechanism, 370–372
net primary production, 341
Biotechnology, 685–686
Biotic factors, 14
Biotic potential, 87, 89–91, 89f, 605
Birch, L. C., 114
Bird(s)
antiparasitic defenses, 220
beak size, 290, 291f
behavior, 148
breeding, 58, 127–129
dispersal, 34–35, 36t
eusociality, 154–155
extinction, 166
fire-dependent, 72
flightless, 525
frugivorous, 185
habitat selection, 33, 33t
habitat size and, 623–625, 625t
longevity, 86, 87t
migration, 27–28, 59, 278–279, 627–629
niches, 280t
parasitism in, 206–207f
photoperiod and, 58, 59
productive energy in, 27
reproductive rates, 138–141, 141t, 142t, 147
sex ratios, 130
spring arrival dates, 277t
territoriality, 110, 111–112
warning calls, 151
Birth rate, 81–82, 102–103, 104t
crude, 82, 115f
human, 115–116
Bison, 526, 529f, 605–606
Biswell, H. H., 490
Blackbird, Brewer's, 266
Blackbird, red-winged, 128, 129, 266, 526
Blackman, F. F., 20
Bladderworts, 594
Blenny, saw tooth, 583
Bloaters, 668, 668f
Blowflies, 383t
Blowouts, 391f, 392–393
Blueberries, 462f
Bluestem, big, 77t, 379, 519, 520f, 531f
Bluestem, little, 396
Bobwhite, northern, 75, 103–104, 103f, 171, 190
Bog(s), 450, 590–596
arctic, 461
as boreal islands, 595–596, 597f
formation, 591f
habitat conditions, 590
hummock-and-hollow cycle, 374–375, 375f
minerotrophic, 596
plants, 592t, 594
spagnum, 419–420, 590, 592t
structure, 271
succession, 590, 591f
Bog forest, 594–595
Booth, D. E., 504
Boreal coniferous forest, 469–476
animals, 472–475, 475t, 499t
biogeographical relations, 475–476
distribution and climate, 469
ecotones, 476
plants, 470–472, 470f
soil, 470
Borgstrom, Georg, 701
Boring, 177
Bormann, F. H., 403
Bornemissza, G., 226
Bottleneck, population, 617
Boulder fields, 481f
Boulding, Kenneth, 117
Bounty system, 607
Braun, E. Lucy, 414, 435, 436, 437
Braun-Blanquet, J., 47
Breeder reactor, 691
Breeding-bird surveys (BBS), 628–629
Brewer, R., 147, 639
Broadleaf forest, 61f, 350–352
Bromfield, Louis, 5
Brown, James H., 231
Brown's Ferry nuclear accident, 659
Browsers, 177, 312, 387
Bt, 202–205
Budworm, spruce, 471, 473, 487
Buffalo wallows, 346, 347f
Buffer zones, 632
Bumblebee, 11, 172, 172t, 174, 256
Burdock, 35f
Bureau of Indian Affairs, 620
Buried seed strategy, 145
Bush, creosote, 375, 376f, 535–536
Butterflies, 256, 257t
Butterfly, cabbage, 181
Butterfly, *Heliconius*, 202
Butterfly, monarch, 181f, 199
Butterfly, viceroy, 199
Butterfly-fish, banded, 583f
Buttress roots, 544, 548f

C_3 plants, 24, 51
C_4 plants, 24, 51
Cacti, 50, 147, 536–537
Cactus, semaphore, 454f
Cainism, 203
Calcification, 62t
Calciphiles, 14
Calciphobes, 14
Caliche, 532
California chaparral, 74–75, 77, 248, 505–510, 506f
Calorie, 23
Camel, 54, 538
Canavinine, 180
Canine distemper, 620
Canopy
changes in, 403
rain forest, 547–548
replacement, 375–377
seed storage, 145
shading, 349
Camouflage, 30, 72, 199–200, 465, 547
Cancers, 694–695, 695f
pollution-related, 655
radiation-induced, 656, 659, 684
Cannibalism, 203–204, 238
Capitalization, 260, 262
Captive breeding, 611f, 619, 620
Carbofuran, 652
Carbon cycle, 355–358
Carbon dioxide, 355–358, 357f
in greenhouse effect, 679–680, 680t, 682, 682–683f
poisoning, 564
Carcinogenicity, 661, 695
Cardinal, northern, 38, 446f
Caribou, 465, 466f, 467
Carl, E. A., 125
Carlquist, Sherman, 302
Carnivores, 309, 312, 661–662
Carpenter, S. R., 332
Carrion, 223–225, 383
Carrying capacity (K), 91, 96, 106–108, 107f, 137–138
Carson, Rachel, 5, 651
Carter, Jimmy, 691
Cascade Range, 490–494
Cascading trophic interactions, 267, 332
Cassia, big, 182
Cassowaries, 551
Catastrophes, 619
Caterpillar phacelia, 508f
Caterpillars, 199, 201f
Cation exchange capacity, 60
Cattle, 226, 334, 619
Cattle plague, 555
Cauliflory, 545, 549f
Caves, 599–600
Cecropia, 441f
Cedar glades, 437
Chamaephytes, 47t, 48f
Chaney, Ralph W., 411
Chaos, 194–197

Chaparral
Arizona, 509f, 510
California, 74–75, 77, 248, 505–510, 506f
fire adaptations, 505–507, 508, 510
Chapman, H. H., 75
Chapman, Royal, 4
Character displacement, 290–292, 292f
Character release, 291–292
Chavez, Cesar, 656
Cheatgrass, 534, 535t
Cheetah, 557, 560f
Chemical defenses, 179–181, 193, 268–269
Chemical ecology, 268–270
Chemical weathering, 59, 344
Chemicals, 647–648
banned or restricted, 652
coevolution and, 160
toxic, 660–661, 663
Chemolithoautotrophs, 24, 251, 329–332, 359, 370
Chernobyl nuclear accident, 659
Cherry, sand, 392, 393
Chestnut, American, 170, 217
Chestnut blight, 170, 217
Chickadee, black-capped, 229f, 233, 236f
Chickadee, brown-capped, 473f
Chickadee, Carolina, 233, 236f
Chickadees, 47, 151–152, 227
Chicken, Attwater's prairie, 527
Chicken, greater prairie, 527, 637
Child, C. M., 428
Chipmunk, eastern, 283, 283f, 395
Chipmunk, least, 234–235
Chipmunk, yellow pine, 234–235
Chlorofluorocarbons (CFCs), 673, 679, 680t, 683–684
Cholla, Christmas tree, 375
Christensen, Martha, 232
Chub, creek, 15f
Chukar, 609
Cinquefoil, 480f, 597f
Circadian rhythms, 58–59, 273
Circannual rhythm, 59
Clam, giant, 582
Clay, 63f
Clean Air Acts, 678
Cleaning symbiosis, 250f, 259, 582–583
Clearcutting, 182, 183f, 439, 495, 626f, 705
Cleavers, 147
Clements, Frederic E., 2, 12, 21, 64, 266, 286, 353, 382, 383, 384, 389, 400, 427, 428–430, 431, 436
Climate
climax communities and, 400, 403
dispersal and, 38–39
diversity and, 299
fire and, 72–73, 74f
future, 424
greenhouse effect and, 679–683
physiognomy and, 427
Pleistocene, 419
rain forest destruction and, 552
soil development and, 62t
stable or predictable, 299
Climatic equability, 415
Climax communities, 379, 380, 398–403
allelopathy in, 248
plant species, 144–145
stability, 139
steady state, 319, 398
Climax pattern concept, 400
Climograph, 54f
Cloud seeding, 485
Clover, red, 11
Coactions, 12f, 167
classification, 168t
community structure and, 284–285
coral reef, 582–584
noncompetitive, 387
role in succession, 387
Coastal marine ecosystems, 329–331
Coccidian, barbershop, 220, 220t, 265
Coefficient of relatedness, 153
Coevolution, 160–165, 260, 270–293, 500
Coexistence, 234, 238–240, 242, 243f, 287
Cohort, 83
Cole, LaMont, C., 89, 141, 143
Colonial invertebrates, 126
Coloration
aposematic, 200
in desert animals, 539
protective, 170, 199
warning, 200, 201f, 273, 273f
Commensalism, 205f, 222–223
Commoner, Barry, 11, 264
Communication, social group, 124
Communities, 263–306
competition in, 285–293
diversity in, 296–306
ecosystem and, 263–267
estimated replacement times, 641t
integrated versus individualistic, 293–296
in landscape ecology, 431
niche as, 279–284
reasons for, 284–285
structure, 267–279, 284–285, 307–310
supersaturated, 296
unsaturated, 296
Community change, 373–398
directional, 373
fluctuation as, 377–379
nondirectional, 373
replacement, 374–377
succession as, 379–398
Community ecology, 168, 264, 411–424
Competition
apparent, 235–238, 265
assembly rules, 288–289
coevolution and, 160–161
coexistence and, 234, 238–240, 242, 243f
in communities, 285–293
contest, 109, 109f
diffuse, 244, 266
extinction and, 239, 239f
in geographical/habitat distribution, 232–235
habitat distribution and, 232–238
interference, 109, 238
interspecific, 209, 229–246
intraspecific, 109–111
Lotka-Volterra model, 240–244
outcomes, 229–230
scramble, 109, 109f
Competitive exclusion, 230, 231–238, 239f, 287–289
Competitive replacement, 231–232
Competitors, 146
Condor, California, 610, 611f
Coneflower, prairie, 518f, 520
Coniferous forest, 61f, 62t, 498–505
Connell, Joseph H., 286, 289, 290, 386
Connor, E. F., 288
Conservancies, 622
Conservation, 166, 603, 640–645
Conservation groups, 553–554, 639
Conservation biology, 8, 603–620
Conservation Biology—an Evolutionary-Ecological Perspective, 603
Conservation easement, 622
Constancy, 21, 270
Consumers, 308–309, 309f
in carbon cycle, 355
efficiencies of, 326–327
primary, 308
secondary, 309
tertiary, 309
Continuum, 299–296
Cooper, W. S., 389, 402–403
Cooperative breeding, 154–156
Co-opted production, 341
Coral reefs, 252, 259, 354, 580–585, 580f
atolls, 581f, 582
barrier, 581–582, 581f
development, 582f, 583
fringing, 581, 581f
zooxanthellae and, 583–584
The Coral Seas, 580
Cormorant, double-crested, 123f, 653f, 669
Corn, 270, 334, 335f, 337t, 338f, 339, 693
Corridors, 636

Cottonwood, 351f, 392, 393, 394, 394t
Coulter, J. M., 389
Cowbird, brown-headed, 611–612, 626
Cowles, Henry Chandler, 2, 14, 285, 389, 402–403, 428, 436
Cox, G. W., 347
Crab, ghost, 578–579
Craig, Wallace, 515
Craighead, F. C., 637
Craighead, John, 409
Cranberry, 594f, 595f
Crane, whooping, 610, 612f
Crassulacean acid metabolism (CAM), 50
Crepuscular organism, 272
Crickets, 165
Cronquist, Arthur, 466
Cuckoldry, 129
Cummins, K. W., 571
Curtis, J. T., 294
Cyclic replacement systems, 374–375
Cyclomorphosis, 566
Cypress, bald, 453f

Daphnia, 252–253
Darwin, Charles, 2, 4–5, 29, 133, 230, 264, 287, 346, 387, 582
Davenport, C. B., 428
Davidson, J., 112
Day-neutral plants, 57–58
DDE, 653f, 662
DDT
  ban on usage, 652
  in Great Lakes, 667–669, 668f
  nontarget organisms and, 651, 653, 653f, 661
  resistance, 133, 133f, 654–655
  salt water and, 21
Death rate, 81–82, 102–103, 104t
  cancer, 695f
  childhood, 696, 696f
  crude, 82, 115f
  human, 115–116
  income and, 696, 696f
DeBach, P., 232
Debt-for-nature swaps, 553–554
Deciduous forest edge, 442–445, 446t
Deciduous Forests of Eastern North America, 436
Decomposers, 309, 309f
Deer, 183, 183f, 184t
Deer, mule, 512, 512f
Deer, white-tailed, 98, 152, 178f, 217, 218f, 219f, 222, 472, 608f, 627, 627f
Deevey, E. S., 84
Defenses
  anti-parasitic, 220
  associational, 182, 265
  chemical, 179–181, 193, 268–269
  herbivory, 178–182
  morphological, 178–179
Defoliation, 325f, 587
Deforestation, 626, 627–629
Dehydration, 54
Demographic transition, 115
Demography, 117
Denitrifiers, 359
Density-dependent factors, 102–105, 108–112, 197–198, 216–220
Desert, 48, 531–543
  animals, 538–542, 539t
  cold, 534–535
  defined, 427
  distribution and climate, 531–532
  human relations, 542
  plants, 533–534, 533f
  soil, 61f, 62t, 532–533
  species richness, 535t
  topography, 534f
  warm, 535–538
Desert ephemerals, 51
Desert scrub, 535–538
Desertification, 543
Design with Nature, 703
Desmid, copperhead, 253
Deterministic extinction, 613–616
Diabetes, 694
Diamond, Jared, 288
Dickcissels, 27t, 627, 627f, 628
Dieldrin, 651–652, 653, 667, 668
Difference equation, 196
Diffusion, 37
Dinoflagellate, 583–584
Dioecious organism, 256
Dioxin, 655, 667
Diplodocus, 408–409
Dipper, American, 501–502
Disaffiliation, 259–260
Disease(s)
  cannibalism and, 204
  causing extinction, 217
  childhood, 214
  deficiency, 694
  health and, 222, 694–698
  human, 213–216
  iatrogenic, 697
  infectious, 211–213
  insect-borne, 697
  metabolic, 694
  microparasitic, 213–214
  radiation-related, 656–657
  pollution-related, 656
  respiratory, 221
  sexually transmitted, 215–216
  stress-related, 695
Dispersal, 33–39
  animal, 222
  environmental, 36
  jump, 37
  Kettle pattern, 35f
  as mutualism, 251–252
  plant, 34t, 144–145
Dissolved organic matter (DOM), 228, 588
Disturbance, 299–300, 299f, 377, 399, 400, 402, 403, 471, 484, 546–547
Diurnalism, 272–273
Diversity
  alpha, 300
  beta, 300–301
  biological, 8, 604
  determinants, 302–306
  ecological, 296–306
  gamma, 300–301
  genetic, 9, 604, 618, 637–638
  indices, 301–302, 302t
  of islands, 302–306
  influences, 297–300
  preservation of, 604
  of rain forest, 545–546
  regional, 300–301
  resource, 298–299
  richness, 296
  stability, 404–405
DNA, 260, 684–685
DNA fingerprinting, 9, 129
Dodders, 208
Doll's-eyes, 433f
Domatia, 251
Dominance, 268, 341
Doppler effect, 163, 163f
Dove, mourning, 399, 444
Downstream drift, 571
Drip tips, 545, 550f
Drought, 378–379, 403, 543, 694
Drude, Oscar, 427
Dubos, René, 222
Duckweed, 453f, 615, 615f
Dumond, D. E., 115
Dung, 225–226
Dust Bowl, 378
Dust storm, 379f
Dutch elm disease, 37, 170

Eagle, bald, 148, 662, 662f
Earth, history of, 369–370
Earth First!, 705
Earth Summit, 8
Earthkeeping: Christian Stewardship of Natural Resources, 644
Earthwolf, 395
Earthworms, 345–346
Ecdysone, 180
Echolocation, 162
Ecological counterparts, 281
Ecological diversity. *See* Diversity

Ecological efficiencies, 324–325
Ecological equivalents, 281
Ecological Society of America, 8, 79, 428, 436, 622, 643, 660
Ecological Systems and the Environment, 702
The Ecological Theater and the Evolutionary Play, 160
Ecology, 1
  global change and, 688–689
  history, 2–8
  as a system, 11
  See also specific kinds of ecology
The Ecology of Invasions by Animals and Plants, 404
The Ecology of North America, 428
Economics
  of pesticide use, 655–656
  of plant extinction, 642
  of pollution, 648–649
Ecosabotage, 705
Ecosystem, 264–306
  energy flow, 311–313
  preservation, 621
  productivity, 322t
  interrelationships, 12f, 264f
Ecosystem ecology, 264
Ecotone, 298, 472, 476
Ecotypes, 31
Ectomycorrhizae, 254
Ectoparasites, 208
Edge effect, 625–627, 626f
Effective population size, 616–617, 631–632
Efficiencies, 324–327
  assimilation, 326
  consumer, 326–327
  ecological growth, 326
  harvesting, 325–326
  Lindeman's, 327
  producer-level, 324–325
  tissue growth, 326
  trophic-level production, 327
Egler, F. E., 382, 389
Egret, cattle, 37
Ehrlich, P. R., 160, 699
Eibl-Eibesfeldt, I., 583
El Niño, 194
Electrophoresis, 129
Elk, 71
Elton, Charles, 2, 191, 224, 231, 404, 431
Emergents, 544, 548f
Emigration
  genetics and, 135
  lemming, 157, 158
  role in population, 81–82, 102–103, 104t
Empiricism, 10
Emu, 525, 557
Enclosure, 387
Endangered species, 148, 456, 542, 610–613
Endangered Species Act, 610
Endomycorrhizae, 254
Endoparasites, 208, 209
Energetics, 4
Energy, 307–342, 689–694
  in agriculture, 334, 341
  assimilated, 25, 317, 318f
  biomass as, 315–316
  compartment approach to, 329
  cultural, 336, 337t
  defined, 22
  ecosystem, 307–342
  excretory, 25
  existence, 25–26
  human expropriation of, 341–346
  kinetic, 23
  net, 689
  in organisms, 23–28
  photovoltaic, 691
  potential, 23
  productive, 25f, 26–27
  quality, 332–334
  regulatory processes, 332–334
  solar, 310–311, 311f, 691
  subsidies, 28
  work and, 22–23
Energy balance, 22–28
Energy conservation, 671, 690, 691
Energy flow, 310–313, 311f, 312f, 313f, 316f, 317f, 319t, 327–328, 464f, 528f, 567, 571–573
Energy intake, 171–172, 173t, 176
Entiat fire, 71
Environment, 14, 17–18
  health and, 222, 694–698
  modification, 125
  species composition, 267, 270–271
Environment, Power, and Society, 689
Environmental crime, 704
Environmental manipulation, 609–610
Environmental movement, 651, 704–705
Environmental planning, 702–703
Environmental protection, 648–649
Environmental Protection Agency (EPA), 652, 655, 671
Environmental resistance, 91
Environmental science, 8
Environmental studies, 8, 648–649
Epidemic, 212–213
Epilimnion, 563, 567
Epiparasite, 255
Epiphytism, 222, 272, 448, 532, 545, 550f
Equilibrium landscape, 402–403, 431
Erosion, 337, 354, 485
Errington, Paul, 103, 112, 171, 190
Eskimos, 467, 468
Estuaries, 579–580
Ethanol, 690
Ethology, 28–29
Eucalyptus, 74
Eucaryote cell, 260–262
Eucaryotes, 260
Euler-Lotka equation, 100
Eulittoral zone, 577
Euonymus, 32–33
Eusociality, 126–127, 154–156
Eutrophication, 361, 363f, 567–569, 568f, 666
Evaporation, 363f, 364
Evaporative power of the air, 48, 49f
Everglades, 450–451, 453f
Evolution, 8, 133–148
  of altruism, 151–156
  of behavior, 148–160
  competition and, 290–293
  of the earth's atmosphere, 369–372
  of eucaryote cell, 260–262
  fitness and, 29
  of foraging strategy, 171–176
  of interference, 111
  of life history traits, 137–148
  of mating systems, 127–129
  of mimicry, 198–199
  of mutualism, 259–260
  parasite-host, 220–222
  of plant defenses, 178–180
  and optimization, 176–177
  and succession, 386–387
Evolutionary ecology, 4
Exclosures, 182–184, 387
Exotic species, 638–639
Exploitation, 109
Extirpation, 610, 637
Extinction, 165–166, 342, 405, 610–620
  case study, 619–620
  competition-caused, 238, 239f
  deterministic causes, 613–614, 614f
  versus immigration, 303
  in island communities, 303–306
  parasite-caused, 217, 219, 220
  of Pleistocene megafauna, 421–423
  population size and, 629–632, 633f, 634f
  process of, 614f
  stochastic causes, 613–619, 614f
  threshold, 619
Exxon Valdez oil spill, 467, 669–671, 669f, 670f

Falcon, peregrine, 662
Falsification, 10
Feeny, P. P., 268
Fenmen, 580, 648
Fens, 579–580, 596–597
Fermentation, 24, 647

Ferret, black-footed, 602f, 619–620, 620t, 635
Fescue, sheep's, 31
Fever, 30–31
Fidelity, 21, 270
Fieldfare, 265
Fig, strangler, 545, 551f
Figs, 160, 185, 258–259
Finches, 290, 291f
Fir, alpine, 482f
Fir, Douglas, 488, 490f, 491f, 495, 496, 498f, 498t
Fir, subalpine, 489f
Fire(s), 65–79
  adaptations, 510
  animals affected by, 71–72
  in chaparral, 505–507, 508f, 510
  crown, 66–67, 490
  cue, 248
  as ecological management, 75–79, 532, 532f, 611–612
  frequency, 72–73
  grassland, 517–518
  ground, 66
  in jack-pine, 76–77
  in longleaf pine ecosystem, 69, 75
  physical environment and, 70–71
  plant maintenance and, 67–69
  in ponderosa pine, 490
  in shrub-steppe, 534
  slash, 439, 495
  suppression, 78
  surface, 66
  in Yellowstone National Park, 77–79
Fire protection, 446–447, 490, 505
Fireflies, 200–202
Fish
  coral reef, 582, 583f
  mercury-tainted, 665, 695
  parasitism in, 206–207f
  pesticide poisoning in, 667–668, 668f
  rheotaxis in, 570
  sex ratios, 130
Fisher, 472
Fisher, R. A., 130, 137
Fisheries, 609
Fitness, 29, 105, 135–136
  absolute, 135
  cannibalism and, 203–204
  foraging and, 176
  relative, 135
Flatwood community, 449, 455
Flies, 199, 200f, 224, 225, 225f
Floristic relays, 382–385, 385f
Fluctuation, 377–379
Fluke, 209, 221
Fly, buffalo, 225
Fly, caddis, 570
Fly, carrion, 257t, 259
Flycatcher, Acadian, 624f
Flycatcher, great crested, 440f
Flycatcher, olive-sided, 500f
Flycatcher, pied, 33t, 129
Foin, T. C., Jr., 702
Food chain, 307–310, 327–328, 333f
  detritus, 314–315, 315f, 316f, 317f
  grazing, 314, 315f
  oceanic, 589
  of phytotelmata, 600
  of prairie, 528f
  of streams, 571–575
  of tundra, 463–464
Food industry, 692
Food web, 307–308, 308f, 327–328, 333f
Foraging
  central-place, 175, 176
  efficiency, 121–124
  energy and, 245
  optimal, 171–177, 171f, 328
  saltatory, 188
Forbes, Stephen A., 2, 79, 428
Forbs, 378, 379, 518f, 519f, 520, 521, 530
Forensics, 382
Forest, defined, 427
Forest edge communities, 442–445, 446t
Forget-me-not, alpine, 483f
Foreman, Dave, 705
Formation, 427, 430
The Formation of Vegetable Mould through the Action of Worms, 346
Fossil fuel, 355–356, 689–690
  pollution, 672–673
  subsidies, 336, 339t
  sulfuric acid and, 361
Fossil(s), 409
  evolution and, 133
  occurrences, 424f
  pollen sites, 414–415
  study of, 37, 407–408
Fox, arctic, 465
Fox, red, 170, 276f
Frass, 201f
Fragmentation, 443, 622–627, 624f
Freeze susceptibility, 43–44
Freeze tolerance, 43–44
Fretwell, Stephen D., 105, 332
Fricke, H. W., 580, 582
Frog, poison arrow, 200
Frugality, 687, 692
Frugivory, 160, 184–185
Fundamentals of Ecology, 4
Fungi, 223, 226–227
  ascomycete, 251
  in lichens, 250–251
  in mycorrhizae, 253–255
  in orchids, 255
Fungicides, 654
Fusion reactor, 691
Gaia hypothesis, 372
Gall formation, 177, 178f, 209
Galleta, 520, 523f
Game animals, 399, 604–606, 605f
Game farming or stocking, 608–609
Game management
  hunting as, 604
  practices, 606–610, 609f
Game Management, 191, 625, 643
Gamma-aminobutyric acid (GABA), 32
Gap phase replacement, 376–377
Gaseous cycles, 365
Gause, G. F., 108, 190, 191–192, 287
Gause's rule, 231
Gazelle, 152
Gene pool, 135, 152f, 616
Gene splicing, 182
General circulation models (GCMs), 680
Generalist species, 165
Genetic death, 136
Genetic diversity, 9, 604, 618, 637–638
Genetic drift, 134–135, 616–618
Genetic engineering, 203, 684–685
Genetic load, 136
Genetics, 9, 133
Genotypes, 18f, 19f, 133, 151f
Geographic Information System (GIS), 702–703
Geological timetable, 412t
Geophytes, 47t, 48f
Gerenuk, 557, 558f
Ghost of competition past, 291
Gila monster, 541f
Gilpin, Michael, 633
Giving-up time, 174
Glaciation, 414, 415–416, 415f, 485, 591f
Gleason, H. A., 284, 294, 381, 466
Glei, 461
Gleization, 62t
GLEMEDS, 667
Global change, 688–689
Global warming, 355–358, 424, 679–683, 688
Goat, mountain, 485f
Goldenrod, 178f
Goldfish, 14, 15f
Gopher, pocket, 347–348, 457
Gorbachev, Mikhail, 660
Gorilla, eastern mountain, 118
Goshawk, 121, 123f
The Grapes of Wrath, 378
Grass, blackoats, 70f
Grass, dune, 389–392, 393
Grass, grama, 522f
Grass, marram, 389–392, 390f, 392f, 393, 394, 394t
Grass, poverty, 386, 396
Grass, wire, 69f, 76
Grasshoppers, 71, 183, 279, 395

Grassland, 514–531
  abiotic factors in, 14, 15f
  animals, 522–526
  defined, 427
  desert, 521–522
  distribution and climate, 514–515, 515t
  energy flow, 528f
  fire and, 71, 525
  oak openings, 528
  plants, 515–519, 516f, 517f
  productivity, 322t
  restoration, 528–531
  soil, 61f, 62t, 515
  types of, 519–522
  variability, 139
Gravel, 63f
Graves, H. S., 500–501
Grazers, 177, 266, 312, 318, 387, 555
Great Drought, 378–379
Great Lakes, 170, 665–669, 666f
Greater ecosystem concept, 637
The Green Hills of Africa, 559
Green revolution, 693, 694
Greenhouse effect, 39, 355–358, 424, 679–683
Greenpeace, 8, 704, 704f
Grime, J. P., 146, 147
Grinnell, Joseph, 231
Grisebach, August, 427
Grosbeak, evening, 474f
Gross energy intake, 25
Gross primary production (GPP), 316, 324–325
Gross secondary production, 316–317
Groundwater, 364, 671–672
Group selection, 157–160, 159f
  intrademic, 158
Groupers, 250, 583
Grouse, red, 111–112
Guilds, 231, 267, 272
Guano, 246–247, 266–267
Gull, Bonaparte's, 473f
Gull, laughing, 28
Gum, tupelo, 453f
Guthrie, Woody, 378

Habitat, 14
  behavioral changes, 17
  distribution, 105, 106f, 232–238
  extreme, 298
  fragmentation, 443, 622–627, 624f
  loss, 166
  modification, 266–267
  patchiness, 35
  physiological changes, 16–17
  selection, 32–33
Habitat island effect, 622
Hadal zone, 574, 575f, 587
Haeckel, Ernst, 1
Haldane, J. B. S., 153–154, 369
Halley, Edmund, 214
Halophytes, 52–53, 585
Hamilton, W. D., 151–153
Hammocks, 447–448, 449f
Haplodiploidy, 126, 154
Hardin, Garrett, 156, 231
Hardy, G. H., 133
Hardy-Weinberg Law, 133–135, 617
Hare, arctic, 237–238, 237t
Hare, snowshoe, 191, 192f, 237–238, 237t, 472
Harper, J. L., 32
Hartebeest, 556f, 557
Haskell, E. F., 168
Haskell classification, 168, 247
Haustoria, 208
Hawks, 151–152, 273
Health, 222, 694–698
Heat gain, 41–42
Heat loss, 41–42
  by conduction, 42
  convectional, 41f, 42
  evaporative, 41f, 42
  infrared radiation, 42
Heather, pink mountain, 483f
Hediger, H., 583
Heinrich, Bernd, 172
Hemicryptophytes, 47t, 48f
Hemiepiphytes, 545, 551f
Hemingway, Ernest, 559
Hemlock, western, 496, 497f, 498f, 498t
Hemlock-white pine-northern hardwoods forest, 394t, 395–396, 438–439, 438f, 439f, 447–448
Hen, heath, 93–95, 94f, 95f, 166
Henry, Alexander, 525–526
Herbicides, 655
Herbivores, 177–178, 308, 311–312, 661
Herbivory, 177–186
  aquatic, 252
  defenses against, 178–182
  harvesting efficiency in, 324–326
  plant distribution and, 182–184
  types of, 177–178
Herding, 522
Heritability, 148–150
Herons, 203
Heterosis, 617
Heterotherms, 46
Heterotrophs, 23, 24, 370
Heterozygosity, 616–617, 617f
Hibernation, 17, 44, 46, 475
Hiemal aspect, 275, 276f
High plateaus, 488–490
Hildén, O., 32
Hodgson, J. R., 332
Holdfast, 577
Holistic medicine, 696–697
Holt, Robert, 183
Homeostasis, 16
Homeotherms, 42–45
Homoiotherms, 42–45, 207
Honeycreeper, 217
Hopkins, A. D., 275
Hopkins' bioclimatic law, 275–276, 279t
Hornaday, W. T., 606
Hornfly, 225
Horowitz, N. H., 370
Horses, 112, 414, 519
Horseweed, 396, 396f
Hosts, 222
Howarth, Robert, 330, 361
Hubbard Brook, 366–367, 367f
Human Immunodeficiency Virus (HIV), 215–216
Humboldt, Alexander von, 427
Humidity, 48–49, 352, 353
Humus, 59, 64, 344–345, 344f
  mor, 345, 346t
  mull, 345, 346t
Hunter-gatherers, 334, 467
Hunting, 467, 604–606
  big game, 559
  in game management, 604, 606–607
  ivory, 559
  legislative control, 604, 606, 607
  limits and closed seasons, 607–608
  as surrogate predation, 604
Hurricanes, 547
Hutchinson, G. E., 98, 160, 231, 281, 286, 287, 327, 567
Hutton, E., 407
Huxley, Julian, 199, 287
Hydric communities, 450–452
Hydrological cycle, 363–365, 363f
Hydroperiod, 52
Hydrophytes, 47t, 48f, 50, 52, 53f
Hydroseres, 353, 381
Hydrothermal vents, 251, 331–332
Hyena, 224, 557, 558f
Hymenoptera, 154, 155f
Hynes, H. B. N., 573
Hypertonic organisms, 54
Hypolimnion, 563–564, 567
Hypotheses, 10
Hypothetico-deductive approach, 286
Hypotonic organisms, 54
Hypsithermal interval, 417–418

Iatrogenic illness, 697
Ice nucleation, 44, 365
Ideal free distribution, 105
Iguana, desert, 30, 31f
Illness. *See* Disease(s)
Iltis, H. H., 642
Immigration
  versus extinction, 303–306

genetics and, 135
human, 700–701
illegal, 116
in island communities, 303–306
role in population, 81–82, 102–103, 104t, 116–117
Immune response, 694
Immunity, 212
Immunization, 216
Inbreeding depression, 617
Indians, American
buffalo hunting, 606
forest-dwelling, 442, 444t
view of nature, 644
Indicator species, 21–22
Indicators, 21–22
Indigo, wild, 258f
Indirect benefit, 266
Indirect effects, 265–267
Individual ecology, 13–39
Infectiousness, 212
Information transfer centers, 124
Infrared radiation, 42
Initial floristic composition model, 382–385, 385t
Injection wells, 671
Insect(s)
as carrion eaters, 224–225
DDT-resistant, 133, 133f, 654–655
eusocial, 126–127, 154, 326
food habits, 177t
pests, 502
pollination, 256, 257t, 258–259
season-affected, 230
Insectivorous plants, 594, 595f
Integrated pest management (IPM), 203
Interactions, 12f, 21, 167–202, 205–228, 229–249, 250–262
cascading trophic, 332
nontrophic, 169
parasite-host, 211–216, 220–221
trophic, 169–171, 267, 267f
Intercompensation, 112, 219
Intermediate disturbance hypothesis, 299–300
International Biological Program, 4
Intrinsic rate of natural increase (*r*), 87, 89–91, 96, 98–100, 137–144
Inuits, 467, 468
IQ, 150, 151f
Island biogeography theory, 300–306, 633
Island habitat, 166, 302–306, 304f, 306f
Isotonic organisms, 54
Iteroparity, 141

Jaeger, long-tailed, 467f
Janzen, Daniel H., 161, 187
Jasmine, rock, 483f
Jay, blue, 447f
Jay, pinyon, 512
Jay, scrub, 154
Jay, Steller's, 499, 401f
Jewell, Minna E., 674
Johnson, M. D., 383t
Joshi, A. C., 586
Joshua tree, 536f
Joule, 23
Juglone, 247
Jungle, 544

K selection, 137–145, 140f, 141t
Keddy, P. A., 300
Keith, Lloyd, 237
Kelp, 577
Kendeigh, S. C., 27, 34, 289, 431, 530
Kenoyer, Leslie A., 545
Kestrel, Seychelles, 232
Kettle curve, 35
Kettle lake bog, 591f, 593f
Keystone species, 265, 268, 629
Kilocalorie, 23
Kin selection, 152–156
Kitchell, J. F., 332
Kittiwake, 125, 125f, 246f
Klamath weed, 182, 202
Knotweed, 147
Krebs cycle, 260
Krill, 469, 588
Krummholz, 481
Kurtén, Bjorn, 409
Kuru, 204

Laboratory and Field Manual of Ecology, 11, 102, 302, 377
Lack, David, 287, 289, 290
Lacunae, 52, 52f
Laessle, A. M., 447
Lagg, 593–594
Lake(s), 561–569
acid, 674–676, 675t
algal growth, 20
beaches, 388, 394
chemolithoautotrophism, 329–331
desert, 542
dimictic, 564
energy flow, 567
eutrophic, 361, 567–569, 666
habitat groups, 565–567
meromictic, 564
origins of, 562
physical factors, 562–564
regions, 564–565
rehabilitation, 334
structure, 271
types of, 562
water levels, 394
zones, 562
Lake Michigan, 388f, 392, 393
Lambkin, Clark's, 219
Lamprey, sea, 38, 170, 208
Land ethic, 644–645
Land preservation. *See* Nature preserves
Land trusts, 622
Land use, 701–703
Landfills, 671, 686
Landscape ecology, 402–403, 431, 637
Landscape preservation, 621, 636–637
Larch, alpine, 489f
Laterization, 62t
Lead poisoning, 79
Leaf mining, 177, 209
Leaf-area index, 324
Leak, W. B., 367, 400
Leatherleaf, 592t, 593f, 596f
Leech, northern, 147
Leek, wild, 434f
Legumes, 380
Lek, 127, 185
Lemmings, 463, 465–466
emigration, 157, 158
population, 193, 194
Leopold, Aldo, 5, 47, 191, 444, 529, 534, 579, 606, 607, 625–626, 643, 644–645
Leslie, P. H., 99
Leslie matrix, 93, 101–102
Let-burn policy, 77–78
Levins, Richard, 114
Lianas, 545
Lichens, 250–251, 251f, 260
air pollution and, 14, 672f, 673
Life, origin of, 369
Life expectation, 83
Life form system, 47–48, 47t, 48f
Life tables, 82–86
age-specific, 83
composite, 83
Pleistocene cave bear, 410
time-specific, 83
U.S., 84–85t
Lignin, 344, 345f
Light, 54–59, 55f
Light intensity, 348–351, 349t
Lightning, 594
Likens, G. E., 320, 403
Lily, swamp, 201f
Lily, trout, 433f
Lily, water, 53f
Liming, 679
Limiting factors, 20–21, 171, 216–220
Limiting similarity, 246, 292–293
Limnetic zone, 564, 565f
Limnology, 561
Limpet, 570
Lincoln, Abraham, 4, 5
Lindeman, Raymond L., 4, 327

Lindeman's efficiency, 327
Lindsey, A. A., 621
Lion, 557, 559f
Litter, 228, 367
  fire and, 70f, 72–75
  in soil formation, 344–345, 344f, 345t
Littoral zone, 564, 565f, 574–578
Lizard, 43, 539, 541t
Loam, 64
Lobster, American, 21
Locoweeds, 519
Locusts, 191
Loevinsohn, M. E., 650
Logan, K. T., 55
Logging, 217, 368, 491, 498, 504, 639
Logistic curve model, 92–93
Long-day plants, 57
Longevity, 83, 87t
Lorenz, Konrad, 28
Lotka, A. J., 89, 238
Lotka-Volterra interspecific competition model, 240–244
Lotka-Volterra predator/prey model, 188–190, 191
Louck, O., 399
Lovelock, J. E., 372
Lovins, Amory, 690
Lovins, Hunter, 690
Lucas, H. L., Jr., 105
Luminescence, 251
Lunarphobic organisms, 272
Lyell, C., 407
Lynx, 191, 192f, 472

MacArthur, Robert, 137, 172, 174, 190, 285–287, 303, 404
MacArthur-Wilson biogeography model, 303–306
MacDougall, W. B., 260
Mack Lake fire, 78
Macroparasites, 208
Madrid Protocol, 469
Magnolia, 449f
Mahogany, mountain, 505, 506f
Malabar Farm, 5
Malaria, 697
Malathion, 653, 654
Malnutrition, 221
Malthus, Thomas Robert, 117
Malyngamides, 181
Management indicator species, 504
Mangroves, 52, 53, 354, 585–587, 586f
Mannakin, 549
Mantis, praying, 165
Maple, sugar, 295, 295f
Maple-basswood forest, 438
Marginal value theorem, 173, 174f
Marl, 353
Marmot, 483, 483f
Marshes, 450, 562, 579–580, 589–590, 590f
Martin, P. S., 422
Mating systems, 127–129, 134
Matrices, projection, 101, 377
May, J. M., 697
May, Robert, 195–196
May-apple, 640–642
McHarg, Ian, 392, 703
McPeek, G. A., 639
Meat production, 334, 337
Mechanical weathering, 59
Medicine, 642, 696–697
Mediterranean climate, 505
Melanism, 133
Mercury, 660, 665, 673, 695–696
Mesic forest, 48, 280t, 295f
Mesophyte, 50, 50f, 52
Mesophytic forest, 435–437
Mesquite, 521–522
Metabolic rate
  standard, 25, 26f
  temperature and, 44f, 45f
Metalimnion, 563–564
Metapopulation, 114, 632–633
Methane, 679, 680t
Microhabitat, 14, 45
Microparasites, 208, 212f
Microsere, 224–227, 382, 383t
  carrion, 382, 383t
Midden, 408, 421
Midge, Vanderplank's, 145
Midgefly, 601
Migration
  bird, 27, 59, 278–279, 627–629
  due to changed environments, 17
  neotropical, 627–629
  post-deglaciation, 420f
  of savanna fauna, 557–559
Mikesell, Thomas, 274
Milkweed, 181f
Miller, S. L., 369
Millipede, 170
Mima mounds, 347
Mimicry, 198–202, 259, 583
Minamata disease, 695
Mineralization, 344–345
Minerals, 254
Minimum viable population (MVP), 629–631
Mining, 484, 485
Minnesota Study of Twins Reared Apart, 148
Missing production, 341
Mistletoe, 208
Mites, 43, 220, 223, 225, 226, 251
Mitigation, 639
Mitochondria, 260, 261f
Mitosis, 261
Moa, 166
Model making, 93
Mohr, C. O., 382
Moisture, 48–54
Monera, 260
Mongoose, 202
Monkey, howler, 551, 552f
Monkey, New World, 547, 551, 552f
Monkey, proboscis, 547
The Monkey Wrench Gang, 705
Monoclimax hypothesis, 400
Monogamy, 127, 129
Montane forest, 485–486, 501t
Montreal Protocol, 684
Moose, 217–219, 472, 472f
Morphology, 407
Morris, D., 472
Mortality rate, 82
Mosquito, anopheles, 697
Mosquito, Asian tiger, 600
Mosquito, night-flying, 217
Mosquito, treehole, 219–220
Mosquito, yellow fever, 32
Mosquito fish, 542
Mosquitoes, 209, 223, 269, 475, 600
Moss, peat, 64, 592, 596f
Moss, Spanish, 222
Moth, gypsy, 654
Moth, peppered, 133
Moth, polyphemus, 269, 269f
Moth, plume, 289–290, 289t
Moth, sphinx, 209
Moth, tiger, 164, 165
Moth, yucca, 160
Moths
  bats and, 162–165
  phenotypic plasticity in, 18
  as pollinators, 256, 257t, 273
Mountain zonation, 477–478, 478f, 554
Mouse, golden, 283, 283f
Mouse, white-footed, 221, 227, 283, 283f
Muir, John, 521
Muller, C. H., 248
Müller, F., 199
Müllerian mimicry, 199
Murie, A., 473
Murre, McKinley, 82–83, 82t
Murrelet, marbled, 502, 502f
Mushroom, stinkhorn, 224
Muskrat, 171, 590
Mussel, zebra, 37f
Mussels, 576f, 577
Mustard, 270
Mutation, 133, 135
Mutch, Robert W., 73
Mutualism, 250–262
  alga-fungus, 250–251
  ant-plant, 182
  blowfly-beetle, 22

evolution of, 259–260
exploitation in, 256–259
grouper-wrasse, 583
in nitrogen fixation, 359
nonsymbiotic, 251–253
nutcracker-whitebark pine, 500
symbiotic, 250–251
Mycorrhizae, 161, 253–255
Myrtle, 474f

Natality rate, 82
National Forest Management Act, 504
Natural area
management, 637–640
defined, 621
preserved, 621–637
restoration, 639–640
Natural burn policy, 77
Natural Resources Defense Council, 8
Natural selection, 29, 133, 135–148
competition-based, 239, 240f, 249
directional, 136f, 137
disruptive, 136f, 137
Fisher's fundamental theorem, 137
seasonal change and, 278
stabilizing, 136–137
Naturalist's Guide to the Americas, 428
Nature centers, 622
Nature Conservancy, 8, 428, 622, 639, 704
Nature preserves, 622–637
buffer zones, 632
corridors, 636
habitat fragmentation, 622–627
landscapes, 636–637
management, 637–639
multiple, 632–636
restoration and, 639–640
size determination, 622–632
Needleleaf forest, 350
Negative externalities, 648–649
Negative feedback, 103
Nekton, 566–567, 588
Nelson, E. W., 500–501
Neotropical migrants, 552, 627–629, 628f
Neritic zone, 575, 575f, 587–588
Net primary production (NPP), 316, 317–318, 320–321, 322t, 324–325, 341t
nutrient limitation of, 361–363
Net reproductive rate (R_0), 100, 100t, 211–216, 635f
Neuston, 567
Neutralism, 248–249
Niches, 230–231, 267, 279–284
fundamental, 281–282
as a hypervolume, 281–283
Nichols, F. H., 579
Nitrification, 359
Nitrogen cycle, 358–359, 358f
Nitrogen fixation, 251, 252f, 253, 359, 362–363, 593, 594
Nocturnalism, 272–273
Non-equilibrium processes, 112–114, 194–197, 299–300, 402–403
Northern hardwood forest, 394t, 395–396, 438–439, 447–448
Nuclear accidents, 659
Nuclear membrane, 260–261
Nuclear power, 657–659, 690–691
Nuclear Regulatory Commission, 691
Nuclear war, 659–660
Nuclear waste, 657–659, 664
Nuclear winter, 660
Nucleation particles, 365
Nurse plants, 537, 538f
Nutcracker, Clark's, 500
Nuthatch, pigmy, 272
Nuthatch, red-breasted, 272
Nuthatch, white-breasted, 271–272
Nutrient cycling, 354–363

Oak, black, 395
Oak, burr, 528, 530f
Oak, shrub live, 510, 510f
Oak, white, 186f
Oak openings, 528
Oak-chestnut forest, 437
Oak-hickory forest, 437
Oak-pine forest, 438
Oaks, 68f, 178, 179f, 556f
Oats, 338f
Observation
direct, 10–11
field, 14
Oceanic zone, 574, 575f, 587–588
Oceanography, 574
Oceans, 574–589
history of, 369
in hydrological cycle, 363–364
life zones, 574–575f
productivity, 588–589
reactions, 353–354
salinity, 574
shores, 577–579
as steady-state control, 371
vent communities, 331–332
Ocotillo, 536
Odum, Eugene P., 4, 231, 279, 324, 365, 580, 692
Odum, H. T., 319, 333, 689, 692
Oil industry, 671, 689–690
Oil Producing Exporting Countries (OPEC), 689–690
Oil spill, 467, 669–671, 669f, 670f
Old-field ecosystem, 320f
Old-field succession, 373f, 381, 386, 396–398, 397f, 398t
Old-growth forest, 343f, 402–403, 496, 498, 502–505, 640
Oligotrophication, 363f
Ombrotrophism, 594
On the Origin of Species, 2, 5, 11, 137
Oozes, calcareous and siliceous, 354
Oparin, A. I., 369
Open-mindedness, 11
Opossum, 37f
Optimization, 176–177
Optimum range, 15–16
Orchids, 255, 259
Organic farming, 692, 693
Orians, G. H., 128
Orphans, 152
Ortstreue, 34–35
Osmotic concentration, 54
Osprey, 167f
Ostrich, 123, 525, 557
Our Vanishing Wild Life, 606
Outbreaks, 405
Ovenbird, 628f
Overgrazing, 21–22, 512, 521, 534, 560
Overhunting, 166, 444
Overkill hypothesis, 423
Overpopulation (human), 111, 698–701
Overshooting, 607
Overturn, of lakes, 563–564
Owl, barn, 232
Owl, great gray, 475
Owl, northern spotted, 503–505
Owl, snowy, 465
Owl, spotted, 77
Oxygen, 370–372
Oxygen depletion, 564
Ozone, 673, 680t, 683–684
Ozone hole, 469, 683–684
Ozone layer, 55, 673

Paleocommunity, 407–408
Paleoecology, 406–425
Palm, cabbage, 449f
Palmetto, 454f
Paloverde, 536, 538f
Palynology, 285
Paradigm shift, 10
Paradigms, 10
Parakeet, Carolina, 638
Paramecium, 108t, 190, 191–192
Parasite-host interactions, 211–216, 220–221, 265–266
Parasitism, 112, 168, 206–222
cowbird, 626
ecology, 209–210
effect on host numbers, 216–220
harm to host, 221–222
helminth, 206f
of humans, 211t
as mutualism, 252, 259–260

Parasitism (*cont.*)
prevalence, 206–208
rates of increase, 211
types of, 208–209
Parasitoids, 208–209, 270
Park, O., 273
Park, Thomas, 4, 220, 238
Parker, M. A., 183
Parkinson's disease, 696
Parrotfish, 582, 582f
Particulate organic matter (POM), 228
Partitioning, 160–161
Partridge, 609
Passionflower, 202
Pathogens, 212, 663
Patterned ground, 461
Peach, 32–33
Pearl, Raymond, 4
Pearse, A. S., 222, 577
Peeper, spring, 13f
Pelagic zone, 588
Percentage base saturation, 60
Perennials, 50, 141–144
Periphyton, 565
Permafrost, 70–71, 461, 467, 470, 476f
Permanent wilting percentage, 63–64
Personal distance, 120
Pest control, biological, 640
Pesticide(s), 202, 649–656
agricultural use, 203, 650, 655–656
ecosystem effects, 654
herbivory and, 182
human toxicity, 650–651
nontarget organisms and, 651–652
persistence and accumulation, 652
predator loss and, 607
resistance, 654–655
spreading, 652
sublethal effects, 652–653, 696
synergistic effects, 653–654
types of, 650
Peyote, 537
pH, 14, 674–675, 675t
Phainopepla, 47
Phanerophytes, 47t, 48f
Pharmaceuticals, 642
Pheasant, ring-necked, 88, 99–100, 526, 609
Phenology, 202, 275–279
Phenotypes, 18–19, 133, 151f
Phenotypic plasticity, 18–19, 147
Pheromones, 203, 268, 269
Philopatry, 34–35
Phlox, Drummond's, 86f
Phoresy, 223
Phosphorus, radioactive, 662–663
Phosphorus cycle, 358–361, 360f
Photic zone, 574
Photochemical smog, 672
Photoperiod, 30, 57–58
Photorespiration, 24
Photosynthesis, 23–24
bacterial, 24
sunlight and, 324
temperature and, 15, 16f
in vent ecosystem, 331
Phreatophytes, 51, 364, 555
Phycomycetes, 254
Phylogenetic trees, 9
Physiognomy, 426–427
Physiological ecology, 40
Phytogeography of Nebraska, 2, 384
Phytoplankton, 565, 587–588
Phytotelmata, 600–601
Pigeon, passenger, 148, 166, 399, 444, 605, 619, 644
Pig, wild, 346–347, 609
Pika, 483
Pimentel, David, 336, 656
Pimm, Stuart, 327
Pine, jack, 69f
community, 394t, 395, 439, 634–635
in dune community, 393
fire dependent, 72
Kirtland's warbler and, 76–77, 105, 439, 611–612
migration, 421f
pollination, 256f
Pine, loblolly, 447, 448t
Pine, lodgepole, 66f, 77, 490
Pine, longleaf, 75, 76f
ecosystem, 449–450, 451f, 629, 639
fire and, 69f
Pine, pinyon, 512
Pine, ponderosa, 479f, 491
Pine, red, 421f
Pine, sand, 121t, 452
Pine, slash, 449f, 452f
Pine, white, 179f, 422f, 438–439, 639
Pine, whitebark, 489f, 500
Pinyon-juniper woodland, 510–513, 511f
Pioneer community, 144–145, 379–380
Pitcherplant, 223, 450f, 594, 595f, 600
Plankton, 565–566, 571, 587–588
Plant distribution, 182–184
Plant Ecology, 21, 384
Playas, 532, 534f
Pleistocene, 414–423
cave bear, 409–411, 410t
climate, 419
glaciation, 414, 415–416, 415f
megafauna extinction, 421–423
Pneumatophores, 585
Poaching, 604, 640f
Podzolization, 62t, 470
Poikilotherms, 42–45, 207
Poisons, 695–696
pesticides as, 650–651
protective, 170
Pollination, 251, 255–256, 257t, 258f
animal, 224, 255, 256
wind, 255–256
Pollution, 79–80, 646–687
biological magnification, 661–663
environmental, 663–686
as global problem, 647–649
human contribution to, 647–649
indicators, 22
types of, 649–661
Pollution control
devices, 673
legislation, 678
technology, 673, 678
Polyandry, 127
Polybrominated biphenyl (PBB), 660–661, 692
Polychlorinated biphenyls (PCBs), 660, 667
Polyclimax concept, 400
Polygamy, 127–129
Polygyny, 127–128, 185, 525
Polygyny threshold model, 128, 128f
Pond zonation, 569
Ponds, 1f, 353, 542, 561–569
Pools, 570–571, 573
Popper, Karl, 10
Population(s)
bottleneck, 617
carrying capacity, 91, 106–108
change factors, 81–82, 82f
cyclic, 190–194, 193t
dispersal and, 36
energy balance and, 27–28
evolution of, 133
global, 114–117, 701
immigration and, 700–701
intercompensatory, 112
local and national, 698–700
optimal, 698
production and, 342
projection, 116f
The Population Bomb, 699
Population density, 102–108
density-dependent factors, 102–105, 193–194
extrinsic, 108
fluctuations, 190–198
habitat distribution and, 105, 106f
intrinsic, 108
nonequilibrium, 112–114
social pressure and, 111
Population ecology, 4, 28, 168, 409–411
Population equivalents (PE), 663
Population growth, 86–102
age structure, 131
Allee effect in, 93–95, 100–101
biotic potential, 87, 89–91
exponential, 86–89, 96

human, 114–117
intrinsic rate of natural increase, 89–91, 90f, 91t, 99–100
logistic, 91–93, 96–100
mathematical treatments, 95–102
net reproduction rate, 100, 100f
suppressed, 132
U.S., 111–117
Population organization, 118–133
age structure, 130–133
mating systems, 127–129
sex ratio, 129–130
of social organisms, 126–127
sociality, 120–126
spacing, 118–120
Population regulation, 102, 108–109, 157–158
human, 114–117
Population viability assessment, 629–630
Population vulnerability analysis (PVA), 629–630
Porcupines, 176, 472
Poulton, E. B., 200
Pound, Roscoe, 384
Power, Mary E., 267
Prairie(s)
bunchgrass, 521
California, 521
climate, 515t
coastal, 520, 527
fires and, 75–76, 77t
grass types, 516f, 517t
hill, 437
life forms, 48
mixed, 520, 527t, 528
Palouse, 521
potholes, 527
restoration and reconstruction, 528–531, 532f
shortgrass, 520, 522f, 523
structure, 271, 271t
stump, 439
tallgrass, 519–520, 521t, 527, 528–529
Prairie dog, black-tailed, 526f
Prairie dog, white-tailed, 619–620
Predation, 187–198
actively-moving, 188
as biological control, 202–203
as diversity factor, 299
evolution and, 146
fire and, 71
foraging and, 176
functional response, 197, 198f
intraguild, 204
keystone, 265
Lotka-Volterra model, 188–190
numerical response, 197, 198f
plant, 188
population regulation and, 108, 112, 171
prey and, 170–171
responses, 197–198
seed, 185–187
sit-and-wait, 188, 188f
surrogate, 604
Predator control, 607, 608f
Predator protection, 154, 155
advantages, 121–124
confusion, 121
geometrical effects, 121
herding, 522
mobbing, 121
mutual vigilance, 121
physical, 121
seed production and, 187
Preserves, 608. *See also* Nature preserves
Prevernal aspect, 274
Prisere, 382
Producers, 308, 309f, 356f, 360f
Production/Biomass (P/B) ratio, 321–324
Production rates, 316
Productivity, 316–324
biomass and, 321–324, 322t, 323t
human expropriation of, 341–342
grassland, 322t
major ecosystem, 322t
oak-pine forest, 323t
savanna, 323t
stress and, 146
Profundal zone, 564, 565f
Promiscuity, 127, 185
Pronghorn, 514f, 526, 538
Protective devices, 170, 198–202
Proximate factors, 30
Pseudomonas syringae, 685
Ptarmigan, 465
Public planning, 701
Pursuit time, 174–175
Pyramid of biomass, 310, 310f, 313, 315
Pyramid of numbers, 310, 310f, 313

Quail, 608

r selection, 138–145
Rabenold, K., 154
Radiation
background, 656–657
electromagnetic, 54–55, 55f
high-level, 657
infrared and heat, 41f, 42
low-level, 657
sickness, 657
solar, 41f, 42, 54–55, 194, 310–311, 348–352
ultraviolet, 54, 194, 369, 469, 683, 684
Radical environmentalism, 703–705
Radioactive materials, 691
doses and effects, 658t
sources, 656–660
Radioactive waste, 657–659
Radioisotopes, 662–663
Radiophosphorus studies, 255, 360
Radon, 657
Rafflesia, 208
Ragweed, 396, 396f
Rail, Sandwich Island, 166
Rail, Virginia, 590f
Rain shadow, 478, 490, 492f
Rainfall, 352, 352f, 353, 364, 364f, 365
Ramensky, L. G., 294
Range expansion, 37–39
Ranson, R. M., 99
Rare species, 610
Rat, naked mole, 127
Raunkiaer, C., 47–48
Raven, P. R., 160
Razorbill, 246
Reaction norms, 18, 150–151
Reactions, 12f, 343–372
on air, 348–353
in fresh water, 353
on land, 344–348
in the ocean, 353–354
role in succession, 382–386
Reading the Landscape of America, 22
Reciprocity, 156
Recombinant DNA, 684–685
Recreation, 484
Recreational vehicles, 543
Recycling, 686–687
Red spruce decline, 488
Redfield, A. C., 371
Redwood, coastal, 411, 494–495, 495f, 496–497, 499f
Redwood, dawn, 406f, 411
Refuges, 608
Reiners, W. A., 368
Relatedness. *See* Kin selection
Releasers, 29
Remmert, Hermann, 165
Remora, 222
Replacement, 374–377
cyclic, 374–375
of forest canopy, 375–377
Reproduction, 29
Reproductive rates, 138–141
Rescue effect, 633
Resource, 229–230, 244–246, 249, 298–299
Respiration, 24, 309
aerobic, 24, 370
anaerobic, 24, 370
autotrophic, 324–325
versus production, 321f
temperature and, 15
Restoration, 639–640

Restoration ecology, 639
Retroviruses, 216
Reznicek, A. A., 300
Rhea, 525, 557
Rheotaxis, 570
Rhinoceros, black, 640f
Richards, P. W., 274, 586
Richardson, R. E., 79
Riffles, 570
Rinderpest, 555, 559
Riparian woodland, 534, 541–542
Rivers, 572–573
Roadrunner, greater, 539, 540f
Robertson, W. B., Jr., 451
Robin, American, 148, 307f
Rocky Mountains, 488–490
Rodents, 231, 266
Roosevelt, Theodore, 603
Root, R. B., 183
Root eating, 177
Rosenthal, G. A., 180
Rosenweig, Michael, 190
Roundie, Schroeter's, 252–253
Roundworm, 209
Ruderals, 146
Rumination, 524

Sacred Sands: the Struggle for Community in the Indiana Dunes, 388
Saddled prominent, 201
Safe site, 32
Sagebrush, 534, 535f, 535t
Sahara desert, 543
Saguaro, 258f, 536–537, 537f, 538f, 629
St.-John's-wort, 182
Salamander, tiger, 203, 204
Salination, 664
Salinity, 52–53, 574
Sampson, Arthur, 510
Sand, 63f, 64
Sand County Almanac, 47, 643, 644
Sand dropseed, 20f
Sand dune(s)
- animal succession, 394–396
- complex, 393–394
- embryonic, 389, 390f, 392
- fore-dune, 389, 391f, 392
- formation, 389–392
- succession, 387–394, 391f, 394t
- vegetation, 393–394
- wandering, 391f, 392–393
- water levels and, 394

Sandhill community, 448, 455–457
Sap sucking, 177, 209
Saprobism, 205f, 223–228
Satiation, 187, 197
Savanna. *See* Tropical savanna
Sawfly, European pine, 197
Sawgrass, 450, 453f
Scale, California red, 232
Scavengers, 71, 223–225
Scepticism, 10–11
Schaffer, W. M., 197
Schimper, A. F. W., 52, 427
Schistosomiasis, 209, 210f, 211, 266
Schoener, T. W., 289
Schulenberg, R. F., 530
Scientific inference, 9–11
Scientific method, 9–11
Sciophytes, 351
Scott, Robert Falcon, 468
Scrub, sand-pine, 452, 454f
The Sea Around Us, 651
Sea hares, 181
Sea rocket, 14, 15f, 390f
Search images, 175
Sears, Paul, 5
Seasonal affective disorder (SAD), 57
Seasonal change, 274–279
- in deciduous forest, 274–275
- in Sonora Desert, 277t
- study of, 275–279

Seasonal forest, 554, 555f
Secondary compounds, 180–181
Secretary bird, 557
Sedge, 450, 453f
Sedge, hairy-fruited, 592f
Sedimentary cycles, 365
Seed bank, 17, 145, 145t, 398
Seed dispersal, 144–145
- by fire, 69, 145

Seed predation, 185–187
Selfish herd, 121, 121f
Selfishness, 151–152
Semelparity, 141, 144, 147
Sequential comparison index (SCI), 301–302, 302t
Sequoia, giant, 411, 491–494, 494f, 495f, 496f
Seral stages, 380–381
Sere, 380–382
Serotinal aspect, 274–275
Serotiny, 69, 69f, 145, 439, 471–472, 490
Seton, Ernest Thompson, 193, 475, 605
Sewage, 79f, 663
Sex ratios, 129–130
Sexual dimorphism, 291–292
Sexually transmitted diseases, 215–216
Shadbush, 387
Shade tolerance, 55–56t
Shading, 348–352, 440
Shaffer, M. L., 629–630
Shannon-Wiener index, 302
Shantz, H. L., 21
Sheathbill, wattled, 469
Sheep, mountain (bighorn), 484f
Shelford, Victor E., 2, 93, 191, 194, 266, 285, 353, 382, 388, 395, 428, 430, 431, 472, 530, 573
Shootable surplus, 604
Short-day plants, 57
Shredders, 571–572, 571f
Shrew, short-tailed, 283, 283f
Shrub-carr, 597–598
Shrublands, 505–513
Shrub-steppe, 521, 524f, 534
Siblicide, 203
Sierra Club, 704
Sierra Nevada, 490–494
Sigmoid growth curve, 91
Silent Spring, 5, 651
Silt, 63f
Silverside, Atlantic, 130
Simberloff, Daniel, 288, 304, 505, 632
Simpson index, 302
Sink habitat, 114, 634–635, 635f
Siphonophores, 126
Site tenacity, 34–35
Skinks, 232, 269
Skunks, 273, 273f
Skutch, Alexander, 154
Slagsvold, T., 265
Slayter, R. O., 386
Slider, pond, 408, 418
Sloth, 549
Smith, J. S., 102
Smog, 672
Snags, 227
Snakeroot, white, 31
Snakes, 298, 539, 541t
Snakeweed, broom, 183
Soapweed, 522f
Social facilitation, 123
Social organisms, 126–127
Social pressure, 111
Sociality, 120–126
Society, defined, 120
Society for Conservation Biology, 603
Sociobiology, 150–157
Sociobiology, 157
Sodium chloride, 21, 21f
Soil, 59–65
- acidification, 676
- erosion, 337
- field capacity, 63, 64
- formation, 59–60, 62t, 344–348
- light, 60
- moisture, 48, 348
- names, 64–65
- orders, 65t
- profile, 59, 60t, 61f
- saline, 52
- structure, 59–60
- texture and fertility, 60–63, 64f

Soil series, 64–65
Soil water, 63–64, 360
- available, 64
- capillary, 63

gravitational, 63
hydroscopic, 63
Solar energy, 310–311, 311f, 691
Solar radiation, 42, 54–55, 194, 310–311, 348–352
Solid waste, 686–687
Solifluction, 481
Sonar, 162, 164, 165, 599
Sonora desert, 277t, 537, 629
Soulé, Michael, 603, 632
Source habitat, 114, 634–635, 635f
Southeastern evergreen forest, 446–458, 448f
animals, 452–458, 455t
climate, 432t
hardwoods, 447–448
pinelands, 448–450
vegetation, 447–452
Spaceship Earth analogy, 705–706
Spatial diversity, 192
Spacing, 118–120, 125
Sparrow, grasshopper, 525f
Sparrow, house, 37, 202
Sparrow, tree, 477f
Specialization, 165
Species composition, 267, 270–271, 287–289, 320f
Species diversity, 296–306
Louck's model of, 399f
Species-area relationship, 304, 305f, 622, 623f
Species/genus ratio, 287–288, 288t
Specific dynamic action, 25–26
Sphagnum, 590, 592–593, 592t, 593f, 594f
Spider, garden, 188
Spider, vagabond, 395
Spider, web-weaving, 395
Spiking, 705
Spittlebug, 289–290, 289t
Sponges, 222, 248
Spring Ephemetals, 285, 432–434
Spruce, black, 462f, 476f
Spruce, red, 487, 488
Spruce, white, 474f
Spruce forest, 416, 417
Spruce-fir forest, 487
Sprugel, Douglas, 403, 504
Squirrel, fox, 443, 445f
Squirrel, gray, 176, 227f, 235–236, 443, 445f
Squirrel, ground, 125–126
Squirrel, red, 235–236
Squirrel, southern flying, 440f, 456
Stability, 399, 403–405
diversity and, 404–405
resilience, 403
resistance, 403
Stand history hypothesis, 488
Standard metabolic rate (SMR), 25, 26f
Starfish, crown of thorns, 584f, 585
Starlings, 37
Starvation, 170
Steady state, 315, 318–320, 320f, 398
Steady state control, 371, 371f
Steinbeck, John, 378
Steppes, 48
Sterilization, 129
Stevenson, Adlai, 705
Stewardship, 644
Stickleback, three-spined, 29, 136–137
Stilt roots, 585, 585f, 586
Stochastic extinction, 613, 614f, 615–619
demographic, 615f, 616
environmental, 618–619
genetic, 616–618
models, 615–616
Stochasticity, 196, 615–616, 632
Stoddard, Herbert, 75
Stokes' law, 566
Stomata, 49–50
Storehouse effect, 634–635
Stork, dry-land, 557–559
Stotting, 152
Stratosphere, 683–684
Stream(s), 486f, 569–574
acid, 674
downstream drift, 571
energy, 317f, 319, 571–575
food chains, 314–315
intermittent, 573
physiographic succession, 573–574
pollution, 663
pools, 570–571, 573
riffles, 570
swift, 573
technological fix for, 673
Stress
disease and, 695
population and, 193
productivity and, 146
Stress tolerators, 146
Subalpine forests, 486–505
animals, 498–505
characteristic birds, 501t
Sublittoral zone, 574, 575f
Subtidal zone, 576, 577
Subsere, 382
Subtropical forests, 554
Succession, 379–398
allogenic factors, 382
animal, 394–396
autogenic factors, 382
bog, 590–594
causes, 382–387
defined, 379
evolutionary view, 386–387
models, 386
old-field, 373f, 381, 386, 396–398
physiographic (streams), 573
primary, 382
rain forest, 546t
sand dune, 387–396
secondary, 382
Succulents, 50
Sulfur dioxide, 14, 672f, 674–679
Sulfur cycle, 361, 362f
Summer kills, 564
Sundby, R. A., 232
Sundew, round-leaved, 595f
Sunflecks, 349–350, 350f
Sunlight, 324–325
Supralittoral fringe, 576
Supratidal zone, 576, 578
Suppressants, 269
Survivorship, 29
Survivorship curves, 84–86, 214f
Susceptibles, 212, 213f
Sustainable agriculture, 693–694
Sustainable Biosphere Initiative, 8, 610, 688
Svärdson, G., 32, 187
Swallow, barn, 122f
Swallow, cliff, 124, 202
Swamps, 451–452, 453f, 590, 598
Swan, L. W., 601
Swander, L., 147
Swift, Vaux's, 502–503
Swordfish, 45–46
Symbiosis, 205–228, 250–262
cleaning, 250f, 259, 582–583
in nitrogen fixation, 359
Sympatric area, 233–234
Synergism, 21, 653–654
Synusia, 272

Taiga, 682f
Tail flagging, 162
Tamarack, 593f
Tamarisk, 52
Tannins, 179–180
Tansley, A. G., 264
Tapeworm, 209, 221
Taxol, 642
Taylor, W. P., 19
Technological fix, 182, 673, 673f, 678, 697
Temperate deciduous forest, 431–442, 448f
animals, 439–442, 443t
food plants, 444t
fragmentation of, 412–414, 443
history of, 412, 416–417
human occupation, 442, 444t
life forms, 48
location and climate, 431, 431t
old-growth, 343f
plants and vegetation, 48, 432–434
seasonal changes, 274–275

Temperature deciduous forest (*cont.*)
soil, 62t, 432
structure, 271
subdivisions, 434–439
Temperate grassland, 514–531
Temperate rain forests, 494–498, 497f, 498t
Temperature
as abiotic factor, 41–48
adaptations, 30–31f
altitude and, 478
global warming, 679–683
heat sum, 279
hydrothermal vent, 331, 332
range of, 15–17, 16f
reactions, 352, 353
seasonal change and, 278, 279
seed germination and, 19–20
sex determination and, 130
wind and, 46–47
Termites, 126, 549, 553f, 555–557, 557f
Terrain modification, 486
Terrestrial habitat, 14
Territoriality, 110, 111–112
Territories, 109–111, 128
Terrorism, 659
Tertiary, 412t
early, 411–412
later, 412–414
Text Book of Botany, 389
Thermal stratification, 563, 563f
Thermocline, 563
Thermoperiodism, 16, 20
Thermoregulation, 26, 43, 125, 538
Therophytes, 47t, 48f
Thicket, 427
Third World countries, 552–554, 684
Thistle, Russian, 379f
Thistle, yellow beach (Pitcher's), 392
Thorson, G., 578
Threadworm, 112
Threatened species, 148, 610
Three Mile Island nuclear accident, 659
Thrips, 112, 113f
Thrushes, 283–284, 284f
Tilapia, blue, 253
Timber industry, 439, 498, 504, 705
Timberlines, 477, 481, 481f
Time lags, 97–98, 193–194, 196
Tinbergen, Niko, 28
Tit, marsh, 233–234, 237t
Tit, willow, 233–234, 237t
Toad, American, 442f
Toad, Couch's spadefoot, 539
Tolerance Range, 14–15
Topminnow, Sonoran, 542
Topography, 346–348
Tortoise, gopher, 268, 457, 457f, 629
Tortoise, Hermann's, 43
Toucan, 549, 553f
Touchstone formula, 646–647
Towhee, Abert's, 58f
Toxic chemicals. *See* Chemicals
Toxic Substances Control Act, 660
Trait selection, 137–145, 141t
Trans-Alaska pipeline, 467, 468f
Transpiration, 49–50, 50f, 361, 364f
Tree(s)
acidification and, 767
age structure, 131–132
canopy replacement, 375–377
dead, 226–228
distribution, 295f, 294
fire adaptations, 75t
fire-affected, 67–69, 68f, 69f
habitat selection and, 32–33
longevity, 146–147
mycorrhizal unions, 255
regeneration, 183f, 184t
seed production, 186–187
shade and, 440
shade-tolerant, 55, 56t
Tree limit, 481
Trilobites, 165
Trivers, Robert, 156
Troglobites, 599–600
Trophic interactions, 169–171, 267, 267f
Trophic levels, 307–310, 328–329, 333f
Trophosome, 331
Tropical biomes, 543–560
Tropical rain forest, 543–554
animals, 547–551
destruction of, 551–554
extinction and, 166
life forms, 48
location and climate, 543, 544, 545f
plants, 544–547
soil, 62t, 543–544
successional communities, 546t
Tropical savanna, 554–560, 556f
animals and plants, 555–559
defined, 427
location and climate, 554–555
productivity, 323t
soil, 62t, 555
Tropopause, 478
Trout, brook, 15f
Trout, lake, 170
Trout, rainbow, 667t
Trumpeters, 549, 551
Tumblebugs, 226
Tundra. *See* Alpine tundra; Arctic tundra
Twin studies, 148–150, 149f
Twister, Roscoff, 251, 252f

Ultimate factors, 30
Ultraviolet radiation, 54, 194, 369, 469, 683, 684
Undercrowding, 93–95
Understory species, 300t, 349–350, 448t
Uniformitarianism, 406–407
United Nations Conference on Environment and Development, 8, 701
U.S. Corps of Engineers, 451
U.S.D.A. Comprehensive System, 64, 65t, 432, 470, 487, 511, 515, 533, 544
U.S. Fish and Wildlife Service, 589, 619
U.S. Forest Service
endangered and threatened species and, 504
enviromentalists and, 705
fire and, 65, 77
habitat conversions, 510
herbicide use, 655
timber practices, 456, 639
watershed studies, 366–367
Utida, Syunro, 230

Vaccines, 216
Vagility, 36
Van Beneden, P. J., 222
VanKat, John, 385–386
Vapor pressure, 49
Vapor pressure gradient, 49
Vectors, 265–266
Veery, 283–284, 284f
Vent ecosystem, 331–332
Verification, 10
Vestal, A. G., 50
Vigilance time, 123, 124f
Vireo, Bell's, 626
Vireo, black-whiskered, 586
Vireo, red-eyed, 102, 426f
Vitousek, Peter M., 361, 368
Volcanoes, 601
Vole, prairie, 418, 525
Vole, red tree, 503, 503f
Voles, 193, 194, 293f, 463, 465
Volterra, Vito, 238
Von Frisch, Karl, 28, 124
Vulture, bearded, 224
Vulture, turkey, 224, 266
Vulture, white-backed, 205f

Walnut, black, 247
Warbler, Bachman's, 628
Warbler, bay-breasted, 147
Warbler, Kirtland's, 72, 76–77, 78, 95, 105, 610–612, 612f, 613f, 626, 635
Warbler, Swainson's, 105
Warbler, yellow-rumped, 474f
Warren, Edward K., 640
Warren Woods, 433f, 434t, 640
Wasp, digger, 395
Wasp, fig, 160, 258–259
Wasp, horntail, 226

Wasp, otherwise-spotted, 230, 232, 233f, 234f, 235f
Wasp, scalesucker, 232
Waste disposal, 661, 671, 686
Waste heat, 663
Waste management, 686–687
Water
  agricultural and industrial use, 671
  of bog and fen, 593t
  density, 574
  drinking, 671–672
  gravitational, 364
  as a habitat, 562–563
  in hydrological cycle, 362–365
  lentic, 562
  lotic, 570
  reactions, 353
  salinity, 574
  soil, 63–64
Water balance, 49–54
Water pennies, 570
Water pollution, 451, 663–672
  nonpoint sources, 664–665
  pathogens, 663
  salination, 664
  sewage, 663
  waste heat, 663, 664
Water strider, 561f, 567
Water table, 70, 348, 364, 365f
Water vapor, 49
Water-flea, 94f
Waterlily, 592f
Watershed studies, 365–368, 366f
Watt, K. E. F., 606
Watterson, A., 651
Watts, May T., 22, 574
Weasels, 291–293, 293t
Weather, 112. *See also* Climate
Weathering, 59, 344
Weaver, John E., 21, 64, 271, 378, 379, 384, 402, 403, 530
Wechsler Adult Intelligence Scale, 150
Weevil, bean, 230
Weinberg, Alvin, 659
Weinberg, W., 133
Westlake, D. F., 598
Wetlands, 52, 450, 451, 579–580, 589–598, 639
Wheat, 338f
Wheatgrass, western, 51f, 379
Wheeler, William Morton, 198–199
Whittaker, Robert H., 268, 300, 320, 400
Whole-tree harvest, 368
Wilcox, Bruce, 603
Wilderness Society, 643, 704
Wildlife management, 19, 604–610, 643
Wildlife sanctuaries, 559
Wildlife Society, 643
Williams, A. B., 102
Williams, G. C., 157, 158
Willow, dwarf, 463f
Wilson, E. O., 137, 150, 157, 303, 304
Wind
  dunes and, 392, 393
  pollination, 255–256
  reactions, 353
  temperature and, 46–47
  tundra and, 481–482
  velocity, 46f
Wind-chill index, 46–47
Windthrow, 347, 348f
Winter depression, 57
Winter kills, 564
Wolverine, 472–473
Wolves, 608f
Woodlands, 505–513
Woodpecker, black-backed, 72, 73f
Woodpecker, downy, 124f
Woodpecker, ivory-billed, 166, 455
Woodpecker, pileated, 625
Woodpecker, red-cockaded, 75, 77, 455–456, 457f, 629
Woodpecker, red-bellied, 455
Woodpigeons, 121, 123f
Work, 22–23
World Health Organization, 216, 653
World Wildlife Fund, 553
The World We Live In, 442
Worm, brain, 217–219, 222
Worm, orange-striped oak, 181f
Worm, Pompeii, 332
Worm, sludge, 22
Worm, spiny-headed, 221–222
Worm, tobacco, 170
Worm, tube, 331
Wrasse, 583
Wren, house, 34–35, 36t
Wren, stripe-backed, 154–155, 156f
Wren-tit, 110f, 508, 508f
Wynne-Edwards, V. C., 139, 157–158

Xerophytes, 50–52, 51f
  true, 50
Xerosere, 381

Yak, 484
Yarnell, R. A., 444t
Yeasts, 647
Yellowstone National Park
  fire management, 78–79
  greater ecosystem plan, 637, 638
  1988 fires, 66f, 71, 72f, 77–78, 78f
Yew, Canada, 626–627
Yew, Pacific, 642
Yucca, 160, 536f, 537

Zebra, 556f, 557
Zeitgeber, 59
Zooplankton, 565, 566, 584, 588
Zooxanthellae, 584
Zurich-Montpellier fidelity, 270